Quantum Theory and Measurement

Edited by
John Archibald Wheeler
and Wojciech Hubert Zurek

Princeton Series in Physics

Princeton University Press
Princeton, New Jersey 1983

Published by Princeton University Press, 41 William Street,
Princeton, New Jersey
In the United Kingdom:
Princeton University Press, Guildford, Surrey

Library of Congress Cataloging in Publication Data will be found on the last printed page of this book

Clothbound editions of Princeton University Press books are printed on acid-free paper, and binding materials are chosen for strength and durability

Printed in the United States of America
by Princeton University Press,
Princeton, New Jersey

ADDITIONAL COPYRIGHT INFORMATION WILL BE FOUND BEGINNING ON P. XXI

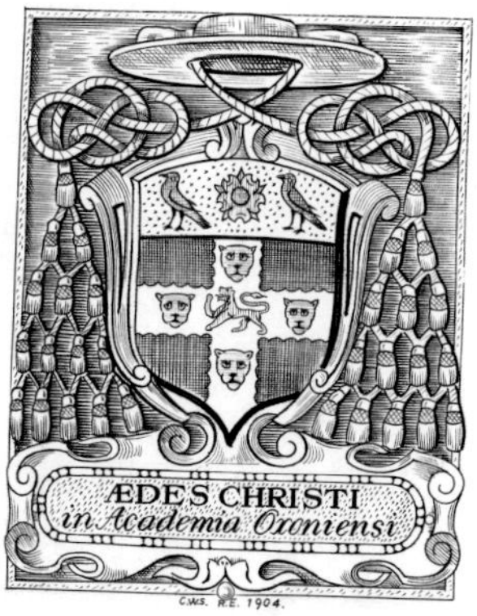
ÆDES CHRISTI
in Academia Oxoniensi
C.W.S. R.E. 1904.

Quantum Theory
and Measurement

Princeton Series in Physics

edited by Arthur S. Wightman
and Philip W. Anderson

Quantum Mechanics for Hamiltonians Defined as Quadratic Forms *by Barry Simon*

Lectures on Current Algebra and Its Applications *by Sam B. Treiman, Roman Jackiw, and David J. Gross*

Physical Cosmology *by P. J. E. Peebles*

The Many-Worlds Interpretation of Quantum Mechanics *edited by B. S. DeWitt and N. Graham*

The $P(\Phi)_2$ Euclidean (Quantum) Field Theory *by Barry Simon*

Homogeneous Relativistic Cosmologies *by Michael P. Ryan, Jr., and Lawrence C. Shepley*

Studies in Mathematical Physics: Essays in Honor of Valentine Bargmann *edited by Elliott H. Lieb, B. Simon, and A. S. Wightman*

Convexity in the Theory of Lattice Gases *by Robert B. Israel*

Surprises in Theoretical Physics *by Rudolf Peierls*

The Large-Scale Structure of the Universe *by P. J. E. Peebles*

Quantum Theory and Measurement *edited by John Archibald Wheeler and Wojciech Hubert Zurek*

Bohr and Einstein in Dialogue

According to the assumption considered here, when a light ray starting from a point is propagated, the energy is not continuously distributed over an ever increasing volume, but it consists of a finite number of energy quanta, localised in space, which move without being divided and which can be absorbed or emitted only as a whole.

EINSTEIN (1905)

During the elementary process of radiative loss, the molecule suffers a recoil of magnitude $h\nu/c$ in a direction which is only determined by 'chance', according to the present state of the theory

EINSTEIN (1916b)

I am studying your great works and—when I get stuck anywhere—now have the pleasure of seeing your friendly young face before me smiling and explaining.

EINSTEIN LETTER OF MAY 2, 1920, AFTER MEETING NIELS BOHR

You believe in a dice-playing God and I in perfect laws in the world of things existing as real objects, which I try to grasp in a wildly speculative way.

EINSTEIN LETTER, AS QUOTED IN SCHILPP (1949), p. 176

. . . time and space are modes by which we think and not conditions in which we live.

EINSTEIN AS QUOTED BY FORSEE (1963), p. 81

Out yonder there was this huge world, which exists independently of us human beings and which stands before us like a great, eternal riddle, at least partially accessible to our inspection.

EINSTEIN IN SCHILPP (1949), p. 5

(1) The dynamical equilibrium of the systems in the stationary states can be discussed by help of the ordinary mechanics, while the passing of the systems between different stationary states cannot be treated on that basis.
(2) The latter process is followed by the emission of a homogeneous radiation, for which the relation between the frequency and the amount of energy emitted is the one given by Planck's theory. BOHR (1913a), p. 7

. . . any observation necessitates an interference with the course of the phenomena, [and requires] a final renunciation of the classical ideal of causality and a radical revision of our attitude towards the problem of physical reality.

1st HALF, BOHR (1934), p. 115; 2nd HALF, BOHR (1935b), p. 697

. . . every atomic phenomenon is closed in the sense that its observation is based on registrations obtained by means of suitable amplification devices with irreversible functioning such as, for example, permanent marks on the photographic plate, caused by the penetration of electrons into the emulsion. . . . the quantum-mechanical formalism permits well-defined applications referring only to such closed phenomena and must be considered a rational generalization of classical physics. BOHR (1958), pp. 73 and 90

. . . the concepts of space and time by their very nature acquire a meaning only because of the possibility of neglecting the interaction with the means of measurement. BOHR (1934), p. 99

. . . we must be prepared for the necessity of an ever extending abstraction from our customary demands for a directly visualizable description of nature. Above all, we may expect new surprises in the domain where the quantum theory meets with the theory of relativity and where unsolved difficulties still stand.

BOHR (1934), p. 115

In tears, Ehrenfest said that he had to make a choice between Bohr's and Einstein's position and that he could not but agree with Bohr. [Samuel Goudsmit's account of a 1927 conversation with Ehrenfest.] PAIS (1979), p. 900

October 1927. The fifth Solvay Conference convenes [in Brussels]. All the founders of the quantum theory were there, from Planck, Einstein, and Bohr to de Broglie, Heisenberg, Schrödinger, and Dirac. During the sessions "Einstein said hardly anything beyond presenting a very simple objection to the probability interpretation. . . . Then he fell back into silence" [de Broglie, 1962, p. 150].

However, the formal meetings were not the only place of discussion. All participants were housed in the same hotel and there, in the dining room, Einstein was much livelier. Otto Stern has given this firsthand account [to Res Jost]: "Einstein came down to breakfast and expressed his misgivings about the new quantum theory, every time he had invented some beautiful experiment from which one saw that it did not work. . . . Pauli and Heisenberg who were there did not react to these matters, "ach was, das stimmt schon, das stimmt schon," ah well, it will be alright, it will be alright. Bohr on the other hand reflected on it with care and in the evening, at dinner, we were all together and he cleared up the matter in detail." PAIS (1979), p. 901

At the sixth Solvay Conference, in 1930, Einstein thought he had found a counterexample to the uncertainty principle. "It was quite a shock for Bohr. . . . he did not see the solution at once. During the whole evening he was extremely unhappy, going from one to the other and trying to persuade them that it couldn't be true, that it would be the end of physics if Einstein were right; but he couldn't produce any refutation. I shall never forget the vision of the two antagonists leaving the club [of the Fondation Universitaire]: Einstein a tall majestic figure, walking quietly, with a somewhat ironical smile, and Bohr trotting near him, very excited. . . . The next morning came Bohr's triumph."

ROSENFELD (1968), p. 232

The photographs on the preceding pages were taken by Paul Ehrenfest in his house in Leyden where Niels Bohr and Albert Einstein were staying as guests. The restoration of the negatives and production of the prints were done by William R. Whipple. Courtesy of the American Institute of Physics Niels Bohr Library.

CONTENTS

PREFACE

A textbook on quantum theory and measurement does not exist, nor is this intended to be one. This is a reference book, containing key papers on quantum theory as it relates to measurement. They are arranged in such a way, and accompanied by such supplementary references, that the collection can be used as a source book for a university course or seminar on the subject. It was so used by us in 1979–1980 and in 1980–1981 at the University of Texas. We have found that these materials are of interest to students and colleagues of all ages, not only in physics, but also in astronomy, philosophy, and mathematics.

We suppose that the reader already has some familiarity with quantum theory, preferably at least the equivalent of an undergraduate course in the subject.

Quantum theory has turned out to be the overarching principle of twentieth-century physics. It would be difficult to find a single subject among the physical sciences which is not affected in its foundations or in its applications by quantum theory. We may feel lost, to be sure, in the beginning stages of the study of the subject. Our supposed knowledge of a particle with its definite track through space and time dissolves into a wave, definiteness becomes indeterminism, and predictability of place is replaced by a predictability of the properties of nuclei, atoms, molecules, solids, liquids, and gases. We soon find ourselves armed with wonderful new tools. The more we use them, the more applications we find; and the more applications we find, the more uses of quantum theory we make.

Why there is no textbook on the measurement side of quantum theory is clear to anyone who participates in a seminar on the subject, and even clearer to one who gives a course on it: puzzlement! Beyond the probability interpretation of quantum mechanics, beyond all the standard analysis of idealized experiments, beyond the principle of indeterminacy and the limits it imposes, lie deep issues on which full agreement has not yet been reached in the physics community. They include questions like these: Does observation demand an irreversible act of amplification such as takes place in a grain of photographic emulsion or in the electron avalanche of a Geiger counter? And if so, what does one mean by "amplification"? And by "irreversible"? Does the quantum theory of observation apply in any meaningful way to the "whole universe"? Or is it restricted, even in principle, to the light cone? And if so, whose light cone? How are the observations made by different observers to be fitted into a single consistent picture in space-time? If these are some of the issues, they lead to other still deeper questions: What is the most productive meaning to assign to the term "reality"?

How are we to look at the subject, so mixed in its character, partly well-understood—and, as such, the unshakable foundation for all of modern physics—and partly still uncaptured frontier territory? What else is it but an unfamiliar animal, confined to an animal house? And how else can one better capture its newness than by walking around, looking at it through one window after another, seeking to combine fragmentary views into a total picture? We here take two tours around the subject. The first tour begins with Niels Bohr's account of his famous 28-year dialog with Albert Einstein. That account is followed by a great trio: Max Born's epochal paper on the probability interpretation of quantum mechanics, Werner Heisenberg's on the principle of indeterminism, and Niels Bohr's Como lecture on what quantum mechanics permits one to know. These are followed by H. P. Robertson's general formulation of indeterminism or "uncertainty," then by Nevill Mott's analysis of the meaning of an α-ray track, which illustrates the principles expounded by Bohr and Born and others and shows the quantum theory of observation in action.

Had quantum mechanics stopped here, its deepest lesson would have escaped attention: "No elementary quantum phenomenon is a phenomenon until it is a registered (observed) phenomenon." The door to this insight—and all the questions that go with it—was opened a crack by the paper of Albert Einstein, Richard Tolman, and Boris Podolsky, and opened wider by the idealized experiment proposed by Einstein, Podolsky, and Nathan Rosen. This EPR paper reasoned that quantum mechanics is incompatible with any reasonable idea of reality. In the next two papers, Bohr replies, in effect, that the EPR concept of reality is too limited. In contrast to Bohr, Erwin Schrödinger, in his famous paper on the "cat paradox," and Eugene Wigner, in a subsequent paper, tried to connect the concept "observation" as it is employed in quantum mechanics with "consciousness." This first tour around the animal house, looking through the various windows, concludes with "law without law," an attempt to assess the situation as it stands today and to evaluate the place of quantum mechanics in the larger scheme of physics.

The second tour of inspection (Sections II–VI) looks into the old windows afresh and into some new ones. It is designed for the more advanced student of the subject. It too gives a view of what is clear and well-established, but also glimpses of what is problematic and mysterious. Section II begins with the classic treatise of Fritz Wolfgang London and Edmond Bauer on the quantum theory of observation and includes notes of Wigner's 1976 lectures on interpretation of quantum mechanics, not previously published. The reader who has mastered this material or the equivalent has in hand the solid foundation of the subject. This section ends with three papers on the problem of measurement within the framework of quantum theory: Hugh Everett's "Relative State" or "Many Worlds" interpretation, and papers of Wigner and H. D. Zeh.

The Einstein-Podolsky-Rosen experiment (Section III) deals with a system which, once united in a definite quantum state, splits into two well-separated systems. It considers the correlation between the state observed for one system and the state observed for the other. It asks: Does the predicted correlation exist? And if so, how does it come about? There is an enormous literature on these questions. Out of the many possible papers we have selected thirteen for this part of our second tour around the subject. They deal with "hidden variables," John Bell's inequality designed to rule out "local hidden variables," the experiments themselves, and generalizations of such experiments.

Up to this point, the statistical interpretation of quantum mechanics and the principle of indeterminacy have been applied to a particle or system of finitely many degrees of freedom. The next group of papers (Section IV), including one by Lev Landau and Rudolf Peierls, and two by Bohr and Leon Rosenfeld, deal with the measurement of the electromagnetic field, a system with infinitely many degrees of freedom.

What is the connection between entropy, information, ergodicity, irreversibility, thermodynamics, and quantum mechanics? The next seven selections (Section V) deal in one way or another with these topics.

The second tour concludes with Section VI, eight papers on the accuracy achievable in measurement as it is affected by quantum limitations. The first (N. F. Mott and H. S. Massey) deals with the impossibility of measuring the spin of an electron that is free rather than bound in an atom or in some experimental combination of electric and magnetic fields. Even when a particle or an atom or a molecule admits of a measurement of its angular momentum, the measuring equipment may be so light in mass that by this very reason its orientation is uncertain (papers of Huzihiro Araki and Mutsuo Yanase, and Yanase). Or the measurement of the time at which an interaction takes place may limit the accuracy with which one can determine the transfer of energy in that interaction (paper of David Bohm and Yakir Aharonov). Or a message may be sent through an amplifier or sent down a communication line and have quantum effects introduced into it along the way (papers of H. Heffner, H. A. Haus and J. A. Mullen, and John Pierce). Or in a measurement of a weak effect, like a gravitational wave from a supernova, can one circumvent quantum indeterminacy limits on the sensitivity of the measuring device? (paper of V. B. Braginsky, Y. I. Vorontsov and K. S. Thorne).

Anyone asking for the practical bearing of the quantum theory of measurement will think of measuring devices, their sensitivity, and the improvements in this sensitivity that can only be achieved by exploiting modern insights to the fullest. In no way do the advances of physics spread more widely to the community than in new and improved measuring devices, whose uses range from biology to medicine, from chemistry to manufacturing. Not otherwise can one understand why the number of kinds of devices listed in the annual "Guide to Scientific Instruments" issued by the American Association for the Advancement of Science (AAAS, 1961, 1970, 1980) has steadily increased. There is at present a heavy push to measure at the Earth the fantastically weak gravitational wave produced by a supernova in the Virgo cluster of galaxies. Central to this enterprise are considerations of quantum theory that have already led to new ideas of optimal design. Achievements on this front will surely make their way from physics to many another field and supply important areas of science and technology with measuring devices of a sensitivity previously unattainable.

We regard as a welcome forerunner to the present collection an earlier collection of about half as many titles and a quarter as many pages issued for the exclusive use of members of the Physical Society of Japan. We have not entered into various schemes for the axiomatization of quantum theory and so-called "quantum logic," and therefore do not reproduce selections from such books as George W. Mackey's *Mathematical Foundations of Quantum Mechanics* (1963) and Josef M. Jauch's *Foundations of Quantum Mechanics* (1968), nor, for example, the work of Günther Ludwig and his collaborators (1968) or P. Mittelstaedt (1978), or Beltrametti and Cassinelli (1981). Neither do we pretend to a proper historical perspective, such as surely someday will be possible thanks to the wonderful collection of tape recordings, notebooks, and other materials inventoried in Kuhn, Heilbron, Forman, and Allen, *Sources for the History of Quantum Physics* (1967). The literature on quantum theory is so vast that one could fill a book this size with bibliography alone. In the annotated bibliographies for each section (located near the end of the book) we have had to content ourselves with a few especially important or representative works, seeking to provide in this way points of entry into the literature.

The reader will look in vain in these selections for any detailed exposition of the wealth of mechanisms that make physics such a rich subject. We leave out such topics as photoelectric effect, scattering, chemical binding, particle theory, the Lamb-Retherford shift, the building up of atoms, angular momentum isomers, superconductivity, and dozens of others equally important for understanding what quantum theory means for the world of today, because we assume that the reader already knows something of "quantum mechanics in action."

If one already knows so much about the applications of quantum theory, what more is to be learned by the study of the quantum theory of measurement? Willis Lamb (1969), some years after receiving the Nobel Prize, wrote, ". . . A discussion of the interpretation of quantum mechanics on any level beyond this almost inevitably becomes rather vague. The major difficulty involves the concept of 'measurement,' which in quantum mechanics means determining the value of a physical observable for a dynamical system with as much precision as is possible.

"I have taught graduate courses in quantum mechanics for over 20 years at Columbia, Stanford, Oxford and Yale, and for almost all of them have dealt with measurement in the following manner. On beginning the lectures I told the students, 'You must first learn the rules of calculation in quantum mechanics, and then I will tell you about the theory of measurement and discuss the meaning of the subject.' Almost invariably, the time allotted to the course ran out before I had to fulfill my promise."

No one who reads among the present selections can escape some contact with the deeper meaning of the subject—and with some of the issues. Is it true that "no elementary quantum phenomenon is a phenomenon until it is a recorded phenomenon"? If so, what does "recording" demand? "Irreversibility"? If so, what does one mean by "irreversibility"? If the "arrow of time" is absent from Schrödinger's equation and from quantum theory generally, what brings it into the act of measurement along with all its ideological connectives, from statistical mechanics to ergodic theory, and from information theory to thermodynamics? Is it true that the result of a measurement must be expressed in classical terms, because only in such terms can one speak in plain language to oneself—and to others? What part does communication play in creating what is called "knowledge"? And from what deeper principle arises the necessity of the quantum in a construction of the world?

With a good conscience we limit ourselves here to the measurement side of quantum theory, because the formalism of the subject is so well treated in so many outstanding texts. Moreover, historic papers in the development of that formalism have been collected in the wonderfully useful book by B. L. Van der Waerden (1967). We wish to express here our indebtedness to those books and above all to the book of Max Jammer (1974) which provides a wealth of historical commentary on the development of the subject to its present state. See also DeWitt and Graham (1971).

We take this opportunity to thank most heartily the many colleagues who have given us their advice in the preparation of this book and to acknowledge our indebtedness to the authors and publishers cited at the front for permission to reprint the selections. We could not have put the collection together without calling on the intelligence and helpfulness of Ruth Bentley, Zelda Davis, Adrienne Harding, Colleen Kieke, Jean F. Otto, Rebecca Stadtner, and Gloria Talcove-Woodward. We appreciate the care given to the project by the staff of Princeton University Press, including especially Alice Calaprice and Judith May. We thank many a favorable chance, many a kind act of hospitality, and thank, too, the University of Texas, the Center for Theoretical Physics, the Center for Statistical Mechanics and Thermodynamics, and the National Science Foundation (Grant PHY 7826592) for the colleagues that they have brought our way who share our concern with the measurement side of quantum theory, among them L. Bartell, J. D. Bekenstein, D. Bohm, A. Bohr, V. B. Braginsky, E. Caianiello, P. Candelas, P. C. W. Davies, B. d'Espagnat, D. Deutsch, B. S. DeWitt, R. H. Dicke, F. Dyson, H. Everett III, L. Z. Fang, R. P. Feynman, E. S. Fry, U. Gerlach, A. M. Gleason, L. P. Horwitz, A. Jaffe, F. Jenc, J. Kalckar, J. R. Klauder, D. Kondepudi, and K. Kuchař, G. Ludwig, G. Mackey, L. Michel, C. W. Misner, P. Mittelstaedt, Y. Ne'eman, A. Peres, C. Piron, I. Prigogine, A. Qadir, L. Radicati, D. Sciama, A. Shimony, H. P. Stapp, E. C. G. Sudarshan, C. Teitelboim, K. S. Thorne, F. Tipler, W. Unruh, E. P. Wigner, W. K. Wootters, and H. D. Zeh.

John Archibald Wheeler
Wojciech Hubert Zurek
Austin, Texas
January 20, 1982

ACKNOWLEDGMENTS AND COPYRIGHT INFORMATION

The editors gratefully acknowledge the permission of the Niels Bohr Library of the American Institute of Physics to reproduce here Ehrenfest's photographs of Bohr and Einstein, and the permissions received from authors, translators, and publishers to reprint the following items.

I.1 Commentary Based on Petersen (1968)
Extracts reprinted by permission from *Quantum Physics and The Philosophical Tradition* by Aage Petersen. Copyright © 1968, Massachusetts Institute of Technology Press.

I.1 Commentary of Heisenberg (1967)
Extracts from "Quantum theory and its interpretations," by W. Heisenberg. Reprinted by permission from *Niels Bohr: His Life and Work as Seen by His Friends and Colleagues*, pp. 94–108, edited by S. Rozental. Copyright © 1967, North Holland Publishing Company.

I.2 Commentary of Rosenfeld (1971a)
Extracts from "Men and ideas in the history of atomic theory," by L. Rosenfeld. From *Selected Papers of Léon Rosenfeld*, pp. 266–296, edited by R. S. Cohen and J. J. Stachel. Copyright © 1979, D. Reidel Publishing Company, reprinted by permission of the publisher.

I.3 Commentary of Heisenberg (1967)
Extracts from "Quantum theory and its interpretations," by W. Heisenberg. Reprinted by permission from *Niels Bohr: His Life and Work as Seen by His Friends and Colleagues*, pp. 94–108, edited by S. Rozental. Copyright © 1967, North Holland Publishing Company.

I.3 Commentary of Rosenfeld (1971a)
Extracts from "Men and ideas in the history of atomic theory," by L. Rosenfeld. From *Selected Papers of Léon Rosenfeld*, pp. 266–296, edited by R. S. Cohen and J. J. Stachel. Copyright © 1979, D. Reidel Publishing Company, reprinted by permission of the publisher.

I.4 Commentary of Rosenfeld (1963)
Extracts from "Niels Bohr's contribution to epistemology," by L. Rosenfeld. From *Selected Papers of Léon Rosenfeld*,

pp. 522–535, edited by R. S. Cohen and J. J. Stachel. Copyright © 1979, D. Reidel Publishing Company, reprinted by permission of the publisher.

I.4 Commentary of Rosenfeld (1971a)
Extracts from "Men and ideas in the history of atomic theory," by L. Rosenfeld. From *Selected Papers of Léon Rosenfeld*, pp. 266–296, edited by R. S. Cohen and J. J. Stachel. Copyright © 1979, D. Reidel Publishing Company, reprinted by permission of the publisher.

I.4 Commentary of Heisenberg (1967)
Extracts from "Quantum theory and its interpretations," by W. Heisenberg. Reprinted by permission from *Niels Bohr: His Life and Work as Seen by His Friends and Colleagues*, pp. 94–108, edited by S. Rozental. Copyright © 1967, North Holland Publishing Company.

I.8 Commentary of Rosenfeld (1967)
Extracts from "Niels Bohr in the thirties: consolidation and extension of the conception of complementarity," by L. Rosenfeld. Reprinted by permission from *Niels Bohr: His Life and Work as Seen by His Friends and Colleagues*, pp. 114–136, edited by S. Rozental. Copyright © 1967, North Holland Publishing Company.

I.9 Commentary of Rosenfeld (1967)
Extracts from "Niels Bohr in the thirties: consolidation and extension of the conception of complementarity," by L. Rosenfeld. Reprinted by permission from *Niels Bohr: His Life and Work as Seen by His Friends and Colleagues*, pp. 114–136, edited by S. Rozental Copyright © 1967, North Holland Publishing Company.

IV.2 Commentary of Rosenfeld (1955)
Extracts from "On quantum electrodynamics," by L. Rosenfeld. Reprinted by permission from *Niels Bohr and the Development of Physics*, edited by W. Pauli. Copyright 1955, Pergamon, New York.

VI.1 Commentary of Rosenfeld (1971b)
Extracts from "Quantum theory in 1929: recollections from the first Copenhagen conference," by L. Rosenfeld. Reprinted by permission from *Selected Papers of Léon Rosenfeld*, pp. 305–307, edited by R. S. Cohen and J. J. Stachel. Copyright © 1979, D. Reidel Publishing Co.

I.1 "Discussion with Einstein on epistemological problems in atomic physics," by Niels Bohr. Reprinted by permission from *Albert Einstein, Philosopher Scientist*, pp. 200–241, edited by Paul Arthur Schilpp. Copyright © 1949, currently handled jointly by Open Court Publishing Company and the American Friends of The Hebrew University, Inc.

I.2 "On the quantum mechanics of collisions," by M. Born. By permission from *Zeitschrift für Physik 37*, pp. 863–867. Copyright © 1926, Springer-Verlag.

I.3 "The physical content of quantum kinematics and mechanics," by W. Heisenberg. Reprinted by permission from *Zeitschrift für Physik 43*, pp. 172–198. Copyright © 1927 by Springer-Verlag. Reprinted in *Dokumente der Naturwissenschaft*, Vol. 4, pp. 9–35. Copyright © 1935.

I.4 "The quantum postulate and the recent development of atomic theory," by Niels Bohr. Reprinted by permission from *Nature*, Vol. 121, pp. 580–590. Copyright © 1928, Macmillian Journals Limited.

I.5 "The uncertainty principle," by H. P. Robertson. Reprinted by permission from *Physical Review 34*, pp. 163–164. Copyright © 1929, American Physical Society.

I.6 "The wave mechanics of α-ray tracks," by N. F. Mott. Reprinted by permission from *Proceedings A126*, pp. 79–84. Copyright © 1929, The Royal Society, London.

I.7 "Knowledge of past and future in quantum mechanics," by A. Einstein, R. C. Tolman, and B. Podolsky. Reprinted by permission from *Physical Review 37*, pp. 780–781. Copyright © 1931, currently handled jointly by American Physical Society and the American Friends of the Hebrew University, Inc.

I.8 "Can quantum-mechanical description of physical reality be considered complete?" by A. Einstein, B. Podolsky and N. Rosen. Reprinted by permission from *Physical Review 47*, pp. 777–780. Copyright © 1935, currently handled jointly by American Physical Society and the American Friends of The Hebrew University, Inc.

I.9 "Quantum mechanics and physical reality," by Niels Bohr. Reprinted by permission from *Nature*, Vol. 136, pp. 65. Copyright © 1935, Macmillian Journals Limited.

I.10 "Can quantum-mechanical description of physical reality be considered complete?" by Niels Bohr. Reprinted by permission from *Physical Review 48*, pp. 696–702. Copyright © 1935, American Physical Society.

I.11 "Die gegenwärtige Situation in der Quantenmechanik," by E. Schrödinger. By permission from *Die Naturwissenschaften 23*, pp. 807–812, 823–828, 844–849. Copyright © 1935, Springer-Verlag New York, Inc. English translation: "The present situation in quantum mechanics: a translation of Schrödinger's 'cat paradox' paper," by John D. Trimmer. Reprinted by permission from *Proceedings of the American Philosophical Society*, Vol. 124, pp. 323–338. Copyright © 1980, American Philosophical Society.

I.12 "Remarks on the mind-body question," pp. 171–184, from *Symmetries and Reflections*, science essay by E. P. Wigner. Copyright © 1967, Indiana University Press, reprinted by permission of the publisher.

I.13 "Law without law," excerpts as follows:

"Frontiers of time" by John A. Wheeler. Reprinted by permission from *Problems in the Foundations of Physics*, Proceedings of the International School of Physics "Enrico Fermi," Course 72, pp. 395–407. Copyright © 1979, Italian Physical Society.

"Beyond the Black Hole," from H. Woolf, ed., *Some Strangeness in the Proportion: A Centennial Symposium to Celebrate the Achievements of Albert Einstein*, Copyright © 1980, Addison-Wesley, Reading, Massachusetts, pp. 362–363. Reprinted with permission.

II.1 "The theory of observation in quantum mechanics," by F. W. London and E. Bauer. No. 775 of *Actualités scientifiques et industrielles: Exposés de physique générale*, edited by Paul Langevin, Hermann et Compagnie, 1939. English translations done independently by Abner Shimony, by J. A. Wheeler and W. H. Zurek, and by James H. McGrath and Susan McLean McGrath; the editors have drawn on all three translations to produce the version here.

II.2 "Interpretation of quantum mechanics," by Eugene P. Wigner. By permission from lecture notes of Eugene P. Wigner, revised and printed here for the first time.

II.3 "Relative state formulation of quantum mechanics," by Hugh Everett III. Reprinted by permission from *Reviews of Modern Physics*, Vol. 29, pp. 454–462. Copyright © 1957, American Physical Society.

II.4 "The problem of measurement," by Eugene P. Wigner. Reprinted by permission from *American Journal of Physics*, Vol. 31, pp. 6–15, edited by John S. Rigden.

Copyright © 1963, American Association of Physics Teachers.

II.5 "On the interpretation of measurement in quantum theory," by H. D. Zeh. Reprinted by permission from *Foundations of Physics*, Vol. 1, pp. 69–76. Copyright © 1970, Plenum Publishing Corporation.

III.1 "Polyelectrons" by John A. Wheeler. Reprinted by permission from *Annals of the New York Academy of Sciences*, Vol. 48, pp. 219–238, edited by Bill M. Boland. Copyright © 1946, New York Academy of Science.

III.2 "The paradox of Einstein, Rosen and Podolsky," by David Bohm. Reprinted by permission from *Quantum Theory*, Chapter 22, Sections 15–19. Copyright © 1951, Prentice-Hall.

III.3 "A suggested interpretation of the quantum theory in terms of 'hidden' variables, I and II," by David Bohm. Reprinted by permission from *Physical Review 85*, pp. 166–179. Copyright © 1952, American Physical Society.

III.4 "On the problem of hidden variables in quantum mechanics," by John S. Bell. Reprinted by permission from *Reviews of Modern Physics*, Vol. 38, pp. 447–452. Copyright © 1966, American Physical Society.

III.5 "On the Einstein Podolsky Rosen paradox," by John S. Bell. Reprinted by permission of J. S. Bell. Originally appeared in *Physics I*, pp. 195–200. Copyright © 1964, Physics Publishing Co.

III.6 "Proposed experiment to test local hidden-variable theories," by John F. Clauser, Michael A. Horne, Richard A. Holt, and Abner Shimony. Reprinted by permission from *Physical Review Letters 23*, pp. 880–884. Copyright © 1969, American Physical Society.

III.7 "Experimental test of local hidden-variable theories," by Stuart J. Freedman and John F. Clauser. Reprinted by permission from *Physical Review Letters 28*, pp. 938–941. Copyright © 1972, American Physical Society.

III.8 "Experimental test of local hidden-variable theories," by Edward S. Fry and Randall C. Thompson. Reprinted by permission from *Physical Review Letters 37*, pp. 465–468. Copyright © 1976, American Physical Society.

III.9 "Quantum mechanics and hidden variables: A test of Bell's inequality by the measurement of the spin correlation in low-energy proton-proton scattering." by M. Lamehi-Rachti and W. Mittig. Reprinted by permission from

Physical Review D14, pp. 2543–2555. Copyright © 1976, American Physical Society.

III.10 "Proposed experiment to test the nonseparability of quantum mechanics," by Alain Aspect. Reprinted by permission from *Physical Review D14*, pp. 1944–1951. Copyright © 1976, American Physical Society.

III.11 "Complementarity in the double-slit experiment: Quantum nonseparability and a quantitative statement of Bohr's principle," by W. K. Wootters and W. H. Zurek. Reprinted by permission from *Physical Review D19*, pp. 473–484. Copyright © 1979, American Physical Society.

III.12 "Complementarity in the double-slit experiment: On simple realizable systems for observing intermediate particle-wave behavior," by L. S. Bartell. Reprinted by permission from *Physical Review D21*, pp. 1698–1699. Copyright © 1980, American Physical Society.

III.13 "A 'delayed-choice' quantum mechanics experiment," by William C. Wickes, C. O. Alley, and O. Jakubowicz. Printed here for the first time by permission of the authors.

IV.1 "Erweiterung des Umbestimmtheitsprinzips für die relativistische Quantentheorie," by L. Landau and Rudolf Peierls. By permission from *Zeitschrift für Physik 69*, p. 56. Copyright © 1931 Springer-Verlag. English translation: "Extension of the uncertainty principle to relativistic quantum theory," translated and edited by Dirk ter Haar. Reprinted by permission from *Collected Papers of L. D. Landau*, pp. 4–51. Copyright © 1965, Gordon and Breach.

IV.2 "On the question of the measurability of electromagnetic field quantities," by Niels Bohr and Léon Rosenfeld. By permission from *Matematisk-Fysiske Meddelelser Kongelige Danske Videnskabernes Selskab 12*, No. 8. Copyright © 1933, Det Kongelige Danske Videnskabernes Selskab. English translation by Aage Petersen. Reprinted by permission from *Selected Papers of Léon Rosenfeld*, pp. 357–412, edited by R. S. Cohen and J. J. Stachel. Copyright © 1979, D. Reidel Publishing Co.

IV.3 "Field and charge measurements in quantum electrodynamics," by Niels Bohr and Léon Rosenfeld. Reprinted by permission from *Physical Review 78*, pp. 794–798. Copyright © 1950, American Physical Society.

V.1 "Uber die Entropieverminderung in einen thermodynamischen System bei Eingriffen intelligenter Wesen," by Leo Szilard. By permission from *Zeitschrift für Physik 53*, pp. 840–856. Copyright © 1929, Springer-Verlag. English translation: "On the decrease of entropy in a thermodynamic system by the intervention of intelligent beings," by Anatol Rapoport and Mechthilde Knoller, revised by Carl Eckart. Reprinted by permission from *Behavioral Science 9*, pp. 301–310. Copyright © 1964, Society for General Systems Research.

V.2 "Measurement and Reversibility" *and* "The Measuring Process," by John von Neumann. By permission from *Mathematische Grundlagen der Quantenmechanik*, Chapters V and VI. Copyright © 1932, Springer-Verlag. English translation reprinted by permission from J. von Neumann, *Mathematical Foundations of Quantum Mechanics*, trans. Robert T. Beyer, pp. 347–445. Copyright © 1955, Princeton University Press.

V.3 "The ergodic behaviour of quantum many-body systems," by Léon Van Hove. Reprinted by permission from *Physica 25*, pp. 268–276. Copyright © 1959, North Holland Publishing Company.

V.4 "Quantum theory of measurement and ergodicity conditions," by A. Daneri, A. Loinger, and G. M. Prosperi. Reprinted by permission from *Nuclear Physics 33*, pp. 297. Copyright © 1962, North Holland Publishing Company.

V.5 "Time symmetry in the quantum process of measurement," by Yakir Aharonov, Peter G. Bergmann, and Joel L. Lebowitz. Reprinted by permission from *Physical Review 134*, pp. B1410. Copyright © 1964, American Physical Society.

V.6 "Lyapounov variable: Entropy and measurement in quantum mechanics," by B. Misra, I. Prigogine, and M. Courbage. Reprinted by permission from *Proceedings of The National Academy of Sciences 76*, pp. 4768–4772. Copyright © 1979, The National Academy of Sciences.

V.7 "Can we undo quantum measurements?" by Asher Peres. Reprinted by permission from *Physical Review D22*, pp. 879–883. Copyright © 1980, American Physical Society.

VI.1 "Magnetic moment of the electron," by N. F. Mott and

H. S. W. Massey. Reprinted by permission of Oxford University Press from *The Theory of Atomic Collisions*, 3rd ed., pp. 214–219, Section 2, Chapter IX. Copyright © 1965, Oxford University Press.

VI.2 "Measurement of quantum mechanical operators," by H. Araki and M. M. Yanase. Reprinted by permission from *Physical Review 120*, pp. 622–626. Copyright © 1960, American Physical Society.

VI.3 "Optimal measuring apparatus," by M. M. Yanase. Reprinted by permission from *Physical Review 123*, pp. 666–668. Copyright © 1961, American Physical Society.

VI.4 "Time in the quantum theory and the uncertainty relation for time and energy," by Y. Aharonov and D. Bohm. Reprinted by permission from *Physical Review 122*, pp. 1649–1658. Copyright © 1961, American Physical Society.

VI.5 "The fundamental noise limit of linear amplifiers," by J. Heffner. Reprinted from *Proceedings of the Institute of Radio Engineers 50*, pp. 1604–1608. Copyright © 1962, The Institute of Radio Engineers.

VI.6 "Quantum noise in linear amplifiers," by Herman A. Haus and James A. Mullen. Reprinted by permission from *Physical Review 128*, pp. 2407–2413. Copyright © 1962, American Physical Society.

VI.7 "Optical channels: Practical limits with photon counting," by J. R. Pierce. Reprinted from the *Institute of Electrical and Electronics Engineers Transactions on Communications Com-26*, pp. 1819–1821. Copyright © 1978, Institute of Electrical and Electronics Engineers.

VI.8 "Quantum nondemolition measurements," by V. B. Braginsky, Y. I. Vorontsov, and K. S. Thorne. Reprinted from *Science 209*, pp. 547–557. Copyright © 1980, The American Association for the Advancement of Science.

I
Questions of Principle

I.1 THE BOHR-EINSTEIN DIALOGUE

BOHR IN BRIEF

Complementarity: any given application of classical concepts precludes the simultaneous use of other classical concepts which in a different connection are equally necessary for the elucidation of the phenomena. BOHR (1934), p. 10

The discovery of the quantum of action shows us, not only the natural limitation of classical physics, but, by throwing a new light upon the old philosophical problem of the objective existence of phenomena independently of our observations, confronts us with a situation hitherto unknown in natural science. As we have seen, any observation necessitates an interference with the course of the phenomena, [and] of such a nature that it deprives us of the foundation underlying the causal mode of description. The limit, which nature herself has thus imposed upon us, of the possibility of speaking about phenomena as existing objectively finds its expression, as far as we can judge, just in the formulation of quantum mechanics.

BOHR (1934), p. 115

. . . atomic phenomena under different experimental conditions, must be termed complementary in the sense that each is well defined and that together they exhaust all definable knowledge about the objects concerned. The quantum-mechanical formalism . . . gives . . . an exhaustive complementary account of a very large domain of experience. BOHR (1958), p. 90

. . . one sometimes speaks of "disturbance of phenomena by observation" or "creation of physical attributes to atomic objects by measurements." Such phrases, however, are apt to cause confusion, since words like phenomena and observation, just as attributes and measurements, are here used in a way incompatible with common language and practical definition. On the lines of objective description, [I advocate using] the word *phenomenon* to refer only to observations obtained under circumstances whose description includes an account of the whole experimental arrangement. In such terminology, the observational problem in quantum physics is deprived of any special intricacy and we are, moreover, directly reminded that every atomic phenomenon is closed in the sense that its observation is based on registrations obtained by means of suitable amplification devices with irreversible functioning such as, for example, permanent marks on a photographic plate, caused by the penetration of electrons into the emulsion. In this connection, it is important to realize that the quantum-mechanical formalism permits well-defined applications referring only to such closed phenomena.

BOHR (1958), p. 73

. . . the finite magnitude of the quantum of action prevents altogether a sharp distinction being made between a phenomenon and the agency by which it is observed . . . . BOHR (1934), p. 11

. . . it . . . can make no difference, as regards observable effects obtainable by a definite experimental arrangement, whether our plans for constructing or handling the instruments are fixed beforehand or whether we prefer to postpone the completion of our planning until a later moment when the particle is already on its way from one instrument to another. BOHR IN SCHILPP (1949), p. 230

. . . a subsequent measurement to a certain degree deprives the information given by a previous measurement of its significance for predicting the future course of the phenomena. Obviously, these facts not only set a limit to the *extent* of the information obtainable by measurements, but they also set a limit to the *meaning* which we may attribute to such information. We meet here in a new light the old truth that in our description of nature the purpose is not to disclose the real essence of the phenomena but only to track down, so far as it is possible, relations between the manifold aspects of our experience. BOHR (1934), p. 18

The experimental conditions can be varied in many ways, but the point is that in each case we must be able to communicate to others what we have done and what we have learned, and that therefore the functioning of the measuring instruments must be described within the framework of classical physical ideas.

BOHR (1958), p. 89

. . . the conscious analysis of any concept stands in a relation of exclusion to its immediate application. BOHR (1934), p. 96

I am quite prepared to talk of the spiritual life of an electronic computer; to say that it is considering or that it is in a bad mood.

What really matters is the unambiguous description of its behaviour, which is what we observe. The question as to whether the machine *really* feels, or whether it merely looks as though it did, is absolutely as meaningless as to ask whether light is "in reality" waves or particles. We must never forget that "reality" too is a human word just like "wave" or "consciousness." Our task is to learn to use these words correctly—that is, unambiguously and consistently.

BOHR AS QUOTED BY KALCKAR (1967), p. 234

We are suspended in language in such a way that we cannot say what is up and what is down.

BOHR AS QUOTED BY PETERSEN (1968), p. 188

COMMENTARY BASED ON PETERSEN (1968)

Aage Petersen was already working on his book, his Copenhagen doctoral thesis, while he was assisting Bohr (†18 November 1962) in preparing some of his last lectures. The thesis is philosophical in character. It is concerned with ideas. It is not intended to be a professional history of science, nor a documentation of stages in Bohr's thinking. However, some of the sections allow one to get an impression of stages in Bohr's development of the concepts of "complementarity," "closure," and "phenomenon." The book states more sharply than Bohr does in his writings the points on which Bohr disagreed with others, explaining, for example, why Bohr introduced the term "complementarity" when Heisenberg had already employed the term "indeterminism."

Petersen (pp. 110–111 and 145) notes that Bohr insisted upon an analysis of the "possibilities of definition" over and above those "possibilities of observation" that he and Heisenberg together had previously considered: "One of the most important issues in the measurement analysis is the question of the nature and origin of the uncertainties involved in the determination of conjugate variables. According to Heisenberg, these uncertainties were due to discontinuous changes, imposed by the quantum on one such variable during the measurement of the other. However, as Bohr pointed out, 'a discontinuous change of energy and momentum during observation could not prevent us from ascribing accurate values to the space-time coordinates, as well as to the momentum-energy components before and after the process.' To clarify the issue it is necessary to consider closely the possibilities of definition. . . . Bohr pointed out that the conditions of description in quantum physics not only 'set a limit to the *extent* of the information obtainable by measurement, but they also set a limit to the *meaning* which we may attribute to such information.' More specifically, the reciprocal uncertainties are '. . . essentially an outcome of the limited accuracy with which changes in energy and momentum can be defined.' "

Petersen goes on to say (p. 120), "Einstein's criticism showed the need for a more rigorous formulation of the Copenhagen interpretation, and Bohr's many attempts to improve the terminology illuminate the development of his attitude to the interpretation question. [How could one] make explicit the conditions set by the formalism for applying the physical concepts unambiguously in the quantum domain? [For this purpose the] shift from 'intuitive understanding' to 'unambiguous communication' [was] an important step."

Petersen adds (pp. 120 and 122–23), "Terminologically, the principal result of Bohr's analysis of Einstein's imaginary experiments was the concept of a quantum phenomenon [which first appeared in Bohr (1939)]. Bohr came to regard it as the basic element of the quantal description. . . . To specify a phenomenon it is not enough to state the initial characteristics of the object, like the momentum with which it emerges from the source. The predictions depend on the whole experimental arrangement and are only well defined if the whole arrangement is specified. To be able to predict the interference pattern we must be given the whole geometry of the optical bench. In other words, 'all unambiguous interpretation of the quantum mechanical formalism involves the fixation of the external conditions, defining the initial state of the atomic system concerned and the character of the possible predictions as regards subsequent observable properties of that system. Any measurement in quantum theory can in fact only refer either to a fixation of the initial state or to the test of such predictions, and it is . . . the combination of measurements of both kinds which constitutes a well-defined phenomenon . . . .' [A phenomenon] is 'indivisible.' In electron interference, the physical 'process' starting at the electron's emergence from the gun and ending at its impact on the plate has no definable course. It cannot be broken up into physically well-defined steps. Unlike a classical phenomenon, a quantum phenomenon is not a sequence of physical events, but a new kind of individual entity."

Petersen (p. 164) recalls Bohr's statement (1934, pp. 19–20): "An interesting example of ambiguity in our use of language is provided by the phrase used to express the failure of the causal mode of description, namely, that one speaks of a free choice on the part of nature. Indeed, properly speaking, such a phrase requires the idea of an external chooser, the existence of which, however, is denied already by the use of the word nature." Petersen adds (p. 172) ". . . Bohr often stressed in discussions that 'reality' is a word in our language and that this word is no different from other words in that we must learn to use it correctly . . . ." Petersen continues (p. 173), "Indivisibility and closure are the two principal characteristics of a quantum phenomenon. . . . The phenomenon's 'interior' is . . . physically inscrutable. . . . In a classical physical process each infinitesimal step is 'closed', i.e. it is a definite physical event. . . . In quantum physics the object's 'behavior' is not a sequence of 'closed' steps." Here Petersen might have quoted from Bohr (1958, p. 73): "... Every atomic phenomenon is closed in the sense that its observation is based on registrations obtained by means of suitable amplification devices with irreversible functioning such as, for example, permanent marks on a photographic plate, caused by the penetration of electrons into the emulsion." But Petersen adds (p. 177–79), ". . . It is clear that Bohr considered the closure of fundamental significance not only in quantum physics but in the whole description of nature. Classical physics did not call attention to the role of this concept because classical processes have, so to say, maximal closure. In quantum mechanics the physically describable aspects of a phenomenon are closed, but the phenomenon's physically inscrutable 'interior' is not. . . . [The] question suggests itself as to whether it is possible to dispense with the classical concepts in the quantum domain or at least supplement them with new physical concepts that are less directly tied to the structure of classical theories and more adapted to the typical quantal parts of quantum mechanics. Bohr gave a negative answer to this question. He held that 'it would be a misconception to believe that the difficulties of the atomic theory may be evaded by eventually replacing the concepts of classical physics by new conceptual forms.'"

"Bohr was remarkably categorical about the question at issue. 'It lies in the nature of physical observation . . . that all experience must ultimately be expressed in terms of classical concepts . . . the unambiguous interpretation of any measurement must be essentially framed in terms of the classical physical theories, and we may say that in this sense the language of Newton and Maxwell will remain the language of physicists for all time.' 'Even when the phenomena transcend the scope of classical physical theories, the account of the experimental arrangement and the recording of observations must be given in plain language, suitably supplemented by technical physical terminology. This is a clear logical demand, since the very word *experiment* refers to a situation where we can tell others what we have done and what we have learned.'"

Commentary of Heisenberg (1967)

The Solvay Conference in Brussels in the autumn of 1927 closed this marvellous period in the history of atomic theory. Planck, Einstein, Lorentz, Bohr, de Broglie, Born, and Schrödinger, and from the younger generation Kramers, Pauli, and Dirac, were gathered here and the discussions were soon focussed to a duel between Einstein and Bohr on the question as to what extent atomic theory in its present form could be considered to be the final solution of the difficulties which had been discussed for several decades. We generally met already at breakfast in the hotel, and Einstein began to describe an ideal experiment in which he thought the inner contradictions of the Copenhagen interpretation were particularly clearly visible. Einstein, Bohr and I walked together from the hotel to the congress building, and I listened to the lively discussion between these two people whose philosophical attitudes were so different, and from time to time I added a remark on the structure of the mathematical formalism. During the meeting and particularly in the pauses we younger people, mostly Pauli and I, tried to analyze Einstein's experiment, and at lunch time the discussions continued between Bohr and the others from Copenhagen. Bohr had usually finished the complete analysis of the ideal experiment late in the afternoon and showed it to Einstein at the supper table. Einstein had no good objection to this analysis, but in his heart he was not convinced. Bohr's friend Ehrenfest, who was also a close friend of Einstein, said to him, "I'm ashamed of you, Einstein. You put yourself here just in the same position as your opponents in their futile attempts to refute your relativity theory." These discussions continued even at the next Solvay meeting in 1930, and it was probably on this occasion that Einstein at breakfast proposed the famous experiment (discussed in Bohr's article on the occasion of Einstein's 70th birthday) in which the colour of a light quantum is to be determined by weighing the source before and after the quantum's emission. As this problem involved gravity, one had to include the theory of gravity, in other words, general relativity theory in the analysis. It was a particular triumph for Bohr that he was able to show that evening, by using just Einstein's own formulae from general relativity, that even in this experiment the uncertainty relations are valid, and that Einstein's objections were unfounded. With this the Copenhagen interpretation of quantum theory seemed from now on to stand on solid ground.

Commentary of Einstein (1936)

There is no doubt that quantum mechanics has seized hold of a beautiful element of truth, and that it will be a test stone for any future theoretical basis, in that it must be deducible as a limiting case from that basis, just as electrostatics is deducible from the Maxwell equations of the electromagnetic field or as thermodynamics is deducible from classical mechanics. However, I do not believe that quantum mechanics will be the *starting point* in the search for this basis, just as, vice versa, one could not go from thermodynamics (resp. statistical mechanics) to the foundations of mechanics.

Commentary of Einstein (before 1953)

That the Lord should play with dice, all right; but that He should gamble according to definite rules, that is beyond me.

I.1 DISCUSSION WITH EINSTEIN ON EPISTEMOLOGICAL PROBLEMS IN ATOMIC PHYSICS

NIELS BOHR

WHEN invited by the Editor of the series, "Living Philosophers," to write an article for this volume in which contemporary scientists are honouring the epoch-making contributions of Albert Einstein to the progress of natural philosophy and are acknowledging the indebtedness of our whole generation for the guidance his genius has given us, I thought much of the best way of explaining how much I owe to him for inspiration. In this connection, the many occasions through the years on which I had the privilege to discuss with Einstein epistemological problems raised by the modern development of atomic physics have come back vividly to my mind and I have felt that I could hardly attempt anything better than to give an account of these discussions which, even if no complete concord has so far been obtained, have been of greatest value and stimulus to me. I hope also that the account may convey to wider circles an impression of how essential the open-minded exchange of ideas has been for the progress in a field where new experience has time after time demanded a reconsideration of our views.

From the very beginning the main point under debate has been the attitude to take to the departure from customary principles of natural philosophy characteristic of the novel development of physics which was initiated in the first year of this century by Planck's discovery of the universal quantum of action. This discovery, which revealed a feature of atomicity in the laws of nature going far beyond the old doctrine of the limited divisibility of matter, has indeed taught us that the classical theories

Originally published in *Albert Einstein: Philosopher-Scientist*, P. A. Schilpp, ed., pp. 200-41, The Library of Living Philosophers, Evanston (1949).

of physics are idealizations which can be unambiguously applied only in the limit where all actions involved are large compared with the quantum. The question at issue has been whether the renunciation of a causal mode of description of atomic processes involved in the endeavours to cope with the situation should be regarded as a temporary departure from ideals to be ultimately revived or whether we are faced with an irrevocable step towards obtaining the proper harmony between analysis and synthesis of physical phenomena. To describe the background of our discussions and to bring out as clearly as possible the arguments for the contrasting viewpoints, I have felt it necessary to go to a certain length in recalling some main features of the development to which Einstein himself has contributed so decisively.

As is well known, it was the intimate relation, elucidated primarily by Boltzmann, between the laws of thermodynamics and the statistical regularities exhibited by mechanical systems with many degrees of freedom, which guided Planck in his ingenious treatment of the problem of thermal radiation, leading him to his fundamental discovery. While, in his work, Planck was principally concerned with considerations of essentially statistical character and with great caution refrained from definite conclusions as to the extent to which the existence of the quantum implied a departure from the foundations of mechanics and electrodynamics, Einstein's great original contribution to quantum theory (1905) was just the recognition of how physical phenomena like the photo-effect may depend directly on individual quantum effects.[1] In these very same years when, in developing his theory of relativity, Einstein laid a new foundation for physical science, he explored with a most daring spirit the novel features of atomicity which pointed beyond the whole framework of classical physics.

With unfailing intuition Einstein thus was led step by step to the conclusion that any radiation process involves the emission or absorption of individual light quanta or "photons" with energy and momentum

$$E = h\nu \quad \text{and} \quad P = h\sigma \qquad (1)$$

[1] A. Einstein, *Ann. d. Phys.*, 17, 132, (1905).

respectively, where h is Planck's constant, while ν and σ are the number of vibrations per unit time and the number of waves per unit length, respectively. Notwithstanding its fertility, the idea of the photon implied a quite unforeseen dilemma, since any simple corpuscular picture of radiation would obviously be irreconcilable with interference effects, which present so essential an aspect of radiative phenomena, and which can be described only in terms of a wave picture. The acuteness of the dilemma is stressed by the fact that the interference effects offer our only means of defining the concepts of frequency and wavelength entering into the very expressions for the energy and momentum of the photon.

In this situation, there could be no question of attempting a causal analysis of radiative phenomena, but only, by a combined use of the contrasting pictures, to estimate probabilities for the occurrence of the individual radiation processes. However, it is most important to realize that the recourse to probability laws under such circumstances is essentially different in aim from the familiar application of statistical considerations as practical means of accounting for the properties of mechanical systems of great structural complexity. In fact, in quantum physics we are presented not with intricacies of this kind, but with the inability of the classical frame of concepts to comprise the peculiar feature of indivisibility, or "individuality," characterizing the elementary processes.

The failure of the theories of classical physics in accounting for atomic phenomena was further accentuated by the progress of our knowledge of the structure of atoms. Above all, Rutherford's discovery of the atomic nucleus (1911) revealed at once the inadequacy of classical mechanical and electromagnetic concepts to explain the inherent stability of the atom. Here again the quantum theory offered a clue for the elucidation of the situation and especially it was found possible to account for the atomic stability, as well as for the empirical laws governing the spectra of the elements, by assuming that any reaction of the atom resulting in a change of its energy involved a complete transition between two so-called stationary quantum states and that, in particular, the spectra were emitted by a step-like process in which each transition is accompanied by the emission of a monochromatic light quantum of an energy just equal to that of an Einstein photon.

These ideas, which were soon confirmed by the experiments of Franck and Hertz (1914) on the excitation of spectra by impact of electrons on atoms, involved a further renunciation of the causal mode of description, since evidently the interpretation of the spectral laws implies that an atom in an excited state in general will have the possibility of transitions with photon emission to one or another of its lower energy states. In fact, the very idea of stationary states is incompatible with any directive for the choice between such transitions and leaves room only for the notion of the relative probabilities of the individual transition processes. The only guide in estimating such probabilities was the so-called correspondence principle which originated in the search for the closest possible connection between the statistical account of atomic processes and the consequences to be expected from classical theory, which should be valid in the limit where the actions involved in all stages of the analysis of the phenomena are large compared with the universal quantum.

At that time, no general self-consistent quantum theory was yet in sight, but the prevailing attitude may perhaps be illustrated by the following passage from a lecture by the writer from 1913:[2]

> I hope that I have expressed myself sufficiently clearly so that you may appreciate the extent to which these considerations conflict with the admirably consistent scheme of conceptions which has been rightly termed the classical theory of electrodynamics. On the other hand, I have tried to convey to you the impression that—just by emphasizing so strongly this conflict—it may also be possible in course of time to establish a certain coherence in the new ideas.

Important progress in the development of quantum theory was made by Einstein himself in his famous article on radiative equilibrium in 1917,[3] where he showed that Planck's law for thermal radiation could be simply deduced from assumptions

[2] N. Bohr, *Fysisk Tidsskrift*, *12*, 97, (1914). (English version in *The Theory of Spectra and Atomic Constitution*, Cambridge, University Press, 1922).

[3] A. Einstein, *Phys. Zs.*, *18*, 121, (1917).

conforming with the basic ideas of the quantum theory of atomic constitution. To this purpose, Einstein formulated general statistical rules regarding the occurrence of radiative transitions between stationary states, assuming not only that, when the atom is exposed to a radiation field, absorption as well as emission processes will occur with a probability per unit time proportional to the intensity of the irradiation, but that even in the absence of external disturbances spontaneous emission processes will take place with a rate corresponding to a certain *a priori* probability. Regarding the latter point, Einstein emphasized the fundamental character of the statistical description in a most suggestive way by drawing attention to the analogy between the assumptions regarding the occurrence of the spontaneous radiative transitions and the well-known laws governing transformations of radioactive substances.

In connection with a thorough examination of the exigencies of thermodynamics as regards radiation problems, Einstein stressed the dilemma still further by pointing out that the argumentation implied that any radiation process was "unidirected" in the sense that not only is a momentum corresponding to a photon with the direction of propagation transferred to an atom in the absorption process, but that also the emitting atom will receive an equivalent impulse in the opposite direction, although there can on the wave picture be no question of a preference for a single direction in an emission process. Einstein's own attitude to such startling conclusions is expressed in a passage at the end of the article (*loc. cit.*, p. 127 f.), which may be translated as follows:

> These features of the elementary processes would seem to make the development of a proper quantum treatment of radiation almost unavoidable. The weakness of the theory lies in the fact that, on the one hand, no closer connection with the wave concepts is obtainable and that, on the other hand, it leaves to chance (*Zufall*) the time and the direction of the elementary processes; nevertheless, I have full confidence in the reliability of the way entered upon.

When I had the great experience of meeting Einstein for the first time during a visit to Berlin in 1920, these fundamental questions formed the theme of our conversations. The discussions, to which I have often reverted in my thoughts, added to all my admiration for Einstein a deep impression of his detached attitude. Certainly, his favoured use of such picturesque phrases as "ghost waves (*Gespensterfelder*) guiding the photons" implied no tendency to mysticism, but illuminated rather a profound humour behind his piercing remarks. Yet, a certain difference in attitude and outlook remained, since, with his mastery for co-ordinating apparently contrasting experience without abandoning continuity and causality, Einstein was perhaps more reluctant to renounce such ideals than someone for whom renunciation in this respect appeared to be the only way open to proceed with the immediate task of co-ordinating the multifarious evidence regarding atomic phenomena, which accumulated from day to day in the exploration of this new field of knowledge.

In the following years, during which the atomic problems attracted the attention of rapidly increasing circles of physicists, the apparent contradictions inherent in quantum theory were felt ever more acutely. Illustrative of this situation is the discussion raised by the discovery of the Stern-Gerlach effect in 1922. On the one hand, this effect gave striking support to the idea of stationary states and in particular to the quantum theory of the Zeeman effect developed by Sommerfeld; on the other hand, as exposed so clearly by Einstein and Ehrenfest,[4] it presented with unsurmountable difficulties any attempt at forming a picture of the behaviour of atoms in a magnetic field. Similar paradoxes were raised by the discovery by Compton (1924) of the change in wave-length accompanying the scattering of X-rays by electrons. This phenomenon afforded, as is well known, a most direct proof of the adequacy of Einstein's view regarding the transfer of energy and momentum in radiative processes; at the same time, it was equally clear that no simple picture of a corpuscular collision could offer an exhaustive description of the phenomenon. Under the impact of such difficulties, doubts

[4] A. Einstein and P. Ehrenfest, *Zs. f. Phys.*, *11*, 31, (1922).

were for a time entertained even regarding the conservation of energy and momentum in the individual radiation processes;[5] a view, however, which very soon had to be abandoned in face of more refined experiments bringing out the correlation between the deflection of the photon and the corresponding electron recoil.

The way to the clarification of the situation was, indeed, first to be paved by the development of a more comprehensive quantum theory. A first step towards this goal was the recognition by de Broglie in 1925 that the wave-corpuscle duality was not confined to the properties of radiation, but was equally unavoidable in accounting for the behaviour of material particles. This idea, which was soon convincingly confirmed by experiments on electron interference phenomena, was at once greeted by Einstein, who had already envisaged the deep-going analogy between the properties of thermal radiation and of gases in the so-called degenerate state.[6] The new line was pursued with the greatest success by Schrödinger (1926) who, in particular, showed how the stationary states of atomic systems could be represented by the proper solutions of a wave-equation to the establishment of which he was led by the formal analogy, originally traced by Hamilton, between mechanical and optical problems. Still, the paradoxical aspects of quantum theory were in no way ameliorated, but even emphasized, by the apparent contradiction between the exigencies of the general superposition principle of the wave description and the feature of individuality of the elementary atomic processes.

At the same time, Heisenberg (1925) had laid the foundation of a rational quantum mechanics, which was rapidly developed through important contributions by Born and Jordan as well as by Dirac. In this theory, a formalism is introduced, in which the kinematical and dynamical variables of classical mechanics are replaced by symbols subjected to a non-commutative algebra. Notwithstanding the renunciation of orbital pictures, Hamilton's canonical equations of mechanics are kept unaltered and

[5] N. Bohr, H. A. Kramers and J. C. Slater, *Phil. Mag.*, 47, 785, (1924).

[6] A. Einstein, *Berl. Ber.*, (1924), 261, and (1925), 3 and 18.

Planck's constant enters only in the rules of commutation

$$qp - pq = \sqrt{-1}\,\frac{h}{2\pi} \qquad (2)$$

holding for any set of conjugate variables q and p. Through a representation of the symbols by matrices with elements referring to transitions between stationary states, a quantitative formulation of the correspondence principle became for the first time possible. It may here be recalled that an important preliminary step towards this goal was reached through the establishment, especially by contributions of Kramers, of a quantum theory of dispersion making basic use of Einstein's general rules for the probability of the occurrence of absorption and emission processes.

This formalism of quantum mechanics was soon proved by Schrödinger to give results identical with those obtainable by the mathematically often more convenient methods of wave theory, and in the following years general methods were gradually established for an essentially statistical description of atomic processes combining the features of individuality and the requirements of the superposition principle, equally characteristic of quantum theory. Among the many advances in this period, it may especially be mentioned that the formalism proved capable of incorporating the exclusion principle which governs the states of systems with several electrons, and which already before the advent of quantum mechanics had been derived by Pauli from an analysis of atomic spectra. The quantitative comprehension of a vast amount of empirical evidence could leave no doubt as to the fertility and adequacy of the quantum-mechanical formalism, but its abstract character gave rise to a widespread feeling of uneasiness. An elucidation of the situation should, indeed, demand a thorough examination of the very observational problem in atomic physics.

This phase of the development was, as is well known, initiated in 1927 by Heisenberg,[7] who pointed out that the knowledge obtainable of the state of an atomic system will always involve a peculiar "indeterminacy." Thus, any measurement of the position of an electron by means of some device,

[7] W. Heisenberg, *Zs. f. Phys.*, 43, 172, (1927).

like a microscope, making use of high frequency radiation, will, according to the fundamental relations (1), be connected with a momentum exchange between the electron and the measuring agency, which is the greater the more accurate a position measurement is attempted. In comparing such considerations with the exigencies of the quantum-mechanical formalism, Heisenberg called attention to the fact that the commutation rule (2) imposes a reciprocal limitation on the fixation of two conjugate variables, q and p, expressed by the relation

$$\Delta q \cdot \Delta p \approx h, \tag{3}$$

where Δq and Δp are suitably defined latitudes in the determination of these variables. In pointing to the intimate connection between the statistical description in quantum mechanics and the actual possibilities of measurement, this so-called indeterminacy relation is, as Heisenberg showed, most important for the elucidation of the paradoxes involved in the attempts of analyzing quantum effects with reference to customary physical pictures.

The new progress in atomic physics was commented upon from various sides at the International Physical Congress held in September 1927, at Como in commemoration of Volta. In a lecture on that occasion,[8] I advocated a point of view conveniently termed "complementarity," suited to embrace the characteristic features of individuality of quantum phenomena, and at the same time to clarify the peculiar aspects of the observational problem in this field of experience. For this purpose, it is decisive to recognize that, *however far the phenomena transcend the scope of classical physical explanation, the account of all evidence must be expressed in classical terms.* The argument is simply that by the word "experiment" we refer to a situation where we can tell others what we have done and what we have learned and that, therefore, the account of the experimental arrangement and of the results of the observations must be expressed in unambiguous language with suitable application of the terminology of classical physics.

This crucial point, which was to become a main theme of the

[8] Atti del Congresso Internazionale dei Fisici, Como, Settembre 1927 (reprinted in *Nature*, *121*, 78 and 580, 1928).

discussions reported in the following, implies the *impossibility of any sharp separation between the behaviour of atomic objects and the interaction with the measuring instruments which serve to define the conditions under which the phenomena appear.* In fact, the individuality of the typical quantum effects finds its proper expression in the circumstance that any attempt of subdividing the phenomena will demand a change in the experimental arrangement introducing new possibilities of interaction between objects and measuring instruments which in principle cannot be controlled. Consequently, evidence obtained under different experimental conditions cannot be comprehended within a single picture, but must be regarded as *complementary* in the sense that only the totality of the phenomena exhausts the possible information about the objects.

Under these circumstances an essential element of ambiguity is involved in ascribing conventional physical attributes to atomic objects, as is at once evident in the dilemma regarding the corpuscular and wave properties of electrons and photons, where we have to do with contrasting pictures, each referring to an essential aspect of empirical evidence. An illustrative example, of how the apparent paradoxes are removed by an examination of the experimental conditions under which the complementary phenomena appear, is also given by the Compton effect, the consistent description of which at first had presented us with such acute difficulties. Thus, any arrangement suited to study the exchange of energy and momentum between the electron and the photon must involve a latitude in the space-time description of the interaction sufficient for the definition of wave-number and frequency which enter into the relation (1). Conversely, any attempt of locating the collision between the photon and the electron more accurately would, on account of the unavoidable interaction with the fixed scales and clocks defining the space-time reference frame, exclude all closer account as regards the balance of momentum and energy.

As stressed in the lecture, an adequate tool for a complementary way of description is offered precisely by the quantum-mechanical formalism which represents a purely symbolic scheme permitting only predictions, on lines of the correspondence principle, as to results obtainable under conditions specified by means of classical concepts. It must here be remembered that even in the indeterminacy relation (3) we are dealing with an implication of the formalism which defies unambiguous expression in words suited to describe classical physical pictures. Thus, a sentence like "we cannot know both the momentum and the position of an atomic object" raises at once questions as to the physical reality of two such attributes of the object, which can be answered only by referring to the conditions for the unambiguous use of space-time concepts, on the one hand, and dynamical conservation laws, on the other hand. While the combination of these concepts into a single picture of a causal chain of events is the essence of classical mechanics, room for regularities beyond the grasp of such a description is just afforded by the circumstance that the study of the complementary phenomena demands mutually exclusive experimental arrangements.

The necessity, in atomic physics, of a renewed examination of the foundation for the unambiguous use of elementary physical ideas recalls in some way the situation that led Einstein to his original revision on the basis of all application of space-time concepts which, by its emphasis on the primordial importance of the observational problem, has lent such unity to our world picture. Notwithstanding all novelty of approach, causal description is upheld in relativity theory within any given frame of reference, but in quantum theory the uncontrollable interaction between the objects and the measuring instruments forces us to a renunciation even in such respect. This recognition, however, in no way points to any limitation of the scope of the quantum-mechanical description, and the trend of the whole argumentation presented in the Como lecture was to show that the viewpoint of complementarity may be regarded as a rational generalization of the very ideal of causality.

At the general discussion in Como, we all missed the presence of Einstein, but soon after, in October 1927, I had the opportunity to meet him in Brussels at the Fifth Physical Conference of the Solvay Institute, which was devoted to the theme "Electrons and Photons." At the Solvay meetings, Einstein had from their beginning been a most prominent figure, and several of us came to the conference with great anticipations to learn his reaction to the latest stage of the development which, to our view, went far in clarifying the problems which he had himself from the outset elicited so ingeniously. During the discussions, where the whole subject was reviewed by contributions from many sides and where also the arguments mentioned in the preceding pages were again presented, Einstein expressed, however, a deep concern over the extent to which causal account in space and time was abandoned in quantum mechanics.

To illustrate his attitude, Einstein referred at one of the sessions[9] to the simple example, illustrated by Fig. 1, of a particle (electron or photon) penetrating through a hole or a narrow slit in a diaphragm placed at some distance before a photographic plate. On account of the diffraction of the wave con-

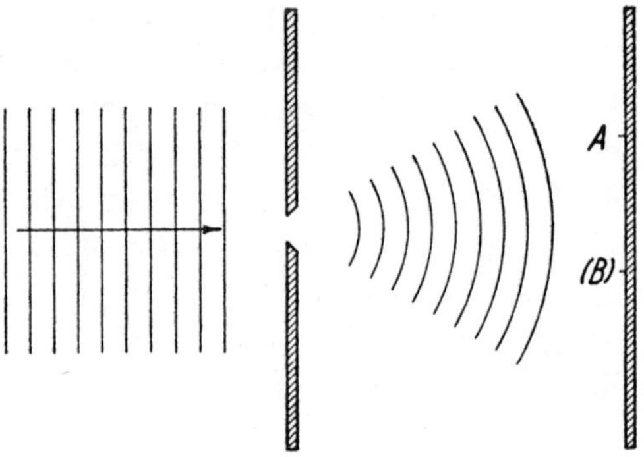

FIG. 1

nected with the motion of the particle and indicated in the figure by the thin lines, it is under such conditions not possible to predict with certainty at what point the electron will arrive at the photographic plate, but only to calculate the probability that, in an experiment, the electron will be found within any given region of the plate. The apparent difficulty, in this description, which Einstein felt so acutely, is the fact that, if in the experiment the electron is recorded at one point A of the plate,

[9] Institut International de Physique Solvay, *Rapport et discussions* du 5[e] Conseil, Paris 1928, 253ff.

then it is out of the question of ever observing an effect of this electron at another point (B), although the laws of ordinary wave propagation offer no room for a correlation between two such events.

Einstein's attitude gave rise to ardent discussions within a small circle, in which Ehrenfest, who through the years had been a close friend of us both, took part in a most active and helpful way. Surely, we all recognized that, in the above example, the situation presents no analogue to the application of statistics in dealing with complicated mechanical systems, but rather recalled the background for Einstein's own early conclusions about the unidirection of individual radiation effects which contrasts so strongly with a simple wave picture (cf. p. 205). The discussions, however, centered on the question of whether the quantum-mechanical description exhausted the possibilities of accounting for observable phenomena or, as Einstein maintained, the analysis could be carried further and, especially, of whether a fuller description of the phenomena could be obtained by bringing into consideration the detailed balance of energy and momentum in individual processes.

To explain the trend of Einstein's arguments, it may be illustrative here to consider some simple features of the momentum and energy balance in connection with the location of a particle in space and time. For this purpose, we shall examine the simple case of a particle penetrating through a hole in a diaphragm without or with a shutter to open and close the hole, as indicated in Figs. 2a and 2b, respectively. The equidistant parallel lines to the left in the figures indicate the train of plane waves corresponding to the state of motion of a particle which, before reaching the diaphragm, has a momentum P related to the wave-number σ by the second of equations (1). In accordance with the diffraction of the waves when passing through the hole, the state of motion of the particle to the right of the diaphragm is represented by a spherical wave train with a suitably defined angular aperture ϑ and, in case of Fig. 2b, also with a limited radial extension. Consequently, the description of this state involves a certain latitude Δp in the momentum component of the particle parallel to the diaphragm and, in the case of a diaphragm with a shutter, an additional latitude ΔE of the kinetic energy.

Since a measure for the latitude Δq in location of the particle in the plane of the diaphragm is given by the radius a of the hole, and since $\vartheta \approx (1/\sigma a)$, we get, using (1), just $\Delta p \approx \vartheta P \approx (h/\Delta q)$, in accordance with the indeterminacy relation (3). This result could, of course, also be obtained directly by noticing that, due to the limited extension of the wave-field at the place of the slit, the component of the wave-number parallel to the plane of the diaphragm will involve a latitude $\Delta\sigma \approx (1/a) \approx (1/\Delta q)$. Similarly, the spread of the frequencies

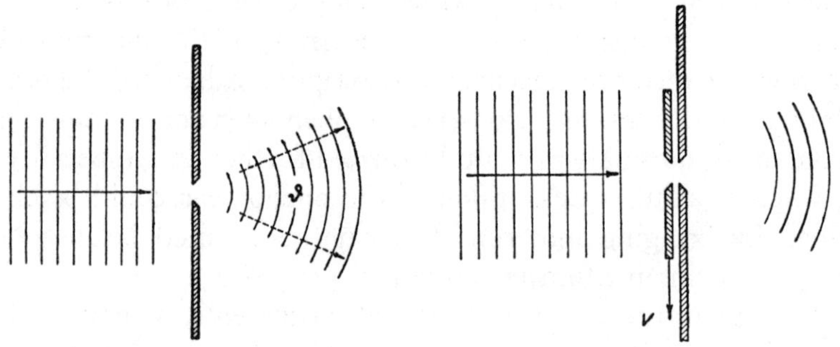

FIG. 2a FIG. 2b

of the harmonic components in the limited wave-train in Fig. 2b is evidently $\Delta\nu \approx (1/\Delta t)$, where Δt is the time interval during which the shutter leaves the hole open and, thus, represents the latitude in time of the passage of the particle through the diaphragm. From (1), we therefore get

$$\Delta E \cdot \Delta t \approx h, \tag{4}$$

again in accordance with the relation (3) for the two conjugated variables E and t.

From the point of view of the laws of conservation, the origin of such latitudes entering into the description of the state of the particle after passing through the hole may be traced to the possibilities of momentum and energy exchange with the diaphragm or the shutter. In the reference system considered in Figs. 2a and 2b, the velocity of the diaphragm may be disregarded and only a change of momentum Δp between the particle and the diaphragm needs to be taken into consideration. The shutter, however, which leaves the hole opened during the time Δt, moves with a considerable velocity $v \approx (a/\Delta t)$, and a momentum transfer Δp involves therefore an energy exchange with the particle, amounting to $v\Delta p \approx (1/\Delta t)\ \Delta q\ \Delta p \approx (h/\Delta t)$, being just of the same order of magnitude as the latitude ΔE given by (4) and, thus, allowing for momentum and energy balance.

The problem raised by Einstein was now to what extent a control of the momentum and energy transfer, involved in a location of the particle in space and time, can be used for a further specification of the state of the particle after passing through the hole. Here, it must be taken into consideration that the position and the motion of the diaphragm and the shutter have so far been assumed to be accurately co-ordinated with the space-time reference frame. This assumption implies, in the description of the state of these bodies, an essential latitude as to their momentum and energy which need not, of course, noticeably affect the velocities, if the diaphragm and the shutter are sufficiently heavy. However, as soon as we want to know the momentum and energy of these parts of the measuring arrangement with an accuracy sufficient to control the momentum and energy exchange with the particle under investigation, we shall, in accordance with the general indeterminacy relations, lose the possibility of their accurate location in space and time. We have, therefore, to examine how far this circumstance will affect the intended use of the whole arrangement and, as we shall see, this crucial point clearly brings out the complementary character of the phenomena.

Returning for a moment to the case of the simple arrangement indicated in Fig. 1, it has so far not been specified to what use it is intended. In fact, it is only on the assumption that the diaphragm and the plate have well-defined positions in space that it is impossible, within the frame of the quantum-mechanical formalism, to make more detailed predictions as to the point of the photographic plate where the particle will be recorded. If, however, we admit a sufficiently large latitude in the knowledge of the position of the diaphragm it should, in principle, be possible to control the momentum transfer to the diaphragm and, thus, to make more detailed predictions as to the direction of the electron path from the hole to the recording point. As regards the quantum-mechanical description, we have to deal here with a two-body system consisting of the diaphragm as well as of the particle, and it is just with an explicit application of conservation laws to such a system that we are concerned in the Compton effect where, for instance, the observation of the recoil of the electron by means of a cloud chamber allows us to predict in what direction the scattered photon will eventually be observed.

The importance of considerations of this kind was, in the course of the discussions, most interestingly illuminated by the examination of an arrangement where between the diaphragm with the slit and the photographic plate is inserted another

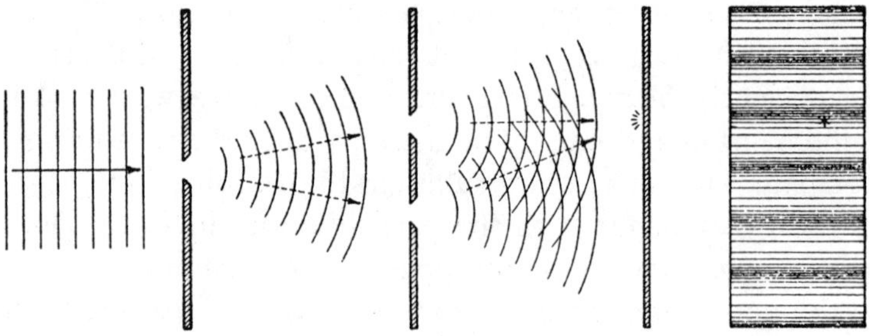

FIG. 3

diaphragm with two parallel slits, as is shown in Fig. 3. If a parallel beam of electrons (or photons) falls from the left on the first diaphragm, we shall, under usual conditions, observe on the plate an interference pattern indicated by the shading of the photographic plate shown in front view to the right of the figure. With intense beams, this pattern is built up by the accumulation of a large number of individual processes, each giving rise to a small spot on the photographic plate, and the distribution of these spots follows a simple law derivable from the wave analysis. The same distribution should also be found in the statistical account of many experiments performed with beams so faint that in a single exposure only one electron (or photon) will arrive at the photographic plate at some spot shown in the figure as a small star. Since, now, as indicated by the broken arrows, the momentum transferred to the first diaphragm ought to be different if the electron was assumed to pass through the upper or the lower slit in the second diaphragm, Einstein suggested that a control of the momentum transfer would permit a closer analysis of the phenomenon and, in particular, to decide through which of the two slits the electron had passed before arriving at the plate.

A closer examination showed, however, that the suggested control of the momentum transfer would involve a latitude in the knowledge of the position of the diaphragm which would exclude the appearance of the interference phenomena in question. In fact, if ω is the small angle between the conjectured paths of a particle passing through the upper or the lower slit, the difference of momentum transfer in these two cases will, according to (1), be equal to $h\sigma\omega$ and any control of the momentum of the diaphragm with an accuracy sufficient to measure this difference will, due to the indeterminacy relation, involve a minimum latitude of the position of the diaphragm, comparable with $1/\sigma\omega$. If, as in the figure, the diaphragm with the two slits is placed in the middle between the first diaphragm and the photographic plate, it will be seen that the number of fringes per unit length will be just equal to $\sigma\omega$ and, since an uncertainty in the position of the first diaphragm of the amount of $1/\sigma\omega$ will cause an equal uncertainty in the positions of the fringes, it follows that no interference effect can appear. The same result is easily shown to hold for any other placing of the second diaphragm between the first diaphragm and the plate, and would also be obtained if, instead of the first diaphragm, another of these three bodies were used for the control, for the purpose suggested, of the momentum transfer.

This point is of great logical consequence, since it is only the circumstance that we are presented with a choice of *either* tracing the path of a particle *or* observing interference effects, which allows us to escape from the paradoxical necessity of concluding that the behaviour of an electron or a photon should depend on the presence of a slit in the diaphragm through which it could be proved not to pass. We have here to do with a typical example of how the complementary phenomena appear under mutually exclusive experimental arrangements (cf. p. 210) and are just faced with the impossibility, in the analysis of quantum effects, of drawing any sharp separation between an independent behaviour of atomic objects and their interaction with the measuring instruments which serve to define the conditions under which the phenomena occur.

Our talks about the attitude to be taken in face of a novel situation as regards analysis and synthesis of experience touched naturally on many aspects of philosophical thinking, but, in spite of all divergencies of approach and opinion, a most humorous spirit animated the discussions. On his side, Einstein mockingly asked us whether we could really believe that the providential authorities took recourse to dice-playing (". . . *ob der liebe Gott würfelt*"), to which I replied by pointing at the great caution, already called for by ancient thinkers, in ascribing attributes to Providence in every-day language. I remember also how at the peak of the discussion Ehrenfest, in his affectionate manner of teasing his friends, jokingly hinted at the apparent similarity between Einstein's attitude and that of the opponents of relativity theory; but instantly Ehrenfest added that he would not be able to find relief in his own mind before concord with Einstein was reached.

Einstein's concern and criticism provided a most valuable incentive for us all to reexamine the various aspects of the situation as regards the description of atomic phenomena. To me it was a welcome stimulus to clarify still further the rôle played by the measuring instruments and, in order to bring into strong relief the mutually exclusive character of the experimental conditions under which the complementary phenomena appear, I tried in those days to sketch various apparatus in a pseudo-realistic style of which the following figures are examples. Thus, for the study of an interference phenomenon of the type indicated in Fig. 3, it suggests itself to use an experimental arrangement like that shown in Fig. 4, where the solid parts of the apparatus, serving as diaphragms and plate-holder, are

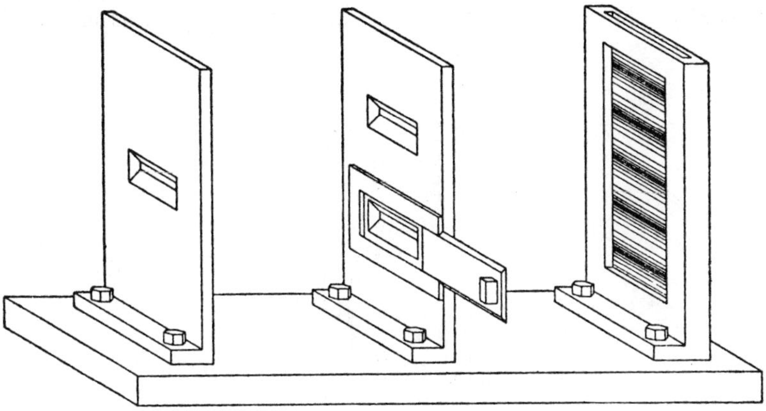

FIG. 4

firmly bolted to a common support. In such an arrangement, where the knowledge of the relative positions of the diaphragms and the photographic plate is secured by a rigid connection, it is obviously impossible to control the momentum exchanged between the particle and the separate parts of the apparatus. The only way in which, in such an arrangement, we could insure that the particle passed through one of the slits in the second diaphragm is to cover the other slit by a lid, as indicated in the figure; but if the slit is covered, there is of course no question of any interference phenomenon, and on the plate we shall simply observe a continuous distribution as in the case of the single fixed diaphragm in Fig. 1.

In the study of phenomena in the account of which we are dealing with detailed momentum balance, certain parts of the whole device must naturally be given the freedom to move independently of others. Such an apparatus is sketched in Fig. 5, where a diaphragm with a slit is suspended by weak springs from a solid yoke bolted to the support on which also other immobile parts of the arrangement are to be fastened. The scale on the diaphragm together with the pointer on the bearings of the yoke refer to such study of the motion of the diaphragm, as may be required for an estimate of the momentum transferred to it, permitting one to draw conclusions as to the deflection suffered by the particle in passing through the slit. Since, however, any reading of the scale, in whatever way performed, will

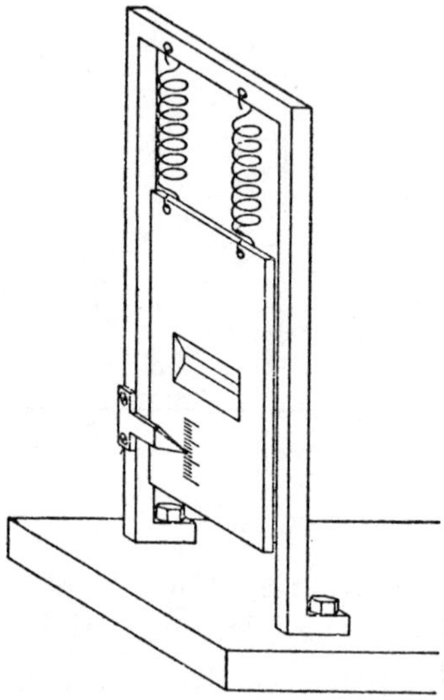

FIG. 5

involve an uncontrollable change in the momentum of the diaphragm, there will always be, in conformity with the indeterminacy principle, a reciprocal relationship between our knowledge of the position of the slit and the accuracy of the momentum control.

In the same semi-serious style, Fig. 6 represents a part of an arrangement suited for the study of phenomena which, in contrast to those just discussed, involve time co-ordination explicitly. It consists of a shutter rigidly connected with a robust clock resting on the support which carries a diaphragm and on which further parts of similar character, regulated by the same clock-work or by other clocks standardized relatively to it, are also to be fixed. The special aim of the figure is to underline that a clock is a piece of machinery, the working of which can completely be accounted for by ordinary mechanics and will be affected neither by reading of the position of its hands nor by the interaction between its accessories and an atomic particle. In securing the opening of the hole at a definite moment, an apparatus of this type might, for instance, be used for an accurate measurement of the time an electron or a photon takes to come from the diaphragm to some other place, but evidently, it would leave no possibility of controlling the energy transfer to

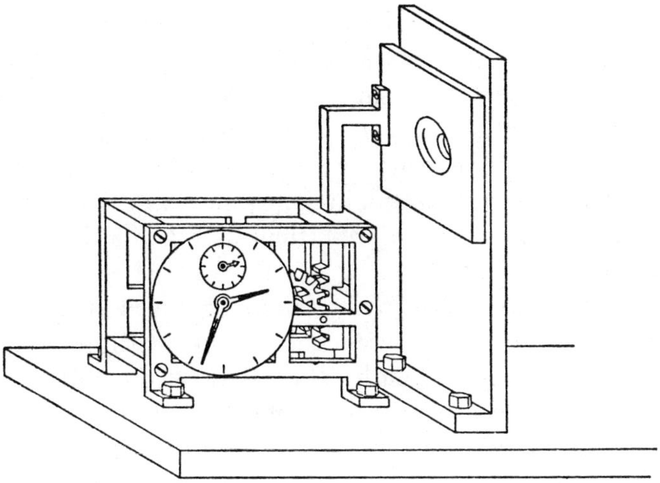

FIG. 6

the shutter with the aim of drawing conclusions as to the energy of the particle which has passed through the diaphragm. If we are interested in such conclusions we must, of course, use an arrangement where the shutter devices can no longer serve as accurate clocks, but where the knowledge of the moment when the hole in the diaphragm is open involves a latitude connected with the accuracy of the energy measurement by the general relation (4).

The contemplation of such more or less practical arrangements and their more or less fictitious use proved most instructive in directing attention to essential features of the problems. The main point here is the distinction between the *objects* under investigation and the *measuring instruments* which serve to define, in classical terms, the conditions under which the phenomena appear. Incidentally, we may remark that, for the illustration of the preceding considerations, it is not relevant that experiments involving an accurate control of the momentum or energy transfer from atomic particles to heavy bodies like diaphragms and shutters would be very difficult to perform, if practicable at all. It is only decisive that, in contrast to the proper measuring instruments, these bodies together with the particles would in such a case constitute the system to which the quantum-mechanical formalism has to be applied. As regards the specification of the conditions for any well-defined application of the formalism, it is moreover essential that the *whole experimental arrangement* be taken into account. In fact, the introduction of any further piece of apparatus, like a mirror, in the way of a particle might imply new interference effects essentially influencing the predictions as regards the results to be eventually recorded.

The extent to which renunciation of the visualization of atomic phenomena is imposed upon us by the impossibility of their subdivision is strikingly illustrated by the following example to which Einstein very early called attention and often has reverted. If a semi-reflecting mirror is placed in the way of a photon, leaving two possibilities for its direction of propagation, the photon may either be recorded on one, and only one, of two photographic plates situated at great distances in the two directions in question, or else we may, by replacing the plates by mirrors, observe effects exhibiting an interference between the two reflected wave-trains. In any attempt of a pictorial representation of the behaviour of the photon we would, thus, meet with the difficulty: to be obliged to say, on the one hand, that the photon always chooses *one* of the two ways and, on the other hand, that it behaves as if it had passed *both* ways.

It is just arguments of this kind which recall the impossibility of subdividing quantum phenomena and reveal the ambiguity in ascribing customary physical attributes to atomic objects. In particular, it must be realized that—besides in the account of the placing and timing of the instruments forming the experimental arrangement—all unambiguous use of space-time concepts in the description of atomic phenomena is confined to the recording of observations which refer to marks on a photographic plate or to similar practically irreversible amplification effects like the building of a water drop around an ion in a cloud-chamber. Although, of course, the existence of the quantum of action is ultimately responsible for the properties of the materials of which the measuring instruments are built and on which the functioning of the recording devices depends, this circumstance is not relevant for the problems of the adequacy and completeness of the quantum-mechanical description in its aspects here discussed.

These problems were instructively commented upon from different sides at the Solvay meeting,[10] in the same session where Einstein raised his general objections. On that occasion an interesting discussion arose also about how to speak of the appearance of phenomena for which only predictions of statistical character can be made. The question was whether, as to the occurrence of individual effects, we should adopt a terminology proposed by Dirac, that we were concerned with a choice on the part of "nature" or, as suggested by Heisenberg, we should say that we have to do with a choice on the part of the "observer" constructing the measuring instruments and reading their recording. Any such terminology would, however, appear dubious since, on the one hand, it is hardly reasonable to endow nature with volition in the ordinary sense, while, on the other hand, it is certainly not possible for the observer to influence the events which may appear under the conditions he has arranged. To my mind, there is no other alternative than to admit that, in this field of experience, we are dealing with individual phenomena and that our possibilities of handling the measuring instruments allow us only to make a choice between the different complementary types of phenomena we want to study.

The epistemological problems touched upon here were more explicitly dealt with in my contribution to the issue of *Naturwissenschaften* in celebration of Planck's 70th birthday in 1929. In this article, a comparison was also made between the lesson derived from the discovery of the universal quantum of action

[10] *Ibid.*, 248ff.

and the development which has followed the discovery of the finite velocity of light and which, through Einstein's pioneer work, has so greatly clarified basic principles of natural philosophy. In relativity theory, the emphasis on the dependence of all phenomena on the reference frame opened quite new ways of tracing general physical laws of unparalleled scope. In quantum theory, it was argued, the logical comprehension of hitherto unsuspected fundamental regularities governing atomic phenomena has demanded the recognition that no sharp separation can be made between an independent behaviour of the objects and their interaction with the measuring instruments which define the reference frame.

In this respect, quantum theory presents us with a novel situation in physical science, but attention was called to the very close analogy with the situation as regards analysis and synthesis of experience, which we meet in many other fields of human knowledge and interest. As is well known, many of the difficulties in psychology originate in the different placing of the separation lines between object and subject in the analysis of various aspects of psychical experience. Actually, words like "thoughts" and "sentiments," equally indispensable to illustrate the variety and scope of conscious life, are used in a similar complementary way as are space-time co-ordination and dynamical conservation laws in atomic physics. A precise formulation of such analogies involves, of course, intricacies of terminology, and the writer's position is perhaps best indicated in a passage in the article, hinting at the mutually exclusive relationship which will always exist between the practical use of any word and attempts at its strict definition. The principal aim, however, of these considerations, which were not least inspired by the hope of influencing Einstein's attitude, was to point to perspectives of bringing general epistemological problems into relief by means of a lesson derived from the study of new, but fundamentally simple physical experience.

At the next meeting with Einstein at the Solvay Conference in 1930, our discussions took quite a dramatic turn. As an objection to the view that a control of the interchange of momentum and energy between the objects and the measuring instruments was excluded if these instruments should serve their purpose of defining the space-time frame of the phenomena, Einstein brought forward the argument that such control should be possible when the exigencies of relativity theory were taken into consideration. In particular, the general relationship between energy and mass, expressed in Einstein's famous formula

$$E = mc^2 \tag{5}$$

should allow, by means of simple weighing, to measure the total energy of any system and, thus, in principle to control the energy transferred to it when it interacts with an atomic object.

As an arrangement suited for such purpose, Einstein proposed the device indicated in Fig. 7, consisting of a box with

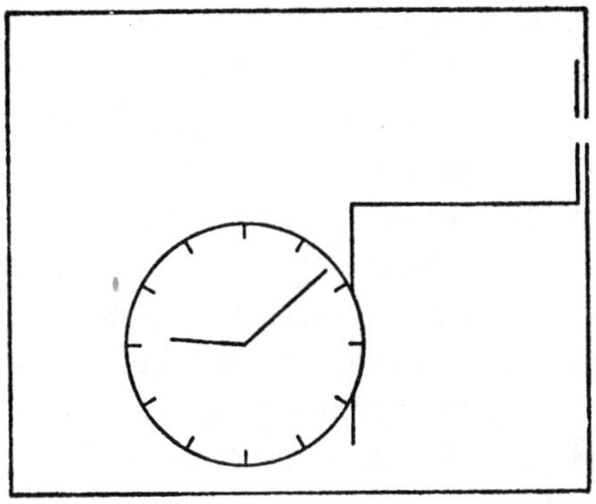

FIG. 7

a hole in its side, which could be opened or closed by a shutter moved by means of a clock-work within the box. If, in the beginning, the box contained a certain amount of radiation and the clock was set to open the shutter for a very short interval at a chosen time, it could be achieved that a single photon was released through the hole at a moment known with as great accuracy as desired. Moreover, it would apparently also be possible, by weighing the whole box before and after this event, to measure the energy of the photon with any accuracy wanted, in definite contradiction to the reciprocal indeterminacy of time and energy quantities in quantum mechanics.

This argument amounted to a serious challenge and gave rise to a thorough examination of the whole problem. At the outcome of the discussion, to which Einstein himself contributed effectively, it became clear, however, that this argument could not be upheld. In fact, in the consideration of the problem, it was found necessary to look closer into the consequences of the identification of inertial and gravitational mass implied in the application of relation (5). Especially, it was essential to take into account the relationship between the rate of a clock and its position in a gravitational field—well known from the red-shift of the lines in the sun's spectrum—following from Einstein's principle of equivalence between gravity effects and the phenomena observed in accelerated reference frames.

Our discussion concentrated on the possible application of an apparatus incorporating Einstein's device and drawn in Fig. 8 in the same pseudo-realistic style as some of the preceding figures. The box, of which a section is shown in order to exhibit its interior, is suspended in a spring-balance and is furnished with a pointer to read its position on a scale fixed to the balance support. The weighing of the box may thus be performed with any given accuracy Δm by adjusting the balance to its zero position by means of suitable loads. The essential point is now that any determination of this position with a given accuracy Δq will involve a minimum latitude Δp in the control of the momentum of the box connected with Δq by the relation (3). This latitude must obviously again be smaller than the total impulse which, during the whole interval T of the balancing procedure, can be given by the gravitational field to a body with a mass Δm, or

$$\Delta p \approx \frac{h}{\Delta q} < T \cdot g \cdot \Delta m, \tag{6}$$

where g is the gravity constant. The greater the accuracy of the reading q of the pointer, the longer must, consequently, be the balancing interval T, if a given accuracy Δm of the weighing of the box with its content shall be obtained.

Now, according to general relativity theory, a clock, when displaced in the direction of the gravitational force by an amount of Δq, will change its rate in such a way that its reading

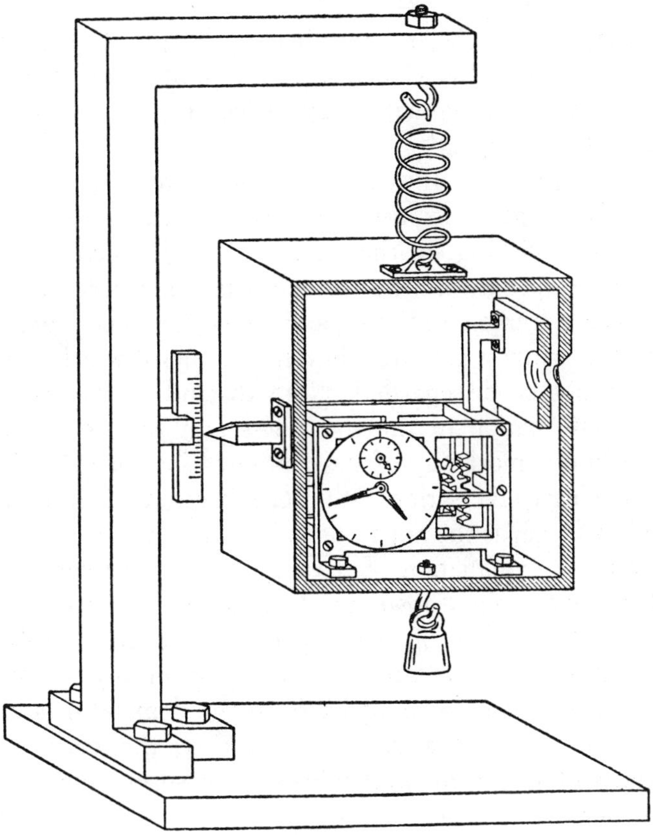

FIG. 8

in the course of a time interval T will differ by an amount ΔT given by the relation

$$\frac{\Delta T}{T} = \frac{1}{c^2} g \Delta q. \qquad (7)$$

By comparing (6) and (7) we see, therefore, that after the weighing procedure there will in our knowledge of the adjustment of the clock be a latitude

$$\Delta T > \frac{h}{c^2 \Delta m}.$$

Together with the formula (5), this relation again leads to

$$\Delta T \cdot \Delta E > h,$$

in accordance with the indeterminacy principle. Consequently, a use of the apparatus as a means of accurately measuring the energy of the photon will prevent us from controlling the moment of its escape.

The discussion, so illustrative of the power and consistency of relativistic arguments, thus emphasized once more the necessity of distinguishing, in the study of atomic phenomena, between the proper measuring instruments which serve to define the reference frame and those parts which are to be regarded as objects under investigation and in the account of which quantum effects cannot be disregarded. Notwithstanding the most suggestive confirmation of the soundness and wide scope of the quantum-mechanical way of description, Einstein nevertheless, in a following conversation with me, expressed a feeling of disquietude as regards the apparent lack of firmly laid down principles for the explanation of nature, in which all could agree. From my viewpoint, however, I could only answer that, in dealing with the task of bringing order into an entirely new field of experience, we could hardly trust in any accustomed principles, however broad, apart from the demand of avoiding logical inconsistencies and, in this respect, the mathematical formalism of quantum mechanics should surely meet all requirements.

The Solvay meeting in 1930 was the last occasion where, in common discussions with Einstein, we could benefit from the stimulating and mediating influence of Ehrenfest, but shortly before his deeply deplored death in 1933 he told me that Einstein was far from satisfied and with his usual acuteness had discerned new aspects of the situation which strengthened his critical attitude. In fact, by further examining the possibilities for the application of a balance arrangement, Einstein had perceived alternative procedures which, even if they did not allow the use he originally intended, might seem to enhance the paradoxes beyond the possibilities of logical solution. Thus, Einstein had pointed out that, after a preliminary weighing of the box with the clock and the subsequent escape of the photon, one was still left with the choice of either repeating the weighing or opening the box and comparing the reading of the clock with the standard time scale. Consequently, we are at this stage still free to choose whether we want to draw conclusions either about the energy of the photon or about the moment when it left the box. Without in any way interfering with the photon between its escape and its later interaction with other suitable measuring instruments, we are, thus, able to make accurate predictions pertaining *either* to the moment of its arrival *or* to the amount of energy liberated by its absorption. Since, however, according to the quantum-mechanical formalism, the specification of the state of an isolated particle cannot involve both a well-defined connection with the time scale and an accurate fixation of the energy, it might thus appear as if this formalism did not offer the means of an adequate description.

Once more Einstein's searching spirit had elicited a peculiar aspect of the situation in quantum theory, which in a most striking manner illustrated how far we have here transcended customary explanation of natural phenomena. Still, I could not agree with the trend of his remarks as reported by Ehrenfest. In my opinion, there could be no other way to deem a logically consistent mathematical formalism as inadequate than by demonstrating the departure of its consequences from experience or by proving that its predictions did not exhaust the possibilities of observation, and Einstein's argumentation could be directed to neither of these ends. In fact, we must realize that in the problem in question we are not dealing with a *single* specified experimental arrangement, but are referring to *two* different, mutually exclusive arrangements. In the one, the balance together with another piece of apparatus like a spectrometer is used for the study of the energy transfer by a photon; in the other, a shutter regulated by a standardized clock together with another apparatus of similar kind, accurately timed relatively to the clock, is used for the study of the time of propagation of a photon over a given distance. In both these cases, as also assumed by Einstein, the observable effects are expected to be in complete conformity with the predictions of the theory.

The problem again emphasizes the necessity of considering the *whole* experimental arrangement, the specification of which is imperative for any well-defined application of the quantum-mechanical formalism. Incidentally, it may be added that paradoxes of the kind contemplated by Einstein are encountered also in such simple arrangements as sketched in Fig. 5. In fact, after a preliminary measurement of the momentum of the diaphragm, we are in principle offered the choice, when an electron or photon has passed through the slit, either to repeat the momentum measurement or to control the position of the diaphragm and, thus, to make predictions pertaining to alternative subsequent observations. It may also be added that it obviously can make no difference as regards observable effects obtainable by a definite experimental arrangement, whether our plans of constructing or handling the instruments are fixed beforehand or whether we prefer to postpone the completion of our planning until a later moment when the particle is already on its way from one instrument to another.

In the quantum-mechanical description our freedom of constructing and handling the experimental arrangement finds its proper expression in the possibility of choosing the classically defined parameters entering in any proper application of the formalism. Indeed, in all such respects quantum mechanics exhibits a correspondence with the state of affairs familiar from classical physics, which is as close as possible when considering the individuality inherent in the quantum phenomena. Just in helping to bring out this point so clearly, Einstein's concern had therefore again been a most welcome incitement to explore the essential aspects of the situation.

The next Solvay meeting in 1933 was devoted to the problems of the structure and properties of atomic nuclei, in which field such great advances were made just in that period due to the experimental discoveries as well as to new fruitful applications of quantum mechanics. It need in this connection hardly be recalled that just the evidence obtained by the study of artificial nuclear transformations gave a most direct test of Einstein's fundamental law regarding the equivalence of mass and energy, which was to prove an evermore important guide for researches in nuclear physics. It may also be mentioned how Einstein's intuitive recognition of the intimate relationship between the law of radioactive transformations and the probability rules governing individual radiation effects (cf. p. 205) was confirmed by the quantum-mechanical explanation of spontaneous nuclear disintegrations. In fact, we are here dealing with a typical example of the statistical mode of description, and the complementary relationship between energy-momentum conservation and time-space co-ordination is most strikingly exhibited in the well-known paradox of particle penetration through potential barriers.

Einstein himself did not attend this meeting, which took place at a time darkened by the tragic developments in the political world which were to influence his fate so deeply and add so greatly to his burdens in the service of humanity. A few months earlier, on a visit to Princeton where Einstein was then guest of the newly founded Institute for Advanced Study to which he soon after became permanently attached, I had, however, opportunity to talk with him again about the epistemological aspects of atomic physics, but the difference between our ways of approach and expression still presented obstacles to mutual understanding. While, so far, relatively few persons had taken part in the discussions reported in this article, Einstein's critical attitude towards the views on quantum theory adhered to by many physicists was soon after brought to public attention through a paper[11] with the title "Can Quantum-Mechanical Description of Physical Reality Be Considered Complete?," published in 1935 by Einstein, Podolsky and Rosen.

The argumentation in this paper is based on a criterion which the authors express in the following sentence: "If, without in any way disturbing a system, we can predict with certainty (i.e., with probability equal to unity) the value of a physical quantity, then there exists an element of physical reality correspond-

[11] A. Einstein, B. Podolsky and N. Rosen, *Phys. Rev.*, 47, 777, (1935).

ing to this physical quantity." By an elegant exposition of the consequences of the quantum-mechanical formalism as regards the representation of a state of a system, consisting of two parts which have been in interaction for a limited time interval, it is next shown that different quantities, the fixation of which cannot be combined in the representation of one of the partial systems, can nevertheless be predicted by measurements pertaining to the other partial system. According to their criterion, the authors therefore conclude that quantum mechanics does not "provide a complete description of the physical reality," and they express their belief that it should be possible to develop a more adequate account of the phenomena.

Due to the lucidity and apparently incontestable character of the argument, the paper of Einstein, Podolsky and Rosen created a stir among physicists and has played a large rôle in general philosophical discussion. Certainly the issue is of a very subtle character and suited to emphasize how far, in quantum theory, we are beyond the reach of pictorial visualization. It will be seen, however, that we are here dealing with problems of just the same kind as those raised by Einstein in previous discussions, and, in an article which appeared a few months later,[12] I tried to show that from the point of view of complementarity the apparent inconsistencies were completely removed. The trend of the argumentation was in substance the same as that exposed in the foregoing pages, but the aim of recalling the way in which the situation was discussed at that time may be an apology for citing certain passages from my article.

Thus, after referring to the conclusions derived by Einstein, Podolsky and Rosen on the basis of their criterion, I wrote:

Such an argumentation, however, would hardly seem suited to affect the soundness of quantum-mechanical description, which is based on a coherent mathematical formalism covering automatically any procedure of measurement like that indicated. The apparent contradiction in fact discloses only an essential inadequacy of the customary viewpoint of natural philosophy for a rational account of physical phenomena of the type with which we are concerned in quantum mechanics. Indeed the *finite interaction between object and measuring agencies* conditioned

[12] N. Bohr, *Phys. Rev.*, *48*, 696, (1935).

by the very existence of the quantum of action entails—because of the impossibility of controlling the reaction of the object on the measuring instruments, if these are to serve their purpose—the necessity of a final renunciation of the classical ideal of causality and a radical revision of our attitude towards the problem of physical reality. In fact, as we shall see, a criterion of reality like that proposed by the named authors contains—however cautious its formulation may appear—an essential ambiguity when it is applied to the actual problems with which we are here concerned.

As regards the special problem treated by Einstein, Podolsky and Rosen, it was next shown that the consequences of the formalism as regards the representation of the state of a system consisting of two interacting atomic objects correspond to the simple arguments mentioned in the preceding in connection with the discussion of the experimental arrangements suited for the study of complementary phenomena. In fact, although any pair q and p, of conjugate space and momentum variables obeys the rule of non-commutative multiplication expressed by (2), and can thus only be fixed with reciprocal latitudes given by (3), the difference $q_1 - q_2$ between two space-co-ordinates referring to the constituents of the system will commute with the sum $p_1 + p_2$ of the corresponding momentum components, as follows directly from the commutability of q_1 with p_2 and q_2 with p_1. Both $q_1 - q_2$ and $p_1 + p_2$ can, therefore, be accurately fixed in a state of the complex system and, consequently, we can predict the values of either q_1 or p_1 if either q_2 or p_2, respectively, are determined by direct measurements. If, for the two parts of the system, we take a particle and a diaphragm, like that sketched in Fig. 5, we see that the possibilities of specifying the state of the particle by measurements on the diaphragm just correspond to the situation described on p. 220 and further discussed on p. 230, where it was mentioned that, after the particle has passed through the diaphragm, we have in principle the choice of measuring either the position of the diaphragm or its momentum and, in each case, to make predictions as to subsequent observations pertaining to the particle. As repeatedly stressed, the principal point is here that such measurements demand mutually exclusive experimental arrangements.

The argumentation of the article was summarized in the following passage:

From our point of view we now see that the wording of the above-mentioned criterion of physical reality proposed by Einstein, Podolsky, and Rosen contains an ambiguity as regards the meaning of the expression 'without in any way disturbing a system.' Of course there is in a case like that just considered no question of a mechanical disturbance of the system under investigation during the last critical stage of the measuring procedure. But even at this stage there is essentially the question of *an influence on the very conditions which define the possible types of predictions regarding the future behaviour of the system.* Since these conditions constitute an inherent element of the description of any phenomenon to which the term "physical reality" can be properly attached, we see that the argumentation of the mentioned authors does not justify their conclusion that quantum-mechanical description is essentially incomplete. On the contrary, this description, as appears from the preceding discussion, may be characterized as a rational utilization of all possibilities of unambiguous interpretation of measurements, compatible with the finite and uncontrollable interaction between the objects and the measuring instruments in the field of quantum theory. In fact, it is only the mutual exclusion of any two experimental procedures, permitting the unambiguous definition of complementary physical quantities, which provides room for new physical laws, the coexistence of which might at first sight appear irreconcilable with the basic principles of science. It is just this entirely new situation as regards the description of physical phenomena that the notion of *complementarity* aims at characterizing.

Rereading these passages, I am deeply aware of the inefficiency of expression which must have made it very difficult to appreciate the trend of the argumentation aiming to bring out the essential ambiguity involved in a reference to physical attributes of objects when dealing with phenomena where no sharp distinction can be made between the behaviour of the objects themselves and their interaction with the measuring instruments. I hope, however, that the present account of the discussions with Einstein in the foregoing years, which contributed so greatly to make us familiar with the situation in quantum physics, may give a clearer impression of the necessity of a radical revision of basic principles for physical explanation in order to restore logical order in this field of experience.

Einstein's own views at that time are presented in an article "Physics and Reality," published in 1936 in the *Journal of the Franklin Institute.*[13]." Starting from a most illuminating exposition of the gradual development of the fundamental principles in the theories of classical physics and their relation to the problem of physical reality, Einstein here argues that the quantum-mechanical description is to be considered merely as a means of accounting for the average behaviour of a large number of atomic systems and his attitude to the belief that it should offer an exhaustive description of the individual phenomena is expressed in the following words: "To believe this is logically possible without contradiction; but it is so very contrary to my scientific instinct that I cannot forego the search for a more complete conception."

Even if such an attitude might seem well-balanced in itself, it nevertheless implies a rejection of the whole argumentation exposed in the preceding, aiming to show that, in quantum mechanics, we are not dealing with an arbitrary renunciation of a more detailed analysis of atomic phenomena, but with a recognition that such an analysis is *in principle* excluded. The peculiar individuality of the quantum effects presents us, as regards the comprehension of well-defined evidence, with a novel situation unforeseen in classical physics and irreconcilable with conventional ideas suited for our orientation and adjustment to ordinary experience. It is in this respect that quantum theory has called for a renewed revision of the foundation for the unambiguous use of elementary concepts, as a further step in the development which, since the advent of relativity theory, has been so characteristic of modern science.

In the following years, the more philosophical aspects of the situation in atomic physics aroused the interest of ever larger circles and were, in particular, discussed at the Second International Congress for the Unity of Science in Copenhagen in July 1936. In a lecture on this occasion,[14] I tried especially to

[13] A. Einstein, *Journ. Frankl. Inst.*, *221*, 349, (1936).

[14] N. Bohr, *Erkenntnis*, *6*, 293, (1937), and *Philosophy of Science*, *4*, 289, (1937).

stress the analogy in epistemological respects between the limitation imposed on the causal description in atomic physics and situations met with in other fields of knowledge. A principal purpose of such parallels was to call attention to the necessity in many domains of general human interest to face problems of a similar kind as those which had arisen in quantum theory and thereby to give a more familiar background for the apparently extravagant way of expression which physicists have developed to cope with their acute difficulties.

Besides the complementary features conspicuous in psychology and already touched upon (cf. p. 224), examples of such relationships can also be traced in biology, especially as regards the comparison between mechanistic and vitalistic viewpoints. Just with respect to the observational problem, this last question had previously been the subject of an address to the International Congress on Light Therapy held in Copenhagen in 1932,[15] where it was incidentally pointed out that even the psycho-physical parallelism as envisaged by Leibniz and Spinoza has obtained a wider scope through the development of atomic physics, which forces us to an attitude towards the problem of explanation recalling ancient wisdom, that when searching for harmony in life one must never forget that in the drama of existence we are ourselves both actors and spectators.

Utterances of this kind would naturally in many minds evoke the impression of an underlying mysticism foreign to the spirit of science; at the above mentioned Congress in 1936 I therefore tried to clear up such misunderstandings and to explain that the only question was an endeavour to clarify the conditions, in each field of knowledge, for the analysis and synthesis of experience.[14] Yet, I am afraid that I had in this respect only little success in convincing my listeners, for whom the dissent among the physicists themselves was naturally a cause of scepticism as to the necessity of going so far in renouncing customary demands as regards the explanation of natural phenomena. Not least through a new discussion with Einstein in Princeton in 1937, where we did not get beyond a humourous contest con-

[15] IIe Congrès international de la Lumière, Copenhague 1932 (reprinted in *Nature*, *131*, 421 and 457, 1933).

cerning which side Spinoza would have taken if he had lived to see the development of our days, I was strongly reminded of the importance of utmost caution in all questions of terminology and dialectics.

These aspects of the situation were especially discussed at a meeting in Warsaw in 1938, arranged by the International Institute of Intellectual Co-operation of the League of Nations.[16] The preceding years had seen great progress in quantum physics due to a number of fundamental discoveries regarding the constitution and properties of atomic nuclei as well as due to important developments of the mathematical formalism taking the requirements of relativity theory into account. In the last respect, Dirac's ingenious quantum theory of the electron offered a most striking illustration of the power and fertility of the general quantum-mechanical way of description. In the phenomena of creation and annihilation of electron pairs we have in fact to do with new fundamental features of atomicity, which are intimately connected with the non-classical aspects of quantum statistics expressed in the exclusion principle, and which have demanded a still more far-reaching renunciation of explanation in terms of a pictorial representation.

Meanwhile, the discussion of the epistemological problems in atomic physics attracted as much attention as ever and, in commenting on Einstein's views as regards the incompleteness of the quantum-mechanical mode of description, I entered more directly on questions of terminology. In this connection I warned especially against phrases, often found in the physical literature, such as "disturbing of phenomena by observation" or "creating physical attributes to atomic objects by measurements." Such phrases, which may serve to remind of the apparent paradoxes in quantum theory, are at the same time apt to cause confusion, since words like "phenomena" and "observations," just as "attributes" and "measurements," are used in a way hardly compatible with common language and practical definition.

As a more appropriate way of expression I advocated the ap-

[16] *New Theories in Physics* (Paris 1938), 11.

plication of the word *phenomenon* exclusively to refer to the observations obtained under specified circumstances, including an account of the whole experimental arrangement. In such terminology, the observational problem is free of any special intricacy since, in actual experiments, all observations are expressed by unambiguous statements referring, for instance, to the registration of the point at which an electron arrives at a photographic plate. Moreover, speaking in such a way is just suited to emphasize that the appropriate physical interpretation of the symbolic quantum-mechanical formalism amounts only to predictions, of determinate or statistical character, pertaining to individual phenomena appearing under conditions defined by classical physical concepts.

Notwithstanding all differences between the physical problems which have given rise to the development of relativity theory and quantum theory, respectively, a comparison of purely logical aspects of relativistic and complementary argumentation reveals striking similarities as regards the renunciation of the absolute significance of conventional physical attributes of objects. Also, the neglect of the atomic constitution of the measuring instruments themselves, in the account of actual experience, is equally characteristic of the applications of relativity and quantum theory. Thus, the smallness of the quantum of action compared with the actions involved in usual experience, including the arranging and handling of physical apparatus, is as essential in atomic physics as is the enormous number of atoms composing the world in the general theory of relativity which, as often pointed out, demands that dimensions of apparatus for measuring angles can be made small compared with the radius of curvature of space.

In the Warsaw lecture, I commented upon the use of not directly visualizable symbolism in relativity and quantum theory in the following way:

Even the formalisms, which in both theories within their scope offer adequate means of comprehending all conceivable experience, exhibit deepgoing analogies. In fact, the astounding simplicity of the generalization of classical physical theories, which are obtained by the use of multidimensional geometry and non-commutative algebra, respectively, rests in both cases essentially on the introduction of the conventional symbol $\sqrt{-1}$. The abstract character of the formalisms concerned is indeed, on closer examination, as typical of relativity theory as it is of quantum mechanics, and it is in this respect purely a matter of tradition if the former theory is considered as a completion of classical physics rather than as a first fundamental step in the thoroughgoing revision of our conceptual means of comparing observations, which the modern development of physics has forced upon us.

It is, of course, true that in atomic physics we are confronted with a number of unsolved fundamental problems, especially as regards the intimate relationship between the elementary unit of electric charge and the universal quantum of action; but these problems are no more connected with the epistemological points here discussed than is the adequacy of relativistic argumentation with the issue of thus far unsolved problems of cosmology. Both in relativity and in quantum theory we are concerned with new aspects of scientific analysis and synthesis and, in this connection, it is interesting to note that, even in the great epoch of critical philosophy in the former century, there was only question to what extent *a priori* arguments could be given for the adequacy of space-time co-ordination and causal connection of experience, but never question of rational generalizations or inherent limitations of such categories of human thinking.

Although in more recent years I have had several occasions of meeting Einstein, the continued discussions, from which I always have received new impulses, have so far not led to a common view about the epistemological problems in atomic physics, and our opposing views are perhaps most clearly stated in a recent issue of *Dialectica*,[17] bringing a general discussion of these problems. Realizing, however, the many obstacles for mutual understanding as regards a matter where approach and background must influence everyone's attitude, I have welcomed this opportunity of a broader exposition of the development by which, to my mind, a veritable crisis in physical science has been overcome. The lesson we have hereby received would seem to have brought us a decisive step further in the never-

[17] N. Bohr, *Dialectica*, *1*, 312 (1948).

ending struggle for harmony between content and form, and taught us once again that no content can be grasped without a formal frame and that any form, however useful it has hitherto proved, may be found to be too narrow to comprehend new experience.

Surely, in a situation like this, where it has been difficult to reach mutual understanding not only between philosophers and physicists but even between physicists of different schools, the difficulties have their root not seldom in the preference for a certain use of language suggesting itself from the different lines of approach. In the Institute in Copenhagen, where through those years a number of young physicists from various countries came together for discussions, we used, when in trouble, often to comfort ourselves with jokes, among them the old saying of the two kinds of truth. To the one kind belong statements so simple and clear that the opposite assertion obviously could not be defended. The other kind, the so-called "deep truths," are statements in which the opposite also contains deep truth. Now, the development in a new field will usually pass through stages in which chaos becomes gradually replaced by order; but it is not least in the intermediate stage where deep truth prevails that the work is really exciting and inspires the imagination to search for a firmer hold. For such endeavours of seeking the proper balance between seriousness and humour, Einstein's own personality stands as a great example and, when expressing my belief that through a singularly fruitful co-operation of a whole generation of physicists we are nearing the goal where logical order to a large extent allows us to avoid deep truth, I hope that it will be taken in his spirit and may serve as an apology for several utterances in the preceding pages.

The discussions with Einstein which have formed the theme of this article have extended over many years which have witnessed great progress in the field of atomic physics. Whether our actual meetings have been of short or long duration, they have always left a deep and lasting impression on my mind, and when writing this report I have, so-to-say, been arguing with Einstein all the time even when entering on topics apparently far removed from the special problems under debate at our meetings. As regards the account of the conversations I am, of course, aware that I am relying only on my own memory, just as I am prepared for the possibility that many features of the development of quantum theory, in which Einstein has played so large a part, may appear to himself in a different light. I trust, however, that I have not failed in conveying a proper impression of how much it has meant to me to be able to benefit from the inspiration which we all derive from every contact with Einstein.

NIELS BOHR

UNIVERSITETETS INSTITUT
FOR TEORETISK FYSIK
COPENHAGEN, DENMARK

I.2 BORN'S PROBABILISTIC INTERPRETATION

Commentary of Rosenfeld (1971a)

... Schrödinger made no secret of his intention to substitute simple classical pictures for the strange conceptions of quantum mechanics, for whose abstract character he expressed deep "aversion"; he was conscious that this last sentiment was shared by all the older generation of physicists, who had not accepted the necessity of giving up their habitual ways of thinking when dealing with phenomena on the atomic scale. Significantly, he turned towards the chevroned peers of the classical era—Lorentz, Planck, Einstein—who did not grudge him praise and encouragement, and shunned the founders of quantum mechanics. The latter, however, who had taken no notice of de Broglie, looked into Schrödinger's ideas most critically; he had indeed enforced the claim to be taken seriously by solving the wave equation for the hydrogen atom and obtaining Rydberg's formula by a calculation much simpler and more elegant than Pauli's quantum algebra. This could hardly be a fortuitous coincidence, and indeed both Pauli and Schrödinger soon discovered that the two theories so different in conception, were mathematically equivalent: for instance, the quantum amplitude governing the transition between two stationary states could be readily computed with the help of the wave amplitudes corresponding to these states. Would this formal equivalence clinch the issue in favor of Schrödinger's contention that the proper quantal concepts can be altogether dispensed with? Far from it, Heisenberg had at once seen that this contention was untenable: Schrödinger's way of treating the charge density as a classical source of radiation would even prevent him from obtaining Planck's law for the distribution of thermal radiation. How could such an elementary, but fatal objection have escaped, not only Schrödinger himself, but above all the creators of the quantum theory of radiation, who gave him such uncritical support? Almost regretting to have ushered in ideas whose revolutionary consequences they had not foreseen, these great masters desperately clung to the "sound philosophy" of classical physics, without realizing its limitations; an emotional resistance dimmed their judgment. The pioneers of quantum mechanics, on their side, were now confronted with a new challenge: Schrödinger's wave theory of matter clearly provided another formal approach to the consistent description of atomic phenomena they were striving for, but its physical content still eluded them.

It did not last long, however, before they realized that wave mechanics was just the appropriate technique they wanted for dealing with the aperiodic phenomena intractable by the quantum algebra. This was explicitly demonstrated by Born, who showed how to treat atomic collisions by a transposition of the mathematical methods applied to the analogous classical problem of the scattering of light waves by a polarizable medium. Born's argument, moreover, embodied an essentially new feature, of decisive importance, with regard to the physical interpretation of the intensity of the matter waves. The optical analogy suggested a comparison of this intensity with that of a classical light wave, which the quantum theory of radiation interprets as the density of the statistical distribution of the associated photons. Born pointed out that the usual particle description of the atomic collision process could be maintained if one adopted a similar statistical relationship between the matter waves and the associated atomic particles. He accordingly proposed to interpret the wave intensity, not as the density of an actual distribution of matter, as Schrödinger imagined, but as a density of probability for the presence of a particle. Thus, the formal equivalence of wave mechanics and Heisenberg's quantum mechanics became physically meaningful: it established a complete harmony between the statistical meaning of the wave intensity and the statistical character of the rules of quantum algebra for the calculation of transition probabilities.

I.2 ON THE QUANTUM MECHANICS OF COLLISIONS

[Preliminary communication]†

MAX BORN

Through the investigation of collisions it is argued that quantum mechanics in the Schrödinger form allows one to describe not only stationary states but also quantum jumps.

Heisenberg's quantum mechanics has so far been applied exclusively to the calculation of stationary states and vibration amplitudes associated with transitions (I purposely avoid the word "transition probabilities"). In this connection the formalism, further developed in the meantime, seems to be well validated. However, questions of this kind deal with only one aspect of quantum theory. Beside them there shows up as equally important the question of the nature of the "transitions" themselves. On this point opinions seem to be divided. Many assume that the problem of transitions is not encompassed by quantum mechanics in its present form, but that here new conceptual developments will be necessary. I myself, impressed with the closed character of the logical nature of quantum mechanics, came to the presumption that this theory is complete and that the problem of transitions must be contained in it. I believe that I have now succeeded in proving this.

Bohr has already directed attention to the fact that all difficulties of principle associated with the quantum approach which meet us in the emission and absorption of light by atoms also occur in the interaction of atoms at short distances and consequently in collision processes. In collisions one deals not with mysterious wave fields, but exclusively with systems of material particles, subject to the formalism of quantum mechanics. I therefore attack the problem of investigating more closely the interaction of the free particle (α-ray or electron) and an arbitrary atom and of determining whether a description of a collision is not possible within the framework of existing theory.

Of the different forms of the theory only Schrödinger's has proved suitable for this process, and exactly for this reason I might regard it as the deepest formulation of the quantum laws. The course of my reasoning is the following.

If one wishes to calculate quantum mechanically the interaction of two systems,

† This report was originally intended for *die Naturwissenschaften*, but could not be accepted there for lack of space. I hope that its publication in this journal [*Zeitschrift für Physik*] does not seem out of place [M.B.].

Originally published under the title, "Zur Quantenmechanik der Stossvorgänge," *Zeitschrift für Physik*, *37*, 863–67 (1926); reprinted in *Dokumente der Naturwissenschaft*, *1*, 48–52 (1962) and in M. Born (1963); translation into English by J.A.W. and W.H.Z., 1981.

then, as is well known, one cannot, as in classical mechanics, pick out a state of the one system and determine how this is influenced by a state of the other system, since all states of both systems are coupled in a complicated way. This is true also in an aperiodic process, such as a collision, where a particle, let us say an electron, comes in from infinity and then goes off to infinity. There is no escape from the conclusion that, as well before as after collision, when the electron is far enough away and the coupling is small enough, a definite state must be specifiable for the atom and likewise a definite rectilinear motion for the electron. The problem is to formulate mathematically this asymptotic behavior of the coupled particles. I did not succeed in doing this with the matrix form of quantum mechanics, but did with the Schrödinger formulation.

According to Schrödinger, the atom in its nth quantum state is a vibration of a state function of fixed frequency W_n^0/h spread over all of space. In particular, an electron moving in a straight line is such a vibratory phenomenon which corresponds to a plane wave. When two such waves interact, a complicated vibration arises. However, one sees immediately that one can determine it through its asymptotic behavior at infinity. Indeed one has nothing more than a "diffraction problem" in which an incoming plane wave is refracted or scattered at an atom. In place of the boundary conditions which one uses in optics for the description of the diffraction diaphragm, one has here the potential energy of interaction between the atom and the electron.

The task is clear. We have to solve the Schrödinger wave equation for the system atom-plus-electron subject to the boundary condition that the solution in a preselected direction of electron space goes over asymptotically into a plane wave with exactly this direction of propagation (the arriving electron). In a thus selected solution we are further interested principally in a behavior of the "scattered" wave at infinity, for it describes the behavior of the system after the collision. We spell this out a little further. Let $\psi_1^0(q_k)$, $\psi_2^0(q_k)$, . . . be the eigenfunctions of the unperturbed atom (we assume that there is only a discrete spectrum). The unperturbed electron, in straight-line motion, corresponds to eigenfunctions $\sin(2\pi/\lambda)(\alpha x + \beta y + \gamma z + \delta)$, a continuous manifold of plane waves. Their wavelength, according to de Broglie, is connected with the energy of translation τ by the relation $\tau = h^2/(2\mu\lambda^2)$. The eigenfunction of the unperturbed state in which the electron arrives from the $+z$ direction, is thus

$$\psi_{n,\tau}^0(q_k, z) = \psi_n^0(q_k)\sin(2\pi z/\lambda).$$

Now let $V(x, y, z; q_k)$ be the potential energy of interaction of the atom and the electron. One can then show with the help of a simple perturbation calculation that there is a uniquely determined solution of the Schrödinger equation with a potential V, which goes over asymptotically for $z \to +\infty$ into the above function.

The question is now how this solution behaves "after the collision."

The calculation gives this result: The scattered wave created by this perturbation has asymptotically at infinity the form:

$$\psi^1_{n\tau}(x, y, z; q_k) = \sum_m \iint_{\alpha x + \beta y + \gamma z > 0} d\omega\, \Phi_{n_\tau m}(\alpha, \beta, \gamma) \sin k_{n_\tau m}(\alpha x + \beta y + \gamma z + \delta)\psi^0_m(q_k).$$

This means that the perturbation, analyzed at infinity, can be regarded as a superposition of solutions of the unperturbed problem. If one calculates the energy belonging to the wavelength $\lambda_{n_\tau m}$ according to the de Broglie formula, one finds

$$W_{n_\tau m} = h\nu^0_{nm} + \tau,$$

where the ν^0_{nm} are the frequencies of the unperturbed atom.

If one translates this result into terms of particles, only one interpretation is possible. $\Phi_{n_\tau m}(\alpha, \beta, \gamma)$ gives the probability* for the electron, arriving from the z-direction, to be thrown out into the direction designated by the angles α, β, γ, with the phase change δ. Here its energy τ has increased by one quantum $h\nu^0_{nm}$ at the cost of the energy of the atom (collision of the first kind for $W^0_n < W^0_m$, $h\nu^0_{nm} < 0$; collision of the second kind $W^0_n > W^0_m$, $h\nu^0_{nm} > 0$).

Schrödinger's quantum mechanics therefore gives quite a definite answer to the question of the effect of the collision; but there is no question of any causal description. One gets no answer to the question, "what is the state after the collision," but only to the question, "how probable is a specified outcome of the collision" (where naturally the quantum mechanical energy relation must be fulfilled).

Here the whole problem of determinism comes up. From the standpoint of our quantum mechanics there is no quantity which in any individual case causally fixes the consequence of the collision; but also experimentally we have so far no reason to believe that there are some inner properties of the atom which condition a definite outcome for the collision. Ought we to hope later to discover such properties (like phases or the internal atomic motions) and determine them in individual cases? Or ought we to believe that the agreement of theory and experiment—as to the impossibility of prescribing conditions for a causal evolution—is a pre-established harmony founded on the nonexistence of such conditions? I myself am inclined to give up determinism in the world of atoms. But that is a philosophical question for which physical arguments alone are not decisive.

In practical terms indeterminism is present for experimental as well as for theoretical physicists. The "yield function" Φ so much investigated by experimentalists is now also sharply defined theoretically. One can determine it from the potential energy of interaction, $V(x, y, z; q_k)$. However, the calculations required

* Addition in proof: More careful consideration shows that the probability is proportional to the square of the quantity $\Phi_{n_\tau m}$.

for this purpose are too complicated to communicate here. I will only clarify briefly the meaning of the function $\Phi_{n_\tau m}$. If, for example, the atom before the collision is in the normal state $n = 1$, then it follows from the equation

$$\tau + h\nu_{1m}^0 = \tau - h\nu_{m1}^0 = W_{1_\tau m} > 0,$$

that, for an electron with less energy than the lowest excitation energy of the atom, the final state is also necessarily $m = 1$, or that $W_{1_\tau 1}$ must be equal to τ. Then we have "elastic reflection" of the electron with the yield function $\Phi_{1_\tau 1}$. If τ increases beyond the first excitation level, then there occurs, besides reflection, also excitation with the yield $\Phi_{1_\tau 2}$, etc. If the target atom is in the excited state $n = 2$ and $\tau < h\nu_{21}^0$, then there occur reflection with yield $\Phi_{2_\tau 2}$ and collisions of the second kind with the yield $\Phi_{2_\tau 1}$. If the kinetic energy $\tau > h\nu_{21}^0$, then further excitation is also possible.

The formulas thus reproduce completely the qualitative character of collisions. The quantitative predictions of the formulas for particular cases require extensive investigation.

I do not exclude the possibility that the strict connection of mechanics and statistics as it comes to light here will demand a revision of basic ideas of thermodynamics and statistical mechanics.

I also believe that the problem of radiation of light—and irradiation—has to be handled in a way entirely analogous to the "boundary value problem" of the wave equation, and will lead to a rational theory of radiation damping and line-breadths in agreement with the theory of light quanta.

An extended treatment will appear shortly in this journal.

I.3 THE PRINCIPLE OF INDETERMINACY

COMMENTARY OF HEISENBERG (1967)

In July [1926] I visited my parents in Munich and on this occasion I heard a lecture given by Schrödinger for the physicists in Munich about his work on wave mechanics. It was thus that I first became acquainted with the interpretation Schrödinger wanted to give his mathematical formalism of wave mechanics, and I was very disturbed about the confusion with which I believed this would burden atomic theory. Unfortunately, nothing came of my attempt during the discussion to put things in order. My argument that one could not even understand Planck's radiation law on the basis of Schrödinger's interpretation convinced no one. Wilhelm Wien, who held the chair of experimental physics at the University of Munich, answered rather sharply that one must really put an end to quantum jumps and the whole atomic mysticism, and the difficulties I had mentioned would certainly soon be solved by Schrödinger. I no longer remember whether or not I wrote to Bohr of this encounter in Munich. Be that as it may, Bohr shortly afterwards invited Schrödinger to Copenhagen and asked him not only to lecture on his wave mechanics, but also to stay in Copenhagen so long that there would be adequate time to discuss the interpretation of quantum theory.

As far as I remember these discussions took place in Copenhagen around September 1926 and in particular they left me with a very strong impression of Bohr's personality. For though Bohr was an unusually considerate and obliging person, he was able in such a discussion, which concerned epistemological problems which he considered to be of vital importance, to insist fanatically and with almost terrifying relentlessness on complete clarity in all arguments. He would not give up, even after hours of struggling, before Schrödinger had admitted that this interpretation was insufficient, and could not even explain Planck's law. Every attempt from Schrödinger's side to get round this bitter result was slowly refuted point by point in infinitely laborious discussions. It was perhaps from over-exertion that after a few days Schrödinger became ill and had to lie abed as a guest in Bohr's home. Even here it was hard to get Bohr away from Schrödinger's bed and the phrase, "But, Schrödinger, you must at least admit that . . ." could be heard again and again. Once Schrödinger burst out almost desperately, "If one has to go on with these damned quantum jumps, then I'm sorry that I ever started to work on atomic theory." To which Bohr answered, "But the rest of us are so grateful that you did, for you have thus brought atomic physics a decisive step forward." Schrödinger finally left Copenhagen rather discouraged, while we at Bohr's Institute felt that at least Schrödinger's interpretation of quantum theory, an interpretation rather too hastily arrived at using the classical wave-theories as models, was now disposed of, but that we still lacked some important ideas before we could really reach a full understanding of quantum mechanics.

From now on the discussions between Bohr and his co-workers in Copenhagen became more and more concentrated on the central problem in quantum theory: how the mathematical formalism was to be applied to each individual problem, and thus how the frequently discussed paradoxes, such as e.g. the apparent contradiction between the wave and particle models, could be cleared up. Ever new imaginary experiments were thought up, each displaying the paradoxes in a more clear-cut way than its predecessors, and we tried to guess what answer nature would probably give to each experiment.

....

Left alone in Copenhagen [when Bohr went on vacation at the end of February 1927] I too was able to give my thoughts freer play, and I decided to make the above uncertainty the central point in the interpretation. Remembering a discussion I had had long before with a fellow student in Göttingen, I got the idea of investigating the possibility of determining the position of a particle with the aid of a gamma-ray microscope, and in this way soon arrived at an interpretation which I believed to be coherent and free of contradictions. I then wrote a long letter to Pauli, more or less the draft of a paper, and Pauli's answer was decidedly positive and encouraging. When Bohr returned from Norway, I was already able to present him with the first version of a paper along with the letter from Pauli. At first Bohr was rather dissatisfied. He pointed out to me that certain statements in this first version were still incorrectly founded, and as he always insisted on relentless clarity in every detail, these points offended him deeply. Further, he had probably already grown familiar, while he was in Norway, with the concept of complementarity which would make it possible to take the dualism between the wave and particle picture as a suitable starting point for an interpretation. This concept of complementarity fitted well the fundamental philosophical attitude which he had always had, and in which the limitations of our means of expressing ourselves entered as a central philosophical problem. He therefore took objection to the fact that I had not started from the dualism between particles and waves. After several weeks of discussion, which were not devoid of stress, we soon concluded, not least thanks to Oskar Klein's participation, that we really meant the same, and that the uncertainty relations were just a special case of the more general complementarity principle. Thus, I sent my improved paper to the printer and Bohr prepared a detailed publication on complementarity.

Commentary of Rosenfeld (1971a)

Bohr was now impatient to come to grips with the outstanding epistemological issue. He was convinced that in the quantum theory of matter just as in that of radiation, one faced an irreducible particle-wave dualism, and it seemed to him that the way was clear to the elucidation of its epistemological significance. The emphasis must be laid here on the role of a scientific theory as a means of unambiguous communication of experience. In atomic theory it is found convenient to use for this purpose, according to the circumstances, either the language appropriate to the description of wave phenomena, or the language of particle mechanics, both modes of description being subject to essential limitations: it is a necessary task of the theory to formulate these limitations so as to fix the conditions of validity of each type of idealized description. Night after night in the first months of 1927, Bohr and Heisenberg pondered together over these questions; again, they were at loggerheads over the strategy: Bohr expected the answer would follow from a direct analysis of the definition of the idealized concepts, Heisenberg argued that the answer was hidden in the formal structure of the theory and that a closer scrutiny of this structure should bring it to light.

Towards the end of February, they parted, after a last fruitless discussion, Bohr having decided to seek a much needed recreation in the snowy hills of Norway. On the same night, however, Heisenberg had an inspiration that put him on the right track. He vividly remembered a discussion he had with Einstein a year before about the new quantum mechanics; his attempt at arguing that a good theory ought only to operate with observable quantities had elicited from Einstein the pointed retort, which had made a strong impression on him: "Only the theory itself can decide what is and is not observable." This remark, he thought, was the key to the whole problem: all one had to do was to investigate, for each given phenomenon, which conclusions one could draw from its theoretical analysis, according to the principles of quantum mechanics, about possible limitations to its observability.

....

Heisenberg based his argumentation on very general "indeterminacy relations" which he had shown to be deeply rooted in the formal structure of quantum mechanics. The mechanical variables are always associated in "conjugate" pairs . . . .

Heisenberg naturally wanted to clinch the argument by analyzing particular examples of idealized experimental situations which could illustrate the application of the indeterminacy relations. He was anxious, in particular, to demonstrate the impossibility of assigning a trajectory to a moving electron: let us imagine we should attempt to localize the electron by means of a microscope: in order to achieve sufficient resolution, we should have to illuminate it with radiation of very high frequency, in the range of the gamma-rays emitted by radioactive substances; the scattering of such photons by the electron, according to the theory of the Compton effect, would throw the electron completely out of the path prescribed by its previous velocity. Bohr's reaction to these considerations was very characteristic: he at once realized that Heisenberg again had forged the weapon that was needed to master the problem, but was soon dissatisfied with the smart way in which he was wielding it. The idea of bringing out the origin of the limitations of classical description by analyzing imaginary processes of observation was of the type that would appeal to Bohr, and he took it up eagerly, but in his own infinitely patient manner, examining the argument from all sides, leaving no point in the shadow. He was not long to discover that Heisenberg's discussion of the gamma-ray microscope touched only one aspect of the question—the role of the frequency of the radiation, but did not really go to the heart of the matter: the Compton recoil of the electron could only be a source of indeterminacy for its momentum to the extent to which the scattering process itself is not completely determined. Now, he pointed out, the formation of an image in a microscope requires a bundle of rays of finite aperture, and it is this latitude in the direction of the photons impinging on the electron that implies a corresponding latitude in the electron recoil. On the other hand a large aperture increases the accuracy of the electron's localization; and a simple computation shows that the two contrary effects of the aperture on the indeterminacies of position and momentum are such as to impose on their product the lower limit given by the relation (2), independently of the aperture and of the frequency of the radiation.

. . . Bohr looked upon such imaginary experiments in a rather different spirit from Heisenberg. The latter was satisfied with the knowledge that he had now established a complete correspondence between the mathematical structure of the theory and the usual physical concepts describing the various aspects of experience; the indeterminacy relations were part of this correspondence, their physical content could be exhibited by the discussion of appropriate experiments. Bohr wanted to pursue the epistemological analysis one step further, and in particular to understand the logical nature of the mutual exclusion of the two aspects opposed in the particle-wave dualism. From this point of view the indeterminacy relations appear in a new light. The conjugate variables for which they hold refer, respectively, to the two mutually exclusive classical pictures centered on the idealized concepts of particle and wave: to the former are attached the momentum and energy variables, to the latter the coordinates of space and time. The particle concept is used to describe the exchanges of momentum and energy between atomic systems, and between such systems and radiation: it is the physical support, so to speak, of the conservation laws governing such exchanges, irrespectively of any space-time localization. The wave concept, or rather the derived density function, describes the localization of atomic systems and the distribution of radiation in space and time; it excludes the specification of momentum and energy, since only a wave of infinite spatial and temporal extension can have a sharply defined periodicity, necessary for a sharp definition of momentum and energy. The odd situation we meet in atomic theory is the necessity of making use of these two conflicting pictures in order to deal with all the conditions under which we can observe the atomic phenomena. The indeterminacy relations are therefore essential to ensure the consistency of the theory, by assigning the limits within which the use of classical concepts belonging to the two extreme pictures may be applied without contradiction. For this novel logical relationship, which called in Bohr's mind echoes of his philosophical meditations over the duality of our mental activity, he proposed the name "complementarity," conscious that he was here breaking new ground in epistemology.

The solemnity of the occasion did not dawn immediately upon Heisenberg. He was not well-disposed toward the idea of a particle-wave dualism, and Bohr was himself still struggling to find the right expression for the new ideas that were taking shape in his mind. However eager he was to let Heisenberg share his elation at the prospects they disclosed, his eagerness was not sufficiently matched by clarity to make impression on a reluctant interlocutor. The memory of the fruitless discussions with Schrödinger still lingered in Heisenberg's mind, and he regarded with deep suspicion any attempt to make more than formal use of the concept of matter wave. He stressed, quite rightly, that according to the correspondence principle, it is the particle concept that has a direct physical meaning with respect to atomic constituents, whereas matter waves are merely mathematical auxiliaries. To this Bohr would oppose the case of radiation, where the correspondence principle, on the contrary, points to the electromagnetic waves as the fundamental classical concept, and leaves only a symbolic part to the photons. Moreover, Oskar Klein, who was then Bohr's assistant, had carefully discussed the relation of matter waves to the correspondence principle, and shown how the latter gave as reliable a guidance for the physical interpretation of wave mechanics as for that of the quantum algebra: but Klein's paper had not found grace with Heisenberg. He stuck obstinately to the view that his own scheme of quantum mechanics formed a sufficient basis for a complete description of the phenomena.

Matters did not improve when Bohr was shown the hastily written paper in which Heisenberg had developed his arguments. Bohr started handling the manuscript as he would have done one of his own, namely as a rough draft which could eventually lead to an acceptable text; and with his usual optimism, he expected Heisenberg to welcome this scrutiny: the latter was simply annoyed to find Bohr raising so many objections to the inaccuracies and careless statements of the paper, and proposing new formulations with which he was in no mood to agree. There is no telling how things would have ended if Pauli's arrival on the scene had not relieved the tension. Heisenberg had informed Pauli of his ideas and received from him approval and encouragement; now, Pauli was in a position both to lend Bohr a dispassionate ear and to gain a hearing from Heisenberg. He was able to explain to the latter that there was no disagreement between his and Bohr's ways of analyzing the physical content of atomic theory, but that Bohr had gone further and deeper in the analysis of its logical structure. Even so, all Heisenberg could be persuaded to do about his paper was to append to it a "remark added on the proof," in which he declared in substance that he had missed essential points, whose clarification would be found in a forthcoming paper by Bohr. This addendum must have puzzled many readers: it is not often that the announcement of a decisive progress in our insight into the workings of nature is qualified by such a warning.

As to Bohr's "forthcoming" publication, more than a year elapsed before it appeared in print . . . .

I.3 THE PHYSICAL CONTENT OF QUANTUM KINEMATICS AND MECHANICS

WERNER HEISENBERG

First we define the terms *velocity*, *energy*, etc. (for example, for an electron) which remain valid in quantum mechanics. It is shown that canonically conjugate quantities can be determined simultaneously only with a characteristic indeterminacy (§1). This indeterminacy is the real basis for the occurrence of statistical relations in quantum mechanics. Its mathematical formulation is given by the Dirac-Jordan theory (§2). Starting from the basic principles thus obtained, we show how microscopic processes can be understood by way of quantum mechanics (§3). To illustrate the theory, a few special *gedankenexperiments* are discussed (§4).

We believe we understand the physical content of a theory when we can see its qualitative experimental consequences in all simple cases and when at the same time we have checked that the application of the theory never contains inner contradictions. For example, we believe that we understand the physical content of Einstein's concept of a closed 3-dimensional space because we can visualize consistently the experimental consequences of this concept. Of course these consequences contradict our everyday physical concepts of space and time. However, we can convince ourselves that the possibility of employing usual space-time concepts at cosmological distances can be justified neither by logic nor by observation. The physical interpretation of quantum mechanics is still full of internal discrepancies, which show themselves in arguments about continuity versus discontinuity and particle versus wave. Already from this circumstance one might conclude that no interpretation of quantum mechanics is possible which uses ordinary kinematical and mechanical concepts. Of course, quantum mechanics arose exactly out of the attempt to break with all ordinary kinematic concepts and to put in their place relations between concrete and experimentally determinable numbers. Moreover, as this enterprise seems to have succeeded, the mathematical scheme of quantum mechanics needs no revision. Equally unnecessary is a revision of space-time geometry at small distances, as we can make the quantum-mechanical laws approximate the classical ones arbitrarily closely by choosing sufficiently great masses, even when arbitrarily small distances and times come into question. But that a revision of kinematical and mechanical concepts is necessary

Originally published under the title, "Über den anschaulichen Inhalt der quantentheoretischen Kinematik und Mechanik," *Zeitschrift für Physik*, *43*, 172–98 (1927); reprinted in *Dokumente der Naturwissenschaft*, *4*, 9–35 (1963); translation into English by J.A.W. and W.H.Z., 1981.

seems to follow directly from the basic equations of quantum mechanics. When a definite mass m is given, in our everyday physics it is perfectly understandable to speak of the position and the velocity of the center of gravity of this mass. In quantum mechanics, however, the relation $\mathbf{pq} - \mathbf{qp} = -i\hbar$ between mass, position, and velocity is believed to hold. Therefore we have good reason to become suspicious every time uncritical use is made of the words "position" and "velocity." When one admits that discontinuities are somehow typical of processes that take place in small regions and in short times, then a contradiction between the concepts of "position" and "velocity" is quite plausible. If one considers, for example, the motion of a particle in one dimension, then in continuum theory one will be able to draw (Fig. 1) a worldline $x(t)$ for the track of the particle (more precisely, its center of gravity), the tangent of which gives the velocity at every instant. In contrast, in a theory based on discontinuity there might be in place of this curve a series of points at finite separation (Fig. 2). In this case it is clearly meaningless to speak about one velocity at one position (1) because one velocity can only be defined by two positions and (2), conversely, because any one point is associated with two velocities.

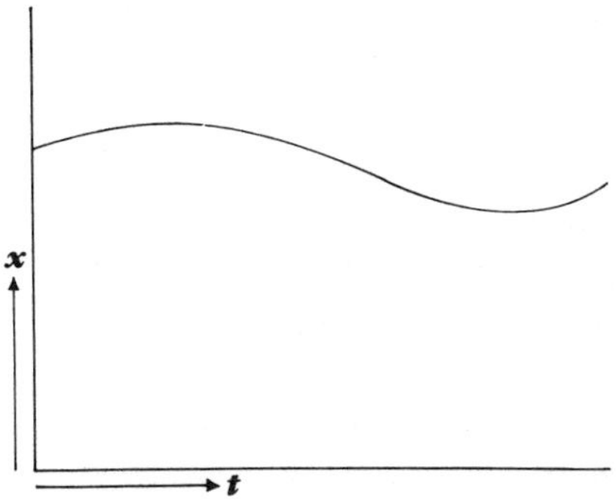

FIGURE 1

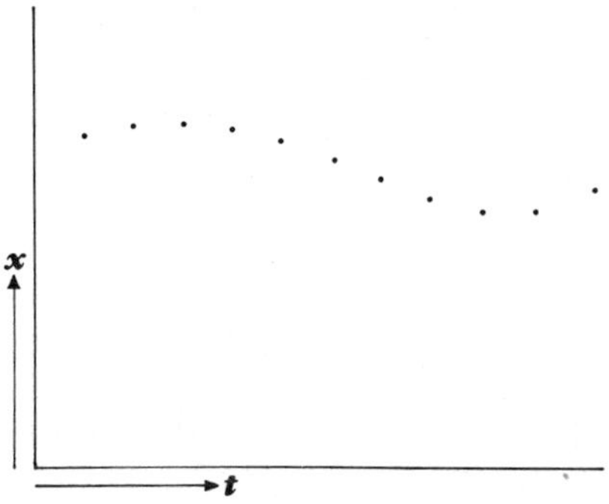

FIGURE 2

The question therefore arises whether, through a more precise analysis of these kinematic and mechanical concepts, it might be possible to clear up the contradictions evident up to now in the physical interpretations of quantum mechanics and to arrive at a physical understanding of the quantum-mechanical formulas.*

* The present work has arisen from efforts and desires to which other investigators have already given clear expression, before the development of quantum mechanics. I call attention here especially to Bohr's papers on the basic postulates of quantum theory (for example, *Zeits. f. Physik, 13*, 117 [1923]) and Einstein's discussions on the connection between wave field and light quanta. The problems dealt with here are discussed most clearly in recent times, and the problems arising are partly answered, by W. Pauli ("Quantentheorie," *Handbuch der Physik*, Vol. XXIII, cited hereafter as *l.c.*); quantum mechanics has changed only slightly the formulation of these problems as given by Pauli. It is also a special pleasure to thank here Herrn Pauli for the repeated stimulus I have received from our oral and written discussions, which have contributed decisively to the present work.

§1. Concepts: Position, Path, Velocity, Energy

In order to be able to follow the quantum-mechanical behavior of any object one has to know the mass of this object and its interactions with any fields and other objects. Only then can the Hamiltonian function be written down for the quantum-mechanical system. (The following considerations ordinarily refer to nonrelativistic quantum mechanics, as the laws of quantum electrodynamics are still very incompletely known.)* About the "Gestalt" (construction) of the object any further assumption is unnecessary; one most usefully employs the word "Gestalt" to designate the totality of these interactions.

When one wants to be clear about what is to be understood by the words "position of the object," for example of the electron (relative to a given frame of reference), then one must specify definite experiments with whose help one plans to measure the "position of the electron"; otherwise this word has no meaning. There is no shortage of such experiments, which in principle even allow one to determine the "position of the electron" with arbitrary accuracy. For example, let one illuminate the electron and observe it under a microscope. Then the highest attainable accuracy in the measurement of position is governed by the wavelength of the light. However, in principle one can build, say, a γ-ray microscope and with it carry out the determination of position with as much accuracy as one wants. In this measurement there is an important feature, the Compton effect. Every observation of scattered light coming from the electron presupposes a photoelectric effect (in the eye, on the photographic plate, in the photocell) and can therefore also be so interpreted that a light quantum hits the electron, is reflected or scattered, and then, once again bent by the lens of the microscope, produces the photoeffect. At the instant when position is determined—therefore, at the moment when the photon is scattered by the electron—the electron undergoes a discontinuous change in momentum. This change is the greater the smaller the wavelength of the light employed—that is, the more exact the determination of the position. At the instant at which the position of the electron is known, its momentum therefore can be known up to magnitudes which correspond to that discontinuous change. Thus, the more precisely the position is determined, the less precisely the momentum is known, and conversely. In this circumstance we see a direct physical interpretation of the equation $\mathbf{pq} - \mathbf{qp} = -i\hbar$. Let q_1 be the precision with which the value q is known (q_1 is, say, the mean error of q), therefore here the wavelength of the light. Let p_1 be the precision with which the value p is determinable; that is, here, the discontinuous change of p in the Compton effect. Then, according to the elementary laws of the Compton effect p_1 and q_1 stand in the relation

* Quite recently, however, great advances in this domain have been made in the papers of P. Dirac [*Proc. Roy. Soc. A114*, 243 (1927) and papers to appear subsequently].

$$p_1 q_1 \sim h. \tag{1}$$

That this relation (1) is a straightforward mathematical consequence of the rule $\mathbf{pq} - \mathbf{qp} = -i\hbar$ will be shown below. Here we can note that equation (1) is a precise expression for the facts which one earlier sought to describe by the division of phase space into cells of magnitude h. For the determination of the position of the electron one can also do other experiments—for example, collision experiments. A precise measurement of the position demands collisions with very fast particles, because for slow electrons the diffraction phenomena—which, according to Einstein, are consequences of de Broglie waves (as, for example, in the Ramsauer effect)—prevent a sharp specification of location. In a precise measurement of position the momentum of the electron again changes discontinuously. An elementary estimate of the precision using the formulas for de Broglie waves leads once more to relation (1).

Throughout this discussion the concept of "position of the electron" seems well enough defined, and only a word need be added about the "size" of the electron. When two very fast particles hit the electron one after the other within a very short time interval Δt, then the positions of the electron defined by the two particles lie very close together at a distance Δl. From the regularities which are observed for α-particles we conclude that Δl can be pushed down to a magnitude of the order of 10^{-12} cm if only Δt is sufficiently small and particles are selected with sufficiently great velocity. This is what we mean when we say that the electron is a corpuscle whose radius is not greater than 10^{-12} cm.

We turn now to the concept of "path of the electron." By path we understand a series of points in space (in a given reference system) which the electron takes as "positions" one after the other. As we already know what is to be understood by "position at a definite time," no new difficulties occur here. Nevertheless, it is easy to recognize that, for example, the often used expression, the "1s orbit of the electron in the hydrogen atom," from our point of view has no sense. In order to measure this 1s "path" we have to illuminate the atom with light whose wavelength is considerably shorter than 10^{-8} cm. However, a single photon of such light is enough to eject the electron completely from its "path" (so that only a single point of such a path can be defined). Therefore here the word "path" has no definable meaning. This conclusion can already be deduced, without knowledge of the recent theories, simply from the experimental possibilities.

In contrast, the contemplated measurement of position can be carried out on many atoms in a 1s state. (In principle, atoms in a given "stationary" state can be selected, for example, by the Stern-Gerlach experiment.) There must therefore exist for a definite state—for example, the 1s state—of the atom a probability function for the location of the electron which corresponds to the mean value for the classical orbit, averaged over all phases, and which can be determined through the measurement with an arbitrary precision. According to Born,* this function is given by $\psi_{1s}(q)\bar{\psi}_{1s}(q)$ where $\psi_{1s}(q)$ designates the Schrödinger wave function belonging to the 1s state. With a view to later generalizations I should like to say—with Dirac and Jordan—that the probability is given by $S(1s, q)\bar{S}(1s, q)$, where $S(1s, q)$ designates that column of the matrix $S(E, q)$ of transformation from E to q that belongs to the energy $E = E_{1s}$.

In the fact that in quantum theory only the probability distribution of the position of the electrons can be given for a definite state, such as 1s, one can recognize, with Born and Jordan, a characteristically statistical feature of quantum theory as contrasted to classical theory. However, one can say, if one will, with Dirac, that the statistics are brought in by our experiments. For plainly *even in classical theory* only the probability of a definite position for the electron can be given as long as we do not know the phase of [the motion of the electron in] the atom. The distinction between classical and quantum mechanics consists rather in this: classically we can always think of the phase as determined through suitable experiments. In reality, however, this is impossible, because every experiment for the determination of phase perturbs or changes the atom. In a definite stationary "state" of the atom, the phases are in principle indeterminate, as one can see as a direct consequence of the familiar equations

$$\mathbf{Et} - \mathbf{tE} = -i\hbar \qquad \text{or} \qquad \mathbf{Jw} - \mathbf{wJ} = -i\hbar,$$

where $\mathbf{J}$ is the action variable and $\mathbf{w}$ is the angle variable.

The word "velocity" can easily be defined for an object by measurements when the motion is free of force. For example, one can illuminate the object with red light and by way of the Doppler effect in the scattered light determine the velocity of the particle. The determination of the velocity is the more exact the longer the wavelength of the light that is used, as then the change in velocity of the particle, per light quantum, by way of the Compton effect is so much less. The determination of position becomes correspondingly inexact, in agreement with equation (1). If one wants to measure the velocity of the electron in the atom at a definite instant, then, for example, one will let the nuclear charge and the forces arising

* The statistical interpretation of de Broglie waves was first formulated by A. Einstein (*Sitzungsber. d. preussische Akad. d. Wiss.*, p. 3 [1925]). This statistical feature of quantum mechanics then played an essential role in M. Born, W. Heisenberg, and P. Jordan, Quantenmechanik II (*Zeits. f. Physik*, *35*, 557 [1926]), especially chapter 4, §3, and P. Jordan (*Zeits. f. Physik*, *37*, 376 [1926]). It was analyzed mathematically in a seminal paper of M. Born (*Zeits. f. Physik*, *38*, 803 [1926]) and used for the interpretation of collision phenomena. One finds how to base the probability picture on the theory of the transformation of matrices in the following papers: W. Heisenberg (*Zeits. f. Physik*, *40*, 501 [1926]), P. Jordan (*Zeits. f. Physik*, *40*, 661 [1926]), W. Pauli (remark in *Zeits. f. Physik*, *41*, 81 [1927]), P. Dirac (*Proc. Roy. Soc. A113*, 621 [1926]), and P. Jordan (*Zeits. f. Physik*, *40*, 809 [1926]). The statistical side of quantum mechanics is discussed more generally in P. Jordan (*Naturwiss.*, *15*, 105 [1927]) and M. Born (*Naturwiss.*, *15*, 238 [1927]).

from the other electrons suddenly be taken away, so that the motion from then on is force-free, and one will then carry out the measurement described above. As above, one can again easily convince oneself that a [momentum] function $p(t)$ cannot be defined for a given state—such as the 1s-state—of an atom. On the contrary, there is again a probability function for p in this state which according to Dirac and Jordan has the value $S(1s, p)\bar{S}(1s, p)$. Here $S(1s, p)$ again designates that column of the matrix $S(E, p)$—that transforms from **E** to **p**—which belongs to $E = E_{1s}$.

Finally we come to experiments which allow one to measure the energy or the value of the action variable J. Such experiments are especially important because only with their help can we define what we mean when we speak of the discontinuous change of the energy and of J. The Franck-Hertz collision experiments allow one to base the measurement of the energy of the atom on the measurement of the energy of electrons in rectilinear motion, because of the validity of the law of conservation of energy in quantum theory. This measurement in principle can be carried out with arbitrary accuracy if only one forgoes the simultaneous determination of the position of the electron or its phase (see the determination of p, above), corresponding to the relation $\mathbf{Et} - \mathbf{tE} = -i\hbar$. The Stern-Gerlach experiment allows one to determine the magnetic or an average electric moment of the atom, and therefore to measure quantities which depend only on the action variable J. The phases remain undetermined in principle. It makes as little sense to speak of the frequency of the light wave at a definite instant as of the energy of an atom at a definite moment. Correspondingly, in the Stern-Gerlach experiment the accuracy of the energy measurement decreases as we shorten the time during which the atom is under the influence of the deflecting field.* Specifically, an upper bound is given for the deviating force through the circumstance that the potential energy of that deflecting force can at most vary inside the beam by an amount which is considerably smaller than the differences in energy of the stationary states. Only then will a determination of the energy of the stationary states be at all possible. Let E_1 be an amount of energy which satisfies this condition (E_1 also fixes the precision of the energy measurement). Then E_1/d specifies the highest allowable value for the deflecting force, if d is the breadth of the beam (measurable through the spacing of the slits employed). The angular deviation of the atomic beam is then $E_1 t_1/dp$, where we designate by t_1 the time during which the atoms are under the influence of the deflecting field, and by p the momentum of the atoms in the direction of the beam. This deflection must be of at least the same order of magnitude as the natural broadening of the beam brought about by the diffraction by the slits, if any measurement is to be possible. The diffraction angle is roughly λ/d if λ denotes the de Broglie wavelength; thus,

* In this connection see W. Pauli, *l.c.*, p. 61.

$$\lambda/d \sim E_1 t_1/dp,$$

or, as $\lambda = h/p$,

$$E_1 t_1 \sim h. \tag{2}$$

This equation corresponds to equation (1) and shows how a precise determination of energy can only be obtained at the cost of a corresponding uncertainty in the time.

§2. The Dirac-Jordan Theory

We might summarize and generalize the results of the preceding section in this statement: *All concepts which can be used in classical theory for the description of a mechanical system can also be defined exactly for atomic processes in analogy to the classical concepts.* The experiments which provide such a definition themselves suffer an indeterminacy introduced purely by the observational procedures we use when we ask of them the simultaneous determination of two canonically conjugate quantities. The magnitude of this indeterminacy is given by relation (1) (generalized to any canonically conjugate quantities whatsoever). It is natural in this respect to compare quantum theory with special relativity. According to relativity, the word "simultaneous" cannot be defined except through experiments in which the velocity of light enters in an essential way. If there existed a "sharper" definition of simultaneity, as, for example, signals which propagate infinitely fast, then relativity theory would be impossible. However, because there are no such signals, or, rather, because already in the definition of simultaneity the velocity of light appears, there is room left for the postulate of the constancy of the speed of light so that this postulate does not contradict any meaningful use of the words "position, velocity, time." We find a similar situation with the definition of the concepts of "position of an electron" and "velocity" in quantum theory. All experiments which we can use for the definition of these terms necessarily contain the uncertainty implied by equation (1), even though they permit one to define exactly the concepts p and q taken in isolation. If there existed experiments which allowed simultaneously a "sharper" determination of p and q than equation (1) permits, then quantum mechanics would be impossible. Thus only the uncertainty which is specified by equation (1) creates room for the validity of the relations which find their most pregnant expression in the quantum-mechanical commutation relations,

$$\mathbf{pq} - \mathbf{qp} = -i\hbar.$$

That uncertainty makes possible this equation without requiring that the physical meaning of the quantities p and q be changed.

For those physical phenomena whose quantum-mechanical formulation is still unknown (for example, electrodynamics), equation (1) makes a demand which may be useful for the discovery of the new laws. For quantum mechanics equation (1) can be derived from the Dirac-Jordan formulation by a slight generalization. If, for any definite state variable η we determine the position q of the electron as q' with an uncertainty q_1, then we can express this fact by a probability amplitude $S(\eta, q)$ which differs appreciably from zero only in a region of spread q_1 near q'. For example, one can write

$$S(\eta, q) \text{ proportional to } \exp[-(q - q')^2/2q_1^2 - (ip'/\hbar)(q - q')], \tag{3a}$$

with therefore

$$S\bar{S}' \text{ proportional to } \exp[-(q - q')^2/q_1^2]. \tag{3b}$$

Then for the probability amplitude for any given value of p we have

$$S(\eta, p) = \int S(\eta, q)S(q, p)\, dq. \tag{4}$$

For $S(q, p)$, according to Jordan, we can write

$$S(q, p) = \exp(ipq/\hbar). \tag{5}$$

Then, according to (4), $S(\eta, p)$ differs appreciably from zero only for values of p for which $(p - p')q_1/\hbar$ is not significantly greater than 1. Specifically, employing (3), we find $S(\eta, p)$ is proportional to

$$\int \exp[i(p - p')q/\hbar - (q - q')^2/2q_1^2]\, dq;$$

that is, proportional to

$$\exp[-(p - p')^2/2p_1^2 + iq'(p - p')/\hbar];$$

and thus

$$S\bar{S} \text{ is proportional to } \exp[-(p - p')^2/p_1^2],$$

where

$$p_1 q_1 = \hbar. \tag{6}$$

The assumption (3) for $S(\eta, q)$ corresponds therefore to the experimental fact that the value p' is measured for p and the value q' for q [with the limit (6) on the precision].

From the purely mathematical point of view it is characteristic of the Dirac-

Jordan formulation of quantum mechanics that the relations between **p**, **q**, **E**, etc. can be described as equations connecting very general matrices in such a way that any predetermined quantum-theoretic quantity appears as a diagonal matrix. The possibility of writing things in such a way is evident when one pictures the matrices as tensors (for example, moment-of-inertia tensors) in a multidimensional space between which there are mathematical connections. One can always pick the axis of the coordinate system in which one expresses these relations along the principal axes of one of these tensors. Finally, one also can always characterize the mathematical relation between two tensors A and B through the transformation equations which take a coordinate system oriented along the principal axes of A over into another oriented along the principal axes of B. This latter formulation corresponds to the Schrödinger theory. In contrast, one will view Dirac's q-number formulation as the formulation of quantum mechanics that is really "invariant" and independent of all coordinate systems. When we want to derive physical results from that mathematical framework, then we have to associate numbers with the quantum-theoretical magnitudes—that is, with the matrices (or "tensors" in multidimensional space). This task is to be understood in these terms: In that multidimensional space a definite direction is arbitrarily prescribed (by the nature of the experimental setup); and it is asked what is the "value" of the matrix (for example, in that picture, what is the value of the moment of inertia) in this given direction. This question only has a well-defined meaning when the given direction coincides with the direction of one of the principal axes of that matrix. In this case there is an exact answer for the question. But also when the prescribed direction differs only little from one of the principal axes of the matrix one can still speak of the "value" of the matrix in the prescribed direction up to a definite uncertainty determined by the angle between the two directions. One can therefore say that associated with every quantum-theoretical quantity or matrix is a number which gives its "value" within a certain definite statistical error. The statistical error depends on the coordinate system. For every quantum-theoretical quantity there exists a coordinate system in which the statistical error for this quantity is zero. Therefore a definite experiment can never give exact information on all quantum-theoretical quantities. Rather, it divides physical quantities into "known" and "unknown" (or more and less accurately known quantities) in a way characteristic of the experiment in question. The results of two experiments can be derived exactly one from the other only then when the two experiments divide the physical quantities in the same way into "known" and "unknown" (that is, when the tensors in that multidimensional space frequently invoked—for ease of visualization—are "looked at" in both experiments from the same direction). When two experiments use different divisions into "known" and "unknown," then their results can be related only statistically.

For a more detailed discussion of this statistical connection let a *gedanken-experiment* be considered. Let a Stern-Gerlach atomic beam be sent first through a field F_1 which is so strongly inhomogeneous in the direction of the beam that it induces many transitions by sudden reversal in the force on the spin. Then let the atomic beam run free up to a definite distance from F_1. But there let a second field F_2 begin, as inhomogeneous as F_1. Between F_1 and F_2 let it be possible to measure the number of atoms in the different stationary states through an optionally applied magnetic field. Let all radiation by the atoms be neglected. If we know that an atom was in a state of energy E_n before it passed F_1, then we can express this fact by ascribing to the atom a wave function—for example, in p-space—with the definite energy E_n and the undetermined phase β_n,

$$S(E_n, p) = \psi(E_n, p) \exp[-i(E_n/\hbar)(\alpha + \beta_n)].$$

After passage through the field F_1, this function is changed into*

$$S(E_n, p) \xrightarrow{F_1} \sum_m c_{nm}\psi(E_m, p) \exp[-i(E_m/\hbar)(\alpha + \beta_m)]. \tag{7}$$

Here we can make some arbitrary determination of the β_m so that the c_{nm} are uniquely determined by F_1. The matrix c_{nm} transforms the energy values before the transition through F_1 to the values after the transition. If after F_1 we carry out a determination of the stationary state, say, by use of an inhomogeneous magnetic field, then we will find that the atom has jumped from the nth state to the mth state with a probability $c_{nm}\bar{c}_{nm}$. When we find experimentally that an atom has indeed jumped to the mth state, then we have to ascribe to it in all calculations thereafter, not the function $\sum_m c_{nm}S_m$, but simply the function S_m with an undetermined phase. Through the experimental determination, "mth state," we select out of the multitude of different possibilities (c_{nm}) a definite one, m. However, at the same time we disturb everything that was still contained in the phase relations between the quantities c_{nm}, as detailed below. In the transition of the atomic beam through F_2, what happened at F_1 repeats itself. Let d_{nm} be the coefficients of the transformation matrix which transform the energies before F_2 to the energies after F_2. If no determination of the state is carried out between F_1 and F_2, then the eigenfunction is transformed according to the following scheme,

$$S(E_n, p) \xrightarrow{F_1} \sum_m c_{nm}S(E_m, p) \xrightarrow{F_2} \sum_m \sum_l c_{nm}d_{ml}S(E_l, p). \tag{8}$$

Let $\sum_m c_{nm}d_{ml}$ be called e_{nl}. If the stationary state of the atom is determined beyond F_2, then one will find the state l with the probability $e_{nl}\bar{e}_{nl}$. In contrast, if between F_1 and F_2 one determines the state—and finds for it the value E_m—then

* See P. Dirac (*Proc. Roy. Soc. A112*, 661 [1926]) and M. Born (*Zeits. f. Physik*, *40*, 167 [1926]).

the probability for "l" beyond F_2 is given by $d_{ml}\bar{d}_{ml}$. In many repetitions of the entire experiment (in which each time the state is determined between F_1 and F_2) one will therefore observe the state l, beyond F_2, with the relative frequency $Z_{nl} = \sum_m c_{nm}\bar{c}_{nm}d_{ml}\bar{d}_{ml}$. This expression does not agree at all with $e_{nl}\bar{e}_{nl}$. For this reason Jordan (*l.c.*) has spoken of an "interference of probabilities." I cannot agree. The two kinds of experiments which lead respectively to $e_{nl}\bar{e}_{nl}$ and Z_{nl} are physically distinct. In one case the atom experiences no disturbance between F_1 and F_2. In the other case it is perturbed by the apparatus which determines its stationary state. This apparatus has as a consequence that the "phase" of the atom changes by an amount that is in principle uncheckable, as the momentum of an electron likewise changes with a determination of its position (see §1). The magnetic field for the determination of the state between F_1 and F_2 will separate the eigenvalues E. In the observation of the path of the atomic beam the atoms are slowed down by statistically different and uncheckable amounts (I think here, say, of Wilson cloud-chamber pictures). This has as a consequence that the final transformation matrix (from the energy value before entry into F_1 to the energy after exit from F_2) is no longer given by $\sum_m c_{nm}d_{ml}$, but every term in this sum has additionally an unknown phase factor. No expectation is therefore open to us, except that the mean value of $e_{nl}\bar{e}_{nl}$ averaged over all these expected phase alterations is equal to Z_{nl}. A simple calculation confirms that this is the case. We can therefore deduce from one experiment the possible results of another by definite statistical rules. The other experiment itself selects out of the plenitude of all possibilities a quite definite one, and thereby limits the possibilities for all later experiments. Such an interpretation of the equation for the transformation matrix S or the Schrödinger wave equation is only possible because the sum of solutions is again a solution. In this circumstance we see the deep significance of the linearity of Schrödinger wave equations. On that account they can be understood only as equations for waves in phase space; and on that account we may regard as hopeless every attempt to replace these equations by nonlinear equations, for example in the relativistic case (for more than one electron).

§3. The Transition from Micro- to Macromechanics

It seems to me that the concepts of kinematics and mechanics in quantum theory are sufficiently clarified by the analysis of the words "position of an electron," "velocity," "energy," etc., in the preceding sections that physical understanding of macroscopic processes from the standpoint of quantum mechanics must also be possible. The transition from micro- to macromechanics has already been treated by Schrödinger,* but I do not believe that Schrödinger's considerations

* E. Schrödinger, *Naturwiss.*, *14*, 664 (1926).

get to the heart of the problem, and this is why: According to Schrödinger, in a high state of excitation a sum of eigenfunctions ought to be able to give a wave packet of limited extent which—through periodic changes in its form—will carry out the periodic motions of the classical "electron." There is an argument against this outlook: If the wave packet had such properties as ascribed to it by this view, then the radiation sent out by the atom would be representable as a Fourier series in which the frequencies of the higher vibrations were integer multiples of the basic frequency. The frequencies of the spectral lines sent out by the atom are, however, according to quantum mechanics, never integer multiples of the basic frequency—except in the special case of the harmonic oscillator. Thus Schrödinger's reasoning is only viable for the case of the harmonic oscillator treated by him; in all other cases a wave packet spreads out in the course of time over the whole immediate neighborhood of the atom. The higher the state of excitation of the atom, the slower is that spreading of the wave packet. However, if one waits long enough it happens. The reasoning developed above about the radiation sent out by the atom might at first sight be used against all experiments which look for a direct transition from quantum to classical mechanics at high quantum numbers. For that reason the attempt was made earlier to circumvent such reasoning by referring to the natural radiation broadening of stationary states—certainly wrongly, first of all because this way out is blocked already in the case of the hydrogen atom on account of the weakness of the radiation for high states, and secondly, because the transition from quantum to classical mechanics ought to be understandable without calling on electrodynamics. Bohr* has already referred many times to these well-known difficulties which stand in the way of a direct connection between quantum and classical theory. We have spelled them out again here so explicitly only because in recent times they seem to be forgotten.

I believe that one can fruitfully formulate the origin of the classical "orbit" in this way: the "orbit" comes into being only when we observe it. For example, let an atom be given in a state of excitation $n = 1000$. The dimensions of the orbit in this case are already relatively large so that, in accordance with §1, it is enough to use light of relatively low wavelength to determine the position of the electron. If the position determination is not to be too fuzzy then the Compton recoil will put the atom in some state of excitation between, say, 950 and 1050. Simultaneously, the momentum of the electron can be determined from the Doppler effect with a precision given by (1). One can characterize the experimental finding by a wave-packet, or, better, a probability-amplitude packet, in q-space of a spread given by the wavelength of the light used, and built up primarily out of eigenfunctions between the 950th and 1050th eigenfunction—and by a corresponding packet in p-space. Let a new determination of position be made after some time with the same precision. Its result, according to §2, can be predicted only statistically. All

* N. Bohr, "Grundpostulate der Quantentheorie," *l.c.*

positions count as likely (with calculable probability) which lie within the bounds of the now broadened wavepacket. The situation would be no different in classical theory, for there too the result of the second position measurement would be predictable only statistically because of the uncertainty in the first measurement. Also, the orbits of classical theory would spread out like the wavepacket. However, statistical laws themselves are different in quantum mechanics and in classical theory. The second determination of the position selects a definite "q" from the totality of possibilities and limits the options for all subsequent measurements. After the second position determination the results of later measurements can only be calculated when one again ascribes to the electron a "smaller" wavepacket of extension λ (wavelength of the light used in the observation). Thus every position determination reduces the wavepacket back to its original extension λ. The "values" of the quantities **p**, **q** are known throughout all the experiments with a certain precision. The values of **p** and **q** stay within the precision limits fixed by the classical equations of motion. This can be seen directly from the quantum-mechanical equations,

$$\dot{\mathbf{p}} = -\partial\mathbf{H}/\partial\mathbf{q}; \qquad \dot{\mathbf{q}} = \partial\mathbf{H}/\partial\mathbf{p}. \tag{9}$$

However, the orbit, as noted earlier, can be calculated only statistically from the initial conditions, a circumstance that one can consider as a consequence of the fundamental indeterminism of the initial conditions. The statistical laws are different for quantum mechanics and classical theory. This distinction, under appropriate circumstances, can give rise to gross macroscopic differences between classical and quantum theory. Before I discuss an example, I should like to indicate how the transition, discussed above, to classical theory is formulated mathematically for a simple mechanical system, the force-free motion of a particle. In one dimension the equations of motion run

$$\mathbf{H} = \mathbf{p}^2/2m; \qquad \dot{\mathbf{q}} = \mathbf{p}/m; \qquad \dot{\mathbf{p}} = 0. \tag{10}$$

As time can be treated as a parameter (or "c-number") when there are no time-dependent external forces, the solution of this equation is

$$\mathbf{q} = \mathbf{p}_0 t/m + \mathbf{q}_0; \qquad \mathbf{p} = \mathbf{p}_0, \tag{11}$$

where $\mathbf{p}_0$ and $\mathbf{q}_0$ are the momentum and the position at the time $t = 0$. At $t = 0$ the value $q_0 = q'$ is measured with accuracy q_1, and $p_0 = p'$ with accuracy p_1 [see equations (3) to (6)]. In order to draw conclusions from the "values" of $\mathbf{p}_0$ and $\mathbf{q}_0$ about the "values" of $\mathbf{q}$ at the time t, one must find—according to Dirac and Jordan—that transformation function which transforms all matrices in the representation* in which $\mathbf{q}_0$ is diagonal to that representation in which $\mathbf{q}$ is diagonal. In the matrix scheme in which $\mathbf{q}_0$ is diagonal, $\mathbf{p}_0$ can be replaced by the operator $-i\hbar\,\partial/\partial q_0$. According to Dirac [*l.c.* equation (11)] the desired transformation amplitude $S(q_0, q)$ satisfies the differential equation,

$$\{-i(\hbar t/m)\,\partial/\partial q_0 + q_0\}S(q_0, q) = qS(q_0, q) \tag{12}$$

or

$$-i(\hbar t/m)\,\partial S/\partial q_0 = (q_0 - q)S(q_0, q),$$

[with the solution]

$$S(q_0, q) = \text{const}\,\exp\left[(im/\hbar t)\int(q - q_0)\,dq_0\right]. \tag{13}$$

$S\bar{S}$ is therefore independent of q_0. In other words, if at the time $t = 0$ we know q_0 exactly, then at any time $t > 0$ all values of q are equally probable; that is, the probability that q lies in any finite region is quite nil. Physically this is already clear without further investigation. Thus the exact determination of q_0 leads to an infinitely large Compton recoil. The same would naturally apply for an arbitrary mechanical system. If, however, q_0 is known at the time $t = 0$ only within the range q_1, and p_0 in the range p_1 [see equation (3)], then

$$S(\eta, q_0) = \text{const}\,\exp[-(q_0 - q')^2/2q_1^2 - (i/\hbar)p'(q_0 - q')],$$

and the probability [amplitude] function for q is to be calculated according to the formula,

$$S(\eta, q) = \int S(\eta, q_0)S(q_0, q)\,dq_0.$$

The result is

$$S(\eta, q) = \text{const}\int\exp\{(im/\hbar t)[q_0(q - tp'/m) - q_0^2/2] - (q' - q_0)^2/2q_1^2\}\,dq_0. \tag{14}$$

With the abbreviation

$$\beta = t\hbar/mq_1^2, \tag{15}$$

the exponent in (14) becomes

* The word "representation," not employed by Heisenberg himself, is introduced here for clarity. He uses the phrase "matrix scheme." (Translators' note.)

$$-\{q_0^2(1 + i/\beta) - 2q_0[q' + i(q - tp'/m)/\beta] + q'^2\}/2q_1^2.$$

The term with q'^2 can be taken into the constant (q-independent factor) and the integration gives

$$\begin{aligned} S(\eta, q) &= \text{const}\exp\{[q' + i(q - tp'/m)/\beta]^2/[2(1 + i/\beta)q_1^2]\} \\ &= \text{const}\exp - \{[(q - tp'/m - i\beta q')^2(1 - i/\beta)]/[2q_1^2(1 + \beta^2)]\}. \quad (16) \end{aligned}$$

It follows that

$$S(\eta, q)\bar{S}(\eta, q) = \text{const}\exp - \{[q - tp'/m - q']^2/q_1^2(1 + \beta^2)\}. \quad (17)$$

Thus the electron is located at the time t at the position $tp'/m + q'$ with a spread $q_1(1 + \beta^2)^{1/2}$. The "wavepacket" or, better, "probability packet" has expanded by the factor $(1 + \beta^2)^{1/2}$. According to (15), β is proportional to the time t, inversely proportional to the mass, as is entirely plausible, and inversely proportional to q_1^2. Too much precision in q_0 produces great uncertainty in p_0 and thus leads to a large uncertainty in q. The parameter η which we have brought in above for formal reasons might be left out here in all formulas, as it does not enter the calculation. To illustrate that the difference between the classical and the quantum statistical laws leads to gross macroscopic differences between the results of the two theories, let the reflection of a beam of electrons at a grating be discussed briefly. When the spacing of the rulings is of the order of the de Broglie wavelength of the electrons, then reflection occurs in definite, discrete directions like the reflection of light at a grating. What classical theory gives is grossly and macroscopically different. Nevertheless, from the orbit of an individual electron we can in no way find a contradiction with a classical theory. We might if we could, direct the electron, say, to a definite point on a grating ruling, and then verify that the reflection there violates classical theory. However, when we want to determine the position of the electron so precisely that we can say at what point on a grating ruling it hits, then the electron acquires through this position determination a large velocity, and the de Broglie wavelength of the electron becomes so much shorter that now the reflection really can and will take place approximately as predicted classically, without violating the laws of the quantum theory.

§4. Discussion of a Few Special Idealized Experiments

According to the physical interpretation of quantum theory aimed at here, the time of transitions or "quantum jumps" must be as concrete and determinable by measurement as, say, energies in stationary states. The spread within which such an instant is specifiable is given according to equation (2) by $h/\Delta E$, if ΔE designates the change of energy in a quantum jump.* We consider, for example, the following experiment. An atom, at time $t = 0$ in state $n = 2$ may transit, via radiation, to the ground state, $n = 1$. Then, in analogy to equation (7), we can ascribe to the atom the eigenfunction

$$S(t, p) = e^{-\alpha t}\psi(E_2, p)e^{-iE_2t/\hbar} + (1 - e^{-2\alpha t})^{1/2}\psi(E_1, p)e^{-iE_1t/\hbar} \qquad (18)$$

if we assume that the radiation damping is expressed in a factor of the form $e^{-\alpha t}$ in the eigenfunction (the real dependence is perhaps not so simple). This atom is sent through an inhomogeneous magnetic field for the determination of its energy level, as is usual in the Stern-Gerlach experiment; yet we also have the inhomogeneous field follow the atomic beam over a long stretch of its path. The consequent acceleration we will measure, say, in this way: we divide the entire stretch that the atomic beam pursues in the magnetic field into short sections, at the end of each of which we determine the deviation of the beam. Depending on the velocity of the atomic beam, the division into intervals of space corresponds for the atom to division into small time intervals Δt. According to §1, equation (2), there corresponds to a time interval Δt a spread in energy of $h/\Delta t$. The probability of measuring the definite energy E can be directly deduced from $S(p, E)$ and is therefore calculated for the interval from $n\,\Delta t$ to $(n + 1)\,\Delta t$ *as*,

$$\underset{n\,\Delta t \to (n+1)\,\Delta t}{S(p, E)} = \int_{n\,\Delta t}^{(n+1)\,\Delta t} S(p, t)e^{iEt/\hbar}\,dt.$$

If the determination "state $n = 2$" is made at the time $(n + 1)\,\Delta t$, then for everything later one must ascribe to the atom not the eigenfunction (18) but one which results from (18) when t is replaced by $t - (n + 1)\,\Delta t$. If, on the contrary, one finds "state $n = 1$," then from that point on one has to attribute to the atom the eigenfunction

$$\psi(E_1, p)e^{-iE_1t/\hbar}.$$

Thus one will first find for a series of intervals Δt "state $n = 2$," then steadily "state $n = 1$." In order that a distinction between the two states will still be possible, Δt cannot be shrunk below $h/\Delta E$. Thus the instant of the transition is determinable within this spread. We imply an experiment of the kind just sketched quite in the spirit of the old formulation of quantum theory founded by Planck, Einstein, and Bohr when we speak of the discontinuous change of the energy. As such an experiment can in principle be carried out, an agreement about its outcome must be possible.

In Bohr's basic postulates of quantum theory, the energy of an atom has the

* Compare W. Pauli, *l.c.*, p. 12.

advantage—just as do the values of the action variables J—over other determinants of the motion (position of the electron, etc.) in that its numerical value can always be given. This preferred position which the energy has over other quantum-mechanical quantities it owes only to the circumstance that it represents an integral of the equations of motion for closed systems (the energy matrix $\mathbf{E}$ is a constant). For open systems, in contrast, the energy is not singled out over any other quantum-mechanical quantity. In particular, one will be able to devise experiments in which the phases, w, of the atom are precisely measurable, but in which then the energy remains in principle undetermined, corresponding to the relation $\mathbf{Jw} - \mathbf{wJ} = -i\hbar$ or $J_1 w_1 \sim h$. Resonance fluorescence is such an experiment. If one irradiates an atom with an eigenfrequency, say $\nu_{12} = (E_2 - E_1)/h$, then the atom vibrates in phase with the external radiation. Then, even in principle, it makes no sense to ask in which state, E_1 or E_2, the atom is thus vibrating. The phase relation between atom and external radiation may be determined, for example, by the phase relations of large numbers of atoms with one another ([R. W.] Wood's experiments). If one prefers to avoid experiments with radiation then one can also measure the phase relation by carrying out exact position determinations on the electron in the sense of §1 at different times relative to the phase of the light impinging (on many atoms). A "wave function," say, of the form,

$$S(q, t) = c_2 \psi_2(E_2, q) e^{-i(E_2 t + \beta)/\hbar} + (1 - c_2^2)^{1/2} \psi_1(E_1, q) e^{-iE_1 t/\hbar}, \tag{19}$$

can be ascribed to the individual atom. Here c_2 depends on the strength and β on the phase of the incident light. The probability of a definite position q is thus

$$\begin{aligned} S(q, t)\bar{S}(q, t) = {} & c_2^2 \psi_2 \bar{\psi}_2 + (1 - c_2^2)\psi_1 \bar{\psi}_1 \\ & + c_2(1 - c_2^2)^{1/2} \{\psi_2 \bar{\psi}_1 e^{-i[(E_2 - E_1)t + \beta]/\hbar} + \bar{\psi}_2 \psi_1 e^{+i[(E_2 - E_1)t + \beta]/\hbar}\}. \end{aligned} \tag{20}$$

The periodic term in (20) is experimentally distinguishable from the unperiodic ones, as the determinations of position can be carried out for different phases of the incident light.

In a well-known idealized experiment proposed by Bohr, the atoms of a Stern-Gerlach atomic beam are first excited to a resonance fluorescence at a definite state by incident radiation. After a little way they go through an inhomogeneous magnetic field. The radiation emerging from the atoms can be observed during the whole path, before and after the magnetic field. Before the atoms enter the magnetic field, ordinary resonance radiation takes place; that is, as in dispersion theory, it must be assumed that all atoms send out spherical waves in phase with the incident light. The latter view at first sight contradicts the result that a crude application of the quantum theory of light or the basic rules of quantum theory would give. Thus from this view one would conclude that only a few atoms are raised to the "upper state" through absorption of the light quantum, and that therefore the entire resonance radiation arises from a few intensively radiating centers. It therefore seemed natural in earlier times to say that the concept of light quanta ought to be brought in here only to account for the balance of energy and momentum, and that "in reality" all atoms in the ground state radiate weak and coherent spherical waves. After the atoms have passed the magnetic field, however, there can hardly be any doubt that the atomic beam has divided into two beams of which one corresponds to atoms in the upper state, the other in the lower. If now the atoms in the lower state were to radiate, then we would have a gross violation of the law of conservation of energy. For all the energy of excitation resides in the beam with atoms in the upper state. Still less can there be any doubt that past the magnetic field only the "upper state" beam sends out light, and indeed incoherent light, from the few intensively radiating atoms in the upper state. As Bohr has shown, this idealized experiment makes it especially clear that care is often needed in applying the concept of "stationary state." The formulation of quantum theory developed here allows a discussion of the Bohr experiment to be carried through without any difficulty. In the external radiation field the phases of the atoms are determined. Therefore it is meaningless to speak of the "energy of the atom." Also, after the atom has left the radiation field one is not entitled to say that it is in a definite stationary state, insofar as one enquires about the coherence properties of the radiation. However, one can set up an experiment to find out in what state the atom is. The result of this experiment can be stated only statistically. Such an experiment is really performed by the inhomogeneous magnetic field. Beyond the magnetic field the energies of the atoms are well determined and therefore the phases are indeterminate. The resulting radiation is incoherent and comes only from atoms in the upper state. The magnetic field determines the energies and therefore destroys the phase relation. Bohr's idealized experiment is a very beautiful illustration of the fact that the energy of the atom "in reality" is not a number but a matrix. The conservation law holds for the energy matrix and therefore also for the value of the energy as precisely as it can be measured. In mathematical terms the lifting of the phase relation can be traced out as follows, for example. Let Q be the coordinates of the center of gravity of the atom, so that one ascribes to the atom, instead of (19), the eigenfunction

$$S(Q, t)S(q, t) = S(Q, q, t). \tag{21}$$

Here $S(Q, t)$ is a function that, like $S(\eta, q)$ in (16), differs from zero only in a small neighborhood of a point in Q-space and propagates with the velocity of the atoms in the direction of the beam. The probability of a relative amplitude q regardless of Q is given by the integral of

$$S(Q, q, t)\bar{S}(Q, q, t)$$

over Q—that is, through (20). However, the eigenfunction (21) will change by a calculable amount in a magnetic field and, on account of the different deviation of atoms in the upper and lower state, will have changed beyond the magnetic field into

$$S(Q, q, t) = c_2 S_2(Q, t)\psi_2(E_2, q)e^{-i(E_2 t + \beta)/\hbar} + (1 - c_2^2)^{1/2} S_1(Q, t)\psi_1(E_1, q)e^{-iE_1 t/\hbar}. \quad (22)$$

Here $S_1(Q, t)$ and $S_2(Q, t)$ will be functions in Q-space which differ from zero only in the small neighborhood of a point; but this point is different for S_1 and S_2. The product $S_1 S_2$ is therefore zero everywhere. The probability of a relative amplitude q and a definite value Q is therefore

$$S(Q, q, t)\bar{S}(Q, q, t) = c_2^2 S_2 \bar{S}_2 \psi_2 \bar{\psi}_2 + (1 - c_2^2) S_1 \bar{S}_1 \psi_1 \bar{\psi}_1. \quad (23)$$

The periodic term of (20) has disappeared and with it the possibility of measuring a phase relation. The statistical result of position determinations will always be the same, whatever the phase of the incident radiation. We may assume that experiments with radiation, the theory of which has not yet been developed, will give the same results about phase relations between atoms and incident radiation.

Finally let us examine the connection* of equation (2), $E_1 t_1 \sim h$, with a complex of problems which Ehrenfest and other investigators have discussed on the basis of Bohr's correspondence principle in two important papers.† Ehrenfest and Tolman speak of "weak quantization" when the quantized periodic motion is interrupted through quantum jumps or rather perturbations in intervals of time which can be regarded as not very long compared to the periods of the system. These cases should reveal not only the exact quantum energy values but also energy values which do not differ too much from the quantum values, and these with a smaller and qualitatively predictable *a priori* probability. In quantum mechanics this behavior is to be interpreted in these terms. As the energy is really changed by external perturbations or quantum jumps, every energy measurement, insofar as it is to be unique, must be done in the time between two perturbations. In this way an upper bound is specified for t_1 in the sense of §1. Therefore we measure the energy value E_0 of a quantized state also only within a spread $E_1 \sim h/t_1$. Here the question is meaningless even in principle whether the system "really" takes on with the correspondingly lower statistical weight such energy values E as deviate from E_0, or whether their experimental realization is to be attributed only to the inaccuracy of the measurement. If t_1 is smaller than the period of the system then it is no longer meaningful to speak of discrete stationary states or discrete energy values.

* W. Pauli drew my attention to this connection.

† P. Ehrenfest and G. Breit (*Zeits. f. Physik*, *9*, 207 [1922]) and P. Ehrenfest and R. C. Tolman (*Phys. Rev.*, *24*, 287 [1924]). See also the discussion in N. Bohr, *Grundpostulate der Quantentheorie*, *l.c.*

Ehrenfest and Breit in a similar connection draw attention to the following paradox. A rotator, which we will visualize as a gear-wheel, is provided with an attachment which after f revolutions of the wheel exactly reverses the direction of its rotation. For example, let the gear-wheel mesh with a toothed sliding member which moves on a straight line between two stops. The slider hits a stop after a definite number of rotations and in that way reverses the rotation of the gear-wheel. The true period T of the system is long in comparison with the rotation period t of the wheel. The discrete energy levels are densely packed—and the denser the packing, the greater the value of T. From the standpoint of consistent quantum theory all stationary states have the same statistical weight. Therefore, for sufficiently great T, practically all energy values occur with equal frequency, in opposition to what would be expected for the rotator. We may sharpen this paradox a little before we treat it from our standpoint. Thus, in order to determine whether the system takes on the discrete energy values belonging to the pure rotator exclusively or particularly often, or whether it assumes with equal probability all possible values (that is, values which correspond to the small energy interval h/T), a time t_1 suffices which is small relative to T (but $\gg t$). In other words, although the long period plays no part at all in such measurements, it appears to express itself in the fact that all possible energy values can occur. We are of the view that, in reality also, such experiments for the determination of the total energy of the system would give all possible energy values with equal probability. The factor responsible for this outcome is not the big period T, but the sliding member. Even if the system sometimes happens to have an energy identical with the quantized energy value of the simple rotator, it can be modified easily—by external forces acting on the stop—to states which do not correspond to the quantization of the simple rotator.* The coupled system, rotator-plus-slider, indeed shows a periodicity entirely different from that of the rotator. The solution of the paradox lies rather in a different circumstance. When we want to measure the energy of the rotator alone, we must first break the coupling between the rotator and the slider. In classical theory, when the mass of the slider is sufficiently small, the coupling can be broken without a change in energy; and there, consequently, the energy of the entire system can be equated to that of the rotator (for small slider mass). In quantum mechanics the energy of interaction between slider and rotator is at least of the same order of magnitude as the level spacing of the rotator (even for small slider mass there is a high zero point energy associated with the elastic interaction between rotator and slider). On decoupling, the slider and the rotator individually take their characteristic quantum energies. Consequently, insofar as we are able to measure the energy values of the rotator alone we always find the quantum energy values with experimental accuracy. Even for

* According to Ehrenfest and Breit this cannot happen, or can happen only rarely, through forces which act on the gear-wheel.

vanishing mass of the slider, however, the energy of the coupled system is different from the energy of the rotator. The energy of the coupled system can take on all possible values (consistent with the T-quantization) with equal probability.

Quantum kinematics and mechanics show far-reaching differences from the ordinary theory. The applicability of classical kinematics and mechanical concepts, however, can be justified neither from our laws of thought nor from experiment. The basis for this conclusion is relation (1), $p_1 q_1 \sim h$. As momentum, position, energy, etc. are precisely defined concepts, one does not need to complain that the basic equation (1) contains only qualitative predictions. Moreover, as we can think through qualitatively the experimental consequences of the theory in all simple cases, we will no longer have to look at quantum mechanics as unphysical and abstract.* Of course we would also like to be able to derive, if possible, the quantitative laws of quantum mechanics directly from the physical foundations—that is, essentially, from relation (1). On this account Jordan has sought to interpret the equation,

$$S(q, q'') = \int S(q, q')S(q', q'')\,dq',$$

as a probability relation. However, we cannot accept this interpretation (§2). We believe, rather, that for the time being the quantitative laws can be derived out of the physical foundations only by use of the principle of maximum simplicity. If, for example, the X-coordinate of the electron is no longer a "number," as can be concluded experimentally, according to equation (1), then the simplest assumption conceivable [that does not contradict (1)] is that this X-coordinate is a diagonal term of a matrix whose nondiagonal terms express themselves in an uncertainty or—by transformation—in other ways (see for example §4). The prediction that, say, the velocity in the X-direction is "in reality" not a number but the diagonal term of the matrix, is perhaps no more abstract and no more unvisualizable than the statement that the electric field strengths are "in reality" the time part of an antisymmetric tensor located in space-time. The phrase "in reality" here is as much and as little justified as it is in any mathematical description of natural processes. As soon as one accepts that all quantum-theoretical quantities are "in reality" matrices, the quantitative laws follow without difficulty.

If one assumes that the interpretation of quantum mechanics is already correct

* Schrödinger describes quantum mechanics as a formal theory of frightening, indeed repulsive, abstractness and lack of visualizability. Certainly one cannot overestimate the value of the mathematical (and to that extent physical) mastery of the quantum-mechanical laws that Schrödinger's theory has made possible. However, as regards questions of physical interpretation and principle, the popular view of wave mechanics, as I see it, has actually deflected us from exactly those roads which were pointed out by the papers of Einstein and de Broglie on the one hand and by the papers of Bohr and by quantum mechanics on the other hand.

in its essential points, it may be permissible to outline briefly its consequences of principle. We have not assumed that quantum theory—in opposition to classical theory—is an essentially statistical theory in the sense that only statistical conclusions can be drawn from precise initial data. The well-known experiments of Geiger and Bothe, for example, speak directly against such an assumption. Rather, in all cases in which relations exist in classical theory between quantities which are really all exactly measurable, the corresponding exact relations also hold in quantum theory (laws of conservation of momentum and energy). But what is wrong in the sharp formulation of the law of causality, "When we know the present precisely, we can predict the future," is not the conclusion but the assumption. Even in principle we cannot know the present in all detail. For that reason everything observed is a selection from a plenitude of possibilities and a limitation on what is possible in the future. As the statistical character of quantum theory is so closely linked to the inexactness of all perceptions, one might be led to the presumption that behind the perceived statistical world there still hides a "real" world in which causality holds. But such speculations seem to us, to say it explicitly, fruitless and senseless. Physics ought to describe only the correlation of observations. One can express the true state of affairs better in this way: Because all experiments are subject to the laws of quantum mechanics, and therefore to equation (1), it follows that quantum mechanics establishes the final failure of causality.

Addition in Proof

After the conclusion of the foregoing paper, more recent investigations of Bohr have led to a point of view which permits an essential deepening and sharpening of the analysis of quantum-mechanical correlations attempted in this work. In this connection Bohr has brought to my attention that I have overlooked essential points in the course of several discussions in this paper. Above all, the uncertainty in our observation does not arise exclusively from the occurrence of discontinuities, but is tied directly to the demand that we ascribe equal validity to the quite different experiments which show up in the corpuscular theory on one hand, and in the wave theory on the other hand. In the use of an idealized gamma-ray microscope, for example, the necessary divergence of the bundle of rays must be taken into account. This has as one consequence that in the observation of the position of the electron the direction of the Compton recoil is only known with a spread which then leads to relation (1). Furthermore, it is not sufficiently stressed that the simple theory of the Compton effect, strictly speaking, only applies to free electrons. The consequent care needed in employing the uncertainty relation is, as Professor Bohr has explained, essential, among other things, for a comprehensive discussion of the transition from micro- to macromechanics. Finally, the discussion of resonance fluorescence is not entirely correct because the connection between the phase of the light and that of the electronic motion is not so simple as was assumed. I owe great thanks to Professor Bohr for sharing with me at an early stage the results of these more recent investigations of his—to appear soon in a paper on the conceptual structure of quantum theory—and for discussing them with me.

Copenhagen, Institute for Theoretical Physics of the University.

I.4 COMPLEMENTARITY

Commentary of Rosenfeld (1963)

Complementarity is no system, no doctrine with ready-made precepts. There is no via regia to it; no formal definition of it can even be found in Bohr's writings, and this worries many people. The French are shocked by this breach of the Cartesian rules; they blame Bohr for indulging in "clair-obscur" and shrouding himself in "les brumes du Nord." The Germans in their thoroughness have been at work distinguishing several forms of complementarity and studying, in hundreds of pages, their relations to Kant. Pragmatic Americans have dissected complementarity with the scalpel of symbolic logic and undertaken to define this gentle art of the correct use of words without using any words at all. Bohr was content to teach by example. He often evoked the thinkers of the past who had intuitively recognized dialectical aspects of existence and endeavored to give them poetical or philosophical expression; our only advantage over these great men, he would observe, is that in physics we have been presented with such a simple and clear case of complementarity that we are able to study it in detail and thus arrive at a precise formulation of a logical relationship of universal scope. The nature of this relation he regarded as sufficiently illustrated by his analyses of the limits of validity of classical physical concepts.

Commentary of Rosenfeld (1971a; continued from §I. 3 above)

As to Bohr's "forthcoming" publication, more than a year elapsed before it appeared in print: it was a much furbished version of a lecture he delivered, shortly after the events just retraced, at a physicists' conference in Como. He had the greatest misgivings about presenting his conception of complementarity to the community of physicists in a state which he judged immature; but he yielded to the advice of his more practically-minded brother Harald. Upon the latter's urging, he even consented to write up a brief account, that could be promptly published in *Nature*, as a letter to the editor: but this letter never reached its destination. With the help of Klein, he actually managed to complete it just on the night of his reluctant departure for Como. However, when Klein came up to the Institute the next morning, and enquired whether the letter had been sent off, he learned that there had been a double hitch, of a kind to delight Freudians. Bohr had missed the night train, because he could not find his passport (which lay on his desk); he had departed by the next train, taking with him the famous letter.

Commentary of Heisenberg (1967; continued from §I. 3 above)

How closely the idea of complementarity was in accord with Bohr's older philosophical ideas became apparent through an episode, which, if I remember correctly, took place on a sailing trip from Copenhagen to Svendborg on the island Fyn. At that time Bohr and a colleague and friend owned a sailing boat, the captain of which was the brilliant and extremely charming chemist Bjerrum. The distinguished surgeon Chievitz kept spirits high even in stormy weather, and the other friends contributed each in his way to this happy and untroubled existence. Bohr was full of the new interpretation of quantum theory, and as the boat took us full sail southwards in sunshine, there was plenty of time to tell of this scientific event and to reflect philosophically on the nature of atomic theory. Bohr began by talking of the difficulties of language, of the limitations of all our means of expressing ourselves, which one had to take into account from the very beginning if one wants to practice science, and he explained how satisfying it was that this limitation had already been expressed in the foundation of atomic theory in a mathematically lucid way. Finally, one of the friends remarked drily, "But, Niels, this is not really new, you said exactly the same ten years ago."

I.4 THE QUANTUM POSTULATE AND THE RECENT DEVELOPMENT OF ATOMIC THEORY

NIELS BOHR

Although it is with great pleasure that I follow the kind invitation of the presidency of the congress to give an account of the present state of the quantum theory in order to open a general discussion on this subject, which takes so central a position in modern physical science, it is with a certain hesitation that I enter on this task. Not only is the venerated originator of the theory present himself, but among the audience there will be several who, due to their participation in the remarkable recent development, will surely be more conversant with details of the highly developed formalism than I am. Still I shall try, by making use only of simple considerations and without going into any details of technical mathematical character, to describe to you a certain general point of view which I believe is suited to give an impression of the general trend of the development of the theory from its very beginning and which I hope will be helpful in order to harmonize the apparently conflicting views taken by different scientists. No subject indeed may be better suited than the quantum theory to mark the development of physics in the century passed since the death of the great genius, whom we are here assembled to commemorate. At the same time, just in a field like this where we are wandering on new paths and have to rely

Originally published in *Nature, 121*, 580-90 (1928). Reprinted in Niels Bohr (1934), *Atomic Theory and the Description of Nature*, Cambridge University Press, pp. 52-91, from which book this paper is reproduced here.

upon our own judgment in order to escape from the pitfalls surrounding us on all sides, we have perhaps more occasion than ever at every step to be remindful of the work of the old masters who have prepared the ground and furnished us with our tools.

I. QUANTUM POSTULATE AND CAUSALITY

The quantum theory is characterized by the acknowledgment of a fundamental limitation in the classical physical ideas when applied to atomic phenomena. The situation thus created is of a peculiar nature, since our interpretation of the experimental material rests essentially upon the classical concepts. Notwithstanding the difficulties which, hence, are involved in the formulation of the quantum theory, it seems, as we shall see, that its essence may be expressed in the so-called quantum postulate, which attributes to any atomic process an essential discontinuity, or rather individuality, completely foreign to the classical theories and symbolized by Planck's quantum of action.

This postulate implies a renunciation as regards the causal space-time co-ordination of atomic processes. Indeed, our usual description of physical phenomena is based entirely on the idea that the phenomena concerned may be observed without disturbing them appreciably. This appears, for example, clearly in the theory of relativity, which has been so fruitful for the elucidation of the classical theories. As emphasized by Einstein, every observation or measurement ultimately rests on the coincidence of two independent events at the same space-time point. Just these coincidences will not be affected by any differences which the space-time co-ordination of different observers otherwise may exhibit. Now, the quantum postulate implies that any observation of atomic phenomena will involve an interaction with the agency of observation not to be neglected. Accordingly, an independent reality in the ordinary physical sense can neither be ascribed to the phenomena nor to the agencies of observation. After all, the concept of observation is in so far arbitrary as it depends upon which objects are included in the system to be observed. Ultimately, every observation can, of course, be reduced to our sense perceptions. The circumstance, however, that in interpreting observations use has always to be made of theoretical notions entails that for every particular case it is a question of convenience at which point the concept of observation involving the quantum postulate with its inherent "irrationality" is brought in.

This situation has far-reaching consequences. On one hand, the definition of the state of a physical system, as ordinarily understood, claims the elimination of all external disturbances. But in that case, according to the quantum postulate, any observation will be impossible, and, above all, the concepts of space and time lose their immediate sense. On the other hand, if in order to make observation possible we permit certain interactions with suitable agencies of measurement, not belonging to the system, an unambiguous definition of the state of the system is naturally no longer possible, and there can be no question of causality in the ordinary sense of the word. The very nature of the quantum theory thus forces us to regard the space-time co-ordination and the claim of causality, the union of which characterizes the classical theories, as complementary but exclusive features of the description, symbolizing the idealization of observation and definition respectively. Just as the relativity theory has taught us that the convenience of distinguishing sharply between space and time rests solely on the smallness of the velocities ordinarily met with compared to the velocity of light, we learn from the quantum theory that the appropriateness of our usual causal space-time description depends entirely upon the small value of the quantum of action as compared to the actions involved in ordinary sense perceptions. Indeed, in the description of atomic phenomena, the quantum postulate presents us with the task of developing a "complementarity" theory the consistency of which can be judged only by weighing the possibilities of definition and observation.

This view is already clearly brought out by the much-discussed question of the nature of light and the ultimate constituents of matter. As regards light, its propagation in space and time is adequately expressed by the electromagnetic theory. Especially the interference phenomena *in vacuo* and the optical properties of material media are completely governed by the wave theory superposition principle. Nevertheless, the conservation of energy and momentum during the interaction between radiation and matter, as evident in the photo-electric and Compton effect, finds its adequate expression just in the light quantum idea put forward by Einstein. As is well known, the doubts regarding the validity of the superposition principle, on one hand, and of the conservation laws, on the other, which were suggested by this apparent contradiction, have been definitely disproved through direct experiments. This situation would seem clearly to indicate the impossibility of a causal space-time description of the light phenomena. On one hand, in attempting to trace the laws of the time-spatial propagation of light according to the quantum postulate, we are confined to statistical considerations. On the other hand, the fulfilment of the claim of causality for the individual light processes, characterized by the quantum of action, entails a renunciation as regards the space-time description. Of course, there can be no question of a quite independent application of the ideas of space and time and of causality. The two views of the nature of light are rather to be considered as different attempts at an interpretation of experimental evidence in which the limitation of the classical concepts is expressed in complementary ways.

The problem of the nature of the constituents of matter presents us with an analogous situation. The individuality of the elementary electrical corpuscles is forced upon us by general evidence. Nevertheless, recent experience, above all the discovery of the selective reflection of electrons from metal crystals, requires the use of the wave theory superposition principle in accordance with the original ideas of L. de Broglie. Just as in the case of light, we have consequently in the question of the nature of matter, so far as we adhere to classical concepts, to face an inevitable dilemma which has to be regarded as the very expression of experimental evidence. In fact, here again we are not dealing with contradictory but with complementary pictures of the phenomena, which only together offer a natural generalization of the classical mode of description. In the discussion of these questions, it must be kept in mind that, according to the view taken above, radiation in free space as well as isolated material particles are abstractions, their properties on the quantum theory being definable and observable only through their interaction with other systems. Nevertheless, these abstractions are, as we shall see, indispensable for a description of experience in connection with our ordinary space-time view.

The difficulties with which a causal space-time description is confronted in the quantum theory, and which have been the subject of repeated discussions, are now placed into the foreground by the recent development of the symbolic methods. An important contribution to the problem of a consistent application of these methods has been made lately by Heisenberg. In particular, he has stressed the peculiar reciprocal uncertainty which affects all measurements of atomic quantities. Before we enter upon his results, it will be advantageous to show how the complementary nature of the description appearing in this uncertainty is unavoidable already in an analysis of the most elementary concepts employed in interpreting experience.

2. QUANTUM OF ACTION AND KINEMATICS

The fundamental contrast between the quantum of action and the classical concepts is immediately apparent from the simple formulae which form the common foundation of the theory of light quanta and of the wave theory of material particles. If Planck's constant be denoted by h, as is well known,

$$E\tau = I\lambda = h, \qquad \ldots\ldots(1)$$

where E and I are energy and momentum respectively, τ and λ the corresponding period of vibration and wavelength. In these formulae the two notions of light and also of matter enter in sharp contrast. While energy and momentum are associated with the concept of particles, and, hence, may be characterized according to the classical point of view by definite space-time co-ordinates, the period of vibration and wave-length refer to a plane harmonic wave train of unlimited extent in space and time. Only with the aid of the superposition principle does it become possible to attain a connection with the ordinary mode of description. Indeed, a limitation of the extent of the wave-fields in space and time can always be regarded as resulting from the interference of a group of elementary harmonic waves. As shown by de Broglie, the translational velocity of the individuals associated with the waves can be represented by just the so-called group-velocity. Let us denote a plane elementary wave by

$$A \cos 2\pi\,(\nu t - x\sigma_x - y\sigma_y - z\sigma_z + \delta),$$

where A and δ are constants determining respectively the amplitude and the phase. The quantity $\nu = 1/\tau$ is the frequency, σ_x, σ_y, σ_z the wave numbers in the direction of the co-ordinate axes, which may be regarded as vector components of the wave number $\sigma = 1/\lambda$ in the directions of propagation. While the wave or phase velocity is given by ν/σ, the group-velocity is defined by $d\nu/d\sigma$. Now according to the relativity theory we have for a particle with the velocity v:

$$I = \frac{v}{c^2} E \text{ and } v dI = dE,$$

where c denotes the velocity of light. Hence by equation (1) the phase velocity is c^2/v and the group-velocity v. The circumstance that the former is in general greater than the velocity of light emphasizes the symbolic character of these considerations. At the same time, the possibility of identifying the velocity of the particle with the group-velocity indicates the field of application of space-time pictures in the quantum theory. Here the complementary character of the description appears, since the use of wave-groups is necessarily accompanied by a lack of sharpness in the definition of period and wave-length, and hence also in the definition of the corresponding energy and momentum as given by relation (1).

Rigorously speaking, a limited wave-field can only be obtained by the superposition of a manifold of elementary waves corresponding to all values of ν and σ_x, σ_y, σ_z. But the order of magnitude of the mean difference between these values for two elementary waves in the group is given in the most favourable case by the condition

$$\Delta t \, \Delta\nu = \Delta x \, \Delta\sigma_x = \Delta y \, \Delta\sigma_y = \Delta z \, \Delta\sigma_z = 1,$$

where Δt, Δx, Δy, Δz denote the extension of the wave-field in time and in the directions of space corresponding to the co-ordinate axes. These relations—well known from the theory of optical instruments, especially from Rayleigh's investigation of the resolving power of spectral apparatus—express the condition that the wave-trains extinguish each other by interference at the space-time boundary of the wave-field. They may be regarded also as signifying that the group as a whole has no phase in the same sense as the elementary waves. From equation (1) we find thus:

$$\Delta t \, \Delta E = \Delta x \, \Delta I_x = \Delta y \, \Delta I_y = \Delta z \, \Delta I_z = h, \quad \ldots\ldots(2)$$

as determining the highest possible accuracy in the

definition of the energy and momentum of the individuals associated with the wave-field. In general, the conditions for attributing an energy and a momentum value to a wave-field by means of formula (1) are much less favourable. Even if the composition of the wave-group corresponds in the beginning to the relations (2), it will in the course of time be subject to such changes that it becomes less and less suitable for representing an individual. It is this very circumstance which gives rise to the paradoxical character of the problem of the nature of light and of material particles. The limitation in the classical concepts expressed through relation (2), is, besides, closely connected with the limited validity of classical mechanics, which in the wave theory of matter corresponds to the geometrical optics in which the propagation of waves is depicted through "rays". Only in this limit can energy and momentum be unambiguously defined on the basis of space-time pictures. For a general definition of these concepts we are confined to the conservation laws, the rational formulation of which has been a fundamental problem for the symbolical methods to be mentioned below.

In the language of the relativity theory, the content of the relations (2) may be summarized in the statement that according to the quantum theory a general reciprocal relation exists between the maximum sharpness of definition of the space-time and energy-momentum vectors associated with the individuals. This circumstance may be regarded as a simple symbolical expression for the complementary nature of the space-time description and the claims of causality. At the same time, however, the general character of this relation makes it possible to a certain extent to reconcile the conservation laws with the space-time co-ordination of observations, the idea of a coincidence of well-defined events in a space-time point being replaced by that of unsharply defined individuals within finite space-time regions.

This circumstance permits us to avoid the well-known paradoxes which are encountered in attempting to describe the scattering of radiation by free electrical particles as well as the collision of two such particles. According to the classical concepts, the description of the scattering requires a finite extent of the radiation in space and time, while in the change of the motion of the electron demanded by the quantum postulate one seemingly is dealing with an instantaneous effect taking place at a definite point in space. Just as in the case of radiation, however, it is impossible to define momentum and energy for an electron without considering a finite space-time region. Furthermore, an application of the conservation laws to the process implies that the accuracy of definition of the energy-momentum vector is the same for the radiation and the electron. In consequence, according to relation (2), the associated space-time regions can be given the same size for both individuals in interaction.

A similar remark applies to the collision between two material particles, although the significance of the quantum postulate for this phenomenon was disregarded before the necessity of the wave concept was realized. Here, this postulate does, indeed, represent the idea of the individuality of the particles which, transcending the space-time description, meets the claim of causality. While the physical content of the light-quantum idea is wholly connected with the conservation theorems for energy and momentum, in the case of the electrical particles the electric charge has to be taken into account in this connection. It is scarcely necessary to mention that for a more detailed description of the interaction between individuals we cannot restrict ourselves to the facts expressed by formulae (1) and (2), but must resort to a procedure which allows us to take into account the coupling of the individuals, characterizing the interaction in question, where just the importance of the electric charge appears. As we shall see, such a procedure necessitates a further departure from visualization in the usual sense.

3. MEASUREMENTS IN THE QUANTUM THEORY

In his investigations already mentioned on the consistency of the quantum-theoretical methods, Heisenberg has given the relation (2) as an expression for the maximum precision with which the space-time co-ordinates and momentum-energy components of a particle can be measured simultaneously. His view was based on the following consideration: On one hand, the co-ordinates of a particle can be measured with any desired degree of accuracy by using, for example, an optical instrument, provided radiation of sufficiently short wave-length is used for illumination. According to the quantum theory, however, the scattering of radiation from the object is always connected with a finite change in momentum, which is the larger the smaller the wave-length of the radiation used. The momentum of a particle, on the other hand, can be determined with any desired degree of accuracy by measuring, for example, the Doppler effect of the scattered radiation, provided the wave-length of the radiation is so large that the effect of recoil can be neglected, but then the determination of the space co-ordinates of the particle becomes correspondingly less accurate.

The essence of this consideration is the inevitability of the quantum postulate in the estimation of the possibilities of measurement. A closer investigation of the possibilities of definition would still seem necessary in order to bring out the general complementary character of the description. Indeed, a discontinous change of energy and momentum during observation could not prevent us from ascribing accurate values to the space-time co-ordinates, as well as to the momentum-energy components before and after the process. The reciprocal uncertainty which always affects the values of these quantities is, as will be clear from the preceding analysis, essentially an outcome of the limited accuracy with which changes in energy and momentum can be defined, when the wave-fields used for the determination of the space-time co-ordinates of the particle are sufficiently small.

In using an optical instrument for determinations of position, it is necessary to remember that the formation of the image always requires a convergent beam of light. Denoting by λ the wave-length of the radiation used, and by ϵ the so-called numerical aperture, that is, the sine of half the angle of convergence, the resolving power of a microscope is given by the well-known expression $\lambda/2\epsilon$. Even if the object is illuminated by parallel light, so that the momentum h/λ of the incident light quantum is known both as regards magnitude and direction, the finite value of the aperture will prevent an exact knowledge of the recoil accompanying the scattering. Also, even if the momentum of the particle were accurately known before the scattering process, our knowledge of the component of momentum parallel to the focal plane after the observation would be affected by an uncertainty amounting to $2\epsilon h/\lambda$. The product of the least inaccuracies with which the positional co-ordinate and the component of momentum in a definite direction can be ascertained is therefore just given by formula (2). One might perhaps expect that in estimating the accuracy of determining the position, not only the convergence but also the length of the wave-train has to be taken into account, because the particle could change its place during the finite time of illumination. Due to the fact, however, that the exact knowledge of the wave-length is immaterial for the above estimate, it will be realized that for any value of the aperture the wave-train can always be taken so short that a change of position of the particle during the time of observation may be neglected in comparison to the lack of sharpness inherent in the determination of position due to the finite resolving power of the microscope.

In measuring momentum with the aid of the Doppler effect—with due regard to the Compton effect—one will employ a parallel wave-train. For the accuracy, however, with which the change in wave-length of the scattered radiation can be measured the extent of the wave-train in the direction of propagation is essential. If we assume that the directions of the incident and scattered radiation are parallel and opposite, respectively, to the direction of the position co-ordinate and momentum component to be measured, then $c\lambda/2l$ can be taken as a measure of the accuracy in the determination of the velocity, where l

denotes the length of the wave-train. For simplicity, we here have regarded the velocity of light as large compared to the velocity of the particle. If m represents the mass of the particle, then the uncertainty attached to the value of the momentum after observation is $cm\lambda/2l$. In this case the magnitude of the recoil, $2h/\lambda$, is sufficiently well defined in order not to give rise to an appreciable uncertainty in the value of the momentum of the particle after observation. Indeed, the general theory of the Compton effect allows us to compute the momentum components in the direction of the radiation before and after the recoil from the wave-lengths of the incident and scattered radiation. Even if the positional co-ordinates of the particle were accurately known in the beginning, our knowledge of the position after observation nevertheless will be affected by an uncertainty. Indeed, on account of the impossibility of attributing a definite instant to the recoil, we know the mean velocity in the direction of observation during the scattering process only with an accuracy $2h/m\lambda$. The uncertainty in the position after observation hence is $2hl/mc\lambda$. Here, too, the product of the inaccuracies in the measurement of position and momentum is thus given by the general formula (2).

Just as in the case of the determination of position, the time of the process of observation for the determination of momentum may be made as short as is desired, if only the wave-length of the radiation used is sufficiently small. The fact that the recoil then gets larger does not, as we have seen, affect the accuracy of measurement. It should further be mentioned, that in referring to the velocity of a particle as we have here done repeatedly, the purpose has only been to obtain a connection with the ordinary space-time description convenient in this case. As it appears already from the considerations of de Broglie mentioned above, the concept of velocity must always in the quantum theory be handled with caution. It will also be seen that an unambiguous definition of this concept is excluded by the quantum postulate. This is particularly to be remembered when comparing the results of successive observations. Indeed, the position of an individual at two given moments can be measured with any desired degree of accuracy; but if, from such measurements, we would calculate the velocity of the individual in the ordinary way, it must be clearly realized that we are dealing with an abstraction, from which no unambiguous information concerning the previous or future behaviour of the individual can be obtained.

According to the above considerations regarding the possibilities of definition of the properties of individuals, it will obviously make no difference in the discussion of the accuracy of measurements of position and momentum of a particle if collisions with other material particles are considered instead of scattering of radiation. In both cases, we see that the uncertainty in question equally affects the description of the agency of measurement and of the object. In fact, this uncertainty cannot be avoided in a description of the behaviour of individuals with respect to a co-ordinate system fixed in the ordinary way by means of solid bodies and unperturbable clocks. The experimental devices—opening and closing of apertures, etc.—are seen to permit only conclusions regarding the space-time extension of the associated wave-fields.

In tracing observations back to our sensations, once more regard has to be taken to the quantum postulate in connection with the perception of the agency of observation, be it through its direct action upon the eye or by means of suitable auxiliaries such as photographic plates, Wilson clouds, etc. It is easily seen, however, that the resulting additional statistical element will not influence the uncertainty in the description of the object. It might even be conjectured that the arbitrariness in what is regarded as object and what as agency of observation would open up a possibility of avoiding this uncertainty altogether. In connection with the measurement of the position of a particle, one might, for example, ask whether the momentum transmitted by the scattering could not be determined by means of the conservation theorem from a measurement of the change of momentum of the microscope—including light source and photographic plate—during the process of observation. A closer investigation shows, however, that such a measurement is impossible, if at the same time one wants to know the position of the microscope with sufficient accuracy. In fact, it follows from the experiences which have found expression in the wave theory of matter that the position of the centre of gravity of a body and its total momentum can only be defined within the limits of reciprocal accuracy given by relation (2).

Strictly speaking, the idea of observation belongs to the causal space-time way of description. Due to the general character of relation (2), however, this idea can be consistently utilized also in the quantum theory, if only the uncertainty expressed through this relation is taken into account. As remarked by Heisenberg, one may even obtain an instructive illustration of the quantum-theoretical description of atomic (microscopic) phenomena by comparing this uncertainty with the uncertainty, due to imperfect measurements, inherently contained in any observation as considered in the ordinary description of natural phenomena. He remarks on that occasion that even in the case of macroscopic phenomena we may say, in a certain sense, that they are created by repeated observations. It must not be forgotten, however, that in the classical theories any succeeding observation permits a prediction of future events with ever-increasing accuracy, because it improves our knowledge of the initial state of the system. According to the quantum theory, just the impossibility of neglecting the interaction with the agency of measurement means that every observation introduces a new uncontrollable element. Indeed, it follows from the above considerations that the measurement of the positional co-ordinates of a particle is accompanied not only by a finite change in the dynamical variables, but also the fixation of its position means a complete rupture in the causal description of its dynamical behaviour, while the determination of its momentum always implies a gap in the knowledge of its spatial propagation. Just this situation brings out most strikingly the complementary character of the description of atomic phenomena which appears as an inevitable consequence of the contrast between the quantum postulate and the distinction between object and agency of measurement, inherent in our very idea of observation.

4. CORRESPONDENCE PRINCIPLE AND MATRIX THEORY

Hitherto we have only regarded certain general features of the quantum problem. The situation implies, however, that the main stress has to be laid on the formulation of the laws governing the interaction between the objects which we symbolize by the abstractions of isolated particles and radiation. Points of attack for this formulation are presented in the first place by the problem of atomic constitution. As is well known, it has been possible here, by means of an elementary use of classical concepts and in harmony with the quantum postulate, to throw light on essential aspects of experience. For example, the experiments regarding the excitation of spectra by electronic impacts and by radiation are adequately accounted for on the assumption of discrete stationary states and individual transition processes. This is primarily due to the circumstance that in these questions no closer description of the space-time behaviour of the processes is required.

Here the contrast with the ordinary way of description appears strikingly in the circumstance that spectral lines, which on the classical view would be ascribed to the same state of the atom, will, according to the quantum postulate, correspond to separate transition processes, between which the excited atom has a choice. Notwithstanding this contrast, however, a formal connection with the classical ideas could be obtained in the limit where the relative difference in the properties of neighbouring stationary states vanishes asymptotically and where in statistical applications the discontinuities may be disregarded. Through this connection it was possible to a large extent to interpret the regularities of spectra on the basis of our ideas about the structure of the atom.

The aim of regarding the quantum theory as a rational generalization of the classical theories led to the formulation of the so-called correspondence principle. The utilization of this principle for the interpretation of spectroscopic results was based on a symbolical application of classical electrodynamics, in which the individual transition processes were each associated with a harmonic in the motion of the atomic particles to be expected according to ordinary mechanics. Except in the limit mentioned, where the relative difference between adjacent stationary states may be neglected, such a fragmentary application of the classical theories could only in certain cases lead to a strictly quantitative description of the phenomena. Especially the connection developed by Ladenburg and Kramers between the classical treatment of dispersion and the statistical laws governing the radiative transition processes formulated by Einstein should be mentioned here. Although it was just Kramers' treatment of dispersion that gave important hints for the rational development of correspondence considerations, it is only through the quantum-theoretical methods created in the last few years that the general aims laid down in the principle mentioned have obtained an adequate formulation.

As is known, the new development was commenced in a fundamental paper by Heisenberg, where he succeeded in emancipating himself completely from the classical concept of motion by replacing from the very start the ordinary kinematical and mechanical quantities by symbols which refer directly to the individual processes demanded by the quantum postulate. This was accomplished by substituting for the Fourier development of a classical mechanical quantity a matrix scheme, the elements of which symbolize purely harmonic vibrations and are associated with the possible transitions between stationary states. By requiring that the frequencies ascribed to the elements must always obey the combination principle for spectral lines, Heisenberg could introduce simple rules of calculation for the symbols which permit a direct quantum-theoretical transcription of the fundamental equations of classical mechanics. This ingenious attack on the dynamical problem of atomic theory proved itself from the beginning to be an exceedingly powerful and fertile method for interpreting quantitatively the experimental results. Through the work of Born and Jordan, as well as of Dirac, the theory was given a formulation which can compete with classical mechanics as regards generality and consistency. Especially, the element characteristic of the quantum theory, Planck's constant, appears explicitly only in the algorithms to which the symbols, the so-called matrices, are subjected. In fact, matrices, which represent canonically conjugated variables in the sense of the Hamiltonian equations, do not obey the commutative law of multiplication, but two such quantities, q and p, have to fulfil the exchange rule

$$pq - qp = \sqrt{-1}\,\frac{h}{2\pi}. \qquad \ldots\ldots(3)$$

Indeed, this exchange relation expresses strikingly the symbolical character of the matrix formulation of the quantum theory. The matrix theory has often been called a calculus with directly observable quantities. It must be remembered, however, that the procedure described is limited just to those problems, in which in applying the quantum postulate the space-time description may largely be disregarded, and the question of observation in the proper sense therefore placed in the background.

In pursuing further the correspondence of the quantum laws with classical mechanics, the stress placed on the statistical character of the quantum-theoretical description, which is brought in by the quantum postulate, has been of fundamental importance. Here the generalization of the symbolical method made by Dirac and Jordan represented a great progress by making possible the operation with matrices, which are not arranged according to the stationary states, but where the possible values of any set of variables may appear as indices of the matrix elements. In analogy to the interpretation considered in the original form of the theory of the "diagonal elements" connected only with a single stationary state, as time averages of the quantity to be represented, the general transformation theory of matrices permits the representation of such averages of a mechanical quantity, in the calculation of which any set of variables characterizing the "state" of the system has given values, while the canonically conjugated variables are allowed to take all possible values. On the basis of the procedure developed by these authors and in close connection with ideas of Born and Pauli, Heisenberg has in the paper already cited above attempted a closer analysis of the physical content of the quantum theory, especially in view of the apparently paradoxical character of the exchange relation (3). In this connection he has formulated the relation

$$\Delta q \, \Delta p \sim h \qquad \ldots\ldots(4)$$

as the general expression for the maximum accuracy with which two canonically conjugated variables can simultaneously be observed. In this way Heisenberg has been able to elucidate many paradoxes appearing in the application of the quantum postulate, and to a large extent to demonstrate the consistency of the symbolic method. In connection with the complementary nature of the quantum-theoretical description, we must, as already mentioned, constantly keep the possibilities of definition as well as of observation before the mind. For the discussion of just this question the method of wave mechanics developed by Schrödinger has, as we shall see, proved of great help. It permits a general application of the principle of superposition also in the problem of interaction, thus offering an immediate connection with the above considerations concerning radiation and free particles. Below we shall return to the relation of wave mechanics to the general formulation of the quantum laws by means of the transformation theory of matrices.

5. WAVE MECHANICS AND THE QUANTUM POSTULATE

Already in his first considerations concerning the wave theory of material particles, de Broglie pointed out that the stationary states of an atom may be visualized as an interference effect of the phase wave associated with a bound electron. It is true that this point of view at first did not, as regards quantitative results, lead beyond the earlier methods of quantum theory, to the development of which Sommerfeld has contributed so essentially. Schrödinger, however, succeeded in developing a wave-theoretical method which has opened up new aspects, and has proved to be of decisive importance for the great progress in atomic physics during the last years. Indeed, the proper vibrations of the Schrödinger wave equation have been found to furnish a representation of the stationary states of an atom meeting all requirements. The energy of each state is connected with the corresponding period of vibration according to the general quantum relation (1). Furthermore, the number of nodes in the various characteristic vibrations gives a simple interpretation to the concept of quantum number which was already known from the older methods, but at first did not seem to appear in the matrix formulation. In addition, Schrödinger could associate with the solutions of the wave equation a continuous distribution of charge and current which, if applied to a characteristic vibration, represents the electrostatic and magnetic properties of an atom in the corresponding stationary state. Similarly, the superposition of two characteristic solutions corresponds to a continuous vibrating distribution of electrical charge, which on classical electrodynamics would give rise to an emission of radiation, illustrating instructively the consequences of the quantum postulate and the correspondence requirement regarding the transition process between two stationary states formulated in matrix mechanics. Another application of the method of Schrödinger, important for the further development, has been made by Born in his investigation of the problem of collisions between atoms and free electric particles. In this connection he succeeded in obtaining a statistical interpretation of the wave functions, allowing a calculation of the probability of the individual transition processes required by the quantum postulate. This includes a wave-mechanical formulation of the adiabatic principle of Ehrenfest, the fertility of which appears strikingly in the promising investigations of Hund on the problem of the formation of molecules.

In view of these results, Schrödinger has expressed the hope that the development of the wave theory will eventually remove the irrational element expressed by the quantum postulate and open the way for a complete description of atomic phenomena along the line of the classical theories. In support of this view, Schrödinger, in a recent paper, emphasizes the fact that the discontinuous exchange of energy between atoms required by the quantum postulate, from the point of view of the wave theory, is replaced by a simple resonance phenomenon. In particular, the idea of individual stationary states would be an illusion and its applicability only an illustration of the resonance mentioned. It must be kept in mind, however, that just in the resonance problem mentioned we are concerned with a closed system which, according to the view presented here, is not accessible to observation. In fact, wave mechanics, just as the matrix theory, on this view represents a symbolic transcription of the problem of motion of classical mechanics adapted to the requirements of quantum theory and only to be interpreted by an explicit use of the quantum postulate. Indeed, the two formulations of the interaction problem might be said to be complementary in the same sense as the wave and particle idea in the description of the free individuals. The apparent contrast in the utilization of the energy concept in the two theories is just connected with this difference in the starting-point.

The fundamental difficulties opposing a space-time description of a system of particles in interaction appear at once from the inevitability of the superposition principle in the description of the behaviour of individual particles. Already for a free particle the knowledge of energy and momentum excludes, as we have seen, the exact knowledge of its space-time co-ordinates. This implies that an immediate utilization of the concept of energy in connection with the classical idea of the potential energy of the system is excluded. In the Schrödinger wave equation these difficulties are avoided by replacing the classical expression of the Hamiltonian by a differential operator by means of the relation

$$p = \sqrt{-1}\,\frac{h}{2\pi}\,\frac{\delta}{\delta q}, \qquad \ldots\ldots(5)$$

where p denotes a generalized component of momentum and q the canonically conjugated variable. Here the negative value of the energy is regarded as conjugated to the time. So far, in the wave equation, time and space as well as energy and momentum are utilized in a purely formal way.

The symbolical character of Schrödinger's method appears not only from the circumstance that its simplicity, similarly to that of the matrix theory, depends essentially upon the use of imaginary arithmetic quantities. But above all there can be no question of an immediate connection with our ordinary conceptions because the "geometrical" problem represented by the wave equation is associated with the so-called co-ordinate space, the number of dimensions of which is equal to the number of degrees of freedom of the system, and, hence, in general greater than the number of dimensions of ordinary space. Further, Schrödinger's formulation of the interaction problem, just as the formulation offered by matrix theory, involves a neglect of the finite velocity of propagation of the forces claimed by relativity theory.

On the whole, it would scarcely seem justifiable, in the case of the interaction problem, to demand a visualization by means of ordinary space-time pictures. In fact, all our knowledge concerning the internal properties of atoms is derived from experiments on their radiation or collision reactions, such that the interpretation of experimental facts ultimately depends on the abstractions of radiation in free space, and free material particles. Hence, our whole space-time view of physical phenomena, as well as the definition of energy and momentum, depends ultimately upon these abstractions. In judging the applications of these auxiliary ideas, we should only demand inner consistency, in which connection special regard has to be paid to the possibilities of definition and observation.

In the characteristic vibrations of Schrödinger's wave equation we have, as mentioned, an adequate representation of the stationary states of an atom allowing an unambiguous definition of the energy of the system by means of the general quantum relation (1). This entails, however, that in the interpretation of observations a fundamental renunciation regarding the space-time description is unavoidable. In fact, the consistent application of the concept of stationary states excludes, as we shall see, any specification regarding the behaviour of the separate particles in the atom. In problems where a description of this behaviour is essential, we are bound to use the general solution of the wave equation which is obtained by superposition of characteristic solutions. We meet here with a complementarity of the possibilities of definition quite analogous to that which we have considered earlier in connection with the properties of light and free material particles. Thus, while the definition of energy and momentum of individuals is attached to the idea of a harmonic elementary wave, every space-time feature of the description of phenomena is, as we have seen, based on a consideration of the interferences taking place inside a group of such elementary waves. Also in the present case the agreement between the possibilities of observation and those of definition can be directly shown.

According to the quantum postulate any observation regarding the behaviour of the electron in the atom will be accompanied by a change in the state of the atom. As stressed by Heisenberg, this change will, in the case of atoms in stationary states of low quantum number, consist in general in the ejection of the electron from the atom. A description of the "orbit" of the electron in the atom with the aid of subsequent observations is, hence, impossible in such a case. This is connected with the circumstance that from characteristic vibrations with only a few nodes no wave packages can be built up which would even approximately represent the "motion" of a particle. The complementary nature of the description, however, appears particularly in that the use of observations concerning the behaviour of particles in the atom rests on the possibility of neglecting, during the process of observation, the interaction between the particles, thus regarding them as free. This requires, however, that the duration of the process is short compared with the natural periods of the atom, which again means that the uncertainty in the knowledge of the energy transferred in the process is large compared to the energy differences between neighbouring stationary states.

In judging the possibilities of observation it must, on the whole, be kept in mind that the wave-mechanical solutions can be visualized only in so far as they can be described with the aid of the concept of free particles. Here the difference between classical mechanics and the quantum-theoretical treatment of the problem of interaction appears most strikingly. In the former such a restriction is unnecessary because the "particles" are here endowed with an immediate "reality", independently of their being free or bound. This situation is particularly important in connection with the consistent utilization of Schrödinger's electric density as a measure of the probability for electrons being present within given space regions of the atom. Remembering the restriction mentioned, this interpretation is seen to be a simple consequence of the assumption that the probability of the presence of a free electron is expressed by the electric density associated with the wave-field in a similar way to that by which the probability of the presence of a light quantum is given by the energy density of the radiation.

As already mentioned, the means for a general consistent utilization of the classical concepts in the quantum theory have been created through the transformation theory of Dirac and Jordan, by the aid of which Heisenberg has formulated his general uncertainty relation (4). In this theory also the Schrödinger wave equation has obtained an instructive application. In fact, the characteristic solutions of this equation appear as auxiliary functions which define a transformation from matrices with indices representing the energy values of the system to other matrices, the indices of which are the possible values of the space co-ordinates. It is also of interest in this connection to mention that Jordan and Klein have recently arrived at the formulation of the problem of interaction expressed by the Schrödinger wave equation, taking as starting-point the wave representation of individual particles and applying a symbolic method closely related to the deep-going treatment of the radiation problem developed by Dirac from the point of view of the matrix theory, to which we shall return below.

6. REALITY OF STATIONARY STATES

In the conception of stationary states we are, as mentioned, concerned with a characteristic application of the quantum postulate. By its very nature this conception means a complete renunciation as regards a time description. From the point of view taken here, just this renunciation forms the necessary condition for an unambiguous definition of the energy of the atom. Moreover, the conception of a stationary state involves, strictly speaking, the exclusion of all interactions with individuals not belonging to the system. The fact that such a closed system is associated with a particular energy value may be considered as an immediate expression for the claim of causality contained in the theorem of conservation of energy. This circumstance justifies the assumption of the supra-mechanical stability of the stationary states, according to which the atom, before as well as after an external influence, always will be found in a well-defined state, and which forms the basis for the use of the quantum postulate in problems concerning atomic structure.

In a judgment of the well-known paradoxes which this assumption entails for the description of collision and radiation reactions, it is essential to consider the limitations of the possibilities of definition of the reacting free individuals, which is expressed by relation (2). In fact, if the definition of the energy of the reacting individuals is to be accurate to such a degree as to entitle us to speak of conservation of energy during the reaction, it is necessary, according to this relation, to co-ordinate to the reaction a time interval long compared to the vibration period associated with the transition process, and connected with the energy difference between the stationary states according to relation (1). This is particularly to be remembered when considering the passage of swiftly moving particles through an atom. According to the ordinary kinematics, the effective duration of such a passage would be very small as compared with the natural periods of the atom, and it seemed impossible to reconcile the principle of conservation of energy with the assumption of the stability of stationary states. In the wave representation, however, the time of reaction is immediately connected with the accuracy of the knowledge of the energy of the colliding particle, and hence there can never be the possibility of a contradiction with the law of conservation. In connection with the discussion of paradoxes of the kind mentioned, Campbell suggested the view that the conception of time itself may be essentially statistical in nature. From the view advanced here, according to which the foundation of space-time description is offered by the abstraction of free individuals, a fundamental distinction between time and space, however, would seem to be excluded by the relativity requirement. The singular position of the time in problems concerned with stationary states is, as we have seen, due to the special nature of such problems.

The application of the conception of stationary states demands that in any observation, say by means of collision or radiation reactions, permitting a distinction between different stationary states, we are entitled to disregard the previous history of the atom. The fact that the symbolical quantum theory methods ascribe a particular phase to each stationary state the value of which depends upon the previous history of the atom, would for the first moment seem to contradict the very idea of stationary states. As soon as we are really concerned with a time problem, however, the consideration of a strictly closed system is excluded. The use of simply harmonic proper vibrations in the interpretation of observations means, therefore, only a suitable idealization which in a more rigorous discussion must always be replaced by a group of harmonic vibrations, distributed over a finite frequency interval. Now, as already mentioned, it is a general consequence of the superposition principle that it has no sense to co-ordinate a phase value to the group as a whole, in the same manner as may be done for each elementary wave constituting the group.

This inobservability of the phase, well known from the theory of optical instruments, is brought out in a particularly simple manner in a discussion of the Stern-Gerlach experiment, so important for the investigation of the properties of single atoms. As pointed out by Heisenberg, atoms with different orientation in the field may only be separated if the deviation of the beam is larger than the diffraction at the slit of the de Broglie waves representing the translational motion of the atoms. This condition means, as a simple calculation shows, that the product of the time of passage of the atom through the field, and the uncertainty due to the finite width of the beam of its energy in the field, is at least equal to the quantum of action. This result was considered by Heisenberg as a support of relation (2) as regards the reciprocal uncertainties of energy and time values. It would seem, however, that here we are not simply dealing with a measurement of the energy of the atom at a given time. But since the period of the proper vibrations of the atom in the field is connected with the total energy by relation (1), we realize that the condition for separability mentioned just means the loss of the phase. This circumstance removes also the apparent contradictions, arising in certain problems concerning the coherence of resonance radiation, which have been discussed frequently, and were also considered by Heisenberg.

To consider an atom as a closed system, as we have done above, means to neglect the spontaneous emission of radiation which even in the absence of external influences puts an upper limit to the lifetime of the stationary states. The fact that this neglect is justified in many applications is connected with the circumstance that the coupling between the atom and the radiation field, which is to be expected on classical electrodynamics, is in general very small compared to the coupling between the particles in the atom. It is, in fact, possible in a description of the state of an atom to a considerable extent to neglect the reaction of radiation, thus disregarding the unsharpness in the energy values connected with the lifetime of the stationary states according to relation (2). This is the reason why it is possible to draw conclusions concerning the properties of radiation by using classical electrodynamics.

The treatment of the radiation problem by the new quantum-theoretical methods meant, to begin with, just a quantitative formulation of this correspondence consideration. This was the very starting-point of the original considerations of Heisenberg. It may also be mentioned that an instructive analysis of Schrödinger's treatment of the radiation phenomena from the point of view of the correspondence principle has been recently given by Klein. In the more rigorous form of the theory developed by Dirac, the radiation field itself is included in the closed system under consideration. Thus it became possible in a rational way to take account of the individual character of radiation demanded by the quantum theory and to build up a dispersion theory, in which the finite width of the spectral lines is taken into consideration. The renunciation regarding space-time pictures characterizing this treatment would seem to offer a striking indication of the complementary character of the quantum theory. This is particularly to be borne in mind in judging the radical departure from the causal description of Nature met with in radiation phenomena, to which we have referred above in connection with the excitation of spectra.

In view of the asymptotic connection of atomic properties with classical electrodynamics, demanded by the correspondence principle, the reciprocal exclusion of the conception of stationary states and the description of the behaviour of individual particles in the atom might be regarded as a difficulty. In fact, the connection in question means that in the limit of large quantum numbers where the relative difference between adjacent stationary states vanishes asymptotically, mechanical pictures of electronic motion may be rationally utilized. It must be emphasized, however, that this connection cannot be regarded as a gradual transition towards classical theory in the sense that the quantum postulate would lose its significance for high quantum numbers. On the contrary, the conclusions obtained from the correspondence principle with the aid of classical pictures depend just upon the assumptions that the conception of stationary states and of individual transition processes are maintained even in this limit.

This question offers a particularly instructive example for the application of the new methods. As shown by Schrödinger, it is possible, in the limit mentioned, by superposition of proper vibrations to construct wave-groups small in comparison to the "size" of the atom, the propagation of which indefinitely approaches the classical picture of moving material particles, if the quantum numbers are chosen sufficiently large. In the special case of a simple harmonic vibrator, he was able to show that such wave-groups will keep together even for any length of time, and will oscillate to and fro in a manner corresponding to the classical picture of the motion. This circumstance Schrödinger has regarded as a support of his hope of constructing a pure wave theory without referring to the quantum postulate. As emphasized by Heisenberg, the simplicity of the case of the oscillator, however, is exceptional and intimately connected with the harmonic nature of the corresponding classical motion. Nor is there in this example any possibility for an asymptotical approach towards the problem of free particles. In general, the wave-group will gradually spread over the whole region of the atom, and the "motion" of a bound electron can only be followed during a number of periods, which is of the order of magnitude of the quantum numbers associated with the proper vibrations. This question has been more closely investigated in a recent paper by Darwin which contains a number of instructive examples of the behaviour of wave groups. From the viewpoint of the matrix theory a treatment of analogous problems has been carried out by Kennard.

Here again we meet with the contrast between the wave-theory superposition principle and the assumption of the individuality of particles with which we have been concerned already in the case of free particles. At the same time the asymptotical connection with the classical theory, to which a distinction between free and bound particles is unknown, offers the possibility of a particularly simple illustration of the above considerations regarding the consistent utilization of the concept of stationary states. As we have seen, the identification of a stationary state by means of collision or radiation reactions implies a gap in the time description, which is at least of the order of magnitude of the periods associated with transitions between stationary states. Now, in the limit of high quantum numbers these periods may be interpreted as periods of revolution. Thus we see at once that no causal connection can be obtained between observations leading to the fixation of a stationary state and earlier observations on the behaviour of the separate particles in the atom.

Summarizing, it might be said that the concepts of stationary states and individual transition processes within their proper field of application possess just as much or as little "reality" as the very idea of individual particles. In both cases we are concerned with a demand of causality complementary to the space-time description, the adequate application of which is limited only by the restricted possibilities of definition and of observation.

7. THE PROBLEM OF THE ELEMENTARY PARTICLES

When due regard is taken of the complementary feature required by the quantum postulate, it seems, in fact, possible with the aid of the symbolic methods to build up a consistent theory of atomic phenomena, which may be considered as a rational generalization of the causal space-time description of classical physics. This view does not mean, however, that classical electron theory may be regarded simply as the limiting case of a vanishing quantum of action. Indeed, the connection of the latter theory with experience is based on assumptions which can scarcely be separated from the group of problems of the quantum theory. A hint in this direction was already given by the well-known difficulties met with in the attempts to account for the individuality of ultimate electrical particles on general mechanical and electrodynamical principles. In this respect also, the general relativity theory of gravitation has not fulfilled expectations. A satisfactory solution of the problems touched upon would seem to be possible only by means of a rational quantum-theoretical transcription of the general field theory, in which the ultimate quantum of electricity has found its natural position as an expression of the feature of individuality characterizing the quantum theory. Recently Klein has directed attention to the possibility of connecting this problem with the five-dimensional unified representation of electromagnetism and gravitation proposed by Kaluza. In fact, the conservation of electricity appears in this theory as an analogue to the conservation theorems for energy and momentum. Just as these concepts are complementary to the space-time description, the appropriateness of the ordinary four-dimensional description as well as its symbolical utilization in the quantum theory would, as Klein emphasizes, seem to depend essentially on the circumstance that in this description electricity always appears in well-defined units, the conjugated fifth dimension being as a consequence not open to observation.

Quite apart from these unsolved deep-going problems, the classical electron theory up to the present time has been the guide for a further development of the correspondence description in connection with the idea first advanced by Compton that the ultimate electrical particles, besides their mass and charge, are endowed with a magnetic moment due to an angular momentum determined by the quantum of action. This assumption, introduced with striking success by Goudsmit and

Uhlenbeck into the discussion of the origin of the anomalous Zeeman effect, has proved most fruitful in connection with the new methods, as shown especially by Heisenberg and Jordan. One might say, indeed, that the hypothesis of the magnetic electron, together with the resonance problem elucidated by Heisenberg, which occurs in the quantum-theoretical description of the behaviour of atoms with several electrons, have brought the correspondence interpretation of the spectral laws and the periodic system to a certain degree of completion. The principles underlying this attack have even made it possible to draw conclusions regarding the properties of atomic nuclei. Thus Dennison, in connection with ideas of Heisenberg and Hund, has succeeded recently in a very interesting way in showing how the explanation of the specific heat of hydrogen, hitherto beset with difficulties, can be harmonized with the assumption that the proton is endowed with a moment of momentum of the same magnitude as that of the electron. Due to its larger mass, however, a magnetic moment much smaller than that of the electron must be associated with the proton.

The insufficiency of the methods hitherto developed as concerns the problem of the elementary particles appears in the questions just mentioned from the fact that they do not allow of an unambiguous explanation of the difference in the behaviour of the electric elementary particles and the "individuals" symbolized through the conception of light quanta expressed in the so-called exclusion principle formulated by Pauli. In fact, we meet in this principle, so important for the problem of atomic structure as well as for the recent development of statistical theories, with one among several possibilities, each of which fulfils the correspondence requirement. Moreover, the difficulty of satisfying the relativity requirement in quantum theory appears in a particularly striking light in connection with the problem of the magnetic electron. Indeed, it seemed not possible to bring the promising attempts made by Darwin and Pauli in generalizing the new methods to cover this problem naturally, in connection with the relativity kinematical consideration of Thomas so fundamental for the interpretation of experimental results. Quite recently, however, Dirac has been able successfully to attack the problem of the magnetic electron through a new ingenious extension of the symbolical method and so to satisfy the relativity requirement without abandoning the agreement with spectral evidence. In this attack not only the imaginary complex quantities appearing in the earlier procedures are involved, but his fundamental equations themselves contain quantities of a still higher degree of complexity that are represented by matrices.

Already the formulation of the relativity argument implies essentially the union of the space-time co-ordination and the demand of causality characterizing the classical theories. In the adaptation of the relativity requirement to the quantum postulate, we must therefore be prepared to meet with a renunciation as to visualization in the ordinary sense going still further than in the formulation of the quantum laws considered here. Indeed, we find ourselves here on the very path taken by Einstein of adapting our modes of perception borrowed from the sensations to the gradually deepening knowledge of the laws of Nature. The hindrances met with on this path originate above all in the fact that, so to say, every word in the language refers to our ordinary perception. In the quantum theory we meet this difficulty at once in the question of the inevitability of the feature of irrationality characterizing the quantum postulate. I hope, however, that the idea of complementarity is suited to characterize the situation, which bears a deep-going analogy to the general difficulty in the formation of human ideas, inherent in the distinction between subject and object.

I.5 THE UNCERTAINTY PRINCIPLE

H. P. ROBERTSON

The uncertainty principle is one of the most characteristic and important consequences of the new quantum mechanics. This principle, as formulated by Heisenberg for two conjugate quantum-mechanical variables, states that the accuracy with which two such variables can be measured simultaneously is subject to the restriction that the product of the uncertainties in the two measurements is at least of order h (Planck's constant). Condon* has remarked that an uncertainty relation of this type can not hold in the general case where the two variables under consideration are not conjugate, and has stressed the desirability of obtaining a general formulation of the principle. It is the purpose of the present letter to give such a general formulation, and to apply it in particular to the case of angular momentum.

We define the "mean value" A_0 of an (Hermitean) operator A in a system whose state is described by the (normal) function ψ as

$$A_0 = \int \bar{\psi} A \psi d\tau$$

where the integral is extended over the entire coordinate space. The Hermitean character of A (i.e.

$$\int \bar{\phi} A \psi d\tau = \int \bar{\psi} A \phi d\tau$$

for arbitrary ϕ, ψ) insures the reality of A_0. The "uncertainty" ΔA in the value of A is then defined, in accordance with statistical usage, as the root mean square of the deviation of A from this mean, i.e.

$$(\Delta A)^2 = \int \bar{\psi} (A - A_0)^2 \psi d\tau.$$

The uncertainty principle for two such variables A, B, whose commutator $AB - BA = hC/2\pi i$, is expressed by

$$\Delta A \cdot \Delta B \geqq h |C_0| / 4\pi$$

i.e. the product of the uncertainties in A, B is not less than half the absolute value of the mean of their commutator.

We here confine ourselves to sketching the proof of this principle for a one-particle system and for quantum mechanical variables $A(q, p)$, $B(q, p)$ which are linear in the momenta (p_x, p_y, p_z).[1] (The proof for the general case in which the operators can be expanded in powers of the momenta can be made along exactly the same lines.) Writing

$$A = a + a_x p_x + a_y p_y + a_z p_z$$

where $p_x = (h/2\pi i)\partial/\partial x$, etc. and the a's are functions of position, the Hermitean character of A requires that these functions be real and that div $(a_x, a_y, a_z) = 0$. The expression for $(\Delta A)^2$ may be written, on integrating once by parts, using the fact that div $(a) = 0$ and discarding the resulting surface integral, in the form

$$(\Delta A)^2 = \int |(A - A_0)\psi|^2 d\tau.$$

We are now in a position to apply the Schwarzian inequality[2]

$$\left[\int (f_1\bar{f}_1 + f_2\bar{f}_2) d\tau\right]\left[\int (g_1\bar{g}_1 + g_2\bar{g}_2) d\tau\right] \geqq \left|\int (f_1 g_1 + f_2 g_2) d\tau\right|^2$$

Taking

$$\bar{f}_1 = (A - A_0)\psi = f_2, \quad g_1 = (B - B_0)\psi = -\bar{g}_2$$

and reducing the integral on the right hand side by integration by parts we find

$$\Delta A \cdot \Delta B \geqq \tfrac{1}{2} \left| \int \bar{\psi} (AB - BA) \psi d\tau \right|,$$

the required result.

We obviously obtain Heisenberg's result if the two variables are conjugate, for then C, and consequently C_0, are ± 1. As a further illustration of the principle, we apply it to the case of angular momentum. Here we have

$$M_x = y p_z - z p_y, \quad M_x M_y - M_y M_x = -h M_z / 2\pi i$$

so *the product of the uncertainties in two of the components of angular momentum is not less than $h/4\pi$ times the mean value of the third component in the state under consideration.*

* E. U. Condon "Remarks on Uncertainty Principles" Science LXIX, p. 573 (May 31, 1929), and in conversations with the writer on this topic.

[1] Cf. proof of special case $A = p$, $B = q$ in H. Weyl "Gruppentheorie und Quantenmechanik" pp. 66, 272.

[2] Weyl, l. c. p. 272.

Originally published in *Physical Review, 34*, 163-64 (1929).

Consider in particular the state, treated by Condon, defined by

$$\psi = f(r)e^{im\phi}P_l{}^m(\cos\theta)$$

where the pole of the spherical coordinates lies on the z-axis. Then M_z, $M^2(=M_x{}^2+M_y{}^2+M_z{}^2)$ have the definite values

$$M_z = M_{z0} = mh/2\pi, \quad M^2 = l(l+1)(h/2\pi)^2$$

the mean values of M_x, M_y are zero and the uncertainties are given by

$$(\Delta M_x)^2 = (\Delta M_y)^2 = \tfrac{1}{2}[l(l+1)-m^2](h/2\pi)^2, \quad \Delta M_z = 0.$$

Now from the uncertainty principle for M_x, M_y we find

$$l(l+1) \geqq m(m+1)$$

which is in fact the case. This example shows that for $m=l$ the equality holds; the inequality is consequently the most restrictive one that can be deduced for angular momenta, for we have here a case in which the ultimate limit has (in principle) been reached.

H. P. Robertson

Palmer Physical Laboratory,
Princeton, N. J.,
June 18, 1929.

I.6 THE WAVE MECHANICS OF α-RAY TRACKS

NEVILL F. MOTT

The present note is suggested by a recent paper by Prof. Darwin,* and is intended to show how one of the most typically particle-like properties of matter can be derived from the wave mechanics. In the theory of radioactive disintegration, as presented by Gamow, the α-particle is represented by a spherical wave which slowly leaks out of the nucleus. On the other hand, the α-particle, once emerged, has particle-like properties, the most striking being the ray tracks that it forms in a Wilson cloud chamber. It is a little difficult to picture how it is that an outgoing spherical wave can produce a straight track ; we think intuitively that it should ionise atoms at random throughout space. We could consider that Gamow's outgoing spherical wave should give the probability of disintegration, but that, when the particle is outside the nucleus, it should be represented by a wave packet moving in a definite direction, so as to produce a straight track. But it ought not to be necessary to do this. The wave mechanics unaided ought to be able to predict the possible results of any observation that we could make on a system, without invoking, until the moment at which the observation is made, the classical particle-like properties of the electrons or α-particles forming that system. If we consider the α-ray alone as the system under consideration, then the gas of the Wilson chamber must be considered as the means by which we observe the particle ; so in this case we must consider the α-ray as a particle as soon as it is outside the nucleus, because that is the moment at which the observation is made. If, however, we consider the α-particle and the gas together as one system, then it is ionised atoms that we observe ; interpreting the wave function should give us simply the probability that such and such an atom is ionised. Until this final interpretation is made, no mention should be made of the α-ray being a particle at all.

The difficulty that we have in picturing how it is that a spherical wave can produce a straight track arises from our tendency to picture the wave as existing in ordinary three dimensional space, whereas we are really dealing with wave functions in the multispace formed by the co-ordinates both of the α-particle and of every atom in the Wilson chamber.

* 'Roy. Soc. Proc.,' A, vol. 124, p. 375 (1929).

Originally published in *Proceedings of the Royal Society, London, A126*, 79-84 (1929).

For our purpose it will be sufficient to consider only two atoms ; for simplicity we shall suppose that they are hydrogen atoms. The position of the nuclei of the atoms we shall treat as parameters ; this is legitimate, since the nuclei are many times heavier than the electrons, and move very much more slowly than the α-particles ; therefore, during the whole time of formation of a track, they may be considered effectively at rest.* We shall then show that the atoms cannot both be ionised unless they lie in a straight line with the radioactive nucleus.

Let $\psi_J(\mathbf{r})$ be the wave function of an excited hydrogen atom, referred to axes that pass through its nucleus. We shall denote by $\psi_0(\mathbf{r})$ the wave function corresponding to the normal state. We shall take axes such that the nucleus of the radioactive atom lies at the origin, and the two hydrogen atoms at the points $\mathbf{a}_1$, $\mathbf{a}_2$. Then the wave functions of the two hydrogen atoms are, in these co-ordinates

$$\Psi^{\mathrm{I}}(\mathbf{r}) = \psi(\mathbf{r} - \mathbf{a}_1)$$
$$\Psi^{\mathrm{II}}(\mathbf{r}) = \psi(\mathbf{r} - \mathbf{a}_2).$$

Let $\mathbf{R}$ be the co-ordinate of the α-particle and $\mathbf{r}_1$, $\mathbf{r}_2$ the co-ordinates of the two electrons. Let $\mathrm{F}(\mathbf{R}, \mathbf{r}_1, \mathbf{r}_2)\, e^{i\mathrm{E}t/h}$ be a periodic wave function of the α-particle and of the two atomic electrons. We can expand F in a series of wave functions of the two atoms, of the form

$$\mathrm{F}(\mathbf{R}, \mathbf{r}_1, \mathbf{r}_2) = \sum_{\mathrm{J}_1\mathrm{J}_2} f_{\mathrm{J}_1\mathrm{J}_2}(\mathbf{R})\, \Psi_{\mathrm{J}_1}^{\mathrm{I}}(\mathbf{r}_1)\, \Psi_{\mathrm{J}_2}^{\mathrm{II}}(\mathbf{r}_2). \tag{1}$$

We can now see what form the wave function must have, in order that we shall obtain straight tracks. Interpreting our wave function, we see that $|f_{\mathrm{J}_1\mathrm{J}_2}(\mathbf{R})|^2\, d\mathrm{V}$ is the probability that we shall find the α-particle in the volume element $d\mathrm{V}$, and at the same time the first atom in the excited (or ionised) state J_1 and the second in the state J_2. To obtain a consistent theory of the straight tracks, we must have $f_{00}(\mathbf{R})$ representing an outgoing spherical wave, at any rate for $|\mathbf{R}|$ less than either $|\mathbf{a}_1|$ or $|\mathbf{a}_2|$. $f_{\mathrm{J}_1 0}(\mathbf{R})$ will represent the probable positions of an α-particle that has excited the first atom, but not the second. It should therefore be independent of $\mathbf{a}_2$, and should represent a

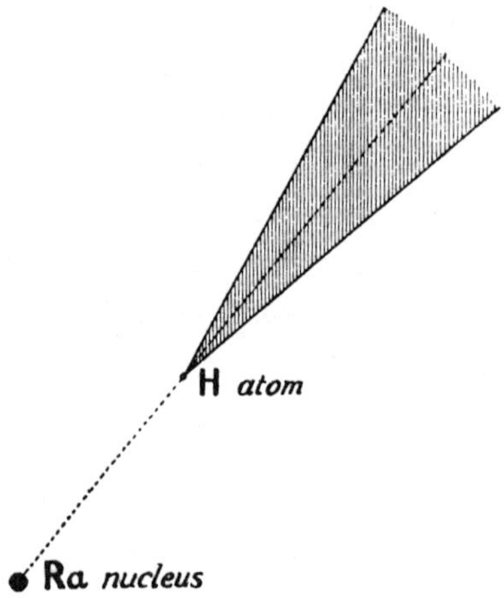

FIG. 1.

* We do not consider the possibility of a collision between the α-particle and the nucleus.

wave diverging from the point $\mathbf{a}_1$, whose amplitude vanishes except inside a small cone, pointing away from the origin. Finally, $f_{J_1J_2}(\mathbf{R})$ will give the probability that the α-particle excites both atoms.

$f_{J_1J_2}(\mathbf{R})$ should vanish, therefore, unless the line joining $\mathbf{a}_1$ and $\mathbf{a}_2$ passes near the origin. We shall obtain a solution with these properties.

The wave equation is

$$\left\{\frac{h^2}{2M}\nabla_R{}^2 + \frac{h^2}{2m}(\nabla_1{}^2 + \nabla_2{}^2) + E + \frac{\varepsilon^2}{|\mathbf{r}_1 - \mathbf{a}_1|} + \frac{\varepsilon^2}{|\mathbf{r}_2 - \mathbf{a}_2|} + \frac{2\varepsilon^2}{|\mathbf{R} - \mathbf{r}_1|} + \frac{2\varepsilon^2}{|\mathbf{R} - \mathbf{r}_2|}\right\} F = 0. \quad (2)$$

where $\nabla_R{}^2$ refers to the co-ordinates of the α-particle, and $\nabla_1{}^2$, $\nabla_2{}^2$ to the co-ordinates of the electrons. We have left out the terms giving the interaction between the α-particle and the nuclei of the atoms, which would in fact produce deviations from the straight track, and are irrelevant to our purpose. We treat the interaction of the atoms and the α-particle as a perturbation, and solve by a method of successive approximations, similar to that used by Born* in his work on collisions. We set

$$F = F^{(0)} + F^{(1)} + F^{(2)} + \dots$$

with

$$F^{(0)} = e^{ik|\mathbf{R}|}/|\mathbf{R}| \,.\, \Psi_0{}^{I}(\mathbf{r}_1)\,\Psi_0{}^{II}(\mathbf{r}_2),$$

where

$$k = \sqrt{2M(E - 2E_0)}/h,$$

representing an outgoing wave for the α-particle, and the normal states of both atoms. Then F is a solution of (2) if

$$\left\{\frac{h^2}{2M}\nabla_R{}^2 + \frac{h^2}{2m}(\nabla_1{}^2 + \nabla_2{}^2) + E + \frac{\varepsilon^2}{|\mathbf{r}_1 - \mathbf{a}_1|} + \frac{\varepsilon^2}{|\mathbf{r}_2 - \mathbf{a}_2|}\right\} F^{(n)} = \left\{\frac{-2\varepsilon^2}{|\mathbf{R} - \mathbf{r}_1|} + \frac{-2\varepsilon^2}{|\mathbf{R} - \mathbf{r}_2|}\right\} F^{(n-1)}. \quad (3)$$

Let us solve first for $F^{(1)}$. We expand $F^{(1)}$ as a series of the form (1), namely

$$F^{(1)}(\mathbf{R}, \mathbf{r}_1, \mathbf{r}_2) = \sum_{J_1J_2} f^{(1)}_{J_1J_2}(\mathbf{R})\,\Psi^{I}_{J_1}(\mathbf{r}_1)\,\Psi^{II}_{J_2}(\mathbf{r}_2). \quad (4)$$

If we insert (4) in (3), multiply by $\Psi_{J_1}{}^{I}(\mathbf{r}_1)\,\Psi_{J_2}{}^{II}(\mathbf{r}_2)$ and integrate over all $\mathbf{r}_1\,\mathbf{r}_2$, we obtain a differential equation satisfied by $f^{(1)}_{J_1J_2}(\mathbf{R})$, namely

$$\left\{\frac{h^2}{2M}\nabla^2 + E - E_{J_1} - E_{J_2}\right\} f^{(1)}_{J_1J_2}(\mathbf{R}) = K(\mathbf{R}), \quad (5)$$

* 'Z. Physik,' vol. 38, p. 803 (1926).

where

$$\begin{aligned} K(\mathbf{R}) &= V_{0J}(\mathbf{R}-\mathbf{a}_1)\, e^{ik|\mathbf{R}|}/|\mathbf{R}| && \text{if } J_1 = J,\ J_2 = 0 \\ &= V_{0J}(\mathbf{R}-\mathbf{a}_2)\, e^{ik|\mathbf{R}|}/|\mathbf{R}| && \text{if } J_2 = J,\ J_1 = 0 \\ &= 0. && \text{otherwise.} \end{aligned}$$

In these formulæ

$$V_{0J}(\mathbf{R}) = \int 2\varepsilon^2/|\mathbf{R}-\mathbf{r}| \,.\, \psi_0(\mathbf{r})\, \psi_J(\mathbf{r})\, d\mathbf{r}.$$

Now, the most general solution of (5) is*

$$f^{(1)}_{J_1J_2}(\mathbf{R}) = \frac{1}{4\pi}\int \frac{2M}{h^2} K(\mathbf{R}') \,.\, \frac{e^{\pm ik'|\mathbf{R}-\mathbf{R}'|}}{|\mathbf{R}-\mathbf{R}'|}\, d\mathbf{R}' + G(\mathbf{R}), \tag{6}$$

where

$$\sqrt{2M(E - E_{J_1} - E_{J_2})} = k'h$$

and $G(\mathbf{R})$ is the most general solution of

$$(\nabla^2 + k'^2)\, G = 0.$$

Now, it is clear that our starting conditions, namely that both atoms are in the normal state before collision, require that

$$G(\mathbf{R}) = 0\,;$$

for $G(\mathbf{R})$ represents streams of particles fired at already excited atoms. We see, therefore, that if neither J_1 nor J_2 define the normal state of an atom, $f^{(1)}_{J_1J_2}(\mathbf{R})$ vanishes, and therefore, to this approximation, there is *no* probability that both atoms will be excited. This is to be expected; we have treated the probability that one atom will be excited as a small quantity of the first order; the probability of both being excited will therefore be a small quantity of the second order.

If, say, J_1 represents the normal state, then $K(\mathbf{R})$ vanishes except in the neighbourhood of $\mathbf{a}_2$. Except in the neighbourhood of $\mathbf{a}_2$, therefore, $f^{(1)}_{0J_2}(\mathbf{R})$ is given by the asymptotic formula

$$f^{(1)}_{0J_2}(\mathbf{R}) \sim \frac{e^{ik'|\mathbf{R}-\mathbf{a}_2|}}{|\mathbf{R}-\mathbf{a}_2|}\, \mathfrak{I}(\mathbf{l})$$

where

$$\mathbf{l} = (\mathbf{R}-\mathbf{a}_2)/|\mathbf{R}-\mathbf{a}_2|$$

and

$$\mathfrak{I}(\mathbf{l}) = \frac{1}{4\pi}\int \frac{2m}{h^2} V_{0J_2}(\mathbf{R}')\, e^{-ik'(\mathbf{l}\mathbf{R}') + ik|\mathbf{R}'+\mathbf{a}_2|}\, d\mathbf{R}'. \tag{7}$$

* *Cf.* Courant-Hilbert, "Methoden der Mathematischen Physik," chap. 5, § 10.

We have taken the positive value of ik' in (6), so that $f^{(1)}_{0J_2}$ shall represent a wave diverging from $\mathbf{a}_2$; $\mathfrak{J}(\mathbf{R}-\mathbf{a}_2/|\mathbf{R}-\mathbf{a}_2|)$ gives the amplitude of the wave in any direction.

We can easily see that $\mathfrak{J}(\mathbf{l})$ is very small except in the neighbourhood of

$$\mathbf{l} = \mathbf{a}_2/|\mathbf{a}_2| \tag{8}$$

that is to say, except in a small cone with its vertex at $\mathbf{a}_2$, pointing away from the radioactive nucleus.

For we can see from (5) that $V_{0J}(\mathbf{R})$ must become very small except in the neighbourhood of $\mathbf{R}=0$. The exponentials oscillate very rapidly in the region where $\mathbf{R}$ does not vanish. $\mathfrak{J}(\mathbf{l})$, therefore, will have a strong maximum when the two exponential terms are in phase at the origin. Since $k-k'$ is negligible compared to k, this will be the case when (8) is satisfied.

It is interesting to work out the function $\mathfrak{J}(\mathbf{l})$ for a particular case. Born* has calculated the function V_{0J} for certain simple cases, and also the functions $\mathfrak{J}(\mathbf{l})$ using a plane wave instead of our spherical wave. His results are applicable to our case, however, since the integrand in (7) vanishes except in a small region, in which the spherical wave may be considered plane. We have then†

$$\Sigma\,|\,\mathfrak{J}(\mathbf{l})\,|^2 = 0\cdot9866.\ \ a_H{}^2\,\frac{M^2}{m^2}\,\frac{k'}{k}\,\frac{1}{z^2(1+z^2)^5}$$

where a_H is the radius of the normal orbit of the hydrogen atom, and

$$z = \tfrac{2}{3}a_H\,|\,k\mathbf{l} - k^1\mathbf{a}_2/|\,\mathbf{a}_2\,|\,|$$

and the summation is over all the three states with principal quantum number two. It is clear that this function is only appreciable in the neighbourhood (8), since ka_H is a very large number.

In order to find the probability that *both* atoms should be excited, we shall now consider $F^{(2)}$. If we expand in a series of the form (4), as before, and insert in (3), then we obtain, analogously to (3), the following differential equation satisfied by $f^{(2)}_{J_1J_2}(\mathbf{R})$

$$\left\{\frac{h^2}{2M}\nabla^2 + E - E_{J_1} - E_{J_2}\right\} f^{(2)}_{J_1J_2}(\mathbf{R})$$

$$= \iint F^{(1)}(\mathbf{R}\,\mathbf{r}_1\mathbf{r}_2)\,\Psi_{J_1}{}^{I}(\mathbf{r}_1)\,\Psi_{J_2}{}^{II}(\mathbf{r}_2)\,d\mathbf{r}_1\,d\mathbf{r}_2$$

$$= f^{(1)}_{J_10}(\mathbf{R})\,V_{0J_2}(\mathbf{R}-\mathbf{a}_2). \tag{9}$$

* 'Gott. Nachr.,' p. 146 (1926).

† *Loc. cit.*, equation (32).

Now, if the point $\mathbf{a}_2$ does not lie very near the straight line joining the origin to $\mathbf{a}_1$, then the right-hand side of (9) will vanish for all $\mathbf{R}$; for $f^{(1)}_{J_1 0}(\mathbf{R})$ vanishes except in the neighbourhood of this line, as we have shown, and $V_{0J_2}(\mathbf{R}-\mathbf{a}_2)$ vanishes except in the neighbourhood of $\mathbf{a}_2$. Therefore in this case, it follows, as before, that the only solution of (9) satisfying our initial conditions is

$$f^{(2)}_{J_1 J_2}(\mathbf{R}) = 0,$$

and there is therefore no probability of both atoms being excited. If, on the other hand, the line joining $\mathbf{a}_1$, $\mathbf{a}_2$ does pass through the origin, then we can obtain as before a solution representing a wave diverging from the point $\mathbf{a}_2$. The amplitude of this wave gives the probability that both atoms are excited, and that the particle is moving in a given direction after exciting both.

In conclusion, the author would like to express his thanks to Prof. Darwin, who has contributed a great deal to the course of the development of this paper.

I.7 KNOWLEDGE OF PAST AND FUTURE IN QUANTUM MECHANICS

ALBERT EINSTEIN,
RICHARD C. TOLMAN,
AND BORIS PODOLSKY

It is well known that the principles of quantum mechanics limit the possibilities of exact prediction as to the future path of a particle. It has sometimes been supposed, nevertheless, that the quantum mechanics would permit an exact description of the past path of a particle.

The purpose of the present note is to discuss a simple ideal experiment which shows that the possibility of describing the past path of one particle would lead to predictions as to the future behaviour of a second particle of a kind not allowed in the quantum mechanics. It will hence be concluded that the principles of quantum mechanics actually involve an uncertainty in the description of past events which is analogous to the uncertainty in the prediction of future events. And it will be shown for the case in hand, that this uncertainty in the description of the past arises from a limitation of the knowledge that can be obtained by measurement of momentum.

Consider a small box B, as shown in the figure, containing a number of identical particles in thermal agitation, and provided with two small openings which are closed by the shutter S. The shutter is arranged to open automatically for a short time and then close again, and the number of particles in the box is so chosen that cases arise in which one particle leaves the box and travels over the direct path SO to an observer at O, and a second particle travels over the longer path SRO through elastic reflection at the ellipsoidal reflector R.

The box is accurately weighed before and after the shutter has opened in order to determine the total energy of the particles which have left, and the observer at O is provided with means for observing the arrival of particles, a clock for measuring their time of arrival, and some apparatus for measuring momentum. Furthermore the distances SO and SRO are accurately measured beforehand, —the distance SO being sufficient so that the rate of the clock at O is not disturbed by the gravitational effects involved in weighing the box, and the distance SRO being very long in order to permit an accurate reweighing of the box before the arrival of the second particle.

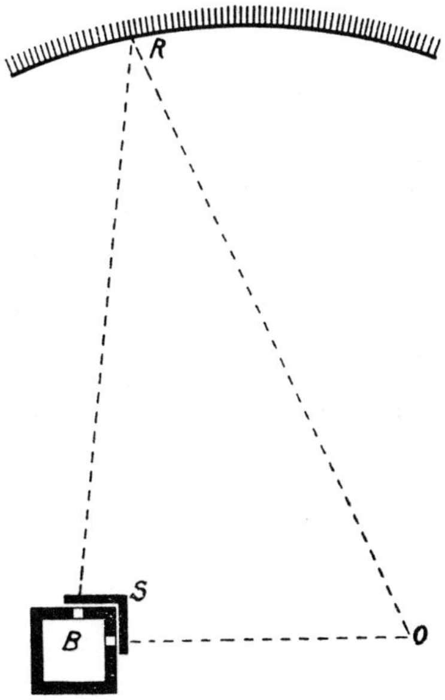

Fig. 1.

Let us now suppose that the observer at O measures the momentum of the first particle as it approaches along the path SO, and then measures its time of arrival. Of course the latter observation, made for example with the help of gamma-ray illumination, will change the momentum in an unknown manner. Nevertheless, knowing the momentum of the particle in the past, and hence also its past velocity and energy, it would seem possible to

Originally published in *Physical Review, 37*, 780-81 (1931).

calculate the time when the shutter must have been open from the known time of arrival of the first particle, and to calculate the energy and velocity of the second particle from the known loss in the energy content of the box when the shutter opened. It would then seem possible to predict beforehand both the energy and the time of arrival of the second particle, a paradoxical result since energy and time are quantities which do not commute in quantum mechanics.

The explanation of the apparent paradox must lie in the circumstance that the past motion of the first particle cannot be accurately determined as was assumed. Indeed, we are forced to conclude that there can be no method for measuring the momentum of a particle without changing its value. For example, an analysis of the method of observing the Doppler effect in the reflected infrared light from an approaching particle shows that, although it permits a determination of the momentum of the particle both before and after collision with the light quantum used, it leaves an uncertainty as to the time at which the collision with the quantum takes place. Thus in our example, although the velocity of the first particle could be determined both before and after interaction with the infrared light, it would not be possible to determine the exact position along the path SO at which the change in velocity occurred as would be necessary to obtain the exact time at which the shutter was open.

It is hence to be concluded that the principles of the quantum mechanics must involve an uncertainty in the description of past events which is analogous to the uncertainty in the prediction of future events. It is also to be noted that although it is possible to measure the momentum of a particle and follow this with a measurement of position, this will not give sufficient information for a complete reconstruction of its past path, since it has been shown that there can be no method for measuring the momentum of a particle without changing its value. Finally, it is of special interest to emphasize the remarkable conclusion that the principles of quantum mechanics would actually impose limitations on the localization in time of a macroscopic phenomenon such as the opening and closing of a shutter.

Albert Einstein
Richard C. Tolman
Boris Podolsky
California Institute of Technology,
February 26, 1931.

I.8 THE EINSTEIN-PODOLSKY-ROSEN PAPER

Commentary of Rosenfeld (1967)

[Einstein] attended [Bohr's 1933 Solvay] lecture and followed the argument with the closest attention; he made no direct comment on it, but put at once the discussion on the general theme of the meaning of quantum theory. He had no longer any doubt about the logic of Bohr's argumentation; but he still felt the same uneasiness as before ("Unbehagen" was his word) when confronted with the strange consequences of the theory. "What would you say of the following situation?" he asked me. "Suppose two particles are set in motion towards each other with the same, very large, momentum, and that they interact with each other for a very short time when they pass at known positions. Consider now an observer who gets hold of one of the particles, far away from the region of interaction, and measures its momentum; then, from the conditions of the experiment, he will obviously be able to deduce the momentum of the other particle. If, however, he chooses to measure the position of the first particle, he will be able to tell where the other particle is. This is a perfectly correct and straightforward deduction from the principles of quantum mechanics; but is it not very paradoxical? How can the final state of the second particle be influenced by a measurement performed on the first, after all physical interaction has ceased between them?"

I had not the impression that Einstein at that time saw in this case, cleverly presented with all the appearances of a paradox, anything else than an illustration of the unfamiliar features of quantum phenomena. Two years later, however, he gave it a much more prominent role in a paper written jointly with Podolsky and Rosen; combined with a "criterion of reality," it was now used with the intention to expose an essential imperfection of quantum theory. Any attribute of a physical system that can be accurately determined without disturbing the system, thus went the argument, is an "element of physical reality," and a description of the system can only be regarded as complete if it embodies all the elements of reality which can be attached to it. Now, the example of the two particles shows that the position and the momentum of a given particle can be obtained by appropriate measurements performed on another particle without disturbing the first, and are therefore elements of reality in the sense indicated. Since quantum theory does not allow both to enter into the description of the state of the particle, such a description is incomplete. The paradox which this incomplete description presents, by suggesting an unaccountable influence of the measurement on the state of the particle, would of course not appear in a complete theory.

This onslaught came down upon us as a bolt from the blue. . . .

I.8 CAN QUANTUM-MECHANICAL DESCRIPTION OF PHYSICAL REALITY BE CONSIDERED COMPLETE?

ALBERT EINSTEIN, BORIS PODOLSKY, AND NATHAN ROSEN

In a complete theory there is an element corresponding to each element of reality. A sufficient condition for the reality of a physical quantity is the possibility of predicting it with certainty, without disturbing the system. In quantum mechanics in the case of two physical quantities described by non-commuting operators, the knowledge of one precludes the knowledge of the other. Then either (1) the description of reality given by the wave function in quantum mechanics is not complete or (2) these two quantities cannot have simultaneous reality. Consideration of the problem of making predictions concerning a system on the basis of measurements made on another system that had previously interacted with it leads to the result that if (1) is false then (2) is also false. One is thus led to conclude that the description of reality as given by a wave function is not complete.

1.

ANY serious consideration of a physical theory must take into account the distinction between the objective reality, which is independent of any theory, and the physical concepts with which the theory operates. These concepts are intended to correspond with the objective reality, and by means of these concepts we picture this reality to ourselves.

In attempting to judge the success of a physical theory, we may ask ourselves two questions: (1) "Is the theory correct?" and (2) "Is the description given by the theory complete?" It is only in the case in which positive answers may be given to both of these questions, that the concepts of the theory may be said to be satisfactory. The correctness of the theory is judged by the degree of agreement between the conclusions of the theory and human experience. This experience, which alone enables us to make inferences about reality, in physics takes the form of experiment and measurement. It is the second question that we wish to consider here, as applied to quantum mechanics.

Whatever the meaning assigned to the term *complete*, the following requirement for a complete theory seems to be a necessary one: *every element of the physical reality must have a counterpart in the physical theory*. We shall call this the condition of completeness. The second question is thus easily answered, as soon as we are able to decide what are the elements of the physical reality.

The elements of the physical reality cannot be determined by *a priori* philosophical considerations, but must be found by an appeal to results of experiments and measurements. A comprehensive definition of reality is, however, unnecessary for our purpose. We shall be satisfied with the following criterion, which we regard as reasonable. *If, without in any way disturbing a system, we can predict with certainty (i.e., with probability equal to unity) the value of a physical quantity, then there exists an element of physical reality corresponding to this physical quantity.* It seems to us that this criterion, while far from exhausting all possible ways of recognizing a physical reality, at least provides us with one

Originally published in *Physical Review, 47*, 777-80 (1935).

such way, whenever the conditions set down in it occur. Regarded not as a necessary, but merely as a sufficient, condition of reality, this criterion is in agreement with classical as well as quantum-mechanical ideas of reality.

To illustrate the ideas involved let us consider the quantum-mechanical description of the behavior of a particle having a single degree of freedom. The fundamental concept of the theory is the concept of *state*, which is supposed to be completely characterized by the wave function ψ, which is a function of the variables chosen to describe the particle's behavior. Corresponding to each physically observable quantity A there is an operator, which may be designated by the same letter.

If ψ is an eigenfunction of the operator A, that is, if

$$\psi' \equiv A\psi = a\psi, \tag{1}$$

where a is a number, then the physical quantity A has with certainty the value a whenever the particle is in the state given by ψ. In accordance with our criterion of reality, for a particle in the state given by ψ for which Eq. (1) holds, there is an element of physical reality corresponding to the physical quantity A. Let, for example,

$$\psi = e^{(2\pi i/h)p_0 x}, \tag{2}$$

where h is Planck's constant, p_0 is some constant number, and x the independent variable. Since the operator corresponding to the momentum of the particle is

$$p = (h/2\pi i)\partial/\partial x, \tag{3}$$

we obtain

$$\psi' = p\psi = (h/2\pi i)\partial\psi/\partial x = p_0\psi. \tag{4}$$

Thus, in the state given by Eq. (2), the momentum has certainly the value p_0. It thus has meaning to say that the momentum of the particle in the state given by Eq. (2) is real.

On the other hand if Eq. (1) does not hold, we can no longer speak of the physical quantity A having a particular value. This is the case, for example, with the coordinate of the particle. The operator corresponding to it, say q, is the operator of multiplication by the independent variable. Thus,

$$q\psi = x\psi \neq a\psi. \tag{5}$$

In accordance with quantum mechanics we can only say that the relative probability that a measurement of the coordinate will give a result lying between a and b is

$$P(a, b) = \int_a^b \bar{\psi}\psi dx = \int_a^b dx = b - a. \tag{6}$$

Since this probability is independent of a, but depends only upon the difference $b-a$, we see that all values of the coordinate are equally probable.

A definite value of the coordinate, for a particle in the state given by Eq. (2), is thus not predictable, but may be obtained only by a direct measurement. Such a measurement however disturbs the particle and thus alters its state. After the coordinate is determined, the particle will no longer be in the state given by Eq. (2). The usual conclusion from this in quantum mechanics is that *when the momentum of a particle is known, its coordinate has no physical reality.*

More generally, it is shown in quantum mechanics that, if the operators corresponding to two physical quantities, say A and B, do not commute, that is, if $AB \neq BA$, then the precise knowledge of one of them precludes such a knowledge of the other. Furthermore, any attempt to determine the latter experimentally will alter the state of the system in such a way as to destroy the knowledge of the first.

From this follows that either (1) *the quantum-mechanical description of reality given by the wave function is not complete* or (2) *when the operators corresponding to two physical quantities do not commute the two quantities cannot have simultaneous reality.* For if both of them had simultaneous reality—and thus definite values—these values would enter into the complete description, according to the condition of completeness. If then the wave function provided such a complete description of reality, it would contain these values; these would then be predictable. This not being the case, we are left with the alternatives stated.

In quantum mechanics it is usually assumed that the wave function *does* contain a complete description of the physical reality of the system in the state to which it corresponds. At first sight this assumption is entirely reasonable, for the information obtainable from a wave function seems to correspond exactly to what can be measured without altering the state of the system. We shall show, however, that this assumption, together with the criterion of reality given above, leads to a contradiction.

2.

For this purpose let us suppose that we have two systems, I and II, which we permit to interact from the time $t=0$ to $t=T$, after which time we suppose that there is no longer any interaction between the two parts. We suppose further that the states of the two systems before $t=0$ were known. We can then calculate with the help of Schrödinger's equation the state of the combined system I+II at any subsequent time; in particular, for any $t>T$. Let us designate the corresponding wave function by Ψ. We cannot, however, calculate the state in which either one of the two systems is left after the interaction. This, according to quantum mechanics, can be done only with the help of further measurements, by a process known as the *reduction of the wave packet*. Let us consider the essentials of this process.

Let $a_1, a_2, a_3, \cdots$ be the eigenvalues of some physical quantity A pertaining to system I and $u_1(x_1), u_2(x_1), u_3(x_1), \cdots$ the corresponding eigenfunctions, where x_1 stands for the variables used to describe the first system. Then Ψ, considered as a function of x_1, can be expressed as

$$\Psi(x_1, x_2) = \sum_{n=1}^{\infty} \psi_n(x_2) u_n(x_1), \tag{7}$$

where x_2 stands for the variables used to describe the second system. Here $\psi_n(x_2)$ are to be regarded merely as the coefficients of the expansion of Ψ into a series of orthogonal functions $u_n(x_1)$. Suppose now that the quantity A is measured and it is found that it has the value a_k. It is then concluded that after the measurement the first system is left in the state given by the wave function $u_k(x_1)$, and that the second system is left in the state given by the wave function $\psi_k(x_2)$. This is the process of reduction of the wave packet; the wave packet given by the infinite series (7) is reduced to a single term $\psi_k(x_2)u_k(x_1)$.

The set of functions $u_n(x_1)$ is determined by the choice of the physical quantity A. If, instead of this, we had chosen another quantity, say B, having the eigenvalues $b_1, b_2, b_3, \cdots$ and eigenfunctions $v_1(x_1), v_2(x_1), v_3(x_1), \cdots$ we should have obtained, instead of Eq. (7), the expansion

$$\Psi(x_1, x_2) = \sum_{s=1}^{\infty} \varphi_s(x_2) v_s(x_1), \tag{8}$$

where φ_s's are the new coefficients. If now the quantity B is measured and is found to have the value b_r, we conclude that after the measurement the first system is left in the state given by $v_r(x_1)$ and the second system is left in the state given by $\varphi_r(x_2)$.

We see therefore that, as a consequence of two different measurements performed upon the first system, the second system may be left in states with two different wave functions. On the other hand, since at the time of measurement the two systems no longer interact, no real change can take place in the second system in consequence of anything that may be done to the first system. This is, of course, merely a statement of what is meant by the absence of an interaction between the two systems. Thus, *it is possible to assign two different wave functions* (in our example ψ_k and φ_r) *to the same reality* (the second system after the interaction with the first).

Now, it may happen that the two wave functions, ψ_k and φ_r, are eigenfunctions of two noncommuting operators corresponding to some physical quantities P and Q, respectively. That this may actually be the case can best be shown by an example. Let us suppose that the two systems are two particles, and that

$$\Psi(x_1, x_2) = \int_{-\infty}^{\infty} e^{(2\pi i/h)(x_1-x_2+x_0)p} dp, \tag{9}$$

where x_0 is some constant. Let A be the momentum of the first particle; then, as we have seen in Eq. (4), its eigenfunctions will be

$$u_p(x_1) = e^{(2\pi i/h)px_1} \tag{10}$$

corresponding to the eigenvalue p. Since we have here the case of a continuous spectrum, Eq. (7) will now be written

$$\Psi(x_1, x_2) = \int_{-\infty}^{\infty} \psi_p(x_2) u_p(x_1) dp, \tag{11}$$

where

$$\psi_p(x_2) = e^{-(2\pi i/h)(x_2 - x_0)p}. \tag{12}$$

This ψ_p however is the eigenfunction of the operator

$$P = (h/2\pi i)\partial/\partial x_2, \tag{13}$$

corresponding to the eigenvalue $-p$ of the momentum of the second particle. On the other hand, if B is the coordinate of the first particle, it has for eigenfunctions

$$v_x(x_1) = \delta(x_1 - x), \tag{14}$$

corresponding to the eigenvalue x, where $\delta(x_1 - x)$ is the well-known Dirac delta-function. Eq. (8) in this case becomes

$$\Psi(x_1, x_2) = \int_{-\infty}^{\infty} \varphi_x(x_2) v_x(x_1) dx, \tag{15}$$

where

$$\varphi_x(x_2) = \int_{-\infty}^{\infty} e^{(2\pi i/h)(x - x_2 + x_0)p} dp$$

$$= h\delta(x - x_2 + x_0). \tag{16}$$

This φ_x, however, is the eigenfunction of the operator

$$Q = x_2 \tag{17}$$

corresponding to the eigenvalue $x + x_0$ of the coordinate of the second particle. Since

$$PQ - QP = h/2\pi i, \tag{18}$$

we have shown that it is in general possible for ψ_k and φ_r to be eigenfunctions of two noncommuting operators, corresponding to physical quantities.

Returning now to the general case contemplated in Eqs. (7) and (8), we assume that ψ_k and φ_r are indeed eigenfunctions of some noncommuting operators P and Q, corresponding to the eigenvalues p_k and q_r, respectively. Thus, by measuring either A or B we are in a position to predict with certainty, and without in any way disturbing the second system, either the value of the quantity P (that is p_k) or the value of the quantity Q (that is q_r). In accordance with our criterion of reality, in the first case we must consider the quantity P as being an element of reality, in the second case the quantity Q is an element of reality. But, as we have seen, both wave functions ψ_k and φ_r belong to the same reality.

Previously we proved that either (1) the quantum-mechanical description of reality given by the wave function is not complete or (2) when the operators corresponding to two physical quantities do not commute the two quantities cannot have simultaneous reality. Starting then with the assumption that the wave function does give a complete description of the physical reality, we arrived at the conclusion that two physical quantities, with noncommuting operators, can have simultaneous reality. Thus the negation of (1) leads to the negation of the only other alternative (2). We are thus forced to conclude that the quantum-mechanical description of physical reality given by wave functions is not complete.

One could object to this conclusion on the grounds that our criterion of reality is not sufficiently restrictive. Indeed, one would not arrive at our conclusion if one insisted that two or more physical quantities can be regarded as simultaneous elements of reality *only when they can be simultaneously measured or predicted.* On this point of view, since either one or the other, but not both simultaneously, of the quantities P and Q can be predicted, they are not simultaneously real. This makes the reality of P and Q depend upon the process of measurement carried out on the first system, which does not disturb the second system in any way. No reasonable definition of reality could be expected to permit this.

While we have thus shown that the wave function does not provide a complete description of the physical reality, we left open the question of whether or not such a description exists. We believe, however, that such a theory is possible.

I.9 BOHR'S REPLY

COMMENTARY OF ROSENFELD (1967)

This onslaught came down upon us as a bolt from the blue. Its effect on Bohr was remarkable. We were then in the midst of groping attempts at exploring the implications of the fluctuations of charge and current distributions, which presented us with riddles of a kind we had not met in electrodynamics. A new worry could not come at a less propitious time. Yet, as soon as Bohr had heard my report of Einstein's argument, everything else was abandoned: we had to clear up such a misunderstanding at once. We should reply by taking up the same example and showing the right way to speak about it. In great excitement, Bohr immediately started dictating to me the outline of such a reply. Very soon, however, he became hesitant: "No, this won't do, we must try all over again . . . we must make it quite clear . . . ." So it went on for a while, with growing wonder at the unexpected subtlety of the argument. Now and then, he would turn to me: "What *can* they mean? Do *you* understand it?" There would follow some inconclusive exegesis. Clearly, we were farther from the mark than we first thought. Eventually, he broke off with the familiar remark that he "must sleep on it." The next morning he at once took up the dictation again, and I was struck by a change in the tone of the sentences: there was no trace in them of the previous day's sharp expressions of dissent. As I pointed out to him that he seemed to take a milder view of the case, he smiled: "That's a sign," he said, "that we are beginning to understand the problem." And indeed, the real work now began in earnest: day after day, week after week, the whole argument was patiently scrutinized with the help of simpler and more transparent examples. Einstein's problem was reshaped and its solution reformulated with such precision and clarity that the weakness in the critics' reasoning became evident, and their whole argumentation, for all its false brilliance, fell to pieces. "They do it 'smartly,'" Bohr commented, "but what counts is to do it right."

The refutation of Einstein's criticism does not add any new element to the conception of complementarity, but it is of great importance in laying bare a very deep-lying opposition between Bohr's general philosophical attitude and the still widespread habits of thought belonging to a glorious but irrevocably bygone stage in the evolution of science. Physical concepts, Einstein used to say, are "free creations of the mind." In the case under debate, the "criterion of reality" he proposed has very much this character, and it turns out to yield a striking illustration of the pitfalls to which one may be exposed by such arbitrary constructions of concepts. In spite of its apparent clarity, the criterion in question contains in fact a very essential ambiguity, hidden in the seemingly harmless restriction "without disturbing the system." To disclose this ambiguity, however, it is necessary to renounce any pretension to impose upon nature our own preconceived notion of what "elements of reality" ought to be, and humbly take guidance, as Bohr exhorts us to do, in what we can learn from nature herself.

....

When one realizes the fundamental nature of the issue at stake, it becomes easier to understand the state of exaltation in which Bohr accomplished this work. The writing of his reply, its typing, polishing, retyping and sending off to print did not take more than six weeks—an astonishing speed when one knows how slow his usual pace was. It was impressive to watch him thus at the height of his powers, in utmost concentration and unrelenting effort to attain clarity through painstaking scrutiny of every detail—true as ever to his favourite Schiller aphorism "Nur die Fülle führt zur Klarheit." He was particularly well served on this occasion by his uncommon ability to go into the opponent's views, dissect his arguments and turn them to the advantage of the truth. In this, however, he always proceeded with complete openmindedness, and only rejoiced in victory if in winning it he had also deepened his own insight into the problem.

The contest about the completeness of the quantal description of physical phenomena was the last clash between the two giants. The confrontation of their diverging conceptions of the nature of scientific knowledge had now reached the limits set by confining it to the problems of the physical world. That there was no hope of carrying it further was soon made clear by Einstein himself, who commented on Bohr's position that it was logically possible, but "so very contrary to my scientific instinct that I cannot forego the search for a more complete conception." Bohr was very unhappy about this deadlock, for he admired Einstein precisely for the way in which he had laid stress on the epistemological aspects of classical physics and, at an early stage, of quantum theory also. In fact, Einstein's approach to these problems had been so closely similar to his own, and such a source of inspiration to him, that he found Einstein's later lack of understanding doubly disheartening. On the other hand, he had good reason to look back with satisfaction on a controversy which had put to such severe test his own conception of the complementarity of physical phenomena, and even, in this last dispute about an alleged "criterion of reality," the underlying general ideas he had formed of the most fundamental aspects of human knowledge and man's position in the universe.

I.9 QUANTUM MECHANICS AND PHYSICAL REALITY

NIELS BOHR

IN a recent article by A. Einstein, B. Podolsky and N. Rosen, which appeared in the *Physical Review* of May 15, and was reviewed in NATURE of June 22, the question of the completeness of quantum mechanical description has been discussed on the basis of a "criterion of physical reality", which the authors formulate as follows: "If, without in any way disturbing a system, we can predict with certainty the value of a physical quantity, then there exists an element of physical reality corresponding to this physical quantity".

Since, as the authors show, it is always possible in quantum theory, just as in classical theory, to predict the value of any variable involved in the description of a mechanical system from measurements performed on other systems, which have only temporarily been in interaction with the system under investigation; and since in contrast to classical mechanics it is never possible in quantum mechanics to assign definite values to both of two conjugate variables, the authors conclude from their criterion that quantum mechanical description of physical reality is incomplete.

I should like to point out, however, that the named criterion contains an essential ambiguity when it is applied to problems of quantum mechanics. It is true that in the measurements under consideration any direct mechanical interaction of the system and the measuring agencies is excluded, but a closer examination reveals that the procedure of measurements has an essential influence on the conditions on which the very definition of the physical quantities in question rests. Since these conditions must be considered as an inherent element of any phenomenon to which the term "physical reality" can be unambiguously applied, the conclusion of the above-mentioned authors would not appear to be justified. A fuller development of this argument will be given in an article to be published shortly in the *Physical Review*.

N. BOHR.

Institute of Theoretical Physics,
Copenhagen.
June 29.

Originally published in *Nature, 136*, 65 (1935).

I.10 CAN QUANTUM-MECHANICAL DESCRIPTION OF PHYSICAL REALITY BE CONSIDERED COMPLETE?

NIELS BOHR

It is shown that a certain "criterion of physical reality" formulated in a recent article with the above title by A. Einstein, B. Podolsky and N. Rosen contains an essential ambiguity when it is applied to quantum phenomena. In this connection a viewpoint termed "complementarity" is explained from which quantum-mechanical description of physical phenomena would seem to fulfill, within its scope, all rational demands of completeness.

IN a recent article[1] under the above title A. Einstein, B. Podolsky and N. Rosen have presented arguments which lead them to answer the question at issue in the negative. The trend of their argumentation, however, does not seem to me adequately to meet the actual situation with which we are faced in atomic physics. I shall therefore be glad to use this opportunity to explain in somewhat greater detail a general viewpoint, conveniently termed "complementarity," which I have indicated on various previous occasions,[2] and from which quantum mechanics within its scope would appear as a completely rational description of physical phenomena, such as we meet in atomic processes.

The extent to which an unambiguous meaning can be attributed to such an expression as "physical reality" cannot of course be deduced from *a priori* philosophical conceptions, but—as the authors of the article cited themselves emphasize—must be founded on a direct appeal to experiments and measurements. For this purpose they propose a "criterion of reality" formulated as follows: "If, without in any way disturbing a system, we can predict with certainty the value of a physical quantity, then there exists an element of physical reality corresponding to this physical quantity." By means of an interesting example, to which we shall return below, they next proceed to show that in quantum mechanics, just as in classical mechanics, it is possible under suitable conditions to predict the value of any given variable pertaining to the description of a mechanical system from measurements performed entirely on other systems which previously have been in interaction with the system under investigation. According to their criterion the authors therefore want to ascribe an element of reality to each of the quantities represented by such variables. Since, moreover, it is a well-known feature of the present formalism of quantum mechanics that it is never possible, in the description of the state of a mechanical system, to attach definite values to both of two canonically conjugate variables, they consequently deem this formalism to be incomplete, and express the belief that a more satisfactory theory can be developed.

Such an argumentation, however, would hardly seem suited to affect the soundness of quantum-mechanical description, which is based on a coherent mathematical formalism covering automatically any procedure of measurement like that indicated.* The apparent contradiction in

[1] A. Einstein, B. Podolsky and N. Rosen, Phys. Rev. **47**, 777 (1935).

[2] Cf. N. Bohr, *Atomic Theory and Description of Nature*, I (Cambridge, 1934).

* The deductions contained in the article cited may in this respect be considered as an immediate consequence of the transformation theorems of quantum mechanics, which perhaps more than any other feature of the formalism contribute to secure its mathematical completeness and its rational correspondence with classical mechanics. In fact, it is always possible in the description of a mechanical system, consisting of two partial systems (1) and (2), interacting or not, to replace any two pairs of canonically conjugate variables (q_1p_1), (q_2p_2) pertaining to systems (1) and (2), respectively, and satisfying the usual commutation rules

$$[q_1p_1]=[q_2p_2]=ih/2\pi,$$
$$[q_1q_2]=[p_1p_2]=[q_1p_2]=[q_2p_1]=0,$$

by two pairs of new conjugate variables (Q_1P_1), (Q_2P_2) related to the first variables by a simple orthogonal transformation, corresponding to a rotation of angle θ in the planes (q_1q_2), (p_1p_2)

$$q_1=Q_1\cos\theta-Q_2\sin\theta \qquad p_1=P_1\cos\theta-P_2\sin\theta$$
$$q_2=Q_1\sin\theta+Q_2\cos\theta \qquad p_2=P_1\sin\theta+P_2\cos\theta.$$

Since these variables will satisfy analogous commutation rules, in particular

$$[Q_1P_1]=ih/2\pi, \qquad [Q_1P_2]=0,$$

it follows that in the description of the state of the combined system definite numerical values may not be assigned to both Q_1 and P_1, but that we may clearly assign

Originally published in *Physical Review, 48*, 696-702 (1935).

fact discloses only an essential inadequacy of the customary viewpoint of natural philosophy for a rational account of physical phenomena of the type with which we are concerned in quantum mechanics. Indeed the *finite interaction between object and measuring agencies* conditioned by the very existence of the quantum of action entails—because of the impossibility of controlling the reaction of the object on the measuring instruments if these are to serve their purpose—the necessity of a final renunciation of the classical ideal of causality and a radical revision of our attitude towards the problem of physical reality. In fact, as we shall see, a criterion of reality like that proposed by the named authors contains—however cautious its formulation may appear—an essential ambiguity when it is applied to the actual problems with which we are here concerned. In order to make the argument to this end as clear as possible, I shall first consider in some detail a few simple examples of measuring arrangements.

Let us begin with the simple case of a particle passing through a slit in a diaphragm, which may form part of some more or less complicated experimental arrangement. Even if the momentum of this particle is completely known before it impinges on the diaphragm, the diffraction by the slit of the plane wave giving the symbolic representation of its state will imply an uncertainty in the momentum of the particle, after it has passed the diaphragm, which is the greater the narrower the slit. Now the width of the slit, at any rate if it is still large compared with the wave-length, may be taken as the uncertainty Δq of the position of the particle relative to the diaphragm, in a direction perpendicular to the slit. Moreover, it is simply seen from de Broglie's relation between momentum and wave-length that the uncertainty Δp of the momentum of the particle in this direction is correlated to Δq by means of Heisenberg's general principle

$$\Delta p \Delta q \sim h,$$

which in the quantum-mechanical formalism is a direct consequence of the commutation relation for any pair of conjugate variables. Obviously the uncertainty Δp is inseparably connected with the possibility of an exchange of momentum between the particle and the diaphragm; and the question of principal interest for our discussion is now to what extent the momentum thus exchanged can be taken into account in the description of the phenomenon to be studied by the experimental arrangement concerned, of which the passing of the particle through the slit may be considered as the initial stage.

Let us first assume that, corresponding to usual experiments on the remarkable phenomena of electron diffraction, the diaphragm, like the other parts of the apparatus,—say a second diaphragm with several slits parallel to the first and a photographic plate,—is rigidly fixed to a support which defines the space frame of reference. Then the momentum exchanged between the particle and the diaphragm will, together with the reaction of the particle on the other bodies, pass into this common support, and we have thus voluntarily cut ourselves off from any possibility of taking these reactions separately into account in predictions regarding the final result of the experiment,—say the position of the spot produced by the particle on the photographic plate. The impossibility of a closer analysis of the reactions between the particle and the measuring instrument is indeed no peculiarity of the experimental procedure described, but is rather an essential property of any arrangement suited to the study of the phenomena of the type concerned, where we have to do with a feature of *individuality* completely foreign to classical physics. In fact, any possibility of taking into account the momentum exchanged between the particle and the separate parts of the apparatus would at once permit us to draw conclusions regarding the "course" of such phenomena,—say through what particular slit of the second diaphragm the particle passes on its way to the photographic plate—which would be quite incompatible with the fact that the probability of the particle reaching a given element of area on this plate is determined not by the presence of any particular slit, but by the positions of all the slits of the second diaphragm within reach

---

such values to both Q_1 and P_2. In that case it further results from the expressions of these variables in terms of (q_1p_1) and (q_2p_2), namely

$$Q_1 = q_1 \cos\theta + q_2 \sin\theta, \qquad P_2 = -p_1 \sin\theta + p_2 \cos\theta,$$

that a subsequent measurement of either q_2 or p_2 will allow us to predict the value of q_1 or p_1 respectively.

of the associated wave diffracted from the slit of the first diaphragm.

By another experimental arrangement, where the first diaphragm is not rigidly connected with the other parts of the apparatus, it would at least in principle* be possible to measure its momentum with any desired accuracy before and after the passage of the particle, and thus to predict the momentum of the latter after it has passed through the slit. In fact, such measurements of momentum require only an unambiguous application of the classical law of conservation of momentum, applied for instance to a collision process between the diaphragm and some test body, the momentum of which is suitably controlled before and after the collision. It is true that such a control will essentially depend on an examination of the space-time course of some process to which the ideas of classical mechanics can be applied; if, however, all spatial dimensions and time intervals are taken sufficiently large, this involves clearly no limitation as regards the accurate control of the momentum of the test bodies, but only a renunciation as regards the accuracy of the control of their space-time coordination. This last circumstance is in fact quite analogous to the renunciation of the control of the momentum of the fixed diaphragm in the experimental arrangement discussed above, and depends in the last resort on the claim of a purely classical account of the measuring apparatus, which implies the necessity of allowing a latitude corresponding to the quantum-mechanical uncertainty relations in our description of their behavior.

The principal difference between the two experimental arrangements under consideration is, however, that in the arrangement suited for the control of the momentum of the first diaphragm, this body can no longer be used as a measuring instrument for the same purpose as in the previous case, but must, as regards its position relative to the rest of the apparatus, be treated, like the particle traversing the slit, as an object of investigation, in the sense that the quantum-mechanical uncertainty relations regarding its position and momentum must be taken explicitly into account. In fact, even if we knew the position of the diaphragm relative to the space frame before the first measurement of its momentum, and even though its position after the last measurement can be accurately fixed, we lose, on account of the uncontrollable displacement of the diaphragm during each collision process with the test bodies, the knowledge of its position when the particle passed through the slit. The whole arrangement is therefore obviously unsuited to study the same kind of phenomena as in the previous case. In particular it may be shown that, if the momentum of the diaphragm is measured with an accuracy sufficient for allowing definite conclusions regarding the passage of the particle through some selected slit of the second diaphragm, then even the minimum uncertainty of the position of the first diaphragm compatible with such a knowledge will imply the total wiping out of any interference effect—regarding the zones of permitted impact of the particle on the photographic plate—to which the presence of more than one slit in the second diaphragm would give rise in case the positions of all apparatus are fixed relative to each other.

In an arrangement suited for measurements of the momentum of the first diaphragm, it is further clear that even if we have measured this momentum before the passage of the particle through the slit, we are after this passage still left with a *free choice* whether we wish to know the momentum of the particle or its initial position relative to the rest of the apparatus. In the first eventuality we need only to make a second determination of the momentum of the diaphragm, leaving unknown forever its exact position when the particle passed. In the second eventuality we need only to determine its position relative to the space frame with the inevitable loss of the knowledge of the momentum exchanged between the diaphragm and the particle. If the diaphragm is sufficiently massive in comparison with the particle, we may even arrange the procedure of measurements in such a way that the diaphragm after the first determination of its momentum will remain at rest in some unknown position relative to the

* The obvious impossibility of actually carrying out, with the experimental technique at our disposal, such measuring procedures as are discussed here and in the following does clearly not affect the theoretical argument, since the procedures in question are essentially equivalent with atomic processes, like the Compton effect, where a corresponding application of the conservation theorem of momentum is well established.

instrument rigidly fixed to the support which defines the space frame of reference. Under the experimental conditions described such a measurement will therefore also provide us with the knowledge of the location, otherwise completely unknown, of the diaphragm with respect to this space frame when the particles passed through the slits. Indeed, only in this way we obtain a basis for conclusions about the initial position of the other particle relative to the rest of the apparatus. By allowing an essentially uncontrollable momentum to pass from the first particle into the mentioned support, however, we have by this procedure cut ourselves off from any future possibility of applying the law of conservation of momentum to the system consisting of the diaphragm and the two particles and therefore have lost our only basis for an unambiguous application of the idea of momentum in predictions regarding the behavior of the second particle. Conversely, if we choose to measure the momentum of one of the particles, we lose through the uncontrollable displacement inevitable in such a measurement any possibility of deducing from the behavior of this particle the position of the diaphragm relative to the rest of the apparatus, and have thus no basis whatever for predictions regarding the location of the other particle.

From our point of view we now see that the wording of the above-mentioned criterion of physical reality proposed by Einstein, Podolsky and Rosen contains an ambiguity as regards the meaning of the expression "without in any way disturbing a system." Of course there is in a case like that just considered no question of a mechanical disturbance of the system under investigation during the last critical stage of the measuring procedure. But even at this stage there is essentially the question of *an influence on the very conditions which define the possible types of predictions regarding the future behavior of the system.* Since these conditions constitute an inherent element of the description of any phenomenon to which the term "physical reality" can be properly attached, we see that the argumentation of the mentioned authors does not justify their conclusion that quantum-mechanical description is essentially incomplete. On the contrary this description, as appears from the preceding discussion, may be characterized as a rational utilization of all possibilities of unambiguous interpretation of measurements, compatible with the finite and uncontrollable interaction between the objects and the measuring instruments in the field of quantum theory. In fact, it is only the mutual exclusion of any two experimental procedures, permitting the unambiguous definition of complementary physical quantities, which provides room for new physical laws, the coexistence of which might at first sight appear irreconcilable with the basic principles of science. It is just this entirely new situation as regards the description of physical phenomena, that the notion of *complementarity* aims at characterizing.

The experimental arrangements hitherto discussed present a special simplicity on account of the secondary role which the idea of time plays in the description of the phenomena in question. It is true that we have freely made use of such words as "before" and "after" implying time-relationships; but in each case allowance must be made for a certain inaccuracy, which is of no importance, however, so long as the time intervals concerned are sufficiently large compared with the proper periods entering in the closer analysis of the phenomenon under investigation. As soon as we attempt a more accurate time description of quantum phenomena, we meet with well-known new paradoxes, for the elucidation of which further features of the interaction between the objects and the measuring instruments must be taken into account. In fact, in such phenomena we have no longer to do with experimental arrangements consisting of apparatus essentially at rest relative to one another, but with arrangements containing moving parts,—like shutters before the slits of the diaphragms,—controlled by mechanisms serving as clocks. Besides the transfer of momentum, discussed above, between the object and the bodies defining the space frame, we shall therefore, in such arrangements, have to consider an eventual exchange of energy between the object and these clock-like mechanisms.

The decisive point as regards time measurements in quantum theory is now completely analogous to the argument concerning measurements of positions outlined above. Just as the transfer of momentum to the separate parts of other parts of the apparatus, and the subsequent fixation of this position may therefore simply consist in establishing a rigid connection between the diaphragm and the common support.

My main purpose in repeating these simple, and in substance well-known considerations, is to emphasize that in the phenomena concerned we are not dealing with an incomplete description characterized by the arbitrary picking out of different elements of physical reality at the cost of sacrifying other such elements, but with a rational discrimination between essentially different experimental arrangements and procedures which are suited either for an unambiguous use of the idea of space location, or for a legitimate application of the conservation theorem of momentum. Any remaining appearance of arbitrariness concerns merely our freedom of handling the measuring instruments, characteristic of the very idea of experiment. In fact, the renunciation in each experimental arrangement of the one or the other of two aspects of the description of physical phenomena,—the combination of which characterizes the method of classical physics, and which therefore in this sense may be considered as *complementary* to one another,—depends essentially on the impossibility, in the field of quantum theory, of accurately controlling the reaction of the object on the measuring instruments, i.e., the transfer of momentum in case of position measurements, and the displacement in case of momentum measurements. Just in this last respect any comparison between quantum mechanics and ordinary statistical mechanics,—however useful it may be for the formal presentation of the theory,—is essentially irrelevant. Indeed we have in each experimental arrangement suited for the study of proper quantum phenomena not merely to do with an ignorance of the value of certain physical quantities, but with the impossibility of defining these quantities in an unambiguous way.

The last remarks apply equally well to the special problem treated by Einstein, Podolsky and Rosen, which has been referred to above, and which does not actually involve any greater intricacies than the simple examples discussed above. The particular quantum-mechanical state of two free particles, for which they give an explicit mathematical expression, may be reproduced, at least in principle, by a simple experimental arrangement, comprising a rigid diaphragm with two parallel slits, which are very narrow compared with their separation, and through each of which one particle with given initial momentum passes independently of the other. If the momentum of this diaphragm is measured accurately before as well as after the passing of the particles, we shall in fact know the sum of the components perpendicular to the slits of the momenta of the two escaping particles, as well as the difference of their initial positional coordinates in the same direction; while of course the conjugate quantities, i.e., the difference of the components of their momenta, and the sum of their positional coordinates, are entirely unknown.* In this arrangement, it is therefore clear that a subsequent single measurement either of the position or of the momentum of one of the particles will automatically determine the position or momentum, respectively, of the other particle with any desired accuracy; at least if the wave-length corresponding to the free motion of each particle is sufficiently short compared with the width of the slits. As pointed out by the named authors, we are therefore faced at this stage with a completely free choice whether we want to determine the one or the other of the latter quantities by a process which does not directly interfere with the particle concerned.

Like the above simple case of the choice between the experimental procedures suited for the prediction of the position or the momentum of a single particle which has passed through a slit in a diaphragm, we are, in the "freedom of choice" offered by the last arrangement, just concerned with a *discrimination between different experimental procedures which allow of the unambiguous use of complementary classical concepts.* In fact to measure the position of one of the particles can mean nothing else than to establish a correlation between its behavior and some

* As will be seen, this description, apart from a trivial normalizing factor, corresponds exactly to the transformation of variables described in the preceding footnote if (q_1p_1), (q_2p_2) represent the positional coordinates and components of momenta of the two particles and if $\theta = -\pi/4$. It may also be remarked that the wave function given by formula (9) of the article cited corresponds to the special choice of $P_2 = 0$ and the limiting case of two infinitely narrow slits.

the apparatus,—the knowledge of the relative positions of which is required for the description of the phenomenon,—has been seen to be entirely uncontrollable, so the exchange of energy between the object and the various bodies, whose relative motion must be known for the intended use of the apparatus, will defy any closer analysis. Indeed, it is *excluded in principle to control the energy which goes into the clocks without interfering essentially with their use as time indicators.* This use in fact entirely relies on the assumed possibility of accounting for the functioning of each clock as well as for its eventual comparison with other clocks on the basis of the methods of classical physics. In this account we must therefore obviously allow for a latitude in the energy balance, corresponding to the quantum-mechanical uncertainty relation for the conjugate time and energy variables. Just as in the question discussed above of the mutually exclusive character of any unambiguous use in quantum theory of the concepts of position and momentum, it is in the last resort this circumstance which entails the complementary relationship between any detailed time account of atomic phenomena on the one hand and the unclassical features of intrinsic stability of atoms, disclosed by the study of energy transfers in atomic reactions on the other hand.

This necessity of discriminating in each experimental arrangement between those parts of the physical system considered which are to be treated as measuring instruments and those which constitute the objects under investigation may indeed be said to form a *principal distinction between classical and quantum-mechanical description of physical phenomena.* It is true that the place within each measuring procedure where this discrimination is made is in both cases largely a matter of convenience. While, however, in classical physics the distinction between object and measuring agencies does not entail any difference in the character of the description of the phenomena concerned, its fundamental importance in quantum theory, as we have seen, has its root in the indispensable use of classical concepts in the interpretation of all proper measurements, even though the classical theories do not suffice in accounting for the new types of regularities with which we are concerned in atomic physics. In accordance with this situation there can be no question of any unambiguous interpretation of the symbols of quantum mechanics other than that embodied in the well-known rules which allow to predict the results to be obtained by a given experimental arrangement described in a totally classical way, and which have found their general expression through the transformation theorems, already referred to. By securing its proper correspondence with the classical theory, these theorems exclude in particular any imaginable inconsistency in the quantum-mechanical description, connected with a change of the place where the discrimination is made between object and measuring agencies. In fact it is an obvious consequence of the above argumentation that in each experimental arrangement and measuring procedure we have only a free choice of this place within a region where the quantum-mechanical description of the process concerned is effectively equivalent with the classical description.

Before concluding I should still like to emphasize the bearing of the great lesson derived from general relativity theory upon the question of physical reality in the field of quantum theory. In fact, notwithstanding all characteristic differences, the situations we are concerned with in these generalizations of classical theory present striking analogies which have often been noted. Especially, the singular position of measuring instruments in the account of quantum phenomena, just discussed, appears closely analogous to the well-known necessity in relativity theory of upholding an ordinary description of all measuring processes, including a sharp distinction between space and time coordinates, although the very essence of this theory is the establishment of new physical laws, in the comprehension of which we must renounce the customary separation of space and time ideas.*

* Just this circumstance, together with the relativistic invariance of the uncertainty relations of quantum mechanics, ensures the compatibility between the argumentation outlined in the present article and all exigencies of relativity theory. This question will be treated in greater detail in a paper under preparation, where the writer will in particular discuss a very interesting paradox suggested by Einstein concerning the application of gravitation theory to energy measurements, and the solution of which offers an especially instructive illustration of the generality of the argument of complementarity. On the same occasion a more thorough discussion of space-time measurements in quantum theory will be given with all necessary mathematical developments and diagrams of experimental

The dependence on the reference system, in relativity theory, of all readings of scales and clocks may even be compared with the essentially uncontrollable exchange of momentum or energy between the objects of measurements and all instruments defining the space-time system of reference, which in quantum theory confronts us with the situation characterized by the notion of complementarity. In fact this new feature of natural philosophy means a radical revision of our attitude as regards physical reality, which may be paralleled with the fundamental modification of all ideas regarding the absolute character of physical phenomena, brought about by the general theory of relativity.

arrangements, which had to be left out of this article, where the main stress is laid on the dialectic aspect of the question at issue.

I.11 THE PRESENT SITUATION IN QUANTUM MECHANICS: A TRANSLATION OF SCHRÖDINGER'S "CAT PARADOX" PAPER

ERWIN SCHRÖDINGER (TRANS. JOHN D. TRIMMER*)

INTRODUCTION

This is a translation of Schrödinger's three-part 1935 paper [1] in *Die Naturwissenschaften.* Earlier that same year there had appeared the Einstein, Podolsky, Rosen paper [2] (also famous in "paradoxology") which, Schrödinger says, in a footnote, motivated his offering. Along with this article in German, Schrödinger had two closely related English-language publications.[3] But the German, aside from its one-paragraph presentation of the famous cat, covers additional territory and gives many fascinating insights into Schrödinger's thought. The translator's goal has been to adhere to the logical and physical content of the original, while at the same time trying to convey something of its semi-conversational, at times slightly sardonic flavor.

TRANSLATION

1. *The Physics of Models*

In the second half of the previous century there arose, from the great progress in kinetic theory of gases and in the mechanical theory of heat, an ideal of the exact description of nature that stands out as the reward of centuries-long search and the fulfillment of millennia-long hope, and that is called classical. These are its features.

Of natural objects, whose observed behavior one might treat, one sets up a representation—based on the experimental data in one's possession but without handcuffing the intuitive imagination—that is worked out in all details exactly, *much* more exactly than any experience, considering its limited extent, can ever authenticate. The representation in its absolute determinacy resembles a mathematical concept or a geometric figure which can be completely calculated from a number of *determining parts;* as, e.g., a triangle's one side and two adjoining angles, as determining parts, also determine the third angle, the other two sides, the three altitudes, the radius of the inscribed circle, etc. Yet the representation differs intrinsically from a geometric figure in this important respect, that also in *time* as fourth dimension it is just as sharply determined as the figure is in the three space dimensions. Thus it is a question (as is self-evident) always of a concept that changes with time, that can assume different *states;* and if a state becomes known in the necessary number of determining parts, then not only are all other parts also given for this moment (as illustrated for the triangle above), but likewise all parts, the complete state, for any given later time; just as the character of a triangle on its base determines its character at the apex. It is part of the inner law of the concept that it should change in a given manner, that is, if left to itself in a given initial state, that it should continuously run through a given sequence of states, each one of which it reaches at a fully determined time. That is its nature, that is the hypothesis, which, as I said above, one builds on a foundation of intuitive imagination.

Of course one must not think so literally, that in this way one learns how things go in the real world. To show that one does not think this, one calls the precise thinking aid that one has created, an *image* or a *model.* With its hindsight-free clarity, which cannot be attained without arbitrariness, one has merely insured that a fully determined hypothesis can be tested for its consequences, without admitting further arbitrariness during the tedious calculations required for deriving these consequences. Here one has explicit marching orders and actually works out only what a clever fellow could have told directly from the data! At least one then knows where the arbitrariness lies and where improvement must be made in case of disagreement with experience: in the initial hypothesis or model. For this one must always be prepared. If in many various experiments the natural object behaves like the model, one is happy and thinks that the image fits the reality in essential features. If it fails to agree, under novel experiments or with refined measuring techniques, it is not said that one should *not* be happy. For basically this is the means of gradually bringing our picture, i.e., our thinking, closer to the realities.

The classical method of the precise model has as principal goal keeping the unavoidable arbitrariness

* Box 79, Route 1, Millington, Md. 21651.

[1] E. Schrödinger, "Die gegenwärtige Situation in der Quantenmechanik," *Naturwissenschaften* **23**: pp. 807–812; 823–828; 844–849 (1935).

[2] A. Einstein, B. Podolsky, and N. Rosen, Phys. Rev. **47**: p. 777 (1935).

[3] E. Schrödinger, *Proc. Cambridge Phil. Soc.* **31**: p. 555 (1935); *ibid.*, **32**: p. 446 (1936).

This translation was originally published in *Proceedings of the American Philosophical Society, 124*, 323-38 (1980).

neatly isolated in the assumptions, more or less as body cells isolate the nucleoplasm, for the historical process of adaptation to continuing experience. Perhaps the method is based on the belief that *somehow* the initial state *really* determines uniquely the subsequent events, or that a *complete* model, agreeing with reality in *complete exactness* would permit predictive calculation of outcomes of all experiments with complete exactness. Perhaps on the other hand this belief is based on the method. But it is quite probable that the adaptation of thought to experience is an infinite process and that "complete model" is a contradiction in terms, somewhat like "largest integer."

A clear presentation of what is meant by classical *model*, its *determining parts*, its *state*, is the foundation for all that follows. Above all, a *determinate model* and a *determinate state of the same* must not be confused. Best consider an example. The Rutherford model of the hydrogen atom consists of two point masses. As determining parts one could for example use the two times three rectangular coordinates of the two points and the two times three components of their velocities along the coordinate axes—thus twelve in all. Instead of these one could also choose: the coordinates and velocity components of the *center of mass*, plus the *separation* of the two points, *two angles* that establish the direction in space of the line joining them, and the speeds (= time derivatives) with which the separation and the two angles are changing at the particular moment; this again adds up of course to twelve. It is *not* part of the concept "R-model of the H-atom" that the determining parts should have particular numerical values. Such being assigned to them, one arrives at a *determinate state* of the model. The clear view over the totality of possible states—yet without relationship among them—constitutes "the model" or "the model in *any* state *whatsoever*." But the concept of the model then amounts to more than merely: the two points in certain positions, endowed with certain velocities. It embodies also knowledge for *every* state how it will change with time in absence of outside interference. (Information on how one half of the determining parts will change with time is indeed given by the other half, but how this other half will change must be independently determined.) *This* knowledge is implicit in the assumptions: the points have the masses m, M and the charges −e, +e and therefore attract each other with force e^2/r^2, if their separation is r.

These results, with definite numerical values for m, M, and e (but of course *not* for r), belong to the description *of the model* (not first and only to that of a definite state). m, M, and e are *not* determining parts. By contrast, separation r is one. It appears as the seventh in the second "set" of the example introduced above. And if one uses the first, r is not an independent thirteenth but can be calculated from the 6 rectangular coordinates:

$$r = [(x_1 - x_2)^2 + (y_1 - y_2)^2 + (z_1 - z_2)^2]^{\frac{1}{2}}.$$

The number of determining parts (which are often called *variables* in contrast to *constants of the model* such as m, M, e) is unlimited. Twelve conveniently chosen ones determine all others, or the *state*. No twelve have the privilege of being *the* determining parts. Examples of other especially important determining parts are: the energy, the three components of angular momentum relative to center of mass, the kinetic energy of center of mass motion. These just named have, however, a special character. They are indeed *variable*, i.e., they have different values in different states. But in every *sequence* of states, that is actually passed through in the course of time, they retain the same value. So they are also called *constants of the motion*—differing from constants of the model.

2. *Statistics of Model Variables in Quantum Mechanics*

At the pivot point of contemporary quantum mechanics (Q.M.) stands a doctrine, that perhaps may yet undergo many shifts of meaning but that will not, I am convinced, cease to be the pivot point. It is this, that models with determining parts that uniquely determine each other, as do the classical ones, cannot do justice to nature.

One might think that for anyone believing this, the classical models have played out their roles. But this is not the case. Rather one uses precisely *them*, not only to express the negative of the new doctrine, but also to describe the diminished mutual determinacy remaining afterwards as though obtaining among the same variables of the same models as were used earlier, as follows:

A. The classical concept of *state* becomes lost, in that at most a well-chosen *half* of a complete set of variables can be assigned definite numerical values; in the Rutherford example for instance the six rectangular coordinates *or* the velocity components (still other groupings are possible). The other half then remains completely indeterminate, while supernumerary parts can show highly varying degrees of indeterminacy. In general, of a complete set (for the R-model twelve parts) *all* will be known only uncertainly. One can best keep track of the degree of uncertainty by following classical mechanics and choosing variables arranged *in pairs* of so-called canonically-conjugate ones. The simplest example is a space coordinate x of a point mass and the component p_x along the same direction, its linear momentum (i.e., mass times velocity). Two such constrain each other in the precision with which they may be simultaneously known, in that the product of their tolerance- or variation-widths (customarily designated by putting a Δ ahead of the quantity) cannot fall *below* the magnitude of a certain universal constant,[4] thus

$$\Delta x \cdot \Delta p_x \geqq \hbar.$$

(Heisenberg uncertainty relation.)

B. If even at any given moment not all variables are determined by some of them, then of course neither are they all determined for a later moment by data obtainable earlier. This may be called a break with causality, but in view of *A.* it is nothing essentially new. If a classical state does not exist at any moment, it can hardly change causally. What do change are the *statistics* or *probabilities, these* moreover causally. Individual variables meanwhile may become more, or less, uncertain. Overall it may be said that the total precision of the description does not change with time, because the principle of limitations described under *A.* remains the same at every moment. - - - -

Now what is the meaning of the terms "uncertain," "statistics," "probability"? Here Q.M. gives the following account. It takes over unquestioningly from the classical model the entire infinite roll call of imaginable variables or determining parts and proclaims each part to be *directly measurable*, indeed measurable to arbitrary precision, so far as it alone is concerned. If through a well-chosen, constrained set of measurements one has gained that maximal knowledge of an object which is just possible according to *A.*, then the mathematical apparatus of the new theory provides means of assigning, for the same or for any later instant of time, a fully determined *statistical distribution* to *every* variable, that is, an indication of the fraction of cases it will be found at this or that value, or within this or that small interval (which is also called probability.) The doctrine is that this is in fact the probability of encountering the relevant variable, if one measures it at the relevant time, at this or that value. By a single trial the correctness of this *probability prediction* can be given at most an approximate test, namely in the case that it is comparatively sharp, i.e., declares possible only a small range of values. To test it thoroughly one must repeat the entire trial *ab ovo* (i.e., including the orientational or preparatory measurements) *very* often and may use only those cases in which the *preparatory* measurements gave exactly the same results. For these cases, then, the statistics of a particular variable, reckoned forward from the preparatory measurements, is to be confirmed by measurement—this is the doctrine.

One must guard against criticizing this doctrine because it is so difficult to express; this is a matter of language. But a different criticism surfaces. Scarcely a single physicist of the classical era would have dared to believe, in thinking about a model, that its determining parts are measurable on the natural object. Only much remoter consequences of the picture were actually open to experimental test. And all experience pointed toward one conclusion: long before the advancing experimental arts had bridged the broad chasm, the model would have substantially changed through gradual adaptation to new facts.—Now while the new theory calls the classical model incapable of specifying all details of the *mutual interrelationship of the determining parts* (for which its creators intended it), it nevertheless considers the model suitable for guiding us as to just which measurements can in principle be made on the relevant natural object. This would have seemed to those who thought up the picture a scandalous extension of their thought-pattern and an unscrupulous proscription against future development. Would it not be pre-established harmony of a peculiar sort if the classical-epoch researchers, those who, as we hear today, had no idea of what *measuring* truly is, had unwittingly gone on to give us as legacy a guidance scheme revealing just what is fundamentally measurable for instance about a hydrogen atom!?

I hope later to make clear that the reigning doctrine is born of distress. Meanwhile I continue to expound it.

3. Examples of Probability Predictions

All of the foregoing pertains to determining parts of a classical model, to positions and velocities of point masses, to energies, angular momenta, etc. The only unclassical feature is that only probabilities are predicted. Let us have a closer look. The orthodox treatment is always that, by way of certain measurements performed *now* and by way of their resulting prediction of results to be expected of other measurements following thereafter either immediately or at some given time, one gains the best possible probability estimates permitted by nature. Now how does the matter really stand? In important and typical cases as follows.

If one measures the energy of a Planck oscillator, the probability of finding for it a value between E and E′ cannot possibly be other than zero unless between E and E′ there lies at least one value from the series $3\pi\hbar\nu$, $5\pi\hbar\nu$, $7\pi\hbar\nu$, $9\pi\hbar\nu$, . . . For any interval containing none of these values the probability is zero. In plain English: other measurement results are ex-

[4] $h = 1.041 \cdot 10^{-27}$ erg sec. Usually in the literature the 2π-fold of this ($6.542 \cdot 10^{-27}$ erg sec) is designated as h and for *our* h an h with a cross-bar is written. [Transl. Note: In conformity with the now universal usage, ħ is used in the translation in place of h.]

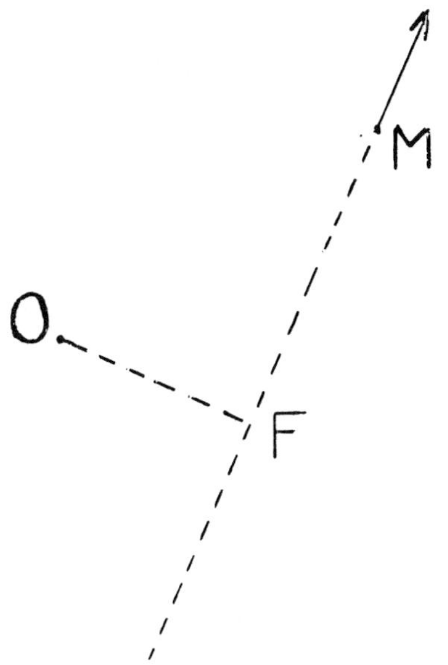

FIG. 1. Angular momentum. M is a material point, O a geometric reference point. The vector arrow represents the momentum (= mass times velocity) of M. Then the *angular momentum* is the product of the length of the arrow by the length OF.

cluded. The values are odd multiples of the *constant of the model* $\pi\hbar\nu$

$$\text{(Planck constant)}/2\pi,$$

$$\nu = \text{frequency of the oscillator}.$$

Two points stand out. First, no account is taken of preceding measurements—these are quite unnecessary. Second, the statement certainly doesn't suffer an excessive lack of precision—quite to the contrary it is sharper than any actual measurement could ever be.

Another typical example is magnitude of angular momentum. In Fig. 1 let M be a moving point mass, with the vector representing, in magnitude and direction, its momentum (mass times velocity). O is any arbitrary fixed point in space, say the origin of coordinates; thus not a physically significant point, but rather a geometric reference point. As magnitude of the angular momentum of M about O classical mechanics designates the product of the length of the momentum vector by the length of the *normal OF*. In Q.M. the magnitude of angular momentum is governed much as the energy of the oscillator. Again the probability is zero for any interval not containing some value(s) from the following series

$$\hbar\,(2)^{\frac{1}{2}},\ \hbar\,(2\times 3)^{\frac{1}{2}},\ \hbar\,(3\times 4)^{\frac{1}{2}},\ \hbar\,(4\times 5)^{\frac{1}{2}},\ \ldots;$$

that is, only one of these values is allowed. Again this is true without reference to preceding measurements. And one readily conceives how important is this precise statement, *much* more important than knowing which of these values, or what probability for each of them, would actually pertain to a given case. Moreover it is also noteworthy here that there is no mention of the reference point: however it is chosen one will get a value from the series. This assertion seems unreasonable for the model, because the normal OF changes *continuously* as the point O is displaced, if the momentum vector remains unchanged. In this example we see how Q.M. does indeed use the model to read off those quantities which one can measure and for which it makes sense to predict results, but finds the classical model inadequate for explicating relationships among these quantities. Now in both examples does one not get the feeling that the essential content of what is being said can only with some difficulty be forced into the Spanish boot of a prediction of probability of finding this or that measurement result for a variable of the classical model? Does one not get the impression that here one deals with fundamental properties of *new* classes of characteristics, that keep only the name in common with classical ones? And by no means do we speak here of exceptional cases, rather it is precisely the truly valuable statements of the new theory that have this character. There are indeed problems more nearly of the type for which the mode of expression is suitable. But they are by no means equally important. Moreover of no importance whatever are those that are naively set up as class exercises. "Given the position of the electron in the hydrogen atom at time t = 0, find the statistics of its position at a later time." No one cares about that.

The big idea seems to be that all statements pertain to the intuitive model. But the useful statements are scarcely intuitive to it, and its intuitive aspects are of little worth.

4. *Can One Base the Theory on Ideal Ensembles?*

The classical model plays a Protean role in Q.M. Each of its determining parts can under certain circumstances become an object of interest and achieve a certain reality. But never all of them together—now it is these, now those, and indeed always at most *half* of the complete set of variables allowed by a full picture of the momentary state. Meantime, how about the others? *Have* they then no reality, perhaps (pardon the expression) a blurred reality; or are all of them always real and is it merely, according to Theorem *A*. of Sect. 2., that simultaneous *knowledge* of them is ruled out?

The second interpretation is especially appealing to those acquainted with the *statistical viewpoint* that came up in the second half of the preceding century;

the more so, considering that on the eve of the new century quantum theory was born *from it*, from a central problem in the statistical theory of heat (Max Planck's *Theory of Heat Radiation*, December, 1899). The essence of this line of thought is precisely this, that one practically never knows all the determining parts of the system, but rather *much* fewer. To describe an actual body at a given moment one relies therefore not on *one* state of the model but on a so-called *Gibbs ensemble*. By this is meant an ideal, that is, merely imagined ensemble of states, that accurately reflects our limited knowledge of the actual body. The body is then considered to behave as though in a single state *arbitrarily chosen from this ensemble*. This interpretation had the most extensive results. Its highest triumphs were in those cases for which *not* all states appearing in the ensemble led to *the same* observable behavior. Thus the body's conduct is now this way, now that, just as foreseen (thermodynamic fluctuations). At first thought one might well attempt likewise to refer back the always uncertain statements of Q.M. to an ideal ensemble of states, of which a quite specific one applies in any concrete instance—but one does not know which one.

That this won't work is shown by the *one* example of angular momentum, as one of many. Imagine in Fig. 1 the point M to be situated at various positions relative to O and fitted with various momentum vectors, and all these possibilities to be combined into an ideal ensemble. Then one can indeed so choose these positions and vectors that in every case the product of vector length by length of normal OF yields one or the other of the acceptable values—relative to the particular point O. But for an arbitrary different point O′, of course, unacceptable values occur. Thus appeal to the ensemble is no help at all. —Another example is the oscillator energy. Take the case that it has a sharply determined value, e.g., the lowest, $3\pi\hbar\nu$. The separation of the two point masses (that constitute the oscillator) then appears very *unsharp*. To be able to refer this statement to a statistical collective of states would require the distribution of separations to be sharply limited, at least toward large values, by that separation for which the *potential energy* alone would equal or exceed the value $3\pi\hbar\nu$. But that's not the way it is—arbitrarily large separations occur, even though with markedly reduced probability. And this is no mere secondary calculation result, that might in some fashion be circumvented, without striking at the heart of the theory: along with many others, the quantum mechanical treatment of radioactivity (Gamow) rests on this state of affairs.—One could go on indefinitely with more examples. One should note that there was no question of any time-dependent changes. It would be of no help to permit the model to vary quite "unclassically," perhaps to "jump." Already for the single instant things go wrong. At no moment does there exist an ensemble of classical states of the model that squares with the totality of quantum mechanical statements of this moment. The same can also be said as follows: if I wish to ascribe to the model at each moment a definite (merely not exactly known to me) state, or (which is the same) to *all* determining parts definite (merely not exactly known to me) numerical values, then there is no supposition as to these numerical values *to be imagined* that would not conflict with some portion of quantum theoretical assertions.

That is not quite what one expects, on hearing that the pronouncements of the new theory are always uncertain compared to the classical ones.

5. *Are the Variables Really Blurred?*

The other alternative consisted of granting reality only to the momentarily sharp determining parts—or in more general terms to each variable a sort of realization just corresponding to the quantum mechanical statistics of this variable at the relevant moment.

That it is in fact not impossible to express the degree and kind of blurring of *all* variables in *one* perfectly *clear* concept follows at once from the fact that Q.M. as a matter of fact has and uses such an instrument, the so-called wave function or ψ-function, also called system vector. Much more is to be said about it further on. That it is an abstract, unintuitive mathematical construct is a scruple that almost always surfaces against new aids to thought and that carries no great message. At all events it is an imagined entity that images the blurring of all variables at every moment just as clearly and faithfully as the classical model does its sharp numerical values. Its equation of motion too, the law of its time variation, so long as the system is left undisturbed, lags not one iota, in clarity and determinacy, behind the equations of motion of the classical model. So the latter could be straight-forwardly replaced by the ψ-function, so long as the blurring is confined to atomic scale, not open to direct control. In fact the function has provided quite intuitive and convenient ideas, for instance the "cloud of negative electricity" around the nucleus, etc. But serious misgivings arise if one notices that the uncertainty affects macroscopically tangible and visible things, for which the term "blurring" seems simply wrong. The state of a radioactive nucleus is presumably blurred in such degree and fashion that neither the instant of decay nor the direction, in which the emitted α-particle leaves the nucleus, is well-established. Inside the nucleus, blurring doesn't bother us. The emerging particle is described, if one wants to explain intuitively, as a spherical wave that continuously emanates in all directions from the nucleus and that impinges continuously on a surrounding luminescent screen over its full expanse. The screen however does not show a more or less constant uniform surface glow, but rather lights up at *one* instant at *one* spot—or, to honor the truth, it lights up now here, now there, for it is impossible to do the experiment with only a single radioactive atom. If in place of the luminescent screen one uses a spatially extended detector, perhaps a gas that is ionised by the α-particles, one finds the ion pairs arranged along rectilinear columns,[5] that project backwards on to the bit of radioactive matter from which the α-radiation comes (C.T.R. Wilson's cloud chamber tracks, made visible by drops of moisture condensed on the ions).

One can even set up quite ridiculous cases. A cat is penned up in a steel chamber, along with the following diabolical device (which must be secured against direct interference by the cat): in a Geiger counter there is a tiny bit of radioactive substance, *so* small, that *perhaps* in the course of one hour one of the atoms decays, but also, with equal probability, perhaps none; if it happens, the counter tube discharges and through a relay releases a hammer which shatters a small flask of hydrocyanic acid. If one has left this entire system to itself for an hour, one would say that the cat still lives *if* meanwhile no atom has decayed. The first atomic decay would have poisoned it. The ψ-function of the entire system would express this by having in it the living and the dead cat (pardon the expression) mixed or smeared out in equal parts.

It is typical of these cases that an indeterminacy originally restricted to the atomic domain becomes transformed into macroscopic indeterminacy, which can then be *resolved* by direct observation. That prevents us from so naively accepting as valid a "blurred model" for representing reality. In itself it would not embody anything unclear or contradictory. There is a difference between a shaky or out-of-focus photograph and a snapshot of clouds and fog banks.

6. *The Deliberate About-face of the Epistemological Viewpoint*

In the fourth section we saw that it is not possible smoothly to take over models and to ascribe, to the momentarily unknown or not exactly known variables, nonetheless determinate values, that we simply don't know. In Sect. 5. we saw that the indeterminacy is not even an actual blurring, for there are always cases where an easily executed observation provides the missing knowledge. So what is left? From this very hard dilemma the reigning doctrine rescues itself or us by having recourse to epistemology. We are told that no distinction is to be made between the state of a natural object and what I know about it, or perhaps better, what I can know about it if I go to some trouble. Actually—so they say—there is intrinsically only awareness, observation, measurement. If through them I have procured at a given moment the best knowledge of the state of the physical object that is possibly attainable in accord with natural laws, then I can turn aside as *meaningless* any further questioning about the "actual state," inasmuch as I am convinced that no further observation can extend my knowledge of it—at least, not without an equivalent diminution in some other respect (namely by changing the state, see below).

Now this sheds some light on the origin of the proposition that I mentioned at the end of Sect. *2.* as something very far-reaching: that all model quantities are measurable in principle. One can hardly get along without this article of belief if one sees himself constrained, in the interests of physical methodology, to call in as dictatorial help the above-mentioned philosophical principle, which no sensible person can fail to esteem as the supreme protector of all empiricism.

Reality resists imitation through a model. So one lets go of naive realism and leans directly on the indubitable proposition that *actually* (for the physicist) after all is said and done there is only observation, measurement. Then all our physical thinking thenceforth has as sole basis and as sole object the results of measurements which can in principle be carried out, for we must now explicitly *not* relate our thinking any longer to any other kind of reality or to a model. All numbers arising in our physical calculations must be interpreted as measurement results. But since we didn't just now come into the world and start to build up our science from scratch, but rather have in use a quite definite scheme of calculation, from which in view of the great progress in Q.M. we would less than ever want to be parted, we see ourselves forced to dictate from the writing-table which measurements are in principle possible, that is, must be possible in order to support adequately our reckoning system. This allows a sharp value for each single variable of the model (indeed for a whole "half set") and so each single variable must be measurable to arbitrary exactness. We cannot be satisfied with less, for we have lost our naively realistic innocence. We have nothing but our reckoning scheme to specify where Nature draws the ignorabimus-line, i.e., what is a *best possible* knowledge of the object. And if we couldn't do that, then indeed would our measurement reality become highly dependent on the diligence or laziness of the experi-

[5] For illustration see Fig. 5 or 6 on p. 375 of the 1927 volume of this journal; or Fig. 1, p. 734 of the preceding year's volume (1934), though these are proton tracks.

menter, how much trouble he takes to inform himself. We must go on to tell him how far he could go if only he were clever enough. Otherwise it would be seriously feared that just there, where we forbid further questions, there might well still be something worth knowing that we might ask about.

7. The ψ-function as Expectation-catalog

Continuing to expound the official teaching, let us turn to the already (Sect. *5*) mentioned ψ-function. It is now the means for predicting probability of measurement results. In it is embodied the momentarily-attained sum of theoretically based future expectation, somewhat as laid down in a *catalog*. It is the relation- and-determinacy-bridge between measurements and measurements, as in the classical theory the model and its state were. With this latter the ψ-function moreover has much in common. It is, in principle, determined by a finite number of suitably chosen measurements on the object, half as many as were required in the classical theory. Thus the catalog of expectations is initially compiled. From then on it changes with time, just as the state of the model of classical theory, in constrained and unique fashion ("causally")—the evolution of the ψ-function is governed by a partial differential equation (of first order in time and solved for $\partial\psi/\partial t$). This corresponds to the undisturbed motion of the model in classical theory. But this goes on only until one again carries out any measurement. For each measurement one is required to ascribe to the ψ-function (= the prediction-catalog) a characteristic, quite sudden change, which *depends on the measurement result obtained*, and so *cannot be foreseen;* from which alone it is already quite clear that this second kind of change of the ψ-function has nothing whatever in common with its orderly development *between* two measurements. The abrupt change by measurement ties in closely with matters discussed in Sect. *5.* and will occupy us further at some length; it is the most interesting point of the entire theory. It is precisely *the* point that demands the break with naive realism. For *this* reason one can *not* put the ψ-function directly in place of the model or of the physical thing. And indeed not because one might never dare impute abrupt unforeseen changes to a physical thing or to a model, but because in the realism point of view observation is a natural process like any other and cannot *per se* bring about an interruption of the orderly flow of natural events.

8. Theory of Measurement, Part One

The rejection of realism has logical consequences. In general, a variable *has* no definite value before I measure it; then measuring it does *not* mean ascertaining the value that it *has*. But then what does it mean? There must still be some criterion as to whether a measurement is true or false, a method is good or bad, accurate, or inaccurate—whether it deserves the name of measurement process at all. Any old playing around with an indicating instrument in the vicinity of another body, whereby at any old time one then takes a reading, can hardly be called a measurement on this body. Now it is fairly clear; if reality does not determine the measured value, then at least the measured value must determine reality—it must actually be present *after* the measurement in *that* sense which alone will be recognized again. That is, the desired criterion can be merely this: repetition of the measurement must give the same result. By many repetitions I can prove the accuracy of the procedure and show that I am not just playing. It is agreeable that this program matches exactly the method of the experimenter, to whom likewise the "true value" is not known beforehand. We formulate the essential point as follows:

The systematically arranged interaction of two systems (measured object and measuring instrument) is called a measurement on the first system, if a directly-sensible variable feature of the second (pointer position) is always reproduced within certain error limits when the process is immediately repeated (on the same object, which in the meantime must not be exposed to any additional influences).

This statement will require considerable added comment: it is by no means a faultless definition. Empirics is more complicated than mathematics and is not so easily captured in polished sentences.

Before the first measurement there might have been an arbitrary quantum-theory prediction *for* it. *After* it the prediction *always* runs: within error limits again the same result. The expectation-catalog (= ψ-function) is therefore changed by the measurement in respect to the variable being measured. If the measurement procedure is known from beforehand to be *reliable*, then the first measurement at once reduces the theoretical expectation within error limits on to the value found, regardless of whatever the prior expectation may have been. This is the typical abrupt change of the ψ-function discussed above. But the expectation-catalog changes in unforeseen manner not only for the measured variable itself, but also for others, in particular for its "canonical conjugate." If for instance one has a rather sharp prediction for the *momentum* of a particle and proceeds to measure its *position* more exactly than is compatible with Theorem A of Sec. *2.*, then the *momentum* prediction must change. The quantum mechanical reckoning scheme moreover takes care of this automatically; there is no ψ-function whatsoever that would contradict Theorem A when one deduces from it the combined expectations.

Since the expectation-catalog changes radically during measurement, the object is then no longer suited for testing, in their full extent, the statistical predictions made earlier; at the very least for the measured variable itself, since for it now the (nearly) same value would occur over and over again. *That* is the reason for the prescription already given in Sect. *2.*: one can indeed test the probability predictions completely, but for this one must repeat the entire experiment *ab ovo*. One's prior treatment of the measured object (or one identical to it) must be exactly the same as that given the first time, in order that the same expectation-catalog ($= \psi$-function) should be valid as before the first measurement. Then one "repeats" it. (This repeating now means of course something quite other than earlier!) All this one must do not twice but very often. Then the predicted statistics are established—that is the doctrine.

One should note the difference between the error limits and the error distribution of the *measurement*, on the one hand, and the theoretically predicted statistics, on the other hand. They have nothing to do with each other. They are established by the two quite different types of *repetition* just discussed.

Here there is opportunity to deepen somewhat the above-attempted delimitation of *measuring*. There are measuring instruments that remain fixed on the reading given by the measurement just made. Or the pointer could remain stuck because of a defect. One would then repeatedly make exactly the same reading, and according to our instruction that would be a spectacularly accurate measurement. Moreover that would be true not merely for the object but also for the instrument itself! As a matter of fact there is still missing from our exposition an important point, but one which could not readily be stated earlier, namely what it is that truly makes the difference between *object* and *instrument* (that it is the latter on which the reading is made, is more or less superficial). We have just seen that the instrument under certain circumstances, as required, must be set back to its neutral initial condition before any control measurement is made. This is well known to the experimentalist. Theoretically the matter may best be expressed by prescribing that on principle the instrument should be subjected to the identical prior treatment before each measurement, so that *for it* each time the same expectation-catalog ($= \psi$-function) applies, as it is brought up to the object. For the object it is just the other way around, any interference being forbidden when a control measurement is to be made, a "repetition of the first kind" (that leads to *error* statistics). That is the characteristic difference between object and instrument. It disappears for a "repetition of the second kind" (that serves for checking the quantum predictions). Here the difference between the two is actually rather insignificant.

From this we gather further that for a second measurement one may use another similarly built and similarly prepared instrument—it need not necessarily be *the same one;* this is in fact sometimes done, as a check on the first one. It may indeed happen that two quite differently built instruments are so related to each other that if one measures with them one after the other (repetition of the first kind!) their two indications are in one-to-one correlation with each other. They then measure on the object essentially the same variable—i.e., the same for suitable calibration of the scales.

9. The ψ-function as Description of State

The rejection of realism also imposes obligations. From the standpoint of the classical model the momentary statement content of the ψ-function is far from complete; it comprises only about 50 per cent of a complete description. From the new standpoint it must be complete for reasons already touched upon at the end of Sect. *6*. It must be impossible to add on to it additional correct statements, without otherwise changing it; else one would not have the right to call meaningless all questions extending beyond it.

Thence it follows that two different catalogs, that apply to the same system under different circumstances or at different times, may well partially overlap, but never so that the one is entirely contained within the other. For otherwise it would be susceptible to completion through additional correct statements, namely through those by which the other one exceeds it.—The mathematical structure of the theory automatically satisfies this condition. There is no ψ-function that furnishes exactly the same statements as another and in addition several more.

Therefore if a system changes, whether by itself or because of measurements, there must always be statements missing from the new function that were contained in the earlier one. In the catalog not just new entries, but also deletions, must be made. Now knowledge can well be *gained*, but not *lost*. So the deletions mean that the previously correct statements have now become incorrect. A correct statement can become incorrect only if the *object* to which it applies changes. I consider it acceptable to express this reasoning sequence as follows:

Theorem 1: If different ψ-functions are under discussion the system is in different states.

If one speaks only of systems for which a ψ-function is in general available, then the inverse of this theorem runs:

Theorem 2: For the same ψ-function the system is in the same state.

This inverse does not follow from Theorem 1 but independently of it, directly from *completeness* or *maximality*. Whoever for the same expectation-catalog would yet claim a difference is possible, would be admitting that it (the catalog) does not give information on all justifiable questions. —The language usage of almost all authors implies validity of the above two theorems. Of course, they set up a kind of new reality—in entirely legitimate fashion, I believe. Moreover they are not trivially tautological, not mere verbal interpretations of "state." Without presupposed maximality of the expectation-catalog, change of the ψ-function could be brought about by mere collecting of new information.

We must face up to yet another objection to the derivation of Theorem 1. One can argue that each individual statement or item of knowledge, under examination there, is after all a probability statement, to which the category of *correct*, or *incorrect* does not apply in any relation to an individual case, but rather in relation to a collective that comes into being from one's preparing the system a thousand times in identical fashion (in order then to allow the same measurement to follow; cf. Sect. *8*.). That makes sense, but we must specify all members of this collective to be identically prepared, since to each the same ψ-function, the same statement-catalog applies and we dare not specify differences that are not expressed in the catalog (*cf.* the foundation of Theorem 2). Thus the collective is made up of identical individual cases. If a statement is wrong for *it,* then the individual case must have changed, or else the collective too would again be the same.

10. Theory of Measurement, Part Two

Now it was previously stated (Sect. *7*) and explained (Sect. *8*) that any *measurement* suspends the law that otherwise governs continuous time-dependence of the ψ-function and brings about in it a quite different change, not governed by any law but rather dictated by the result of the measurement. But laws of nature differing from the usual ones cannot apply during a measurement, for objectively viewed it is a natural process like any other, and it cannot interrupt the orderly course of natural events. Since it does interrupt that of the ψ-function, the latter—as we said in Sect. *7*—can *not* serve, like the classical model, as an experimentally verifiable representation of an objective reality. And yet in the last Section something like that has taken shape.

So, using catchwords for emphasis, I try again to contrast: 1.) The discontinuity of the expectation-catalog due to measurement is *unavoidable,* for if measurement is to retain any meaning at all then the *measured value,* from a good measurement, *must* obtain. 2.) The discontinuous change is certainly *not* governed by the otherwise valid causal law, since it depends on the measured value, which is not predetermined. 3.) The change also definitely includes (because of "maximality") some *loss* of knowledge, but knowledge cannot be lost, and so the object *must* change—*both* along with the discontinuous changes and *also,* during these changes, in an unforeseen, *different* way.

How does this add up? Things are not at all simple. It is the most difficult and most interesting point of the theory. Obviously we must try to comprehend objectively the interaction between measured object and measuring instrument. To that end we must lay out a few very abstract considerations.

This is the point. Whenever one has a complete expectation-catalog—a maximum total knowledge—a ψ-function—for two completely separated bodies, or, in better terms, for each of them singly, then one obviously has it also for the two bodies together, i.e., if one imagines that neither of them singly but rather the two of them together make up the object of interest, of our questions about the future.[6]

But the converse is not true. *Maximal knowledge of a total system does not necessarily include total knowledge of all its parts, not even when these are fully separated from each other and at the moment are not influencing each other at all.* Thus it may be that some part of what one knows may pertain to relations or stipulations between the two subsystems (we shall limit ourselves to two), as follows: if a particular measurement on the first system yields *this* result, then for a particular measurement on the second the valid expectation statistics are such and such; but if the measurement in question on the first system should have *that* result, then some other expectation holds for that on the second; should a third result occur for the first, then still another expectation applies to the second; and so on, in the manner of a complete disjunction of all possible measurement results which the one specifically contemplated measurement on the first system can yield. In this way, any measurement process at all or, what amounts to the same, any variable at all of the second system can be tied to the not-yet-known value of any variable at all of the first, and of course *vice versa* also. If that is the case, if such conditional statements occur in the combined catalog, *then it can not possibly be maximal in regard to the individual systems.* For the content of two maximal individual catalogs would by itself suffice for a maximal combined catalog; the conditional statements could not be added on.

---

[6] Obviously. We cannot fail to have, for instance, statements on the relation of the two to each other. For that would be, at least for one of the two, something in addition to its Ψ-function. And such there cannot be.

These conditioned predictions, moreover, are not something that has suddenly fallen in here from the blue. They are in every expectation-catalog. If one knows the ψ-function and makes a particular measurement and this has a particular result, then one again knows the ψ-function, *voilà tout*. It's just that for the case under discussion, because the combined system is supposed to consist of two fully separated parts, the matter stands out as a bit strange. For thus it becomes meaningful to distinguish between measurements on the one and measurements on the other subsystem. This provides to each full title to a private maximal catalog; on the other hand it remains possible that a portion of the attainable combined knowledge is, so to say, squandered on conditional statements, that operate between the subsystems, so that the private expectancies are left unfulfilled—even though the combined catalog is maximal, that is even though the ψ-function of the combined system is known.

Let us pause for a moment. This result in its abstractness actually says it all: Best possible knowledge of a whole does not necessarily include the same for its parts. Let us translate this into terms of Sect. *9*: The whole is in a definite state, the parts taken individually are not.

"How so? Surely a system must be in some sort of state." "No. State is ψ-function, is maximal sum of knowledge. I didn't necessarily provide myself with this, I may have been lazy. Then the system is in no state."

"Fine, but then too the agnostic prohibition of questions is not yet in force and in our case I can tell myself: the subsystem is already in some state, I just don't know which."

"Wait. Unfortunately no. There is no 'I just don't know'. For as to the total system, maximal knowledge is at hand . . ."

The insufficiency of the ψ-function as model replacement rests solely on the fact that one doesn't always have it. If one does have it, then by all means let it serve as description of the state. But sometimes one does not have it, in cases where one might reasonably expect to. And in that case, one dare not postulate that it "is actually a particular one, one just doesn't know it"; the above-chosen standpoint forbids this. "It" is namely a sum of knowledge; and knowledge, that no one knows, is none. ——

We continue. That a portion of the knowledge should float in the form of disjunctive conditional statements *between* the two systems can certainly not happen if we bring up the two from opposite ends of the world and juxtapose them without interaction. For then indeed the two "know" nothing about each other. A measurement on one cannot possibly furnish any grasp of what is to be expected of the other. Any "entanglement of predictions" that takes place can obviously only go back to the fact that the two bodies at some earlier time formed in a true sense *one* system, that is were interacting, and have left behind *traces* on each other. If two separated bodies, each by itself known maximally, enter a situation in which they influence each other, and separate again, then there occurs regularly that which I have just called *entanglement* of our knowledge of the two bodies. The combined expectation-catalog consists initially of a logical sum of the individual catalogs; during the process it develops causally in accord with known law (there is no question whatever of measurement here). The knowledge remains maximal, but at its end, if the two bodies have again separated, it is not again split into a logical sum of knowledges about the individual bodies. What still remains *of that* may have become less than maximal, even very strongly so.—One notes the great difference over against the classical model theory, where of course from known initial states and with known interaction the individual end states would be exactly known.

The measuring process described in Sect. *8* now fits neatly into this general scheme, if we apply it to the combined system, measured object + measuring instrument. As we thus construct an objective picture of this process, like that of any other, we dare hope to clear up, if not altogether to avoid, the singular jump of the ψ-function. So now the one body is the measured object, the other the instrument. To suppress any interference from outside we arrange for the instrument by means of built-in clockwork to creep up automatically to the object and in like manner creep away again. The reading itself we postpone, as our immediate purpose is to investigate whatever may be happening "objectively"; but for later use we let the result be recorded automatically in the instrument, as indeed is often done these days.

Now how do things stand, after automatically completed measurement? We possess, afterwards same as before, a maximal expectation-catalog for the total system. The recorded measurement result is of course not included therein. As to the instrument the catalog is far from complete, telling us nothing at all about where the recording pen left its trace. (Remember that poisoned cat!) What this amounts to is that our knowledge has evaporated into conditional statements: *if* the mark is at line 1, *then* things are thus and so for the measured object, *if* it is at line 2, then such and such, if at 3, then a third, etc. Now has the ψ-function of the measured *object* made a leap? Has it developed further in accord with natural law (in accord with the partial differential equation)? No to both questions. It is no more. It has become snarled up, in accord with the causal law for the *combined* ψ-function, with that of the measuring instrument. *The expectation-catalog of the object has split into a conditonal disjunction of expectation-*

catalogs—like a Baedeker that one has taken apart in the proper manner. Along with each section there is given also the probability that it proves correct—transcribed from the original expectation-catalog of the object. But which one proves right—which section of the Baedeker should guide the ongoing journey—that can be determined only by actual inspection of the record.

And what if we *don't* look? Let's say it was photographically recorded and by bad luck light reaches the film before it is developed. Or we inadvertently put in black paper instead of film. Then indeed have we not only not learned anything new from the miscarried measurement, but we have suffered loss of knowledge. This is not surprising. It is only natural that outside interference will almost always spoil the knowledge that one has of a system. The interference, if it is to allow the knowledge to be gained back afterwards, must be circumspect indeed.

What have we won by this analysis? *First,* the insight into the disjunctive splitting of the expectation-catalog, which still takes place quite continuously and is brought about through embedment in a combined catalog for instrument and object. From this amalgamation the object can again be separated out only by the living subject actually taking cognizance of the result of the measurement. Some time or other this must happen if that which has gone on is actually to be called a measurement—however dear to our hearts it was to prepare the process throughout as objectively as possible. And that is the *second* insight we have won: ***not until this inspection,*** which determines the disjunction, does anything discontinuous, or leaping, take place. One is inclined to call this a *mental* action, for the object is already out of touch, is no longer physically affected; what befalls it is already past. But it would not be quite right to say that the ψ-function of the object which changes *otherwise* according to a partial differential equation, independent of the observer, should *now* change leap-fashion because of a mental act. For it had disappeared, it was no more. Whatever is not, no more can it change. It is born anew, is reconstituted, is separated out from the entangled knowledge that one has, through an act of perception, which as a matter of fact is not a physical effect on the measured object. From the form in which the ψ-function was last known, to the new in which it reappears, runs no continuous road—it ran indeed through annihilation. Contrasting the two forms, the thing looks like a leap. In truth something of importance happens in between, namely the influence of the two bodies on each other, during which the object possessed no private expectation-catalog nor had any claim thereto, because it was not independent.

11. Resolution of the "Entanglement." Result Dependent on the Experimenter's Intention

We return to the general case of "entanglement," without having specificially in view the special case, just considered, of a measurement process. Suppose the expectation-catalogs of two bodies A and B have become entangled through transient interaction. Now let the bodies be again separated. Then I can take one of them, say B, and by successive measurements bring my knowledge of it, which had become less than maximal, back up to maximal. I maintain: just as soon as I succeed in this, and not before, then first, the entanglement is immediately resolved and, second, I will also have acquired maximal knowledge of A through the measurements on B, making use of the conditional relations that were in effect.

For in the first place the knowledge of the total system remains always maximal, being in no way damaged by good and exact measurements. In the second place: conditional statements of the form "if for A. . , then for B. ." can no longer exist, as soon as we have reached a maximal catalog on B. For it is ***not*** conditional and to it nothing at all can be added on relevant to B. Thirdly: conditional statements in the inverse sense (if for B. . , then for A. .) can be transformed into statements about A alone, because all probabilities for B are already known unconditionally. The entanglement is thus completely put aside, and since the knowledge of the total system has remained maximal, it can only mean that along with the maximal catalog for B came the same thing for A.

And it cannot happen the other way around, that A becomes maximally known indirectly, through measurements on B, before B is. For then all conclusions work in the reversed direction—that is, B is too. The systems become simultaneously maximally known, as asserted. Incidentally, this would also be true if one did not limit the measurement to just one of the two systems. But the interesting point is precisely this, that one *can* limit it to one of the two; that thereby one reaches his goal.

Which measurements on B and in what sequence they are undertaken, is left entirely to the arbitrary choice of the experimenter. He need not pick out specific variables, in order to be able to use the conditional statements. He is free to formulate a plan that would lead him to maximal knowledge of B, even if he should know nothing at all about B. And it can do no harm if he carries through this plan to the end. If he asks himself after each measurement whether he has perhaps already reached his goal, he does so only to spare himself from further, superfluous labor.

What sort of A-catalog comes forth in this indirect way depends obviously on the measured values that are found in B (before the entanglement is entirely resolved; not on more, on any later ones, in case the measuring goes on superfluously). Suppose now that in this way I derived an A-catalog in a particular case. Then I can look back and consider whether I might perhaps have found a *different one* if I had put into action a *different* measuring plan for B. But since after all I neither have actually touched the system A, nor in the imagined other case would have touched it, the statements of the other catalog, whatever it might be, must all *also* be correct. They must therefore be entirely contained within the first, since the first is maximal. But so is the second. So it must be identical with the first.

Strangely enough, the mathematical structure of the theory by no means satisfies this requirement automatically. Even worse, examples can be set up where the requirement is necessarily violated. It is true that in any experiment one can actually carry out only *one* group of measurements (always on B), for once that has happened the entanglement is resolved and one learns nothing more about A from further measurements on B. But there are cases of entanglement in which *two definite programs* are specifiable, of which each 1) must lead to resolution of the entanglement, and 2) must lead to an A-catalog to which the *other can* not possibly lead—whatsoever measured values may turn up in one case or the other. It is simply like this, that the *two series* of A-catalogs, that can possibly arise from the one or the other of the programs, are sharply separated and have in common not a single term.

These are especially pointed cases, in which the conclusion lies so clearly exposed. In general one must reflect more carefully. If two programs of measurement on B are proposed, along with the two series of A-catalogs to which they can lead, then it is by no means sufficient that the two series have one or more terms in common in order for one to be able to say: well now, surely one of these will always turn up—and so to set forth the requirements as "presumably fulfilled." That's not enough. For *indeed one knows* the probability of every measurement on B, considered as measurement on the total system, and under many ab-ovo-repetitions each one must occur with the frequency assigned to it. Therefore the two series of A-catalogs would have to agree, member by member, and furthermore the probabilities in each series would have to be the same. And that not merely for these two programs but also for each of the infinitely many that one might think up. But this is utterly out of the question. The requirement that the A-catalog that one gets should always be the same, regardless of what measurements on B bring it into being, this requirement is plainly and simply never fulfilled.

Now we wish to discuss a simple "pointed" example.

12. *An Example* [7]

For simplicity, we consider two systems with just *one* degree of freedom. That is, each of them shall be specified through a *single* coordinate q and its canonically conjugate momentum p. The classical picture would be a point mass that could move only along a straight line, like the spheres of those playthings on which small children learn to calculate. p is the product of mass by velocity. For the second system we denote the two determining parts by capital Q and P. As to whether the two are "threaded on the same wire" we shall not be at all concerned, in our abstract consideration. But even if they are, it may in that case be convenient not to reckon q and Q from the same reference point. The equation $q = Q$ thus does not necessarily mean coincidence. The two systems may in spite of this be fully separated.

In the cited paper it is shown that between these two systems an entanglement can arise, which at a particular moment, to which everything following is referred, can be compactly shown in the two equations: $q = Q$ and $p = -P$. That means: *I know, if* a measurement of q on the system yields a certain value, that a Q-measurement performed immediately thereafter on the second will give the *same* value, and vice versa; *and I know*, if a p-measurement on the first system yields a certain value, that a P-measurement performed immediately thereafter will give the opposite value, and vice versa.

A single measurement of *q or p or Q or P* resolves the entanglement and makes both systems maximally known. A second measurement on the same system modifies only the statements about *it*, but teaches nothing more about the other. So one cannot check both equations in a single experiment. But one can repeat the experiment *ab ovo* a thousand times; each time set up the same entanglement; according to whim check one or the other of the equations; and find confirmed that one which one is momentarily pleased to check. We assume that all this has been done.

If for the thousand-and-first experiment one is then seized by the desire to give up further checking and instead measure q on the first system and P on the second, and one obtains

$$q = 4; \quad P = 7;$$

can one then doubt that

$$q = 4; \quad p = -7$$

would have been a correct prediction for the first system, or

$$Q = 4; \quad P = 7$$

[7] A. Einstein, B. Podolsky, and N. Rosen, *Phys. Rev.* **47**: 777 (1935). The appearance of this work motivated the present—shall I say lecture or general confession?

a correct prediction for the second? Quantum predictions are indeed not subject to test as to their full content, ever, in a single experiment; yet they are correct, in that whoever possessed them suffered no disillusion, whichever half he decided to check.

There's no doubt about it. Every measurement is for its system the first. Measurements on separated systems cannot directly influence each other—that would be magic. Neither can it be by chance, if from a thousand experiments it is established that virginal measurements agree.

The prediction-catalog $q = 4$, $p = -7$ would of course be hypermaximal.

13. Continuation of the Example: All Possible Measurements Are Entangled Unequivocally

Now a *prediction* of this extent is thus utterly impossible according to the teaching of Q.M., which we here follow out to its last consequences. Many of my friends remain reassured in this and declare: what answer a system *would have given* to the experimenter *if* . . . ,—has nothing to do with an actual measurement and so, from our epistemological standpoint, does not concern us.

But let us once more make the matter very clear. Let us focus attention on the system labeled with small letters p, q and call it for brevity the "small" one. Then things stand as follows. I can direct *one* of two questions to the small system, either that about q or that about p. Before doing so I can, if I choose, procure the answer to *one* of these questions by a measurement on the fully separated other system (which we shall regard as auxiliary apparatus), or I may intend to take care of this afterwards. My small system, like a schoolboy under examination, *cannot possibly know* whether I have done this or for which questions, or whether and for which I intend to do it later. From arbitrarily many pretrials I know that the pupil will correctly answer the first question that I put to him. From that it follows that in every case he *knows* the answer to *both* questions. That the answering of the first question, that it pleases me to put to him, so tires or confuses the pupil that his further answers are worthless, changes nothing at all of this conclusion. No school principal would judge otherwise, if this situation repeated itself with thousands of pupils of similar provenance, however much he might wonder *what* makes all the scholars so dimwitted or obstinate after the answering of the first question. He would not come to think that his, the teacher's, consulting a textbook first suggests to the pupil the correct answer, or even, in the cases when the teacher chooses to consult it only after ensuing answers by the pupil, that the pupil's answer has changed the text of the notebook in the pupil's favor.

Thus my small system holds a quite definite answer to the q-question and to the p-question in readiness for the case that one or the other is the first to be put directly to it. Of this preparedness not an iota can be changed if I should perhaps measure the Q on the auxiliary system (in the analogy: if the teacher looks up one of the questions in his notebook and thereby indeed ruins with an inkblot *the* page where the other answer stands). The quantum mechanician maintains that after a Q-measurement on the auxiliary system my small system has a ψ-function in which "q is fully sharp, but p fully indeterminate." And yet, as already mentioned, not an iota is changed of the fact that my small system also has ready an answer to the p-question, and indeed the same one as before.

But the situation is even worse yet. Not only to the q-question and to the p-question does my clever pupil have a definite answer ready, but rather also to a thousand others, and indeed without my having the least insight into the memory technique by which he is able to do it. p and q are not the only variables that I can measure. Any combination of them whatsoever, for example

$$p^2 + q^2$$

also corresponds to a fully definite measurement according to the formulation of Q.M. Now it may be shown [8] that also for this the answer can be obtained by a measurement on the auxiliary system, namely by measurement of $P^2 + Q^2$, and indeed the answers are just the same. By general rules of Q.M. this sum of squares can only take on a value from the series

$$\hbar,\ 3\hbar,\ 5\hbar,\ 7\hbar,\ \ldots .$$

The answer that my small system has ready for the $(p^2 + q^2)$-question (in case this should be the first it must face) must be a number from this series.—It is very much the same with measurement of

$$p^2 + a^2q^2$$

where a is an arbitrary positive constant. In this case the answer must be, according to Q.M., a number from the following series

$$a\hbar,\ 3a\hbar,\ 5a\hbar,\ 7a\hbar,\ \ldots .$$

For each numerical value of a one gets a different question, and to each my small system holds ready an answer from the series (formed with the a-value in question).

Most astonishing is this: these answers cannot possibly be related to each other in the way given by the formulas! For let q' be the answer held ready for the q-question, and p' that for the p-question, then the relation

$$(p'^2 + a^2q'^2)/(a\hbar) = \text{an odd integer}$$

---

[8] E. Schrödinger, *Proc. Cambridge Phil. Soc.* (in press).

cannot possibly hold for given numerical values q′ and p′ and for *any positive number a*. This is by no means an operation with imagined numbers, that one can not really ascertain. One can in fact get two of the numbers, e.g., q′ and p′, the one by direct, the other by indirect measurement. And then one can (pardon the expression) convince himself that the above expression, formed with the numbers q′ and p′ and an arbitrary a, is not an odd integer.

The lack of insight into the relationships among the various answers held in readiness (into the "memory technique" of the pupil) is a total one, a gap not to be filled perhaps by a new kind of algebra of Q.M. The lack is all the stranger, since on the other hand one can show: the entanglement is already uniquely determined by the requirements $q = Q$ and $p = -P$. If we know that the coordinates are equal and the momenta equal but opposite, then there follows by quantum mechanics a *fully determined* one-to-one arrangement of *all possible* measurements on both systems. For *every* measurement on the "small" one the numerical result can be procured by a suitably arranged measurement on the "large" one, and each measurement on the large stipulates the result that a particular measurement on the small would give or has given. (Of course in the same sense as always heretofore: only the virgin measurement on each system counts.) As soon as we have brought the two systems into the situation where they (briefly put) coincide in coordinate and momentum, then they (briefly put) coincide also in regard to all other variables.

But as to how the numerical values of all these variables of *one* system relate to each other we know nothing at all, even though for each the system must have a quite specific one in readiness, for if we wish we can learn it from the auxiliary system and then find it always confirmed by direct measurement.

Should one now think that because we are so ignorant about the relations among the variable-values held ready in *one* system, that none exists, that far-ranging arbitrary combination can occur? That would mean that such a system of "*one* degree of freedom" would need not merely *two* numbers for adequately describing it, as in classical mechanics, but rather many more, perhaps infinitely many. It is then nevertheless strange that two systems always agree in *all* variables if they agree in two. Therefore one would have to make the second assumption, that this is due to our awkwardness; would have to think that as a practical matter we are not competent to bring two systems into a situation such that they coincide in reference to two variables, without *nolens volens* bringing about coincidence also for all other variables, even though that would not in itself be necessary. One would have to make these *two* assumptions in order not to perceive as a great dilemma the complete lack of insight into the interrelationship of variable-values within one system.

14. Time-dependence of the Entanglement. Consideration of the Special Role of Time

It is perhaps not superfluous to recall that everything said in sections 12 and 13 pertains to a single instant of time. The entanglement is not constant in time. It does continue to be a one-to-one entanglement of *all* variables, but the arrangement changes. That means the following. At a later time t one can very well again learn the values of q or of p that *then* obtain, by a measurement on the auxiliary system, but the measurements, that one must undertake thereto on the auxiliary system, are *different*. Which ones they should be, one can easily see in simple cases. It now of course becomes a question of the forces at work within each of the two systems. Let us assume that no forces are working. For simplicity we will set the mass of each to be the same and call it m. Then in the classical model the momenta p and P would remain constant, since they are still the masses multiplied by the velocities; and the coordinates at time t, which we shall distinguish by giving them subscripts t, (q_t, Q_t), would be calculated from the initial ones, which henceforth we designate q,Q, thus:

$$q_t = q + (p/m)t$$
$$Q_t = Q + (P/m)t$$

Let us first talk about the small system. The most natural way of describing it classically at time t is in terms of coordinate and momentum *at this time*, i.e., in terms of q_t and p. But one may do it differently. In place of q_t one could specify q. It too is a "determining part at time t," and indeed at every time t, and in fact one that does not change with time. This is similar to the way in which I can specify a certain determining part of my own person, namely my *age*, either through the number 48, which changes with time and in the system corresponds to specifying q_t, or through the number 1887, which is usual in documents and corresponds to specifying q. Now according to the foregoing:

$$q = q_t - (p/m)t$$

Similarly for the second system. So we take as determining parts

for the first system $q_t - (p/m)t$ and p.
for the second system $Q_t - (P/m)t$ and P.

The advantage is that *among these the same entanglement goes on indefinitely:*

$$q_t - (p/m)t = Q_t - (P/m)t$$
$$p = -P$$

or solved:

$$q_t = Q_t - (2\,t/m)P; \quad p = -P.$$

So that what changes with time is just this: the coordinate of the "small" system is not ascertained simply by a coordinate measurement on the auxiliary system, but rather by a measurement of the aggregate

$$Q_t - (2\,t/m)P.$$

Here however, one must not get the idea that maybe he measures Q_t *and* P, because that just won't go. Rather one must suppose, as one always must suppose in Q.M., that there is a direct measurement procedure for this aggregate. Except for this change, *everything* that was said in Sections *12* and *13* applies at any point of time; in particular there exists at all times the one-to-one entanglement of *all* variables together with its evil consequences.

It is just this way too, if within each system a force works, except that then q_t and p are entangled with variables that are more complicated combinations of Q_t and P.

I have briefly explained this in order that we may consider the following. That the entanglement should change with time makes us after all a bit thoughtful. Must perhaps all measurements, that were under discussion, be completed in very short time, actually *instantaneously*, in zero time, in order that the unwelcome consequences be vindicated? Can the ghost be banished by reference to the fact that measurements take time? No. For each single experiment one needs just *one* measurement on each system; only the virginal one matters, further ones apart from this would be without effect. How long the measurement lasts need not therefore concern us, since we have no second one following on. One must merely be able to so arrange the two virgin measurements that they yield variable-values for the same definite *point* of time, known to us in advance—known in advance, because after all we must direct the measurements at a pair of variables that are entangled at just this point of time.

"Perhaps it is not possible so to direct the measurements?"

"Perhaps. I even presume so. Merely: *today's* Q.M. must require this. For it is now so set up that its predictions are always made for a *point* of time. Since they are supposed to relate to measurement results, they would be entirely without content if the relevant variables were not measurable *for* a definite point of time, whether the measurement itself lasts a long or a short while."

When we *learn* the result is of course quite immaterial. Theoretically that has as little weight as for instance the fact that one needs several months to integrate the differential equations of the weather for the next three days.—The drastic analogy with the pupil examination misses the mark in a few points of the law's letter, but it fits the spirit of the law. The expression "the system knows" will perhaps no longer carry the meaning that the answer comes forth from an instantaneous situation; it may perhaps derive from a succession of situations, that occupies a finite length of time. But even if it be so, it need not concern us so long as the system somehow brings forth the answer from within itself, with no other help than that we tell it (through the experimental arrangement) *which* question we would like to have answered; and so long as the answer itself is uniquely tied to a *moment* of time; which for better or for worse must be presumed for every measurement to which contemporary Q.M. speaks, for otherwise the quantum mechanical predictions would have no content.

In our discussion, however, we have stumbled across a possibility. If the formulation could be so carried out that the quantum mechanical predictions did not or did not always pertain to a quite sharply defined point of time, then one would also be freed from requiring this of the measurement results. Thereby, since the entangled variables change with time, setting up the antinomical assertions would become much more difficult.

That prediction for sharply-defined time is a blunder, is probable also on other grounds. The numerical value of time is like any other the result of observation. Can one make exception just for measurement with a clock? Must it not like any other pertain to a variable that in general has no sharp value and in any case cannot have it simultaneously with *any* other variable? If one predicts the value of *another* for a particular *point of time*, must one not fear that both can never be sharply known together? Within contemporary Q.M. one can hardly deal with this apprehension. For time is always considered a priori as known precisely, although one would have to admit that every look-at-the-clock disturbs the clock's motion in uncontrollable fashion.

Permit me to repeat that we do not possess a Q.M. whose statements should *not* be valid for sharply fixed points of time. It seems to me that this lack manifests itself directly in the former antinomies. Which is not to say that it is the only lack which manifests itself in them.

15. Natural Law or Calculating Device?

That "sharp time" is an anomaly in Q.M. and that besides, more or less independent of that, the special treatment of time forms a serious hindrance to adapting Q.M. to the *relativity principle*, is something that in recent years I have brought up again and again, unfortunately without being able to make the shadow of a useful counterproposal.[9] In an overview of the entire contemporary situation, such as I have tried to sketch here, there comes up, in addition, a quite different kind of remark in relation to the so ardently sought, but not yet actually attained, "relativisation" of Q.M.

The remarkable theory of measurement, the apparent jumping around of the ψ-function, and finally the "antinomies of entanglement," all derive from the simple manner in which the calculation methods of quantum mechanics allow two separated systems conceptually to be combined together into a single one; for which the methods seem plainly predestined. When two systems interact, their ψ-functions, as we have seen, do not come into interaction but rather they immediately cease to exist and a single one, for the combined system, takes their place. It consists, to mention this briefly, at first simply of the *product* of the two individual functions; which, since the one function depends on quite different variables from the other, is a function of all these variables, or "acts in a space of much higher dimension number" than the individual functions. As soon as the systems begin to influence each other, the combined function ceases to be a product and moreover does not again divide up, after they have again become separated, into factors that can be assigned individually to the systems. Thus one disposes provisionally (until the entanglement is resolved by an actual observation) of only a *common* description of the two in that space of higher dimension. This is the reason that knowledge of the individual systems can decline to the scantiest, even to zero, while that of the combined system remains continually maximal. Best possible knowledge of a whole does *not* include best possible knowledge of its parts—and that is what keeps coming back to haunt us.

Whoever reflects on this must after all be left fairly thoughtful by the following fact. The conceptual joining of two or more systems into *one* encounters great difficulty as soon as one attempts to introduce the principle of special relativity into Q.M. Already seven years ago P.A.M. Dirac found a startlingly simple and elegant relativistic solution to the problem of a single electron.[10] A series of experimental confirmations, marked by the key terms electron spin, positive electron, and pair creation, can leave no doubt as to the basic correctness of the solution. But in the first place it does nevertheless very strongly transcend the conceptual plan of Q.M. (that which I have attempted to picture *here*),[11] and in the second place one runs into stubborn resistance as soon as one seeks to go forward, according to the prototype of non-relativistic theory, from the Dirac solution to the problem of several electrons. (This shows at once that the solution lies outside the general plan, in which, as mentioned, the combining together of subsystems is extremely simple.) I do not presume to pass judgment on the attempts which have been made in this direction.[12] That they have reached their goal, I must doubt first of all because the authors make no such claim.

Matters stand much the same with another system, the electromagnetic field. Its laws are "relativity theory personified," a *non*-relativistic treatment being in general impossible. Yet it was this field, which in terms of the classical model of heat radiation provided the first hurdle for quantum theory, that was the first system to be "quantized." That this could be successfully done with simple means comes about because here one has things a bit easier, in that the photons, the "atoms of light," do not in general interact directly with each other,[13] but only via the charged particles. Today we do not as yet have a truly unexceptionable quantum theory of the electromagnetic field.[14] One can go a long way with *building up out of subsystems* according to the pattern of the non-relativistic theory (Dirac's theory of light[15]), yet without quite reaching the goal.

The simple procedure provided for this by the non-relativistic theory is perhaps after all only a convenient calculational trick, but one that today, as we have seen, has attained influence of unprecedented scope over our basic attitude toward nature.

My warmest thanks to Imperial Chemical Industries, London, for the leisure to write this article.

---

[9] Berl. Ber. 16 Apr. 1931; *Annales de l'Institut Henri Poincare*, p. 269 (Paris, 1931); *Cursos de la Universidad Internacional de Verano en Santander*, **1**: p. 60 (Madrid, Signo, 1935).

[10] *Proc. Roy. Soc. Lond.* **A117**: p. 610 (1928).

[11] P. A. M. Dirac, *The principles of quantum mechanics*, 1st ed. p. 239; 2nd ed. p. 252. Oxford: Clarendon Press, 1930 or 1935.

[12] Herewith a few of the more important references: G. Breit, *Phys. Rev.* **34**: p. 553 (1929) and **39**: p. 616 (1932); C. Møller, *Z. Physik* **70**: p. 786 (1931); P. A. M. Dirac, *Proc. Roy. Soc. Lond.* **A136**: p. 453 (1932) and *Proc. Cambridge Phil. Soc.* **30**: p. 150 (1934); R. Peierls, *Proc. Roy. Soc. Lond.* **A146**: p. 420 (1934); W. Heisenberg, *Z. Physik.* **90**: p. 209 (1934).

[13] But this holds, probably, only approximately. See M. Born and L. Infeld, *Proc. Roy. Soc. Lond.* **A144**: p. 425 and **A147**: p. 522 (1934); **A150**: p. 141 (1935). This is the most recent attempt at a quantum electrodynamics.

[14] Here again the most important works, partially assignable, according to their contents, also to the penultimate citation: P. Jordan and W. Pauli, *Z. Physik* **47**: p. 151 (1928); W. Heisenberg and W. Pauli, *Z. Physik* **56**: p. 1 (1929); **59**: p. 168 (1930); P. A. M. Dirac, V. A. Fock, and B. Podolsky, *Physik. Z. Sowjetunion* **6**: p. 468 (1932); N. Bohr and L. Rosenfeld, *Danske. Videns. Selsk.* (math.-phys.) **12**: p. 8 (1933).

[15] An excellent reference: E. Fermi, *Rev. Mod. Phys.* **4**: p. 87 (1932).

I.12 REMARKS ON THE MIND-BODY QUESTION

EUGENE P. WIGNER

Introductory Comments

F. Dyson, in a very thoughtful article,[1] points to the everbroadening scope of scientific inquiry. Whether or not the relation of mind to body will enter the realm of scientific inquiry in the near future—and the present writer is prepared to admit that this is an open question—it seems worthwhile to summarize the views to which a dispassionate contemplation of the most obvious facts leads. The present writer has no other qualification to offer his views than has any other physicist and he believes that most of his colleagues would present similar opinions on the subject, if pressed.

Until not many years ago, the "existence" of a mind or soul would have been passionately denied by most physical scientists. The brilliant successes of mechanistic and, more generally, macroscopic physics and of chemistry overshadowed the obvious fact that thoughts, desires, and emotions are not made of matter, and it was nearly universally accepted among physical scientists that there is nothing besides matter. The epitome of this belief was the conviction that, if we knew the positions and velocities of all atoms at one instant of time, we could compute the fate of the universe for all future. Even today, there are adherents to this

[1] F. J. Dyson, *Scientific American,* 199, 74 (1958). Several cases are related in this article in which regions of inquiry, which were long considered to be outside the province of science, were drawn into this province and, in fact, became focuses of attention. The best-known example is the interior of the atom, which was considered to be a metaphysical subject before Rutherford's proposal of his nuclear model, in 1911.

Originally published in *The Scientist Speculates*, I. J. Good, ed., pp. 284-302, Heinemann, London (1961); Basic Books, New York (1962). Reprinted in E. Wigner (1967), *Symmetries and Reflections*, Indiana University Press, Bloomington, pp. 171-84 from which book this paper is reproduced here.

view[2] though fewer among the physicists than – ironically enough – among biochemists.

There are several reasons for the return, on the part of most physical scientists, to the spirit of Descartes's "*Cogito ergo sum*," which recognizes the thought, that is, the mind, as primary. First, the brilliant successes of mechanics not only faded into the past; they were also recognised as partial successes, relating to a narrow range of phenomena, all in the macroscopic domain. When the province of physical theory was extended to encompass microscopic phenomena, through the creation of quantum mechanics, the concept of consciousness came to the fore again: it was not possible to formulate the laws of quantum mechanics in a fully consistent way without reference to the consciousness.[3] All that quantum mechanics purports to provide are probability connections between subsequent impressions (also called "apperceptions") of the consciousness, and even though the dividing line between the observer, whose consciousness is being affected, and the observed physical object can be shifted towards the one or the other to a considerable degree,[4] it cannot be eliminated. It may be premature to believe that the present philosophy of quantum mechanics will remain a permanent feature of future physical theories; it will remain remarkable, in whatever way our future concepts may develop, that the very study of the external world led to the conclusion that the content of the consciousness is an ultimate reality.

It is perhaps important to point out at this juncture that the question concerning the existence of almost anything (even the whole external world) is not a very relevant question. All of us recognize at once how meaningless the query concerning the existence of the electric field in vacuum would be. All that is relevant is that the concept of the electric

[2] The book most commonly blamed for this view is E. F. Haeckel's *Welträtsel* (1899). However, the views propounded in this book are less extreme (though more confused) than those of the usual materialistic philosophy.

[3] W. Heisenberg expressed this most poignantly [*Daedalus*, 87, 99 (1958)]: "The laws of nature which we formulate mathematically in quantum theory deal no longer with the particles themselves but with our knowledge of the elementary particles." And later: "The conception of objective reality . . . evaporated into the . . . mathematics that represents no longer the behavior of elementary particles but rather our knowledge of this behavior." The "our" in this sentence refers to the observer who plays a singular role in the epistemology of quantum mechanics. He will be referred to in the first person and statements made in the first person will always refer to the observer.

[4] J. von Neumann, *Mathematische Grundlagen der Quantenmechanik* (Berlin: Julius Springer, 1932), Chapter VI; English translation (Princeton, N.J.: Princeton University Press, 1955).

field is useful for communicating our ideas and for our own thinking. The statement that it "exists" means only that: (*a*) it can be measured, hence uniquely defined, and (*b*) that its knowledge is useful for understanding past phenomena and in helping to foresee further events. It can be made part of the *Weltbild*. This observation may well be kept in mind during the ensuing discussion of the quantum mechanical description of the external world.

The Language of Quantum Mechanics

The present and the next sections try to describe the concepts in terms of which quantum mechanics teaches us to store and communicate information, to describe the regularities found in nature. These concepts may be called the language of quantum mechanics. We shall not be interested in the regularities themselves, that is, the contents of the book of quantum mechanics, only in the language. It may be that the following description of the language will prove too brief and too abstract for those who are unfamiliar with the subject, and too tedious for those who are familiar with it.[5] It should, nevertheless, be helpful. However, the knowledge of the present and of the succeeding section is not necessary for following the later ones, except for parts of the section on the Simplest Answer to the Mind-Body Question.

Given any object, all the possible knowledge concerning that object can be given as its wave function. This is a mathematical concept the exact nature of which need not concern us here—it is composed of a (countable) infinity of numbers. If one knows these numbers, one can foresee the behavior of the object as far as it *can* be foreseen. More precisely, the wave function permits one to foretell with what probabilities the object will make one or another impression on us if we let it interact with us either directly, or indirectly. The object may be a radiation field, and its wave function will tell us with what probability we shall see a

[5] The contents of this section should be part of the standard material in courses on quantum mechanics. They are given here because it may be helpful to recall them even on the part of those who were at one time already familiar with them, because it is not expected that every reader of these lines had the benefit of a course in quantum mechanics, and because the writer is well aware of the fact that most courses in quantum mechanics do not take up the subject here discussed. See also, in addition to references 3 and 4, W. Pauli, *Handbuch der Physik*, Section 2.9, particularly page 148 (Berlin: Julius Springer, 1933). Also F. London and E. Bauer, *La Théorie de l'observation en mécanique quantique* (Paris: Hermann and Co., 1939). The last authors observe (page 41), "Remarquons le rôle essential que joue la conscience de l'observateur. . . ."

flash if we put our eyes at certain points, with what probability it will leave a dark spot on a photographic plate if this is placed at certain positions. In many cases the probability for one definite sensation will be so high that it amounts to a certainty—this is always so if classical mechanics provides a close enough approximation to the quantum laws.

The information given by the wave function is communicable. If someone else somehow determines the wave function of a system, he can tell me about it and, according to the theory, the probabilities for the possible different impressions (or "sensations") will be equally large, no matter whether he or I interact with the system in a given fashion. In this sense, the wave function "exists."

It has been mentioned before that even the complete knowledge of the wave function does not permit one always to foresee with certainty the sensations one may receive by interacting with a system. In some cases, one event (seeing a flash) is just as likely as another (not seeing a flash). However, in most cases the impression (e.g., the knowledge of having or not having seen a flash) obtained in this way permits one to foresee later impressions with an increased certainty. Thus, one may be sure that, if one does not see a flash if one looks in one direction, one surely does see a flash if one subsequently looks in another direction. The property of observations to increase our ability for foreseeing the future follows from the fact that all knowledge of wave functions is based, in the last analysis, on the "impressions" we receive. In fact, the wave function is only a suitable language for describing the body of knowledge—gained by observations—which is relevant for predicting the future behaviour of the system. For this reason, the interactions which may create one or another sensation in us are also called observations, or measurements. One realises that *all* the information which the laws of physics provide consists of probability connections between subsequent impressions that a system makes on one if one interacts with it repeatedly, i.e., if one makes repeated measurements on it. The wave function is a convenient summary of that part of the past impressions which remains relevant for the probabilities of receiving the different possible impressions when interacting with the system at later times.

An Example

It may be worthwhile to illustrate the point of the preceding section on a schematic example. Suppose that all our interactions with the system consist in looking at a certain point in a certain direction at times

$t_0, t_0 + 1, t_0 + 2, \cdots$, and our possible sensations are seeing or not seeing a flash. The relevant law of nature could then be of the form: "If you see a flash at time t, you will see a flash at time $t + 1$ with a probability $1/4$, no flash with a probability $3/4$; if you see no flash, then the next observation will give a flash with the probability $3/4$, no flash with a probability $1/4$; there are no further probability connections." Clearly, this law can be verified or refuted with arbitrary accuracy by a sufficiently long series of observations. The wave function in such a case depends only on the last observation and may be ψ_1 if a flash has been seen at the last interaction, ψ_2 if no flash was noted. In the former case, that is for ψ_1, a calculation of the probabilities of flash and no flash after unit time interval gives the values $1/4$ and $3/4$; for ψ_2 these probabilities must turn out to be $3/4$ and $1/4$. This agreement of the predictions of the law in quotation marks with the law obtained through the use of the wave function is not surprising. One can either say that the wave function was invented to yield the proper probabilities, or that the law given in quotation marks has been obtained by having carried out a calculation with the wave functions, the use of which we have learned from Schrödinger.

The communicability of the information means, in the present example, that if someone else looks at time t, and tells us whether he saw a flash, we can look at time $t + 1$ and observe a flash with the same probabilities as if we had seen or not seen the flash at time t ourselves. In other words, he can tell us what the wave function is: ψ_1 if he did, ψ_2 if he did not see a flash.

The preceding example is a very simple one. In general, there are many types of interactions into which one can enter with the system, leading to different types of observations or measurements. Also, the probabilities of the various possible impressions gained at the next interaction may depend not only on the last, but on the results of many prior observations. The important point is that the impression which one gains at an interaction may, and in general does, modify the probabilities with which one gains the various possible impressions at later interactions. In other words, the impression which one gains at an interaction, called also *the result of an observation*, modifies the wave function of the system. The modified wave function is, furthermore, in general unpredictable before the impression gained at the interaction has entered our consciousness: it is the entering of an impression into our consciousness which alters the wave function because it modifies our appraisal of the probabilities for different impressions which we expect to receive in the future. It is at this point that the consciousness enters the theory unavoidably and unalterably. If one speaks in terms of the wave function, its changes are coupled with the entering of impressions into our consciousness. If one formulates the laws of quantum mechanics in terms of probabilities of impressions, these are *ipso facto* the primary concepts with which one deals.

It is natural to inquire about the situation if one does not make the observation oneself but lets someone else carry it out. What is the wave function if my friend looked at the place where the flash might show at time t? The answer is that the information available about the *object* cannot be described by a wave function. One could attribute a wave function to the joint system: friend plus object, and this joint system would have a wave function also after the interaction, that is, after my friend has looked. I can then enter into interaction with this joint system by asking my friend whether he saw a flash. If his answer gives me the impression that he did, the joint wave function of friend + object will change into one in which they even have separate wave functions (the total wave function is a product) and the wave function of the object is ψ_1. If he says no, the wave function of the object is ψ_2, i.e., the object behaves from then on as if I had observed it and had seen no flash. However, even in this case, in which the observation was carried out by someone else, the typical change in the wave function occurred only when some information (the *yes* or *no* of my friend) entered *my* consciousness. It follows that the quantum description of objects is influenced by impressions entering my consciousness.[6] Solipsism may be logically consistent with present quantum mechanics, monism in the sense of materialism is not. The case against solipsism was given at the end of the first section.

The Reasons for Materialism

The principal argument against materialism is not that illustrated in the last two sections: that it is incompatible with quantum theory. The

[6] The essential point is not that the states of objects cannot be described by means of position and momentum co-ordinates (because of the uncertainty principle). The point is, rather, that the valid description, by means of the wave function, is influenced by impressions entering our consciousness. See in this connection the remark of London and Bauer, quoted above, and S. Watanabe's article in *Louis de Broglie, Physicien et Penseur* (Paris: Albin Michel, 1952), p. 385.

principal argument is that thought processes and consciousness are the primary concepts, that our knowledge of the external world is the content of our consciousness and that the consciousness, therefore, cannot be denied. On the contrary, logically, the external world could be denied—though it is not very practical to do so. In the words of Niels Bohr,[7] "The word consciousness, applied to ourselves as well as to others, is indispensable when dealing with the human situation." In view of all this, one may well wonder how materialism, the doctrine[8] that "life could be explained by sophisticated combinations of physical and chemical laws," could so long be accepted by the majority of scientists.

The reason is probably that it is an emotional necessity to exalt the problem to which one wants to devote a lifetime. If one admitted anything like the statement that the laws we study in physics and chemistry are limiting laws, similar to the laws of mechanics which exclude the consideration of electric phenomena, or the laws of macroscopic physics which exclude the consideration of "atoms," we could not devote ourselves to our study as wholeheartedly as we have to in order to recognise any new regularity in nature. The regularity which we are trying to track down must appear as the all-important regularity—if we are to pursue it with sufficient devotion to be successful. Atoms were also considered to be an unnecessary figment before macroscopic physics was essentially complete—and one can well imagine a master, even a great master, of mechanics to say: "Light may exist but I do not need it in order to explain the phenomena in which I am interested." The present biologist uses the same words about mind and consciousness; he uses them as an expression of his disbelief in these concepts. Philosophers do not need these illusions and show much more clarity on the subject. The same is true of most truly great natural scientists, at least in their years of maturity. It is now true of almost all physicists—possibly, but not surely, because of the lesson we learned from quantum mechanics. It is also possible that we learned that the principal problem is no longer the fight with the adversities of nature but the difficulty of understanding ourselves if we want to survive.

[7] N. Bohr, *Atomic Physics and Human Knowledge*, section on "Atoms and Human Knowledge," in particular p. 92 (New York: John Wiley & Sons, 1960).

[8] The quotation is from William S. Beck, *The Riddle of Life, Essay in Adventures of the Mind* (New York: Alfred A. Knopf, 1960), p. 35. This article is an eloquent statement of the attitude of the open-minded biologists toward the questions discussed in the present note.

Simplest Answer to the Mind-Body Question

Let us first specify the question which is outside the province of physics and chemistry but is an obviously meaningful (because operationally defined) question: Given the most complete description of my body (admitting that the concepts used in this description change as physics develops), what are my sensations? Or, perhaps, with what probability will I have one of the several possible sensations? This is clearly a valid and important question which refers to a concept—sensations—which does not exist in present-day physics or chemistry. Whether the question will eventually become a problem of physics or psychology, or another science, will depend on the development of these disciplines.

Naturally, I have direct knowledge only of my own sensations and there is no strict logical reason to believe that others have similar experiences. However, everybody believes that the phenomenon of sensations is widely shared by organisms which we consider to be living. It is very likely that, if certain physico-chemical conditions are satisfied, a consciousness, that is, the property of having sensations, arises. This statement will be referred to as our first thesis. The sensations will be simple and undifferentiated if the physico-chemical substrate is simple; it will have the miraculous variety and colour which the poets try to describe if the substrate is as complex and well organized as a human body.

The physico-chemical conditions and properties of the substrate not only create the consciousness, they also influence its sensations most profoundly. Does, conversely, the consciousness influence the physico-chemical conditions? In other words, does the human body deviate from the laws of physics, as gleaned from the study of inanimate nature? The traditional answer to this question is, "No": the body influences the mind but the mind does not influence the body.[9] Yet at least two reasons can be given to support the opposite thesis, which will be referred to as the second thesis.

The first and, to this writer, less cogent reason is founded on the

[9] This writer does not profess to a knowledge of all, or even of the majority of all, metaphysical theories. It may be significant, nevertheless, that he never found an affirmative answer to the query of the text—not even after having perused the relevant articles in the earlier (more thorough) editions of the *Encyclopaedia Britannica.*

quantum theory of measurements, described earlier in sections 2 and 3. In order to present this argument, it is necessary to follow my description of the observation of a "friend" in somewhat more detail than was done in the example discussed before. Let us assume again that the object has only two states, ψ_1 and ψ_2. If the state is, originally, ψ_1, the state of object plus observer will be, after the interaction, $\psi_1 \times \chi_1$; if the state of the object is ψ_2, the state of object plus observer will be $\psi_2 \times \chi_2$ after the interaction. The wave functions χ_1 and χ_2 give the state of the observer; in the first case he is in a state which responds to the question "Have you seen a flash?" with "Yes"; in the second state, with "No." There is nothing absurd in this so far.

Let us consider now an initial state of the object which is a linear combination $\alpha \psi_1 + \beta \psi_2$ of the two states ψ_1 and ψ_2. It then *follows* from the linear nature of the quantum mechanical equations of motion that the state of object plus observer is, after the interaction, $\alpha (\psi_1 \times \chi_1) + \beta (\psi_2 \times \chi_2)$. If I now ask the observer whether he saw a flash, he will with a probability $|\alpha|^2$ say that he did, and in this case the object will also give to me the responses as if it were in the state ψ_1. If the observer answers "No"—the probability for this is $|\beta|^2$—the object's responses from then on will correspond to a wave function ψ_2. The probability is zero that the observer will say "Yes," but the object gives the response which ψ_2 would give because the wave function $\alpha (\psi_1 \times \chi_1) + \beta (\psi_2 \times \chi_2)$ of the joint system has no $(\psi_2 \times \chi_1)$ component. Similarly, if the observer denies having seen a flash, the behavior of the object cannot correspond to χ_1 because the joint wave function has no $(\psi_1 \times \chi_2)$ component. All this is quite satisfactory: the theory of measurement, direct or indirect, is logically consistent so long as I maintain my privileged position as ultimate observer.

However, if after having completed the whole experiment I ask my friend, "What did you feel about the flash before I asked you?" he will answer, "I told you already, I did [did not] see a flash," as the case may be. In other words, the question whether he did or did not see the flash was already decided in his mind, before I asked him.[10] If we accept this, we are driven to the conclusion that the proper wave func-

[10] F. London and E. Bauer (*op. cit.*, reference 5) on page 42 say, "Il [l'observateur] dispose d'une faculté caractéristique et bien familière, que nous pouvons appeler la 'faculté d'introspection': il peut se rendre compte de manière immédiate de son propre état."

tion immediately after the interaction of friend and object was already either $\psi_1 \times \chi_1$ or $\psi_1 \times \chi_2$ and not the linear combination $\alpha (\psi_1 \times \chi_1) + \beta (\psi_2 \times \chi_2)$. This is a contradiction, because the state described by the wave function $\alpha (\psi_1 \times \chi_1) + \beta (\psi_2 \times \chi_2)$ describes a state that has properties which neither $\psi_1 \times \chi_1$ nor $\psi_2 \times \chi_2$ has. If we substitute for "friend" some simple physical apparatus, such as an atom which may or may not be excited by the light-flash, this difference has observable effects and *there is no doubt that* $\alpha (\psi_1 \times \chi_1) + \beta (\psi_2 \times \chi_2)$ *describes the properties of the joint system correctly, the assumption that the wave function is either* $\psi_1 \times \chi_1$ or $\psi_2 \times \chi_2$ *does not.* If the atom is replaced by a conscious being, the wave function $\alpha (\psi_1 \times \chi_1) + \beta (\psi_2 \times \chi_2)$ (which also follows from the linearity of the equations) appears absurd because it implies that my friend was in a state of suspended animation before he answered my question.[11]

It follows that the being with a consciousness must have a different role in quantum mechanics than the inanimate measuring device: the atom considered above. In particular, the quantum mechanical equations of motion cannot be linear if the preceding argument is accepted. This argument implies that "my friend" has the same types of impressions and sensations as I—in particular, that, after interacting with the object, he is not in that state of suspended animation which corresponds to the wave function $\alpha (\psi_1 \times \chi_1) + \beta (\psi_2 \times \chi_2)$. It is not necessary to see a contradiction here from the point of view of orthodox quantum mechanics, and there is none if we believe that the alternative is meaningless, whether my friend's consciousness contains either the impression of having seen a flash or of not having seen a flash. However, to deny the existence of the consciousness of a friend to this extent is surely an

[11] In an article which will appear soon [*Werner Heisenberg und die Physik unserer Zeit* (Braunschweig: Friedr. Vieweg, 1961)] G. Ludwig discusses the theory of measurements and arrives at the conclusion that quantum mechanical theory cannot have unlimited validity (see, in particular, Section IIIa, also Ve). This conclusion is in agreement with the point of view here represented. However, Ludwig believes that quantum mechanics is valid only in the limiting case of microscopic systems, whereas the view here represented assumes it to be valid for all inanimate objects. At present, there is no clear evidence that quantum mechanics becomes increasingly inaccurate as the size of the system increases, and the dividing line between microscopic and macroscopic systems is surely not very sharp. Thus, the human eye can perceive as few as three quanta, and the properties of macroscopic crystals are grossly affected by a single dislocation. For these reasons, the present writer prefers the point of view represented in the text even though he does not wish to deny the possibility that Ludwig's more narrow limitation of quantum mechanics may be justified ultimately.

unnatural attitude, approaching solipsism, and few people, in their hearts, will go along with it.

The preceding argument for the difference in the roles of inanimate observation tools and observers with a consciousness—hence for a violation of physical laws where consciousness plays a role—is entirely cogent so long as one accepts the tenets of orthodox quantum mechanics in all their consequences. Its weakness for providing a specific effect of the consciousness on matter lies in its total reliance on these tenets—a reliance which would be, on the basis of our experiences with the ephemeral nature of physical theories, difficult to justify fully.

The second argument to support the existence of an influence of the consciousness on the physical world is based on the observation that we do not know of any phenomenon in which one subject is influenced by another without exerting an influence thereupon. This appears convincing to this writer. It is true that under the usual conditions of experimental physics or biology, the influence of any consciousness is certainly very small. "We do not need the assumption that there is such an effect." It is good to recall, however, that the same may be said of the relation of light to mechanical objects. Mechanical objects influence light—otherwise we could not see them—but experiments to demonstrate the effect of light on the motion of mechanical bodies are difficult. It is unlikely that the effect would have been detected had theoretical considerations not suggested its existence, and its manifestation in the phenomenon of light pressure.

More Difficult Questions

Even if the two theses of the preceding section are accepted, very little is gained for science as we understand science: as a correlation of a body of phenomena. Actually, the two theses in question are more similar to existence theorems of mathematics than to methods of construction of solutions and we cannot help but feel somewhat helpless as we ask the much more difficult question: how could the two theses be verified experimentally? i.e., how could a body of phenomena be built around them. It seems that there is no solid guide to help in answering this question and one either has to admit to full ignorance or to engage in speculations.

Before turning to the question of the preceding paragraph, let us note in which way the consciousnesses are related to each other and to the physical world. The relations in question again show a remarkable similarity to the relation of light quanta to each other and to the material bodies with which mechanics deals. Light quanta do not influence each other directly[12] but only by influencing material bodies which then influence other light quanta. Even in this indirect way, their interaction is appreciable only under exceptional circumstances. Similarly, consciousnesses never seem to interact with each other directly but only via the physical world. Hence, any knowledge about the consciousness of another being must be mediated by the physical world.

At this point, however, the analogy stops. Light quanta can interact directly with virtually any material object but each consciousness is uniquely related to some physico-chemical structure through which alone it receives impressions. There is, apparently, a correlation between each consciousness and the physico-chemical structure of which it is a captive, which has no analogue in the inanimate world. Evidently, there are enormous gradations between consciousnesses, depending on the elaborate or primitive nature of the structure on which they can lean: the sets of impressions which an ant or a microscopic animal or a plant receives surely show much less variety than the sets of impressions which man can receive. However, we can, at present, at best, guess at these impressions. Even our knowledge of the consciousness of other men is derived only through analogy and some innate knowledge which is hardly extended to other species.

It follows that there are only two avenues through which experimentation can proceed to obtain information about our first thesis: observation of infants where we may be able to sense the progress of the awakening of consciousness, and by discovering phenomena postulated by the second thesis, in which the consciousness modifies the usual laws of physics. The first type of observation is constantly carried out by millions of families, but perhaps with too little purposefulness. Only very crude observations of the second type have been undertaken in the past, and all these antedate modern experimental methods. So far as it is known, all of them have been unsuccessful. However, every phenomenon is unexpected and most unlikely until it has been discovered—and some of them remain unreasonable for a long time after they have been discovered. Hence, lack of success in the past need not discourage.

[12] This statement is certainly true in an approximation which is much better than is necessary for our purposes.

Non-linearity of Equations as Indication of Life

The preceding section gave two proofs—they might better be called indications—for the second thesis, the effect of consciousness on physical phenomena. The first of these was directly connected with an actual process, the quantum mechanical observation, and indicated that the usual description of an indirect observation is probably incorrect if the primary observation is made by a being with consciousness. It may be worthwhile to show a way out of the difficulty which we encountered.

The simplest way out of the difficulty is to accept the conclusion which forced itself on us: to assume that the joint system of friend plus object cannot be described by a wave function after the interaction—the proper description of their state is a mixture.[13] The wave function is $(\psi_1 \times \chi_1)$ with a probability $|\alpha|^2$; it is $(\psi_2 \times \chi_2)$ with a probability $|\beta|^2$. It was pointed out already by Bohm[14] that, if the system is sufficiently complicated, it may be in practice impossible to ascertain a difference between certain mixtures, and some pure states (states which *can* be described by a wave function). In order to exhibit the difference, one would have to subject the system (friend plus object) to very complicated observations which cannot be carried out in practice. This is in contrast to the case in which the flash or the absence of a flash is registered by an atom, the state of which I can obtain precisely by much simpler observations. This way out of the difficulty amounts to the postulate that the equations of motion of quantum mechanics cease to be linear, in fact that they are grossly non-linear if conscious beings enter the picture.[15] We saw that the linearity condition led uniquely to the

[13] The concept of the mixture was put forward first by L. Landau, Z. *Physik*, 45, 430 (1927). A more elaborate discussion is found in J. von Neumann's book (footnote 4), Chapter IV. A more concise and elementary discussion of the concept of mixture and its characterisation by a statistical (density) matrix is given in L. Landau and E. Lifshitz, *Quantum Mechanics* (London: Pergamon Press, 1958), pp. 35-38.

[14] The circumstance that the mixture of the states $(\psi_1 \times \chi_1)$ and $(\psi_2 \times \chi_2)$, with weights $|\alpha|^2$ and $|\beta|^2$, respectively, cannot be distinguished in practice from the state $\alpha(\psi_1 \times \chi_1) + \beta(\psi_2 \times \chi_2)$, if the states χ are of great complexity, has been pointed out already in Section 22.11 of D. Bohm's *Quantum Theory* (New York: Prentice Hall, 1951). The reader will also be interested in Sections 8.27, 8.28 of this treatise.

[15] The non-linearity is of a different nature from that postulated by W. Heisenberg in his theory of elementary particles [cf., e.g., H. P. Dürr, W. Heisenberg, H. Mitter, S. Schlieder, K. Yamazaki, Z. *Naturforsch.*, 14, 441 (1954)]. In our case the equations giving the time variation of the state vector (wave function) are postulated to be non-linear.

unacceptable wave function $\alpha(\psi_1 \times \chi_1) + \beta(\psi_2 \times \chi_2)$ for the joint state. Actually, in the present case, the final state is uncertain even in the sense that it cannot be described by a wave function. The statistical element which, according to the orthodox theory, enters only if I make an observation enters equally if my friend does.

It remains remarkable that there is a continuous transition from the state $\alpha(\psi_1 \times \chi_1) + \beta(\psi_2 \times \chi_2)$ to the mixture of $\psi_1 \times \chi_1$ and $\psi_2 \times \chi_2$, with probabilities $|\alpha|^2$ and $|\beta|^2$, so that every member of the continuous transition has all the statistical properties demanded by the theory of measurements. Each member of the transition, except that which corresponds to orthodox quantum mechanics, is a mixture, and must be described by a statistical matrix. The statistical matrix of the system friend-plus-object is, after their having interacted $(|\alpha|^2 + |\beta|^2 = 1)$,

$$\left\| \begin{matrix} |\alpha|^2 & \alpha\beta^* \cos\delta \\ \alpha^*\beta \cos\delta & |\beta|^2 \end{matrix} \right\|$$

in which the first row and column corresponds to the wave function $\psi_1 \times \chi_1$, the second to $\psi_2 \times \chi_2$. The $\delta = 0$ case corresponds to orthodox quantum mechanics; in this case the statistical matrix is singular and the state of friend-plus-object can be described by a wave function, namely, $\alpha(\psi_1 \times \chi_1) + \beta(\psi_2 \times \chi_2)$. For $\delta = \frac{1}{2}\pi$, we have the simple mixture of $\psi_1 \times \chi_1$ and $\psi_2 \times \chi_2$, with probabilities $|\alpha|^2$ and $|\beta|^2$, respectively. At intermediate δ, we also have mixtures of two states, with probabilities $\frac{1}{2} + (\frac{1}{4} - |\alpha\beta|^2 \sin\delta)^{\frac{1}{2}}$ and $\frac{1}{2} - (\frac{1}{4} - |\alpha\beta|^2 \sin^2\delta)^{\frac{1}{2}}$. The two states are $\alpha(\psi_1 \times \chi_1) + \beta(\psi_2 \times \chi_2)$ and $-\beta^*(\psi_1 \times \chi_1) + \alpha^*(\psi^2 \times \chi^2)$ for $\delta = 0$ and go over continuously into $\psi_1 \times \chi_1$ and $\psi_2 \times \chi_2$ as δ increases to $\frac{1}{2}\pi$.

The present writer is well aware of the fact that he is not the first one to discuss the questions which form the subject of this article and that the surmises of his predecessors were either found to be wrong or unprovable, hence, in the long run, uninteresting. He would not be greatly surprised if the present article shared the fate of those of his predecessors. He feels, however, that many of the earlier speculations on the subject, even if they could not be justified, have stimulated and helped our thinking and emotions and have contributed to re-emphasize the ultimate scientific interest in the question, which is, perhaps, the most fundamental question of all.

I.13 LAW WITHOUT LAW

JOHN ARCHIBALD WHEELER

* * * * *

The second phase of the dialog began in Europe but continued in America from Einstein's arrival at Princeton in October, 1933, to his death there in April, 1955. Here Einstein tried to show that quantum theory — in making what happens depend upon what the observer chooses to measure — is incompatible with any reasonable idea of reality.[18] Bohr's reply[19] briefly summarized was this: Your concept of reality is too limited.

THE BEAM SPLITTER

Of all the idealized experiments taken up by the two friends in their effort to win agreement, none is simpler than the beam splitter of fig. 4. With the final half-silvered mirror in place the photodetector at the lower right goes click-click as the successive photons arrive but the adjacent counter registers nothing. This is evidence of interference between beams 4a and 4b; or, in photon language, evidence that each arriving light quantum has arrived by both routes, A *and* B. In such experiments,[20] Einstein originally argued, it is unreasonable for a single photon to travel simultaneously two routes. Remove the half-silvered mirror, as at the lower left, and one will find that the one counter goes off, or the other. Thus the photon has traveled only *one* route. It travels only one route, but it travels both routes; it travels both routes, but it travels only one route. What nonsense! How obvious it is that quantum theory is inconsistent!

[18] A. Einstein, B. Podolsky and N. Rosen, "Can quantum-mechanical description of physical reality be considered complete?" *Physical Review* **47:** pp. 777-780 (1935).

[19] N. Bohr, "Can quantum-mechanical description of physical reality be considered complete?" *Physical Review* **48:** pp. 696-702 (1935).

[20] The center of discussion in the Bohr-Einstein dialog was more often the so-called double-slit experiment than the beam splitter depicted in figure 4. The latter is made the focus of attention here because it presents the central point without getting into the physics of interference patterns.

The first section between stars (* * * * *) appeared in Wheeler, 1981a; the next section between stars from Wheeler, 1979; the following from Wheeler, 1980; and the final section (a single paragraph) from Wheeler, 1981b. Preparation for publication of all four items was assisted by The University of Texas Center for Theoretical Physics and by NSF Grant PHY78-26592.

Bohr emphasized that there is no inconsistency. We are dealing with two different experiments. The one with the half-silvered mirror removed tells which route. The one with the half-silvered mirror in place provides evidence that the photon traveled both routes. But it is impossible to do both experiments at once. One can observe one feature of nature, or the complementary feature of nature but not both features simultaneously. What we choose to measure has an irretrievable consequence for what we will find.

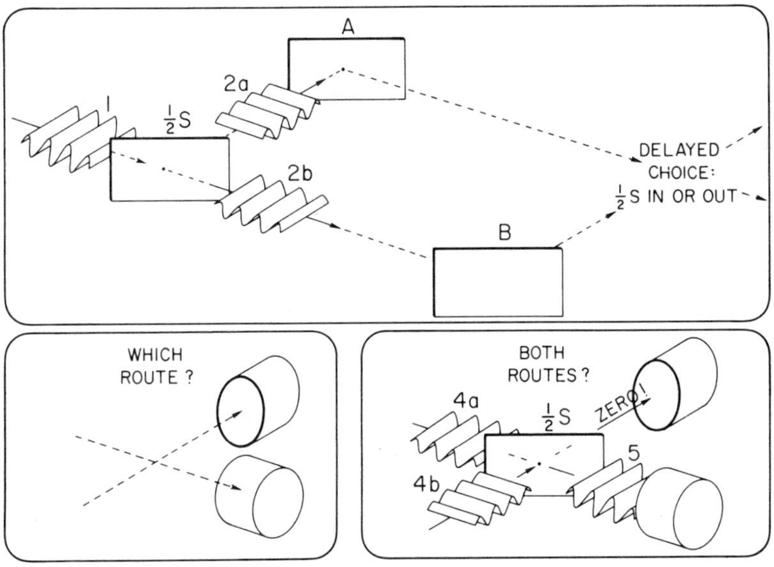

Fig. 4. Beam splitter (above) and its use in a delayed-choice experiment (below). An electromagnetic wave comes in at 1 and encounters the half-silvered mirror marked "½S" which splits it into two beams, 2a and 2b, of equal intensity which are reflected by mirrors A and B to a crossing point at the right. Counters (lower left) located past the point of crossing tell by which route an arriving photon has come. In the alternative arrangement at the lower right, a half-silvered mirror is inserted at the point of crossing. On one side it brings beams 4a and 4b into destructive interference, so that the counter located on that side never registers anything. On the other side the beams are brought into constructive interference to reconstitute a beam, 5, of the original strength, 1. Every photon that enters at 1 is registered in that second counter in the idealized case of perfect mirrors and 100 per cent photodetector efficiency. In the one arrangement (lower left) one finds out by *which* route the photon came. In the other arrangement (lower right) one has evidence that the arriving photon came by both routes. In the new "delayed-choice" version of the experiment one decides whether to put in the half-silvered mirror or take it out at the very last minute. Thus one decides whether the photon "shall have come by one route, or by both routes" after it has "*already done* its travel."

THE DELAYED-CHOICE EXPERIMENT

In our own day we have learned to state the point even more sharply by way of a so-called delayed-choice experiment.[21] There we make the decision whether to put the final half-silvered mirror in place or to take it out at the very last picosecond, after the photon has already accomplished its travel. In this sense, we have a strange inversion of the normal order of time. We, now, by moving the mirror in or out have an unavoidable effect on what we have a right to say about the *already* past history of that photon.

"PHENOMENON"

The dependence of what is observed upon the choice of experimental arrangement made Einstein unhappy. It conflicts with the view that the universe exists "out there" independent of all acts of observation. In contrast Bohr stressed that we confront here an inescapable new feature of nature, to be welcomed because of the understanding it gives us. In struggling to make clear to Einstein the central point as he saw it, Bohr found himself forced to introduce the word "phenomenon."[7] In today's words Bohr's point — and the central point of quantum theory — can be put into a single, simple sentence. "No elementary phenomenon is a phenomenon until it is a registered (observed) phenomenon."[8] It is wrong to speak of the "route" of the photon in the experiment of the beam splitter. It is wrong to attribute a tangibility to the photon in all its travel from the point of entry to its last instant of flight. A phenomenon is not yet a phenomenon

[21] J.A. Wheeler, "The 'past' and the 'delayed-choice' double-slit experiment," in A.R. Marlow, ed., *Mathematical Foundations of Quantum Theory* (Academic Press, New York, 1978), pp. 9-48.

[22] "Closed by irreversible amplification", p. 73; "irreversible amplification," p. 88: N. Bohr, *Atomic Physics and Human Knowledge* (Wiley, New York, 1958).

[23] A homely illustration of this idea is provided by the old parlor game of Twenty Questions in the "surprise version" described by the author in several places, most recently in "Beyond the black hole," a chapter in H. Woolf, ed., *Some Strangeness in the Proportions: An Einstein Centenary Celebration* (Addison-Wesley, Reading, Mass., 1980).

until it has been brought to a close by an irreversible act of amplification such as the blackening of a grain of silver bromide emulsion or the triggering of a photodetector.[22] In broader terms, we find that nature at the quantum level is not a machine that goes its inexorable way. Instead what answer we get depends on the question we put, the experiment we arrange, the registering device we choose. We are inescapably involved in bringing about that which appears to be happening.[23]

CONCERN ABOUT OBSERVER-PARTICIPANCY TODAY

Most applications of quantum theory deal with stationary states of elementary particles, of atomic nuclei, atoms, molecules and larger systems, and with processes of collision between one quantum system and another. Only in recent years has increasing attention come back to the point of central concern of Bohr and Einstein, the elementary quantum phenomenon, the process of measurement, the involvement of the registering device in bringing about that which appears to be happening, the strangest part of a strange subject. How can one contemplate indeterminism, complementarity and "phenomenon" without being reminded of the words of Gertrude Stein about modern art? "It looks strange and it looks strange and it looks very strange; and then suddenly it doesn't look strange at all and you can't understand what made it look strange in the first place." Many investigators, believing that the greatest insights are to be won from nature's strangest features, are—in research papers, review articles and books—giving fresh coverage of the strange "observer-participancy" forced to our attention by the quantum.[24]

[24] See for example B. d'Espagnat, ed., *Foundations of Quantum Mechanics* (Academic Press, New York, 1971); E.P. Wigner, "Interpretation of quantum mechanics," 93 pages of mimeographed notes of lectures delivered at Princeton University in 1976 on deposit in Fine Library, Princeton University, Princeton, N.J.; M.M. Yanase, M. Namiki and S. Machida, eds., *Theory of Measurement in Quantum Mechanics* (Physical Society of Japan, Tokyo, 1980); J.A. Wheeler, "Frontiers of time," in N. Toraldo di Francia, ed., *Problems in the Foundations of Physics*, Rendiconti della Scuola Internazionale di Fisica 'Enrico Fermi', LXXII Corso (North-Holland, Amsterdam, 1979).

MANY QUANTA VERSUS ONE QUANTUM

How does quantum mechanics today differ from what Bishop George Berkeley told us two centuries ago, *"Esse est percipi,"* to be is to be perceived?[25] Does the tree not exist in the forest unless there is someone there to see it? Do Bohr's conclusions about the role of the observer differ from those of Berkeley? Yes, and in an important way. Bohr deals with the individual quantum process. Berkeley — like all of us under everyday circumstances — deals with multiple quantum processes.

Pondering the difference between the individual quantum phenomenon and the tree that falls, unobserved, in the forest, we walk through the art gallery on our way to visit again a favorite picture. We pass by the painting "Impressions," first shown by Claude Monet in 1863 at the Salon des Refusés. From a tiny dab of color on that canvas in the single second of our passage the pupil of our eye receives 50,000 photons. Each is accidental in its direction and time of arrival. The quanta in that hail of information are so numerous that they give the impression of perfect steadiness of illumination. What one of us busy mortals has the time to count them all? We rely instead on some gross and handier measure of intensity, such as the eye so aptly passes to the brain. There is no place in that message for the qualifying words, "with a root mean square fluctuation of 224 relative to an average number of photons of 50,000." Who needs to know about quanta to know the dot of color is there?

Unexpectedly the power blacks out. A guard with electric torch pointed at the floor guides our return. Our eye receives no photons from the dab of paint on the canvas. However, a touch of the hand as we pass the painting in the dark is enough to comfort us that it is still there. It would outreach any on-the-spot bookkeeping to count the 10^{16} atomic points of contact between the fingers and the picture frame, or the even more numerous quantum processes that impinge from the frame onto the fingertips. The message is still clear. How-

[25] G. Berkeley (1685-1783) in M.W. Calkins, ed., *Berkeley: Essays, Principles, Dialogs, with Selections from Other Writings* (Scribner, New York, 1929, as reprinted in 1957), pp. 125-126.

ever, we now go through a longer chain of theory and interpretation in reaching the conclusion that the dab of paint is still there. Or was the luminous dot of color an illusion created by trick illumination from a concealed lamp? That was conceivable when we passed it first but highly unlikely given the integrity of the museum and the difficulty of the undertaking. During the exit through the dark it is more difficult to check against deception but the best indirect evidence one has says that the painting is still there with all its dots of color. Moreover, one is free to stop and extend the investigation and transform questionable evidence into convincing evidence.

When we emerge from the gallery and start thinking again of the tree, we recognize that this problem differs from the case of the picture only in degree, not in kind. The supposition that it fell we can check more and more conclusively according to the amount of effort we are willing to put into investigating impact points, ground dislocations and acoustic records. Anything macroscopic that happened in the past makes, we know, a rich fallout of consequences in the present. But whether we deal with the fall of the tree or the evidence for the dab of paint on the canvas or the motion of the moon through the sky, the number of quanta that come into play is so enormous that the unseen quantum individuality of the act of observation can hardly be said to influence the event observed.

In contrast the choice of question asked has a decisive consequence for[26] the elementary quantum phenomenon. For illustration it is enough to recall the inquiry of fig. 4 about the "track" of the photon, or a similar inquiry about the "path" of an electron through a beam splitter or the "motion" of an electron in an atom. In each of these examples, moreover, at least one of the available choices of question to be asked (which route for the photon or electron; or what position or momentum does the electron have in the atom) has a

[26] Why not change "has a decisive consequence for . . ." to "makes all the difference in the elementary quantum phenomenon"? The word "difference" is not allowable. We can do the one experiment or the other experiment but the two experiments simply will not fit into one place at one time. We are dealing with one phenomenon, one "act of creation." The very individuality of the quantum phenomenon leaves no place for comparing what is with what might have been.

completely unpredictable answer. We can send a million photons through the beam splitter when it is operated in the "which route" configuration at the lower left of fig. 4. Then we can be assured half a million photons, more or less (statistical variations of the order of magnitude ± 500) will be recorded by each counter. However, when via the same arrangement we deal with a single photon we have not the slightest possibility to tell in advance which of the two counters it will strike.

QUANTUM OUTCOME: GOVERNED BY HIDDEN VARIABLES?

Is there not some underground machinery beneath the working of the world which one can ferret out to secure an advance indication of the outcome? Some secret determiner, some "hidden variable"? Every attempt, theoretical or observational, to defend such a hypothesis has been struck down.[27] Not the slightest hard evidence has ever been found that would throw doubt on the plain, straightforward prediction of quantum mechanics, the prediction that no prediction is possible. Probability? Yes. A definite forecast? No. Einstein could be unhappy that "God plays dice"; but Bohr could tell him jokingly, "Einstein, stop telling God what to do."[28]

QUANTUM OUTCOME: ALLAH WILLED IT?

If no identifiable machinery is at hand to tell the lone photon which way to go then why not simply say of the route it actually takes, Allah willed it? And willed the outcome of every other individual quantum process?

To strike down a proposal of this kind, it has been

[27] For a review of relevant experiments, see especially F.M. Pipkin, "Atomic physics tests of the basic concepts in quantum mechanics," pp. 281-340 in *Advances in Atomic and Molecular Physics* (Academic Press, New York, 1978).

[28] N. Bohr as quoted by J. Bronowski, *The Ascent of Man* (Little, Brown and Co., Boston/Toronto, 1973), p. 122.

pointed out more than once,[29] is beyond the power of logic. One has to appeal instead to pragmatism. In the struggle for survival, other things being equal, that way of life will go under that takes all that comes in a blindly fatalistic spirit. To evade danger and to seize opportunity every faculty has to be mobilized to predict what lies ahead of peril and promise. Society charges science with the task of prediction. Science makes some progress with the task. In the individual quantum process, however, prediction comes to the end of the road. Science does not have to be ashamed of its finding. It has only to be honest about it. Why demand of science a cause when cause there is none?

QUANTUM OUTCOME: ELEMENTARY ACT OF CREATION?

How did the universe come into being? Is that some strange, far-off process, beyond hope of analysis? Or is the mechanism that came into play one which all the time shows itself?

Of the signs that testify to "quantum phenomenon" as being the elementary act of creation, none is more striking than its untouchability. In the delayed-choice version of the split-beam experiment, for example, we have no right to say what the photon is doing in all its long course from point of entry to point of detection. Until the act of detection the phenomenon-to-be is not yet a phenomenon. We could have intervened at some point along the way with a different measuring device; but then regardless whether it is the new registering device or the previous one that happens to be triggered we have a new phenomenon. We have come no closer than before to penetrating to the untouchable interior of the phenomenon. For a process of creation that can and does operate anywhere, that reveals itself and yet hides itself, what could one have dreamed up out of pure imagination more magic —and more fitting—than this?

[29] For a discussion of this point I am indebted to Professor Andrew Gleason.

DELAYED CHOICE AT THE COSMOLOGICAL SCALE

Of all the features of the "act of creation" that is the elementary quantum phenomenon, the most startling is that seen in the delayed-choice experiment. It reaches back into the past in apparent opposition to the normal order of time. The distance of travel in a laboratory split-beam experiment might be thirty meters and the time a tenth of a microsecond; but the distance could as well have been billions of light years and the time billions of years. Thus the observing device in the here and now, according to its last minute setting one way or the other, has an irretrievable consequence for what one has the right to say about a photon that was given out long before there was any life in the universe.

Two astronomical objects, known as 0957+561A,B (fig. 5), once considered to be two distinct quasistellar objects or "quasars" because they are separated by six seconds of arc, are considered now by many observers to be two distinct images of one quasar.[30] Evidence has been found for an intervening galaxy, roughly a quarter of the way from us to the quasar. Calculations indicate[31] that a normal galaxy at such a distance has the power to take two light rays, spread apart by

[30] D. Walsh, R.F. Carswell and R.J. Weymann, "0957+561A,B: twin quasistellar objects or gravitational lens?" *Nature* **279**: pp. 381-384 (1979); R.J. Weymann, F.H. Chaffee Jr., M. Davis, N.P. Carleton, D. Walsh, and R.F. Carswell, "Multiple-mirror observations of the twin QSO 0957+561A, B," *Astrophysical Journal* **233**, L43-L46 (1979); P.J. Young, W.L.W. Sargent, J.A. Kristian and J.A. Westphal, "CCD photometry of the nuclei of three supergiant elliptical galaxies: evidence for a supermassive object in the center of the radiogalaxy NGC6251," *Astrophysical Journal* **234**: pp. 76-85 (1979); D.H. Roberts, P.E. Greenfield and B.F. Burke, "The double quasar 0957+561: a radio study at 6 centimeters wavelength," *Science* **205**: pp. 894-896 (1979); G.G. Pooley, I. Browne, E.J. Daintree, P.K. Moore, R.G. Noble and D. Walsh, "Radio studies of the double QSO 0957+561A,B," *Nature* **280**: pp. 461-464 (1979); P.E. Greenfield, D.H. Roberts and B.F. Burke, "The double quasar 0957+561: examination of the gravitational lens hypothesis using the very large array," *Science* **208**: pp. 495-497 (1980); P.J. Young, J.E. Gunn, J.A. Kristian, J.B. Oke and J.A. Westphal, "Q0957+561A,B: a gravitational lens formed by a galaxy at $z = 0.39$," *Astrophysical Journal*, in press (1980); B. Wills and D. Wills, "Spectrophotometry of the double QSO 0957+561," *Astrophysical Journal* **238**: pp. 1-9 (1980); B.T. Soifer, G. Neugebauer, K. Matthews, E.E. Becklin, C.G. Wynn-Williams and R. Capps, "IR observations of the double quasar 0957+561A,B and the intervening galaxy," *Nature* **285**: pp. 91-93 (1980).

[31] C.C. Dyer and R.C. Roeder, "Possible multiple imaging by spherical galaxies," *Astrophysical Journal* **238**, L67-L70 (1980); C.C. Dyer and R.C. Roeder, "A range of time delays for the double quasar 0957+561A,B," *Astrophysical Journal*, submitted for publication June 16, 1980.

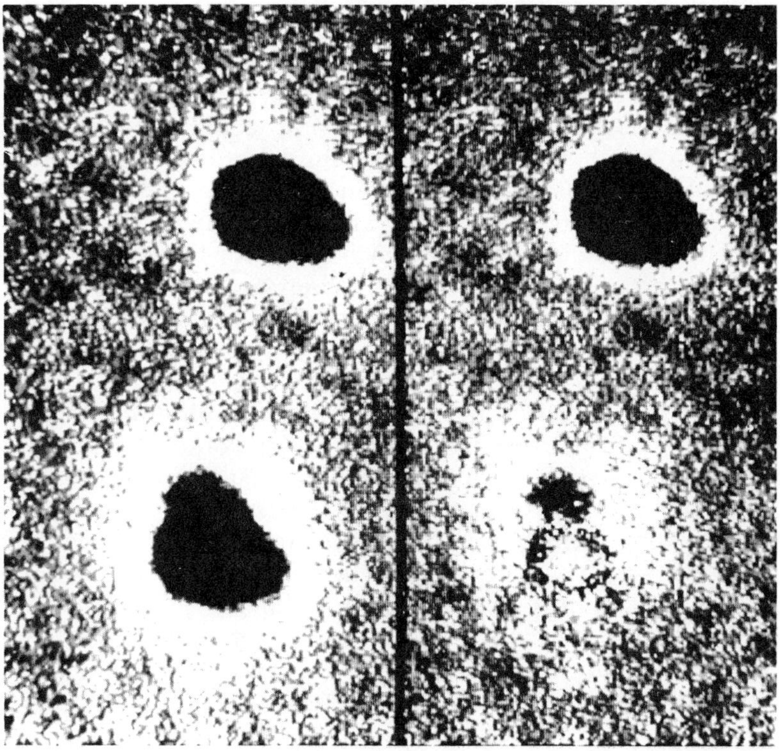

Fig. 5. Left, the double quasistellar object ("quasar": red shift z = 1.41), identified by its right ascension and declination as 0957 + 561A,B, and suspected to be the two images—produced by gravitational lens action—of one and the same quasar. This photograph, made at the University of Hawaii telescope by Alan Stockton and kindly communicated and discussed by Derek Wills of the University of Texas at Austin, is the digital sum of five one-minute exposures in red light (5700 to 7000Å). The stellar images appear elongated because of a telescope tracking problem. Right, the same digital photographic record after a stellar profile has been subtracted from the southern image (B), the residual being compatible with the existence near B of a lensing galaxy (G-1). Evidence has been found by Young, Gunn, Kristian, Oke and Westphal at Caltech for such a galaxy (0.02″ to the West and 0.8″ North of B; red shift z = 0.39), much closer to B than to A (which is 1.2″ to the West and 6″ North of B), and for its membership *in* a cluster of perhaps 1000 to 10,000 galaxies (centered 2″ to the West and 15″ North of B).

fifty thousand light years on their way out from the quasar, and bring them back together at the Earth. This circumstance, and evidence for a new case of gravitational lensing,[32] make it reasonable to promote the split-beam experiment in the delayed-choice version from the laboratory level to the cosmological scale as illustrated in fig. 6.

We get up in the morning and spend the day in meditation whether to observe by "which route" or to observe interference between "both routes." When night comes and the telescope is at last usable we leave the half-silvered mirror out or put it in, according to our choice. The monochromatizing filter placed over the telescope makes the counting rate low. We may have to wait an hour for the first photon. When it triggers a counter, we discover "by which route" it came with the one arrangement; or by the other, what the relative phase is of the waves associated with the passage of the photon from source to receptor "by both routes"—perhaps 50,000 light years apart as they pass the lensing galaxy G-1. But the photon has already *passed* that galaxy billions of years before we made our decision. This is the sense in which, in a loose way of speaking, we decide what the photon *shall have done* after it has *already* done it. In actuality it is wrong to talk of the "route" of the photon. For a proper way of speaking we recall once more that it makes no sense to talk of the phenomenon until it has been brought to a close by an irreversible act of amplification: "No elementary phenomenon is a phenomenon until it is a registered (observed) phenomenon."

[32] R.J. Weymann, D. Latham, J.R.P. Angel, R.F. Green, J.W. Liebert, D.A. Turnshek, D.E. Turnshek and J.A. Tyson, "The triple QSO PG1115+08: another probable gravitational lens," *Nature* **205**: pp. 641-643 (1980).

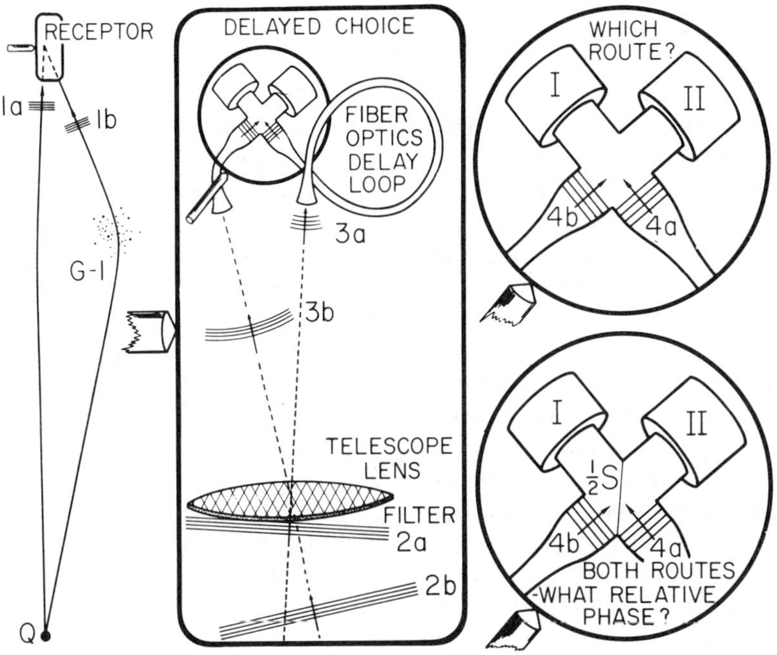

Fig. 6. Proposed delayed-choice experiment extending over a cosmological reach of space and time. Left, quasar Q recorded at receptor as two quasars by reason of the gravitational lens action of the intervening galaxy G-1. Middle, schematic design of receptor for delayed-choice experiment: (a) filter to pass only wave lengths in a narrow interval, corresponding to a long wave train, suitable for interference experiments; (b) lens to focus the two apparent sources onto the acceptor faces of two optic fibers; (c) delay loop in one of these fibers of such length, and of such rate of change of length with time, as to bring together the waves traveling the two very different routes with the same, or close to the same, phase. Right, the choice. Upper diagram, nothing is interposed in the path of the two waves at the crossing of the optic fibers. Wave 4a goes into counter I, and 4b into counter II. Whichever of these photodetectors goes off, that—in a bad way of speaking—signals "by *which* route, a or b, the photon in question traveled from the quasar to the receptor." Lower diagram, a half-silvered mirror, ½S, is interposed as indicated at the crossing of the two fibers. Let the delay loop be so adjusted that the two arriving waves have the same phase. Then there is never a count in I. All photons are recorded in II. This result, again in a misleading phraseology, says that "the photons in question come by *both* routes." However, at the time the choice was made whether to put in ½S or leave it out, the photon in question had *already* been on its way for billions of years. It is not right to attribute to it a route. No elementary phenomenon is a phenomenon until it is a registered phenomenon.

THE "PAST" IN THE LIGHT OF THE DELAYED-CHOICE EXPERIMENT

To use other language, we are dealing with an elementary act of creation. It reaches into the present from billions of years in the past. It is wrong to think of that past as "already existing" in all detail. The "past" is theory. The past has no existence except as it is recorded in the present. By deciding what questions our quantum registering equipment shall put in the present we have an undeniable choice in what we have the right to say about the past.

What we call reality consists (fig. 7) of a few iron posts of observation between which we fill in by an elaborate papier-maché construction of imagination and theory.[33]

Spacetime in the prequantum dispensation was a great record parchment. This sheet, this continuum, this carrier of all that is, was and shall be, had its definite structure with its curves, waves and ripples; and on this great page every event, like a glued down grain of sand, had its determinate place. In this frozen picture a far-reaching modification is forced by the quantum. What we have the right to say of past spacetime, and past events, is decided by choices — of what measurements to carry out — made in the near past and now. The phenomena called into being by these decisions reach backward in time in their consequences as indicated in fig. 8, back even to the earliest days of the universe. Registering equipment operating in the here and now has an undeniable part in bringing about that which appears to have happened. Useful as it is under everyday circumstances to say that the world exists "out there" independent of us, that view can no longer be upheld. There is a strange sense in which this is a "participatory universe."

[33] In this connection see especially E.H. Gombrich, *Art and Illusion: A Study in the Psychology of Pictorial Representation* (Princeton University Press, Princeton, N.J., 1961, 2nd edition, revised), pp. 273, 329 and 394.

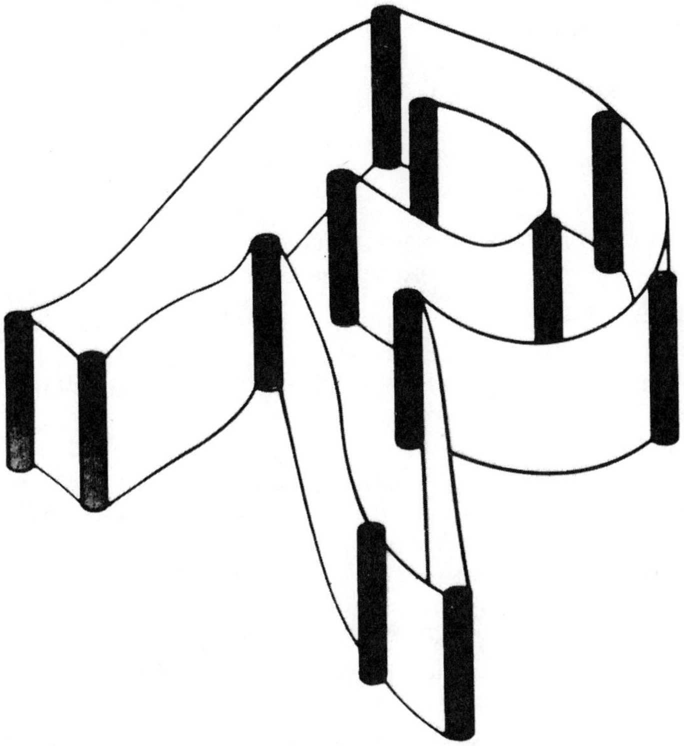

Fig. 7. What we call "reality," symbolized by the letter "R" in the diagram, consists of an elaborate papier-maché construction of imagination and theory fitted in between a few iron posts of observation.

FROM MEASUREMENT TO MEANING

We cannot speak in these terms without a caution and a question. The caution: "Consciousness" has nothing whatsoever to do with the quantum process. We are dealing with an event that makes itself known by an irreversible act of amplification, by an indelible record,[34] an act of registration. Does that record subsequently enter into the "consciousness" of some person, some animal or some computer? Is that the first step in translating the measurement into "meaning"—meaning regarded as "the joint product of all the evidence that is available to those who communicate"?[35] Then that is a separate part of the story, important but not to be confused with "quantum phenomenon."

IS THE UNIVERSE CONSTRUCTED OUT OF ELEMENTARY PHENOMENA?

From this caution we turn to the question: If the elementary quantum process is an act of creation, is an act of creation of any other kind required to bring into being all that is?

At first sight no question could seem more ridiculous. How fantastic the disproportion seems between the microscopic scale of the typical quantum phenomenon and the gigantic reach of the universe! Disproportion, however, we have learned, does not give us the right to dismiss. Else how would we have discovered that the heat of the carload of molten pig iron goes back for its explanation to the random motions of billions of microscopic atoms and the shape of the elephant to the message on a microscopic strand of DNA? Is the term "big bang" merely a shorthand way to describe the

[34] F.J. Belinfante, *Measurements and Time Reversal in Objective Quantum Theory* (Oxford University Press, Oxford, 1975); terminology "indelible," p. 39.

[35] D. Føllesdal, "Meaning and experience" in S. Guttenplan, ed., *Mind and Language* (Clarendon Press, Oxford, 1975), pp. 254. Føllesdal's article, the other articles in this book and the references they make to the still larger literature of meaning, a central topic of philosophy in Britain and America in recent decades, will indicate the representative character of this statement.

cumulative consequence of billions upon billions of elementary acts of observer-participancy reaching back into the past, as symbolized in fig. 8?

An old legend describes a dialog between Abraham and Jehovah. Jehovah chides Abraham, "You would not even exist if it were not for me!" "Yes, Lord, that I know," Abraham replies, "but also You would not be known if it were not for me."[36]

In our time the participants in the dialog have changed. They are the universe and man. The universe, in the words of some who would aspire to speak for it, says, "I am a giant machine. I supply the space and time for your existence. There was no before before I came into being, and there will

[36] Thanks are expressed here to Professors Lawrence P. Horwitz, Zvi Kurzweil, Yuval Ne'eman, Asher Peres, Shmuel Sambursky, Lawrence Schulman and Elie Wiesel, each for his part in leading the author to this legend and documenting it, as follows: (i) H. Freedman and M. Simon, translators and eds., *Midrash Rabbah, Genesis I* (Soncino Press, London, 1939), p. 238, commentary on "Noah walked with God": "*The God before whom my fathers Abraham and Isaac did walk,* etc. (Genesis 48:15). R. Berekiah in R. Johanan's name and Resh Lakish gave two illustrations of this. R. Johanan said: It was as if a shepherd stood and watched his flocks. Resh Lakish said: It was as if a prince walked along while the elders preceded him [Footnote: As an escort, to make known his coming. Similarly, Abraham and Isaac walked before God, spreading His knowledge]. On R. Johanan's view: We need His proximity. On the view of Resh Lakish: He needs us to glorify Him [Footnote: By propagating the knowledge of His greatness]." (ii) *Ibid,* p. 357, commentary on, "And he blessed him, and said: blessed be Abram of the God most high, who has acquired [Koneh = maker of] heaven and earth" (Genesis 14:19): "From whom then did He acquire them?— Said R. Abba: [Acquired is attributive,] as one says, So-and-so has [Koneh = in possession of] beautiful eyes and hair. R. Isaac said: Abraham used to entertain wayfarers, and after they had eaten he would say to them, 'Say a blessing,' 'What shall we say?' they asked. 'Blessed be the God of the Universe of Whose bounty we have eaten,' replied he. Then the Holy One, blessed be He, said to him: 'My Name was not known among My creatures, and thou hast made it known among them: I will regard thee as though thou wast associated with Me in the creation of the world'. . . . " (iii) Deuteronomy 32:10: "He found him [Jacob] in a desert land, and in the waste howling wilderness; he led him about, he instructed him, he kept him as the apple of his eye," as commented on in *Sifrei* [analogous to the *Midrash* of (i) and (ii) but contains in addition to the Aggadic or legend of the Midrash the Halakhic or law; ed. in the Holy Land before the end of the 4th century A.D.] §313, "he led him about": "This is related to Genesis 12:1, 'Get thee out of thy country' . . . .; 'he instructed him': . . . . before our father Abraham came into this world it seemed as if the Lord, Blessed Be He, reigned only in Heaven, since it is said, 'The Lord, God of Heaven, which took me from my father's house' (Genesis 24:7). But once *Abraham* had come into the world [= was born], he Abraham [thereby] enthroned Him over Heaven *and Earth*" (translation from the Hebrew by Y. Ne'eman). (iv) Isaiah 43:10: "Ye are my witnesses, saith the Lord, and my servant whom I have chosen; that ye may know and believe me, and understand that I am he: before me there was no God formed, neither shall there be after me."

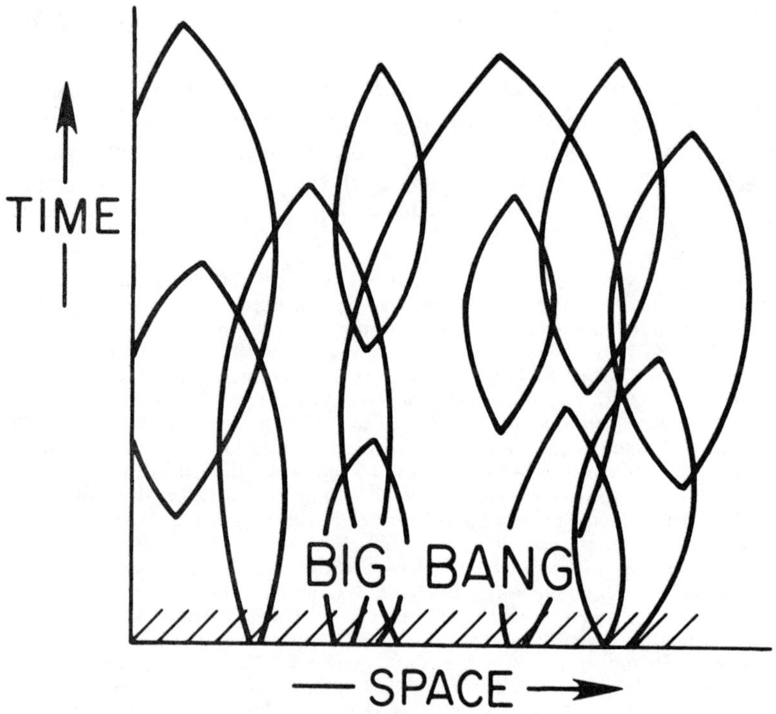

Fig. 8. Symbolic description how all that "has happened" in the past is influenced by choices made in the present as to what to observe. The upper tip of each "leaf" stands for the elementary act of registration. The lower end of each leaf stands for the beginning of the elementary phenomenon being investigated by the observational means at hand. Is anything else required to make up space and time and all their burden of physical content except the information carried in the elementary quantum acts thus symbolized? [Details in the original publication.]

be no after after I cease to exist. You are an unimportant bit of matter located in an unimportant galaxy."

How shall we reply? Shall we say, "Yes, oh universe, without you I would not have been able to come into being. Yet you, great system, are made of phenomena; and every phenomenon rests on an act of observation. You could never even exist without elementary acts of registration such as mine"?

Are elementary quantum phenomena, those untouchable, indivisible acts of creation, indeed the building material of all that is? Beyond particles, beyond fields of force, beyond geometry, beyond space and time themselves, is the ultimate constituent, the still more ethereal act of observer-participancy? For Dr. Samuel Johnson the stone was real enough when he kicked it. The subsequent discovery that the matter in that rock is made of positive and negative electric charges and more than 99.99 per cent empty space does not diminish the pain that it inflicts on one's toe. If the stone is someday revealed to be altogether emptiness, "reality" will be none the worse for the finding.

Roland M. Frye, in reminding us[37] of Shakespeare and of ways of seeing, gives us opportunity to recall those words of almost four hundred years ago,

> And as imagination bodies forth
> The form of things unknown, the poet's pen
> Turns them to shapes, and gives to airy nothing
> A local habitation and a name.

Are billions upon billions of acts of observer-participancy the foundation of everything? We are about as far as we can be today from knowing enough about the deeper machinery of the universe to answer this question. Increasing knowledge about detail has brought an increasing ignorance about plan. The very fact that we can ask such a strange question shows how uncertain we are about the deeper foundations of the quantum and its ultimate implications.

[37] R.M. Frye, "Ways of seeing: unities and disunities in Shakespeare and Elizabethan painting," *infra*, pp.43 ff.

THE QUANTUM: ITS USES — AND ITS USE

To encounter the quantum is to feel like an explorer from a faraway land who has come for the first time upon an automobile. It is obviously meant for use, and an important use, but what use? One opens the door, cranks the window up and down, flashes the lights on and off, and perhaps even turns over the starter, all the time without knowing the central point of the thing. The quantum is the automobile. We use the quantum in a transistor to control machinery, in a molecule to design an anesthetic, in a superconductor to make a magnet. Could it be that all the time we have been missing the central point, the use of the quantum phenomenon in the construction of the universe itself?

We have turned over the starter. We haven't got the engine going.

* * * * *

1. – Law without law.

species will never vary, and have remained the same since the creation of each species.

Charles LYELL [1], writing almost three decades before *The Origin of Species*

[The astronomer Sir John Frederick William] *Herschel says my book is 'the law of higgledy-piggledy'.*

Charles DARWIN [2], 18 days after the November 24, 1859 publication of *The Origin of Species*

Are the laws of physics eternal and immutable? or are these laws, like species, mutable [3] and of « higgledy-piggledy » origin?

The hierarchical speciation of plant and animal life, we now know, arises out of the blind accidents of genetic mutation and natural selection [5, 6]. Likewise the gas laws, the pressure-volume-temperature relation for water and for other substances, and the laws of thermodynamics take their origin in the chaos of molecular collisions. But as for the molecules themselves, the particles of which they are made and the fields of force that couple them, is it conceivable that they too derive their way of action, their structure and even their existence from multitudinous accidents?

Such questions about the « plan » of physics we would hardly raise if we had the skeleton of it in hand. But we don't. Now and then we meet a colleague in another realm of thought who still thinks physics is in possession of this plan. He cites the words of Laplace [7] and reiterates the Laplacean vision as he understands it: the laws are definite, the initial co-ordinates and momenta are definite, and therefore the future is definite. The Universe is a machine.

No, we have to tell him; that is a cracked paradigm. Quantum mechanics allows us to know a co-ordinate, or a momentum, but not both. Of the initial-value data that LAPLACE needed, the principle of complementarity [8] or indeterminacy [9] says half do not and cannot exist.

. . . .

You tell me what isn't the plan of physics, our friend rejoins. If you understand quantum mechanics so well, why don't you tell we what *is* the plan of physics?

No one knows, we reply. We have clues, clues most of all in the writings of Bohr [23-25], but no answer. That he did not propose an answer, not philosophize, not go an inch beyond the soundest fullest statement of the inescapable lessons of quantum mechanics, was his way to build a clean pier for some later day's bridge to the future.

What kind of a « plan of physics » do you think BOHR had in mind, our colleague asks. I know Einstein's words [26], « Physics is an attempt to grasp reality as it is thought independently of its being observed ». I know Bohr's reply [28], « These conditions [of measurement] constitute an inherent element of any phenomenon to which the term ' physical reality ' can be attached.... [This requires] a final renunciation of the classical ideal of causality and a radical revision of our attitude towards the problem of physical reality ». But if I could have asked BOHR, how did he think the Universe came into being, and what is its substance, what would he have said?

It is too late to ask. The plan is up to us to find.

The Universe can't be Laplacean. It may be higgledy-piggledy. But have hope. Surely someday we will see the necessity of the quantum in its construction. Would you like a little story along this line?

Of course! About what?

About the game of twenty questions. You recall how it goes—one of the after-dinner party sent out of the living room, the others agreeing on a word, the one fated to be questioner returning and starting his questions. « Is it a living object? » « No. » « Is it here on earth? » « Yes. » So the questions go from respondent to respondent around the room until at length the word emerges: victory if in twenty tries or less; otherwise, defeat.

Then comes the moment when we are fourth to be sent from the room. We are locked out unbelievably long. On finally being readmitted, we find a smile on everyone's face, sign of a joke or a plot. We innocently start our questions. At first the answers come quickly. Then each question begins to take longer in the answering—strange, when the answer itself is only a simple « yes » or « no ». At length, feeling hot on the trail, we ask, « Is the word ' cloud '? » « Yes », comes the reply, and everyone bursts out laughing. When we were out of the room, they explain, they had agreed not to agree in advance on any word at all. Each one around the circle could respond « yes » or « no » as he pleased to whatever question we put to him. But however he replied he had to have a word in mind compatible with his own reply—and with all the replies that went before. No wonder some of those decisions between « yes » and « no » proved so hard!

And the point of your story?

Compare the game in its two versions with physics in its two formulations, classical and quantum. First, we thought the word already existed « out there » as physics once thought that the position and momentum of the electron existed « out there », independent of any act of observation. Second, in actuality the information about the word was brought into being step by step through the questions we raised, as the information about the electron is brought into being, step by step, by the experiments that the observer chooses to make. Third, if we had chosen to ask different questions we would have ended up with a different word—as the experimenter would have ended up with a different story for the doings of the electron if he had measured different quantities or the same quantities in a different order. Fourth, whatever power we had in bringing the particular word « cloud » into being was partial only. A major part of the selection—unknowing selection—lay in the « yes » or « no » replies of the colleagues around the room. Similarly, the experimenter has some substantial influence on what will happen to the electron by the choice of experiments he will do on it; but he knows there is much impredictability about what any given one of his measurements will disclose. Fifth, there was a « rule of the game » that required of every participator that his choice of yes or no should be compatible with *some* word. Similarly, there is a consistency about the observations made in physics. One person must be able to tell another in plain language what he finds and the second person must be able to verify the observation.

Go on!

That is difficult! Interesting though our comparison is between the world of physics and the world of the game, there is an important point of difference. The game has few participants and terminates after a few steps. In contrast, the making of observations is a continuing process. Moreover, it is extraordinarily difficult to state sharply and clearly where the community of observer-participators begins and where it ends.

This comparison between the world of quantum observations and the game of twenty questions misses much, but it makes the vital central point. In the real world of quantum physics, *no elementary phenomenon is a phenomenon until it is an observed phenomenon.* In the surprise version of the game no word is a word until that word is promoted to reality by the choice of questions asked and answers given. « Cloud » sitting there waiting to be found as we entered the room? Pure delusion! Momentum, $p_x = 1.4 \cdot 10^{-19}$ gcm/s, or position, $x = 0.31 \cdot 10^{-8}$ cm, of the electron waiting to be found as we start to probe the atom? Pure fantasy! MANN may be going too far when he suggests [29] that « ... we are actually bringing about what seems to be happening to us ». However, it is undeniable that each of us, as observer, is also *one* of the participators in bringing « reality » into being.

. . . .

To say « no elementary phenomenon is a phenomenon until it is an observed phenomenon » is to make no small change in our traditional view that something has « already happened » before we observe it. The word « cloud », we mistakenly thought, already existed in the room before we « uncovered » it. The photons of the primordial cosmic fireball radiation that enter our telescope today, we customarily assume, already had an existence in the very earliest days of the Universe, long before life evolved. However, not until we catch a particular one of those photons in a particular state with particular parameters, not until the elementary phenomenon is an observed phenomenon, do we have the right even to call it a phenomenon. This is the sense, the limited sense, but the inescapable sense, in which we, here, now, have a part in bringing about that which « had already happened » at a time when no observers existed.

But what about the unbelievably more numerous relict photons that escape our telescope? Surely you do not deny them « reality »?

Of course not; but their « reality » is of a paler and more theoretic hue. The vision of the Universe that is so vivid in our minds is framed by a few iron posts of true observation—themselves also resting on theory for their meaning—but most of the walls and towers in the vision are of papier-maché, plastered in between those posts by an immense labor of imagination and theory. In this labor, « ... we can never neatly separate what we see from what we know ... what we call seeing is invariably coloured and shaped by our knowledge (or belief) of what we see » [61]. « Without some initial system, without a first guess to which we can stick unless it is disproved, we could ... make no 'sense' of the milliards of ambiguous stimuli that reach us from our environment. In order to learn, we must make mistakes ... the simplicity hypothesis cannot be learned. It is ... the only condition under which we could learn at all » [62]. « ... our mind will still react to the challenge of this conundrum [of what we 'see'] by throwing out a random answer, making ready to test it in terms of consistent possible worlds. It is these answers that will transform the ambiguous stimulus pattern into the image of something 'out there' » [63].

What keeps these images of something « out there » from degenerating into separate and private universes: one observer, one universe; another observer, another universe?

That is prevented by the very solidity of those iron posts, the elementary acts of observership-participancy. That is the importance of Bohr's point that no observation is an observation unless we can communicate the results of that observation to others in plain language [49].

. . . .

The only thing harder to understand than a law of statistical origin would be a law that is not of statistical origin, for then there would be no way for it—or its progenitor principles—to come into being. On the other hand, when we view each of the laws of physics—and no laws are more magnificent in scope or better tested—as at bottom statistical in character, then we are at last able to forego the idea of a law that endures from everlasting to everlasting.

Individual events. Events beyond law. Events so numerous and so uncoordinated that, flaunting their freedom from formula, they yet fabricate firm form.

« Fabricate form »? Do you suggest that even the 4-dimensional space-time manifold is only a fabrication, only a theory—irreplaceable convenience though that theory is?

Yes! Compare space-time with cloth. Each it is useful under everyday circumstances to call a manifold. Yet each is exactly then most obviously not a manifold where it comes to an end, whether in the selvedge made by the loom, or in the geodesic terminations made by one of the « gates of time »—big bang or big crunch [31, 32] or black hole [33]. Nowhere more clearly than in the ending of space-time are we warned that time is not an ultimate category in the description of Nature [34].

Aren't you being extreme? I see the lesson of the game of twenty questions. I begin to believe with you that no elementary phenomenon is a phenomenon until it is an observed phenomenon. I accept that events of observer-participancy, as you call them, occupy a special place in the scheme of things. I agree that that word « cloud » was brought into being entirely through such elementary events. But that such events, however numerous, should be the *sole* blocks for building the laws of physics—and space and time themselves—seems to me preposterous. You surely have been involved enough in times past with nuts-and-bolts physics to know the difference between science and poetry; yet if I appreciate the drift of what you say, you might as well be quoting SHAKESPEARE [35],

> ... These our actors,
> As I foretold you, were all spirits and
> Are melted into air, into thin air:
> And, like the baseless fabric of this vision,
> The cloud-capp'd towers, the gorgeous palaces,
> The solemn temples, the great globe itself,
> Yea, all which it inherit, shall dissolve
> And, like this insubstantial pageant faded,
> Leave not a rack behind. We are such stuff
> As dreams are made on ...

I can't believe any such dreamlike vision of the physical world. As Samuel JOHNSON used to say, I have only to kick a stone to find it real enough.

Why do you say « preposterous »? Perhaps SHAKESPEARE understood this universe of ours better than we do ourselves! You have known for years that the atom is more than 99.99 percent emptiness. If matter turns out in the end to be altogether ephemeral, what difference can that make in the pain you feel when you kick the rock? And how can matter—and space-time—be anything but mutable, coming into being at one gate of time and fading out of existence at the other? No physics before the big bang, or after the big crunch? No! The lesson of Einstein's standard closed-space cosmology is different and stronger. It denies all meaning to such terms as « before the big bang » and « after the big crunch ».

Particles or fields or mathematics won't do for ultimate building blocks. They can't come into being or fade out of existence [30].

Yes, I appreciate the reasons given [36] against believing in any « magic particle » or any « magic field » or [37] any « magic mathematics » as the foundation of physics; but isn't it even more difficult to think of acts of observer-participancy as the magic ingredient?

Difficult, yes; inconceivable, no.

Go on!

No, we have to stop here. It is beyond the power of today to fit together the pieces of the puzzle.

Don't stop! You've carried me halfway into an exciting mystery story. You can't leave me without the traditional half-way-point review of the important clues and first try at a working hypothesis.

Review? A proper review would be impossibly ambitious. And how can one advance a working hypothesis that will not be wrong tomorrow and ridiculous the day after?

I appeal to you to go on. You have told me more than once that science advances only by making all possible mistakes; that the main thing is to make the mistakes as fast as possible—and recognize them. You like to quote the motto of that engine inventor, John Kris: « Start her up and see why she don't run ». You point to Einstein's definition of a scientist, « An unscrupulous opportunist ». If you believe all this, and are a true colleague of mine, you must go on.

You leave no escape!

Good!

Then let us agree to go on; but let us replace the comprehensive review of clues that you wanted by something more modest. How would it do, for example, to survey some of the lessons we have learned from the study of time, and how those lessons bear on « observer-participancy »?

I accept, and with many thanks. But first tell me the central point as you see it.

The absolute central point would seem to be this: The Universe had to have a way to come into being out of nothingness, with no prior laws, no Swiss watchworks, no nucleus of crystallization to help it—as on a more modest level, we believe, life came into being out of lifeless matter with no prior life to guide the process [5, 6, 38].

When we say « out of nothingness » we do not mean out of the vacuum of physics. The vacuum of physics is loaded with geometrical structure and vacuum fluctuations and virtual pairs of particles. The Universe is already in existence when we have such a vacuum. No, when we speak of nothingness we mean nothingness: neither structure, nor law, nor plan.

A conception more clearly impossible I never heard!

Preposterous we have to agree is the idea that everything is produced out of nothing—as preposterous, but perhaps also as inescapable, as the view that life had its origin in lifeless matter.

But how?

« Omnibus ex nihil ducendis sufficit unum », LEIBNIZ told us [39]; for producing everything out of nothing one principle is enough. Of all principles that might meet this requirement of Leibniz nothing stands out more strikingly in this era of the quantum than the necessity to draw a line between the observer-participator and the system under view. Without that demarcation it would make no sense to do quantum mechanics, no sense to speak of quantum theory of measurement, no sense to say that « No elementary phenomenon is a phenomenon until it is an observed phenomenon ». The necessity for that line of separation is the most mysterious feature of the quantum. We take that demarcation as being, if not the central principle, the clue to the central principle in constructing out of nothing everything.

Let me ask if your reasoning couldn't be turned around. You talk of the observer-participancy of quantum theory as the mechanism for the Universe to come into being. If that is a proper way of speaking, would the converse not also hold: The strange necessity of the quantum as we see it everywhere in the scheme of physics comes from the requirement that—via observer-participancy—the Universe should have a way to come into being?

Your point is exciting indeed. If true—and it is attractive—it should provide someday a means to *derive* quantum mechanics from the requirement that the Universe must have a way to come into being [40].

I know that in that empty courtyard many a game cannot be a game until a line has been drawn—it does not matter where—to separate one side from the other. I know that no Gaussian flux integral can be a flux integral until the 2-surface over which it runs—bumpy and rippled though we make it and deform it as we will—has been extended to closure. But how much arbitrariness is there in this more ethereal kind of demarcation, the line between « system » and « observing device »?

Much arbitrariness! BOHR stresses [42] that the stick we hold can itself be an object of investigation, as when we run our fingers over its surface. The same stick, when grasped firmly and used to explore something else, becomes an extension of the observer or—when we depersonalize—a part of the measuring equipment. As we withdraw the stick from the one role, and recast it in the other role, we transpose the line of demarcation from one end of it to the other. The distinction between the probed and the probe, so evident at this scale of the everyday, is the without-which-nothing of every elementary phenomenon, of every « closed » quantum process.

Do we possess today any mathematical or legalistic formula for what the line is or where it is to be drawn?

No.

Then what is important about this demarcation?

Existence, yes; position, no. It is the mark of an observation to leave an « indelible » record, according to BELINFANTE [43]. WIGNER argues that an observation is only then an observation when it becomes part of « the consciousness of the observer » [44] and points to « the impressions which the observer receives as the basic entities between which quantum mechanics postulates correlations » [45]. For BOHR the central point is not « consciousness », not even an « observer », but an experimental device—grain of silver bromide, Geiger counter, retina of the eye—capable of an « irreversible act of amplification » [47]. This act brings the measuring process to a « close » [48]. Only then, he emphasized, is one person able « to describe the result of the measurement to another in plain language » [49]. He adds that « all departures from common language and ordinary logic are entirely avoided by reserving the word 'phenomenon' solely for reference to unambiguously communicable information » [50].

I would have felt very uncomfortable if BOHR had used the term « consciousness » in defining the elemental act of observation. I would not have known what he meant. However, I am beginning to understand and accept the terms he actually adopts, « brought to a close by an irreversible act of amplification » and « communicable in plain language ». What *was* his position on consciousness?

We have asked Jørgen KALCKAR, who collaborated with BOHR in his last months, and he has kindly replied [51], « During work on the preparation of some lecture, to define the phenomenon of consciousness, BOHR used a phrase somewhat like this: a behaviour so complex that an adequate account would require references to the organism's 'self-awareness'. I objected jokingly that with this definition he would soon have to ascribe a consciousness to the highly developed electronic computers. This did not worry BOHR. 'I am absolutely prepared', said he, 'to talk of the spiritual life of an electronic computer; to state that it is reflecting or that it is in a bad mood .... The question whether the machine *really* feels or ponders, or whether it merely looks as though it did, is of course absolutely meaningless'. »

Other outstanding thinkers have argued otherwise. For them « consciousness » makes an unclimbable difference of principle between even the most powerful imaginable computer and the brain [52].

Do you agree with that argument?

How can we possibly accept such a difference of principle?

Do we not believe that brain function itself will someday be explained entirely in terms of physical chemistry and electrochemical potentials? What escape is there from the reasoning of von Neumann [53] and Bohr and many active present-day investigators? When one of the three discoverers of the mechanism of superconductivity today gives us, chapter by chapter and verse by verse, an entirely cellular account of the mechanism of memory [54-56], who can dismiss it?

When a distinguished computer expert and student of the structure of society details, one by one, the distinctions proposed in times past between « consciousness » and the computer, and painstakingly analyzes each down to nothingness [57], what case can anyone possibly maintain for *any* distinction of principle between the computer and the brain?

I am happy not to have to delve today into the term « consciousness ». I find it hard enough to know what to make of « irreversible act of amplification ». Never have I heard of an act of amplification that was not characterized by an amplification factor, or an equivalent quantity; and never an amplification factor that was not a finite number.

Between infinity and a finite number there may be a difference of principle; but between one finite number and another there is only a difference of degree. How big does the grain of silver bromide have to be, or the avalanche of electrons in the Geiger counter, before we count the measuring process as brought to a close by an irreversible act of amplification?

According as I specify one or another number as the critical level of amplification, don't I make all the difference between rating or not rating a given process as an « elementary phenomenon »?

According as the closed Gaussian surface encloses a given elementary charge or not, we find an unmistakable difference in the surface integral of the electric flux. Nevertheless, we know enough about the relevant invariance principle never to question the correctness of always identifying flux with enclosed charge. About « elementary quantum phenomenon » we have not today learned, but have a deep obligation someday to learn, enough to display a similar covariance with respect to where we draw the line. That is what « complementarity » is all about.

Even if neither you nor I know how to define that line, I like the idea that the « game » in the empty courtyard is only then possible when *a* line is drawn. May I question you now about the game itself? How would you describe it if forced to commit yourself?

* * * * *

From "nothingness ruled out as meaningless,"[106] to the line of distinction that rules it out; from this dividing line to "phenomenon"; from one phenomenon to many; from the statistics of many to regularity and structure: these considerations lead us at the end to ask if the universe is not best conceived as a self-excited circuit[107] (Fig. 22.13): Beginning with the big bang, the universe expands and cools. After eons of dynamic development it gives rise to observership. Acts of observer-participancy — via the mechanism of the delayed-choice experiment — in turn give tangible "reality" to the universe not only now but back to the beginning. To speak of the universe as a self-excited circuit is to imply once more a participatory universe.

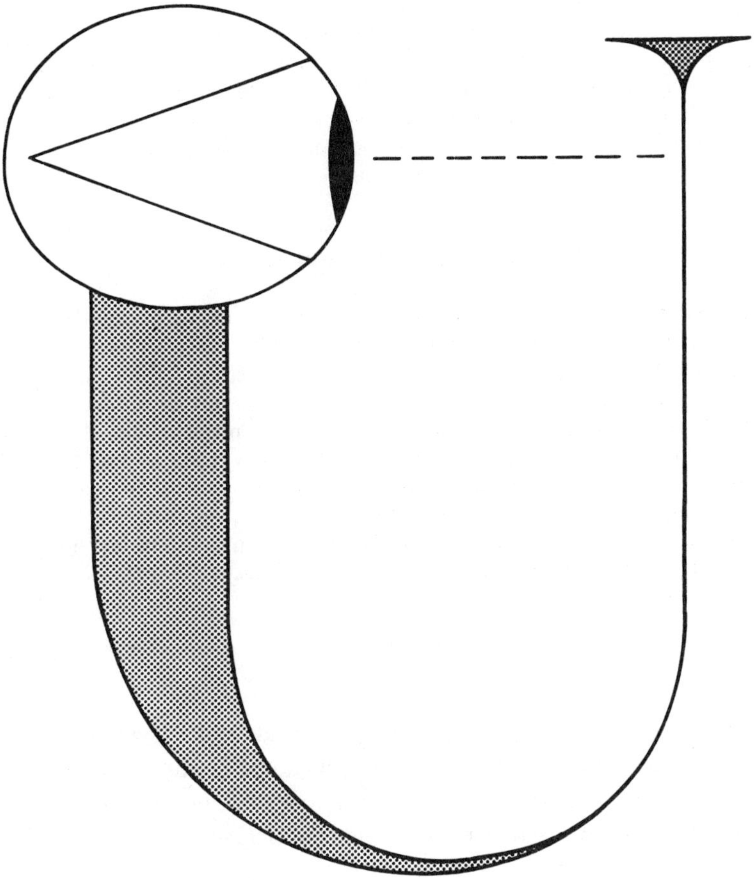

Fig. 22.13 The universe viewed as a self-excited circuit. Starting small (thin U at *upper right*), it grows (loop of U) and in time gives rise (*upper left*) to obsever-participancy — which in turn imparts "tangible reality" (cf. the delayed-choice experiment of Fig. 22.9) to even the earliest days of the universe.

[106, 107] See the original publication for these rather long references.

If the views that we are exploring here are correct, one principle, observer-participancy, suffices to build everything. The picture of the participatory universe will flounder, and have to be rejected, if it cannot account for the building of law; and space-time as part of law; and out of law substance. It has no other than a higgledy-piggledy way to build law: out of the statistics of billions upon billions of acts of observer-participancy each of which by itself partakes of utter randomness.

Two Tests

No test of these views looks more like being someday doable, nor more interesting and more instructive, than a *derivation* of the structure of quantum theory from the requirement that everything have a way to come into being[108] — as the word "cloud" was brought into being in the surprise version of the game of twenty questions. No prediction lends itself to a more critical test than this, that every law of physics, pushed to the extreme, will be found to be statistical and approximate, not mathematically perfect and precise.

The Challenge of "Law without Law"

We can ask ourselves if it is not absolutely preposterous to put into a formula anything at first sight so vague as law without law and substance without substance. How can we hope to move forward with no solid ground at all under our feet? Then we remember that Einstein had to perform the same miracle. He had to reexpress all of physics in a new language. His curved space seemed to take all definite structure away from anything we can call solidity. In the end physics, after being moved bodily over onto the new underpinnings, shows itself as clear and useful as ever. We have to demand no less here. We have to move the imposing structure of science over onto the foundation of elementary acts of observer-participancy.[109] No one who has lived through the revolutions made in our time by relativity and quantum mechanics — not least through the work of Einstein himself — can doubt the power of theoretical physics to grapple with this still greater challenge.

* * * * *

Recent decades have taught us that physics is a magic window. It shows us the illusion that lies behind reality—and the reality that lies behind illusion. Its scope is immensely greater than we once realized. We are no longer satisfied with insights only into particles, or fields of force, or geometry, or even space and time. Today we demand of physics some understanding of existence itself.

[108, 109] See the original publication for these rather long references.

REFERENCES

[1] C. LYELL: *Principles of Geology, being an attempt to explain the former changes of the earth's surfaces, by reference to cause now in operation,* Vol. II (London, 1830-1833).

[2] C. DARWIN: as quoted in H. WARD: *Darwin: The Man and His Warfare* (Indianapolis, Ind., 1927), p. 297.

[3] J. A. WHEELER: *From relativity to mutability,* in ref. [4], pp. 202-247.

[4] J. MEHRA, editor: *The Physicists' Conception of Nature* (Dordrecht, 1973).

[5] M. EIGEN: *The origin of biological information,* in ref. [4], pp. 594-632.

[6] M. EIGEN and R. WINKLER: *Das Spiel: Naturgesetze steuern den Zufall* (München, 1975).

[7] P. S. LAPLACE: *Essai philosophique sur les probabilités* (Paris, 1814), 2nd edition, pp. 3-4.

[8] N. BOHR: *Das Quantenpostulat und die neuere Entwicklung der Atomistik, Naturwiss.*, **16**, 245-257 (1928).

[9] W. HEISENBERG: *Über den auschaulichen Inhalt der quantentheoretischen Kinematik und Mechanik, Zeits. Phys.*, **43**, 172-198 (1927).

[10] J. BURCKHARDT: *Die Kultur der Renaissance in Italien* (Leipzig, 1860); authorized English translation from the 15th edition, by S. G. C. MIDDLEMORE: *The Civilization of the Renaissance in Italy* (New York, N. Y., 1929).

[11] P. GAY: *Style in History* (New York, N. Y., 1974).

[12] H. EVERETT III: *The theory of the universal wave function,* doctoral dissertation, Princeton University (1957); published in abbreviated form in ref. [13]; published in full for the first time in ref. [14].

[13] H. EVERETT III: *« Relative state » formulation of quantum mechanics, Rev. Mod. Phys.*, **29**, 454-462 (1957); reprinted in ref. [14].

[14] B. S. DEWITT and N. GRAHAM, editors: *The Many-Worlds Interpretation of Quantum Mechanics* (Princeton, N. J., 1973).

[15] J. VON NEUMANN: *Mathematischen Grundlagen der Quantenmechanik* (Berlin, 1932); translated as *Mathematical Foundations of Quantum Mechanics* (Princeton, N. J., 1955).

[16] J. A. WHEELER: *Assessment of Everett's « relative state » formulation of quantum mechanics, Rev. Mod. Phys.*, **29**, 463-465 (1974); reprinted in ref. [14].

[17] E. P. WIGNER: *Epistemological perspective on quantum theory,* in *Contemporary Research in the Foundations and Philosophy of Quantum Theory,* edited by C. A. HOOKER (Dordrecht, 1973), pp. 369-385.

[18] E. P. WIGNER: *Remarks on the mind-body question,* in *The Scientist Speculates,* edited by I. J. GOOD (London, 1962), pp. 284-302.

[19] E. P. WIGNER: *Are we machines?, Proc. Amer. Phil. Soc.*, **113**, 95-101 (1969).

[20] E. P. WIGNER: *Physics and the explanation of life, Found. Phys.*, **1**, 35-45 (1970).

[21] C. F. VON WEIZSÄCKER: *Classical and quantum decriptions,* in ref. [4], pp. 635-667.

[22] J. A. WHEELER: *Include the observer in the wave function?,* in *Quantum Mechanics, a Half Century Later,* edited by J. LEITE LOPES and M. PATY (Dordrecht, 1977).

[23] N. BOHR: *Atomic Theory and the Description of Nature* (Cambridge, 1934).

[24] N. BOHR: *Atomic Physics and Human Knowledge* (New York, N. Y., 1958).

[25] N. BOHR: *Essays 1958-1962 on Atomic Physics and Human Knovledge* (New York, N. Y., 1963).

[26] A. EINSTEIN: *Autobiological Notes,* in ref. [27], quotation from p. 81.

[27] P. A. Schilpp, editor: *Albert Einstein: Philosopher-Scientist* (Evanston, Ill., 1949), and subsequent paperback editions elsewhere.

[28] N. Bohr: *Can quantum-mechanical description of physical reality be considered complete?*, *Phys. Rev.*, **48**, 696-702 (1935); the quotation comes from p. 697.

[29] T. Mann: *Freud, Goethe, Wagner* (New York, N. Y., 1937), p. 20; translated by H. T. Lowe-Porter from *Freud und die Zukunft* (Vienna, 1936); the cited words were included in the lecture given at the 80th birthday celebration for Sigmund Freud, 8 May 1936.

[30] C. S. Peirce: *The Philosophy of Peirce: Selected Writings*, edited by J. Buchler (London, 1940); paperback reprint, *Philosophical Writings of Peirce* (New York, N. Y., 1955); « fallibilist », p. 358.

[31] B. K. Harrison, M. Wakano and J. A. Wheeler: *Matter-energy at high density; end point of thermonuclear evolution*, in *La structure et l'évolution de l'univers, Onzième conseil de physique Solvay* (Bruxelles, 1958), pp. 124-146; terminology « crushing points », pp. 134-136.

[32] J. R. Gott III, J. E. Gunn, D. N. Schramm and B. M. Tinsley: *Will the Universe expand forever*, *Sci. Amer.*, **234**, 62-79 (March 1976); terminology « big crunch », p. 69.

[33] J. A. Wheeler: *Our universe: the known and the unknown*, address before the American Association for the Advancement of Science, New York, December 29, 1967; *Amer. Scholar*, **37**, 248-274 (1968); terminology « black hole », pp. 258-262.

[34] J. A. Wheeler: *Superspace and the nature of quantum geometrodynamics*, in *Battelle Rencontres: 1967 Lectures in Mathematics and Physics*, edited by C. DeWitt and J. A. Wheeler (New York, N. Y., 1968), pp. 242-307; « no before, no after », p. 253.

[35] W. Shakespeare: *The Tempest*, London, about 1610; Prospero in Act IV, Scene I, lines 148-158.

[36] J. A. Wheeler: ref. [3]; no magic particle, no magic field, p. 235.

[37] C. M. Patton and J. A. Wheeler: *Is physics legislated by cosmogony?*, in *Quantum Gravity*, edited by C. J. Isham, R. Penrose and D. W. Sciama (Oxford, 1975); no magic mathematics, pp. 589-591. © Oxford University Press 1975, by permission of Oxford University Press. Appreciation also expressedl to Charles Patton for permission to quote cited passages.

[38] A. I. Oparin, editor: *Evolutionary Biochemistry, Proceedings of the V International Congress on Biochemistry* (Moscow, 1961; London, 1963), pp. 12-51.

[39] G. W. Leibniz: source of quotation not traced.

[40] J. A. Wheeler: *Genesis and Observership*, ref. [41], pp. 3-33; see p. 29.

[41] R. E. Butts and K. J. Hintikka: *Foundational Problems in the Special Sciences* (Dordrecht, 1977).

[42] N. Bohr: ref. [23]; Bohr's stick, p. 99.

[43] F. J. Belinfante: *Measurements and Time Reversal in Objective Quantum Theory* (Oxford, 1975); terminology « indelible », p. 39.

[44] E. P. Wigner: *Are we machines?*, *Proc. Amer. Phil. Soc.*, **113**, 95-101 (1969); quotation from p. 97.

[45] E. P. Wigner: *The philosophical problem*, in ref. [46], pp. 1-3; quotation from p. 3.

[46] B. d'Espagnat, editor: *Foundations of Quantum Mechanics* (New York, N. Y., 1971).

[47] N. Bohr: ref. [24]; irreversible amplification, p. 88.

[48] N. Bohr: ref. [24]; closed by irreversible amplification, p. 73.

[49] N. Bohr: ref. [25]; plain language, p. 3.

[50] N. Bohr: ref. [25]; unambiguously communicable, pp. 5, 6.

[51] J. KALCKAR: letter to J. A. WHEELER dated June 10, 1977. Appreciation is expressed here to Dr. J. KALCKAR both for the letter and for subsequent permission to quote from it. He adds that he cannot guarantee to have correct wording; and that Bohr's reply, as so often, was a joke with a definite « point »—hence to be taken seriously, but not *quite* seriously.

[52] K. R. POPPER and J. C. ECCLES: *The Self and Its Brain* (Berlin, New York, N. Y., and London, 1977); see especially pp. 207-208, 438-440 and 515.

[53] J. VON NEUMANN: *The Computer and the Brain* (New Haven, Conn., 1958); see especially pp. 60-61.

[54] L. N. COOPER: *A possible organization of animal memory and learning*, in *Nobel Symposium on the Collective Properties of Physical Systems*, edited by B. LUNDQVIST and S. LUNDQVIST (New York, N. Y., 1973), pp. 252-264.

[55] M. M. NASS and L. N. COOPER: *A theory for the development of feature-detecting cells in visual cortex*, *Biol. Cyber.*, **19**, 1-18 (1975).

[56] L. N. COOPER: *A theory for the acquisition of animal memory*, in *Lepton and Hadron Structure (1974 International School of Subnuclear Physics, Erice, Trapani, Sicily: July 14-31, 1974)*, edited by A. ZICHICHI (New York, N. Y., 1975), pp. 808-839.

[57] G. E. PUGH: *On the Origin of Human Values* (New York, N. Y., 1976), chapter « Human values, free will, and the conscious mind »; preprinted in *Zygon*, **11**, 2-24 (1976).

[58] INTERNATIONAL UNION OF PURE AND APPLIED PHYSICS: *Report of the Commission on Symbols, Units and Nomenclature* (Amsterdam, 1948).

[59] J. A. WHEELER: *The « past » and the « delayed-choice » double-slit experiment*, in ref. [60].

[60] A. R. MARLOW, editor: *Mathematical Foundations of Quantum Theory* (New York, N. Y., 1978).

[61] E. H. GOMBRICH: *Art and Illusion: A Study in the Psychology of Pictorial Representation* (Princeton, N. J., 1961), 2nd edition, revised, p. 394. Appreciation is expressed to Prof. E. H. GOMBRICH and Princeton University Press for permission to quote cited passages in the text.

[62] E. H. GOMBRICH: ref. [61], p. 273.

[63] E. H. GOMBRICH: ref. [61], p. 329.

[64] A. EINSTEIN, B. PODOLSKY and N. ROSEN: *Can quantum-mechanical description of physical reality be considered complete?*, *Phys. Rev.*, **47**, 777-780 (1935).

[65] R. M. F. HOUTAPPEL, H. VAN DAM and E. P. WIGNER: *The conceptual basis and use of the geometric invariance principles*, *Rev. Mod. Phys.*, **37**, 595-632 (1965); see especially §§ 4.1-4.5 on pp. 610-616.

[66] C. N. YANG and R. L. MILLS: *Conservation of isotopic spin and isotopic gauge invariance*, *Phys. Rev.*, **96**, 191-195 (1954).

[67] S. HOJMAN, K. KUCHAŘ and C. TEITELBOIM: *New approach to general relativity*, *Nature Phys. Sci.*, **245**, 97-98 (1973).

[68] S. A. HOJMAN, K. KUCHAŘ and C. TEITELBOIM: *Geometrodynamics regained*, *Ann. of Phys.*, **76**, 88-135 (1976).

[69] C. TEITELBOIM: *Surface deformations, space-time structure and gauge invariance*, in *Relativity, Fields, Strings and Gravity: Proceedings of the Second Latin American Symposium on Relativity and Gravitation SILARG II* held in Caracas, December 1975, Universidad Simon Bolivar, edited by C. ARAGONE (Caracas, 1976).

[70] J. E. NELSON and C. TEITELBOIM: *Hamiltonian for the Einstein-Dirac field*, *Phys. Lett.*, **69** B, 81-84 (1977).

[71] J. A. WHEELER: ref. [3]; machinery hidden, pp. 236-240.

[72] J. A. WHEELER: *Beyond the end of time*, Marchon lecture, University of Newcastle Upon Tyne, May 18, 1971, and Nuffield lecture, Cambridge University,

II

Interpretations of the Act of Measurement

II.1 THE THEORY OF OBSERVATION IN QUANTUM MECHANICS

FRITZ LONDON AND EDMOND BAUER

PREFACE OF PAUL LANGEVIN

Quantum physics has brought an essential advance to science, the finding that in every experiment or measurement there inescapably enters the duality between subject and object, the action and reaction of observer and system observed, the observer and the measuring apparatus being viewable as one entity.

The classical view, disregarding the necessarily limited character of our knowledge and the retroactive effect of the measurement on the system observed, always postulated the possibility of an infinitely precise knowledge of the simultaneous values of all the parameters used for the description of the system. Heisenberg, in giving concrete significance to his principle of indeterminism, has shown how the

Originally published as *La théorie de l'observation en mécanique quantique*, No. 775 of *Actualités scientifiques et industrielles: Exposés de physique générale*, publiés sous la direction de Paul Langevin, Hermann, Paris (1939). English translations—including a new paragraph by Professor Fritz London—done independently by A. Shimony, and by J. A. Wheeler and W. H. Zurek, and by J. McGrath and S. McLean McGrath; reconciled in 1982.

very existence of quanta excludes the possibility of knowing precisely at the same time all the quantities which might be the object of our measurement.

The form in which quantum mechanics is presented today provides an admirable translation of this new situation. The wave function it uses to describe the object no longer depends solely on the object, as was the case in the classical representation, but, above all, states what the observer knows and what, in consequence, are his possibilities for predictions about the evolution of the object. For a given object, this function, consequently, is modified in accordance with the information possessed by the observer. The introduction of the wave function at the very foundation of our representation common-sensibly recognizes what is almost ignored by classical physics, that our possiblities for prediction depend, above all, on our information. It also expresses quite exactly the fact that certain quantities, called noncommutable, cannot be known simultaneously with complete certainty. It characterizes the system by a certain number of observable quantities, different forms of "maximum knowledge" corresponding to different so-called "pure cases."

The present work, where the authors expand lectures given by one of them at the Sorbonne, demonstrates the precision and clarity with which the formalism of quantum theory expresses this representation by the wave function of the information acquired by the observer, and the manner in which each new measurement intervenes to modify this representation.

The act of observation is analyzed here in a particularly penetrating way. The essential character of the new physics emerges with complete clarity in the two stages of change of the wave function: by coupling of the system observed with the measuring device; and by the intervention of the observer, who becomes aware of the result of the measurement and thus determines the new wave function—following the observation—by using the new datum to reconstitute his information bank.

This treatise does a valuable service. It brings out the important finding of the new physics: how we express our knowledge of the external world.

Authors' Preface

The majority of introductions to quantum mechanics follow a rather dogmatic path from the moment that they reach the statistical interpretation of the theory. In general, they are content to show by more or less intuitive considerations how the actual measuring devices always introduce an element of indeterminism, as this interpretation demands. However, care is rarely taken to verify explicitly that the formalism of the theory, applied to that special process which constitutes the measurement, truly implies a transition of the system under study to a state of affairs less fully determined than before. A certain uneasiness arises. One does not see exactly with what right and up to what point one may, in spite of this loss of determinism, attribute to the system an appropriate state of its own. Physicists are to some extent sleepwalkers, who try to avoid such issues and are accustomed to concentrate on concrete problems. But it is exactly these questions of principle which nevertheless interest nonphysicists and all who wish to understand what modern physics says about the analysis of the act of observation itself.

Although these problems have already been the subject of deep discussions (see especially von Neumann, 1932), there does not yet exist a treatment both concise and simple. This gap we have tried to fill.

Paris, June 1939

INTRODUCTION

It is well known that theoretical physics has been transformed since the beginning of the century into an *essentially statistical* doctrine and that the *discovery of quanta* made this revolution inevitable.

The principal aim of this study will be the *statistical interpretation* of the formalism of quantum theory. Although these questions of interpretation were systematized about ten years ago (Heisenberg, 1927; von Neumann, 1927; Dirac, 1927), one still often meets rather fuzzy ideas about what it means that probabilities appear in modern physics.

According to some, this statistical character shows that our knowledge of laws at the atomic level is still *incomplete*: that there remain to be found some *hidden parameters*, determining those processes which, *provisionally*, we are content to describe in a statistical language. To believe them, one might hope some day to recast the theory in a deterministic mold.

Others would have it understood that the *action of the observer* is involved. They sometimes consider that this would be an action that is *causal*, but *incompletely known*, because one never knows the exact state of the observer. From this circumstance would arise the statistical spread of measurements, the exact results of which might be predictable if one could take better account of the intervention of the observer.

It has also been said that the law of *causality* may be *correct* but *inapplicable* because there is never any way to reproduce identical conditions.

The discussion of these questions is not at all a matter of speculation. It is a definite problem. To treat it one ought to apply quantum theory to—and thereby extract the central features of—the very process of measurement. One can convince oneself that statistical distributions, such as are given by quantum mechanics and verified by experiment, have such a structure that they cannot be reproduced by hidden parameters. It is not, as often claimed, a question of philosophical interpretation; quantum mechanics ought to be *testably false*, if atomic processes in fact were deterministic and only incompletely known. It would be necessary to change the theory fundamentally and give up some battle-tested results, if one wished to reconstitute it on a deterministic basis. Causality is *no longer applicable*, it is true; but the reason for this fact is not the impossibility, in the last analysis, of reproducing identically the conditions of an experiment. The heart of the matter is the difficulty of separating the object and the observer.

Modern physics often advances only by sacrificing some of our traditional philosophical convictions. The case of quantum mechanics is especially instructive. In all innocence one sought to construct a theory which would contain only relations between the "observable" quantities of Bohr's theory, such as the frequencies and the intensities of spectral lines. Heisenberg followed this route and in this way succeeded in obtaining a formalism which would resolve this problem. But, as often happens in theoretical physics, the formalism of the theory, once established, carried one further than one expected. It implied more relations than its founders had started with, relations between quantities altogether disconnected from the original spectroscopy, but themselves also observable (coordinates, momenta, etc.). One was led quite naturally in this way to try—after the initial shock—to interpret these relations which had been exposed automatically by the theory. In this way the discussion of this formalism taught us that the apparent philosophical point of departure of the theory, the idea of an observable world, totally independent of the observer, was a vacuous idea. Without intending to set up a theory of knowledge, although they were guided by a rather questionable philosophy, physicists were so to speak trapped in spite of themselves into discovering that the formalism of quantum mechanics already implies a well-defined theory of the relation between the object and the observer, a relation quite different from that implicit in naive realism, which had seemed, until then, one of the indispensable foundation stones of every natural science.

To discuss the process of measurement it is necessary to consider at least two systems, the observer and the object. It is therefore necessary to apply the quantum theory of the many-body system. This exists at present only in the nonrelativistic approximation. We are therefore forced to limit ourselves to this approximation, which still neglects all effects of the time delay in the propagation of forces.

It is not possible to give here a detailed introduction to quantum mechanics. We will limit ourselves (§1 and 2) to recalling briefly, and a bit dogmatically, the definitions and the laws that we will need. For a more detailed exposition of quantum theory see, for example, de Broglie, 1930; Bloch, 1930; Kemble, 1937; Dushman, 1938.

§1. Resumé of the Principles of Quantum Physics

In atomic physics the use of statistical concepts came far earlier than wave mechanics. The first step in this direction was probably made at the moment when one described spontaneous radioactive decay by the laws of probability. Of course, at the beginning we thought that this was a provisional approach, forced by our ignorance of what is going on inside the nucleus. But when Bohr, obviously guided by an analogy with statistical concepts, constructed his model of the atom with its spontaneous quantum jumps, and, above all, when Einstein gave his famous demonstration of Planck's radiation law on the basis of the idea of spontaneous and stimulated *transition probabilities* (Einstein's A and B coefficients), one already had the strong feeling that these probabilities ought to be something basic and primordial. In a world of discontinuous phenomena, the appearance of a statistical form for the elementary laws would seem almost inescapable. The theory of Bohr, although it does not yet furnish a mathematical scheme that is complete and coherent, has already allowed us to state questions of principle to which quantum theory must address itself. In a physics that deals with magnitudes, whose domains of variation are not necessarily continuous, one wants to know:

(1) *What are the possible values* of a physical quantity;
(2) *With what probabilities* are they realized in a given system and under given circumstances.

Quantum mechanics furnishes us with a precise scheme which allows a quantitative treatment of questions of this kind. We can summarize it in the following way:

"The state" of a system—given in classical mechanics at each instant t by the $2f$ values of the variables $q_1(t), q_2(t) \ldots q_f(t), p_1(t), p_2(t) \ldots p_f(t)$—is represented in quantum mechanics by a *complex function* of the f variables $q_1, q_2 \ldots q_f$ and of t, the "Schrödinger wave function"

$$\psi(q_1, q_2 \ldots q_f; t),$$

which is normalized in such a way that*

$$\int \psi\psi^* \, dq = 1. \tag{1}$$

The evolution of the system in time is governed in classical mechanics by a "Hamiltonian function" $H(q, p)$, characteristic of the system in question. This function of the coordinates $q_1, q_2 \ldots q_f$ and of the momenta $p_1, p_2 \ldots p_f$, which is nothing other than the energy, permits one to write the Hamiltonian equations of motion. This is essentially† the same function $H(q, p)$ which in quantum mechanics also gives the law of evolution of the ψ that represents the state of the system: one forms the operator‡ $H(q, -i\hbar\partial/\partial q)$ by replacing p_k in the Hamiltonian

* In what follows the integrals $\int dq$ are always taken over the entire configuration space $q_1, q_2 \ldots q_f$, and ψ^* is the complex conjugate of ψ.

† Completed by a term referring to "spin."

‡ Here the symbol $h/2\pi$ employed by London and Bauer has been replaced for convenience, and in accord with modern practice, by the Dirac notation for the quantum of angular momentum, $\hbar$. —Eds.

by the differential operator $-i\hbar\partial/\partial q_k$ and q_k by the operation of multiplication by q_k. Here a certain ambiguity enters in the ordering of factors, because the operations q_k and $\partial/\partial q_k$ are not commutable. It is enough for us to know that there are prescriptions which ordinarily suffice to determine uniquely the order of the operators, but we will not go into this in detail here. The operator $H(q, -i\hbar\partial/\partial q)$, once given, allows us to write the equation of evolution:[§]

$$H(q, -i\hbar\partial/\partial q)\psi = i\hbar\psi\partial/\partial t. \tag{2}$$

This equation, discovered by Schrödinger, has the important property of leaving invariant the integral $\int \psi\psi^* \, dq$, which is necessary for our normalization $\int \psi\psi^* \, dq = 1$ to be possible. Thus ψ, once normalized, always retains its normalization.

$\psi(q, t_0)$ represents a "state" of the system at an instant t_0. Here we take this term in a sense completely analogous to that which it has in classical mechanics, where one says that the data $q_1(t_0) \cdots q_f(t_0)$, $p_1(t_0) \cdots p_f(t_0)$ "represent a state." The knowledge of the representative of the state at a given instant is *necessary* and *sufficient* for an unambiguous calculation, with the aid of the dynamic law, of the representative of the state at every subsequent moment. We cannot forgo a part of these data without losing the ability to calculate the future. Neither can we add to them supplementary data without introducing useless tautologies or contradictions of the data already in hand.

The *stationary states* of Bohr correspond to special solutions of the Schrödinger equation, solutions exactly periodic in time and of the form:

$$\psi = \exp(-iEt/\hbar)u(q). \tag{3}$$

In consequence of (1) and (2) the "amplitude" $u(q)$ obeys a time-independent equation,

$$Hu = Eu, \tag{4}$$

and the condition of normalization,

$$\int uu^* \, dq = 1. \tag{5}$$

The pair of equations (4) and (5) ordinarily does not have a solution for every value of E. They present a "proper value" or "eigenvalue" or "characteristic value" problem. Only for a "spectrum" of special values $E_1, E_2, E_3 \ldots$, possibly also

[§] In accordance with present-day practice we omit the square brackets in which the authors enclose the left-hand side of this and subsequent similar equations; and we use on the right-hand side the opposite sign of i from that which they use whenever they write the time-evolution operator. —Eds.

containing continuous intervals,* can condition (5) be fulfilled. For other values of E the linear equation (4) of course also possesses solutions u containing an arbitrary factor; but they are not square integrable, ruling out the normalization (5).

The allowed values,

$$E_1, E_2, \ldots E_k, \ldots$$

are called the "eigenvalues" of the operator H. The corresponding solutions u are called "eigenfunctions" and designated by corresponding indices,

$$u_1, u_2, \ldots u_k, \ldots.$$

It was the fundamental idea of Schrödinger to identify the spectrum of eigenvalues, $E_1, E_2, \ldots,$ which have the dimensions of energy, with the allowed values of the energy in Bohr's theory, and the success of this ingenious idea is well known.

One knows also that the founders of wave mechanics were guided in the beginning by the conviction that it was necessary to *get rid of discontinuities*, or rather to derive them from an essentially *continuous* substructure, from a field theory, and thus overturn the basically statistical picture of Bohr and Einstein. But this program did not turn out to be realizable.

The statistical interpretation of quantum mechanics may be considered to be a particularly conservative attempt to maintain the picture worked out by Bohr and Einstein and to embody it in a coherent theoretical system.

Now that we know how to interpret the "monochromatic" solutions of the wave equation (2) such as

$$\psi = \exp(-iE_k t/\hbar)u_k(q)$$

as describing a state of energy E_k, we have to *discover the meaning of the most general solutions.* One can show that if one limits oneself to square integrable functions (1), *the most general solution* of equation (2) is written in the form:

$$\psi = \sum_k c_k \exp(-iE_k t/\hbar)u_k(q), \tag{6}$$

where the c_k are complex constants.

Born, replying to this question, formulated the foundations of the statistical interpretation of quantum theory. He postulated that the quantity $|c_k|^2$ gives the

* In the case of a continuous spectrum condition, (5) has to be imposed on a function u associated with a small *interval of E* of the continuum ("proper differential"). In what follows we will not take account of these sophistications, which do not concern questions of principle, and we will write all formulas as if there were nothing but discrete spectra.

probability of finding the value E_k of the energy when the system in the state (6) is subjected to a measurement of energy.

If we introduce the abbreviation* $\psi_k = c_k \exp(-iE_k t/\hbar)$, we can write

$$\psi = \sum_k \psi_k(t) u_k(q). \tag{6'}$$

The coefficients ψ_k depend only on the time, and, as $|\psi_k|^2 = |c_k|^2$, we can also interpret the quantity $|\psi_k(t)|^2$ as the probability of finding the value E_k for the energy in the state ψ.

The coefficients ψ_k may, moreover, be calculated very easily with the help of two basic properties of eigenfunctions:

(1) "*Orthogonality*":

$$\int u_k^* u_l \, dq = \delta_{kl} \tag{7}$$

where

$$\delta_{kl} \equiv \begin{Bmatrix} 1 \text{ if } k = l \\ 0 \text{ if } k \neq l \end{Bmatrix};$$

(2) "*Completeness*": for an arbitrary square integrable function $f(q)$ (that is, $\int f^* f \, dq =$ finite) one has the identity:

$$\sum_k \left| \int f u_k \, dq \right|^2 = \int f^* f \, dq. \tag{8}$$

Orthogonality immediately gives the form of the coefficients ψ_k of the expansion (6′). Multiplying (6′) by u_l^* and integrating, one obtains:

$$\int \psi(q) u_l^*(q) \, dq = \psi_l. \tag{9}$$

Completeness guarantees that the series thus defined converges to the function $\psi(q)$. Thus, the integral of the square of the absolute value of the difference,

$$\psi(q) - \sum \psi_l u_l(q),$$

goes to zero,

* Not a customary abbreviation and one possibly confusing to the reader, but well adapted to the purpose of the next few sections. —Eds.

$$\int \left|\psi(q)-\sum_l \psi_l u_l(q)\right|^2 dq = \int \psi\psi^* \, dq - \sum_l \psi_l \int \psi^* u_l \, dq - \sum_l \psi_l^* \int \psi u_l^* \, dq + \sum_l \psi_l \psi_l^*$$

$$= \int \psi\psi^* \, dq - \sum_l \left|\int \psi u_l^* \, dq\right|^2 = 0,$$

from which also follows the result

$$\sum_l \psi_l \psi_l^* = \int \psi\psi^* \, dq = 1.$$

In other words, the sum of the probabilities of finding the various values of the energy is unity. Thus if we ask what is the energy, we can be assured of always finding some value. Naturally, it is necessary that things be so if the definitions of Born are to make sense.

Thus we see how, for energy at least, quantum theory answers the two questions which show up (first paragraph of §1) in every theory of the discontinuous:

1. The *possible values of the energy* are the eigenvalues $E_1, E_2, E_3, \ldots$ of the operator $H(q, -i\hbar\partial/\partial q)$.

2. The *probability* of finding the value E_k of the energy in the state represented by ψ is given by

$$Prob_\psi(E_k) = \left|\int \psi u_k^* \, dq\right|^2, \tag{10}$$

where u_k is the eigenfunction associated with the value E_k of the energy.

In particular, if by chance this state is represented by an eigenfunction of the energy; that is, if

$$\psi = u_k,$$

then equation (10) gives the probabilities 1 for the eigenvalue E_k and 0 for all other eigenvalues of the energy.

In the original theory of Bohr one was occupied above all with the energy. However, in our present formalism energy does not play an exceptional part except for the time-evolution of the state as represented by its ψ function (equation 2). If ψ is given at a certain instant, we can also look for statistical predictions for an arbitrary physical quantity $F(q, p)$ at that moment.

The *generalization* of our definitions *for other quantities* $F(q, p)$ (as, for example, $xp_y - yp_x$, x, p_x, etc.) is fully specified as follows.

We form the operator $F(q, -i\hbar\partial/\partial q)$ and define, with the help of the equations of the eigenvalue problem,

$$Fv(q) = fv(q)$$

$$\int vv^* \, dq = 1, \tag{11}$$

the eigenvalues $f_1, f_2, \ldots f_\rho \ldots$ of the operator F and the corresponding eigenfunctions $v_1, v_2, \ldots v_\rho \ldots$. Mathematicians have shown that these eigenfunctions, too, under rather general conditions, form a *complete system* of orthogonal functions; that is, that we have the relations

$$\int v_\rho v_\sigma^* \, dq = \delta_{\rho\sigma} \tag{12}$$

and that we can expand the function ψ at a given moment t_0 in a convergent series of these functions v_ρ

$$\psi(q, t_0) = \sum_\rho \psi_\rho(t_0) v_\rho(q),$$

where

$$\psi_\rho(t_0) = \int \psi(q, t_0) v_\rho^*(q) \, dq.$$

The generalization of our previous definitions is immediate:

1. The *possible values of the physical quantity* F are given by the eigenvalues of the operator $F(q, -i\hbar\partial/\partial q)$.
2. The *probability* for finding the eigenvalue f_ρ of the quantity $F(q, p)$ for the state represented by ψ is given by

$$Prob_\psi(f_\rho) = \left| \int \psi(q, t_0) v_\rho^*(q) \, dq \right|^2 = |\psi_\rho|^2. \tag{13}$$

In particular, if by chance the state is represented by the eigenfunction v_ρ of F; that is, if

$$\psi = v_\rho,$$

one obtains the probability 1 for finding the value f_ρ of the quantity F and 0 for the probability for every other eigenvalue.

From these definitions immediately follows the *mean value of* F *in the state* ψ,

$$Mean_\psi(F) = \sum_\rho |\psi_\rho|^2 f_\rho. \tag{14}$$

This expression can be written in a more convenient form, which allows one to calculate immediately the mean value without having to go back to an explicit evaluation of the individual eigenvalues f_ρ or the expansion of ψ as a series in the eigenfunctions v_ρ. One easily checks that

$$Mean_\psi(F) = \int \psi^* F \psi \, dq \tag{14'}$$

because, as consequence of (11) and (12) this expression lets itself be written as

$$\int \sum_{\rho\sigma} \psi_\rho^* v_\rho^* F \psi_\sigma v_\sigma \, dq = \sum_{\rho\sigma} \psi_\rho^* \psi_\sigma \int v_\rho^* f_\sigma v_\sigma \, dq = \sum_{\rho\sigma} \psi_\rho^* \psi_\sigma f_\sigma \delta_{\rho\sigma}.$$

§2. Vector Notation

Our definitions are now complete. We have only to introduce a slightly more convenient notation, that of the language of vectors.

Vectors. We will say that the function $\psi(q)$, representing the instantaneous state of a system, is a "vector" in a space of an infinite number of dimensions, the function space of Hilbert. The integral of the product of the two functions ψ^* and ϕ taken over all coordinates $q_1, q_2 \ldots q_f$ will be called the "*scalar product of* ψ *and* ϕ,"

$$(\psi, \phi) = \int \psi^*(q)\phi(q) \, dq = \sum_\rho \psi_\rho^* \phi_\rho. \tag{1}$$

The quantity $(\psi, \psi) = \int \psi^* \psi \, dq = \sum_\rho |\psi_\rho|^2$ will be called the *square of the length of the vector.*

$(\psi, \phi) = 0$ means that the vectors ψ and ϕ are "orthogonal."

The eigenfunctions $v_1, v_2 \ldots v_\rho \ldots$ of an operator F satisfy the relation (§1, equation 12),

$$(v_\rho, v_\sigma) = \begin{Bmatrix} 1 \text{ if } \rho = \sigma \\ 0 \text{ if } \rho \neq \sigma \end{Bmatrix}.$$

They form therefore a system of orthogonal unit vectors that define an "*orthogonal coordinate* system" with the help of which one can represent any vector ψ whatsoever in the form,

$$\psi(q) = \sum_\rho \psi_\rho v_\rho(q). \tag{1)*}$$

The "components" ψ_ρ of the vector ψ are defined by the "projection"

* Both equations are numbered (1), presumably an oversight. —Eds.

$$\psi_\rho = (v_\rho, \psi)$$

of ψ in the direction of the unit vector v_ρ.

This decomposition into components is quite analogous to the resolution of a vector in ordinary space into projections. We can consider the set of ψ_ρ as equivalent to the function $\psi(q)$ itself; it is one particular decomposition of the vector ψ into orthogonal components. The coefficients

$$\psi_k = (u_k, \psi)$$

give an analogous decomposition of the same vector with respect to another system of orthogonal axes $u_1, u_2 \ldots u_k, \ldots$.

The representation of ψ by itself, that is, the function $\psi(q)$, may be regarded as a *special case* of representation in terms of orthogonal components—specifically, the orthogonal system composed of the eigenfunctions of the particular operator $F = q$. The eigenvalue problem for this operator has the form,

$$qv_\alpha(q) = q_\alpha v_\alpha(q),$$

or

$$(q - q_\alpha)v_\alpha(q) = 0.$$

The solutions are the "limiting" or symbolic functions of Dirac,

$$v_\alpha(q) = \delta(q - q_\alpha).$$

Such a function by definition vanishes for $q \neq q_\alpha$ but for $q = q_\alpha$ is so singular that

$$\int \delta(q - q_\alpha)\, dq = 1.$$

In terms of these special eigenfunctions one obtains for $\psi(q)$ the trivial expansion,

$$\psi(q) = \sum_\alpha \psi(q_\alpha)\delta(q - q_\alpha),$$

where the

$$\psi(q_\alpha) = \int \psi(q)\delta(q - q_\alpha)$$

are the coefficients in the expansion.

From our general definitions thus follows the particular result,

$$|\psi(q)|^2\, dq = \text{(probability to find } q \text{ in the interval from } q \text{ to } q + dq).$$

Tensors. In the present picture the typical operator $F(q, -i\hbar\partial/\partial q)$, representing a physical quantity, is a *tensor*; that is, a linear transformation of *vectors*. Applied to a vector ψ, it transforms it into another vector,

$$\psi' = F\psi.$$

It is *linear* because, for every combination of multiplications and differentiations, one always has the distributive relation,

$$F(\psi + \chi) = F\psi + F\chi, \tag{2}$$

and because, for every constant c, one has

$$Fc\psi = cF\psi.$$

As it represents a *real* (noncomplex) physical quantity, it has one more important property,

$$(\phi, F\psi) = (\psi, F\phi)^*. \tag{3}$$

Such an operator is termed "*Hermitian.*" Relation (3) is easily demonstrated by an integration by parts that takes account of the facts that every operation of differentiation contained in F brings in a factor i, and that F, considered as a function of q and $-i\hbar\partial/\partial q$, is a real function.

When we are using for ψ a representation in terms of the ψ_k—that is, when we refer the state vector to a system of coordinates $u_1, u_2 \ldots u_k \ldots$, we must also decompose the operator F into components referred to the same coordinates. In this way F evidently finds itself expressed as the linear transformation brought about by the matrix

$$F_{kl} = \int u_k^* F u_l \, dq = (u_k, Fu_l). \tag{4}$$

Thus if one applies this transformation to a vector ψ_l,

$$\sum_l F_{kl}\psi_l = (u_k, F \sum_l \psi_l u_l) = (u_k, F\psi),$$

one obtains the kth component of the function $F\psi$.

From (3) immediately follows the relation

$$F_{kl} = F_{lk}^* \tag{5}$$

of "*Hermitian matrices.*"

There is a system of coordinates in which the matrix representing the operator

F shows an especially simple form. These are the coordinates defined by the eigenfunctions $v_1 \cdots v_\rho \cdots$ of this operator itself. In terms of them one finds,

$$F_{\rho\sigma} = (v_\rho, Fv_\sigma) = f_\sigma(v_\rho, v_\sigma) = f_\sigma \delta_{\rho\sigma} = \begin{Bmatrix} f_\sigma \text{ for } \sigma = \rho \\ 0 \text{ for } \sigma \neq \rho \end{Bmatrix}.$$

In its "eigen"-basis the matrix F therefore takes a diagonal form and its diagonal elements are the eigenvalues of F,

$$F_{\rho\sigma} = \begin{Vmatrix} f_1 & 0 & 0 & 0 & \cdots \\ 0 & f_2 & 0 & 0 & \cdots \\ 0 & 0 & f_3 & 0 & \cdots \\ 0 & 0 & 0 & f_4 & \cdots \\ & \cdots & \cdots & \cdots & \\ & \cdots & \cdots & \cdots & \end{Vmatrix} \tag{6}$$

In the system of coordinates ψ_ρ used to expand ψ in a series of eigenfunctions $v_1 \cdots v_\rho \cdots$ of an arbitrary operator F, the Hamiltonian operator H takes the form $H_{\rho\sigma} = (v_\rho, Hv_\sigma)$ and the Schrödinger equation becomes

$$\sum_\rho H_{\rho\sigma}\psi_\sigma = i\hbar\dot{\psi}_\rho, \tag{7}$$

the discontinuous form in which Heisenberg, Born, and Jordan found the equations of quantum mechanics in the first place.

Invariants. We use capital letters H, F, etc. for tensors, and Greek letters ψ, ϕ, etc. for vectors. We can then suppress indices (or arguments) associated with any special decomposition into components and write (7) in an *invariant* form, independent of the system of coordinates, or rather encompassing all possible systems (see §1, equation 2),

$$H\psi = i\hbar\dot{\psi}.$$

Two distinct representations of the same vector ψ, for example $\psi_k = (u_k, \psi)$ and $\psi_\rho = (v_\rho, \psi)$, are related by a linear transformation,

$$\psi_k = \sum_\rho S_{k\rho}\psi_\rho \quad \text{with} \quad S_{k\rho} = (u_k, v_\rho). \tag{8}$$

Thus, as $\psi = \sum_\rho v_\rho(v_\rho, \psi)$, one has,

$$\psi_k = (u_k, \psi) = \sum_\rho (u_k, v_\rho)(v_\rho, \psi) = \sum_\rho S_{k\rho}\psi_\rho.$$

For the coefficients $S_{k\rho}$ one easily finds the relations,

$$\sum_{\rho} S_{k\rho} S_{l\rho}^{*} = \delta_{kl} \quad \text{or} \quad S\tilde{S}^{*} = 1, \tag{9}$$

where $\tilde{S}_{k\rho} = S_{\rho k}$.

These relations characterize the transformation (8) as a "*unitary*" *transformation.*

Two different representations of the same tensor F, for example $F_{kl} = (u_k, Fu_l)$ and $F_{\rho\sigma} = (u_\rho, Fu_\sigma)$, are related, as one easily verifies, by the relations,

$$\mathrm{F}_{kl} = \sum_{\rho\sigma} \mathrm{S}_{k\rho} \mathrm{F}_{\rho\sigma} \tilde{\mathrm{S}}_{\sigma l}^{*}.$$

Thus, as $u_k = \sum_{\rho} v_\rho (v_\rho u_k) = \sum_{\rho} S_{k\rho}^{*} v_\rho$, one has

$$F_{kl} = (u_k, Fu_l) = \int \sum_{\rho} S_{k\rho} v_\rho^{*} F \sum_{\sigma} S_{l\sigma}^{*} v_\sigma \, dq = \sum_{\rho\sigma} S_{k\rho} \int v_\rho^{*} F v_\sigma \, dq \, S_{l\sigma}^{*}$$

$$= \sum_{\rho\sigma} S_{k\rho} F_{\rho\sigma} \tilde{S}_{\sigma l}^{*}.$$

Physically significant numerical values naturally ought to be scalars *invariant* under these unitary transformations. The only scalar invariants that we will meet are the "*scalar product*" of two vectors ψ and ϕ,

$$(\psi, \phi) = \int \psi^{*}(q) \phi(q) \, dq = \sum_{k} \psi_k^{*} \phi_k = \sum_{\rho} \psi_\rho^{*} \phi_\rho,$$

and the "trace" of a tensor F_{kl},

$$\mathit{Trace}(F) = \sum_{k} F_{kk} = \sum_{\rho} F_{\rho\rho}.$$

Thus, for example, mean value of a physical quantity F in a state ψ is given in invariant form by the scalar product

$$\mathit{Mean}_{\psi}(F) = \sum_{k} F_{kk} = \sum_{\rho} F_{\rho\rho}.$$

The other results of the theory can also be expressed in invariant form. We will come back to this in §5.

The scalar product can also be considered as the trace of a special matrix $(\psi \times \phi)$ defined by

$$(\psi \times \phi)_{k\rho} = \psi_k^{*} \phi_\rho, \tag{10}$$

termed the "direct product" of the vectors ψ and ϕ.

§3. Statistics and Objectivity

Already in the classic memoir (Born, 1926) where he proposed foundations for the statistical interpretation of quantum mechanics, Born remarked that the probabilities which he introduced there must have a strange character, quite different from what one normally understands when one speaks of probability.

This feature he expressed in a form a bit paradoxical: "Although the movements of particles are not determined, except by probabilities, these probabilities themselves evolve according to a causal law." What he understands here by "causal law" is a connection between the "states" at different moments, such that a knowledge of the initial state at an arbitrary instant uniquely implies a knowledge of the state at every subsequent time. A "state," on the other hand, is a well-defined collection of data on the system in question at a given moment.

Naturally, there is no way of predicting *a priori* whether, in a given domain of science, there exist causal laws as so defined, nor what are the necessary and sufficient conditions for giving rise to such laws. If one does not end up with unique predictions, if one finds oneself forced to be satisfied with *probabilities*, that may be either because our knowledge of the "state" is not yet complete or because causality does not hold. But conversely, when one has succeeded in establishing causal laws, that is evidently a criterion for deciding that one has attained a complete knowledge of the object in question and thus, in some measure, a maximum description.

But the Schrödinger equation has all the features of a causal connection. If the ψ function is known at a given moment, it is determined at every subsequent time. It therefore seems difficult to believe that this function nevertheless contains a statistical collection. At first sight it seems impossible to avoid the following dilemma.

1. One might imagine that the ψ function has the character of the ordinary probability function such as one uses, for example, to describe Brownian motion. A function of this type contains certain statistical predictions that we can test. We then verify which of the possibilities foreseen in the theory is realized in fact in a given case. After this observation we are naturally entitled to use, for the subsequent predictions, the knowledge thus obtained and to replace our original probability function by a function of the same type, but better tuned. Evidently this is only possible by virtue of the enrichment of our knowledge, which is always partial. Of course we do not claim that the object itself has changed its state as a consequence of our observation. All that has changed is the discrepancy between our knowledge and the object. In this case the ψ function will therefore represent the state of our *partial knowledge* of the object and not the state of the object itself.

2. Imagine, on the contrary, that the ψ function has an "objective" character, as, for example the wave functions of optics. It then claims to represent, in an idealized and simplified form, something *complete*, a *maximum* picture of the state

of the *object*. But if this is the situation, it seems difficult to understand how this ψ function implies a statistic. If one checks experimentally predictions that can be made from it, and if one observes which of the possible outcomes is realized—an outcome predicted by the theory, but only with a certain probability—*by what right can we add this new knowledge* to the supposedly complete knowledge that we already had?

Heisenberg found the solution to this dilemma. He emphasized that it is the *process of measurement* itself which introduces the element of indeterminacy in the state of the object.

Thus the statistical feature would not show up except on the occasion of a measurement. If the ψ function gives us probabilities, it does so only *in anticipation of an eventual measurement*. Thus these are only, so to speak, "*potential*" *probabilities* which come into force only on the occasion of an actual measurement. They do not affect the precision with which the state of the system is currently known; thus it is already maximal when the ψ function is given.

Of course it may happen that there is an additional uncertainty in the state of the system—that is, in the ψ function itself. In this case it is a question of *probabilities* in the *ordinary* sense of the word. They arise from an incomplete knowledge of the state of the object. It is necessary to distinguish clearly between these probabilities and the "potential" probabilities furnished by ψ functions.

§4. Mixtures and Pure States

That an essential distinction is in question can be seen most clearly by an example.

I. Let us consider first the case where the system is represented by a wave function,

$$\psi = \sum_k \psi_k u_k(q),$$

where $u_1, u_2 \cdots u_k \cdots$ are, for example, eigenfunctions of the energy. We know that the quantity $|\psi_k|^2$ gives us the probability of finding the value E_k of the energy when an energy measurement is made on a system in a state ψ.

II. It has often been considered that this case corresponds straightforwardly to a *virtual ensemble* of identical systems in different states possessing respectively energies $E_1, E_2 \ldots E_k \ldots$, each one of them being contained in an ensemble with a relative weight $p_k = |\psi_k|^2$.

However, this latter case (II), which will interest us also, is basically different from the *pure state* (I) represented by a single function ψ. It is a *mixture* of several distinct pure states, each represented by *its wave function*,*

* In what follows the upper indices $(\overset{(1)}{\psi}, \overset{(2)}{\psi}, \ldots \overset{(n)}{\psi})$ always designate distinct pure states.

$$\overset{(1)}{\psi}(q) = u_1(q)$$

$$\overset{(2)}{\psi}(q) = u_2(q)$$

$$\cdots\cdots\cdots\cdots\cdots$$

$$\overset{(k)}{\psi}(q) = u_k(q)$$

$$\cdots\cdots\cdots\cdots\cdots$$

with the respective fractional abundances

$$p_1 = |\psi_1|^2, \qquad p_2 = |\psi_2|^2, \ldots\ p_k = |\psi_k|^2, \ldots$$

One easily verifies that case I and case II give two completely different statistical distributions. Of course they are identical for energy, because we have arranged that their statistics should be the same. But let us consider another physical quantity, F for example, not having the same eigenfunctions as energy. In case I the mean value of F takes the form (see §1),

$$Mean_{\mathrm{I}}(F) = (\psi, F\psi) = \sum_{k,l} \psi_k^* \psi_l F_{kl}, \tag{I}$$

while in case II each component $\overset{(k)}{\psi} = u_k$ gives a contribution of this kind:

$$\overset{(k)}{Mean}(F) = (\overset{(k)}{\psi}, F\overset{(k)}{\psi}) = (u_k, Fu_k) = F_{kk}.$$

When this component appears with the probability p_k in the mixture, one obtains altogether

$$Mean(F) = \sum_k p_k(\overset{(k)}{\psi}, F\overset{(k)}{\psi}) = \sum p_k F_{kk}. \tag{II}$$

If in particular we consider the case of a mixture where $p_k = |\psi_k|^2$, we find,

$$Mean_{\mathrm{II}}(F) = \sum_k |\psi_k|^2 F_{kk}. \tag{II'}$$

Thus, provided that F_{kl} is not by a chance a diagonal matrix (that is, F does not have the same eigenfunctions as H), the two cases I and II are completely different.

It is evident that in case II our knowledge of the system is much more restricted than in case I. If our knowledge is limited to the statistics of energy—that is, if we have only the equations

$$p_1 = |\psi_1|^2, \qquad p_2 = |\psi_2|^2 \cdots p_k = |\psi_k|^2—$$

we do not know the coefficients ψ_k themselves, but only their absolute values. Given that the ψ_k are normally complex quantities, we can write them in the form

$$\psi_k = \sqrt{p_k}\, \exp(i\alpha_k), \tag{1}$$

where the phases α_k are still indeterminate. One easily verifies that the difference between cases II and I arises from this ignorance about the phases α_k. Thus, in introducing (1) in the expression $Mean_{\text{I}}(F)$, we find

$$Mean_{\text{I}}(F) = \sum_{kl} \sqrt{p_k p_l} F_{kl} \exp[i(\alpha_l - \alpha_k)].$$

If we then average over the unknown phases we find that all of the terms with $k \neq l$ drop out and we get precisely $Mean_{\text{II}}(F)$.

One thus sees that it is necessary to make a careful distinction between:

I. A *pure state* described by a single wave function ψ that represents, we see, something irreducible, the probabilities that it implies being only "potential" probabilities;

II. A *mixture*, composed of different pure states

$$\overset{(1)}{\psi},\ \overset{(2)}{\psi} \cdots \overset{(n)}{\psi},$$

realized with probabilities $p_1, p_2 \cdots p_n \ldots$. These latter probabilities are understood in the ordinary sense of the word. Naturally, they are all non-negative. We suppose them to be normalized:

$$\sum_n p_n = 1.$$

§5. The Statistical Operator

It will be useful to introduce here a concise notation to describe statistical ensembles in all generality. We will consider a mixture such as we have just defined (§4, case II). The mean value of a physical quantity G in this mixture is

$$Mean(G) = \sum_n p_n(\overset{(n)}{\psi}, G\overset{(n)}{\psi}),$$

or, referred to a concrete coordinate system,

$$Mean(G) = \sum_n p_n \sum_{\rho\sigma} \overset{(n)}{\psi}_\sigma \overset{(n)}{\psi}{}^*_\rho G_{\rho\sigma},$$

which can be written

$$\begin{aligned} Mean(G) &= Trace(PG) \\ &= Trace(GP), \end{aligned} \tag{I}$$

when one introduces a Hermitian matrix P, the *statistical matrix*, defined by

$$P = \sum_n p_n(\overset{(n)}{\psi} \times \overset{(n)}{\psi}); \tag{1}$$

that is, in some chosen system of coordinates,

$$P_{\sigma\rho} = \sum_n p_n \overset{(n)}{\psi}{}^*_\rho \overset{(n)}{\psi}_\sigma. \tag{1'}$$

The case of a pure state $\psi = \sum_\rho \psi_\rho u_\rho$ is included in these formulas as the special case of a mixture where all the p_n are zero except for the single one, which is equal to unity. Its statistical matrix takes the form

$$\overset{(\psi)}{P}_{\sigma\rho} = \psi^*_\rho \psi_\sigma, \quad \text{or} \quad \overset{(\psi)}{P} = (\psi \times \psi). \tag{2}$$

Let us call the statistical matrix for a pure state an *elementary matrix* (Einzelmatrix). The matrix P for a general mixture can thus be considered as a linear superposition of "elementary matrices,"

$$P = \sum_n p_n \overset{(n)}{P}.$$

The elementary matrix for a pure state ψ takes an especially simple form in a system of coordinates $v_1, v_2 \cdots v_\rho$ in which the wave function ψ is identical with one of the axes. For example, let ψ be equal to v_2. Then one has

$$\overset{(\psi)}{P}_{\rho\sigma} = \delta_{2\rho} \cdot \delta_{2\sigma} = \begin{Vmatrix} 0 & 0 & 0 & 0 & \cdots \\ 0 & 1 & 0 & 0 & \cdots \\ 0 & 0 & 0 & 0 & \cdots \\ 0 & 0 & 0 & 0 & \cdots \\ \cdots & \cdots & \cdots & \cdots & \cdots \end{Vmatrix} \tag{2'}$$

Let us now calculate the probability of finding the value g_α of G in the ensemble characterized by the statistical matrix P.

Let us take as axes the eigenfunctions v_α of the operator G. For the pure component

$$\overset{(n)}{\psi} = \sum \overset{(n)}{\psi}_\alpha v_\alpha$$

with the index n we have (§1, equation 13)

$$\overset{(n)}{Prob}(g_\alpha) = |\overset{(n)}{\psi}_\alpha|^2 = \overset{(n)}{P}_{\alpha\alpha}.$$

When this component occurs in the ensemble with the probability p_n one has altogether

$$Prob(g_\alpha) = \sum_n p_n |\overset{(n)}{\psi}_\alpha|^2 = P_{\alpha\alpha}.$$

If we make use of the elementary matrix $\overset{(\alpha)}{P}$ for the pure state in which $G = g_\alpha$ (see 2′),

$$\overset{(\alpha)}{P}_{\beta\gamma} = \delta_{\alpha\beta}\delta_{\alpha\gamma},$$

we can express *the probability of finding the value* g_α *in the mixture* P in the invariant form,

$$Prob_P(g_\alpha) = P_{\alpha\alpha} = Trace(\overset{(\alpha)}{P} P). \tag{II}$$

In particular, if $P = \overset{(n)}{P}$ designates the case of the pure state where an arbitrary physical quantity F has the eigenvalue f_n, one sees that

$$Trace(\overset{(n)}{P}\ \overset{(\alpha)}{P}) = Prob(f_n, g_\alpha) \qquad \text{(III)}$$

is the *probability of finding the value* $G = g_\alpha$ *for the pure state in which* $F = f_n$. One notes that this expression is completely symmetric between F and G. The same expression (III) also gives the *probability of finding* $F = f_n$ *for the pure state in which* $G = g_\alpha$.

The statistical matrices P therefore present an evident advantage. They permit us to express all our definitions so that the very form (I, II, III) already indicates the invariant character.

§6. Some Mathematical Properties of Statistical Matrices

(a) Let us calculate the *trace* of an arbitrary statistical matrix P,

$$Trace\, P = \sum_\alpha P_{\alpha\alpha} = \sum_n p_n \sum_\alpha |\overset{(n)}{\psi_\alpha}|^2 = \sum_n p_n = 1.$$

In this way we obtain the relation

$$Trace\, P = 1, \qquad (1)$$

which expresses in brief form the normalization of probabilities.

(b) In particular, the *elementary* statistical matrices P for *pure states possess* the additional property

$$\overset{(\psi)}{P^2} = \overset{(\psi)}{P}, \qquad (2)$$

which results immediately from the definition (§5, equation 2) of these matrices,

$$(\overset{(\psi)}{P^2})_{\sigma\rho} = \sum_\tau \psi_\rho^* \psi_\tau \psi_\tau^* \psi_\sigma = \psi_\rho^* \psi_\sigma = \overset{(\psi)}{P}_{\sigma\rho}.$$

Relation (2) is in any case evident if one recalls the diagonal representation (§5, equation 2′) of elementary matrices.

One perceives immediately that the converse is also true. From $P^2 = P$ and from $Trace\ P = 1$ it follows that P is an elementary matrix. Thus when P is written in its diagonal form, P^2 is likewise diagonal, and $P = P^2$ implies $p_i = p_i^2$. The eigenvalues are therefore zero or unity. From the equation $Trace\, P = \sum_i p_i = 1$

it follows, finally, that a single one of p_i is equal to unity, and that all the others vanish. The relation $P^2 - P = 0$ is therefore *necessary and sufficient* for the statistical matrix P to be the matrix for a pure state.

(c) We have not yet placed any restrictions on the choice of the pure states which constitute the mixture. In particular, we have not assumed that they were represented by orthogonal wave functions ψ (later to be distinguished from one another by an index (n) written directly above the ψ). But one can show very easily that an arbitrary mixture composed of arbitrary pure states can always be written in the form of a mixture of *orthogonal* pure states with *non-negative* relative probabilities p_i.

Let us first verify that a matrix P is a *semi-definite* matrix; that is, that for our arbitrary vector ξ one always has

$$(\xi, P\xi) \geq 0. \tag{3}$$

Thus the definition (§5, equation 1′) of P gives

$$(\xi, P\xi) = \sum_{n\alpha\beta} p_n \xi_\alpha^* \overset{(n)}{\psi_\alpha} \cdot \xi_\beta \overset{(n)}{\psi_\beta^*} = \sum_n p_n |(\xi, \overset{(n)}{\psi})|^2,$$

an expression which cannot be negative because

$$p_n \geq 0 \quad \text{and} \quad |(\xi, \overset{(n)}{\psi})|^2 \geq 0.$$

But P is a Hermitian matrix. Therefore, there exists an *orthogonal* coordinate system $v_1, v_2 \ldots v_\rho$ in which P takes its diagonal form,

$$P_{\rho\sigma} = \begin{Vmatrix} p_1' & 0 & 0 & 0 & \cdots \\ 0 & p_2' & 0 & 0 & \cdots \\ 0 & 0 & p_3' & 0 & \cdots \\ 0 & 0 & 0 & p_4' & \cdots \\ \cdots & \cdots & \cdots & \cdots & \cdots \end{Vmatrix}. \tag{4}$$

The values of the diagonal elements $p_1', p_2' \ldots p_\rho'$ cannot be negative. Thus, again one has in these coordinates

$$(\xi, P\xi) = \sum_\rho p_\rho' |\xi_\rho|^2 \geq 0,$$

which is not possible for an arbitrary vector ξ unless $p'_\rho \geq 0$ for every ρ. The matrix P therefore can always be written in the form

$$P = \sum_\rho p'_\rho \overset{(\rho)}{P}, \tag{4'}$$

where the $\overset{(\rho)}{P}$ are the elementary matrices for a certain system of *orthogonal* pure states $v_1, v_2 \ldots v_\rho$. The $p'_1, p'_2 \ldots p'_\rho \geq 0$ are the relative probabilities for these states.

As *Trace* $P = 1$, we also have

$$\sum p'_\rho = 1,$$

and $p'_\rho \geq 0$ implies

$$0 \leq p'_\rho \leq 1.$$

(d) From this inequality follows another,

$$p'_\rho - p'^2_\rho \geq 0.$$

From this equation we conclude that for an arbitrary vector ξ,

$$(\xi, (P - P^2)\xi) = \sum_\rho |\xi_\rho|^2(p'_\rho - p'^2_\rho) \geq 0. \tag{5}$$

The matrix $P - P^2$ is therefore likewise semidefinite.

In particular for a pure state the quantity $P - P^2$ vanishes (see equation 2).

Elementary statistical operators like $P = (\psi \times \psi)$ can be considered as "projectors" or "projection operators." Applied to an arbitrary state vector ξ, P singles out the projection of this vector in the direction of the vector ψ,

$$\sum_\sigma P_{\sigma\rho}\xi_\sigma = \sum_\sigma \psi_\rho \psi^*_\sigma \xi_\sigma = \psi_\rho(\psi, \xi).$$

The magnitude of this vector is (ψ, ξ) and its direction is that of the unit vector ψ. The iteration of the projection produces no further change: $P^2 = P$.

§7. The Statistical Operator and Thermodynamics

The operator P describes an ensemble of identical systems which are distributed in an arbitrary way among the different states. It plays the role analogous to that of a distribution function in ordinary statistical mechanics. Therefore there ought to exist a connection between operator P and the macroscopic thermodynamic quantities. It will be enough for us to point out briefly the main features of this connection without stopping for proofs (see von Neumann, 1927, p. 273). Everything is summarized in the definition of entropy, S:

$$S = -kN \; Trace(P \ln P), \tag{1}$$

where

$$k = \textit{Boltzmann's constant}$$

and

$$N = \textit{the total number of systems.}$$

This relation becomes quite plausible when one recalls that in statistical mechanics one has

$$S = -k \sum_{\alpha} (n_\alpha \ln n_\alpha - N \ln N),$$

where n_α is the number of systems in the state α. Therefore if $p_\alpha = n_\alpha/N$ represents the probability for a system to be in the state α, one has again

$$S = -kN\left[\sum_{\alpha} p_\alpha(\ln p_\alpha + \ln N) - \ln N\right],$$

which is identical with (1) in a system of coordinates in which P is diagonal (§6, equation 4). Our definition of entropy is therefore the entirely straightforward generalization of the usual definition. One sees immediately that for a pure state the entropy thus defined is zero. Thus, if one represents P in diagonal form, the entropy becomes

$$S = -kN \sum_{\alpha} p_\alpha \ln p_\alpha.$$

Every term $p_\alpha \ln p_\alpha$ of the sum vanishes, because in a pure state the p_α are all zero except for a single one which is unity. One also sees immediately that the entropy of a mixture is always positive.

To maximize the entropy for a given total energy E is to impose on P the following conditions:

$$-kN \; Trace(P \ln P) \rightarrow Max;$$

$$Trace\, P = 1;$$

$$N \; Trace(PH) = E.$$

The solution of this extremum problem is represented by the matrix,

$$P = e^{-\beta H}/Trace(e^{-\beta H}) = e^{-\beta H}/Z(\beta), \tag{2}$$

where

$$Z(\beta) = Trace(e^{-\beta H}).$$

The Lagrange factor β is determined, as always, by equilibrium with a perfect gas; thus, $\beta = 1/kT$, where T is the absolute temperature.

One thus obtains for the entropy of the most probable distribution,

$$\begin{aligned} S &= kN\ Trace[e^{-\beta H}(\beta H + \ln Z)/Z(\beta)] \\ &= -kN[\beta\partial \ln Z/\partial\beta - \ln Z], \end{aligned}$$

and for the energy,

$$E = -N\partial \ln Z/\partial\beta,$$

from which one easily gets all the other thermodynamic quantities, for example the free energy

$$F = E - TS = -kNT \ln Z,$$

et cetera.

§8. The Irreducibility of the Pure Case

Our definitions would be worthless if pure cases were not characterized by some kind of irreducibility. We have to show that *it is not possible to represent a pure case in the form of a mixture.* For this purpose we will show that a statistical matrix P, obtained as a mixture of two statistical matrices Q and R,

$$P = \alpha Q + \beta R, \tag{1}$$

with

$$\alpha + \beta = 1, \qquad \alpha \geq 0, \qquad \beta \geq 0,$$

cannot be an elementary statistical matrix (such that $P = P^2$) except if $Q = R = P$. Let us calculate

$$\begin{aligned} P^2 &= \alpha^2 Q^2 + \beta^2 R^2 + \alpha\beta(QR + RQ) \\ &= \alpha^2 Q^2 + \beta^2 R^2 + \alpha\beta[Q^2 + R^2 - (Q - R)^2] \\ &= \alpha Q^2 + \beta R^2 - \alpha\beta(Q - R)^2, \end{aligned}$$

where we used the condition $\alpha + \beta = 1$. Thus

$$P - P^2 = \alpha(Q - Q^2) + \beta(R - R^2) + \alpha\beta(Q - R)^2.$$

We now recall that the matrices $Q - Q^2$ and $R - R^2$, as well as $(Q - R)^2$, are always semidefinite (§6, equation 5). Therefore it is necessary that all these quantities should vanish for P to be an elementary matrix ($P = P^2$). In particular, we have $(Q - R)^2 = 0$, from which we conclude

$$Q = R,$$

because the square of a Hermitian matrix cannot vanish unless the original matrix vanishes,

$$\left(\sum_k A_{ik}A_{ki} = \sum_k A_{ik}A_{ik}^* = \sum_k |A_{ik}|^2 = 0 \text{ implies } A_{ik} = 0\right).$$

From $Q = R$ and from (1) it follows that $Q = R = P$.

Statistical operators thus form an ensemble of a characteristic structure, called *convex ensemble*. Its boundary is formed by the operators for pure cases. No pure case can be constructed by linear superposition with positive coefficients—that is, by mixture—of two nonidentical pure cases.

Although the statistical operator P of a pure case cannot be decomposed, one might imagine that there perhaps exists some other means to reduce directly the corresponding statistics.

In actuality this problem does not differ from the problem that we have just discussed. However, it would perhaps be useful to consider it explicitly.

Of course we always take as our foundation the statistical distributions as predicted by the formalism of the theory and verified by so many experiments. Therefore we will not discuss the validity of these statistical predictions. We will ask ourselves rather if, once assumed, they might not be represented by mixtures of arbitrary form, but belonging to systems that are well defined in the ordinary sense of classical mechanics.

Let us take a concrete example that is as simple as possible, that of the statistical description associated with a "spin." Let us consider an atom of angular momentum $\hbar/2$ and let us focus attention on the component of this angular momentum in an arbitrary direction. For any given direction there are only two values possible for this component of the spin: $+\hbar/2$ and $-\hbar/2$. Let us fix in space the axis of a system of spherical polar coordinates. Let u_+ and u_- be the eigenfunctions associated with the two possible values for the component along this axis. To be concrete, suppose that

$$\psi = u_+$$

is the wave function of the state in question—that this state is a "pure case."

Let us now consider, for the same state ψ, the component of the spin in another direction, oblique to the polar axis and characterized, for example, by the polar coordinates θ, ϕ. The only possible values for the component of the spin in this direction are again $+\hbar/2$ and $-\hbar/2$. To evaluate the probabilities for finding each of these two values it is necessary to represent ψ in terms of the two eigenfunctions associated with those two possibilities. We will call them u'_+ and u'_-. A calculation gives

$$u_+ = c_+ u'_+ + c_- u'_-,$$

with

$$c_+ = e^{i\phi/2} \cos \theta/2,$$

$$c_- = e^{-i\phi/2} \sin \theta/2,$$

The squares of the absolute values of the coefficients u'_+ and u'_- represent the probabilities for finding the one or the other of the two possible components in the direction θ, ϕ; thus,

$$p'_+ = |c_+|^2 = \cos^2 \theta/2,$$

$$p'_- = |c_-|^2 = \sin^2 \theta/2.$$

But, in the same state, the probabilities of the two values for the component along the original polar axis are, respectively,

$$p_+ = 1$$

and

$$p_- = 0.$$

Evidently it is impossible to decompose these statistics into a mixture of definitely oriented spins. In such a mixture it would be necessary that a fraction $\cos^2 \theta/2$ of the atoms should have a component $\hbar/2$ in the direction θ, ϕ and that a fraction $\sin^2 \theta/2$ should have the opposite orientation. This, by itself, would be possible. But we ought, in addition, to have 100% of the atoms with the same component $\hbar/2$ along the direction of the polar axis, and yet a fraction $\cos^2 \theta'/2$ along any other direction θ'. That would be a juggling trick rather difficult to bring off!

Evidently it is impossible to arrange a virtual ensemble of oriented atoms that meets simultaneously all of these statistical requirements. The mathematics of the probability calculations already precludes this possibility.

We will use the same simple example in section §12 to study in detail how, by his own intervention, the observer succeeds in doing the juggling trick. However, to keep the discussion clear, it is first necessary for us to bring out an aspect of quantum mechanics which we have not yet mentioned, but which contains the very essence of the theory, the feature responsible for the appearance of probabilities.

§9. Statistics of a System Composed of Two Subsystems

At first sight the mathematical formalism of quantum mechanics seems entirely analogous to that of the theories of "classical" physics: a differential equation uniquely prescribes the evolution of the wave function ψ that describes the state of the system. Therefore it seems as though our task, faced with Schrödinger's equation, is no different from that of Laplace faced with the equations of Newton. The state of a closed system, perhaps the entire universe, is completely determined for all time if it is known at a given instant. According to the Schrödinger equation, a pure case represented by a ψ function remains always a pure case. One does not immediately see any occasion for the introduction of probabilities, and our statistical definitions might appear in the theory as a foreign structure.

We will see that that is not the case. It is true that the state of a closed system, once given pure, always remains pure. But let us study what happens when one puts into contact *two systems*, both originally in pure states, and afterwards separates them.

Let us therefore consider two systems, I and II, originally separated. Let x be the ensemble of the coordinates of I and y the coordinates of II. *Each of the two systems* is assumed to be in a *pure state* given by its wave function:

$$\psi(x) = \sum_k \psi_k u_k(x) \qquad (\textit{system I}),$$

$$\phi(y) = \sum_\rho \phi_\rho v_\rho(y) \qquad (\textit{system II}).$$

We have expressed the functions ψ and ϕ as series built on the orthogonal functions $u_k(x)$ and $v_\rho(y)$. The coefficients ψ_k and ϕ_ρ depend only on time.

Although the two systems are originally taken to be separated, we can nevertheless describe them by a combined wave function $\Psi(x, y)$, whose evolution is governed by a combined Hamiltonian. The fact that the two systems are isolated from each other is expressed in the form of the Hamiltonian. It is the *sum* of two terms, each depending only upon the coordinates of one of the two systems; thus,

$$H(x, p_x, y, p_y) = H_I(x, p_x) + H_{II}(y, p_y).$$

It is easily verified that the combined wave function $\Psi(x, y)$, which unites the statistics contained in $\psi(x)$ and $\phi(y)$, is the simple product of the two separate wave functions. It obeys the wave equation with Hamiltonian $H_I + H_{II}$. One has therefore *before contact*

$$\Psi(x, y) = \psi(x)\phi(y) = \sum_{k,\rho} \psi_k \phi_\rho u_k(x) v_\rho(y). \tag{1}$$

Let us now bring the two systems into contact. That makes it necessary to add to H an *interaction* term, $H_i(x, p_x, y, p_y)$, containing the two sets of variables x, p_x and y, p_y in a form that is not simply additive. It is evident that the combined function Ψ for the pair of systems will no longer in general keep the form of a product of two functions each depending on a single set of variables. However, it naturally can be expanded *at each instant as a series of products* $u_k(x)v_\rho(y)$ with coefficients that will depend on the instant chosen. Thus, if $u_k(x)$ and $v_\rho(y)$ form *complete* systems of orthogonal functions in their own domains, x and y, the products $u_k(x)v_\rho(y)$ also form an orthogonal system that is complete in the space of functions on the domain (x, y) of the ensemble. *During or after the contact* the wave function will be written in every case in the general form,

$$\Psi(x, y) = \sum_{k,\rho} \Psi_{k\rho} u_k(x) v_\rho(y), \tag{2}$$

where the coefficients $\Psi_{k\rho}$ in the generic case will not have the special form of a product, $\psi_k \phi_\rho$. As we are always dealing with a unique wave function evolving in accord with a Schrödinger equation, we have *for the combined system a constantly pure case.* Its statistical matrix is an elementary matrix,

$$P_{k\rho,l\sigma} = \Psi_{k\rho} \Psi^*_{l\sigma}.$$

As one pair of indices, k, ρ, is needed here to characterize a state of the total system, the elements of the statistical matrix of this system will evidently depend on two pairs of indices, k, ρ and l, σ.

Let us now focus attention on the system I. What is its statistical matrix?

Let F be a function solely of the variables of system I and F_{kl} its representation in the coordinates $u_k(x)$. The mean or expectation value of F in the state Ψ will be given by

$$(\Psi, F\Psi) = \sum_{kl,\rho} \Psi^*_{k\rho} \Psi_{l\rho} F_{kl} = Trace(P^I F).$$

Therefore the matrix

$$P^I_{lk} = \sum_\rho \Psi_{l\rho} \Psi^*_{k\rho} \tag{3a}$$

plays the role of statistical matrix for system I. Similarly one obtains for system II the statistical matrix

$$P^{II}_{\sigma\rho} = \sum_k \Psi_{k\sigma}\Psi^*_{k\rho}. \tag{3b}$$

P^I and P^{II} are evidently no longer elementary matrices. We are dealing with mixtures. What are their components? And what are the relative probabilities or concentrations of these components?

We note that the $\Psi_{k\rho}$ are normalized in the space of the indices k, ρ; thus,

$$\sum_{k,\rho} |\Psi_{k\rho}|^2 = 1.$$

Therefore the magnitude $\overset{(\rho)}{\psi}$ with the components

$$\overset{(\rho)}{\psi}_k = \Psi_{k\rho}\bigg/\left[\sum_l |\Psi_{l\rho}|^2\right]^{1/2}$$

is a normalized state vector for system I and represents a pure case.

We can therefore give to P^I the following form,

$$P^I = \sum_\rho p_\rho(\overset{(\rho)}{\psi} \times \overset{(\rho)}{\psi}), \tag{3a}$$

where

$$p_\rho = \sum_l |\Psi_{l\rho}|^2 \tag{4a}$$

is the concentration with which the pure case $\overset{(\rho)}{\psi}$ is contained in the mixture I. In the same way one has

$$P^{II} = \sum_k p_k(\overset{(k)}{\phi} \times \overset{(k)}{\phi}), \tag{3b$'$}$$

where

$$\overset{(k)}{\phi}_\rho = \Psi_{k\rho}\bigg/\left[\sum_\sigma |\Psi_{k\sigma}|^2\right]^{1/2}$$

is a unit vector, representing a pure case in II; and

$$p_k = \sum_\sigma |\Psi_{k\sigma}|^2 \tag{4b}$$

is the concentration with which $\overset{(k)}{\phi}$ is contained in the mixture II.

While the combined system I + II, which we suppose isolated from the rest of the world, is and remains in a pure state, we see that during the interaction systems I and II individually transform themselves from pure cases into mixtures.

This is a rather strange result. In classical mechanics we are not astonished by the fact that a *maximal knowledge of a composite system* implies a *maximal knowledge of all its parts*. We see that this equivalence, which might have been considered trivial, does not take place in quantum mechanics. There a maximal knowledge of a composite system ordinarily implies only mixtures for the component parts—that is, a knowledge that is not maximal.

The mixtures represented by P^I and P^{II} naturally cannot express all that it is possible to know about the combined system I + II. It is evident that the elementary combination of statistical mixtures for two individual systems cannot by itself reproduce a pure case for the combined system. Thus the function $\Psi(x, y)$ for the combined system contains still other relations, to wit, *statistical correlations between* the components of the two mixtures I and II.

The fact that the description we obtain for each of the two individual systems does not have the character of a pure case warns us that we are *renouncing part of the knowledge contained in* $\Psi(x, y)$ when we calculate probabilities for each of the two individual systems separately. This renunciation expresses itself by the summation over the index ρ in the definition of P^I, where we abstract away from what might be known about the state ρ of system II and about its connection with system I. This loss of knowledge expresses itself by the appearance of probabilities, now understood in the ordinary sense of the word, as expression of the fact that our knowledge about the combined system is not maximal.

§10. Reversible and Irreversible Evolution

It is evidently necessary to make a characteristic distinction between two essentially different modes of evolution of an individual system, a distinction which has no analog in classical mechanics.

I. *Reversible* or "*causal*" transformations. These take place when the system is isolated. They can be described by the change with time of a ψ function (or of a *certain number* of distinct ψ functions when one deals with a mixture). If $\psi(t_0)$ represents a pure case at the instant t_0, its evolution can be written in the form

$$\psi(t) = T\psi(t_0), \tag{1}$$

with the operator

$$T = \exp[-iH(t - t_0)/\hbar] = \sum_n (1/n!)[-iH(t - t_0)/\hbar]^n.$$

In the case of a mixture, the time dependence of the statistical operator

$$P(t) = \sum_\rho p_\rho \left(\overset{(\rho)}{\psi}(t) \times \overset{(\rho)}{\psi}(t) \right) = \sum_\rho p_\rho \left(T\overset{(\rho)}{\psi}(t_0) \times T\overset{(\rho)}{\psi}(t_0) \right)$$

is thus given by the equation

$$P(t) = TP(t_0)\tilde{T}^*,$$

where $\tilde{T}_{kl}$ is the abbreviation for the matrix T_{lk}. One verifies that

$$\tilde{T}^* = \sum_n (1/n!)[iH(t - t_0)/\hbar]^n$$

and that

$$T\tilde{T}^* = 1.$$

It is therefore a *unitary transformation* that characterizes a causal evolution. It transforms a pure case into a pure case.

A unitary transformation, keeping invariant—as it does—the trace of a tensor, does not change the value of the entropy; the quantity *Trace* $(P, \ln P)$ stays constant.

II. *Irreversible* transformations, which one might also call "*acausal.*" These take place only when the system in question (I) makes physical contact with another system (II). The total system, comprising the two systems (I + II), again in this case undergoes a reversible transformation so long as the combined system I + II is isolated. But if we fix our attention on system I, this system will undergo an irreversible transformation. If it was in a pure state before the contact, it will ordinarily be transformed into a mixture. If it was already a mixture, it will be transformed into another mixture, the entropy of which (§7, equation 1) will be increased. Once thus degraded, the system has no chance in and by itself ever to regain its initial degree of determination.

We shall see specifically that measurement processes bring about an irreversible transformation of the state of the measured object, such that the initial statistical operator for a pure case

$$[\psi(q) = \sum_n \psi_n u_n(q)],$$

$$P = (\psi \times \psi),$$

is transformed, by interaction with the measuring apparatus, into the mixture,

$$P' = \sum |\psi_n|^2 (u_n \times u_n),$$

the u_n being the eigenfunctions of the quantity measured. The transition from P to P' clearly cannot be represented by a unitary transformation. It is associated with an increase of the entropy from 0 to $-k \sum_n |\psi_n|^2 \ln |\psi_n|^2$, which cannot come about by a unitary transformation.

The distinction between these two modes of evolution has no analog in classical mechanics, where it is always possible to give a maximal description of the state of an object by its $2f$ coordinates and momenta $q_1, q_2 \cdots p_f$, whatever the interactions between systems.

§11. Measurement and Observation. The Act of Objectification

We are now ready to analyze what happens in the act of measurement. We will first outline a protocol for this process and then verify in the following section that it describes properly the typical course of a measurement.

Suppose that we want to measure the quantity $F(x, p_x)$ of a system ("the object") given to be in the state $\psi = \sum_k \psi_k u_k(x)$ where u_k is an eigenfunction corresponding to the value f_k of F. We couple it with an apparatus capable of measuring F.

Let $G(y, p_y)$ be the coordinate specifying the position of the "needle" of the measuring device, and $g_0, g_1 \ldots g_\rho$ its eigenvalues, with eigenfunctions $v_0(y)$, $v_1(y) \ldots v_\rho(y)$. The state $v_0(y)$ corresponds to the zero of this apparatus.

Before the coupling we will attribute to the combined system a collective wave function of the form,

$$\Psi(x, y) = v_0(y) \sum_k \psi_k u_k(x). \tag{1}$$

This is a pure case for each of the two individual systems. *After the interaction* the wave function will be of a more general character,

$$\Psi(x, y) = \sum_{k,\rho} \Psi_{k\rho} u_k(x) v_\rho(y). \tag{2}$$

But an arbitrary interaction does not provide a measurement. In order for it to do so, it is necessary that it disturb the state of the object as little as possible and in addition that it should let one deduce from a g_ρ the corresponding f_k. Thus the values of the measurement scale g_ρ should be coordinated one-to-one with the values f_k of the quantity under consideration, so that one can inscribe directly onto the g_ρ scale the corresponding values of F. That occurs by replacing the index $\rho_{(k)}$ by k. Thus $g_k = g_{\rho(k)}$ will correspond to f_k.

We will see (§12) that after a measurement of the quantity the wave function takes the special form,

$$\Psi(x, y) = \sum \psi_k u_k(x) v_k(y). \tag{2'}$$

According to the preceding section, this function represents a state of the combined system that has for each separate system, object, and apparatus, the character of a *mixture*. According to §9, equation (4), the quantity $p_k = |\psi_k|^2$ gives the probability of finding the object in the pure state u_k with $F = f_k$; and the same quantity $p_k = |\psi_k|^2$ also gives the probability of finding the apparatus indication $G = g_k$. Moreover, we have a correlation between the two mixtures: specifically, we know with certainty that if $G = g_k$, then $F = f_k$. But of course quantum mechanics does not allow us to predict *which* value will actually be found in the measurement. The interaction with the apparatus does not put the object into a new pure state. Alone, it does not confer on the object a new wave function. On the contrary, it actually gives nothing but a statistical mixture: It leads to one mixture for the object and one mixture for the apparatus. For either system regarded individually there results uncertainty, incomplete knowledge. Yet nothing prevents our reducing this uncertainty by further observation. And this is our opportunity.

So far we have only coupled one apparatus with one object. But a coupling, even with a measuring device, is not yet a measurement. A measurement is achieved only when the position of the pointer has been *observed*. It is precisely this increase of knowledge, acquired by observation, that gives the observer the right to choose among the different components of the mixture predicted by theory, to reject those which are not observed, and to attribute thenceforth to the object a new wave function, that of the pure case which he has found.

We note the essential role played by the consciousness of the observer in this transition from the mixture to the pure case. Without his effective intervention, one would never obtain a new ψ function. In order to see this point clearly, let us consider the ensemble of three systems, (*object x*) + (*apparatus y*) + (*observer z*), as a combined and unique system. We will describe it by a global wave function with a form analogous to (2),

$$\Psi(x, y, z) = \sum_k \psi_k u_k(x) v_k(y) w_k(z),$$

where the w_k represent the different states of the observer.

"Objectively"—that is, *for us* who consider as "object" the combined system x, y, z—the situation seems little changed compared to what we just met when we were considering only apparatus and object. We now have three mixtures, one for each system, with those statistical correlations between them that are tied to a pure case for the combined system. Thus the function $\Psi(x, y, z)$ represents a maximal description of the combined "object," consisting of the actual object x, the apparatus y, and the observer z; and nevertheless we do not know in what state the object x is.

The observer has a completely different impression. For him it is only the object x and the apparatus y that belong to the external world, to what he calls "objectivity." By contrast he has *with himself* relations of a very special character. He possesses a characteristic and quite familiar faculty which we can call the "faculty of introspection." He can keep track from moment to moment of his own state. By virtue of this "immanent knowledge" he attributes to himself the right to create his own objectivity—that is, to cut the chain of statistical correlations summarized in $\sum_k \psi_k u_k(x) v_k(y) w_k(z)$ by declaring, "I am in the state w_k" or more simply, "I see $G = g_k$," or even directly, "$F = f_k$."

Accordingly, we will label this creative action as "making objective." By it the observer establishes his own framework of objectivity and acquires a new piece of information about the object in question.*

Thus it is not a mysterious interaction between the apparatus and the object that produces a new ψ for the system during the measurement. It is only the consciousness of an "I" who can separate himself from the former function $\Psi(x, y, z)$ and, by virtue of his observation, *set up a new objectivity* in attributing to the object henceforward a new function $\psi(x) = u_k(x)$.

Neither is it some ignorance as to the state of the observer that creates quantum indeterminacy. *On the contrary*, in assuming a pure case for the combined system, we have implicitly presupposed an equally perfect knowledge of the initial state $w_0(z)$ of the observer and of the apparatus $v_0(y)$, that is, maximal information. Moreover, we have assumed that the observer can keep track perfectly of his own state.

Of course there might also be restrictions on the immanent knowledge of the observer. But these, if they existed, would in any case have nothing to do with quantum indeterminism; they would be additional restrictions of a completely different character. Moreover, it is not ordinarily required for a discussion of the measuring process that one should have an *all-encompassing* knowledge of the state of the observer; for example, there is little chance of making a big mistake if one does not know his age.

§12. An Example of Measurement

It only remains for us now to verify the protocol for measurement that we have just discussed. Let us take as a typical example the determination of the value of

* This paragraph is new. We have translated it from a typed addition inserted by Professor Fritz London in his own copy of the printed book, kindly sent to us October 24, 1980, by Mrs. Fritz London.—Eds.

one component of the magnetic moment of an atom by the method of Stern and Gerlach. The formulas that we will get can be generalized without difficulty to an arbitrary measurement.

This measurement is made, as is well known, by observing the motion of an atom through a nonuniform magnetic field. The field points in the direction along which one wants to determine the component of the magnetic moment of the atom. The coordinates, y, of the center of gravity of the atom play the role of the pointer reading $G(y, \beta_y)$. The internal coordinates of the atom, relative to the center of gravity, serve as object coordinates x. Specifically, we are concerned with the component, $M = M(x, p_x)$, of the magnetic moment in the direction of the field.

Let us write the wave equation for this problem in the form

$$\{-(\hbar^2/2m)\Delta_y + H_0(x, \partial/\partial x) + [M(x, p_x), F(y)]\}\Psi(x, y) = i\hbar\dot{\Psi}(x, y). \quad (1)$$

Here H_0 is the Hamiltonian operator for the field-free atom after one has separated off the center-of-gravity variables, $-(\hbar^2/2m)\Delta_y$ is the operator of the "apparatus" corresponding to the kinetic energy of the center of gravity, (M, F) is the contribution arising from the magnetic field F, and M is the operator for the magnetic moment of the atom in the direction of the field.

So long as the field F is constant—that is, so long as it does not depend on the coordinates y—the variables x and y can be separated in equation (1). For the different states of the "object" we must deal with the eigenvalue problem,

$$(H_0 + MF)u_k(x) = E_k u_k(x). \quad (2)$$

Let us limit ourselves to the lowest state of the atom, a state which we assume to be degenerate, with its components splitting in the field in proportion to the field strength F,

$$E_k = E_0 + (k\mu/j)F. \quad (3)$$

Here μ is the magnetic moment of the atom and k is the magnetic quantum number, $k = j, j-1, j-2, j-3 \ldots -j$, where $j\hbar$ represents the total magnetic moment of the atom. When the field is no longer constant, $F = F(y)$, equation (2) contains the coordinates y as *parameters*. Consequently the eigenvalues of (2),

$$E_k(y) = E_0 + (k\mu/j)F(y), \quad (3a)$$

will also depend on these parameters y; likewise the eigenfunctions $u_k(x)$, which will be written more appropriately as $u_k(x, y)$. In practical terms the perturbation of the u_k by the nonuniformity of the field is so slight that we can forget this dependence of the u_k on the parameter y.

Let us now develop $\Psi(x, y)$ in a series in terms of the functions $u_k(x)$,*

$$\Psi(x, y) = \sum_k v_k(t, y)u_k(x),$$

and let us introduce this development into the Schrödinger equation (1). After multiplication by u_k and integration over x we find for the coefficients $v_k(t, y)$ the following equations:

$$\{-(\hbar^2/2m)\Delta_y + E_k(y)\}v_k(y) = i\hbar\dot{v}_k(y). \tag{4}$$

They are still of the Schrödinger type but now refer solely to the motion of the center of gravity. In a typical such equation, the eigenvalue (3a) of (2), $E_k(y)$, plays the role of the potential. Let us now consider an atomic beam and let us develop $E_k(y)$ in the neighborhood of this beam,

$$E_k = E_0 + (k\mu/j)[F_0 + (y, grad\ F) + \cdots].$$

In first approximation the "potential" $E_k(y)$ varies linearly; thus it behaves like the potential of gravity near the earth's surface. Consequently equation (4) is nothing other than the equation of free fall. But in it the acceleration is proportional to the quantum number k. Thus the acceleration depends on the value of the component of the magnetic moment in the direction of the field, and can be *positive* or *negative*. Therefore it is easy to foresee in general terms and without detailed calculation the shape of the various trajectories which come from a well-collimated and initially monokinetic source and then travel through the nonuniform magnetic field. A single beam v_{00} splits into separate beams belonging to different values of k. The cross-hatched regions in Figure 1 show where the functions v_k are appreciably different from zero.

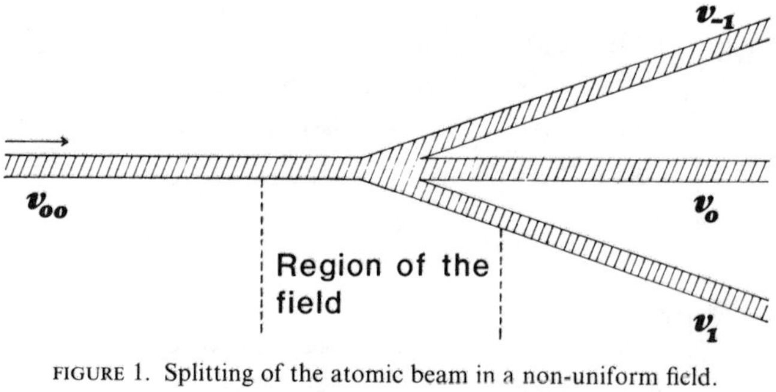

FIGURE 1. Splitting of the atomic beam in a non-uniform field.

* This is certainly permissible. The $v_k(y, t)$ show up in this development, not as given eigenfunctions, but as the still unknown coefficients of the functions $u_k(x)$.

So far we have been discussing the not-yet-normalized solutions of the auxiliary equations (4). A complete solution of the Schrödinger equation (1) will have the form

$$\Psi(x, y) = \sum_k c_k v_k(y) u_k(x). \tag{5}$$

Let

$$\psi = \sum_k \psi_k u_k(x)$$

with

$$\sum_k |\psi_k|^2 = 1$$

be the state of the atom *before the measurement* in the region at the left of the figure, where we can still separate off the motion of the center of gravity and where all the functions $v_k(y)$ are identical $[= v_{00}(y)]$.

The total wave function of object-plus-apparatus *before entry into the region of the field* is

$$\Psi(x, y) = v_{00}(y) \sum_k \psi_k u_k(x).$$

On the other side of the field region, according to (5) and the principle of continuity, it is

$$\Psi'(x, y) = \sum_k \psi_k u_k(x) v_k(y).$$

A measurement of the y coordinate of the center of gravity of an atom that has crossed the field is equivalent to a determination of k because each $v_k(y)$ differs from zero only in a limited region, fixed by k, and because a determination of k is equivalent to a knowledge of the component of the magnetic moment parallel to F. This is exactly the type of statistical link that we presumed in §11.

Let us take, for example, an atom of total angular momentum 1, for which k is restricted to the values 1, 0, -1. The effect of the measurement shows up in the transition of the coefficients $\Psi_{k\rho}$ of the wave function of the combined system from

$$\Psi_{k\rho} = \psi_k \delta_{0\rho} = \begin{Vmatrix} 0 & 0 & 0 \\ \psi_{-1} & \psi_0 & \psi_1 \\ 0 & 0 & 0 \end{Vmatrix} \quad \textit{(before the measurement)}$$

to

$$\Psi'_{k\rho} = \psi_k \delta_{k\rho} = \begin{Vmatrix} \psi_{-1} & 0 & 0 \\ 0 & \psi_0 & 0 \\ 0 & 0 & \psi_1 \end{Vmatrix} \quad \textit{(after the measurement)}.$$

Before the measurement the statistical matrices P^I and P^{II} have the form of two pure cases,

$$P^I_{kl} = \begin{Vmatrix} \psi_{-1}\psi^*_{-1} & \psi_{-1}\psi^*_0 & \psi_{-1}\psi^*_1 \\ \psi_0\psi^*_{-1} & \psi_0\psi^*_0 & \psi_0\psi^*_1 \\ \psi_1\psi^*_{-1} & \psi_1\psi^*_0 & \psi_1\psi^*_1 \end{Vmatrix}$$

and

$$P^{II}_{\rho\sigma} = \begin{Vmatrix} 0 & 0 & 0 \\ 0 & 1 & 0 \\ 0 & 0 & 0 \end{Vmatrix}.$$

The action of the magnetic field transforms them into mixtures,

$$(P^I_{kl})' = \begin{Vmatrix} |\psi_{-1}|^2 & 0 & 0 \\ 0 & |\psi_0|^2 & 0 \\ 0 & 0 & |\psi_1|^2 \end{Vmatrix}$$

and

$$(P^{II}_{\rho\sigma})' = \begin{Vmatrix} |\psi_{-1}|^2 & 0 & 0 \\ 0 & |\psi_0|^2 & 0 \\ 0 & 0 & |\psi_1|^2 \end{Vmatrix}.$$

Here the components have a one-to-one correspondence, implying the associated correlations. For the combined system the transition $\Psi \rightarrow \Psi'$ is a unitary transformation,

$$\Psi'_{k\rho} = \sum_{l,\sigma} T_{k\rho,l\sigma} \Psi_{l\sigma},$$

where

$$T_{k\rho,l\sigma} = \delta_{kl}\delta_{\rho,l+\sigma},$$

it being understood that

$$\rho = \pm 2 \quad \text{is equivalent to} \quad \rho = \mp 1.$$

§13. Individuality and Pure Case

A measuring arrangement, such as we have just finished describing in the previous section, can be used to "filter" through objects possessing a prescribed value of a certain physical quantity. It is enough to make a suitable slit through the screen on which the atoms fall to transform the Stern-Gerlach apparatus into a source of identically oriented atoms. It thus becomes a good set-up for producing pure cases.

Does not this conclusion contradict what we just learned (§11), to the effect that a pure case is brought about only by an "act of objectifying," accomplished by the appropriate intervention of an observer? It is necessary, however, to be more precise: the filter never puts any *individual* object whatsoever into a new pure state. It can only put it into a mixed state. This is what always happens when one couples one system with another. By contrast, we can evidently say that the states of atoms which have gone through the slit have the desired property. We can attribute to them the wave function of the pure case in question. But this attribution only works out, so to speak, at the expense of the individuality of the object, as one does not know in advance *which* are the atoms that have the property in question. We can easily attribute to the objects that get through the slit the ψ function of the pure case, but we cannot say *which object*, that is, which variable is the argument of this ψ function. Without a supplementary check by an observer, it is not possible to guarantee whether a given atom has gone through the filter or been caught in it. The filter alone thus truly produces pure cases, but in an absolutely *anonymous* form. Of course we can attribute to these cases afresh some names of their own, for example by numbering in sequence the atoms that really get through the slit. But that is no different from a true measurement, and we would be led back in that way to what we have already discussed.

Moreover, anonymous objects are precisely the focus of one's interest in many experiments. The majority of the measurements in atomic physics really do not deal with an *individual system*; rather, they seek to find out the general properties of an *entire species* of atoms—or of molecules, or of elementary particles. Thus for example the Stern-Gerlach set-up just discussed is ordinarily used, not to measure a component of the spin of an *individual atom*, but rather to determine the spin *of the silver atom.*

Quantum mechanics, truly a "theory of species," is perfectly adapted to this experimental task. But given that every measurement contains a macroscopic process, unique and separate, we can hardly escape asking ourselves to what extent and within what limits the everyday concept of an individual object is still recognizable in quantum mechanics.

§14. Scientific Community and Objectivity

At first sight it would appear that in quantum mechanics the concept of scientific objectivity has been strongly shaken. Since the classic period, the idea has become familiar that a physical object is something real, existing outside of the observer, independent of him, and in particular independent of whether or not the object has been subjected to measurement. The situation is not the same in quantum mechanics. Far from it being possible to attribute to a system at every instant its measurable properties, one cannot even claim that to attribute to it so much as a wave function has a well-defined meaning, unless referring explicitly to a definite measurement. Moreover, it looks as if the result of a measurement is intimately linked to the consciousness of the person making it, and as if quantum mechanics thus drives us toward complete solipsism.

Actually, however, we know that the relations between physicists have undergone practically no change since the discovery of quantum mechanics. No physicist has retired into a solipsistic isolation. Physicists use the same means of scientific exchange as in the past and are capable of cooperation in studying the same object. Thus there really exists something like a community of scientific perception, an agreement on what constitutes the object of the investigation, and it is this that still has to be looked into.

First of all, it is easy to recognize that the act of observation, that is, the coupling between the measuring apparatus and the observer (see our example in §11), is truly a *macroscopic* action and not basically quantal. Consequently one always has the right to neglect the effect on the apparatus of the "scrutiny" of the observer. Tracing things back in time, one will obtain definite conclusions about the state of the apparatus (or the photographic plate) and consequently the state of the object before the observation (*but of course after the coupling* is turned off). Moreover, nothing prevents another observer from looking at the same apparatus; and one can predict that, barring errors, his observations will be the same. The possibility of abstracting away from the individuality of the observer and of creating a collective scientific perception therefore in no way comes seriously into question.

It might appear that the scientific community thus created is a kind of *spiritualistic society* which studies imaginary phenomena—that the objects of physics are phantoms produced by the observer himself. In classical physics, one can picture a system at every instant in a unique and continuous way by the set of all of its measurable properties, even when it is not subjected to observation. It is exactly the possibility of this continuity of connection between properties and object that has ordinarily been considered as proof that physics deals with something "real," that is, having in principle an existence "independent of all observers." In quantum mechanics an object is the carrier, not of a definite set of measurable properties, but only of a set of "potential" probability distributions or statistics

(§3) referring to measurable properties, statistics which only come into force on the occasion of an effective, well-defined measurement. If one abstracts away from all acts of measurement, it is meaningless to claim these measurable properties as realized; the very mathematical form of the statistics does not allow it (see §8).

But that does not keep us from predicting or interpreting experimental results. Theory fixes the rules. It teaches us first of all how to filter an object to get a pure case—that is, reproducible conditions; then it suggests how to make measurements, either to check theoretical predictions or to discover new empirical regularities. The theory adapts itself truly marvelously to the realities of experiment. It gives answers on all desired details and is silent on hypothetical questions without experimental meaning.

In present physics the concept of "objectivity" is a little more abstract than the classical idea of a material object. Is it not a guarantee of "*the objectivity*" *of an object* that one can at least formally attribute measurable properties to it in a *continuous manner* even at times when it is not under observation? The answer is No, as this new theory shows by its internal consistency and by its impressive applications. It is enough, evidently, that the properties of the object should be present at the moment they are measured and that they should be predicted by theory in agreement with experiment.

In the limiting case of macroscopic phenomena, quantum theory rejoins classical theory. Thus it justifies the use of the "naive" concept of "objectivity" and at the same time specifies the limitations of this concept.

What has just been said relates to an important philosophical problem that we cannot enter into here: the determination of the necessary and sufficient conditions for an object of thought to possess objectivity and to be an object of science. This problem was perhaps posed for the first time in any general way by such mathematicians as Malebranche, Leibniz, and especially by B. Bolzano. More recently Husserl (1901, 1913; see also the rather similar ideas in Cassirer, 1910, 1936) has systematically studied such questions and has thus created a new method of investigation called "Phenomenology."

Physics insofar as it is an empirical science cannot enter into such problems in all their generality. It is satisfied to use philosophical concepts *sufficient* for its needs; but on occasion it can recognize that some of the concepts that once served it have become *quite unnecessary*, that they contain elements that are useless and even incorrect, actual obstacles to progress. One can doubt the possibility of establishing philosophical truths by the methods of physics, but it is surely not outside the competence of physicists to demonstrate that *certain statements which pretend to have a philosophical validity do not*. And sometimes these "negative" philosophical discoveries by physicists are no less important, no less revolutionary for philosophy than the discoveries of recognized philosophers.

II.2 INTERPRETATION OF QUANTUM MECHANICS

EUGENE P. WIGNER

§1. PROBLEMS RAISED BY QUANTUM THEORY BEFORE THE ADVENT OF QUANTUM MECHANICS

The conceptual problems generated by the generally accepted interpretation of quantum mechanics overshadow in philosophical depth those generated by the older quantum theory (see for example the collections of basic papers edited by ter Haar, 1967, and Kangro, 1972) to such an extent that one is likely to forget the latter. Nevertheless, these were very real also, though more concrete and lying more within physics proper than those generated by quantum mechanics.

The idea of the quantum emission and absorption of radiation was conceived by M. Planck (1900a,b) in order to explain the finite energy density of the black body radiation. Classical electrodynamic theory gave an infinite density for this radiation. The energy density per unit frequency, as calculated on the basis of this theory, was

$$8\pi\nu^2 kT/c^3$$

and the integral of this over the frequency ν is clearly infinite. Partly on the basis of experimental information, partly on intuition, and partly to maintain conformance with Wien's displacement law, Planck replaced this by

$$\frac{8\pi h\nu^3/c^3}{e^{h\nu/kT} - 1},$$

Lectures originally given in the Physics Department of Princeton University during 1976, as revised for publication, 1981.

and in order to "explain" it, postulated that both emission and absorption of radiation occur instantaneously and in finite quanta. It may be worth mentioning that, somewhat later, realizing how drastic this assumption was, he modified it somewhat, postulating that the absorption is a continuous process. [An account of the history of Planck's black body theory and of its influence on the development of quantum mechanics can be found in a book by Kuhn (1978) which contains also an extensive list of references. —Eds.]

Bohr's postulate of the quantum condition for the electronic orbits was the second building block of pre-quantum-mechanical quantum theory. This also was highly successful, explaining the spectrum of atomic hydrogen clearly and also, in a qualitative, but only qualitative way, the periodic system and hence some basic properties of all atoms (Bohr, 1913a,b,c; 1914; 1915a,b). The latter considerations were qualitative and it remains true that, in spite of many efforts, no mathematically consistent rules could be formulated specifying the orbits of electrons in systems with more than a single electron. In spite of this, the picture, analogous to the present Hartree-Fock picture, was quite successful. Nevertheless, the absence of mathematically consistent rules on the basis of which the electronic orbits, and hence the energy levels, could be determined was greatly disturbing. In addition, it was, of course, a mystery how the electron jumps from one precisely defined orbit to another. The problems which were most deeply felt were, however, different, and were concerned principally with the flow of the conserved quantities, such as energy.

The large cross section of atoms for the absorption of light, which, for instance, for the Na resonance radiation is around 10^{-9} cm^2, was in agreement with classical electromagnetic theory but difficult to understand if light consisted of quanta with a point-like structure, since the radii of atoms are around 3×10^{-8} cm. On the other hand, the uniform spread of the radiation energy over the area of the light beam, as postulated by classical theory, was in apparent contradiction with the fact that the emission of the photoelectrons was not delayed by a decrease of the intensity of the light beam—not even if the average energy incident on an atom on the surface of the electron emitter was less than a thousandth of the energy needed to liberate the electron. This did indicate a concentration of the energy in point-like quanta, so that the striking of an atom by a quantum could furnish the energy needed to liberate the electron. In order to reconcile the two phenomena, Bohr, Kramers, and Slater (1924) postulated that the conservation law of energy is valid only statistically. However, the experiments of Bothe and Geiger (1924), and of Compton and Simon (1952a,b,c), refuted this assumption.

Another phenomenon which was difficult to explain was the Stern-Gerlach effect (Gerlach and Stern, 1921; 1922). A beam of silver atoms is split, by an inhomogeneous magnetic field, into two beams, such that the angular momentum of the atoms in one of the beams has a definite direction, and that of the atoms in the other beam has the opposite direction. Originally, evidently, the angular momenta were randomly oriented. How could they all assume one of two definite directions? Surely the final state must depend continuously on the intial state. It was also perturbing that the classical picture, developed by Ehrenfest, could not account for the transfer of the angular momentum to those atoms whose angular momentum included, originally, a considerable angle with both of the two final momentum directions (Einstein and Ehrenfest, 1922).

As the last example of the difficulties, chemical association reactions may be mentioned. An illustration is $2NO_2 \rightarrow N_2O_4$. For such a reaction to take place, the two associating molecules must collide with an energy which corresponds to one of the energy levels of the compound molecule, N_2O_4 in the example cited. This has a small probability but since the levels of the compound molecule have a certain width—this was already recognized before the advent of quantum mechanics—it is not entirely impossible. However, the angular momentum of the compound is strictly quantized and the probability of a collision with a sharply defined angular momentum surely has zero probability. In spite of this, association reactions do take place. How this happens was at least as much of a mystery as how the electron from one orbit of the hydrogen atom jumps into another, equally sharply defined, orbit. An explanation was proposed by Polanyi and Wigner (1925): that the angular momentum (as considered from the coordinate system in which the total center of mass is at rest) is increased, as a result of the collision, to the nearest integer multiple of $\hbar$—an assumption leading to very nearly correct results but surely in contradiction to a basic conservation theorem.

All the phenomena mentioned were very puzzling—so puzzling indeed that many physicists doubted that a rational explanation of quantum phenomena would ever be found. It may be worthwhile, therefore, to mention Einstein's suggestion for overcoming the paradox mentioned in connection with the photoelectric effect. He believed in the concentration of the energy in quanta and that these quanta have structures similar to particles. However, their motion is governed by what he called *Führungsfeld*—that is, "guiding field"—and this obeys the equations of electrodynamics. In this way the existence of interference phenomena could be reconciled with the concentration of energy in very small—perhaps infinitesimally small—volumes. However, the picture could not be reconciled with the conservation laws for energy and momentum, and Einstein firmly believed in these. As a result Einstein never published the *Führungsfeld* idea. Schrödinger's theory (1926) reconciled the two postulates: his *Führungsfeld* the Schrödinger wave, moved not in ordinary space, but in configuration space and referred not to single particles, but to the change of the configuration of the whole system, i.e., the motion of all particles. However, his theory was not a pre-quantum-mechanical theory but a fascinating reformulation and even reinterpretation of the original Heisenberg-Born-Jordan (Heisenberg, 1925; Born and Jordan, 1925; Born, Heisenberg, and Jordan, 1926) quantum mechanics which, originally, attempted only the calculation of energy levels and transition probabilities.

It may be useful, at this point, to recall a few dates:

1900–(Dec. 14) M. Planck announces his quantum theory (Planck, 1900a,b).
1905–Einstein proposes his law of the photoelectric effect (actually, the Nobel prize was awarded to Einstein for this, not for his theory of relativity) (Einstein, 1905).
1913–N. Bohr's theory of the H atom (Bohr, 1913a,b,c).
1914–The experiment of Franck and Hertz (1914).
1916–Einstein's derivation of Planck's black-body-radiation formula (Einstein, 1916a,b; 1917a).
1921–Stern-Gerlach experiment (Gerlach and Stern, 1921).
1923–L. de Broglie suggests that matter also has a wave nature (de Broglie, 1923a,b,c; 1924a,b,c,d,e; 1925; 1926).
1925–W. Heisenberg's article which was to form the basis of matrix mechanics (Heisenberg, 1925).
1925–M. Born and P. Jordan establish matrix mechanics (Born and Jordan, 1925).
1926–E. Schrödinger proposes his equations of wave mechanics (Schrödinger, 1926, 1930).
1927–Davisson and Germer verify the wave nature of matter by their interference experiment (Davisson and Germer, 1927).
1927–W. Heisenberg's article "Über den anschaulichen Inhalt der quantentheoretischen Kinematik und Mechanik"—the uncertainty principle (Heisenberg, 1927).

At first, it was difficult for the community of physicists to accept Planck's quantum idea of absorption and emission of radiation. In fact he himself proposed a modification of it which, he expected, would make it more palatable: he suggested that only the emission process is instantaneous, while the absorption of light is a continuous process (Kuhn, 1978). Einstein was the first to accept Planck's original idea at face value and his proposal of the law of photoelectric emission was a result of this. As to the interpretation of the Schrödinger waves as *Führungsfeld*, this became generally accepted as a result of Heisenberg's 1927 paper. This paper also convinced most physicists that it is not meaningful to attribute a definite orbit, or a definite path, to particles—that the concepts in terms of which classical mechanics characterizes the states of a particle are not applicable in the microscopic domain.

Actually, Heisenberg's argument was not entirely rigorous—it was based on the analysis of measurements by means of a γ-ray microscope, but the analysis was not complete in all details. Nevertheless, the article had a profound influence. One can substitute for Heisenberg's γ-ray microscope the measurement of position by means of a light quantum sent out toward the object at time t_1, reflected by it, and received at the position of the emission at time t_2 (Figure 1). The collision of the light quantum with the object occurred at time

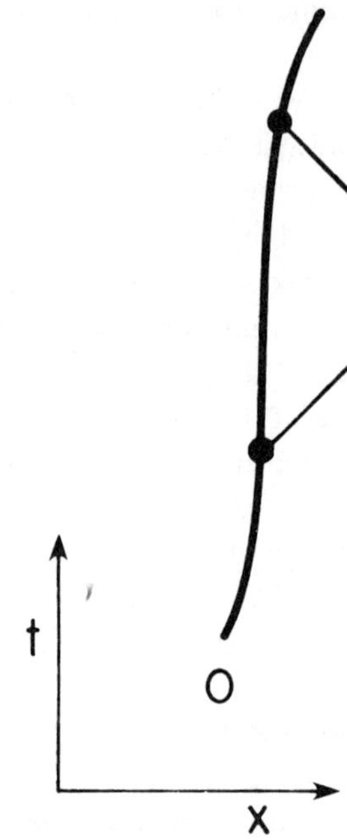

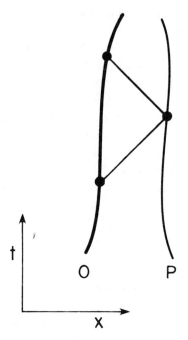

FIGURE 1. The uncertainty principle, illustrated for the particle, P. The observer, O, sends out a pulse of light at t_1 and receives this pulse, reflected, at t_2. The distance of the particle is determined to be $(t_2 - t_1)/2$, but only with the accuracy of the length of the pulse. The momentum of the particles is uncertain to the extent of the kick the photon has given to it—i.e., at least $\sim h/$[the length of the pulse].

$$t_{collision} = \tfrac{1}{2}(t_1 + t_2) \tag{1}$$

and the object's position at that time was

$$x_{collision} = \tfrac{1}{2}c(t_2 - t_1). \tag{1a}$$

However, the times t_1 and t_2 cannot be measured exactly because the light quantum's field had to have a finite extension, to be denoted by Δx. Its frequency, hence, was indeterminate to the extent $c/\Delta x$, its energy had an uncertainty $hc/\Delta x$, and its momentum $h/\Delta x$. It imparted twice its momentum to the particle; thus even if the momentum of this was accurately known before the measurement, after the measurement it was uncertain to about $\Delta p = 2h/\Delta x$. Hence the measurement described permitted the determination of the position only with an accuracy of the order $\Delta x \approx h/\Delta p$—the conclusion arrived at by Heisenberg. The point is that the measurement described could be made accurate only if one had a light wave of sharply defined frequency which was at the same time accurately localized in space. The two requirements cannot be met simultaneously.

Let us now give a mathematical proof of the uncertainty relation. We will give two different proofs, the second one of which is due to H. P. Robertson (1929).

We consider a system of units in which $\hbar\ (\equiv h/2\pi) = 1$. Then we have

$$\bar{x} = \int |\psi(x)|^2 x\, dx, \qquad \bar{p} = \int |\phi(p)|^2 p\, dp, \tag{2a}$$

with

$$\phi(p) = \frac{1}{\sqrt{2\pi}} \int \psi(x) e^{-ipx}\, dx. \tag{2b}$$

Now it will simplify the calculation a bit to set $\bar{x} = 0$ and $\bar{p} = 0$. This represents no loss of generality as it merely amounts to a change of the integration variable as far as x is concerned and a multiplication of ψ by $\exp(-i\bar{p}x)$ to annul the new $\bar{p}$. The uncertainties Δx, Δp are then defined by

$$(\Delta x)^2 = \int |\psi(x)|^2 x^2\, dx, \tag{3a}$$

and

$$(\Delta p)^2 = \int |\phi(p)|^2 p^2\, dp = \int |\partial\psi/\partial x|^2\, dx. \tag{3b}$$

We want to prove that for every state

$$(\Delta x)^2 (\Delta p)^2 \geq \tfrac{1}{4}.$$

We can restrict ourselves without loss of generality to the case where the two uncertainties are equal. This is so because the substitution of $\psi_{new}(x) = \sqrt{\alpha}\psi(\alpha x)$ (the $\sqrt{\alpha}$ is inserted to keep ψ_{new} normalized) decreases Δx by α and increases Δp by α. It does not change $\Delta x\ \Delta p$, but makes it possible to adjust them so that Δx and Δp become equal. Hence, we can assume that this is the case and we can set

$$\begin{aligned} \Delta x\, \Delta p &= \tfrac{1}{2}[(\Delta x)^2 + (\Delta p)^2] \\ &= \tfrac{1}{2} \int [x^2|\psi(x)|^2 + |\partial\psi/\partial x|^2]\, dx. \end{aligned} \tag{4}$$

This is the functional that gives the energy of the harmonic oscillator. It is minimized by $\psi = c \exp(-x^2/2)$ (c is the normalization constant, $\pi^{-1/4}$). The two terms inside the integral of (4) are then indeed equal, and equal to $\frac{1}{2}$, and add to unity. Hence we find, in this case,

$$\Delta x\, \Delta p \geq \tfrac{1}{2} \tag{5}$$

and, by virtue of the remark before (4), this holds also generally, as we wanted to show.

The second proof goes as follows. One writes

$$\int \psi^* \left(-x \frac{\partial}{\partial x} \right) \psi \, dx = a, \qquad \int \psi^* \left(\frac{\partial}{\partial x} x \right) \psi \, dx = b, \tag{6}$$

the star denoting transition to the conjugate complex. Then, because of the normalization of ψ, we know that

$$\int \psi^* \left(-x \frac{\partial}{\partial x} + \frac{\partial}{\partial x} x \right) \psi \, dx = \int \psi^* \psi \, dx = \int |\psi|^2 \, dx = 1,$$

or

$$a + b = 1. \tag{6a}$$

Now, by Schwartz's inequality, we have

$$|a|^2 = \left| \int \psi^* x \frac{\partial \psi}{\partial x} \, dx \right|^2 \leq \int |x\psi(x)|^2 \, dx \int \left| \frac{\partial \psi}{\partial x} \right|^2 dx, \tag{7}$$

so (again assuming for simplicity $\bar{x} = \bar{p} = 0$)

$$|a|^2 \leq (\Delta x)^2 (\Delta p)^2. \tag{8}$$

Next one recognizes by partial integration that

$$b = a^*. \tag{9}$$

This means that if we set

$$a = u + iv \tag{9a}$$

we must have, by (6a),

$$2u = 1. \tag{9b}$$

This result in turn implies

$$|a|^2 = u^2 + v^2 \geq \tfrac{1}{4}; \tag{10}$$

and this together with (7) again proves (5).

It is worthwhile to remark here that the above "proof" for the uncertainty relation is fundamentally different from Heisenberg's. Like equations (3a) and (3b), it is based on the statistical interpretation of (nonrelativistic) quantum mechanics. Those equations do not refer to the actual ways x and p could be measured—the subjects of Heisenberg's article. Thus, they avoid the basic question: how is a measurement to be carried out, how do we prepare the object on which the measurement is taking place? They assume that the probabilities of the various outcomes of the measurement of x and p are correctly given by the present quantum mechanics. The conclusion we obtained is of a statistical nature: There is no state of any system such that the results of the measurements of x and p would be predictable with greater accuracies Δx and Δp than are indicated by (5). Evidently, the verification of this statement requires, first, the repeated production of the same state many times so that the statistical distribution of the outcomes of the measurements of x and of p can be ascertained. This is in itself a difficult problem; one can never be absolutely sure that one has produced the same state of the system. At the first preparation somebody may have looked on—can we be sure that this did not affect the state of the system produced? Similar remarks apply to the measurements. There is also the question: When is the measurement completed? We will see that, if we adhere strictly to the principles of quantum mechanics, the measurement is completed only when we have observed its outcome, i.e., have read the recording of the measuring apparatus. Apparently, this is not a precisely repeatable process. Could the fluctuations of the outcome of the measurements of x and p come from these sources? Evidently we believe that this is not the case; but it would be difficult to explain this belief convincingly to a fully detached person. In spite of this, we are all fully convinced of all this and do not believe that any refinement of the preparation of the system the x and p of which is to be measured, or any improvement of the measuring technique, would lead to a violation of (5).

Before going into a more detailed discussion of these questions, we will review the mathematical structure of quantum mechanics, its description of states, and its calculation of the probabilities of the outcomes of measurements. The equations used to derive (5) will naturally appear as special cases of the general theory.

§2. The Mathematical Formalism of Quantum Mechanics

Hilbert Space

In quantum mechanics, as in classical physics, we postulate the existence of *isolated systems.* In both theories, if a complete description of an isolated system is given at one time, a complete description for any other time is uniquely determined as long as the system remains isolated—i.e., is not influenced by any other system. In this sense, both theories are deterministic. The means of description of the state of the system have, however, undergone drastic changes over the course of the history of physics. In the original Newtonian mechanics, the state of the system was described by the positions and velocities of its constituents. In the later field theories, it was described by the field strengths at all points of space—that is, by one or more—in the case of electrodynamics six—functions of three real variables, the latter characterizing the points of ordinary space. The usual quantum-mechanical description of the state of a system is much more abstract; it is given by a vector (or, more accurately, by a ray) in an abstract complex Hilbert space. This vector is usually called the state vector. Since the Hilbert space has infinitely many dimensions, this amounts to the specification of the state by an infinite set of complex numbers, $a_1, a_2, a_3, \ldots$, the components of the state vector in Hilbert space, which however, satisfy the condition that the sum

$$\sum |a_i|^2 < \infty \tag{1}$$

is finite. This means that the *length* of the vectors in Hilbert space is finite. Furthermore, it is postulated that two state vectors which have the same direction, i.e., the components of which differ only by a common factor, characterize the same state. The set of vectors the components of which differ only by a common factor, e.g., the vectors with the components $a_1, a_2, a_3 \ldots$ and the vectors $ca_1, ca_2, ca_3 \ldots$ for any $c \neq 0$, are said to form a *ray*. We can choose, therefore, one of these vectors to describe the state, and we usually choose one that is normalized, i.e., one for which

$$\sum |a_i|^2 = 1. \tag{1a}$$

The different normalized vectors describing the same state differ only by a factor ω of modulus 1.

Schrödinger's original formulation of his "wave mechanics" was, of course, not in terms of the Hilbert space. Wave mechanics characterized the states of systems in terms of complex valued functions, actually functions in configuration space. However, if one introduces an orthonormal set of functions $u_n(x_1, x_2 \ldots)$ in any space,

$$\int \cdots \int u_n(x_1, x_2 \ldots)^* u_m(x_1, x_2 \ldots)\, dx_1\, dx_2 \cdots = \delta_{nm}, \tag{2}$$

one can expand the wave function ψ of any state in terms of this set. Thus,

$$\psi(x_1, x_2 \ldots) = \sum a_n u_n(x_1, x_2 \ldots), \tag{2a}$$

$$a_n = \iint u_n(x_1, x_2 \ldots)^* \psi(x_1, x_2 \ldots)\, dx_1\, dx_2 \ldots; \tag{2b}$$

and the numbers a_n can be considered to be the components of a vector in Hilbert space as long as

$$\int \cdots \int |\psi(x_1, x_2 \ldots)|^2\, dx_1\, dx_2 \cdots = \sum |a_n|^2 \tag{2c}$$

is finite—which we assume to be the case for Schrödinger's wave functions.

It may be worthwhile to observe at this point that the correspondence between functions ψ and the corresponding vector a in Hilbert space is not one to one. Two functions ψ and ψ' which differ only on a set of measure 0 (for instance only on a denumerable set of points) correspond to the same vector in Hilbert space. However, this observation rarely plays an important role in actual calculations.

The most important derivative concept in Hilbert space is that of the scalar product. The scalar product of two vectors a and b is defined as

$$(a, b) = \sum a_n^* b_n. \tag{3}$$

In terms of the wave functions this is

$$\begin{aligned}(\psi, \phi) &= \int \cdots \int \psi(x_1, x_2 \ldots)^* \phi(x_1, x_2 \ldots) dx_1\, dx_2 \cdots \\ &= \sum a_n^* b_n,\end{aligned} \tag{3a}$$

the last part of the equation being valid if the a are, according to (2a), the expansion coefficients of ψ and the b are, similarly, the expansion coefficients of ϕ. That the sum in (3) is finite follows from (1), and the similar restriction follows for b, by means of Schwartz's inequality. It follows from (3) or (3a) also that

$$(a, b) = (b, a)^*, \tag{3b}$$

the star denoting, as before, the conjugate complex.

If a and b are normalized according to (1a), the absolute square $|(a, b)|^2$ of the scalar product (a, b) is called, for reasons which will appear soon, the transition probability from the state a into the state b—or conversely, because of (3b). Otherwise, the transition probability is

$$\frac{|(a, b)|^2}{(a, a)(b, b)}. \tag{4}$$

It may be worth remarking that (a, a) is real and also positive unless $a = 0$, and that the scalar product is linear in the second factor,

$$(a, \beta b + \beta' b') = \beta(a, b) + \beta'(a\ b'), \tag{5}$$

and antilinear in the first factor,

$$(\alpha a + \alpha' a', b) = \alpha^*(a, b) + \alpha'^*(a', b), \tag{5a}$$

where α, α', β, and β' are arbitrary complex numbers. These equations are immediate consequences of the definition (3) of the scalar product.

Linear Operators in Hilbert Space

An operator in Hilbert space transforms a vector in Hilbert space into another (or the same) vector in the same space. One should really write $A(\phi)$ for the vector into which the operator A transforms the vector ϕ, but, at least for linear operators A, one writes $A\phi$ instead. An operator A is called linear if for any two vectors ϕ, ψ and any two numbers α, β

$$A(\alpha\phi + \beta\psi) = \alpha A\phi + \beta A\psi \tag{6}$$

holds. More generally, one can write

$$A(\sum a_n\psi_n) = \sum a_n A\psi_n. \tag{6a}$$

Hence if we assume that the ψ_n in (6a) form a complete orthonormal set of vectors, and if we expand the $A\psi_n$ in terms of these vectors, we can write

$$A\psi_n = \sum_m A_{mn}\psi_m. \tag{7}$$

Then (6a) gives for the transformation properties of the coefficients a_n,

$$a_n \to \sum_m A_{nm}a_m. \tag{7a}$$

The operation A acts on the expansion coefficients as the matrix (A_{nm}), the matrix elements of which are, because of (7)

$$A_{mn} = (\psi_m, A\psi_n). \tag{7b}$$

The last two equations are valid if the ψ_m form a complete orthonormal set; but (6a) is valid as a consequence of (6) for any set of vectors ψ_m.

Two special kinds of linear operators play particularly important roles in quantum mechanics. The invariance transformations, such as the time-displacement transformation, are mediated by unitary transformations U which leave the scalar products, and hence the transition probabilities (4), between any two vectors unchanged:

$$(U\phi, U\psi) = (\phi, \psi). \tag{8}$$

They will play a lesser role in the considerations which follows. The other special kind of linear operators of basic significance for what follows are the self-adjoint operators. In order to define them, one first defines the adjoint $A^\dagger$ to an operator A. This is so defined that for any two vectors ϕ and ψ,

$$(\phi, A\psi) = (A^\dagger\phi, \psi) \tag{9}$$

is valid. One concludes from (9) and (3b) that

$$(A\psi, \phi) = (\psi, A^\dagger\phi), \tag{9a}$$

so that A is the adjoint of $A^\dagger$ if $A^\dagger$ is the adjoint of A. The matrix elements of $A^\dagger$ are, as one can infer from (7b),

$$(A^\dagger)_{nm} = A^*_{mn}, \tag{9b}$$

i.e., the matrices of A and $A^\dagger$ are what we call hermitian adjoints. The second type of operators of basic significance can now be defined. They are those which are equal to their adjoints,

$$A = A^\dagger. \tag{10}$$

In the corresponding matrices the matrix elements which lie symmetrically with respect to the main diagonal are the conjugate complexes of each other.

For the sake of accuracy it should be pointed out that the existence of the adjoint $A^\dagger$ to an arbitrary linear operator is by no means obvious, and requires, if all the questions of convergence are to be treated rigorously, a reasonably elaborate proof (von Neumann, 1932). Once its existence is established, i.e., once it is shown that there is a ϕ' independent of ψ satisfying the equation

$$(\phi, A\psi) = (\phi', \psi) \tag{11}$$

(ϕ' depending only on A and ϕ), the linear dependence of this $\phi' = A^\dagger\phi$ on ϕ is easily proved. We have on the one hand,

$$(\alpha\phi_1 + \beta\phi_2, A\psi) = [A^\dagger(\alpha\phi_1 + \beta\phi_2), \psi], \tag{12}$$

and on the other,

$$\begin{aligned}(\alpha\phi_1 + \beta\phi_2, A\psi) &= \alpha^*(\phi_1, A\psi) + \beta^*(\phi_2, A\psi)\\ &= \alpha^*(A^\dagger\phi_1, \psi) + \beta^*(A^\dagger\phi_2, \psi)\\ &= (\alpha A^\dagger\phi_1 + \beta A^\dagger\phi_2, \psi).\end{aligned} \tag{12a}$$

Hence the right sides of (12) and (12a) are equal and, since this is true for every ψ,

$$A^\dagger(\alpha\phi_1 + \beta\phi_2) = \alpha A^\dagger\phi_1 + \beta A^\dagger\phi_2, \tag{12b}$$

i.e., $A^\dagger$ is linear.

Normal Form of Self-Adjoint Operators

Except for the proof of the existence of the hermitian adjoint $A^\dagger$, the preceding discussion is straightforward and easy. This is not the case for the following discussion, particularly not if the concept of the operator is extended so that it encompasses also unbounded operators. The following discussion gives only the results. The detailed proofs can be found, for instance, in the book of J. von Neumann (1932) or in the book of G. R. Mackey (1963).

Let us consider first bounded self-adjoint operators, i.e., operators such that $(A\psi, A\psi)$ has an upper limit. The usual procedure to obtain the normal form is to look for the characteristic vectors ψ_ν of A

$$A\psi_\nu = \lambda_\nu\psi_\nu. \tag{13}$$

One easily sees that the characteristic values λ_ν have to be real and the characteristic vectors ψ_ν, which are solutions for different λ_ν, are orthogonal. We assume that they are also normalized, i.e., that $(\psi_\nu, \psi_\nu) = 1$. If the characteristic vectors ψ_ν form a complete orthonormal set, we say that A has only a point spectrum, this consisting of the λ_ν. The effect of A on an arbitrary vector ϕ

$$\phi = \sum (\psi_\nu, \phi)\psi_\nu \tag{14}$$

is then, because of its linear character,

$$\begin{aligned} A\phi &= A\sum (\psi_\nu, \phi)\psi_\nu = \sum (\psi_\nu, \phi)A\psi_\nu \\ &= \sum (\psi_\nu, \phi)\lambda_\nu\psi_\nu. \end{aligned} \tag{14a}$$

The measurement theory then postulates that the measurement of A on a system in the state ϕ gives one of the values λ and that it gives the value λ_ν with the probability

$$p_\nu = |(\psi_\nu, \phi)|^2. \tag{15}$$

It is assumed here that ϕ and the ψ_ν are normalized,

$$(\phi, \phi) = (\psi_\nu, \psi_\nu) = 1. \tag{15a}$$

It is also reasonable to assume that the system which was originally in the state ϕ is, after the measurement—if the result of that is λ_ν—in the state ψ_ν. Naturally it would be good to justify this postulate of measurement theory by means of the equations of motion of quantum mechanics—that is, to decribe the measurement process. Such an analysis would have to treat in quantum-mechanical language the apparatus used for the measurement of A. We will discuss later the extent to which this has so far proved to be possible, and the examples which show that for certain A it is impossible.

The preceding remarks apply to a self-adjoint operator which has a point spectrum only. One can well say that most operators do not have such a spectrum. A self-adjoint operator may have a continuous spectrum also—or only a continuous spectrum. In other words the solutions ψ_ν of (13)—with a finite length so that (15a) can be made valid—may not form a complete set. In fact, there may be no such solution at all; that is, there may be no vector ψ_ν in Hilbert space which satisfies (13) for some λ_ν. In the general case the preceding equations have to be replaced by much more complicated ones. In order to illustrate this point, it is helpful to rewrite (14) and (15) somewhat by decomposing A into projection operators P_ν. A projection operator is defined by

$$P_\nu\phi = (\psi_\nu, \phi)\psi_\nu. \tag{16}$$

It then follows from (14) that

$$\sum P_\nu = 1, \tag{17}$$

and from (14a),

$$\sum \lambda_\nu P_\nu = A. \tag{17a}$$

One easily verifies that the P_ν are self-adjoint, identical with their squares, and more generally satisfy the equations,

$$P_\nu P_\mu = \delta_{\nu\mu}P_\nu. \tag{17b}$$

In fact,

$$\begin{aligned} P_\nu P_\mu\phi &\equiv P_\nu(P_\mu\phi) = P_\nu(\psi_\mu, \phi)\psi_\mu = (\psi_\mu, \phi)P_\nu\psi_\mu \\ &= (\psi_\mu, \phi)(\psi_\nu, \psi_\mu)\psi_\nu = \delta_{\nu\mu}(\psi_\mu, \phi)\psi_\nu = \delta_{\nu\mu}P_\nu\phi. \end{aligned} \tag{17c}$$

The expression for the transition probability into ψ_ν—which is also the probability that the outcome of the measurement of A on ϕ will be λ_ν—becomes, in terms of P_ν,

$$p_\nu = (\phi, P_\nu\phi). \tag{18}$$

One can also write this in the form

$$p_\nu = (P_\nu\phi, P_\nu\phi) \tag{18a}$$

as, because of the self-adjoint nature of P_v and (17),

$$(P_v\phi, P_v\phi) = (\phi, P_v^2\phi) = (\phi, P_v\phi). \tag{18b}$$

The derivation so far has supposed that A has only a discrete spectrum, a condition for the validity of equations (17).

We now proceed to the more general case that A may also have a continuous spectrum. In this case one has to admit that the measurement will not yield a mathematically precise value—no one asks whether the outcome of this measurement is a rational or irrational number. It is more reasonable to ask, for instance, whether the outcome is smaller than a number λ. The probability for a "yes" answer to this question can be written, in the case of a discrete spectrum, as

$$p(\lambda) = [\phi, P(\lambda)\phi]. \tag{19}$$

Here

$$P(\lambda) = \sum_{\lambda_v < \lambda} P_v. \tag{19a}$$

It then follows from (18) that the probability for an outcome of the measurement between λ' and $\lambda > \lambda'$ is

$$p(\lambda) - p(\lambda') = \{\phi, [P(\lambda) - P(\lambda')]\phi\}. \tag{20}$$

It follows from the theory of self-adjoint operators in Hilbert space that the operators $P(\lambda)$ can also be defined if the spectrum is not exclusively discrete. In other words, the $P(\lambda)$ can be defined for every self-adjoint operator, for any real λ. One clearly has

$$P(\lambda) \to 0 \quad \text{for} \quad \lambda \to -\infty; \tag{21a}$$

and the analogs of (17) and (17b) are

$$P(\lambda) \to 1 \quad \text{for} \quad \lambda \to \infty \tag{21b}$$

and

$$P(\lambda)P(\lambda') = P(\lambda')P(\lambda) = P(\lambda') \quad \text{for} \quad \lambda' \leq \lambda. \tag{21c}$$

These equations, of course, do not determine the $P(\lambda)$ since they do not involve A. In order fully to define the $P(\lambda)$, one has to write the analog of (17a). This becomes, in the general case, a somewhat complicated expression—a limit of a sum with increasingly many terms. One has to form series of increasingly many λ values which cover the real line with increasing density. If the density is $1/N$, one has the series

$$A = \sum_{n=-\infty}^{\infty} \sum_{m=1}^{N} (n + m/N)[P(n + m/N) - P(n + (m - 1)/N)], \tag{22}$$

and this is valid in the limit $N = \infty$. Naturally, this is a special form for A in which the distance between successive λ values is uniformly $1/N$. There are infinitely many other ways to increase the density of the λ. The mathematician writes a Stieltjes integral for (22) and, in fact, (22) is the definition of the Stieltjes integral. The expression for A is written in the elementary form (22) because it does not presuppose familiarity with that integral. The point is that (22), together with the equations (21), fully defines the λ and hence determines the mathematical expression (20) for the probability of the outcome of the measurement of A lying between λ' and λ.

A few remarks are needed to complete the discussion. First, the possibility of finding a mathematical expression for the outcome of the measurement of an operator obviously does not guarantee the possibility of such a measurement. Second, the theory postulates measurements corresponding to operators, not to classical quantities. Originally, the attempt was made to define operators to correspond to classical quantities. The present theory does not strive for such a coordination. Thus, one does not try to choose, for instance, between $(px^2 + x^2p)/2$ and xpx as operators to correspond to the classical quantity, "product of the square of the coordinate and first power of the momentum." Instead, one asks if an experimental device can be constructed that will realize a measurement of the operator $(px^2 + x^2p)/2$ and whether another can be built that will measure xpx.

On the mathematical side: if A is bounded, $P(\lambda)$ remains 0 up to the lower bound of A and becomes 1 at the upper bound. Hence that part of the sum in (22) vanishes for which $n + m/N$ is below the lower bound; and so does that which is above the upper bound. However, and this theorem is due to von Neumann (1932), the preceding equations as written are valid also for general self-adjoint operators, not only for bounded ones. It is worthwhile, nevertheless, to add a few words relating to the former, since most common operators are not bounded. There are vectors in Hilbert space to which an unbounded operator cannot be applied. Both multiplication by x and $i\partial/\partial x$ are unbounded operators. For instance, multiplication of the state vector $1/(i + x)$ by x does not produce a state vector, since $\psi(x) = x/(i + x)$ is not square integrable; for it the expression (2c) is infinite. For an unbounded operator it is postulated only that there be an everywhere dense set of vectors to which it can be applied—i.e., that in the neighborhood of any chosen vector there be vectors arbitrarily close to it to which the operator can be applied. In the case of $1/(i + x)$ such vectors are, for instance $e^{-\varepsilon x^2}/(i + x)$ with decreasing values of ε. In fact, the square of the difference vector,

$$\int \left| \frac{1}{i + x} - \frac{e^{-\varepsilon x^2}}{i + x} \right|^2 dx, \tag{23}$$

goes to zero as ε goes to zero. Moreover, multiplication with x leaves $e^{-\varepsilon x^2}/(i + x)$ within the Hilbert space no matter how small ε is, as long as it remains positive.

It is remarkable that the probability of the outcome of measurements, given by (20), is well defined even for unbounded operators, even in the case of state vectors ϕ to which the unbounded operator cannot be applied. The intrinsic reason for this is that the measurement hardly refers to the operator and its spectrum—what it really purports to define is the transition probability into the state $[P(\lambda) - P(\lambda')]\phi$ if the original state was ϕ. The projection operators $P(\lambda)$ could be given other labels, such as $\lambda/(\lambda^2 + 1)^{1/2}$, and the measurement would still be the same, only its outcome would be called differently. To repeat, however, all this is theory, and there is no mathematical guarantee that even a transition probability into an arbitrary state can be measured. In order to guarantee the possibility of such a measurement one would have to describe a way to do it. The mathematical theory of the measurement, as formulated first by von Neumann but now generally accepted, ingenious as it is, does not do that.

It may be worthwhile to observe, nevertheless, that the self-adjoint nature of an unbounded operator imposes rather rigorous criteria. First, the operator and its adjoint must give the same vector if applied to a vector which lies in the intersection of the domains of definition of the two operators. Second, the two domains of definition must be identical (von Neumann called them hypermaximal). Fortunately, this criterion apparently plays no practical role in the theory.

§3. Direct Product and Quantum Mechanical Description of the Measuring Process

The Direct Product of Hilbert Spaces

Starting from two Hilbert spaces, one can construct a single larger Hilbert space, a sort of union of the two. If the axes of the first Hilbert space are specified by Greek indices ν, and of the second by Latin indices such as n, then the axes of the direct product are specified by a double index, such as νn. Thus νn specifies a single axis of the Hilbert space which is the direct product of the original two Hilbert spaces. The number of axes νn is still denumerable even though it appears to be greater than the number of axes of the factors: one can order pairs of ordered numbers into a single series such as 11; 12, 21; 13, 22, 31; 14, 23, and so on. The components $\Psi_{\nu n}$ of a vector Ψ in the space of the direct product have two indices, and the scalar product of two vectors Ψ and Φ in that space is

$$(\Psi, \Phi) = \sum_{\nu n} \Psi^*_{\nu n}\Phi_{\nu n}. \tag{24}$$

The Hilbert space which is the direct product of two Hilbert spaces H_1 and H_2 is usually denoted by $H_1 \otimes H_2$.

One also defines the direct product of two vectors, one of which, ϕ, is in H_1, and the other, ψ, in H_2. This direct product, denoted by $\phi \otimes \psi$, is in the space $H_1 \otimes H_2$. Its νn component is

$$(\phi \otimes \psi)_{\nu n} = \phi_\nu \psi_n. \tag{25}$$

The direct product is linear in its two factors. Thus

$$(\alpha\phi + \alpha'\phi' \otimes \psi) = \alpha(\phi \otimes \psi) + \alpha'(\phi' \otimes \psi), \tag{25a}$$

and a similar equation applies if ψ is replaced by $\beta\psi + \beta\psi'$. In order to verify (25a), one compares the νn components of the two sides. These are $(\alpha\phi + \alpha'\phi')_\nu\psi_n$ and $\alpha\phi_\nu\psi_n + \alpha'\phi'_\nu\psi_n$ and the two are equal.

The square length of $\phi \otimes \psi$ is

$$\begin{aligned}(\phi \otimes \psi, \phi \otimes \psi) &= \sum_{\nu n} |(\phi \otimes \psi)_{\nu n}|^2 = \sum_{\nu n} |\phi_\nu \psi_n|^2 \\ &= \sum_\nu |\phi_\nu|^2 \sum_n |\psi_n|^2,\end{aligned} \tag{25a}$$

and is the product of the squares of the lengths of the two factors ϕ and ψ. The scalar product of two direct product vectors $\phi \otimes \psi$ and $\phi' \otimes \psi'$ in the new Hilbert space becomes, similarly,

$$\begin{aligned}(\phi \otimes \psi, \phi' \otimes \psi') &= \sum_{\nu n} (\phi \otimes \psi)^*_{\nu n}(\phi' \otimes \psi')_{\nu n} = \sum_{\nu n} (\phi_\nu\psi_n)^*\phi'_\nu\psi'_n \\ &= \sum_\nu \phi^*_\nu\phi'_\nu \sum_n \psi^*_n\psi'_n = (\phi, \phi')(\psi, \psi'),\end{aligned} \tag{25b}$$

that is, the product of the scalar products of the two factors.

It is important to remark, finally, that the direct product of two factors ϕ and ψ is the same vector in the new Hilbert space no matter which coordinate systems are used for its definition in (25). If the unit vectors in the direction of the coordinate axes used in (25) for the two Hilbert spaces H_1 and H_2 are denoted by e_ν and f_n, the νn component of $\phi \otimes \psi$ is, by (25),

$$(\phi \otimes \psi)_{\nu n} = \phi_\nu\psi_n = (e_\nu, \phi)(f_n, \psi) = (e_\nu \otimes f_n, \phi \otimes \psi), \tag{26}$$

This equation remains valid for any other choice of the coordinate axes in the original Hilbert spaces. One can verify this statement by explicit calculation. It follows from (26) that

$$\phi \otimes \psi = \sum_{\nu n} (e_\nu, \phi)(f_n, \psi)e_\nu \otimes f_n. \tag{26a}$$

Hence, the component of the state vector in the new $e'_\mu \otimes f'_m$ direction becomes

$$(e'_\mu \otimes f'_m, \phi \otimes \psi) = \sum_{\nu n} (e_\nu, \phi)(f_n, \psi)(e'_\mu \otimes f'_m, e_\nu \otimes f_n)$$

$$= \sum_{\nu n} (e_\nu, \phi)(f_n, \psi)(e'_\mu, e_\nu)(f'_m, f_n). \tag{26b}$$

But we have the equality,

$$\sum_\nu (e'_\mu, e_\nu)(e_\nu, \phi) = \sum_\nu ((e_\nu, e'_\mu)e_\nu, \phi) = (e'_\mu, \phi); \tag{26c}$$

and a similar equation applies for the f. Therefore, we have verified explicitly that

$$(e'_\mu \otimes f'_m, \phi \otimes \psi) = (e'_\mu, \phi)(f'_m, \psi). \tag{26d}$$

This result proves that the equation (25) defining the direct product of the two vectors is independent of the coordinate systems used in the original Hilbert spaces.

The direct product of two Hilbert spaces, and the direct product of vectors in them, is introduced in order to describe the joining of two systems into a single system. This is important if one wants to describe the interaction of two systems which were originally separated—in our case the interaction between the measuring apparatus and the system on which the measurement is undertaken. If it is possible to describe the two systems in separate Hilbert spaces, their union can indeed be most easily characterized in the direct product of these Hilbert spaces. In this "product space" the state vector of the union is the direct product of the state vectors of the components. Indeed, if one wants to calculate the probability that the first system is in state ϕ' and the second is in state ψ' when their actual state vectors are ϕ and ψ, one obtains by (25b),

$$|(\phi' \otimes \psi', \phi \otimes \psi)|^2 = |(\phi', \phi)|^2 |(\psi', \psi)|^2, \tag{27}$$

which is the product of the transition probabilities from ϕ into ϕ' and from ψ into ψ'. Since the two system are assumed to be independent, this is the expected result.

Direct Products of Operators

In order to obtain the more general expressions corresponding to (18), etc. for the joint system, it is useful to introduce the concept of the direct product of two operators A and B, acting in the two original Hilbert spaces H_1 and H_2. The action of A is described by the equation

$$(A\phi)_\nu = \sum_{\nu'} A_{\nu'\nu}\phi_{\nu'}; \tag{28a}$$

that of B by

$$(B\psi)_n = \sum_{n'} (B_{n'n}\psi_{n'}. \tag{28b}$$

Their direct product, to be denoted by $A \otimes B$, will then transform Ψ with the components $\Psi_{\nu n}$ into $A \otimes B\Psi$, the components of which are

$$((A \otimes B)\Psi)_{\nu n} = \sum_{\nu' n'} A_{\nu'\nu}B_{n'n}\Psi_{\nu' n}. \tag{28}$$

This is the definition of $A \otimes B$. Clearly, if Ψ is a direct product, $\Psi = \phi \otimes \psi$. The action of $A \otimes B$ on this will give

$$(A \otimes B)\phi \otimes \psi = A\phi \otimes B\psi, \tag{29}$$

the direct product of the results of the actions of A and of B in their respective Hilbert spaces.

The concept of the direct product of two operators enables us to generalize the expression for the transition probability (27) as the similar expression (15) for a single system was generalized, first in (18) for the case of a discrete spectrum, and then in (19) and (20) to the case where a continuous spectrum may also be present. First, if both operators A and B, to be measured on the two systems, have only a discrete spectrum, then let us denote the projection operators which correspond to the characteristic values α_ν of A and b_n of B with P_ν and Q_n, respectively. The probability that the outcome is a_ν for the measurement of A on ϕ is then $(\phi, P_\nu\phi)$. The probability that the outcome is b_n for the measurement of B on ψ is $(\psi, Q_n\psi)$. The probability that the two outcomes on the joint system be a_ν and b_n is then given by

$$P_{\nu n} = (\phi \otimes \psi, (P_\nu \otimes Q_n)(\phi \otimes \psi)) = (\phi, P_\nu\phi)(\psi, Q_n\psi), \tag{30}$$

as follows from (29). We conclude that $P_\nu \otimes Q_n$ is the projection operator for the outcome a_ν of A and b_n of B. Similarly, we can generalize (19): the probability for A giving a result smaller than λ *and* B giving a result smaller than l is

$$p(\lambda, l) = \{\phi \otimes \psi, [P(\lambda) \otimes Q(l)]\phi \otimes \psi\} = [\phi, P(\lambda)\phi][\psi, Q(l)\psi]. \tag{31}$$

Here $P(\lambda)$ is defined for operator A as in (19a) and $Q(l)$ is defined similarly for B. Equation (31) is valid even when there is a continuous spectrum. The formula for the probability that the outcomes will fall in the intervals λ, λ' and l, l', respectively, is obtained equally easily as

$$[p(\lambda', l') - p(\lambda, l')] - [p(\lambda', l) - p(\lambda, l)]$$
$$= (\phi \otimes \psi, [P(\lambda') - P(\lambda)][Q(l') - Q(l)]\phi \otimes \psi), \tag{32}$$

in generalization of (20). The first two terms provide the probability that B will fall below l' and A between λ and λ'. The square bracket gives the probability for the same interval of A and that B will fall below l. Hence (32), the difference between the two, gives the probability that A will fall between λ and λ' and B between l and l'. The right side is a symmetric expression for this probability. Actually, these formulae are given only for the sake of completeness; they will not be used. The same applies to the general formula,

$$(A \otimes B)(A' \otimes B') = (AA' \otimes BB'). \tag{33}$$

It is worth remembering, though, that the projection operator in the direct product space for the probability of the outcome a_ν for A without specification of the outcome for any measurement on the second system is $P_\nu \otimes 1$; and conversely, it is $1 \otimes Q_n$ for the outcome b_n of B as measured on the second system if no measurement on the first system is undertaken. This concludes our mathematical discussion of the direct product concept.

It should be admitted, however, that the exclusion principle's requirement of the antisymmetrization of the state vector puts a certain restriction on the postulate that the states of the two noninteracting systems be described in separate Hilbert spaces. As a result of the antisymmetrization, this separation is not actually directly possible. A similar remark applies for particles obeying Bose statistics and demanding a symmetrized state vector. It does not seem, though, that this restriction is truly relevant in the case to be considered: the two Hilbert spaces may describe not the two different objects but the conditions in two distinct parts of space, in our case that of the object and that of the apparatus. It seems that field theories might not encounter this difficulty, but since actually no one seems to believe that a more precise discussion, taking the symmetric or antisymmetric nature of the Schrödinger wave functions into account, would alter any of the conclusions, such a precise discussion of this point does not appear in the literature. It may be true that it would be good to provide such a discussion for the sake of accuracy.

The Quantum Mechanical Description of Measurement

The concept of the direct product greatly facilitates the quantum mechanical description of the measurement process. This consists of a temporary interaction between the object on which the measurement is undertaken and the apparatus which performs the measurement. Let us consider, first, a measurement on an object which is in a state for which the outcome is surely determined—i.e., a state, the state vector σ_κ of which is a characteristic vector of the quantity to be measured.

If we denote the initial state vector of the apparatus by a_0, then the initial state of apparatus plus object is $a_0 \otimes \sigma_\kappa$. After the measurement, the apparatus has assumed a state which shows the outcome of the measurement; we denote its state vector by a_κ. Hence, if the object did not change its state as a result of the measurement—this assumption will be discussed further later—the interaction between apparatus and object transforms $a_0 \otimes \sigma_\kappa$ into

$$a_0 \otimes \sigma_\kappa \rightarrow a_\kappa \otimes \sigma_\kappa, \tag{34}$$

and this is assumed to be valid for all κ. Thus, the art of measuring the quantity with the characteristic vectors σ_κ consists in producing an apparatus the interaction of which with the object has the result indicated by (34).

It now follows from the linear nature of the quantum-mechanical equations of motion that if the initial state of the object is a linear combination of the σ_κ, say $\sum \alpha_\kappa \sigma_\kappa$, the final state of object plus apparatus will be given by

$$a_0 \otimes \sum \alpha_\kappa \sigma_\kappa = \sum \alpha_\kappa (a_0 \otimes \sigma_\kappa) \rightarrow \sum \alpha_\kappa (a_\kappa \otimes \sigma_\kappa). \tag{35}$$

The second member of (35) follows from the linearity of the direct product in terms of its factors, an immediate consequence of (25); the last member, from the linear nature of the time-development operator.

Do the processes postulated in (34) and (35) fully describe the measurement? In the case of (34) this is true: the apparatus assumes a definite state which indicates the state of the object. In the general case, described by (35), this is not the case: the apparatus is not, with the desired probability $|\alpha_\kappa|^2$, in the state a_κ. In fact, the joint state of object and apparatus appears quite complicated. This could not have been expected to be otherwise: the transformation indicated by the arrow is a consequence of the quantum-mechanical equation of motion and this is deterministic. The outcome of the measurement in the general case of (35) has a probabilistic nature. What the final state of (35) does show is that a statistical correlation between the state of the apparatus and that of the object has been established: the probability of finding both in the state v is, according to the standard postulate of quantum mechanics,

$$\begin{aligned} \left| a_v \otimes \sigma_v, \sum \alpha_\kappa (a_\kappa \otimes \sigma_\kappa) \right|^2 &= \left| \sum_\kappa \alpha_\kappa (a_v, a_\kappa)(\sigma_v, \sigma_\kappa) \right|^2 \\ &= \left| \sum_\kappa \alpha_\kappa \delta_{v\kappa} \right|^2 = |\alpha_v|^2. \end{aligned} \tag{36}$$

Both (a_v, a_κ) and $(\sigma_v, \sigma_\kappa)$ have been set equal to $\delta_{\kappa\bar{v}}$—the former because the states a_v and a_κ are supposed to be distinguishable even at the macroscopic level, the latter because the σ are normalized characteristic vectors of the self-adjoint operator which is being measured. A calculation entirely similar to (36) then shows also that the probability is zero of finding the apparatus in state a_λ and the object in the state σ_κ with $\kappa \neq \lambda$—i.e., that indeed the right side of (35) represents a joint state of object and apparatus with the statistical correlation between the states as indicated.

It is evident, therefore, that interesting and suggestive as (35) may be (it is due, essentially, to von Neumann, 1932) it does not completely describe the quantum-mechanical measurement. In order to give (35) the meaning just outlined, we must assume that a measurement on the state of the right side of (35) is possible and that it gives the different possible values with the probabilities postulated by the usual interpretation. Actually, in order to obtain the state of the object, it is necessary only to measure the state of the apparatus. However, the quantum-mechanical description of this measurement suffices no more than before to pick out one definite value of κ from many. The equations, being deterministic, lead always from a superposition to a superposition. Nevertheless, if (1) one measures the state of the apparatus a by a second apparatus b, and if (2) the states a_κ of a definitely put the apparatus b into the state b_κ; in other words, if

$$b_0 \otimes a_\kappa \to b_\kappa \otimes a_\kappa, \tag{37}$$

then (3) the interaction of this apparatus with the state (34) resulting from the measurement on σ_κ by a, will give

$$b_0 \otimes (a_\kappa \otimes \sigma_\kappa) \to b_\kappa \otimes a_\kappa \otimes \sigma_\kappa. \tag{37a}$$

This is true, at least, if b interacts only with a, not with the object. Hence, in this case, the interaction of b_0 with the result of the measurement on the general state (cf. [35]) will give

$$b_0 \otimes \sum_\kappa \alpha_\kappa(a_\kappa \otimes \sigma_\kappa) \to \sum_\kappa \alpha_\kappa(b_\kappa \otimes a_\kappa \otimes \sigma_\kappa). \tag{37b}$$

Thus, a correlation between the states of all three systems—object, apparatus a, and apparatus b—is established; but, naturally, no choice between the different states σ_κ is made. A similar statement applies if a fourth apparatus is used to "measure" the state of b, and so on. The measurement process, as far as it can be described by standard quantum mechanics, only establishes statistical correlations between the states of the apparata and those of the object. No choice for a definite state emerges, except if the object was, to begin with, in a state in which the outcome of the measurement is unique, as it is in the case considered in (34).

Before discussing some further problems arising from the assumption embodied in (34) and (35), and before proposing a possible resolution of the problem here encountered, it may be worth pointing out the obvious fact that while it may be possible to envisage an interaction of the form (34) between apparatus and object, this in no way guarantees that an apparatus with this interaction with the object can be found. This also will be discussed further later.

§4. The Physics of the Measurement Description (35)

The description of the measurement indicated by (34) and (35) has one very happy consequence: it shows that only self-adjoint operators are measurable. This follows from the unitary nature of the transformation indicated by the arrow. If one writes (34) for two different indices, say κ and λ, the final states of the corresponding transformations are orthogonal.

We have

$$(a_\kappa \otimes \sigma_\kappa, a_\lambda \otimes \sigma_\lambda) = (a_\kappa, a_\lambda)(\sigma_\kappa, \sigma_\lambda). \tag{38}$$

Moreover, a_κ and a_λ are two clearly distinguishable states—distinguishable, as a rule, even macroscopically. Therefore $(a_\kappa, a_\lambda) = 0$. Hence, it follows from the unitary nature of the transformation indicated by the arrow that the initial states were also orthogonal:

$$(a_0 \otimes \sigma_\kappa)(a_0 \otimes \sigma_\lambda) = (a_0, a_0)(\sigma_\kappa, \sigma_\lambda) = 0. \tag{38a}$$

In addition we know that $(a_0, a_0) = 1$. We conclude that $(\sigma_\kappa, \sigma_\lambda) = 0$ if a_κ and a_λ indicate two different outcomes of the measurement. If we further postulate that the outcomes are described by real numbers, i.e., that the characteristic values of the operator Q of which the a_κ are characteristic vectors are real,

$$Qa_\kappa = q_\kappa a_\kappa, \quad \text{with} \quad q_\kappa = q_\kappa^*, \tag{38b}$$

then the operator Q is necessarily self-adjoint. This means that for any two functions $\sum \alpha_\kappa a_\kappa$ and $\sum \beta_\lambda a_\lambda$ we have

$$\begin{aligned}(\textstyle\sum \alpha_\kappa a_\kappa, Q \sum \beta_\lambda a_\lambda) &= \sum_\kappa \alpha_\kappa^*(a_\kappa, \textstyle\sum \beta_\lambda Q a_\lambda) = \sum_{\kappa\lambda} \alpha_\kappa^* \beta_\lambda (a_\kappa, q_\lambda a_\lambda)\\ &= \sum_{\kappa\lambda} \alpha_\kappa^* \beta_\lambda q_\lambda \delta_{\kappa\lambda} = \sum_\kappa \alpha_\kappa^* \beta_\kappa q_\kappa\end{aligned} \tag{38c}$$

and one obtains the same expression for $(Q \sum \alpha_\kappa a_\kappa, \sum \beta_\lambda a_\lambda)$. This is satisfactory. It may be worth observing that all the preceding calculations seem to imply a discrete set of outcomes of the measurement—the last proof implies that the measured quantity Q has a discrete spectrum. This is both realistic and essentially unavoidable: the outcome of a measurement is restricted to a discrete set; in the case of the measurement of a continuous quantity, such as the momentum, the finite accuracy of the measuring device still guarantees that the different outcomes realized form only a discrete set, which in practice is even finite. If one does not want to accept this argument, it is ncessary to rewrite the preceding discussion, essentially in terms of Stieltjes integrals, and this is quite possible, though, in the opinion of this writer, unnecessary.

This is on the favorable side. One has to admit, on the other hand, that (35) is a highly idealized description of the measurement. It does not specify the duration of the measuring process. In fact, most writers, including von Neumann, at least imply that the transition indicated by the arrow in (35) is instantaneous. Of course, even if one accepts this idealization,unless the quantity to be measured commutes with the Hamiltonian of the system, its value will change after the measurement, and this applies also for the microscopic, that is, quantum-mechanical, description of the apparatus. However, these changes in the system and in the apparatus do not introduce basic problems. Thus, if the σ_κ are orthogonal to each other immediately after the measurement, orthogonality will continue also after the measurement as long as the system remains isolated.

The fact that the measurement is of finite duration introduces a more serious problem. If the operator of the quantity which is being measured does not commute with the Hamiltonian of the system, as is the case, for instance, when position is measured, it will change in the course of the measurement. To which position at which time does the measurement then refer? This issue is unclear and is rarely discussed. The existence of this issue reemphasizes that the quantum-mechanical description of the measurement, embodied in (34) and (35), is a highly idealized description—unless, as was mentioned before, the quantity to be measured is stationary.

In view of all these reservations, it is worthwhile to give a practical example of a measurement (Wigner, 1963). The example most often given is the measurement of the operator s_z—that is, the component of the spin of a particle in a fixed direction. This is a quantity which, for the free particle, is stationary; it commutes with the Hamiltonian. The measurement is called the Stern-Gerlach experiment and was discussed above in a cursory fashion. A particle with spin is passed through an inhomogeneous magnetic field. It is easiest to discuss the case of total spin $\frac{1}{2}$. When the magnetic moment of the particle points in the direction of increase of the field, the particle experiences a force—and is deflected—in the direction of decrease of the field, and conversely. The incoming beam is split into two beams. In one, the spin variable in the field direction has one value, and in the other beam, the other value. Hence, a statistical correlation is established between the position of the particle and its spin direction. The "a" of (34) or (35) is, in this case, the position of the particle. The two σ_κ are the two states of the spin, one parallel, the other antiparallel to the magnetic field. The right side of (35) resembles an expression such as

$$\alpha_+ e^{-x^2-y^2-(z-c)^2}\delta(s_z, \tfrac{1}{2}) + \alpha_- e^{-x^2-y^2-(z+c)^2}\delta(s_z, -\tfrac{1}{2}). \tag{39}$$

Of course, even if the beam is split, one does not yet know what the spin of the particle is; (39) is still a consequence of the quantum-mechanical equation of motion. In order to have a particle with a definite spin direction one must make an additional measurement, determining whether it is in one beam or the other—whether its z coordinate is positive or negative. This is the process indicated in equations (37a) and (37b)—and it is much more difficult to describe quantum mechanically because the quantum-mechanical description of the position measurement is not at all unique. Nevertheless, this example is instructive, and the expression for the state vector after the original measurements, indicated in (39), is a relatively easy consequence of the quantum-mechanical equations of motion. This is not the case for the measurements of most quantities—including the position of a particle.

Conceptual Problems of the Measurement Description

It was pointed out at the end of §3—and also sometime ago, particularly by von Neumann—that the description given is incomplete. Even if we accept the validity of the equations postulated, (34) and (35), we can account only for the establishment of a statistical correlation between the states of the object and those of the apparatus, not for the fact that the measurement gives one result once, another result another time—i.e., not for the statistical nature of the outcome of the measurement process. As was also pointed out earlier, this is hardly surprising because the deterministic nature of the equations of motion prevents them from accounting for a probabilistic result. The present section will deal with the problem resulting from this circumstance.

Naturally, one way out of this difficulty would be to postulate that the equations are not fully correct, or, at least, that they do not fully describe the actual situation. This has often been suggested, by various writers, and this possibility will be discussed in some detail later—on the whole with a rather negative, but not completely negative result. A more natural explanation would be that the statistical nature of the measurement outcome arises because the initial state of the apparatus—of a macroscopic apparatus—is not unique. Some initial states of the apparatus give one, others another result, and the statistical nature of the measurement outcome is due to the statistical nature of the initial state of the apparatus. Of course this proposed explanation would not apply to all processes of measurement. It does not apply to the Stern-Gerlach experiment insofar as it leads to a state like that given by (39). However, the final process of observation always involves some macroscopic apparatus, and the statistical explantion could be imagined to apply to that phase of the measurement. We shall see, however, that this proposed way out is also unacceptable. If the outcome of the measurement obeys the statistics postulated by quantum-mechanical theory, this correctness of prediction cannot be explained by the statistical nature of the initial state of the apparatus if we continue to believe in the unlimited validity of the present quantum-mechanical equations.

Let us consider, for the sake of simplicity, a measurement which can have, as in the Stern-Gerlach experiment discussed before, only two outcomes. For σ_+ it will have the outcome 1, for σ_- the outcome -1, for $\alpha\sigma_+ + \beta\sigma_-$ (with $|\alpha|^2 + |\beta|^2 = 1$) the outcome 1 with probability $|\alpha|^2$, the outcome -1 with probability $|\beta|^2$. The possible initial states of the apparatus will be denoted by a_0', a_0'', a_0''', and so on. If the object's initial state is, for instance, $(\sigma_+ + \sigma_-)/\sqrt{2}$, we would expect on this view that half of the states a_0', a_0'', a_0''' $\cdots$ will give the result 1, half of them -1.

This expectation, however, is at odds with the linear nature of the equations. We are supposing that each of the apparatus states a_0 gives, with σ_+, the result 1, with σ_- the result -1. This supposition means that

$$a_0 \otimes \sigma_+ \to a_+ \otimes \sigma_+ \quad \text{and} \quad a_0 \otimes \sigma_- \to a_- \otimes \sigma_- \tag{40}$$

or, in other words, that a_+ shows a result $+1$, and a_- the result -1. It now follows from the linear nature of the interaction operator that

$$a_0 \otimes (\sigma_+ + \sigma_-)/\sqrt{2} \to a_+ \otimes \sigma_+/\sqrt{2} + a_- \otimes \sigma_-/\sqrt{2}; \tag{40a}$$

that is, that none of the a_0', a_0'', a_0''' ... can give definitely $a = +1$ or definitely $a = -1$ as a result. We believe that each apparatus will give the correct result whenever the state of the object is definitely σ_+, and also whenever the state of the object is definitely σ_-. But we see from (40a) that as long as we maintain this belief, we cannot blame the statistical nature of the outcome of the measurement for the state $(\sigma_+ + \sigma_-)/\sqrt{2}$ on the uncertain initial state of the apparatus. Thus, if (40) holds, none of the states a_0 can give either purely $a_+ \otimes \sigma_+$ or purely $a_- \otimes \sigma_-$. All give the linear combination (40a) of these two states.

This situation suggests a drastic reformulation of the basic concepts of quantum mechanics. It appears that the statistical nature of the outcome of a measurement is a basic postulate, that the function of quantum mechanics is not to describe some "reality," whatever this term means, but only to furnish statistical correlations between subsequent observations. This assessment reduces the state vector to a calculational tool, an important and useful tool, but not a representation of "reality." The statistical correlations can be calculated with the aid of the state vector. If the first observation tells us that the state vector is σ_κ, we can calculate the probability of the outcome λ of an experiment the characteristic functions of the operator of which are $b_1, b_2 \ldots$. This probability is $|(a_\kappa, b_\lambda)|^2$. The probability that the next measurement of the quantity the characteristic functions of which are

$c_1, c_2 \ldots$ will give the result μ is then $|(b_\lambda, c_\mu)|^2$ and the probability of both outcomes b_λ and c_μ is

$$|(a_\kappa, b_\lambda)(b_\lambda, c_\mu)|^2 = (a_\kappa, b_\lambda)(b_\lambda, c_\mu)(c_\mu, b_\lambda)(b_\lambda, a_\kappa). \tag{41}$$

The generalization to more measurements is obvious (Houtappel, Van Dam, and Wigner, 1963).

It is of some interest that (41) can be given a more concise form. If the state vectors in (41) can be written as functions of a variable x, then (41) can be given the form,

$$\int \cdots \int a_\kappa(x)^* b_\lambda(x) b_\lambda(x')^* c_\mu(x') c_\mu(x'')^* \cdot b_\lambda(x'') b_\lambda(x''')^* a_\kappa(x''') dx \cdots dx'''. \tag{41a}$$

However, $b_\lambda(x)b_\lambda(x'')^*$ is the kernel of the projection operator P defined in (16) and equations (17) for the operator A. Hence, if we make $a_\kappa(x''')$ of (41a) the first factor, the factors $a_\kappa(x''')a_\kappa(x)^*$, $b_\lambda(x)b_\lambda(x'')^*$, and so on, are the kernels of the projection operators P_κ for the first quantity measured, P'_λ for the second one, P''_μ for the third, and again P'_λ for the second. The integrations over x, x', x'' give the product of these projection operators, and the integration over x''' the trace of the product. Hence, the first measurement having given the outcome κ, the probability that the second one gives λ, the third one μ etc. becomes

$$\frac{Trace\ P_\kappa P'_\lambda P''_\mu P'''_\nu \cdots P'''_\nu P''_\mu P'_\lambda}{Trace\ P_\kappa}, \tag{42}$$

where P, P', P'' ... are the proper projection operators for the first, second, third, etc. measurement, respectively. The *Trace* P_κ in the denominator appears because the characteristic value κ of the operator of the first measurement may not be simple—in which case one has to divide with the multiplicity of this characteristic value. It may be worth remarking that one can add a last factor P_κ in the numerator of (42) to make it more symmetric. This does not change its value because, with $P_\kappa^2 = P_\kappa$, one can insert another factor P_κ on the left side of the expression the trace of which appears in the numerator. This extra factor P_κ can then be shifted to the right end of the numerator since, quite generally, *Trace* AB = *Trace* BA or, in our case, *Trace* $P_\kappa Q$ = *Trace* QP_κ where Q is the expression in the numerator of (42). The insertion of the P_κ factor on the right therefore does not change the value of (42); it only makes it appear more symmetric.

The preceding derivation of (42) is incomplete, because the possibility of multiple characteristic values of the measured operators is not explicitly taken into account. This has little significance for (35)—it does not matter whether or not the same "pointer position" corresponds to several a_κ. As far as the primed P are concerned (that is, as far as the measurements after the initial one are concerned), the possibility of multiple characteristic values can easily be taken into account and the result is correctly represented by (42). However, (42) also contains the assumption that, if P_κ corresponds to a multiple characteristic value, that is, if *Trace* $P_\kappa \geq 2$, the state produced by this initial measurement contains the different states of this characteristic value with equal probabilities. This is an assumption which cannot be fully justified.

It should be observed also that if the later measurements, those to which the primed P correspond, take place at later times, the corresponding P must be modified so that they are in the "Heisenberg picture." This means that if the measurement to which $P_\nu^{(n)}$ corresponds takes place at time t_n after the first measurement, the $P_\nu^{(n)}$ in (42) can be obtained from the $P_\nu^{(n)}$ applicable if the measurement n had taken place immediately after the first one by the formula,

$$P_\nu^{(n)}(t_n) = \exp(-iHt_n/\hbar)P_\nu^{(n)}(0)\exp(iHt_n/\hbar). \tag{43}$$

(Naturally $t_1 \leq t_2 \leq t_3 \ldots$.)

The preceding interpretation of the quantum-mechanical formalism reflects its concern with observations, and with giving probabilities for the outcomes of these observations. It does not eliminate the difficulty which the finite length of the measurement time creates, the difficulty mentioned at the end of the preceding section. Nor does it alter the fact that we have not specified how the measurements are to be carried out. A discussion of limitations on the possibility of measuring certain operators will be given in the next chapter. In spite of all these reservations, it remains essentially correct to say that the basic statement of quantum mechanics can be given in a formula as simple as (42).

§5. Other Proposed Resolutions of the Measurement Paradox

The measurement paradox referred to in the title of this section is the contradiction between the deterministic nature of the quantum-mechanical equations of motion and the probabilistic outcome of the measurements—processes which should be describable by the quantum-mechanical equations of motion. Section 4 above proposed a resolution of the paradox: the quantum-mechanical equations of motion do not describe the measurement process; they only help in the calculation of the probabilities of the different outcomes. These probabilities form the real content of quantum-mechanical theory. The formalism of state vectors, equations of motion, etc., are only means to calculate these probabilities. The observation results are the true "reality" which underlie quantum mechanics. The state vector does not represent "reality." It is a calculational tool. It should be mentioned that von Neumann's idea was not truly different from this. He postulated that the state vector varies in two different ways. As long as the system is isolated, its state vector is subject to the quantum-mechanical equation of motion and its behavior is deterministic. When an observation takes place, there is a second type of change of the state vector. Its change then has a probabilistic nature. It jumps discontinuously. It becomes one of the characteristic vectors of the operator which is being measured. If the initial state vector (normalized) was ϕ, it jumps with the probability $|(\psi_\nu, \phi)|^2$ into the state ψ_ν which is one of the (normalized) characteristic vectors of the operator which is being measured. Naturally, the sum of these probabilities must be 1 and this is the consequence of the normalized nature of ϕ. The second type of change of the state vector, the jump from ϕ into one of the ψ_ν, according to von Neumann's picture, is not described by the quantum-mechanical equations of motion.

The other resolutions of the measurement paradox propose more "physical" pictures. The most popular of these is the picture of hidden parameters—embraced particularly by D. Bohm (1952a,b), Y. Aharonov (Bohm and Aharonov, 1957), J. Bub (1969), (see also Bohm and Bub, 1966a,b), J. P. Vigier (1951; 1956), and Bohm and Vigier (1954), but also by many others. The story of this picture is given in some detail by F. Belinfante (1973) as well as by M. Jammer in his book, *The Philosophy of Quantum Mechanics* (Jammer, 1974) which is worth reading for other reasons—it encompasses an amazing amount of information on the history of our general subject. The other attempt at the resolution of the paradox is due to H. Everett (1957), B. S. DeWitt (1970), and J. A. Wheeler (1957) (see also DeWitt and Graham, 1973). This is the "relative state" theory, postulating that, as a result of an observation, the world splits into several new worlds, existing independently of each other. Both these pictures will be discussed in more detail, and arguments against them will be presented. It should perhaps be added now that J. A. Wheeler (1977) no longer supports this view.

Theory of "Hidden Variables"

The idea of "hidden variables" postulates that the description of states, by the quantum-mechanical state vector, is incomplete; that there is a more detailed description, by means of variables now "hidden," which would be complete and the knowledge of which would permit one to foresee the actual outcomes of observations—observations about whose outcomes present-day quantum mechanics makes only probabilistic statements. The relation of the postulated theory of hidden variables to present quantum mechanics would be similar to the relation of classical microscopic physics to macroscopic physics. The former uses the positions and velocities of the atoms as variables, while macroscopic physics, such as, for instance, hydrodynamics, describes only the average velocities of the atoms situated in volumes which are large on the microscopic scale and contain many atoms.

There is no clear specification of the nature of the "hidden variables"—they are hidden. The best known replacement for Schrödinger's equation is Bohm's

(1952a,b), in which he reverses Schrödinger's analysis. Schrödinger derived the wave equation named after him by viewing the classical Hamilton-Jacobi equation as giving an incomplete and approximate description of this wave: incomplete, in the sense that it deals only with the phase, $S(x, y, z, t)/\hbar$, of this wave, $\psi(x\ y, z, t)$; approximate, in the sense that the equation for the Hamilton-Jacobi function S is nonlinear, whereas by demanding linearity Schrödinger got the right equation for ψ. Bohm turns this reasoning around. He assumes that Schrödinger's equation for

$$\psi = Re^{-iS/\hbar} \tag{44}$$

is valid (R and S being real and functions of the position coordinates) and obtains equations for R and S. The equation for S,

$$\frac{\partial S}{\partial t} = \sum_j \frac{1}{2m_j}\left[\left(\frac{\partial S}{\partial x_j}\right)^2 + \left(\frac{\partial S}{\partial y_j}\right)^2 + \left(\frac{\partial S}{\partial z_j}\right)^2\right] + V + Q, \tag{44a}$$

differs from the classical Hamilton-Jacobi equation by containing an extra "quantum term" Q. All the other terms are obtained from the Hamilton equation for the energy, $E = H(x, p)$, by replacing E by $-\partial S/\partial t$ and p_{x_j} by $\partial S/\partial x_j$, etc. The additional term, Q, is the "quantum potential,"

$$Q = -\sum_j \frac{\hbar^2}{2m_j R}\left(\frac{\partial^2}{\partial x_j^2} + \frac{\partial^2}{\partial y_j^2} + \frac{\partial^2}{\partial z_j^2}\right)R, \tag{44b}$$

in the equation for S. The equation for R is the classical one:

$$\frac{\partial R}{\partial t} = \sum_j \frac{1}{m_j}\left(\frac{\partial R}{\partial x_j}\frac{\partial S}{\partial x_j} + \frac{\partial R}{\partial y_j}\frac{\partial S}{\partial y_j} + \frac{\partial R}{\partial z_j}\frac{\partial S}{\partial z_j}\right) + \frac{R}{2m_j}\left(\frac{\partial^2 S}{\partial x_j^2} + \frac{\partial^2 S}{\partial y_j^2} + \frac{\partial^2 S}{\partial z_j^2}\right). \tag{44c}$$

One can obtain these equations by introducing (44) into Schrödinger's time-dependent equation for ψ and separating real and imaginary parts.

The interpretation of these equations from the point of view of hidden variables theory is not so simple. It seems to be agreed that R^2 gives, as in the quantum interpretation, the probability of the configuration indicated by its variables, i.e., that it refers to an ensemble. For S, the classical interpretation is postulated, but it is difficult to understand then how the properties of the ensemble, described by R^2, can influence the motion of an individual system, the system which S is supposed to describe. Yet Q depends on R and R describes the *ensemble* containing the system the motion of which should be described by S. Is it that a system's behavior is different depending on the set of systems of which it may be a part?

This dependence of the individual on the ensemble is an objection against a specific theory of hidden variables, not a general objection against all such theories; i.e., it is not an argument against the existence of a deterministic theory of the motion of atomic objects and constituents. Von Neumann was convinced that no such deterministic theory is compatible with quantum mechanics. The reason is easily illustrated by the Stern-Gerlach experiment, or rather, by an indefinite number of repetitions of that experiment. One may consider the measurement of the spin component first in the z direction, then in the x direction, then again in the z direction, then again in the x direction, and so on. If the total spin of the particle on which the measurements are undertaken is $\frac{1}{2}$, all measurements succeeding the first one give the two possible results with a probability $\frac{1}{2}$. If these results are, fundamentally, all determined by the initial values of the hidden parameters, the outcome of each measurement should give some information on the initial values of these parameters. Eventually, it would seem, the values of the "hidden parameters" which determine the outcomes of the first N measurements would be in such a narrow range that they would determine, if N is large enough, the outcomes of all later measurements. Yet this is in contradiction to the quantum-mechanical prediction.

The preceding argument can be made more convincing by substituting other measurements, such as a position and momentum measurements, for the measurements of spin direction. Yet this argument apparently cannot be made mathematically rigorous and it was not published by von Neumann. The proof he published (see p. 173 of von Neumann, 1932; pp. 326–28 in English translation), though it was made much more convincing later on by Kochen and Specker (1967), still uses assumptions which, in my opinion, can quite reasonably be questioned.

Bell's Argument

In my opinion, the most convincing argument against the theory of hidden variables was presented by J. S. Bell (1964). His argument, in its simplest form, starts with a system of two particles with spins $\frac{1}{2}$, these spins being antiparallel, i.e., forming a singlet state. The spin part of the state vector is

$$\sigma_+(1)\sigma_-(2) - \sigma_-(1)\sigma_+(2). \tag{45}$$

Here $\sigma_+(1)$ is the state vector of particle 1, with positive component in the z direction. The meaning of the other symbols should then be obvious. Actually, the state (45) is spherically symmetric. It is the only antisymmetric combination possible for the two spin functions. Thus the z direction mentioned above can be replaced by any other direction without changing the value of (45) (as can be verified, of course, by actual calculation). The state (45) is called the spin singlet state. The spin parts of the state vectors of the two electrons of the He atom, or of the H_2 molecule, are actually in that state. In fact, the spin state remains essentially unchanged even if the electrons are torn away from the atom or molecule by dipole radiation, so that their spins remain, to a very good approximation, in the state (45) even when they are widely separated from each other, as is demanded for some applications of Bell's argument. Even for that case, (45) represents a state of the two spins which is not only theoretically possible but also, to a good approximation, experimentally realizable.

For the state (45), there is, according to quantum mechanics, a statistical correlation for the measurement of the spin components in two directions, $\mathbf{e}_1$ and $\mathbf{e}_2$, which enclose an angle θ_{12}. The probability that both components are positive is

$$P_{++} = (\tfrac{1}{2}) \sin^2(\theta_{12}/2) = P_{--} \tag{46}$$

and this is also the probability that the measurement of the spin components of the two particles, in the $\mathbf{e}_1$ and $\mathbf{e}_2$ directions respectively, will have a negative outcome for both. The probability that one component will be positive the other negative, or for the converse, is

$$P_{+-} = P_{-+} = (\tfrac{1}{2}) \cos^2(\theta_{12}/2). \tag{46a}$$

If $\theta_{12} = 0$, i.e., if the two directions are parallel, the probability (46) of both having the same component is zero. This vanishing of P_{++} is what makes the aggregate spin 0, i.e., makes the total state a singlet. The expressions (46) and (46a) for the probabilities will not be proved explicitly. They are contained, at least implicitly, in the usual textbooks on quantum mechanics.

Bell's inequality, based on the assumption that the "hidden variables" of the two particles uniquely determine the outcomes of the measurements of spin components in all directions, will be shown to be in conflict with equations (46) and (46a). Actually, it suffices to consider three directions, $\mathbf{e}_1$, $\mathbf{e}_2$, $\mathbf{e}_3$, and measurements of the spin components in these directions. Let us denote $(+-+;--+)$ the probability-weighted integral over the values of the hidden variables, an integral taken over that domain of these variables which ensures that the first particle's spin has a positive component in the $\mathbf{e}_1$ and $\mathbf{e}_3$ directions, and a negative component in the $\mathbf{e}_2$ direction, while the second particle's spin has a negative component in the $\mathbf{e}_1$ and $\mathbf{e}_2$ directions, and a positive component in the $\mathbf{e}_3$ direction. The meaning of the other combinations of six $+$ and $-$ signs is similar; thus $(+++;---)$ gives the probability that the spin component of the first particle is positive in all three directions, that of the second negative. Since the hidden variables are supposed to determine completely the properties of the objects to which they refer, all these quantities are fully defined and are, surely, non-negative. It appears that there are $2^6 = 64$ of them.

It can be noted, next, that most of the 64 symbols are 0. Thus $(++-;+-+)$ must vanish because it corresponds to such values of the hidden variables as give, for both particles, a positive component of the spin as measured in the $\mathbf{e}_1$ direction. Because of (46), the probability for this spin configuration vanishes. That is, no value of the hidden variables produces a positive spin component for both particles in the $\mathbf{e}_1$ direction, at least not if quantum mechanics, and hence (46), properly gives the probabilities of the outcomes of measurements. Similarly, if quantum mechanics is correct, all symbols vanish in which the same sign appears at the same position both before and after the semicolon. This then leaves 8 symbols such as $(+--;-++)$ which can have non-zero values.

Next, we write down the probabilities for those outcomes of the measurements which indicate a positive spin component for both particles if these are measured in two different directions, and equate these with the corresponding quantum-mechanical expressions. First, if the spin components are measured in the $\mathbf{e}_1$ and $\mathbf{e}_2$ directions, respectively, the probability of finding positive components is, according to the theory of hidden variables (the first sign before, and the second sign after the semicolon must be $+$),

$$(+-+;-+-)+(+--;-++)=(\tfrac{1}{2})\sin^2(\theta_{12}/2). \tag{47}$$

All other bracket symbols vanish in which the first sign before the semicolon and the second sign after the semicolon are $+$. Similarly, if the measurements are made in the $\mathbf{e}_2$ and $\mathbf{e}_3$ directions, respectively, we have

$$(++-;--+)+(-+-;+-+)=(\tfrac{1}{2})\sin^2(\theta_{23}/2). \tag{47a}$$

Likewise we have

$$(++-;--+)+(+--;-++)=(\tfrac{1}{2})\sin^2(\theta_{13}/2). \tag{47b}$$

It follows from equations (47) that

$$\begin{aligned}\sin^2(\theta_{12}/2)+\sin^2(\theta_{23}/2)&=\sin^2(\theta_{13}/2)+2(+-+;-+-)+2(-+-;+-+)\\&\geq\sin^2(\theta_{13}/2).\end{aligned} \tag{48}$$

This is the Bell inequality. It should be fulfilled if the theory of hidden variables is correct, and if, in addition, quantum mechanics correctly predicts—as it apparently does—the probabilities of all conceivable outcomes of a measurement.

It is easy to find directions $\mathbf{e}_1$, $\mathbf{e}_2$, $\mathbf{e}_3$, however, for which (48) is not valid. It is not valid, in particular, if the three directions are in the same plane and $\mathbf{e}_2$ is between $\mathbf{e}_1$ and $\mathbf{e}_3$. Specifically, if $\theta_{12}=\theta_{23}=\pi/3$, $\theta_{13}=2\pi/3$, then the left side of (48) is $\frac{1}{4}+\frac{1}{4}=\frac{1}{2}$, the right side $\frac{3}{4}$. Hence, the hidden variable theory, as applied here, leads to a contradiction.

There are then two questions. First, is (46) correct, i.e., can it be confirmed experimentally? This will be discussed later. Second, can one modify the theory of hidden variables in such a way that Bell's inequality (48) does not follow? The answer to this second question is "yes"; but the modified theory implies the use of hidden variables which specify not only the state of the two particles but also the states of the measuring devices, and postulates correlations between the state of the particles and the directions in which the spin is going to be measured. The theory that assumes that there are no such correlations is often called "local" hidden variable theory. The preceding argument shows that any theory of hidden variables conforming with the postulate of locality is in conflict with quantum mechanics. On the other hand, if one admits hidden variables which establish correlations between the measuring devices and the objects, i.e., describe the states of both together, one cannot see the limit of the complex that has to be described jointly. It seems highly questionable whether a theory which does not permit the specification of the states of isolated objects can be in any way useful.

Of course, if future experiments were to demonstrate that Bell's inequalities, in particular (48), are correct and the quantum-mechanical equations, in particular (46) and (46a), incorrect, this finding would constitute strong evidence in favor of some theory of hidden variables. As was mentioned before, the present status of the experimental research will be discussed later. As of 1981 it appears to be definitely established that what is called "Bell's inequality" is strongly violated, whereas the quantum-mechanical predictions appear to be well supported.

Many-World Theories

"Many-world" theories are much more difficult to discuss than theories of hidden variables. They postulate—as mentioned earlier—that if a measurement with a probabilistic outcome is undertaken, the world splits into several worlds, and each possible outcome of the observations appears in the fraction of the new worlds given by the quantum-mechanical probability of that outcome. This re-establishes the determinism which the laws of nature are expected to exhibit.

It is, of course, difficult to see the meaning of the statement that there are other worlds with which we never will have any contact, which have no influence on us, and which we cannot influence or perceive in any way. From a positivistic point of view, the statement that there are such worlds, and that they are constantly created in large numbers, is entirely meaningless. It can be neither confirmed nor refuted.

Let us admit, however, in conclusion, that the weakness of the theories which have been proposed to replace the standard interpretation of quantum mechanics, and which form the subject of most of these notes, does not establish the full validity of quantum theory. The weaknesses of quantum theory, though not as marked as the deficiencies of the theories discussed in the present chapter, are real nevertheless. They will be discussed subsequently.

§6. Experimental Tests of Bell's Inequality

The conflict between quantum theory and the theory of hidden variables, as manifested by Bell's inequality, was discussed in the last section. The original version of the present section was kindly provided by S. J. Freedman, then at Princeton University. It dealt, principally, with the experimental work then available and aimed at deciding which of the two formulations—Bell's inequality or the conclusions of quantum mechanics—was correct. Since that time (1976) the experimental information has been greatly extended, and also some errors in the earlier work have been detected. As a result, it would not be reasonable to repeat Freedman's analysis fully. Instead we refer the reader to Pipkin's (1978) summary of the experimental findings. These findings disagree with Bell's inequality and seem to agree with the consequences of quantum theory. This is a very important point.

In spite of the fact that the more recent experimental results are of great relevance and are, naturally, not contained in Freedman's original analysis, his discussion will be presented below, though in much abbreviated form, as being of some historical interest. It is also good to realize that experimental results can be in error. It is even better to realize that these errors can be corrected.

Freedman also presented some remarks on three general, largely theoretical, questions on our subject. (1) He characterized the structure of a general hidden variable model. One of his "hidden variables" was the wave function (and, in many cases, this also is hidden). But he introduced a set of other variables which *always* remain hidden and which are supposed to give a complete description of the state of the system. (2) He presented a special model in order to prove that it is conceivable that hidden variables could be found which would determine the future of the system completely. He admitted, though, at this point, that in order to eliminate the difficulty (for hidden variables!) of the invalidity of Bell's inequality one has to abandon the postulate of the local character of the hidden variables. These must be so constituted, he said, that they establish statistical correlations between the state of the object on which the measurement will be undertaken, and the state of the apparatus which will carry out the measurement. These correlations, he supposed, exist before the measurement takes place. (3) Freedman also presented a modification of Bell's inequality which is more easily and more directly subject to an experimental test than is the original inequality.

Perhaps the present writer will be permitted to voice his opinion on theories of hidden variables of the type which were crudely described under (2) and which are often proposed. In my opinion, it is not very meaningful to introduce variables, the magnitude of which cannot be determined. These variables could be defined as giving the state of the system for the entire future. Such a description would require the introduction of one additional variable, most naturally denoted by t, to describe the system. However, it is the purpose of physics to give information on the future based on facts ascertainable at present. The unqualified existence of "hidden variables" would nullify this purpose.

Let us go over to a short description of the experimental information available at the time Freedman made his remarks. Actually, the first two experiments were not carried out with spin-$\frac{1}{2}$ particles in the singlet state of (45) but with two photons emitted in succession by an atom. Their polarizations then give correlations similar to those of the two spin-$\frac{1}{2}$ particles discussed in the preceding chapter.

The first set of experiments was carried out on Ca by Freedman in collaboration with Clauser (1972). The result was a correlation factor of 0.300 ± 0.009. This agrees with the quantum-mechanical value of 0.301 but contradicts Bell's inequality which postulates a value smaller than 0.25.

The next set of experiments was done by Holt and Pipkin (1974) with isotopically pure Hg^{198} (zero nuclear spin). Their result was 0.216 ± 0.013. This agrees with Bell's inequality but differs grossly from the quantum-mechanical value of 0.301. Perhaps fortunately, it was demonstrated not long after this section was originally written that an experimental error had crept into the original value (see Clauser, 1976).

Still other experiments deal with the decay of positronium, in a ground state of spin zero, into two photons. Experiments on the correlation of the polarization of these two photons were originally suggested by Wheeler (1946). The early experiments, carried out before 1970, gave excellent agreement with quantum mechanics (Wu and Shaknov, 1950; Kasday, Ullman, and Wu, 1970). Their results did not really contradict Bell's inequality, but it is hard to believe that the laws of quantum mechanics are violated in any area under consideration if they are so well obeyed in the region investigated. But more recent work disagrees with these results, referring to conditions in which quantum mechanics and Bell's inequality are in contradiction (Faraci, Gutkowski, Notarrigo, and Pennisi, 1974). This creates a confusing situation.

The last experiment referred to by Freedman related to the spin-spin correlation when a low energy proton is scattered by a proton at rest. Low energy scattering is dominated by proton pairs of zero orbital momentum. The space part of the wave function is therefore symmetric on interchange of the two particles. Consequently the spin part must be antisymmetric, of the form (45), corresponding to zero total spin. The original experiments did not seem to confirm quantum mechanics, but the later, improved ones, did, and were in contradiction with Bell's inequality (Lamehi-Rachti and Mittig, 1976).

This will conclude the somewhat abbreviated review of the experimental material available in 1975 on the validity of Bell's inequality and hence on the possibility of explaining the statistical nature of the outcomes of measurements with hidden variables of a local and reasonable nature. As was mentioned before, the subject was treated in more detail in the original version of these lecture notes as given by Freedman. As was also mentioned, more recent experimental data definitely deny the possibility that Bell's inequality is generally valid, and hence rule out any possibility of eliminating the statistical nature of the measurement process by the introduction of local hidden variables.

The next section, the last one, will deal with problems of the standard interpretation. That interpretation assumes that the principle of determinism does not apply to the measurement process. It also shows weaknesses. Nevertheless, no current experimental information contradicts it. Its principal weakness is that it gives no clear and simple rule for the way the measurement can be carried out nor for the limits of the accuracy of the measurement process. The preceding discussion was confined to measuring the spin component or the state of polarization, both of which clearly appear to be determinable with high accuracy.

§7. Problems of the Standard Interpretation: Unmeasurable Quantities

Most physicists working on or with quantum mechanics were quite surprised that the experimental confirmation of some of its simple consequences, such as the violation of Bell's inequality, was as ambiguous as described in the preceding sections. Most of us are convinced, nevertheless, that the consequences of quantum theory, which were tested in the experiments described, will be unambiguously confirmed by further and perhaps more precise experiments.

The present section will be devoted to a discussion of internal problems of the standard interpretation as discussed in §3. Let us admit that these problems do mar the mathematical beauty of the theory by demonstrating the difficulty of making measurements and the unavoidable limitations on the accuracy of most of them. These difficulties and limitations indicate, as do the remarks of §4, that quantum mechanics shares a degree of incompleteness with all other theories of physics.

It is a reassuring feature of quantum theory, however, that the earliest problem of measurement—a problem which generated a great deal of discussion—was not really a problem. The apparent problem was to construct the proper quantum-mechanical operator to correspond to a given classical expression. For example, what is the operator which corresponds to the classical expression xp? Surely xp, where p stands for $(\hbar/i)\partial/\partial x$, has to be rejected because it is not self-adjoint. It is natural to choose, instead,

$$xp \rightarrow \frac{1}{2}(xp + px) = \frac{1}{2}\frac{\hbar}{i}\left(x\frac{\partial}{\partial x} + \frac{\partial}{\partial x}x\right) = \frac{\hbar}{i}\left(x\frac{\partial}{\partial x} + \frac{1}{2}\right). \tag{49}$$

However, in somewhat more complicated cases, such as x^2p^2, the choice is not unique. Two possible choices are $\frac{1}{2}(x^2p^2 + p^2x^2)$ and xp^2x (p here stands again for $(\hbar/i)\partial/\partial x$), both of which are self-adjoint but not equal to each other. It should perhaps be mentioned that Weyl (1927) has proposed a definite operator to correspond to any classical function of position and momentum. But as long as no experimental prescription is provided for the measurement of that operator, the meaning of the Weyl prescription is not clear.

The quandary just described has been resolved by considering the operator to be defined by the quantity which is measured. Thus one does not ask for the operator which corresponds to the classical expression x^2p^2 but asks for ways in which either $\frac{1}{2}(x^2p^2 + p^2x^2)$ or xp^2x $(= px^2p)$ can be measured. We will be concerned henceforth with questions of this nature. Unfortunately, as we shall see, there are serious limitations on the measurability of an arbitrary quantity. They blur the mathematical elegance of von Neumann's original postulate that all self-adjoint operators are measurable. Von Neumann does produce an expression for the interaction between object and apparatus which would lead to equation (34) of our §3. However, the ensuing considerations show that for many if not most operators, this expression—or any other expression which might lead to that equation—contradicts some of the basic principles of quantum theory. What then are the limitations of measurability?

Only Quantities Which Commute with All Additive Conserved Quantities Are Precisely Measurable

This theorem, dating back to 1952 (Wigner, 1952; see also Araki and Yanase, 1960; Yanase, 1961), will not be proved generally. Only a characteristic example will be given. The quantity to be measured is the component of the spin in the x direction. The "additive conserved quantity" will be the angular momentum in the z direction. By "additive" we mean that the magnitude of the quantity for object-plus-apparatus is the sum of this quantity for object and for apparatus, both before and after the measurement.

In the case considered, the argument is very simple. Let us denote the spin states with positive and negative z-component by α and β respectively. The state vector associated with the positive x-component spin is then $2^{-1/2}(\alpha + \beta)$, and that of the negative one, $2^{-1/2}(\alpha - \beta)$. Hence equation (34) of §3 reads in this case

$$a \otimes (\alpha + \beta) \to a^+ \otimes (\alpha + \beta), \tag{50a}$$

$$a \otimes (\alpha - \beta) \to a^- \otimes (\alpha - \beta). \tag{50b}$$

Let us decompose the apparatus states a, a^+, and a^- into state vectors each having a definite angular momentum in the z direction,

$$a = \sum a_m; \qquad a^+ = \sum a_m^+; \qquad a^- = \sum a_m^-. \tag{51}$$

It then follows from the addition and subtraction of equations (51) that

$$2\sum a_m \otimes \alpha \to \sum a_m^+ \otimes (\alpha + \beta) + \sum a_m^- \otimes (\alpha - \beta), \tag{51a}$$

$$2\sum a_m \otimes \beta \to \sum a_m^+ \otimes (\alpha + \beta) - \sum a_m^- \otimes (\alpha - \beta). \tag{51b}$$

It simplifies these equations to introduce the symbols

$$2b_m = a_m^+ + a_m^- \quad \text{and} \quad 2c_m = a_m^+ - a_m^-. \tag{52}$$

Hence (51a) and (51b) become

$$\sum a_m \otimes \alpha \to \sum (b_m \otimes \alpha + c_m \otimes \beta), \tag{52a}$$

$$\sum a_m \otimes \beta \to \sum (b_m \otimes \beta + c_m \otimes \alpha). \tag{52b}$$

The angular momentum of $a_m \otimes \alpha$ in the z direction is $m + \frac{1}{2}$, that of $a_m \otimes \beta$ is $m - \frac{1}{2}$. Since the angular momentum in the z direction (or any other direction) does not change as a result of the interaction, it follows that the interaction of the state a_m alone with α will yield

$$a_m \otimes \alpha \to b_m \otimes \alpha + c_{m+1} \otimes \beta, \tag{53a}$$

and of a_m alone with β will give

$$a_m \otimes \beta \to b_m \otimes \beta + c_{m-1} \otimes \alpha. \tag{53b}$$

Now the lengths of the vectors on the left and right sides of (53a) must be equal. Moreover, the two terms on the right side are orthogonal. Therefore we have

$$(a_m, a_m) = (b_m, b_m) + (c_{m+1}, c_{m+1}); \tag{54a}$$

and similarly, from (53b)

$$(a_m, a_m) = (b_m, b_m) + (c_{m-1}, c_{m-1}). \tag{54b}$$

Hence, we have

$$(c_{m+1}, c_{m+1}) = (c_{m-1}, c_{m-1}), \tag{54}$$

which is a contradiction. Thus (54) means either (1) that the c_m are all 0—in which case the $a_m^+ = a_m^-$ so that $a^+ = a^-$ whereas they should be orthogonal—or (2) that

$$(c, c) = \sum (c_m, c_m) \tag{55}$$

is infinite unless all (c_m, c_m) vanish. But this is impossible. Thus from (52) $(c, c) = \frac{1}{2}$ because a^+ and a^- are of the same unit length and are orthogonal. However, (55) already indicates how this difficulty will be overcome. We will recognize that (52a) and (52b) can be only approximately valid (see equations 56a and 56b). In other words, all (c_m, c_m) will be very small and they will not be absolutely equal (Gaussian dependence on m, with a very large spread in m-values). This means that the measurement is not absolutely accurate, as (52a) and (52b) would imply. However, the magnitude of the inaccuracy, i.e., the magnitude of the terms η in (56a) and (56b), will be minimized.

The conclusion thus arrived at seems to deny the possibility of measuring the spin component precisely in an arbitrary direction, as described in §3. However, two points have to be remembered. First, the measuring equipment uses an external magnetic field. If that field is considered to be "external," it invalidates the law of conservation of angular momentum. The only way to remedy this difficulty is to consider the magnet as part of the apparatus—as it is. The thus enlarged apparatus becomes quite macroscopic. Second, as equation (39) indicates, the measurement is not perfect. The two beams, with positive and negative s_z, overlap to a certain extent. These two points suggest first, that the measurement of a quantity not commuting with additive conserved quantities requires a large apparatus and, second, that the limitations on the possible accuracy of the measurement can decrease with increasing "size" of the apparatus. "Size," we shall see, means the amount of the additive conserved quantity that is contained in the apparatus.

Instead of (50a) and (50b) we now write

$$a \otimes (\alpha + \beta) \to a^+ \otimes (\alpha + \beta) + \eta^+ \otimes (\alpha - \beta), \tag{56a}$$

$$a \otimes (\alpha - \beta) \to a^- \otimes (\alpha - \beta) + \eta^- \otimes (\alpha + \beta). \tag{56b}$$

Here the terms η^+ and η^- express the error in the measurement. We will try to make η^+ and η^- as small as possible. We write, similar to (51),

$$\eta^+ = \sum \eta_m^+, \qquad \eta^- = \sum \eta_m^- \tag{57}$$

and, in analogy to (52), we set

$$\eta_m^+ + \eta_m^- = 2\sigma_m \quad \text{and} \quad \eta_m^+ - \eta_m^- = 2\delta_m. \tag{58}$$

Hence, (56a) and (56b) give as a result of the conservation of the z angular momentum the relations

$$a_m \otimes \alpha \to (b_m + \sigma_m) \otimes \alpha + (c_{m+1} - \delta_{m+1}) \otimes \beta, \tag{58a}$$

$$a_m \otimes \beta \to (b_m - \sigma_m) \otimes \beta + (c_{m-1} + \delta_{m-1}) \otimes \alpha. \tag{58b}$$

These are the analogs of (52a) and (52b). The arrow here, as there, denotes a unitary transformation. Can this requirement be satisfied? Previously, using only part of the unitary restrictions, we arrived at the contradiction (54). Now we have to take into account all the consequences of unitarity.

The left sides of (58a) and (58b) are orthogonal to each other and orthogonal to all similar expressions with other m-values. Their unitary transforms must likewise be orthogonal. This they are automatically when they have different values of the conserved quantity, "total z angular momentum." When the left sides, say $a_m \otimes \alpha$ and $a_{m+1} \otimes \beta$, have the same value for this quantity, orthogonality of the right sides gives

$$(b_m + \sigma_m, c_m + \delta_m) + (c_{m+1} - \delta_{m+1}, b_{m+1} - \sigma_{m+1}) = 0. \tag{59}$$

Since the lengths of the a_m vectors are free, the only further relation imposed by the unitarity condition is that the lengths of the vectors on the right sides of (58a) and (58b) be equal. This gives, again for all m,

$$(a_m, a_m) = (b_m + \sigma_m, b_m + \sigma_m) + (c_{m+1} - \delta_{m+1}, c_{m+1} - \delta_{m+1}), \tag{60a}$$

$$(a_m, a_m) = (b_m - \sigma_m, b_m - \sigma_m) + (c_{m-1} + \delta_{m-1}, c_{m-1} + \delta_{m-1}). \tag{60b}$$

As when (54) was derived, so here we take the difference of these two equations. The (b_m, b_m) term drops out to give

$$\begin{aligned}(c_{m+1}, c_{m+1}) &- (c_{m-1}, c_{m-1}) \\ &= 2Re[(c_{m+1}, \delta_{m+1}) + (c_{m-1}, \delta_{m-1}) - 2(b_m, \sigma_m) \\ &\quad + (\delta_{m-1}, \delta_{m-1}) - (\delta_{m+1}, \delta_{m+1})].\end{aligned} \tag{60}$$

In the case of an ideal measurement, the left side vanished; we had $\sigma = \delta = 0$; and we encountered a contradiction. The sums of (60a) and (60b) for definite m-values need not be considered; they only give the probabilities (a_m, a_m) of the specified angular momenta of the measuring device in terms of the other quantities that appear in the state vector of the system after the measurement has taken place. The sum of these probabilities for all m-values, however, does give the normalization condition for the apparatus. Because $(\alpha, \alpha) = (\beta, \beta) = 1$, this condition becomes

$$\begin{aligned}\sum (a_m, a_m) = \sum \{&(b_m, b_m) + (\sigma_m, \sigma_m) + \tfrac{1}{2}(\delta_{m-1}, \delta_{m-1}) + \tfrac{1}{2}(\delta_{m+1}, \delta_{m+1}) \\ &+ \tfrac{1}{2}(c_{m+1}, c_{m+1}) + \tfrac{1}{2}(c_{m-1}, c_{m-1}) + Re[(c_{m-1}, \delta_{m-1}) - (c_{m+1}, \delta_{m+1})]\} = 1.\end{aligned} \tag{61}$$

The second member of (61) gives (a_m, a_m) as the average of the two expressions given by (60). We have a final condition. We require that at least the first terms on the right sides of (50a) and (50b) should represent different states—i.e., that $(a^+, a^-) = 0$. In terms of the b_m and c_m this condition says,

$$\sum_m (b_m + c_m, b_m - c_m) = 0. \tag{62}$$

Equations (59), (60) and (61), (62) represent the conditions which b, c, σ, and δ must satisfy in order to guarantee the unitary nature of the transformation indicated by (56a) and (56b). In equations (61) and (62) there is summation over m. They give the normalization condition and the requirement that the transformations (56a) and (56b) represent a measurement at least in the approximate sense.

Next, we define the error ε implicit in the measurement process indicated by these expressions,

$$\varepsilon = \sum [(\eta_m^+, \eta_m^+) + (\eta_m^-, \eta_m^-)] = 2 \sum [(\sigma_m, \sigma_m) + (\delta_m, \delta_m)]. \tag{63}$$

The "size," M, of the apparatus which permits an error so small will be defined by

$$\begin{aligned} M^2 &= \sum m^2 (a_m, a_m) \\ &= \tfrac{1}{2} \sum m^2 [2(b_m, b_m) + (c_{m+1}, c_{m+1}) + (c_{m-1}, c_{m-1}) \\ &\quad + 2Re(c_{m-1}, \delta_{m-1}) - 2Re(c_{m+1}, \delta_{m+1}) + (\delta_{m+1}, \delta_{m+1}) + (\delta_{m-1}, \delta_{m-1})]. \end{aligned} \tag{64}$$

Again, the average of the right sides of (60a) and (60b) was substituted for (a_m, a_m). The problem then is to find expressions for b_m, c_m, σ_m, and δ_m satisfying equations (59) to (62) which, for a given error ε, make M as small as possible; or, equivalently, for a given "size" of the apparatus, M, make the error ε as small as possible.

This minimum problem will not be solved exactly but only under the assumption that the error ε is small as compared with 1—i.e., that the σ_m and δ_m are small as compared with b_m and c_m. This means that (b_m, b_m) and (c_m, c_m) are appreciable for a rather wide range of m-values. Therefore they will be assumed to depend continuously on m. Even though this idealization is not mathematically rigorous, we adopt it. We rewrite in terms of this continuum approximation all of the equations from (59) to (64) except for (60), which only gives (a_m, a_m) in terms of the other quantities:

$$(b_m, c_m) = 0; \tag{59a}$$

$$2 \frac{d}{dm} (c_m, c_m) = 2Re[(c_m, \delta_m) - (b_m, \sigma_m)]; \tag{60c}$$

$$\sum [(b_m, b_m) + (c_m, c_m)] = 1; \tag{61a}$$

$$\sum (b_m, b_m) = \sum (c_m, c_m), \qquad Im \sum (b_m, c_m) = 0; \tag{62a}$$

$$\varepsilon = 2 \sum [(\sigma_m, \sigma_m) + (\delta_m, \delta_m)]; \tag{63a}$$

$$M^2 = \sum m^2[(b_m, b_m) + (c_m, c_m)]. \tag{64a}$$

It now follows that the Hilbert space vectors δ_m and σ_m are best assumed to be parallel to c_m and b_m, respectively, and indeed the proportionality constant between them is real. A part of δ_m which would be orthogonal to c_m would not change any of the equations, except by adding to the error ε. The same applies to σ_m with respect to b_m. The same argument shows that the proportionality factor between δ_m and c_m, and between σ_m and b_m, has to be real. Next, we assume that the lengths of b_m and c_m, and of δ_m and σ_m, are equal, but of course, wherever (c_m, δ_m) is positive, (b_m, σ_m) is negative, and conversely. Otherwise the derivative of (c_m, c_m) would become 0 and we would face the same difficulty as in the case of the idealized measurement (cf. equation 54). The equalities assumed can be justified, but this will not be done here. Capitalizing on these observations, and writing $c(m)$ for the length of c_m and $c(m)\delta(m)$ for (c_m, δ_m), we can restate the preceding set of equations:

$$4c(m)\frac{dc(m)}{dm} = 4c(m)\delta(m); \tag{60d}$$

$$\int c(m)^2\, dm = \tfrac{1}{2}; \tag{61b}$$

$$\varepsilon = 4 \int \sigma(m)\, dm; \tag{63b}$$

$$M^2 = 2 \int m^2 c(m)^2\, dm. \tag{64b}$$

The other equations are automatically satisfied—except that the vectors b_m and c_m must be assumed to be orthogonal.

It follows from (60a) that $\delta(m)$ is the derivative of $c(m)$, so that the error becomes

$$\varepsilon = 4 \int \left(\frac{dc(m)}{dm}\right)^2 dm. \tag{63c}$$

The problem, therefore, reduces either to minimizing the size M (or its square) given by (64b) while fixing the error ε of (63c) and the normalization (61b)—or to minimizing the error but fixing the size and taking the normalization into account. The Lagrange equation of the minimum problem reduces in both cases to a linear relation between $c(m)$, $m^2c(m)$, and $d^2c(m)/dm^2$—i.e., the quantum-mechanical equation of the oscillator. The solution is, therefore, that $c(m)$ is a Gauss error curve

$$c(m) = \alpha e^{-(\beta/2)m^2}, \tag{65}$$

the constants of which are determined by (61b) and either (63c) or (64b). The relation between error ε and size naturally is the same no matter which procedure is used:

$$\varepsilon = 1/(2M^2). \tag{66}$$

This is an interesting result, due in this form to Araki and Yanase (1960). It shows that most measurements cannot be made mathematically precise. Moreover, the more accuracy we demand, the greater must be the "size" (equation 64b) of the measuring apparatus.

Are there any quantities where the preceding size-accuracy correlation does not apply, quantities which, as far as this argument goes, may be measured precisely with a small apparatus? For elementary particles, only the mass and the magnitude of the spin are such precisely measurable quantities; none of the quantities specifying the actual state is measurable with arbitrary precision. If one has a more complex system, the preceding quantities do depend on the state because the degree of excitation is determined by the Minkowski length of the four-dimensional momentum vector (the rest mass). In addition, if the system contains several particles, the scalar product of the momentum vectors and many other similar quantities do commute with all additive conserved quantities. Nevertheless, it is clear that the limitation on the measurability of many quantities with a small apparatus is very severe (Yanase, 1961).

The preceding discussion, leading to equation (66), is incomplete—and this not only from the point of view of mathematical rigor. We considered only a system of spin $\frac{1}{2}$ and only one conservation law, conservation of the angular momentum in one of the directions perpendicular to that in which the spin component is to be measured. It is true that the components of the linear momentum do commute with the spin components; but the other components of the angular momentum do not. It would be interesting to generalize the result (66) both as to the system considered, and as to the inclusion of all additive conservation laws. It should also be repeated that (66) represents only a necessary condition for the apparatus to measure the spin component, and even such a generalization as is suggested above would only give a *necessary* condition. To demonstrate the actual measurability, one would have to describe the measuring apparatus and its functioning—as was done to some degree in §3 with respect to the measurement here considered, but not generally. In fact we shall derive in what follows several other constraints on measurability.

It has been observed that, in order to define a position coordinate, one has to have a coordinate system with a definite position in space and equipped with a clock with well-defined zero of time. The former requires that the apparatus which defines the coordinate system have a spread in momentum which corresponds to the position-momentum uncertainty relation, and a similar remark, to be discussed later, applies to the clock. Hence, in the case of the measurement of the position coordinate, the need to have an apparatus with a large momentum uncertainty, that is, momentum spread, is obvious. The same applies for many other measurements, but not with the generality established in the preceding discussion. In particular, the argument in this section also applies to the conservation laws for electric charge, for baryon number, etc., and already shows the difficulty of measuring any operator not commuting with these quantities. This last point will be taken up again when the superselection rules are discussed.

The Boson-Fermion Superselection Rule

The preceding section dealt with difficulties inherent in the measurement of a great many, if not most, operators, but the difficulties considered there could be overcome to an increasing extent by an increase in the "size" of the measuring equipment. The superselection rules absolutely exclude the measurability of certain operators. This exclusion is demonstrated in the first case, the boson-fermion superselection rule, on the basis of the postulate of the rotational invariance of the theory. The demonstration in the second case, the charge, baryon-number, etc. superselection rules, is more intricate and touches on a deeper philosophical problem.

The boson-fermion superselection rule tells us that no operator is measurable which has a finite matrix element between two states, one of which has an integer, the other a half-integer, angular momentum. The former states are associated with bosons, the latter with fermions but, of course, in the sense here considered, a hydrogen atom is a boson—the half-integer spins of the electron and of the proton combine to an integer spin. If an operator with a finite matrix element is measured, and the measurement results in characteristic vectors of the operator being produced, some of these characteristic vectors will have components with both integer and half integer spin: $b + f$.

From here on, the proof of the inconsistency with rotational invariance can proceed in many ways. A rather natural way to carry out the demonstration starts from the decompositions of b and f in terms of (1) total angular momentum and (2) z-component of the angular momentum, as viewed from some arbitrary coordinate system:

$$b = \sum_{Jm} b_m^{Jm}, \tag{67}$$

$$f = \sum_{Jm} f_m^{Jm}. \tag{67a}$$

Naturally, the J and m in (67) are integers, in (67a) half integers—this is the definition of b and f. The index m appears in (67) and (67a) twice in order to remind us that the terms in (67) or (67a) are not partners in the sense of invariance theory. Rather, the effect of a rotation R on these is given by the formulae

$$O_R b_m^{Jm} = \sum_{m'} D^{(J)}(R)_{m'm} b_{m'}^{Jm} \tag{68}$$

and the same formula applies to the $f_{m'}^{Jm}$, the $b_{m'}^{Jm}$ being the "partners" of b_m^{Jm}, the $f_{m'}^{Jm}$ partners of f_m^{Jm}. If the rotation R is applied on $b_{m'}^{Jm}$, we have

$$O_R b_{m'}^{Jm} = \sum_{m''} D^{(J)}(R)_{m''m'} b_{m''}^{Jm}, \tag{69}$$

and a similar formula applies to the rotation of the $f_{m'}^{Jm}$. The $D^{(J)}(R)$ are the matrices of the representations of the rotation group in our three-dimensional space (Wigner, 1959).

It now follows that the state vector of the state which looks like $b + f$ from the point of view of a coordinate system obtained from the original one by a rotation R is

$$\begin{aligned} O_R(b + f) &= \sum_{Jm} O_R b_m^{Jm} + O_R f_m^{Jm} \\ &= \sum_{Jm} \sum_{m'} D^{(J)}(R)_{m'm} b_{m'}^{Jm} + \omega_R \sum_{Jm} \sum_{m'} D^{(J)}(R)_{m'm} f_{m'}^{Jm}. \end{aligned} \tag{70}$$

The ω_R, with absolute value 1, had to be introduced in (70) because the effect of a transformation is always indefinite within a factor of absolute value 1—the state determines the state vector only up to such a factor. Actually, the first term of (70) should also be provided with such a factor. However, one factor of this nature can be eliminated from all such equations. One has only to postulate that the state vector be one of the state vectors representing the state in question. The others can be obtained from it by multiplication with an arbitrary phase factor, i.e., a factor of modulus 1. Actually, the omission of this phase factor simplifies the calculation relatively little.

It may be well to observe at this point that if the operator projecting into the state $b + f$ is observable—and we have assumed it is—then the operator projecting into $O_R(b + f)$ is likewise, by means of the same apparatus rotated by R. We shall take for R, first a rotation by π about z, to be denoted by Z. We then have, using the usual formulae for $D^{(J)}$ (remembering, though, that the m of b is an integer, that of f a half integer),

$$O_Z(b + f) = \sum_{Jm} i^{2m}(b_m^{Jm} + \omega_Z f_m^{Jm}) \tag{71}$$

and

$$O_Z^2(b + f) = \sum_{Jm} i^{4m}[b_m^{Jm} + \omega_Z(\omega_Z f_m^{Jm})]. \tag{71a}$$

We recall that O_Z^2 is the unit operation. Therefore (71a) should differ only by a factor from $b + f$. In the first term m is an integer and the factor in question indeed has the value unity. Hence $i^{4m}\omega_Z^2$ must also be 1 for the half integer m of the second term. It follows that $\omega_Z^2 = -1$, or

$$\omega_Z = \pm i. \tag{71b}$$

Let us observe that, in order to derive this result, it was necessary to assume that neither b nor f is a null-vector.

We consider next a rotation by π about the y axis. The standard formulae give

$$O_Y(b + f) = \sum_{Jm} i^{2J-2m}(b^{Jm}_{-m} + \omega_Y f^{Jm}_{-m}). \tag{72}$$

Calculation of $O_Y^2(b + f)$ yields this time:

$$\begin{aligned} O_Y^2(b + f) &= \sum_{Jm} i^{2J-2m} O_Y(b^{Jm}_{-m} + \omega_Y f^{Jm}_{-m}) \\ &= \sum_{Jm} i^{2J-2m} i^{2J+2m}(b^{Jm}_{m} + \omega_Y^2 f^{Jm}_{m}). \end{aligned} \tag{72a}$$

The fact that the second ω_Z factor in (71a) and the second ω_Y factor in (72a) must be the same as the first follows from the unitary nature of the transformations O_Z and O_Y. Otherwise the scalar products $[b + f, O_Z(b + f)]$ and $[O_Z(b + f), O_Z^2(b + f)]$ would not be equal. A similar remark applies to O_Y. Now $O_Y^2(b + f)$ can again differ from $b + f$ only in a factor. Moreover, the first term shows that this factor is 1. Also we have $i^{4J} = -1$ for the J of the second term. Therefore we arrive at a result similar to (71b),

$$\omega_Y = \pm i. \tag{72b}$$

A similar calculation yields

$$\omega_X = \pm i, \tag{73}$$

as was to be expected. However, the product of rotations by π about z and by π about y is a rotation by π about x. Therefore the vector $O_Y O_Z(b + f)$ can differ from $O_X(b + f)$ only by a factor. Contrary to this requirement we find

$$\begin{aligned} O_Y O_Z(b + f) &= O_Y \sum_{Jm} i^{2m}(b^{Jm}_{m} + \omega_Z f^{Jm}_{m}) \\ &= \sum_{Jm} i^{2m} i^{2J-2m}(b^{Jm}_{-m} + \omega_Y \omega_Z f^{Jm}_{-m}). \end{aligned} \tag{74}$$

Comparison of this expression with that for $O_X(b + f)$ shows that the factor in the first term is 1. Comparison of the second terms then shows that

$$\omega_X = \omega_Y \omega_Z, \tag{74a}$$

which is in contradiction to (71b), (72b), and (73). This disagreement shows that the existence of a state $b + f$, with finite b and finite f, is incompatible with the principle of rotational invariance. As was mentioned above, this "boson-fermion superselection rule" can be established in many different ways, each of which, however, relies to some degree on the theory of rotational invariance, in particular on the difference between the $D^{(J)}$ which apply to bosons and to fermions.

Superselection Rules for Charge, etc.

The creation of a state $b + f$ with both b (boson integer spin) and f (fermion, half-integer spin) finite, or, in other words, "any violation of the boson-fermion superselection rule," would conflict, as we have just seen, with the postulate of the rotational invariance of the theory, provided that the basic ideas of quantum mechanics are valid. This deduction may be a blemish on the very general principles of quantum theory, but it does not affect any of its practical applications or conceivable experimental conclusions.

According to the "charge superselection rule," it is also impossible to produce states

$$\sum c_n, \tag{75}$$

with c_n representing a state with electric charge number n and more than one of the vectors c_n finite. According to the "baryon superselection rule," the same prohibition applies to the baryon number. In contrast to the situation with the state vectors $b + f$, no conflict with any of the fundamental principles has been derived from the possibility of producing a state with the state vector (75). Rather, the validity of the somewhat controversial superselection rules for charge, etc. is based on the physical impossibility of producing states such as (75). We do not know of any macroscopic object that would be in a state such as (75). Even if we do not know the exact electric charge of an object—and in the case of a macroscopic object this is difficult to know—we cannot distinguish between (75) and other states, such as

$$\sum e^{i\phi_n} c_n, \tag{75a}$$

the exponentials in (75a) being phase factors of modulus 1. In the mathematical terminology of quantum mechanics this ambiguity is expressed by the statement that the actual state is not a superposition of the states c_n but a mixture of them. The state then cannot be characterized by a state vector. The system can be in any of the states c_n. The probability that a measurement of the charge will give the value n is (c_n, c_n). The theory of mixtures has been given an elegant mathematical form, but it will not be elaborated here.

State vectors with different charges do not interact with each other. Thus the time development of every c_n of (75) is the same as if it were present alone. Therefore the mixture character of the state remains preserved in time. Similarly, if two systems are united to form a joint system, and if each of the two is a mixture of states with definite charge numbers, the same will be true of the united system. It follows from this reasoning that unless nature supplies some superposition of different charge states, such as (75), no such superposition state will ever be found. Then the superselection rule for charge will be valid. The same applies to the superselection rules for baryon number, etc.

The law of conservation of electric charge, the independence of the time development of the states with different charges, and our observation on the union of two systems, remain equally valid if "linear momentum" is substituted for "electric charge." It is natural to ask, therefore, why no superselection rule applies for the components of linear momentum or angular momentum. How is it that one can produce a state which is a superposition of states with different linear momenta? And how is it that one can ascertain the coefficients $e^{i\phi_n}$ in the expression corresponding to (75a) with n now referring to a component of linear momentum (n a continuous variable) or to a component of the angular momentum? The answer to this question, it must be admitted, cannot be read out of the basic equations of quantum mechanics. We really do not know how we acquired the ability to see light signals and to feel objects, and why there are no similar phenomena in connection with electric charges. Superconductivity is often claimed to provide such signals. Indeed, the usual theory does use a description of the superconducting state which is a definite superposition of states with different electric charges. However, it is evident that the conclusions of the theory, describing currents, magnetic fields, and similar observable phenomena, would not change in any way if the different charge states were multiplied with different phase factors as they are in (75a). None of the observables used has matrix elements connecting states with different charges. Only the finiteness of such matrix elements could permit one to distinguish (75a) from (75). As far as linear momentum is concerned, the situation is entirely different. Any position operator, for instance, has major matrix elements between states with different momenta.

The question naturally arises whether and how phase relations between different charge states could manifest themselves. Surely the fact that the observed phenomena of superconductivity do not prove the existence of such phase relations does not prove their absence. Naturally, the existence of such phase relations, that is, the breakdown of the charge superselection rule, would be very surprising. Present quantum mechanics would do as little to explain such a breakdown as to explain how the states of different energy in a thermodynamic ensemble could manifest phase relations. Nevertheless, it may be interesting to find an experimental criterion to tell whether such phase relations exist—that is, to tell whether superpositions such as (75) or (75a) with definite ψ_n are distinguishable.

The simplest experiment would consist in the uniting of two systems. Let each system be imagined to be in a superposition of states with the charges n and m. If these are superpositions such as (75) or (75a), their state vectors could be denoted by

$$\psi_n + e^{i\phi}\psi_m \quad \text{and} \quad \psi'_n + e^{i\phi'}\psi'_m, \tag{76}$$

with definite ϕ and ϕ'. Their union would then have the state vector

$$\psi_n \otimes \psi'_n + [e^{i\phi'}\psi_n \otimes \psi'_m + e^{i\phi}\psi_m \otimes \psi'_n] + e^{i\phi + i\phi'}\psi_m \otimes \psi'_m. \tag{76a}$$

This formula indicates that the component with charge $n + m$, given in the brackets of (76a), would have a definite structure and might well be distinguished by conceivable experiments from a mixture of the two states $\psi_n \otimes \psi'_m$ and $\psi_m \otimes \psi'_n$. The factor $\exp[i(\phi - \phi')]$ measuring the difference in phase between the two states of equal charge would be definite. There is no reason to believe that that factor would be unobservable. Needless to say, (76) and hence (76a) could be greatly generalized and, equally obvious, the experiment alluded to here is far from concrete. Moreover, it may be impossible to carry it out.

The Measurement of Position

As was mentioned at the beginning of this section, the early interest in the correspondence of a quantum-mechanical operator to classical expressions, such as x^2p^2, has largely ceased. Nevertheless, there are classical quantities, such as linear and angular momentum, energy, and position, to which one would like to coordinate an operator. It appears, at least superficially, that these quantities have a simple meaning, and can be measured, and that it should be possible to find the proper operator for them.

Whether easily measurable or not, the coordination of operators to energy, linear and angular momentum components (and the three other quantities associated with them in relativistic theory) can easily be made. The operators are the infinitesimal operators of the (special) relativistic invariance group, also called the Poincaré group. Position is different. Quite apart from the difficulties of measuring position soon to be discussed, we find great problems in coordinating an operator to position. These difficulties have been most elegantly demonstrated by Hegerfeldt (1974) (see also Pryce, 1948; Møller, 1972; Wightman, 1962; and Fleming, 1965). The following discussion is based on his remark. It is more technical than the rest of these notes, being based on the theory of the Poincaré invariance of quantum mechanics, and may be, for some, difficult to follow.

In order to simplify the demonstration, we shall use a spacetime with only one spacelike dimension. The generalization to three spacelike dimensions is not difficult but would make the discussion a good deal longer. The spirit of the demonstration is not impaired by the restriction to one spacelike dimension.

Let us assume that there is a position operator and consider its projection operators $P(x)$ as defined, for an arbitrary operator, by equations (21) of §2 ("all values up to x"). Let us then produce out of some initial state ϕ, by the operator $P(x_2) - P(x_1)$ (with $x_2 > x_1$), a state ψ for which the position is confined to the x_1, x_2 interval. We express the state vector of such a state in the momentum representation, as is usual in the quantum-mechanical invariance theory. Thus we write $\psi = \psi(p, \sigma)$, where p is the momentum vector (one-dimensional for our world with one spacelike dimension) and σ is a discrete variable, characterizing the other variables, such as spin component in one direction, etc. Because of the confinement of the position to the x_1, x_2 interval, that is, because of

$$\psi(p, \sigma) = [P(x_2) - P(x_1)]\phi,$$

we have

$$P(x)\psi(p, \sigma) = 0 \qquad \text{for all} \quad x \leq x_1, \tag{77}$$

and

$$P(x)\psi(p, \sigma) = \psi(p, \sigma) \qquad \text{for all} \quad x \geq x_2. \tag{77a}$$

The scalar product of two vectors ϕ and ϕ', expressed in the (p, σ)-representation, is

$$(\phi, \phi') = \sum_{\sigma} \int \phi(p, \sigma)^* \phi'(p, \sigma)\, dp/p_0. \tag{78}$$

Here cp_0 is the energy associated with the momentum p,

$$p_0 = (m^2c^2 + p^2)^{1/2}. \tag{78a}$$

The p_0 in the denominator of (78) will play no significant role. It is introduced in the usual definition for the scalar product because it simplifies the expressions for Lorentz transformations. We shall not make use of general transformations but only of displacements in space and time. The operator for a displacement in space by a and displacement in time by t is simply multiplication by

$$e^{-ip_0t + ipa}. \tag{79}$$

If $|a| > x_2 - x_1$, a space displacement by a moves the original confinement into a new, nonoverlapping one so that $e^{ipa}\psi(p, \sigma)$ will be orthogonal to $\psi(p, \sigma)$,

$$(e^{ipa}\psi(p, \sigma), \psi(p, \sigma)) = \int e^{-ipa} \sum_{\sigma} |\psi(p, \sigma)|^2 \, dp/p_0 = 0, \tag{80}$$

for $|a| > x_2 - x_1$. This can be proved also more formally by noting that

$$P(x - a)e^{ipa} = e^{ipa}P(x) \tag{79a}$$

and making use of (21c) of §2. From the validity of (79), and the confinement of x for $\psi(p, \sigma)$ to an interval of the width $x_2 - x_1$, (80) should be evident anyway.

Equation (80) expresses the fact that the Fourier transform of

$$\sum_{\sigma} \psi(p, \sigma)^* \psi(p, \sigma)/p_0 \tag{80a}$$

is confined to a region of width $2(x_2 - x_1)$. It then follows from the expression of (80a) in terms of its Fourier transform—an integral over a region of width $2(x_2 - x_1)$—that the expression (80a) is a meromorphic function of p; that is, it has no singularity in the finite complex plane. The scalar product of a space displaced $\psi(p, \sigma)$ and a time displaced $\psi(p, \sigma)$,

$$(e^{ipa}\psi(p, \sigma), e^{-ip_0t}\psi(p, \sigma)) = \int e^{-ipa - ip_0t} \sum_{\sigma} |\psi(p, \sigma)|^2 \, dp/p_0, \tag{81}$$

evidently is the Fourier coefficient of the function

$$e^{-ip_0t} \sum_{\sigma} \psi(p, \sigma)^* \psi(p, \sigma)/p_0. \tag{81a}$$

This function, for finite t, is not a meromorphic function of p. To be sure, (80a) *is* such a function; but the p_0 in the exponent has singularities at $\pm imc$. Hence, there is no finite interval of a outside of which its Fourier transform, (81), will vanish. This means that the time displacement spreads the position, originally confined to the interval x_1, x_2, over all space. Otherwise there could be no finite transition probability to functions $e^{ipa}\psi(p, \sigma)$ for arbitrarily large a, since the latter state is confined to the interval $x_1 - a, x_2 - a$. No matter how one defines the position, one has to conclude that the velocity, defined as the ratio of two subsequent position measurements divided by the time interval between them, has a finite probability of assuming an arbitrarily large value, exceeding c. One either has to accept this, or deny the possibility of measuring the position precisely or even giving significance to this concept: a very difficult choice! And the author of these notes will admit that since writing them, he has devoted a good deal of attention to the problem (Ahmad and Wigner, 1975; O'Connell and Wigner, 1977, 1978).

Summary

The present section starts with the observation that the measured quantities correspond to (self-adjoint) operators, not to the quantities of classical physics. It then points to the demonstrable difficulties of certain measurements, i.e., where it can be shown that it is difficult to find an apparatus the interaction of which with the object has the result indicated by the basic equation (34) of §3.

It is shown, first, that no precise measurement of any quantity is possible unless this commutes with all conserved additive quantities. In some cases this conclusion can be derived, at least qualitatively, with ease. In other cases, such as operators which do not commute with the operator of electric charge, the situation is less obvious. It is shown then that a measurement which has a small probability of giving an incorrect result surely needs a large apparatus, i.e., one in a state in which the additive conserved quantities, not commuting with the quantity to be measured, have probabilities for values spread over a wide spectrum.

The second type of case discussed refers to quantities which cannot be measured at all. Examples are, first, operators which have a finite matrix element between a state with integer and another state with half-integer angular momentum. The so-called boson-fermion superselection rule shows that it is impossible to produce states which are superpositions of states with integer and half-integer angular momenta. A similar statement is then made for the superposition of states with different electric charges and also with different baryon numbers, and perhaps other similar descriptors.

Finally, we had to recognize, every attempt to provide a precise definition of a position coordinate stands in direct contradiction with special relativity.

All these are concrete and clearly demonstrated limitations on the measurability of operators. They should not obscure the other, perhaps even more fundamental weakness of the standard theory, that it postulates the measurability of operators but does not give directions as to how the measurement should be carried out. This problem has already been emphasized at the end of §3, the description of the quantum theory of measurement.

All the preceding discussion is based on the usual quantum-mechanical theory, as taught in classes, not on the more advanced axiomatic quantum field theory. The question therefore arises whether quantum field theory avoids the difficulties mentioned. Surely, as far as the superselection rules are concerned, it does not. As to the rest, my opinion is negative also. The best known discussion of the measurement of field strengths, that given by Bohr and Rosenfeld, postulates an electric test charge with arbitrarily large charge and arbitrarily small size (Bohr and Rosenfeld, 1933, 1950). Naturally, the fact that the quantum field theory does not resolve our problems, though regrettable, should not be considered as an argument against that theory.

Acknowledgments

The author would like to express his appreciation to all who helped him to write up his lectures, and this includes not only S. J. Freedman but several others who attended the lectures. He also wishes to record his gratitude to W. H. Zurek for reviewing the material and, even more, for assembling the references.

II.3 "RELATIVE STATE" FORMULATION OF QUANTUM MECHANICS*

HUGH EVERETT III

1. INTRODUCTION

THE task of quantizing general relativity raises serious questions about the meaning of the present formulation and interpretation of quantum mechanics when applied to so fundamental a structure as the space-time geometry itself. This paper seeks to clarify the foundations of quantum mechanics. It presents a reformulation of quantum theory in a form believed suitable for application to general relativity.

The aim is not to deny or contradict the conventional formulation of quantum theory, which has demonstrated its usefulness in an overwhelming variety of problems, but rather to supply a new, more general and complete formulation, from which the conventional interpretation can be *deduced*.

The relationship of this new formulation to the older formulation is therefore that of a metatheory to a theory, that is, it is an underlying theory in which the nature and consistency, as well as the realm of applicability, of the older theory can be investigated and clarified.

The new theory is not based on any radical departure from the conventional one. The special postulates in the old theory which deal with observation are omitted in the new theory. The altered theory thereby acquires a new character. It has to be analyzed in and for itself before any identification becomes possible between the quantities of the theory and the properties of the world of experience. The identification, when made, leads back to the omitted postulates of the conventional theory that deal with observation, but in a manner which clarifies their role and logical position.

We begin with a brief discussion of the conventional formulation, and some of the reasons which motivate one to seek a modification.

2. REALM OF APPLICABILITY OF THE CONVENTIONAL OR "EXTERNAL OBSERVATION" FORMULATION OF QUANTUM MECHANICS

We take the conventional or "external observation" formulation of quantum mechanics to be essentially the following[1]: A physical system is completely described by a state function ψ, which is an element of a Hilbert space, and which furthermore gives information only to the extent of specifying the probabilities of the results of various observations which can be made *on* the system *by* external observers. There are two fundamentally different ways in which the state function can change:

Process 1: The discontinuous change brought about by the observation of a quantity with eigenstates $\phi_1, \phi_2, \cdots$, in which the state ψ will be changed to the state ϕ_j with probability $|(\psi,\phi_j)|^2$.

Process 2: The continuous, deterministic change of state of an isolated system with time according to a wave equation $\partial\psi/\partial t = A\psi$, where A is a linear operator.

This formulation describes a wealth of experience. No experimental evidence is known which contradicts it.

Not all conceivable situations fit the framework of this mathematical formulation. Consider for example an isolated system consisting of an observer or measuring apparatus, plus an object system. Can the change with time of the state of the *total* system be described by Process 2? If so, then it would appear that no discontinuous probabilistic process like Process 1 can take place. If not, we are forced to admit that systems which contain observers are not subject to the same kind of quantum-mechanical description as we admit for all other physical systems. The question cannot be ruled out as lying in the domain of psychology. Much of the discussion of "observers" in quantum mechanics has to do with photoelectric cells, photographic plates, and similar devices where a mechanistic attitude can hardly be contested. For the following one can *limit himself to this class of problems*, if he is unwilling to consider observers in the more familiar sense on the same mechanistic level of analysis.

What mixture of Processes 1 and 2 of the conventional formulation is to be applied to the case where only an approximate measurement is effected; that is, where an apparatus or observer interacts only weakly and for a limited time with an object system? In this case of an

* Thesis submitted to Princeton University March 1, 1957 in partial fulfillment of the requirements for the Ph.D. degree. An earlier draft dated January, 1956 was circulated to several physicists whose comments were helpful. Professor Niels Bohr, Dr. H. J. Groenewald, Dr. Aage Peterson, Dr. A. Stern, and Professor L. Rosenfeld are free of any responsibility, but they are warmly thanked for the useful objections that they raised. Most particular thanks are due to Professor John A. Wheeler for his continued guidance and encouragement. Appreciation is also expressed to the National Science Foundation for fellowship support.

[1] We use the terminology and notation of J. von Neumann, *Mathematical Foundations of Quantum Mechanics*, translated by R. T. Beyer (Princeton University Press, Princeton, 1955).

Originally published in *Reviews of Modern Physics, 29*, 454-62 (1957).

approximate measurement a proper theory must specify (1) the new state of the object system that corresponds to any particular reading of the apparatus and (2) the probability with which this reading will occur. von Neumann showed how to treat a special class of approximate measurements by the method of projection operators.[2] However, a general treatment of all approximate measurements by the method of projection operators can be shown (Sec. 4) to be impossible.

How is one to apply the conventional formulation of quantum mechanics to the space-time geometry itself? The issue becomes especially acute in the case of a closed universe.[3] There is no place to stand outside the system to observe it. There is nothing outside it to produce transitions from one state to another. Even the familiar concept of a proper state of the energy is completely inapplicable. In the derivation of the law of conservation of energy, one defines the total energy by way of an integral extended over a surface large enough to include all parts of the system and their interactions.[4] But in a closed space, when a surface is made to include more and more of the volume, it ultimately disappears into nothingness. Attempts to define a total energy for a closed space collapse to the vacuous statement, zero equals zero.

How are a quantum description of a closed universe, of approximate measurements, and of a system that contains an observer to be made? These three questions have one feature in common, that they all inquire about the *quantum mechanics* that is *internal to an isolated system.*

No way is evident to apply the conventional formulation of quantum mechanics to a system that is not subject to *external* observation. The whole interpretive scheme of that formalism rests upon the notion of external observation. The probabilities of the various possible outcomes of the observation are prescribed exclusively by Process 1. Without that part of the formalism there is no means whatever to ascribe a physical interpretation to the conventional machinery. But Process 1 is out of the question for systems not subject to external observation.[5]

3. QUANTUM MECHANICS INTERNAL TO AN ISOLATED SYSTEM

This paper proposes to regard pure wave mechanics (Process 2 only) as a complete theory. It postulates that a wave function that obeys a linear wave equation everywhere and at all times supplies a complete mathematical model for every isolated physical system without exception. It further postulates that every system that is subject to external observation can be regarded as part of a larger isolated system.

The wave function is taken as the basic physical entity with *no a priori interpretation.* Interpretation only comes *after* an investigation of the logical structure of the theory. Here as always the theory itself sets the framework for its interpretation.[5]

For any interpretation it is necessary to put the mathematical model of the theory into correspondence with experience. For this purpose it is necessary to formulate abstract models for observers that can be treated within the theory itself as physical systems, to consider isolated systems containing such model observers in interaction with other subsystems, to deduce the changes that occur in an observer as a consequence of interaction with the surrounding subsystems, and to interpret the changes in the familiar language of experience.

Section 4 investigates representations of the state of a composite system in terms of states of constituent subsystems. The mathematics leads one to recognize the concept of the *relativity of states,* in the following sense: a constituent subsystem cannot be said to be in any single well-defined state, independently of the remainder of the composite system. To any arbitrarily chosen state for one subsystem there will correspond a unique *relative state* for the remainder of the composite system. This relative state will usually depend upon the choice of state for the first subsystem. Thus the state of one subsystem does not have an independent existence, but is fixed only by the state of the remaining subsystem. In other words, the states occupied by the subsystems are not independent, but *correlated.* Such correlations between systems arise whenever systems interact. In the present formulation all measurements and observation processes are to be regarded simply as interactions between the physical systems involved—interactions which produce strong correlations. A simple model for a measurement, due to von Neumann, is analyzed from this viewpoint.

Section 5 gives an abstract treatment of the problem of observation. This uses only the superposition principle, and general rules by which composite system states are formed of subsystem states, in order that the results shall have the greatest generality and be applicable to any form of quantum theory for which these principles hold. Deductions are drawn about the state of the observer relative to the state of the object system. It is found that experiences of the observer (magnetic tape memory, counter system, etc.) are in full accord with predictions of the conventional "external observer" formulation of quantum mechanics, based on Process 1.

Section 6 recapitulates the "relative state" formulation of quantum mechanics.

---

[2] Reference 1, Chap. 4, Sec. 4.

[3] See A. Einstein, *The Meaning of Relativity* (Princeton University Press, Princeton, 1950), third edition, p. 107.

[4] L. Landau and E. Lifshitz, *The Classical Theory of Fields,* translated by M. Hamermesh (Addison-Wesley Press, Cambridge, 1951), p. 343.

[5] See in particular the discussion of this point by N. Bohr and L. Rosenfeld, Kgl. Danske Videnskab. Selskab, Mat.-fys. Medd. **12,** No. 8 (1933).

4. CONCEPT OF RELATIVE STATE

We now investigate some consequences of the wave mechanical formalism of composite systems. If a composite system S, is composed of two subsystems S_1 and S_2, with associated Hilbert spaces H_1 and H_2, then, according to the usual formalism of composite systems, the Hilbert space for S is taken to be the *tensor product* of H_1 and H_2 (written $H=H_1 \otimes H_2$). This has the consequence that if the sets $\{\xi_i^{S_1}\}$ and $\{\eta_j^{S_2}\}$ are complete orthonormal sets of states for S_1 and S_2, respectively, then the general state of S can be written as a superposition:

$$\psi^S = \sum_{i,j} a_{ij} \xi_i^{S_1} \eta_j^{S_2}. \tag{1}$$

From (3.1) although S is in a definite state ψ^S, the subsystems S_1 and S_2 do not possess anything like definite states independently of one another (except in the special case where all but one of the a_{ij} are zero).

We *can*, however, for any choice of a state in one subsystem, *uniquely* assign a corresponding *relative* state in the other subsystem. For example, if we choose ξ_k as the state for S_1, while the composite system S is in the state ψ^S given by (3.1), then the corresponding *relative state* in S_2, $\psi(S_2;\, \text{rel}\xi_k, S_1)$, will be:

$$\psi(S_2;\, \text{rel}\xi_k, S_1) = N_k \sum_j a_{kj} \eta_j^{S_2} \tag{2}$$

where N_k is a normalization constant. This relative state for ξ_k is *independent* of the choice of basis $\{\xi_i\}$ $(i \neq k)$ for the orthogonal complement of ξ_k, and is hence determined uniquely by ξ_k alone. To find the relative state in S_2 for an arbitrary state of S_1 therefore, one simply carries out the above procedure using any pair of bases for S_1 and S_2 which contains the desired state as one element of the basis for S_1. To find states in S_1 relative to states in S_2, interchange S_1 and S_2 in the procedure.

In the conventional or "external observation" formulation, the relative state in S_2, $\psi(S_2;\, \text{rel}\phi, S_1)$, for a state ϕ^{S_1} in S_1, gives the conditional probability distributions for the results of all measurements in S_2, given that S_1 has been measured and found to be in state ϕ^{S_1}—i.e., that ϕ^{S_1} is the eigenfunction of the measurement in S_1 corresponding to the observed eigenvalue.

For any choice of basis in S_1, $\{\xi_i\}$, it is always possible to represent the state of S, (1), as a *single* superposition of pairs of states, each consisting of a state from the basis $\{\xi_i\}$ in S_1 and its relative state in S_2. Thus, from (2), (1) can be written in the form:

$$\psi^S = \sum_i \frac{1}{N_i} \xi_i^{S_1} \psi(S_2;\, \text{rel}\xi_i, S_1). \tag{3}$$

This is an important representation used frequently.

Summarizing: There does not, in general, exist anything like a single state for one subsystem of a composite system. Subsystems do not possess states that are independent of the states of the remainder of the system, so that the subsystem states are generally correlated *with one another. One can arbitrarily choose a state for one subsystem, and be led to the relative state for the remainder. Thus we are faced with a fundamental* relativity of states, *which is implied by the formalism of composite systems. It is meaningless to ask the absolute state of a subsystem—one can only ask the state relative to a given state of the remainder of the subsystem.*

At this point we consider a simple example, due to von Neumann, which serves as a model of a measurement process. Discussion of this example prepares the ground for the analysis of "observation." We start with a system of only one coordinate, q (such as position of a particle), and an apparatus of one coordinate r (for example the position of a meter needle). Further suppose that they are initially independent, so that the combined wave function is $\psi_0^{S+A} = \phi(q)\eta(r)$ where $\phi(q)$ is the initial system wave function, and $\eta(r)$ is the initial apparatus function. The Hamiltonian is such that the two systems do not interact except during the interval $t=0$ to $t=T$, during which time the total Hamiltonian consists only of a simple interaction,

$$H_I = -i\hbar q(\partial/\partial r). \tag{4}$$

Then the state

$$\psi_t^{S+A}(q,r) = \phi(q)\eta(r-qt) \tag{5}$$

is a solution of the Schrödinger equation,

$$i\hbar(\partial\psi_t^{S+A}/\partial t) = H_I \psi_t^{S+A}, \tag{6}$$

for the specified initial conditions at time $t=0$.

From (5) at time $t=T$ (at which time interaction stops) there is no longer any definite independent apparatus state, nor any independent system state. The apparatus therefore does not indicate any definite object-system value, and nothing like process 1 has occurred.

Nevertheless, we *can* look upon the total wave function (5) as a *superposition* of pairs of subsystem states, each element of which has a definite q value and a correspondingly displaced apparatus state. Thus after the interaction the state (5) has the form:

$$\psi_T^{S+A} = \int \phi(q')\delta(q-q')\eta(r-qT)dq', \tag{7}$$

which is a superposition of states $\psi_{q'} = \delta(q-q')\eta(r-qT)$. Each of these elements, $\psi_{q'}$, of the superposition describes a state in which the system has the definite value $q=q'$, and in which the apparatus has a state that is displaced from its original state by the amount $q'T$. These elements $\psi_{q'}$ are then superposed with coefficients $\phi(q')$ to form the total state (7).

Conversely, if we transform to the representation where the *apparatus* coordinate is definite, we write (5) as

$$\psi_T^{S+A} = \int (1/N_{r'})\xi^{r'}(q)\delta(r-r')dr',$$

where

$$\xi^{r'}(q) = N_{r'}\phi(q)\eta(r'-qT) \quad (8)$$

and

$$(1/N_{r'})^2 = \int \phi^*(q)\phi(q)\eta^*(r'-qT)\eta(r'-qT)dq.$$

Then the $\xi^{r'}(q)$ are the relative system state functions[6] for the apparatus states $\delta(r-r')$ of definite value $r=r'$.

If T is sufficiently large, or $\eta(r)$ sufficiently sharp (near $\delta(r)$), then $\xi^{r'}(q)$ is nearly $\delta(q-r'/T)$ and the relative system states $\xi^{r'}(q)$ are nearly eigenstates for the values $q=r'/T$.

We have seen that (8) is a superposition of states $\psi_{r'}$, *for each of which* the apparatus has recorded a definite value r', and the system is left in approximately the eigenstate of the measurement corresponding to $q=r'/T$. The discontinuous "jump" into an eigenstate is thus only a relative proposition, dependent upon the mode of decomposition of the total wave function into the superposition, and relative to a particularly chosen apparatus-coordinate value. So far as the complete theory is concerned all elements of the superposition exist simultaneously, and the entire process is quite continuous.

von Neumann's example is only a special case of a more general situation. Consider any measuring apparatus interacting with any object system. As a result of the interaction the state of the measuring apparatus is no longer capable of independent definition. It can be defined only *relative* to the state of the object system. In other words, there exists only a correlation between the states of the two systems. It seems as if nothing can ever be settled by such a measurement.

This indefinite behavior seems to be quite at variance with our observations, since physical objects always appear to us to have definite positions. Can we reconcile this feature wave mechanical theory built purely on Process 2 with experience, or must the theory be abandoned as untenable? In order to answer this question we consider the problem of observation itself within the framework of the theory.

5. OBSERVATION

We have the task of making deductions about the appearance of phenomena to observers which are considered as purely physical systems and are treated within the theory. To accomplish this it is necessary to identify some present properties of such an observer with features of the past experience of the observer.

[6] This example provides a model of an approximate measurement. However, the relative system states after the interaction $\xi^{r'}(q)$ cannot ordinarily be generated from the original system state ϕ by the application of *any* projection operator, E. Proof: Suppose on the contrary that $\xi^{r'}(q) = NE\phi(q) = N'\phi(q)\eta(r'-qt)$, where N, N' are normalization constants. Then

$$E(NE\phi(q)) = NE^2\phi(q) = N''\phi(q)\eta^2(r'-qt)$$

and $E^2\phi(q) = (N''/N)\phi(q)\eta^2(r'-qt)$. But the condition $E^2=E$ which is necessary for E to be a projection implies that $N'/N''\eta(q) = \eta^2(q)$ which is generally false.

Thus, in order to say that an observer 0 has observed the event α, it is necessary that the state of 0 has become changed from its former state to a new state which is dependent upon α.

It will suffice for our purposes to consider the observers to possess memories (i.e., parts of a relatively permanent nature whose states are in correspondence with past experience of the observers). In order to make deductions about the past experience of an observer it is sufficient to deduce the present contents of the memory as it appears within the mathematical model.

As models for observers we can, if we wish, consider automatically functioning machines, possessing sensory apparatus and coupled to recording devices capable of registering past sensory data and machine configurations. We can further suppose that the machine is so constructed that its present actions shall be determined not only by its present sensory data, but by the contents of its memory as well. Such a machine will then be capable of performing a sequence of observations (measurements), and furthermore of deciding upon its future experiments on the basis of past results. If we consider that current sensory data, as well as machine configuration, is immediately recorded in the memory, then the actions of the machine at a given instant can be regarded as a function of the memory contents only, and all relavant experience of the machine is contained in the memory.

For such machines we are justified in using such phrases as "the machine has perceived A" or "the machine is aware of A" if the occurrence of A is represented in the memory, since the future behavior of the machine will be based upon the occurrence of A. In fact, all of the customary language of subjective experience is quite applicable to such machines, and forms the most natural and useful mode of expression when dealing with their behavior, as is well known to individuals who work with complex automata.

When dealing with a system representing an observer quantum mechanically we ascribe a state function, ψ^0, to it. When the state ψ^0 describes an observer whose memory contains representations of the events $A, B, \cdots, C$ we denote this fact by appending the memory sequence in brackets as a subscript, writing:

$$\psi^0{}_{[A,B,\cdots,C]}. \quad (9)$$

The symbols $A, B, \cdots, C$, which we assume to be ordered time-wise, therefore stand for memory configurations which are in correspondence with the past experience of the observer. These configurations can be regarded as punches in a paper tape, impressions on a magnetic reel, configurations of a relay switching circuit, or even configurations of brain cells. We require only that they be capable of the interpretation "The observer has experienced the succession of events $A, B, \cdots, C$." (We sometimes write dots in a memory sequence, $\cdots A, B, \cdots, C$, to indicate the possible presence of previous memories which are irrelevant to the case being considered.)

The mathematical model seeks to treat the interaction of such observer systems with other physical systems (observations), within the framework of Process 2 wave mechanics, and to deduce the resulting memory configurations, which are then to be interpreted as records of the past experiences of the observers.

We begin by defining what constitutes a "good" observation. A good observation of a quantity A, with eigenfunctions ϕ_i, for a system S, by an observer whose initial state is ψ^0, consists of an interaction which, in a specified period of time, transforms each (total) state

$$\psi^{S+0}=\phi_i\psi^0{}_{[\ldots]} \tag{10}$$

into a new state

$$\psi^{S+0'}=\phi_i\psi^0{}_{[\ldots\alpha_i]} \tag{11}$$

where α_i characterizes[7] the state ϕ_i. (The symbol, α_i, might stand for a recording of the eigenvalue, for example.) That is, we require that the system state, *if it is an eigenstate*, shall be unchanged, and (2) that the observer state shall change so as to describe an observer that is "aware" of which eigenfunction it is; that is, some property is recorded in the memory of the observer which characterizes ϕ_i, such as the eigenvalue. The requirement that the eigenstates for the system be unchanged is necessary if the observation is to be significant (repeatable), and the requirement that the observer state change in a manner which is different for each eigenfunction is necessary if we are to be able to call the interaction an observation at all. How closely a general interaction satisfies the definition of a good observation depends upon (1) the way in which the interaction depends upon the dynamical variables of the observer system—including memory variables—and upon the dynamical variables of the object system and (2) the initial state of the observer system. Given (1) and (2), one can for example solve the wave equation, deduce the state of the composite system after the end of the interaction, and check whether an object system that was originally in an eigenstate is left in an eigenstate, as demanded by the repeatability postulate. This postulate is satisfied, for example, by the model of von Neumann that has already been discussed.

From the definition of a good observation we first deduce the result of an observation upon a system which is *not* in an eigenstate of the observation. We know from our definition that the interaction transforms states $\phi_i\psi^0{}_{[\cdots]}$ into states $\phi_i\psi^0{}_{[\cdots\alpha_i]}$. Consequently these solutions of the wave equation can be superposed to give the final state for the case of an arbitrary initial system state. Thus if the initial system state is not an eigenstate, but a general state $\sum_i a_i\phi_i$, the final total state will have the form:

$$\psi^{S+0'}=\sum a_i\phi_i\psi^0{}_{[\cdots\alpha_i]}. \tag{12}$$

This superposition principle continues to apply in the presence of further systems which do not interact during the measurement. Thus, if systems $S_1, S_2, \cdots, S_n$ are present as well as 0, with original states $\psi^{S_1}, \psi^{S_2}, \cdots, \psi^{S_n}$, and the only interaction during the time of measurement takes place between S_1 and 0, the measurement will transform the initial total state:

$$\psi^{S_1+S_2+\cdots+S_n+0}=\psi^{S_1}\psi^{S_2}\cdots\psi^{S_n}\psi^0{}_{[\cdots]} \tag{13}$$

into the final state:

$$\psi'^{S_1+S_2+\cdots+S_n+0}=\sum_i a_i\phi_i{}^{S_1}\psi^{S_2}\cdots\psi^{S_n}\psi^0{}_{[\cdots\alpha_i]} \tag{14}$$

where $a_i=(\phi_i{}^{S_1},\psi^{S_1})$ and $\phi_i{}^{S_1}$ are eigenfunctions of the observation.

Thus we arrive at the general rule for the transformation of total state functions which describe systems within which observation processes occur:

Rule 1: The observation of a quantity A, with eigenfunctions $\phi_i{}^{S_1}$, in a system S_1 by the observer 0, transforms the total state according to:

$$\psi^{S_1}\psi^{S_2}\cdots\psi^{S_n}\psi^0{}_{[\cdots]} \rightarrow\sum_i a_i\phi_i{}^{S_1}\psi^{S_2}\cdots\psi^{S_n}\psi^0{}_{[\cdots\alpha_i]} \tag{15}$$

where

$$a_i=(\phi_i{}^{S_1},\psi^{S_1}).$$

If we next consider a *second* observation to be made, where our total state is now a superposition, we can apply Rule 1 separately to each element of the superposition, since each element separately obeys the wave equation and behaves independently of the remaining elements, and then superpose the results to obtain the final solution. We formulate this as:

Rule 2: Rule 1 may be applied separately to each element of a superposition of total system states, the results being superposed to obtain the final total state. Thus, a determination of B, with eigenfunctions $\eta_j{}^{S_2}$, on S_2 by the observer 0 transforms the total state

$$\sum_i a_i\phi_i{}^{S_1}\psi^{S_2}\cdots\psi^{S_n}\psi^0{}_{[\cdots\alpha_i]} \tag{16}$$

into the state

$$\sum_{i,j} a_i b_j\phi_i{}^{S_1}\eta_j{}^{S_2}\psi^{S_3}\cdots\psi^{S_n}\psi^0{}_{[\cdots\alpha_i,\beta_j]} \tag{17}$$

where $b_j=(\eta_j{}^{S_2},\psi^{S_2})$, which follows from the application of Rule 1 to each element $\phi_i{}^{S_1}\psi^{S_2}\cdots\psi^{S_n}\psi^0{}_{[\cdots\alpha_i]}$, and then superposing the results with the coefficients a_i.

These two rules, which follow directly from the superposition principle, give a convenient method for determining final total states for any number of observation processes in any combinations. We now seek the *interpretation* of such final total states.

[7] It should be understood that $\psi^0{}_{[\cdots\alpha_i]}$ is a *different* state for each i. A more precise notation would write $\psi^0{}_i{}_{[\cdots\alpha_i]}$, but no confusion can arise if we simply let the $\psi_i{}^0$ be indexed only by the index of the memory configuration symbol.

Let us consider the simple case of a single observation of a quantity A, with eigenfunctions ϕ_i, in the system S with initial state ψ^S, by an observer 0 whose initial state is $\psi^0_{[\cdots]}$. The final result is, as we have seen, the superposition

$$\psi'^{S+0}=\sum_i a_i\phi_i\psi^0_{[\cdots\alpha_i]}. \tag{18}$$

There is no longer any independent system state or observer state, although the two have become correlated in a one-one manner. However, in each *element* of the superposition, $\phi_i\psi^0_{[\cdots\alpha_i]}$, the object-system state is a particular eigenstate of the observation, and *furthermore the observer-system state describes the observer as definitely perceiving that particular system state.* This correlation is what allows one to maintain the interpretation that a measurement has been performed.

We now consider a situation where the observer system comes into interaction with the object system for a second time. According to Rule 2 we arrive at the total state after the second observation:

$$\psi''^{S+0}=\sum_i a_i\phi_i\psi^0_{[\cdots\alpha_i,\alpha_i]}. \tag{19}$$

Again, each element $\phi_i\psi^0_{[\cdots\alpha_i,\alpha_i]}$ describes a system eigenstate, but this time also describes the observer as having obtained the *same result* for each of the two observations. Thus for every separate state of the observer in the final superposition the result of the observation was repeatable, even though different for different states. This repeatability is a consequence of the fact that after an observation the *relative* system state for a particular observer state is the corresponding eigenstate.

Consider now a different situation. An observer-system 0, with initial state $\psi^0_{[\cdots]}$, measures the *same* quantity A in a number of separate, identical, systems which are initially in the same state, $\psi^{S_1}=\psi^{S_2}=\cdots=\psi^{S_n}=\sum_i a_i\phi_i$ (where the ϕ_i are, as usual, eigenfunctions of A). The initial total state function is then

$$\psi_0^{S_1+S_2+\cdots+S_n+0}=\psi^{S_1}\psi^{S_2}\cdots\psi^{S_n}\psi^0_{[\cdots]}. \tag{20}$$

We assume that the measurements are performed on the systems in the order $S_1, S_2, \cdots S_n$. Then the total state after the first measurement is by Rule 1,

$$\psi_1^{S_1+S_2+\cdots+S_n+0}=\sum_i a_i\phi_i^{S_1}\psi^{S_2}\cdots\psi^{S_n}\psi^0_{[\cdots\alpha_i^1]} \tag{21}$$

(where α_i^1 refers to the first system, S_1).

After the second measurement it is, by Rule 2,

$$\psi_2^{S_1+S_2+\cdots+S_n+0}=\sum_{i,j}a_ia_j\phi_i^{S_1}\phi_j^{S_2}\psi^{S_3}\cdots\psi^{S_n}\psi^0_{[\cdots\alpha_i^1,\alpha_j^2]} \tag{22}$$

and in general, after r measurements have taken place ($r\leq n$), Rule 2 gives the result:

$$\psi_r=\sum_{i,j,\ldots k}a_ia_j\cdots a_k\phi_i^{S_1}\phi_j^{S_2}\cdots\phi_k^{S_r}\psi^{S_{r+1}}\cdots\psi^{S_n}\psi^0_{[\cdots\alpha_i^1,\alpha_j^2,\cdots\alpha_k^r]} \tag{23}$$

We can give this state, ψ_r, the following interpretation. It consists of a superposition of states:

$$\psi'_{ij\ldots k}=\phi_i^{S_1}\phi_j^{S_2}\cdots\phi_k^{S_r}\times\psi^{S_{r+1}}\cdots\psi^{S_n}\psi^0_{[\alpha_i^1,\alpha_j^2,\cdots\alpha_k^r]} \tag{24}$$

each of which describes the observer with a definite memory sequence $[\alpha_i^1,\alpha_j^2,\cdots\alpha_k^r]$. Relative to him the (observed) system states are the corresponding eigenfunctions $\phi_i^{S_1}, \phi_j^{S_2},\cdots, \phi_k^{S_r}$, the remaining systems, $S_{r+1},\cdots, S_n$, being unaltered.

A typical element $\psi'_{ij\ldots k}$ of the final superposition describes a state of affairs wherein the observer has perceived an apparently random sequence of definite results for the observations. Furthermore the object systems have been left in the corresponding eigenstates of the observation. At this stage suppose that a redetermination of an earlier system observation (S_l) takes place. Then it follows that every element of the resulting final superposition will describe the observer with a memory configuration of the form $[\alpha_i^1,\cdots\alpha_j^l,\cdots\alpha_k^r,\alpha_j^l]$ in which the earlier memory coincides with the later—i.e., the memory states are *correlated.* It will thus *appear* to the observer, as described by a typical element of the superposition, that each initial observation on a system caused the system to "jump" into an eigenstate in a random fashion and thereafter remain there for subsequent measurements on the same system. Therefore—disregarding for the moment quantitative questions of relative frequencies—the probabilistic assertions of Process 1 *appear* to be valid to the observer described by a typical element of the final superposition.

We thus arrive at the following picture: Throughout all of a sequence of observation processes there is only one physical system representing the observer, yet there is no single unique *state* of the observer (which follows from the representations of interacting systems). Nevertheless, there is a representation in terms of a *superposition*, each element of which contains a definite observer state and a corresponding system state. Thus with each succeeding observation (or interaction), the observer state "branches" into a number of different states. Each branch represents a different outcome of the measurement and the *corresponding* eigenstate for the object-system state. All branches exist simultaneously in the superposition after any given sequence of observations.‡

‡ *Note added in proof.*—In reply to a preprint of this article some correspondents have raised the question of the "transition from possible to actual," arguing that in "reality" there is—as our experience testifies—no such splitting of observer states, so that only one branch can ever actually exist. Since this point may occur to other readers the following is offered in explanation.

The whole issue of the transition from "possible" to "actual" is taken care of in the theory in a very simple way—there is no such transition, nor is such a transition necessary for the theory to be in accord with our experience. From the viewpoint of the theory *all* elements of a superposition (all "branches") are "actual," none any more "real" than the rest. It is unnecessary to suppose that all but one are somehow destroyed, since all the

The "trajectory" of the memory configuration of an observer performing a sequence of measurements is thus not a linear sequence of memory configurations, but a branching tree, with all possible outcomes existing simultaneously in a final superposition with various coefficients in the mathematical model. In any familiar memory device the branching does not continue indefinitely, but must stop at a point limited by the capacity of the memory.

In order to establish quantitative results, we must put some sort of measure (weighting) on the elements of a final superposition. This is necessary to be able to make assertions which hold for almost all of the observer states described by elements of a superposition. We wish to make quantitative statements about the relative frequencies of the different possible results of observation—which are recorded in the memory—for a typical observer state; but to accomplish this we must have a method for selecting a typical element from a superposition of orthogonal states.

We therefore seek a general scheme to assign a measure to the elements of a superposition of orthogonal states $\sum_i a_i\phi_i$. We require a positive function m of the complex coefficients of the elements of the superposition, so that $m(a_i)$ shall be the measure assigned to the element ϕ_i. In order that this general scheme be unambiguous we must first require that the states themselves always be normalized, so that we can distinguish the coefficients from the states. However, we can still only determine the *coefficients*, in distinction to the states, up to an arbitrary phase factor. In order to avoid ambiguities the function m must therefore be a function of the amplitudes of the coefficients alone, $m(a_i)=m(|a_i|)$.

We now impose an additivity requirement. We can regard a subset of the superposition, say $\sum_{i=1}^{n} a_i\phi_i$, as a single element $\alpha\phi'$:

$$\alpha\phi' = \sum_{i=1}^{n} a_i\phi_i. \tag{25}$$

We then demand that the measure assigned to ϕ' shall be the sum of the measures assigned to the ϕ_i (i from 1 to n):

$$m(\alpha) = \sum_{i=1}^{n} m(a_i). \tag{26}$$

Then we have already restricted the choice of m to the square amplitude alone; in other words, we have $m(a_i)=a_i^*a_i$, apart from a multiplicative constant.

To see this, note that the normality of ϕ' requires that $|\alpha| = (\sum a_i^*a_i)^{\frac{1}{2}}$. From our remarks about the dependence of m upon the amplitude alone, we replace the a_i by their amplitudes $u_i=|a_i|$. Equation (26) then imposes the requirement,

$$m(\alpha)=m(\textstyle\sum a_i^*a_i)^{\frac{1}{2}}=m(\sum u_i^2)^{\frac{1}{2}} \\ =\textstyle\sum m(u_i)=\sum m(u_i^2)^{\frac{1}{2}}. \tag{27}$$

Defining a new function $g(x)$

$$g(x)=m(\sqrt{x}) \tag{28}$$

we see that (27) requires that

$$g(\textstyle\sum u_i^2)=\sum g(u_i^2). \tag{29}$$

Thus g is restricted to be linear and necessarily has the form:

$$g(x)=cx \quad (c \text{ constant}). \tag{30}$$

Therefore $g(x^2)=cx^2=m(\sqrt{x^2})=m(x)$ and we have deduced that m is restricted to the form

$$m(a_i)=m(u_i)=cu_i^2=ca_i^*a_i. \tag{31}$$

We have thus shown that the only choice of measure consistent with our additivity requirement is the square amplitude measure, apart from an arbitrary multiplicative constant which may be fixed, if desired, by normalization requirements. (The requirement that the total measure be unity implies that this constant is 1.)

The situation here is fully analogous to that of classical statistical mechanics, where one puts a measure on trajectories of systems in the phase space by placing a measure on the phase space itself, and then making assertions (such as ergodicity, quasi-ergodicity, etc.) which hold for "almost all" trajectories. This notion of "almost all" depends here also upon the choice of measure, which is in this case taken to be the Lebesgue measure on the phase space. One could contradict the statements of classical statistical mechanics by choosing a measure for which only the exceptional trajectories had nonzero measure. Nevertheless the choice of Lebesgue measure on the phase space can be justified by the fact that it is the only choice for which the "conservation of probability" holds, (Liouville's theorem) and hence the only choice which makes possible any reasonable statistical deductions at all.

In our case, we wish to make statements about "trajectories" of observers. However, for us a trajectory is constantly branching (transforming from state to superposition) with each successive measurement. To

---

separate elements of a superposition individually obey the wave equation with complete indifference to the presence or absence ("actuality" or not) of any other elements. This total lack of effect of one branch on another also implies that no observer will ever be aware of any "splitting" process.

Arguments that the world picture presented by this theory is contradicted by experience, because we are unaware of any branching process, are like the criticism of the Copernican theory that the mobility of the earth as a real physical fact is incompatible with the common sense interpretation of nature because we feel no such motion. In both cases the argument fails when it is shown that the theory itself predicts that our experience will be what it in fact is. (In the Copernican case the addition of Newtonian physics was required to be able to show that the earth's inhabitants would be unaware of any motion of the earth.)

have a requirement analogous to the "conservation of probability" in the classical case, we demand that the measure assigned to a trajectory at one time shall equal the sum of the measures of its separate branches at a later time. This is precisely the additivity requirement which we imposed and which leads uniquely to the choice of square-amplitude measure. Our procedure is therefore quite as justified as that of classical statistical mechanics.

Having deduced that there is a unique measure which will satisfy our requirements, the square-amplitude measure, we continue our deduction. This measure then assigns to the $i, j, \cdots k$th element of the superposition (24),

$$\phi_i^{S_1}\phi_j^{S_2}\cdots\phi_k^{S_r}\psi^{S_{r+1}}\cdots\psi^{S_n}\psi^0{}_{[\alpha_i{}^1,\alpha_j{}^2,\cdots\alpha_k{}^r]} \tag{32}$$

the measure (weight)

$$M_{ij\ldots k}=(a_ia_j\cdots a_k)^*(a_ia_j\cdots a_k) \tag{33}$$

so that the observer state with memory configuration $[\alpha_i{}^1,\alpha_j{}^2,\cdots,\alpha_k{}^r]$ is assigned the measure $a_i^*a_ia_j^*a_j\cdots a_k^*a_k=M_{ij\ldots k}$. We see immediately that this is a product measure, namely,

$$M_{ij\ldots k}=M_iM_j\cdots M_k \tag{34}$$

where

$$M_l=a_l^*a_l$$

so that the measure assigned to a particular memory sequence $[\alpha_i{}^1, \alpha_j{}^2, \cdots, \alpha_k{}^r]$ is simply the product of the measures for the individual components of the memory sequence.

There is a direct correspondence of our measure structure to the probability theory of random sequences. *If we regard* the $M_{ij\ldots k}$ as probabilities for the sequences then the sequences are equivalent to the random sequences which are generated by ascribing to each term the *independent* probabilities $M_l=a_l^*a_l$. Now probability theory is equivalent to measure theory mathematically, so that we can make use of it, while keeping in mind that all results should be translated back to measure theoretic language.

Thus, in particular, if we consider the sequences to become longer and longer (more and more observations performed) *each* memory sequence of the final superposition will satisfy any given criterion for a randomly generated sequence, generated by the independent probabilities $a_l^*a_l$, except for a set of total measure which tends toward zero as the number of observations becomes unlimited. Hence all averages of functions over *any* memory sequence, including the special case of frequencies, can be computed from the probabilities $a_i^*a_i$, except for a set of memory sequences of measure zero. We have therefore shown that the statistical assertions of Process 1 will appear to be valid to the observer, *in almost all* elements of the superposition (24), in the limit as the number of observations goes to infinity.

While we have so far considered only sequences of observations of the same quantity upon identical systems, the result is equally true for arbitrary sequences of observations, as may be verified by writing more general sequences of measurements, and applying Rules 1 and 2 in the same manner as presented here.

We can therefore summarize the situation when the sequence of observations is arbitrary, when these observations are made upon the same or different systems in any order, and when the number of observations of each quantity in each system is very large, with the following result:

> Except for a set of memory sequences of measure nearly zero, the averages of any functions over a memory sequence can be calculated approximately by the use of the independent probabilities given by Process 1 for each initial observation, on a system, and by the use of the usual transition probabilities for succeeding observations upon the same system. In the limit, as the number of all types of observations goes to infinity the calculation is exact, and the exceptional set has measure zero.

This prescription for the calculation of averages over memory sequences by probabilities assigned to individual elements is precisely that of the conventional "external observation" theory (Process 1). Moreover, these predictions hold for almost all memory sequences. Therefore all predictions of the usual theory will appear to be valid to the observer in amost all observer states.

In particular, the uncertainty principle is never violated since the latest measurement upon a system supplies all possible information about the relative system state, so that there is no direct correlation between any earlier results of observation on the system, and the succeeding observation. Any observation of a quantity B, between two successive observations of quantity A (all on the same system) will destroy the one-one correspondence between the earlier and later memory states for the result of A. Thus for alternating observations of different quantities there are fundamental limitations upon the correlations between memory states for the same observed quantity, these limitations expressing the content of the uncertainty principle.

As a final step one may investigate the consequences of allowing several observer systems to interact with (observe) the same object system, as well as to interact with one another (communicate). The latter interaction can be treated simply as an interaction which correlates parts of the memory configuration of one observer with another. When these observer systems are investigated, in the same manner as we have already presented in this section using Rules 1 and 2, one finds that in *all elements* of the final superposition:

1. When several observers have separately observed the same quantity in the object system and then communicated the results to one another they find that they

are in agreement. This agreement persists even when an observer performs his observation *after* the result has been communicated to him by another observer who has performed the observation.

2. Let one observer perform an observation of a quantity A in the object system, then let a second perform an observation of a quantity B in this object system which does not commute with A, and finally let the first observer repeat his observation of A. Then the memory system of the first observer will *not* in general show the same result for both observations. The intervening observation by the other observer of the noncommuting quantity B prevents the possibility of any one to one correlation between the two observations of A.

3. Consider the case where the states of two object systems are correlated, but where the two systems do not interact. Let one observer perform a specified observation on the first system, then let another observer perform an observation on the second system, and finally let the first observer repeat his observation. Then it is found that the first observer always gets the same result both times, and the observation by the second observer has no effect whatsoever on the outcome of the first's observations. Fictitious paradoxes like that of Einstein, Podolsky, and Rosen[8] which are concerned with such correlated, noninteracting systems are easily investigated and clarified in the present scheme.

Many further combinations of several observers and systems can be studied within the present framework. The results of the present "relative state" formalism agree with those of the conventional "external observation" formalism in all those cases where that familiar machinery is applicable.

In conclusion, the continuous evolution of the state function of a composite system with time gives a complete mathematical model for processes that involve an idealized observer. When interaction occurs, the result of the evolution in time is a superposition of states, each element of which assigns a different state to the memory of the observer. Judged by the state of the memory in almost all of the observer states, the probabilistic conclusion of the usual "external observation" formulation of quantum theory are valid. In other words, pure Process 2 wave mechanics, without any initial probability assertions, leads to all the probability concepts of the familiar formalism.

6. DISCUSSION

The theory based on pure wave mechanics is a conceptually simple, causal theory, which gives predictions in accord with experience. It constitutes a framework in which one can investigate in detail, mathematically, and in a logically consistent manner a number of sometimes puzzling subjects, such as the measuring process itself and the interrelationship of several observers. Objections have been raised in the past to the conventional or "external observation" formulation of quantum theory on the grounds that its probabilistic features are postulated in advance instead of being derived from the theory itelf. We believe that the present "relative-state" formulation meets this objection, while retaining all of the content of the standard formulation.

While our theory ultimately justifies the use of the probabilistic interpretation as an aid to making practical predictions, it forms a broader frame in which to understand the consistency of that interpretation. In this respect it can be said to form a *metatheory* for the standard theory. It transcends the usual "external observation" formulation, however, in its ability to deal logically with questions of imperfect observation and approximate measurement.

The "relative state" formulation will apply to all forms of quantum mechanics which maintain the superposition principle. It may therefore prove a fruitful framework for the quantization of general relativity. The formalism invites one to construct the formal theory first, and to supply the statistical interpretation later. This method should be particularly useful for interpreting quantized unified field theories where there is no question of ever isolating observers and object systems. They all are represented in a *single* structure, the field. Any interpretative rules can probably only be deduced in and through the theory itself.

Aside from any possible practical advantages of the theory, it remains a matter of intellectual interest that the statistical assertions of the usual interpretation do not have the status of independent hypotheses, but are deducible (in the present sense) from the pure wave mechanics that starts completely free of statistical postulates.

---

[8] Einstein, Podolsky, and Rosen, Phys. Rev. **47**, 777 (1935). For a thorough discussion of the physics of observation, see the chapter by N. Bohr in *Albert Einstein, Philosopher-Scientist* (The Library of Living Philosophers, Inc., Evanston, 1949).

II.4 THE PROBLEM OF MEASUREMENT

EUGENE P. WIGNER

Introduction

The last few years have seen a revival of interest in the conceptual foundations of quantum mechanics.[1] This revival was stimulated by the attempts to alter the probabilistic interpretation of quantum mechanics. However, even when these attempts turned out to be less fruitful than its protagonists had hoped,[2] the interest continued. Hence, after the subject had been dormant for more than two decades, we again hear discussions on the basic principles of quantum theory and the epistemologies that are compatible with it. As is often the case under similar

Reprinted by permission from the *American Journal of Physics,* Vol. 31, No. 1 (January, 1963).

[1] Some of the more recent papers on the subject are: Y. Aharonov and D. Bohm, *Phys. Rev.*, 122, 1649 (1961); *Nuovo Cimento*, 17, 964 (1960); B. Bertotti, *Nuovo Cimento Suppl.*, 17, 1 (1960); L. de Broglie, *J. Phys. Radium*, 20, 963 (1959); J. A. de Silva, *Ann. Inst. Henri Poincaré*, 16, 289 (1960); A. Datzeff, *Compt. Rend.*, 251, 1462 (1960); *J. Phys. Radium*, 21, 201 (1960); 22, 101 (1961); J. M. Jauch, *Helv. Phys. Acta*, 33, 711 (1960); A. Landé, *Z. Physik*, 162, 410 (1961); 164, 558 (1961); *Am. J. Phys.*, 29, 503 (1961); H. Margenau and R. N. Hill, *Progr. Theoret. Phys.*, 26, 727 (1961); A. Peres and P. Singer, *Nuovo Cimento*, 15, 907 (1960); H. Putnam, *Phil. Sci.*, 28, 234 (1961); M. Renninger, *Z. Physik*, 158, 417 (1960); L. Rosenfeld, *Nature*, 190, 384 (1961); F. Schlögl, *Z. Physik*, 159, 411 (1960); J. Schwinger, *Proc. Natl. Acad. Sci. U.S.*, 46, 570 (1960); J. Tharrats, *Compt. Rend.*, 250, 3786 (1960); H. Wakita, *Progr. Theoret. Phys.*, 23, 32 (1960); 27, 139 (1962); W. Weidlich, *Z. Naturforsch.*, 15a, 651 (1960); J. P. Wesley, *Phys. Rev.*, 122, 1932 (1961). See also the articles of E. Teller, M. Born, A. Landé, F. Bopp, and G. Ludwig in *Werner Heisenberg und die Physik unserer Zeit* (Braunschweig: Friedrich Vieweg und Sohn, 1961).

[2] See the comments of V. Fock in the *Max Planck Festschrift* (Berlin: Deutscher Verlag der Wissenschaften, 1958), p. 177, particularly Sec. II.

Originally published in *American Journal of Physics, 31*, 6-15 (1963). Reprinted in E. Wigner (1967), *Symmetries and Reflections*, Indiana University Press, Bloomington, pp. 153-170, from which book this paper is reproduced here.

circumstances, some of the early thinking had been forgotten; in fact, a small fraction of it remains as yet unrediscovered in the modern literature. Equally naturally, some of the language has changed but, above all, new ideas and new attempts have been introduced. Having spoken to many friends on the subject which will be discussed here, it became clear to me that it is useful to review the standard view of the late "Twenties," and this will be the first task of this article. The standard view is an outgrowth of Heisenberg's paper in which the uncertainty relation was first formulated.[3] The far-reaching implications of the consequences of Heisenberg's ideas were first fully appreciated, I believe, by von Neumann,[4] but many others arrived independently at conclusions similar to his. There is a very nice little book, by London and Bauer,[5] which summarizes quite completely what I shall call the orthodox view.

The orthodox view is very specific in its epistemological implications. This makes it desirable to scrutinize the orthodox view very carefully and to look for loopholes which would make it possible to avoid the conclusions to which the orthodox view leads. A large group of physicists finds it difficult to accept these conclusions and, even though this does not apply to the present writer, he admits that the far-reaching nature of the epistemological conclusions makes one uneasy. The misgivings, which are surely shared by many others who adhere to the orthodox view, stem from a suspicion that one cannot arrive at valid epistemological conclusions without a careful analysis of the *process of the acquisition of knowledge*. What will be analyzed, instead, is only the type of information which we can acquire and possess concerning the external inanimate world, according to quantum-mechanical theory.

We are facing here the perennial question whether we physicists do not go beyond our competence when searching for philosophical truth.

[3] W. Heisenberg, Z. *Physik*, 43, 172 (1927); also his article in *Niels Bohr and the Development of Physics* (London: Pergamon Press, 1955); N. Bohr, *Nature*, 121, 580 (1928); *Naturwissen.*, 17, 483 (1929) and particularly *Atomic Physics and Human Knowledge* (New York: John Wiley & Sons, Inc., 1958).

[4] See J. von Neumann, *Mathematische Grundlagen der Quantenmechanik* (Berlin: Verlag Julius Springer, 1932), English translation (Princeton, N.J.: Princeton University Press, 1955). See also P. Jordan, *Anschauliche Quantentheorie* (Berlin: Julius Springer, 1936), Chapter V.

[5] F. London and E. Bauer, *La Théorie de l'observation en mécanique quantique* (Paris: Hermann et Cie., 1939); or E. Schrödinger, *Naturwissen.*, 23, 807 ff. (1935); *Proc. Cambridge Phil. Soc.*, 31, 555 (1935).

I believe that we probably do.[6] Nevertheless, the ultimate implications of quantum theory's formulation of the laws of physics appear interesting even if one admits that the conclusions to be arrived at may not be the ultimate truth.

The Orthodox View

The possible states of a system can be characterized, according to quantum-mechanical theory, by state vectors. These state vectors—and this is an almost verbatim quotation of von Neumann—change in two ways. As a result of the passage of time, they change continuously, according to Schrödinger's time-dependent equation—this equation will be called the equation of motion of quantum mechanics. The state vector also changes discontinuously, according to probability laws, if a measurement is carried out on the system. This second type of change is often called the reduction of the wavefunction. It is this reduction of the state vector which is unacceptable to many of our colleagues.

The assumption of two types of changes of the state vector is a strange dualism. It is good to emphasize at this point that the dualism in question has little to do with the oft-discussed wave-versus-particle dualism. This latter dualism is only part of a more general pluralism or even "infinitesilism" which refers to the infinity of noncommuting measurable quantities. One can measure the position of the particles, or one can measure their velocity, or, in fact, an infinity of other observables. The dualism here discussed is a true dualism and refers to the *two* ways in which the state vector changes. It is also worth noting, though only parenthetically, that the probabilistic aspect of the theory is almost diametrically opposite to what ordinary experience would lead one to expect. The place where one expects probability laws to prevail is the change of the system with time. The interaction of the particles, their collisions, are the events which are ordinarily expected to be governed by statistical laws. This is not at all the case here: the uncertainty in the behavior of a system does not increase in time if the system is left alone, that is, if it is not subjected to measurements. In this case, the properties of the system, as described by its state vector,

[6] This point is particularly well expressed by H. Margenau, in the first two sections of the article in *Phil. Sci.*, 25, 23 (1958).

change causally, no matter what the period of time is during which it is left alone. On the contrary, the phenomenon of chance enters when a measurement is carried out on the system, when we try to check whether its properties did change in the way our causal equations told us they would change. However, the extent to which the results of all possible measurements on the system can be predicted does not decrease, according to quantum-mechanical theory, with the time during which the system was left alone; it is as great right after an observation as it is a long time thereafter. The uncertainty of the result, so to say, increases with time for some measurements just as much as it decreases for others. The Liouville theorem is the analog for this in classical mechanics. It tells us that, if the point which represents the system in phase space is known to be in a finite volume element at one given time, an equally large volume element can be specified for a given later time which will then contain the point representing the state of the system. Similarly, the uncertainty in the result of the measurement of Q, at time 0, is exactly equal to the uncertainty of the measurement of $Q_t = \exp(-iHt/\hbar)Q_0 \exp(iHt/\hbar)$ at time t. The information which is available at a later time may be less valuable than the information which was available on an earlier state of the system (this is the cause of the increase of the entropy); in principle, the amount of information does not change in time.

Consistency of the Orthodox View

The simplest way that one may try to reduce the two kinds of changes of the state vector to a single kind is to describe the whole process of measurement as an event in time, governed by the quantum-mechanical equations of motion. One might think that, if such a description is possible, there is no need to assume a second kind of change of the state vector; if it is impossible, one might conclude, the postulate of the measurement is incompatible with the rest of quantum mechanics. Unfortunately, the situation will turn out not to be this simple.

If one wants to describe the process of measurement by the equations of quantum mechanics, one will have to analyze the interaction between object and measuring apparatus. Let us consider a measurement from the point of view of which the "sharp" states are $\sigma^{(1)}, \sigma^{(2)}, \cdots$. For these states of the object the measurement will surely yield the values $\lambda_1, \lambda_2,$

· · ·, respectively. Let us further denote the initial state of the apparatus by a; then, if the initial state of the system was $\sigma^{(\nu)}$, the total system—apparatus plus object—will be characterized, before they come into interaction, by $a \times \sigma^{(\nu)}$. The interaction should not change the state of the object in this case and hence will lead to

$$a \times \sigma^{(\nu)} \to a^{(\nu)} \times \sigma^{(\nu)}. \tag{1}$$

The state of the object has not changed, but the state of the apparatus has and will depend on the original state of the object. The different states $a^{(\nu)}$ may correspond to states of the apparatus in which the pointer has different positions, which indicate the state of the object. The state $a^{(\nu)}$ of the apparatus will therefore be called also "pointer position ν." The state vectors $a^{(1)}, a^{(2)}, \cdots$ are orthogonal to each other —usually the corresponding states can be distinguished even macroscopically. Since we have considered, so far, only "sharp" states, for each of which the measurement in question surely yields one definite value, no statistical element has yet entered into our considerations.[7]

Let us now see what happens if the initial state of the object is not sharp, but an arbitrary linear combination $\alpha_1\sigma^{(1)} + \alpha_2\sigma^{(2)} + \cdots$. It then *follows* from the linear character of the quantum-mechanical equation of motion (as a result of the so-called superposition principle) that the state vector of object-plus-apparatus after the measurement becomes the right side of

$$a \times [\sum \alpha_\nu \sigma^{(\nu)}] \to \sum \alpha_\nu [a^{(\nu)} \times \sigma^{(\nu)}]. \tag{2}$$

Naturally, there is no statistical element in this result, as there cannot be. However, in the state (2), obtained by the measurement, there is a statistical correlation between the state of the object and that of the apparatus: the simultaneous measurement on the system—object-plus-apparatus--of the two quantities, one of which is the originally measured quantity of the object and the second the position of the pointer of the apparatus, always leads to concordant results. As a result, one of these measurements is unnecessary: The state of the object can be ascertained by an observation on the apparatus. This is a consequence of the special form of the state vector (2), of not containing any $a^{(\nu)} \times \sigma^{(\mu)}$ term with $\nu \neq \mu$.

[7] The self-adjoint (Hermitean) character of every observable can be derived from Eq. (1) and the unitary nature of the transformation indicated by the arrow. Cf. E. Wigner, Z. *Physik*, 133, 101 (1952), footnote 2 on p. 102.

It is well known that statistical correlations of the nature just described play a most important role in the structure of quantum mechanics. One of the earliest observations in this direction is Mott's explanation of the straight track left by the spherical wave of outgoing α particles.[8] In fact, the principal conceptual difference between quantum mechanics and the earlier Bohr-Kramers-Slater theory is that the former, by its use of configuration space rather than ordinary space for its waves, allows for such statistical correlations.

Returning to the problem of measurement, we see that we have not arrived either at a conflict between the theory of measurement and the equations of motion, nor have we obtained an explanation of that theory in terms of the equations of motion. The equations of motion permit the description of the process whereby the state of the object is mirrored by the state of an apparatus. The problem of a measurement on the object is thereby transformed into the problem of an observation on the apparatus. Clearly, further transfers can be made by introducing a second apparatus to ascertain the state of the first, and so on. However, the fundamental point remains unchanged and a full description of an observation must remain impossible since the quantum-mechanical equations of motion are causal and contain no statistical element, whereas the measurement does.

It should be admitted that when the quantum theorist discusses measurements, he makes many idealizations. He assumes, for instance, that the measuring apparatus will yield some result, no matter what the initial state of the object was. This is clearly unrealistic since the object may move away from the apparatus and never come into contact with it. More importantly, he has appropriated the word "measurement" and used it to characterize a special type of interaction by means of which information can be obtained on the state of a definite object. Thus, the measurement of a physical constant, such as cross section, does not fall into the category called "measurement" by the theorist. His measurements answer only questions relating to the ephemeral state of a physical system, such as, "What is the x component of the momentum of this atom?" On the other hand, since he is unable to follow the path of the information until it enters his, or the observer's, mind, he considers the measurement completed as soon as a statistical relation has been established between the quantity to be measured and the state of some

[8] N. F. Mott, *Proc. Roy. Soc.* (London), 126, 79 (1929).

idealized apparatus. He would do well to emphasize his rather specialized use of the word "measurement."

This will conclude the review of the orthodox theory of measurement. As was mentioned before, practically all the foregoing is contained, for instance, in the book of London and Bauer.

Critiques of the Orthodox Theory

There are attempts to modify the orthodox theory of measurement by a complete departure from the picture epitomized by Eqs. (1) and (2). The only attempts of this nature which will be discussed here presuppose that the result of the measurement is not a state vector, such as (2), but a so-called mixture, namely, *one* of the state vectors

$$a^{(\mu)} \times \sigma^{(\mu)}, \tag{3}$$

and that this particular state will emerge from the interaction between object and apparatus with the probability $|\alpha_\mu|^2$. If this were so, the state of the system would not be changed when one ascertains—in some unspecified way—which of the state vectors (3) corresponds to the actual state of the system; one would merely "ascertain which of various possibilities has occurred." In other words, the final observation only increases our knowledge of the system; it does not change anything. This is not true if the state vector, after the interaction between object and apparatus, is given by (2) because *the state represented by the vector (2) has properties which neither of the states (3) has.* It may be worthwhile to illustrate this point, which is fundamental though often disregarded, by an example.

The example is the Stern-Gerlach experiment,[9] in which the projection of the spin of an incident beam of particles, into the direction which is perpendicular to the plane of the drawing, is measured. (See Fig. 1.) The index ν has two values in this case; they correspond to the two possible orientations of the spin. The "apparatus" is that positional coordinate of the particle which is also perpendicular to the plane of the drawing. If this coordinate becomes, in the experiment illustrated, positive, the spin is directed toward us; if it is negative, the spin is directed away from us. The experiment illustrates the statistical correla-

[9] The same experiment was discussed recently from another point of view by H. Wakita, *Progr. Theoret. Phys.*, 27, 139 (1962).

tion between the state of the "apparatus" (the position coordinate) and the state of the object (the spin) which we have discussed. The ordinary use of the experiment is to obtain the spin direction, by observing the position, i.e., the location of the beam. The measurement is, therefore, as far as the establishment of a statistical correlation is concerned, complete when the particle reaches the place where the horizontal spin arrows are located.

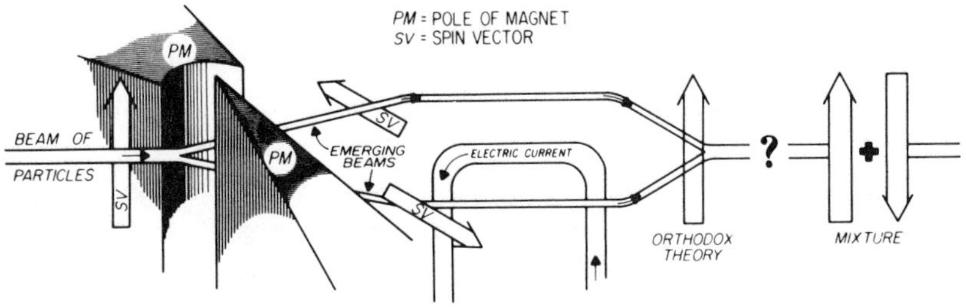

Fig. 1

What is important for us, however, is the right side of the drawing. This shows that the state of the system—object-plus-apparatus (spin and positional coordinates of the particle, i.e., the whole state of the particle)—shows characteristics which neither of the separated beams alone would have. If the two beams are brought together by the magnetic field due to the current in the cable indicated, the two beams will interfere and the spin will be vertical again. This could be verified by letting the united beam pass through a second magnet which is, however, not shown on the figure. If the state of the system corresponded to the beam toward us, its passage through the second magnet would show that it has equal probabilities to assume its initial and the opposite directions. The same is true of the second beam which was deflected away from us. Even though the experiment indicated would be difficult to perform, there is little doubt that the behavior of particles and of their spins conforms to the equations of motion of quantum mechanics under the conditions considered. Hence, the properties of the system, object-plus-apparatus, are surely correctly represented by an expression of the form (2) which gives, *in this case,* properties which are different from those of *either* alternative (3).

In the case of the Stern-Gerlach experiment, one can thus point to a specific and probably experimentally realizable way to distinguish between the state vector (2), furnished by the orthodox theory, and the more easily visualizable mixture of the states (3) which one would offhand expect. There is little doubt that in this case the orthodox theory is correct. It remains remarkable how difficult it is, even in this very simple case, to distinguish between the two, and this raises two questions. The first of these is whether there is, in more complicated cases, a principle which makes the distinction between the state vector (2), and the mixture of the states (3), impossible. As far as is known to the present writer, this question has not ever been posed seriously heretofore, and it will be considered in the present discussion also only obliquely. The second question is whether there is a continuous transition between (2) and the mixture of states (3) so that in simpler cases (2) is the result of the interaction between object and measuring apparatus, but in more complicated and more realistic cases the actual state of object-plus-apparatus more nearly resembles a mixture of the states (3). Again, this question can be investigated within the framework of quantum mechanics, or one can postulate deviations from the quantum-mechanical equations of motion, in particular from the superposition principle.

"More complicated" and "more realistic" mean in the present context that the measuring apparatus, the state of which is to be correlated with the quantity to be measured, is of such a nature that it is easy to measure *its* states, i.e., correlate it with the state of another "apparatus." If this is done, the state of that second "apparatus" will be correlated also to the state of the object. The case of establishing correlations between the state of the apparatus which came into direct contact with the object and another "apparatus" is usually greatest if the first one is of macroscopic nature, i.e., complicated from the quantum-mechanical point of view. The ease with which the secondary correlations can be established is a direct measure of how realistically one can say that the measurement has been completed. Clearly, if the state of the apparatus which carried out the primary measurement is just as difficult to ascertain as the state of the object, it is not very realistic to say that the establishment of a correlation between its and the object's state is a fully completed measurement. Nevertheless, it is so regarded by the orthodox theory. The question which we pose is, therefore, whether it is consistent with the principles of quantum mechanics to assume that at the end of a realistic measurement the state of object-plus-apparatus is not a wavefunction, as given by (2), but a mixture of the states (3). We shall see that the answer is negative. Hence, the modification of the orthodox theory of measurement mentioned at the beginning of this section is not consistent with the principles of quantum mechanics.

Let us now proceed with the calculation. Even though this point is not usually emphasized, it is clear that, in order to obtain a mixture of states as a result of the interaction, the initial state must have been a mixture already.[10] This follows from the general theorem that the characteristic values of the density matrix are constants of motion. The assumption that the initial state of the system, object-plus-apparatus, is a mixture, is indeed a very natural one because the state vector of the apparatus, which is under the conditions now considered usually a macroscopic object, is hardly ever known. Let us assume, therefore, that the initial state of the apparatus is a mixture of the states $A^{(1)}$, $A^{(2)}$, $\cdots$, the probability of $A^{(\rho)}$ being p_ρ. The vectors $A^{(\rho)}$ can be assumed to be mutually orthogonal. The equations of motion will yield, for the state $A^{(\rho)}$ of the apparatus and the state $\sigma^{(\nu)}$ of the object, a final state

$$A^{(\rho)} \times \sigma^{(\nu)} \to A^{(\rho\nu)} \times \sigma^{(\nu)}. \tag{4}$$

Every state $A^{(1\nu)}$, $A^{(2\nu)}$, $\cdots$ will indicate the same state $\sigma^{(\nu)}$ of the object; the position of the pointer is ν for all of these. For different ν, however, the position of the pointer is also different. It follows that the $A^{(\rho\nu)}$, for different ν, are orthogonal, even if the ρ are also different. On the other hand, $A^{(\rho\nu)}$ and $A^{(\sigma\nu)}$, for $\rho \neq \sigma$, are also orthogonal because

[10] This point is disregarded by several authors who have rediscovered von Neumann's description of the measurement, as given by (1) and (2). These authors assume that it follows from the macroscopic nature of the measuring apparatus that if several values of the "pointer position" have finite probabilities [as is the case if the state vector is (2)], the state is necessarily a *mixture* (rather than a linear combination) of the states (3)—that is, of states in each of which the pointer position is definite (sharp). The argument given is that classical mechanics applies to macroscopic objects, and states such as (2) have no counterpart in classical theory. *This argument is contrary to present quantum-mechanical theory.* It is true that the motion of a macroscopic body can be adequately described by the classical equations of motion if its state has a classical description. That this last premise is, according to present theory, not always fulfilled, is clearly, though in an extreme fashion, demonstrated by Schrödinger's cat-paradox (cf. reference 5). Further, the discussion of the Stern-Gerlach experiment, given in the text, illustrates the fact that there are, in principle, observable differences between the state vector given by the right side of (2), and the *mixture* of the states (3), each of which has a definite position. Proposals to modify the quantum-mechanical equations of motion so as to permit a mixture of the states (3) to be the result of the measurement even though the initial state was a state vector, will be touched upon later.

$A^{(\rho\nu)} \times \sigma^{(\nu)}$ and $A^{(\sigma\nu)} \times \sigma^{(\nu)}$ are obtained by a unitary transformation from two orthogonal states, $A^{(\rho)} \times \sigma^{(\nu)}$ and $A^{(\sigma)} \times \sigma^{(\nu)}$ and the scalar product of $A^{(\rho\nu)} \times \sigma^{(\nu)}$ with $A^{(\sigma\nu)} \times \sigma^{(\nu)}$ is $(A^{(\rho\nu)}, A^{(\sigma\nu)})$. Hence, the $A^{(\rho\nu)}$ form an orthonormal (though probably not complete) system

$$(A^{(\rho\nu)}, A^{(\sigma\mu)}) = \delta_{\rho\sigma}\delta_{\nu\mu}. \tag{5}$$

It again follows from the linear character of the equation of motion that, if the initial state of the object is the linear combination $\sum\alpha_\nu\sigma^{(\nu)}$, the state of object-plus-apparatus will be, after the measurement, a mixture of the states

$$A^{(\rho)} \times \sum \alpha_\nu\sigma^{(\nu)} \rightarrow \sum_\nu \alpha_\nu[A^{(\rho\nu)} \times \sigma^{(\nu)}] = \Phi^{(\rho)}, \tag{6}$$

with probabilities p_ρ. This same mixture should then be, according to the postulate in question, equivalent to a mixture of orthogonal states

$$\Psi^{(\mu k)} = \sum_\rho x_\rho^{(\mu k)}[A^{(\rho\mu)} \times \sigma^{(\mu)}]. \tag{7}$$

These are the most general states for which the originally measured quantity has a definite value, namely λ_μ, and in which this state is coupled with some state (one of the states $\sum_\rho x_\rho^{(\mu\kappa)}A^{(\rho\mu)}$) with a pointer position μ. Further, if the probability of the state $\Psi^{(\mu k)}$ is denoted by $P_{\mu k}$, we must have

$$\sum_k P_{\mu k} = |\,\alpha_\mu\,|^2. \tag{7a}$$

The $x_\rho^{(\mu k)}$ will naturally depend on the α.

It turns out, however, that a mixture of the states $\Phi^{(\rho)}$ cannot be, at the same time, a mixture of the states $\Psi^{(\mu k)}$ (unless only one of the α is different from zero). A necessary condition for this would be that the $\Psi^{(\mu k)}$ are linear combinations of the $\Phi^{(\rho)}$, so that one should be able to find coefficients u so that

$$\begin{aligned}\sum_\rho x_\rho^{(\mu k)}\,[A^{(\rho\mu)} \times \sigma^{(\mu)}] = \Psi^{(\mu k)} &= \sum_\rho u_\rho\Phi^{(\rho)} \\ &= \sum_{\rho\nu} u_\rho\alpha_\nu[A^{(\rho\nu)} \times \sigma^{(\nu)}].\end{aligned} \tag{8}$$

From the linear independence of the $A^{(\rho\nu)}$ it then follows that

$$u_\rho\alpha_\nu = \delta_{\nu\mu}x_\rho^{(\mu k)}, \tag{8a}$$

which cannot be fulfilled if more than one α is finite. It follows that it is not compatible with the equations of motion of quantum mechanics to assume that the state of object-plus-apparatus is, after a measurement, a mixture of states each with one definite position of the pointer.

It must be concluded that *measurements which leave the system object-plus-apparatus in one of the states with a definite position of the pointer cannot be described by the linear laws of quantum mechanics.* Hence, if there are such measurements, quantum mechanics has only limited validity. This conclusion must have been familiar to many even though the detailed argument just given was not put forward before. Ludwig, in Germany, and the present writer have independently suggested that the equations of motion of quantum mechanics must be modified so as to permit measurements of the aforementioned type.[11] These suggestions will not be discussed in detail because they are suggestions and do not have convincing power at present. Even though either may well be valid, one must conclude that the only known theory of measurement which has a solid foundation is the orthodox one and that this implies the dualistic theory concerning the changes of the state vector. It implies, in particular, the so-called reduction of the state vector. However, to answer the question posed earlier: yes, there is a continuous transition between the state vector (2), furnished by orthodox theory, and the requisite mixture of the states (3), postulated by a more visualizable theory of measurement.[11]

What Is the State Vector?

The state vector concept plays such an important part in the formulation of quantum-mechanical theory that it is desirable to discuss its role and the ways to determine it. Since, according to quantum mechanics, all information is obtained in the form of the results of measurements, the standard way to obtain the state vector is also by carrying out measurements on the system.[12]

In order to answer the question proposed, we shall first obtain a for-

[11] See G. Ludwig's article "Solved and Unsolved Problems in the Quantum Mechanics of Measurement" (reference 1) and the present author's article "Remarks on the Mind-Body Question" in *The Scientist Speculates,* edited by I. J. Good (London: William Heinemann, 1962), p. 284, reprinted in this volume.

[12] There are, nevertheless, other procedures to bring a system into a definite state. These are based on the fact that a small system, if it interacts with a large system in a definite and well-known state, may assume itself a definite state with almost absolute certainty. Thus, a hydrogen atom, in some state of excitation, if placed into a large container with no radiation in it, will almost surely transfer all its energy to the radiation field of the container and go over into its normal state. This method of preparing a state has been particularly stressed by H. Margenau.

mula for the probability that successive measurements carried out on a system will give certain specified results. This formula will be given both in the Schrödinger and in the Heisenberg picture. Let us assume that n successive measurements are carried out on the system, at times $t_1, t_2, \cdots, t_n$. The operators of the quantities which are measured are, in the Schrödinger picture, $Q_1, Q_2, \cdots, Q_n$. The characteristic vectors of these will all be denoted by ψ with suitable upper indices. Similarly, the characteristic values will be denoted by q so that

$$Q_j \psi_\kappa{}^{(j)} = q_\kappa{}^{(j)} \psi_\kappa{}^{(j)} \tag{9}$$

The Heisenberg operators which correspond to these quantities, if measured at the corresponding times, are

$$Q_j{}^H = e^{iHt_j} Q_j e^{-iHt_j} \tag{10}$$

and the characteristic vectors of these will be denoted by $\varphi_\kappa{}^{(j)}$, where

$$\varphi_\kappa{}^{(j)} = e^{iHt_j} \psi_\kappa{}^{(j)} \qquad Q_j{}^H \varphi_\kappa{}^{(j)} = q_\kappa{}^{(j)} \varphi_\kappa{}^{(j)}. \tag{10a}$$

If the state vector is originally Φ, the probability for the sequence $q_\alpha{}^{(1)}$, $q_\beta{}^{(2)}, \ldots, q_\mu{}^{(n)}$ of measurement-results is the absolute square of

$$(e^{-iHt_1}\Phi, \psi_\alpha{}^{(1)})\,(e^{-iH(t_2-t_1)}\psi_\alpha{}^{(1)}, \psi_\beta{}^{(2)}) \cdots (e^{-iH(t_n-t_{n-1})}\psi_\lambda{}^{(n-1)}, \psi_\mu{}^{(n)}). \tag{11}$$

The same expression in terms of the characteristic vectors of the Heisenberg operators is simpler,

$$(\Phi, \varphi_\alpha{}^{(1)})\,(\varphi_\alpha{}^{(1)}, \varphi_\beta{}^{(2)}) \cdots (\varphi_\lambda{}^{(n-1)}, \varphi_\mu{}^{(n)}). \tag{11a}$$

It should be noted that the probability is not determined by the n Heisenberg operators $Q_j{}^H$ and their characteristic vectors: the *time order* in which the measurements are carried out enters into the result essentially. Von Neumann already derived these expressions as well as their generalizations for the case in which the characteristic values $q_\alpha{}^{(1)}$, $q_\beta{}^{(2)}, \cdots$ have several characteristic vectors. In this case, it is more appropriate to introduce projection operators for every characteristic value $q^{(j)}$ of every Heisenberg operator $Q_j{}^H$. If the projection operator in question is denoted by $P_{j\kappa}$, the probability for the sequence $q_\alpha{}^{(1)}, q_\beta{}^{(2)}, \cdots$, $q_\mu{}^{(n)}$ of measurement-results is

$$(P_{n\mu} \cdots P_{2\beta} P_{1\alpha} \Phi, P_{n\mu} \cdots P_{2\beta} P_{1\alpha} \Phi). \tag{12}$$

The expressions (11) or (11a) can be obtained also by postulating that the state vector became $\psi_\kappa^{(j)}$ when the measurement of $Q^{(j)}$ gave the result $q_\kappa^{(j)}$. Indeed, the statement that the state vector is $\psi_\kappa^{(j)}$ is only a short expression for the fact that the last measurement on the system, of the quantity $Q^{(j)}$, just carried out, gave the result $q_\kappa^{(j)}$. In the case of simple characteristic values the state vector depends only on the result of the last measurement and the future behavior of the system is independent of the more distant past history thereof. This is not the case if the characteristic value $q^{(j)}$ is multiple.

The most simple expression for the Heisenberg state vector, when the jth measurement gave the value $q_\kappa^{(j)}$, is, in this case,

$$P_{j\kappa} \cdots P_{2\beta} P_{1\alpha} \Phi, \tag{12a}$$

properly normalized. If, after normalization, the expression (12a) is independent of the original state vector Φ, the number of measurements has sufficed to determine the state of the system completely and a pure state has been produced. If the vector (12a) still depends on the original state vector Φ, and if this was not known to begin with, the state of the system is a mixture, a mixture of all the states (12a), with all possible Φ. Evidently, the measurement of a single quantity Q, the characteristic values of which are all nondegenerate, suffices to bring the system into a pure state though it is not in general foreseeable which pure state will result.

We recognize, from the preceding discussion, that the state vector is only a shorthand expression of that part of our information concerning the past of the system which is relevant for predicting (as far as possible) the future behavior thereof. The density matrix, incidentally, plays a similar role except that it does not predict the future behavior as completely as does the state vector. We also recognize that *the laws of quantum mechanics only furnish probability connections between results of subsequent observations carried out on a system.* It is true, of course, that the laws of classical mechanics can also be formulated in terms of such probability connections. However, they can be formulated also in terms of objective reality. The important point is that the laws of quantum mechanics can be expressed only in terms of probability connections.

Problems of the Orthodox View

The incompatibility of a more visualizable interpretation of the laws of quantum mechanics with the equations of motion, in particular the superposition principle, may mean that the orthodox interpretation is here to stay; it may also mean that the superposition principle will have to be abandoned. This may be done in the sense indicated by Ludwig, in the sense proposed by me, or in some third, as yet unfathomed sense. The dilemma which we are facing in this regard makes it desirable to review any possible conceptual weaknesses of the orthodox interpretation and the present, last, section will be devoted to such a review.

The principal conceptual weakness of the orthodox view is, in my opinion, that it merely abstractly postulates interactions which have the effect of the arrows in (1) or (4). For some observables, in fact for the majority of them (such as xyp_z), nobody seriously believes that a measuring apparatus exists. It can even be shown that no observable which does not commute with the additive conserved quantities (such as linear or angular momentum or electric charge) can be measured precisely, and in order to increase the accuracy of the measurement one has to use a very large measuring apparatus. The simplest form of the proof heretofore was given by Araki and Yanase.[13] On the other hand, most quantities which we believe to be able to measure, and surely all the very important quantities such as position, momentum, fail to commute with all the conserved quantities, so that their measurement cannot be possible with a microscopic apparatus. This raises the suspicion that the macroscopic nature of the apparatus is necessary in principle and reminds us that our doubts concerning the validity of the superposition principle for the measurement process were connected with the macroscopic nature of the apparatus. The oint state vector (2), resulting from a measurement with a very large apparatus, surely *cannot be distinguished* as *simply from a mixture* as was the state vector obtained in the Stern-Gerlach experiment which we discussed.[14]

A second, though probably less serious, difficulty arises if one tries to

[13] H. Araki and M. Yanase, *Phys. Rev.*, 120, 666 (1961); cf. also E. P. Wigner, Z. *Physik,* 131, 101 (1952).

[14] This point was recognized already by D. Bohm. See Section 22.11 of his *Quantum Theory* (Englewood Cliffs, New Jersey: Prentice-Hall, Inc., 1951).

calculate the probability that the interaction between object and apparatus be of such nature that there exist states $\sigma^{(\nu)}$ for which (1) is valid. We recall that an interaction leading to this equation was simply postulated as the type of interaction which leads to a measurement. When I talk about the probability of a certain interaction, I mean this in the sense specified by Rosenzweig or by Dyson, who have considered ensembles of possible interactions and defined probabilities for definite interactions.[15] If one adopts their definition (or any similar definition) the probability becomes zero for the interaction to be such that there are states $\sigma^{(\nu)}$ satisfying (1). The proof for this is very similar to that[16] which shows that the probability is zero for finding reproducing systems—in fact, according to (1), each $\sigma^{(\nu)}$ is a reproducing system. The resolution of this difficulty is presumably that if the system with the state vector a—that is, the apparatus—is very large, (1) can be satisfied with a very small error. Again, the large size of the apparatus appears to be essential for the possibility of a measurement.

The simplest and least technical summary of the conclusions which we arrived at when discussing the orthodox interpretation of the quantum laws is that these laws merely provide probability connections between the results of several consecutive observations on a system. This is not at all unreasonable and, in fact, this is what one would naturally strive for once it is established that there remains some inescapable element of chance in our measurements. However, there is a certain weakness in the word "consecutive," as this is not a relativistic concept. Most observations are not local and one will assume, similarly, that they have an irreducible extension in time, that is, duration. However, the "observables" of the present theory are instantaneous, and hence unrelativistic, quantities. The only exceptions from this are the local field operators and we know, from the discussion of Bohr and Rosenfeld, how many extreme abstractions have to be made in order to describe their measurement.[17] This is not a reassuring state of affairs.

[15] C. E. Porter and N. Rosenzweig, *Suomalaisen Tiedeakatemian Toimotuksia*, VI, No. 44 (1960); *Phys. Rev.*, 120, 1698 (1960); F. Dyson, *J. Math. Phys.*, 3, 140, 157, 166 (1962). See also E. P. Wigner, *Proceedings of the Fourth Canadian Mathematics Congress* (Toronto: University of Toronto Press, 1959), p. 174, reprinted in this volume.

[16] Cf. the writer's article in *The Logic of Personal Knowledge* (London: Routledge and Kegan Paul, 1961), p. 231, reprinted in this volume.

[17] N. Bohr and L. Rosenfeld, *Kgl. Danske Videnskab. Selskab, Mat.-fys. Medd.*, 12, No. 8 (1933); *Phys. Rev.*, 78, 194 (1950); E. Corinaldesi, *Nuovo Cimento*, 8, 494 (1951); B. Ferretti, *ibid.*, 12, 558 (1954).

The three problems just discussed—or at least two of them—are real. It may be useful, therefore, to re-emphasize that they are problems of the formal mathematical theory of measurement, and of the description of measurements by macroscopic apparatus. They do not affect the conclusion that a "reduction of the wave packet" (however bad this terminology may be) takes place in some cases. Let us consider, for instance, the collision of a proton and a neutron and let us imagine that we view this phenomenon from the coordinate system in which the center of mass of the colliding pair is at rest. The state vector is then, if we disregard the unscattered beam, in very good approximation (since there is only S-scattering present),

$$\psi(r_p,r_n) = r^{-1}\, e^{ikr} w(r), \tag{13}$$

where $r = |r_p - r_n|$ is the distance of the two particles and $w(r)$ some very slowly varying damping function which vanishes for $r < r_0 - \frac{1}{2}c$ and $r > r_0 + \frac{1}{2}c$, where r_0 is the mean distance of the two particles at the time in question and c the coherence length of the beam. If a measurement of the momentum of one of the particles is carried out—the possibility of this is never questioned—and gives the result $\mathbf{p}$, the state vector of the other particle suddenly becomes a (slightly damped) plane wave with the momentum $-\mathbf{p}$. This statement is synonymous with the statement that a measurement of the momentum of the second particle would give the result $-\mathbf{p}$, as follows from the conservation law for linear momentum. The same conclusion can be arrived at also by a formal calculation of the possible results of a joint measurement of the momenta of the two particles.

One can go even further[18]: instead of measuring the linear momentum of one particle, one can measure its angular momentum about a fixed axis. If this measurement yields the value $m\hbar$, the state vector of the other particle suddenly becomes a cylindrical wave for which the same component of the angular momentum is $-m\hbar$. This statement is again synonymous with the statement that a measurement of the said component of the angular momentum of the second particle certainly would give the value $-m\hbar$. This can be inferred again from the conservation law of the angular momentum (which is zero for the two

[18] See, in this connection, the rather similar situation discussed by A. Einstein, B. Podolsky, and N. Rosen, *Phys. Rev.*, 47, 777 (1935).

particles together) or by means of a formal analysis. Hence, a "contraction of the wave packet" took place again.

It is also clear that it would be wrong, in the preceding example, to say that even before any measurement, the state was a mixture of plane waves of the two particles, traveling in opposite directions. For no such pair of plane waves would one expect the angular momenta to show the correlation just described. This is natural since plane waves are not cylindrical waves, or since (13) is a state vector with properties different from those of any mixture. The statistical correlations which are clearly postulated by quantum mechanics (and which can be shown also experimentally, for instance in the Bothe-Geiger experiment) demand in certain cases a "reduction of the state vector." The only possible question which can yet be asked is whether such a reduction must be postulated also when a measurement with a macroscopic apparatus is carried out. The considerations around Eq. (8) show that even this is true *if* the validity of quantum mechanics is admitted for all systems.

II.5 ON THE INTERPRETATION OF MEASUREMENT IN QUANTUM THEORY

H. D. ZEH

Received September 19, 1969

It is demonstrated that neither the arguments leading to inconsistencies in the description of quantum-mechanical measurement nor those "explaining" the process of measurement by means of thermodynamical statistics are valid. Instead, it is argued that the probability interpretation is compatible with an objective interpretation of the wave function.

1. INTRODUCTION

The problem of measurement in quantum theory and the related problem of how to describe classical phenomena in the framework of quantum theory have received increased attention during recent years. The various contributions express very different viewpoints, and may roughly be classified as follows:

1. Those emphasizing contradictions obtained when the process of measurement is itself described in terms of quantum theory.(1)

2. Those claiming that measurement may well be explained by quantum theory in the sense that "quantum-mechanical noncausality" can be derived from statistical uncertainties inherent in the necessarily macroscopic apparatus of measurement.(2)

3. Those introducing new physical concepts like hidden variables.(3)

Suggestions of the third group are usually based on the first viewpoint, and are meaningful only if they lead to experimental consequences. These have not been confirmed so far.

Originally published in *Foundations of Physics, 1*, 69-76 (1970).

A measurement in quantum theory is axiomatically described by means of a Hermitian operator. If the eigenstates of this operator are φ_n, and the state of the measured system S is $\varphi = \sum c_n \varphi_n$, then, according to the axiom, the result of the measurement will, with probability $|c_n|^2$, be the corresponding eigenvalue a_n represented physically by a "pointer position," i.e., by an appropriate state of the measuring device M. For the most frequent class of measurements, it is furthermore predicted that any following measurement can be described by assuming S to be in the state φ_n after the measurement.

When describing the process of measurement as a whole in the framework of quantum theory, it is assumed that the apparatus M can be described by a wave function ϕ_α, the state of the total system $M + S$ obeying the Schrödinger equation,

$$\psi(t) = e^{iHt}\phi_\alpha \sum_n c_n \varphi_n = \sum_{n,m,\beta} c_n U_{\alpha\beta}^{nm}(t)\, \phi_\beta \varphi_m \tag{1}$$

with $U_{\alpha\beta}^{nm}(0) = \delta_{nm}\,\delta_{\alpha\beta}$. As the state of a macroscopic apparatus can be determined only incompletely, there must be a large set of states $\{\phi\}_0$ compatible with the knowledge about M. If this set of states is assumed to be independent of the state of S before measurement, a condition on the coefficients $U_{\alpha\beta}^{nm}(t)$ can be derived from the requirement that the axiom of measurement be fulfilled in the case $c_n = \delta_{nn_0}$, i.e., $\varphi = \varphi_{n_0}$. The interaction must be of the von Neumann type[4]

$$U_{\alpha\beta}^{nm}(t) = \delta_{nm} u_{\alpha\beta}^{n}(t) \tag{2}$$

for all but a negligible measure of states of the set $\{\phi\}_0$, and for times t larger than the duration of the measurement. Furthermore, practically all states $\sum_\beta u_{\alpha\beta}^{n}(t)\, \phi_\beta$ must be members of a set $\{\phi\}_n$ corresponding to a "pointer position n" of M.

In the case of a general state φ, the final total state now takes the form

$$\psi(t) = \sum_{n,\beta} c_n u_{\alpha\beta}^{n}(t)\, \phi_\beta \varphi_n \tag{3}$$

It represents a superposition of different pointer positions. This result is said to be in contradiction to the axiom of measurement, because the latter states that the result of the measurement is one of the states $\sum_\beta u_{\alpha\beta}^{n}(t)\, \phi_\beta \varphi_n$. It is of course very unsatisfactory to assume that the laws of nature change according to whether or not a physical process is a measurement.

The difficulties arising when a macroscopic system is described by quantum theory can be seen more directly by applying the main axiom of quantum theory, i.e., the superposition principle. If there are two possible pointer positions $\{\phi\}_{n_1}$ and $\{\phi\}_{n_2}$, any superposition $c_1\phi_{n_1} + c_2\phi_{n_2}$ must be a possible state. As such superpositions have never been observed (see Wigner[1]) one should at least find dynamical causes for their nonoccurrence. Although recent work[5] has shown that dynamical stability conditions in the original sense of Schrödinger's[6] have a much wider field of applicability than previously expected, the process of measurement does not, because of the above arguments, belong to this class of phenomena.

2. CRITICISM OF STATISTICAL INTERPRETATIONS

Results apparently in contradiction to those of the preceding section have been derived in a series of papers[2] which try to make use of the uncertainties in the microscopic properties of the apparatus of measurement. The mathematical concept used in these theories is the density matrix formalism.

A simple example may illustrate such theories. If the density matrix describing M is $\sum_\alpha p_\alpha \phi_\alpha \phi_\alpha{}^*$, the total system is described by

$$\rho(0) = \sum_{\alpha,n,n'} p_\alpha c_n c_{n'}^* \phi_\alpha \phi_\alpha{}^* \varphi_n \varphi_{n'}^* \tag{4}$$

For a von Neumann interaction, one obtains

$$\rho(t) = e^{iHt} \rho(0)\, e^{-iHt} = \sum_{\alpha n n' \beta \beta'} p_\alpha c_n c_{n'}^* u_{\alpha\beta}^n(t)\, u_{\alpha\beta'}^{*n'}(t)\, \phi_\beta \phi_{\beta'}^* \varphi_n \varphi_{n'}^* \tag{5}$$

Provided the coefficients $u_{\alpha\beta}^n(t)$ possess arbitrarily distributed phases guaranteeing that

$$\sum_\alpha p_\alpha u_{\alpha\beta}^n(t)\, u_{\alpha\beta'}^{*n'}(t) \approx \delta_{nn'} q_{\beta\beta'}^n(t) \tag{6}$$

(the diagonality in $\beta\beta'$ is not needed), $\rho(t)$ becomes

$$\rho(t) \approx \sum_n | c_n |^2 \varphi_n \varphi_n{}^* \sum_{\beta\beta'} q_{\beta\beta'}^n(t)\, \phi_\beta \phi_{\beta'}^* \tag{7}$$

This density matrix describes exactly the situation postulated by the axiom of measurement.[4]

It is tempting to interpret this result by saying that the statistical uncertainty inherent in the macroscopic apparatus is transferred by means of the interaction to the system S. This means that the outcome of a measurement, i.e., the pointer position, should be exactly predictable if we knew the microscopic state of M. Equation (3) demonstrates that this interpretation is wrong.[1]

The contradiction between Eqs. (3) and (7) is—aside from the dubious nature of the statistical assumption—due to a circular argument. The density matrix formalism is itself based upon the axiom of measurement. In order to see this, consider the case of a set of states $\psi^{(i)} = \sum_n c_n^{(i)} \psi_n$ prepared with probabilities $p^{(i)}$. The probability of finding the eigenvalue a_n is then

$$w_n = \sum_i p^{(i)} | c_n^{(i)} |^2 = \mathrm{tr}\{P_n \rho\} \tag{8}$$

[1] The above example is not identical with any of the theories of Ref. 2. It does not, however, use any additional assumptions. As it leads to a contradiction, one of the assumptions used must be wrong. Some of these theories do not start with an ensemble for the initial state of the apparatus, but assume instead that the "pointer position" is represented by some time average. The latter is then transformed into an ensemble average by means of the ergodic theorem. Interpreted rigorously, these theories would prove that the pointer position fluctuates in time.

if $P_n = \psi_n\psi_n{}^*$, and $\rho = \sum_i p^{(i)}\psi^{(i)}\psi^{(i)*}$. The states $\psi^{(i)}$ will in general not be linearly independent, although ρ may of course be expanded quadratically in terms of a complete orthogonal set. The reason for the usefulness of ρ is that, according to the axiom of measurement, all observable quantities can be expressed as linear-antilinear functionals of the wave function.

For example, the statistical ensemble consisting of equal probabilities of neutrons with spin up and spin down in the x direction cannot be distinguished by measurement from the analogous ensemble having the spins parallel or antiparallel to the y direction. Both ensembles, however, can be easily prepared by appropriate versions of the Stern–Gerlach experiment. One is justified in describing both ensembles by the same density matrix as long as the axiom of measurement is accepted. However, the density matrix formalism cannot be a complete description of the ensemble, as the ensemble cannot be rederived from the density matrix. The discrepancy between Eqs. (3) and (7) arises since, on the one hand, Eq. (3) must hold for all but a negligible number of members of the ensemble, whereas Eq. (7) is interpreted as describing an ensemble of states $\varphi_n \sum_\beta u^n_{\alpha\beta}(t)\, \phi_\beta$, i.e., each state being essentially different from (3). Only if the measurement axiom is accepted can these ensembles not be distinguished by subsequent observations.

The circularity is more obvious in some versions which avoid the density matrix formalism (e.g., Rosenfeld[2] who made repeated use of the probability interpretation although the latter is to be derived). In such cases, the circular argument is considered a "proof of consistency." This viewpoint cannot be accepted, as it would mean that the secondary observation of the pointer position (by a conscious observer or a second apparatus) is a measurement in the axiomatic sense. It corresponds to the interpretation of measurement due to Heisenberg and von Neumann[4] (claiming the arbitariness of the position of the "*Heisenbergscher Schnitt*"), and does not require any contribution from thermodynamics. Bohm's analysis of the process of measurement,[7] however, shows the importance of the amplification of the result of a measurement up to the macroscopic scale, thus leading to a natural position of the "*Heisenbergscher Schnitt.*" (Relative phases between microscopically realized pointer positions could still be measured.)

The secondary (macroscopic) observation is significantly different from the primary (microscopic) one, for the physical situation between these two observations is described by the reduced wave function. The macroscopic observation can thus be performed in a reversible way, in contrast to the microscopic one, which must result in the reduction. It is implicitly assumed in applying the density matrix formalism that the macroscopic measurement is accompanied by a reduction of the wave function.

3. CONSEQUENCES OF A UNIVERSALLY VALID QUANTUM THEORY

The arguments presented so far were based on the assumption that a macroscopic system (the apparatus of measurement) can be described by a wave function ϕ. It appears that this assumption is not valid, for dynamical reasons:

If two systems are described in terms of basic states $\phi^{(1)}_{k_1}$ and $\phi^{(2)}_{k_2}$, the wave function of the total system can be written as $\phi = \sum_{k_1 k_2} c_{k_1 k_2} \phi^{(1)}_{k_1} \phi^{(2)}_{k_2}$. The case where the subsystems are in definite states ($\phi = \phi^{(1)}\phi^{(2)}$) is therefore an exception. Any sufficiently effective interaction will induce correlations. The effectiveness may be measured by the ratios of the interaction matrix elements and the separation of the corresponding unperturbed energy levels. Macroscopic systems possess extremely dense energy spectra. The level distances, for example, of a rotator with moment of inertia 1 g cm^2 are of the order 10^{-42} eV, which value may be compared with the interaction between two electric dipoles of 1 e $\times$ cm at distance R, e$^2 \times$ cm$^2/R^3 \approx 10^{-7}(\text{cm}/R)^3$ eV. It must be concluded that macroscopic systems are always strongly correlated in their microscopic states. They still do have uncorrelated macroscopic properties, however, if the summations over k_1 and k_2 are each essentially limited to macroscopically equivalent states.[8] Since the interactions between macroscopic systems are effective even at astronomical distances, the only "closed system" is the universe as a whole. The assumption of a closed system $M + S$ is hence unrealistic on a microscopic scale.

The arguments leading to Eq. (3) can be accepted only if the states ϕ_α are interpreted as those of the "remainder of the universe" including the apparatus of measurement, instead of those of the latter alone. It is of course very questionable to describe the universe by a wave function that obeys a Schrödinger equation. Otherwise, however, there is no inconsistency in measurement, as there is no theory. This assumption is referred to as that of "universal validity of quantum theory." It leads—as is demonstrated below—to some unusual consequences, but is able to avoid the discrepancies of quantum theory.

The nonexistence of the microscopic states of macroscopic subsystems of the universe leads to severe difficulties in the interpretation of observation or measurement in terms of information transfer between systems. In particular, since no microscopic state of an organism exists, the principle of "psychophysical parallelism"[4] does not apply.

In order to understand Eq. (3), the meaning of superpositions of macroscopically different states has to be investigated. Consider, for the moment, a right-handed sugar molecule with wave function φ_R. This is different from an eigenstate of its Hamiltonian H_S, $\varphi_R \pm \varphi_L$. In contrast to the analogous situation for an ammonia molecule, the tunneling time from φ_R to φ_L is much larger than the age of the universe. The interaction matrix element $\langle \varphi_R \mid H_S \mid \varphi_L \rangle$ is extremely small, as H_S can at most change the state of two particles. Assume now that an eigenstate $\varphi_R \pm \varphi_L$ had been prepared. The two components would then interact in different ways with their environment,

$$e^{iHt}\phi(\varphi_R \pm \varphi_L) \approx \phi^{(R)}(t)\, \varphi_R \pm \phi^{(L)}(t)\, \varphi_L \equiv \psi^{(R)}(t) \pm \psi^{(L)}(t) \tag{9}$$

(Destruction of the sugar molecule is neglected, and excitations may be taken into ϕ.) With respect to the parity quantum number, the sugar molecule behaves like a macroscopic object—the energy difference between the eigenstates is extremely small. The two world components $\psi^{(R)}$ and $\psi^{(L)}$ will behave practically independently after they have been prepared, since $\langle \psi^{(R)} \mid H \mid \psi^{(L)} \rangle$ becomes even smaller with increasing time. There are no transitions between them any more. The "handedness" of the sugar is dynamically stable, whereas one component of the oriented ammonia molecule would emit a photon.

Such a dynamical decoupling of components is even more extreme if φ_R and φ_L represent two states of a pointer corresponding to different positions. Each state will now produce macroscopically correlated states: different images on the retina, different events in the brain, and different reactions of the observer. The different components represent two completely decoupled worlds. This decoupling describes exactly the "reduction of the wave function." As the "other" component cannot be observed any more, it serves only to save the consistency of quantum theory. Omitting this component is justified pragmatically, but leads to the discrepancies discussed above.

This interpretation, corresponding to a "localization of consciousness" not only in space and time, but also in certain Hilbert-space components, has been suggested by Everett[(9)] in connection with the quantization of general relativity, and called the "relative state interpretation" of quantum theory. It amounts to a reformulation of the "psychophysical parallelism" which has in any case become necessary as a consequence of the above discussion of dynamical correlations between states of macroscopic systems.[2] A theory of measurement must necessarily be empty if it does not have a substitute for psychophysical parallelism. Everett's relative state interpretation is ambiguous, however, since the dynamical stability conditions[3] are not considered. This ambiguity is present in the orthodox interpretation of quantum theory as well, where it has always been left to intuition which property of a system is measured "automatically" (e.g., handedness for the sugar, but parity for the ammonia molecule). The dynamical stability appears also to be the cause why microscopic oscillators are observed in energy eigenstates, whereas macroscopic ones occur in "coherent states."[(5)]

According to the twofold localization of consciousness, there are two kinds of subjectivity: The result of a measurement is subjective in that it depends on the world component of the observer; it is objective in the sense that all observers of this world component observe the same result. The question of whether the other components still "exist" after the measurement is as meaningless as asking about the existence of an object while it is not being observed. It *is* meaningful, however, to ask whether or not the assumption of this existence (i.e., of an objective world) leads to a contradiction.

The probability postulate of quantum theory can be formulated in the following way: Suppose a sequence of equivalent measurements have been performed, each creating an equivalent "branching of the universe." The observer can explain the results by assuming that his final branch has been "chosen randomly" if the components are weighted by their norm. The irreversibility connected with this branching is different from that due to thermodynamical statistics, and thus cannot be explained in terms of the latter. Instead, the effect of branching, i.e., measurement, should be of importance for the foundation of thermodynamics. It seems to be partly taken into account by using the density matrix formalism.[4]

[2] Another suggestion of Wigner's,[(10)] which postulates an active role of consciousness, would require corrections to the equations of motion.

[3] The importance of stability for organic systems has been emphazised by Elsasser.[(11)]

[4] This may indeed be the reason why the foundation of quantum-mechanical thermodynamics appears simpler than that of classical thermodynamics. Proofs of the master equation would, however, be circular again if the process of measurement and hence the density matrix formalism were themselves based on thermodynamics.

The famous paradox of Einstein, Rosen, and Podolski[12] is solved straightforwardly: A particle of vanishing spin is assumed to decay into two spin-$\frac{1}{2}$ particles. As a consequence, and according to the axiom of measurement, each particle possesses spin projections of equal probability with respect to any direction in space. After measuring the spin of one particle, however, the spin of the other one is determined. According to Einstein *et al.*, this cannot be true if quantum theory is complete, as there is no interaction with the second particle. The interpretation is that the measurement corresponds to the transformation

$$e^{iHt}\phi(\varphi_1^+\varphi_2^- - \varphi_1^-\varphi_2^+) = \varphi_1^+\varphi_2^-\phi^{(+)}(t) - \varphi_1^-\varphi_2^+\phi^{(-)}(t) \tag{10}$$

where $\phi^{(+)}$ and $\phi^{(-)}$ are dynamically decoupled after a short time. Hence, there is one world component in which the experimentalists observe φ_1^+ and φ_2^-, another one in which they observe φ_1^- and φ_2^+. As these components cannot "communicate," the result is in accord with the axiom of measurement.

This interpretation of measurement may also explain certain "superselection rules"[13] which state, for example, that superpositions of states with different charge cannot occur. It is very plausible that any measurement performed with such a system must necessarily also be a measurement of the charge. Superpositions of states with different charge therefore cannot be observed for similar reasons as those valid for superpositions of macroscopically different states: They cannot be dynamically stable because of the significantly different interaction of their components with their environment, in analogy to the different handedness components of a sugar molecule.

If experimental evidence verifies a spontaneous symmetry-breaking of the vacuum as predicted by many field theories[14] this would not prove an asymmetry of the world. One may formally construct invariant wave functions $\Psi = \int d\Omega \, U_\Omega \phi$ from symmetry-violating wave functions ϕ (as done for microscopic systems[15]). The former cannot be distinguished from its components $U_\Omega \phi$ if the relative state interpretation is accepted.

It appears that the objective interpretation of quantum theory does not contradict the probability interpretation. It has to be admitted, however, that the "relative state wave function" describes only part of the universe. There is no information on other components except for those which have been created by branching in the past. No estimate can therefore be made on the probability of an inverse branching process, i.e., the spontaneous occurrence of components by accidental overlap.

ACKNOWLEDGMENT

I wish to thank Prof. E. P. Wigner for encouraging a more detailed version of the third section, and Dr. M. Böhning for several valuable remarks.

REFERENCES

1. E. P. Wigner, *Am. J. Phys.* **31**, 6 (1963); B. d'Espagnat, *Nuovo Cimento (Suppl.)* **4**, 828 (1966); T. Earman and A. Shimony, *Nuovo Cimento* **54B**, 332 (1968); J. M. Jauch, E. P. Wigner, and

M. M. Yanase, *Nuovo Cimento* **48B**, 144 (1967); G. Ludwig, in *Werner Heisenberg und die Physik unserer Zeit* (Braunschweig, 1961).
2. G. Ludwig, *Die Grundlagen der Quantenmechanik* (Berlin, 1954), p. 122 ff.; *Z. Physik* **135**, 483 (1953); A. Danieri, A. Loinger, and G. M. Prosperi, *Nucl. Phys.* **33**, 297 (1962); *Nuovo Cimento* **44B**, 119 (1966); L. Rosenfeld, *Progr. Theoret. Phys. (Suppl.)* p. 222 (1965); W. Weidlich, *Z. Physik* **205**, 199 (1967).
3. J. S. Bell, *Rev. Mod. Phys.* **38**, 447 (1966); D. Bohm and J. Bub, *Rev. Mod. Phys.* **38**, 453 (1966).
4. J. von Neumann, *Mathematische Grundlagen der Quantenmechanik* (Springer, Berlin, 1932) [English translation: Mathematical Foundations of Quantum Mechanics (Princeton University Press, Princeton, N. J., 1955)].
5. R. J. Glauber, *Phys. Rev.* **131**, 2766 (1963); P. Caruthers and M. M. Nieto, *Am. J. Phys.* **33**, 537 (1965); B. Jancovici and D. Schiff, *Nucl. Phys.* **58**, 678 (1964); C. L. Mehta and E. C. G. Sudarshan, *Phys. Rev.* **138**, B274 (1965).
6. E. Schrödinger, *Z. Physik* **14**, 664 (1926).
7. D. Bohm, *Quantum Theory* (Prentice-Hall, Englewood Cliffs, N.J., 1951).
8. J. M. Jauch, *Helv. Phys. Acta* **33**, 711 (1960).
9. H. Everett, *Rev. Mod. Phys.* **29**, 454 (1957); J. A. Wheeler, *Rev. Mod. Phys.* **29**, 463 (1957).
10. E. P. Wigner, in *The Scientist Speculates*, L. J. Good, ed. (Heinemann, London, 1962), p. 284.
11. W. M. Elsasser, *The Physical Foundation of Biology* (Pergamon Press, New York and London, 1958).
12. A. Einstein, N. Rosen, and B. Podolski, *Phys. Rev.* **47**, 777 (1935); D. Bohm and Y. Aharonov, *Phys. Rev.* **108**, 1070 (1957).
13. G. C. Wick, A. S. Wightman, and E. P. Wigner, *Phys. Rev.* **88**, 101 (1952); E. P. Wigner and M. M. Yanase, *Proc. Natl. Acad. Sci. (US)* **49**, 910 (1963).
14. W. Heisenberg, *Rev. Mod. Phys.* **29**, 269 (1957); Y. Nambu and G. Jona-Lasinio, *Phys. Rev.* **122**, 345 (1961).
15. H. D. Zeh, *Z. Physik* **202**, 38 (1967).

III

"Hidden Variables" versus "Phenomenon" and Complementarity

III.1 POLYELECTRONS

JOHN ARCHIBALD WHEELER

SUMMARY

Theoretical evidence for the existence of entities composed entirely of electrons and positrons is presented in the following article, together with a discussion of their properties.[1] The simplest of these entities consists of one electron and one positron, bound together in a structure similar to that of the hydrogen atom. The next higher entity is composed of two positrons and one electron or of two electrons and one positron. The bi-electron system is stable by 6.77 ev against dissociation. Against annihilation, it has a life time of 1.24×10^{-10} sec., when the spins of the two particles are parallel, and a life several orders of magnitude greater, when the spins are antiparallel. The tri-electron system also has a radioactive mean life of the order of 10^{-10} sec., and is calculated to be stable by at least 0.19 ev against dissociation into a bi-electron and a free electron or positron. The production of a bi-electron, by interaction of an energetic gamma ray with the field of force of an atomic nucleus, is calculated to occur with a probability about 10^{-6} times less than that for production of an electron-positron pair. The possibilities are discussed for observing atomic and molecular spectra in which the positron plays the role of an especially light hydrogen ion. An experiment is suggested which can be used to check the theory of the perpendicular polarization of the gamma rays given off in the annihilation process. The similarities and distinct differences between polyelectrons and cosmic ray mesons are discussed.

. . . .

POSSIBILITY OF TESTING SOME OF THE PRESENT PREDICTIONS OF PAIR THEORY

Quite apart from the possibly questionable direct bearing of polyelectrons upon cosmic ray phenomena, it is natural to ask if there are any experimental implications which may be examined in the laboratory as new tests of the validity of pair theory itself.

One of the most interesting possibilities for a test is suggested by the existence of excited energy levels in the polyelectron, P^{+-}. Radiative transition of the system between these energy levels generates an optical

Originally published in *Annals of the New York Academy of Sciences, 48*, 219-38 (1946).

spectrum which differs from that of hydrogen, in its major features, only through the displacement of all lines to the red by a displacement factor of two. To observe any well-defined spectrum of such a character would, of course, appear to call, in the first place, for a gaseous emitter. In addition, the securing of slow polyelectrons requires that a slow positron should be able easily to detach an electron from an atom of the gas. This condition requires that the first ionization potential of the substance of the gas should be close to 6.7 volts. Finally, the means of observation must be capable of picking up over the background spectral lines which have been considerably broadened by the Doppler effect, inevitable in systems which have only twice the electronic mass and which are in thermal equilibrium near room temperature.

The difficulties about Doppler effect and choice of substance with suitable ionization potential are considerably alleviated by renouncing, in the beginning, the study of the particular entity, P^{+-}, and looking at the problem of the test of pair theory in a broader light. The essential point is to find an atomic or molecular system which contains a positron and which possesses several optically combining energy levels. In looking for such a system, it is only necessary to remember that the positron may be regarded as a superlight isotope of the highly reactive hydrogen ion. Consequently, one can look among such compounds as e^+Cl^- for systems which may possess the desired type of energy levels and which will be free of objectionable Doppler effect, on account of their substantial mass.

The actual experiment would consist in irradiation of a suitable gas with slow positrons, the radiative capture of some of these positrons into excited states of entities having somewhat the character of molecules, the transition of these entities to lower levels with the emission of characteristic spectral lines lying in, or near, the visible region, the observation of this spectrum, and the annihilation of each positron by an electron of the corresponding molecule-like entity. The gamma rays given off in this experiment would be of no direct concern to the problem at issue. The test of the pair theory would come in the comparison of the observed and calculated positions of the spectral lines. Obviously, the experiment, though interesting, is difficult.

A second and somewhat simpler experiment would seem to offer a means to check on one of the details of the annihilation process itself. We have already remarked that by far the dominating type of annihilation is that in which the positron combines with an electron whose spin forms a singlet state with respect to the spin of the positron. Associated with this selection of pairs which have zero relative angular momentum, before the annihilation process, is an analogous polarization phenomenon in the two quanta which are left at the end of the process. According to the pair theory, if one of these photons is linearly polarized in one plane, then the photon which goes off in the opposite direction with equal momentum is linearly polarized in the perpendicular plane.

To test this prediction, the following experimental arrangement suggests itself: A radioactive source of slow positrons is covered with a foil thick enough to guarantee annihilation of all the positrons. A sphere of lead centered on this source prevents the escape of any of the annihilation quanta, except through a relatively narrow hole drilled through the sphere along one of its diameters. When a photon of energy mc^2 comes out of one end of this channel, we expect a photon also of energy mc^2 to emerge simultaneously from the other end. At each end, a carbon scatterer is placed. Photons scattered by one of these blocks through approximately ninety degrees and into the proper azimuth pass through a gamma ray counter. The scattering process gives a preference to the recording of photons with a selected polarization. A similar arrangement applies at the other end of the channel. The relative azimuth of the two counters may be varied at will. Coincidences between the two counters are recorded, (a) when the azimuths of the two counters are identical, (b) when the azimuths differ by a right angle. The observed ratio of (b) to (a) is compared with the computed ratio, as a check on the theory of the annihilation process. The calculated ratio for the case of ideal geometry is 1.080, when the arrangement requires the photons to be scattered through 90°. The theoretically most favorable ratio of 1.100 is obtained when the scattering angle is reduced to 74°30′.

[To be corrected to 2.60 at 90° and maximum of 2.85 at 82°: Pryce and Ward (1947), Snyder, Pasternack, and Hornbostel (1948)—Eds.]

Another possible means of studying this scattering is to use the knock-on electrons, instead of the recoil photons. The polarization is, obviously, as complete for the particles as for the radiation. In case this arrangement is employed, the detecting counters are set to catch electrons knocked on at an angle of about 30°, with respect to the annihilation radiation. The efficiency of counting is increased by this alteration in the plan of the experiment.

Evidently, it is possible, by means of a reasonable experimental procedure, to obtain information bearing most closely upon the problem of the intimate interaction of an electron and a positron.

. . . .

REFERENCES AND NOTES

1. On October 5, four days after the present paper was submitted to the New York Academy of Sciences, the author learned from Professor **Arthur Ruark** that he had previously envisaged the existence of the particular entity composed of one electron and one positron. Dr. Ruark has discussed the optical spectrum and the life time of this two-particle system in a note dated September 23, 1945, which he intends to submit for publication to the *Physical Review* in the form of a "Letter to the Editor." A reference to unpublished work by **L. Landau** on the properties of the bi-electron has been made by **Alichanian, A., & T. Asatiani.** 1945. J. Phys. U.S.S.R. **9:** 56.

III.2 THE PARADOX OF EINSTEIN, ROSEN, AND PODOLSKY

DAVID BOHM

15. The Paradox of Einstein, Rosen, and Podolsky. In an article in the Physical Review,† Einstein, Rosen, and Podolsky raise a serious criticism of the validity of the generally accepted interpretation of quantum theory. This objection is raised in the form of a paradox to which they are led on the basis of their analysis of a certain hypothetical experiment, which we shall discuss in detail later. Their criticism has, in fact, been shown to be unjustified,‡ and based on assumptions concerning the nature of matter which implicitly contradict the quantum theory at the outset. Nevertheless, these implicit assumptions seem, at first sight, so natural and inevitable that a careful study of the points which the authors raised affords deep and penetrating insight into the difference between classical and quantum concepts of the nature of matter.

The authors first undertook to define criteria for a complete physical

† *Phys. Rev.*, **47**, 777 (1935).
‡ N. Bohr, *Phys. Rev.* **48**, 696 (1935); W. H. Furry, *Phys. Rev.* **49**, 393, 476 (1936).

Originally published as sections 15-19, Chapter 22 of *Quantum Theory*, David Bohm, pp. 611-23, Prentice-Hall, Englewood Cliffs (1951).

theory. It seemed to them that a necessary requirement for a *complete* physical theory was the following:

(1) Every element of physical reality must have a counterpart in a *complete* physical theory.

As to what actually constituted the correct elements in terms of which physical theory should be expressed, they felt that this question can be decided finally only by recourse to experiments and observations. They nevertheless suggested the following criterion for recognizing an element of reality, which seemed to them a *sufficient* criterion:

(2) If, without in any way disturbing the system, we can predict with certainty (i.e., with probability equal to unity) the value of a physical quantity, then there exists an element of reality corresponding to this physical quantity.

The authors agreed that elements of physical reality might well be recognized in other ways also, but they intended to show that even if one restricted oneself to elements that could be recognized by means of *this* criterion alone, quantum theory as now interpreted led to contradictory results.

The use of the above explicit criteria rests, however, on certain implicit assumptions, which are an integral part of the treatment given by the authors, but which are never explicitly stated. These assumptions are:

(3) The world can correctly be analyzed in terms of distinct and separately existing "elements of reality,"

(4) Every one of these elements must be a counterpart of a *precisely* defined mathematical quantity appearing in a *complete* theory.*

We shall temporarily accept the above criteria and assumptions, in order to permit the further development of the arguments given by the authors, but in Sec. 18 we shall show that these criteria should not be applied at the quantum level of accuracy.

Now, let us recall that in the present quantum theory, one assumes that all relevant physical information about a system is contained in its wave function, so that when two systems have wave functions which differ by at most a constant phase factor, they are said to be in the same quantum state.† What the authors wished to do with their criteria for reality was to show that the above interpretation of the present quantum theory is untenable and that the wave function cannot possibly contain a complete description of all physically significant factors (or "elements of reality") existing within a system. If their contention could be proved, then one would be led to search for a more complete theory,

* This criterion is essentially a strengthened form of (1). Einstein, Rosen, and Podolsky do not restrict themselves to the assumption (1), that every element of reality always has a counterpart in a complete theory, but they also assume implicitly that this counterpart must always be precisely definable.

† See Chap. 9, Sec. 4.

perhaps containing something like hidden variables,* in terms of which the present quantum theory would be a limiting case.

Let us now consider an arbitrary observable A having a set of eigenfunctions, ψ_a, belonging to a series of eigenvalues which are denoted by a. When the wave function is ψ_a, then the system is said to be in a quantum state in which the observable A has the definite value a. In this situation, ERP would say that there is in the system an element of reality corresponding to the observable, A. But now let us consider another observable B which does not commute with A, so that there exists no wave function for which A and B have simultaneously definite values. Now if we adopt the implicit assumption (4) that every element of reality must be a counterpart of a *precisely* defined mathematical quantity appearing in a *complete* theory, then the usual assumption that the wave function provides a *complete* description of reality leads to the conclusion that A and B cannot exist simultaneously.† This follows from the fact that the supposedly complete wave theory contains no *precisely* defined mathematical elements corresponding to the simultaneous existence of A and B. From this point of view, we must also assume, however, that when B is measured and obtains a definite value, the elements corresponding to A are destroyed (since we have assumed that they cannot exist together with those corresponding to B). It seems natural to suppose that this destruction is brought about by the quanta that are transferred from the measuring apparatus to the system under observation. It is clear, however, that in such an interpretation of the noncommutativity of two observables, it is essential that in every measurement there shall actually be a disturbance arising from the apparatus that destroys all elements of reality corresponding to observables that do not commute with the measured variable. For if there were no such disturbance, then one could take a system initially having a definite value of A and then measure B without in any way altering the elements corresponding to A, thus obtaining a system in which the elements of reality corresponding to A and B exist together at the same time. Now, in the next section, we shall discuss a type of hypothetical experiment suggested by ERP that actually permits us to measure a given observable without in any way disturbing the associated system. With the aid of this type of hypothetical experiment, they are then able to obtain a contradiction between the assumption that the quantum theory provides a complete description of reality and the assumption that their criteria for reality must necessarily apply in any complete theory. If one accepts their

* Chap. 2, Sec. 5; Chap. 5, Sec. 3.

† In Sec. 18, we shall make the alternative assumption that elements of reality exist in a roughly defined form and do not necessarily have to be counterparts of precisely defined mathematical quantities appearing in a complete theory. Thus, we shall give up the implicit assumptions (3) and (4).

criteria, one is left with a single remaining alternative, viz., that quantum theory does not provide a complete description of reality. This is the conclusion that they originally set out to obtain.

16. The Hypothetical Experiment of Einstein, Rosen, and Podolsky. We shall now describe the hypothetical experiment of Einstein, Rosen, and Podolsky. We have modified the experiment somewhat, but the form is conceptually equivalent to that suggested by them, and considerably easier to treat mathematically.

Suppose that we have a molecule containing two atoms in a state in which the total spin is zero and that the spin of each atom is $\hbar/2$. Roughly speaking, this means that the spin of each particle points in a direction exactly opposite to that of the other, insofar as the spin may be said to have any definite direction at all. Now suppose that the molecule is disintegrated by some process that does not change the total angular momentum. The two atoms will begin to separate and will soon cease to interact appreciably. Their combined spin angular momentum, however, remains equal to zero, because by hypothesis, no torques have acted on the system.

Now, if the spin were a classical angular momentum variable, the interpretation of this process would be as follows: While the two atoms were together in the form of a molecule, each component of the angular momentum of each atom would have a definite value that was always opposite to that of the other, thus making the total angular momentum equal to zero. When the atoms separated, each atom would continue to have every component of its spin angular momentum opposite to that of the other. The two spin-angular-momentum vectors would therefore be correlated. These correlations were originally produced when the atoms interacted in such a way as to form a molecule of zero total spin, but after the atoms separate, the correlations are maintained by the deterministic equations of motion of each spin vector separately, which bring about conservation of each component of the separate spin-angular-momentum vectors.

Suppose now that one measures the spin angular momentum of any one of the particles, say No. 1. Because of the existence of correlations, one can immediately conclude that the angular-momentum vector of the other particle (No. 2) is equal and opposite to that of No. 1. In this way, one can measure the angular momentum of particle No. 2 indirectly by measuring the corresponding vector of particle No. 1.

Let us now consider how this experiment is to be described in the quantum theory. Here, the investigator can measure either the x, y, or z component of the spin of particle No. 1, but not more than one of these components, in any one experiment. Nevertheless, it still turns out as we shall see that whichever component is measured, the results are correlated, so that if the same component of the spin of atom No. 2 is measured, it will always turn out to have the opposite value. This means that a measurement of any component of the spin of atom No. 1 provides, as in classical theory, an indirect measurement of the same component of the spin of atom No. 2. Since, by hypothesis, the two particles no longer interact, we have obtained a way of measuring an arbitrary component of the spin of particle No. 2 without in any way disturbing that particle. If we accept the definition of an element of reality (2) suggested by ERP, it is clear that after we have measured σ_z for particle 1, then σ_z for particle 2 must be regarded as an element of reality, existing separately in particle No. 2 alone. If this is true, however, this element of reality must have existed in particle No. 2 even before the measurement of σ_z for particle No. 1 took place. For since there is no interaction with particle No. 2, the process of measurement cannot have affected this particle in any way. But now let us remember that, in each case, the observer is always free to reorient the apparatus in an arbitrary direction while the atoms are still in flight, and thus to obtain a definite (but unpredictable) value of the spin component in any direction that he chooses. Since this can be accomplished without in any way disturbing the second atom, we conclude that if criterion (2) of ERP is applicable, precisely defined elements of reality must exist in the second atom, corresponding to the simultaneous definition of all three components of its spin. Because the wave function can specify, at most, only one of these components at a time with complete precision, we are then led to the conclusion that the wave function does not provide a complete description of all elements of reality existing in the second atom.

If this conclusion were valid, then we should have to look for a new theory in terms of which a more nearly complete description was possible. We shall see, however, in Sec. 18, that the analysis given by ERP involves in an integral way the implicit assumptions (3) and (4) that the world is actually made up of separately existing and precisely defined "elements of reality." Quantum theory, however, implies a quite different picture of the structure of the world at the microscopic level. This picture leads, as we shall see, to a perfectly rational interpretation of the hypothetical experiment of ERP within the present framework of the theory.

17. Mathematical Analysis of Experiment According to Quantum Theory. Before discussing the physical interpretation that the present quantum theory gives to the hypothetical experiment of Einstein, Rosen and Podolsky, we shall first show how this experiment is to be described in mathematical terms.

The system containing the spin of two atoms has four basic wave functions, from which an arbitrary wave function can be constructed.*

* The complete wave function for the system is then obtained by multiplying the spin wave functions by appropriate space wave functions, which depend on the space co-ordinates of both particles.

These are

$$\psi_a = u_+(1)u_+(2) \qquad \psi_c = u_+(1)u_-(2)$$
$$\psi_b = u_-(1)u_-(2) \qquad \psi_d = u_-(1)u_+(2)$$

where u_+ and u_- are the one-particle spin wave functions representing, respectively, a spin $\hbar/2$ and $-\hbar/2$, and the argument (1) or (2) refers, respectively, to the particle which has this spin. Now ψ_c and ψ_d represent the two possible situations in which each particle has a definite z component of the spin in a direction which is opposite to that of the other. The wave function for a system of total spin zero is the following linear combination of ψ_c and ψ_d (see Chap. 17, Sec. 9):

$$\psi_0 = \frac{1}{\sqrt{2}} (\psi_c - \psi_d) \tag{26}$$

The particular sign with which ψ_c and ψ_d are combined is of crucial importance in determining the combined spin, for if they are combined with a + sign, one obtains an angular momentum of $\hbar$ (but with a zero value of the z component of the angular momentum). We denote this result below:

$$\psi_1 = \frac{1}{\sqrt{2}} (\psi_c + \psi_d) \tag{27}$$

It is clear, then, that the total angular momentum is an interference property of ψ_c and ψ_d. On the other hand, the only states in which each particle has a definite spin opposite to that of the other are represented either by ψ_c or by ψ_d separately. Thus, in any state in which the value of σ_z for each particle is definite, the total angular momentum must be indefinite. Vice versa, whenever the total angular momentum is definite, then neither atom can correctly be regarded as having a definite value of its own spin, for if it did, there could be no interference between ψ_c and ψ_d, and it is just this interference which is required to produce a definite total angular momentum.

Besides leading to a definite value of the combined spin, however, definite phase relations between ψ_c and ψ_d have additional physical meaning, for they also imply that if the same component of the spin of each atom is measured, the results will be correlated. Such correlations can be demonstrated, for example, in a process in which the z component of the spin of each atom is measured by allowing each atom to pass through a separate Stern-Gerlach apparatus (see Fig. 1). For the sake of simplicity, we can suppose that both spins are measured at the same time, although no results will depend significantly on this assumption. The Hamiltonian at the time of measurement is then [see eqs. (10a) and (10b)]:

$$W = \mu(\mathcal{H}_0 + z\mathcal{H}_0')\sigma_{1,z} + \mu(\mathcal{H}_0 + z_2\mathcal{H}_0')\sigma_{2,z}$$

where z_1 is the z co-ordinate of the first atom and z_2 is the z co-ordinate of the second atom. (We assume that both pieces of apparatus are identical in construction.)

We now expand the spin wave function during the course of the measurement in terms of the four basic functions, ψ_a, ψ_b, ψ_c, and ψ_d. Since this measurement does not change σ_z, it will remain true that only ψ_c and ψ_d are needed during the course of the measurement.* Thus, we write

$$\psi = f_c\psi_c + f_d\psi_d$$

In our case, the initial value of f_c is $1/\sqrt{2}$, and the initial value of f_d is $-1/\sqrt{2}$. By methods similar to those leading to equation (13b), one derives

$$i\hbar \frac{\partial f_c}{\partial t} = \mu f_c[(\mathcal{H}_0 + \mathcal{H}_0' z_1) - (\mathcal{H}_0 + \mathcal{H}_0' z_2)]$$

$$i\hbar \frac{\partial f_d}{\partial t} = -\mu f_d[(\mathcal{H}_0 + \mathcal{H}_0' z_1) - (\mathcal{H}_0 + \mathcal{H}_0' z_2)]$$

The solution for f_c and f_d with the proper boundary conditions yields for the wave function just after the particles leave the magnetic field

$$f_c = \frac{1}{\sqrt{2}} e^{-i\frac{\mu\mathcal{H}_0'}{\hbar}(z_1 - z_2)\,\Delta t} \qquad f_d = -\frac{1}{\sqrt{2}} e^{i\frac{\mu\mathcal{H}_0'}{\hbar}(z_1 - z_2)\,\Delta t}$$

where we have inserted $t = \Delta t$ = time of interaction between atoms and the inhomogeneous magnetic field.

This wave function implies that the two results represented, respectively, by ψ_c and by ψ_d are equally probable. In the first possible result, atom No. 1 has a positive value of σ_z, while atom number 2 has a negative value. The factor $e^{-i\mu\mathcal{H}_0'\Delta t(z_1 - z_2)/\hbar}$ represents the fact that in the Stern-Gerlach experiment, each atom obtains an opposite momentum corresponding to its opposite spin. Similarly, in the second possible result, atom No. 2 has a negative value of σ_z, whereas atom No. 1 has a positive value. As in Secs. 9 and 11, we can show that because the apparatus is classically describable, the apparatus wave functions (which depend on z_1 and z_2), multiply the spin wave function by uncontrollable phase factors, so that we finally obtain

$$\psi = \frac{1}{\sqrt{2}} (\psi_c\, e^{i\alpha_c} + \psi_d\, e^{i\alpha_d})$$

where α_c and α_d are separate and uncontrollable phase factors.

This result shows that if the value of σ_z is measured for each atom, the result will come out a definite number for each, which is always

* Note that these are the only terms present initially.

opposite to that of the other. In this way, we prove that correlations resembling those of classical theory will also be obtained in the quantum theory. After the measurement is over, however, the system has been transformed from one that had a definite combined angular momentum and an indefinite value of σ_z for each particle to one which has a definite value of σ_z for each particle, but an indefinite combined angular momentum. Moreover, the precise value of σ_z which will be obtained for each particle is not related deterministically to the state of the system before the measurement, but only statistically.

Let us now describe the process of measurement of σ_x. The results are very similar, because the wave function for a system of zero total spin is the same when expressed in terms of v_+, v_- (the eigenfunctions of σ_x) as in terms of u_+, u_-. Thus, we obtain

$$\psi_0 = \frac{1}{\sqrt{2}} [v_+(1)v_-(2) - v_-(1)v_+(2)]$$

One can now describe the measurement of σ_x for each particle in exactly the same way as was done with σ_z, and after the interaction with the measuring apparatus, one obtains

$$\psi = \frac{1}{\sqrt{2}} [v_+(1)v_-(2)\, e^{i\alpha_1} + v_-(1)v_+(2)\, e^{i\alpha_2}]$$

where α_1 and α_2 are separate uncontrollable phase factors.

We conclude that the value of σ_x for each particle is also correlated to that of the other in such a way that the sum of the two is zero. Moreover, it is readily verified that if one had taken the function

$$\psi_1 = \frac{1}{\sqrt{2}} (\psi_a + \psi_b)$$

then with the substitution, $v_+ = \frac{1}{\sqrt{2}}(u_+ + u_-)$ and $v_- = \frac{1}{\sqrt{2}}(u_+ - u_-)$, one would have the wave function

$$\psi_1 = \frac{1}{\sqrt{2}} [v_+(1)v_+(2) + v_-(1)v_-(2)]$$

This represents a situation in which measurement of σ_x will disclose that both particles have a positive value together, or that both particles have a negative value together. We see therefore that the type of correlation of σ_x which can develop depends on the sign with which ψ_c and ψ_d are added, and therefore also on the combined angular momentum.

One more significant point arises in connection with this experiment; namely, that the existence of correlations does not imply that the behavior of either atom is affected in any way at all by what happens to the other, after the two have ceased to interact. To prove this statement, we first evaluate the mean value of any function $g(\boldsymbol{\sigma}_2)$ of the spin variables of particle No. 2 alone. With the wave function before a measurement took place, we obtain

$$\bar{g}_0(\boldsymbol{\sigma}_2) = \tfrac{1}{2}(\psi_c^* - \psi_d^*)g(\boldsymbol{\sigma}_2)(\psi_c - \psi_d) = \tfrac{1}{2}[\psi_c^* g(\boldsymbol{\sigma}_2)\psi_c + \psi_d^* g(\boldsymbol{\sigma}_2)\psi_d]$$

(By virtue of the orthogonality of ψ_c and $g(\boldsymbol{\sigma}_2)\psi_d$.) After the spin of the first particle is measured, the average of $g(\boldsymbol{\sigma}_2)$ becomes

$$\bar{g}_f(\boldsymbol{\sigma}_2) = \tfrac{1}{2}(\psi_c^* e^{-i\alpha_c} - \psi_d^* e^{-i\alpha_d})g(\boldsymbol{\sigma}_2)(\psi_c e^{i\alpha_c} - \psi_d e^{i\alpha_d})$$
$$= \tfrac{1}{2}[\psi_c^* g(\boldsymbol{\sigma}_2)\psi_c + \psi_d^* g(\boldsymbol{\sigma}_2)\psi_d]$$

This is the same as what was obtained without a measurement of the spin variables of particle No. 1. The behavior of the two spins is, however, correlated despite the fact that each behaves in a way that does not depend on what actually happens to the other after interaction has ceased.

18. Physical Description of Origin of Correlations. We have deduced mathematically that in a system of two atoms having a total spin of zero, the spin components of each atom in an arbitrary direction will be correlated, despite the fact that according to our present interpretation of quantum theory these spin components cannot all exist simultaneously in precisely defined forms. We wish to show now that the paradoxical results obtained by ERP in the interpretation of this fact will not be obtained if one avoids making their implicit assumptions (3) and (4); viz., that the world can correctly be analyzed into elements of reality, each of which is a counterpart of a precisely defined mathematical quantity appearing in a complete theory. These assumptions, which are at the root of all classical theory, might perhaps be called the hypothesis that reality is built upon a mathematical plan, for it is required that every element appearing in the real world shall correspond *precisely* to some term appearing in a complete set of mathematical equations. Although such a hypothesis seems quite natural to us at this time, it is by no means inescapable.† In fact, in quantum theory, one makes a quite different, but equally plausible, hypothesis concerning the fundamental nature of matter. Here, we assume that the one-to-one correspondence between mathematical theory and well-defined "elements of reality" exists only at the classical level of accuracy. For at the quantum level, the mathematical description provided by the wave function is certainly not in a one-to-one correspondence with the actual behavior of

† Historically speaking, it is a comparatively new idea, having arisen in connection with the great success of mathematical analysis in mechanics and electrodynamics during the period between the sixteenth and early twentieth centuries (see Chap. 8, Secs. 2 to 10).

the system under description, but only in a statistical correspondence.* Yet, we assert that the wave function (in principle) can provide the most complete possible description of the system that is consistent with the actual structure of matter. How can we reconcile these two aspects of the wave function? We do so in terms of the assumption that the properties of a given system exist, in general, only in an imprecisely defined form, and that on a more accurate level, they are not really well-defined properties at all, but instead only potentialities,† which are more definitely realized in interaction with an appropriate classical system, such as a measuring apparatus. For example, consider two noncommuting observables, such as momentum and position of an electron. We say that, in general, neither exists in a given system in a *precisely* defined form, but that both exist together in a roughly defined form, such that the uncertainty principle is not violated.‡ Either variable is potentially capable of becoming better defined at the expense of the degree of definition of the other, in interaction with a suitable measuring apparatus. We see then that the properties of position and momentum are not only incompletely defined and opposing potentialities, but also that in a very accurate description, they cannot be regarded as belonging to the electron alone; for the realization of these potentialities depends just as much on the systems with which it interacts as on the electron itself.§ This means that there are actually no precisely defined "elements of reality" belonging to the electron. Thus, we contradict the assumptions (3) and (4) of Einstein, Rosen, and Podolsky.

Quantum-mechanical spin variables must be interpreted in a similar way. Whereas ERP would say that the only existing component of the spin is the one which may happen to be defined precisely by the wave function, we say that, in general, all three components exist simultaneously in roughly defined forms, and that any one component has the potentiality for becoming better defined at the expense of the others if the associated atom interacts with a suitable measuring apparatus. The probability for the development of a definite value of any spin component in a suitable process of measurement is proportional to the square of the amplitude of the coefficient of the part of the wave function corresponding to this component. We must, however, recall that the complete spin wave function for a given atom can be expanded in terms of the eigenfunctions, u_+ and u_-, of the spin variables in *any* direction. Thus $\psi = a_+u_+ + a_-u_-$. In such an expansion, the phase relations between u_+ and u_- help determine the distribution over spin components in other directions.‖ (Thus, if u_+, u_- represent eigenfunctions of σ_z, then

* See Chap. 6, Sec. 4.

† See Chap. 6, Secs. 9 and 13, Chap. 8, Secs. 14 and 15, Chap. 22, Sec. 13.

‡ See discussion of complementarity in Chap. 8, Sec. 15.

§ Chap. 6, Sec. 13, Chap. 8, Sec. 16.

‖ See Chap. 17, Secs. 6 and 7.

an eigenfunction of σ_x is obtained when $\psi = \frac{1}{\sqrt{2}}(u_+ \pm u_-)$.) This means that as long as definite phase relations exist between u_+ and u_-, one cannot categorize (or classify) the system as having a spin which corresponds either *entirely* to u_+ or *entirely* to u_-, with respective probabilities,* $|a_+|^2$ and $|a_-|^2$. Instead, we must say that the system cuts across this method of classification, and in some sense, covers both states at once in a poorly defined way.† Thus, we must give up the classical picture of a precisely defined spin variable associated with each atom, and replace it by our quantum concept of a potentiality, the probability of whose development is given by the wave function. It is only when the wave function is an eigenfunction of a given spin component that the system is certain (in interaction with a suitable apparatus) to develop a predictable value of that spin component.

Now, when we come to our system of two atoms having a total spin of zero, we see from eq. (26) that because the wave function

$$\psi_0 = \frac{1}{\sqrt{2}}(\psi_c - \psi_d)$$

has definite phase relations between ψ_c and ψ_d, the system must cover the states corresponding to ψ_c and ψ_d simultaneously. Thus, for a given atom, *no* component of the spin of a given variable exists with a precisely defined value, until interaction with a suitable system, such as a measuring apparatus, has taken place. But as soon as either atom (say, No. 1) interacts with an apparatus measuring a given component of the spin, definite phase relations between ψ_c and ψ_d are destroyed. This means that the system then acts as if it is either in the state ψ_c or ψ_d. Thus, in every instance in which particle No. 1 develops a definite spin component in, for example, the z direction, the wave function of particle No. 2 will automatically take such a form that it guarantees the development of the opposite value of σ_z if this particle also interacts with an apparatus which measures the same component of the spin. The wave function therefore describes the propagation of correlated potentialities. Because the expansion of the wave function ψ_0 takes the same form when expanded in terms of the eigenfunctions of an arbitrary component of the spin, we conclude that similar correlations will be obtained if the same component of the spin of each atom in any direction is measured. Moreover, because the potentialities for development of a definite spin component are not realized irrevocably until interaction with the apparatus actually takes place, there is no inconsistency in the statement that while the atoms are still in flight, one can rotate the apparatus into an arbitrary

* Chap. 6, Sec. 4, Chap. 22, Sec. 10.
† Chap. 16, Sec. 25, and Chap. 8, Sec. 15.

direction, and thus choose to develop definite and correlated values for any desired spin component of each atom.

Finally, it is perhaps interesting to consider in a new light the fact that the mathematical description provided by the wave function is not in a one-to-one correspondence with the actual behavior of matter. From this fact, we are led to conclude that, contrary to general opinion, quantum theory is less mathematical in its philosophical basis than is classical theory, for, as we have seen, it does not assume that the world is constructed according to a precisely defined mathematical plan. Instead, we have come to the point of view that the wave function is an abstraction, providing a mathematical reflection of certain aspects of reality, but not a one-to-one mapping. To obtain a description of *all* aspects of the world, one must, in fact, supplement the mathematical description with a physical interpretation in terms of incompletely defined potentialities.* Moreover, the present form of quantum theory implies that the world cannot be put into a one-to-one correspondence with any conceivable kind of precisely defined mathematical quantities, and that a complete theory will always require concepts that are more general than that of analysis into precisely defined elements. We may probably expect that even the more general types of concepts provided by the present quantum theory will also ultimately be found to provide only a partial reflection of the infinitely complex and subtle structure of the world. As science develops, we may therefore look forward to the appearance of still newer concepts, which are only faintly foreshadowed at present, but there is no strong reason to suppose that these new concepts are likely to lead to a return to the comparatively simple idea of a one-to-one correspondence between the real world and precisely defined mathematical abstractions.

19. Proof that Quantum Theory Is Inconsistent with Hidden Variables. We can now use some of the results of the analysis of the paradox of Einstein, Rosen, and Podolsky to help prove that quantum theory is inconsistent with the assumption of hidden causal variables. (See Chap. 2, Sec. 5 and Chap. 5, Sec. 3.) We first note that the assumption that there are separately existing and precisely defined elements of reality would be at the base of any precise causal description in terms of hidden variables; for without such elements there would be nothing to which a precise causal description could apply. Similarly, as we saw in Chap. 8, Sec. 20, the existence of separate elements requires a precise causal theory of the relationships between these elements for its consistent application. Thus, the analysis of the world into pre-

* See Chap. 23 for a fuller discussion of how the wave function must be supplemented with its interpretation in terms of potentialities for the production of various classically describable results.

cisely defined elements and the synthesis of these elements according to precise causal laws must stand or fall together.

Now, from the reasoning of ERP we conclude that if the world can be explained in terms of such precisely defined elements, then the correct interpretation of two noncommuting variables, such as momentum and position, would be that they correspond to simultaneously existing elements of reality. To interpret the uncertainty principle, we would then have to assume that we are simply unable to measure the values of the two simultaneously with complete precision. But we saw in Chap. 6, Sec. 11, that any such assumption would lead to a contradiction with the uncertainty principle, which is one of the most fundamental deductions of the quantum theory. We conclude then that no theory of mechanically determined hidden variables can lead to *all* of the results of the quantum theory. Such a mechanical theory might conceivably be so ingeniously framed that it would agree with quantum theory for a wide range of predicted experimental results.* But the hypothetical experiment suggested in Chap. 6, Sec. 11 would then be an example of a crucial test of the theory. If, in this experiment, we were able to violate the uncertainty principle, then the theory of mechanically determined underlying variables would be strongly indicated, whereas if we were not able to violate the uncertainty principle, we should obtain a fairly convincing proof that no correct mechanical theory could ever be found. Unfortunately, such an experiment is still far beyond present techniques, but it is quite possible that it could some day be carried out. Until and unless some such disagreement between quantum theory and experiment is found, however, it seems wisest to assume that quantum theory is substantially correct, because it is a self-consistent theory yielding agreement with such a wide range of experiments not correctly treated by any other known theory.

* We do not wish to imply here that anyone has ever produced a concrete and successful example of such a theory, but only state that such a theory is, as far as we know, conceivable.

III.3 A SUGGESTED INTERPRETATION OF THE QUANTUM THEORY IN TERMS OF "HIDDEN" VARIABLES, I AND II

DAVID BOHM

The usual interpretation of the quantum theory is self-consistent, but it involves an assumption that cannot be tested experimentally, *viz.*, that the most complete possible specification of an individual system is in terms of a wave function that determines only probable results of actual measurement processes. The only way of investigating the truth of this assumption is by trying to find some other interpretation of the quantum theory in terms of at present "hidden" variables, which in principle determine the precise behavior of an individual system, but which are in practice averaged over in measurements of the types that can now be carried out. In this paper and in a subsequent paper, an interpretation of the quantum theory in terms of just such "hidden" variables is suggested. It is shown that as long as the mathematical theory retains its present general form, this suggested interpretation leads to precisely the same results for all physical processes as does the usual interpretation. Nevertheless, the suggested interpretation provides a broader conceptual framework than the usual interpretation, because it makes possible a precise and continuous description of all processes, even at the quantum level. This broader conceptual framework allows more general mathematical formulations of the theory than those allowed by the usual interpretation. Now, the usual mathematical formulation seems to lead to insoluble difficulties when it is extrapolated into the domain of distances of the order of 10^{-13} cm or less. It is therefore entirely possible that the interpretation suggested here may be needed for the resolution of these difficulties. In any case, the mere possibility of such an interpretation proves that it is not necessary for us to give up a precise, rational, and objective description of individual systems at a quantum level of accuracy.

1. INTRODUCTION

THE usual interpretation of the quantum theory is based on an assumption having very far-reaching implications, *viz.*, that the physical state of an individual system is completely specified by a wave function that determines only the probabilities of actual results that can be obtained in a statistical ensemble of similar experiments. This assumption has been the object of severe criticisms, notably on the part of Einstein, who has always believed that, even at the quantum level, there must exist precisely definable elements or dynamical variables determining (as in classical physics) the actual behavior of each individual system, and not merely its probable behavior. Since these elements or variables are not now included in the quantum theory and have not yet been detected experimentally, Einstein has always regarded the present form of the quantum theory as incomplete, although he admits its internal consistency.[1–5]

Most physicists have felt that objections such as those raised by Einstein are not relevant, first, because the present form of the quantum theory with its usual probability interpretation is in excellent agreement with an extremely wide range of experiments, at least in the domain of distances[6] larger than 10^{-13} cm, and, secondly, because no consistent alternative interpretations have as yet been suggested. The purpose of this paper (and of a subsequent paper hereafter denoted by II) is, however, to suggest just such an alternative interpretation. In contrast to the usual interpretation, this alternative interpretation permits us to conceive of each individual system as being in a precisely definable state, whose changes with time are determined by definite laws, analogous to (but not identical with) the classical equations of motion. Quantum-mechanical probabilities are regarded (like their counterparts in classical statistical mechanics) as only a practical necessity and not as a manifestation of an inherent lack of complete determination in the properties of matter at the quantum level. As long as the present general form of Schroedinger's equation is retained, the physical results obtained with our suggested alternative interpretation are precisely the same as those obtained with the usual interpretation. We shall see, however, that our alternative interpretation permits modifications of the mathematical formulation which could not even be described in terms of the usual interpretation. Moreover, the modifications can quite easily be formulated in such a way that their effects are insignificant in the atomic domain, where the present quantum theory is in such good agreement with experiment, but of crucial importance in the domain of dimensions of the order of 10^{-13} cm, where, as we have seen, the present theory is totally inadequate. It is thus entirely possible that some of the modifications describable in terms of our suggested alternative interpretation, but

* Now at Universidade de São Paulo, Faculdade de Filosofia, Ciencias, e Letras, São Paulo, Brasil.

[1] Einstein, Podolsky, and Rosen, Phys. Rev. **47**, 777 (1933).

[2] D. Bohm, *Quantum Theory* (Prentice-Hall, Inc., New York, 1951), see p. 611.

[3] N. Bohr, Phys. Rev. **48**, 696 (1935).

[4] W. Furry, Phys. Rev. **49**, 393, 476 (1936).

[5] Paul Arthur Schilp, editor, *Albert Einstein, Philosopher-Scientist* (Library of Living Philosophers, Evanston, Illinois, 1949). This book contains a thorough summary of the entire controversy.

[6] At distances of the order of 10^{-13} cm or smaller and for times of the order of this distance divided by the velocity of light or smaller, present theories become so inadequate that it is generally believed that they are probably not applicable, except perhaps in a very crude sense. Thus, it is generally expected that in connection with phenomena associated with this so-called "fundamental length," a totally new theory will probably be needed. It is hoped that this theory could not only deal precisely with such processes as meson production and scattering of elementary particles, but that it would also systematically predict the masses, charges, spins, etc., of the large number of so-called "elementary" particles that have already been found, as well as those of new particles which might be found in the future.

Originally published in *Physical Review*, *85*, 166-93 (1952).

not in terms of the usual interpretation, may be needed for a more thorough understanding of phenomena associated with very small distances. We shall not, however, actually develop such modifications in any detail in these papers.

After this article was completed, the author's attention was called to similar proposals for an alternative interpretation of the quantum theory made by de Broglie[7] in 1926, but later given up by him partly as a result of certain criticisms made by Pauli[8] and partly because of additional objections raised by de Broglie[7] himself.† As we shall show in Appendix B of Paper II, however, all of the objections of de Broglie and Pauli could have been met if only de Broglie had carried his ideas to their logical conclusion. The essential new step in doing this is to apply our interpretation in the theory of the measurement process itself as well as in the description of the observed system. Such a development of the theory of measurements is given in Paper II,[9] where it will be shown in detail that our interpretation leads to precisely the same results for all experiments as are obtained with the usual interpretation. The foundation for doing this is laid in Paper I, where we develop the basis of our interpretation, contrast it with the usual interpretation, and apply it to a few simple examples, in order to illustrate the principles involved.

2. THE USUAL PHYSICAL INTERPRETATION OF THE QUANTUM THEORY

The usual physical interpretation of the quantum theory centers around the uncertainty principle. Now, the uncertainty principle can be derived in two different ways. First, we may start with the assumption already criticized by Einstein,[1] namely, that a wave function that determines only probabilities of actual experimental results nevertheless provides the most complete possible specification of the so-called "quantum state" of an individual system. With the aid of this assumption and with the aid of the de Broglie relation, $\mathbf{p}=\hbar\mathbf{k}$, where $\mathbf{k}$ is the wave number associated with a particular fourier component of the wave function, the uncertainty principle is readily deduced.[10] From this derivation, we are led to interpret the uncertainty principle as an inherent and irreducible limitation on the precision with which it is correct for us even to conceive of momentum and position as simultaneously defined quantities. For if, as is done in the usual interpretation of the quantum theory, the wave intensity is assumed to determine only the probability of a given position, and if the $\mathbf{k}$th Fourier component of the wave function is assumed to determine only the probability of a corresponding momentum, $\mathbf{p}=\hbar\mathbf{k}$, then it becomes a contradiction in terms to ask for a state in which momentum and position are simultaneously and precisely defined.

A second possible derivation of the uncertainty principle is based on a theoretical analysis of the processes with the aid of which physically significant quantities such as momentum and position can be measured. In such an analysis, one finds that because the measuring apparatus interacts with the observed system by means of indivisible quanta, there will always be an irreducible disturbance of some observed property of the system. If the precise effects of this disturbance could be predicted or controlled, then one could correct for these effects, and thus one could still in principle obtain simultaneous measurements of momentum and position, having unlimited precision. But if one could do this, then the uncertainty principle would be violated. The uncertainty principle is, as we have seen, however, a necessary consequence of the assumption that the wave function and its probability interpretation provide the most complete possible specification of the state of an individual system. In order to avoid the possibility of a contradiction with this assumption, Bohr[3,5,10,11] and others have suggested an additional assumption, namely, that the process of transfer of a single quantum from observed system to measuring apparatus is inherently unpredictable, uncontrollable, and not subject to a detailed rational analysis or description. With the aid of this assumption, one can show[10] that the same uncertainty principle that is deduced from the wave function and its probability interpretation is also obtained as an inherent and unavoidable limitation on the precision of all possible measurements. Thus, one is able to obtain a set of assumptions, which permit a self-consistent formulation of the usual interpretation of the quantum theory.

The above point of view has been given its most consistent and systematic expression by Bohr,[3,5,10] in terms of the "principle of complementarity." In formulating this principle, Bohr suggests that at the atomic level we must renounce our hitherto successful practice of conceiving of an individual system as a unified and precisely definable whole, all of whose aspects are, in a manner of speaking, simultaneously and

---

[7] L. de Broglie, *An Introduction to the Study of Wave Mechanics* (E. P. Dutton and Company, Inc., New York, 1930), see Chapters 6, 9, and 10. See also Compt. rend. **183**, 447 (1926); **184**, 273 (1927); **185**, 380 (1927).

[8] *Reports on the Solvay Congress* (Gauthiers-Villars et Cie., Paris, 1928), see p. 280.

† *Note added in proof.*—Madelung has also proposed a similar interpretation of the quantum theory, but like de Broglie he did not carry this interpretation to a logical conclusion. See E. Madelung, Z. f. Physik **40**, 332 (1926), also G. Temple, *Introduction to Quantum Theory* (London, 1931).

[9] In Paper II, Sec. 9, we also discuss von Neumann's proof [see J. von Neumann, *Mathematische Grundlagen der Quantenmechanik* (Verlag, Julius Springer, Berlin, 1932)] that quantum theory cannot be understood in terms of a statistical distribution of "hidden" causal parameters. We shall show that his conclusions do not apply to our interpretation, because he implicitly assumes that the hidden parameters must be associated only with the observed system, whereas, as will become evident in these papers, our interpretation requires that the hidden parameters shall also be associated with the measuring apparatus.

[10] See reference 2, Chapter 5.

[11] N. Bohr, *Atomic Theory and the Description of Nature* (Cambridge University Press, London, 1934).

unambiguously accessible to our conceptual gaze. Such a system of concepts, which is sometimes called a "model," need not be restricted to pictures, but may also include, for example, mathematical concepts, as long as these are supposed to be in a precise (i.e., one-to-one) correspondence with the objects that are being described. The principle of complementarity requires us, however, to renounce even mathematical models. Thus, in Bohr's point of view, the wave function is in no sense a conceptual model of an individual system, since it is not in a precise (one-to-one) correspondence with the behavior of this system, but only in a statistical correspondence.

In place of a precisely defined conceptual model, the principle of complementarity states that we are restricted to complementarity pairs of inherently imprecisely defined concepts, such as position and momentum, particle and wave, etc. The maximum degree of precision of definition of either member of such a pair is reciprocally related to that of the opposite member. This need for an inherent lack of complete precision can be understood in two ways. First, it can be regarded as a consequence of the fact that the experimental apparatus needed for a precise measurement of one member of a complementary pair of variables must always be such as to preclude the possibility of a simultaneous and precise measurement of the other member. Secondly, the assumption that an individual system is completely specified by the wave function and its probability interpretation implies a corresponding unavoidable lack of precision in the very conceptual structure, with the aid of which we can think about and describe the behavior of the system.

It is only at the classical level that we can correctly neglect the inherent lack of precision in all of our conceptual models; for here, the incomplete determination of physical properties implied by the uncertainty principle produces effects that are too small to be of practical significance. Our ability to describe classical systems in terms of precisely definable models is, however, an integral part of the usual interpretation of the theory. For without such models, we would have no way to describe, or even to think of, the result of an observation, which is of course always finally carried out at a classical level of accuracy. If the relationships of a given set of classically describable phenomena depend significantly on the essentially quantum-mechanical properties of matter, however, then the principle of complementarity states that no single model is possible which could provide a precise and rational analysis of the connections between these phenomena. In such a case, we are not supposed, for example, to attempt to describe in detail how future phenomena arise out of past phenomena. Instead, we should simply accept without further analysis the fact that future phenomena do in fact somehow manage to be produced, in a way that is, however, necessarily beyond the possibility of a detailed description. The only aim of a mathematical theory is then to predict the statistical relations, if any, connecting these phenomena.

3. CRITICISM OF THE USUAL INTERPRETATION OF THE QUANTUM THEORY

The usual interpretation of the quantum theory can be criticized on many grounds.[5] In this paper, however, we shall stress only the fact that it requires us to give up the possibility of even conceiving precisely what might determine the behavior of an individual system at the quantum level, without providing adequate proof that such a renunciation is necessary.[9] The usual interpretation is admittedly consistent; but the mere demonstration of such consistency does not exclude the possibility of other equally consistent interpretations, which would involve additional elements or parameters permitting a detailed causal and continuous description of all processes, and not requiring us to forego the possibility of conceiving the quantum level in precise terms. From the point of view of the usual interpretation, these additional elements or parameters could be called "hidden" variables. As a matter of fact, whenever we have previously had recourse to statistical theories, we have always ultimately found that the laws governing the individual members of a statistical ensemble could be expressed in terms of just such hidden variables. For example, from the point of view of macroscopic physics, the coordinates and momenta of individual atoms are hidden variables, which in a large scale system manifest themselves only as statistical averages. Perhaps then, our present quantum-mechanical averages are similarly a manifestation of hidden variables, which have not, however, yet been detected directly.

Now it may be asked why these hidden variables should have so long remained undetected. To answer this question, it is helpful to consider as an analogy the early forms of the atomic theory, in which the existence of atoms was postulated in order to explain certain large-scale effects, such as the laws of chemical combination, the gas laws, etc. On the other hand, these same effects could also be described directly in terms of existing macrophysical concepts (such as pressure, volume, temperature, mass, etc.); and a correct description in these terms did not require any reference to atoms. Ultimately, however, effects were found which contradicted the predictions obtained by extrapolating certain purely macrophysical theories to the domain of the very small, and which could be understood correctly in terms of the assumption that matter is composed of atoms. Similarly, we suggest that if there are hidden variables underlying the present quantum theory, it is quite likely that in the atomic domain, they will lead to effects that can also be described adequately in the terms of the usual quantum-mechanical concepts; while in a domain associated with much smaller dimensions, such as the level associated with the "fundamental length" of the order of 10^{-13} cm, the hidden variables may lead to completely new effects not consistent with the extrapolation of the present quantum theory down to this level.

If, as is certainly entirely possible, these hidden variables are actually needed for a correct description at small distances, we could easily be kept on the wrong track for a long time by restricting ourselves to the usual interpretation of the quantum theory, which excludes such hidden variables as a matter of principle. It is therefore very important for us to investigate our reasons for supposing that the usual physical interpretation is likely to be the correct one. To this end, we shall begin by repeating the two mutually consistent assumptions on which the usual interpretation is based (see Sec. 2):

(1) The wave function with its probability interpretation determines the most complete possible specification of the state of an individual system.

(2) The process of transfer of a single quantum from observed system to measuring apparatus is inherently unpredictable, uncontrollable, and unanalyzable.

Let us now inquire into the question of whether there are any experiments that could conceivably provide a test for these assumptions. It is often stated in connection with this problem that the mathematical apparatus of the quantum theory and its physical interpretation form a consistent whole and that this combined system of mathematical apparatus and physical interpretation is tested adequately by the extremely wide range of experiments that are in agreement with predictions obtained by using this system. If assumptions (1) and (2) implied a unique mathematical formulation, then such a conclusion would be valid, because experimental predictions could then be found which, if contradicted, would clearly indicate that these assumptions were wrong. Although assumptions (1) and (2) do limit the possible forms of the mathematical theory, they do not limit these forms sufficiently to make possible a unique set of predictions that could in principle permit such an experimental test. Thus, one can contemplate practically arbitrary changes in the Hamiltonian operator, including, for example, the postulation of an unlimited range of new kinds of meson fields each having almost any conceivable rest mass, charge, spin, magnetic moment, etc. And if such postulates should prove to be inadequate, it is conceivable that we may have to introduce nonlocal operators, nonlinear fields, S-matrices, etc. This means that when the theory is found to be inadequate (as now happens, for example, at distances of the order of 10^{-13} cm), it is always possible, and, in fact, usually quite natural, to assume that the theory can be made to agree with experiment by some as yet unknown change in the mathematical formulation alone, not requiring any fundamental changes in the physical interpretation. This means that as long as we accept the usual physical interpretation of the quantum theory, we cannot be led by any conceivable experiment to give up this interpretation, even if it should happen to be wrong. The usual physical interpretation therefore presents us with a considerable danger of falling into a trap, consisting of a self-closing chain of circular hypotheses, which are in principle unverifiable if true. The only way of avoiding the possibility of such a trap is to study the consequences of postulates that contradict assumptions (1) and (2) at the outset. Thus, we could, for example, postulate that the precise outcome of each individual measurement process is in principle determined by some at present "hidden" elements or variables; and we could then try to find experiments that depended in a unique and reproducible way on the assumed state of these hidden elements or variables. If such predictions are verified, we should then obtain experimental evidence favoring the hypothesis that hidden variables exist. If they are not verified, however, the correctness of the usual interpretation of the quantum theory is not necessarily proved, since it may be necessary instead to alter the specific character of the theory that is supposed to describe the behavior of the assumed hidden variables.

We conclude then that a choice of the present interpretation of the quantum theory involves a real physical limitation on the kinds of theories that we wish to take into consideration. From the arguments given here, however, it would seem that there are no secure experimental or theoretical grounds on which we can base such a choice because this choice follows from hypotheses that cannot conceivably be subjected to an experimental test and because we now have an alternative interpretation.

4. NEW PHYSICAL INTERPRETATION OF SCHROEDINGER'S EQUATION

We shall now give a general description of our suggested physical interpretation of the present mathematical formulation of the quantum theory. We shall carry out a more detailed description in subsequent sections of this paper.

We begin with the one-particle Schroedinger equation, and shall later generalize to an arbitrary number of particles. This wave equation is

$$i\hbar\partial\psi/\partial t=-(\hbar^2/2m)\nabla^2\psi+V(\mathbf{x})\psi. \tag{1}$$

Now ψ is a complex function, which can be expressed as

$$\psi=R\exp(iS/\hbar), \tag{2}$$

where R and S are real. We readily verify that the equations for R and S are

$$\frac{\partial R}{\partial t}=-\frac{1}{2m}[R\nabla^2 S+2\nabla R\cdot\nabla S], \tag{3}$$

$$\frac{\partial S}{\partial t}=-\left[\frac{(\nabla S)^2}{2m}+V(\mathbf{x})-\frac{\hbar^2}{2m}\frac{\nabla^2 R}{R}\right]. \tag{4}$$

It is convenient to write $P(\mathbf{x})=R^2(\mathbf{x})$, or $R=P^{\frac{1}{2}}$ where $P(\mathbf{x})$ is the probability density. We then obtain

$$\frac{\partial P}{\partial t}+\nabla\cdot\left(P\frac{\nabla S}{m}\right)=0, \tag{5}$$

$$\frac{\partial S}{\partial t}+\frac{(\nabla S)^2}{2m}+V(\mathbf{x})-\frac{\hbar^2}{4m}\left[\frac{\nabla^2 P}{P}-\frac{1}{2}\frac{(\nabla P)^2}{P^2}\right]=0. \tag{6}$$

Now, in the classical limit ($\hbar\rightarrow 0$) the above equations are subject to a very simple interpretation. The function $S(\mathbf{x})$ is a solution of the Hamilton-Jacobi equation. If we consider an ensemble of particle trajectories which are solutions of the equations of motion, then it is a well-known theorem of mechanics that if all of these trajectories are normal to any given surface of constant S, then they are normal to all surfaces of constant S, and $\nabla S(\mathbf{x})/m$ will be equal to the velocity vector, $\mathbf{v}(\mathbf{x})$, for any particle passing the point $\mathbf{x}$. Equation (5) can therefore be re-expressed as

$$\partial P/\partial t+\nabla\cdot(P\mathbf{v})=0. \tag{7}$$

This equation indicates that it is consistent to regard $P(\mathbf{x})$ as the probability density for particles in our ensemble. For in that case, we can regard $P\mathbf{v}$ as the mean current of particles in this ensemble, and Eq. (7) then simply expresses the conservation of probability.

Let us now see to what extent this interpretation can be given a meaning even when $\hbar\neq 0$. To do this, let us assume that each particle is acted on, not only by a "classical" potential, $V(\mathbf{x})$ but also by a "quantum-mechanical" potential,

$$U(\mathbf{x})=\frac{-\hbar^2}{4m}\left[\frac{\nabla^2 P}{P}-\frac{1}{2}\frac{(\nabla P)^2}{P^2}\right]=\frac{-\hbar^2}{2m}\frac{\nabla^2 R}{R}. \tag{8}$$

Then Eq. (6) can still be regarded as the Hamilton-Jacobi equation for our ensemble of particles, $\nabla S(\mathbf{x})/m$ can still be regarded as the particle velocity, and Eq. (5) can still be regarded as describing conservation of probability in our ensemble. Thus, it would seem that we have here the nucleus of an alternative interpretation for Schroedinger's equation.

The first step in developing this interpretation in a more explicit way is to associate with each electron a particle having precisely definable and continuously varying values of position and momentum. The solution of the modified Hamilton-Jacobi equation (4) defines an ensemble of possible trajectories for this particle, which can be obtained from the Hamilton-Jacobi function, $S(\mathbf{x})$, by integrating the velocity, $\mathbf{v}(\mathbf{x})=\nabla S(\mathbf{x})/m$. The equation for S implies, however, that the particles moves under the action of a force which is not entirely derivable from the classical potential, $V(\mathbf{x})$, but which also obtains a contribution from the "quantum-mechanical" potential, $U(\mathbf{x})=(-\hbar^2/2m)\times\nabla^2 R/R$. The function, $R(\mathbf{x})$, is not completely arbitrary, but is partially determined in terms of $S(\mathbf{x})$ by the differential Eq. (3). Thus R and S can be said to codetermine each other. The most convenient way of obtaining R and S is, in fact, usually to solve Eq. (1) for the Schroedinger wave function, ψ, and then to use the relations,

$$\psi=U+iW=R[\cos(S/\hbar)+i\sin(S/\hbar)],$$
$$R^2=U^2+V^2;\quad S=\hbar\tan^{-1}(W/U).$$

Since the force on a particle now depends on a function of the absolute value, $R(\mathbf{x})$, of the wave function, $\psi(\mathbf{x})$, evaluated at the actual location of the particle, we have effectively been led to regard the wave function of an individual electron as a mathematical representation of an objectively real field. This field exerts a force on the particle in a way that is analogous to, but not identical with, the way in which an electromagnetic field exerts a force on a charge, and a meson field exerts a force on a nucleon. In the last analysis, there is, of course, no reason why a particle should not be acted on by a ψ-field, as well as by an electromagnetic field, a gravitational field, a set of meson fields, and perhaps by still other fields that have not yet been discovered.

The analogy with the electromagnetic (and other) field goes quite far. For just as the electromagnetic field obeys Maxwell's equations, the ψ-field obeys Schroedinger's equation. In both cases, a complete specification of the fields at a given instant over every point in space determines the values of the fields for all times. In both cases, once we know the field functions, we can calculate force on a particle, so that, if we also know the initial position and momentum of the particle, we can calculate its entire trajectory.

In this connection, it is worth while to recall that the use of the Hamilton-Jacobi equation in solving for the motion of a particle is only a matter of convenience and that, in principle, we can always solve directly by using Newton's laws of motion and the correct boundary conditions. The equation of motion of a particle acted on by the classical potential, $V(\mathbf{x})$, and the "quantum-mechanical" potential, Eq. (8), is

$$md^2\mathbf{x}/dt^2=-\nabla\{V(\mathbf{x})-(\hbar^2/2m)\nabla^2 R/R\}. \tag{8a}$$

It is in connection with the boundary conditions appearing in the equations of motion that we find the only fundamental difference between the ψ-field and other fields, such as the electromagnetic field. For in order to obtain results that are equivalent to those of the usual interpretation of the quantum theory, we are required to restrict the value of the initial particle momentum to $\mathbf{p}=\nabla S(\mathbf{x})$. From the application of Hamilton-Jacobi theory to Eq. (6), it follows that this restriction is consistent, in the sense that if it holds initially, it will hold for all time. Our suggested new interpretation of the quantum theory implies, however, that this restriction is not inherent in the conceptual structure. We shall see in Sec. 9, for example, that it is quite consistent in our interpretation to contemplate modifications in the theory, which permit an arbitrary relation between $\mathbf{p}$ and $\nabla S(\mathbf{x})$. The law of force on the particle can, however, be so chosen that in the atomic domain, $\mathbf{p}$ turns out to be very nearly equal to $\nabla S(\mathbf{x})/m$, while in processes involving very small distances, these two quantities may be very different. In this way, we can improve the analogy between the ψ-field and the electromagnetic field (as well as between quantum mechanics and classical mechanics).

Another important difference between the ψ-field and the electromagnetic field is that, whereas Schroedinger's equation is homogeneous in ψ, Maxwell's equations are inhomogeneous in the electric and magnetic fields. Since inhomogeneities are needed to give rise to radiation, this means that our present equations imply that the ψ-field is not radiated or absorbed, but simply changes its form while its integrated intensity remains constant. This restriction to a homogeneous equation is, however, like the restriction to a homogeneous equation is, however, like the restriction to $\mathbf{p}=\nabla S(\mathbf{x})$, not inherent in the conceptual structure of our new interpretation. Thus, in Sec. 9, we shall show that one can consistently postulate inhomogeneities in the equation governing ψ, which produce important effects only at very small distances, and negligible effects in the atomic domain. If such inhomogeneities are actually present, then the ψ-field will be subject to being emitted and absorbed, but only in connection with processes associated with very small distances. Once the ψ-field has been emitted, however, it will in all atomic processes simply obey Schroedinger's equation as a very good approximation. Nevertheless, at very small distances, the value of the ψ-field would, as in the case of the electromagnetic field, depend to some extent on the actual location of the particle.

Let us now consider the meaning of the assumption of a statistical ensemble of particles with a probability density equal to $P(\mathbf{x})=R^2(\mathbf{x})=|\psi(\mathbf{x})|^2$. From Eq. (5), it follows that this assumption is consistent, provided that ψ satisfies Schroedinger's equation, and $\mathbf{v}=\nabla S(\mathbf{x})/m$. This probability density is numerically equal to the probability density of particles obtained in the usual interpretation. In the usual interpretation, however, the need for a probability description is regarded as inherent in the very structure of matter (see Sec. 2), whereas in our interpretation, it arises, as we shall see in Paper II, because from one measurement to the next, we cannot in practice predict or control the precise location of a particle, as a result of corresponding unpredictable and uncontrollable disturbances introduced by the measuring apparatus. Thus, in our interpretation, the use of a statistical ensemble is (as in the case of classical statistical mechanics) only a practical necessity, and not a reflection of an inherent limitation on the precision with which it is correct for us to conceive of the variables defining the state of the system. Moreover, it is clear that if in connection with very small distances we are ultimately required to give up the special assumptions that ψ satisfies Schroedinger's equation and that $\mathbf{v}=\nabla S(\mathbf{x})/m$, then $|\psi|^2$ will cease to satisfy a conservation equation and will therefore also cease to be able to represent the probability density of particles. Nevertheless, there would still be a true probability density of particles which is conserved. Thus, it would become possible in principle to find experiments in which $|\psi|^2$ could be distinguished from the probability density, and therefore to prove that the usual interpretation, which gives $|\psi|^2$ only a probability interpretation must be inadequate. Moreover, we shall see in Paper II that with the aid of such modifications in the theory, we could in principle measure the particle positions and momenta precisely, and thus violate the uncertainty principle. As long as we restrict ourselves to conditions in which Schroedinger's equation is satisfied, and in which $\mathbf{v}=\nabla S(\mathbf{x})/m$, however, the uncertainty principle will remain an effective practical limitation on the possible precision of measurements. This means that at present, the particle positions and momenta should be regarded as "hidden" variables, since as we shall see in Paper II, we are not now able to obtain experiments that localize them to a region smaller than that in which the intensity of the ψ-field is appreciable. Thus, we cannot yet find clear-cut experimental proof that the assumption of these variables is necessary, although it is entirely possible that, in the domain of very small distances, new modifications in the theory may have to be introduced, which would permit a proof of the existence of the definite particle position and momentum to be obtained.

We conclude that our suggested interpretation of the quantum theory provides a much broader conceptual framework than that provided by the usual interpretation, for all of the results of the usual interpretation are obtained from our interpretation if we make the following three special assumptions which are mutually consistent:

(1) That the ψ-field satisfies Schroedinger's equation.

(2) That the particle momentum is restricted to $\mathbf{p}=\nabla S(\mathbf{x})$.

(3) That we do not predict or control the precise location of the particle, but have, in practice, a statistical ensemble with probability density $P(\mathbf{x})=|\psi(\mathbf{x})|^2$. The use of statistics is, however, not inherent in the conceptual structure, but merely a consequence of our ignorance of the precise initial conditions of the particle.

As we shall see in Sec. 9, it is entirely possible that a better theory of phenomena involving distances of the order of 10^{-13} cm or less would require us to go beyond the limitations of these special assumptions. Our principal purpose in this paper (and in Paper II) is to show, however, that if one makes these special assumptions, our interpretation leads in all possible experiments to the same predictions as are obtained from the usual interpretation.[9]

It is now easy to understand why the adoption of the

usual interpretation of the quantum theory would tend to lead us away from the direction of our suggested alternative interpretation. For in a theory involving hidden variables, one would normally expect that the behavior of an individual system should not depend on the statistical ensemble of which it is a member, because this ensemble refers to a series of similar but disconnected experiments carried out under equivalent initial conditions. In our interpretation, however, the "quantum-mechanical" potential, $U(\mathbf{x})$, acting on an individual particle depends on a wave intensity, $P(\mathbf{x})$, that is also numerically equal to a probability density in our ensemble. In the terminology of the usual interpretation of the quantum theory, in which one tacitly assumes that the wave function has only one interpretation; namely, in terms of a probability, our suggested new interpretation would look like a mysterious dependence of the individual on the statistical ensemble of which it is a member. In our interpretation, such a dependence is perfectly rational, because the wave function can consistently be interpreted both as a force and as a probability density.[12]

It is instructive to carry our analogy between the Schroedinger field and other kinds of fields a bit further. To do this, we can derive the wave Eqs. (5) and (6) from a Hamiltonian functional. We begin by writing down the expression for the mean energy as it is expressed in the usual quantum theory:

$$\bar{H}=\int\psi^*\left(-\frac{\hbar^2}{2m}\nabla^2+V(\mathbf{x})\right)\psi d\mathbf{x}$$

$$=\int\left\{\frac{\hbar^2}{2m}|\nabla\psi|^2+V(\mathbf{x})|\psi|^2\right\}d\mathbf{x}.$$

Writing $\psi=P^{\frac{1}{2}}\exp(iS/\hbar)$, we obtain

$$\bar{H}=\int P(\mathbf{x})\left\{\frac{(\nabla S)^2}{2m}+V(\mathbf{x})+\frac{\hbar^2}{8m}\frac{(\nabla P)^2}{P^2}\right\}d\mathbf{x}. \quad (9)$$

We shall now reinterpret $P(\mathbf{x})$ as a field coordinate, defined at each point, $\mathbf{x}$, and we shall tentatively assume that $S(\mathbf{x})$ is the momentum, canonically conjugate to $P(\mathbf{x})$. That such an assumption is appropriate can be verified by finding the Hamiltonian equations of motion for $P(\mathbf{x})$ and $S(\mathbf{x})$, under the assumption that the Hamiltonian functional is equal to $\bar{H}$ (See Eq. (9)). These equations of motion are

$$\dot{P}=\frac{\delta\bar{H}}{\delta S}=-\frac{1}{m}\nabla\cdot(P\nabla S),$$

$$\dot{S}=-\frac{\delta\bar{H}}{\delta P}=-\left[\frac{(\nabla S)^2}{2m}+V(\mathbf{x})-\frac{\hbar^2}{4m}\left(\frac{\nabla^2P}{P}-\frac{1}{2}\frac{(\nabla P)^2}{P^2}\right)\right].$$

These are, however, the same as the correct wave Eqs. (5) and (6).

We can now show that the mean particle energy averaged over our ensemble is equal to the usual quantum mechanical mean value of the Hamiltonian, $\bar{H}$. To do this, we note that according to Eqs. (3) and (6), the energy of a particle is

$$E(\mathbf{x})=-\frac{\partial S(\mathbf{x})}{\partial t}=\left[\frac{(\nabla S)^2}{2m}+V(\mathbf{x})-\frac{\hbar^2}{2m}\frac{\nabla^2R}{R}\right]. \quad (10)$$

The mean particle energy is found by averaging $E(\mathbf{x})$ with the weighting function, $P(\mathbf{x})$. We obtain

$$\langle E\rangle_{\substack{\text{ensemble}\\\text{average}}}=\int P(\mathbf{x})E(\mathbf{x})d\mathbf{x}$$

$$=\int P(\mathbf{x})\left[\frac{(\nabla S)^2}{2m}+V(\mathbf{x})\right]d\mathbf{x}-\frac{\hbar^2}{2m}\int R\nabla^2R d\mathbf{x}.$$

A little integration by parts yields

$$\langle E\rangle_{\substack{\text{ensemble}\\\text{average}}}=\int P(\mathbf{x})\left[\frac{(\nabla S)^2}{2m}+V(\mathbf{x})+\frac{\hbar^2}{8m}\frac{(\nabla P)^2}{P^2}\right]d\mathbf{x}=\bar{H}. \quad (11)$$

5. THE STATIONARY STATE

We shall now show how the problem of stationary states is to be treated in our interpretation of the quantum theory.

The following seem to be reasonable requirements in our interpretation for a stationary state:

(1) The particle energy should be a constant of the motion.

(2) The quantum-mechanical potential should be independent of time.

(3) The probability density in our statistical ensemble should be independent of time.

It is easily verified that these requirements can be satisfied with the assumption that

$$\psi(\mathbf{x},t)=\psi_0(\mathbf{x})\exp(-iEt/\hbar)$$
$$=R_0(\mathbf{x})\exp[i(\Phi(\mathbf{x})-Et)/\hbar]. \quad (12)$$

From the above, we obtain $S=\Phi(\mathbf{x})-Et$. According to the generalized Hamilton-Jacobi Eq. (4), the particle energy is given by

$$\partial S/\partial t=-E.$$

Thus, we verify that the particle energy is a constant of the motion. Moreover, since $P=R^2=|\psi|^2$, it follows that P (and R) are independent of time. This means that both the probability density in our ensemble and the quantum-mechanical potential are also time independent.

[12] This consistency is guaranteed by the conservation Eq. (7). The questions of why an arbitrary statistical ensemble tends to decay into an ensemble with a probability density equal to $\psi^*\psi$ will be discussed in Paper II, Sec. 7.

The reader will readily verify that no other form of solution of Schroedinger's equation will satisfy all three of our criteria for a stationary state.

Since ψ is now being regarded as a mathematical representation of an objectively real force field, it follows that (like the electromagnetic field) it should be everywhere finite, continuous, and single valued. These requirements will guarantee in all cases that occur in practice that the allowed values of the energy in a stationary state, and the corresponding eigenfunctions are the same as are obtained from the usual interpretation of the theory.

In order to show in more detail what a stationary state means in our interpretation, we shall now consider three examples of stationary states.

Case 1: "s" State

The first case that we shall consider is an "s" state. In an "s" state, the wave function is

$$\psi = f(r)\exp[i(\alpha - Et)/\hbar], \tag{13}$$

where α is an arbitrary constant and r is the radius taken from the center of the atom. We conclude that the Hamilton-Jacobi function is

$$S = \alpha - Et.$$

The particle velocity is

$$\mathbf{v} = \nabla S = 0.$$

The particle is therefore simply standing still, wherever it may happen to be. How can it do this? The absence of motion is possible because the applied force, $-\nabla V(\mathbf{x})$, is balanced by the "quantum-mechanical" force, $(\hbar^2/2m)\nabla(\nabla^2 R/R)$, produced by the Schroedinger ψ-field acting on its own particle. There is, however, a statistical ensemble of possible positions of the particle, with a probability density, $P(\mathbf{x}) = (f(r))^2$.

Case 2: State with Nonzero Angular Momentum

In a typical state of nonzero angular momentum, we have

$$\psi = f_n^{\,l}(r)P_l^{\,m}(\cos\theta)\exp[i(\beta - Et + \hbar m\phi)/\hbar], \tag{14}$$

where θ and ϕ are the colatitude and azimuthal polar angles, respectively, $P_l^{\,m}$ is the associated Legendre polynomial, and β is a constant. The Hamilton-Jacobi function is $S = \beta - Et + \hbar m\phi$. From this result it follows that the z component of the angular momentum is equal to $\hbar m$. To prove this, we write

$$L_z = xp_y - yp_x = x\partial S/\partial y - y\partial S/\partial x = \partial S/\partial\phi = \hbar m. \tag{15}$$

Thus, we obtain a statistical ensemble of trajectories which can have different forms, but all have the same "quantized" value of the z component of the angular momentum.

Case 3: A Scattering Problem

Let us now consider a scattering problem. Because it is comparatively easy to analyze, we shall discuss a hypothetical experiment, in which an electron is incident in the z direction with an initial momentum, p_0, on a system consisting of two slits.[13] After the electron passes through the slit system, its position is measured and recorded, for example, on a photographic plate.

Now, in the usual interpretation of the quantum theory, the electron is described by a wave function. The incident part of the wave function is $\psi_0 \sim \exp(ip_0 z/\hbar)$; but when the wave passes through the slit system, it is modified by interference and diffraction effects, so that it will develop a characteristic intensity pattern by the time it reaches the position measuring instrument. The probability that the electron will be detected between $\mathbf{x}$ and $\mathbf{x}+d\mathbf{x}$ is $|\psi(\mathbf{x})|^2 d\mathbf{x}$. If the experiment is repeated many times under equivalent initial conditions, one eventually obtains a pattern of hits on the photographic plate that is very reminiscent of the interference patterns of optics.

In the usual interpretation of the quantum theory, the origin of this interference pattern is very difficult to understand. For there may be certain points where the wave function is zero when both slits are open, but not zero when only one slit is open. How can the opening of a second slit prevent the electron from reaching certain points that it could reach if this slit were closed? If the electron acted completely like a classical particle, this phenomenon could not be explained at all. Clearly, then the wave aspects of the electron must have something to do with the production of the interference pattern. Yet, the electron cannot be identical with its associated wave, because the latter spreads out over a wide region. On the other hand, when the electron's position is measured, it always appears at the detector as if it were a localized particle.

The usual interpretation of the quantum theory not only makes no attempt to provide a single precisely defined conceptual model for the production of the phenomena described above, but it asserts that no such model is even conceivable. Instead of a single precisely defined conceptual model, it provides, as pointed out in Sec. 2, a pair of complementary models, *viz.*, particle and wave, each of which can be made more precise only under conditions which necessitate a reciprocal decrease in the degree of precision of the other. Thus, while the electron goes through the slit system, its position is said to be inherently ambiguous, so that if we wish to obtain an interference pattern, it is meaningless to ask through which slit an individual electron actually passed. Within the domain of space within which the position of the electron has no meaning we can use the wave model and thus describe the subsequent production of interference. If, however, we

[13] This experiment is discussed in some detail in reference 2, Chapter 6, Sec. 2.

tried to define the position of the electron as it passed the slit system more accurately by means of a measurement, the resulting disturbance of its motion produced by the measuring apparatus would destroy the interference pattern. Thus, conditions would be created in which the particle model becomes more precisely defined at the expense of a corresponding decrease in the degree of definition of the wave model. When the position of the electron is measured at the photographic plate, a similar sharpening of the degree of definition of the particle model occurs at the expense of that of the wave model.

In our interpretation of the quantum theory, this experiment is described causally and continuously in terms of a single precisely definable conceptual model. As we have already shown, we must use the same wave function as is used in the usual interpretation; but instead we regard it as a mathematical representation of an objectively real field that determines part of the force acting on the particle. The initial momentum of the particle is obtained from the incident wave function, $\exp(ip_0z/\hbar)$, as $p=\partial s/\partial z=p_0$. We do not in practice, however, control the initial location of the particle, so that although it goes through a definite slit, we cannot predict which slit this will be. The particle is at all times acted on by the "quantum-mechanical" potential, $U=(-\hbar^2/2m)\nabla^2R/R$. While the particle is incident, this potential vanishes because R is then a constant; but after it passes through the slit system, the particle encounters a quantum-mechanical potential that changes rapidly with position. The subsequent motion of the particle may therefore become quite complicated. Nevertheless, the probability that a particle shall enter a given region, $d\mathbf{x}$, is as in the usual interpretation, equal to $|\psi(\mathbf{x})|^2d\mathbf{x}$. We therefore deduce that the particle can never reach a point where the wave function vanishes. The reason is that the "quantum-mechanical" potential, U, becomes infinite when R becomes zero. If the approach to infinity happens to be through positive values of U, there will be an infinite force repelling the particle away from the origin. If the approach is through negative values of U, the particle will go through this point with infinite speed, and thus spend no time there. In either case, we obtain a simple and precisely definable conceptual model explaining why particles can never be found at points where the wave function vanishes.

If one of the slits is closed, the "quantum-mechanical" potential is correspondingly altered, because the ψ-field is changed, and the particle may then be able to reach certain points which it was unable to reach when both slits were open. The slit is therefore able to affect the motion of the particle only indirectly, through its effect on the Schroedinger ψ-field. Moreover, as we shall see in Paper II, if the position of the electron is measured while it is passing through the slit system, the measuring apparatus will, as in the usual interpretation, create a disturbance that destroys the interference pattern. In our interpretation, however, the necessity for this destruction is not inherent in the conceptual structure; and as we shall see, the destruction of the interference pattern could in principle be avoided by means of other ways of making measurements, ways which are conceivable but not now actually possible.

6. THE MANY-BODY PROBLEM

We shall now extend our interpretation of the quantum theory to the problem of many bodies. We begin with the Schroedinger equation for two particles. (For simplicity, we assume that they have equal masses, but the extension of our treatment to arbitrary masses will be obvious.)

$$i\hbar\frac{\partial\psi}{\partial t}=-\frac{\hbar^2}{2m}(\nabla_1^2\psi+\nabla_2^2\psi)+V(\mathbf{x}_1,\mathbf{x}_2)\psi.$$

Writing $\psi=R(\mathbf{x}_1,\mathbf{x}_2)\exp[iS(\mathbf{x}_1,\mathbf{x}_2)/\hbar]$ and $R^2=P$, we obtain

$$\frac{\partial P}{\partial t}+\frac{1}{m}[\nabla_1\cdot P\nabla_1S+\nabla_2\cdot P\nabla_2S]=0, \tag{16}$$

$$\frac{\partial S}{\partial t}+\frac{(\nabla_1S)^2+(\nabla_2S)^2}{2m}+V(\mathbf{x}_1,\mathbf{x}_2)-\frac{\hbar^2}{2mR}[\nabla_1^2R+\nabla_2^2R]=0. \tag{17}$$

The above equations are simply a six-dimensional generalization of the similar three-dimensional Eqs. (5) and (6) associated with the one-body problem. In the two-body problem, the system is described therefore by a six-dimensional Schroedinger wave and by a six-dimensional trajectory, specifying the actual location of each of the two particles. The velocity of this trajectory has components, ∇_1S/m and ∇_2S/m, respectively, in each of the three-dimensional surfaces associated with a given particle. $P(\mathbf{x}_1,\mathbf{x}_2)$ then has a dual interpretation. First, it defines a "quantum-mechanical" potential, acting on each particle

$$U(\mathbf{x}_1,\mathbf{x}_2)=-(\hbar^2/2mR)[\nabla_1^2R+\nabla_2^2R].$$

This potential introduces an additional effective interaction between particles over and above that due to the classically inferrable potential $V(\mathbf{x})$. Secondly, the function $P(\mathbf{x}_1,\mathbf{x}_2)$ can consistently be regarded as the probability density of representative points $(\mathbf{x}_1,\mathbf{x}_2)$ in our six-dimensional ensemble.

The extension to an arbitrary number of particles is straightforward, and we shall quote only the results here. We introduce the wave function, $\psi=R(\mathbf{x}_1,\mathbf{x}_2,\cdots\mathbf{x}_n)\exp[iS(\mathbf{x}_1,\mathbf{x}_2\cdots\mathbf{x}_n)/\hbar]$ and define a $3n$-dimensional trajectory, where n is the number of particles, which describes the behavior of every particle in the system. The velocity of the ith particle is $\mathbf{v}_i=\nabla_iS(\mathbf{x}_1,\mathbf{x}_2\cdots\mathbf{x}_n)/m$. The function $P(\mathbf{x}_1,\mathbf{x}_2\cdots\mathbf{x}_n)=R^2$ has two interpretations. First, it defines a "quantum-mechanical" potential

$$U(\mathbf{x}_1, \mathbf{x}_2\cdots\mathbf{x}_n) = -\frac{\hbar^2}{2mR}\sum_{s=1}^{n}\nabla_s^2 R(\mathbf{x}_1, \mathbf{x}_2\cdots\mathbf{x}_n). \quad (18)$$

Secondly, $P(\mathbf{x}_1, \mathbf{x}_2\cdots\mathbf{x}_n)$ is equal to the density of representative points $(\mathbf{x}_1, \mathbf{x}_2\cdots\mathbf{x}_n)$ in our $3n$-dimensional ensemble.

We see here that the "effective potential," $U(\mathbf{x}_1, \mathbf{x}_2, \cdots\mathbf{x}_n)$, acting on a particle is equivalent to that produced by a "many-body" force, since the force between any two particles may depend significantly on the location of every other particle in the system. An example of the effects of such a force is given by the exclusion principle. Thus, if the wave function is antisymmetric, we deduce that the "quantum-mechanical" forces will be such as to prevent two particles from ever reaching the same point in space, for in this case, we must have $P=0$.

7. TRANSITIONS BETWEEN STATIONARY STATES—THE FRANCK-HERTZ EXPERIMENT

Our interpretation of the quantum theory describes all processes as basically causal and continuous. How then can it lead to a correct description of processes such as the Franck-Hertz experiment, the photoelectric effect, and the Compton effect, which seem to call most strikingly for an interpretation in terms of discontinuous and incompletely determined transfers of energy and momentum? In this section, we shall answer this question by applying our suggested interpretation of the quantum theory in the analysis of the Franck-Hertz experiment. Here, we shall see that the apparently discontinuous nature of the process of transfer of energy from the bombarding particle to the atomic electron is brought about by the "quantum-mechanical" potential, $U=(-\hbar^2/2m)\nabla^2 R/R$, which does not necessarily become small when the wave intensity becomes small. Thus, even if the force of interaction between two particles is very weak, so that a correspondingly small disturbance of the Schroedinger wave function is produced by the interaction of these particles, this disturbance is capable of bringing about very large transfers of energy and momentum between the particles in a very short time. This means that if we view only the end results, this process presents the aspect of being discontinuous. Moveover, we shall see that the precise value of the energy transfer is in principle determined by the initial position of each particle and by the initial form of the wave function. Since we cannot in practice predict or control the initial particle positions with complete precision, we are also unable to predict or control the final outcome of such an experiment, and can, in practice, predict only the probability of a given outcome. Because the probability that the particles will enter a region with coordinates, $\mathbf{x}_1$, $\mathbf{x}_2$, is proportional to $R^2(\mathbf{x}_1, \mathbf{x}_2)$, we conclude that although a Schroedinger wave of low intensity can bring about large transfers of energy, such a process is (as in the usual interpretation) highly improbable.

In Appendix A of Paper II, we shall see that similar possibilities arise in connection with the interaction of the electromagnetic field with charged matter, so that electromagnetic waves can very rapidly transfer a full quantum of energy (and momentum) to an electron, even after they have spread out and fallen to a very low intensity. In this way, we shall explain the photoelectric effect and the Compton effect. Thus, we are able in our interpretation to understand by means of a causal and continuous model just those properties of matter and light which seem most convincingly to require the assumption of discontinuity and incomplete determinism.

Before we discuss the process of interaction between two particles, we shall find it convenient to analyze the problem of an isolated single particle that happens to be in a nonstationary state. Because the field function ψ is a solution of Schroedinger's equation, we can linearly suppose stationary-state solutions of this equation and in this way obtain new solutions. As an illustration, let us consider a superposition of two solutions

$$\psi = C_1\psi_1(\mathbf{x})\exp(-iE_1t/\hbar) + C_2\psi_2(\mathbf{x})\exp(-iE_2t/\hbar),$$

where C_1, C_2, ψ_1, and ψ_2 are real. Thus we write $\psi_1=R_1$, $\psi_2=R_2$, and

$$\psi = \exp[-i(E_1+E_2)t/2\hbar]\{C_1R_1\exp[-i(E_1-E_2)t/2\hbar] + C_2R_2\exp[i(E_1-E_2)t/2\hbar]\}.$$

Writing $\psi=R\exp(iS/\hbar)$, we obtain

$$R^2 = C_1^2R_1^2(\mathbf{x}) + C_2^2R_2^2(\mathbf{x}) + 2C_1C_2R_1(\mathbf{x})R_2(\mathbf{x})\cos[(E_1-E_2)t/2\hbar], \quad (19)$$

$$\tan\left\{\frac{S+(E_1-E_2)t/2}{\hbar}\right\} = \frac{C_2R_2(\mathbf{x})-C_1R_1(\mathbf{x})}{C_2R_2(\mathbf{x})+C_1R_1(\mathbf{x})}\tan\left\{\frac{(E_1-E_2)t}{2\hbar}\right\}. \quad (20)$$

We see immediately that the particle experiences a "quantum-mechanical" potential, $U(\mathbf{x})=(-\hbar/2m)\nabla^2R/R$, which fluctuates with angular frequency, $w=(E_1-E_2)/\hbar$, and that the energy of this particle, $E=-\partial S/\partial t$, and its momentum $\mathbf{p}=\nabla S$, fluctuate with the same angular frequency. If the particle happens to enter a region of space where R is small, these fluctuations can become quite violent. We see then that, in general, the orbit of a particle in a nonstationary state is very irregular and complicated, resembling Brownian motion more closely than it resembles the smooth track of a planet around the sun.

If the system is isolated, these fluctuations will continue forever. The result is quite reasonable, since as is well known, a system can make a transition from one stationary state to another only if it can exchange energy with some other system. In order to treat the problem of transition between stationary states, we must therefore introduce another system capable of exchanging energy with the system of interest. In this section, we shall discuss the Franck-Hertz experiment, in which this other system consists of a bombarding particle. For the sake of illustration, let us suppose that we have hydrogen atoms of energy E_0 and wave function, $\psi_0(\mathbf{x})$, which are bombarded by particles that can be scattered inelastically, leaving the atom with energy E_n and wave function, $\psi_n(\mathbf{x})$.

We begin by writing down the initial wave function, $\Psi_i(\mathbf{x}, \mathbf{y}, t)$. The incident particle, whose coordinates are represented by $\mathbf{y}$ must be associated with a wave packet, which can be written as

$$f_0(\mathbf{y}, t)=\int e^{i\mathbf{k}\cdot\mathbf{y}}f(\mathbf{k}-\mathbf{k}_0)\exp(-i\hbar k^2t/2m)d\mathbf{k}. \quad (21)$$

The center of this packet occurs where the phase has an extremum as a function of $\mathbf{k}$, or where $\mathbf{y}=\hbar\mathbf{k}_0t/m$.

Now, as in the usual interpretation, we begin by writing the incident wave function for the combined system as a product

$$\Psi_i=\psi_0(\mathbf{x})\exp(-iE_0t/\hbar)f_0(\mathbf{y}, t). \quad (22)$$

Let us now see how this wave function is to be understood in our interpretation of the theory. As pointed out in Sec. 6, the wave function is to be regarded as a mathematical representation of a six-dimensional but objectively real field, capable of producing forces that act on the particles. We also assume a six-dimensional representative point, described by the coordinates of the two particles, $\mathbf{x}$ and $\mathbf{y}$. We shall now see that when the combined wave function takes the form (22) involving a product of a function of $\mathbf{x}$ and a function of $\mathbf{y}$, the six-dimensional system can correctly be regarded as being made up of two independent three-dimentional subsystems. To prove this, we write

$$\psi_0(\mathbf{x})=R_0(\mathbf{x})\exp[iS_0(\mathbf{x})/\hbar] \quad \text{and}$$
$$f_0(\mathbf{y}, t)=M_0(\mathbf{y}, t)\exp[iN_0(\mathbf{y}, t)/\hbar].$$

We then obtain for the particle velocities

$$d\mathbf{x}/dt=(1/m)\nabla S_0(\mathbf{x});\quad d\mathbf{y}/dt=(1/m)\nabla N_0(y, t), \quad (23)$$

and for the "quantum-mechanical" potential

$$U=-\frac{\hbar^2\{(\nabla_x^2+\nabla_y^2)R(\mathbf{x}, \mathbf{y})\}}{2mR(\mathbf{x}, \mathbf{y})}$$
$$=\frac{-\hbar^2}{2m}\left\{\frac{\nabla^2R_0(\mathbf{x})}{R_0(\mathbf{x})}+\frac{\nabla^2M_0(\mathbf{y}, t)}{M_0(\mathbf{y}, t)}\right\}. \quad (24)$$

Thus, the particle velocities are independent and the "quantum-mechanical" potential reduces to a sum of terms, one involving only $\mathbf{x}$ and the other involving only $\mathbf{y}$. This means that the particles move independently. Moreover, the probability density, $P=R_0^2(\mathbf{x})\times M_0^2(\mathbf{y}, t)$, is a product of a function of $\mathbf{x}$ and a function of $\mathbf{y}$, indicating that the distribution in $\mathbf{x}$ is statistically independent of that in $\mathbf{y}$. We conclude, then, that whenever the wave function can be expressed as a product of two factors, each involving only the coordinates of a single system, then the two systems are completely independent of each other.

As soon as the wave packet in $\mathbf{y}$ space reaches the neighborhood of the atom, the two systems begin to interact. If we solve Schroedinger's equation for the combined system, we obtain a wave function that can be expressed in terms of the following series:

$$\Psi=\Psi_i+\sum_n\psi_n(\mathbf{x})\exp(-iE_nt/\hbar)f_n(\mathbf{y}, t), \quad (25)$$

where the $f_n(\mathbf{y}, t)$ are the expansion coefficients of the complete set of functions, $\psi_n(\mathbf{x})$. The asymptotic form of the wave function is[14]

$$\Psi=\Psi_i(\mathbf{x}, \mathbf{y})+\sum_n\psi_n(\mathbf{x})\exp\left(-\frac{iE_nt}{\hbar}\right)\int f(\mathbf{k}-\mathbf{k}_0)$$
$$\times\frac{\exp[ik_n\cdot\mathbf{r}-(\hbar k_n^2/2n)t]}{r}g_n(\theta, \phi, \mathbf{k})d\mathbf{k}, \quad (26)$$

where

$$\hbar^2k_n^2/2m=(\hbar^2k_0^2/2m)+E_0-E_n \quad \text{(conservation of energy)}. \quad (27)$$

The additional terms in the above equation represent outgoing wave packets, in which the particle speed, $\hbar k_n/m$, is correlated with the wave function, $\psi_n(\mathbf{x})$, representing the state in which the hydrogen atom is left. The center of the nth packet occurs at

$$r_n=(\hbar k_n/m)t. \quad (28)$$

It is clear that because the speed depends on the hydrogen atom quantum number, n, every one of these packets will eventually be separated by distances which are so large that this separation is classically describable.

When the wave function takes the form (25), the two particles system must be described as a single six-dimensional system and not as a sum of two independent three-dimensional subsystems, for at this time, if we try to express the wave function as $\psi(\mathbf{x}, \mathbf{y})=R(\mathbf{x}, \mathbf{y})\times\exp[iS(\mathbf{x}, \mathbf{y})/\hbar]$, we find that the resulting expressions for R and S depend on $\mathbf{x}$ and $\mathbf{y}$ in a very complicated way. The particle momenta, $\mathbf{p}_1=\nabla_2S(\mathbf{x}, \mathbf{y})$ and $\mathbf{p}_2=\nabla_yS(\mathbf{x}, \mathbf{y})$, therefore become inextricably interdependent. The "quantum-mechanical" potential,

$$U=-\frac{\hbar^2}{2mR(\mathbf{x}, \mathbf{y})}(\nabla_x^2R+\nabla_y^2R)$$

ceases to be expressible as the sum of a term involving $\mathbf{x}$ and a term involving $\mathbf{y}$. The probability density,

[14] N. F. Mott and H. S. W. Massey, *The Theory of Atomic Collisions* (Clarendon Press, Oxford, 1933).

$R^2(\mathbf{x}, \mathbf{y})$ can no longer be written as a product of a function of $\mathbf{x}$ and a function of $\mathbf{y}$, from which we conclude that the probability distributions of the two particles are no longer statistically independent. Moreover, the motion of the particle is exceedingly complicated, because the expressions for R and S are somewhat analogous to those obtained in the simpler problem of a nonstationary state of a single particle [see Eqs. (19) and (20)]. In the region where the scattered waves $\psi_n(\mathbf{x})f_n(\mathbf{y}, t)$ have an amplitude comparable with that of the incident wave, $\psi_0(\mathbf{x})f_0(\mathbf{y}, t)$, the functions R and S, and therefore the "quantum-mechanical" potential and the particle momenta, undergo rapid and violent fluctuations, both as functions of position and of time. Because the quantum-mechanical potential has $R(\mathbf{x}, \mathbf{y}, t)$ in the denominator, these fluctuations may become very large in this region where R is small. If the particles happen to enter such a region, they may exchange very large quantities of energy and momentum in a very short time, even if the classical potential, $V(\mathbf{x}, \mathbf{y})$ is very small. A small value of $V(\mathbf{x}, \mathbf{y})$ implies, however, a correspondingly small value of the scattered wave amplitudes, $f_n(\mathbf{y}, t)$. Since the fluctuations become large only in the region where the scattered wave amplitude is comparable with the incident wave amplitude and since the probability that the particles shall enter a given region of $\mathbf{x}$, $\mathbf{y}$ space is proportional to $R^2(\mathbf{x}, \mathbf{y})$, it is clear that a large transfer of energy is improbable (although still always possible) when $V(\mathbf{x}, \mathbf{y})$ is small.

While interaction between the two particles takes place then, their orbits are subject to wild fluctuations. Eventually, however, the behavior of the system quiets down and becomes simple again. For after the wave function takes its asymptotic form (26), and the packets corresponding to different values of n have obtained classically describable separations, we can deduce that because the probability density is $|\psi|^2$, the outgoing particle must enter one of these packets and stay with that packet thereafter (since it does not enter the space between packets in which the probability density is negligibly different from zero). In the calculation of the particle velocities, $\mathbf{V}_1=\nabla_x S/m$, $\mathbf{V}_2=\nabla_y S/m$, and of the quantum-mechanical potential, $U=(-\hbar^2/2mR)(\nabla_x^2 R+\nabla_y^2 R)$, we can therefore ignore all parts of the wave function other than the one actually containing the outgoing particle. It follows that the system acts as if it had the wave function

$$\Psi_n=\psi_n(\mathbf{x})\exp\left(\frac{iE_n t}{\hbar}\right)\int f(\mathbf{k}-\mathbf{k}_0)\times\frac{\exp\{i[\mathbf{k}_n\cdot\mathbf{r}-(\hbar k_n^2 t/2m)t]\}}{r}g_n(\theta, \phi, \mathbf{k})d\mathbf{k}, \quad (29)$$

where n denotes the packet actually containing the outgoing particle. This means that for all practical purposes the complete wave function (26) of the system may be replaced by Eq. (29), which corresponds to an atomic electron in its nth quantum state, and to an outgoing particle with a correlated energy, $E_n'=\hbar^2 k_n^2/2m$. Because the wave function is a product of a function of $\mathbf{x}$ and a function of $\mathbf{y}$, each system once again acts independently of the other. The wave function can now be renormalized because the multiplication of Ψ_n by a constant changes no physically significant quantity, such as the particle velocity or the "quantum-mechanical" potential. As shown in Sec. 5, when the electronic wave function is $\psi_n(\mathbf{x})\exp(-iE_n t/\hbar)$, its energy must be E_n. Thus, we have obtained a description of how it comes about that the energy is always transferred in quanta of size E_n-E_0.

It should be noted that while the wave packets are still separating, the electron energy is not quantized, but has a continuous range of values, which fluctuate rapidly. It is only the final value of the energy, appearing after the interaction is over that must be quantized. A similar result is obtained in the usual interpretation if one notes that because of the uncertainty principle, the energy of either system can become definite only after enough time has elapsed to complete the scattering process.[15]

In principle, the actual packet entered by the outgoing particle could be predicted if we knew the initial position of both particles and, of course, the initial form of the wave function of the combined system.[16] In practice, however, the particle orbits are very complicated and very sensitively dependent on the precise values of these initial positions. Since we do not at present know how to measure these initial positions precisely, we cannot actually predict the outcome of such an interaction process. The best that we can do is to predict the probability that an outgoing particle enters the nth packet within a given range of solid angle, $d\Omega$, leaving the hydrogen atom in its nth quantum state. In doing this, we use the fact that the probability density in $\mathbf{x}$, $\mathbf{y}$ space is $|\psi(\mathbf{x}, \mathbf{y})|^2$ and that as long as we are restricted to the nth packet, we can replace the complete wave function (26) by the wave function (29), corresponding to the packet that actually contains the particle. Now, by definition, we have $\int|\psi_n(\mathbf{x})|^2 d\mathbf{x}=1$. The remaining integration of

$$\left|\int f(\mathbf{k}-\mathbf{k}_0)\frac{\exp\{i[k_n r-(\hbar k_n^2/2m)t]\}}{r}g_n(\theta, \phi, k)d\mathbf{k}\right|^2$$

over the region of space corresponding to the nth outgoing packet leads, however, to precisely the same probability of scattering as would have been obtained by applying the usual interpretation. We conclude, then, that if ψ satisfies Schroedinger's equation, that if $\mathbf{v}=\nabla S/m$, and that if the probability density of particles is $P(\mathbf{x}, \mathbf{y})=R^2(\mathbf{x}, \mathbf{y})$, we obtain in every respect

[15] See reference 2, Chapter 18, Sec. 19.

[16] Note that in the usual interpretation one assumes that *nothing* determines the precise outcome of an individual scattering process. Instead, one assumes that all descriptions are inherently and unavoidably statistical (see Sec. 2).

exactly the same physical predictions for this problem as are obtained when we use the usual interpretation.

There remains only one more problem; namely, to show that if the outgoing packets are subsequently brought together by some arrangement of matter that does not act on the atomic electron, the atomic electron and the scattered particle will continue to act independently.[17] To show that these two particles will continue to act independently, we note that in all practical applications, the outgoing particle soon interacts with some classically describable system. Such a system might consist, for example, of the host of atoms of the gas with which it collides or of the walls of a container. In any case, if the scattering process is ever to be observed, the outgoing particle must interact with a classically describable measuring apparatus. Now all classically describable systems have the property that they contain an enormous number of internal "thermodynamic" degrees of freedom that are inevitably excited when the outgoing particle interacts with the system. The wave function of the outgoing particle is then coupled to that of these internal thermodynamic degrees of freedom, which we represent as $y_1, y_2, \cdots y_s$. To denote this coupling, we write the wave function for the entire system as

$$\Psi = \sum_n \psi_n(\mathbf{x}) \exp(-iE_n t/\hbar) f_n(\mathbf{y}, y_1, y_2 \cdots y_s). \quad (30)$$

Now, when the wave function takes this form, the overlapping of different packets in $\mathbf{y}$ space is not enough to produce interference between the different $\psi_n(\mathbf{x})$. To obtain such interference, it is necessary that the packets $f_n(\mathbf{y}, y_1, y_2, \cdots y_s)$ overlap in every one of the $S+3$ dimensions, $\mathbf{y}, y_1, y_2 \cdots y_s$. The reader will readily convince himself, by considering a typical case such as a collision of the outgoing particle with a metal wall, that it is overwhelmingly improbable that two of the packets $f_n(\mathbf{y}_1, y_1, y_2 \cdots y_s)$ will overlap with regard to every one of the internal thermodynamic coordinates, $y_1, y_2, \cdots y_s$, even if they are successfully made to overlap in $\mathbf{y}$ space. This is because each packet corresponds to a different particle velocity and to a different time of collision with the metal wall. Because the myriads of internal thermodynamic degrees of freedom are so chaotically complicated, it is very likely that as each of the n packets interacts with them, it will encounter different conditions, which will make the combined wave packet $f_n(\mathbf{y}, y_1, \cdots y_s)$ enter very different regions of $y_1, y_2 \cdots y_s$ space. Thus, for all practical purposes, we can ignore the possibility that if two of the packets are made to cross in $\mathbf{y}$ space, the motion either of the atomic electron or of the outgoing particle will be affected.[18]

8. PENETRATION OF A BARRIER

According to classical physics, a particle can never penetrate a potential barrier having a height greater than the particle kinetic energy. In the usual interpretation of the quantum theory, it is said to be able, with a small probability, to "leak" through the barrier. In our interpretation of the quantum theory, however, the potential provided by the Schroedinger ψ-field enables it to "ride" over the barrier, but only a few particles are likely to have trajectories that carry them all the way across without being turned around.

We shall merely sketch in general terms how the above results can be obtained. Since the motion of the particle is strongly affected by its ψ-field, we must first solve for this field with the aid of "Schroedinger's equation." Initially, we have a wave packet incident on the potential barrier; and because the probability density is equal to $|\psi(\mathbf{x})|^2$, the particle is certain to be somewhere within this wave packet. When the wave packet strikes the repulsive barrier, the ψ-field undergoes rapid changes which can be calculated[19] if desired, but whose precise form does not interest us here. At this time, the "quantum-mechanical" potential, $U = (-\hbar^2/2m)\nabla^2 R/R$, undergoes rapid and violent fluctuations, analogous to those described in Sec. 7 in connection with Eqs. (19), (20), and (25). The particle orbit then becomes very complicated and, because the potential is time dependent, very sensitive to the precise initial relationship between the particle position and the center of the wave packet. Ultimately, however, the incident wave packet disappears and is replaced by two packets, one of them a reflected packet and the other a transmitted packet having a much smaller intensity. Because the probability density is $|\psi|^2$, the particle must end up in one of these packets. The other packet can, as shown in Sec. 7, subsequently be ignored. Since the reflected packet is usually so much stronger than the transmitted packet, we conclude that during the time when the packet is inside the barrier, most of the particle orbits must be turned around, as a result of the violent fluctuations in the "quantum-mechanical" potential.

9. POSSIBLE MODIFICATIONS IN MATHEMATICAL FORMULATION LEADING TO EXPERIMENTAL PROOF THAT NEW INTERPRETATION IS NEEDED

We have already seen in a number of cases and in Paper II we shall prove in general, that as long as we

[17] See reference 2, Chapter 22, Sec. 11, for a treatment of a similar problem.

[18] It should be noted that exactly the same problem arises in the usual interpretation of the quantum theory for (reference 16), for whenever two packets overlap, then even in the usual interpretation, the system must be regarded as, in some sense, covering the states corresponding to both packets simultaneously. See reference 2, Chapter 6 and Chapter 16, Sec. 25. Once two packets have obtained classically describable separations, then, both in the usual interpretation and in our interpretation the probability that there will be significant interference between them is so overwhelmingly small that it may be compared to the probability that a tea kettle placed on a fire will happen to freeze instead of boil. Thus, we may for all practical purposes neglect the possibility of interference between packets corresponding to the different possible energy states in which the hydrogen atom may be left.

[19] See, for example, reference 2, Chapter 11, Sec. 17, and Chapter 12, Sec. 18.

assume that ψ satisfies Schroedinger's equation, that $\mathbf{v}=\nabla S(\mathbf{x})/m$, and that we have a statistical ensemble with a probability density equal to $|\psi(\mathbf{x})|^2$, our interpretation of the quantum theory leads to physical results that are identical with those obtained from the usual interpretation. Evidence indicating the need for adopting our interpretation instead of the usual one could therefore come only from experiments, such as those involving phenomena associated with distances of the order of 10^{-13} cm or less, which are not now adequately understood in terms of the existing theory. In this paper we shall not, however, actually suggest any specific experimental methods of distinguishing between our interpretation and the usual one, but shall confine ourselves to demonstrating that such experiments are conceivable.

Now, there are an infinite number of ways of modifying the mathematical form of the theory that are consistent with our interpretation and not with the usual interpretation. We shall confine ourselves here, however, to suggesting two such modifications, which have already been indicated in Sec. 4, namely, to give up the assumption that $\mathbf{v}$ is necessarily equal to $\nabla S(\mathbf{x})/m$, and to give up the assumption that ψ must necessarily satisfy a homogeneous linear equation of the general type suggested by Schroedinger. As we shall see, giving up either of those first two assumptions will in general also require us to give up the assumption of a statistical ensemble of particles, with a probability density equal to $|\psi(\mathbf{x})|^2$.

We begin by noting that it is consistent with our interpretation to modify the equations of motion of a particle (8a) by adding any conceivable force term to the right-hand side. Let us, for the sake of illustration, consider a force that tends to make the difference, $\mathbf{p}-\nabla S(\mathbf{x})$, decay rapidly with time, with a mean decay time of the order of $\tau=10^{-13}/c$ seconds, where c is the velocity of light. To achieve this result, we write

$$m\frac{d^2\mathbf{x}}{dt^2}=-\nabla\left\{V(\mathbf{x})-\frac{h^2}{2m}\frac{\nabla^2 R}{R}\right\}+\mathbf{f}(\mathbf{p}-\nabla S(\mathbf{x})), \quad (31)$$

where $\mathbf{f}(\mathbf{p}-\nabla S(\mathbf{x}))$ is assumed to be a function which vanishes when $\mathbf{p}=\nabla S(\mathbf{x})$ and more generally takes such a form that it implies a force tending to make $\mathbf{p}-\nabla S(\mathbf{x})$ decrease rapidly with the passage of time. It is clear, moreover, that f can be so chosen that it is large only in processes involving very short distances (where $\nabla S(\mathbf{x})$ should be large).

If the correct equations of motion resembled Eq. (31), then the usual interpretation would be applicable only over times much longer than τ, for only after such times have elapsed will the relation $\mathbf{p}=\nabla S(\mathbf{x})$ be a good approximation. Moreover, it is clear that such modifications of the theory cannot even be described in the usual interpretation, because they involve the precisely definable particle variables which are not postulated in the usual interpretation.

Let us now consider a modification that makes the equation governing ψ inhomogeneous. Such a modification is

$$ih\partial\psi/\partial t=H\psi+\xi(\mathbf{p}-\nabla S(\mathbf{x}_i)). \quad (32)$$

Here, H is the usual Hamiltonian operator, $\mathbf{x}_i$, represents the actual location of the particle, and ξ is a function that vanishes when $\mathbf{p}=\nabla S(\mathbf{x}_i)$. Now, if the particle equations are chosen, as in Eq. (31), to make $\mathbf{p}-\nabla S(\mathbf{x}_i)$ decay rapidly with time, it follows that in atomic processes, the inhomogeneous term in Eq. (32) will become negligibly small, so that Schroedinger's equation is a good approximation. Nevertheless, in processes involving very short distances and very short times, the inhomogeneities would be important, and the ψ-field would, as in the case of the electromagnetic field, depend to some extent on the actual location of the particle.

It is clear that Eq. (32) is inconsistent with the usual interpretation of the theory. Moreover, we can contemplate further generalizations of Eq. (32), in the direction of introducing nonlinear terms that are large only for processes involving small distances. Since the usual interpretation is based on the hypothesis of linear superposition of "state vectors" in a Hilbert space, it follows that the usual interpretation could not be made consistent with such a nonlinear equation for a one-particle theory. In a many-particle theory, operators can be introduced, satisfying a nonlinear generalization of Schroedinger's equation; but these must ultimately operate on wave functions that satisfy a linear homogeneous Schroedinger equation.

Finally, we repeat a point already made in Sec. 4, namely, that if the theory is generalized in any of the ways indicated here, the probability density of particles will cease to equal $|\psi(\mathbf{x})|^2$. Thus, experiments would become conceivable that distinguish between $|\psi(\mathbf{x})|^2$ and this probability; and in this way we could obtain an experimental proof that the usual interpretation, which gives $|\psi(\mathbf{x})|^2$ *only* a probability interpretation, must be inadequate. Moreover, we shall show in Paper II that modifications like those suggested here would permit the particle position and momentum to be measured simultaneously, so that the uncertainty principle could be violated.

ACKNOWLEDGMENT

The author wishes to thank Dr. Einstein for several interesting and stimulating discussions.

II

In this paper, we shall show how the theory of measurements is to be understood from the point of view of a physical interpretation of the quantum theory in terms of "hidden" variables, developed in a previous paper. We find that in principle, these "hidden" variables determine the precise results of each individual measurement process. In practice, however, in measurements that we now know how to carry out, the observing apparatus disturbs the observed system in an unpredictable and uncontrollable way, so that the uncertainty principle is obtained as a practical limitation on the possible precision of measurements. This limitation is not, however, inherent in the conceptual structure of our interpretation. We shall see, for example, that simultaneous measurements of position and momentum having unlimited precision would in principle be possible if, as suggested in the previous paper, the mathematical formulation of the quantum theory needs to be modified at very short distances in certain ways that are consistent with our interpretation but not with the usual interpretation.

We give a simple explanation of the origin of quantum-mechanical correlations of distant objects in the hypothetical experiment of Einstein, Podolsky, and Rosen, which was suggested by these authors as a criticism of the usual interpretation.

Finally, we show that von Neumann's proof that quantum theory is not consistent with hidden variables does not apply to our interpretation, because the hidden variables contemplated here depend both on the state of the measuring apparatus and the observed system and therefore go beyond certain of von Neumann's assumptions.

In two appendixes, we treat the problem of the electromagnetic field in our interpretation and answer certain additional objections which have arisen in the attempt to give a precise description for an individual system at the quantum level.

1. INTRODUCTION

IN a previous paper,[1] to which we shall hereafter refer as I, we have suggested an interpretation of the quantum theory in terms of "hidden" variables. We have shown that although this interpretation provides a conceptual framework that is broader than that of the usual interpretation, it permits of a set of three mutually consistent special assumptions, which lead to the same physical results as are obtained from the usual interpretation of the quantum theory. These three special assumptions are: (1) The ψ-field satisfies Schroedinger's equation. (2) If we write $\psi = R\exp(is/\hbar)$, then the particle momentum is restricted to $\mathbf{p} = \nabla S(\mathbf{x})$. (3) We have a statistical ensemble of particle positions, with a probability density, $P = |\psi(\mathbf{x})|^2$. If the above three special assumptions are not made, then one obtains a more general theory that cannot be made consistent with the usual interpretation. It was suggested in Paper I that such generalizations may actually be needed for an understanding of phenomena associated with distances of the order of 10^{-13} cm or less, but may produce changes of negligible importance in the atomic domain.

In this paper, we shall apply the interpretation of the quantum theory suggested in Paper I to the development of a theory of measurements in order to show that as long as one makes the special assumptions indicated above, one is led to the same predictions for all measurements as are obtained from the usual interpretation. In our interpretation, however, the uncertainty principle is regarded, not as an inherent limitation on the precision with which we can correctly conceive of the simultaneous definition of momentum and position, but rather as a practical limitation on the precision with which these quantities can simultaneously be measured, arising from unpredictable and uncontrollable disturbances of the observed system by the measuring apparatus. If the theory needs to be generalized in the ways suggested in Paper I, Secs. 4 and 9, however, then these disturbances could in principle either be eliminated, or else be made subject to prediction and control, so that their effects could be corrected for. Our interpretation therefore demonstrates that measurements violating the uncertainty principle are at least conceivable.

2. QUANTUM THEORY OF MEASUREMENTS

We shall now show how the quantum theory of measurements is to be expressed in terms of our suggested interpretation of the quantum theory.[2]

In general, a measurement of any variable must always be carried out by means of an interaction of the system of interest with a suitable piece of measuring apparatus. The apparatus must be so constructed that any given state of the system of interest will lead to a certain range of states of the apparatus. Thus, the interaction introduces correlations between the state of the observed system and the state of the apparatus. The range of indefiniteness in this correlation may be called the uncertainty, or the error, in the measurement.

Let us now consider an observation designed to measure an arbitrary (hermitian) "observable" Q, associated with an electron. Let $\mathbf{x}$ represent the position of the electron, y that of the significant apparatus coordinate (or coordinates if there are more than one). Now, one can show[2] that it is enough to consider an impulsive measurement, i.e., a measurement utilizing a very strong interaction between apparatus and system

* Now at Universidade de São Paulo, Faculdade de Filosofia, Ciencias e Letras, São Paulo, Brasil.

[1] D. Bohm, Phys. Rev. **84**, 166 (1951).

[2] For a treatment of how the theory of measurements can be carried out with the usual interpretation, see D. Bohm, *Quantum Theory* (Prentice-Hall, Inc., New York, 1951), Chapter 22.

under observation, which lasts for so short a time that the changes of the apparatus and the system under observation that would have taken place in the absence of interaction can be neglected. Thus, at least while the interaction is taking place, we can neglect the parts of the Hamiltonian associated with the apparatus alone and with the observed system alone, and we need retain only the part of the Hamiltonian, H_I, representing the interaction. Moreover, if the Hamiltonian operator is chosen to be a function only of quantities that commute with Q, then the interaction process will produce no uncontrollable changes in the observable, Q, but only in observables that do not commute with Q. In order that the apparatus and the system under observation shall be coupled, however, it is necessary that H_I shall also depend on operators involving y.

For the sake of illustration of the principles involved, we shall consider the following interaction Hamiltonian:

$$H_I = -aQp_y, \tag{1}$$

where a is a suitable constant and p_y is the momentum conjugate to y.

Now, in our interpretation, the system is to be described by a four-dimensional but objectively real wave field that is a function of $\mathbf{x}$ and y and by a corresponding four-dimensional representative point, specified by the coordinates, $\mathbf{x}$, of the electron and the coordinate, y, of the apparatus. Since the motion of the representative point is in part determined by forces produced by the ψ-field acting on both electron and apparatus variables, our first step in solving this problem is to calculate the ψ-field. This is done by solving Schroedinger's equation, with the appropriate boundary conditions on ψ.

Now, during interaction, Schroedinger's equation is approximated by

$$i\hbar\partial\Psi/\partial t = -aQp_y\Psi = (ia/\hbar)Q\partial\Psi/\partial y. \tag{2}$$

It is now convenient to expand Ψ in terms of the complete set $\psi_q(\mathbf{x})$ of eigenfunctions of the operator, Q, where q denotes an eigenvalue of Q. For the sake of simplicity, we assume that the spectrum of Q is discrete, although the results are easily generalized to a continuous spectrum. Denoting the expansion coefficients by $f_q(y, t)$, we obtain

$$\Psi(\mathbf{x}, y, t) = \sum_q \psi_q(\mathbf{x}) f_q(y, t). \tag{3}$$

Noting that $Q\psi_q(\mathbf{x}) = q\psi_q(\mathbf{x})$, we readily verify that Eq. (2) can now be reduced to the following series of equations for $f_q(y, t)$:

$$i\hbar\partial f_q(y, t)/\partial t = (ia/\hbar)qf_q(y, t). \tag{4}$$

If the initial value of $f_q(y, t)$ was $f_q{}^0(y)$, we obtain as a solution

$$f_q(y, t) = f_q{}^0(y - aqt/\hbar^2), \tag{5}$$

and

$$\Psi(\mathbf{x}, y, t) = \sum_q \psi_q(\mathbf{x}) f_q{}^0(y - aqt/\hbar^2). \tag{6}$$

Now, initially the apparatus and the electron were independent. As shown in Paper I, Sec. 7, in our interpretation (as in the usual interpretation), independent systems must have wave fields $\Psi(\mathbf{x}, y, t)$ that are equal to a product of a function of $\mathbf{x}$ and a function of y. Initially, we therefore have

$$\Psi_0(\mathbf{x}, y) = \psi_0(\mathbf{x}) g_0(y) = g_0(y) \sum_q c_q \psi_q(\mathbf{x}), \tag{7}$$

where the c_q are the (unknown) expansion coefficients of $\psi_q(\mathbf{x})$, and $g_0(y)$ is the initial wave function of the apparatus coordinate, y. The function $g_0(y)$ will take the form of a packet. For the sake of convenience, we assume that this packet is centered at $y=0$ and that its width is Δy. Normally, because the apparatus is classically describable, the definition of this packet is far less precise than that allowed by the limits of precision set by the uncertainty principle.

From Eqs. (7) and (3), we shall readily deduce that $f_q{}^0(y) = c_q g_0(y)$. When this value of $f_q{}^0(y)$ is inserted into Eq. (6), we obtain

$$\Psi(\mathbf{x}, y, t) = \sum_q c_q \psi_q(\mathbf{x}) g_0(y - aqt/\hbar^2). \tag{8}$$

Equation (8) indicates already that the interaction has introduced a correlation between q and the apparatus coordinate, y. In order to show what this correlation means in our interpretation of the quantum theory, we shall use some arguments that have been developed in more detail in Paper I, Sec. 7, in connection with a similar problem involving the interaction of two particles in a scattering process. First we note that while the electron and the apparatus are interacting, the wave function (8) becomes very complicated, so that if it is expressed as

$$\Psi(\mathbf{x}, y, t) = R(\mathbf{x}, y, t) \exp[iS(\mathbf{x}, y, t)/\hbar],$$

then R and S undergo rapid oscillations both as a function of position and of time. From this we deduce that the "quantum-mechanical" potential,

$$U = (-\hbar^2/2mR)(\nabla_x{}^2R + \partial^2R/\partial y^2),$$

undergoes violent fluctuations, especially where R is small, and that the particle momenta, $\mathbf{p} = \nabla_x S(\mathbf{x}, y, t)$ and $p_y = \partial S(\mathbf{x}, y, t)/\partial y$, also undergo corresponding violent and extremely complicated fluctuations. Eventually, however, if the interaction continues long enough, the behavior of the system will become simpler because the packets $g_0(y - aqt/\hbar^2)$, corresponding to different values of q, will cease to overlap in y space. To prove this, we note that the center of the qth packet in y space is at

$$y = aqt/\hbar^2; \quad \text{or} \quad q = \hbar^2 y/at. \tag{9}$$

If we denote the separation of adjacent values of q by δq, we then obtain for the separation of the centers of adjacent packets in y space

$$\delta y = at\delta q/\hbar^2. \tag{10}$$

It is clear that if the product of the strength of interaction a, and the duration of interaction, t, is large enough, then δy can be made much larger than the width Δy of the packet. Then packets corresponding to different values of q will cease to overlap in y space and will, in fact, obtain separations large enough to be classically describable.

Because the probability density is equal to $|\Psi|^2$, we deduce that the apparatus variable, y, must finally enter one of the packets and remain with that packet thereafter (since it does not enter the intermediate space between packets in which the probability density is practically zero). Now, the packet entered by the apparatus variable y determines the actual result of the measurement, which the observer will obtain when he looks at the apparatus. The other packets can (as shown in Paper I, Sec. 7) be ignored, because they affect neither the quantum-mechanical potential acting on the particle coordinates $\mathbf{x}$ and y, nor the particle momenta, $\mathbf{p}_x = \nabla_x S$ and $p_y = \partial S/\partial y$. Moreover, the wave function can also be renormalized without affecting any of the above quantities. Thus, for all practical purposes, we can replace the complete wave function, Eq. (8), by a new renormalized wave function

$$\Psi(\mathbf{x}, y) = \psi_q(\mathbf{x}) g_0(y - aqt/\hbar^2), \tag{11}$$

where q now corresponds to the packet actually containing the apparatus variable, y. From this wave function, we can deduce, as shown in Paper I, Sec. 7, that the apparatus and the electron will subsequently behave independently. Moreover, by observing the approximate value of the apparatus coordinate within an error $\Delta y \ll \delta y$, we can deduce with the aid of Eq. (9) that since the electron wave function can for all practical purposes be regarded as $\psi_q(\mathbf{x})$, the observable, Q, must have the definite value, q. However, if the product, "$at\delta q/\hbar^2$," appearing in Eqs. (8), (9), (10), and (11), had been less than Δy, then no clear measurement of Q would have been possible, because packets corresponding to different q would have overlapped, and the measurement would not have had the requisite accuracy.[3]

Finally, we note that even if the apparatus packets are subsequently caused to overlap, none of those conclusions will be altered. For the apparatus variable y will inevitably be coupled to a whole host of internal thermodynamic degrees of freedom, $y_1, y_2, \cdots y_s$, as a result of effects such as friction and brownian motion. As shown in Paper I, Sec. 7, interference between packets corresponding to different values of q would be possible only if the packets overlapped in the space of $y_1, y_2, \cdots y_s$, as well as in y space. Such an overlap, however, is so improbable that for all practical purposes, we can ignore the possibility that it will ever occur.

3. THE ROLE OF PROBABILITY IN MEASUREMENTS—THE UNCERTAINTY PRINCIPLE

In principle, the final result of a measurement is determined by the initial form of the wave function of the combined system, $\Psi^0(\mathbf{x}, y)$, and by the initial position of the electron particle, $\mathbf{x}_0$, and the apparatus variable, y_0. In practice, however, as we have seen, the orbit fluctuates violently while interaction takes place, and is very sensitive to the precise initial values of $\mathbf{x}$ and y, which we can neither predict nor control. All that we can predict in practice is that in an ensemble of similar experiments performed under equivalent initial conditions, the probability density is $|\Psi(\mathbf{x}, y)|^2$. From this information, however, we are able to calculate only the probability that in an individual experiment, the result of a measurement of Q will be a specific number q. To obtain the probability of a given value of q, we need only integrate the above probability density over all $\mathbf{x}$ and over all values of y in the neighborhood of the qth packet. Because the packets do not overlap, the Ψ-field in this region is equal to $c_q\psi_q(\mathbf{x})g_0(y - aqt/\hbar^2)$ [see Eq. (8)]. Since, by definition, $\psi_q(\mathbf{x})$ and $g_0(y)$ are normalized, the total probability that a particle is in the qth packet is

$$P_q = |c_q|^2. \tag{12}$$

The above is, however, just what is obtained from the usual interpretation. We conclude then that our interpretation is capable of leading in all possible experiments to identical predictions with those obtained from the usual interpretation (provided, or course, that we make the special assumptions indicated in the introduction).

Let us now see what a measurement of the observable, Q, implies with regard to the state of the electron particle and its Ψ-field. First, we note that the process of interaction with an apparatus designed to measure the observable, Q, effectively transforms the electron ψ-field from whatever it was before the measurement took place into an eigenfunction $\psi_q(\mathrm{x})$ of the operator Q. The precise value of q that comes out of this process is as we have seen, not, in general, completely predictable or controllable. If, however, the same measurement is repeated after the ψ-field has been transformed into $\psi_q(\mathbf{x})$, we can then predict that (as in the usual interpretation), the same value of q, and therefore the same wave function, $\psi_q(\mathbf{x})$, will be obtained again. If, however, we measure an observable "P" that does not commute with Q, then the results of this measurement are not, in practice, predictable or controllable. For as shown in Eq. (8), the Ψ-field after interaction with the measuring apparatus is now transformed into

$$\Psi(\mathbf{x}, z, t) = \sum_p a_{p,q}\phi_p(\mathbf{x})g_0(z - apt/\hbar^2), \tag{13}$$

where $\phi_p(\mathbf{x})$ is an eigenfunction of the operator, P, belonging to an eigenvalue, p, and where $a_{p,q}$ is an expansion coefficient defined by

$$\psi_q(\mathbf{x}) = \sum_p a_{p,q}\phi_p(\mathbf{x}). \tag{14}$$

Since the packets corresponding to different p ultimately become completely separate in z space, we deduce, as in the case of the measurement of Q, that

[3] A similar requirement is obtained in the usual interpretation. See reference 2, Chapter 22, Sec. 8.

for all practical purposes, this wave function may be replaced by

$$\Psi = a_{pq}\phi_p(\mathbf{x})g_0(z - apt/\hbar^2),$$

where p now represents the packet actually entered by the apparatus coordinate, y. As in the case of measurement of Q, we readily show that the precise value of p that comes out of this experiment cannot be predicted or controlled and that the probability of a given value of p is equal to $|a_{pq}|^2$. This is, however, just what is obtained in the usual interpretation of this process.

It is clear that if two "observables," P and Q, do not commute, one cannot carry out a measurement of both simultaneously on the same system. The reason is that each measurement disturbs the system in a way that is incompatible with carrying out the process necessary for the measurement of the other. Thus, a measurement of P requires that wave field, ψ, shall become an eigenfunction of P, while a measurement of Q requires that it shall become an eigenfunction of Q. If P and Q do not commute, then by definition, no ψ-function can be simultaneously an eigenfunction of both. In this way, we understand in our interpretation why measurements, of complementary quantities, must (as in the usual interpretation) necessarily be limited in their precision by the uncertainty principle.

4. PARTICLE POSITIONS AND MOMENTA AS "HIDDEN VARIABLES"

We have seen that in measurements that can now be carried out, we cannot make precise inferences about the particle position, but can say only that the particle must be somewhere in the region in which $|\psi|$ is appreciable. Similarly, the momentum of a particle that happens to be at the point, $\mathbf{x}$, is given by $\mathbf{p} = \nabla S(\mathbf{x})$, so that since $\mathbf{x}$ is not known, the precise value of $\mathbf{p}$ is also not, in general, inferrable. Hence, as long as we are restricted to making observations of this kind, the precise values of the particle position and momentum must, in general, be regarded as "hidden," since we cannot at present measure them. They are, however, connected with real and already observable properties of matter because (along with the ψ-field) they determine in principle the actual result of each individual measurement. By way of contrast, we recall here that in the usual interpretation of the theory, it is stated that although each measurement admittedly leads to a definite number, nothing determines the actual value of this number. The result of each measurement is assumed to arise somehow in an inherently indescribable way that is not subject to a detailed analysis. Only the statistical results are said to be predictable. In our interpretation, however, we assert that the at present "hidden" precisely definable particle positions and momenta determine the results of each individual measurement process, but in a way whose precise details are so complicated and uncontrollable, and so little known, that one must for all practical purposes restrict oneself to a statistical description of the connection between the values of these variables and the directly observable results of measurements. Thus, we are unable at present to obtain direct experimental evidence for the existence of precisely definable particle positions and momenta.

5. "OBSERVABLES" OF USUAL INTERPRETATION ARE NOT A COMPLETE DESCRIPTION OF SYSTEM IN OUR INTERPRETATION

We have seen in Sec. 3 that in the measurement of an "observable," Q, we cannot obtain enough information to provide a complete specification of the state of an electron, because we cannot infer the precisely defined values of the particle momentum and position, which are, for example, needed if we wish to make precise predictions about the future behavior of the electron. Moreover, the process of measuring an observable does not provide any unambiguous information about the state that existed before the measurement took place; for in such a measurement, the ψ-field is transformed into an in practice unpredictable and uncontrollable eigenfunction, $\psi_q(\mathbf{x})$, of the measured "observable" Q. This means that the measurement of an "observable" is not really a measurement of any physical property belonging to the observed system alone. Instead, the value of an "observable" measures only an incompletely predictable and controllable potentiality belonging just as much to the measuring apparatus as to the observed system itself.[4] At best, such a measurement provides unambiguous information only at a classical level of accuracy, where the disturbance of the ψ-field by the measuring apparatus can be neglected. The usual "observables" are therefore not what we ought to try to measure at a quantum level of accuracy. In Sec. 6, we shall see that it is conceivable that we may be able to carry out new kinds of measurements, providing information not about "observables" having a very ambiguous significance, but rather about physically significant properties of a system, such as the actual values of the particle position and momentum.

As an example of the rather indirect and ambiguous significance of the "observable," we may consider the problem of measuring the momentum of an electron. Now, in the usual interpretation, it is stated that one can always measure the momentum "observable" without changing the value of the momentum. The result is said, for example, to be obtainable with the aid of an impulsive interaction involving only operators which commute with the momentum operator, $\mathbf{p}_x$. To represent such a measurement, we could choose $H_I = -ap_xp_y$ in Eq. (1). In our interpretation, however, we cannot in general conclude that such an interaction will enable us to measure the actual particle momentum without changing its value. In fact, in our interpreta-

[4] Even in the usual interpretation, an observation must be regarded as yielding a measure of such a potentiality. See reference 2, Chapter 6, Sec. 9.

tion, a measurement of particle momentum that does not change the value of this momentum is possible only if the ψ-field initially takes the special form, $\exp(i\mathbf{p}\cdot\mathbf{x}/\hbar)$. If, however, ψ initially takes its most general possible form,

$$\psi=\sum_p a_1\mathbf{p}\exp(i\mathbf{p}\cdot\mathbf{x}/\hbar), \tag{15}$$

then as we have seen in Secs. 2 and 3, the process of measuring the "observable" p_x will effectively transform the ψ-field of the electron into

$$\exp(ip_x/\hbar) \tag{16}$$

with a probability $|a_p|^2$ that a given value of p_x will be obtained. When the ψ-field is altered in this way, large quantities of momentum can be transferred to the particle by the changing ψ-field, even though the interaction Hamiltonian, H_I, commutes with the momentum operator, $\mathbf{p}$.

As an example, we may consider a stationary state of an atom, of zero angular momentum. As shown in Paper I, Sec. 5, the ψ-field for such a state is real, so that we obtain

$$\mathbf{p}=\nabla S=0.$$

Thus, the particle is at rest. Nevertheless, we see from Eqs. (14) and (15) that if the momentum "observable" is measured, a large value of this "observable" may be obtained if the ψ-field happens to have a large fourier coefficient, $a_{\mathbf{p}}$, for a high value of $\mathbf{p}$. The reason is that in the process of interaction with the measuring apparatus, the ψ-field is altered in such a way that it can give the electron particle a correspondingly large momentum, thus transforming some of the potential energy of interaction of the particle with its ψ-field into kinetic energy.

A more striking illustration of the points discussed above is afforded by the problem of a "free" particle contained between two impenetrable and perfectly reflecting walls, separated by a distance L. For this case, the spatial part of the ψ-field is

$$\psi=\sin(2\pi nx/L),$$

where n is an integer and the energy of the electron is

$$E=(1/2m)(nh/L)^2.$$

Because the ψ-field is real, we deduce that the particle is at rest.

Now, at first sight, it may seem puzzling that a particle having a high energy should be at rest in the empty space between two walls. Let us recall, however, that the space is not really empty, but contains an objectively real ψ-field that can act on the particle. Such an action is analogous to (but of course not identical with) the action of an electromagnetic field, which could create non-uniform motion of the particle in this apparently "empty" enclosure. We observe that in our problem, the ψ-field is able to bring the particle to rest and to transform the entire kinetic energy into potential energy of interaction with the ψ-field. To prove this, we evaluate the "quantum-mechanical potential" for this ψ-field

$$U=\frac{-\hbar^2}{2m}\frac{\nabla^2 R}{R}=\frac{-\hbar^2}{2m}\frac{\nabla^2\psi}{\psi}=\frac{1}{2m}\left(\frac{nh}{L}\right)^2$$

and note that it is precisely equal to the total energy, E.

Now, as we have seen, any measurement of the momentum "observable" must change the ψ-field in such a way that in general some (and in our case, all) of this potential energy is transformed into kinetic energy. We may use as an illustration of this general result a very simple specific method of measuring the momentum "observable," namely, to remove the confining walls suddenly and then to measure the distance moved by the particle after a fairly long time. We can compute the momentum by dividing this distance by the time of transit. If (as in the usual interpretation of the quantum theory) we assume that the electron is "free," then we conclude that the process of removing the walls should not appreciably change the momentum if we do it fast enough, for the probability that the particle is near a wall when this happens can then in principle be made arbitrarily small. In our interpretation, however, the removal of the walls alters the particle momentum indirectly, because of its effect on the ψ-field, which acts on the particle. Thus, after the walls are removed, two wave packets moving in opposite directions begin to form, and ultimately they become completely separate in space. Because the probability density is $|\psi|^2$, we deduce that the particle must end up in one packet or the other. Moreover, the reader will readily convince himself that the particle momentum will be very close to $\pm nh/L$, the sign depending on which packet the particle actually enters. As in Sec. (2), the packet not containing the particle can subsequently be ignored. In principle, the final particle momentum is determined by the initial form of the ψ-field and by the initial particle position. Since we do not in practice know the latter, we can at best predict a probability of $\frac{1}{2}$ that the particle ends up in either packet. We conclude then that this measurement of the momentum "observable" leads to the same result as is predicted in the usual interpretation. However, the actual particle momentum existing before the measurement took place is quite different from the numerical value obtained for the momentum "observable," which, in the usual interpretation, is called the "momentum."

6. ON THE POSSIBILITY OF MEASUREMENTS OF UNLIMITED PRECISION

We have seen that the so-called "observables" do not measure any very readily interpretable properties of a system. For example, the momentum "observable" has in general no simple relation to the actual particle momentum. It may therefore be fruitful to consider how we might try to measure properties which, according to our interpretation, are (along with the ψ-field) the physically significant properties of an electron, namely, the actual particle position and momentum. In connection with this problem, we shall show that if, as suggested in Paper I, Secs. 4 and 9, we give up the three mutually consistent special assumptions leading to the same results as those of the usual interpretation of the quantum theory, then in our interpretation, the particle position and momentum can in principle be measured simultaneously with unlimited precision.

Now, for our purposes, it will be adequate to show that precise predictions of the future behavior of a system are in principle possible. In our interpretation, a sufficient condition for precise predictions is as we have seen that we shall be able to prepare a system in a state in which the ψ-field and the initial particle position and momentum are precisely known. We have shown that it is possible, by measuring the "observable," Q, with the aid of methods that are now available, to prepare a state in which the ψ-field is effectively transformed into a known form, $\psi_q(\mathbf{x})$; but we cannot in general predict or control the precise position and momentum of the particle. If we could now measure the position and momentum of the particle without altering the ψ-field, then precise predictions would be possible. However, the results of Secs. 2, 3, and 4 prove that as long as the three special assumptions indicated above are valid, we cannot measure the particle position more accurately without effectively transforming the ψ-function into an incompletely predictable and controllable packet that is much more localized than $\psi_q(\mathbf{x})$. Thus, efforts to obtain more precise definition of the state of the system will be defeated. But it is clear that the difficulty originates in the circumstance that the potential energy of interaction between electron and apparatus, $V(\mathbf{x}, y)$, plays two roles. For it not only introduces a direct interaction between the two particles, proportional in strength to $V(\mathbf{x}, y)$ itself, but it introduces an indirect interaction between these particles, because this potential also appears in the equation governing the ψ-field. This indirect interaction may involve rapid and violent fluctuations, even when $V(\mathbf{x}, y)$ is small. Thus, we are led to lose control of the effects of this interaction, because no matter how small $V(\mathbf{x}, y)$ is, very large and chaotically complicated disturbances in the particle motion may occur.

If, however, we give up the three special assumptions mentioned previously, then it is not inherent in our conceptual structure that every interaction between particles must inevitably also produce large and uncontrollable changes in the ψ-field. Thus, in Paper I, Eq. (31), we give an example in which we postulate a force acting on a particle that is not necessarily accompanied by a corresponding change in the ψ-field. Equation (31), Paper I, is concerned only with a one-particle system, but similar assumptions can be made for systems of two or more particles. In the absence of any specific theory, our interpretation permits an infinite number of kinds of such modifications, which can be chosen to be important at small distances but negligible in the atomic domain. For the sake of illustration, suppose that it should turn out that in certain processes connected with very small distances, the force acting on the apparatus variable is

$$F_y = ax,$$

where a is a constant. Now if "a" is made large enough so that the interaction is impulsive, we can neglect all changes in y that are brought about by the forces that would have been present in the absence of this interaction. Moreover, for the sake of illustration of the principles involved, we are permitted to make the assumption, consistent with our interpretation, that the force on the electron is zero. The equation of motion of y is then

$$\ddot{y} = ax/m.$$

The solution is

$$y - y_0 = (axt^2/2m) + \dot{y}_0 t,$$

where $\dot{y}_0$ is the initial velocity of the apparatus variable and y_0 its initial position. Now, if the product, at^2, is large enough, then $y - y_0$ can be made much larger than the uncertainty in y arising from the uncertainty of y_0, and the uncertainty of $\dot{y}_0$. Thus, $y - y_0$ will be determined primarily by the particle position, x. In this way, it is conceivable that we could obtain a measurement of x that does not significantly change x, $\dot{x}$, or the ψ-function. The particle momentum can then be obtained from the relation, $p = \nabla S(\mathbf{x})$, where $S/\hbar$ is the phase of the ψ-function. Thus, precise predictions would in principle be possible.

7. THE ORIGIN OF THE STATISTICAL ENSEMBLE IN THE QUANTUM THEORY

We shall now see that even if, because of a failure of the three special assumptions mentioned in Secs. 1 and 6, we are able to determine the particle positions and momenta precisely, we shall nevertheless ultimately obtain a statistical ensemble again at the atomic level, with a probability density equal to $|\psi|^2$. The need for such an ensemble arises from the chaotically complicated character of the coupling between the electron and classical systems, such as volumes of gas, walls of containers, pieces of measuring apparatus, etc., with which this particle must inevitably in practice interact. For as we have seen in Sec. 2, and in Paper I, Sec. 7, during the course of such an interaction, the "quantum-mechanical" potential undergoes violent and rapid fluctuations, which tend to make the particle orbit wander over the whole region in which the ψ-field is appreciable. Moreover, these fluctuations are further complicated by the effects of molecular chaos in the very large number of internal thermodynamic degrees of freedom of these classically describable systems, which are inevitably excited in any interaction process.

Thus, even if the initial particle variables were well defined, we should soon in practice lose all possibility of following the particle motion and would be forced to have recourse to some kind of statistical theory. The only question that remains is to show why the probability density that ultimately comes about should be equal to $|\psi|^2$ and not to some other quantity.

To answer this question, we first note that a statistical ensemble with a probability density $|\psi(\mathbf{x})|^2$ has the property that under the action of forces which prevail at the atomic level, where our three special assumptions are satisfied, it will be preserved by the equations of motion of the particles, once it comes into existence. There remains only the problem of showing that an arbitrary deviation from this ensemble tends, under the action of the chaotically complicated forces described in the previous paragraph, to decay into an ensemble with a probability density of $|\psi(\mathbf{x})|^2$. This problem is very similar to that of proving Boltzmann's H theorem, which shows in connection with a different but analogous problem that an arbitrary ensemble tends as a result of molecular chaos to decay into an equilibrium Gibbs ensemble. We shall not carry out a detailed proof here, but we merely suggest that it seems plausible that one could along similar lines prove that in our problem, an arbitrary ensemble tends to decay into an ensemble with a density of $|\psi(\mathbf{x})|^2$. These arguments indicate that in our interpretation, quantum fluctuations and classical fluctuations (such as the Brownian motion) have basically the same origin; *viz.*, the chaotically complicated character of motion at the microscopic level.

8. THE HYPOTHETICAL EXPERIMENT OF EINSTEIN, PODOLSKY, AND ROSEN

The hypothetical experiment of Einstein, Podolsky, and Rosen[5] is based on the fact that if we have two particles, the sum of their momenta, $p=p_1+p_2$, commutes with the difference of their positions, $\xi=x_1-x_2$. We can therefore define a wave function in which p is zero, while ξ has a given value, a. Such a wave function is

$$\psi=\delta(x_1-x_2-a). \tag{17}$$

In the usual interpretation of the quantum theory, p_1-p_2 and x_1+x_2 are completely undetermined in a system having the above wave function.

The whole experiment centers on the fact that an observer has a choice of measuring either the momentum or the position of any one of the two particles. Whichever of these quantities he measures, he will be able to infer a definite value of the corresponding variable in the other particle, because of the fact that the above wave function implies correlations between variables belonging to each particle. Thus, if he obtains a position x_1 for the first particle, he can infer a position of $x_2=a-x_1$ for the second particle; but he loses all possibility of making any inferences about the momenta of either particle. On the other hand, if he measures the momentum of the first particle and obtains a value of p_1, he can infer a value of $p_2=-p_1$ for the momentum of the second particle; but he loses all possibility of making any inferences about the position of either particle. Now, Einstein, Podolsky, and Rosen believe that this result is itself probably correct, but they do not believe that quantum theory as usually interpreted can give a complete description of how these correlations are propagated. Thus, if these were classical particles, we could easily understand the propagation of correlations because each particle would then simply move with a velocity opposite to that of the other. But in the usual interpretation of quantum theory, there is no similar conceptual model showing in detail how the second particle, which is not in any way supposed to interact with the first particle, is nevertheless able to obtain either an uncontrollable disturbance of its position or an uncontrollable disturbance of its momentum depending on what kind of measurement the observer decided to carry out on the first particle. Bohr's point of view is, however, that no such model should be sought and that we should merely accept the fact that these correlations somehow manage to appear. We must note, of course, that the quantum-mechanical description of these processes will always be consistent, even though it gives us no precisely definable means of describing and analyzing the relationships between the classically describable phenomena appearing in various pieces of measuring apparatus.

In our suggested new interpretation of the quantum theory, however, we can describe this experiment in terms of a single precisely definable conceptual model, for we now describe the system in terms of a combination of a six-dimensional wave field and a precisely definable trajectory in a six-dimensional space (see Paper I, Sec. 6). If the wave function is initially equal to Eq. (17), then since the phase vanishes, the particles are both at rest. Their possible positions are, however, described by an ensemble, in which $x_1-x_2=a$. Now, if we measure the position of the first particle, we introduce uncontrollable fluctuations in the wave function for the entire system, which, through the "quantum-mechanical" forces, bring about corresponding uncontrollable fluctuations in the momentum of each particle. Similarly, if we measure the momentum of the first particle, uncontrollable fluctuations in the wave function for the system bring about, through the "quantum-mechanical" forces, corresponding uncontrollable changes in the position of each particle. Thus, the "quantum-mechanical" forces may be said to transmit uncontrollable disturbances instantaneously from one particle to another through the medium of the ψ-field.

What does this transmission of forces at an infinite rate mean? In nonrelativistic theory, it certainly causes

[5] Einstein, Podolsky, and Rosen, Phys. Rev. **47**, 777 (1935).

no difficulties. In a relativistic theory, however, the problem is more complicated. We first note that as long as the three special assumptions mentioned in Sec. 2 are valid, our interpretation can give rise to no inconsistencies with relativity, because it leads to precisely the same predictions for all physical processes as are obtained from the usual interpretation (which is known to be consistent with relativity). The reason why no contradictions with relativity arise in our interpretation despite the instantaneous transmission of momentum between particles is that no signal can be carried in this way. For such a transmission of momentum could constitute a signal only if there were some practical means of determining precisely what the second particle would have done if the first particle had not been observed; and as we have seen, this information cannot be obtained as long as the present form of the quantum theory is valid. To obtain such information, we require conditions (such as might perhaps exist in connection with distances of the order of 10^{-13} cm) under which the usual form of the quantum theory breaks down (see Sec. 6), so that the positions and momenta of the particles can be determined simultaneously and precisely. If such conditions should exist, then there are two ways in which contradictions might be avoided. First, the more general physical laws appropriate to the new domains may be such that they do not permit the transmission of controllable aspects of interparticle forces faster than light. In this way, Lorentz covariance could be preserved. Secondly, it is possible that the application of the usual criteria of Lorentz covariance may not be appropriate when the usual interpretation of quantum theory breaks down. Even in connection with gravitational theory, general relativity indicates that the limitation of speeds to the velocity of light does not necessarily hold universally. If we adopt the spirit of general relativity, which is to seek to make the properties of space dependent on the properties of the matter that moves in this space, then it is quite conceivable that the metric, and therefore the limiting velocity, may depend on the ψ-field as well as on the gravitational tensor $g^{\mu,\nu}$. In the classical limit, the dependence on the ψ-field could be neglected, and we would get the usual form of covariance. In any case, it can hardly be said that we have a solid experimental basis for requiring the same form of covariance at very short distances that we require at ordinary distances.

To sum up, we may assert that wherever the present form of the quantum theory is correct, our interpretation cannot lead to inconsistencies with relativity. In the domains where the present theory breaks down, there are several possible ways in which our interpretation could continue to treat the problem of covariance consistently. The attempt to maintain a consistent treatment of covariance in this problem might perhaps serve as an important heuristic principle in the search for new physical laws.

9. ON VON NEUMANN'S DEMONSTRATION THAT QUANTUM THEORY IS INCONSISTENT WITH HIDDEN VARIABLES

Von Neumann[6] has studied the following question: "If the present mathematical formulation of the quantum theory and its usual probability intepretation are assumed to lead to absolutely correct results for every experiment that can ever be done, can quantum-mechanical probabilities be understood in terms of any conceivable distribution over hidden parameters?" Von Neumann answers this question in the negative. His conclusions are subject, however, to the criticism that in his proof he has implicitly restricted himself to an excessively narrow class of hidden parameters and in this way has excluded from consideration precisely those types of hidden parameters which have been proposed in this paper.

To demonstrate the above statements, we summarize Von Neumann's proof briefly. This proof (which begins on p. 167 of his book), shows that the usual quantum-mechanical rules of calculating probabilities imply that there can be no "dispersionless states," i.e., states in which the values of all possible observables are simultaneously determined by physical parameters associated with the observed system. For example, if we consider two noncommuting observables, p and q, then Von Neumann shows that it would be inconsistent with the usual rules of calculating quantum-mechanical probabilities to assume that there were in the observed system a set of hidden parameters which simultaneously determined the results of measurements of position and momentum "observables." With this conclusion, we are in agreement. However, in our suggested new interpretation of the theory, the so-called "observables" are, as we have seen in Sec. 5, not properties belonging to the observed system alone, but instead potentialities whose precise development depends just as much on the observing apparatus as on the observed system. In fact, when we measure the momentum "observable," the final result is determined by hidden parameters in the momentum-measuring device as well as by hidden parameters in the observed electron. Similarly, when we measure the position "observable," the final result is determined in part by hidden parameters in the position-measuring device. Thus, the statistical distribution of "hidden" parameters to be used in calculating averages in a momentum measurement is different from the distribution to be used in calculating averages in a position measurement. Von Neumann's proof (see p. 171 in his book) that no single distribution of hidden parameters could be consistent with the results of the quantum theory is therefore irrelevant here, since in our interpretation of measurements of the type that can now be carried out, the distribution of hidden parameters varies in accordance with the different mutually exclusive experimental

[6] J. von Neumann, *Mathematische Grundlagen der Quantenmechanik* (Verlag. Julius Springer, Berlin, 1932).

arrangements of matter that must be used in making different kinds of measurements. In this point, we are in agreement with Bohr, who repeatedly stresses the fundamental role of the measuring apparatus as an inseparable part of the observed system. We differ from Bohr, however, in that we have proposed a method by which the role of the apparatus can be analyzed and described in principle in a precise way, whereas Bohr asserts that a precise conception of the details of the measurement process is as a matter of principle unattainable.

Finally, we wish to stress that the conclusions drawn thus far refer only to the measurement of the so-called "observables" carried out by the methods that are now available. If the quantum theory needs to be modified at small distances, then, as suggested in Sec. 6, precise measurements can in principle be made of the actual position and momentum of a particle. Here, it should be noted that Von Neumann's theorem is likewise irrelevant, this time because we are going beyond the assumption of the unlimited validity of the present general form of quantum theory, which plays an integral part in his proof.

10. SUMMARY AND CONCLUSIONS

The usual interpretation of the quantum theory implies that we must renounce the possibility of describing an individual system in terms of a single precisely defined conceptual model. We have, however, proposed an alternative interpretation which does not imply such a renunciation, but which instead leads us to regard a quantum-mechanical system as a synthesis of a precisely definable particle and a precisely definable ψ-field which exerts a force on this particle. An experimental choice between these two interpretations cannot be made in a domain in which the present mathematical formulation of the quantum theory is a good approximation; but such a choice is conceivable in domains, such as those associated with dimensions of the order of 10^{-13} cm, where the extrapolation of the present theory seems to break down and where our suggested new interpretation can lead to completely different kinds of predictions.

At present, our suggested new interpretation provides a consistent alternative to the usual assumption that no objective and precisely definable description of reality is possible at the quantum level of accuracy. For, in our description, the problem of objective reality at the quantum level is at least in principle not fundamentally different from that at the classical level, although new problems of measurement of the properties of an individual system appear, which can be solved only with the aid of an improvement in the theory, such as the possible modifications in the nuclear domain suggested in Sec. 6. In this connection, we wish to point out that what we can measure depends not only on the type of apparatus that is available, but also on the existing theory, which determines the kind of inference that can be used to connect the directly observable state of the apparatus with the state of the system of interest. In other words, our epistemology is determined to a large extent by the existing theory. It is therefore not wise to specify the possible forms of future theories in terms of purely epistomological limitations deduced from existing theories.

The development of the usual interpretation of the quantum theory seems to have been guided to a considerable extent by the principle of not postulating the possible existence of entities which cannot now be observed. This principle, which stems from a general philosophical point of view known during the nineteenth century as "positivism" or "empiricism" represents an extraphysical limitation on the possible kinds of theories that we shall choose to take into consideration.[7] The word "extraphysical" is used here advisedly, since we can in no way deduce, either from the experimental data of physics, or from its mathematical formulation, that it will necessarily remain forever impossible for us to observe entities whose existence cannot now be observed. Now, there is no reason why an extraphysical general principle is necessarily to be avoided, since such principles could conceivably serve as useful working hypotheses. The particular extraphysical principle described above cannot, however, be said to be a good working hypothesis. For the history of scientific research is full of examples in which it was very fruitful indeed to assume that certain objects or elements might be real, long before any procedures were known which would permit them to be observed directly. The atomic theory is just such an example. For the possibility of the actual existence of individual atoms was first postulated in order to explain various macrophysical results which could, however, also be understood directly in terms of macrophysical concepts without the need for assuming the existence of atoms. Certain nineteenth-century positivists (notably Mach) therefore insisted on purely philosophical grounds that it was incorrect to suppose that individual atoms actually existed, because they had never been observed as such. The atomic theory, they thought, should be regarded only as an interesting way of calculating various observable large-scale properties of matter. Nevertheless, evidence for the existence of individual atoms was ultimately discovered by people who took the atomic hypothesis seriously enough to suppose that individual atoms might actually exist, even though no one had yet observed them. We may have here, perhaps, a close analogy to the usual interpretation of the quantum theory, which avoids considering the possibility that the wave function of an individual system may represent objective reality, because we cannot observe it with the aid of existing experiments and theories.

[7] A leading nineteenth-century exponent of the positivist point of view was Mach. Modern positivists appear to have retreated from this extreme position, but its reflection still remains in the philosophical point of view implicitly adopted by a large number of modern theoretical physicists.

Finally, as an alternative to the positivist hypothesis of assigning reality only to that which we can now observe, we wish to prevent here another hypothesis, which we believe corresponds more closely to conclusions that can be drawn from general experience in actual scientific research. This hypothesis is based on the simple assumption that the world as a whole is objectively real and that, as far as we now know, it can correctly be regarded as having a precisely describable and analyzable structure of unlimited complexity. The pattern of this structure seems to be reflected completely but indirectly at every level, so that from experiments done at the level of size of human beings, it is very probably possible ultimately to draw inferences concerning the properties of the whole structure at all levels. We should never expect to obtain a complete theory of this structure, because there are almost certainly more elements in existence than we possibly can be aware of at any particular stage of scientific development. Any specified element, however, can in principle ultimately be discovered, but never all of them. Of course, we must avoid postulating a new element for each new phenomenon. But an equally serious mistake is to admit into the theory only those elements which can now be observed. For the purpose of a theory is not only to correlate the results of observations that we already know how to make, but also to suggest the need for new kinds of observations and to predict their results. In fact, the better a theory is able to suggest the need for new kinds of observations and to predict their results correctly, the more confidence we have that this theory is likely to be good representation of the actual properties of matter and not simply an empirical system especially chosen in such a way as to correlate a group of already known facts.

APPENDIX A. PHOTOELECTRIC AND COMPTON EFFECTS

In this appendix, we shall show how the electromagnetic field is to be described in our new interpretation, with the purpose of making possible a treatment of the photoelectric and Compton effects. For our purposes, it is adequate to restrict ourselves to a gauge in which $\operatorname{div}\mathbf{A}=0$, and to consider only the transverse part of the electromagnetic field, for in this gauge, the longitudinal part of the field can be expressed through Poisson's equation entirely in terms of the charge density. The Fourier analysis of the vector potential is then

$$\mathbf{A}(\mathbf{x})=(4\pi/V)^{\frac{1}{2}}\textstyle\sum_{\mathbf{k},\mu}\epsilon_{\mathbf{k},\mu}q_{\mathbf{k},\mu}e^{i\mathbf{k}\cdot\mathbf{x}} \qquad \text{(A1)}$$

with

$$q_{\mathbf{k},\mu}{}^{*}=q_{-\mathbf{k},\mu}.$$

The $q_{\mathbf{k},\mu}$ are coordinates of the electromagnetic field, associated with oscillations of wave number, $\mathbf{k}$, and polarization direction, μ, where $\boldsymbol{\varepsilon}_{\mathbf{k},\mu}$ is a unit vector normal to $\mathbf{k}$ and μ runs over two indices, corresponding to a pair of orthogonal directions of polarization. V is the volume of the box, which is assumed to be very large.

We also introduce the momenta $\Pi_{\mathbf{k},\mu}=\partial q_{\mathbf{k},\mu}{}^{*}/\partial t$, canonically conjugate[8] to the $q_{\mathbf{k},\mu}$. We have for the transverse part of the electric field

$$\mathfrak{E}(\mathbf{x})=-\frac{1}{c}\frac{\partial\mathbf{A}(\mathbf{x})}{\partial t}=-\left(\frac{4\pi}{Vc^{2}}\right)^{\frac{1}{2}}\sum_{\mathbf{k},\mu}\epsilon_{\mathbf{k},\mu}\Pi_{\mathbf{k},\mu}{}^{*}e^{i\mathbf{k}\cdot\mathbf{x}} \qquad \text{(A2)}$$

and for the magnetic field

$$\mathfrak{H}(\mathbf{x})=\nabla\times\mathbf{A}=-(4\pi/V)^{\frac{1}{2}}i\textstyle\sum_{\mathbf{k},\mu}(\mathbf{k}\times\epsilon_{\mathbf{k},\mu})q_{\mathbf{k},\mu}e^{i\mathbf{k}\cdot\mathbf{x}}. \qquad \text{(A3)}$$

The Hamiltonian of the radiation field corresponds to a collection of independent harmonic oscillators, each with angular frequency, $\omega=kc$. This Hamiltonian is

$$H^{(R)}=\textstyle\sum_{\mathbf{k},\mu}(\Pi_{\mathbf{k},\mu}\Pi_{\mathbf{k},\mu}{}^{*}+k^{2}c^{2}q_{\mathbf{k},\mu}q_{\mathbf{k},\mu}{}^{*}). \qquad \text{(A4)}$$

Now, in our interpretation of the quantum theory, the quantity $q_{\mathbf{k},\mu}$ is assumed to refer to the actual value of the $\mathbf{k}$, μ Fourier component of the vector potential. As in the case of the electron, however, there is present an objectively real superfield that is a function of all the electromagnetic field coordinates $q_{\mathbf{k},\mu}$. Thus, we have

$$\Psi^{(R)}=\Psi^{(R)}(\cdots q_{\mathbf{k},\mu}\cdots). \qquad \text{(A5)}$$

Writing $\Psi^{(R)}=R\exp(iS/\hbar)$, we obtain (in analogy with Paper I, Sec. 4)

$$\partial q_{\mathbf{k},\mu}/\partial t=\Pi_{\mathbf{k},\mu}{}^{*}=\partial s/\partial q_{\mathbf{k},\mu}{}^{*}. \qquad \text{(A6)}$$

The function $R(\cdots q_{\mathbf{k},\mu}\cdots)$ has two interpretations. First, it defines an additional quantum-mechanical term appearing in Maxwell's equations. To see the origin of this term, let us write the generalized Hamilton-Jacobi equation of the electromagnetic field, analogous to Paper I, Eq. (4),

$$\frac{\partial s}{\partial t}+\sum_{\mathbf{k},\mu}\frac{\partial s}{\partial q_{\mathbf{k},\mu}}\frac{\partial s}{\partial q_{\mathbf{k},\mu}{}^{*}}+\sum_{\mathbf{k},\mu}(kc)^{2}q_{\mathbf{k},\mu}q_{\mathbf{k},\mu}{}^{*}-\frac{\hbar^{2}}{2R}\sum_{\mathbf{k},\mu}\frac{\partial^{2}R(\cdots q_{\mathbf{k},\mu}\cdots)}{\partial q_{\mathbf{k},\mu}\partial q_{\mathbf{k},\mu}{}^{*}}=0. \qquad \text{(A7)}$$

The equation of motion of $q_{\mathbf{k},\mu}$, derived from the Hamiltonian implied by Eq. (A7) becomes

$$\ddot{q}_{\mathbf{k},\mu}+k^{2}c^{2}q_{\mathbf{k},\mu}=\frac{\partial}{\partial q_{\mathbf{k},\mu}{}^{*}}\left(\frac{\hbar^{2}}{2R}\sum_{\mathbf{k}',\mu'}\frac{\partial^{2}R}{\partial q_{\mathbf{k}',\mu'}\partial q_{\mathbf{k}',\mu'}{}^{*}}\right). \qquad \text{(A8)}$$

Since Maxwell's equations for empty space follow when the right-hand side is zero, we see that the "quantum-mechanical" terms can profoundly modify the behavior of the electromagnetic field. In fact, it is this modification which will contribute to the explana-

[8] See G. Wentzel, *Quantum Theory of Fields* (Interscience Publishers, Inc., New York, 1948).

tion of the ability of an oscillator, $q_{\mathbf{k},\mu}$, to transfer large quantities of energy and momentum rapidly even when $q_{\mathbf{k},\mu}$ is very small, for when $q_{\mathbf{k},\mu}$ is small, the right-hand side of Eq. (A8) may become very large.

The second interpretation of R is that as in Paper I, Eq. (5), it defines a conserved probability density that each of the $q_{\mathbf{k},\mu}$ has a certain specified value. From this fact, we see that although large transfers of energy and momentum to a radiation oscillator can occur in a short time when R is small, the probability of such a process is (as was also shown in Paper I, Sec. 7) very small.

In the lowest state (when no quanta are present) every oscillator is in the ground state. The super wave fields is then

$$\Psi_0^{(R)}=\exp[-\textstyle\sum_{\mathbf{k},\mu}(kcq_{\mathbf{k},\mu}q_{\mathbf{k},\mu}^*+\frac{1}{2}ikct)]. \quad (A9)$$

If the $\mathbf{k}'$, μ' oscillator is excited to the nth quantum state, the super wave field is

$$\Psi^{(R)}=h_n(q_{\mathbf{k}',\mu'})e^{-ink'ct}\Psi_0^{(R)}, \quad (A10)$$

where h_n is the nth hermite polynomial. As shown in Paper I, Sec. 5, the stationary states of such a system correspond to a quantized energy equal to the same value, $E_n=(n+\frac{1}{2})\hbar kc$, obtained from the usual interpretation. In nonstationary states, however, Eqs. (A7) and (A8) imply that the energy of each oscillator may fluctuate violently, as was also true of nonstationary states of the hydrogen atom (see Paper I, Sec. 7).

A nonstationary state of particular interest in the photoelectric and Compton effects is a state corresponding to the presence of an electromagnetic wave packet containing a single quantum. The super wave field for such a state is

$$\Psi_P^{(R)}=\textstyle\sum_{\mathbf{k},\mu}f_\mu(\mathbf{k}-\mathbf{k}_0)q_{\mathbf{k},\mu}e^{-ikct}\Psi_0^{(R)}, \quad (A11)$$

where $f_\mu(\mathbf{k}-\mathbf{k}_0)$ is a function that is large only near $\mathbf{k}=\mathbf{k}_0$ and the first hermite polynomial is represented by $q_{\mathbf{k},\mu}$, to which it is proportional.

To prove that Eq. (A11) represents an electromagnetic wave packet, we can evaluate the difference

$$\langle\Delta W\rangle_{\text{Av}}=\langle W\rangle_{\text{Av}}-\langle W_0\rangle_{\text{Av}}, \quad (A12)$$

where $\langle W(\mathbf{x})\rangle_{\text{Av}}$ is the actual mean energy density present (averaged over the ensemble), and $\langle W_0(\mathbf{x})\rangle_{\text{Av}}$ is the mean energy that would be present even in the ground state, because of zero-point fluctuations. We have

$$\langle W(\mathbf{x})\rangle_{\text{Av}}=\int\int\cdots\int\Psi_P^{*(R)}(\cdots q_{\mathbf{k},\mu}\cdots)\times\frac{[\mathfrak{E}^2(\mathbf{x})+\mathfrak{H}^2(\mathbf{x})]}{8\pi}\Psi_P^{(R)}(\cdots q_{\mathbf{k},\mu}\cdots)\times(\cdots dq_{\mathbf{k},\mu}\cdots), \quad (A13)$$

$$\langle W_0(\mathbf{x})\rangle_{\text{Av}}=\int\int\cdots\int\Psi_0^{*(R)}(\cdots q_{\mathbf{k},\mu}\cdots)\times\frac{[\mathfrak{E}^2(\mathbf{x})+\mathfrak{H}^2(\mathbf{x})]}{8\pi}\Psi_0^{(R)}(\cdots q_{\mathbf{k},\mu}\cdots)\times(\cdots dq_{\mathbf{k},\mu}\cdots). \quad (A14)$$

Obtaining $\mathfrak{E}(\mathbf{x})$ from Eq. (A2), $\mathfrak{H}(\mathbf{x})$ from Eq. (A3), $\Psi_P^{(R)}$ from Eq. (A10), $\Psi_0^{(R)}$ from Eq. (A9), we readily show that

$$\langle\Delta W(\mathbf{x})\rangle_{\text{Av}}=\textstyle\sum_{\mathbf{k},\mu}\sum_{\mathbf{k}',\mu'}f_\mu(\mathbf{k}-\mathbf{k}_0)f_{\mu'}(\mathbf{k}'-\mathbf{k}_0)\times e^{i(\mathbf{k}+\mathbf{k}')\cdot\mathbf{x}}\epsilon_{\mathbf{k},\mu}\cdot\epsilon_{\mathbf{k}',\mu'}. \quad (A15)$$

This means that the wave packet implies an excess over zero-point energy that is localized within a region in which the packet function, $\mathfrak{g}(\mathbf{x})$ is appreciable, where

$$\mathfrak{g}(\mathbf{x})=\textstyle\sum_{\mathbf{k},\mu}f_\mu(\mathbf{k}-\mathbf{k}_0)e^{i\mathbf{k}\cdot\mathbf{x}}\epsilon_{\mathbf{k},\mu}. \quad (A16)$$

We are now ready to treat the photoelectric and Compton effects. The entire treatment is so similar to that of the Franck-Hertz experiment (Paper I, Sec. 7) that we need merely sketch it here. We begin by adding to the radiation Hamiltonian, $H^{(R)}$, the particle Hamiltonian,

$$H^{(P)}=(1/2m)[\mathbf{p}-(e/c)\mathbf{A}(\mathbf{x})]^2. \quad (A17)$$

(We restrict ourselves here to nonrelativistic treatment.) The photoelectric effect corresponds to the transition of a radiation oscillator from an excited state to the ground state, while the atomic electron is ejected, with an energy $E=h\nu-I$, where I is the ionization potential of the atom. The initial super wave field, corresponding to an incident packet containing only one quantum, plus an atom in the ground state is (see Eq. (A11))

$$\Psi_i=\psi_0(\mathbf{x})\exp(-iE_0t/\hbar)\Psi_0^{(R)}(\cdots q_{\mathbf{k},\mu}\cdots)\times\textstyle\sum_{\mathbf{k},\mu}f_\mu(\mathbf{k}-\mathbf{k}_0)q_{\mathbf{k},\mu}e^{-ikct}. \quad (A18)$$

By solving Schroedinger's equation for the combined system, we obtain an asymptotic wave field analogous to Paper I, Eq. (26), containing terms corresponding to the photoelectric effect. These terms, which must be added to Ψ_i, to yield the complete superfield, are (asymptotically)

$$\delta\Psi_a=\Psi_0^{(R)}(\cdots q_{\mathbf{k},\mu}\cdots)\sum_{\mathbf{k},\mu}f_\mu(\mathbf{k}-\mathbf{k}_0)\times\frac{\exp[i\mathbf{k}'\cdot\mathbf{r}-i\hbar(k'^2/2m)t]}{r}g_\mu(\theta,\phi,k'), \quad (A19)$$

where the energy of the outgoing electron is $E=\hbar^2k'^2/2m=\hbar kc+E_0$. The function $g_\mu(\theta,\phi,k')$ is the amplitude associated with the ψ-field of the outgoing electron. This quantity can be calculated from the matrix element of the interaction term, $-(e/c)\mathbf{p}\cdot\mathbf{A}(\mathbf{x})$, by methods that are easily deducible from the usual perturbation theory.[6]

The outgoing electron packet has its center at $r=(\hbar k'/m)t$. Eventually, this packet will become completely separated from the initial electron wave function, $\psi_0(\mathbf{x})$. If the electron happens to enter the outgoing packet, the initial wave function can subsequently be ignored. The system then acts for all practical purposes as if its wave field were given by Eq. (A9), from which we conclude that the radiation field is in the ground state, while the electron has been liberated. It is readily shown that, as in the usual interpretation, the probability that the electron appears in the direction θ, ϕ can be calculated from $|g_\mu(\theta, \phi, k')|^2$ (see Paper I, Sec. 7).

To describe the Compton effect, we need only add to the super wave field the term corresponding to the appearance of an outgoing electromagnetic wave, as well as an outgoing electron. This part is asymptotically

$$\delta\Psi_b=\Psi_0^{(R)}(\cdots q_{\mathbf{k},\mu}\cdots)\sum_{\mathbf{k}',\mu'} f_\mu(\mathbf{k}-\mathbf{k}_0)$$
$$\times c_{\mathbf{k}',\mu'}{}^{\mathbf{k},\mu} q_{k',\mu'} g_{\mathbf{k}',\mu'}{}^{\mathbf{k},\mu}(\theta,\phi)\frac{e^{ik''r}}{r}$$
$$\times \exp\left(-ik'ct-\frac{i\hbar k''^2 t}{2m}\right), \quad \text{(A20)}$$

where

$$(\hbar^2k''^2/2m)+\hbar k'c=\hbar kc+E_0.$$

The quantity, $c_{\mathbf{k}',\mu'}{}^{\mathbf{k},\mu}$ is proportioned to the matrix element for a transition in which the $\mathbf{k}$, μ-radiation oscillator falls from the first excited state, to the ground state, while the $\mathbf{k}'$, μ'-oscillator rises from the ground state to the first excited state. This matrix element is determined mainly by the term $(e^2/8mc^2)A^2(\mathbf{x})$ in the hamiltonian.

It is easily seen that the outgoing electron packet eventually becomes completely separated both from the initial wave field, $\Psi_i(\mathbf{x}, \cdots q_{\mathbf{k},\mu}\cdots)$, and from the packet for the photoelectric effect, $\delta\Psi_a$ [defined in Eq. (A19)]. If the electron should happen to enter this packet, then the others can be ignored, and the system acts for all practical purposes like an outgoing electron, plus an independent outgoing light quantum. The reader will readily verify that the probability that the light quantum $\mathbf{k}'$, μ' appears along with an electron with angles θ, ϕ is precisely the same as in the usual interpretation.

APPENDIX B. A DISCUSSION OF INTERPRETATIONS OF THE QUANTUM THEORY PROPOSED BY DE BROGLIE AND ROSEN

After this article had been prepared, the author's attention was called to two papers in which an interpretation of the quantum theory similar to that suggested here was proposed, first by L. de Broglie,[9] and later by N. Rosen.[10] In both of these papers, it was suggested that if one writes $\psi=R\exp(is/\hbar)$, then one can regard R^2 as a probability density of particles having a velocity, $\mathbf{v}=\nabla s/m$. De Broglie regarded the ψ-field as an agent "guiding" the particle, and therefore referred to ψ as a "pilot wave." Both of these authors came to the conclusion that this interpretation could not consistently be carried through in those cases in which the field contained a linear combination of stationary state wave functions. As we shall see in this appendix, however, the difficulties encountered by the above authors could have been overcome by them, if only they had carried their ideas to a logical conclusion.

De Broglie's suggestions met strong objections on the part of Pauli,[11] in connection with the problem of inelastic scattering of a particle by a rigid rotator. Since this problem is conceptually equivalent to that of inelastic scattering of a particle by a hydrogen atom, which we have already treated in Paper I, Sec. 7, we shall discuss the objections raised by Pauli in terms of the latter example.

Now, according to Pauli's argument, the initial wave function in the scattering problem should be $\Psi=\exp(i\mathbf{p}_0\cdot\mathbf{y}/\hbar)\psi_0(\mathbf{x})$. This corresponds to a stationary state for the combined system, in which the particle momentum is $\mathbf{p}_0$, while the hydrogen atom is in its ground state, with a wave function, $\psi_0(\mathbf{x})$. After interaction between the incident particle and the hydrogen atom, the combined wave function can be represented as

$$\Psi=\sum_n f_n(\mathbf{y})\psi_n(\mathbf{x}), \quad \text{(B1)}$$

where $\psi_n(\mathbf{x})$ is the wave function for the nth excited state of the hydrogen atom, and $f_n(\mathbf{y})$ is the associated expansion coefficient. It is easily shown[12] that asymptotically, $f_n(\mathbf{y})$ takes the form of an outgoing wave, $f_n(\mathbf{y})\sim g_n(\theta,\phi)e^{ik_nr}/r$, where $(\hbar k_n)^2/2m=[(\hbar k_0)^2/2m]+E_n-E_0$. Now, if we write $\psi=R\exp(iS/\hbar)$, we find that the particle momenta, $p_x=\nabla_x S(\mathbf{x},\mathbf{y})$ and $\mathbf{p}_y=\nabla_y S(\mathbf{x},\mathbf{y})$, fluctuate violently in a way that depends strongly on the position of each particle. Thus, neither atom nor the outgoing particle ever seem to approach a stationary energy. On the other hand, we know from experiment that both the atom and the outgoing particle do eventually obtain definite (but presumably unpredictable) energy values. Pauli therefore concluded that the interpretation proposed by de Broglie was untenable. De Broglie seems to have agreed with the conclusion, since he subsequently gave up his suggested interpretation.[9]

---

[9] L. de Broglie, *An Introduction to the Study of Wave Mechanics* (E. P. Dutton and Company, Inc., New York, 1930), see Chapters 6, 9, and 10.

[10] N. Rosen, J. Elisha Mitchel Sci. Soc. **61**, Nos. 1 and 2 (August, 1945).

[11] *Reports on the 1927 Solvay Congress* (Gauthiers-Villars et Cie., Paris, 1928), see p. 280.

[12] N. F. Mott and H. S. W. Massey, *The Theory of Atomic Collisions* (Clarendon Press, Oxford, 1933).

Our answer to Pauli's objection is already contained in Paper I, Sec. 7, as well as in Sec. 2 of this paper. For as is well known, the use of an incident plane wave of infinite extent is an excessive abstraction, not realizable in practice. Actually, both the incident and outgoing parts of the ψ-field will always take the form of bounded packets. Moreover, as shown in Paper I, Sec. 7, all packets corresponding to different values of n will ultimately obtain classically describable separations. The outgoing particle must enter one of these packets, and it will remain with that particular packet thereafter, leaving the hydrogen atom in a definite but correlated stationary state. Thus, Pauli's objection is seen to be based on the use of the excessively abstract model of an infinite plane wave.

Although the above constitutes a complete answer to Pauli's specific objections to our suggested interpretation, we wish here to amplify our discussion somewhat, in order to anticipate certain additional objections that might be made along similar lines. For at this point, one might argue that even though the wave packet is bounded, it can nevertheless in principle be made arbitrarily large in extent by means of a suitable adjustment of initial conditions. Our interpretation predicts that in the region in which incident and outgoing ψ-waves overlap, the momentum of each particle will fluctuate violently, as a result of corresponding fluctuations in the "quantum-mechanical" potential produced by the ψ-field. The question arises, however, as to whether such fluctuations can really be in accord with experimental fact, especially since in principle they could occur when the particles were separated by distances much greater than that over which the "classical" interaction potential, $V(\mathbf{x}, \mathbf{y})$, was appreciable.

To show that these fluctuations are not in disagreement with any experimental facts now available, we first point out that even in the usual interpretation the energy of each particle cannot correctly be regarded as definite under the conditions which are assumed here, namely, that the incident and outgoing wave packets overlap. For as long as interference between two stationary state wave function is possible, the system acts as if it, in some sense, covered both states simultaneously.[13] In such a situation, the usual interpretation implies that a precisely defined value for the energy of either particle is meaningless. From such a wave function, one can predict only the probability that if the energy is measured, a definite value will be obtained. On the other hand, the very experimental conditions needed for measuring the energy play a key role in making a definite value of the energy possible because the effect of the measuring apparatus is to destroy interference between parts of the wave function corresponding to different values of the energy.[14]

[13] Reference 2, Chapter 16, Sec. 25.
[14] Reference 2, Chapter 6, Secs. 3 to 8; Chapter 22, Secs. 8 to 10.

In our interpretation, the overlap of incident and outgoing wave packets signifies not that the precise value of the energy of either particle can be given no meaning, but rather that this value fluctuates violently in an, in practice, unpredictable and uncontrollable way. When the energy of either particle is measured, however, then our interpretation predicts, in agreement with the usual interpretation, that the energy of each particle will become definite and constant, as a result of the effects of the energy-measuring apparatus on the observed system. To show how this happens, let us suppose that the energy of the hydrogen atom is measured by means of an interaction in which the "classical" potential, V, is a function only of the variables associated with the electron in the hydrogen atom and with the apparatus, but is not a function of variables associated with the outgoing particle. Let z be the coordinate of the measuring apparatus. Then as shown in Sec. 2, interaction with an apparatus that measures the energy of the hydrogen atom will transform the Ψ-function (B1), into

$$\Psi=\sum_n f_n(\mathbf{y})\psi_n(\mathbf{x})g_0(z-aEnt/\hbar^2). \tag{B2}$$

Now, we have seen that if the product at is large enough to make a distinct measurement possible, packets corresponding to different values of n will ultimately obtain classically describable separations in z space. The apparatus variable, z, must enter one of these packets; and, thereafter, all other packets can for practical purposes be ignored. The hydrogen atom is then left in a state having a definite and constant energy, while the outgoing particle has a correspondingly definite but correlated constant value for its energy. Thus, we find that as with the usual interpretation, our interpretation predicts that whenever we measure the energy of either particle by methods that are now available, a definite and constant value will always be obtained. Nevertheless, under conditions in which incident and outgoing wave packets overlap, and in which neither particle interacts with an energy-measuring device, our interpretation states unambiguously that real fluctuations in the energy of each particle will occur. These fluctuations are moreover, at least in principle, observable (for example, by methods discussed in Sec. 6). Meanwhile, under conditions in which we are limited by present methods of observation, our interpretation leads to predictions that are precisely the same as those obtained from the usual interpretation, so that no experiments supporting the usual interpretation can possibly contradict our interpretation.

In his book,[9] de Broglie raises objections to his own suggested interpretation of the quantum theory, which are very similar to those raised by Pauli. It is therefore not necessary to answer de Broglie's objections in detail here, since the answer is essentially the same as that which has been given to Pauli. We wish, however, to add one point. De Broglie assumes that not only electrons, but also light quanta, are associated with particles. A consistent application of the interpretation suggested here requires, however, as shown in Appendix A, that light quanta be described as electromagnetic wave packets. The only precisely definable quantities in such a packet are the Fourier components, $q_{\mathbf{k},\mu}$, of the vector potential and the corresponding canonically conjugate momenta, $\prod_{\mathbf{k},\mu}$. Such packets have many particle-like properties, including the ability to transfer rapidly a full quantum of energy at great distances. Nevertheless, it would not be consistent to assume the existence of a "photon" particle, associated with each light quantum.

We shall now discuss Rosen's paper briefly.[10] Rosen gave up his suggested interpretation of the quantum theory, because of difficulties arising in connection with the interpretation of standing waves. In the case of the stationary states of a free particle in a box, which we have already discussed in Sec. 8, our interpretation leads to the conclusion that the particle is standing still. Rosen did not wish to accept this conclusion, because it seemed to disagree with the statement of the usual interpretation that in such a state the electron is moving with equal probability that the motion is in either direction. To answer Rosen's objections, we need merely point out again that the usual interpretation can give no meaning to the motion of particles in a stationary state; at best, it can only predict the probability that a given result will be obtained, if the velocity is measured. As we saw in Sec. 8, however, our interpretation leads to precisely the same predictions as are obtained from the usual interpretation, for any process which could actually provide us with a measurement of the velocity of the electron. One must remember, however, that the value of the momentum "observable" as it is now "measured" is not necessarily equal to the particle momentum existing before interaction with the measuring apparatus took place.

We conclude that the objections raised by Pauli, de Broglie, and Rosen, to interpretations of the quantum theory similar to that suggested here, can all be answered by carrying every aspect of our suggested interpretation to its logical conclusion.

III.4 ON THE PROBLEM OF HIDDEN VARIABLES IN QUANTUM MECHANICS*

JOHN S. BELL†

The demonstrations of von Neumann and others, that quantum mechanics does not permit a hidden variable interpretation, are reconsidered. It is shown that their essential axioms are unreasonable. It is urged that in further examination of this problem an interesting axiom would be that mutually distant systems are independent of one another.

I. INTRODUCTION

To know the quantum mechanical state of a system implies, in general, only statistical restrictions on the results of measurements. It seems interesting to ask if this statistical element be thought of as arising, as in classical statistical mechanics, because the states in question are averages over better defined states for which individually the results would be quite determined. These hypothetical "dispersion free" states would be specified not only by the quantum mechanical state vector but also by additional "hidden variables"—"hidden" because if states with prescribed values of these variables could actually be prepared, quantum mechanics would be observably inadequate.

Whether this question is indeed interesting has been the subject of debate.[1,2] The present paper does not contribute to that debate. It is addressed to those who do find the question interesting, and more particularly to those among them who believe that[3] "the question concerning the existence of such hidden variables received an early and rather decisive answer in the form of von Neumann's proof on the mathematical impossibility of such variables in quantum theory." An attempt will be made to clarify what von Neumann and his successors actually demonstrated. This will cover, as well as von Neumann's treatment, the recent version of the argument by Jauch and Piron,[3] and the stronger result consequent on the work of Gleason.[4] It will be urged that these analyses leave the real question untouched. In fact it will be seen that these demonstrations require from the hypothetical dispersion free states, not only that appropriate ensembles thereof should have all measurable properties of quantum mechanical states, but certain other properties as well. These additional demands appear reasonable when results of measurement are loosely identified with properties of isolated systems. They are seen to be quite unreasonable when one remembers with Bohr[5] "the impossibility of any sharp distinction between the behavior of atomic objects and the interaction with the measuring instruments which serve to define the conditions under which the phenomena appear."

The realization that von Neumann's proof is of limited relevance has been gaining ground since the 1952 work of Bohm.[6] However, it is far from universal. Moreover, the writer has not found in the literature any adequate analysis of what went wrong.[7] Like all authors of noncommissioned reviews, he thinks that he can restate the position with such clarity and simplicity that all previous discussions will be eclipsed.

II. ASSUMPTIONS, AND A SIMPLE EXAMPLE

The authors of the demonstrations to be reviewed were concerned to assume as little as possible about quantum mechanics. This is valuable for some purposes, but not for ours. We are interested only in the possibility of hidden variables in ordinary quantum me-

* Work supported by U.S. Atomic Energy Commission.

† Permanent address: CERN, Geneva.

[1] The following works contain discussions of and references on the hidden variable problem: L. de Broglie, *Physicien et Penseur* (Albin Michel, Paris, 1953); W. Heisenberg, in *Niels Bohr and the Development of Physics*, W. Pauli, Ed. (McGraw-Hill Book Co., Inc., New York, and Pergamon Press, Ltd., London, 1955); *Observation and Interpretation*, S. Körner, Ed. (Academic Press Inc., New York, and Butterworths Scientific Publ., Ltd., London, 1957); N. R. Hansen, *The Concept of the Positron* (Cambridge University Press, Cambridge, England, 1963). See also the various works by D. Bohm cited later, and Bell and Nauenberg.[8] For the view that the possibility of hidden variables has little interest, see especially the contributions of Rosenfeld to the first and third of these references, of Pauli to the first, the article of Heisenberg, and many passages in Hansen.

[2] A. Einstein, *Philosopher Scientist*, P. A. Schilp, Ed. (Library of Living Philosophers, Evanston, Ill., 1949). Einstein's "Autobiographical Notes" and "Reply to Critics" suggest that the hidden variable problem has some interest.

[3] J. M. Jauch and C. Piron, Helv. Phys. Acta **36**, 827 (1963).

[4] A. M. Gleason, J. Math. & Mech. **6**, 885 (1957). I am much indebted to Professor Jauch for drawing my attention to this work.

[5] N. Bohr, in Ref. 2.

[6] D. Bohm, Phys. Rev. **85**, 166, 180 (1952).

[7] In particular the analysis of Bohm[6] seems to lack clarity, or else accuracy. He fully emphasizes the role of the experimental arrangement. However, it seems to be implied (Ref. 6, p. 187) that the circumvention of the theorem *requires* the association of *hidden* variables with the apparatus as well as with the system observed. The scheme of Sec. II is a counter example to this. Moreover, it will be seen in Sec. III that if the essential additivity assumption of von Neumann were granted, hidden variables wherever located would not avail. Bohm's further remarks in Ref. 16 (p. 95) and Ref. 17 (p. 358) are also unconvincing. Other critiques of the theorem are cited, and some of them rebutted, by Albertson [J. Albertson, Am. J. Phys. **29**, 478 (1961)].

Originally published in *Reviews of Modern Physics, 38*, 447-52 (1966).

chanics and will use freely all the usual notions. Thereby the demonstrations will be substantially shortened.

A quantum mechanical "system" is supposed to have "observables" represented by Hermitian operators in a complex linear vector space. Every "measurement" of an observable yields one of the eigenvalues of the corresponding operator. Observables with commuting operators can be measured simultaneously.[8] A quantum mechanical "state" is represented by a vector in the linear state space. For a state vector ψ the statistical expectation value of an observable with operator O is the normalized inner product $(\psi, O\psi)/(\psi, \psi)$.

The question at issue is whether the quantum mechanical states can be regarded as ensembles of states further specified by additional variables, such that given values of these variables together with the state vector determine precisely the results of individual measurements. These hypothetical well-specified states are said to be "dispersion free."

In the following discussion it will be useful to keep in mind as a simple example a system with a two-dimensional state space. Consider for definiteness a spin $-\frac{1}{2}$ particle without translational motion. A quantum mechanical state is represented by a two-component state vector, or spinor, ψ. The observables are represented by 2×2 Hermitian matrices

$$\alpha+\boldsymbol{\beta}\cdot\boldsymbol{\sigma}, \tag{1}$$

where α is a real number, $\boldsymbol{\beta}$ a real vector, and $\boldsymbol{\sigma}$ has for components the Pauli matrices; α is understood to multiply the unit matrix. Measurement of such an observable yields one of the eigenvalues,

$$\alpha\pm|\boldsymbol{\beta}|, \tag{2}$$

with relative probabilities that can be inferred from the expectation value

$$\langle\alpha+\boldsymbol{\beta}\cdot\boldsymbol{\sigma}\rangle=(\psi, [\alpha+\boldsymbol{\beta}\cdot\boldsymbol{\sigma}]\psi).$$

For this system a hidden variable scheme can be supplied as follows: The dispersion free states are specified by a real number λ, in the interval $-\frac{1}{2}\leq\lambda\leq\frac{1}{2}$, as well as the spinor ψ. To describe how λ determines which eigenvalue the measurement gives, we note that by a rotation of coordinates ψ can be brought to the form

$$\psi=\begin{pmatrix}1\\0\end{pmatrix}.$$

Let β_x, β_y, β_z, be the components of $\boldsymbol{\beta}$ in the new coordinate system. Then measurement of $\alpha+\boldsymbol{\beta}\cdot\boldsymbol{\sigma}$ on the state specified by ψ and λ results with certainty in the eigenvalue

$$\alpha+|\boldsymbol{\beta}|\operatorname{sign}(\lambda|\boldsymbol{\beta}|+\tfrac{1}{2}|\beta_z|)\operatorname{sign}X, \tag{3}$$

where

$$\begin{aligned} X&=\beta_z && \text{if } \beta_z\neq 0\\ &=\beta_x && \text{if } \beta_z=0,\ \beta_x\neq 0\\ &=\beta_y && \text{if } \beta_z=0, \text{ and } \beta_x=0 \end{aligned}$$

and

$$\begin{aligned} \operatorname{sign}X&=+1 && \text{if } X\geq 0\\ &=-1 && \text{if } X<0. \end{aligned}$$

The quantum mechanical state specified by ψ is obtained by uniform averaging over λ. This gives the expectation value

$$\langle\alpha+\boldsymbol{\beta}\cdot\boldsymbol{\sigma}\rangle$$

$$=\int_{-\frac{1}{2}}^{\frac{1}{2}} d\lambda\{\alpha+|\boldsymbol{\beta}|\operatorname{sign}(\lambda|\boldsymbol{\beta}|+\tfrac{1}{2}|\beta_z|)\operatorname{sign}X\}=\alpha+\beta_z$$

as required.

It should be stressed that no physical significance is attributed here to the parameter λ and that no pretence is made of giving a complete reinterpretation of quantum mechanics. The sole aim is to show that at the level considered by von Neumann such a reinterpretation is not excluded. A complete theory would require for example an account of the behavior of the hidden variables during the measurement process itself. With or without hidden variables the analysis of the measurement process presents peculiar difficulties,[8] and we enter upon it no more than is strictly necessary for our very limited purpose.

III. VON NEUMANN

Consider now the proof of von Neumann[9] that dispersion free states, and so hidden variables, are impossible. His essential assumption[10] is: *Any real linear combination of any two Hermitian operators represents an observable, and the same linear combination of expecta-*

---

[8] Recent papers on the measurement process in quantum mechanics, with further references, are: E. P. Wigner, Am. J. Phys. **31**, 6 (1963); A. Shimony, *ibid.* **31**, 755 (1963); J. M. Jauch, Helv. Phys. Acta **37**, 293 (1964); B. d'Espagnat, *Conceptions de la physique contemporaine* (Hermann & Cie., Paris, 1965); J. S. Bell and M. Nauenberg, in *Preludes in Theoretical Physics, In Honor of V. Weisskopf* (North-Holland Publishing Company, Amsterdam, 1966).

[9] J. von Neumann, *Mathematische Grundlagen der Quantenmechanik* (Julius Springer-Verlag, Berlin, 1932) [English transl.: Princeton University Press, Princeton, N.J., 1955]. All page numbers quoted are those of the English edition. The problem is posed in the preface, and on p. 209. The formal proof occupies essentially pp. 305–324 and is followed by several pages of commentary. A self-contained exposition of the proof has been presented by J. Albertson (see Ref. 7).

[10] This is contained in von Neumann's **B′** (p. 311), **1** (p. 313), and **11** (p. 314).

tion values is the expectation value of the combination. This is true for quantum mechanical states; it is required by von Neumann of the hypothetical dispersion free states also. In the two-dimensional example of Sec. II, the expectation value must then be a linear function of α and $\boldsymbol{\beta}$. But for a dispersion free state (which has no statistical character) the expectation value of an observable must equal one of its eigenvalues. The eigenvalues (2) are certainly not linear in $\boldsymbol{\beta}$. Therefore, dispersion free states are impossible. If the state space has more dimensions, we can always consider a two-dimensional subspace; therefore, the demonstration is quite general.

The essential assumption can be criticized as follows. At first sight the required additivity of expectation values seems very reasonable, and it is rather the nonadditivity of allowed values (eigenvalues) which requires explanation. Of course the explanation is well known: A measurement of a sum of noncommuting observables cannot be made by combining trivially the results of separate observations on the two terms—it requires a quite distinct experiment. For example the measurement of σ_x for a magnetic particle might be made with a suitably oriented Stern Gerlach magnet. The measurement of σ_y would require a different orientation, and of $(\sigma_x+\sigma_y)$ a third and different orientation. But this explanation of the nonadditivity of allowed values also establishes the nontriviality of the additivity of expectation values. The latter is a quite peculiar property of quantum mechanical states, not to be expected *a priori*. There is no reason to demand it individually of the hypothetical dispersion free states, whose function it is to reproduce the *measurable* peculiarities of quantum mechanics *when averaged over*.

In the trivial example of Sec. II the dispersion free states (specified λ) have additive expectation values only for commuting operators. Nevertheless, they give logically consistent and precise predictions for the results of all possible measurements, which when averaged over λ are fully equivalent to the quantum mechanical predictions. In fact, for this trivial example, the hidden variable question as posed informally by von Neumann[11] in his book is answered in the affirmative.

Thus the formal proof of von Neumann does not justify his informal conclusion[12]: "It is therefore not, as is often assumed, a question of reinterpretation of quantum mechanics—the present system of quantum mechanics would have to be objectively false in order that another description of the elementary process than the statistical one be possible." It was not the objective measurable predictions of quantum mechanics which ruled out hidden variables. It was the arbitrary assumption of a particular (and impossible) relation between the results of incompatible measurements either of which *might* be made on a given occasion but only one of which can in fact be made.

[11] Reference 9, p. 209.
[12] Reference 9, p. 325.

IV. JAUCH AND PIRON

A new version of the argument has been given by Jauch and Piron.[3] Like von Neumann they are interested in generalized forms of quantum mechanics and do not assume the usual connection of quantum mechanical expectation values with state vectors and operators. We assume the latter and shorten the argument, for we are concerned here only with possible interpretations of ordinary quantum mechanics.

Consider only observables represented by projection operators. The eigenvalues of projection operators are 0 and 1. Their expectation values are equal to the probabilities that 1 rather than 0 is the result of measurement. For any two projection operators, a and b, a third $(a\cap b)$ is defined as the projection on to the intersection of the corresponding subspaces. The essential axioms of Jauch and Piron are the following:

(A) Expectation values of *commuting* projection operators are additive.

(B) If, for some state and two projections a and b,

$$\langle a\rangle=\langle b\rangle=1,$$

then for that state

$$\langle a\cap b\rangle=1.$$

Jauch and Piron are led to this last axiom (4° in their numbering) by an analogy with the calculus of propositions in ordinary logic. The projections are to some extent analogous to logical propositions, with the allowed value 1 corresponding to "truth" and 0 to "falsehood," and the construction $(a\cap b)$ to (a "and" b) In logic we have, of course, if a is true and b is true then (a and b) is true. The axiom has this same structure.

Now we can quickly rule out dispersion free states by considering a 2-dimensional subspace. In that the projection operators are the zero, the unit operator, and those of the form

$$\tfrac{1}{2}+\tfrac{1}{2}\hat{\boldsymbol{\alpha}}\cdot\boldsymbol{\sigma},$$

where $\hat{\alpha}$ is a unit vector. In a dispersion free state the expectation value of an operator must be one of its eigenvalues, 0 or 1 for projections. Since from A

$$\langle\tfrac{1}{2}+\tfrac{1}{2}\hat{\boldsymbol{\alpha}}\cdot\boldsymbol{\sigma}\rangle+\langle\tfrac{1}{2}-\tfrac{1}{2}\hat{\boldsymbol{\alpha}}\cdot\boldsymbol{\sigma}\rangle=1,$$

we have that for a dispersion free state either

$$\langle\tfrac{1}{2}+\tfrac{1}{2}\hat{\boldsymbol{\alpha}}\cdot\boldsymbol{\sigma}\rangle=1 \quad\text{or}\quad \langle\tfrac{1}{2}-\tfrac{1}{2}\hat{\boldsymbol{\alpha}}\cdot\boldsymbol{\sigma}\rangle=1.$$

Let $\boldsymbol{\alpha}$ and $\hat{\boldsymbol{\beta}}$ be any noncollinear unit vectors and

$$a=\tfrac{1}{2}\pm\tfrac{1}{2}\hat{\boldsymbol{\alpha}}\cdot\boldsymbol{\sigma},\qquad b=\tfrac{1}{2}\pm\tfrac{1}{2}\hat{\boldsymbol{\beta}}\cdot\boldsymbol{\sigma},$$

with the signs chosen so that $\langle a\rangle=\langle b\rangle=1$. Then B requires

$$\langle a\cap b\rangle=1.$$

But with $\hat{\alpha}$ and $\hat{\beta}$ noncollinear, one readily sees that

$$a \cap b = 0,$$

so that

$$\langle a \cap b \rangle = 0.$$

So there can be no dispersion free states.

The objection to this is the same as before. We are not dealing in B with logical propositions, but with measurements involving, for example, differently oriented magnets. The axiom holds for quantum mechanical states.[13] But it is a quite peculiar property of them, in no way a necessity of thought. Only the quantum mechanical averages over the dispersion free states need reproduce this property, as in the example of Sec. II.

V. GLEASON

The remarkable mathematical work of Gleason[4] was not explicitly addressed to the hidden variable problem. It was directed to reducing the axiomatic basis of quantum mechanics. However, as it apparently enables von Neumann's result to be obtained without objectionable assumptions about noncommuting operators, we must clearly consider it. The relevant corollary of Gleason's work is that, if the dimensionality of the state space is greater than two, the additivity requirement for expectation values of *commuting operators* cannot be met by dispersion free states. This will now be proved, and then its significance discussed. It should be stressed that Gleason obtained more than this, by a lengthier argument, but this is all that is essential here.

It suffices to consider projection operators. Let $P(\Phi)$ be the projector on to the Hilbert space vector Φ, i.e., acting on any vector ψ

$$P(\Phi)\psi = (\Phi, \Phi)^{-1}(\Phi, \psi)\Phi.$$

If a set Φ_i are complete and orthogonal,

$$\sum_i P(\Phi_i) = 1.$$

Since the $P(\Phi_i)$ commute, by hypothesis then

$$\sum_i \langle P(\Phi_i) \rangle = 1. \tag{4}$$

Since the expectation value of a projector is nonnegative (each measurement yields one of the allowed values 0 or 1), and since any two orthogonal vectors can be regarded as members of a complete set, we have:

(A) If with some vector Φ, $\langle P(\Phi) \rangle = 1$ for a given state, then for that state $\langle P(\psi) \rangle = 0$ for any ψ orthogonal on Φ.

If ψ_1 and ψ_2 are another orthogonal basis for the subspace spanned by some vectors Φ_1 and Φ_2, then from (4)

$$\langle P(\psi_1) \rangle + \langle P(\psi_2) \rangle = 1 - \sum_{i \neq 1, i \neq 2} \langle P(\Phi_i) \rangle$$

or

$$\langle P(\psi_1) \rangle + \langle P(\psi_2) \rangle = \langle P(\Phi_1) \rangle + \langle P(\Phi_2) \rangle.$$

Since ψ_1 may be any combination of Φ_1 and Φ_2, we have:

(B) If for a given state

$$\langle P(\Phi_1) \rangle = \langle P(\Phi_2) \rangle = 0$$

for some pair of orthogonal vectors, then

$$\langle P(\alpha\Phi_1 + \beta\Phi_2) \rangle = 0$$

for all α and β.

(A) and (B) will now be used repeatedly to establish the following. Let Φ and ψ be some vectors such that for a given state

$$\langle P(\psi) \rangle = 1, \tag{5}$$

$$\langle P(\Phi) \rangle = 0. \tag{6}$$

Then Φ and ψ cannot be arbitrarily close; in fact

$$| \Phi - \psi | > \tfrac{1}{2} | \psi |. \tag{7}$$

To see this let us normalize ψ and write Φ in the form

$$\Phi = \psi + \epsilon\psi',$$

where ψ' is orthogonal to ψ and normalized and ϵ is a real number. Let ψ'' be a normalized vector orthogonal to both ψ and ψ' (it is here that we need three dimensions at least) and so to Φ. By (A) and (5),

$$\langle P(\psi') \rangle = 0, \qquad \langle P(\psi'') \rangle = 0.$$

Then by (B) and (6),

$$\langle P(\Phi + \gamma^{-1}\epsilon\psi'') \rangle = 0,$$

where γ is any real number, and also by (B),

$$\langle P(-\epsilon\psi' + \gamma\epsilon\psi'') \rangle = 0.$$

The vector arguments in the last two formulas are orthogonal; so we may add them, again using (B):

$$\langle P(\psi + \epsilon(\gamma + \gamma^{-1})\psi'') \rangle = 0.$$

Now if ϵ is less than $\frac{1}{2}$, there are real γ such that

$$\epsilon(\gamma + \gamma^{-1}) = \pm 1.$$

Therefore,

$$\langle P(\psi + \psi'') \rangle = \langle P(\psi - \psi'') \rangle = 0.$$

The vectors $\psi \pm \psi''$ are orthogonal; adding them and again using (B),

$$\langle P(\psi) \rangle = 0.$$

[13] In the two-dimensional case $\langle a \rangle = \langle b \rangle = 1$ (for some quantum mechanical state) is possible only if the two projectors are identical ($\hat{\alpha} = \hat{\beta}$). Then $a \cap b = a = b$ and $\langle a \cap b \rangle = \langle a \rangle = \langle b \rangle = 1$.

This contradicts the assumption (5). Therefore,

$$\epsilon > \tfrac{1}{2},$$

as announced in (7).

Consider now the possibility of dispersion free states. For such states each projector has expectation value either 0 or 1. It is clear from (4) that both values must occur, and since there are no other values possible, there must be arbitrarily close pairs ψ, Φ with different expectation values 0 and 1, respectively. But we saw above such pairs could *not* be arbitrarily close. Therefore, there are no dispersion free states.

That so much follows from such apparently innocent assumptions leads us to question their innocence. Are the requirements imposed, which are satisfied by quantum mechanical states, reasonable requirements on the dispersion free states? Indeed they are not. Consider the statement (B). The operator $P(\alpha\Phi_1+\beta\Phi_2)$ commutes with $P(\Phi_1)$ and $P(\Phi_2)$ only if either α or β is zero. Thus in general measurement of $P(\alpha\Phi_1+\beta\Phi_2)$ requires a quite distinct experimental arrangement. We can therefore reject (B) on the grounds already used: it relates in a nontrivial way the results of experiments which cannot be performed simultaneously; the dispersion free states need not have this property, it will suffice if the quantum mechanical averages over them do. How did it come about that (B) was a consequence of assumptions in which only commuting operators were explicitly mentioned? The danger in fact was not in the explicit but in the implicit assumptions. It was tacitly assumed that measurement of an observable must yield the same value independently of what other measurements may be made simultaneously. Thus as well as $P(\Phi_3)$ say, one might measure *either* $P(\Phi_2)$ *or* $P(\psi_2)$, where Φ_2 and ψ_2 are orthogonal to Φ_3 but not to one another. These different possibilities require different experimental arrangements; there is no *a priori* reason to believe that the results for $P(\Phi_3)$ should be the same. The result of an observation may reasonably depend not only on the state of the system (including hidden variables) but also on the complete disposition of the apparatus; see again the quotation from Bohr at the end of Sec. I.

To illustrate these remarks, we construct a very artificial but simple hidden variable decomposition. If we regard all observables as functions of commuting projectors, it will suffice to consider measurements of the latter. Let $P_1, P_2, \cdots$ be the set of projectors measured by a given apparatus, and for a given quantum mechanical state let their expectation values be $\lambda_1, \lambda_2-\lambda_1, \lambda_3-\lambda_2, \cdots$. As hidden variable we take a real number $0<\lambda\leq 1$; we specify that measurement on a state with given λ yields the value 1 for P_n if $\lambda_{n-1}<\lambda\leq\lambda_n$, and zero otherwise. The quantum mechanical state is obtained by uniform averaging over λ. There is no contradiction with Gleason's corollary, because the result for a given P_n depends also on the choice of the others. Of course it would be silly to let the result be affected by a mere permutation of the other P's, so we specify that the same order is taken (however defined) when the P's are in fact the same set. Reflection will deepen the initial impression of artificiality here. However, the example suffices to show that the implicit assumption of the impossibility proof was essential to its conclusion. A more serious hidden variable decomposition will be taken up in Sec. VI.[14]

VI. LOCALITY AND SEPARABILITY

Up till now we have been resisting arbitrary demands upon the hypothetical dispersion free states. However, as well as reproducing quantum mechanics on averaging, there *are* features which can reasonably be desired in a hidden variable scheme. The hidden variables should surely have some spacial significance and should evolve in time according to prescribed laws. These are prejudices, but it is just this possibility of interpolating some (preferably causal) space-time picture, between preparation of and measurements on states, that makes the quest for hidden variables interesting to the unsophisticated.[2] The ideas of space, time, and causality are not prominent in the kind of discussion we have been considering above. To the writer's knowledge the most successful attempt in that direction is the 1952 scheme of Bohm for elementary wave mechanics. By way of conclusion, this will be sketched briefly, and a curious feature of it stressed.

Consider for example a system of two spin $-\frac{1}{2}$ particles. The quantum mechanical state is represented by a wave function,

$$\psi_{ij}(\mathbf{r}_1, \mathbf{r}_2),$$

where i and j are spin indices which will be suppressed. This is governed by the Schrödinger equation,

$$\partial\psi/\partial t = -i(-(\partial^2/\partial\mathbf{r}_1^2)-(\partial^2/\partial\mathbf{r}_2^2)+V(\mathbf{r}_1-\mathbf{r}_2)$$
$$+a\boldsymbol{\sigma}_1\cdot\mathbf{H}(\mathbf{r}_1)+b\boldsymbol{\sigma}_2\cdot\mathbf{H}(\mathbf{r}_2))\psi, \quad (8)$$

where V is the interparticle potential. For simplicity we have taken neutral particles with magnetic moments, and an external magnetic field $\mathbf{H}$ has been allowed to represent spin analyzing magnets. The hidden variables are then two vectors $\mathbf{X}_1$ and $\mathbf{X}_2$, which give directly the results of position measurements. Other measurements are reduced ultimately to position measurements.[15] For example, measurement of a spin component means observing whether the particle emerges with an upward or downward deflection from a Stern–

[14] The simplest example for illustrating the discussion of Sec. V would then be a particle of spin 1, postulating a sufficient variety of spin–external-field interactions to permit arbitrary complete sets of spin states to be spacially separated.

[15] There are clearly enough measurements to be interesting that can be made in this way. We will not consider whether there are others.

Gerlach magnet. The variables $\mathbf{X}_1$ and $\mathbf{X}_2$ are supposed to be distributed in configuration space with the probability density,

$$\rho(\mathbf{X}_1, \mathbf{X}_2) = \sum_{ij} | \psi_{ij}(\mathbf{X}_1, \mathbf{X}_2) |^2,$$

appropriate to the quantum mechanical state. Consistently, with this $\mathbf{X}_1$ and $\mathbf{X}_2$ are supposed to vary with time according to

$$d\mathbf{X}_1/dt = \rho(\mathbf{X}_1, \mathbf{X}_2)^{-1} \operatorname{Im} \sum_{ij} \psi_{ij}^*(\mathbf{X}_1, \mathbf{X}_2)(\partial/\partial\mathbf{X}_1)\psi_{ij}(\mathbf{X}_1, \mathbf{X}_2),$$

$$d\mathbf{X}_2/dt = \rho(\mathbf{X}_1, \mathbf{X}_2)^{-1} \operatorname{Im} \sum_{ij} \psi_{ij}^*(\mathbf{X}_1, \mathbf{X}_2)(\partial/\partial\mathbf{X}_2)\psi_{ij}(\mathbf{X}_1, \mathbf{X}_2). \quad (9)$$

The curious feature is that the trajectory equations (9) for the hidden variables have in general a grossly nonlocal character. If the wave function is factorable before the analyzing fields become effective (the particles being far apart),

$$\psi_{ij}(\mathbf{X}_1, \mathbf{X}_2) = \Phi_i(\mathbf{X}_1)\chi_j(\mathbf{X}_2),$$

this factorability will be preserved. Equations (8) then reduce to

$$d\mathbf{X}_1/dt = [\sum_i \Phi_i^*(\mathbf{X}_1)\Phi_i(X_1)]^{-1} \times \operatorname{Im} \sum_i \Phi_i^*(\mathbf{X}_1)(\partial/\partial\mathbf{X}_1)\Phi_i(X_1),$$

$$d\mathbf{X}_2/dt = [\sum_j \chi_j^*(\mathbf{X}_2)\chi_j(X_2)]^{-1} \times \operatorname{Im} \sum_j \chi_j^*(\mathbf{X}_2)(\partial/\partial\mathbf{X}_2)\chi(X_2).$$

The Schrödinger equation (8) also separates, and the trajectories of $\mathbf{X}_1$ and $\mathbf{X}_2$ are determined separately by equations involving $\mathbf{H}(\mathbf{X}_1)$ and $\mathbf{H}(\mathbf{X}_2)$, respectively. However, in general, the wave function is not factorable. The trajectory of 1 then depends in a complicated way on the trajectory and wave function of 2, and so on the analyzing fields acting on 2—however remote these may be from particle 1. So in this theory an explicit causal mechanism exists whereby the disposition of one piece of apparatus affects the results obtained with a distant piece. In fact the Einstein–Podolsky–Rosen paradox is resolved in the way which Einstein would have liked least (Ref. 2, p. 85).

More generally, the hidden variable account of a given system becomes entirely different when we remember that it has undoubtedly interacted with numerous other systems in the past and that the total wave function will certainly not be factorable. The same effect complicates the hidden variable account of the theory of measurement, when it is desired to include part of the "apparatus" in the system.

Bohm of course was well aware[6,16–18] of these features of his scheme, and has given them much attention. However, it must be stressed that, to the present writer's knowledge, there is no *proof* that *any* hidden variable account of quantum mechanics *must* have this extraordinary character.[19] It would therefore be interesting, perhaps,[1] to pursue some further "impossibility proofs," replacing the arbitrary axioms objected to above by some condition of locality, or of separability of distant systems.

ACKNOWLEDGMENTS

The first ideas of this paper were conceived in 1952. I warmly thank Dr. F. Mandl for intensive discussion at that time. I am indebted to many others since then, and latterly, and very especially, to Professor J. M. Jauch.

---

[16] D. Bohm, *Causality and Chance in Modern Physics* (D. Van Nostrand Co., Inc., Princeton, N.J., 1957).

[17] D. Bohm, in *Quantum Theory*, D. R. Bates, Ed. (Academic Press Inc., New York, 1962).

[18] D. Bohm and Y. Aharonov, Phys. Rev. **108**, 1070 (1957).

[19] Since the completion of this paper such a proof has been found [J. S. Bell, Physics **1**, 195 (1965)].

III.5 ON THE EINSTEIN PODOLSKY ROSEN PARADOX*

JOHN S. BELL†

I. Introduction

THE paradox of Einstein, Podolsky and Rosen [1] was advanced as an argument that quantum mechanics could not be a complete theory but should be supplemented by additional variables. These additional variables were to restore to the theory causality and locality [2]. In this note that idea will be formulated mathematically and shown to be incompatible with the statistical predictions of quantum mechanics. It is the requirement of locality, or more precisely that the result of a measurement on one system be unaffected by operations on a distant system with which it has interacted in the past, that creates the essential difficulty. There have been attempts [3] to show that even without such a separability or locality requirement no "hidden variable" interpretation of quantum mechanics is possible. These attempts have been examined elsewhere [4] and found wanting. Moreover, a hidden variable interpretation of elementary quantum theory [5] has been explicitly constructed. That particular interpretation has indeed a grossly non-local structure. This is characteristic, according to the result to be proved here, of any such theory which reproduces exactly the quantum mechanical predictions.

II. Formulation

With the example advocated by Bohm and Aharonov [6], the EPR argument is the following. Consider a pair of spin one-half particles formed somehow in the singlet spin state and moving freely in opposite directions. Measurements can be made, say by Stern-Gerlach magnets, on selected components of the spins $\vec{\sigma}_1$ and $\vec{\sigma}_2$. If measurement of the component $\vec{\sigma}_1 \cdot \vec{a}$, where $\vec{a}$ is some unit vector, yields the value $+1$ then, according to quantum mechanics, measurement of $\vec{\sigma}_2 \cdot \vec{a}$ must yield the value -1 and vice versa. Now we make the hypothesis [2], and it seems one at least worth considering, that if the two measurements are made at places remote from one another the orientation of one magnet does not influence the result obtained with the other. Since we can predict in advance the result of measuring any chosen component of $\vec{\sigma}_2$, by previously measuring the same component of $\vec{\sigma}_1$, it follows that the result of any such measurement must actually be predetermined. Since the initial quantum mechanical wave function does *not* determine the result of an individual measurement, this predetermination implies the possibility of a more complete specification of the state.

Let this more complete specification be effected by means of parameters λ. It is a matter of indifference in the following whether λ denotes a single variable or a set, or even a set of functions, and whether the variables are discrete or continuous. However, we write as if λ were a single continuous parameter. The result A of measuring $\vec{\sigma}_1 \cdot \vec{a}$ is then determined by $\vec{a}$ and λ, and the result B of measuring $\vec{\sigma}_2 \cdot \vec{b}$ in the same instance is determined by $\vec{b}$ and λ, and

---

*Work supported in part by the U.S. Atomic Energy Commission

†On leave of absence from SLAC and CERN

Originally published in *Physics, 1*, 195-200 (1964).

$$A(\vec{a}, \lambda) = \pm 1, \; B(\vec{b}, \lambda) = \pm 1. \tag{1}$$

The vital assumption [2] is that the result B for particle 2 does not depend on the setting $\vec{a}$, of the magnet for particle 1, nor A on $\vec{b}$.

If $\rho(\lambda)$ is the probability distribution of λ then the expectation value of the product of the two components $\vec{\sigma}_1 \cdot \vec{a}$ and $\vec{\sigma}_2 \cdot \vec{b}$ is

$$P(\vec{a}, \vec{b}) = \int d\lambda \, \rho(\lambda) A(\vec{a}, \lambda) B(\vec{b}, \lambda) \tag{2}$$

This should equal the quantum mechanical expectation value, which for the singlet state is

$$< \vec{\sigma}_1 \cdot \vec{a} \; \vec{\sigma}_2 \cdot \vec{b} > = -\vec{a} \cdot \vec{b} \, . \tag{3}$$

But it will be shown that this is not possible.

Some might prefer a formulation in which the hidden variables fall into two sets, with A dependent on one and B on the other; this possibility is contained in the above, since λ stands for any number of variables and the dependences thereon of A and B are unrestricted. In a complete physical theory of the type envisaged by Einstein, the hidden variables would have dynamical significance and laws of motion; our λ can then be thought of as initial values of these variables at some suitable instant.

III. Illustration

The proof of the main result is quite simple. Before giving it, however, a number of illustrations may serve to put it in perspective.

Firstly, there is no difficulty in giving a hidden variable account of spin measurements on a single particle. Suppose we have a spin half particle in a pure spin state with polarization denoted by a unit vector $\vec{p}$. Let the hidden variable be (for example) a unit vector $\vec{\lambda}$ with uniform probability distribution over the hemisphere $\vec{\lambda} \cdot \vec{p} > 0$. Specify that the result of measurement of a component $\vec{\sigma} \cdot \vec{a}$ is

$$\text{sign } \vec{\lambda} \cdot \vec{a}' \; , \tag{4}$$

where $\vec{a}'$ is a unit vector depending on $\vec{a}$ and $\vec{p}$ in a way to be specified, and the sign function is $+1$ or -1 according to the sign of its argument. Actually this leaves the result undetermined when $\lambda \cdot a' = 0$, but as the probability of this is zero we will not make special prescriptions for it. Averaging over $\vec{\lambda}$ the expectation value is

$$< \vec{\sigma} \cdot \vec{a} > = 1 - 2\theta'/\pi \, , \tag{5}$$

where θ' is the angle between $\vec{a}'$ and $\vec{p}$. Suppose then that $\vec{a}'$ is obtained from $\vec{a}$ by rotation towards $\vec{p}$ until

$$1 - \frac{2\theta'}{\pi} = \cos \theta \tag{6}$$

where θ is the angle between $\vec{a}$ and $\vec{p}$. Then we have the desired result

$$< \vec{\sigma} \cdot \vec{a} > = \cos \theta \tag{7}$$

So in this simple case there is no difficulty in the view that the result of every measurement is determined by the value of an extra variable, and that the statistical features of quantum mechanics arise because the value of this variable is unknown in individual instances.

Secondly, there is no difficulty in reproducing, in the form (2), the only features of (3) commonly used in verbal discussions of this problem:

$$\left. \begin{aligned} P(\vec{a}, \vec{a}) &= -P(\vec{a}, -\vec{a}) = -1 \\ P(\vec{a}, \vec{b}) &= 0 \ \text{if} \ \vec{a} \cdot \vec{b} = 0 \end{aligned} \right\} \tag{8}$$

For example, let λ now be unit vector $\vec{\lambda}$, with uniform probability distribution over all directions, and take

$$\left. \begin{aligned} A(\vec{a}, \vec{\lambda}) &= \text{sign } \vec{a} \cdot \vec{\lambda} \\ B(a, b) &= -\text{sign } \vec{b} \cdot \vec{\lambda} \end{aligned} \right\} \tag{9}$$

This gives

$$P(\vec{a}, \vec{b}) = -1 + \frac{2}{\pi} \theta , \tag{10}$$

where θ is the angle between a and b, and (10) has the properties (8). For comparison, consider the result of a modified theory [6] in which the pure singlet state is replaced in the course of time by an isotropic mixture of product states; this gives the correlation function

$$-\frac{1}{3} \vec{a} \cdot \vec{b} \tag{11}$$

It is probably less easy, experimentally, to distinguish (10) from (3), than (11) from (3).

Unlike (3), the function (10) is not stationary at the minimum value -1 (at $\theta = 0$). It will be seen that this is characteristic of functions of type (2).

Thirdly, and finally, there is no difficulty in reproducing the quantum mechanical correlation (3) if the results A and B in (2) are allowed to depend on $\vec{b}$ and $\vec{a}$ respectively as well as on $\vec{a}$ and $\vec{b}$. For example, replace $\vec{a}$ in (9) by $\vec{a}'$, obtained from $\vec{a}$ by rotation towards $\vec{b}$ until

$$1 - \frac{2}{\pi} \theta' = \cos \theta ,$$

where θ' is the angle between $\vec{a}'$ and $\vec{b}$. However, for given values of the hidden variables, the results of measurements with one magnet now depend on the setting of the distant magnet, which is just what we would wish to avoid.

IV. Contradiction

The main result will now be proved. Because ρ is a normalized probability distribution,

$$\int d\lambda \rho(\lambda) = 1 , \tag{12}$$

and because of the properties (1), P in (2) cannot be less than -1. It can reach -1 at $\vec{a} = \vec{b}$ only if

$$A(\vec{a}, \lambda) = -B(\vec{a}, \lambda) \tag{13}$$

except at a set of points λ of zero probability. Assuming this, (2) can be rewritten

$$P(\vec{a}, \vec{b}) = -\int d\lambda \rho(\lambda) A(\vec{a}, \lambda) A(\vec{b}, \lambda) . \tag{14}$$

It follows that $\vec{c}$ is another unit vector

$$P(\vec{a}, \vec{b}) - P(\vec{a}, \vec{c}) = -\int d\lambda\, \rho(\lambda)\, [A(\vec{a}, \lambda)\, A(\vec{b}, \lambda) - A(\vec{a}, \lambda)\, A(\vec{c}, \lambda)]$$
$$= \int d\lambda\, \rho(\lambda)\, A(\vec{a}, \lambda)\, A(\vec{b}, \lambda)\, [A(\vec{b}, \lambda)\, A(\vec{c}, \lambda) - 1]$$

using (1), whence

$$|P(\vec{a}, \vec{b}) - P(\vec{a}, \vec{c})| \leq \int d\lambda\, \rho(\lambda)\, [1 - A(\vec{b}, \lambda)\, A(\vec{c}, \lambda)]$$

The second term on the right is $P(\vec{b}, \vec{c})$, whence

$$1 + P(\vec{b}, \vec{c}) \geq |P(\vec{a}, \vec{b}) - P(\vec{a}, \vec{c})| \tag{15}$$

Unless P is constant, the right hand side is in general of order $|\vec{b}-\vec{c}|$ for small $|\vec{b}-\vec{c}|$. Thus $P(\vec{b}, \vec{c})$ cannot be stationary at the minimum value (-1 at $\vec{b} = \vec{c}$) and cannot equal the quantum mechanical value (3).

Nor can the quantum mechanical correlation (3) be arbitrarily closely approximated by the form (2). The formal proof of this may be set out as follows. We would not worry about failure of the approximation at isolated points, so let us consider instead of (2) and (3) the functions

$$\bar{P}(\vec{a}, \vec{b}) \text{ and } \overline{-\vec{a} \cdot \vec{b}}$$

where the bar denotes independent averaging of $P(\vec{a}', \vec{b}')$ and $-\vec{a}' \cdot \vec{b}'$ over vectors $\vec{a}'$ and $\vec{b}'$ within specified small angles of $\vec{a}$ and $\vec{b}$. Suppose that for all $\vec{a}$ and $\vec{b}$ the difference is bounded by ϵ:

$$|\bar{P}(\vec{a}, \vec{b}) + \vec{a} \cdot \vec{b}| \leq \epsilon \tag{16}$$

Then it will be shown that ϵ cannot be made arbitrarily small.

Suppose that for all a and b

$$|\overline{\vec{a} \cdot \vec{b}} - \vec{a} \cdot \vec{b}| \leq \delta \tag{17}$$

Then from (16)

$$|\bar{P}(\vec{a}, \vec{b}) + \vec{a} \cdot \vec{b}| \leq \epsilon + \delta \tag{18}$$

From (2)

$$\bar{P}(\vec{a}, \vec{b}) = \int d\lambda\, \rho(\lambda)\, \bar{A}(\vec{a}, \lambda)\, \bar{B}(\vec{b}, \lambda) \tag{19}$$

where

$$|\bar{A}(\vec{a}, \lambda)| \leq 1 \text{ and } |\bar{B}(\vec{b}, \lambda)| \leq 1 \tag{20}$$

From (18) and (19), with $\vec{a} = \vec{b}$,

$$d\lambda\, \rho(\lambda)\, [\bar{A}(\vec{b}, \lambda)\, \bar{B}(\vec{b}, \lambda) + 1] \leq \epsilon + \delta \tag{21}$$

From (19)

$$\bar{P}(\vec{a}, \vec{b}) - \bar{P}(\vec{a}, \vec{c}) = \int d\lambda\, \rho(\lambda)\, [\bar{A}(\vec{a}, \lambda)\, \bar{B}(\vec{b}, \lambda) - \bar{A}(\vec{a}, \lambda)\, \bar{B}(\vec{c}, \lambda)]$$
$$= \int d\lambda\, \rho(\lambda)\, \bar{A}(\vec{a}, \lambda)\, \bar{B}(\vec{b}, \lambda)\, [1 + \bar{A}(\vec{b}, \lambda)\, \bar{B}(\vec{c}, \lambda)]$$
$$- \int d\lambda\, \rho(\lambda)\, \bar{A}(\vec{a}, \lambda)\, \bar{B}(\vec{c}, \lambda)\, [1 + \bar{A}(\vec{b}, \lambda)\, \bar{B}(\vec{b}, \lambda)]$$

Using (20) then

$$|\bar{P}(\vec{a}, \vec{b}) - \bar{P}(\vec{a}, \vec{c})| \leq \int d\lambda\, \alpha(\lambda)\, [1 + \bar{A}(\vec{b}, \lambda)\, \bar{B}(\vec{c}, \lambda)] + \int d\lambda\, \rho(\lambda)\, [1 + \bar{A}(\vec{b}, \lambda)\, \bar{B}(\vec{b}, \lambda)]$$

Then using (19) and 21)

$$|\bar{P}(\vec{a}, \vec{b}) - \bar{P}(\vec{a}, \vec{c})| \leq 1 + \bar{P}(\vec{b}, \vec{c}) + \epsilon + \delta$$

Finally, using (18),

$$|\vec{a} \cdot \vec{c} - \vec{a} \cdot \vec{b}| - 2(\epsilon + \delta) \leq 1 - \vec{b} \cdot \vec{c} + 2(\epsilon + \delta)$$

or

$$4(\epsilon + \delta) \geq |\vec{a} \cdot \vec{c} - \vec{a} \cdot \vec{b}| + \vec{b} \cdot \vec{c} - 1 \tag{22}$$

Take for example $\vec{a} \cdot \vec{c} = 0$, $\vec{a} \cdot \vec{b} = \vec{b} \cdot \vec{c} = 1/\sqrt{2}$ Then

$$4(\epsilon + \delta) \geq \sqrt{2} - 1$$

Therefore, for small finite δ, ϵ cannot be arbitrarily small.

Thus, the quantum mechanical expectation value cannot be represented, either accurately or arbitrarily closely, in the form (2).

V. Generalization

The example considered above has the advantage that it requires little imagination to envisage the measurements involved actually being made. In a more formal way, assuming [7] that any Hermitian operator with a complete set of eigenstates is an "observable", the result is easily extended to other systems. If the two systems have state spaces of dimensionality greater than 2 we can always consider two dimensional subspaces and define, in their direct product, operators $\vec{\sigma}_1$ and $\vec{\sigma}_2$ formally analogous to those used above and which are zero for states outside the product subspace. Then for at least one quantum mechanical state, the "singlet" state in the combined subspaces, the statistical predictions of quantum mechanics are incompatible with separable predetermination.

VI. Conclusion

In a theory in which parameters are added to quantum mechanics to determine the results of individual measurements, without changing the statistical predictions, there must be a mechanism whereby the setting of one measuring device can influence the reading of another instrument, however remote. Moreover, the signal involved must propagate instantaneously, so that such a theory could not be Lorentz invariant.

Of course, the situation is different if the quantum mechanical predictions are of limited validity. Conceivably they might apply only to experiments in which the settings of the instruments are made sufficiently in advance to allow them to reach some mutual rapport by exchange of signals with velocity less than or equal to that of light. In that connection, experiments of the type proposed by Bohm and Aharonov [6], in which the settings are changed during the flight of the particles, are crucial.

I am indebted to Drs. M. Bander and J. K. Perring for very useful discussions of this problem. The first draft of the paper was written during a stay at Brandeis University; I am indebted to colleagues there and at the University of Wisconsin for their interest and hospitality.

References

1. A. EINSTEIN, N. ROSEN and B. PODOLSKY, *Phys. Rev.* **47**, 777 (1935); see also N. BOHR, *Ibid.* **48**, 696 (1935), W. H. FURRY, *Ibid.* **49**, 393 and 476 (1936), and D. R. INGLIS, *Rev. Mod. Phys.* **33**, 1 (1961).
2. "But on one supposition we should, in my opinion, absolutely hold fast: the real factual situation of the system S_2 is independent of what is done with the system S_1, which is spatially separated from the former." A. EINSTEIN in *Albert Einstein, Philosopher Scientist,* (Edited by P. A. SCHILP) p. 85, Library of Living Philosophers, Evanston, Illinois (1949).
3. J. VON NEUMANN, *Mathematishe Grundlagen der Quanten-mechanik.* Verlag Julius-Springer, Berlin (1932), [English translation: Princeton University Press (1955)]; J. M. JAUCH and C. PIRON, *Helv. Phys. Acta* **36**, 827 (1963).
4. J. S. BELL, to be published.
5. D. BOHM, *Phys. Rev.* **85**, 166 and 180 (1952).
6. D. BOHM and Y. AHARONOV, *Phys. Rev.* **108**, 1070 (1957).
7. P. A. M. DIRAC, *The Principles of Quantum Mechanics* (3rd Ed.) p. 37. The Clarendon Press, Oxford (1947).

III.6 PROPOSED EXPERIMENT TO TEST LOCAL HIDDEN-VARIABLE THEORIES

JOHN F. CLAUSER, MICHAEL A. HORNE, ABNER SHIMONY, AND RICHARD A. HOLT

A theorem of Bell, proving that certain predictions of quantum mechanics are inconsistent with the entire family of local hidden-variable theories, is generalized so as to apply to realizable experiments. A proposed extension of the experiment of Kocher and Commins, on the polarization correlation of a pair of optical photons, will provide a decisive test between quantum mechanics and local hidden-variable theories.

Einstein, Podolsky, and Rosen (EPR) in a classic paper[1] presented a paradox which led them to infer that quantum mechanics is not a complete theory. They concluded that the quantum mechanical description of a physical system should be supplemented by postulating the existence of "hidden variables," the specification of which would predetermine the result of measuring any observable of the system. They believed the predictions of quantum mechanics to be correct, but only as consequences of statistical distributions of the hidden variables. Bohr[2] argued in reply that no paradox can be derived from the assumption of completeness if one recognizes that quantum mechanics concerns only the interaction of microsystems with experimental apparatus and not their intrinsic character.

There is an extensive literature purporting to prove the inconsistency of hidden-variable theories with the statistical predictions of quantum mechanics. These proofs, though mathematically valid, rest upon physically unrealistic postulates.[3] Bell[4] succeeded in replacing these postulates by a physically reasonable condition of locality. He showed that in a Gedankenexperiment of Bohm[5] (a variant of that of EPR) no local hidden-variable theory can reproduce all of the statistical predictions of quantum mechanics. This result is somewhat ironical in view of Einstein's convictions[6] that quantum mechanical predictions concerning spatially separated systems are incompatible with his conditions for locality unless hidden variables exist.

Bell's theorem has profound implications in that it points to a decisive experimental test of the entire family of local hidden-variable theories. The aim of this paper is to propose explicitly such an

Originally published in *Physical Review Letters, 23*, 880-84 (1969).

experiment. For this purpose, we first present a generalization of Bell's theorem which applies to realizable experiments. Second, we indicate that neither of the experimental realizations[7] of Bohm's Gedankenexperiment has produced evidence against local hidden-variable theories, even though the results of both are compatible with quantum mechanical predictions. Third, we show that a simple extension of one of these experiments can provide a decisive test.

Generalization of Bell's theorem.—Consider an ensemble of correlated pairs of particles moving so that one enters apparatus I_a and the other apparatus II_b, where a and b are adjustable apparatus parameters. In each apparatus a particle must select one of two channels labeled +1 and −1. Let the results of these selections be represented by $A(a)$ and $B(b)$, each of which equals ±1 according as the first or second channel is selected.

Suppose now that a statistical correlation of $A(a)$ and $B(b)$ is due to information carried by and localized within each particle, and that at some time in the past the particles constituting one pair were in contact and communication regarding this information. The information, which emphatically is not quantum mechanical, is part of the content of a set of hidden variables, denoted collectively by λ. The results of the two selections are then to be deterministic functions $A(a,\lambda)$ and $B(b,\lambda)$. Locality reasonably requires $A(a,\lambda)$ to be independent of the parameter b and $B(b,\lambda)$ to be likewise independent of a, since the two selections may occur at an arbitrarily great distance from each other. Finally, since the pair of particles is generally emitted by a source in a manner physically independent of the adjustable parameters a and b, we assume that the normalized probability distribution $\rho(\lambda)$ characterizing the ensemble is independent of a and b.

Defining the correlation function $P(a,b) \equiv \int_\Gamma A(a,\lambda)B(b,\lambda)\rho(\lambda)d\lambda$, where Γ is the total λ space, we have

$$|P(a,b)-P(a,c)| \leq \int_\Gamma |A(a,\lambda)B(b,\lambda)-A(a,\lambda)B(c,\lambda)|\rho(\lambda)d\lambda = \int_\Gamma |A(a,\lambda)B(b,\lambda)|[1-B(b,\lambda)B(c,\lambda)]\rho(\lambda)d\lambda$$
$$= \int_\Gamma [1-B(b,\lambda)B(c,\lambda)]\rho(\lambda)d\lambda = 1-\int_\Gamma B(b,\lambda)B(c,\lambda)\rho(\lambda)d\lambda.$$

Suppose that for some b' and b we have $P(b',b)=1-\delta$, where $0\leq\delta\leq 1$. Experimentally interesting cases will have δ close to but not equal to zero. Here we avoid Bell's experimentally unrealistic restriction that for some pair of parameters b' and b there is perfect correlation (i.e., $\delta=0$). Dividing Γ into two regions Γ_+ and Γ_- such that $\Gamma_\pm = \{\lambda | A(b',\lambda) = \pm B(b,\lambda)\}$, we have $\int_{\Gamma_-}\rho(\lambda)d\lambda = \frac{1}{2}\delta$. Hence,

$$\int_\Gamma B(b,\lambda)B(c,\lambda)\rho(\lambda)d\lambda = \int_\Gamma A(b',\lambda)B(c,\lambda)\rho(\lambda)d\lambda - 2\int_{\Gamma_-} A(b',\lambda)B(c,\lambda)\rho(\lambda)d\lambda$$
$$\geq P(b',c) - 2\int_{\Gamma_-}|A(b',\lambda)B(c,\lambda)|\rho(\lambda)d\lambda = P(b',c)-\delta,$$

and therefore

$$|P(a,b)-P(a,c)| \leq 2-P(b',b)-P(b',c). \tag{1a}$$

In the experiment proposed below $P(a,b)$ depends only on the parameter difference $b-a$. Defining $\alpha \equiv b-a$, $\beta \equiv c-b$, and $\gamma \equiv b-b'$, we have

$$|P(\alpha)-P(\alpha+\beta)| \leq 2-P(\gamma)-P(\beta+\gamma). \tag{1b}$$

In principle entire measuring devices, each consisting of a filter followed by a detector, could be used for I_a and II_b, and the values ±1 of $A(a)$ and $B(b)$ would denote detection or nondetection of the particles. Inequalities (1) would then apply directly to experimental counting rates. Unfortunately, if the particles are optical photons (as in the experiment proposed below) no practical tests of (1) can presently be performed in this way, because available photoelectric efficiencies are rather small. We shall therefore henceforth interpret $A(a)=\pm1$ and $B(b)=\pm1$ to mean emergence or nonemergence of the photons from the respective filters. Also the filters will be taken to be linear polarization filters, and a and b will represent their orientations. It will be convenient to introduce an exceptional value ∞ of the parameter a (and likewise of b) to represent the removal of a polarizer; clearly, $A(\infty)$ and $B(\infty)$ necessarily equal +1. Since $P(a,b)$ is an emergence correlation function, in order to derive an experimental prediction from (1) an additional assumption[8] must be made: that if a pair of photons emerges from I_a, II_b the probability of their joint detection is independent of a and b. Then if the flux into I_a, II_b is a constant independent of a and b, the rate of coincidence detection $R(a,b)$ will be proportional to $w[A(a)_+, B(b)_+]$, where $w[A(a)_\pm, B(b)_\pm]$ is the probability that $A(a)=\pm1$ and $B(b)=\pm1$. Letting $R_0 = R(\infty,$

∞), $R_1(a) = R(a, \infty)$, and $R_2(b) = R(\infty, b)$, making use of the evident formulas

$$P(a, b) = w[A(a)_+, B(b)_+] - w[A(a)_+, B(b)_-] - w[A(a)_-, B(b)_+] + w[A(a)_-, B(b)_-]$$

and

$$w[A(a)_+, B(\infty)_+] = w[A(a)_+, B(b)_+] + w[A(a)_+, B(b)_-],$$

and of similar formulas for $w[A(\infty)_+, B(b)_+]$ and $w[A(\infty)_+, B(\infty)_+]$, we obtain

$$P(a, b) = \frac{4R(a, b)}{R_0} - \frac{2R_1(a)}{R_0} - \frac{2R_2(b)}{R_0} + 1.$$

We can now express (1) in terms of experimental quantities, namely coincidence rates with both polarizers in, and with one and then the other removed. If $R_1(a)$ and $R_2(b)$ are found experimentally to be constants R_1 and R_2, the result is

$$|R(a, b) - R(a, c)| + R(b', b) + R(b', c) - R_1 - R_2 \leq 0. \tag{2a}$$

In the special case in which $P(a, b) = P(a - b)$, (2a) becomes

$$|R(\alpha) - R(\alpha + \beta)| + R(\gamma) + R(\beta + \gamma) - R_1 - R_2 \leq 0. \tag{2b}$$

Existing experimental results.—Bohm's Gedankenexperiment, involving correlated spin-$\frac{1}{2}$ particles, has not been performed, nor does it appear to be easily realizable. Two related experiments have been performed on polarization correlation of photons. Wu and Shaknov (WS)[7] examined polarization correlation of 0.5-MeV γ rays emitted during positronium annihilation.[9] Although the polarization state of the pair is suitable for our purposes, their high energy requires the use of Compton polarimeters. Thus, instead of directly examining the polarization correlation, WS examined the polarization-dependent joint distribution for Compton scattering of the pair. Inequality (2) cannot be immediately applied to such a scattering experiment because neither photon is forced to make a binary decision. However, a suitable binary result may be imposed by partitioning the scattering sphere into two arbitrary regions, denoted, respectively, by ±1, and by letting the adjustable parameter a (or b) designate the particular mode of partitioning. But as one can see[10] by examining the joint scattering distribution,[11] no such partitioning can yield a correlation in violation of (2). The essential difficulty is that the direction of Compton scattering of a photon is a statistically weak index of its linear polarization.

The other experiment, that of Kocher and Commins (KC),[7] involved polarization correlation of photon pairs emitted in the $6^1S_0 \rightarrow 4^1P_1 \rightarrow 4^1S_0$ cascade of calcium. Since the two photons are in the visible, filters of the Polaroid type could be used. The photons impinged normally upon a pair of these polarizers whose planes were parallel, and the polarization correlation was measured with standard coincidence techniques. With this arrangement inequality (2b) is applicable upon assumption of a local hidden-variable theory. However, the data obtained by KC do not suffice to test (2b), because their polarizers were not highly efficient and were placed only in the relative orientations 0° and 90°.

Proposed experiment.—A decisive test can be obtained by modifying the KC experiment to include observations at two appropriate relative orientations of the polarizers, and also with one and then the other removed. For realizable apparatus, quantum mechanics predicts violation of inequality (2b).

Define $\epsilon_M{}^i$ as the efficiency of the polarizer i (i = I, II) for light polarized parallel to the polarizer axis and $\epsilon_m{}^i$ as that for light perpendicularly polarized. Consider a point source and filter-detector assemblies, each of which gathers the photons emitted into a cone of half-angle θ. Then for a $J=0 \rightarrow J=1 \rightarrow J=0$ electric-dipole cascade (0-1-0) the quantum mechanical predictions for the counting rates are[10]

$$R(\varphi)/R_0 = \tfrac{1}{4}(\epsilon_M{}^{\mathrm{I}} + \epsilon_m{}^{\mathrm{I}})(\epsilon_M{}^{\mathrm{II}} + \epsilon_m{}^{\mathrm{II}}) + \tfrac{1}{4}(\epsilon_M{}^{\mathrm{I}} - \epsilon_m{}^{\mathrm{I}})(\epsilon_M{}^{\mathrm{II}} - \epsilon_m{}^{\mathrm{II}})F_1(\theta)\cos 2\varphi,$$
$$R_1/R_0 = \tfrac{1}{2}(\epsilon_M{}^{\mathrm{I}} + \epsilon_m{}^{\mathrm{I}}), \quad R_2/R_0 = \tfrac{1}{2}(\epsilon_M{}^{\mathrm{II}} + \epsilon_m{}^{\mathrm{II}}). \tag{3}$$

Here φ is the angle between the polarizer axes,

$$F_1(\theta) = 2G_1{}^2(\theta)[G_2{}^2(\theta) + \tfrac{1}{2}G_3{}^2(\theta)]^{-1},$$

and

$$G_1(\theta)=\tfrac{1}{4}(\tfrac{4}{3}-\cos\theta+\sin^2\theta-\tfrac{1}{3}\cos^3\theta),\quad G_2(\theta)=\tfrac{2}{3}-\tfrac{1}{3}(\sin^2\theta+2)\cos\theta,\quad G_3(\theta)=\tfrac{4}{3}-\cos\theta-\tfrac{1}{3}\cos^3\theta.$$

The predictions for a 0-1-1 electric-dipole cascade (and for a 1-1-0, provided the initial statistical state of the atom is isotropic) are obtained from (3) upon replacing $F_1(\theta)$ with $-F_2(\theta)$ where

$$F_2(\theta)=2G_1^{\,2}(\theta)[2G_2(\theta)G_3(\theta)+\tfrac{1}{2}G_3^{\,2}(\theta)]^{-1}.$$

For sufficiently efficient polarizers one easily sees that there exist sets of relative orientations for which the quantum mechanical counting rates violate Inequality (2b). The greatest violation always occurs at $\alpha=22.5^\circ$, $\beta=45^\circ$, and $\gamma=157.5^\circ$ for the 0-1-0 cascade and at $\alpha=67.5^\circ$, $\beta=135^\circ$, and $\gamma=112.5^\circ$ for the 0-1-1 cascade. Note that in each case the four angles α, $\alpha+\beta$, γ, and $\beta+\gamma$ which occur in Inequality (2b) characterize only two distinct relative orientations of the polarizers, namely 22.5° and 67.5°.

In an actual experiment $F_j(\theta)$ is less than 1, because of finite half-angle θ, and ϵ_M is never unity. Assuming the use of calcite polarizers (for which $\epsilon_m\approx 10^{-5}$), taking $\epsilon_M^{\,\mathrm{I}}=\epsilon_M^{\,\mathrm{II}}=\epsilon_M$ for simplicity, and using the above choices for α, β, and γ, we find that for either type of cascade, the condition for violation of Inequality (2b) is

$$\sqrt{2}\,F_j(\theta)+1>2/\epsilon_M. \tag{4}$$

This is the essential requirement on the design of a decisive experiment. For given $F_j(\theta)$, (4) implies a lower limit on ϵ_M, and vice versa. Since both $F_1(\theta)$ and $F_2(\theta)$ are monotonically decreasing functions, a lower limit on $F_j(\theta)$ implies an upper limit on θ. Condition (4) and numerical evaluation of $F_1(\theta)$ and $F_2(\theta)$ lead to the curves shown in Fig. 1, from which one can directly read combinations of θ and ϵ_M suitable for a decisive experiment. The experiment can be performed with uncoated calcite polarizers ($\epsilon_M\approx 0.92$); however, if the polarizers have antireflection coatings ($\epsilon_M\approx 0.95$), a larger θ and hence a larger counting rate can be achieved.

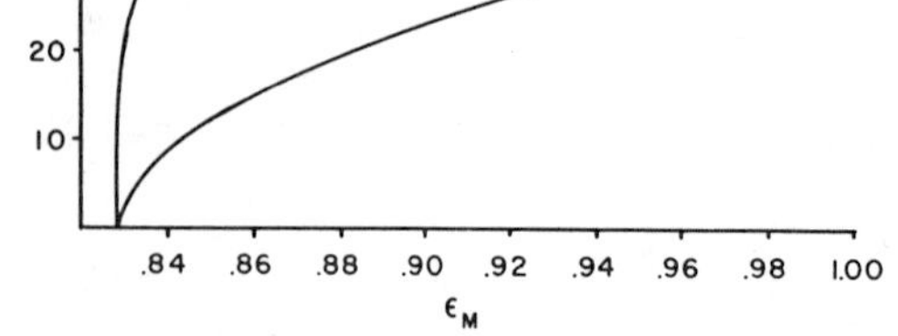

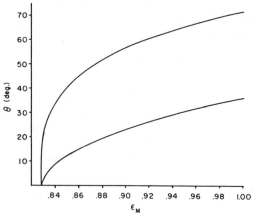

FIG. 1. Upper limits on detector half-angle θ as a function of polarizer efficiency ϵ_M. To test for hidden-variable theories, the experiment must be performed in the region below the appropriate curve—the upper curve for a 0-1-0 cascade, the lower for a 0-1-1.

The authors gratefully acknowledge helpful discussions with Y. Aharonov, M. Jammer, L. Kasday, D. Nartonis, C. Papaliolios, F. Pipkin, D. Pritchard, J. L. Snider, H. Stein, and C. R. Willis.

---

*This paper is an expansion of ideas presented by one of us (J.F.C.) at the Spring 1969 meeting of the American Physical Society [Bull. Am. Phys. Soc. 14, 578 (1969)]. The same conclusions had been reached independently by two of us (A.S. and M.A.H.).

†Present address: Department of Physics, University of California, Berkeley, Calif.

[1]A. Einstein, B. Podolsky, and N. Rosen, Phys. Rev. 47, 777 (1935).

[2]N. Bohr, Phys. Rev. 48, 696 (1935).

[3]For an excellent survey of these proofs see J. S. Bell, Rev. Mod. Phys. 38, 447 (1966).

[4]J. S. Bell, Physics 1, 195 (1965).

[5]D. Bohm, Quantum Theory (Prentice-Hall, Inc., Englewood Cliffs, New Jersey, 1951), p. 614.

[6]Albert Einstein: Philosopher-Scientist, edited by P. Schilpp (Library of Living Philosophers, Evanston, Ill., 1949), pp. 681-682.

[7]C. S. Wu and I. Shaknov, Phys. Rev. 77, 136 (1950); C. A. Kocher and E. D. Commins, Phys. Rev. Letters 18, 575 (1967). See also C. A. Kocher, thesis, University of California, 1967, University of California Radiation Laboratory Report No. UCRL-17587 (unpublished).

[8]It may appear that the assumption can be established experimentally by measuring detection rates when a controlled flux of photons of known polarization impinges on each detector. From the standpoint of hidden-variable theories, however, these measurements are irrelevant, since the distribution of the hidden variables when the fluxes are thus controlled is almost certain to be different from the $\rho(\lambda)$ governing our ensemble. In view of the difficulty of an experimental check, the assumption could be challenged by an advocate of hidden-variable theories in case the outcome of the proposed experiment favors quantum mechanics. However, highly pathological detectors are required to con-

vert hidden-variable emergence rates into quantum mechanical counting rates.

[9]It has been shown by D. Bohm and Y. Aharonov, Phys. Rev. 108, 1070 (1957), that the WS experiment is a decisive refutation of a hypothesis studied by W. H. Furry, Phys. Rev. 49, 393, 476 (1936).

[10]For details see M. A. Horne, thesis, Boston University, 1969 (unpublished).

[11]The distribution is given in H. S. Snyder, S. Pasternack, and J. Hornbostel, Phys. Rev. 73, 440 (1948).

III.7 EXPERIMENTAL TEST OF LOCAL HIDDEN-VARIABLE THEORIES

STUART J. FREEDMAN AND JOHN F. CLAUSER

We have measured the linear polarization correlation of the photons emitted in an atomic cascade of calcium. It has been shown by a generalization of Bell's inequality that the existence of local hidden variables imposes restrictions on this correlation in conflict with the predictions of quantum mechanics. Our data, in agreement with quantum mechanics, violate these restrictions to high statistical accuracy, thus providing strong evidence against local hidden-variable theories.

Since quantum mechanics was first developed, there have been repeated suggestions that its statistical features possibly might be described by an underlying deterministic substructure. Such features, then, arise because a quantum state represents a statistical ensemble of "hidden-variable states." Proofs by von Neumann and others, demonstrating the impossibility of a hid-

Originally published in *Physical Review Letters, 28*, 938-41 (1972).

den-variable substructure consistent with quantum mechanics, rely on various assumptions concerning the character of the hidden variables.[1] Bell has argued that these assumptions are unduly restrictive. However, by considering an idealized case of two spatially separated but quantum-mechanically correlated systems, he was able to show that any hidden-variable theory satisfying only the natural assumption of "locality" also leads to predictions ("Bell's inequality") in conflict with quantum mechanics.[2]

Bell's proof was extended to realizable systems by Clauser, Horne, Shimony, and Holt,[3] who also pointed out that their generalization of Bell's inequality can be tested experimentally, thus testing all local hidden-variable theories, but that existing experimental results were insufficient for this purpose. This Letter reports the results of an experiment which are sufficiently precise to rule out local hidden-variable theories with high statistical accuracy

In the present work we measured the correlation in linear polarization of two photons γ_1 and γ_2 emitted in a $J=0 \to J=1 \to J=0$ atomic cascade. The decaying atoms were viewed by two symmetrically placed optical systems, each consisting of two lenses, a wavelength filter, a rotatable and removable polarizer, and a single-photon detector (see Fig. 1). The following quantities were measured: $R(\varphi)$, the coincidence rate for two-photon detection, as a function of the angle φ between the planes of linear polarization defined by the orientation of the inserted polarizers; R_1, the coincidence rate with polarizer 2 removed; R_2, the coincidence rate with polarizer 1 removed[4]; R_0, the coincidence rate with both polarizers removed. Quantum mechanics predicts that $R(\varphi)$ and R_0 are related as follows[3,5]:

$$R(\varphi)/R_0 = \tfrac{1}{4}(\epsilon_M{}^1 + \epsilon_m{}^1)(\epsilon_M{}^2 + \epsilon_m{}^2) + \tfrac{1}{4}(\epsilon_M{}^1 - \epsilon_m{}^1) \times (\epsilon_M{}^2 - \epsilon_m{}^2) F_1(\theta) \cos 2\varphi, \qquad (1a)$$

while

$$R_1/R_0 = \tfrac{1}{2}(\epsilon_M{}^1 + \epsilon_m{}^1), \qquad (1b)$$

and

$$R_2/R_0 = \tfrac{1}{2}(\epsilon_M{}^2 + \epsilon_m{}^2). \qquad (1c)$$

Here $\epsilon_M{}^i$ $(\epsilon_m{}^i)$ is the transmittance of the ith polarizer for light polarized parallel (perpendicular) to the polarizer axis, and $F_1(\theta)$ is a function of the half-angle θ subtended by the primary lenses. It represents a depolarization due to noncollinearity of the two photons, and approaches unity for infinitesimal detector solid angles. [For this experiment, $\theta \approx 30°$, and $F_1(30°) \approx 0.99$.]

We make the following assumptions for any local hidden-variable theory: (1) The two photons propagate as separated localized particles. (2) A binary selection process occurs for each photon at each polarizer (transmission or no-transmission). This selection does not depend upon the orientation of the distant polarizer.

In addition, we make the following assumption to allow a comparison of the generalization of Bell's inequality with out experiment: (3) All photons incident on a detector have a probability of detection that is independent of whether or not the photon has passed through a polarizer.[6]

The above assumptions constrain the coincidence rates by the following inequalities[7]:

$$-1 \leq \Delta(\varphi) \leq 0, \qquad (2)$$

where

$$\Delta(\varphi) = \frac{3R(\varphi)}{R_0} - \frac{R(3\varphi)}{R_0} - \frac{R_1 + R_2}{R_0}.$$

For sufficiently small detector solid angles and highly efficient polarizers, these inequalities (2) are not satisfied by the quantum-mechanical prediction (1) for a range of values of φ. Maximum violations occur at $\varphi = 22\tfrac{1}{2}°$ $[\Delta(\varphi) > 0]$ and $\varphi = 67\tfrac{1}{2}°$ $[\Delta(\varphi) < -1]$. At these angles of maximum violation, inequalities (2) can be combined into the simpler and more convenient expression

$$\delta = |R(22\tfrac{1}{2}°)/R_0 - R(67\tfrac{1}{2}°)/R_0| - \tfrac{1}{4} \leq 0, \qquad (3)$$

which does not involve R_1 or R_2.

The experimental arrangement was similar to

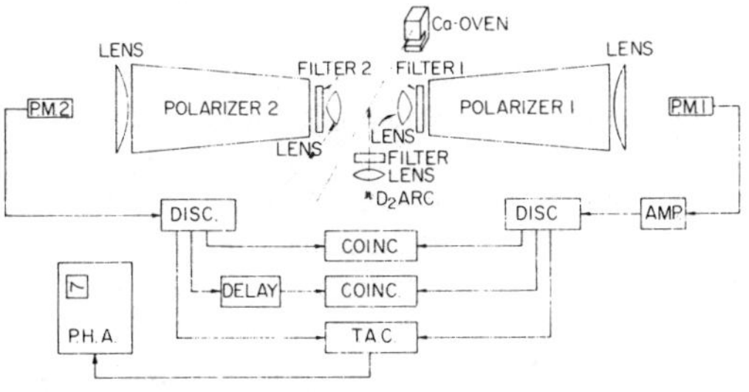

FIG. 1. Schematic diagram of apparatus and associated electronics. Scalers (not shown) monitored the outputs of the discriminators and coincidence circuits during each 100-sec count period. The contents of the scalers and the experimental configuration were recorded on paper tape and analyzed on an IBM 1620-II computer.

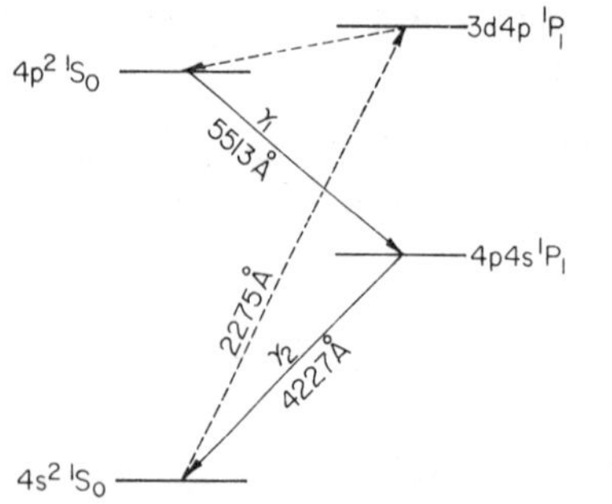

FIG. 2. Level scheme of calcium. Dashed lines show the route for excitation to the initial state $4p^2\,{}^1S_0$.

that of Kocher and Commins.[8] A calcium atomic beam effused from a tantalum oven, as shown in Fig. 1. The continuum output of a deuterium arc lamp (ORIEL C-42-72-12) was passed through an interference filter [250 Å full width at half-maximum (FWHM), 20% transmission at 2275 Å] and focused on the beam. Resonance absorption of a 2275-Å photon excited calcium atoms to the $3d4p\,{}^1P_1$ state. Of the atoms that did not decay directly to the ground state, about 7% decayed to the $4p^2\,{}^1S_0$ state, from which they cascaded through the $4s4p\,{}^1P_1$ intermediate state to the ground state with the emission of two photons at 5513 Å (γ_1) and 4227 Å (γ_2) (see Fig. 2). At the interaction region (roughly, a cylinder 5 mm high and 3 mm in diameter) the density of the calcium was about 1×10^{10} atoms/cm³. To avoid spherical aberrations which would have reduced counter efficiencies, aspheric primary lenses (8.0 cm diam, $f=0.8$) were used. Photons γ_1 were selected by a filter with 10 Å FWHM and 50% transmission, and γ_2 by a filter with 6 Å FWHM and 20% transmission. The requirement for large efficient linear polarizers led us to employ "pile-of-plates" polarizers. Each polarizer consisted of ten 0.3-mm-thick glass sheets inclined nearly at Brewster's angle. The sheets were attached to hinged frames, and could be folded completely out of the optical path. A Geneva mechanism rotated each polarizer through increments of $22\frac{1}{2}°$. The measured transmittances of the polarizers were $\epsilon_M{}^1=0.97\pm0.01$, $\epsilon_m{}^1=0.038\pm0.004$, $\epsilon_M{}^2=0.96\pm0.01$, and $\epsilon_m{}^2=0.037\pm0.004$. The photomultiplier detectors (RCA C31000E, quantum efficiency ≈0.13 at 5513 Å; and RCA 8850, quantum efficiency ≈0.28 at 4227 Å) were cooled, reducing dark rates to 75 and 200 counts/sec, respectively. The measured counter efficiencies with polarizers removed were $\eta_1\approx1.7\times10^{-3}$ and $\eta_2\approx1.5\times10^{-3}$.[9]

A diagram of the electronics is included in Fig. 1. The overall system time resolution was about 1.5 nsec. The short intermediate state lifetime (~5 nsec) permitted a narrow coincidence window (8.1 nsec). A second coincidence channel displaced in time by 50 nsec monitored the number of accidental coincidences, the true coincidence rate being determined by subtraction.[10] A time-to-amplitude converter and pulse-height analyzer measured the time-delay spectrum of the two photons. The resulting exponential gave the intermediate state lifetime.[11]

The coincidence rates depended upon the beam and lamp intensities, the latter gradually decreasing during a run. The typical coincidence rate with polarizers removed ranged from 0.3 to 0.1 countx/sec, and the accidental rate ranged from 0.01 to 0.002 counts/sec. Long runs required by the low coincidence rate necessitated automatic data collections.

The system was cycled with 100-sec counting periods. Periods with one or both polarizers inserted alternated with periods in which both polarizers were removed. Both polarizers rotated according to a prescribed sequence. For a given run, $R(\varphi)/R_0$ was calculated by summing counts for all configurations corresponding to angle φ and dividing by half the sum of the counts in the adjacent periods of the sequence in which both polarizers were moved. Data for R_1/R_0 and R_2/R_0 were analyzed in a similar fashion. The values given here are averages over the orientation of the inserted polarizer. This cycling and averaging procedure minimized the effects of drift and apparatus asymmetry.

The results of the measurements of the correlation $R(\varphi)/R_0$, corresponding to a total integration time of ~200 h, are shown in Fig. 3. All error limits are conservative estimates of 1 standard deviation. Using the values at $22\frac{1}{2}°$ and $67\frac{1}{2}°$, we obtain $\delta=0.050\pm0.008$ in clear violation of inequality (3).[12] Furthermore, we observe no evidence for a deviation from the predictions of quantum mechanics, calculated from the measured polarizer efficiences and solid angles, and shown as the solid curve in Fig. 3. We consider these results to be strong evidence against local hidden-variable theories.

The authors wish to express their sincerest appreciation for guidance and help from Professor Eugene Commins, to Professor Charles Townes for his encouragement of this work, and to M. Sim-

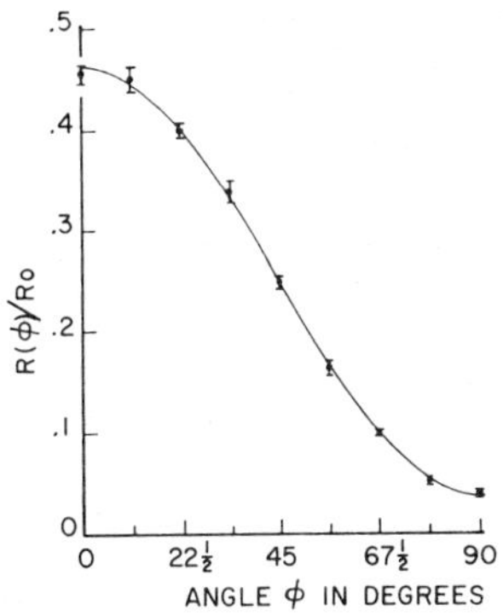

FIG. 3. Coincidence rate with angle φ between the polarizers, divided by the rate with both polarizers removed, plotted versus the angle φ. The solid line is the prediction by quantum mechanics, calculated using the measured efficiencies of the polarizers and solid angles of the experiment.

mons for helpful suggestions.

---

*Work supported by U. S. Atomic Energy Commission.

[1]The best-known proof is by J. von Neumann, *Mathematische Grundlagen der Quantemechanik* (Springer, Berlin, 1932) [*Mathematical Foundations of Quantum Mechanics* (Princeton Univ. Press. Princeton, N. J., 1955)]. For a critical review of this and other proofs see J. S. Bell, Rev. Mod. Phys. 38, 447 (1966).

[2]J. S. Bell, Physics (Long Is. City, N.Y.) 1, 195 (1964).

[3]J. Clauser, M. Horne, A. Shimony, and R. Holt, Phys. Rev. Lett. 23, 880 (1969).

[4]A hidden-variable theory need not require that R_1 and R_2 be independent of the orientation of the inserted polarizer, and we do not assume this independence in our data analysis. However, the results are consistent with R_1 and R_2 being independent of angle, and for simplicity they are so denoted.

[5]M. Horne, Ph. D. thesis, Boston University, 1970 (unpublished). See also A. Shimony, in "Foundations of Quantum Mechanics, Proceedings of the International School of Physics 'Enrico Fermi,' Course IL" (Academic, New York, to be published).

[6]This assumption is much weaker than the assumption made by L. R. Kasday, J. Ullman, and C. S. Wu, Bull. Amer. Phys. Soc. 15, 586 (1970), in their discussion of the two-γ decay of positronium; see L. R. Kasday, in "Foundations of Quantum Mechanics, Proceedings of the International School of Physics 'Enrico Fermi,' Course IL" (Academic, New York, to be published).

[7]The inequality $\Delta(\varphi) \leq 0$ is derived in Refs. 3 and 5. The other forms of the hidden-variable restriction are obtained by similar arguments; see S. Freedman, Ph. D. thesis, University of California, Berkeley, Lawrence Berkeley Laboratory Report No. LBL-391, 1972 (unpublished).

[8]C. A. Kocher and E. D. Commins, Phys. Rev. Lett. 18, 575 (1967); C. A. Kocher, Ph. D. thesis, University of California, Berkeley, Lawrence Berkeley Laboratory Report No. UCRL-17587, 1967 (unpublished).

[9]The counter efficiencies are given by $\eta_i = (\Omega_i/4\pi)T_i \times \epsilon_i L_i$, where Ω_i is the solid angle, T_i is the transmission of the filter, ϵ_i is the quantum efficiency, and L_i accounts for other losses. The measurement of η_2 was made, employing the properties of the calcium cascade, by comparing the coincidence rate and the γ_1 singles rate after suitable background correction; η_1 was then inferred from the known quantum efficiencies and filter transmissions assuming that Ω_i and L_i were the same for both detector systems.

[10]An estimate of the accidental rate was also obtained from the singles rates. The two estimates gave consistent results. In fact, our conclusions are not changed if accidentals are neglected entirely; the signal-to-accidental ratio with polarizer removed is about 40 to 1 for the data presented.

[11]Resonance trapping, encountered at high beam densities, resulted in a lengthening of the observed lifetime and a slight decrease in the polarization correlation amplitude, see J. P. Barrat, J. Phys. Radium 20, 541, 633 (1959). At low beam densities the measured lifetime is consistent with previously measured values. See W. L. Weise, M. W. Smith, and B. M. Miles, *Atomic Transition Probabilities*, U. S. National Bureau of Standards Reference Data Series—22 (U.S. GPO, Washington, D.C., 1969), Vol. 2.

[12]The results that are of interest in comparison with the hidden-variable inequalities are $R_1/R_0 = 0.497 \pm 0.009$, $R_2/R_0 = 0.499 \pm 0.009$, $R(22\frac{1}{2}°)/R_0 = 0.400 \pm 0.007$, and $R(67\frac{1}{2}°)/R_0 = 0.100 \pm 0.003$. We thus obtain $\Delta(22\frac{1}{2}°) = 0.104 \pm 0.026$ and $\Delta(67\frac{1}{2}°) = -1.097 \pm 0.018$, in violation of inequalities (2).

III.8 EXPERIMENTAL TEST OF LOCAL HIDDEN-VARIABLE THEORIES*

EDWARD S. FRY AND RANDALL C. THOMPSON

We have measured the linear polarization correlation between the two photons from the $7^3S_1 \to 6^3P_1 \to 6^1S_0$ cascade of Hg^{200}. The results were used to evaluate Freedman's version of the Bell inequality, $\delta \leq 0$. Our result is $\delta_{exp} = +0.046 \pm 0.014$, in clear violation of the inequality and in excellent agreement with the quantum mechanical prediction, $\delta_{QM} = +0.044 \pm 0.007$. An important feature of the experiment was the explicit measurement of the initial density matrix for the cascading atoms.

We have measured the linear polarization correlation of photon pairs from the 7^3S_1-6^3P_1-6^1S_0 cascade of Hg^{200}. Under appropriate experimental conditions, quantum mechanics (QM) predicts there should be a very strong correlation. The essence of Bell's theorem[1] is that any local hidden variable (LHV) theory restricts the strength of this correlation. This LHV restriction can be put in a form derived by Freedman,[2]

$$\delta = |R(67\tfrac{1}{2}^\circ)/R_0 - R(22\tfrac{1}{2}^\circ)/R_0| - \tfrac{1}{4} \leq 0. \qquad (1)$$

Here the two photons are respectively detected on the $\pm Z$ axes, $R(\varphi)$ is the coincidence rate with an angle φ between the transmission axes of the polarizers, and R_0 is the coincidence rate with polarizers removed. A decisive experimental test of LHV theories can then be obtained by choosing experimental conditions such that inequality (1) is violated by the quantum mechanical predictions.[3] Previously, results have been obtained from three such experiments. The first by Freedman and Clauser[4] used the calcium cascade 6^1S_0-4^1P_1-4^1S_0. Their results violated the inequality and were in agreement with the QM predictions. The second by Holt[5] used the mercury (Hg^{198}) cascade 9^1P_1-7^3S_1-6^3P_0. The results satisfied the inequality and were in apparent disagreement with QM. The third, also by Clauser,[6] was recently completed using the same cascade in mercury (Hg^{202}) and the same excitation technique as Holt. The results violated the inequality and were in agreement with QM.

The present experiment used a different cascade, 7^3S_1-6^3P_1-6^1S_0 in mercury (Hg^{200}) (see Fig. 1). The 7^3S_1 state was populated in a two-step process, i.e., electron bombardment excitation to the 6^3P_2 state followed by absorption of resonant 5461-Å radiation from a laser. An atomic beam was used and the two excitation steps occurred at physically different locations. Consequently, in the interaction region (where the 7^3S_1 state is populated) there are essentially no rapidly decaying states other than the cascade states. Therefore there is a one-to-one correspondence between all 4358-Å and 2537-Å photons. As a result comparatively high data accumulation rates were obtained.

The experimental arrangement is shown in Fig. 2. The mercury atomic beam passes through a solenoid electron gun where atoms are excited to the 6^3P_2 state. Laser beams tuned to the resonant frequency of the 5461-Å transition in Hg^{200} intersect the atomic beam at two locations. The 4358-Å fluorescence from the first location provides a reference to lock the laser cavity onto the Hg resonance. The second location is the interaction region. Its dimensions, defined by the intersection of the two beams, are $0.3 \times 0.8 \times 0.8$ mm^3. The first location is slightly off the atomic beam axis so that atoms which can "see" laser radiation in the first location cannot enter the interaction volume.

The laser radiation incident on the interaction region is polarized parallel to the Z axis. At the interaction region the emitted 4358-Å (2537-Å) photons are collected over a half-angle $\theta = 19.9° \pm 0.3°$, pass through a pile-of-plates polarizer and a filter, and are detected on the $+(-)$ Z axis.

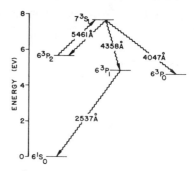

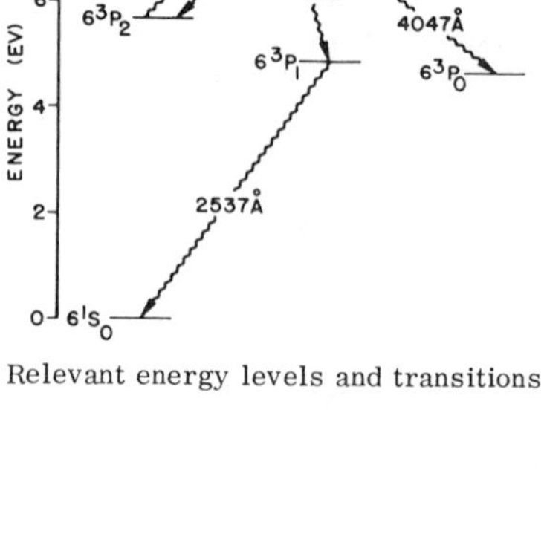

FIG. 1. Relevant energy levels and transitions in mercury.

Originally published in *Physical Review Letters, 37*, 465-68 (1976).

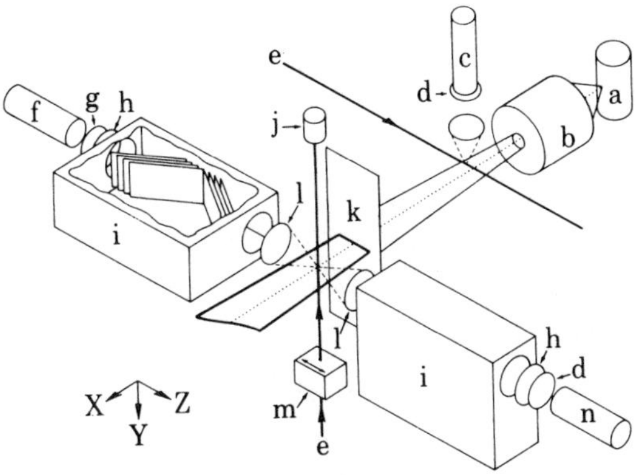

FIG. 2. Schematic of the apparatus. Polarizer plate arrangement is also indicated. Actual polarizers have 14 plates. (a) Hg oven; (b) solenoid electron gun; (c) RCA 8575; (d) 4358-Å filter; (e) 5461-Å laser beam; (f) Amperex 56 DUVP/03; (g) 2537-Å filter; (h) focusing lens; (i) pile-of-plates polarizer; (j) laser beam trap; (k) atomic beam defining slit; (l) light collecting lens; (m) crystal polarizer; (n) RCA 8850.

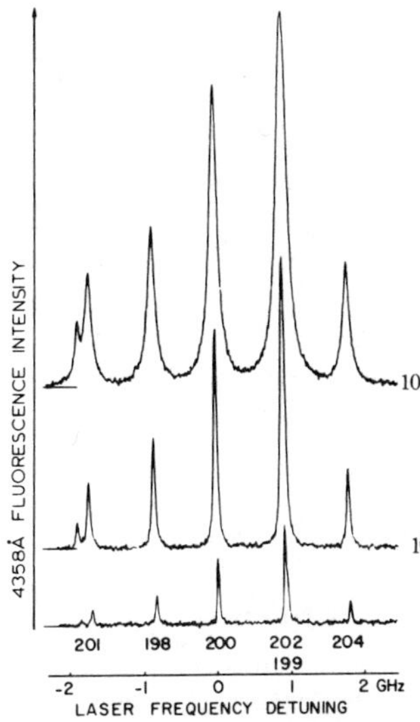

FIG. 3. Fluorescence intensity of 4358-Å radiation versus laser frequency for three intensities of the incident 5461-Å radiation. The lines for the various Hg isotopes are labeled with their mass numbers. The vertical scale is the same for all three scans; the zeroes have been offset for clarity. $I_0 \sim 2.5$ mW/cm^2.

The collection optics are lens pairs whose radii have been adjusted to minimize the Seidel spherical aberration coefficient. Each polarizer consists of two sets of seven plates symmetrically arranged so as to cancel out transverse ray displacements. The magnetic field in the interaction region is zeroed to less than 5 mG in all directions.

A valid test can only be made with zero-spin isotopes. Our beam uses mercury of natural isotopic abundance; but we selectively excite only atoms of the zero-spin isotope, Hg^{200}, to the initial state of the cascade by using 5461-Å radiation from a narrow-linewidth (15 MHz) tunable dye laser. By sweeping the laser frequency and observing the 4358-Å fluorescence we can observe the structure of the 5461-Å absorption line.[7] Figure 3 shows the results for the central portion of the line for three incident laser intensities. The transition is power broadened, but at our operating intensity, $I_0 \sim 2.5$ mW/cm^2, the isotope separation is very clean.

The initial state of our cascade has $J = 1$. Its density matrix is 3×3 and has elements ρ_{ij} where i and j are magnetic quantum numbers. With detectors on the $\pm Z$ axes, the QM prediction for the coincidence rate $S(\varphi)$ shows no dependence on ρ_{10} or ρ_{0-1}. (The coordinate system is indicated in Fig. 2.) When ρ_{1-1} is zero, the normalized coincidence rate is

$$R(\varphi)/R_0 = \tfrac{1}{4}(\epsilon_M{}^1 + \epsilon_m{}^1)(\epsilon_M{}^2 + \epsilon_m{}^2) - \tfrac{1}{4}(\epsilon_M{}^1 - \epsilon_m{}^1)(\epsilon_M{}^2 - \epsilon_m{}^2)F(\theta)\cos 2\varphi. \qquad (2)$$

Here $\epsilon_M{}^i$ ($\epsilon_m{}^i$) is the transmission efficiency of the ith polarizer for light polarized parallel (perpendicular) to the transmission axis, and $F(\theta)$ is given by

$$F(\theta) = \rho' J^2(\theta)[(1+\rho')G^2(\theta) - (1-2\rho')G(\theta)H(\theta) - (2-\rho')H^2(\theta)]^{-1}. \qquad (3)$$

The functions[8] $G(\theta)$, $H(\theta)$, and $J(\theta)$ depend on the half-angle θ subtended by the light collection optics, and ρ' is given by

$$\rho' = \rho_{00}/(\rho_{11} + \rho_{-1-1}). \qquad (4)$$

It is essential to measure ρ'; to check that ρ_{1-1} is zero; and to verify that the QM prediction, Eq. (2), violates inequality (1).

The density matrix for atoms in the 7^3S_1 state is determined experimentally by measuring the polarization of the 4358-Å fluorescence at appropriate angles. It is found that at the high intensities at which the transition is saturated, the off-

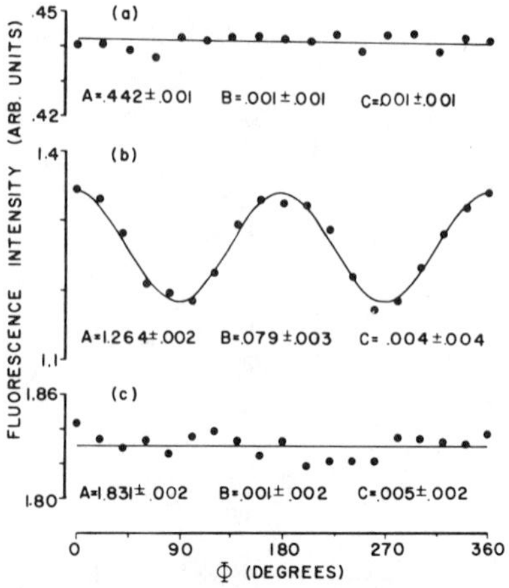

FIG. 4. Linear polarization dependences of the intensities of the cascade photons. Φ is the angle of the polarizer transmission axis with respect to the X axis. The data are least-squares fitted by a function of the form $A + B\cos 2\Phi + C\sin 2\Phi$ and the fitted parameters are given with each curve. (a) 4358-Å intensity on the $+Z$ axis. (b) 4358-Å intensity on the $-Y$ axis. (c) 2537 Å on the $-Z$ axis.

diagonal elements are nonzero and all elements are intensity dependent. With the intensity reduced to $I_0 \sim 2.5$ mW/cm^2, our data (Fig. 4) show that the density matrix has the desired form. Figure 4(a) shows the results for the 4358-Å fluorescence on the $+Z$ axis. The absence of a linear polarization dependence here implies $\rho_{1-1} = 0$. Figure 4(b) shows the polarization measurements of 4358-Å radiation detected on the $-Y$ axis; from the fitted parameters we find $\mathrm{Re}(\rho_{10} - \rho_{0-1}) = 0$, and $\rho' = 0.633 \pm 0.005$. For completeness, Fig. 4(c) shows the polarization measurements of 2537-Å fluorescence on the $-Z$ axis.

The polarizer efficiency parameters are $\epsilon_M{}^1 = 0.98 \pm 0.01$, $\epsilon_M{}^2 = 0.97 \pm 0.01$, $\epsilon_m{}^1 = \epsilon_m{}^2 = 0.02 \pm 0.005$. Hence the QM prediction, Eq. (2), for the normal-

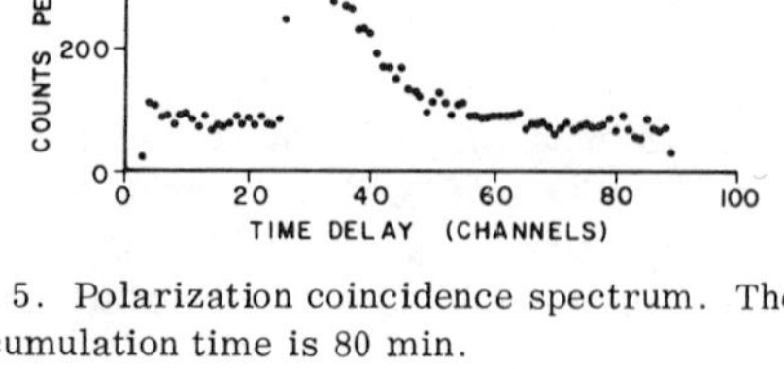

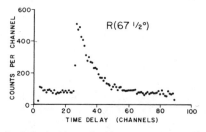

FIG. 5. Polarization coincidence spectrum. The total accumulation time is 80 min.

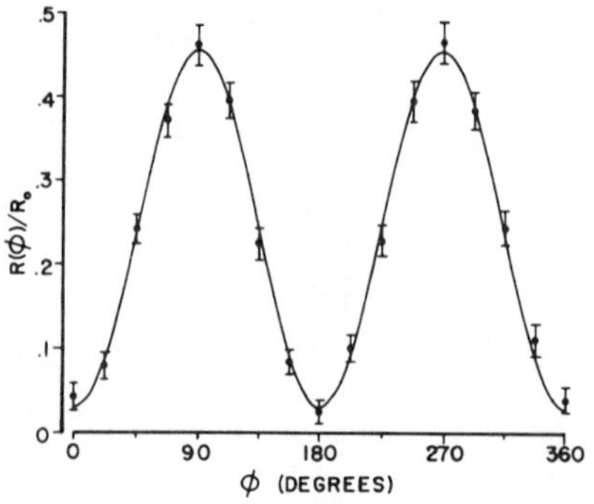

FIG. 6. Normalized polarization coincidence data from 0° to 360°. The datum point at 0° is duplicated at 360°. The smooth curve is a least-squares fit to $R(\varphi)/R_0 = A + B\cos 2\varphi + C\sin 2\varphi$. The fitted parameters are $A = 0.242 \pm 0.003$; $B = -0.212 \pm 0.004$; $C = -0.003 \pm 0.004$.

ized coincidence rate is

$$R(\varphi)/R_0 = (0.248 \pm 0.004) - (0.208 \pm 0.004)\cos 2\varphi \qquad (5)$$

and we also find

$$\delta_{\mathrm{QM}} = 0.044 \pm 0.007. \qquad (6)$$

All errors are ± 1 standard deviation.

Coincidence data were obtained using a time-to-amplitude converter and a pulse-height analyzer. Figure 5 shows the total coincidence spectrum with $67\frac{1}{2}°$ between the polarizer axes. The total accumulation time for this spectrum was 80 min. To obtain the true coincidences, one must subtract out the accidental coincidence background. Consequently, the error depends on the width of the coincidence window. For our data, minimum error is obtained for a window width of 12 to 14 channels (1.3τ). The quality factor, defined by Freedman,[2] was $Q = 1.03$ with the polarizers removed.

Figure 6 shows the polarization data for the full 360° together with a least-squares fit. The fitted parameters are in good agreement with the QM prediction, Eq. (5).

From the R_0, $R(22\frac{1}{2}°)$, and $R(67\frac{1}{2}°)$ data, we find

$$\delta_{\exp} = 0.046 \pm 0.014 \qquad (7)$$

in excellent agreement with the QM prediction, Eq. (6), and in clear violation of the LHV restriction, inequality (1).

The authors wish to thank the many people who have contributed to this work, especially Jim McGuire, Jim Ellis, Norman Alexander, and our instrument shop personnel.

*Work supported by the Robert A. Welch Foundation, Grant No. A437, and by a grant from the Research Corportion.

[1]J. S. Bell, Physics (L. I. City, N. Y.) 1, 195 (1965).
[2]S. J. Freedman, Lawrence Berkeley Laboratory Report No. LBL-391, 1972 (unpublished).
[3]J. F. Clauser and M. A. Horne, Phys. Rev. D 10, 526 (1974).
[4]S. J. Freedman and J. F. Clauser, Phys. Rev. Lett. 28, 938 (1972).
[5]R. A. Holt, Ph.D. thesis, Harvard University, 1973 (unpublished).
[6]J. F. Clauser, Phys. Rev. Lett. 36, 1223 (1976).
[7]J. Blaise and H. Chantrel, J. Phys. Radium 18, 193 (1957).
[8]E. S. Fry, Phys. Rev. A 8, 1219 (1973).

III.9 QUANTUM MECHANICS AND HIDDEN VARIABLES: A TEST OF BELL'S INEQUALITY BY THE MEASUREMENT OF THE SPIN CORRELATION IN LOW-ENERGY PROTON-PROTON SCATTERING

M. LAMEHI-RACHTI AND W. MITTIG

The inequality of Bell has been tested by the measurement of the spin correlation in proton-proton scattering. Measurements were made at $E_p = 13.2$ and 13.7 MeV using carbon analyzers of 18.6 and 29 mg/cm^2, respectively, accumulating a total of 10^4 coincidences. The experimental analyzing power, geometric correlation coefficients, and energy spectra are compared to the result of a Monte Carlo simulation of the apparatus. The results are in good agreement with quantum mechanics and in disagreement with the inequality of Bell if the same additional assumptions are made. The conditions for comparing the results of the experiments to the inequality of Bell are discussed.

I. INTRODUCTION

Since the beginning of quantum mechanics (QM) a number of physicists who contributed the most to the development of this theory had serious doubts about its logical foundations. Most of the problems were illustrated by a number of paradoxes, such as those of Einstein, Podolsky, and Rosen[1] and Schrödinger (namely, the cat paradox).[2] These discussions never died down and even today there is no theory of measurement which satisfies everybody.

One attempt to overcome these difficulties was to suppose that there are some supplementary variables outside the scope of QM ("hidden variables") which determine the result of the individual measurement. A theorem derived by J. von Neumann was taken for a long time as proof that such interpretations are impossible. But Bohm[3] in 1952 developed a model of the hidden-variables theory which was in complete agreement with the predictions of QM, and Bell[4] showed in 1965 why the theorem of von Neumann was not valid as applied to physical systems. Bell showed, too, that all hidden-variables models which give complete agreement with QM must have an undesirable feature. They do not obey the principle of locality as stated by Einstein[5]: "If S_1 and S_2 are two systems that have interacted in the past but are now arbitrarily distant, the real, factual situation of system S_1 does not depend on what is done with system S_2, which is spatially separated from the former."

Developments[6-8] of the argumentation of Bell led for the first time in a more than 30-year-old discussion to the possibility of a critical experimental test which could distinguish among the different interpretations. The consequences of such experimental verifications have more profound implications than just eliminating special models which interpret the measuring process. They will test the validity of a general conception of the foundations of microphysics: the principle of locality or, as written more precisely in Ref. 7, the validity of objective local theories.

II. BELL'S INEQUALITY

The first derivation of the inequality, which later led to an experimental test, was given by Bell.[4] It has been generalized by Clauser *et al.*[6,7] In the meantime various ways of demonstration have been derived and can be found in Ref. 9 together with a description of the actual state of hidden-variables theories. We will follow here a demonstration given by Bell.[8]

To be definite we take the example of two spin-$\frac{1}{2}$ particles which have been coupled in the past in a singlet state and which are now widely separated. The principle of locality as formulated by Einstein means that each of these particles has some properties, which we will denote by λ(λ can be a whole set of variables) which do *not* depend on what is happening to the other particle. The result of the measurement is determined by these properties λ. We denote by A, B the result of the measurement in the direction $\vec{a}$ and $\vec{b}$ of the sign of the spin of the two particles respectively. For a realistic apparatus and/or if the dependence on λ is not strictly deterministic, but only stochastic, one will have

$$|A(\lambda, \vec{a})| \leq 1 \text{ and } |B(\lambda, \vec{b})| \leq 1.$$

The correlation function $P(\vec{a}, \vec{b})$ is defined to be the mean value of the product AB and thus

$$P(\vec{a}, \vec{b}) = \int A(\lambda, \vec{a}) B(\lambda, \vec{b}) \rho(\lambda) d\lambda,$$

where $\rho(\lambda)$ denotes the frequency of the properties λ with the normalization condition $\int \rho(\lambda) d\lambda = 1$. Thus,

Originally published in *Physical Review, D14*, 2543-55 (1976).

$$\begin{aligned}|P(\vec{a},\vec{b})-P(\vec{a},\vec{b}')| &= \Big|\int A(\lambda,\vec{a})B(\lambda,\vec{b})\rho(\lambda)\,d\lambda-\int A(\lambda,\vec{a})B(\lambda,\vec{b}')\rho(\lambda)d\lambda\pm\int A(\lambda,\vec{a})B(\lambda,\vec{b})A(\lambda,\vec{a}')B(\lambda,\vec{b}')\rho(\lambda)d\lambda\\ &\quad\mp\int A(\lambda,\vec{a})B(\lambda,\vec{b}')A(\lambda,\vec{a}')B(\lambda,\vec{b})\rho(\lambda)\,d\lambda\Big|\\ &= \Big|\int A(\lambda,\vec{a})B(\lambda,\vec{b})[1\pm A(\lambda,\vec{a}')B(\lambda,\vec{b}')]\rho(\lambda)d\lambda-\int A(\lambda,\vec{a})B(\lambda,\vec{b}')[1\pm A(\lambda,\vec{a}')B(\lambda,\vec{b})]\rho(\lambda)d\lambda\Big|\\ &\le \int |A(\lambda,\vec{a})B(\lambda,\vec{b})|\,|1\pm A(\lambda,\vec{a}')B(\lambda,\vec{b}')|\rho(\lambda)d\lambda+\int |A(\lambda,\vec{a})B(\lambda,\vec{b}')|\,|1\pm A(\lambda,\vec{a}')B(\lambda,\vec{b})|\rho(\lambda)d\lambda\\ &\le 2[\,P(\vec{a}',\vec{b}')+P(\vec{a}',\vec{b})]\quad\text{using } |AB|\le 1.\end{aligned}$$

For coplanar vectors $\vec{a}$, $\vec{a}'$, $\vec{b}$, $\vec{b}'$ only the angle between these vectors is important, and one can write

$$|P(\theta)-P(\theta+\gamma)|+|P(\theta+\gamma+\phi)+P(\theta+\phi)|\le 2.$$

QM predicts for this correlation function[4, 17, 18]

$$\begin{aligned}P(\vec{a},\vec{b})_{QM} &= \langle\vec{\sigma}_1\cdot\vec{a}\,\vec{\sigma}_2\cdot\vec{b}\rangle\\ &= -\cos(\vec{a},\vec{b})\\ &= -\cos\theta\ .\end{aligned}$$

Putting in special values for the angles θ, γ, ϕ and using the invariance of P by reflection and rotation one gets the upper limits for the absolute value of $P(\theta)$ as compared to predictions of QM in Table I. As can be seen, there is a definite contradiction between the values predicted by QM and the upper limit as implied by the inequality of Bell.

III. DESCRIPTION OF A "PERFECT" EXPERIMENTAL SETUP AND DISCUSSION OF REALIZED EXPERIMENTS

The inequality of Bell can be tested by special experimental devices. We will first describe an example of an ideal experiment and then discuss how the actually performed experiments differ from such an ideal arrangement.

We will take as an example Bohm's[10] version of the Einstein-Podolsky-Rosen paradox. Consider (Fig. 1) a source which prepares two particles of spin $J=\frac{1}{2}$ in an intermediate state $J=0$. This state disintegrates by emitting the two particles with a velocity v in opposite directions. The two possible states, + and −, of the direction of spin are split up for example by Stern-Gerlach magnets and the particles are detected by the detectors d. The vectors $\vec{a}$ and $\vec{b}$ denote the orientations of the magnets. Then we define N_0 as the number of pairs of particles which enter the analyzers in coincidence (preparation of a beam of coincident particles) which would be measured in the case of charged particles of sufficiently high energy for example by the use of thin ΔE counters as entrance collimators of the analyzers without depolarization of the particles. As in Ref. 7 we define $p_1^+(\vec{a},\lambda)$ as the probability of having a count in the counter d_1^+, and in the same way for the other counters. For sufficiently large number N_0 we have

$$p_1^+(\vec{a})=\frac{N_1^+}{N_0}=\int P_1^+(\vec{a},\lambda)\rho(\lambda)d\lambda,$$

and clearly $p_1^+(\vec{a},\lambda)\le 1$ and $-1\le p_1^+(\vec{a},\lambda)-p_1^-(\vec{a},\lambda)\le 1$. For objective local theories one gets

$$\frac{N_{++}}{N_0}=\int p_1^+(\vec{a},\lambda)p_2^+(\vec{b},\lambda)\rho(\lambda)d\lambda,$$

where N_{++} are coincidences between the counters d_1^+ and d_2^+.

Defining the measured correlation function as

$$P'_{\text{meas}}(\vec{a},\vec{b})=\frac{N_{++}+N_{--}-N_{-+}-N_{+-}}{N_0}$$

and using the above relation for N_{++} and an analogous relation for N_{+-} and so on we get

$$P'_{\text{meas}}(\vec{a},\vec{b})=\int[p_1^+(\vec{a},\lambda)-p_1^-(\vec{a},\lambda)][p_2^+(\vec{b},\lambda)-p_2^-(\vec{b},\lambda)]\times\rho(\lambda)d\lambda.$$

Setting

$$p_1^+(\vec{a},\lambda)-p_1^-(\vec{a},\lambda)=A(\vec{a},\lambda)$$

and

$$p_2^+(\vec{b},\lambda)-p_2^-(\vec{b},\lambda)=B(\vec{b},\lambda),$$

we see that this is equivalent to the function $P(\vec{a},\vec{b})$ defined earlier and thus must obey the same inequality, if objective local theories are valid. Quantum mechanics predicts for this correlation

TABLE I. Comparison of predictions of QM for $|P(\phi)|$ with the upper limit given by the inequality of Bell.

ϕ	QM	Upper limit of Bell's inequality
0°	1	≤ 1
30°	$\frac{3}{2}$	$\le\frac{2}{3}$
45°	$1/\sqrt{2}$	$\le\frac{1}{2}$
60°	$\frac{1}{2}$	$[2-P(0°)]/3$
90°	0	0

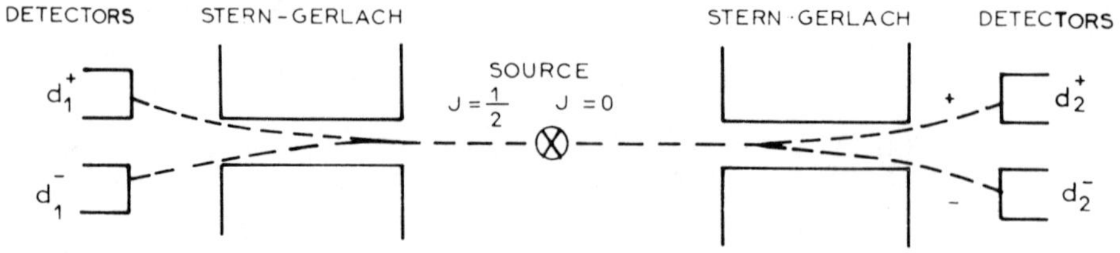

FIG. 1. Schematic experimental setup for the test of the inequality of Bell.

function

$$P'_{\text{meas}}(\vec{a}, \vec{b}) = P_1 P_2 T_1 T_2 \cos(\vec{a}, \vec{b}),$$

where $P_{1,2}$ and $T_{1,2}$ are the analyzing power and transmission of the analyzers. Therefore in order to get a contradiction with the inequality an ideal apparatus should have the property

(a) $|P_1 P_2 T_1 T_2| > 1/\sqrt{2}$, which is a very stringent condition, and supposes yet that one has measured the number of particles which enter the analyzer in coincidence. Apart from this, the ideal apparatus should have the following properties:

(b) The lifetime τ of the intermediate state should be short and the source should be pulsed (or by some other information—coincidence—one should know when the particles enter the analyzers) with a resolution t_r and the restrictions

$$\tau v << D \text{ and } t_r v << D,$$

where v is the velocity of the particles and D is the dimension of the apparatus. This ensures that one knows at each moment with a good precision where to find the particles in the apparatus.

(c) The time t_v between the moment when the particles enter the analyzer and the detection should satisfy the relation

$$t_v c < d,$$

where c is the speed of light and d is the distance between the analyzers. Thus the theory of relativity excludes the coupling of the particles after or during the measurement process.

(d) The orientation $\vec{a}, \vec{b}$ of the two analyzers should be changed in an arbitrary way during the time of flight of the particles, satisfying the relation

$$(t_{\text{ch}} - t_{\text{det}})c < d,$$

where t_{ch} denotes the time of the change of orientation and t_{det} denotes the time of detection of the particles. Thus one analyzer cannot "know," with a speed of exchange of information $\leq c$, what the other one is doing.

If *all* these conditions are satisfied the above-defined correlation can directly be compared to the inequality of Bell without extra assumptions. To see what can happen if there are deviations from the above-described ideal setup let us consider as an example the case where the detection probability is low, $P_{\text{det}} << 100\%$. It is possible to imagine that for a perfect apparatus with $P_{\text{det}} = 100\%$, one would measure a correlation function which agrees with the inequality of Bell. When the detection probability becomes low the properties of the particles, which determine the result of the spin measurement, determine at the same time the probability of joint detection, doing so in such a way that agreement of the counting rates with QM is reestablished. This means that the properties of the particles which determine the result of the spin measurement would determine in a *correlated* way other properties (in this example to be detected or not). One can imagine that for one special experiment there is such a correlation of properties. But it seems difficult to imagine that in very different experimental setups different properties will always be correlated in just such a way to reestablish agreement with QM. This shows that it is necessary to test the inequality of Bell in very different experimental conditions. If the spacelike separation of the particles and of the different parts of the measuring device is not realized, one can imagine some hypothetical coupling or exchange of information. This would mean that the result of the measurement on one particle could depend on what is done with the other particle and agreement with QM could be obtained.

The first experiment done specifically to verify the inequality of Bell was the measurement of the correlation of polarization of positronium annihilation γ rays by Kasday.[11] Agreement with QM was obtained. Somewhat later the correlation of polarization of photons of an atomic cascade was studied by Freedman and Clauser,[12] and again agreement with QM was obtained. The experiment with atomic photons has the advantage that in atomic physics one can build polarization analyzers of nearly 100% transmission and analyzing power, which is not the case for the experiment with annihilation γ rays. Some of this advantage is lost by the low probability of response of photomultipliers used to detect the photons (~10%) and the fact that the second photon is not emitted in a well-defined direction with respect to the first one, but into the whole space. The photon experiment is related to a three-body phenomenon, involving the two photons and the atom which emits the photons. Thus, as in the other experiment, only a very small number of the photons detected in one analyzer is in coincidence with the photons detected in the other analyzer. In the atomic photon experiment the additional assumption necessary concerns the response probability of the photomultiplier ("no enhancement assumption"[7]), whereas in the annihilation γ experiment it concerns the scattering process in the first scatterer.

Both types of experiments used photons. Photons cannot be localized by a Lorentz transformation. One can attribute to photons a length, the coherence length $l=c\tau$, where c is the speed of light and τ is the mean lifetime of the state which produced the photon. For experiments with annihilation γ rays this length is ~17 cm and ~300 cm for the atomic cascade case. These dimensions are comparable to or bigger than the dimension of the apparatus used and therefore it is not clear that the condition of localization is respected. This is why it had been suggested[13] that we use particles with a mass at rest different from zero.

We developed an experimental device to measure the spin correlation of protons after scattering in a singlet state, and this will be described in some detail below. Since the beginning of the development of this device the experimental situation for the photons became more confused. The experiment with annihilation γ rays was repeated by Faraci *et al.*[14] and results in contradiction with QM and in agreement with Bell's inequality were obtained. For another atomic cascade Holt and Pipkin[15] found a result which is also in contradiction with QM and in agreement with Bell's inequality. These experiments will be repeated in other laboratories, and so we can hope that in the not-too-distant future the experimental situation for the photon experiments will be clarified.[28]

IV. MEASUREMENT OF THE SPIN CORRELATION IN LOW-ENERGY PROTON-PROTON SCATTERING

A. The experimental setup

In Fig. 2 the schematic experimental device is shown. A beam of protons, delivered by the Saclay tandem accelerator, hits a target containing hydrogen. After scattering, the two protons enter in kinematical coincidence into the analyzers at $\theta_{\text{lab}}=45°$ ($\theta_{\text{c.m.}}=90°$). In the analyzers the protons are scattered by a carbon foil and the coincidences between the detectors of one analyzer with the detectors of the other are counted. The detectors of one analyzer are in the reaction plane, and the detectors of the other are rotated by an angle θ around the axis defined by the protons entering in the analyzer.

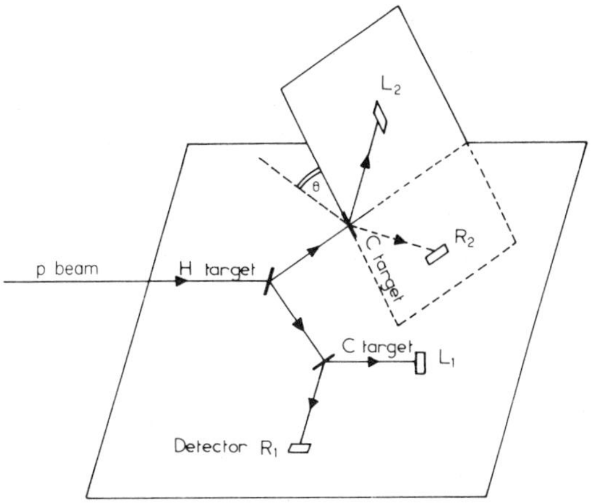

FIG. 2. Schematic experimental setup for the measurement of the spin correlation in proton-proton scattering.

We define the measured correlation function

$$P_{\text{meas}}(a,b)=\frac{N_{LL}+N_{RR}-N_{RL}-N_{LR}}{N_{LL}+N_{RR}+N_{RL}+N_{LR}},$$

where N_{LL} are the coincidences between left counters L_1 and L_2, and so on.

It is not possible to compare this correlation function directly with the inequality of Bell. Some extra assumptions are necessary because our apparatus does not fulfill the conditions for a perfect apparatus. The following assumptions are necessary:

H1: It is possible to construct a perfect apparatus. As in the annihilation γ case, there is no experimental method known which could give nearly 100% analyzing power and transmission, even extrapolating present techniques. A Stern-Gerlach apparatus is not suited for charged particles as can be shown by uncertainty relations.[16] Nonetheless, there does not seem to exist *a priori* an obstacle to constructing such a perfect apparatus.

H2: Our device does not fulfill the conditions of spacelike separation discussed in the preceding section. We assume that this does not affect the result of the measurement. In the experiment the coherence length can be estimated from the lifetime of the intermediate state formed in p-p scattering. Because there are no sharp resonances this lifetime is of the order of $\tau\simeq 10^{-22}$ sec, resulting in a coherence length of $\lambda\simeq 4\times10^{-13}$ cm. This is very small compared to the dimension of the apparatus (5 cm), in distinction with the photon experiments where the coherence length and the dimension of the apparatus are of the same order of magnitude. We tried to minimize the distance between the detectors and the carbon foil, in order to prevent a hypothetical coupling between the particles after the second scattering. In our device the time of flight $t_v = 0.3$ nsec, whereas $d/c = 0.2$ nsec. This means that we were quite near the condition (c) in Sec. III, but we did not yet fulfill it.

The analyzing power of our measuring device is ~0.7 which is very similar to the one obtained in the annihilation γ experiments and, as in the case of annihilation γ's the transmission of the analyzers is very low. Therefore the product $P_1P_2T_1T_2$ does not fulfill the conditions of a perfect device. Examples for which it is not possible to compare with the inequality without additional assumptions have been given.[7,8,11] Consider a pair of particles which enter the analyzers in coincidence. Suppose that a perfect device would give the results $p_{IL}^{id}(\vec{a}, \lambda)$, $p_{IR}^{id}(\vec{a}, \lambda)$, etc. for the probability of having a count in the detectors d_{IL}, d_{IR}, etc. It makes sense to consider such a device because of assumption H1. We now make the following assumption:

H3: The analyzing power and the transmission of the measuring apparatus can be considered as intrinsic constants of the apparatus. This means that the parameters λ which determine the result of the spin measurement do not determine in a correlated way the value of the analyzing power and/or transmission of the analyzers. A similar assumption is necessary in the atomic photon experiment concerning the response probability of the photomultipliers.

Then one can write

$$p_{1L}(\vec{a}, \lambda) + p_{1R}(\vec{a}, \lambda) = T_1[p_{1L}^{id}(\vec{a}, \lambda) + p_{1R}^{id}(\vec{a}, \lambda)] = T_1,$$

$$p_{1L}(\vec{a}, \lambda) - p_{1R}(\vec{a}, \lambda) = T_1P_1[p_{1L}^{id}(\vec{a}, \lambda) - p_{1R}^{id}(\vec{a}, \lambda)] = T_1P_1A(\vec{a}, \lambda),$$

where $p_{IL}(\vec{a}, \lambda)$ is the probability of having a count in the left counter of the analyzer[1] and so on, and P_1 and T_1 are the mean analyzing power and transmission. Analogous equations hold for the apparatus 2.

Then for N_0 particles which enter the analyzers in coincidence, one has

$$N_0(p_{1L} - p_{1R})(p_{2L} - p_{2R}) = N_0P_1P_2T_1T_2 \times \int A(\vec{a}, \lambda)B(\vec{b}, \lambda)\rho(\lambda)d\lambda = N_{LL} + N_{RR} - N_{LR} - N_{RL},$$

$$N_0(p_{1L} + p_{1R})(p_{2L} + p_{2R}) = N_0T_1T_2 = N_{LL} + N_{RR} + N_{RL} + N_{LR}.$$

Using these relations one has

$$P_{\text{meas}}(\vec{a}, \vec{b}) = \frac{N_{LL} + N_{RR} - N_{RL} - N_{LR}}{N_{LL} + N_{RR} + N_{RL} + N_{LR}} = P_1P_2\int A(\vec{a}, \lambda)B(\vec{b}, \lambda)\rho(\lambda)d\lambda.$$

The transmission coefficients for left-right scattering of the analyzers depend slightly on the angle ψ of the first scattering. Replacing N_0 by $dN_0/d\psi$ and introducing the dependence of the transmission coefficient on ψ, one finds after some arithmetic and integration between $\psi_{\min}$ and $\psi_{\max}$ a correction to the preceding formula; and the final result is, to first order in the correction term,

$$P_{\text{meas}}(\vec{a}, \vec{b}) = P_1P_2\int A(\vec{a}, \lambda)B(\vec{b}, \lambda)\rho(\lambda)d\lambda + C_g\left\{1 - \left[P_1P_2\int A(\vec{a}, \lambda)B(\vec{b}, \lambda)\rho(\lambda)d\lambda\right]^2\right\}\cos(\vec{a}, \vec{b}),$$

where

$$C_g = \frac{\int(T_{1L}T_{2L} + T_{1R}T_{2R} - T_{1L}T_{2R} - T_{1R}T_{2L})d\psi}{\int(T_{1L}T_{2L} + T_{1R}T_{2R} + T_{1L}T_{2R} + T_{1R}T_{2L})d\psi}.$$

Defining

$$\int A(\vec{a}, \lambda)B(\vec{b}, \lambda)\rho(\lambda)d\lambda = P_{\exp}(\vec{a}, \vec{b}),$$

we can write

$$P_{\text{meas}}(\vec{a}, \vec{b}) = \frac{N_{LL} + N_{RR} - N_{RL} - N_{LR}}{N_{LL} + N_{RR} + N_{RL} + N_{LR}} = P_1P_2P_{\exp}(\vec{a}, \vec{b}) + C_g[1 - |P_1P_2P_{\exp}(\vec{a}, \vec{b})|^2]\cos(\vec{a}, \vec{b}).$$

Measuring P_1, P_2, and C_g, the experimental counting rates can be used to extract $P_{\exp}(\vec{a}, \vec{b})$.

Now we must remind the reader that in order to derive the inequality of Bell it is necessary to vary $\vec{a}$ as well as $\vec{b}$ (see Sec. II). Because in the laboratory system the two proton directions form

90° and not, as in the center-of-mass system, 180°, when rotating $\vec{a}$ and $\vec{b}$ around the propagation direction of the protons to $\vec{a}'$, $\vec{b}'$, the vectors $\vec{a}$, $\vec{b}$, $\vec{a}'$, and $\vec{b}'$ are not coplanar. The prediction of QM is

$$P_{QM}(\vec{a}, \vec{b}) = C_{nn} \cos(\vec{a}, \vec{b}) + C_{kp} \sin\varphi_1 \sin\varphi_2$$
$$= C_{nn} \cos\varphi_1 \cos\varphi_2 + C_{kp} \sin\varphi_1 \sin\varphi_2,$$

where φ_1, φ_2 are the angles by which the analyzers are turned out of the reaction plane. Using this correlation function directly one sees that it does not violate the inequality of Bell. This comes from the fact that our analyzers are sensitive only to the transversal component of the polarization. Nonetheless, it is possible to obtain a contradiction using the fact that most of the scattering passes through a $J = 0$ intermediate state and the correlation function of $J = 0$ must be invariant with respect to rotation. We can decompose the correlation function into a part $f(\vec{a}, \vec{b}) = f(\varphi)$ that is rotationally invariant, and another part $g(\vec{a}, \vec{b})$ that is not, with

$$P(\vec{a}, \vec{b}) = f(\varphi) + g(\vec{a}, \vec{b}),$$

and therefore

$$P(\vec{a}, \vec{b}) - g(\vec{a}, \vec{b}) = P(\vec{a}', \vec{b}') - g(\vec{a}', \vec{b}')$$

if $(\vec{a}, \vec{b}) = (\vec{a}', \vec{b}')$. We define β such that $|g(\vec{a}, \vec{b})| \leq \beta$ for all $\vec{a}, \vec{b}$. The singlet scattering cannot contribute to the part of $P(\vec{a}, \vec{b})$ that is not rotationally invariant. Therefore an upper limit for β is the probability of triplet scattering. From the measurements of Catillon *et al.*[19] one obtains $\beta \leq 0.02 \pm 0.01$. We make the following assumption:

H4: An upper limit for the contribution of triplet scattering can be obtained from the scattering of polarized protons on polarized protons. This assumption could be eliminated by the use of a 90° (electrostatic) deflector before one of the polarimeters. This would give a 180° correlation and thus the vectors $\vec{a}$, $\vec{b}$, $\vec{a}'$, and $\vec{b}'$ would be coplanar. Introducing the above expressions in the equality gives

$$|P(\varphi) + P(\varphi')| + |P(\varphi+\gamma) - P(\varphi'+\gamma)| \leq 2 + 4\beta \approx 2.08,$$

where φ, φ' and $\varphi+\gamma$, $\varphi'+\gamma$ are the angles that one polarimeter is turned out of the reaction plane, whereas the other remains fixed in the reaction plane.

Quantum mechanics predicts that

$$P_{exp}(\theta) = -C_{nn} \cos\theta,$$

when one analyzer is in the reaction plane and the other is rotated by an angle θ out of it (Fig. 2). Experimental values obtained by Catillon *et al.*[19] can be interpolated to the energies used for this experiment (13.2 and 13.7 MeV) and give $C_{nn} = -0.95 \pm 0.015$. The deviation of C_{nn} from -1 reflects the 2% contribution of triplet scattering. In order to get a contradiction between predictions of quantum mechanics and the inequality of Bell, we need some additional assumptions (H1 – H4).

Because the analyzing power and the transmission are similar to the experiment with annihilation γ's, equivalent assumptions are needed. The main difference for protons is the fact their coherence length is $\sim 10^{-13}$ cm, which is extremely small compared to the dimension of the apparatus (~ 5 cm), whereas for the photons the coherence length is larger than or comparable to the size of the measuring apparatus. Therefore for photons the locality condition seems not as clear as in the present experiment.

Even if in the atomic cascade experiments it is possible to construct polarizers of nearly 100% analyzing power and transmission, the assumption made concerning the response probability of the photomultipliers is qualitatively the same as needed here, only it concerns a more simple effect (photoelectric effect) than the scattering process here, and is better isolated from other effects such as analyzing power.

B. Design of the experimental device

The design of the experimental device is mainly conditioned by the low transmission of polarization analyzers available in nuclear physics. For the analyzers used here it was of the order of 10^{-5}; because of the coincidence between the analyzers only roughly one of 10^{10} pairs of protons which enter in the analyzers is detected in coincidence. To have, in spite of this, high enough counting rates, it is necessary to have a high beam intensity, thick targets, and large solid angles.

As hydrogen targets, polyethylene (CH_2) foils of 9 mg/cm^2 were used. The beam used of 1.5 μA with a spot size of 1.5-mm diameter instantly burned a hole in the target. Therefore it was necessary to construct a target rotating in an eccentric way with respect to the beam with the speed of 1 rev/sec. Thus, mechanically, the targets well withstood the beam, but still a chemical burning resulted in a blackening of the target and, after some hours, the hydrogen content on the beam trace diminished by a factor of 2. Thus every two hours the eccentricity was changed together with the orientations of the analyzers and every six hours the target was changed. Because the detectors were cooled to −20°C, ice buildup by the humidity contained in air had to be avoided and thus the opening was done under argon atmosphere.

The main problem for the analyzer was the reduction of the background. γ rays, mainly from the target, produce a background which rises exponentially for low energies and which completely covers the events of interest if one does not take great care. This background was reduced to essentially zero in the region of interest by a lead protection and by using silicon detectors 300 to 400 μ thick. The use of thinner detectors would have reduced the background even more but they would not have stopped the most energetic protons (~ 6 MeV) and thus the energy spectrum would have been deformed, not allowing a detailed interpretation. A careful design of slits is necessary to reduce the background produced by protons. The protons which are scattered by an angle < 10° leave the analyzer by a tube which serves at the same time as a rotation axis. The mean angle of scattering of the protons detected in the analyzer was fixed to 50°.

The final design of the analyzers is shown in Fig. 3. In an early stage of the experiment, one of the analyzers contained only two detectors. This was changed to have simultaneously a measurement of the geometric correlation (see Sec. IV F). All pieces, including beam-defining slits, were mounted on the cover of the scattering chamber and aligned to better than $\frac{1}{10}$ mm. The most important dimensions are given in Table II.

C. Monte Carlo simulation

Apart from suitable experimental tests, it seemed highly desirable to make exact calculations simulating the experimental device in order to have a supplementary control that the apparatus not show any undesirable and uncontrolled feature. At the same time this provides help for the choice of geometry and for the optimization of the apparatus. A numerical integration is excluded because of the dimension of the integrals involved. Thus a Monte Carlo evaluation of the integrals was programmed for the CII 10020 computer.

The geometry was treated exactly taking into account the three dimensions of the device. The energy loss in the targets was calculated using the tables of Northcliff and Schilling.[20] Angular straggling was taken into account. The carbon cross section and polarization were obtained using the results of phase-shift analyses.[21-24] The

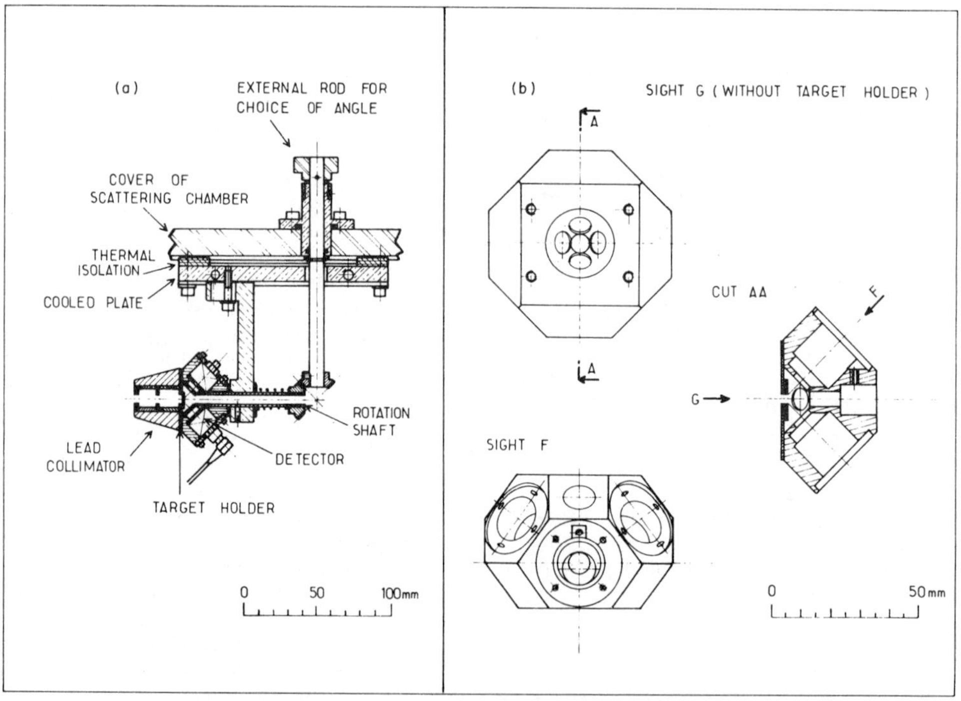

FIG. 3. Design of the polarization analyzers: Part (a) shows the general design, and part (b) shows the details of the analyzer.

TABLE II. Most important dimensions used in the experimental device.

	Diameter (mm)	Distance (mm) from target	Number
beam-defining slit 1	2	180	1
beam-defining slit 2	2	150	1
beam-defining slit 3	1.5	50	1
beam-defining slit 4	2	20	1
entrance slit of the analyzers	4.5	51.5	1
slit before detectors	7	12.5	2

phase shifts were parametrized and resonances at 4808 and 5373 keV (see Ref. 22) were included. Good agreement with published[21-24] cross sections and analyzing powers was obtained. For Ta scatterers pure Rutherford scattering was assumed, which is justified because the energy of the protons is much lower than the Coulomb barrier.

In the following the experimental results will be compared to the results of the Monte Carlo calculation.

D. Measurement of the analyzing power

The analyzing power was measured in a double-scattering arrangement. A beam of protons of 8.07 MeV is scattered by a 2-mg/cm^2 carbon target. The scattered protons enter at $\theta_{lab} = 70°$ in the analyzer and are scattered with a mean angle of 50° a second time by the carbon foil of the analyzer, and one measures the asymmetry of the counting rates of the left-right detectors. Polyethylene foils are put between the first and second target to slow down the protons to the desired energy. The protons after the first scattering are polarized to 100% with an error of less than 2%.[21,22] Thus the analyzing power is directly obtained by the measured asymmetry. Before the measurement of the analyzing power the carbon target of the analyzer was replaced by a gold target (35 mg/cm^2) to be sure that no misalignment affected the results.

The main problems connected with these measurements, once the background problem was solved, were target problems. First we used, for the analyzer, carbon targets prepared with Aquadag (Acheson Company). For proton energies above 6 MeV the experimental analyzing power was much lower than the calculated one, as stated in Ref. 25. Exposing the targets to the beam showed no significant target contamination. Heating them for several days to 200°C gave no improvement. But when these targets were heated in a vacuum to 1500 °C they lost 20% of their weight, demonstrating that they contained still a large amount of the liquid solvent. After heating to 1500 °C the targets were so frail that they could no longer be used. Thus the targets finally used were mechanically worked out of solid carbon (Carbone Lorraine).

In Fig. 4 a typical energy spectrum is shown. The results of the asymmetry measurements for the two targets used, 18.6 and 29 mg/cm^2, are shown in Figs. 5 and 6. A background correction of about 2% was applied with an estimated error of 1%. The transmission of the analyzer, defined as the number of protons detected by the two detectors in the reaction plane divided by the number of protons which enter the analyzer, was 4.4×10^{-5} and 6.5×10^{-5} for the 18.6-mg/cm^2 and the 29-mg/cm^2 targets respectively.

E. Electronics associated with coincidence measurements

A block diagram of the coincidence measurements which will be described below is shown in Fig. 7.

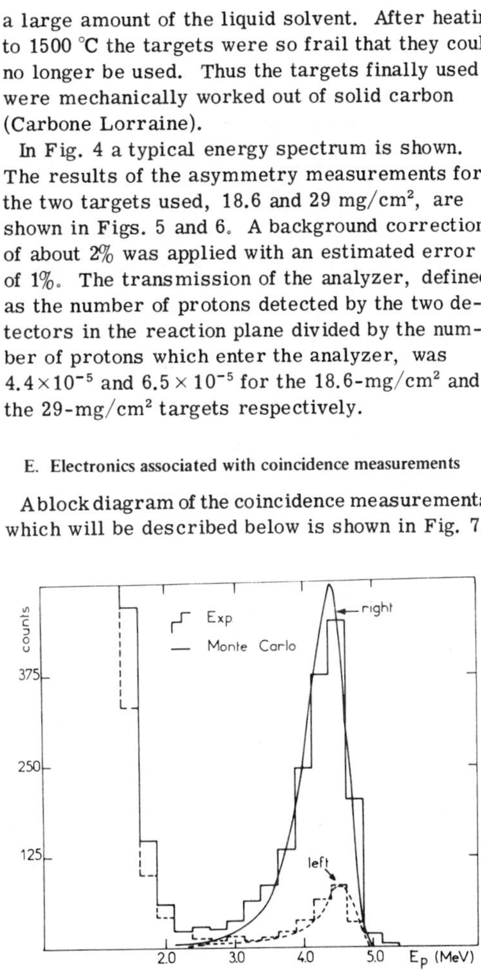

FIG. 4. Typical energy spectrum obtained in the measurement of the analyzing power of the polarimeters by double scattering of the protons, with $E_p = 8.07$ MeV, a first target of carbon of 2 mg/cm^2 and a second carbon target of 18.6 mg/cm^2. The protons entering in the analyzer were slowed down by a 17-mg/cm^2 CH_2 foil to $E_p = 5.95$ MeV.

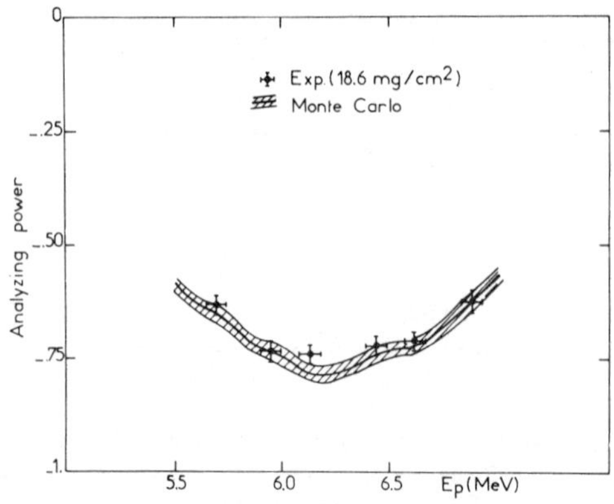

FIG. 5. Analyzing power of the analyzer with a 18.6-mg/cm^2 carbon target, as a function of the energy of the incoming protons. The width of the Monte Carlo calculation shows the uncertainty which was used to calculate the errors of the final results.

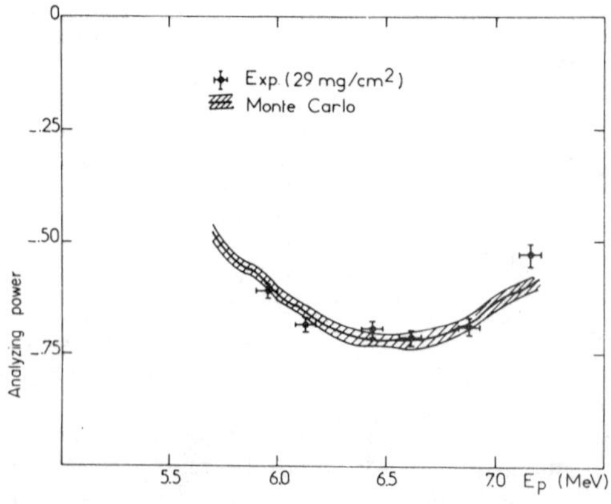

FIG. 6. Same as Fig. 5 for a 29-mg/cm^2 carbon target.

All the electronic devices up the analog-to-digital converters (ADC) used were standard electronics developed at the SPNBE at Saclay,[26] and they provide great versatility and facility toward a fast logic control of coincidence, anticoincidence, dead time, selection of energy domain, and control of counting rates at each level before the numerical analysis.

The four detectors of each analyzer are connected to a device which delivers a rapid pulse if one, and only one of the detector pulses delivered by a rapid proportional amplifier, is above a discriminator level. If there is a coincidence in the time-to-pulse converter, linear gates are opened which let the slower energy pulse of the

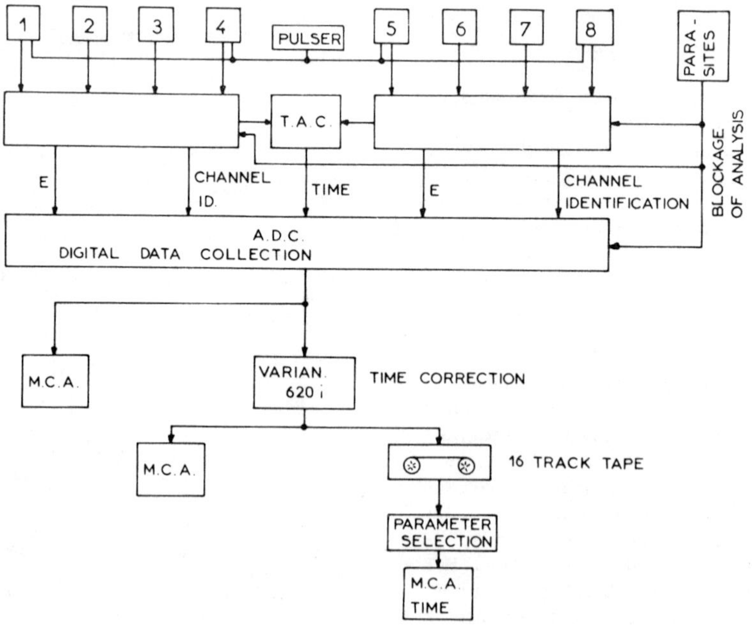

FIG. 7. Block diagram of the electronic setup used for coincidence measurements.

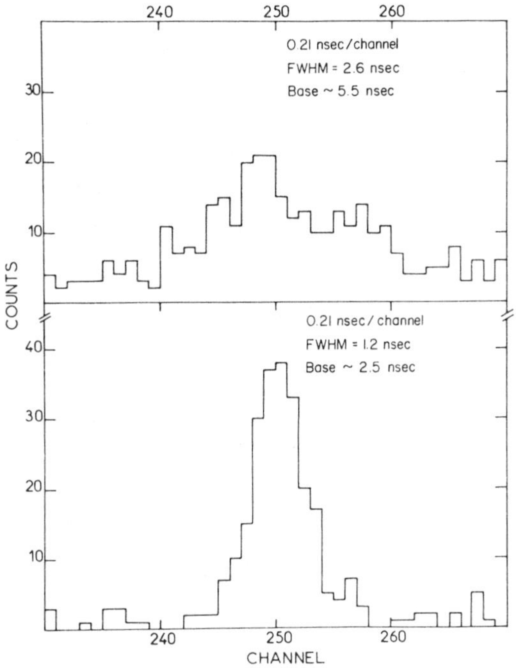

FIG. 8. Typical time spectra before and after rise-time correction by the 620 I Varian computer.

two, and only of these, detectors in coincidence pass, and an order is given to the ADC to analyze the energy and time signals. The results of this analysis together with an identification of the detectors in coincidence are sent to a Varian 620 I computer. This computer had mainly two roles: It coded the spectra following the 16 possibilities of coincidence and made a correction for the finite rise time (~50 nsec) of the rapid pulses. The rise-time correction was done by the formula

$$t_c = t_o + \frac{C_1}{E_1 + D_1} - \frac{C_2}{E_2 + D_2} ,$$

where t_o and t_c are the uncorrected and the corrected times, $E_{1,2}$ is the pulse height (energy) of the signals, and C, D were adjusted to give the best time resolution. A typical uncorrected and corrected time spectrum is shown in Fig. 8. A pulse generator was used continuously to control the electronic setup.

A good protection against electric parasites had to be included to prevent these parasites from simulating coincidences. A first protection is the anticoincidence between detectors which should not be in coincidence. A supplementary protection is given by an amplifying chain mounted in the same conditions as for the detectors. A discriminator is set just above the noise, and if it delivers a pulse, the coincidence circuits at the rapid level are blocked and the analysis of the pulses by the ADC is inhibited for 200 μsec. With these safeguards it was verified that over a period of several days no parasites were analyzed.

The Varian 620 I computer delivers for each event sequences containing the two energies, the corrected time, and the coded coincidence possibility, together with the number of the run. This information is written on magnetic tape. After an appropriate energy selection the 16 time spectra are accumulated in a 4096-channel memory. Because the information was written event-by-event on magnetic tape, various controls could be made out of beam time. For example, it was verified, by repeating the data reduction with different energy windows, that the results were insensitive to the setting of the energy windows within reasonable limits.

F. The geometric correlation

The geometric correlation which arises from the kinematic correlation between the protons after the first scattering and the anisotropy of the carbon (p, p) cross section gives rise to an enhancement of the coincidences between left-left and right-right detectors with respect to the

TABLE III. Results of the measurements of the geometric correlation and comparison with Monte Carlo predictions (see text).

E_p (MeV)	Target (analyzer 1)	Target (analyzer 2)	C_g (Monte Carlo)	C_g (experiment)
13.7	Ta (70 mg/cm^2)	Ta (70 mg/cm^2)	0.26	0.22 $\pm$0.01
	Ta (70 mg/cm^2)	C (27 mg/cm^2)	0.17	0.12 $\pm$0.02
	C (27 mg/cm^2)	C (27 mg/cm^2)	0.11	0.07 $\pm$0.02
13.2	Ta (70 mg/cm^2)	Ta (70 mg/cm^2)	0.26	0.225$\pm$0.01
	Ta (70 mg/cm^2)	C (18.6/mg/cm^2)	0.15	0.10 $\pm$0.03
	C (18.6 mg/cm^2)	C (18.6/mg/cm^2)	0.09	0.05 $\pm$0.02

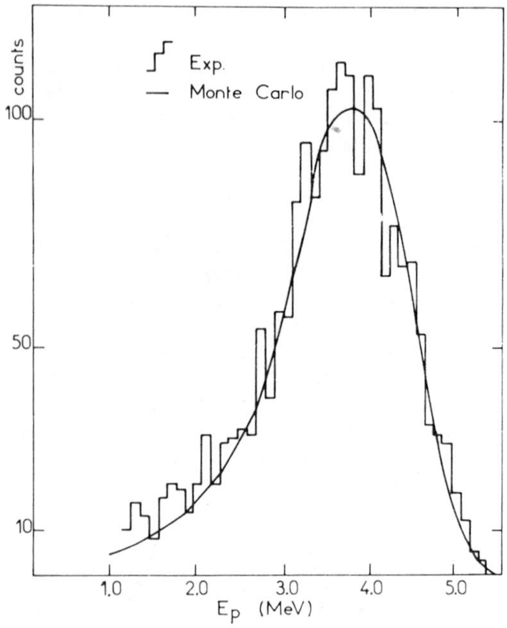

FIG. 9. Typical energy spectrum obtained in coincidence between the two analyzers with a 29-mg/cm^2 carbon target. Accidental coincidences have been subtracted.

right-left coincidences. Two methods were used to determine the coefficient of geometric correlation.

One method consists of first measuring the geometric correlation with both carbon foils in the analyzers replaced by 70-mg/cm^2 tantalum foils The analyzing power of tantalum is zero; thus the measured correlation function is, for one pair of detectors in the reaction plane and the other rotated out of it by an angle θ, $P(\theta)=C_g(\text{Ta-Ta})\cos\theta$, where C_g (Ta-Ta) is the geometric correlation coefficient for this arrangement. At the same

TABLE IV. Final results for the measured correlation function $P_{\text{meas}}(\theta)$ as a function of the angle θ for 18.6-mg/cm^2 and 29-mg/cm^2 targets. Errors are statistical one-standard-deviation errors.

θ	18.6 mg/cm^2	29 mg/cm^2
0°	-0.40 ± 0.05	-0.38 ± 0.025
30°	-0.38 ± 0.04	-0.27 ± 0.025
45°	-0.29 ± 0.04	-0.26 ± 0.023
60°	-0.24 ± 0.04	-0.17 ± 0.025
90°	-0.01 ± 0.03	-0.03 ± 0.04

time this setup permits an easy control of the electronic setup and of the whole apparatus; the coincidence rates (about 1000 counts/hour and coincidence possibility) are about a factor of 100 larger than with the carbon foils. Then one replaces one tantalum foil by a carbon foil and one gets the coefficient C_g(Ta-C). The coefficient for two carbon foils is connected to these coefficients by the relation

$$C_g(\text{C}-\text{C})=\frac{C_g(\text{C-Ta})^2}{C_g(\text{Ta-Ta})} \quad .$$

The second method is provided by the measurement with two carbon foils, one pair of detectors turned out of the reaction plane by 90°, and the other by an angle θ. Then $P_{\text{exp}}(\theta)=0$ (see Sec. III) and $P_{\text{meas}}(\theta)=-C_g\sin\theta$. Because the analyzers contained two detector pairs with 90° between them, simultaneously with the final results the value for the geometric correlations was obtained. Both methods gave within error bars identical results. The results are given in Table III.

As can be seen, the Monte Carlo simulation predicts a somewhat higher geometric correlation.

TABLE V. Final results for $P_{\text{exp}}(\theta)$ as compared to QM and to the limit of Bell. The results are given separately for the 18.6-mg/cm^2 and the 29-mg/cm^2 targets together with their weighted mean. $\langle P_1P_2\rangle$ and C_g are the values of the product of the analyzing power and of the geometric correlation coefficient, respectively, which were used to extract the values of P_{exp} from the values of P_{meas} given in Table IV.

θ	29 mg/cm^2 $\langle P_1P_2\rangle = 0.44 \pm 0.025$ $C_g = +0.07 \pm 0.02$	18.6 mg/cm^2 $\langle P_1P_2\rangle = 0.52 \pm 0.025$ $C_g = 0.05 \pm 0.02$	Mean	Bell's limit for the absolute value	QM
0°	-0.99 ± 0.09	-0.85 ± 0.11	-0.93 ± 0.07	≤ 1	-0.90
30°	-0.74 ± 0.08	-0.81 ± 0.10	-0.77 ± 0.06	≤ 0.69	-0.78
45°	-0.69 ± 0.08	-0.63 ± 0.09	-0.66 ± 0.06	≤ 0.52	-0.64
60°	-0.48 ± 0.07	-0.50 ± 0.10	-0.48 ± 0.06	≤ 0.38	-0.45
90°	$+0.07 \pm 0.10$	-0.01 ± 0.07	$+0.02 \pm 0.05$	≤ 0.02	0

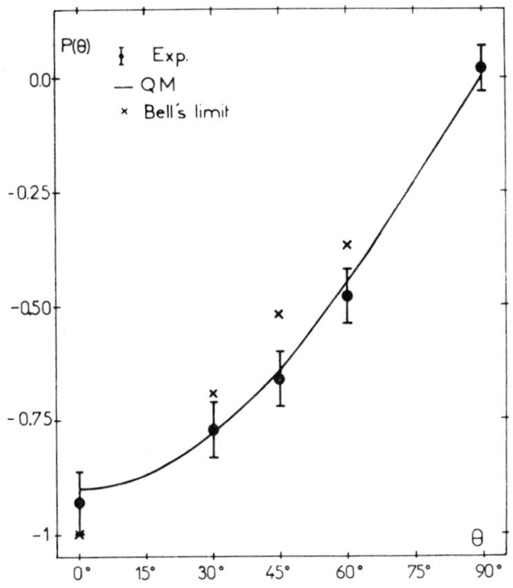

FIG. 10. Experimental results (see Table V) for $P_{exp}(\theta)$ as compared to the limit of Bell and predictions of QM.

The angular straggling mainly in the first target has some influence on the geometric correlation. To be sure that the values from Ref. 27 used in the Monte Carlo code were not the cause of this disagreement, we measured in a special device the angular straggling for the different targets; a good agreement was found.

The Monte Carlo calculation given in Table III was done for a perfect alignment of the apparatus. All misalignment contributes coherently to reduce the geometric correlation. Allowing for misalignment within the experimental uncertainties, $\frac{2}{10}$ mm for the center of gravity of the beam spot and $\frac{1}{10}$ mm for other geometric misalignments, the Monte Carlo code well reproduces the experimental results. These calculations showed, too, as one would expect, that the product of the analyzing powers, $\langle P_1P_2\rangle$, is independent of such small misalignments.

G. Final results

After all these preliminary measurements the final measurements of the coincidences of the two analyzers with carbon scatters could be made. It was verified that the final results for the correlation function did not depend on the lower limits set on the energies of the detected protons. A typical energy spectrum is shown in Fig. 9. Accidental coincidences, which represent about 5% of the total, have been subtracted. For each angle about 30 individual measurements were made with different combinations of orientations of the analyzers. A total of 10^4 true coincidences were accumulated. The final results for $P_{meas}(\theta)$ are given in Table IV. In Table V the results for $P_{exp}(\theta)$ together with the values of the product of the analyzing power and of the geometric correlation used to extract the value of $P_{exp}(\theta)$ are given.

In Table V the weighted mean of these results for $P_{exp}(\theta)$ is also given and compared to predictions of QM and to the limit of Bell. The quoted experimental errors are one-standard-deviation errors containing statistical errors and uncertainties in the analyzing power and in the geometrical-correlations coefficient. In Fig. 10, the values of Table V are shown. As can be seen, the agreement with QM is good, whereas the results are in contradiction with the limit of Bell with a statistical significance of $\frac{7}{10\,000}$. Using the prediction of QM for $P_{exp}(\phi_1, \phi_2) = C_{nn}\cos\phi_1\cos\phi_2$ one obtains

$$C_{nn} = -0.97 \pm 0.05$$

in good agreement with the value of $A_{yy} = C_{nn} = -0.95 \pm 0.015$ of Catillon *et al.*[19]

V. CONCLUSION

The measurement of the spin correlation of protons gave good agreement with QM. To compare with the limit of Bell, as in previous experiments with photons, some extra assumptions are necessary. With these assumptions which seem natural but cannot be tested in our device, a contradiction is obtained with the limit of Bell providing an argument against the validity of this limit, and thus being in favor of nonlocal properties of microphysics.

All experiments performed up to now do not respect all the conditions necessary to permit a direct comparison with the limit of Bell. The conditions for transmission and/or analyzing power and the spacelike separation of the particles and the different parts of the measuring device[29] are not respected and seem very difficult to realize even extrapolating up today's techniques in spin or polarization correlation measurements. Thus it would be interesting to find some other type of experiments where it would be more easy to fulfill these conditions.

ACKNOWLEDGMENTS

We wish to thank J. S. Bell, B. d'Espagnat, and A. Shimony for helpful discussions. The support of this work by E. Cotton, C. Lévi, and L. Papineau was appreciated. We thank the members of the mechanical workshop for very careful machining of all the pieces we needed and S. Valéro for the assistance and discussions during implantation of the Monte Carlo program.

---

*Present address: Univ. de Sao Paulo, Instituto de Fisica, C. Postal 20516, Sao Paulo, SP, Brasil.

[1]A. Einstein, N. Podolsky, and B. Rosen, Phys. Rev. 47, 777 (1935).

[2]E. Schrödinger, Naturwiss. 23, 807 (1935).

[3]D. Bohm, Phys. Rev. 85, 166 (1952); 85, 180 (1952).

[4]J. S. Bell, Rev. Mod. Phys. 38, 447 (1966); Physics (N.Y.) 1, 195 (1964).

[5]A. Einstein, in *A. Einstein Philosopher Scientist*, edited by P. Schilpp (Library of Living Philosophers, Evanston, Ill., 1949).

[6]J. F. Clauser, M. A. Horne, A. Shimony, and R. Holt, Phys. Rev. Lett. 23, 880 (1969).

[7]J. F. Clauser and M. A. Horne, Phys. Rev. D 10, 526 (1974).

[8]J. S. Bell, in *Foundations of Quantum Mechanics*, proceedings of the International School of Physics, Varenna, 1970, "Enrico Fermi," Course 49, edited by B. D'Espagnat (Academic, New York, 1972).

[9]F. J. Belinfante, *A Survey of Hidden Variables Theories* (Pergamon, New York, 1973).

[10]D. Bohm, *Quantum Theory* (Constable, London, England, 1954).

[11]L. R. Kasday, in *Foundations of Quantum Mechanics*, Ref. 8.

[12]S. J. Freedman and J. F. Clauser, Phys. Rev. Lett. 28, 938 (1972).

[13]R. Fox, Lett. Nuovo Cimento 2, 565 (1971).

[14]C. Faraci, D. Gutkowski, S. Notarrigo, and A. R. Pennisi, Lett. Nuovo Cimento 9, 607 (1974).

[15]R. A. Holt and F. M. Pipkin, report (unpublished).

[16]N. F. Mott and F. E. Mayers, *Theory of Atomic Collisions*, 2nd ed. (Oxford Univ. Press, London, 1949).

[17]L. Wolfenstein, Annu. Rev. Nucl. Sci. 6, 43 (1956).

[18]J. Raynal, Nucl. Phys. 28, 220 (1961).

[19]P. Catillon, M. Chapellier, and D. Garreta, Nucl. Phys. B2, 93 (1967).

[20]L. C. Northcliff and R. F. Schilling, Nucl. Data Tables 7, 233 (1970).

[21]S. J. Moss and W. Haeberli, Nucl. Phys. 72, 417 (1965).

[22]C. W. Reich *et al.*, Phys. Rev. 104, 143 (1956).

[23]E. M. Bernstein and G. E. Terrel, Phys. Rev. 173, 937 (1968).

[24]J. S. Duval, A. C. L. Barnard, and J. B. Swint, Nucl. Phys. A93, 164 (1967).

[25]G. Gurd, G. Roy, and H. Leighton, Nucl. Instrum. Meth. 61, 72 (1968).

[26]M. Avril *et al.*, Compte rendu d'activité du Département de Physique Nucléaire, Note No. CEA N-1232, 1968–1969 (unpublished), p. 66, and Note No. CEA N-1390, 1969–1970 (unpublished), p. 74.

[27]J. B. Marion and F. C. Young, *Nuclear Reaction Analysis* (North-Holland, Amsterdam, 1968).

[28]The experiment of Holt and Pipkin has been repeated very recently with a slightly different experimental device by J. F. Clauser [Phys. Rev. Lett. 36, 1223 (1976)], and he obtained agreement with QM. E. S. Fry [Phys. Rev. Lett. 37, 465 (1976)] made a measurement for another atomic cascade in Hg and obtained agreement with QM.

[29]An experiment with atomic cascade is in progress at Orsay, Institut d'Optique, where the orientation of the polarizers will be changed randomly [A. Aspect, Phys. Lett. 54A, 117 (1975)].

III.10 PROPOSED EXPERIMENT TO TEST THE NONSEPARABILITY OF QUANTUM MECHANICS

ALAIN ASPECT

As a criterion between quantum mechanics and local hidden-variable theories, the so-called Einstein-Podolsky-Rosen paradox is mainly tested in the form of the statistical correlation between polarizations of photons issuing from a cascade transition. It has been stated more than once that an improved form of the test would make use of polarizers, the orientation of which would change randomly in a time comparable with the time of flight of the two photons; the Bell locality assumption could then be replaced by a weaker assumption also considered by Bell: The Einstein principle of separability. However, to our knowledge, no workable experimental scheme has yet been proposed, and we believe the one described in this paper to be a workable one. After explaining the difference between the Bell locality assumption and the Einstein principle of separability, we briefly discuss the theoretical implications of the modified experiment. The overall scheme of the apparatus we are proposing is described, and the generalized Bell inequalities, modified for our case, are derived. As in previous experiments, supplementary assumptions are made in order to derive experimentally testable inequalities. Finally, we describe the device we intend to use to carry out the proposed scheme.

I. TEST OF THE PRINCIPLE OF SEPARABILITY

The so-called nonlocality paradox of Einstein, Podolsky, and Rosen[1] has been much discussed. Bell[2] has shown the possibility of bringing the question into the experimental domain. Then, several experiments have been proposed and performed.[3-8] All these experiments are able to discriminate between quantum mechanics and "local" hidden-variable theories that fulfill Bell's condition of locality: The setting of a measuring device does not influence the result obtained with another remote measuring device (nor does it influence the way in which particles are emitted by a distant source). Most of the experiments contradict these local hidden-variable theories,[4,5,7] although conflicting results exist.[6,8]

Although such a condition of locality looks highly reasonable, it is not prescribed by any fundamental physical law. Following a suggestion made by Bell,[2] we are proposing an experiment able to discriminate between quantum mechanics and "separable" hidden-variable theories fulfilling Einstein's principle of separability[9,10] that we can formulate in the following way for the experiments under consideration: The setting of a measuring device at a certain time (event A) does not influence the result obtained with another measuring device (event B) if the event B is not in the forward light cone of event A (nor does it influence the way in which particles are emitted by a source if the emission event is not in the forward light cone of event A).

Any theory fulfilling Bell's condition of locality also obeys Einstein's principle of separability. But one can conceive separable theories that do not fulfill Bell's condition of locality; such theories take into account the possibility of interactions between remote measuring devices (i.e., these theories do not fulfill Bell's condition of locality), but these interactions do not propagate with velocity greater than that of light (i.e., these theories obey Einstein's principle of separability). These theories were not within the reach of previous experiments, but they could be tested with a modification of these experiments.

To discuss this point, let us recall the optical transposition of Bohm's "*Gedankenexperiment*"[11] as performed by Freedman and Clauser[4] (Fig. 1). Letting $N(a_i, b_j)$ be the joint detection rate when the polarizers are in orientations a_i and b_j (the value ∞ represents the removal of the corresponding polarizer), one considers the quantiy

$$S = [1/N(\infty,\infty)][N(a_1, b_1) - N(a_1, b_2) + N(a_2, b_1) + N(a_2, b_2) - N(a_2,\infty) - N(\infty, b_1)], \quad (1)$$

where a_1, a_2, b_1, b_2 denote specific orientations of the polarizers in successive measurements. Local hidden-variable theories fulfilling Bell's condition of locality predict (modulo a supplementary assumption on the detector's efficiency) that S is constrained by the generalized Bell inequalities[3,12-15]

$$-1 \leq S \leq 0. \quad (2)$$

For certain values of the orientation parameters a_1, a_2, b_1, b_2 the quantum-mechanical predictions violate the inequalities (2). Hence an experimental test between the conflicting theories is possible.

It has been emphasized that a crucial point in the derivation of the Bell inequalities is the locality assumption.[16] These inequalities could not be

Originally published in *Physical Review, D14*, 1944-51 (1976).

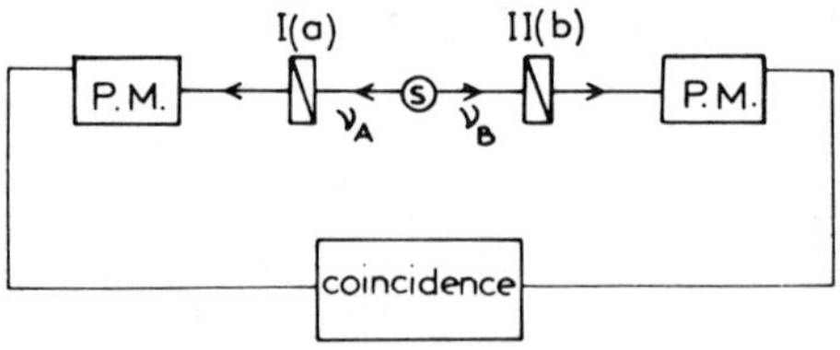

FIG. 1. Optical transposition of Bohm's "*Gedankenexperiment*." The correlated photons ν_A and ν_B, issuing from the cascade source S, impinge upon the linear polarizers I and II in orientations a and b. The rate of joint detections by the photomultipliers is monitored for various couples of orientations (a, b).

proved if the response of one polarizer was depending on the orientation of the other one (or if the emission of the pairs of photons was depending on the orientations of the polarizers). In the previous experiments, such interactions are not precluded by the principle of separability because "the settings of the instruments are made sufficiently in advance to allow them to reach some mutual rapport by exchange of signals with velocity less than or equal to that of light".[2] Therefore the separable theories that do not fulfill Bell's condition of locality cannot be tested by the previous experiments.

To test these theories, it has been proposed[2,13,15,17] to change rapidly, repeatedly, and independently the orientations of the polarizers. Then one finds as a consequence of the principle of separability that the response of one polarizer, when analyzing a photon, cannot be influenced by the orientation of the other polarizer at the same time (when analyzing the coupled photon); likewise, the way in which a pair of photons is emitted cannot be influenced by the orientations of the polarizers when later analyzing this pair. Therefore, for such improved experiments, inequalities (2) can be derived from the principle of separability, with no further locality assumption made. Since inequalities (2) still conflict with the quantum-mechanical predictions, such modified experiments would be able to discriminate between quantum mechanics and separable hidden-variable theories.

A result consonant with the quantum-theory predictions would imply the rejection of separable hidden-variable theories. But as a matter of fact, it would imply more.[18] According to a recent analysis[19] it would constitute an experimental confirmation of the reality of the nonseparability introduced formally in the quantum theory. More generally, d'Espagnat[20] pointed out that such a result would entail consequences practically amounting to a disproof of the principle of separability, and he showed that these consequences (violation of some general assumptions) could be derived without reference to any specific theory (in particular without incorporating in the general assumptions just mentioned any *a priori* hypothesis about the existence of hidden variables[21]). Then, some reinterpretations of the quantum theory would be untenable, while others would be upheld.[22]

II. PROPOSED EXPERIMENTAL SCHEME AND CORRESPONDING GENERALIZED BELL INEQUALITIES

Several authors[2,13,15,17] have already proposed to change rapidly and repeatedly the orientations of the polarizers, but few experimental practical suggestions have been given. One could think of using Kerr or Pockels cells, allowing changes in the polarization orientations in less than one nanosecond. Unfortunately, there are several drawbacks: Only very narrow beams could be transmitted, yielding very low coincidence rates; as these cells heat up, and then become inoperative, long runs would be prohibited. Last, a very sophisticated system would be needed for monitoring the change in time of the orientations; the calibration of the system would thus be exceedingly difficult.

We believe that these difficulties could be overcome by using optical commutators (Fig. 2). During a short time interval, the commutator C_A directs the photons ν_A towards the polarizer I_1; then its state changes and, during the following period, it directs ν_A towards the polarizer I_2. The commutator C_B works similarly with the photons ν_B, independently of C_A. The time intervals between two commutations are taken to be stochastic, so that two states of the commutator, separated by a time longer than the autocorrelation time, are statistically independent. The autocorrelation time

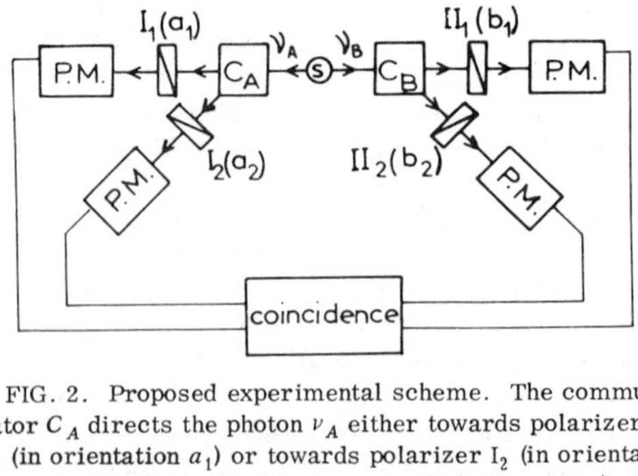

FIG. 2. Proposed experimental scheme. The commutator C_A directs the photon ν_A either towards polarizer I_1 (in orientation a_1) or towards polarizer I_2 (in orientation a_2). Similarly C_B directs ν_B towards II_1 or II_2 (in orientations b_1 and b_2). The two commutators work independently in a stochastic way. The four joint detection rates are monitored, and the orientations a_1, a_2, b_1, and b_2 are not changed for the whole experiment.

of each commutator is taken as short as L/c; L denotes the distance between the commutators and c denotes the speed of light.

One guesses that the separability assumption then leads to inequalities analogous to Bell's, with the arguments the same as in Sec. I[23,24]

Let us proceed to the actual derivation of the modified inequalities. We use the formalism of hidden-variable theories, but, as emphasized in Sec. I, the validity of the result will be more general.

Each emitted pair of photons is assumed to be characterized by a hidden variable λ (λ might stand for any number of parameters). We denote by t_1 the time of emission[25] of a pair of photons in the laboratory frame, and by $\hat{u}(\lambda, t_1)$ the probability distribution of λ. At the time $t_2 = t_1 + L/2c$ the photons ν_A and ν_B impinge upon the commutators. We define "commutation functions" $\hat{\alpha}_i(t)$ and $\hat{\beta}_j(t)$ ($i=1$ or 2 and $j=1$ or 2), the values of which are 1 or 0 according as, respectively, the photons are sent along the corresponding channel or not; $\hat{\alpha}_i(t)$ and $\hat{\beta}_j(t)$ are of course λ independent. We denote by $t_3 > t_2$ the time at which the photons are anlayzed by the polarizers, and by $\hat{\mathfrak{A}}_i(\lambda, t_3)$ and $\hat{\mathfrak{B}}_j(\lambda, t_3)$ the corresponding response functions, the values of which are 1 or 0 according as the corresponding photon does or does not pass the polarizer. Thus $\hat{\alpha}_i(t_2)\hat{\beta}_j(t_2) \times \hat{\mathfrak{A}}_i(\lambda, t_3)\hat{\mathfrak{B}}_j(\lambda, t_3)$ assumes the value 1 if both photons ν_A and ν_B emerge from the polarizers I_i and II_j and 0 otherwise. Finally, the probability that a pair of photons ν_A and ν_B emerge in coincidence from I_i and II_j is

$$P_{ij} = \lim_{T\to\infty} \frac{1}{T}\int_T dt_1 \int d\lambda\, \hat{u}(\lambda, t_1)\hat{\alpha}_i(t_2)\hat{\beta}_j(t_2) \times \hat{\mathfrak{A}}_i(\lambda, t_3)\hat{\mathfrak{B}}_j(\lambda, t_3), \tag{3}$$

where the considered λ is the initial value at the time t_1 of emission.

Following Bell, we can generalize this to the case where the polarizers themselves contain hidden variables contributing to the result. We denote by λ'_{I_i} and λ'_{II_j} the instrumental parameters of each polarizer; then the response of each polarizer is $\hat{\mathfrak{A}}_i(\lambda, \lambda'_{I_i}, t_3)$ and $\hat{\mathfrak{B}}_j(\lambda, \lambda'_{II_j}, t_3)$. This formalism is also appropriate for a stochastic theory,[16] since λ'_{I_i} and λ'_{II_j} can be taken as random variables without any specific interpretation. Then the probability P_{ij} assumes the form

$$P_{ij} = \lim_{T\to\infty} \frac{1}{T}\int_T dt_1 \int d\lambda \int \cdots \int d\lambda'_{I_1} \cdots d\lambda'_{II_2}\hat{u}(\lambda, \lambda'_{I_1}, \ldots, \lambda'_{II_2}, t_1, t_3) \times \hat{\alpha}_i(t_2)\hat{\beta}_j(t_2)\hat{\mathfrak{A}}_i(\lambda, \lambda'_{I_i}, t_3)\hat{\mathfrak{B}}_j(\lambda, \lambda'_{II_j}, t_3), \tag{4}$$

where $\hat{u}(\cdots)$ is the probability distribution of λ at time t_1 and of the instrumental parameters at time t_3.

Correlations may exist between the instrumental parameters λ'_{I_i} and λ'_{II_j}. However, in accordance with the principle of separability, correlations at time t_3 can be produced only by common causes at times $t \leq t_1$. Introducing these common causes in formula (4), one remarks that these common causes are always coupled with λ; then for simplicity these common causes are included as a part of the fully general λ.

In accordance with the principle of separability, the given λ at times $t \leq t_1$ describe in a complete manner all the correlations between the instrumental responses at time t_3; therefore the conditional probability distribution of the instrumental parameters, given a particular value of λ, is factorized.[26] Hence, with $\hat{u}(\lambda, \lambda'_{I_1}, \ldots, \lambda'_{II_2}, t_1, t_3)$ expressed as a function of the probability distribution $\hat{\rho}(\lambda, t_1)$ of λ at time t_1 and of the conditional probability distributions of each instrumental parameter, P_{ij} (formula 4) assumes the form

$$P_{ij} = \lim_{T\to\infty} \frac{1}{T}\int_T dt_1 \int d\lambda\, \hat{\rho}(\lambda, t_1)\hat{\alpha}_i(t_2)\hat{\beta}_j(t_2) \times \hat{A}_i(\lambda, t_3)\hat{B}_j(\lambda, t_3), \tag{5}$$

where $\hat{A}_i(\lambda, t_3)$ and $\hat{B}_j(\lambda, t_3)$ are the average values—over the respective instrumental parameters—of the instrumental response functions at time t_3.

When integrating over time, we must consider the possibility of interactions between the commutators and the other devices. The principle of separability precludes instantaneous interactions but not retarded ones. Hence the response of polarizer I_i (for instance) at time t_3 might depend upon the state of commutator C_A at times $t \leq t_2$ and also upon the state of commutator C_B at times $t \leq t_2 - L/c$. Similarly, the emission at time t_1 might depend upon the states of both commutators at times $t \leq t_1 - L/2c$. Taking into account these possible interactions, it can be shown (see Appendix)—by using the assumed properties of the commutators (stochastic independent workings, autocorrelation times shorter than L/c)—that the probability of coincident emergence factorizes as

$$P_{ij}=\alpha_i\beta_j\int d\lambda\,\rho(\lambda)A_i(\lambda)B_j(\lambda)\,, \tag{6}$$

where α_i and β_j denote the averages over time of the commutation functions $\hat{\alpha}_i(t)$ and $\hat{\beta}_j(t)$, $\rho(\lambda)$ is a function with the properties of a probability distribution, and $A_i(\lambda)$ and $B_j(\lambda)$ are functions related to the responses of polarizers I_i and II_j, constrained by the inequalities

$$\begin{aligned}&0\leqslant A_i(\lambda)\leqslant 1\,,\\&0\leqslant B_j(\lambda)\leqslant 1\,.\end{aligned} \tag{7}$$

With the same notations, the probabilities that one photon emerges from I_i (or II_j) are (respectively)

$$\begin{aligned}&P_{i0}=\alpha_i\int d\lambda\,\rho(\lambda)A_i(\lambda)\,,\\&P_{0j}=\beta_j\int d\lambda\,\rho(\lambda)B_j(\lambda)\,.\end{aligned} \tag{8}$$

Thus, using the principle of separability, we have derived a factorized form similar to the definition of objective local theories.[15] Nevertheless, there is a large difference: The response of one polarizer, which (given λ) depends upon its orientation, might also depend upon the orientations of the other polarizers (since we do not make the Bell locality assumption). Similarly, the way in which the pairs of photons are emitted [and therefore $\rho(\lambda)$] might depend upon the polarizers' orientations. But, in our experiment, all the orientations remain unchanged during the whole course of the experiment; hence, in formula (6), $\rho(\lambda)$ does not depend upon the indices i or j; likewise, $A_i(\lambda)$ does not depend upon the index j [nor does $A_j(\lambda)$ depend upon i].

Therefore, the derivation of Clauser and Horne[15] holds. Inequalities (7) entail

$$-1\leqslant U\leqslant 0\,, \tag{9}$$

with (λ is dropped for simplicity)

$$U\equiv A_1B_1-A_1B_2+A_2B_1+A_2B_2-A_2-B_1\,.$$

After multiplication by $\rho(\lambda)$ and integration, one obtains

$$-1\leqslant S\leqslant 0\,, \tag{10}$$

with

$$S\equiv\frac{P_{11}}{\alpha_1\beta_1}-\frac{P_{12}}{\alpha_1\beta_2}+\frac{P_{21}}{\alpha_2\beta_1}+\frac{P_{22}}{\alpha_2\beta_2}-\frac{P_{20}}{\alpha_2}-\frac{P_{01}}{\beta_1}\,. \tag{11}$$

Inequalities (10) are isomorphic to the generalized Bell inequalities.[15] On the other hand, the quantum-mechanical predictions, which are

$$\begin{aligned}&P_{ij}=\tfrac{1}{2}\alpha_i\beta_j\cos^2(a_i,b_j)\,,\\&P_{i0}=\tfrac{1}{2}\alpha_i\,,\\&\text{and}\\&P_{0j}=\tfrac{1}{2}\beta_j\,,\end{aligned} \tag{12}$$

lead to a violation of the inequalities (10), the maximum of which occurs respectively for

$$(a_1,b_1)=(b_1,a_2)=(a_2,b_2)=\begin{cases}22.5^\circ\\ \text{or}\\ 67.5^\circ\end{cases}$$

and (13)

$$(a_1,b_2)=\begin{cases}67.5^\circ\\ \text{or}\\ 202.5^\circ\,.\end{cases}$$

For these specific orientations, we indeed have

$$S=\begin{cases}0.207\\ \text{or}\\ 1.207\,.\end{cases}$$

All the quantities involved in formula (11) are probabilities of coincident—or single—emergence and could, in principle, be measured. We thus have a test for separable hidden-variable theories. This test will be operational if we find means for measuring these probabilities and designing commutators obeying the assumptions we have made.

III. TESTABLE INEQUALITIES

Although theoretically measurable, the probabilities in formula (11) cannot be measured directly for two reasons.[3,15] First, optical photomultipliers have a low quantum efficiency; hence the rate of joint detection will be lower than the true rate of coincident emergence from the corresponding polarizers. Second, in atomic cascades only a fraction of pairs fly in opposite directions (since we have a three-body decay); hence a photon ν_A (for instance) may impinge upon commutator C_A and be analyzed while the corresponding ν_B is lost; therefore the rate of single detection is not a faithful measurement of the probability that a single photon is transmitted by the corresponding polarizer.

Denoting by N the average rate of emission of processed pairs (i.e., with both photons impinging upon the commutators) and by ϵ_{ij} a numerical factor accounting for the quantum efficiency of the corresponding photomultipliers, the rate of joint detection may be expressed as

$$N_{ij}(a_i,b_j)=\epsilon_{ij}P_{ij}N\,. \tag{14}$$

As in previous work,[3,4,8,12] we assume that ϵ_{ij} does not depend on whether or not the photons have passed through a polarizer.[27] We also assume that the average rate of pair emissions N is not changed when a polarizer is removed [although, as stated in Sec. II, we accept that $\hat{u}(\lambda, t)$ might be changed]. Therefore the rate of joint detection in channels I_i and II_j when the corresponding polarizers are removed will be

$$N_{ij}(\infty, \infty) = \epsilon_{ij}\, \alpha_i \beta_j N, \tag{15}$$

where $\alpha_i \beta_j$ is the probability of a coincident emergence from the commutators C_A and C_B along the corresponding channels.

We thus obtain

$$\frac{P_{ij}}{\alpha_i \beta_j} = \frac{N_{ij}(a_i, b_j)}{N_{ij}(\infty, \infty)} \tag{16}$$

and then the four first terms in formula (11) are put as functions of measurable quantities.

One more assumption is needed to render the two last terms of S measurable: that the probability that a photon emerges out of a polarizer does not depend upon whether or not another polarizer has been removed[28] (although, as stated above, we accept that the elementary response might be changed).

Therefore, if the polarizer II_j is removed, the probability of joint emergence of ν_A from polarizer I_i and of ν_B along channel II_j is $\beta_j P_{i0}$ (see Appendix), and the rate of joint detections, with polarizer II_j removed, is

$$N_{ij}(a_i, \infty) = \epsilon_{ij} \beta_j P_{i0} N. \tag{17}$$

We finally obtain

$$\frac{P_{i0}}{\alpha_i} = \frac{N_{ij}(a_i, \infty)}{N_{ij}(\infty, \infty)}, \qquad \frac{P_{0j}}{\beta_j} = \frac{N_{ij}(\infty, b_j)}{N_{ij}(\infty, \infty)}, \tag{18}$$

and the two last terms in formula (11) can be measured.

On the whole, S in formula (11) is expressed as

$$S = \frac{N_{11}(a_1, b_1)}{N_{11}(\infty, \infty)} - \frac{N_{12}(a_1, b_2)}{N_{12}(\infty, \infty)} + \frac{N_{21}(a_2, b_1)}{N_{21}(\infty, \infty)} + \frac{N_{22}(a_2, b_2)}{N_{22}(\infty, \infty)} - \frac{N_{2j}(a_2, \infty)}{N_{2j}(\infty, \infty)} - \frac{N_{i1}(\infty, b_1)}{N_{i1}(\infty, \infty)}. \tag{19}$$

The four quantities $N_{ij}(a_i, b_j)$ are measured during one single run of the experiment, and this is a very significant difference between our proposal and the previous schemes. The other quantities are measured in auxiliary calibrations.

Concluding this section, we have been able to define a practical scheme for measuring the probabilities, modulo some reasonable assumptions which, however, restrict somewhat the generality of the derivation of Sec. II.

IV. OVERALL EXPERIMENTAL SETUP

Clauser *et al.*[3] have already discussed the case of nonideal polarizers and extended beams. These features of a realistic experiment decrease somewhat the quantum-mechanical violation of the inequalities (10). Significant experiments have nevertheless been carried out to test the locality condition. Our experiment could be built with the same sort of source, polarizers, and detecting devices.

The specificity of our experiment is the presence of two optical commutators. These can consist of acoustic standing waves working as adjustable gratings (Fig. 3). The deviation of a light beam by strong interactions with an acoustic wave has been studied[29] both theoretically and experimentally. We are planning to use commutators with a surface of 4 cm^2 and an angular aperture of $1° \times 30°$. By appropriately adjusting the various parameters, we should be able to obtain one single beam diffracted, i.e., two channels.

The transmitted and diffracted beams will be modulated in opposition at twice the frequency of the sound wave. We expect the modulation rate of the transmitted beam to be over 90%, and polarization independent.

We finally must discuss to what extent our commutators obey the assumptions stated in Sec. II. Since the modulation rate is not exactly 100% and the commutation is not instantaneous, the values of the commutation functions $\hat{\alpha}_i(t)$ and $\hat{\beta}_j(t)$ are not restricted to 0 and 1. Nevertheless, the reasoning in Sec. II still holds if, with $\hat{\alpha}_i(t)$ and $\hat{\beta}_j(t)$ denoting the probability that a photon is directed towards channel I_i or II_j, these commutation func-

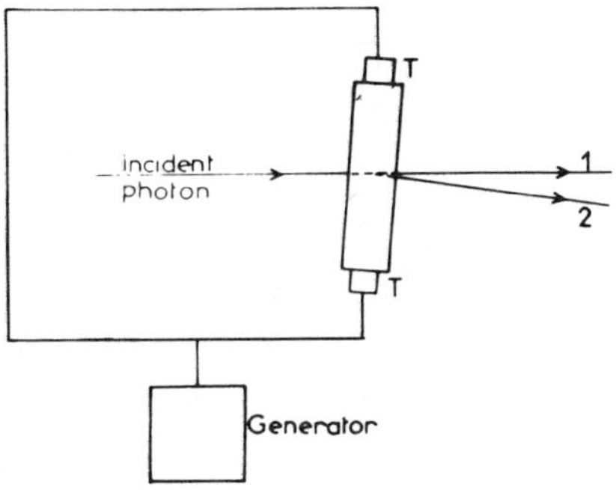

FIG. 3. Optical commutator. The generator supplies two identical transducers T, producing an acoustic standing wave in a crystal. The diffracted beam (channel 2) is modulated at twice the frequency of the standing wave. The transmitted beam (channel 1) can be modulated at a rate greater than 90%; hence the commutation is nearly complete.

tions do not depend on the hidden variable λ. This makes one more assumption. If it is unsatisfied, a fraction of the photons could be directed one way or the other, depending on the value of λ, and a "conspiracy" of the commutators and polarizers could decrease the difference between the quantum-mechanics predictions and the separable hidden-variable theories predictions. However, if the quantum-mechanics predictions were vindicated for various modulation rates, the occurrence of such a conspiracy might appear as *a priori* unlikely. Nevertheless, such a "conspiracy" could be avoided by inhibiting the detectors when the commutation functions assume values significantly different from 0 or 1 (during the rise time).

Another difficulty is that the commutators are operated periodically and not in a truly stochastic way. However, the significant requirement is that we have two independent commutation functions, each with autocorrelation time shorter than L/c—say, shorter than 20 nsec if L is 6 m. It seems that these conditions will be fulfilled if the commutators are separately driven by macroscopic generators whose frequencies deviate independently. We can drive a pseudorandom deviation of the frequency of each commutator, and a direct action of the source upon the driving mechanisms (and possibly upon the operator's descision; see footnote 13 of Ref. 15) seems very unlikely. For instance, the standing-wave frequency can vary between 100 and 125 MHz, and the commutation frequency will vary between 200 and 250 MHz. Then the autocorrelation time, which is of the order of the inverse of the line width, is about 20 nsec. If a sweeping of the frequency over a broad line turned out to be too difficult, a supplementary assumption should be exhibited: The polarizers have no "memory," i.e., they can be influenced by signals received at a certain time from the commutators (with a certain delay) but they cannot store all this information for a long time and extrapolate in the future even if there is some regularity in the working of the commutators. With this very natural supplementary assumption, the experiment would be significant even with periodic commutations.

V. BRIEF CONCLUSION AND ACKNOWLEDGMENTS

We believe that the experimental scheme we are proposing, although it is not an ideal one, is interesting in that it embodies a device for changing the orientations of the analyzers in a time comparable to the time of flight of the photons. Such a feature has been considered a crucial one by quite a few workers in the field, and therefore such experiments are worth making, even if they are not ideal.

The author gratefully acknowledges Professor C. Imbert and Dr. O. Costa de Beauregard for having suggested this study and for many fruitful discussions. He especially thanks Dr. J. S. Bell for his encouragement, and Professor B. d'Espagnat for his thorough consideration and discussion of the theoretical aspects of our scheme.

APPENDIX

The probability P_{ij} in Eq. (11) is the time average of the quantity [see Eq. (5)]

$$\hat{F}_{ij}(\lambda, t_1) = \int d\lambda\, \hat{\rho}(\lambda, t_1)\hat{\alpha}_i(t_2)\hat{\beta}_j(t_2) \times \hat{A}_i(\lambda, t_3)\hat{B}_j(\lambda, t_3),$$

where $t_2 = t_1 + L/2c$ and $t_3 - t_2$ is the time of flight of photons between a commutator and a polarizer.

Since the commutators work in a stationary stochastic way, we may replace the time average by ensemble averages. Denoting by $X(t)$ and $Y(t)$ random variables specifying the states at the time t of commutators C_A and C_B, respectively, we replace the time functions $\hat{}$ by functions $\tilde{}$ of these random variables. Taking care of the various possible interactions (as explained in Sec. II) we obtain

$$\hat{\alpha}_i(t_2) \to \tilde{\alpha}_i[X(t_2)],$$

$$\hat{\beta}_j(t_2) \to \tilde{\beta}_j[Y(t_2)],$$

$$\hat{\rho}(\lambda, t_1) \to \tilde{\rho}[\lambda, X(t_0), Y(t_0)],$$

$$\hat{A}_i(\lambda, t_3) \to \tilde{A}_i[\lambda, X(t_0), X(t_2), X(t_4), Y(t_0)],$$

$$\hat{B}_j(\lambda, t_3) \to \tilde{B}_j[\lambda, Y(t_0), Y(t_2), Y(t_4), X(t_0)],$$

and in general

$$\hat{F}_{ij}(\lambda, t_1) \to \tilde{F}_{ij}[\lambda, X(t_0) \cdots Y(t_4)],$$

where

$$t_0 \leq t_1 - L/2c$$

and

$$t_1 - L/2c < t_4 < t_2 .$$

As the two commutators are working independently, $X(t)$ and $Y(t)$ are independent random variables. As the autocorrelation time of each commutator is shorter than L/c, $X(t_0)$ and $X(t_2)$ are independent random variables, and so are $Y(t_0)$ and $Y(t_2)$.

We average $\tilde{F}_{ij}$, i.e., we integrate after multiplying by the probability distribution

$$\tilde{g}[X(t_0), X(t_2), X(t_4)]\tilde{h}[Y(t_0), Y(t_2), Y(t_4)]$$

in a factorized form since $X(t)$ and $Y(t)$ are independent. Integrating over $X(t_2)$ and $X(t_4)$, we then define

$$A_i[\lambda, X(t_0), Y(t_0)] = \{\alpha_i \bar{g}'[X(t_0)]\}^{-1} \times \iint dX(t_2)\, dX(t_4)\, \bar{g}[X(t_0), X(t_2), X(t_4)]\, \tilde{\alpha}_i[X(t_2)]\, \tilde{A}_i[\lambda, X(t_0), X(t_2), X(t_4)],$$

where

$$\bar{g}'[X(t_0)] = \iint dX(t_2)\, dX(t_4) \bar{g}[X(t_0), X(t_2), X(t_4)]$$

is the probability distribution of $X(t_0)$ and α_i is the average of $\tilde{\alpha}_i\,[X(t_2)]$. Remembering that $\tilde{A}_i$ assumes values 0 or 1 and that all the factors are positive, we obtain

$$0 \leq A_i[\lambda, X(t_0), Y(t_0)] \leq \{\alpha_i \bar{g}'[X(t_0)]\}^{-1} \iint dX(t_2)\, dX(t_4) \bar{g}[X(t_0), X(t_2), X(t_4)]\, \tilde{\alpha}_i[X(t_2)],$$

hence, remarking that $\int dX(t_4) \bar{g}[X(t_0), X(t_2), X(t_4)]$ factorizes, because $X(t_0)$ and $X(t_2)$ are independent random variables, we obtain

$$0 \leq A_i[\lambda, X(t_0), Y(t_0)] \leq 1 .$$

By a completely similar fashion, with $\bar{h}'[Y(t_0)]$ replacing $\bar{g}'[X(t_0)]$, we obtain

$$0 \leq B_j[\lambda, X(t_0), Y(t_0)] \leq 1 .$$

The average of $\tilde{F}_{ij}$ is then

$$P_{ij} = \alpha_i \beta_j \iiint d\lambda\, dX(t_0)\, dY(t_0)\, \bar{g}'[X(t_0)]\, \bar{h}'[Y(t_0)]\, \bar{\rho}[\lambda, X(t_0), Y(t_0)]\, A_i[\lambda, X(t_0), Y(t_0)]\, B_j[\lambda, X(t_0), Y(t_0)].$$

For simplicity we can include $X(t_0)$ and $Y(t_0)$ into the parameter λ, thus obtaining formulas (6) and (7). Similarly, the single probability that a photon emerges from polarizer I_i is the time average of

$$\int d\lambda\, \hat{\rho}(\lambda, t_1) \hat{\alpha}_i(t_2) \hat{A}_i(\lambda, t_3),$$

and, through the same procedure as above, we obtain formula (8).

Finally, the joint probability of photon ν_A emerging from polarizer I_i and photon ν_B emerging from commutator C_B into the channel II_j is the time average of

$$\int d\lambda\, \hat{\rho}(\lambda, t_1) \hat{\alpha}_i(t_2) \hat{\beta}_j(t_2) \hat{A}_i(\lambda, t_3).$$

Remembering that $X(t_2)$ and $Y(t_2)$ are independent, we obtain $\beta_j P_{i0}$ as the expression of this probability; this expression is used in Sec. III [Eq. (17)].

---

[1]A. Einstein, B. Podolsky, and N. Rosen, Phys. Rev. 47, 777 (1935).
[2]J. S. Bell, Physics 1, 195 (1965).
[3]J. F. Clauser, M. A. Horne, A. Shimony, and R. A. Holt, Phys. Rev. Lett. 23, 880 (1969).
[4]S. J. Freedman and J. F. Clauser, Phys. Rev. Lett. 28, 938 (1972).
[5]L. Kasday, in *Foundations of Quantum Mechanics*, Proceedings of the International School of Physics "Enrico Fermi," Course 49, edited by B. d'Espagnat (Academic, New York, 1972).
[6]G. Faraci, D. Gutkowski, S. Notarrigo, and A. R. Pennisi, Lett. Nuovo Cimento 9, 607 (1974).
[7]M. Lamehi-Rachti and W. Mittig, Phys. Rev. D (to be published).
[8]R. A. Holt, Ph.D. thesis, Harvard University, 1973 (unpublished).

[9]The words "locality" and "separability" (and likewise "local" and "separable") are sometimes taken as synonymous. However, the specific definitions given here are natural and, in the context of the present paper, useful.

[10]The expression "principle of separability of Einstein" is used by B. d'Espagnat in *Conceptual Foundations of Quantum Mechanics* (Benjamin, Reading, Mass., 1971). This principle is applied to "the real, factual situation of the system" [A. Einstein, in *Albert Einstein: Philosopher-Scientist*, edited by P. Schilpp (Open Court, La Salle, Ill., 1949)]; it is more stringent than the principle of causality; for instance, in the Einstein-Podolsky-Rosen situation, the principle of separability may be violated, although one cannot send orders

faster than light.

[11]D. J. Bohm, *Quantum Theory* (Prentice-Hall, Englewood Cliffs, N. J., 1951).

[12]M. A. Horne, thesis, Boston Univ., 1970 (unpublished).

[13]A. Shimony, in *Foundations of Quantum Mechanics* (Ref. 5).

[14]E. P. Wigner, Am. J. Phys. 38, 1005 (1970).

[15]J. F. Clauser and M. A. Horne, Phys. Rev. D 10, 526 (1974).

[16]J. S. Bell (in Ref. 5) and Clauser and Horne (Ref. 15) have proved that the hypothesis of determinism is not necessary to derive inequalities conflicting with some quantum-mechanical predictions. The latter have shown that such inequalities hold for objective local theories which are "essentially... stochastic hidden variable theories with a certain local character considered by Bell" (see Ref. 15).

[17]Y. Aharonov and D. Bohm, Phys. Rev. 108, 1970 (1957).

[18]The hypothesis of Furry (that the wave function of a many-body system tends to factorize into a product of localized states when the two parts are separated) would be rejected without the locality assumption (see Ref. 17).

[19]H. P. Stapp, LBL Report No. 3837, 1975 (unpublished).

[20]B. d'Espagnat, Phys. Rev. D 11, 1424 (1975).

[21]The idea that the hypothesis of the existence of a hidden variable would not be necessary to derive Bell's inequalities is in several papers [see for instance L. E. Ballentine, Phys. Today 27 (No. 10), 53 (1974); A. Baracca, C. J. Bohm, and B. J. Hiley, report (unpublished); see also Ref. 15 and a recent paper of J. S. Bell: CERN Report No. TH. 2053, 1975 (unpublished)].

[22]Among the authors who have considered a possible rejection of the principle of separability, Costa de Beauregard considers "the paradoxical scheme accepting an indirect spacelike connection via the two timelike vectors connecting the cascade event to the commutation events (and to the two subsequent counting events)." He goes on to say the following: "The corresponding space-time channels are indeed carrying waves and, moreover, the calculation of the quantal correlation probabilities does imply the phase binding between these waves. In this sense the (indirect) spacelike correlation, or nonseparability, predicted by the quantum theory, can be said to be physical *and* mathematical *and* logical.

"This correlation is nevertheless paradoxical, inasmuch as it asserts that the two (apparently) independent reductions of the wave function occurring in the distant pieces of apparatus are indeed in contact through their common past (and, thus, not really independent).

"This implied in some sense that Einstein's prohibition to telegraph into the past does not hold at the level of the individual quantal stochastic event (transition, or Ψ collapse). This is in line with the intrinsic time-symmetry also postulated in classical statistical mechanics for the individual stochastic event" (private communication).

See also O. Costa de Beauregard, unpublished work and work reported in *Proceedings of the International Conference on Thermodynamics*, edited by P. T. Landsberg (Butterworths, London, 1970), p. 540.

[23]A. Aspect, Phys. Lett. 54A, 117 (1975).

[24]A modification of our scheme has been suggested by J. F. Clauser (private communication). The two detectors I_1 and I_2 (Fig. 2) could be "or-wired", giving the equivalent of a single-variable polarizer with an orientation jumping from a_1 to a_2 and reversely. The time of change of the orientation should be monitored in order that the experiment be significant.

[25]The time of emission is not actually the same for the two photons, but our derivation remains valid when the lifetime of the medium state is shorter than the time of flight of the photons.

[26]A similar argument is used by Bell for a very general derivation of inequalities [J. S. Bell, CERN Report No. TH.2053, 1975 (unpublished)].

[27]Clauser and Horne (Ref. 15) found a weaker assumption for previous experiments. However, the widening of their reasoning to our case is not straightforward.

[28]This hypothesis may appear as strong as the Bell condition of locality. Actually it is quite a different hypothesis, concerning the probability and not the elementary response. It could then be experimentally tested with an ideally efficient photomultiplier. We can practically make several tests; for instance, we can verify that

$$\frac{N_{i1}(a_i,\infty)}{N_{i1}(\infty,\infty)}=\frac{N_{i2}(a_i,\infty)}{N_{i2}(\infty,\infty)}.$$

Such verifications would give good confidence in our hypothesis.

[29]J. M. Bauza, C. Carles, and R. Torguet, Acustica 30, 137 (1974); V. N. Mahajan and Jack D. Gaskill, Optica Acta 21, 893 (1974).

Editors' Note: A letter of 7 July 1982 from Alain Aspect to W.H.Z. reports that the experiment proposed in this paper has been completed by Alain Aspect, Jean Dalibard, and Gérard Roger, and that the results will be submitted to *Physical Review Letters*. They find that with the two commutators separated by the distance $L = c \cdot 4 \cdot 10^{-8}$s, and with the choice of polarization observable switched every 10^{-8}s, the correlation parameter S, defined by Eq. (19) in the paper, is $S_{EXP} = 0.101 \pm 0.020$. "This result violates the generalized Bell's inequality ($-1 \leq S \leq 0$) by 5 standard deviations and is in good agreement with [the quantum-mechanical prediction] $S_{QM} = 0.112$ [estimated for the actual experimental setup]. . . . The commutators did not work completely at random, but of course were driven by two independent generators. . . . The total duration of data accumulation was 7 hours."

III.11 COMPLEMENTARITY IN THE DOUBLE-SLIT EXPERIMENT: QUANTUM NONSEPARABILITY AND A QUANTITATIVE STATEMENT OF BOHR'S PRINCIPLE

WILLIAM K. WOOTTERS AND WOJCIECH H. ZUREK

A detailed analysis of Einstein's version of the double-slit experiment, in which one tries to observe both wave and particle properties of light, is performed. Quantum nonseparability appears in the derivation of the interference pattern, which proves to be surprisingly sharp even when the trajectories of the photons have been determined with fairly high accuracy. An information-theoretic approach to this problem leads to a quantitative formulation of Bohr's complementarity principle for the case of the double-slit experiment. A practically realizable version of this experiment, to which the above analysis applies, is proposed.

I. INTRODUCTION

In Einstein's version of the double-slit experiment,[1,2] one can retain a surprisingly strong interference pattern by not insisting on a 100% reliable determination of the slit through which each photon passes. The analysis leading us to this conclusion involves the following considerations. The plate which receives the kick from each photon can either be stopped and its position measured, or released and its momentum measured. These two options give us two ways of subdividing the original ensemble of photons: (1) according to the measured position of the plate, and (2) according to the measured momentum of the plate. In case (1) each subensemble produces a perfect but differently shifted interference pattern. In case (2) each subensemble produces a smeared out pattern, but also gives us some information about the photons' paths. We will thus be led to study the following questions. Does our choice of what to measure affect the total interference pattern? Do we not violate the complementarity principle by measuring both the fringes and the kick? Can this kind of experiment be performed in practice?

A monochromatic wave passing through the familiar apparatus shown in Fig. 1 will produce a perfect interference pattern on plate 3. This kind of experiment was used by Young to demonstrate the wave nature of light, and by Davisson and Germer to demonstrate the wave nature of electrons. How does one reconcile this result with the fact that these same photons or electrons also have distinctly particlelike features, such as being individually detectable? For if they are indeed particles, then each one should go through a definite slit, and we should see a sum of single-slit diffraction patterns, rather than the observed interference pattern. This apparent contradiction has served as an archetypal example of the wave-particle duality encountered in the microworld. As stated by Feynman, the double-slit experiment is a phenomenon "which has in it the heart of quantum mechanics; in reality it contains the *only* mystery" of the theory.[3]

The traditional resolution of this problem, consistent with the rest of quantum mechanics, states that unless we actually *measure* the path of each photon (for definiteness let us restrict our attention to photons; similar arguments apply to any other particle), we have no right to maintain that any given photon actually follows a definite path. Indeed, according to this view, if we were to succeed in measuring the path of each photon, then the interference pattern would be destroyed. Thus, we can observe either the wave properties or the particle properties of light, but not both simultaneously.

At the Fifth Solvay Congress at Brussels, Einstein devised a modification of the double-slit experiment by which he hoped to show the inconsistency of quantum mechanics.[1,2] He considered an experiment in which the first screen (Fig. 1) is free to move up or down. Photons deflected toward a given slit always impart a characteristic momentum to screen 1. Thus, Einstein hoped that by measuring the momentum imparted one could determine the path of each photon without disturbing the interference pattern. If that were the case, then the complementarity principle would be proven false.

Bohr, in his defense of the consistency of quantum mechanics, pointed out that in order to deduce the slit through which the photon will pass from a knowledge of the final momentum of screen 1, one would also have to know the *initial* momentum of screen 1 to within an uncertainty $\Delta p = (s/L)p = (h/\lambda) \times (s/L)$ (λ is the wavelength of the light), since this is the difference between the momenta

Originally published in *Physical Review, D19*, 473-84 (1979).

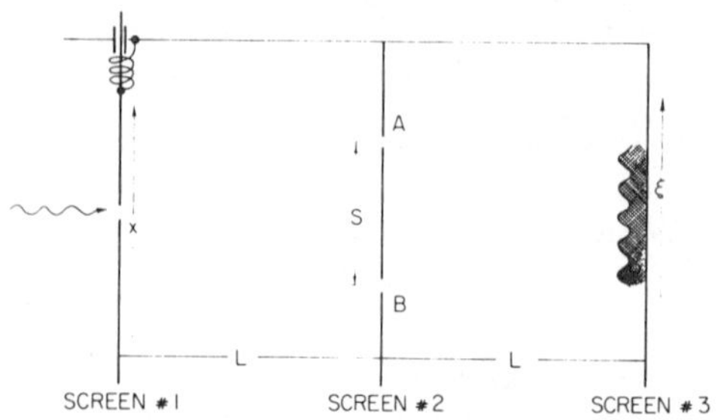

FIG. 1. Path determination in the double-slit experiment, as proposed by Einstein (Ref. 2).

transferred by photons following the two possible paths to the final screen. But then the Heisenberg uncertainty principle requires that the initial *position* of screen 1 be uncertain by at least $\Delta x \sim h/\Delta p = \lambda s/L$. This is exactly what is needed to wash out the interference pattern, since this Δx is the spacing between fringes. In this way Bohr succeeded in defending the consistency of quantum mechanics.[1]

The purpose of this paper is to examine Bohr's idea in detail; that is, we will find out exactly what interference pattern is produced when we attempt to determine the slit through which each photon passes. To do this, we shall represent the plate as a quantum-mechanical harmonic oscillator. The photon and the plate, having once interacted, become nonseparable parts of a single quantum-mechanical system. This forces us to consider the effect of our measurement of the plate on the photon wave function, and, consequently, on the interference pattern produced by these photons.

In the next section we shall derive the interference pattern attenuated as a result of our measurements on the position of the plate. There we shall also obtain the distribution of momentum of plate 1 associated with the relevant photons, that is, those which contribute to the image on the photographic plate.

In Sec. III we shall find what partial interference patterns correspond to ensembles of photons correlated to definite final momenta of the first plate. We shall also show that although these partial interference patterns depend on the measurements performed on the first plate, the total interference pattern (the sum of the partial interference patterns) is always the same.

In Sec. IV we shall find an analogy between this situation and the celebrated Einstein-Rosen-Podolsky paradox.

Section V discusses in terms of Shannon's theory the information one can obtain about the slit through which the photon passed. The complementarity principle is stated in terms of an inequality, which sets the limit on the amount of retrievable information about the photons' paths (photon-particle) for an assumed sharpness of the interference pattern (photon-wave). A practically realizable version of Einstein's double-slit experiment is discussed in Appendix A. Appendix B gives details of the proof of the inequality derived in Sec. V.

II. APPROXIMATE PATH-DETERMINATION

Let us now analyze in more detail the experiment described in Sec. I. Assuming that the Heisenberg uncertainty principle holds, let us ask exactly to what extent the interference pattern is smeared out if we insist on determining the path of each photon with a given accuracy. In the spirit of Bohr's rebuttal, we will represent screen 1 by a quantum-mechanical wave function.

The harmonic-oscillator wave function is a natural choice, since it satisfies (in its lowest-energy state) the equality

$$\Delta x \Delta p = \hbar/2,$$

thus minimizing the cost (disturbance of photon's phase due to Δx) of obtaining information about the photon's momentum with an error Δp.[11] Both in the position (x) and wave-vector (k) representation it is given by a Gaussian:

$$\begin{aligned} \psi(x) &= \pi^{-1/4} a^{-1/2} \exp(-x^2/2a^2), \quad \Delta x^2 = a^2/2, \\ \phi(k) &= \pi^{-1/4} a^{1/2} \exp(-a^2k^2/2), \quad \Delta k^2 = 1/(2a^2), \end{aligned} \tag{1}$$

where $k = p/\hbar$ is the x component of the wave vector of plate 1. [One sees that $(\Delta x)(\Delta k) = \frac{1}{2}$, the minimal irreducible demand of the uncertainty principle.]

The object of this section is to calculate the interference pattern one obtains when plate 1 has this wave function. But first we need to state the assumptions we will be making regarding the photon wave function in a *single slit* experiment. Let us consider for example the wave function at plate 3 of photons which have passed through slit A, slit B being closed for the time being. This wave function is

$$\mu_A(\xi) = f(\xi) \exp\{i(2\pi/\lambda)[L^2 + (\xi - s/2)^2]^{1/2}\},$$

where f is a slowly varying envelope function whose presence is due to the fact that the slit is not infinitesimally narrow. For an ordinary slit, f would be the usual diffraction pattern. Moreover, one could also consider a slit containing one or another type of lens, and this would lead to a different envelope function. For example, the process of apodization can be used to smooth out the diffraction pattern.[12] In any case, all we ask is that there be a region $|\xi| < \xi_0$ in which $f(\xi)$ is essentially constant. We will then restrict our attention to a section of plate 3 where $|\xi| < \xi_0$ and $|\xi| \ll L$. There the wave function can be approximated by

$$\mu_A(\xi) = f(0) e^{i\alpha(\xi)} e^{ik_0\xi},$$

where $\alpha(\xi) = (2\pi/L)[L^2 + (s^2/8) + (\xi^2/2)]$ and $k_0 = (\pi s)/(L\lambda)$. Similarly, the wave function for photons coming only from slit B is

$$\mu_B(\xi) = f(0) e^{i\alpha(\xi)} e^{-ik_0\xi}.$$

When we superimpose these two wave functions, the common factor $e^{i\alpha(\xi)}$ does not contribute to the interference pattern. We will therefore omit this factor throughout the rest of this paper. Finally, coming back to the *double-slit* experiment, since we have abandoned the usual normalization in considering only a part of plate 3, our convention will be to let the average of the interference pattern over one period be 1.

Just after each photon passes, we can measure exactly either the position or momentum of plate 1. Let us assume for now that we measure the position, and let us calculate the resulting interference pattern. In this case we know exactly where each photon starts its journey to plate 3. All photons starting at the same place x form a subensemble whose contribution $I_x(\xi)$ to the total interference pattern is perfect, but shifted by an amount $\Delta\xi = -x$:

$$I_x(\xi) = 1 + \cos 2k_0(\xi + x),$$

where $k_0 = \pi s/(L\lambda)$, and s and L are defined in Fig. 1. The number of photons in the subensemble characterized by the position x is proportional to $|\psi(x)|^2$. Therefore, the total interference pattern $\mathfrak{F}(\xi)$ is just the sum of all the contributions from the various subensembles, weighted by $|\psi(x)|^2$:

$$\mathfrak{F}(\xi) \propto \int dx |\psi(x)|^2 I_x(\xi) = 1 + e^{-k_0^2 a^2} \cos 2k_0\xi. \qquad (2)$$

This is our smeared-out interference pattern. Now how much can we determine about the paths of the photons using the same experimental arrangement, if we choose to measure the momentum of screen 1 rather than its position? Clearly the only photons which concern us are those which succeed in arriving at the final screen. Therefore, we will record only the momenta imparted by these photons. From the geometry of the apparatus we see that for these photons there are only two possible values of the wave vector imparted to screen 1, namely, $\pm\pi s/(L\lambda) = \pm k_0$. (We have assumed here that $a \ll s \ll L$, and that all photons arrive at screen 1 with no initial momentum in the x direction.) The measured momentum of screen 1, however, will not be confined to these two values, even if our measurements are exact. As before, the uncertainty comes from the fact that the *initial* momentum of the plate is not definite, but rather has a distribution given by $|\varphi(k)|^2$. If the plate has initial wave vector k, then its final wave vector κ (after the plate has collided with one of the photons under consideration) can have either of the two values $\kappa = k \pm k_0$. The total recorded distribution $\mathfrak{D}(\kappa)$ of wave numbers of the plate will be the sum of all the partial distributions $D_k(\kappa)$, weighted by $|\phi(k)|^2$:

$$\begin{aligned} D_k(\kappa) &= \tfrac{1}{2}[\delta(\kappa - k + k_0) + \delta(\kappa - k - k_0)], \\ \mathfrak{D}(\kappa) &= \int dk |\varphi(k)|^2 D_k(\kappa) \\ &= [a/(2\pi^{1/2})]\{\exp[-a^2(\kappa + k_0)^2] \\ &\qquad + \exp[-a^2(\kappa - k_0)^2]\}. \end{aligned} \qquad (3)$$

It is convenient to define a dimensionless "smudging parameter" $u = ak_0$. From Eq. (2) it is clear that this is the parameter characterizing the suppression of the interference pattern. It is also a measure of the uncertainty of our determination of the path of each photon. To clarify these points, let us choose a particular value of u, namely, $u = 0.4769$. If we use this value of u to calculate the interference pattern according to Eq. (2), and the distribution of measured momenta according to Eq. (3), then the results are those shown in Fig. 2.

If the measured momentum is positive, then we will guess that the photon passed through slit A; if it is negative, then we will guess that the photon passed through slit B. Clearly some of our guesses will be wrong—there are photons that have positive values of measured momentum even though their actual momentum was negative and they went through slit B. The fraction of photons which misbehave in this way equals the fraction $F(u)$ of our guesses which are wrong. This fraction is just the ratio of the area under the "tails" of the Gaussians (that is, the parts lying on the wrong side of $\kappa = 0$) to the total area under $\mathfrak{D}(\kappa)$:

$$\begin{aligned} F(u) &= (a/\pi^{1/2}) \int_{-\infty}^{0} \exp[-a^2(\kappa - k_0)^2]\, d\kappa \\ &= (1/\pi^{1/2}) \int_{u}^{\infty} e^{-\eta^2} d\eta \\ &= \tfrac{1}{2}[1 - \operatorname{erf}(u)]. \end{aligned}$$

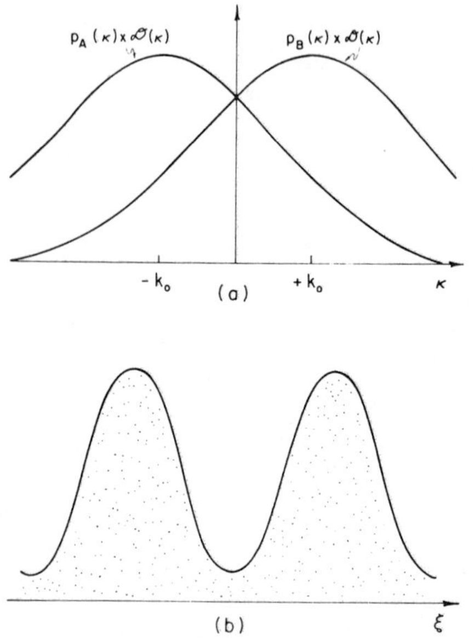

FIG. 2. Results of the approximate momentum measurement in the double-slit experiment, for the smudging parameter $u = 0.4796$. (a) Distributions of measured momenta. (b) The corresponding total interference patterns.

Our particular value of u was chosen so that $F(u) = 0.25$. That is, out of four guesses, one is usually wrong. Thus, we have done a fairly good (but not perfect) job of determining the path of each photon, and therefore we might expect an almost completely obliterated interference pattern. But in fact this is not true. Figure 2(b) shows that we still get an almost perfect interference pattern; the crest-to-valley ratio $R(u)$ of intensities is $R(u) = [1 + \exp(-u^2)]/[1 - \exp(-u^2)] = 8.8$ for our value of u. Have we not then succeeded in observing both particlelike and wavelike properties of the same photons?

One will no doubt object: The interference pattern $\mathfrak{F}(\xi)$ was calculated for an experiment in which the position of screen 1 was measured. The path determination depended on the wave vector k of the photon inferred from the measured momentum of the screen 1. But these two measurements cannot both be performed for the same photons, and so Figs. 2(a) and 2(b) do not refer to the same experiment. Hence, in accord with the Copenhagen interpretation of quantum mechanics, there is no paradox. The complementarity principle does not prevent photons from behaving once as waves and once as particles. It only states that the same photon should not reveal this "split personality" in the same experiment.

The above objection is certainly valid. However, we shall see in the next section that even when we perform the momentum measurements on screen 1, obtaining information about the slit each photon chooses, the same total interference pattern $\mathfrak{F}(\xi)$ forms on the photosensitive screen 3. That is, Figs. 2(a) and 2(b) describe the outcome of the same hypothetical experiment.

III. CORRELATION BETWEEN THE MEASURED MOMENTUM AND THE CORRESPONDING INTERFERENCE PATTERN

In this section we will analyze the experiment in which the momentum of screen 1 is measured rather than the position. From each momentum measurement we will infer the relative probabilities of the photon's passing through one slit or the other. In practice, we cannot actually perform this momentum measurement with the necessary accuracy. However, as is shown in Appendix A, one can perform an experiment which will give the desired probabilities. The analysis in this section is sufficiently general to be applicable to the experiment considered in Appendix A.

In our experiment (Fig. 1), photons entering the apparatus change the momentum of screen 1. Then some of them pass through the slits in plate 2, as announced by a flash of light emitted by a scintillator on plate 3. We wait for these scintillations, and immediately record the momentum of plate 1 and the position of the scintillation. That is, our record contains for each photon two numbers: the measured position on plate 3 and the measured momentum. Of course, we can obtain from this record both the interference pattern $\mathfrak{F}(\xi)$ and the distribution $\mathfrak{D}(\kappa)$ of measured photon momenta. Indeed, these two distributions could have been obtained directly, without keeping a photon-by-photon record. The fact that each scintillation is actually associated with a definite kick to plate 1 can have no effect on the shapes of $\mathfrak{F}(\xi)$ and $\mathfrak{D}(\xi)$. Given that we do have a photon-by-photon record, we find $\mathfrak{F}(\xi)$ simply by counting all the photons that landed in the interval $\xi \to \xi + d\xi$. $\mathfrak{D}(\kappa)$ is clearly constructed in an analogous manner.

But a new question which we have not considered so far comes to our attention. We can look for the distribution of scintillations arising only from those photons which have been associated with a definite measured momentum κ of the plate 1. What kind of interference pattern will *these* photons produce? The first thing to notice is that these partial interference patterns $i_\kappa(\xi)$ cannot be independent of κ; that is, they cannot all have the shape of the total interference pattern. For if we consider only the photons with a high-enough κ, we could improve indefinitely our chances of a correct determination of their paths; that is, by increasing κ we would be able to tell with as small a margin of error as we wish through which of the two slits the photon has passed. In this way we would be able to get around Bohr's resolution of the double-slit paradox.

Therefore, a correlation between κ—the measured value of the plate's momentum—and $i_\kappa(\xi)$—the partial interference pattern—seems inescapable. To find out the shape of $i_\kappa(\xi)$, let us consider the fraction of error $f(\kappa)$ defined as the ratio of the probability of the photon's passing through slit A to the probability of passing through slit B,

$$f(\kappa) = p_A/p_B = \exp[-a^2(\kappa + k_0)^2]/\exp[-a^2(\kappa - k_0)^2]. \tag{4}$$

Here p_A and p_B are defined by this formula together with the relation $p_A + p_B = 1$. This ratio $f(\kappa)$ clearly represents the uncertainty in our knowledge about the two possibilities from which the photon can choose.

The same situation and identical error ratio appear when we consider a double-slit experiment in which the two slits have different areas. The uncertainty ratio f will in this case be expressed by the ratio of the two areas. In this situation the interference pattern can be immediately written:

$$i_f(\kappa) = 1 + 2p_A^{1/2} p_B^{1/2} \cos 2k_0\xi . \tag{5}$$

It is natural to expect that if our correlation measurement is characterized by the same value of f as this unequal-double-slit experiment, then it should yield the same interference pattern. Therefore,

$$i_\kappa(\xi) = 1 + 2p_A^{1/2}(\kappa) p_B^{1/2}(\kappa) \cos 2k_0\xi . \tag{6}$$

These formulas (5) and (6) are certainly compatible with quantum mechanics and with the uncertainty and complementarity principles. What do they predict for a definite value of f? For instance, let us take $f = \frac{99}{1}$; that is, out of 100 photons we expect 99 of them to pass through the more likely slit. An accuracy of 99% is certainly high, and one would expect that such a measurement should destroy the interference pattern. But a straightforward calculation shows that the crest-to-valley ratio R given by $R = (1 + 2p_A^{1/2} p_B^{1/2})/(1 - 2p_A^{1/2} p_B^{1/2})$ equals approximately $\frac{3}{2}$. Despite the fact that we know with 99% certainty the paths of the photons, they still have strong wavelike properties. In a sense we have localized the cause of what appeared to be a violation of the complementarity principle.

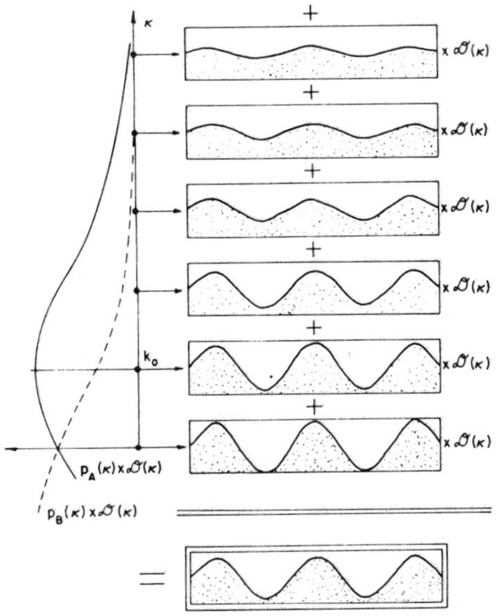

FIG. 3. Partial-interference patterns $i_\kappa(\xi)$ corresponding to the different measured momenta κ and the resulting total interference pattern $\mathfrak{F}(\xi)$, according to formula (7).

To make sure that our present reasoning is consistent and that our apparent paradox survives the objection raised at the end of Sec. I, let us add up all the partial interference patterns, all the differently smudged-out contributions. According to Eqs. (3) and (5), each of them can be written as

$$i_\kappa(\xi) = 1 + \frac{2\exp[-a^2(\kappa^2 + k_0^2)]}{\exp[-a^2(\kappa - k_0)^2] + \exp[-a^2(\kappa + k_0)^2]} \times \cos 2k_0\xi .$$

Weighting them by $\mathfrak{D}(\kappa)$ and adding them together, as represented in Fig. 3, we obtain

$$\mathfrak{F}(\xi) = \int \mathfrak{D}(\kappa) i_\kappa(\xi) d\kappa = 1 + e^{-a^2 k_0^2} \cos 2k_0\xi . \tag{7}$$

This is exactly what we obtained previously [Eq. (1)] by adding perfect, but shifted, interference patterns. This way of obtaining the total interference pattern is illustrated symbolically in Fig. 3.

IV. EINSTEIN-ROSEN-PODOLSKY "PARADOX" IN THE DOUBLE-SLIT EXPERIMENT

Let us recall the two possible measurements one can perform using the experimental set-up of Fig. 1: (1) Measure the position of screen 1—

get *perfect* but *shifted* partial interference patterns. (2) Measure the momentum of screen 1—get *smeared out* but *centered* partial interference patterns.

Notice that in both cases the measurement on screen 1 is performed *after* the photon has interacted with it. How does the photon know in which partial interference pattern to fall? How can it know what we decided to measure when it is already separated from the plate by a large distance?

There is no doubt that our choice affects the wave function of the photon: (1) If we measure plate 1 to have position x, the photon arrives at plate 3 in the state

$$\mu_x(\xi) \propto e^{ik_0(\xi+x)} + e^{-ik_0(\xi+x)}. \tag{8a}$$

(2) If we measure plate 1 to have momentum κ, the photon arrives at plate 3 in the state

$$\nu_\kappa(\xi) \propto p_A^{1/2}(\kappa)\, e^{ik_0\xi} + p_B^{1/2}(\kappa)\, e^{-ik_0\xi}. \tag{8b}$$

This effect of our measurement on the photon's wave function can be understood in the following way. After the photon has interacted with plate 1, it no longer has an individual wave function; there is only the state Ψ of the combined system (plate 1 + photon). This combined state can be expanded in terms of any complete set of orthogonal states of plate 1. For example, if σ_x is the eigenstate of position of plate 1 corresponding to the eigenvalue x, we can always write the combined state as

$$\Psi = \int dx c_x \mu_x \otimes \sigma_x , \tag{9a}$$

where the photon states μ_x are all normalized but not necessarily orthogonal, and $|c_x|^2$ is the probability distribution of the plate's position x. Alternatively, we can use the set of eigenstates of momentum of plate 1, $\{\tau_\kappa\}$, as our basis, in which case Ψ is written as

$$\Psi = \int d\kappa d_\kappa \nu_\kappa \otimes \tau_\kappa . \tag{9b}$$

Thus, each plate eigenstate (σ_x or τ_κ) has its corresponding photon state. When we measure the plate's position and obtain the value x, the photon is forced into the state μ_x, and similarly for momentum. In this way our choice affects the wave function of the photon.

Our object now is to identify the various quantities which appear in Eqs. (9), and thus to check that this idea of the combined state is consistent with the analysis of the previous sections. For definiteness we will write Ψ as a function of the positions of plate 1 and the photon, to which we assign the symbols "z" and "ξ" respectively ("x" is reserved as a lable for eigenfunctions; hence the necessity of introducing "z"). According to our previous analysis, μ_x and ν_κ should be given by Eqs. (8), and the other quantities in Eqs. (9) should be interpreted as follows:

$$c_x = \psi(x) = \frac{1}{\pi^{1/4} a^{1/2}} e^{-x^2/2a^2} \quad \text{[see Eq. (1)]},$$

$$\sigma_x(z) = \delta(z - x),$$

$$d_\kappa = \mathfrak{D}^{1/2}(\kappa) \quad \text{[see Eq. (3)]},$$

$$\tau_\kappa(z) = e^{i\kappa z}.$$

Inserting these identifications into Eqs. (9), we obtain the following expressions for Ψ:

According to Eq. (9a),

$$\Psi = \frac{1}{2^{1/2} a^{1/2} \pi^{1/4}} \int dx\, e^{-x^2/2a^2} [e^{ik_0(\xi+x)} + e^{-ik_0(\xi+x)}] \times \delta(z - x); \tag{10a}$$

according to Eq. (9b),

$$\Psi = \frac{a}{2\pi^{1/4}} \int d\kappa \mathfrak{D}^{1/2}(\kappa) [p_A^{1/2}(\kappa)\, e^{ik_0\xi} + p_B^{1/2}(\kappa)\, e^{-ik_0\xi}] \times e^{i\kappa z}, \tag{10b}$$

where the constant factors have been chosen so that Ψ conforms to our normalization convention; i.e., $\int dz |\Psi|^2$ is a function of ξ whose average value is 1. One can verify that these two expressions for the combined wave function are indeed equal. This is in fact the real justification for our identification of ν_κ [Eq. (8b)] as the wave function of photons corresponding to plate momentum κ, and hence also for our expression (6) for the partial interference pattern.

In Secs. II and III we found that although our choice of what to measure (about plate 1) does affect the photon's wave function, it does not affect the *total* interference pattern. From the point of view of the present section, this result is seen quite easily. The total interference pattern is

$$f(\xi) = \int dz |\Psi(z, \xi)|^2 ,$$

which is independent of any measurements made on plate 1.

The situation described here and the analysis outlined above are essentially the same as in the Einstein-Podolsky-Rosen "paradox."[4,5,6] There too what we choose to measure on one system affects the wave-function of another system far away. But the final distribution of measured values of the far away system is again independent of our choice. Thus in discussing Einstein's double-slit experiment we have encountered the classic nonseparability feature of quantum mechanics.[7]

To make sure that the analogy is complete, one may consider a situation where the photon is first localized on screen 3, and the position (or momentum) of screen 1 is measured only afterwards. In that case the wave function of the photon is simply $\delta(\xi - \xi_0)$. It is readily seen that the total wave function $\Psi(z, \xi)$ can be decomposed into a wave function of a photon localized at ξ_0 and a wave function describing plate 1:

$$\Psi(z, \xi) = \int d\xi_0 \{\delta(\xi - \xi_0)\} \otimes \{(2a)^{-1/2}\pi^{-1/4} \exp(-z^2/(2a^2))[\exp(ik_0(\xi_0 + z)) + \exp(-ik_0(\xi_0 + z))]\}.$$

From this expression a wave function, and, consequently, a corresponding probability distribution correlated to an ensemble of photons landing at the given coordinate value ξ_0 can be obtained.

That is, the measurement can be first performed either on the plate 1 or on the photon, and the obtained information will allow us to determine what is the probability distribution of the other, yet unmeasured part of an unseparable system.

V. INFORMATION AND THE COMPLEMENTARITY PRINCIPLE

In the preceding sections we have presented a a result which, although not paradoxical, was nevertheless surprising (that is, that one can make a fairly precise determination of the slit through which each photon comes with only a slight disturbance of the interference pattern). It is worthwhile to notice that the two limiting cases of Einstein's version of the double-slit experiment do not surprise us at all. These are the cases in which the smudging parameter u takes the values zero (no determination is made of the path of the photon) and infinity (each photon's path is determined completely). As we can see from Eq. (1), in the former case the interference pattern is perfect, while in the latter case it is completely washed out. This result has been known for so long that we have learned how to talk about it, and it is therefore no longer surprising.

The apparent paradox of our present example clearly arises from the fact that we are considering an intermediate situation in which one obtains *some* information about the photons' paths, and still retains an interference pattern having *some* degree of clarity. The problem is that we lack a good way of talking about such a situation, and we have no simple rule which tells us what to expect. The aim of this section is to fill this gap. We will find that information theory provides a good language for dealing with an imprecise path determination, and this will lead us naturally to a rule which defines the extent to which the two complementary aspects of light (wave and particle) may be manifested simultaneously.

Intuitively, one expects that if the interference pattern has a certain sharpness of definition, then there must be some limit on the amount of information which could have been obtained regarding through which slit each photon passed. What we wish to do here is simply to make this statement more precise. For this we need a way to quantify "information." We will use the measure of information discovered by Shannon,[8] which has all the mathematical properties one usually requires of information, and which has been fruitfully applied to a wide variety of problems,[9] including quantum mechanics itself.[10] Its most general definition is the following: If a system can be in one of N possible states, but if we know only the probabilities p_i of its being in each state i, then the amount of information we lack about the system is the positive number

$$H = -\sum_{i=1}^{N} p_i \ln p_i .$$

In our experiment we can consider each photon as a system with two possible states (in the above sense): passing through slit A and passing through slit B. If we determine the probabilities of these two paths to be p_A and p_B (where $p_A + p_B = 1$), then the information we lack about the path of that photon is

$$H = -(p_A \ln p_A + p_B \ln p_B). \tag{11}$$

As an example, we note that if we know nothing about the path of a photon, that is, if each slit is equally likely, then the amount of information we lack is

$$H_0 = -(\tfrac{1}{2}\ln\tfrac{1}{2} + \tfrac{1}{2}\ln\tfrac{1}{2}) = \ln 2 .$$

Now we are ready to formulate the problem.

Suppose we are interested in obtaining an interference pattern with a certain crest-to-valley ratio of intensities, say, 8.8 as in the above example. Notice that there are many ways to make such a pattern. We list a few of them here:

(1) Perform the experiment described in Sec. II, in which the measured momentum of plate 1 is used to judge the path of each photon. In this case we have seen that the paths of some photons will be better determined than the paths of others, depending on the value κ of the measured momentum.

(2) Keep all the plates fixed, but let one of the slits be bigger than the other. We have seen in Sec. III that the resulting pattern will again be partially smeared out, and that the same pattern arises from a subensemble of photons all having the same value of κ. The crucial thing about these experiments is that we cannot differentiate further among the photons. There is only *one* subensemble (the whole ensemble); each photon gives us the same amount of information; the photons are in a pure state.

(3) Allow some of the photons to pass through one slit only (the other being blocked), making a completely smeared out pattern; then open the other slit, allowing the rest of the photons to interfere perfectly. The sum of the two patterns can be arranged to have the desired crest-to-valley ratio.

(4) Make no measurement of the photons' paths, but let the slit in plate 1 be of a size comparable to the spacing between fringes, so that the resulting pattern is smudged.

The point is that the amount of information one obtains about the photons' paths varies from one case to another, even though the "summed-up interference pattern" is the same in each case. For example, in case 4 one loses some of the sharpness of the pattern without gaining any information at all.

The question naturally arises: Of all possible methods of generating this particular interference pattern, which one gives us the most information regarding the photons' paths? If we can find this method, then we will have found the limit on the amount of information which can be obtained, given that we insist on a certain sharpness of definition of the pattern. We will then have to admit that no amount of cleverness can produce an experiment in which the information gained exceeds this limit.

We will find that this question is indeed answerable, and that the best method turns out to be experiment (2) above, in which the photons are in a pure state. To see this, we first need to write down a general expression for information which applies to all the experiments we are considering.

A typical experiment can be described as follows. As in Sec. III, all the photons can be divided into different subensembles, such that the photons of a given subensemble all have the same probability of passing through slit A. In Sec. III we labeled the subensembles by the measured quantity κ, and we called the probabilities of the two paths $p_A(\kappa)$ and $p_B(\kappa)$. In the general case which we are considering now, it will be more convenient simply to label each subensemble by its values of p_A and p_B. To simplify the notation let us define $\gamma = p_A$, from which it follows that $p_B = 1-\gamma$. For our purpose, each experiment is completely characterized once we specify the fraction of the photons in each subensemble. Therefore, let $\rho(\gamma)d\gamma$ be the fraction of the photons whose probabilities of going through slit A are between γ and $\gamma + d\gamma$.

According to Eq. (7) the information we lack about the path of each photon of the subensemble γ is

$$H(\gamma) = -[\gamma \ln\gamma + (1-\gamma)\ln(1-\gamma)] .$$

The total information we lack is the sum of $H(\gamma)$ over all photons. To obtain the average information H we lack per photon, we divide this sum by the total number of photons, arriving at the formula

$$H = \int_0^1 d\gamma\, \rho(\gamma) H(\gamma)$$

$$= -\int_0^1 d\gamma\, \rho(\gamma)[\gamma \ln\gamma + (1-\gamma)\ln(1-\gamma)] . \tag{12}$$

Now we would also like to write an expression for the interference pattern $\mathfrak{F}(\xi)$ in terms of the function $\rho(\gamma)$. As in Sec. III [Eq. (6)], we again associate with each subensemble γ a partial interference pattern $i_\gamma(\xi)$, given by

$$i_\gamma(\xi) \propto 1 + 2\gamma^{1/2}(1-\gamma)^{1/2}\cos 2k_0\xi .$$

Notice that we have assumed the best possible $i_\gamma(\xi)$. One could always be sloppy (as in experiment 4) and obtain a more smudged pattern. Furthermore, we have assumed that all of the contributions i_γ from the various subensembles are not shifted relative to each other; that is, they all have their maxima at the same places. For our purpose here, these assumptions entail no loss of generality; we are trying to get as good a pattern as we can in order to investigate the theoretical limit on the amount of information which can be obtained. We now generalize Eq. (7), and find that the total interference pattern is

$$\mathfrak{F}(\xi) = \int_0^1 d\gamma\, \rho(\alpha)\, i_\gamma(\xi) = 1 + S\cos 2k_0\xi , \tag{13}$$

where

$$S \equiv 2\int_0^1 d\gamma\, \rho(\gamma)\gamma^{1/2}(1-\gamma)^{1/2}. \qquad (14)$$

From Eq. (13) it is clear that the quantity S is a measure of the sharpness of definition of the interference pattern. In terms of S, the main problem of this section can now be stated very simply: For a given value of S, what distribution ρ minimizes H, and what is this minimum value of H?

Our answer, whose proof is outlined in Appendix B, is that the best ρ is

$$\rho_0(\gamma) = \delta(\gamma - \gamma_0), \qquad (15)$$

where γ_0 is determined by the chosen sharpness of the interference pattern. In fact, according to Eqs. (13)–(15),

$$S = 2\gamma_0^{1/2}(1-\gamma_0)^{1/2},$$

or, if we solve this for γ_0,

$$\gamma_0 = \tfrac{1}{2}[1 \pm (1-S^2)^{1/2}]. \qquad (16)$$

What Eq. (15) says is that if we wish to get as much information as we can about the photons' paths, the best experiment we can do is of the kind exemplified by experiment 2, where all photons have the same probability γ_0 of passing through slit A, and are thus in a pure state. (Actually, as is shown in Appendix B, there is a one-parameter family of distributions which are as good as ρ_0; they differ from ρ_0 only in that some of the photons have their probabilities p_A and p_B reversed.) We have thus answered the first part of our question.

For an interference pattern of given sharpness S, the amount of information $H(S)$ one gives up when he uses this "best" method can be found from Eqs. (12) and (15) (and this answers the second part of our question):

$$H(S) = -[\gamma_0 \ln\gamma_0 + (1-\gamma_0)\ln(1-\gamma_0)], \qquad (17)$$

where γ_0 is given by Eq. (16). The main result of this section is that one must forfeit at least this much information about the photons' paths in order to obtain an interference pattern of sharpness S. That is,

$$H \geqslant H(S). \qquad (18)$$

Let us now apply this result to our original experiment (experiment 1), with the smudging parameter u having the same value as before. The interference pattern is [Eq. (2)]

$$\mathfrak{F}(\xi) = 1 + e^{-u^2}\cos 2k_0\xi$$
$$= 1 + (0.796)\cos 2k_0\xi. \qquad (19)$$

That is, $S = 0.796$. The value of $H(S)$, given by Eqs. (17) and (16), is then $H(S) = 0.497$. This number becomes more meaningful if we compare it to the total amount of information $H_0 = \ln 2$ available about each photon:

$$H(S)/H_0 = 0.717.$$

Thus, according to our result, in order to obtain the interference pattern (19), we must sacrifice at least 71.7% of the available information regarding the photons' paths.

This is the *minimum* amount of information we must sacrifice. How much information do we in fact give up when we perform Einstein's experiment? The easiest way to answer this is to return to the κ notation. According to Eq. (4), the probability $p_A(\kappa)$ that a photon in the subensemble κ will pass through slit A is

$$p_A(\kappa) = \frac{\exp[-a^2(\kappa+k_0)^2]}{\exp[-a^2(\kappa+k_0)^2] + \exp[-a^2(\kappa-k_0)^2]},$$

and $p_B(\kappa) = 1 - p_A(\kappa)$. The average information we lack per photon is

$$H = \int_{-\infty}^{\infty} d\kappa\, \mathfrak{D}(\kappa)[p_A(\kappa)\ln p_A(\kappa) + p_B(\kappa)\ln p_B(\kappa)],$$

where the distribution $\mathfrak{D}$ of measured momentum κ is given by Eq. (3). Upon evaluating this integral numerically, we find that

$$H/H_0 = 0.728.$$

This is only slightly greater than the minimum value $H(S)/H_0 = 0.717$, associated with this interference pattern. Thus, Einstein's experiment gives almost as much information as one could possibly get [which, by the way, is not very much in this case—only 28.3% (i.e., 1–71.7%) of the available information; this is perhaps in better agreement with what we might expect for this interference pattern, as compared to the expression "75% accuracy" which we used earlier]. But the point is that the amount of information one can get *is* limited by $H(S)$.

Let us conclude this section with a clear statement of the complementarity principle in the language of information theory, as it applies to the double-slit experiment. The sharpness of the interference pattern can be regarded as a measure of how wavelike the light is, and the amount of information we have obtained about the photons' trajectories can be regarded as a measure of how particlelike it is. Equation (14) can be expressed in words as follows: (Information lost about the photons' paths) $\geqslant$ (information $H(S)$ lost in pure-state experiment giving the same interference pattern). $H(S)$ [given by Eqs. (17) and (16)] increases monotonically as the sharpness increases. The above inequality is thus a precise statement of the following fact: The more clearly we wish to observe the wave nature of light, the more information we must give up about its particle properties.

ACKNOWLEDGMENTS

We would like to thank Professor John Archibald Wheeler, who encouraged us to pursue this project, for many stimulating discussions and for reading the manuscript. We would also like to thank Dilip K. Kondepudi for sharing with us some of his ideas on the double-slit experiment.

APPENDIX A: MULTIPLATE DOUBLE-SLIT EXPERIMENT

As we have discussed in the introduction, Einstein proposed to modify the double-slit experiment in such a way that *in principle* one should be able to determine through which slit the photon passed. In this appendix we shall propose an arrangement where one would be able to carry out an equivalent experiment *in practice*

The major difficulty encountered in an attempt to realize Einstein's proposal is the error in the determination of the lateral kick. This measurement may be performed by releasing any of the three screens involved (we do not consider any causality effects in this appendix), and measuring the momentum of the screen after its interaction with the photon. But regardless of the particular screen, the square average momentum of the random, noiselike Brownian motion is larger than the value of the kick we want to determine. For the case where the photographic plate is released Wheeler[13] recently calculated that the resulting fluctuations in the value of the momentum are 12 orders of magnitude larger than the signal itself. We are consequently forced to look for some alternative indicator of the particle path that would replace the momentum kick.

The direction was all that the kick could have told us. Can we determine the direction of the photon in a straightforward manner? To do it we propose the experimental arrangement envisaged in Fig. 4. There, the usual photographic plate is replaced by the set of nontransparent, thin "photoplates" covered with the photographic emulsion on both sides. In such a "multiplate double-slit" experiment, when the photon blackens a grain of sliver bromide on the top (bottom) of one of the photoplates, we expect that it came from the top (bottom) slit. The analysis of the experiment is simplest when the photoplates are occupying the region $\Delta\xi \ll s$, that is, when they lie close to the optical axis of the double-slit apparatus. We shall assume that this is true in the

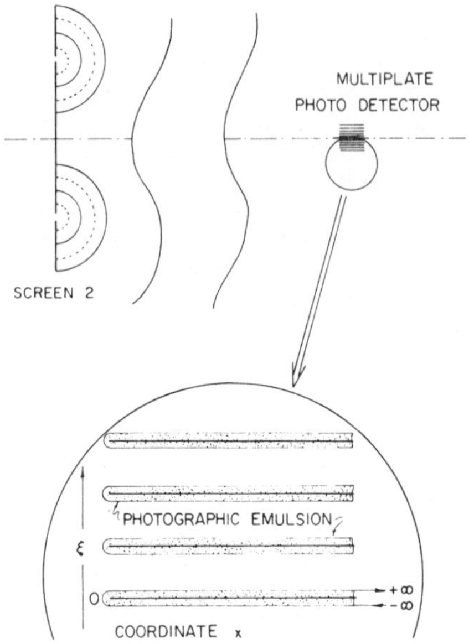

FIG. 4. Multiplate double-slit experiment.

rest of this section unless otherwise stated.

Let us now consider only those plates in the positions where previously, with the help of the normal photographic plate, maxima and minima of the perfect interference pattern

$$\mathcal{F}(\xi) \propto 1 + \cos 2k_0\xi$$

were observed. Will the plates placed at $\xi = 2n\pi/2k_0$ become dark while the ones at $\xi = [(2n+1)\pi]/2k_0$ remain untouched by the radiation? Shall we recover the perfect interference pattern? If so, quantum mechanics would prove inconsistent: The same photon goes through two slits producing an interference pattern and yet it comes from the top (bottom) slit.

What other outcomes of the experiment are conceivable? Let us examine more carefully the possible fate of photons landing on the photoplates to establish if: (1) The interference pattern may disappear—and all the plates will be "grey." (2) The fact that the photon lands on the top (bottom) of the photoplate does not yet allow us to be sure from which slit it came.

While the experiment remains to be done, one really expects (1) to occur. The reason for it in a carefully performed multiplate double-slit experiment is clear enough: The photons coming from two different slits never meet on the same side of the photoplate—they can never interfere. We can conclude this already from the wave optics interpretation, unless the wave representing the electrical field of the photon can get in some fashion from the bottom slit to the top side of the photoplate.

If this "landing on the wrong side" does occur, then we have the situation described by (2). Physical reasons for that effect can be found in the reflections that direct the ray coming from the top slit to the bottom side of the photoplate [see Fig. 5(a)]. Consequently, the probability of detecting a photon from a certain slit changes accordingly. We can use $p_A(x)$ as in Sec. III, to denote the probability that a photon from the top slit will arrive at the position x, where x is defined in Fig. 5(a). As before, we assume $p_A(x) + p_B(x) = 1$. We can also assume that the phases of the photons arriving at the given x are not randomized through the undergone reflection processes.

With this in mind one does not encounter any difficulties in carrying out an analysis similar to the one performed in Sec. III. The only important change is the different shape of $p_A(x)$ and $p_B(x)$, which is qualitatively shown in Fig. 5(b). What we shall measure here can be described in terms of the partial interference pattern and total interference pattern. For when we record the intensity at a certain position x on each plate, we sample at a set of discrete points $\{\xi_i\}$ (which are the positions of the photoplates) the partial interference pattern $i_x(\xi)$. As in Sec. III, from the knowledge of $p_A(x)$ and $p_B(x)$, we have

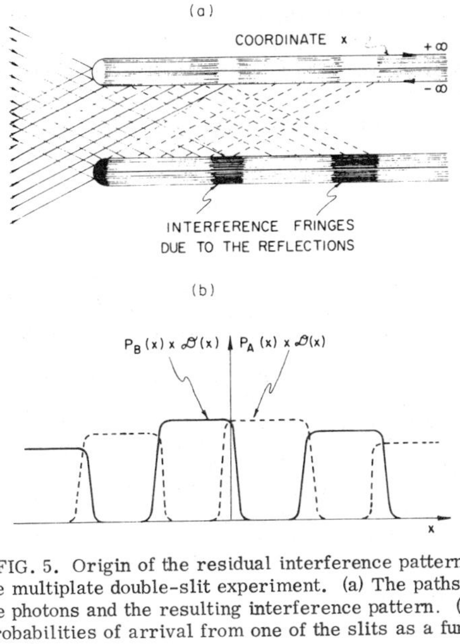

FIG. 5. Origin of the residual interference pattern in the multiplate double-slit experiment. (a) The paths of the photons and the resulting interference pattern. (b) Probabilities of arrival from one of the slits as a function of x.

$$i_x(\xi) = 1 + p_A^{1/2}(x) p_B^{1/2}(x) \cos 2k_0\xi .$$

This formula is true since all the photons landing at a certain x have the identical probabilities $p_A(x)$ and $p_B(x)$, and for that reason constitute one "subensemble" in the sense of Sec. III. Thus, we have succeeded in designing a measurement that gives directly the partial-interference pattern.

To obtain the total interference pattern, $\mathfrak{F}(\xi)$, we use formula (6):

$$\mathfrak{F}(\xi) = \int_{-\infty}^{\infty} dx\, \rho(x) i_x(\xi) . \tag{A1}$$

That is, the total number of grains blackened on the photoplate at a position ξ gives the value of the total interference pattern $\mathfrak{F}(\xi)$. The $\rho(x)$ used in Eq. (A1) is the weight function proportional to the total number of photons that land at a certain x and plays the same role as $\mathfrak{D}(\kappa)$ in formula (6).

To show that the discussed proposal is practical and realizable, let us go through the calculations in an example. The separation of photoplates δ should not be larger than the distance between the maximum and minimum expected in the interference pattern—in fact it is advantageous to sample the pattern only at these points: This yields for δ:

$$\delta = \pi / k_0 .$$

We have assumed previously that $\delta/s \ll 1$. This assumption is not crucial. We have used it solely to ascertain that in the set of photoplates one side of the photoplate can see only one slit. However, even if $\delta \sim s$, we can still make sure that this last statement is true by orienting the photoplates sc that the plane in which each one of them lies always intercepts the double-slit screen half-way between the two slits. Still, it is easier to do with s bigger than δ.

Taking $\lambda = 10\ \mu$m (there are lasers which have wavelength of this order), we get for $L = 100$ cm and $s = 5$ mm the photoplate separation $\delta = 0.85$ mm. This is well within reach of experimental possibilities.

We have seen here an experiment that attempts to determine the slit through which each photon passed while trying to retain the interference pattern—the same purpose for which the experiment described in the introduction was designed. However, no data clarifying the Einstein-Podolsky-Rosen paradox as discussed in Sec. IV can be obtained here, since both the path determination and contribution to the interference pattern follow from a single event—photoinduced transition in a silver-bromide grain. Still, our proposal is the only practical version of Einstein's double-slit experiment we are aware of.

APPENDIX B: OUTLINE OF A PROOF THAT THE PURE STATE MINIMIZES THE LOSS OF INFORMATION

For a given value of S, what distribution ρ minimizes H? This question can be simplified by the following definitions:

$$s(\gamma)=\gamma^{1/2}(1-\gamma)^{1/2},$$

$$h(\gamma)=-[\gamma\ln\gamma+(1-\gamma)\ln(1-\gamma)],$$

$$\hat{\rho}(\gamma)=\rho(\gamma)+\rho(1-\gamma),$$

restricted to the interval $0\leqslant\gamma\leqslant\frac{1}{2}$. Then

$$S=\int_0^{1/2}d\gamma\hat{\rho}(\gamma)s(\gamma), \tag{B1}$$

$$H=\int_0^{1/2}d\gamma\hat{\rho}(\gamma). \tag{B2}$$

What we wish to show is that the distribution $\hat{\rho}_0(\gamma)=\delta(\gamma-\gamma_0)$ minimizes H for a given value of S. This will imply in particular that the pure state $\rho_0=\delta(\gamma-\gamma_0)$ minimizes H. [So does any other ρ which gives the same $\hat{\rho}$. These other ρ's are all of the form

$$\rho(\gamma)=\alpha\delta(\gamma-\gamma_0)+(1-\alpha)\delta(\gamma-(1-\gamma_0)),$$

$$0\leqslant\alpha\leqslant 1.]$$

Step 1. One can prove that s and h are monotonically increasing in the interval $0\leqslant\gamma\leqslant\frac{1}{2}$, and that $s''/s'\leqslant h''/h'$ in this interval. (A prime denotes differentiation with respect to γ.)

Step 2. Let us compare the distribution $\hat{\rho}_0(\gamma)=\delta(\gamma-\gamma_0)$ with the trial distribution $\hat{\rho}_1(\gamma)=\eta_1\delta(\gamma-\gamma_1)+\eta_2\delta(\gamma-\gamma_2)$, in which $\eta_1+\eta_2=1$, and the parameters are assumed to be adjusted so that both distributions give the same value of S; that is [Eq. (B1)],

$$s(\gamma_0)=\eta_1 s(\gamma_1)+\eta_2 s(\gamma_2). \tag{B3}$$

We will show that $\hat{\rho}_0$ gives a smaller value of H than $\hat{\rho}_1$ does; that is [Eq. (B2)],

$$h(\gamma_0)\leqslant\eta_1 h(\gamma_1)+\eta_2 h(\gamma_2). \tag{B4}$$

To prove this, it is helpful to define $\bar{h}(\gamma)$ by

$$\bar{h}(\gamma)=\frac{[s(\gamma_2)-s(\gamma_1)]h(\gamma)+s(\gamma_1)h(\gamma_2)-s(\gamma_2)h(\gamma_1)}{h(\gamma_2)-h(\gamma_1)}.$$

The following things are true about $\bar{h}$:

(i) $\bar{h}(\gamma_1)=s(\gamma_1)$ and $\bar{h}(\gamma_2)=s(\gamma_2)$.

(ii) $\dfrac{\bar{h}''}{\bar{h}'}=\dfrac{h''}{h'}\geqslant\dfrac{s''}{s'}$ for $0\leqslant\gamma\leqslant\frac{1}{2}$.

(iii) By (i) and Eq. (B3), $s(\gamma_0)=\eta_1\bar{h}(\gamma_1)+\eta_2\bar{h}(\gamma_2)$.

(iv) To prove Eq. (B4), it is sufficient to prove that

$$\bar{h}(\gamma_0)\leqslant\eta_1\bar{h}(\gamma_1)+\eta_2\bar{h}(\gamma_2).$$

From (iii) and (iv) it is clear that what we have to prove is that $\bar{h}(\gamma_0)\leqslant s(\gamma_0)$. In fact it is not hard to show from properties (i) and (ii) that $\bar{h}(\gamma)\leqslant s(\gamma)$ for all γ between γ_1 and γ_2, and thus for γ_0 in particular.

Step 3. We have seen in Step 2 that the distribution $\hat{\rho}_0=\delta(\gamma-\gamma_0)$ is better at minimizing H than any sum of two δ functions. This result can be extended by induction to any sum of n δ functions.

Step 4. For the purpose of calculating S and H, any distribution can be approximated arbitrarily closely by a sum of n δ functions. Therefore, the distribution $\hat{\rho}_0=\delta(\gamma-\gamma_0)$ minimizes H with respect to all possible distributions.

---

[1] M. Jammer, *The Philosophy of Quantum Mechanics* (Wiley, New York, 1974), pp. 121–132.

[2] F. J. Belinfante, *Measurement and Time Reversal in Objective Quantum Theory* (Pergamon, New York, 1975), p. 32.

[3] R. P. Feynman, R. B. Leighton, and M. Sands, *The Feynman Lectures on Physics* (Wesley, Reading, Mass., 1965), Vol. 3, p. 1-1.

[4] M. Jammer, Ref. 1, pp. 160–247.

[5] F. J. Belinfante, Ref. 2, p. ix.

[6] A. Einstein, B. Podolsky, and N. Rosen, Phys. Rev. 47, 777 (1935).

[7] B. D'Espagnat, *Conceptual Foundations of Quantum Mechanics* (Benjamin, Reading, Mass., 1976), 2nd edition.

[8] C. E. Shannon, Bell Syst. Techn. J. 27, 379 (1948).

[9] J. R. Pierce, *Symbols, Signals and Noise* (Harper, New York, 1961).

[10] H. Everett III, in *The Many-Worlds Interpretation of Quantum Mechanics*, edited by B. DeWitt and N. Graham (Princeton Univ. Press, Princeton, New Jersey, 1973).

[11] H. Weyl, *The Theory of Groups and Quantum Mechanics* (Methuen, London, 1931).

[12] P. Jacquinot and B. Roizen-Dossier, in *Progress in Optics*, edited by E. Wolf (North-Holland, Amsterdam, 1964), Vol. 3.

[13] J. A. Wheeler, in *Mathematical Foundations of Quantum Theory: Proceedings of the New Orleans Conference on the Mathematical Foundations of Quantum Theory*, edited by A. R. Marlow (Academic, New York, 1978).

III.12 COMPLEMENTARITY IN THE DOUBLE-SLIT EXPERIMENT: ON SIMPLE REALIZABLE SYSTEMS FOR OBSERVING INTERMEDIATE PARTICLE-WAVE BEHAVIOR

LAWRENCE S. BARTELL

In the course of a detailed analysis of observable intermediate particle-wave aspects of light, Wootters and Zurek invented an ingenious but technically difficult practical apparatus for performing Einstein's version of the double-slit experiment. The present note discusses two very simple alternative systems for displaying the Bohr-Einstein particle-wave extremes or, if desired, any intermediate trajectory-interference manifestation associated with the double-slit experiment.

In a recent analysis of Einstein's version of the double-slit experiment, Wootters and Zurek[1] presented a detailed treatment of the potentially observable manifestations of particle and wave characteristics of light. This approach led them to a quantitative expression of Bohr's complementarity principle appropriate for such experiments. Einstein had originally proposed that the lateral kick imparted by a photon to an interference screen could be used to identify which slit the photon traveled through on its way to the screen. Pointing out that thermal noise in practical experiments would utterly obscure such photon-induced recoils, Wheeler[2] recently reformulated the Einstein experiment in such a way that simple optical discrimination replaced measurements of photon kicks. In Wheeler's revised experiment, the operator is free to choose (after the photon has passed the slits, if he wishes) *either* to observe which slit has been traversed *or* to accumulate interference fringes. Wootters and Zurek showed how to modify the double-slit experiment to permit intermediate choices. In their thought experiment strongly visible interference fringes are displayed under conditions allowing "trajectories of the photons" to be "determined with fairly high accuracy." Although their theoretical analysis was developed in terms of measurements of Einstein photon kicks, Wootters and Zurek also proposed an ingenious practical version of the experiment not requiring measurements of recoil. Instead of adapting the Wheeler version, however, they introduced a new type of multiplate detector.

It seems worthwhile to point out that a small modification of Wheeler's apparatus leads to an even simpler demonstration of intermediate "particle-wave" behavior. This modification is illustrated in Fig. 1. In Wheeler's design the interference screen is placed at the focal plane of the lens at a distance L from the slits, where the Fraunhofer pattern can be observed. If, instead, it is desired to observe the trajectory of a particular photon, the screen is turned aside and the photon activates one or the other counter.[3] All that is needed to observe intermediate behavior is to move the interference screen to a new distance $l = L + z$, on either side of the focus, where the Fresnel pattern can be viewed. It is convenient to analyze a system with equivalent slits incorporating Wheeler's Gaussian transmission filters. Such filters circumvent the subsidiary diffraction spots characteristic of slits with abrupt edges. Elementary diffraction theory for a case $l \gg d \gg \lambda$ yields the intensity

$$I(y) = K[F(y)]^2 \cosh(ay)[1 + V(y)\cos by], \tag{1}$$

where $F(y)$ is the slit form factor $\exp(-y^2/2\sigma^2)$, $V(y)$ is the Michelson "fringe visibility" (parameter S of Wootters and Zurek) given by $\operatorname{sech}(ay)$, and K, a, b, and σ are identified in Ref. 4.

Although the fringe visibility damps as y increases, it is easy to choose conditions making the variation of $V(y)$ small across a span of many fringes. As is true also for the Wootters and Zurek system, fringe visibility is related to the slit transmission probabilities $P_A(y)$ and $P_B(y)$ for radiation reaching coordinate y via

$$V(y) = 2[P_A(y)P_B(y)]^{1/2}. \tag{2}$$

Therefore, if the slit discrimination ratio $P_A : P_B$ is selected to be $4:1$, for example, the resultant

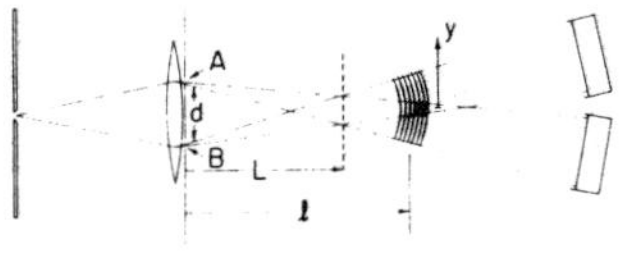

FIG. 1. Modification of Wheeler's version of Einstein's double-slit experiment. Fraunhofer fringes can be accumulated on a screen at L, or counters at right can detect which slit a photon came through. Alternatively, an observation region y at $l \neq L$ can be selected to yield any desired slit transmission probability ratio, $P_A{:}P_B$, and fringe visibility. See text.

Originally published in *Physical Review, D21*, 1698-99 (1980).

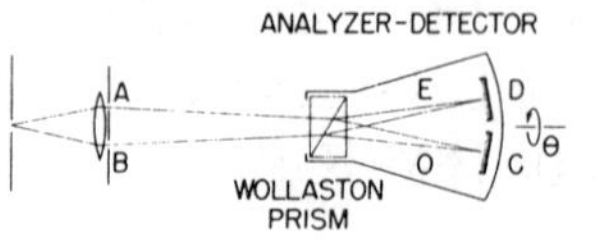

FIG. 2. Double-slit apparatus with controllable trajectory/interference characteristics. Slits A and B transmit light polarized at $+45°$ and $-45°$, respectively. Any desired slit transmission ratio $P_A:P_B$ and corresponding fringe visibility received at the detectors can be chosen by adjusting the analyzer-detector orientation. O and E represent ordinary and extraordinary rays.

fringe visibility is only 20% lower than that of an ideal two-slit pattern. Although such a ratio suggests a fairly high probability of a trajectory through slit A, it cannot, of course, be rigidly interpreted in terms of individual photons going through slit A four times out of five. Equation (2) obviously reduces to the Bohr-Einstein extremes as V, P_A, or P_B approach unity. At intermediate values of these variables, Eq. (2) offers a quantitative expression of complementarity for intermediate cases. Expectation values of V (wave) and P_A (particle) cannot both be unity simultaneously; the sacrifice of knowledge of one for the other is plain.

The extreme simplicity of the foregoing experiment may not be an adequate compensation for its unattractive but characteristic variation of $P_A : P_B$ across the diffraction pattern. Therefore, an alternate apparatus designed to produce patterns with uniform, controllable $P_A : P_B$ ratios is sketched in Fig. 2. In this apparatus, slits A and B transmit light polarized at $+45°$ and $-45°$, respectively. When the analyzer-recorder is oriented as illustrated, two-slit patterns of unit visibility arrive at C and D (one pattern with a light fringe and one with a dark fringe at the center). A rotation of the analyzer system by 45° about the optic axis produces a single-slit pattern from A at C and a corresponding pattern from B at D. Intermediate rotations θ generate intermediate fringe visibilities ($V = \cos 2\theta$) and intermediate transmission probabilities P_A and P_B. Again, the visibility is related to transmission probabilities by Eq. (2) but the ratio $P_A : P_B$ now depends only on the orientation θ. In principle, by adjusting θ, a choice can be made (*after* the photon has passed the slits, if desired) of what fraction of it (89%, say) shall have passed through one slit and what fraction through the other. The experimenter cannot, of course, choose which slit shall have transmitted 89%—that is left to chance—but only that one of the slits shall have transmitted 89%, and one, 11%. The 11%, although small, is quite crucial to the 63% fringe visibility.

In each of the two experiments sketched above, the Fresnel and the polarization, Wootters and Zurek reflectionless multiplate detectors could be introduced to determine directly the $P_A : P_B$ ratio inferred only indirectly in the foregoing. Naturally, such direct observations would render the interference fringes unobservable. Each of the experiments can be adjusted to manifest a variable degree of wave-particle behavior analogous to that discussed at length by Wootters and Zurek. As thought experiments the present designs offer no new lessons. As realizable experiments, however, they offer quite the simplest arrangements possible.

This research was supported by a grant from the National Science Foundation.

---

[1]W. K. Wootters and W. H. Zurek, Phys. Rev. D 19, 473 (1979).

[2]J. A. Wheeler, in *Mathematical Foundations of Quantum Theory: Proceedings of the New Orleans Conference on the Mathematical Foundations of Quantum Theory*, edited by A. R. Marlow (Academic, New York, 1978).

[3]A greater discrimination can be secured by placing a lens at or beyond the interference screen to project images of the slits upon the counters.

[4]Let the slit amplitude transmission function be $\exp(-x^2/2x_0^2)$, where x is the vertical distance from the slit center. Then $\sigma^2 = l^2 x_0^2 [k^{-2} x_0^{-4} + (l^{-1} - L^{-1})^2]$, $a = ld(l^{-1} - L^{-1})/\sigma^2$, and $b = ld/kx_0^2\sigma^2$. The symbol K represents a collection of constants of no concern here.

III.13 A "DELAYED-CHOICE" QUANTUM MECHANICS EXPERIMENT

WILLIAM C. WICKES, CARROLL O. ALLEY, AND OLEG JAKUBOWICZ

We describe an experiment currently being assembled to test the quantum-mechanical assertion that the results of a split-beam interference measurement are unchanged even if the decision to recombine the two light beams is delayed until after each photon interacts with the beam splitter and enters a superposition state.

In a traditional optical double-slit interference experiment (Fig. 1), photons from a distant point-source are incident upon a screen containing two small apertures. With a lens behind the screen, an image of the source is formed on, for example, a photographic plate situated in the focal plane of the lens. The size and shape of the image depend upon the apertures, but the image will be crossed by interference fringes of angular frequency λ/D, where D is the separation of the apertures and λ is the wavelength of the light. Such an experiment isolates the wave character of the incident radiation: the interference fringes can be best understood by describing the energy contained in the radiation as passing through both apertures, even if the intensity is reduced so that usually only one photon at a time enters the system.

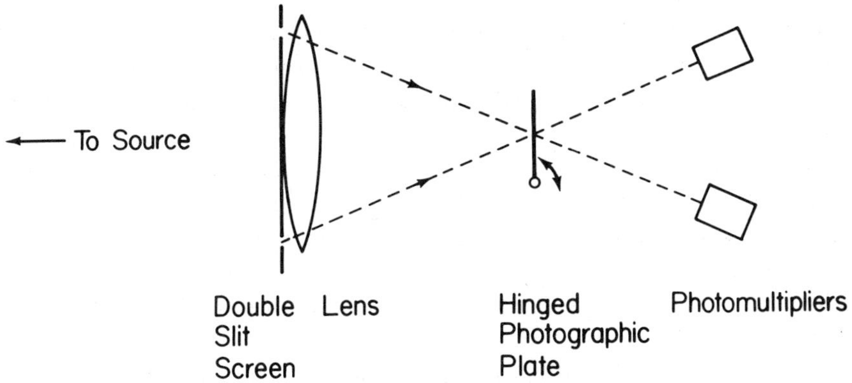

FIGURE 1. Ideal Double-Slit Experiment

If the photographic plate is replaced by a pair of photomultipliers situated so that each one receives light from only one of the apertures, then the experiment emphasizes the particle nature of the light. If, for example, one of the apertures is covered, there is no effect on the count rate observed with the photomultiplier associated with the other aperture. This effect implies that each photon passes through one aperture or the other, never both.

This reasoning suggests that the configuration of the apparatus, and hence, the choice of an experimenter, determines whether the photons exhibit particle-like or wave-like behavior. Acceptance of this apparent paradox is central to the Copenhagen interpretation of quantum mechanics. Bohr (1949) moreover asserts that the outcome of a double-slit experiment will not change even if the apparatus is changed after the photon is already in flight. The experiment we propose will test this prediction.

In an idealized "delayed choice" double-slit experiment (Wheeler, 1978, 1979), the choice of whether the experiment is to demonstrate the wave properties (interference fringes on a photographic plate) or the particle properties (individual events in photomultipliers) is delayed until after a particular photon has passed the plane of the double slit. The light intensity is adjusted so that there is a low probability of more than one quantum being in the apparatus at any time. We might imagine that the photographic plate is hinged to swing rapidly into or out of the photon paths: if the plate intercepts the beams, an interference pattern is recorded, one grain at a time; if the plate is swung out of the way, photons passing through individual apertures are detected. If the interference pattern obtained is the same as obtained with a static plate, the quantum-mechanical prediction is confirmed. In what Wheeler (1979) calls a bad set of words, the behavior of the photons at the time of their interaction with the double slit, i.e., whether they pass through one aperture or both, is determined by our choice of photographic plate position *after* the interaction has taken place. There is no doubt that interference effects are obtained with a fixed system; however, an actual delayed-choice experiment of this type has not been performed. Various tests of hidden variable theories and of the EPR paradox have been proposed (Clauser et al., 1969; Wootters and Zurek, 1979) or carried out (Faraci, 1974; Freedman and Clauser, 1972; Kasday, 1971). Aspect (1975, 1976) is undertaking an experiment to test Einstein separability by studying the polarization of correlated photons separated by spacelike intervals. The delayed-choice experiment considered here looks at the behavior of single photons with a choice of trajectories to a single recombination region in space-time rather than at pairs of photons at distant points.

For any practical realization of a delayed-choice experiment, devices analogous to the hinged photographic plate are not possible because of the high speed of propagation of the photons. It is conceptually equivalent and technologically feasible to change the beam paths, switching among stationary detectors. In our configuration, to facilitate the beam switching the "double slit" is replaced by a beam splitter, resulting in an arrangement similar to the split-beam apparatus described by Wheeler (1978).

In the proposed system (Fig. 2), the detection of both "particles" and "waves" is accomplished by photomultipliers in which detection events can be localized in time. We will use RCA 31024 photomultipliers, which have a pulse jitter of approximately 500 ps for detection of single photons. Discriminated photon

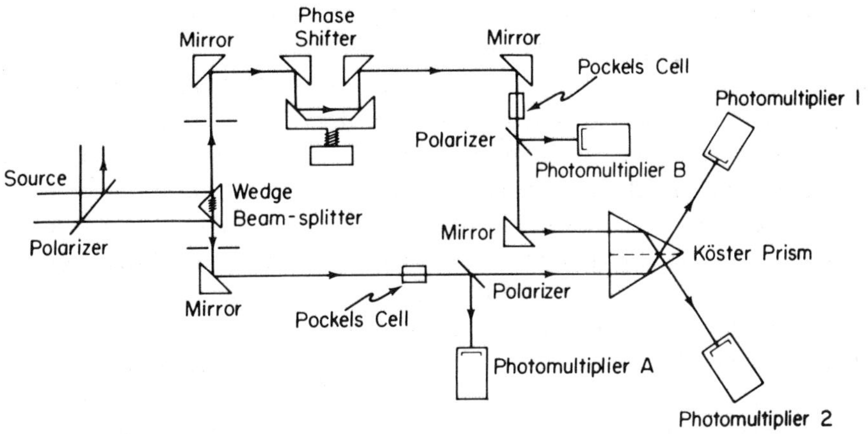

FIGURE 2. The "Delayed Choice" Interferometer

events will be recorded with the University of Maryland Lunar Ranging Program event timer (Currie et al., 1972; Steggerda, 1973), modified to accept inputs from four independent photomultipliers. The event timer will record the epoch of each photon event to an accuracy of 200–300 ps, as well as identify the appropriate detector. The delayed-choice switching is implemented in both arms of the apparatus by a combination of a Pockels cell and a thin-film polarizer. Upon application of a high voltage, the Pockels cells produce a 90° rotation of the plane of polarization of transmitted light. If the light from the original source is plane polarized, then the state of the Pockels cells at the time the photons pass through them will determine whether the photons are transmitted by the thin-film polarizers into the Köster prism, or reflected into photomultipliers A and B, depending upon the relative orientations of the source polarizer and the thin-film polarizers. This type of switching is routinely employed in modern short-pulse laser systems, with a rise time of the Pockels cells switching of about 2–3 ns. The cumulative uncertainty in the overall system timing is thus less than 3 ns; the photon paths from beam splitter to Pockels cells are about 15 ns light travel time.

Photomultipliers A and B play the role of "particle" detectors, since each can receive light from only one arm of the apparatus and thus is isolated from any wave interference effects. Photomultipliers 1 and 2 then serve to detect "wave" properties through the interference of the beams recombined within the Köster prism. In Figure 2, the dotted line in the Köster prism represents a beam splitter surface. The silvering of that surface is adjusted to provide a net 90° phase difference between the reflected and transmitted rays, so that if the two waves from the two arms of the apparatus entering the prism differ in phase by 90°, there will be destructive interference in the two waves leaving the prism on one side, and constructive interference on the other side. The initial 90° phase difference is supplied by the adjustable phase shifter situated in one arm of the system. The presence of interference is thus simply indicated by the relative count rates in photomultipliers 1 and 2. In the absence of interference, for example, if one arm or the other is blocked, the two photomultipliers will exhibit identical count rates. Perfect interference would be indicated by an absence of counts in one detector.

An advantage other than simplicity of using photomultipliers to detect interference effects is that there is no necessity then to control in advance the emission times of the photons entering the apparatus. Since the arrival time of each detected photon is recorded, its distance from the Pockels cells can be computed backwards from that time rather than forwards from an emission time. We may therefore use a low intensity thermal source, emitting photons at random, and may also switch the Pockels cells at random (or periodically, which is more convenient). The recorded events can later be sorted according to the computed position of each detected photon at the time of a Pockels cells switch. The only events of interest are those for which the photons are "found" to have been between the beam splitter and the Pockels cells at the time of the switch; these events are then analyzed for possible delayed-choice effects on their interference properties as reflected in the relative number of these events found with photomultipliers 1 and 2.

As a second approach, we plan also to use a mode-locked neodymium YAG laser as a light source. This emits pulses with a time duration of 100 ps, corresponding to a wave packet with spatial extension of 3 cm. In this case, the time can be known at which a suitable attenuated pulse containing energy equivalent to one photon enters the apparatus.

Any interpretation of the results of this experiment will depend, of course, on whether it is sensible to speak of the "position" of individual photons, especially as regards the extrapolation backwards in time from the detection of photomultiplier pulses. It should be emphasized that there is no requirement to identify the photons entering the Köster prism as having been in one arm or the other of the apparatus; we are only interested in whether the photons, considered as particles, wave packets, etc., were between the initial beam splitter and the Pockels cells at the time when the cells changed state, with no consideration of whether the photons were "in" one arm or the other or both. That such a "one-dimensional" positioning is meaningful is clear from experiments such as the lunar ranging work (Alley, 1972; Bender et al., 1973; Silverberg, 1974; Williams et al., 1976), where a 3 ns pulse of laser radiation is emitted, sent to a distant corner reflector target, and returned. The return detections, which require single photon sensitivity, occur within a well-defined time window corresponding to the emission time plus round trip travel time. We could, in principle, calculate the emission epoch from the reception time with equally consistent results. According to the uncertainty principle, the uncertainty in a photon's position along the direction of propagation is given by $\lambda^2/\Delta\lambda$, where λ and $\Delta\lambda$ are respectively the central wavelength and bandwidth of the optical bandpass filter at the entrance to the delayed-choice apparatus. Choosing $\lambda/\Delta\lambda$ to be $\simeq 50$ gives a position uncertainty of $\simeq 50\lambda$, which is negligible compared to the dimensions of the apparatus while still permitting small path length differences in the two arms without eliminating interference.

All measurements for calibration purposes such as determining the overall photon count rate, relative transmissions of the two arms overall and of individual components, photomultiplier efficiencies and dark rates, and adjustments of the phase shifter, can be made with a static system. The basic split-beam/Köster prism interferometer is the principal part of an existing long baseline stellar interferometer (Liewer, 1979), so that the optical stability of and detection of interference with the system have already been verified. This experiment is further forgiving in that by asking only one simple question (at least initially), i.e., does the delayed-choice switching of the Pockels cells affect the wave/particle properties exhibited by photons in a static system, there is no necessity for ideal behavior of components such as the thin-film polarizers, the initial beam splitter, or the Köster prism. The initial experiment will be to detect gross effects; refinements can follow if the results are interesting.

The most "interesting" result would be an observable delayed-choice effect on the photons' interference. At present, it is safe to say that few physicists would expect such an effect. However, until questions of superluminal connections (Stapp, 1977), hidden variable theories (see d'Espagnat, 1979, for a review), or the continuing problem of interpretation of the wave function (Wheeler, 1977) are satisfactorily answered, we believe that it is desirable to continue to test and probe all fundamental predictions of quantum mechanics as it becomes technologically possible to do so.

The undertaking of this experiment was inspired by an Einstein Centennial Address delivered at the University of Maryland by J. A. Wheeler. We are grateful to A. Shimony and to A. Wightman for their encouragement.

IV
Field Measurements

IV.1 EXTENSION OF THE UNCERTAINTY PRINCIPLE TO RELATIVISTIC QUANTUM THEORY

LEV DAVIDOVICH LANDAU AND RUDOLF PEIERLS

It is shown by considering possible methods of measurement that all the physical quantities occurring in wave mechanics can in general no longer be defined in the relativistic rannge. This is related to the well-known failure of the methods of wave mechanics in that range.

1. INTRODUCTION

IT is known that the application of the methods of wave mechanics to problems in which the speed of light cannot be regarded as infinite leads to absurd results. In the first place, states with negative mass appear in Dirac's relativistic equation[1]. This difficulty arises because the relation between momentum and energy in relativity theory is quadratic, so that two energy states are possible for a given momentum. In contrast to classical ($h = 0$) relativity theory, where the continuous change of all quantities means that transitions between the two kinds of state are impossible, such transitions cannot reasonably be forbidden in quantum theory.

In the second place, the interaction of a charged particle with the field produced by itself is inevitably divergent[2].

The infinite zero-point energy of the radiation field which occurs on quantisation of the field[3] can be avoided by the use of suitable variables[4], but it still has the effect that the energy-density matrix elements become infinite. This is very closely related to the self-energy difficulty mentioned above (see also ref. 5).

This complete failure of the theory suggests that in the range considered the physical requirements for the applicability of the methods of wave mechanics are no longer satisfied. The present paper investigates this problem.†

2‡. THE CONCEPT OF MEASUREMENT IN WAVE MECHANICS

The significance of any physical theory is to derive from the result of an experiment conclusions regarding the results of subsequent experiments. Thus the relations between measurements and the physical states of a system are

L. Landau und R. Peierls, Erweiterung des Unbestimmtheitsprinzips für die relativistische Quantentheorie, *Z. Phys.* **69**, 56 (1931).

† The uncertainty relations on which our conclusions are based are derived mainly from discussions in Copenhagen. Professor Bohr's attitude to these relations will be described in an article to appear shortly in *Nature*.

‡ This section is essentially a development of ideas put forward by N. Bohr in his lecture at Como[6].

Translation originally published in *Collected Papers of Landau*, D. ter Haar, ed., pp. 40-51, Gordon and Breach, New York (1965).

of two kinds. Firstly, the measurement determines the state of the system *after* the measurement is made, and secondly it examines the state of the system which existed *before* the measurement. In classical ($h = 0$) mechanics this distinction is of no importance, since the states of the system before and after the measurement can be regarded as identical.

In wave mechanics, however, the situation is quite different, since the measurement always causes a change in the state of the system, and this change is in principle impossible to determine. If the measurement had no other property, the wave-mechanics description would be neither possible nor meaningful. It is necessary to make use of another physical property of measurements, which is usually described as repeatability. This signifies that, when the same measurement is immediately repeated, the same result is certainly obtained. In this form, however, the hypothesis is physically incorrect in most cases, as will be shown in more detail below; and in this strict form it is not necessary for wave mechanics. The important point is that for any system there should exist *predictable* measurements. This means measurements such that for every result there is a state of the system in which this measurement *certainly* gives that result. For, if this requirement were not fulfilled, the state of the system after a measurement could not be described by a ψ function. This may be seen as follows. We can describe the state of the system and the measuring apparatus together by a wave function which, before the measurement, consists of a product $\psi\,\varphi_0$. Here ψ is the initially arbitrary wave function of the system, and φ_0 the known wave function of the measuring apparatus. After the interaction, the wave function will in general no longer be a product. If we expand it in terms of the eigenfunctions of the measuring apparatus, in the form $\sum \psi_n\,\varphi_n$, then ψ_n describes the state in which the system remains after the measurement. In general, the form of ψ_n depends on that of ψ. If the wave function of the system is to be deduced from an observation of the measuring apparatus, ψ_n must be independent of ψ apart from a constant factor, i.e. $\psi_n = a_n\,u_n$, with u_n normalised to unity. From the linearity of the wave equation it follows that a_n depends linearly on ψ, i.e. can be written in the form $\int \psi\, v_n^*\, d\tau$, with v_n any function dependent on the process of measurement. Then $|a_n|^2$ is the probability that the measurement gives the nth result. The sum of all these probabilities must be equal to unity, i.e. $\sum |a_n|^2 = 1$ independent of ψ (rpovided that ψ is normalised):

$$\sum a_n a_n^* = \int \sum a_n\, v_n\, \psi^*\, d\tau .$$

This expression must therefore always equal unity if $\int \psi\,\psi^*\, d\tau = 1$, i.e. we must have

$$\sum a_n\, v_n = \psi .$$

(The v_n form a complete orthogonal system). From this, however, it follows that the measurement is predictable if we take ψ to have the particular value of one of the v_n; then only one of the a_n is not zero. The repeatability of the measurement would signify that the v_n and the u_n are identical, and this is not in general true.†

If, however, the wave function of the system can not be determined by any measurement, it can have no physical meaning. The use of wave functions would then be as pointless as, for example, the use of the concept of paths in quantum mechanics. Thus the existence of predictable measurements is an absolutely necessary condition for the validity of wave mechanics.

The condition of repeatability cannot in general be satisfied. This is particularly seen if the time necessary for the measurement is taken into account. This time is restricted by the relation $\Delta E \, \Delta t > \hbar$, which has very often been stated, but which has been correctly interpreted only by Bohr[6]. Clearly it does *not* signify that the energy can not be known exactly at a given time (for in that case the concept of energy would have no meaning), nor does it mean that the energy can not be measured with arbitrary accuracy within a short time. We must take into account the change caused by the process of measurement even in the case of a predictable measurement, i.e. of the difference between the result of the measurement (v_n) and the state after the measurement (u_n). The relation then signifies that this difference causes an energy uncertainty of the order of $\hbar/\Delta t$, so that in time Δt no measurement can be performed for which the energy uncertainty in *both* states is less than $\hbar/\Delta t$.

This follows from a consideration of the time evolution of the interaction process. The method of the variation of constants shows that transitions within short times occur not only between states which satisfy the condition $E + \varepsilon = E' + \varepsilon'$ (E and E' being the energy of the system before and after the transition, ε and ε' that of the apparatus). These states are given preference by resonance only after a long time, the corresponding transition probabilities increasing greatly with time. In practice, after a time Δt, only transitions for which $|E + \varepsilon - E' - \varepsilon'| \lesssim \hbar/\Delta t$ are of importance. This fact does not, of course, contradict the strict validity of the law of conservation of energy in wave mechanics, but the energy of interaction between the system and the apparatus is also indeterminate by the same amount. In the most favourable case, where ε and ε' are precisely known, the uncertainty must be $\Delta(E - E') > \hbar/\Delta t$.

This relation has important consequences as regards the measurement of momentum. Any measurement of momentum is made by allowing the body to collide with another. In measuring a component of momentum (most simply done by collision with a plane mirror) the law of conservation of momentum is to be applied rigorously, but that of conservation of energy applies only to within $\hbar/\Delta t$, because of the unknown interaction energy. Thus to determine the

† In a measurement which occupies a short time it can easily be shown that the u_n are identical with the v_n only when the corresponding operator commutes with the energy of interaction between the system and the apparatus. In wave mechanics (neglecting relativity) this interaction energy is always a function of the co-ordinate. The only quantity for which a repeatable measurement is possible is therefore the co-ordinate. Measurements of the co-ordinate always actually have this property. It is also seen that the u_n need not in general be orthogonal, i.e. the measurement does not in general diagonalise an operator. This physical circumstance also is usually overlooked in the presentation of transformation theory.

particle momentum $\boldsymbol{P}$ we have the equations

$$\boldsymbol{p} + \boldsymbol{P} - \boldsymbol{p}' - \boldsymbol{P}' = 0,$$

$$|\varepsilon + E - \varepsilon' - E'| \gtrsim \frac{\hbar}{\Delta t};$$

$\boldsymbol{p}$, $\boldsymbol{p}'$, ε, ε', i.e. the motion of the measuring apparatus before and after the collision, may be regarded as known. Then $\Delta P = \Delta P'$ and, since $\Delta E = v\,\Delta P$,

$$(v - v')\,\Delta P > \frac{\hbar}{\Delta t}. \tag{1}$$

Thus any measurement of momentum necessarily involves a *definite* change of momentum (in addition to the unknown change which restricts the accuracy of the measurement).† This fact was first recognised by Bohr[6]. The non-repeatability of the momentum measurement in a short time is thus shown with particular clarity. Momentum measurements which last a long time, on the other hand, are meaningful only for free particles.

3. Momentum Measurement in the Relativistic Case

We now wish to make use of relativity, i.e. of the finite speed of propagation. There exists as yet no satisfactory relativistic quantum theory, but it is clear that here also we certainly cannot go beyond the limits imposed on the accuracy of measurement by the general principles of wave mechanics.

The scope of the relation just derived for momentum measurement is considerably extended by relativity. In the non-relativistic theory, the definite change of velocity could be made arbitrarily large, and so the momentum could be measured with arbitrary accuracy even in a short time. If, however, we take into account the fact that the velocity cannot exceed c, then $v - v'$ can be at most of the order of c, so that equation (1) gives

$$\Delta P\,\Delta t > \frac{\hbar}{c}. \tag{2}$$

The inequality (2) is particularly easy to derive for the state *after* the measurements. If we assume that the particle had a definite position before the measurement, then after a time Δt, on account of the finite velocity limit, the position is still known with accuracy $c\,\Delta t$. If the momentum after this time were determined more accurately than as given by (2), this would contradict the result $\Delta P\,\Delta q > \hbar$.

† Here an important point is that not every Hamilton's function can actually occur in nature; as already mentioned, the interaction function is always a function of the co-ordinates and so does not commute with the momentum. If the form of the Hamilton's function could be chosen arbitrarily, the momentum could be measured in an arbitrarily short time without change of velocity; this is a trivial deduction from the fact that co-ordinates and momenta are then equivalent.

On account of (2) the concept of momentum has a precise significance only over long times. Thus, in cases where the momentum changes appreciably within such times, the use of the concept of momentum is purposeless.

In the measurement of momentum of a charged body, in addition to the above-mentioned inaccuracy, a further perturbation of the measurement arises because the body will emit radiation in the necessary change of velocity. We shall consider only the case where the velocity of the body before the measurement is certainly small compared with c. In this case it is favourable to conduct the measurement so that after the measurement the velocity is again considerably less than c. For, if the velocity approaches c, the relation (1) gives very little benefit, while the accuracy is greatly reduced by the emission of radiation. Thus the non-relativistic formula for radiation damping can be used. The energy emitted is then

$$\frac{e^2}{c^3}\int \dot{v}^2\,\mathrm{d}t,$$

where e is the charge on the body. This energy evidently has its least value for uniform acceleration, i.e. for $\dot{v} = (v' - v)/\Delta t$, so that the energy emitted is

$$\frac{e^2}{c^3}\frac{(v' - v)^2}{\Delta t}.$$

This unknown change of energy has to be taken into account in the energy balance, and there thus arises in the momentum a further inaccuracy:

$$(v' - v)\,\Delta P > \frac{e^2}{c^3}\frac{(v' - v)^2}{\Delta t},$$

or

$$\Delta P\,\Delta t > \frac{e^2}{c^3}(v' - v). \tag{3}$$

For electrons this inequality gives no new information, since even in the most unfavourable case where $v \sim v' + c$ it gives only $\Delta P\,\Delta t > e^2/c^2$, and this is weaker than (2), since $e^2 < \hbar c$. For macroscopic bodies, however, the relation (3) is significant. Multiplication by (1) gives

$$\Delta P\,\Delta t > \frac{\hbar}{c}\sqrt{\frac{e^2}{\hbar c}}, \tag{4}$$

and in this form we shall make use of it later. The inequality (4) is, of course, valid independently of the method of measurement used, and in particular when the measurement is made by means of the charge on the body, as in the case of the Compton effect, where, in addition to the Compton scattered radiation used in the measurement, there is a further radiation corresponding to that discussed above, obtained when higher approximations are taken into account in the perturbation calculation for the interaction between the radiation and the particle[7]. (In the ordinary Compton effect with electrons this effect is of no importance, on account of the smallness of $e^2/\hbar c$.)

4. Field Measurement

The simplest method of measuring an electric field is to observe the acceleration of a charged test body. In order to avoid interference by magnetic fields, we use a body of very large mass and very small velocity. Let the momentum of the body before the measurement be known, and let the momentum afterwards be measured, again with accuracy ΔP. From this we car deduce the electric field strength with accuracy such that

$$e\,\Delta\,\mathscr{E}\,\Delta t > \Delta P. \tag{5}$$

In addition, however, the condition (4) must be satisfied in the momentum measurement. Multiplication of equations (4) and (5) gives

$$\Delta\,\mathscr{E} > \frac{\sqrt{\hbar\,c}}{(c\,\Delta t)^2}. \tag{6}$$

For the magnetic field strength we easily obtain the same result by considering the motion of a magnetic needle:

$$\Delta\,\mathscr{H} > \frac{\sqrt{\hbar c}}{(c\,\Delta t)^2}. \tag{6a}$$

If it is desired to measure the electric and magnetic field strengths simultaneously, then, in addition to the effects already discussed, we have to take into account the effect on the needle of the magnetic field due to the charged body and vice versa. This magnetic field is, in order of magnitude,

$$\Delta\,\mathscr{H} > \frac{e}{(\Delta l)^2}\cdot\frac{v'}{c}, \tag{7}$$

where Δe is the distance between the test body and the needle. If we multiply this inequality by equations (5) and (1), then (with $v = 0$) we have

$$\Delta\,\mathscr{E}\,\Delta\,\mathscr{H} > \frac{\hbar\,c}{(c\,\Delta t)^2}\cdot\frac{1}{(\Delta e)^2}. \tag{6b}$$

This condition differs from the product of (6) and (6a) in that $c\,\Delta t$ in the denominator is partly replaced by Δe.

If follows from (6), (6a) and (6b) that for $\Delta t = \infty$ the measurement can be made arbitrarily accurate for both $\mathscr{E}$ and $\mathscr{H}$. Thus static fields can be completely defined in the classical sense.†

In wave fields (that is, field which are further than $c/v = \lambda$ from the bodies which produce them), it is sufficient to use (6) and (6a), because as a result of the coupling of the space and time variation nothing is discovered about the

† Our thanks are due to Professor Bohr for pointing out this situation and the significance of time in general.

field if the region of measurement has an extent less than $c\,\Delta t$ for a given Δt. Thus here also the measurements of $\mathscr{E}$ and $\mathscr{H}$ do not interfere, and to the extent that the field strengths can be measured in accordance with (6) and (6a) they can be measured simultaneously. Thus the field strengths are in accordance with the classical theory inasfar as they can be defined at all. In the quantum range, on the other hand, the field strengths are not measurable quantities.†

5. Measurements on Light Quanta

We shall now show that in a radiation field no measurements can be carried out with certainty within a short time, i. e. measurements for which every possible result gives information about the state of the system. (Thus we do not consider such measurements as, for example, a measurement of position by means of a collision which does not occur with probability unity within the period of observation, so that, although a deflection of the test body shows that the body under measurement was at the point considered, the absence of such a deflection shows nothing.) The time necessary for the measurement depends on the state of the system. If the energy of the radiation field is approximately determined as E, we shall show that this time is greater than $\hbar/E$. Since the field consists of light quanta, the greatest frequency occurring in the Fourier resolution of the field can be at most $E/\hbar$; if we carry out the measurements in times small compared with $\hbar/E$, we remain within the period of oscillation, and so the field strength may be regarded as constant during the measurements. All measurements in such short times are therefore field measurements and are subject to the inaccuracy (6). Thus, in order than an effect should be detectable, the field strength must considerably exceed $\sqrt{\hbar\, c}/(c\,\Delta t)^2$. The smallest wavelength occurring, on the other hand, is $\hbar\, c/E$, and so the field strength, if non-zero at any point, must be non-zero in a region of at least this extent. Consequently, the total field energy must be at least of the order $E > \mathscr{E}^2(\hbar\, c/E)^3 > (\hbar\, c)^4/E^3(c\Delta t)^4$, i.e.

$$\Delta t > \frac{\hbar}{E}, \tag{8}$$

in contradiction with our hypothesis. Thus measurements which do not satisfy equation (8) are impossible.

This result applies in particular, of course, when the radiation field consists of a single quantum of light. Within the time given by (8), a quantum of light is therefore undetectable, and in particular its position cannot be determined with any accuracy. In a measurement of position, the time to which that position refers is therefore indeterminate by more than $\hbar/E$. If the measurement of position is to be used to investigate a state, as discussed in section 2, we are interested in the position at a time up to which the state under investigation

† The inaccuracy for the field measurement with an electron found by Jordan and Fock[8] is greater than (6) and therefore proves only that the electron is not a suitable means of measuring the field.

(i.e. the state whose energy was of the order of E) existed. The measurement of position indicates such a quantity with inaccuracy, at best $\Delta q > \hbar c/E$.

It might be thought that the accuracy could be improved by measuring momentum at the same time as position (within the limits given by $\Delta P \Delta q > \hbar$, of course), and seeking to deduce how far and in which direction the light quantum has meanwhile travelled. A closer examination, however, shows that the resulting accuracy can be no better than $\hbar c/E$. Thus in every state it is meaningful to give the probability of presence of the light quantum only for regions large compared with the wavelength, and the position of the light quantum is a meaningful concept only in geometrical optics.

If the number of light quanta varies appreciably within the period of oscillation, the concept of light quanta itself is meaningless.

6. Measurements on Material Particles

Let us now investigate the corresponding relations for material particles. (We shall always speak of electrons, but the arguments of course apply for any kind of material particle.) Such particles can best be detected by means of collision processes, for example using the Compton effect. Thus the presence of an electron is demonstrated by making two momentum measurements on the light quantum and deducing from the change in momentum that a collision has occurred in between. Here, however, the course of the process depends considerably on the length of the time interval between the two measurements. Over long times the Compton effect is obtained, i.e. the momentum of the light quantum changes either not at all or by an amount determined by the initial momenta, which can be made arbitrarily large by using very hard radiation. Over very short times, however, any changes of momentum may occur, provided that the law of conservation of momentum remains valid; the sum of the energies of the electron and the light quantum need be conserved only to within $\hbar/\Delta t$, as shown in section 2. For the same reasons as in measurements on light quanta, the small momentum changes have much the greatest probability. An elementary calculation shows that the second behaviour begins when the time interval is no longer large compared with $\hbar/E$, where E is the approximate energy of the electron before the measurement.

Thus, if the duration of the process of measurement is made shorter than $\hbar/E$, the momentum of the light quantum (and therefore also that of the electron) changes by an arbitrary amount. Hence, from the fact that no measurable change of momentum has occurred, we cannot conclude that no collision has taken place. Physically this signifies that the measurement of the momentum of the light quantum destroys the initial state of the electron. We cannot ensure that the electron is found with probability unity at the first measurement: if it was in a volume δq before the measurement, a time $\delta q/c$ is necessary before we can be sure that the light has reached the electron. Since $\delta q/c > \hbar/c\,\delta P > \hbar/c\,P > \hbar/E$, we should therefore have to make several measurements before being able to detect the electron, and thus completely destroy its state before we find it. Measurements in times less than $\hbar/E$ are therefore useless.

We now again ask how accurately this measurement can be used to derive the position of the electron at a time up to which it was in its former state. To do this, of course, only the knowledge about the velocity which is compatible with the position measurement can be used, and not, for example, the velocity in the state before the measurement. If an exact measurement of position is made, no information is obtained as regards the velocity, which thus remains indeterminate within c. The co-ordinate can therefore be derived only with an error $\Delta q > \hbar c/E$. Elementary considerations show that no higher accuracy can be achieved by measuring the momentum and co-ordinate simultaneously with any accuracies compatible with $\Delta P \Delta q > \hbar$. Thus

$$\Delta q > \frac{\hbar c}{E} \tag{9}$$

represents the limit of the accuracy with which the position of the electron can reasonably be defined. If in particular, if the velocity of the electron is not very close to c, this becomes

$$\Delta q > \frac{\hbar}{m c}. \tag{10a}$$

The derivation of (10a) shows that it is valid only for electrons which are not moving too rapidly. The statement frequently found in the literature that $\hbar/m c$ is a general limit for the accuracy of measurements of position is based on incorrect arguments.

A superficial consideration might suggest that the uncertainty relations derived above are not relativistically invariant. In reality, of course, there can be no contradiction with relativity, which has been taken into account throughout the derivations. The explanation is that the inequalities themselves need not transform in a relativistically invariant manner, since the most favourable possible measurements of a quantity need not be so when they are viewed from a moving system of co-ordinates. Thus we have only to require that the limit of accuracy should not be exceeded when such a measurement is viewed from a moving system of co-ordinates. This requirement is, of course, always satisfied.

Particular care is needed in this respect with position measurements, for here the statement of the problem is itself not relativistically invariant, but distinguishes a time axis, since we ask for the co-ordinate at a *time* up to which the unperturbed state existed.

7. Mathematical Failure of the Methods of Wave Mechanics

The above-stated unmeasurability of all wave-mechanical quantities also appears, of course, in the formalism which results when we attempt to apply the methods of wave mechanics to the relativistic case.

The most fundamental quantity in the theory, both for electrons and for light quanta, is the momentum; this is of course due to the fact that if it remains constant in time it can be defined with arbitrary accuracy, although very long times are needed for its measurement. This latter fact does not, of course, appear in the wave-mechanics formalism, and in consequence the statements of the theory regarding short times have no meaning.

The unmeasurability of the position, on the other hand, is directly expressed in the formalism. For electrons this is because the Dirac equation also allows the physically meaningless solutions with negative energy. The result of a measurement can, of course, in reality only be a wave function composed only of states with positive energy. Such states cannot, however, form an arbitrary wave packet. It is easily seen that the dimensions of a wave packet in general cannot be less than $\hbar/m\,c$. There are, it is true, special wave packets of smaller size (namely those whose centre moves at almost the speed of light†), but the corresponding wave functions do not form a complete system, and the state before the measurement cannot in general be expanded in terms of them. This corresponds to the result shown earlier that in short times a determination of position may sometimes chance to be possible, but a measurement cannot be carried out with certainty.

The conditions for light quanta are still more extreme in that no mathematical expression can be given for the probability density. This is seen from the fact that, on account of polarisation properties, the wave function for a light quantum must be a tensor of rank two[4]. The probability density and current must form a four-vector, which is impossible, because they depend quadratically on the wave function. In geometrical optics it is, of course, possible to construct wave packets in which all effects vanish outside a certain region. But here also these wave functions do not form a complete system.

The unmeasurability of the field strength is shown by the fact that in empty space (no light quanta) the field strength operator[2,4] is not zero, but even the expectation of the square of the field strength is infinite. This is related to the fact that for $\Delta t = 0$ we have from (6) an infinite indeterminacy of the field strength.

8. Conclusions

We have seen that no predictable measurements can exist for the fundamental quantities of wave mechanics (except when these quantities are constant in time, and then an infinitely long time is needed for an exactly predictable measurement). It cannot, of course, be formally demonstrated that there are not in nature some particularly complicated quantities for which predictable measurements are possible, but such a speculation need not be discussed. The assumptions of wave mechanics which have been shown to be necessary in section 2 are therefore not fulfilled in the relativistic range, and the application of wave mechanics methods to this range goes beyond their scope. It is

† Our thanks are due to Professor O. Klein for pointing this out.

therefore not surprising that the formalism leads to various infinities; it would be surprising if the formalism bore any resemblance to reality.

The applicability of wave mechanics is restricted to processes where the state of the system varies sufficiently slowly. In cases where the ordinary Schrödinger's equation is applicable this is of course not always true. For radiation alone wave mechanics is never meaningful, since the limit $c = \infty$ has no meaning.

In the correct relativistic quantum theory (which does not yet exist), there will therefore be no physical quantities and no measurements in the sense of wave mechanics. One can, of course, cause the system to interact with some apparatus and ask what happens to the latter. The theory will give a probability for the result of this experiment, but this cannot be interpreted as the probability of a parameter of the system under investigation, since it can in no way be ensured that the probability of a given result is unity and that of all other results is zero. In addition, it is in principle impossible to make the duration of such an experiment arbitrarily short.

This view is confirmed by the known fact that the β-spectra of radioactive nuclei are continuous, although the uniform lifetime indicates that the nuclei are not in different states. For, if all the β-particles had the same energy, the process could be regarded as a predictable measurement.

This fact presents an insuperable difficulty in wave mechanics because, as Bohr has emphasised, it means that the law of conservation of energy is probably invalid for nuclear electrons. This law is indissolubly connected with the foundations of wave mechanics. In relativistic quantum theory, however, the preceding discussion shows that the concept of energy need not be mechanically definable. It is of course definable in a certain sense in terms of the total mass of the nucleus, because the nucleus in its motion as a whole satisfies wave mechanics, but this does not imply a predictable measurement of quantities related to the internal state of the nucleus.

If the law of conservation of energy is not valid, then in radioactive processes the mass of the whole system will of course change, but this change cannot be followed in the course of time, since the mass cannot be measured in an arbitrarily short time. If we consider the process of measurement of the mass as in section 3, the time needed for the measurement is such that

$$\Delta m \, \Delta t > \frac{\hbar}{c^2}$$

The preceding discussion is not contradicted by the fact that the spectra of protons and α-particles are discrete. On account of their large mass (low velocity) these particles obey wave mechanics even in the nucleus, rather as the nuclei in a molecule can be essentially described in classical terms despite their strong interaction with the electrons, for which classical mechanics fails completely.

One of us (Landau) thanks the Rockefeller Foundation for making it possible for him to work in Copenhagen and in Zürich.

References

1. P. A. M. Dirac. *Proc. Roy. Soc.* **A 117**, 610 (1928).
2. W. Heisenberg and W. Pauli, *Z. Phys.* **56**, 1 (1929); **59**, 168 (1930).
3. P. Jordan and W. Pauli, *Z. Phys.* **47**, 151 (1928).
4. L. Landau and R. Peierls, *Z. Phys.* **62**, 188 (1930); Collected Papers No 3, p. 19.
5. L. Rosenfeld, *Z. Phys.* **65**, 589 (1930).
6. N. Bohr, *Naturwiss.* **16**, 245 (1928).
7. W. Pauli, *Z. Phys.* **18**, 272 (1923).
8. P. Jordan and V. Fock, *Z. Phys.* **66**, 206 (1930).

IV.2 THE MEASURABILITY OF THE ELECTROMAGNETIC FIELD

Commentary of Rosenfeld (1955)

When I arrived at the Institute on the last day of February, 1931, for my annual stay, the first person I saw was Gamow. As I asked him about the news, he replied in his own picturesque way by showing me a neat pen drawing he had just made.* It represented Landau, tightly bound to a chair and gagged, while Bohr, standing before him with upraised forefinger, was saying: "Bitte, bitte, Landau, muss ich† nur ein Wort sagen!" I learned that Landau and Peierls had just come a few days before with some new paper of theirs which they wanted to show Bohr, "but" (Gamow added airily) "he does not seem to agree—and this is the kind of discussion which has been going on all the time." Peierls had left the day before, "in a state of complete exhaustion," Gamow said. Landau stayed for a few weeks longer, and I had the opportunity of ascertaining that Gamow's representation of the situation was only exaggerated to the extent usually conceded to artistic fantasy.

There was indeed reason for excitement, for the point raised by Landau and Peierls was a very fundamental one. They questioned the logical consistency of quantum electrodynamics by contending that the very concept of electromagnetic field is not susceptible, in quantum theory, to any physical determination by means of measurements. The measurement of a field component requires determinations of the momentum of a charged test-body; and the reaction from the field radiated by the test-body in the course of these operations would (except in trivial cases) lead to a limitation of the accuracy of the field measurement, entirely at variance with the premises of the theory. In fact, the quantization of the field only entails reciprocal limitations of the measurements of pairs of components, arising from their non-commutability, but no limitation whatsoever to the definition of any single field component. On the other hand, one had to face another inescapable consequence of the field quantization, the occurrence of irregular fluctuations in the value of any field component; the existence of this fluctuating "zero-field" (as it was called because it persists even in a vacuum) was known to be responsible for one of the divergent contributions to the self-energy of charged particles, but its meaning was very obscure. Landau and Peierls, somewhat illogically, tried to bring it in relation with their alleged limitation of measurability of the field, and this only further confused an already tangled issue . . . .

. . . Bohr's state of mind when he attacked the problem reminded me of an anecdote about Pasteur. When the latter set about investigating the silk-

* I am afraid this work of art has been allowed to disintegrate before its historical value could be realized.

† This is a familiar danicism of Bohr's for "darf ich."

worm sickness, he went to Avignon to consult Fabre. "I should like to see cocoons," he said, "I have never seen any, I know them only by name." Fabre gave him a handful: "he took one, turned it between his fingers, examined it curiously as we would some singular object brought from the other end of the world. He shook it near his ear. 'It rattles,' he said, much surprised, 'there is something inside.'" My first task was to lecture Bohr on the fundamentals of field quantization; the mathematical structure of the commutation relations and the underlying physical assumptions of the theory were subjected to unrelenting scrutiny. After a very short time, needless to say, the roles were inverted and he was pointing out to me essential features to which nobody had as yet paid sufficient attention.

His first remark, which threw decisive light on the problem, was that field components taken at definite space-time points are used in the formalism as idealizations without immediate physical meaning; the only meaningful statements of the theory concern averages of such field components over finite space-time regions. This meant that in studying the measurability of field components we must use as test-bodies finite distributions of charge and current, and not point charges as had been loosely done so far. The consideration of finite test-bodies immediately disposed of Landau and Peierls' argument concerning the perturbation of the momentum measurements by the radiation reaction: it is easily seen that this reaction is so much reduced, for finite test-bodies, as to be always negligible.

On the other hand, the construction and manipulation of extended test-bodies proved a most perplexing affair. . . . in fact, it seemed as if we had hit upon a mode of measurement exactly suited to give the best combined accuracy, for pairs of field determinations, compatible with the theoretical limitation. We little imagined that it would still take us almost two years to reach that goal . . . .

. . . At the end of our laborious inquiry, we had completely vindicated the consistency of quantum electrodynamics, at least in its simplest form. Our increased insight invited a re-assessment of the scope of the analysis we had just completed. We had set out on the suspicion of grave defects in the logical structure of the theory, and used the direct method of testing definitions of concepts by investigating the concrete measuring processes they embody.

IV.2 ON THE QUESTION OF THE MEASURABILITY OF ELECTROMAGNETIC FIELD QUANTITIES*

NIELS BOHR AND LÉON ROSENFELD

1. INTRODUCTION

The question of limitations on the measurability of electromagnetic field quantities, rooted in the quantum of action, has acquired particular interest in the course of the discussion of the still unsolved difficulties in relativistic atomic mechanics. On the basis of exploratory considerations, Heisenberg[1] attempted to demonstrate that the connection between the limitation on the measurability of field quantities and the quantum theory of fields is similar to the relationship between the complementary limitations on the measurability of kinematical and dynamical quantities expressed in the indeterminacy principle and the non-relativistic formalism of quantum mechanics. However, in the course of a critical investigation of the foundations of the relativistic generalization of this formalism, Landau and Peierls[2] came to the conclusion that the measurability of field quantities is subjected to further restrictions which go essentially beyond the presuppositions of quantum field theory, and which therefore deprive this theory of any physical basis.

At first glance, one might see in this contradiction a serious dilemma. On the one hand, the quantum theory of fields surely ought to be considered a consistent re-interpretation of classical electromagnetic theory in the sense of the correspondence argument, just as quantum mechanics represents a reformulation of classical mechanics adapted to the existence of the quantum of action. On the other hand, quantum electrodynamics has essentially increased the difficulties, already encountered in classical electron theory, of a harmonious blending of field theory and atomic theory. However, closer consideration shows that the various problems involved here can be to a large extent separated from each other, since the quantum-electromagnetic formalism in itself is independent of all ideas concerning the atomic constitution of matter. This is evident from the fact that in addition to the velocity of light only the quantum of action

Originally published under the title, "Zur Frage der Messbarkeit der elektromagnetischen Feldgrössen," *Mat.-fys. Medd. Dan. Vid. Selsk., 12*, no. 8 (1933); this translation into English by Aage Petersen is taken from *Selected Papers of Léon Rosenfeld*, Cohen and Stachel, eds. (1979), Reidel, Dordrecht, pp. 357-400.

enters the formalism as a universal constant; for these two constants obviously do not suffice to determine any specific space-time dimensions. In the quantum theory of atomic structure such a determination will be obtained only by the inclusion of the electric charge and the rest masses of the elementary particles.

Just the insufficient distinction between field theory and atomic theory is the principal reason for the conflicting results of previous investigations of the measurability of field quantities in which single electrically charged mass points were used as test bodies. The utilization of classical electron theory, in the sense of the correspondence argument, which underlies current atomic mechanics, rests above all on the smallness of the elementary charge in comparison with the square root of the product of the quantum of action and the velocity of light, which makes it possible to treat all radiation reactions as small compared to the ponderomotive forces exerted on the particles. However, in measurements of field quantities it turns out to be essential to be able to adjust the charge of the test bodies to an extent which would be in conflict with the latter presupposition if one were to consider these bodies as point charges. As we shall see, however, these difficulties disappear if one uses test bodies whose linear extensions are chosen sufficiently large, compared to atomic dimensions, that their charge density can be considered approximately constant over the whole body.

In this connection, it is also of essential importance that the customary description of an electric field in terms of the field components at each space-time point, which characterizes classical field theory and according to which the field should be measurable by means of point charges in the sense of the electron theory, is an idealization which has only a restricted applicability in quantum theory. This circumstance finds its proper expression in the quantum-electromagnetic formalism, in which the field quantities are no longer represented by true point functions but by functions of space-time regions, which formally correspond to the average values of the idealized field components over the regions in question. The formalism only allows the derivation of unambiguous predictions about the measurability of such region-functions, and our task will thus consist in investigating whether the complementary limitations on the measurability of field quantities, defined in this way, are in accordance with the physical possibilities of measurement.

Insofar as we can disregard all restrictions arising from the atomistic structure of the measuring instruments, it is actually possible to demonstrate a complete accord in this respect. Besides a thorough investigation of the construction and handling of the test bodies, this demonstration requires, however, consideration of certain new features of the complementary mode of description, which come to light in the discussion of the measurability question, but which were not included in the customary formulation of the indeterminacy principle in connection with non-relativistic quantum mechanics. Not only is it an essential complication of the problem of field measurements that, when comparing field averages over different space-time regions, we cannot in an unambiguous way speak about a temporal sequence of the measurement processes; but even the interpretation of individual measurement results requires a still greater caution in the case of field measurements than in the usual quantum-mechanical measurement problem.

Characteristic of the latter problem is the possibility of attributing to each individual measurement result a well-defined meaning in the sense of classical mechanics, while the quantum-imposed interaction, uncontrollable in principle, between instrument and object is fully taken into account through the influence of each measuring process on the statistical expectations testable in succeeding measurements. In contrast, in measurements of field quantities, indeed, every measuring result is well defined on the basis of the classical field concept; but, the limited applicability of classical field theory to the description of the unavoidable electromagnetic field effects of the test bodies during the measurement implies, as we shall see, that these field effects to a certain degree influence the measurement result itself in a way which cannot be compensated for. However, a closer investigation of the fundamentally statistical character of the consequences of the quantum-electromagnetic formalism shows that this influence on the object of measurement by the measuring process in no way impairs the possibility of testing such consequences, but rather is to be regarded as an essential feature of the intimate adaptation of quantum field theory to the measurability problem.

Before we turn to a detailed exposition of the considerations indicated above, we want to stress once more that the fundamental difficulties which confront the consistent utilization of field theory in atomic theory remain entirely untouched by the present investigation. Indeed, consideration of the atomistic constitution of all measuring instruments would be essential for an assessment of the connection between these difficulties and the well-known paradoxes of the measurement problem in relativistic quantum mechanics. Also particularly relevant in this context is the limitation which the finite value of the elementary charge compared to the square root of the product of the velocity of light and the quantum of action places on atomic mechanics based on the correspondence argument.[3]

2. Measurability of Fields According to Quantum Theory

The quantum electromagnetic formalism found its starting point in the quantum theory of radiation developed by Dirac, which is characterized by the introduction of a non-commutativity, consistent with the quantum-mechanical commutation relations, of canonically conjugate amplitudes of vibration of a radiation field. On the basis of this theory, Jordan and Pauli set up commutation relations between the electromagnetic field components for the case of charge-free fields, and the formalism was then brought to a certain completion by Heisenberg and Pauli who treated the interaction between field and material charges on correspondence lines. However, the consistent application of the theory to atomic problems is essentially impaired by the occurrence of the well-known paradoxes of the infinite self-energy of the elementary particles, which were not removed by Dirac's proposed modifications in the representation of the formalism.[4] Yet, in our discussion of the limitations of the measurability of field quantities these difficulties play no role, since for this purpose the atomistic structure of matter is not an essential issue. It is true that the measurement of fields requires the use of material charged test bodies, but their unambiguous application as measuring instruments depends exactly on the extent to which we can treat their response to the fields as well as their influence as field sources on the basis of classical electrodynamics.

In these circumstances we may restrict ourselves to the pure field theory, and thus take the commutation relations for charge-free fields as our starting point for the investigation of the consequences of the quantum electromagnetic formalism with respect to the measurability of field quantities. Using the usual notation, $[p, q]=pq-qp$, we thus have the following typical relations between the field components in two space-time points (x_1, y_1, z_1, t_1) and (x_2, y_2, z_2, t_2), from which the remaining commutation relations are obtained by cyclic permutation:[5]

$$(1)\quad \begin{cases} [\mathfrak{E}_x^{(1)}, \mathfrak{E}_x^{(2)}] = [\mathfrak{H}_x^{(1)}, \mathfrak{H}_x^{(2)}] = \sqrt{-1}\,\hbar(A_{xx}^{(12)} - A_{xx}^{(21)}) \\ [\mathfrak{E}_x^{(1)}, \mathfrak{E}_y^{(2)}] = [\mathfrak{H}_x^{(1)}, \mathfrak{H}_y^{(2)}] = \sqrt{-1}\,\hbar(A_{xy}^{(12)} - A_{xy}^{(21)}) \\ [\mathfrak{E}_x^{(1)}, \mathfrak{H}_x^{(2)}] = 0 \\ [\mathfrak{E}_x^{(1)}, \mathfrak{H}_y^{(2)}] = -[\mathfrak{H}_x^{(1)}, \mathfrak{E}_y^{(2)}] = \sqrt{-1}\,\hbar(B_{xy}^{(12)} - B_{xy}^{(21)}). \end{cases}$$

Here, $\mathfrak{E}_x^{(1)}, \mathfrak{E}_y^{(1)}, \mathfrak{E}_z^{(1)}, \mathfrak{H}_x^{(1)}, \mathfrak{H}_y^{(1)}, \mathfrak{H}_z^{(1)}$ are the electric and magnetic field components at the space-time point (x_1, y_1, z_1, t_1) while the following abbreviations have been used:

$$(2)\quad \begin{cases} A_{xx}^{(12)} = -\left(\dfrac{\partial^2}{\partial x_1\,\partial x_2} - \dfrac{1}{c^2}\dfrac{\partial^2}{\partial t_1\,\partial t_2}\right)\left\{\dfrac{1}{r}\,\delta\left(t_2 - t_1 - \dfrac{r}{c}\right)\right\} \\ A_{xy}^{(12)} = -\dfrac{\partial^2}{\partial x_1\,\partial y_2}\left\{\dfrac{1}{r}\,\delta\left(t_2 - t_1 - \dfrac{r}{c}\right)\right\} \\ B_{xy}^{(12)} = -\dfrac{1}{c}\dfrac{\partial}{\partial t_1\,\partial z_2}\left\{\dfrac{1}{r}\,\delta\left(t_2 - t_1 - \dfrac{r}{c}\right)\right\} \end{cases}$$

Further, $\hbar$ is Planck's constant divided by 2π, c the velocity of light, and r the spatial separation between the two points. Finally, δ denotes the symbolic function introduced by Dirac which, as is well known, is characterized by the property

$$(3)\quad \int_{t'}^{t''} \delta(t - t_0)\,dt = \begin{cases} 1 \text{ for } t' < t_0 < t'' \\ 0 \text{ for } t_0 < t' \text{ or } t_0 > t'' \end{cases}$$

and which is formally differentiated like an ordinary function.

The occurrence of the δ-function, defined in (3), in the commutation relations (1) brings to the fore the fact mentioned above that the quantum theoretical field quantities are not to be considered as true point functions but that unambiguous meaning can be attached only to space-time integrals of the field components. With a view to the simplest possibility of testing the formalism we shall consider only averages of field components over simply connected space-time regions whose spatial extension remains constant during a given time interval. Thus, for example, if the volume of such a region G is denoted by V and the corresponding time interval is T, we define the average value of $\mathfrak{E}_x$ by the formula

$$\mathfrak{E}_x^{(G)} = \frac{1}{VT}\int_T dt \int_V \mathfrak{E}_x \, dv. \tag{4}$$

For the average values of two field components over two given space-time regions I and II there exist commutation relations which follow immediately from (1) by integration over the two regions and division by the product of their four-dimensional extensions. Thus, the value of the bracket symbols $[\mathfrak{E}_x^{(\mathrm{I})}, \mathfrak{E}_x^{(\mathrm{II})}], \ldots$ are obtained from (1) simply by replacing the quantities $A^{(12)}$, $B^{(12)}$ by their average values over the two regions

$$\left\{\begin{aligned}
\bar{A}_{xx}^{(\mathrm{I,\,II})} &= -\frac{1}{V_{\mathrm{I}}V_{\mathrm{II}}T_{\mathrm{I}}T_{\mathrm{II}}}\int_{T_{\mathrm{I}}} dt_1 \int_{T_{\mathrm{II}}} dt_2 \int_{V_{\mathrm{I}}} dv_1 \int_{V_{\mathrm{II}}} dv_2 \\
&\qquad \left(\frac{\partial^2}{\partial x_1\,\partial x_2} - \frac{1}{c^2}\frac{\partial^2}{\partial t_1\,\partial t_2}\right)\left\{\frac{1}{r}\,\delta\left(t_2 - t_1 - \frac{r}{c}\right)\right\} \\
\bar{A}_{xy}^{(\mathrm{I,\,II})} &= -\frac{1}{V_{\mathrm{I}}V_{\mathrm{II}}T_{\mathrm{I}}T_{\mathrm{II}}}\int_{T_{\mathrm{I}}} dt_1 \int_{T_{\mathrm{II}}} dt_2 \int_{V_{\mathrm{I}}} dv_1 \int_{V_{\mathrm{II}}} dv_2 \\
&\qquad \frac{\partial^2}{\partial x_1\,\partial y_2}\left\{\frac{1}{r}\,\delta\left(t_2 - t_1 - \frac{r}{c}\right)\right\} \\
\bar{B}_{xy}^{(\mathrm{I,\,II})} &= -\frac{1}{V_{\mathrm{I}}V_{\mathrm{II}}T_{\mathrm{I}}T_{\mathrm{II}}}\int_{T_{\mathrm{I}}} dt_1 \int_{T_{\mathrm{II}}} dt_2 \int_{V_{\mathrm{I}}} dv_1 \int_{V_{\mathrm{II}}} dv_2 \\
&\qquad \frac{1}{c}\frac{\partial^2}{\partial t_1\,\partial z_2}\left\{\frac{1}{r}\,\delta\left(t_2 - t_1 - \frac{r}{c}\right)\right\}
\end{aligned}\right. \tag{5}$$

In exactly the same way as the Heisenberg relation for two canonically conjugate mechanical quantities

$$\Delta p\,\Delta q \sim \hbar, \tag{6}$$

which is the basis for the uncertainty principle, can be derived from the general quantum mechanical commutation relation

$$[q, p] = \sqrt{-1}\,\hbar, \tag{7}$$

one obtains for the products of the complementary uncertainties of the field averages in question the following typical formulae:

$$\Delta\bar{\mathfrak{E}}_x^{(\mathrm{I})}\,\Delta\bar{\mathfrak{E}}_x^{(\mathrm{II})}\sim\hbar|\bar{A}_{xx}^{(\mathrm{I,\,II})}-\bar{A}_{xx}^{(\mathrm{II,\,I})}|,\quad \Delta\bar{\mathfrak{E}}_x^{(\mathrm{I})}\,\Delta\bar{\mathfrak{E}}_y^{(\mathrm{II})}\sim\hbar|\bar{A}_{xy}^{(\mathrm{I,\,II})}-\bar{A}_{xy}^{(\mathrm{II,\,I})}|$$

$$\Delta\bar{\mathfrak{E}}_x^{(\mathrm{I})}\,\Delta\bar{\mathfrak{H}}_x^{(\mathrm{II})}=0,\qquad \Delta\bar{\mathfrak{E}}_x^{(\mathrm{I})}\,\Delta\bar{\mathfrak{H}}_y^{(\mathrm{II})}\sim\hbar|\bar{B}_{xy}^{(\mathrm{I,\,II})}-\bar{B}_{xy}^{(\mathrm{II,\,I})}|. \tag{8}$$

Several results of importance for our problem follow immediately from the expressions (5) and (8). Above all, we see that, in accord with the property of the δ-function expressed in (3), the quantities $\bar{A}^{(\mathrm{I,\,II})}$, $\bar{B}^{(\mathrm{I,\,II})}$ change continuously as a result of a continuous displacement of the boundaries of regions I and II, as long as the extension of these regions, i.e. the values of V_{I}, T_{I}, V_{II}, T_{II}, remain different from zero. In particular, the differences $\bar{A}^{(\mathrm{I,\,II})}-\bar{A}^{(\mathrm{II,\,I})}$ and $\bar{B}^{(\mathrm{I,\,II})}-\bar{B}^{(\mathrm{II,\,I})}$ vanish without discontinuity when the boundaries of the two regions gradually are made to coincide. From this it follows that the averages of all field components over the same space-time region commute, and thus should be exactly measurable, independently of one another. In fact, this consequence of the theory, which goes essentially beyond the presupposition of unrestricted measurability of each single field quantity, appears as a special case of two more general theorems which follow from the symmetry properties of the quantities $\bar{A}^{(\mathrm{I,\,II})}$ and $\bar{B}^{(\mathrm{I,\,II})}$. For the fact that the expressions $A^{(12)}-A^{(21)}$ change their sign when the times t_1 and t_2 are interchanged, implies that the averages of two components of the same kind (i.e. two electric or two magnetic components) over two arbitrary space-time regions always commute if the associated time intervals coincide. Further, the corresponding antisymmetry of the expressions $B^{(12)}$ and $B^{(21)}$ with respect to interchange of the spatial points (x_1, y_1, z_1) and (x_2, y_2, z_2) implies that the averages of two components of different kind, e.g. $\mathfrak{E}_x$ and $\mathfrak{H}_y$, over two arbitrary time intervals commute when the corresponding spatial regions coincide.

At first sight these results might seem incompatible with the commutation relations between averages of field quantities at one and the same time and over finite space regions, which can be derived from the Heisenberg–Pauli representation of the formalism and are discussed in the book by Heisenberg previously cited. While it is stated there too that averages of components of the same kind commute, it is deduced that components of different kind over one and the same region of space do not commute. However, this contradiction is solved easily by noting that Heisenberg's treatment involves a limiting process in which two originally non-coinciding space-time regions are brought into coincidence only after their temporal extension has been contracted to one and the same instant of time. For from the symmetry of the expression (2) for $B_{xy}^{(12)}$ with respect to t_1 and t_2, and from the property of the δ-function stated in (3), we find in the case of coinciding time intervals

$$(9) \qquad \bar{B}_{xy}^{(\mathrm{I},\,\mathrm{II})} - \bar{B}_{xy}^{(\mathrm{II},\,\mathrm{I})} = \frac{2}{V_{\mathrm{I}} V_{\mathrm{II}} T^2} \cdot \frac{1}{c} \int\limits_{V_{\mathrm{I}}} \mathrm{d}v_1 \int\limits_{V_{\mathrm{II}}} \mathrm{d}v_2 \frac{\partial}{\partial z_1}\left(\frac{1}{r}\right),$$

where we have put $T_{\mathrm{I}} = T_{\mathrm{II}} = T$ and where the double integration is over all pairs of points, one in each of the spatial regions, whose separation r is smaller than cT. If we now further assume that the two spatial regions have the same volume, $V_{\mathrm{I}} = V_{\mathrm{II}} = V$, and the same shape, but are displaced relative to each other in the z-direction, then in the limit where cT can be considered negligible compared with the linear extensions of the space regions, the space integral in (9) gives by partial integration $\pm 2\pi c^2 T^2 F$, where F is the area enclosed by the projection on the xy-plane of the curve in which the surfaces of V_{I} and V_{II} intersect, and where the sign is + or − depending on whether region II is displaced relative to region I in the positive or the negative z-direction. Thus, if the regions are displaced continuously through each other the difference $\bar{B}_{xy}^{(\mathrm{I},\,\mathrm{II})} - \bar{B}_{xy}^{(\mathrm{II},\,\mathrm{I})}$ undergoes a discontinuous change of $8\pi c F/V^2$, while both expressions $\bar{B}_{xy}^{(\mathrm{I},\,\mathrm{II})}$ and $\bar{B}_{xy}^{(\mathrm{II},\,\mathrm{I})}$ change their sign. Therefore, the commutation relation between the instantaneous spatial averages of $\mathfrak{E}_x$ and $\mathfrak{H}_y$ in the limiting case under discussion displays an essential ambiguity, which is responsible for the apparent contradiction mentioned above.

Furthermore, as we shall see, the supposed demonstration, on the basis of previous investigations of the physical possibilities of measurement, that there is a complementary limitation on the measurability of field components of different kinds in one and same region of space, also depends entirely on the use of point charges as test bodies, which does not permit a sufficiently sharp delimitation of the region of measurement. As we have already emphasized, only measurements employing test bodies with a charge distribution of finite extension should be considered for the testing of the quantum-electromagnetic formalism, since every well-defined statement of this formalism refers to averages of field components over finite space-time regions. Yet this circumstance in no way prevents us from testing by field measurements all unambiguous conclusions that can be drawn from the Heisenberg–Pauli representation concerning the time dependence of spatial field averages. For this purpose we need only introduce averaging regions whose temporal extension T, multiplied by c, is sufficiently small compared to their linear spatial dimensions, the order of magnitude of which we shall in the following always denote by L.

In fact, the case $L > cT$ is particularly suited for a thorough testing of the consequences of the formalism in the properly quantum-theoretical domain. Of course, already in the domain of validity of classical theory the case $L \leqslant cT$ is of little interest, since all the peculiarities of the wave fields inside the volume V are smoothed out to a large extent by the averaging procedure, because of the propagation during the time interval T. In addition to this smoothing out, there are, in the quantum domain, the peculiar fluctuation phenomena which derive from the basically statistical character of the formalism. As we shall see shortly, these fluctuations, while essentially entering the solutions of problems in the case $L \leqslant ct$, play a comparatively small role in the case $L > cT$.

The fluctuations in question are intimately related to the impossibility, which is characteristic of the quantum theory of fields, of visualizing the concept of light quanta in terms of classical concepts. In particular, they give expression to the mutual exclusiveness of an accurate knowledge of the light quantum composition of an electromagnetic field and of knowledge of the average value of any of its components in a well-defined space-time region. Let the density $\omega_i(\kappa_x, \kappa_y, \kappa_z)$ of light quanta with definite polarization parameter i and given momentum and energy, $\hbar\kappa_x$, $\hbar\kappa_y$, $\hbar\kappa_z$ and $\hbar\nu = \hbar c \sqrt{\kappa_x^2 + \kappa_y^2 + \kappa_z^2}$ be known; then the expectation value of all field averages are indeed zero, but the mean square fluctuation of each field quantity, such as $\mathfrak{E}_x^{(G)}$ defined in (4), is given by the easily derivable formula

$$(10)\quad \left\{ \begin{aligned} & S(G) = \frac{1}{V^2 T^2} \cdot \frac{\hbar}{3} \int_T dt_1 \int_T dt_2 \int_V dv_1 \int_V dv_2 \frac{\partial^2}{\partial t_1 \, \partial t_2} \int_{-\infty}^{\infty} \Big(\sum_i \omega_i + 1\Big) \\ & \cos\left[\kappa_x(x_1 - x_2) + \kappa_y(y_1 - y_2) + \kappa_z(z_1 - z_2) - \nu(t_1 - t_2)\right] \\ & \frac{d\kappa_x \, d\kappa_y \, d\kappa_z}{\nu} \end{aligned} \right.$$

From formula (10) we see that for a given light quantum composition the fluctuations in question can never vanish, since even when $\omega_i = 0$, i.e., in the complete absence of light quanta, they assume a finite positive value, which by an easy calculation can be put in the form

$$(11) \qquad S_0(G) = \frac{2}{3\pi^2}\frac{\hbar c}{V^2}\int_V dv_1 \int_V dv_2 \frac{1}{r^2[(cT)^2 - r^2]}\,.$$

For every other light quantum distribution defined by a given density ω_i, the mean square fluctuation of the average of a field component becomes larger than $S_0(G)$. However, the fluctuations of a field average expected according to the formalism can be arbitrarily small when a direct knowledge of field quantities, obtained by measurement for example, is assumed. In such a case the light quantum density ω_i is obviously not well defined, and we must content ourselves with statistical statements about this density.

Furthermore, it is of decisive importance for the discussion of the measurement possibilities that the expression (11) holds not only for the field fluctuations in a region free of light quanta, but also represents the mean square fluctuation of each field average in the more general case where only classically describable current and charge distributions occur as sources of the field. The state of the field is then uniquely defined by the requirement that the expectation value of every field quantity coincide with the classically computed value, and that the numbers of light quanta of given momentum and polarization be distributed around their mean value n_0, estimated by means of the correspondence argument, according to the probability law valid for independent events

$$(12) \qquad w(n) = \frac{n_0^n\, e^{-n_0}}{n!}$$

An easy calculation shows that the field fluctuations in this state are given just by the expression (11). Moreover, in correspondence with the characteristics of black-body fluctuations, it follows that also in the general case of a field of given light quantum composition, the inclusion of field effects of any classically describable sources will have no influence on the fluctuation phenomena.

The square root of the expression (11) may be regarded as a critical field strength, $\mathfrak{S}$, in the sense that only when considering field averages essentially larger than $\mathfrak{S}$ are we allowed to neglect the corresponding fluctuations. To assess the possibilities of testing the formalism in the properly quantum domain, still another critical field size, $\mathfrak{U}$, is relevant, which equals the square root of the products, given by (8), of the complementary uncertainties of two field averages over space-time regions that only partially coincide, being displaced relative to each other by spatial and temporal distances of order of magnitude L and T, respectively. For when the field strengths are essentially larger than $\mathfrak{U}$ we obviously enter the domain of validity of classical electromagnetic theory, where all quantum mechanical features of the formalism lose their significance. A simple estimate based on formulae (8) and (11), shows that in the case $L \leqslant cT$ both critical expressions $\mathfrak{U}$ and $\mathfrak{S}$ are of the same order of magnitude

$$\mathfrak{U} \sim \mathfrak{S} \sim \frac{\sqrt{\hbar c}}{L \cdot cT} \tag{13}$$

On the other hand, in the case $L > cT$ one has

$$\mathfrak{U} \sim \sqrt{\frac{\hbar}{L^3 T}} \text{ and } \mathfrak{S} \sim \frac{\sqrt{\hbar c}}{L^2}, \tag{14}$$

so that in the limiting case $L \gg cT$ the critical field strength $\mathfrak{U}$ is much larger than $\mathfrak{S}$ and, therefore, in testing the characteristic consequences of the formalism we can to a large extent disregard the field fluctuations.

Before we turn to the comparison of the consequences of the quantum electromagnetic formalism discussed in this section, with the physical measurement possibilities for field quantities, we want to emphasize here once again that the consistent interpretability of this formalism is in no way endangered by such paradoxical features of its mathematical representation as the infinite zero-point energy. In particular, this latter paradox, which moreover can be removed by a formal change in the representation[6] that does not influence the physical interpretation, has no direct connection with the problem of measurability of field quantities. In fact, a field-theoretic determination of the electromagnetic energy in a given space-time domain would require knowledge of the values of the field components at each space-time point of a region, which are inaccessible to measurement. A physical measurement of the field energy can be carried out only by means of a suitable mechanical device that would make it possible to separate the electromagnetic fields in a given region from the rest of the field, so that the energy contained in the region could be measured subsequently by application of the conservation law. However, because of the interaction with the measuring mechanism, any such separation of the fields would be accompanied by an uncontrollable change in the field energy in the region in question, the consideration of which is essential for clarification of the well-known paradoxes that arise in the discussion of energy fluctuations in black-body radiation.[7]

3. Presuppositions for physical field measurements

The measurement of electromagnetic field quantities rests by definition on the transfer of momentum to suitable electric or magnetic test bodies situated in the field. Quite apart from the caution required by quantum theory in applying the customary idealization of field components defined in each space-time point, we are here always concerned with averages of these components over the finite time intervals necessary for the momentum transfer as well as over the spatial domains in which the electric charges or magnetic pole strengths of the test bodies in question are distributed. Obviously, even the assumption of a uniform charge distribution on a test body is an idealization, subject to a certain restriction because of the atomic constitution of all material bodies, but indispensable for the unambiguous definition of field quantities.

In order to have a definite case in mind we consider the measurement of the average of the electrical field component in the x-direction, $\mathfrak{E}_x$, over a space-time domain of volume V and duration T. For this purpose we therefore use a test body whose electric charge is uniformly distributed over the volume V with a density ρ, and determine the values p_x' and p_x'' of this body's momentum components in the x-direction at the beginning t' and at the end t'' of the interval T. The average $\bar{\mathfrak{E}}_x$ we are looking for is then determined by the equation

$$p_x'' - p_x' = \rho \bar{\mathfrak{E}}_x VT, \qquad (15)$$

where it is assumed that the time intervals required for the momentum measurements, whose order of magnitude we shall denote by Δt, can be regarded as negligibly small compared to T, and that we can disregard the displacements suffered by the test body due to the momentum measurements as well as the acceleration given to it during the time interval T by the field that is being measured, in comparison with the linear dimensions L of the spatial domain V.

By choosing a sufficiently heavy test body we can obviously make its acceleration due to the field arbitrarily small. In the momentum measurements, however, we encounter conditions which are independent of the mass of the test body. As a consequence of the indeterminacy principle, any measurement of the momentum component p_x carried out with the accuracy Δp_x is accompanied by a loss Δx in the knowledge of the position of the body in question, the order of magnitude of which is given by the relation contained in (6)

$$(16) \qquad \Delta p_x \, \Delta x \sim \hbar.$$

Nevertheless, in itself this state of affairs does not imply any restriction on the accuracy to be achieved by the field measurement, because we still have at our disposal the value of the charge density. In fact, if we neglect Δt and Δx in comparison with T and L, we get from (15) and (16) for the order of magnitude of the accuracy $\Delta\bar{\mathfrak{E}}_x$ of the field measurement

$$(17) \qquad \Delta\bar{\mathfrak{E}}_x \sim \frac{\hbar}{\rho \Delta x \cdot VT},$$

which for any value of Δx, however small, can be made arbitrarily small by choosing a sufficiently large value of ρ.

Strictly speaking, the accuracy of the field measurement is also dependent on the absolute magnitude of the value of $\bar{\mathfrak{E}}_x$ itself, for with given ranges of Δt and Δx the value of $\bar{\mathfrak{E}}_x$ ascertained from (15), even if Δp_x were zero, would be affected with an uncertainty arising from the latitude in the delineation of the domain of measurement which would surpass any limit as $\bar{\mathfrak{E}}_x$ increases indefinitely. Yet, the latter circumstance merely reflects the general limitation on all physical measurements, for which a knowledge of the order of magnitude of the expected effects is always required in order to choose the appropriate measuring instruments. In our problem an upper limit to the effects that we are interested in is set by the fact that as the magnitude of the field components increases we gradually reach the domain of validity of classical electromagnetic theory. As mentioned in the previous section, in the case $L > cT$, which is particularly suited for testing the quantum electromagnetic formalism, the expression

(18) $$Q=\sqrt{\frac{\hbar}{VT}},$$

which is equivalent to the right-hand side of the first formula (14), represents a critical field size in this respect. Substituting this into (17), the latter relation assumes the form

(19) $$\Delta\mathfrak{E}_x \sim \lambda Q,$$

where

(20) $$\lambda=\frac{Q}{\rho\Delta x}$$

is a dimensionless factor determining the accuracy of the field measurement.

The requirement that λ be small compared to unity and simultaneously Δx be small compared to L means that the total electric charge of the test body must consist of a large number of elementary charges ε. In fact, according to (20) this number is given by

(21) $$N=\frac{\rho V}{\varepsilon}=\frac{QV}{\lambda\varepsilon\Delta x}=\frac{1}{\lambda}\cdot\frac{L}{\Delta x}\cdot\sqrt{\frac{L}{cT}}\cdot\sqrt{\frac{\hbar c}{\varepsilon^2}}$$

and is very large when the above requirements are fulfilled and when $L > cT$ as assumed. The last factor is of course the reciprocal square root of the fine-structure constant whose smallness, as we already mentioned in the Introduction, is an essential presupposition for the correspondence approach to electron theory. As emphasized there, essential restrictions are imposed on a field measurement with an elementary charge as a test body, a fact which is also directly visible from (21) if one puts $N=1$.[8] Moreover, the assumption of a large value of N is a necessary condition for the physical realization of a uniform distribution of the charge of the test body over the volume V; and as long as the linear dimensions of the test body are large compared to the atomic dimensions, its fulfillment obviously presents no difficulties in principle. It need hardly be mentioned that with this presupposition the assumption used above about the mass of the test body, equivalent to the requirement that this mass be very large compared to that of a light quantum of wave length L, always can be satisfied.

Thus far we have completely disregarded the electromagnetic field effects which accompany the acceleration of any test body during the momentum measurement. These effects superpose themselves on the original field and must be included in the field averages defined by equations of type (15). Hence, the main task of the following investigation will be to find a measuring arrangement in which the field effects of the test bodies can be controlled or compensated to the largest possible extent. Yet here we must first of all discuss the question of whether the reaction of the radiation fields produced by the acceleration of the test body in the momentum measurements could impair even the practicability of measuring the values occurring in (15) of the test body's momentum components at the beginning and end of the measuring interval. It was just this possibility that led Landau and Peierls, in the work cited at the beginning, to doubt the reliability of the indeterminacy relation (16) for charged bodies and to conclude that it should be replaced by another even more restrictive relation in which the charge of the test body enters in an essential way. However, they likened the electromagnetic behavior of such a body to that of a point charge e, and consequently used the following expression for estimating the order of magnitude of the test body's momentum change, brought about by radiation recoil, during the time Δt

$$(22) \qquad \delta_e p_x \sim \frac{e^2}{c^3} \frac{\Delta x}{\Delta t^2}.$$

If, however, $\delta_e p_x$ is considered an additional indeterminacy of the momentum measurement, then if one puts $\rho V = e$ and does not distinguish between $\bar{\mathfrak{E}}_x$ and $\mathfrak{E}_x$, one gets instead of (17)

$$(23) \qquad \Delta_e \mathfrak{E}_x \sim \frac{\hbar}{eT\Delta x} + \frac{e\Delta x}{c^3 T \Delta t^2},$$

whose minimum under variation of e is obviously given by

$$(24) \qquad \Delta_m \mathfrak{E}_x \sim \frac{\sqrt{\hbar c}}{c^2 T \Delta t}.$$

If, still following Landau and Peierls, one does not distinguish between T and Δt, this expression agrees with the absolute limit on the measurability of field components that they gave, on which they based their criticism of the foundations of the quantum electromagnetic formalism.

However, the supposed difficulties of the momentum measurement disappear as soon as sufficient account is taken of the finite extension of the test body's electric charge. Using the idealization of a uniform, rigidly displaceable charge distribution, to be more closely examined below, the electric field strengths in the region V during the acceleration of the test body within the time Δt can at most reach a value of the order of magnitude $\rho \Delta x$; since, according to Maxwell's equations, their time derivatives are at most of the same order of magnitude as that of the current density, given by $\rho(\Delta x/\Delta t)$. Hence, any electromagnetic reaction on the body during the measuring interval Δt can only contribute a momentum transfer of the order of magnitude

$$(25) \qquad \delta_\rho p_x \sim \rho^2 V \Delta x \Delta t.$$

Thus, in view of (18) and (20), we get by comparison of (16) and (25)

$$(26) \qquad \delta_\rho p_x \sim \Delta p_x \cdot \lambda^{-2} \frac{\Delta t}{T},$$

which implies that for any desired accuracy of the field measurement, symbolized by a given value of λ, the influence of the electromagnetic reaction on the momentum measurement of the test body can be neglected if only Δt is chosen sufficiently small in comparison with T. It is precisely this circumstance that is decisive for assessing the accuracy of the field measurements; for it turns out to be impossible to directly take into account the influence of the radiation reaction on the momentum and energy balance in the individual momentum measurements. For example, Pauli's proposal[9] to measure subsequently by means of a special device the momentum and energy contained in the emitted radiation would already be impracticable because of the fact that, at least in the case $L > cT$ which is of particular importance for field measurements, the radiation fields that are produced in the momentum measurements at the beginning and at the end of the interval T cannot be separated from each other to the degree sufficient for this purpose. In fact, in the following sections we shall show quite generally that any attempt at such a control of the test body's field effects would essentially impair the realization of the field measurement in question.

Besides, not only for discussing the behavior of an individual test body during the measurement but also for assessing the mutual influence of several test bodies, it is essential to treat these not as point charges but as continuous charge distributions. This is because the customary identification of the position indeterminacy of a test body, considered as a point charge, with the linear dimensions of the domain of measurement is an arbitrary assumption that is foreign to the measurability problem. For this reason, not only do the estimates of the product of the uncertainties of $\mathfrak{E}_x$ and $\mathfrak{H}_y$ inside the same space-time domain, obtained by Heisenberg and by Landau and Peierls by considering point charges, deviate from the predictions of the quantum-electromagnetic formalism as already mentioned; they are in agreement with each other only in the special case $L \sim cT$. In this case, both estimates give the expression Q^2, which corresponds to the order of magnitude, to be expected from the formalism, of the value of the product of the complementary indeterminacies of two field averages in space-time regions that are displaced relative to each other through space-time distances of the same order of magnitude as L and T. Moreover, it is an essential feature of the formalism that the product in question vanishes identically for coinciding regions. The physical meaning of this result becomes obvious as soon as one takes into account the uniform charge distribution of the test body used to measure $\bar{\mathfrak{E}}_x$; since the magnetic field strength which is produced at a point P_2 of the volume V by the displacement of the charge ρdv contained in a volume element situated at the point P_1 is exactly equal and opposite to the magnetic field strength produced at the point P_1 by the same displacement of the charge ρdv at the point P_2, so that the average over the volume V of every magnetic field component produced by the displacement of the test body disappears.

From the foregoing it emerges that in the investigation of the measureability of field quantities it is decisively important to assume that the test bodies to be used behave like uniformly charged rigid bodies whose momenta can be measured, in any given arbitrarily small time interval, with an accuracy, expressed by (16), complementary to the accompanying uncontrollable displacement. Of course, in view of the finite propagation of all forces, we should not think here of the usual mechanical idealization of rigid bodies, but must think of every test body as a system of individual components of sufficiently small dimensions; and think of the measurement of the total momentum of this system as carried out in such a way that, to a sufficient approximation, all the components undergo the same displacement during the momentum measurement. That this requirement can be fulfilled without difficulties of principle, at least insofar as one can disregard the atomic constitution of the test body, is due to the fact that the required momentum measurements can be fully described on a classical basis, irrespective of whether they depend on looking at a collision process between the test body and a suitable material colliding body; or, say, on the study of the Doppler effect involved in the reflection of radiation from the test body. For, if only the mass of the colliding body is sufficiently large or if the packet of radiation that is used to measure the Doppler effect contains a sufficient number of light quanta, then the interaction between the test body and the colliding body can be described classically to any approximation. In fact, the loss of knowledge of the position of the test body, which accompanies the momentum measurement, is due solely to the impossibility of simultaneously fixing the course of the collision process relative to a well-defined space-time reference frame. Indeed, the peculiar complementarity of the mode of description ultimately derives from the fact that any such fixation is bound up with an unavoidable, and in principle uncontrollable, transfer of energy and momentum to the scales and clocks needed to establish the co-ordinate system.[10]

We recall that, according to the indeterminacy principle, the latitude in the time Δt that is left open in any description and the accuracy with which the energy exchanged in the collision process between colliding body and test body is known are connected by the well-known relation

$$\Delta E \cdot \Delta t \sim \hbar. \tag{27}$$

Because of the relation between energy and momentum and velocity components

$$dE = v_x \, dp_x \tag{28}$$

which is valid for both bodies, it follows directly that

$$\Delta p_x |v_x'' - v_x'| \Delta t \sim \hbar. \tag{29}$$

Even though, as noted above, the change of velocity $|v_x'' - v_x'|$ of the test body in the momentum measurement can be considered as arbitrarily well known for a sufficiently heavy test body, the factor

$$|v_x'' - v_x'| \Delta t = \Delta x \tag{30}$$

obviously implies a completely free latitude in the position of the body relative to the fixed frame of reference, in complete agreement with the indeterminacy relation (16). From (30) the condition

(31) $\Delta x < c\Delta t,$

follows immediately, which, because of (16), imposes an absolute limit on the accuracy Δp_x that can be achieved in a momentum measurement with a given upper limit for the time latitude Δt. However, in view of the relativistic invariance of the relations (16) and (27) and in particular of formula (28), this circumstance implies no restriction in the formulation and applicability of the indeterminacy principle. In our problem it is even permissible to disregard all mechanical relativistic effects, for by using sufficiently heavy test bodies we can always arrange that the velocities of all test bodies during the whole measurement process remain small compared to the velocity of light. Consequently, we can even consider any displacement Δx in the momentum measurements very small compared to the corresponding value of $c\Delta t$ which itself must be chosen arbitrarily small.

It is exactly the possibility of accurately tracing the relative space-time course of the process serving as momentum measurement that enables us to measure the total momentum of an extended body within any given time interval with the required accuracy expressed by (16). Thus, we can determine the total momentum of the system of charged material component bodies serving as test body by a single collision process if we make use of a colliding body of special construction which intervenes everywhere in the test body system and gives every component the same acceleration at the same time. It is true that this device imposes severe demands on the construction of the colliding and test bodies, but these demands do not present any difficulties of principle as long as we can disregard the atomic constitution of the bodies. The measurement of the total momentum of the test body would presumably be performed most simply by optical means, i.e. by determination of the Doppler effect; for this purpose one might proceed as follows: imagine that every component body is equipped with a small mirror at right angles to the x-direction, and that a number of other mirrors are placed in fixed positions in such a way that the light path from the radiation source to each component body is the same. If now by means of a suitable device we produce a packet of radiation of duration Δt and containing a number of light quanta sufficiently large compared to the number of component bodies, then all these bodies will suffer a collision simultaneously and undergo an acceleration which for all component bodies can be made equal with arbitrary accuracy.

In order to show that one can in fact measure the total momentum of the test body with an accuracy satisfying relation (16) by means of such an arrangement, we shall consider somewhat more closely the interaction between the test body system and the packet of radiation. In view of the assumed smallness of the velocity of the test body compared to the velocity of light. we have for each component body

$$\left\{\begin{aligned} & m_\tau(v''_{\tau,x} - v'_{\tau,x}) = \frac{\hbar}{c} \sum_{n_\tau} (\nu' + \nu''), \\ & \frac{1}{2} m_\tau(v''^2_{\tau,x} - v'^2_{\tau,x}) = \hbar \sum_{n_\tau} (\nu' - \nu''), \end{aligned}\right. \tag{32}$$

where m_τ denotes the mass of a component body, $v'_{\tau,x}$, $v''_{\tau,x}$ are its velocity before and after the reflection, and where the summation extends over the n_τ light quanta reflected from the component body whose frequency (reciprocal period times 2π) before and after the reflection are denoted by ν' and ν'', respectively. It follows from (32) that the momentum of the component body in question before and after the collision is

$$\left.\begin{aligned} p'_{\tau,x} = m_\tau v'_{\tau,x} = \\ p''_{\tau,x} = m_\tau v''_{\tau,x} = \end{aligned}\right\} m_\tau c \frac{\sum_{n_\tau} (\nu' - \nu'')}{\sum_{n_\tau} (\nu' + \nu'')} \mp \frac{1}{2} \frac{\hbar}{c} \sum_{n_\tau} (\nu' + \nu''), \tag{33}$$

If we now assume that the mean spectral frequency ν_0 of the radiation packet is very large compared to the mean width $(\Delta t)^{-1}$ of its frequency distribution as well as to all frequency changes $\nu' - \nu''$, then we can take the change of velocity of the component body in the collision, to a sufficient approximation, as

$$v''_{\tau,x} - v'_{\tau,x} = \frac{\hbar}{m_\tau c} \sum_{n_\tau} (\nu' + \nu'') = \frac{2 n_\tau \hbar \nu_0}{m_\tau c} \tag{34}$$

and we can assume that it is the same for all component bodies. Thus, in the collision all component bodies suffer displacements which, although uncontrollable, are arbitrarily close to being identical and whose order of magnitude Δx satisfies relation (30) where $|v''_x - v'_x|$ is to be identified with the common velocity change of the whole test body system. Consequently, since according to our assumptions Δx can be considered negligibly small compared to $c\Delta t$, we get from (33) and (34), for the product of Δx and the uncertainty of the total momentum of the test body, approximately

$$(35) \qquad \Delta p_x \Delta x \sim \Delta t \cdot \Delta\left(\sum_{\tau}\sum_{n_\tau} \hbar\nu' - \sum_{\tau}\sum_{n_\tau} \hbar\nu''\right).$$

The quantities in the bracket in (35) are precisely the total energies of the radiation packets impinging on and reflected from the test body. The energy of the latter packet can be measured with arbitrary accuracy, e.g. by spectral analysis of the reflected radiation. For the incoming radiation packet, however, such an analysis would obviously be incompatible with the experimental conditions. Yet the total energy of this radiation can be measured with an accuracy that is complementary to Δt, as given by (27). To do that it suffices to use a purely mechanical device by means of which the packet in question is separated from a radiation field, whose energy before and after the separation can be determined with arbitrary accuracy, e.g. by spectral measurements. Thus, the relation (35) is identical with the usual indeterminacy relation (16). Note further that the demonstration of this identity is essentially dependent on the fact that in accordance with the described arrangement we obtain no information about the momentum of the single component bodies but only about the total momentum of the test body.

The fact that the test body system suffers a common translation during the required momentum measurements is not only important for the calculation of the field effects of the test body which accompany these measurements, but also gives us the possibility to arrange things in such a way that outside the short time intervals occupied by the momentum measurements all the test bodies employed in the field measurement can be considered as being at rest, which greatly simplifies the calculation. For immediately after each momentum measurement, i.e. practically speaking still inside the interval Δt, we can give the test body system a second push in the opposite direction by means of a suitable device, such that the velocity change which every component body suffered in the first collision is cancelled out; this can be done with an arbitrary accuracy, i.e.

an accuracy that is inversely proportional to the mass of the component body, and without losing the desired knowledge of the total momentum of the test body. However, with this arrangement it is impossible to know the time interval between the two collision processes with a latitude smaller than Δt, and so, as required by the indeterminacy principle, the test body is not returned to its original position by the counter-collision, but rather is brought to rest to the required approximation at an unknown position, displaced by a distance of the order Δx.

To assess the complementary limitations on the measurability of field quantities, which will be more closely investigated in the following sections, we must follow the behavior of the test body as accurately as possible during the whole measuring process. It turns out that for this purpose it is necessary first of all to know accurately the position of each test body at all times before and after its use in the measurement. This is achieved most expediently by having the test body firmly attached to a rigid frame serving as a spatial reference system, except in the time interval during which the momentum transfer to the test body from the field is to be determined. At the beginning of this interval the attachment must be disconnected, and the momentum component of the test body in the direction of the field component that is to be determined must be measured. We always assume that by an immediately following counter-impulse, as discussed above, the body is brought back to rest with an accuracy inversely proportional to its mass, at a position which is not accurately predictable. At the end of the time interval and after renewed measurement of the momentum component in question, the firm attachment is re-established; here it turns out to be not unessential that the test body be brought back into exactly the same position as it had originally. If the space-time averaging domains are to be sufficiently sharply defined, these prescriptions alone impose far-reaching demands on the detailed construction of the test body system. For due to the retardation of all forces it is strictly speaking necessary that the severance as well as the re-establishment of the attachment of the test body system to the fixed frame are performed in such a way that all its independent component bodies, whose linear dimensions must be at least as small as the smallest relevant value of $c\Delta t$, are unfastened and fastened simultaneously, i.e. within the time latitude Δt of the momentum measurement, which itself must be chosen sufficiently small compared to the time interval T.

Still more far-reaching demands on idealization with respect to the construction and handling of the test body system are obviously needed to measure field averages over two partially overlapping space-time regions. For in this case we must have test bodies at our disposal which can be displaced inside each other without mutual mechanical influence. In order that the electromagnetic field to be measured be disturbed as little as possible by the presence of the test body system, we shall imagine, moreover, that every electric or magnetic component body is placed adjacent to a neutralization body with exactly the opposite charge. In the case of a magnetic test body system it is to be noted that a uniform pole strength distribution cannot exist on a strictly delimited body. However, one can imagine, at least in principle, that every component body of such a system is connected with the corresponding neutralization body by magnetizable flexible threads. All these neutralization bodies are to remain connected with the fixed frame during the whole measurement process without mechanically influencing the free mobility of the component bodies belonging to the test body system proper. Of course, the idealizations entailed in such presuppositions, as well as in the still needed compensation mechanisms to be introduced below, are justifiable only as long as we can neglect the atomic constitution of the test body. However, as already mentioned, this neglect does not imply any restrictions in principle on the possibility of testing the quantum electromagnetic formalism, since no universal space-time dimensions appear in this formalism. Accordingly, the purpose of the preceding considerations was above all to show that in the purely mechanical problems which are relevant to the field measurements, it is possible to distinguish strictly between the restrictions on the constitution of the test body stemming from the atomic structure of matter and the restrictions on the handling of these bodies that are due to the quantum of action, formulated in particular in the principle of indeterminacy.

4. Calculation of the field effects of the test body

After having investigated the physical presuppositions for the constitution of the test body we now turn to a closer consideration of the electromagnetic field effects of the test body which accompany the measurement of field quantities and which are of decisive importance for the measurability question. In accordance with the discussion above we shall treat each test body as a charge distribution which uniformly fills the spatial averaging domain and which undergoes a simple translation during the momentum measurement. We shall first carry out the calculation of the electromagnetic fields thereby produced on the basis of classical electrodynamics, and only afterwards discuss the restriction on the validity of this treatment due to the quantum of action.

Let us consider two space-time regions, I and II, with volumes V_{I} and V_{II} and durations T_{I} and T_{II}, and let us ask for the electromagnetic field 'which is produced at a point (x_2, y_2, z_2, t_2) of region II by a measurement of the average of $\mathfrak{E}_x$ over the region I. Thus, we assume that in volume V_{I} there are originally two electric charge distributions with the constant densities $+\rho_{\mathrm{I}}$ and $-\rho_{\mathrm{I}}$. In the interval from t'_{I} to $t'_{\mathrm{I}}+\Delta t_{\mathrm{I}}$ the first charge distribution experiences a simple non-uniform translation in the x-direction through a distance $D_x^{(\mathrm{I})}$; in the interval from $t'_{\mathrm{I}}+\Delta t_{\mathrm{I}}$ to t''_{I} it remains at rest at the displaced position; finally, in the interval from t''_{I} to $t''_{\mathrm{I}}+\Delta t_{\mathrm{I}}$ it moves non-uniformly parallel to the x-axis back to its original position, which coincides with that of the neutralization distribution. In accordance with the requirement discussed in the preceding sections, we assume further that Δt_{I} is very small compared to $T_{\mathrm{I}}=t''_{\mathrm{I}}-t'_{\mathrm{I}}$ and that $D_x^{(\mathrm{I})}$ is very small not only compared to the linear dimensions of the spatial averaging region of volume V_{I}, but also small compared with $c\Delta t_{\mathrm{I}}$.

Hence, in the limiting case of vanishingly small Δt_{I}, the sources of the field that we are looking for may be represented as a polarization in the x-direction of constant density, $P_x^{(\mathrm{I})}=\rho_{\mathrm{I}}\, D_x^{(\mathrm{I})}$, existing in the region I during the time interval from t'_{I} to t''_{I}, as well as a current density present only in the immediate vicinity of the times t'_{I} and t''_{I}, which we can write as

$$J_x^{(\mathrm{I})}=\rho_{\mathrm{I}} D_x^{(\mathrm{I})}[\delta(t-t'_{\mathrm{I}})-\delta(t-t''_{\mathrm{I}})], \tag{36}$$

using δ symbol defined by formula (3). By means of the same symbol we can similarly express the polarization at an arbitrary time t by the formula

$$P_x^{(\mathrm{I})}=\rho_{\mathrm{I}} D_x^{(\mathrm{I})}\int_{t'_{\mathrm{I}}}^{t''_{\mathrm{I}}}\delta(t-t_1)\,\mathrm{d}t_1. \tag{37}$$

As is well-known, the components of the fields at the space time point

(x_2, y_2, z_2, t_2) produced by these sources may be calculated from the formulae

$$(38)\quad \begin{cases} E_x^{(\mathrm{I})} = -\dfrac{\partial \varphi^{(\mathrm{I})}}{\partial x_2} - \dfrac{1}{c}\dfrac{\partial \psi_x^{(\mathrm{I})}}{\partial t_2}, & E_y^{(\mathrm{I})} = -\dfrac{\partial \varphi^{(\mathrm{I})}}{\partial y_2}, \quad E_z^{(\mathrm{I})} = -\dfrac{\partial \varphi^{(\mathrm{I})}}{\partial z_2}, \\ H_x^{(\mathrm{I})} = 0, & H_y^{(\mathrm{I})} = \dfrac{\partial \psi_x^{(\mathrm{I})}}{\partial z_2}, \quad H_z^{(\mathrm{I})} = -\dfrac{\partial \psi_x^{(\mathrm{I})}}{\partial y_2}, \end{cases}$$

where we have used latin letters to distinguish these fields from the field components that are to be measured. In (38), $\varphi^{(\mathrm{I})}$ signifies the retarded scalar potential

$$(39)\qquad \varphi^{(\mathrm{I})} = \int\limits_{V_1} \frac{\partial}{\partial x_1}\left[\frac{P_x^{(\mathrm{I})}(t_2 - r/c)}{r}\right] dv_1$$

and $\psi_x^{(\mathrm{I})}$ the retarded vector potential component

$$(40)\qquad \psi_x^{(\mathrm{I})} = \frac{1}{c}\int\limits_{V_1} \frac{J_x^{(\mathrm{I})}(t_2 - r/c)}{r}\, dv_1,$$

where r is the distance between the space points (x_1, y_1, z_1) and (x_2, y_2, z_2). Noting that the expression (36) can also be written in the form

$$(41)\qquad J_x^{(\mathrm{I})} = -\rho_1 D_x^{(\mathrm{I})} \int\limits_{t_1'}^{t_1''} \frac{\partial}{\partial t_1}\,\delta(t - t_1)\, dt_1$$

and taking into account (37) and (41) one sees that the field components given by (38), (39) and (40) can be expressed by the typical formulae

$$(42)\quad \begin{cases} E_x^{(\mathrm{I})} = \rho_1 D_x^{(\mathrm{I})} \displaystyle\int\limits_{V_1} dv_1 \int\limits_{T_1} dt_1 A_{xx}^{(12)}, & E_y^{(\mathrm{I})} = \rho_1 D_x^{(\mathrm{I})} \displaystyle\int\limits_{V_1} dv_1 \int\limits_{T_1} dt_1 A_{xy}^{(12)}, \\ H_x^{(\mathrm{I})} = 0, & H_y^{(\mathrm{I})} = \rho_1 D_x^{(\mathrm{I})} \displaystyle\int\limits_{V_1} dv_1 \int\limits_{T_1} dt_1 B_{xy}^{(12)}, \end{cases}$$

where the abbreviations defined in (2) have been used.

In view of the properties of the symbolic δ-function it is easy to see that the field components given by (42) always remain finite and even cannot surpass a value of the order of magnitude $\rho_{\mathrm{I}} D_x^{(\mathrm{I})}$ at any space-time point (x_2, y_2, z_2, t_2). As already mentioned, the electromagnetic forces which occur during the momentum measurement of the test body in the time interval Δt are just of this order of magnitude (cf. p. 372). The fact that the field intensities do not subsequently increase essentially is solely a consequence of the counter collision taking place right after the momentum measurement which brings the body back to rest, and finds its idealized mathematical expressions in (36) and (37).

The averages of these field components over the region II, which are of particular interest to us, are obtained from (42) by a simple space-time integration and, in accordance with (5), are given by the formulae

$$(43)\qquad \begin{cases} \bar{E}_x^{(\mathrm{I},\,\mathrm{II})} = D_x^{(\mathrm{I})} \rho_{\mathrm{I}} V_{\mathrm{I}} T_{\mathrm{I}} \bar{A}_{xx}^{(\mathrm{I},\,\mathrm{II})}, & \bar{E}_y^{(\mathrm{I},\,\mathrm{II})} = D_x^{(\mathrm{I})} \rho_{\mathrm{I}} V_{\mathrm{I}} T_{\mathrm{I}} \bar{A}_{xy}^{(\mathrm{I},\,\mathrm{II})} \\ \bar{H}_x^{(\mathrm{I},\,\mathrm{II})} = 0, & \bar{H}_y^{(\mathrm{I},\,\mathrm{II})} = D_x^{(\mathrm{I})} \rho_{\mathrm{I}} V_{\mathrm{I}} T_{\mathrm{I}} \bar{B}_{xy}^{(\mathrm{I},\,\mathrm{II})}. \end{cases}$$

As a result of the properties of the expressions $\bar{A}$ and $\bar{B}$, already discussed in Section 2, we see that for a given value of $D_x^{(\mathrm{I})}$ the field averages given by (43) are well-defined continuous functions of the regions I and II. Hence, for decreasing latitudes, Δt and Δx, of the duration of the momentum measurements and of the accompanying unpredictable displacements, these field averages are completely independent of the detailed space-time course of the collision process, and simply proportional to the constant displacement of the test body in the measuring interval T_{I}. As we shall see, this very fact turns out to be decisive for the possibility of an extensive compensation of the uncontrollable field effects of the test bodies.

Thus far, the calculation of the field effects has been carried out on a purely classical basis. Yet for a more detailed comparison of the measurement possibilities and the requirements of the quantum-electromagnetic formalism one must also take into account the limitation imposed on the classical mode of calculation by the quantum-theoretical features of any field effect, symbolized by the concept of light quanta. In order to get a general view of the situation we assume that the averaging regions in question are of the same order of magnitude and spatially displaced relative to each other through distances of the same order of magnitude as their linear dimensions, which we denote by L; and that further the corresponding time intervals of order of magnitude T are smaller than L/c. Under these conditions, the spectral decomposition of the field effects contains essentially only waves of wave length of the same order of magnitude as L. Since, furthermore, in the case under consideration the intensity of the field produced in the momentum measurement is of the order $\rho\Delta x$, and consequently the field energy contained in the volume V is of the order $\rho^2\Delta x^2 V$, then the number of light quanta in question is approximated by the expression

$$(44) \qquad n \sim \rho^2 \Delta x^2 V \frac{L}{\hbar c} = \lambda^{-2} \frac{L}{cT},$$

where λ is the factor, defined in (20), that provides the measure for the accuracy of the measurement. Thus we see that in our case n is always large compared to unity if an accuracy of measurement is required that permits field strengths to be measured which are smaller than the critical field quantity Q.

Evidently, the classically calculated expressions (42) and (43) for the field effects become relatively more exact, the greater the accuracy aimed at for the field measurements. However, it is essential to note that the absolute accuracy of these expressions does not change for increasing values of n. For in our case the statistical range of fluctuation of the field averages is approximately given by

$$\frac{\rho\Delta x}{\sqrt{n}} \sim \sqrt{\frac{\hbar c}{VL}} \sim \frac{\sqrt{\hbar c}}{L^2}.$$

This expression for the range of fluctuations of the field effects of the test body, which depends only on the linear dimensions of the measurement domain and which always remains finite, agrees in fact with the expression (14) for the order of magnitude of the pure black-body fluctuations which was derived from the formalism in the case $L > cT$. Actually, in the above consideration we are dealing merely with an example of the general relation, mentioned in Section 2, between black-body fluctuations and the deviations, only describable statistically, of field averages from field quantities that are calculated according to classical theory from specification of the sources. Furthermore, as was already there pointed out, in the case $L > cT$, especially important for testing the formalism, the black-body fluctuations are always smaller than the field strength Q which is a measure of the complementary measurability of field quantities; and, indeed, so much the smaller, the larger the ratio between L and cT. Thus, in the following comparison between field measurements and formalism we shall always start from the classically calculated expressions (43), and only afterwards discuss the significance of the fluctuation phenomena for the consistency of the formalism.

5. Measurement of single field averages

By definition we base the investigation of the measurement possibilities for field averages on equation (15) which expresses the classically described momentum balance for a test body situated in the field. According to the preceding arguments, each field component, such as $\mathfrak{E}_x$, is in this context to be regarded as the superposition of the fields originating from all field sources, including the test body itself, and the core of the measurement problem is precisely the question of the extent to which these fields can be associated with the various sources. However, we must emphasize here immediately that the strict applicability of the classical field concept in defining field averages is not in itself impaired by the previously discussed limited validity of the classical description of the field effects of the test body. Quite apart from the question of the accuracy attainable in the momentum measurements of the test body at the beginning and end of the measuring interval, which was discussed in Section 3, the unambiguous character of this definition would itself require that the masses of the test bodies be chosen sufficiently large that any modifications of the electromagnetic fields, stemming from their accelerations under the influence of these fields during the measuring interval, may be neglected. If one were inclined to regard this neglect as in contradiction with the atomic character of the momentum transfer between electromagnetic wave fields and material bodies, then it must be recalled that in the measurement problem under consideration there is no question of tracing well-defined elementary processes in the sense of the light quantum concept. In particular, in the measuring arrangement described, an uncontrollable amount of momentum is absorbed by the rigid frame to which every test body is attached before and after the measuring interval. In the limiting case of a classically describable interaction between an electromagnetic wave train and a sufficiently heavy charged body, the momentum transfer just mentioned would obviously exactly compensate the momentum absorbed by the test body in the measuring interval.

As a preparation for the general discussion of the measurement problem, we first consider a single field measurement and, as in Section 3, ask for the average value of $\mathfrak{E}_x$ in a certain space-time domain which we denote, as in the previous sections, by I. Thus, from the fundamental equation (15), we get for the momentum balance of the test body

$$(45) \qquad p_x^{(\mathrm{I})''} - p_x^{(\mathrm{I})'} = \rho_\mathrm{I} V_\mathrm{I} T_\mathrm{I} (\mathfrak{E}_x^{(\mathrm{I})} + \bar{E}_x^{(\mathrm{I},\,\mathrm{I})}),$$

where $\mathfrak{E}_x^{(\mathrm{I})}$ represents the part of the average of $\mathfrak{E}_x$ which would be present in the space-time domain I under consideration if no momentum measurement were made on the test body at time t', while $\bar{E}_x^{(\mathrm{I},\,\mathrm{I})}$ is the part of the field average that arises from this measurement, whose classically estimated expression is given by (43), if regions I and II are set equal.

According to the arguments of Section 3, the sum of the field averages $\mathfrak{E}_x^{(\mathrm{I})}$ and $\bar{E}_x^{(\mathrm{I},\,\mathrm{I})}$ appearing in (45) can be determined with arbitrary accuracy by choosing a sufficiently large value of ρ_I. However, the larger ρ_I is chosen, the larger will be the uncontrollable value of $\bar{E}_x^{(\mathrm{I},\,\mathrm{I})}$; and therefore, the attainable accuracy in the determination of $\mathfrak{E}_\mathrm{I}^{(\mathrm{I})}$ by means of the simple measuring arrangement previously described, which according to (45) is given by

$$(46) \qquad \Delta\mathfrak{E}_x^{(\mathrm{I})} \sim \frac{\Delta p_x^{(\mathrm{I})}}{\rho_\mathrm{I} V_\mathrm{I} T_\mathrm{I}} + \Delta \bar{E}_x^{(\mathrm{I},\,\mathrm{I})},$$

has a limit imposed upon it. Indeed, due to relation (16) and to the fact that the quantity $D_x^{(\mathrm{I})}$ appearing in (43) is predictable only with a latitude Δx_I, we get from (46) the expression

$$(47) \qquad \Delta\mathfrak{E}_x^{(\mathrm{I})} \sim \frac{\hbar}{\rho_\mathrm{I} \Delta x_\mathrm{I} V_\mathrm{I} T_\mathrm{I}} + \rho_\mathrm{I} \Delta x_\mathrm{I} V_\mathrm{I} T_\mathrm{I} |\bar{A}_{xx}^{(\mathrm{I},\,\mathrm{I})}|,$$

for $\Delta\mathfrak{E}_x^{(\mathrm{I})}$, whose minimal value obviously is

$$(48) \qquad \Delta_m \mathfrak{E}_x^{(\mathrm{I})} \sim \sqrt{\hbar |\bar{A}_{xx}^{(\mathrm{I},\,\mathrm{I})}|}$$

and in the case $L_\mathrm{I} > cT_\mathrm{I}$ precisely equals the critical quantity Q_I. It is true that when L_I is large compared to cT_I, (48) is essentially smaller than the expression (24) which was given by Landau and Peierls as the absolute limit on the measurability of field quantities; yet if (48) were to be considered as an unavoidable limit on the accuracy of measurement, we should still arrive at the conclusion, in agreement with the view of these authors, that the quantum-electromagnetic formalism admits of no test in the properly quantum domain, and that therefore physical reality can be ascribed to the entire field theory only in the classical limit.

However, this conclusion cannot be maintained, for the fact that according to (43) the coefficient of the unpredictable displacement $D_x^{(\mathrm{I})}$ in $\bar{E}_x^{(\mathrm{I},\mathrm{I})}$ is a well-defined quantity depending solely on geometrical relations, allows us to so arrange things in the measurements that the effect of the field $E_x^{(\mathrm{I})}$ is completely compensated except for the unavoidable field fluctuations. This is achieved by a measuring arrangement in which the test body is not freely movable, even during the measuring interval T_I, but remains connected with the rigid frame through a spring mechanism whose tension is proportional to $D_x^{(\mathrm{I})}$. If the force in the x-direction exerted by this mechanism on the test body is $-F_x D_x^{(\mathrm{I})}$, then the total momentum transferred from the field $E_x^{(\mathrm{I})}$ to this body will obviously be completely cancelled by the spring if the spring constant is chosen to be

$$(49) \qquad F_\mathrm{I} = \rho_\mathrm{I}^2 V_\mathrm{I}^2 T_\mathrm{I} \bar{A}_{xx}^{(\mathrm{I},\mathrm{I})}.$$

At any rate, this holds when the test body is so heavy that its oscillation period under the influence of the spring is large compared to T_I and thus its displacement due to the spring tension during the time T_I is small compared to $D_x^{(\mathrm{I})}$. Furthermore, the action of the spring, which strictly speaking is classically describable only in the asymptotic limiting case, may be calculated on the basis of classical mechanics with an accuracy which is the greater the larger the mass of the test body. Apart from the limitations due to the atomic structure of all bodies, no objection of principle could exist against such a compensation device. In the first place, by using a mechanical spring all electromagnetic fields are avoided, which would be inseparable from the fields to be measured. Secondly, if the length of the spring is sufficiently small, i.e. small compared to cT_I, one may obviously disregard all retardation effects. In doing so, if the test body system is sufficiently heavy, it is clearly immaterial whether the spring acts only on a component body or whether one uses a system of springs that affects each component body uniformly.

Thus we see that the sharpness of a single field measurement is restricted solely by the limit set for the classical description of the field effects of the test body. However, even in the case $L_I \leqslant cT_I$ this limitation, which is the more insignificant the larger L_I is compared to cT_I, implies no restriction at all on the possibility of testing the consequences of the quantum-electromagnetic formalism. In assessing this question, we must distinguish sharply between the testing of theoretical predictions which presuppose data concerning electric or magnetic forces obtained by field measurements, and of those which depend on knowledge of the state of the field in question obtained on some other basis. As for the former predictions, their testing obviously requires a closer investigation of the mutual relations between several field measurements; thus, to begin with, here it can only be a matter of testing predictions of the latter kind.

Now, as mentioned in Section 2, it is a major result of the quantum theory of fields that all predictions concerning field averages which do not rest on true field measurements, but on the light quantum composition of the field to be investigated or on the knowledge of classically described field sources, must be of an essentially statistical nature. Further, the more detailed argument presented there shows that inclusion of the fluctuations of the test body's field effects around their classically estimated value brings about no change whatsoever in these statistical predictions. Without further correction, the measurement results obtained by means of the experimental arrangement described thus appear as the desired field averages for testing the theoretical statements. Such a view of the measuring results, whose general justification we shall investigate more closely in the following, is also suggested by the fact that all measurements of physical quantities, by definition, must be a matter of the application of classical concepts; and that, therefore, in field measurements any consideration of limitations on the strict applicability of classical electrodynamics would be in contradiction with the measurement concept itself.

Even though, consequently, as already stressed in the Introduction, the measurement concept is to be applied with even greater caution in field measurements than in the usual quantum mechanical measurement problems; nevertheless, as regards the inseparability of phenomenon and measuring process the situation described exhibits a far-reaching analogy to these problems. Indeed, even in a position or momentum measurement on the electron in a hydrogen atom in a given stationary state one can assert with a certain right that the measuring result is produced only by the measurement itself. It is here not a question of a limitation on the sharpness of the measuring result on the basis of classical mechanics, indeed; but only of abandonment of any control over the influence of the measuring process on the state of the atom. In field measurements, this complementary feature of the description, essential for consistency, corresponds to the fact that the knowledge of the light quantum composition of the field is lost through the field effects of the test body; and in fact, according to (44) the more so, the greater the desired accuracy of the measurement. Moreover, it will appear from the following discussion that any attempt to re-establish the knowledge of the light quantum composition of the field through a subsequent measurement by means of any suitable device would at the same time prevent any further utilization of the field measurement in question.

However, the fact that pure black-body fluctuations appear as the common limitation in the demonstration of the correspondence between the testability of the consequences of the quantum-electromagnetic formalism by means of a single field measurement and the interpretability of such a measurement on the basis of classical electrodynamics in no way implies that these fluctuations set an absolute limit for any utilization of field measurements. Indeed, such a general limitation exists neither for the consequences of the formalism regarding relations between averages of a field component over different regions, nor for the testing of such relations through direct field measurements. This will become clear from the considerations in the following sections; and in particular it will be shown that the requirement of repeatability of measurements of kinematical and dynamical quantities, essential for the discussion of the consistency of the usual quantum mechanics, possesses its natural analog in field measurements.

6. Measurability of Two-Fold Averages of a Field Component

In investigating the measurability of two field quantities it is convenient to start with the measurement of the average of one and the same field component over two different regions, I and II. Thus, considering as above the field component $\mathfrak{E}_x$, and disregarding to begin with the limitations of the classical describability of the test bodies' field effects, we have in this case for the momentum balance of the two test bodies, instead of (45):

$$(50)\quad \begin{cases} p_x^{(\mathrm{I})''} - p_x^{(\mathrm{I})'} = \rho_{\mathrm{I}} V_{\mathrm{I}} T_{\mathrm{I}} (\mathfrak{E}_x^{(\mathrm{I})} + \bar{E}_x^{(\mathrm{I},\,\mathrm{I})} + \bar{E}_x^{(\mathrm{II},\,\mathrm{I})}), \\ p_x^{(\mathrm{II})''} - p_x^{(\mathrm{II})'} = \rho_{\mathrm{II}} V_{\mathrm{II}} T_{\mathrm{II}} (\mathfrak{E}_x^{(\mathrm{II})} + \bar{E}_x^{(\mathrm{II},\,\mathrm{II})} + \bar{E}_x^{(\mathrm{I},\,\mathrm{II})}), \end{cases}$$

where $\bar{E}_x^{(\mathrm{I},\,\mathrm{II})}$ is defined by expression (43), and $\bar{E}_x^{(\mathrm{II},\,\mathrm{I})}$ is obtained from this expression by simple interchange of the indices I and II.

According to the considerations in the previous sections, the appearance of the expressions $\bar{E}_x^{(\mathrm{I},\,\mathrm{I})}$ and $\bar{E}_x^{(\mathrm{II},\,\mathrm{II})}$ in equations (50) implies that each of the desired field averages, $\mathfrak{E}_x^{(\mathrm{I})}$ and $\mathfrak{E}_x^{(\mathrm{II})}$, can only be determined with a limited accuracy, given by (48), by means of a simple measuring arrangement. Thus, it is evident from the beginning that a compensation procedure is unavoidable, and for preliminary orientation about the more complicated measuring problem considered here we therefore first use a measuring arrangement in which the reactions, $\rho_{\mathrm{I}} V_{\mathrm{I}} T_{\mathrm{I}} \bar{E}_x^{(\mathrm{I},\,\mathrm{I})}$ and $\rho_{\mathrm{II}} V_{\mathrm{II}} T_{\mathrm{II}}$-$\bar{E}_x^{(\mathrm{II},\,\mathrm{II})}$, are cancelled by means of two springs acting on the test bodies I and II, the spring constants being given by (49) and an analogous expression.

From equations (50), with $\bar{E}_x^{(\mathrm{I},\,\mathrm{I})}$ and $\bar{E}_x^{(\mathrm{II},\,\mathrm{II})}$ omitted, it follows, using (16) and (43), that in this arrangement the uncertainties of the two field measurements, taking into account that the displacements of the test bodies, $D_x^{(\mathrm{I})}$ and $D_x^{(\mathrm{II})}$, appearing in $\bar{E}_x^{(\mathrm{I},\,\mathrm{II})}$ and $\bar{E}_x^{(\mathrm{II},\,\mathrm{I})}$ are completely independent of each other and known only with the latitudes Δx_{I} and Δx_{II}, are given by

$$(51)\quad \begin{cases} \Delta \mathfrak{E}_x^{(\mathrm{I})} \sim \dfrac{\hbar}{\rho_{\mathrm{I}} \Delta x_{\mathrm{I}} V_{\mathrm{I}} T_{\mathrm{I}}} + \rho_{\mathrm{II}} \Delta x_{\mathrm{II}} V_{\mathrm{II}} T_{\mathrm{II}} |\bar{A}_{xx}^{(\mathrm{II},\,\mathrm{I})}| \\ \Delta \mathfrak{E}_x^{(\mathrm{II})} \sim \dfrac{\hbar}{\rho_{\mathrm{II}} \Delta x_{\mathrm{II}} V_{\mathrm{II}} T_{\mathrm{II}}} + \rho_{\mathrm{I}} \Delta x_{\mathrm{I}} V_{\mathrm{I}} T_{\mathrm{I}} |\bar{A}_{xx}^{(\mathrm{I},\,\mathrm{II})}|. \end{cases}$$

By suitable choice of the values of $\rho_{\mathrm{I}} \Delta x_{\mathrm{I}}$ and $\rho_{\mathrm{II}} \Delta x_{\mathrm{II}}$ either one of the quantities $\Delta \mathfrak{E}_x^{(\mathrm{I})}$, $\Delta \mathfrak{E}_x^{(\mathrm{II})}$ can obviously be arbitrarily diminished, but only at the expense of an increase of the other. For according to (51) we get for the product of the two quantities the minimum value

$$(52)\quad \Delta \mathfrak{E}_x^{(\mathrm{I})} \Delta \mathfrak{E}_x^{(\mathrm{II})} \sim \hbar [|\bar{A}_{xx}^{(\mathrm{I},\,\mathrm{II})}| + |\bar{A}_{xx}^{(\mathrm{II},\,\mathrm{I})}|].$$

In spite of the great similarity of relation (52) to the uncertainty relations (8) required by the formalism, there is, nevertheless, a fundamental difference in that the latter contains not the sum of the magnitudes of the quantities $\bar{A}_{xx}^{(\mathrm{I},\mathrm{II})}$ and $\bar{A}_{xx}^{(\mathrm{II},\mathrm{I})}$ but their algebraic difference. It is true that (8) and (52) are in general in agreement as to order of magnitude when the regions I and II are spatially and temporally displaced relative to each other by distances of order of magnitude L and T, in which case they both have the approximate value Q^2. However, as mentioned in Section 2, the difference sign appearing in the indeterminacy relation (8) has the effect that in important cases the product of the complementary uncertainties vanishes identically, even though the quantities $\bar{A}_{xx}^{(\mathrm{I},\mathrm{II})}$ and $\bar{A}_{xx}^{(\mathrm{II},\mathrm{I})}$ each remain different from zero. This happens, for example, when the temporal averaging intervals T_{I} and T_{II} coincide, and in particular when the two averaging regions I and II completely overlap. In the latter case even the limit on the measurability of two field averages given by (52) would be in glaring contradiction to the result of the previous discussion of measurement of a single field average. In general, the two expressions (52) and (8) agree exactly only when at least one of the quantities $\bar{A}_{xx}^{(\mathrm{I},\mathrm{II})}$ or $\bar{A}_{xx}^{(\mathrm{II},\mathrm{I})}$ vanishes which in general requires that one of the expressions $t_1 - t_2 - r/c$ or $t_2 - t_1 - r/c$, appearing as arguments of the δ-function in the integrals (5), remain different from zero for every pair of points (x_1, y_1, z_1, t_1) and (x_2, y_2, z_2, t_2) of regions I and II.

Thus, apart from the last mentioned case in which there exists no correlation, or at any rate only a one-way correlation, between the two field averages, the demonstration of the agreement between measurability and quantum electromagnetic formalism requires a more refined measuring arrangement in which the uncontrollable effects can be compensated to a larger extent. It is true that there appears here, in comparison with the compensation procedure needed already for measuring a single field quantity, the further complication that the displacements of the two test bodies not only must remain unknown but are also completely independent of each other. However, this circumstance implies no fundamental difficulty; only a somewhat more complicated procedure is necessary in order to compensate as much as possible the influence of the relative displacement of the test bodies on the field measurements. For this purpose we select two component bodies ε_{I} and $\varepsilon_{\mathrm{II}}$, one from each test body system I and II, for which the expression $r - c(t_1 - t_2)$ vanishes for two times t_{I}^* and t_{II}^* lying in the time intervals T_{I} and T_{II}, respectively. If such a choice were not possible, then as said above the agreement between measurability and formalism would already be attained without further compensation. To establish the necessary correlation between the test bodies one might at first think of a spring which should connect the bodies ε_{I} and $\varepsilon_{\mathrm{II}}$ directly with each other; however, due to the retardation of the forces one would thereby run into difficulties. But one can manage with a short spring, i.e. small compared to cT, if one adds to the second test body system a neutral component body $\varepsilon_{\mathrm{III}}$ which is situated in the immediate neighborhood of component body ε_{I}, belonging to the first system, and connected with it by a spring.

Like all component bodies of the two test body systems, the body $\varepsilon_{\mathrm{III}}$ is initially to be bound to the rigid frame. Then at time t_{I}', its momentum is to be measured, after severing of this link, with the same accuracy as that of test body system II. It thereby undergoes an unknown displacement $D_x^{(\mathrm{III})}$ in the x-direction which is of the same order of magnitude as Δx_{II}. If now the tension of the spring mounted between $\varepsilon_{\mathrm{III}}$ and ε_{I} is chosen equal to $\frac{1}{2}\rho_{\mathrm{I}}\rho_{\mathrm{II}}V_{\mathrm{I}}V_{\mathrm{II}}T_{\mathrm{II}}(\bar{A}_{xx}^{(\mathrm{I},\mathrm{II})}+\bar{A}_{xx}^{(\mathrm{II},\mathrm{I})})$, then in the time interval T_{I} the momentum

$$(53) \qquad P=\tfrac{1}{2}\rho_{\mathrm{I}}\rho_{\mathrm{II}}V_{\mathrm{I}}V_{\mathrm{II}}T_{\mathrm{I}}T_{\mathrm{II}}(\bar{A}_{xx}^{(\mathrm{I},\mathrm{II})}+\bar{A}_{xx}^{(\mathrm{II},\mathrm{I})})(D_x^{(\mathrm{I})}-D_x^{(\mathrm{III})})$$

will be transferred from $\varepsilon_{\mathrm{III}}$ to ε_{I}, while $\varepsilon_{\mathrm{III}}$ undergoes the momentum change $-P$ during the same time interval. At time t'' the momentum of $\varepsilon_{\mathrm{III}}$ is measured again with the same accuracy. However, before this measurement, and in fact at time t_{II}^*, a short light signal is to be sent from $\varepsilon_{\mathrm{II}}$ to $\varepsilon_{\mathrm{III}}$, by which the relative displacement $D_x^{(\mathrm{III})}-D_x^{(\mathrm{II})}$ of these bodies can be measured with arbitrary accuracy by means of a suitable device. At the emission and absorption of the signal the two bodies undergo momentum changes which indeed remain completely unknown, but cancel each other exactly in the sum of the momentum changes measured on the bodies.

Thus, for the momentum balance of the two test body systems during the measurement we have, if we include the body $\varepsilon_{\mathrm{III}}$ in system II,

$$(54) \qquad \begin{cases} p_x^{(\mathrm{I})''}-p_x^{(\mathrm{I})'}=\rho_{\mathrm{I}}V_{\mathrm{I}}T_{\mathrm{I}}(\mathfrak{E}_x^{(\mathrm{I})}+\bar{E}_x^{(\mathrm{II},\mathrm{I})})+P \\ p_x^{(\mathrm{II})''}-p_x^{(\mathrm{II})'}+p_x^{(\mathrm{III})''}-p_x^{(\mathrm{III})'}=\rho_{\mathrm{II}}V_{\mathrm{II}}T_{\mathrm{II}}(\bar{\mathfrak{E}}_x^{(\mathrm{II})}+\bar{E}_x^{(\mathrm{I},\mathrm{II})})-P. \end{cases}$$

Taking into account (43) and (53), these formulae can be put into the form

$$
(55)\quad \begin{cases}
p_x^{(\mathrm{I})''}-p_x^{(\mathrm{I})'}=\rho_{\mathrm{I}}V_{\mathrm{I}}T_{\mathrm{I}}\mathfrak{E}_x^{(\mathrm{I})} \\
+\tfrac{1}{2}\rho_{\mathrm{I}}\rho_{\mathrm{II}}V_{\mathrm{I}}V_{\mathrm{II}}T_{\mathrm{I}}T_{\mathrm{II}}\{-D_x^{(\mathrm{II})}(\bar{A}_{xx}^{(\mathrm{I},\,\mathrm{II})}-\bar{A}_{xx}^{(\mathrm{II},\,\mathrm{I})}) \\
+(D_x^{(\mathrm{II})}-D_x^{(\mathrm{III})})(\bar{A}_{xx}^{(\mathrm{I},\,\mathrm{II})}+\bar{A}_{xx}^{(\mathrm{II},\,\mathrm{I})})+D_x^{(\mathrm{I})}(\bar{A}_{xx}^{(\mathrm{I},\,\mathrm{II})}+\bar{A}_{xx}^{(\mathrm{II},\,\mathrm{I})})\} \\
p_x^{(\mathrm{II})''}-p_x^{(\mathrm{II})'}+p_x^{(\mathrm{III})''}-p_x^{(\mathrm{III})'}=\rho_{\mathrm{II}}V_{\mathrm{II}}T_{\mathrm{II}}\,\mathfrak{E}_x^{(\mathrm{II})} \\
+\tfrac{1}{2}\rho_{\mathrm{I}}\rho_{\mathrm{II}}V_{\mathrm{I}}V_{\mathrm{II}}T_{\mathrm{I}}T_{\mathrm{II}}\{D_x^{(\mathrm{I})}(\bar{A}_{xx}^{(\mathrm{I},\,\mathrm{II})}-\bar{A}_{xx}^{(\mathrm{II},\,\mathrm{I})}) \\
-(D_x^{(\mathrm{II})}-D_x^{(\mathrm{III})})(\bar{A}_{xx}^{(\mathrm{I},\,\mathrm{II})}+\bar{A}_{xx}^{(\mathrm{II},\,\mathrm{I})})+D_x^{(\mathrm{II})}(\bar{A}_{xx}^{(\mathrm{I},\,\mathrm{II})}+\bar{A}_{xx}^{(\mathrm{II},\,\mathrm{I})})\}
\end{cases}
$$

The last terms in the curly brackets in (55) are proportional to the unknown displacements of the test bodies I and II and can therefore, exactly as the simple reactions of each test body on itself, be cancelled by means of suitable spring connections with the rigid frame. This simply amounts to replacing the expression (49) for the tension of the spring acting on body I by

$$
(56)\qquad F_{\mathrm{I},\,\mathrm{II}}=\rho_{\mathrm{I}}^2 V_{\mathrm{I}}^2 T_{\mathrm{I}}\bar{A}_{xx}^{(\mathrm{I},\,\mathrm{I})}+\tfrac{1}{2}\rho_{\mathrm{I}}\rho_{\mathrm{II}}V_{\mathrm{I}}V_{\mathrm{II}}T_{\mathrm{II}}(\bar{A}_{xx}^{(\mathrm{I},\,\mathrm{II})}+\bar{A}_{xx}^{(\mathrm{II},\,\mathrm{I})})
$$

and similarly changing the tension of the spring between the frame and body II. Furthermore, in the measuring arrangement described, the terms proportional to the relative displacement $D_x^{(\mathrm{III})}-D_x^{(\mathrm{II})}$ are known with arbitrary accuracy and can therefore easily be taken into account in the field measurements. In fact, by means of a somewhat more complicated device one could even obtain the vanishing of the difference $D_x^{(\mathrm{III})}-D_x^{(\mathrm{II})}$ by using $P_x^{(\mathrm{II})}+P_x^{(\mathrm{III})}$ (in analogy to the arrangement for measuring the total momentum of a test body system described in Section 3) to determine one and the same radiation packet and, by means of suitably placed fixed mirrors, by regulating the light path in such a way that in the first momentum measurement the reflections at body $\varepsilon_{\mathrm{III}}$ and at all component bodies of system II occur at the times t_{I}' and t_{II}', respectively, and in the second momentum measurement occur at the times t_{I}'' and t_{II}''.

By means of all these contrivances, whose considerable complexity lies in the nature of the problem, being due solely to the finite propagation of all field effects, we now have actually removed the apparent conflict between the determination of single and two-fold averages of a field component described at the beginning of this section. For from (55) we now obtain for the indeterminacies of $\mathfrak{E}_x^{(\mathrm{I})}$ and $\mathfrak{E}_x^{(\mathrm{II})}$, instead of (51),

$$(57)\quad \begin{cases} \Delta\mathfrak{E}_x^{(\mathrm{I})} \sim \dfrac{\hbar}{\rho_{\mathrm{I}}\Delta x_{\mathrm{I}} V_{\mathrm{I}} T_{\mathrm{I}}} + \dfrac{1}{2}\rho_{\mathrm{II}}\Delta x_{\mathrm{II}} V_{\mathrm{II}} T_{\mathrm{II}} \left|\bar{A}_{xx}^{(\mathrm{I},\,\mathrm{II})} - \bar{A}_{xx}^{(\mathrm{II},\,\mathrm{I})}\right| \\ \Delta\mathfrak{E}_x^{(\mathrm{II})} \sim \dfrac{\hbar}{\rho_{\mathrm{II}}\Delta x_{\mathrm{II}} V_{\mathrm{II}} T_{\mathrm{II}}} + \dfrac{1}{2}\rho_{\mathrm{I}}\Delta x_{\mathrm{I}} V_{\mathrm{I}} T_{\mathrm{I}} \left|\bar{A}_{xx}^{(\mathrm{I},\,\mathrm{II})} - \bar{A}_{xx}^{(\mathrm{II},\,\mathrm{I})}\right|, \end{cases}$$

which immediately yields for the minimal value of the product of the uncertainties

$$(58)\quad \Delta\mathfrak{E}_x^{(\mathrm{I})}\Delta\mathfrak{E}_x^{(\mathrm{II})} \sim \hbar\left|\bar{A}_{xx}^{(\mathrm{I},\,\mathrm{II})} - \bar{A}_{xx}^{(\mathrm{II},\,\mathrm{I})}\right|$$

in agreement with the consequence of the quantum theory of fields expressed by (8).

However, in order to demonstrate the complete accord between the measurability of the averages of a field component over two space-time regions and the requirements of the quantum electromagnetic formalism, we must go somewhat further into the question of the extent to which the assumption of the classical describability of the test body's field effects impairs the possibilities of testing theoretical predictions. For, as already indicated, exactly in the case of measurement of several field averages one might think beforehand that the neglect of the fluctuations of all field effects of the test body, which cannot be followed classically, and are inseparable from the pure black-body fluctuations, in this respect signifies an essential renunciation. At any rate, as long as we are dealing with averaging regions which are displaced spatially and temporally relative to each other by distances of the same order of magnitude as their linear dimensions L and corresponding time intervals T, this neglect is of little significance in the important case where L is large compared to cT. However, if L is of the same order of magnitude or smaller than cT, then the black-body fluctuations, as mentioned in Section 2, are of just the same order of magnitude as the critical field strength $\mathfrak{U}$, which is defined for such displaced regions by means of the indeterminacy relations and is to be regarded as the limit of the classical field description. The neglect in question might appear even more doubtful and seem to imply a complete renunciation of the repeatability of field measurements in the case of two almost coinciding domains, in which the product of the complementary uncertainties of the field averages, given by (8), tends to zero independently of the ratio between L and cT, and where thus the critical field strength $\mathfrak{U}$ can be arbitrarily small compared to the black-body fluctuations.

Nevertheless, a closer consideration shows that we obtain a consistent interpretation of all consequences of the quantum theory of fields, if, in a necessary generalization of the measurement concept, we interpret the measuring results obtained by the arrangement described as the desired field averages. For the classically undescribable fluctuations included in the field effects of all test bodies cannot be separated in any way from the fundamentally statistical features of every theoretical assertion whose conditions do not refer to actual field measurements. Without in any way limiting the given measurement problem we can therefore always regard the fluctuations in question as an integral part of the field to be measured. The situation in multiple field measurements differs in this respect from that obtaining in measurement of a single field average only insofar as the state of the field, with which we are concerned in every single measurement in the general case, is codetermined by the result of the other field measurements.

However, with regard to this state of affairs it may not be superfluous to point out that in the correlation of several field measurements we have to do with a feature of the general complementarity of description which is alien to the usual measuring problem of non-relativistic quantum mechanics. Indeed, the fundamental simplification which we meet in the latter theory lies precisely in the separation made there between spatial coordination and temporal evolution, which makes it possible to order all measuring processes in a simple temporal sequence. On the other hand, it is possible to speak of such a sequence of measuring processes during the measurement of two field averages only when the corresponding time intervals do not overlap at all. In general, in accordance with the formalism, the correlation of the two measurements is also a reciprocal one; and only when one of the quantities $r-c(t_1-t_2)$ and $r-c(t_2-t_1)$ remains different from zero for all pairs of points of the regions I and II, do we encounter conditions similar to those of the usual measurement problem of atomic mechanics, since the result of the one field measurement may then be simply included in the preconditions used in the predictions to be tested by the other measurement.

We meet an instructive example of an intimate reciprocal correlation in measurements of the averages of a field component over two almost coinciding space-time domains. In conformity with the requirement of the repeatability of measurement results, the theory demands in this case that both measurements yield the same result to an arbitrary degree of approximation quite independently of the statistical assertions about the values of the field quantities to be measured which are implied by the preconditions. That this requirement is actually fulfilled in our experimental arrangement follows from the fact that in this case we have to do with two test body systems which occupy almost the same spatial region and are used during almost the same interval of time. Thus, by definition, they are exposed to almost the same field, quite irrespective of the sources producing this field, and of which contribution comes from one or the other test body.

Actually, it follows from the last remark that in the case of coinciding averaging regions we would get exactly identical results of the two measurements even without any compensation. However, on account of the field effects of the test bodies the measurement results so obtained would differ in an unpredictable way from the theoretical predictions to be tested, the more so, the greater the desired measuring accuracy. It is true that by means of the compensation mechanism suitable for single field measurements, which we had retained unaltered at the beginning of this section, these deviations are in general diminished; but at the same time any strict correlation of the measuring results is prevented by the effects of the springs which are proportional to the independent displacements of the test bodies. In the arrangement for two field measurements finally adopted, in which all well-defined differences between measurement results and theoretical predictions are compensated, such a correlation is also re-established just in the case of coinciding regions. For, as one easily sees, quite independently of the relation between the magnitudes of the uncontrollable displacements of the test bodies, the momenta transferred to each test body through the combined effect of all the springs, divided by the corresponding charge densities, are exactly identical in this case.

As far as the consistency of the description is concerned, we might still remark that any attempt to control the change in the light quantum composition of the field caused by the field measurement by investigation of the test body's radiation, as already mentioned several times, would exclude the possibility of utilizing the measurement result for a comparison with a second field measurement. For in order even to be able to speak of such a utilization, there must exist pairs of points from the regions I and

II, respectively, for which one of the expressions $r-c(t_1-t_2)$ or $r-c(t_2-t_1)$ vanishes. But this implies that the radiation fields produced by the test bodies I and II during the measurement cannot be intercepted and analyzed on their way from one test body to the other without at the same time essentially influencing the fields to be measured by these bodies. Only after the completion of all field measurements, when their direct utilization is no longer of concern, is it possible to perform an arbitrarily accurate analysis of the light quantum composition of the entire field without adversely affecting the given measurement problem.

7. Measurability of Two Averages of Different Field Components

As far as measurements of averages of different field components are concerned, only the case of perpendicular, similar or dissimilar components needs closer investigation; for the complete commutativity and independent measurability of averages of parallel dissimilar components required by the quantum electromagnetic formalism finds its direct interpretation in the identical vanishing of the component $H_x^{(\mathrm{I})}$ of the field produced by the measurement of $\mathfrak{E}_x^{(\mathrm{I})}$, as shown by (42). Besides, on the basis of equations (43), the measurement of averages of perpendicular field components allows a method of treatment analogous to the procedure described in the previous section.

Let us consider the measurement of the average of $\mathfrak{E}_x$ over the region I and the average of $\mathfrak{E}_y$ or $\mathfrak{H}_y$ over the region II. If to begin with we use a measuring arrangement in which the field effects of each test body on itself during the measurement are compensated in the manner described in Section 5, we get equations of the following type for the momentum balance of the two test bodies to be used:

$$(59) \qquad \begin{cases} p_x^{(\mathrm{I})''} - p_x^{(\mathrm{I})'} = \rho_{\mathrm{I}} V_{\mathrm{I}} T_{\mathrm{I}} (\bar{\mathfrak{E}}_x^{(\mathrm{I})} + D_y^{(\mathrm{II})} \sigma_{\mathrm{II}} V_{\mathrm{II}} T_{\mathrm{II}} \bar{C}_{xy}^{(\mathrm{II},\,\mathrm{I})}) \\ p_y^{(\mathrm{II})''} - p_y^{(\mathrm{II})'} = \sigma_{\mathrm{II}} V_{\mathrm{II}} T_{\mathrm{II}} \bar{\mathfrak{R}}_y^{(\mathrm{II})} + D_x^{(\mathrm{I})} \rho_{\mathrm{I}} V_{\mathrm{I}} T_{\mathrm{I}} \bar{C}_{xy}^{(\mathrm{I},\,\mathrm{II})}) \end{cases}$$

Depending on whether we are dealing with a measurement of similar or dissimilar components, the letter $\mathfrak{R}$ here represents the usual designation of the field components $\mathfrak{E}$ or $\mathfrak{H}$, while C is written instead of the symbols A or B appearing in (43); further, the designation σ_{II} represents the charge density or the pole strength distribution of the test body II accordingly.

In a manner similar to the derivation of (52) one obtains from (59) the relation

$$(60) \qquad \Delta\mathfrak{E}_x^{(\mathrm{I})}\Delta\mathfrak{H}_y^{(\mathrm{II})} \sim \hbar[|\bar{C}_{xy}^{(\mathrm{I,\,II})}|+|\bar{C}_{xy}^{(\mathrm{II,\,I})}|],$$

which, like (52), does not generally, but only in certain cases, represent an agreement between measurability and quantum electromagnetic formalism. Of such cases we might mention in particular the measurement of dissimilar perpendicular field components inside the same spatial region, for which, as stressed in Section 2, both expressions $\bar{B}_{xy}^{(\mathrm{I,\,II})}$ and $\bar{B}_{xy}^{(\mathrm{II,\,I})}$ vanish. The correctness of interpreting this fact as an arbitrarily accurate independent measurability of the field quantities in question was suggested already in Section 3 by elementary considerations in the case of coinciding space-time regions.

For the general treatment of the measurability problem of perpendicular field components we choose, as in the previous section, two component bodies ε_{I} and $\varepsilon_{\mathrm{II}}$ of the test body systems I and II, respectively, whose separation is $r=c(t_{\mathrm{I}}^*-t_{\mathrm{II}}^*)$, where t_{I}^* and t_{II}^* are within the time intervals T_{I} and T_{II}, respectively. Furthermore, in the immediate neighborhood of ε_{I} we introduce a third body $\varepsilon_{\mathrm{III}}$, whose momentum in the y-direction is measured at the times t_{I}' and t_{I}''; the relative displacements $D_y^{(\mathrm{III})}-D_y^{(\mathrm{II})}$ of the bodies $\varepsilon_{\mathrm{III}}$ and $\varepsilon_{\mathrm{II}}$ are again determined by means of a light signal, as a result of which both bodies undergo equal and opposite momentum changes. Instead of connecting $\varepsilon_{\mathrm{III}}$ directly with ε_{I} by a spring we must, however, in order to make the force transfer through the spring mechanism proportional to $D_y^{(\mathrm{III})}-D_x^{(\mathrm{I})}$, use a device which consists of two springs and an angular level with two equally long mutually perpendicular arms which can rotate on a hinge mounted on the rigid frame, and the arms of which are initially parallel to the x- and y-directions, respectively. A spring parallel to the y-axis is fastened between the first arm and the body $\varepsilon_{\mathrm{III}}$, and a spring parallel to the x-axis acts between the second arm and the body ε_{I}. Let the tension of the springs be so chosen that the force which acts on the body $\varepsilon_{\mathrm{III}}$ in the y-direction and on the body ε_{I} in the x-direction during the time interval T_{I} is given by

$$\tfrac{1}{2}\rho_{\mathrm{I}}\sigma_{\mathrm{II}}V_{\mathrm{I}}V_{\mathrm{II}}T_{\mathrm{II}}(\bar{C}_{xy}^{(\mathrm{I,\,II})}+\bar{C}_{xy}^{(\mathrm{II,\,I})})(D_x^{(\mathrm{I})}-D_y^{(\mathrm{III})}).$$

Thus, the momentum balance of the two test body systems, after suitable rearrangement, may be expressed in the following form, analogous to (55)

$$(61)\quad \begin{cases} p_x^{(\mathrm{I})''} - p_x^{(\mathrm{I})'} = \rho_\mathrm{I} V_\mathrm{I} T_\mathrm{I} \mathfrak{E}_x^{(\mathrm{I})} \\ +\tfrac{1}{2}\rho_\mathrm{I}\sigma_\mathrm{II} V_\mathrm{I} V_\mathrm{II} T_\mathrm{I} T_\mathrm{II}\{-D_y^{(\mathrm{II})}(\bar{C}_{xy}^{(\mathrm{I},\,\mathrm{II})} - \bar{C}_{xy}^{(\mathrm{II},\,\mathrm{I})}) \\ +(D_y^{(\mathrm{II})} - D_y^{(\mathrm{III})})(\bar{C}_{xy}^{(\mathrm{I},\,\mathrm{II})} + \bar{C}_{xy}^{(\mathrm{II},\,\mathrm{I})}) + D_x^{(\mathrm{I})}(\bar{C}_{xy}^{(\mathrm{I},\,\mathrm{II})} + \bar{C}_{xy}^{(\mathrm{II},\,\mathrm{I})})\} \\ p_y^{(\mathrm{II})''} - p_y^{(\mathrm{II})'} + p_y^{(\mathrm{III})''} - p_y^{(\mathrm{III})'} = \sigma_\mathrm{II} V_\mathrm{II} T_\mathrm{II} \mathfrak{H}_y^{(\mathrm{II})} \\ +\tfrac{1}{2}\rho_\mathrm{I}\sigma_\mathrm{II} V_\mathrm{I} V_\mathrm{II} T_\mathrm{I} T_\mathrm{II}\{D_x^{(\mathrm{I})}(\bar{C}_{xy}^{(\mathrm{I},\,\mathrm{II})} - \bar{C}_{xy}^{(\mathrm{II},\,\mathrm{I})}) \\ +(D_y^{(\mathrm{II})} - D_y^{(\mathrm{III})})(\bar{C}_{xy}^{(\mathrm{I},\,\mathrm{II})} + \bar{C}_{xy}^{(\mathrm{II},\,\mathrm{I})}) - D_y^{(\mathrm{II})}(\bar{C}_{xy}^{(\mathrm{I},\,\mathrm{II})} + \bar{C}_{xy}^{(\mathrm{II},\,\mathrm{I})})\}. \end{cases}$$

After compensation of the last terms we thus obtain for the uncertainties of the field averages:

$$(62)\quad \begin{cases} \Delta\mathfrak{E}_x^{(\mathrm{I})} \sim \dfrac{\hbar}{\rho_\mathrm{I}\Delta x_\mathrm{I} V_\mathrm{I} T_\mathrm{I}} + \dfrac{1}{2}\sigma_\mathrm{II}\Delta y_\mathrm{II} V_\mathrm{II} T_\mathrm{II} |\bar{C}_{xy}^{(\mathrm{I},\,\mathrm{II})} - \bar{C}_{xy}^{(\mathrm{II},\,\mathrm{I})}| \\ \Delta\mathfrak{H}_y^{(\mathrm{II})} \sim \dfrac{\hbar}{\sigma_\mathrm{II}\Delta y_\mathrm{II} V_\mathrm{II} T_\mathrm{II}} + \dfrac{1}{2}\rho_\mathrm{I}\Delta x_\mathrm{I} V_\mathrm{I} T_\mathrm{I} |\bar{C}_{xy}^{(\mathrm{I},\,\mathrm{II})} - \bar{C}_{xy}^{(\mathrm{II},\,\mathrm{I})}|, \end{cases}$$

from which the minimal value of their product is obtained as

$$(63)\quad \Delta\mathfrak{E}_x^{(\mathrm{I})}\Delta\mathfrak{H}_y^{(\mathrm{II})} \sim \hbar|\bar{C}_{xy}^{(\mathrm{I},\,\mathrm{II})} - \bar{C}_{xy}^{(\mathrm{II},\,\mathrm{I})}|,$$

again in complete agreement with the quantum electromagnetic formalism.

Furthermore, from the general arguments at the end of the previous section it follows that also in the case considered here the utilization of field measurements for testing the formalism's assertions in no way is impaired by the classical evaluation of the field effects. Besides, in measurements of averages of dissimilar field components the question of repeatability does not arise at all, and the pure black-body fluctuations are included in all theoretical assertions as an unavoidable statistical feature.

8. Concluding remarks

We thus arrive at the conclusion already stated at the beginning, that with respect to the measurability question the quantum theory of fields represents a consistent idealization to the extent that we can disregard all limitations due to the atomic structure of the field sources and the measuring instruments. As already emphasized in the Introduction, this result should properly be regarded as an immediate consequence of the fact that both the quantum electromagnetic formalism and the viewpoints on which the possibilities of testing this formalism are to be assessed have as their common foundation the correspondence argument. Nevertheless, it would seem that the somewhat complicated character of the considerations used to demonstrate the agreement between formalism and measurability are hardly avoidable. For in the first place the physical requirements to be imposed on the measuring arrangement are conditioned by the integral form in which the assertions of the quantum-electromagnetic formalism are expressed, whereby the peculiar simplicity of the classical field theory as a purely differential theory is lost. Furthermore, as we have seen, the interpretation of the measuring results and their utilization by means of the formalism require consideration of certain features of the complementary mode of description which do not appear in the measurement problems of non-relativistic quantum mechanics.

At the completion of this work we should not like to leave unmentioned that we have found much stimulation and help in many discussions of the questions considered with past and present colleagues at the Institute, among them Heisenberg and Pauli as well as Landau and Peierls.

Universitetets Institut for teoretisk Fysik
Copenhagen, April 1933

NOTES

* Translated by Prof. Aage Petersen; revised by RSC and JS.

[1] W. Heisenberg, *The Physical Principles of the Quantum Theory*, transl. by C. Eckart and F. Hoyt (Dover, New York, 1930), pp. 42 ff.

[2] L. Landau and R. Peierls, *Zs. f. Phys.* **69** (1931), 56.

[3] Cf. N. Bohr, 'Atomic Stability and Conservation Laws', *Atti del Congresso di Fisica Nucleare* (1932). Added in the proof: A separate publication to appear shortly will contain a discussion of the consequences for the problems discussed in the cited reference implied by the recent discovery of the occurrence, under special circumstances, of so-called 'positive electrons'; and by the recognition of the connection of this discovery with Dirac's relativistic electron theory.

[4] Cf. L. Rosenfeld, *Zs. f. Phys.* **76** (1932), 729.

[5] Cf. P. Jordan and W. Pauli, *Zs. f. Phys.* **47** (1928), 151 and also W. Heisenberg and W. Pauli, *Zs. f. Phys.* **56** (1929), 33. Apart from an unessential difference in sign resulting from a difference in the choice of time direction in the Fourier decomposition of the field strengths, the formulae above are equivalent in content with those derived in the papers quoted. In particular, the notation used here, where all terms appear as retarded, is a purely formal change which aims at an interpretation of the measurement problems that is as intuitive as possible.

[6] Cf. L. Rosenfeld and J. Solomon, *J. de Physique* **2** (1931), 139 and also W. Pauli, *Handbuch d. Physik*, 2nd edition, Vol. 24/1 (1933), p. 255.

[7] Cf. W. Heisenberg, *Leipziger Berichte* **83** (1931), 1.

[8] Cf. V. Fock and P. Jordan, *Zs. f. Phys.* **66** (1930), 206, where reference is made to such restrictions on field measurements, which are unrelated to the quantum theory of fields. Cf. also J. Solomon, *J. de Physique* **4** (1933), 368.

[9] Cf. W. Pauli, *Handbuch d. Physik*, 2nd edition, Vol. 24/1 (1933), p. 257.

[10] See N. Bohr, *Atomic Theory and the Description of Nature* (Cambridge University Press, Cambridge, England, 1934). In the meantime, this question has been treated in more detail by the author in a guest lecture in Vienna, to appear shortly, in which in particular the paradoxes arising in the interpretation of the indeterminacy principle when account is taken of the requirements of relativity are further discussed.

IV.3 FIELD AND CHARGE MEASUREMENTS IN QUANTUM ELECTRODYNAMICS

NIELS BOHR AND LÉON ROSENFELD

ABSTRACT. A survey is given of the problem of measurability in quantum electrodynamics and it is shown that it is possible in principle, by the use of idealized measuring arrangements, to achieve full conformity with the interpretation of the formalism as regards the determination of field and charge quantities.

INTRODUCTION

Recent important contributions[1] to quantum electrodynamics by Tomonaga. Schwinger and others have shown that the problem of the interaction between charged particles and electromagnetic fields can be treated in a manner satisfying at every step the requirements of relativistic covariance. In this formulation, essential use is made of a representation of the electromagnetic field components on the one hand, and of the quantities specifying the electrified particles on the other, corresponding to a vanishing interaction between field and particles. The account of such interaction is subsequently introduced by an approximation procedure based on an expansion in powers of the nondimensional constant $e^2/\hbar c$. As regards the interpretation of the formalism, this method has the advantage of a clear emphasis on the dualistic aspect of electrodynamics. In fact, an unambiguous definition of the electromagnetic field quantities rests solely on the consideration of the momentum imparted to appropriate test bodies carrying charges or currents, while the charge-current distributions referring to the presence of particles are ultimately defined by the fields to which these distributions give rise.

Just from this point of view the problem of the measurability of field quantities has been discussed by the authors in a previous paper.[2] A similar investigation of the measurability of electric charge density was then also undertaken, but, owing to various circumstances, its publication has been delayed.[3] When recently the work was resumed, it appeared that by making use of the new development as regards the formulation of quantum electrodynamics a more general and exhaustive treatment

Originally published in *Physical Review, 78*, 794-98 (1950); reproduced here from *Selected Papers of Léon Rosenfeld*, Cohen and Stachel, eds. (1979), Reidel, Dordrecht, pp. 401-412.

could be obtained.[4] As these considerations my be helpful in the current discussion of the situation in atomic physics, we shall here give a brief account of the implications of present electron theory for measurements of charge-current densities. For this purpose, it will be convenient to start with a summary of our earlier treatment of the measurability of field quantities.[5]

1. Measurements of electromagnetic fields

Classical electrodynamics operates with the idealization of field components $f_{\mu\nu}(x)$ defined at every point (x) of space-time. Although in the quantum theory of fields these concepts are formally upheld, it is essential to realize that only averages of such field components over finite space-time regions R, like

$$(1) \qquad F_{\mu\nu}(R)=\frac{1}{R}\int_R f_{\mu\nu}(x)\,\mathrm{d}^4x$$

have a well-defined meaning (I, §2). In the initial step of approximation, in which all effects involving $e^2/\hbar c$ are disregarded, these averages obey commutation relations of the general form

$$(2) \qquad [F_{\mu\nu}(R), F_{\kappa\lambda}(R')]=i\hbar c\,[A_{\mu\nu,\kappa\lambda}(R, R')-A_{\kappa\lambda,\mu\nu}(R', R)],$$

where the expressions of the type $A_{\mu\nu,\kappa\lambda}(R, R')$, defined as integrals over the space-time regions R and R' of certain singular functions, have finite values depending on the shapes and relative situation of the regions R and R'.

The measurement of a field average $F_{\mu\nu}(R)$ demands the control of the total momentum transferred within the space-time region R to a system of movable test bodies with an appropriate distribution of charge or current, of density ρ_ν, covering the whole part of space which at any time belongs to the region R. In the case of an electric field component F_{4l}, we shall take a distribution of charge, with constant density ρ_4, and in the case of a magnetic field component F_{mn}, a uniform distribution of current in a perpendicular direction, with density components ρ_m and ρ_n. The field action of such charge-current distribution, so far as it does not originate from the displacements of the test bodies accompanying the momentum control, can in principle be eliminated by the use of fixed auxiliary bodies carrying a charge-current distribution of opposite sign, and constructed in a way which does not hinder the free motion of the test bodies. In the case of a current distribution, such auxiliary bodies are even indispensable in providing closed circuits for the currents by means of some flexible conducting connection with the test bodies. As a result of this compensation, the field sources of the whole measuring arrangement will thus merely be described by a polarization $P_{\mu\nu}$ arising from the uncontrollable displacements of the test bodies in the course of the field measurements.

If the test bodies are chosen sufficiently heavy, we can throughout disregard any latitude in their velocities, but the control of their momentum will of course imply an essential latitude in their position, to the extent demanded by the indeterminacy relation. Still, it is possible, without violating any requirement of quantum mechanics, not only to keep every test body fixed in its original position except during the time interval within the region R corresponding to this position, but also to secure that, during such time intervals, the displacements of all test bodies in the direction of the momentum transfer to be measured, although uncontrollable, are exactly the same. This common displacement D_μ is described, in the case of the measurement of an electric field, by the component D_l parallel to the field component F_{4l}, and when a magnetic field is measured, by the components D_m and D_n perpendicular to F_{mn}. Without imposing any limit on the accuracy of the field measurement, it is, moreover, possible to keep the displacement D_μ arbitrarily small, if only the charge-current density ρ_ν of the test bodies is chosen sufficiently large. By a further refinement of the composite measuring arrangement described in our earlier paper (1, § 3), it is even possible to reduce the measurement of any field average to the momentum control of a single supplementary body, and thus to obtain a still more compendious expression for the ultimate consequences of the general indeterminacy relation.

An essential point in field measurements is, however, the necessity of eliminating so far as possible the uncontrollable contribution to the average field present in R, arising from the displacement of the test bodies in the course of the measurement. In fact, the expectation value of this contribution will vary in inverse proportion to the latitude allowed in the field measurement, since it is proportional to the polarization $P_{\mu\nu} = D_\mu\rho_\nu - D_\nu\rho_\mu$ within the region R. Just this circumstance, however, makes it possible, by a suitable mechanical device, by which a force proportional to their displacement is exerted on the test bodies, to compensate the momentum transferred to these bodies by the uncontrollable field, insofar as the relation of this field to its sources is expressed by classical field theory. With the compensation procedure described, the resulting measurement of $F_{\mu\nu}(R)$ actually fulfils all requirements of the quantum theory of fields as regards the definition of field averages (I, §5). In fact, the incompensable part of the field action of the text bodies due to the essentially statistical character of the elementary processes involving photon emission and absorption, corresponds exactly to the characteristic field fluctuations which in quantum electrodynamics are superposed on all expectation values determined by the field sources.

When the measurement of two field averages $F_{\mu\nu}(R)$ and $F_{\kappa\lambda}(R')$ is considered, it appears (I, §4) that the expectation value of the average field component $\Phi_{\mu\nu,\kappa\lambda}(R, R')$ which the displacement of the test bodies operated in the region R produces in the region R' is equal to the product of $\frac{1}{2}RP_{\mu\nu}$ with the quantity $A_{\mu\nu,\kappa\lambda}(R, R')$ occurring in the commutation relation (2). Likewise, the expectation value of the average component $\Phi_{\kappa\lambda,\mu\nu}(R', R)$ of the field in R due to the test bodies in R' is equal to $\frac{1}{2}R'P'_{\kappa\lambda}A_{\kappa\lambda,\mu\nu}(R', R)$. When optimum compensation of the momenta transferred to the test bodies by these fields is established by suitable devices, making use of a correlation by light signals transmitted between points of the two regions R and R', it can be deduced from the reciprocal indeterminacy of position and momentum control that the only limitations of the measurability of the two field averages considered correspond exactly to the consequences of the commutation rule (2) for such averages (I, §6, 7). In this connection, it must be stressed that the field fluctuations which are inseparable form the incompensable parts of the fields created by the operation of the test bodies, do not imply any restriction in the measurability of a field component in two asymptotically coinciding space-time regions. In fact, we have here to do with a complete analog to the reproducibility of the fixation of observables in quantum mechanics by immediately repeated measurements.

2. Charge-current measurements in initial approximation

In the formalism of quantum electrodynamics, charge-current densities, like field quantities, are introduced by components $j_\nu(x)$ at every space-time point, but, even in the initial approximation in which such symbols are formally commutable, well-defined expressions are only given by integrals of the type

$$(3)\qquad J_\nu(R)=\frac{1}{R}\int_R j_\nu(x)\,\mathrm{d}^4x,$$

representing the average charge-current density within the finite space-time region R. From the fundamental equations of electrodynamics it follows quite generally that

$$(4)\qquad RJ_\nu(R)=\int_R \frac{\partial f_{\nu\mu}}{\partial x_\mu}\,\mathrm{d}^4x=\int_S f_{\nu\mu}\,\mathrm{d}\sigma_\mu,$$

which expresses the definition of the average charge-current density over the region R in terms of the flux of the electromagnetic field through the boundary S of this region. In this four-dimensional representation, such generalized fluxes comprise, of course, besides the ordinary electric field flux defining the average charge density, other expressions pertaining to the average current densities and representing magnetic field circulations and displacement currents.

In the simple special case in which the region R is defined by a fixed spatial extension V and a constant time interval T, the average charge density, in accordance with (4), will be given, in the ordinary vectorial representation, by

$$(5)\qquad J_4(V,T)=\frac{1}{VT}\int_T \mathrm{d}t\int_S \mathbf{E}\mathbf{n}\,\mathrm{d}\sigma,$$

where S is the surface limiting the extension V, and $\mathbf{n}$ the unit vector in the outward normal direction on this surface. In such representation, the average current density will be given by

$$(6)\qquad \mathbf{J}(V,T)=\frac{1}{VT}\int_T \mathrm{d}t\int_S \mathbf{n}\wedge\mathbf{H}\,\mathrm{d}\sigma-\frac{1}{VT}\int_V \mathbf{E}\,\mathrm{d}v\,\Big|_{t_1}^{t_2},$$

where the first term on the right-hand side represents the time integral of the tangential component of the magnetic field integrated over the surface S, while the last term expresses the difference of the volume integrals of the electric field at the beginning and at the end of the time interval T.

The determination of an average charge-current density $J_\nu(R)$ thus demands the measurement of a field flux through the boundary S of the space-time region R. The approach to the problem of such measurement must rationally start from the consideration of the average flux over a thin four-dimensional shell situated at the boundary S, and which for simplicity we shall assume to have a constant thickness in space-time. As in the situation met with in the measurement of an average field component $F_{\mu\nu}(R)$, we shall require for this purpose a system of movable test bodies, filling the space which belongs to the shell at any time with an appropriate uniform charge-current distribution, and whose field actions are ordinarily neutralized by a distribution of opposite sign on fixed, penetrable, auxiliary bodies. For the measurement of an average charge density J_4, it suffices to take a set of test bodies with a uniform charge distribution of density ρ_4, while in the measurement of a current component, J_l, we shall have to use, besides such test bodies, another independent set of freely movable test bodies with a uniform current distribution ρ_l parallel to the current component to be measured.

In the measurement of an average charge density, the estimation of the flux over the shell demands the determination of the algebraic sum of the momenta transferred to the test bodies in the direction of the normal to the instantaneous spatial boundary. The evaluation of this sum, however, does not require independent measurements of the momenta transferred to the individual test bodies within the time intervals during which their positions belong to the space-time shell, but can be obtained by a composite measuring process in which the positions of all test bodies are correlated by suitable devices to secure during these intervals a displacement of every test body in the normal direction by the same amount. By choosing the product of the thickness of the shell and the charge density of the test bodies sufficiently large, it is possible to keep the uncontrollable common displacement D of all the test bodies in the normal direction arbitrarily small, and still to obtain unlimited accuracy for the average flux over the shell. Like in the measurement of a simple field average, it is further possible to achieve an automatic compensation of the uncontrollable contribution to this average flux, due to the fields created by the displacement of the test bodies, and proportional to $D\rho_4$. This compensation will even be complete, in the initial approximation considered, because the field fluctuations, owing to their source-free character, do not give any contribution to the flux. Since these considerations hold for any given thickness of the shell, it is in principle possible, in the asymptotic limit of a sharp boundary, to measure accurately the average charge density within a well-defined space-time region.

In measurements of an average current component J_l, we have to take into account the magnetic circulation as well as the electric field in the space-time shell. Thus, in the special case in which R is defined by a spatial extension V and a time interval T, we have to do, according to (6), not only with a contribution from the time average over T of the magnetic circulation around the direction l within a thin spatial shell on the boundary of V, but also with a contribution representing the difference between the volume integrals over V of the electric field component in the direction l, averaged over two short time invervals at the beginning and at the end of the interval T. The evaluation of these contributions requires measuring procedures of a similar kind as those described above in the case of measurements of simple field averages. While the measurement of the latter contribution demands the control of the momentum in the direction l transferred to a set of test bodies with uniform charge density ρ, the evaluation of the former contribution demands the control of the momentum normal to the spatial boundary transferred to another set of test bodies with uniform current density ρ_l.

Just as in the field or change measurements discussed above, all these operations can be correlated in such a way that the determination of the algebraic sum of the momenta transferred to each test body within the time interval and in the direction required can be reduced to the momentum control of some supplementary body. In such a correlation, all the test bodies of charge density ρ will be subjected during the appropriate time intervals to the same displacement D_l and all the test bodies of current density ρ_l to the same normal displacement D. The interpretation of the current measurement requires further the establishment of a correlation between these two displacement, satisfying the condition $\rho D_l = \rho_l D$. Under such circumstances, it is possible, by choosing ρ and ρ_l sufficiently large, to achieve that the displacements D_l and D be arbitrarily small without imposing any limitation upon the accuracy of the measurement. Moreover, it is possible, by suitable mechanical devices of the kind already mentioned, to obtain a complete automatic elimination of the uncontrollable contributions from the operation of the test bodies to the average current to be measured.

It need hardly be added that the procedure can be extended to quite general space-time regions R, by using an arrangement in which each test body is displaced just in the time interval during which its position belongs to the space-time shell surrounding the region R. In this connection, it may be noted that a compendious four-dimensional description of all the measuring processes pertaining to charge-current components involves a uniform four-vector current distribution in the shell, parallel to the charge-current component to be measured.

Like in charge measurements, all the considerations concerning current measurements are independent of the thickness of the shell, and in principle it is therefore possible, in the initial approximation considered, to determine with unlimited accuracy any average charge-current component $J_\nu(R)$ within a sharply bounded region R. As regards charge-current measurements over two space-time regions, it can easily be seen that, in the limiting case of sharp boundaries, all field actions accompanying the flux measurements will vanish at any point of space-time which does not belong to the boundaries. In conformity with the formalism, there will therefore, to the approximation concerned, be no mutual influence of measurements of average charge-current densities in different space-time regions.

The situation so far described is of course merely an illustration of the compatibility of a consistent mathematical scheme with a strict application of the definition of the physical concepts to which it refers, and is in particular quite independent of the question of the possibility of actually constructing and manipulating test bodies with the required properties. The disregard of all limitations in this respect, which may originate in the atomic constitution of matter, is, however, entirely justified when dealing with quantum electrodynamics in the initial stage of approximation. In fact, at this stage, the formalism is essentially independent of space-time scale, since it contains only the universal constants c and $\hbar$ which alone do not suffice to define any quantity of the dimensions of a length or time interval.

3. Charge-current measurements in pair theory

New aspects of the problem of measurements arise in quantum electrodynamics in the next approximation, in which effects proportional to $e^2/\hbar c$ are taken into consideration, and where we meet with additional features connected with electron pair production induced by the electromagnetic fields. For the commutation rules of the field components, this means in general only a smaller modification expressed by additional terms containing $e^2/\hbar c$. The charge-current quantities, however, will no longer be commutable but will obey commutation relations of the form.

$$(7)\qquad [J_\nu(R), J_\mu(R')] = i\hbar c[B_{\nu\mu}(R, R') - B_{\mu\nu}(R', R)],$$

where the expressions $B_{\nu\mu}(R, R')$ are integrals of singular functions over the regions R and R'. In contrast to the quantities $A_{\mu\nu,\kappa\lambda}(R, R')$ occurring in (2), which depend only on simple spatio-temporal characteristics of the problem, the B's will, however, besides such characteristics, also essentially involve the length $\hbar/mc$ and the period $\hbar/mc^2$, related to the electron mass m.

To approach the problem of the measurability of a charge-current quantity $J_\nu(R)$ in this approximation, we must again consider systems of electrified test bodies operated in a space-time shell on the boundary of the region R, but we shall now have to examine the effect of the charge-current density appearing as a consequence of actual or virtual electron pair production by the field action of the displacement of the test bodies during the measuring process. As we shall see, these effects, which are inseparably connected with the measurements, do not in any way limit the possibilities of testing the theory.[6]

In the first place, the average effect of the polarization of the vacuum by virtual and actual pair production in the measuring process can be eliminated by a compensation arrangement like that previously described. It is true that a direct estimate of these polarization effects in quantum electrodynamics involves divergent expressions which can only be given finite values by some renormalization or regularization procedure.[7] By such a procedure the average polarization effects will give rise to a contribution to the charge current density which is proportional to the common displacement of the test bodies. Thus in the limit of a sharp boundary of the region R we get, denoting the surface polarization on the boundary by P_ν, the expression $RP_\nu B_{\mu\nu}(R, R)$, where the last factor represents the value of $B_{\mu\nu}(R, R')$ in (7) for coinciding space-time extensions.

Moreover, the statistical effects caused by actual production of electron pairs in the measurement process are inseparably connected with the interpretation of the fluctuations of average charge-current densities in quantum electrodynamics. While the mean square deviation of the field component $F_{\mu\nu}(R)$ over a sharply bounded space-time region R has a finite value, finite mean-square fluctuations of charge-current quantities can only be obtained, however, by further averaging over an ensemble of regions R whose boundaries are allowed a certain latitude around some given surface.[8]

This feature finds its exact counterpart in the estimate of the statistical effects of the real pairs which are produced in measurements of charge-current quantities by the indicated procedure. In fact, the mean square fluctuation of an average flux will increase indefinitely with decreasing thickness of the shell in which the test bodies are operated, in just the same way as, according to the formalism, the mean-square fluctuation of the corresponding charge-current density will vary with the latitude of the ensemble of space-time extensions over which the averaging is performed. The appearance of an infinite mean-square fluctuation in a sharply limited space-time region is in no way connected with the divergencies which appear in vacuum polarization effects but is a direct consequence of the fundamental assumptions of the theory, according to which the electrons are regarded as point charges.

In the case of measurements of charge-current averages over two space-time regions, it can be shown that the polarization effects of the manipulation of the test bodies used for the measurement of $J_\nu(R)$ will give rise, in the limit of sharp boundaries, to a contribution to the average charge-current density component of index μ in the region R', equal to the product of the quantity $B_{\mu\nu}(R', R)$ occurring in formula (7) with RP_ν, where P_ν is the surface polarization created on the boundary of R during the measuring process. Conversely, the measurement of $J_\mu(R')$ will give a contribution $R'P'_\mu B_{\nu\mu}(R, R')$ to the average charge-current density of index ν in R. By similar compensation devices as required for two field measurements, it is therefore possible, as readily seen, to obtain an accuracy of measurements of average charge-current densities in two space-time regions subject only to the reciprocal limitation expressed by the commutation relation (7).

4. Concluding remarks

The conformity of the formalism of quantum electrodynamics with the interpretation of idealized field and charge measurements has of course no immediate relation to the question of the scope of the theory and of the actual possibility of measuring the physical quantities with which it deals.

In the present state of atomic physics, the problem of an actual limitation of measurements interpreted by means of the concepts of classical electrodynamics can hardly be fully explored. Still, in view of the great success of quantum electrodynamics in accounting for numerous phenomena, the formal interpretation of which involves space-time coordination of electrons within regions of dimensions far smaller than $\hbar/mc$ and $\hbar/mc^2$, it may be reasonable to assume that measurements within such regions are in principle possible Indeed, the comparatively heavy and highly charged test bodies of such small dimensions and operated over such short time intervals, which would be required for these measurements, might be conceived to be built up of nuclear particles.

Yet, an ultimate limitation of the consistent application of the formalism is indicated by the necessity of introducing forces of short range in nuclear theory, with no analog in classical electrodynamics, and by the circumstance that the ratio between the electron mass and the rest mass of the quanta of the nuclear field has the same order of magnitude as the fundamental parameter $e^2/\hbar c$ of quantum electrodynamics.[9] The further exploration of such problems may, however, demand a radical revision of the foundation for the application of the basic dual concepts of fields and particles.

NOTES

[1] S. Tomonaga, *Prog. Theor. Phys.* **1** (1946), 27; *Phys. Rev.* **74** (1948), 224. J. Schwinger, *Phys. Rev.* **74** (1948), 1439; **75** (1949), 651; **75** (1949), 1912; **76** (1949), 790. F. Dyson, *Phys. Rev.* **75** (1949), 486; **75** (1949), 1736. R. Feynman, *Phys. Rev.* **76** (1949), 749; **76** (1949), 769.

[2] N. Bohr and L. Rosenfeld, *Kgl. Danske Vid. Seis., Math.-fys. Medd.* **12** (1933), No. 8. This paper will be referred to in the following as I. [English translation: this volume, p. 357].

[3] An account of the preliminary results of the investigation, which were discussed at several physical conferences in 1938, has recently been included in the monograph by

A. Pais, *Developments in the Theory of the Electron*, Princeton University Press, Princeton, New Jersey, 1948.

[4] The bearing of this development on the elucidation of the problem of measurability was brought to the attention of the writers in a stimulating correspondence with Professor Pauli.

[5] A more detailed account of the subject with fuller references to the literature will appear later in the Communications of the Copenhagen Academy.

[6] In a paper by Halpern and Johnson, *Phys. Rev.* **59** (1941), 896, arguments are brought forward pointing to a far more restrictive limitation of the field and charge measurements. In these arguments, however, no sufficient separation is made between such actions of the charged test bodies as are directly connected with their use in the measuring procedure and those actions which can be eliminated by appropriate neutralization by auxiliary bodies of opposite charge.

[7] Cf. W. Pauli and F. Villars, *Rev. Mod. Phys.* **21** (1949), 434.

[8] Cf. W. Heisenberg, *Leipziger Ber.* **86** (1934), 317. We are indebted to Drs. Jost and Luttinger for information about their more precise evaluation of charge-current fluctuations, showing that the unlimited increase of the charge-current fluctuations in a space-time region with decreasing latitude in the fixation of its boundary involves only the logarithm of the ratio between the linear dimensions of the region and the width of this latitude. Even a latitude very small compared with $\hbar/mc$ will therefore imply no excessive effect of the charge fluctuations. A situation entirely similar in all such respects to that in electron theory is met with in a quantum electrodynamics dealing with electrical particles of spin zero which obey Bose statistics. We are indebted to Dr. Corinaldesi for the communication of his results regarding the charge-current fluctuations and pair production effects in such a theory.

[9] Cf., e.g., N. Bohr, *Report of the Solvay Council* (1948).

V

Irreversibility and Quantum Theory

V.1 THE DECREASE OF ENTROPY BY INTELLIGENT BEINGS

COMMENTARY OF
BEHAVIORAL SCIENCE (1964)

In memory of Leo Szilard, who passed away on May 30, 1964, we present an English translation of his classical paper *Über die Entropieverminderung in einem thermodynamischen System bei Eingriffen intelligenter Wesen*, which appeared in the *Zeitschrift für Physik*, 1929, 53, 840-856. The publication in this journal of this translation was approved by Dr. Szilard before he died, but he never saw the copy. At Mrs. Szilard's request, Dr. Carl Eckart revised the translation.

This is one of the earliest, if not the earliest paper, in which the relations of physical entropy to information (in the sense of modern mathematical theory of communication) were rigorously demonstrated and in which Maxwell's famous demon was successfully exorcised: a milestone in the integration of physical and cognitive concepts.

V.1 ON THE DECREASE OF ENTROPY IN A THERMODYNAMIC SYSTEM BY THE INTERVENTION OF INTELLIGENT BEINGS

LEO SZILARD

Translated by Anatol Rapoport and Mechthilde Knoller from the original article "Über die Entropieverminderung in einem thermodynamischen System bei Eingriffen intelligenter Wesen." Zeitschrift für Physik, 1929, 53, 840–856.

The objective of the investigation is to find the conditions which apparently allow the construction of a perpetual-motion machine of the second kind, if one permits an intelligent being to intervene in a thermodynamic system. When such beings make measurements, they make the system behave in a manner distinctly different from the way a mechanical system behaves when left to itself. We show that it is a sort of a memory faculty, manifested by a system where measurements occur, that might cause a permanent decrease of entropy and thus a violation of the Second Law of Thermodynamics, were it not for the fact that the measurements themselves are necessarily accompanied by a production of entropy. At first we calculate this production of entropy quite generally from the postulate that full compensation is made in the sense of the Second Law (Equation [1]). Second, by using an inanimate device able to make measurements—however under continual entropy production—we shall calculate the resulting quantity of entropy. We find that it is exactly as great as is necessary for full compensation. The actual production of entropy in connection with the measurement, therefore, need not be greater than Equation (1) requires.

THERE is an objection, already historical, against the universal validity of the Second Law of Thermodynamics, which indeed looks rather ominous. The objection is embodied in the notion of Maxwell's demon, who in a different form appears even nowadays again and again; perhaps not unreasonably, inasmuch as behind the precisely formulated question quantitative connections seem to be hidden which to date have not been clarified. The objection in its original formulation concerns a demon who catches the fast molecules and lets the slow ones pass. To be sure, the objection can be met with the reply that man cannot in principle foresee the value of a thermally fluctuating parameter. However, one cannot deny that we can very well measure the value of such a fluctuating parameter and therefore could certainly gain energy at the expense of heat by arranging our interven-

Translation originally published in *Behavioral Science*, *9*, 301-10 (1964).

tion according to the results of the measurements. Presently, of course, we do not know whether we commit an error by not including the intervening man into the system and by disregarding his biological phenomena.

Apart from this unresolved matter, it is known today that in a system left to itself no "perpetuum mobile" (perpetual motion machine) of the second kind (more exactly, no "automatic machine of continual finite work-yield which uses heat at the lowest temperature") can operate in spite of the fluctuation phenomena. A perpetuum mobile would have to be a machine which in the long run could lift a weight at the expense of the heat content of a reservoir. In other words, if we want to use the fluctuation phenomena in order to gain energy at the expense of heat, we are in the same position as playing a game of chance, in which we may win certain amounts now and then, although the expectation value of the winnings is zero or negative. The same applies to a system where the intervention from outside is performed strictly periodically, say by periodically moving machines. We consider this as established (Szilard, 1925) and intend here only to consider the difficulties that occur when intelligent beings intervene in a system. We shall try to discover the quantitative relations having to do with this intervention.

Smoluchowski (1914, p. 89) writes: "As far as we know today, there is no automatic, permanently effective perpetual motion machine, in spite of the molecular fluctuations, but such a device might, perhaps, function regularly if it were appropriately operated by intelligent beings. . . ."

A perpetual motion machine therefore is possible if—according to the general method of physics—we view the experimenting man as a sort of *deus ex machina*, one who is continuously and exactly informed of the existing state of nature and who is able to start or interrupt the macroscopic course of nature at any moment without expenditure of work. Therefore he would definitely not have to possess the ability to catch single molecules like Maxwell's demon, although he would definitely be different from real living beings in possessing the above abilities. In eliciting any physical effect by action of the sensory as well as the motor nervous systems a degradation of energy is always involved, quite apart from the fact that the very existence of a nervous system is dependent on continual dissipation of energy.

Whether—considering these circumstances—real living beings could continually or at least regularly produce energy at the expense of heat of the lowest temperature appears very doubtful, even though our ignorance of the biological phenomena does not allow a definite answer. However, the latter questions lead beyond the scope of physics in the strict sense.

It appears that the ignorance of the biological phenomena need not prevent us from understanding that which seems to us to be the essential thing. We may be sure that intelligent living beings—insofar as we are dealing with their intervention in a thermodynamic system—can be replaced by nonliving devices whose "biological phenomena" one could follow and determine whether in fact a compensation of the entropy decrease takes place as a result of the intervention by such a device in a system.

In the first place, we wish to learn what circumstance conditions the decrease of entropy which takes place when intelligent living beings intervene in a thermodynamic system. We shall see that this depends on a certain type of coupling between different parameters of the system. We shall consider an unusually simple type of these ominous couplings.[1] For brevity we shall talk about a "measurement," if we succeed in coupling the value of a parameter y (for instance the position co-ordinate of a pointer of a measuring instrument) at one moment with the simultaneous value of a fluctuating parameter x of the system, in such a way that, from the value y, we can draw conclusions about the value that x had at the moment of the "measurement." Then let x and y be uncoupled after the measurement, so that x can change, while y retains its value for some time. Such measurements are not harmless interventions. A system in which such measurements occur shows a sort of memory

[1] The author evidently uses the word "ominous" in the sense that the possibility of realizing the proposed arrangement threatens the validity of the Second Law.—*Translator*

faculty, in the sense that one can recognize by the state parameter y what value another state parameter x had at an earlier moment, and we shall see that simply because of such a memory the Second Law would be violated, if the measurement could take place without compensation. We shall realize that the Second Law is not threatened as much by this entropy decrease as one would think, as soon as we see that the entropy decrease resulting from the intervention would be compensated completely in any event if the execution of such a measurement were, for instance, always accompanied by production of $k \log 2$ units of entropy. In that case it will be possible to find a more general entropy law, which applies universally to all measurements. Finally we shall consider a very simple (of course, not living) device, that is able to make measurements continually and whose "biological phenomena" we can easily follow. By direct calculation, one finds in fact a continual entropy production of the magnitude required by the above-mentioned more general entropy law derived from the validity of the Second Law.

The first example, which we are going to consider more closely as a typical one, is the following. A standing hollow cylinder, closed at both ends, can be separated into two possibly unequal sections of volumes V_1 and V_2 respectively by inserting a partition from the side at an arbitrarily fixed height. This partition forms a piston that can be moved up and down in the cylinder. An infinitely large heat reservoir of a given temperature T insures that any gas present in the cylinder undergoes isothermal expansion as the piston moves. This gas shall consist of a single molecule which, as long as the piston is not inserted into the cylinder, tumbles about in the whole cylinder by virtue of its thermal motion.

Imagine, specifically, a man who at a given time inserts the piston into the cylinder and somehow notes whether the molecule is caught in the upper or lower part of the cylinder, that is, in volume V_1 or V_2. If he should find that the former is the case, then he would move the piston slowly downward until it reaches the bottom of the cylinder. During this slow movement of the piston the molecule stays, of course, above the piston. However, it is no longer constrained to the upper part of the cylinder but bounces many times against the piston which is already moving in the lower part of the cylinder. In this way the molecule does a certain amount of work on the piston. This is the work that corresponds to the isothermal expansion of an ideal gas—consisting of one single molecule—from volume V_1 to the volume $V_1 + V_2$. After some time, when the piston has reached the bottom of the container, the molecule has again the full volume $V_1 + V_2$ to move about in, and the piston is then removed. The procedure can be repeated as many times as desired. The man moves the piston up or down depending on whether the molecule is trapped in the upper or lower half of the piston. In more detail, this motion may be caused by a weight, that is to be raised, through a mechanism that transmits the force from the piston to the weight, in such a way that the latter is always displaced upwards. In this way the potential energy of the weight certainly increases constantly. (The transmission of force to the weight is best arranged so that the force exerted by the weight on the piston at any position of the latter equals the average pressure of the gas.) It is clear that in this manner energy is constantly gained at the expense of heat, insofar as the biological phenomena of the intervening man are ignored in the calculation.

In order to understand the essence of the man's effect on the system, one best imagines that the movement of the piston is performed mechanically and that the man's activity consists only in determining the altitude of the molecule and in pushing a lever (which steers the piston) to the right or left, depending on whether the molecule's height requires a down- or upward movement. This means that the intervention of the human being consists only in the coupling of two position co-ordinates, namely a co-ordinate x, which determines the altitude of the molecule, with another co-ordinate y, which determines the position of the lever and therefore also whether an upward or downward motion is imparted to the piston. It is best to imagine the mass of the piston as large and its speed sufficiently great, so that the thermal agitation of the piston at the temperature in question can be neglected.

In the typical example presented here, we wish to distinguish two periods, namely:

1. The period of *measurement* when the piston has just been inserted in the middle of the cylinder and the molecule is trapped either in the upper or lower part; so that if we choose the origin of co-ordinates appropriately, the x-co-ordinate of the molecule is restricted to either the interval $x > 0$ or $x < 0$;

2. The period of *utilization of the measurement*, "the period of decrease of entropy," during which the piston is moving up or down. During this period the x-co-ordinate of the molecule is certainly not restricted to the original interval $x > 0$ or $x < 0$. Rather, if the molecule was in the upper half of the cylinder during the period of measurement, i.e., when $x > 0$, the molecule must bounce on the downward-moving piston in the lower part of the cylinder, if it is to transmit energy to the piston; that is, the co-ordinate x has to enter the interval $x < 0$. The lever, on the contrary, retains during the whole period its position toward the right, corresponding to downward motion. If the position of the lever toward the right is designated by $y = 1$ (and correspondingly the position toward the left by $y = -1$) we see that during the period of measurement, the position $x > 0$ corresponds to $y = 1$; but afterwards $y = 1$ stays on, even though x passes into the other interval $x < 0$. We see that in the utilization of the measurement the coupling of the two parameters x and y disappears.

We shall say, quite generally, that a parameter y "measures" a parameter x (which varies according to a probability law), if the value of y is directed by the value of parameter x at a given moment. A measurement procedure underlies the entropy decrease effected by the intervention of intelligent beings.

One may reasonably assume that a measurement procedure is fundamentally associated with a certain definite average entropy production, and that this restores concordance with the Second Law. The amount of entropy generated by the measurement may, of course, always be greater than this fundamental amount, but not smaller. To put it precisely: we have to distinguish here between two entropy values. One of them, $\bar{S}_1$, is produced when during the measurement y assumes the value 1, and the other, $\bar{S}_2$, when y assumes the value -1. We cannot expect to get general information about $\bar{S}_1$ or $\bar{S}_2$ separately, but we shall see that *if* the amount of entropy produced by the "measurement" is to compensate the entropy decrease affected by utilization, the relation must always hold good.

$$e^{-\bar{S}_1/k} + e^{-\bar{S}_2/k} \leqq 1 \tag{1}$$

One sees from this formula that one can make one of the values, for instance $\bar{S}_1$, as small as one wishes, but then the other value $\bar{S}_2$ becomes correspondingly greater. Furthermore, one can notice that the magnitude of the interval under consideration is of no consequence. One can also easily understand that it cannot be otherwise.

Conversely, as long as the entropies $\bar{S}_1$ and $\bar{S}_2$, produced by the measurements, satisfy the inequality (1), we can be sure that the expected decrease of entropy caused by the later utilization of the measurement will be fully compensated.

Before we proceed with the proof of inequality (1), let us see in the light of the above mechanical example, how all this fits together. For the entropies $\bar{S}_1$ and $\bar{S}_2$ produced by the measurements, we make the following Ansatz:

$$\bar{S}_1 = \bar{S}_2 = k \log 2 \tag{2}$$

This ansatz satisfies inequality (1) and the mean value of the quantity of entropy produced by a measurement is (of course in this special case independent of the frequencies w_1, w_2 of the two events):

$$\bar{S} = k \log 2 \tag{3}$$

In this example one achieves a decrease of entropy by the isothermal expansion:[2]

$$\begin{aligned} -\bar{s}_1 &= -k \log \frac{V_1}{V_1 + V_2}; \\ -\bar{s}_2 &= -k \log \frac{V_2}{V_1 + V_2}, \end{aligned} \tag{4}$$

[2] The entropy generated is denoted by $\bar{s}_1$, $\bar{s}_2$.

depending on whether the molecule was found in volume V_1 or V_2 when the piston was inserted. (The decrease of entropy equals the ratio of the quantity of heat taken from the heat reservoir during the isothermal expansion, to the temperature of the heat reservoir in question). Since in the above case the frequencies w_1, w_2 are in the ratio of the volumes V_1, V_2, the mean value of the entropy generated is (a negative number):

$$\bar{s} = w_1 \cdot (+\bar{s}_1) + w_2 \cdot (+\bar{s}_2) = \frac{V_1}{V_1 + V_2} k \log \frac{V_1}{V_1 + V_2} + \frac{V_2}{V_1 + V_2} k \log \frac{V_1}{V_1 + V_2} \quad (5)$$

As one can see, we have, indeed

$$\frac{V_1}{V_1 + V_2} k \log \frac{V_1}{V_1 + V_2} + \frac{V_2}{V_1 + V_2} \cdot k \log \frac{V_2}{V_1 + V_2} + k \log 2 \geqq 0 \quad (6)$$

and therefore:

$$\bar{S} + \bar{s} \geqq 0. \quad (7)$$

In the special case considered, we would actually have a full compensation for the decrease of entropy achieved by the utilization of the measurement.

We shall not examine more special cases, but instead try to clarify the matter by a general argument, and to derive formula (1). We shall therefore imagine the whole system—in which the co-ordinate x, exposed to some kind of thermal fluctuations, can be measured by the parameter y in the way just explained—as a multitude of particles, all enclosed in one box. Every one of these particles can move freely, so that they may be considered as the molecules of an ideal gas, which, because of thermal agitation, wander about in the common box independently of each other and exert a certain pressure on the walls of the box—the pressure being determined by the temperature. We shall now consider two of these molecules as chemically different and, in principle, separable by semipermeable walls, if the co-ordinate x for one molecule is in a preassigned interval while the corresponding co-ordinate of the other molecule falls outside that interval. We also shall look upon them as chemically different, if they differ only in that the y co-ordinate is $+1$ for one and -1 for the other.

We should like to give the box in which the "molecules" are stored the form of a hollow cylinder containing four pistons. Pistons A and A' are fixed while the other two are movable, so that the distance BB' always equals the distance AA', as is indicated in Figure 1 by the two brackets. A', the bottom, and B, the cover of the container, are impermeable for all "molecules," while A and B' are semi-permeable; namely, A is permeable only for those "molecules" for which the parameter x is in the preassigned interval, i.e., (x_1, x_2), B' is only permeable for the rest.

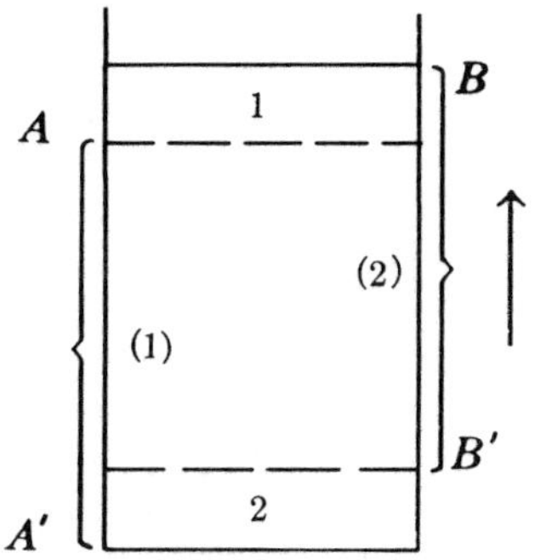

FIG. 1

In the beginning the piston B is at A and therefore B' at A', and all "molecules" are in the space between. A certain fraction of the molecules have their co-ordinate x in the preassigned interval. We shall designate by w_1 the probability that this is the case for a randomly selected molecule and by w_2 the probability that x is outside the interval. Then $w_1 + w_2 = 1$.

Let the distribution of the parameter y be over the values $+1$ and -1 in any proportion but in any event independent of the x-values. We imagine an intervention by an intelligent being, who imparts to y the value 1 for all "molecules" whose x at that moment is in the selected interval. Otherwise the value -1 is assigned. If then, because of thermal fluctuation, for any "molecule," the parameter x should come out of the preassigned interval or, as we also may put it, if the "molecule" suffers a monomolecular chemical reaction with regard to x (by which it is transformed from a species that can pass the semipermeable piston A into a species for which the piston is impermeable), then the parameter y retains its value 1 for the time being, so that the "molecule," because of the value of the parameter y, "remembers" during the whole following process that x originally was in the preassigned interval. We shall see immediately what part this memory may play. After the intervention just discussed, we move the piston, so that we separate the two kinds of molecules without doing work. This results in two containers, of which the first contains only the one modification and the second only the other. Each modification now occupies the same volume as the mixture did previously. In one of these containers, if considered by itself, there is now no equilibrium with regard to the two "modifications in x." Of course the ratio of the two modifications has remained $w_1:w_2$. If we allow this equilibrium to be achieved in both containers independently and at constant volume and temperature, then the entropy of the system certainly has increased. For the total heat release is 0, since the ratio of the two "modifications in x" $w_1:w_2$ does not change. If we accomplish the equilibrium distribution in both containers in a reversible fashion then the entropy of the rest of the world will decrease by the same amount. Therefore the entropy increases by a negative value, and, the value of the entropy increase per molecule is exactly:

$$\bar{s} = k(w_1 \log w_1 + w_2 \log w_2). \qquad (9)$$

(The entropy constants that we must assign to the two "modifications in x" do not occur here explicitly, as the process leaves the total number of molecules belonging to the one or the other species unchanged.)

Now of course we cannot bring the two gases back to the original volume without expenditure of work by simply moving the piston back, as there are now in the container—which is bounded by the pistons BB'—also molecules whose x-co-ordinate lies outside of the preassigned interval and for which the piston A is not permeable any longer. Thus one can see that the calculated decrease of entropy (Equation [9]) does not mean a contradiction of the Second Law. *As long as we do not use the fact that the molecules in the container BB', by virtue of their co-ordinate y, "remember" that the x-co-ordinate for the molecules of this container originally was in the preassigned interval, full compensation exists for the calculated decrease of entropy,* by virtue of the fact that the partial pressures in the two containers are smaller than in the original mixture.

But now we can use the fact that all molecules in the container BB' have the y-co-ordinate 1, and in the other accordingly -1, to bring all molecules back again to the original volume. To accomplish this we only need to replace the semipermeable wall A by a wall A^*, which is semipermeable not with regard to x but with regard to y, namely so that it is permeable for the molecules with the y-co-ordinate 1 and impermeable for the others. Correspondingly we replace B' by a piston B'^*, which is impermeable for the molecules with $y = -1$ and permeable for the others. Then both containers can be put into each other again without expenditure of energy. The distribution of the y-co-ordinate with regard to 1 and -1 now has become statistically independent of the x-values and besides we are able to re-establish the original distribution over 1 and -1. Thus we would have gone through a complete cycle. The only change that we have to register is the resulting decrease of entropy given by (9):

$$\bar{s} = k(w_1 \log w_1 + w_2 \log w_2). \qquad (10)$$

If we do not wish to admit that the Second Law has been violated, we must conclude *that the intervention which establishes the coupling between y and x, the measurement of x by y, must be accompanied by a production of entropy.* If a definite way of achieving this coupling is adopted and if the quantity of entropy that is inevitably produced is designated by S_1 and S_2, where S_1 stands for the mean increase in entropy that occurs when y acquires the value 1, and accordingly S_2 for the increase that occurs when y acquires the value -1, we arrive at the equation:

$$w_1 S_1 + w_2 S_2 = \bar{S} \qquad (11)$$

In order for the Second Law to remain in force, this quantity of entropy must be greater than the decrease of entropy $\bar{s}$, which according to (9) is produced by the utilization of the measurement. Therefore the following unequality must be valid:

$$\bar{S} + \bar{s} \geqq 0$$

$$w_1 S_1 + w_2 S_2 \quad (12)$$

$$+ k(w_1 \log w_1 + w_2 \log w_2) \geqq 0$$

This equation must be valid for any values of w_1 and w_2,[3] and of course the constraint $w_2 + w_2 = 1$ cannot be violated. We ask, in particular, for which w_1 and w_2 and given S-values the expression becomes a minimum. For the two minimizing values w_1 and w_2 the inequality (12) must still be valid. Under the above constraint, the minimum occurs when the following equation holds:

$$\frac{S_1}{k} + \log w_1 = \frac{S_2}{k} + \log w_2 \quad (13)$$

But then:

$$e^{-\bar{S}_1/k} + e^{-\bar{S}_2/k} \leqq 1. \quad (14)$$

This is easily seen if one introduces the notation

$$\frac{S_1}{k} + \log w_1 = \frac{S_2}{k} + \log w_2 = \lambda; \quad (15)$$

then:

$$w_1 = e^{\lambda} \cdot e^{-S_1/k}; \quad w_2 = e^{\lambda} \cdot e^{-S_2/k}. \quad (16)$$

If one substitutes these values into the inequality (12) one gets:

$$\lambda e^{\lambda}(e^{-S_1/k} + e^{-S_2/k}) \geqq 0. \quad (17)$$

Therefore the following also holds:

$$\lambda \geqq 0. \quad (18)$$

If one puts the values w_1 and w_2 from (16) into the equation $w_1 + w_2 = 1$, one gets

$$e^{-S_1/k} + e^{-S_2/k} = e^{-\lambda}. \quad (19)$$

And because $\lambda \geqq 0$, the following holds:

$$e^{-S_1/k} + e^{-S_2/k} \leqq 1. \quad (20)$$

This equation must be universally valid, if thermodynamics is not to be violated.

As long as we allow intelligent beings to perform the intervention, a direct test is not possible. But we can try to describe simple nonliving devices that effect such coupling, and see if indeed entropy is generated and in what quantity. Having already recognized that the only important factor is a certain characteristic type of coupling, a "measurement," we need not construct any complicated models which imitate the intervention of living beings in detail. We can be satisfied with the construction of this particular type of coupling which is accompanied by memory.

In our next example, the position co-ordinate of an oscillating pointer is "measured" by the energy content of a body K. The pointer is supposed to connect, in a purely mechanical way, the body K—by whose energy content the position of the pointer is to be measured—by heat conduction with one of two intermediate pieces, A or B. The body is connected with A as long as the co-ordinate—which determines the position of the pointer—falls into a certain preassigned, but otherwise arbitrarily large or small interval a, and otherwise if the co-ordinate is in the interval b, with B. Up to a certain moment, namely the moment of the "measurement," both intermediate pieces will be thermally connected with a heat reservoir at temperature T_0. At this moment the insertion A will be cooled reversibly to the temperature T_A, e.g., by a periodically functioning mechanical device. That is, after successive contacts with heat reservoirs of intermediate temperatures, A will be brought into contact with a heat reservoir of the temperature T_A. At the same time the insertion B will be heated in the same way to temperature T_B. Then the intermediate pieces will again be isolated from the corresponding heat reservoirs.

We assume that the position of the pointer changes so slowly that all the operations that we have sketched take place while the position of the pointer remains unchanged. If the position co-ordinate of the pointer fell in the preassigned interval, then the body was connected with the insertion A during the above-mentioned operation, and consequently is now cooled to temperature T_A.

In the opposite case, the body is now heated to temperature T_B. Its energy content becomes—according to the position of

[3] The increase in entropy can depend only on the types of measurement and their results but not on how many systems of one or the other type were present.

the pointer at the time of "measurement"—small at temperature T_A or great at temperature T_B and will retain its value, even if the pointer eventually leaves the preassigned interval or enters into it. After some time, while the pointer is still oscillating, one can no longer draw any definite conclusion from the energy content of the body K with regard to the momentary position of the pointer but one can draw a definite conclusion with regard to the position of the pointer at the time of the measurement. Then the measurement is completed.

After the measurement has been accomplished, the above-mentioned periodically functioning mechanical device should connect the thermally isolated insertions A and B with the heat reservoir T_0. This has the purpose of bringing the body K—which is now also connected with one of the two intermediate pieces—back into its original state. The direct connection of the intermediate pieces and hence of the body K—which has been either cooled to T_A or heated to T_B—to the reservoir T_0 consequently causes an increase of entropy. This cannot possibly be avoided, because it would make no sense to heat the insertion A reversibly to the temperature T_0 by successive contacts with the reservoirs of intermediate temperatures and to cool B in the same manner. After the measurement we do not know with which of the two insertions the body K is in contact at that moment; nor do we know whether it had been in connection with T_A or T_B in the end. Therefore neither do we know whether we should use intermediate temperatures between T_A and T_0 or between T_0 and T_B.

The mean value of the quantity of entropy S_1 and S_2, per measurement, can be calculated, if the heat capacity as a function of the temperature $\bar{u}(T)$ is known for the body K, since the entropy can be calculated from the heat capacity. We have, of course, neglected the heat capacities of the intermediate pieces. If the position co-ordinate of the pointer was in the preassigned interval at the time of the "measurement," and accordingly the body in connection with insertion A, then the entropy conveyed to the heat reservoirs during successive cooling was

$$\int_{T_A}^{T_0} \frac{1}{T} \frac{d\bar{u}}{dT} \cdot \qquad (21)$$

However, following this, the entropy withdrawn from the reservoir T_0 by direct contact with it was

$$\frac{\bar{u}(T_0) - \bar{u}(T_A)}{T_0} \cdot \qquad (22)$$

All in all the entropy was increased by the amount

$$S_A = \frac{\bar{u}(T_A) - \bar{u}(T_0)}{T_0} + \int_{T_A}^{T_0} \frac{1}{T} \frac{d\bar{u}}{dT} dT. \quad (23)$$

Analogously, the entropy will increase by the following amount, if the body was in contact with the intermediate piece B at the time of the "measurement":

$$S_B = \frac{\bar{u}(T_B) - \bar{u}(T_0)}{T_0} + \int_{T_B}^{T_0} \frac{1}{T} \frac{d\bar{u}}{dT} dT. \quad (24)$$

We shall now evaluate these expressions for the very simple case, where the body which we use has only two energy states, a lower and a higher state. If such a body is in thermal contact with a heat reservoir at any temperature T, the probability that it is in the lower or upper state is given by respectively:

$$\left.\begin{aligned} p(T) &= \frac{1}{1 + ge^{-u/kT}} \\ q(T) &= \frac{ge^{-u/kT}}{1 + ge^{-u/kT}} \end{aligned}\right\} \qquad (25)$$

Here u stands for the difference of energy of the two states and g for the statistical weight. We can set the energy of the lower state equal to zero without loss of generality. Therefore:[4]

$$\left.\begin{aligned} S_A &= q(T_A)\, k \log \frac{q(T_A)\, p\,(T_0)}{q(T_0)\, p\,(T_A)} \\ &\quad + k \log \frac{p(T_A)}{p(T_0)} \\ S_B &= p(T_B)\, k \log \frac{q(T_0)\, p\,(T_B)}{q(T_B)\, p\,(T_0)} \\ &\quad + k \log \frac{q(T_B)}{q(T_0)} \end{aligned}\right\} . \qquad (26)$$

Here q and p are the functions of T given

[4] See the Appendix.

by equation (25), which are here to be taken for the arguments T_0, T_A, or T_B.

If (as is necessitated by the above concept of a "measurement") we wish to draw a dependable conclusion from the energy content of the body K as to the position co-ordinate of the pointer, we have to see to it that the body surely gets into the lower energy state when it gets into contact with T_B. In other words:

$$p(T_A) = 1,\ q(T_A) = 0; \qquad p(T_B) = 0,\ q(T_B) = 1. \tag{27}$$

This of course cannot be achieved, but may be arbitrarily approximated by allowing T_A to approach absolute zero and the statistical weight g to approach infinity. (In this limiting process, T_0 is also changed, in such a way that $p(T_0)$ and $q(T_0)$ remain constant.) The equation (26) then becomes:

$$S_A = -k \log p(T_0); \qquad S_B = -k \log q(T_0) \tag{28}$$

and if we form the expression $e^{-S_A/k} + e^{-S_B/k}$, we find:

$$e^{-S_A/k} + e^{-S_B/k} = 1. \tag{29}$$

Our foregoing considerations have thus just realized the smallest permissible limiting care. The use of semipermeable walls according to Figure 1 allows a complete utilization of the measurement: inequality (1) certainly cannot be sharpened.

As we have seen in this example, a simple inanimate device can achieve the same essential result as would be achieved by the intervention of intelligent beings. We have examined the "biological phenomena" of a nonliving device and have seen that it generates exactly that quantity of entropy which is required by thermodynamics.

APPENDIX

In the case considered, when the frequency of the two states depends on the temperature according to the equations:

$$p(T) = \frac{1}{1 + ge^{-u/kT}};\ q(T) = \frac{ge^{-u/kT}}{1 + ge^{-u/kT}} \tag{30}$$

and the mean energy of the body is given by:

$$\bar{u}(T) = uq(T) = \frac{uge^{-u/kT}}{1 + ge^{-u/kT}}, \tag{31}$$

the following identity is valid:

$$\frac{1}{T}\frac{d\bar{u}}{dT} = \frac{d}{dT}\left\{\frac{\bar{u}(T)}{T} + k \log\left(1 + e^{-u/kT}\right)\right\}. \tag{32}$$

Therefore we can also write the equation:

$$B_A = \frac{\bar{u}(T_A) - \bar{u}(T_0)}{T_0} + \int_{T_A}^{T_0} \frac{1}{T}\frac{d\bar{u}}{dT}\, dT \tag{33}$$

as

$$S_A = \frac{\bar{u}(T_A) - \bar{u}(T_0)}{T_0} + \left\{\frac{\bar{u}(T)}{T} + k \log(1 + ge^{-u/kT})\right\}_{T_A}^{T_0}, \tag{34}$$

and by substituting the limits we obtain:

$$S_A = \bar{u}(T_A)\left(\frac{1}{T_0} - \frac{1}{T_A}\right) + k \log \frac{1 + ge^{-u/kT_0}}{1 + ge^{-u/kT_A}}. \tag{35}$$

If we write the latter equation according to (25):

$$1 + ge^{-u/kT} = \frac{1}{p(T)} \tag{36}$$

for T_A and T_0, then we obtain:

$$S_A = \bar{u}(T_A)\left(\frac{1}{T_0} - \frac{1}{T_A}\right) + k \log \frac{p(T_A)}{p(T_0)} \tag{37}$$

and if we then write according to (31):

$$\bar{u}(T_A) = uq(T_A) \tag{38}$$

we obtain:

$$S_A = q(T_A)\left(\frac{u}{T_0} - \frac{u}{T_A}\right) + k \log \frac{p(T_A)}{p(T_0)}. \tag{39}$$

If we finally write according to (25):

$$\frac{u}{T} = -k \log \frac{q(T)}{gp(T)} \tag{40}$$

for T_A and T_0, then we obtain:

$$S_A = q(T_A)\, k \log \frac{p(T_0)\, q(T_A)}{q(T_0)\, p(T_A)} + k \log \frac{p(T_A)}{p(T_0)}. \tag{41}$$

We obtain the corresponding equation for S_B, if we replace the index A with B. Then we obtain:

$$S_B = q(T_B)\, k \log \frac{p(T_0)}{q(T_0)} \frac{q((T_B)}{p((T_B)} + k \log \frac{p(T_B)}{p(T_0)}. \quad (42)$$

Formula (41) is identical with (26), given, for S_A, in the text.

We can bring the formula for S_B into a somewhat different form, if we write:

$$q(T_B) = 1 - p(T_B), \quad (43)$$

expand and collect terms, then we get

$$S_B = p(T_B)\, k \log \frac{q(T_0)}{p(T_0)} \frac{p(T_B)}{q(T_B)} + k \log \frac{q(T_B)}{q(T_0)}. \quad (44)$$

This is the formula given in the text for S_B.

REFERENCES

Smoluchowski, F. *Vorträge über die kinetische Theorie der Materie u. Elektrizitat.* Leipzig: 1914.

Szilard, L. Zeitschrift fur Physik, 1925, 32, 753.

V.2 "MEASUREMENT AND REVERSIBILITY" AND "THE MEASURING PROCESS"

JOHN VON NEUMANN

1. MEASUREMENT AND REVERSIBILITY

What happens to a mixture with the statistical operator U, if a quantity $\mathfrak{R}$ with the operator R is measured in it? This operator must be thought of as measuring $\mathfrak{R}$ in each element of the ensemble and collecting the elements that have been thus treated into a new ensemble. We can answer this question -- to the extent to which it admits of an unambiguous answer.

First, let R have a pure discrete, simple spectrum, let $\phi_1, \phi_2, \ldots$ be the complete orthonormal set of eigenfunctions and $\lambda_1, \lambda_2, \ldots$ the corresponding eigenvalues (by assumption, all different from each other). After the measurement, the state of affairs is the following: In the fraction $(U\phi_n, \phi_n)$ of the original ensemble, $\mathfrak{R}$ has the value λ_n $(n = 1, 2, \ldots)$. This fraction then forms an ensemble in which $\mathfrak{R}$ has the value λ_n with certainty (**M.** in IV.3.); it is therefore in the state ϕ_n with the (correctly normalized) statistical operator $P_{[\phi_n]}$. Upon collecting these sub-ensembles, therefore, we obtain a mixture with the statistical operator

$$U' = \sum_{n=1}^{\infty} (U\phi_n, \phi_n) P_{[\phi_n]} \, .$$

Originally published as chapters V and VI of *Mathematische Grundlagen der Quantenmechanik* by John von Neumann, pp. 184-237, Springer, Berlin (1932); translation into English by Robert T. Beyer, *Mathematical Foundations of Quantum Mechanics*, pp. 347-445, Princeton University Press, Princeton (1955).

Second, let R have just a pure discrete spectrum, and let the meaning of $\phi_1, \phi_2, \ldots$ and $\lambda_1, \lambda_2, \ldots$ be as before, except that the eigenvalues λ_n are not all simple -- i.e., among the λ_n there are coincidences.* Then the measuring process of $\mathfrak{R}$ is not uniquely defined (the same was the case, for example, with $\mathfrak{E}$ in IV.3.). Indeed: Let $\mu_1, \mu_2, \ldots$ be distinct real numbers, and S the operator corresponding to the $\phi_1, \phi_2, \ldots$ and $\mu_1, \mu_2, \ldots$. Let $\mathfrak{S}$ be the corresponding quantity. If F(x) is a function with

$$F(\mu_n) = \lambda_n \qquad (n = 1, 2, \ldots)$$

then $F(S) = R$, therefore $F(\mathfrak{S}) = \mathfrak{R}$. Hence the $\mathfrak{S}$ measurement can also be regarded as an $\mathfrak{R}$ measurement. This now changes U into the U' given above, and U' is independent of the (entirely arbitrary) $\mu_1, \mu_2, \ldots$, but not of the $\phi_1, \phi_2, \ldots$. Yet the $\phi_1, \phi_2, \ldots$ are not uniquely determined, because of the multiplicity of the eigenvalues of R . In IV.2., we stated (following II.8.), what can be said regarding the $\phi_1, \phi_2, \ldots$: Let $\lambda', \lambda'', \ldots$ be the different eigenvalues among the $\lambda_1, \lambda_2, \ldots$, let $\mathfrak{M}_{\lambda'}, \mathfrak{M}_{\lambda''}, \ldots$ be the sets of the f with $Rf = \lambda' f$, $Rf = \lambda'' f$, ... respectively. Finally, let $\chi_1', \chi_2', \ldots$; $\chi_1'', \chi_2'', \ldots$; ... , respectively be arbitrary orthonormal sets which span $\mathfrak{M}_{\lambda'}, \mathfrak{M}_{\lambda''}, \ldots$. Then $\chi_1', \chi_2', \ldots, \chi_1'', \chi_2'', \ldots, \ldots$ is the most general $\phi_1, \phi_2, \ldots$ set. Hence U' may be depending upon the choice of $\mathfrak{S}$, i.e., depending upon the actual measuring arrangement, any expression

$$U' = \sum_n (U\chi_n', \chi_n')P_{[\chi_n']} + \sum_n (U\chi_n'', \chi_n'')P_{[\chi_n'']} + \cdots .$$

This expression, however, is unambiguous only in special cases.

We determine this special case. Each individual term must be unambiguous. That is, for each eigenvalue λ ,

* Eds. note: A. S. Wightmann informs us that von Neumann's expression here is incorrect for the density matrix after a measurement in which the observable in question has degenerate eigenvalues (Lüders, 1951; Furry, 1966).

if $\mathfrak{M}_\lambda$ is the set of the f with $Rf = \lambda f$, the sum

$$\sum_n (U\chi_n, \chi_n) P_{[\chi_n]}$$

must have the same value for every choice of the orthonormal set $\chi_1, \chi_2, \ldots$ spanning the manifold $\mathfrak{M}_\lambda$. If we call this sum V, then verbatim repetition of the observations in IV.3. (in which the U_0, U, $\mathfrak{M}$ there are to be replaced by U, V, $\mathfrak{M}_\lambda$) shows that we must have $V = c_\lambda P_{\mathfrak{M}}$ (c_λ constant, > 0), and that this is equivalent to the validity of $(Uf, f) = c_\lambda (f, f)$ for all f of $\mathfrak{M}_\lambda$. Since these f are the same as the $P_{\mathfrak{M}_\lambda} g$ for all g, we require: $(UP_{\mathfrak{M}_\lambda} g, P_{\mathfrak{M}_\lambda} g) = c_\lambda (P_{\mathfrak{M}_\lambda} g, P_{\mathfrak{M}_\lambda} g)$, i.e., $(P_{\mathfrak{M}_\lambda} UP_{\mathfrak{M}_\lambda} g, g) = c_\lambda (P_{\mathfrak{M}_\lambda} g, g)$, i.e.,

$$P_{\mathfrak{M}_\lambda} UP_{\mathfrak{M}_\lambda} = c_\lambda P_{\mathfrak{M}_\lambda}$$

for all eigenvalues λ of R. But if this condition, clearly restricting U sharply, is not satisfied, then different arrangements of measurement for $\mathfrak{R}$ can actually transform U into different U'. (Nevertheless, we shall succeed in V.4. in making some statements about the result of a general $\mathfrak{R}$ measurement, on a thermodynamical basis.

Third, let R have no pure discrete spectrum. Then by III.3. (or IV.3., criterion 1.), it is not measurable with absolute precision, and $\mathfrak{R}$ measurements of limited precision (as we discussed in the case referred to) are equivalent to measurements of quantities with pure discrete spectra.

Another type of intervention in material systems, in contrast to the discontinuous, non-causal and instantaneously acting experiments or measurements, is given by the time dependent Schrödinger differential equation. This describes how the system changes continuously and causally in the course of time, if its total energy is known. For states ϕ, these equations are

$$(\mathbf{T}_1.) \qquad \frac{\partial}{\partial t}\phi_t = -\frac{2\pi i}{h} H\phi_t$$

where H is the energy operator.

For the statistical operator of the state ϕ_t, $U_t = P_{[\phi_t]}$, this means:

$$\left(\frac{\partial}{\partial t}U_t\right)f = \frac{\partial}{\partial t}(U_t f) = \frac{\partial}{\partial t}((f,\phi_t)\cdot\phi_t) = \left(f, \frac{\partial}{\partial t}\phi_t\right)\cdot\phi_t + (f,\phi_t)\cdot\frac{\partial}{\partial t}\phi_t$$

$$= -\left(f, \frac{2\pi i}{h}H\phi_t\right)\cdot\phi_t - (f,\phi_t)\frac{2\pi i}{h}H\phi_t$$

$$= \frac{2\pi i}{h}((Hf,\phi_t)\cdot\phi_t - (f,\phi_t)\cdot H\phi_t) = \frac{2\pi i}{h}(U_t H - H U_t)f,$$

that is:

$$(\mathbf{T}_2.) \qquad \frac{\partial}{\partial t}U_t = \frac{2\pi i}{h}(U_t H - H U_t).$$

Now if U_t is not a state, but a mixture of several states, say $P_{[\phi_t^{(1)}]}, P_{[\phi_t^{(2)}]}, \ldots$ with the respective weights $w_1, w_2, \ldots$, then it must be changed in such a way as results from the changes of the individual $P_{[\phi_t^{(1)}]}, P_{[\phi_t^{(2)}]}, \ldots$. By the addition of the corresponding equations $\mathbf{T}_2.$, we recognize that $\mathbf{T}_2.$ holds for this U_t also. Now since all U are such mixtures, or limiting cases of such (for example, each U with finite $\mathrm{Tr}\, U$ is such a mixture), we can claim the general validity of $\mathbf{T}_2.$

In $\mathbf{T}_2.$, moreover, H may also depend on t, just as in the Schrödinger differential equation $\mathbf{T}_1.$ If that is not the case, then we can even given explicit solutions: For $\mathbf{T}_1.$, as we already know,

$$(\mathbf{T}_1'.)\qquad \phi_t = e^{-\frac{2\pi i}{h} t\cdot \mathrm{H}} \phi_0 ,$$

and for $\mathbf{T}_2$.,

$$(\mathbf{T}_2'.)\qquad U_t = e^{-\frac{2\pi i}{h} t\cdot \mathrm{H}} U_0 e^{\frac{2\pi i}{h} t \mathrm{H}} .$$

(It is easily verified that these are solutions, and also that they follow from each other. It is clear also that there is only one solution with a fixed initial ϕ_0 or U_0 respectively: the differential equations $\mathbf{T}_1$., $\mathbf{T}_2$. are of first order in t.)

We therefore have two fundamentally different types of interventions which can occur in a system $\mathbf{S}$ or in an ensemble $[\mathbf{S}_1, \ldots, \mathbf{S}_N]$. First, the arbitrary changes by measurements which are given by the formula

$$(\mathbf{1}.)\qquad U \longrightarrow U' = \sum_{n=1}^{\infty} (U\phi_n, \phi_n) P_{[\phi_n]}$$

($\phi_1, \phi_2, \ldots$ a complete orthonormal set, cf. supra). Second, the automatic changes which occur with passage of time. These are given by the formula

$$(\mathbf{2}.)\qquad U \longrightarrow U_t = e^{-\frac{2\pi i}{h} t \mathrm{H}} U e^{\frac{2\pi i}{h} t \mathrm{H}}$$

(H is the energy operator, t the time; H is independent of t). If H depends on t, then we may divide the time interval under consideration into small time intervals in each one of which H does not change -- or changes only very slightly, and apply **2.** to these individual intervals. Superposition then gives the final result.

We must now analyze in more detail these two types of intervention, their nature, and their relation one to another.

First of all, it is noteworthy that the time

dependence of H is included in **2.** (in the manner described there), so that one should expect that **2.** would suffice to describe the intervention caused by a measurement: Indeed, a physical intervention can be nothing else than the temporary insertion of a certain energy coupling into the observed system, i.e., the introduction of an appropriate time dependency of H (prescribed by the observer). Why then do we need the special process **1.** for the measurement? The reason is this: In the measurement we cannot observe the system **S** by itself, but must rather investigate the system **S** + **M** , in order to obtain (numerically) its interaction with the measuring apparatus **M**. The theory of the measurement is a statement concerning **S** + **M** , and should describe how the state of **S** is related to certain properties of the state of **M** (namely, the positions of a certain pointer, since the observer reads these). Moreover, it is rather arbitrary whether or not one includes the observer in **M** , and replaces the relation between the **S** state and the pointer positions in **M** by the relations of this state and the chemical changes in the observer's eye or even in his brain (i.e., to that which he has "seen" or "perceived"). We shall investigate this more precisely in VI.1. In any case, therefore, the application of **2.** is of importance only for **S** + **M** . Of course, we must show that this gives the same result for **S** as the direct application of **1.** on **S** . If this is successful, then we have achieved a unified way of looking at the physical world on a quantum mechanical basis. We postpone the discussion of this question until VI.3.

Second, it is to be noted, with regard to **1.**, that we have repeatedly shown that a measurement in the sense of **1.** must be instantaneous, i.e., must be carried through in so short a time that the change of U given by **2.** is not yet noticeable. (If we wanted to correct this by calculating the changed U_t by **2.**, we would still gain nothing, because to apply any U_t , we must first know t , the moment of measurement, exactly, i.e., the time duration of the measurement must be short.) This is now questionable in principle, because it is well-known that there is a quantity which, in classical mechanics, is canonically conjugate with the time: the energy.[180] Therefore it is to be expected that for the canonically conjugate pair time-energy, there must exist indeterminacy relations similar to those of the pair cartesian coordinate-momentum.[181] Note that the special relativity theory shows that a far reaching analogy must exist: the three space coordinates and time form a "four vector" as do the three momentum coordinates and the energy. Such an indeterminacy relation would mean that it is not possible to carry out a very precise measurement of the energy in a very short time. In fact, one would expect for the error of measurement (in the energy) and the time duration τ a relation of the form

$$\epsilon\tau \sim h \ .$$

A physical discussion, similar to that carried out in III.4. for p, q, actually leads to this result.[181] Without going into details, we shall consider the case of a light quantum. Its energy uncertainty ϵ is, because of the Bohr frequency condition, h times the frequency uncertainty: $h\Delta\nu$. But, as we discussed in Note 137, $\Delta\nu$ is at best the reciprocal of the time duration, $1/\tau$, i.e., $\epsilon \gtrsim h/\tau$ -- and in order that the monochromatic nature of the light quantum be established in the entire time interval τ, the measurement must extend over this entire time interval. The case of the light quantum is characteristic,

---

[180]Any textbook of classical (Hamiltonian) mechanics gives an account of these connections.

[181]The uncertainty relations for the pair time-energy have been discussed frequently. Cf. the comprehensive treatment of Heisenberg, Die Physikalischen Prinzipien der Quantentheorie, II.2.d., Leipzig, 1930.

since the atomic energy levels, as a rule, are determined from the frequency of the corresponding spectral lines. Since the energy behaves in such fashion, a relation between the precision of measurement for other quantities $\mathfrak{R}$ and the duration of the measurement is also possible. Then how can our assumption of instantaneous measurements be justified?

First of all we must admit that this objection points at an essential weakness which is, in fact, the chief weakness of quantum mechanics: its non-relativistic character, which distinguishes the time t from the three space coordinates x, y, z, and presupposes an objective simultaneity concept. In fact, while all other quantities (especially those x, y, z closely connected with t by the Lorentz transformation) are represented by operators, there corresponds to the time an ordinary number-parameter t, just as in classical mechanics. Or: a system consisting of 2 particles has a wave function which depends on its $2 \times 3 = 6$ space coordinates, and only upon one time t, although, because of the Lorentz transformation, two times would be desirable. It may be connected with this non-relativistic character of quantum mechanics that we can ignore the natural law of minimum duration of the measurements. This might be a clarification, but not a happy one!

A more detailed investigation of the problem, however, shows that the situation is really not so bad as this. For what we really need is not that the change of t be small, but only that it have little effect in the calculation of the probabilities $(U\phi_n, \phi_n)$, and therefore in the formation of

$$U' = \sum_{n=1}^{\infty} (U\phi_n, \phi_n) P_{[\phi_n]}$$

whether we start out from U itself or from a

$$U_t = e^{-\frac{2\pi i}{h} tH} U e^{\frac{2\pi i}{h} tH} .$$

Because of

$$(U_t\phi_n, \phi_n) = \left(e^{-\frac{2\pi i}{h}tH} Ue^{\frac{2\pi i}{h}tH}\phi_n, \phi_n\right)$$

$$= \left(\overline{U}e^{\frac{2\pi i}{h}tH}\phi_n, e^{\frac{2\pi i}{h}tH}\phi_n\right),$$

this can be accomplished by so changing H by an appropriate perturbation energy that

$$e^{\frac{2\pi i}{h}tH}\phi_n$$

differs from ϕ_n only by a constant factor of absolute value 1. That is, the state ϕ_n should be essentially constant under the influence of **2**., i.e., a stationary state; or equivalently $H\phi_n$ must be equal to a real constant times ϕ_n, i.e., ϕ_n an eigenfunction of H. At first glance, such a change of the energy operator H, which makes the eigenfunctions of R stationary, and therefore eigenfunctions of H (i.e., R, H commutative) may seem implausible. But this is not really the case, and one can even see that the typical arrangements of measurement aim at exactly this sort of effect on H.

In fact, each measurement results in the emission of a light quantum or a mass particle, with a certain energy, in a certain direction. It is then by these characteristics, i.e., by its momentum, that the particle expresses the result of the measurement or, a mass point (for example, a pointer on a scale) comes to rest, and its cartesian coordinates give the result of the measurement. In the case of light quanta, using the terminology of III.6., the desired measurement is thus equivalent to the statement as to which $M_n = 1$ (the rest being $= 0$), i.e., to the enumeration of all $M_1, M_2, \ldots$ values. For a moving (departing) mass point, the statement of its three momentum components P^x, P^y, P^z is the corresponding equivalent; for a mass point at rest (the index point), the statement of its three cartesian coordinates x, y, z, or, using their operators, of the Q^x, Q^y, Q^z. But the measurement is completed only if the light quantum or mass point is actually borne "away," i.e., only when the light quantum is not in danger of absorption; or when the mass point may no longer be deflected by potential energies; or, if the mass point is actually at rest, in which case a large mass is necessary.[182] (This latter is certainly necessary because of the uncertainty relations, since the velocity must be near 0, and therefore its dispersion must be small, although its product with the mass -- the momentum -- has a large dispersion, because of the small dispersion of the coordinates. Ordinarily, the pointers are macroscopic objects, i.e., enormous.) Now the energy operator H, so far as it concerns the light quantum, is (III.6, page 270)

$$\sum_{n=1}^{\infty} h\rho_n \cdot M_n$$

$$+ \sum_{p=1}^{\infty} \sum_{n=1}^{\infty} w_{kj}^{n} \left(\sqrt{M_n + 1} \cdot \left(\begin{matrix} k \\ M_n \end{matrix} \rightrightarrows \begin{matrix} j \\ M_n + 1 \end{matrix} \right) \right.$$

$$\left. + \sqrt{M_n} \cdot \begin{matrix} k \\ M_n \end{matrix} \rightrightarrows \begin{matrix} j \\ M_n - 1 \end{matrix} \right) \Bigg)$$

while for both mass point examples, H is given by

$$\frac{(P^x)^2 + (P^y)^2 + (P^z)^2}{2m} + V(Q^x, Q^y, Q^z)$$

---

[182]All other details of the measuring arrangement aim only at the connection of the quantity $\mathfrak{R}$, which is actually of interest, or of its operator R, with the M_n or the P^x, P^y, P^z or the Q^x, Q^y, Q^z, respectively, that have

(m the mass, V the potential energy). Our criteria say: the w^n_{kj} should vanish, or V should be constant, or m should be very large. But this actually produces the effect that the P^x, P^y, P^z and the Q^x, Q^y, Q^z respectively commute with the H given above.

In conclusion, it should be mentioned that the making stationary of the really interesting states (here the $\phi_1, \phi_2, \ldots$) plays a role elsewhere, too, in theoretical physics. The assumptions on the possibility of the interruption of chemical reactions (i.e., their "poisoning"), which are often unavoidable in physical-chemical "ideal experiments," are of this nature.[183]

The two interventions **1.** and **2.** are fundamentally different from one another. That both are formally unique, i.e., causal, is unimportant; indeed, since we are working in terms of the statistical properties of mixtures, it is not surprising that each change, even if it is statistical, effects a causal change of the probabilities and the expectation values. Indeed, it is precisely for this reason, that one introduces statistical ensembles and probabilities! On the other hand, it is important that **2.** does not increase the statistical uncertainty existing in U, but that **1.** does: **2.** transforms states into states

$$\left(P_{[\phi]} \quad \text{into} \quad P_{[e^{-\frac{2\pi i}{h} tH} \phi]} \right),$$

while **1.** can transform states into mixtures. In this sense, therefore, the development of a state according to **1.** is statistical, while according to **2.** it is causal.

---

been mentioned. Of course, this is the most important practical aspect of the measuring technique.

[183] Cf. e.g., Nernst, Theoretische Chemie, Stuttgart (numerous editions since 1893), Book IV, Discussion of the thermodynamic proof of the "mass action law."

Furthermore, for fixed H and t , **2.** is simply a unitary transformation of all U : $U_t = AUA^{-1}$, $A = e^{-\frac{2\pi i}{h} tH}$ is unitary. That is, $Uf = g$ implies that $U_t(Af) = Ag$, so that U_t results from U by the unitary transformation A of Hilbert space, that is, by an isomorphism which leaves all our basic geometric concepts invariant (cf. the principles set down in I.4.). Therefore it is reversible: it suffices to replace A by A^{-1} -- and this is possible, since A, A^{-1} can be regarded as entirely arbitrary unitary operators because of the far reaching freedom in the choice of H, t . Just as in classical mechanics therefore, **2.** does not reproduce one of the most important and striking properties of the real world, namely its irreversibility, the fundamental difference between the time directions, "future" and "past."

1. behaves in a fundamentally different fashion: the transition

$$U \longrightarrow U' = \sum_{n=1}^{\infty} (U\phi_n, \phi_n) P_{[\phi_n]}$$

is certainly not prima facie reversible. We shall soon see that it is in general irreversible, in the sense that it is not possible in general to come back from a given U' to its U by repeated applications of any processes ., **2.**

Therefore, we have reached a point at which it is desirable to utilize the thermodynamical method of analysis, because it alone makes it possible for us to understand correctly the difference between **1.** and **2.**, into which reversibility questions obviously enter.

2. THERMODYNAMICAL CONSIDERATIONS

We shall investigate the thermodynamics of quantum mechanical ensembles according to two different points of view. First, let us assume the validity of both fundamental laws of thermodynamics, i.e., the impossibility of perpetual motion of the first and second kind (energy law and entropy law),[184] and calculate the entropy for each ensemble from this. In this case, normal methods of the phenomenological thermodynamics are applied, and quantum mechanics plays a role only insofar as our thermodynamical observations relate to such objects whose behavior is regulated by the laws of quantum mechanics (our ensembles, as well as their statistical operators U) -- but the correctness of both laws will be assumed and not proved. Afterwards we shall prove the validity of these fundamental laws in quantum mechanics. Since the energy law holds in any case, only the entropy law has to be considered. That is, we shall show that the interventions **1.**, **2.** never decrease the entropy, as calculated by the first method. This order may seem somewhat unnatural, but it is based on the fact that it is by the phenomenological discussion that we obtain that overall view of the problem which is required for considerations of the second kind.

We therefore begin with the phenomenological consideration, which will also permit us to solve a well-known paradox of classical thermodynamics. First we must emphasize that the unusual character of our "ideal experiments," i.e., their practical infeasibility, does not impair their demonstrative power: In the sense of phenomenological thermodynamics, each conceivable process constitutes valid evidence, provided that it does not conflict with the two fundamental laws of thermodynamics.

---

[184] The phenomenological system of thermodynamics built upon this foundation can be found in numerous texts. For example, Planck, Treatise on Thermodynamics, London, 1927. For the following, the statistical aspect of these laws is of chief importance. This is analyzed in the following treatises: Einstein, Verh. d. dtsch. physik, Ges. 12 (1914); Szilard, Z. Physik 32 (1925).

Our purpose is to determine the entropy of an ensemble $[\mathbf{S}_1, \ldots, \mathbf{S}_N]$ with the statistical operator U, where U is assumed to be correctly normalized, i.e., Tr U = 1. In the terminology of classical statistical mechanics, we are dealing with a Gibbs ensemble: i.e., the application of statistics and thermodynamics will be made not on the (interacting) components of a single, very complicated mechanical system with many (only imperfectly known) degrees of freedom[185] -- but on an ensemble of very many (identical) mechanical systems, each of which may have an arbitrarily large number of degrees of freedom, and each of which is entirely separated from the others, and does not interact with any of them.[186] As a consequence of the complete separation of the systems $\mathbf{S}_1, \ldots, \mathbf{S}_N$, and of the fact that we shall apply to them the ordinary methods of enumeration of the calculus of probability, it is evident that ordinary statistics be used, and that the Bose-Einstein and Fermi-Dirac statistics, which differ from those and which are applicable to certain ensembles of indistinguishable and interacting particles (namely, for light quanta or electrons and protons, cf. III.6., in particular, Note 147), do not enter into the problem.

---

[185]This is the Maxwell-Boltzmann method of statistical mechanics (cf. the review in the article of P. and T. Ehrenfest in Enzykl. d. Math. Wiss., Vol. II.4. D., Leipzig, 1907). In the gas theory for example, the "very complicated" system is the gas which consists of many (interacting) molecules, and the molecules are investigated statistically.

[186]This is the Gibbs method (cf. the reference in Note 185). Here the individual system is the entire gas, and many replicas of the same system (i.e., of the same gas) are considered simultaneously, and their properties are evaluated statistically.

The method introduced by Einstein for the thermodynamical treatment of such ensembles $[\mathbf{s}_1, \ldots, \mathbf{s}_N]$ is the following:[187] Each system $\mathbf{s}_1, \ldots, \mathbf{s}_N$ is confined in a box $\mathbf{K}_1, \ldots, \mathbf{K}_N$, whose walls are impenetrable to all transmission effects -- which is possible for this system because of the lack of interaction. Furthermore, each box must have a very large mass, so that the possible state (and hence energy and mass) changes of the $\mathbf{s}_1, \ldots, \mathbf{s}_N$ affects its mass only slightly. Also, their velocities in the ideal experiments which are to be carried out are thereby kept so small that the calculations may be performed non-relativistically. We then enclose these boxes into a very large box $\overline{\mathbf{K}}$ (i.e., the volume $\mathcal{V}$ of $\overline{\mathbf{K}}$ should be much larger than the sum of the volumes of the $\mathbf{K}_1, \ldots, \mathbf{K}_N$). For simplicity, no force field will be present in $\overline{\mathbf{K}}$ (in particular, it should be free from all gravitational fields, and so large that the masses of the $\mathbf{K}_1, \ldots, \mathbf{K}_N$ have no relevant effects either. We can therefore regard the $\mathbf{K}_1, \ldots, \mathbf{K}_N$ (which contain $\mathbf{s}_1, \ldots, \mathbf{s}_N$ respectively) as the molecules of a gas which is enclosed in the large container $\overline{\mathbf{K}}$. If we now bring $\overline{\mathbf{K}}$ into contact with a very large heat reservoir of temperature $\mathbf{T}$, then the walls of $\overline{\mathbf{K}}$ also take on this temperature, and its (true) molecules assume the corresponding Brownian motion. Therefore they will contribute momentum to the adjacent $\mathbf{K}_1, \ldots, \mathbf{K}_N$, so that these engage in motion, and transfer momentum to the other $\mathbf{K}_1, \ldots, \mathbf{K}_N$. Soon all $\mathbf{K}_1, \ldots, \mathbf{K}_N$ will be in motion and will be exchanging momentum (on the wall of $\overline{\mathbf{K}}$) with the (true) molecules of the wall, and with each other (in the interior of $\overline{\mathbf{K}}$) by collision processes. The stationary equilibrium state of motion is then obtained if the $\mathbf{K}_1, \ldots, \mathbf{K}_n$ have taken on that velocity distribution which is in equilibrium with the Brownian motion of the wall

---

[187] See the reference in Note 184. This was further developed by L. Szilard.

molecules (of temperature $\mathbf{T}$) -- i.e., the Maxwellian velocity distribution of a gas of temperature $\mathbf{T}$, the "molecules" of which are the $\mathbf{K}_1, \ldots, \mathbf{K}_N$.[188] We can then say: the $[\mathbf{S}_1, \ldots, \mathbf{S}_N]$-gas has taken on the temperature $\mathbf{T}$. For brevity, we shall call the ensemble $[\mathbf{S}_1, \ldots, \mathbf{S}_N]$ with the statistical operator U the U-ensemble, and the $[\mathbf{S}_1, \ldots, \mathbf{S}_N]$-gas the U-gas.

The reason that we concern ourselves with such a gas is that we must determine the entropy difference of the U-ensemble and the V-ensemble (U, V definite operators with $\mathrm{Tr}\, U = 1$, $\mathrm{Tr}\, V = 1$, and with the corresponding ensembles $[\mathbf{S}_1, \ldots, \mathbf{S}_N]$ and $[\mathbf{S}'_1, \ldots, \mathbf{S}'_N]$). The determination requires by definition a reversible transformation of the former ensemble into the latter,[189] and this is best accomplished by the aid of the U- and V-gases. That is, we maintain that the entropy difference of the U- and V-ensembles is exactly the same as that of the U- and V-gases -- if both are observed at the same temperature $\mathbf{T}$, but are otherwise arbitrary. If $\mathbf{T}$ is very near 0, then this is obviously the case with arbitrary precision; because the difference between the U-ensemble and the V-gas vanishes at the temperature 0, since the $\mathbf{K}_1, \ldots, \mathbf{K}_N$ of the latter have then no motion of their own, and the

---

[188] The kinetic theory of gases, as is well-known, describes in this way that process in which the walls communicate their temperature to the gas enclosed by them. Cf. the references in Notes 184 and 185.

[189] In this transformation, if the heat quantities $\mathbf{Q}_1, \ldots, \mathbf{Q}_i$ are required at the respective temperatures $\mathbf{T}_1, \ldots, \mathbf{T}_i$, then the entropy difference is equal to

$$\frac{\mathbf{Q}_1}{\mathbf{T}_1} + \frac{\mathbf{Q}_2}{\mathbf{T}_2} + \cdots + \frac{\mathbf{Q}_i}{\mathbf{T}_i} .$$

Cf. the reference in Note 184.

presence of the $\mathbf{K}_1,\ldots,\mathbf{K}_N$, $\overline{\mathbf{K}}$, when they are at rest is thermodynamically unimportant (and likewise for V). Therefore we shall have accomplished our aim if we can show that for a given change of $\mathbf{T}$, the entropy of the U-gas changes just as much as the entropy of the V-gas. The entropy change of a gas which is heated from $\mathbf{T}_1$ to $\mathbf{T}_2$ depends only upon its caloric equation of state, or more precisely, upon its specific heat.[190] Naturally, the gas must not be assumed to be an ideal gas here if, as in our case, $\mathbf{T}_1$ must be chosen near 0.[191] On the other hand, it is certain that both gases (U and V) have the same equation of state and the same specific heats because, by kinetic theory, the boxes $\mathbf{K}_1,\ldots,\mathbf{K}_N$ dominate and cover completely the systems $\mathbf{S}_1,\ldots,\mathbf{S}_N$ and $\mathbf{S}'_1,\ldots,\mathbf{S}'_N$ which are enclosed in them. In this heating process therefore, the difference of U and V is not noticeable, and the two entropy differences coincide, as was maintained. In the following therefore, we shall compare only the U- and V-gases with each other, and we shall choose the temperature $\mathbf{T}$ so high that these can be regarded as ideal gases.[192] In this way, we control its kinetic behavior

---

[190]If $c(\mathbf{T})$ is the specific heat at the temperature $\mathbf{T}$ of the gas quantum under discussion, then in the temperature interval $\mathbf{T}$, $\mathbf{T}+d\mathbf{T}$ it takes on the quantity of heat $c(\mathbf{T})d\mathbf{T}$. By Note 185, the entropy difference is then

$$\int_{\mathbf{T}_1}^{\mathbf{T}_2} \frac{c(\mathbf{T})d\mathbf{T}}{\mathbf{T}}$$

[191]For an ideal gas, $c(\mathbf{T})$ is constant; for very small $\mathbf{T}$, this certainly fails. Cf. for example, the reference in Note 6.

[192]In addition to this, it is required that the volume V of $\overline{\mathbf{K}}$ be large in comparison to the total volume of the

completely, and we can apply ourselves to the real problem: to transform U-gas reversibly into V-gas. In this case, in contrast to the processes used so far, we shall also have to consider the $\mathbf{s}_1,\ldots,\mathbf{s}_N$ found in the interior of the $\mathbf{K}_1,\ldots,\mathbf{K}_N$ i.e., we shall have to "open" the boxes $\mathbf{K}_1,\ldots,\mathbf{K}_N$.

Next, we show that all states $U = P_{[\phi]}$ have the same entropy, i.e., that the reversible transformation of the $P_{[\phi]}$ ensemble into the $P_{[\psi]}$ ensemble is accomplished without the absorption or liberation of heat energy (mechanical energy must naturally be consumed or produced if the expectation value of the energy in $P_{[\phi]}$ is different from that in $P_{[\psi]}$), cf. Note 189. In fact, we shall not even have to refer to the gases just considered. This transformation succeeds even at the temperature 0 , i.e., with the ensembles themselves. It should be mentioned, furthermore, that as soon as this is proved, we shall be able to and shall so normalize the entropies of the U ensembles that all states have the entropy 0 .

Moreover, the transformation of $P_{[\phi]}$ into $P_{[\psi]}$ described above does not need to be reversible: Because if it is not so, then the entropy difference must be $\geq$ the expression given in Note 189 (cf. reference in Note 185), therefore ≥ 0 . Permutation of $P_{[\phi]}$, $P_{[\psi]}$ shows that this value must also be ≤ 0 . Therefore the value is $= 0$.

$\mathbf{K}_1,\ldots,\mathbf{K}_N$; furthermore that the "energy per degree of freedom" $\kappa\mathbf{T}$ (κ = Boltzmann's constant) be large in comparison to

$$h^2 / \mu \mathcal{V}^{2/3}$$

(h = Planck's constant, μ = mass of the individual molecule; this quantity is of the dimensions of energy). Cf. for example, Fermi, Z. Physik, 36 (1926).

The simplest process would be to refer to the time dependent Schrödinger differential equation, i.e., our process **2**., in which an energy operator H and a numerical value of t must be found such that the unitary operator

$$e^{-\frac{2\pi i}{h} tH}$$

transforms ϕ into ψ. Then, in t seconds, $P_{[\phi]}$ would change spontaneously into $P_{[\psi]}$. The process is also reversible, and no mention has been made of the heat (cf. V.1.). However, we prefer to avoid assumptions regarding the possible forms of the energy operators H and to apply the process **1**. alone, i.e., measuring interventions. The simplest such measurement would be to measure the quantity $\mathfrak{R}$ in the ensemble $P_{[\phi]}$, whose operator R has a pure discrete spectrum with simple eigenvalues $\lambda_1, \lambda_2, \ldots$, and in which ψ occurs among the eigenfunctions $\psi_1, \psi_2, \ldots$, say $\psi_1 = \psi$. This measurement transforms ϕ into a mixture of the states $\psi_1, \psi_2, \ldots$, and there $\psi_1 = \psi$ will be present along with the other states ψ_n. However, this procedure is unsuitable, because $\psi_1 = \psi$ occurs only with the probability $|(\phi, \psi)|^2$, while the portion $1 - |(\phi, \psi)|^2$ goes over into other states. In fact, the latter portion is the entire result for orthogonal ϕ, ψ. A different experiment however will accomplish our purpose. By repetition of a great number of different measurements, we shall change $P_{[\phi]}$ into such an ensemble, which differs from $P_{[\psi]}$ by an arbitrarily small amount. That all these operators are (or at least, can be) irreversible is unimportant, as we discussed above.

We assume ϕ, ψ orthogonal, since we could otherwise choose a χ ($||\chi|| = 1$) orthogonal to both, and could go from ϕ to χ, and then from χ to ψ. Now let $k = 1, 2, \ldots$ be a number which is at our disposal, and set $\psi^{(\nu)} = \cos\frac{\pi\nu}{2k}\cdot\phi + \sin\frac{\pi\nu}{2k}\cdot\psi$ $(\nu = 0, 1, \ldots, k)$. Clearly, $\psi^{(0)} = \phi$, $\psi^{(k)} = \psi$, and $||\psi^{(\nu)}|| = 1$. We extend each $\psi^{(\nu)}$ $(\nu = 1,2,\ldots,k)$ to a complete orthonormal set $\psi_1^{(\nu)},\psi_2^{(\nu)},\ldots$ with $\psi_1^{(\nu)} = \psi^{(\nu)}$. Let $R^{(\nu)}$ be an operator with a pure discrete spectrum and different eigenvalues, say $\lambda_1^{(\nu)},\lambda_2^{(\nu)},\ldots$, whose eigenfunctions are the $\psi_1^{(\nu)},\psi_2^{(\nu)},\ldots$, and $\mathfrak{R}^{(\nu)}$ the corresponding quantity. We observe further that

$$(\psi^{(\nu-1)}, \psi^{(\nu)}) = \cos\frac{\pi(\nu-1)}{2k}\cos\frac{\pi\nu}{2k} + \sin\frac{\pi(\nu-1)}{2k}\sin\frac{\pi\nu}{2k}$$

$$= \cos\left(\frac{\pi\nu}{2k} - \frac{\pi(\nu-1)}{2k}\right) = \cos\frac{\pi}{2k}\ .$$

In the ensemble with $U^{(o)} = P_{[\phi^{(o)}]} = P_{[\phi]}$ we now measure the quantity $\mathfrak{R}^{(1)}$, in which case $U^{(1)}$ results. We then measure the quantity $\mathfrak{R}^{(2)}$ on $U^{(1)}$, when $U^{(2)}$ results, etc. We finally measure the quantity $\mathfrak{R}^{(k)}$ on $U^{(k-1)}$ whence $U^{(k)}$ results. That $U^{(k)}$, for sufficiently large k, lies arbitrarily close to $P_{[\psi^{(k)}]} = P_{[\psi]}$ can easily be established. If we measure $\mathfrak{R}^{(\nu)}$ on $\psi^{(\nu-1)}$, then the fraction $|(\psi^{(\nu-1)}, \psi^{(\nu)})|^2 = (\cos\frac{\pi}{2k})^2$ goes over into $\psi^{(\nu)}$, and in the successive measurements of $\mathfrak{R}^{(1)}, \mathfrak{R}^{(2)},\ldots, \mathfrak{R}^{(k)}$ therefore, at least the fraction $(\cos\frac{\pi}{2k})^2$ will go over from $\psi^{(o)} = \phi$ over $\psi^{(1)},\psi^{(2)},\ldots,\psi^{(k-1)}$ into $\psi = \psi^{(k)}$. And since $(\cos\frac{\pi}{2k})^{2k} \longrightarrow 1$ as $k \longrightarrow \infty$, ψ results as nearly exclusively as one may wish, if k is sufficiently large. The exact proof runs as follows. Since the process **1.** does not change the trace, and since $\mathrm{Tr}\,U^{(o)} = \mathrm{Tr}\,P_{[\phi]} = 1$, therefore $\mathrm{Tr}\,U^{(1)} = \mathrm{Tr}\,U^{(2)} = \ldots = \mathrm{Tr}\,U^{(k)} = 1$. On the other hand,

$$(U^{(\nu)}f, f) = \sum_n (U^{(\nu-1)}\psi_n^{(\nu)}, \psi_n^{(\nu)})(P_{[\psi_n^{(\nu)}]}f, f)$$

$$= \sum_n (U^{(\nu-1)}\psi_n^{(\nu)}, \psi_n^{(\nu)})|(\psi_n^{(\nu)}, f)|^2\ .$$

Therefore, for $\nu = 1,\dots,k-1$ and $f = \psi_1^{(\nu+1)} = \psi^{(\nu+1)}$, and for $\nu = k$ and $f = \psi_1^{(k)} = \psi^{(k)} = \psi$, we have:

$$(U^{(\nu)}\psi^{(\nu+1)},\psi^{(\nu+1)}) \geq (U^{(\nu-1)}\psi^{(\nu)}, \psi^{(\nu)})|(\psi^{(\nu)}, \psi^{(\nu+1)})|^2$$

$$= (\cos \frac{\pi}{2k})^2 \cdot (U^{(\nu-1)}\psi^{(\nu)}, \psi^{(\nu)}) ,$$

$$(U^{(k)}\psi^{(k)}, \psi^{(k)}) = (U^{(k-1)}\psi^{(k)}, \psi^{(k)})$$

together with

$$(\overline{U}^{(o)}\psi^{(1)}, \psi^{(1)}) = (P_{[\psi^{(o)}]}\psi^{(1)}, \psi^{(1)}) = |(\psi^{(o)}, \psi^{(1)})|^2$$

$$= (\cos \frac{\pi}{2k})^2 ,$$

this gives

$$(U^{(k)}\psi, \psi) \geq (\cos \frac{\pi}{2k})^{2k} .$$

Since $\mathrm{Tr}\, U^{(k)} = 1$ and $(\cos \frac{\pi}{2k})^{2k} \longrightarrow 1$ as $k \longrightarrow \infty$, we can apply the result obtained in II.11.: $U^{(k)}$ converges to $P_{[\psi]}$. Hence our aim is accomplished.

How far may we use one of the main instruments of "ideal experiments" of phenomenological thermodynamics, namely the so-called semipermeable walls, when dealing with quantum mechanical systems?

In phenomenological thermodynamics, this theorem holds: If I and II are two different states of the same system **S**, then it is permissible to assume the existence of a wall which is completely permeable for I and not permeable for II[193] -- this is, so to speak, the thermodynamical definition of difference, and therefore of

---

[193] Cf. for example, the reference in Note 184.

equality also, for two systems. How far is such an assumption permissible in quantum mechanics?

We first show that if $\phi_1, \phi_2, \ldots, \psi_1, \psi_2, \ldots$ is an orthonormal set, then there is a semi-permeable wall which lets the system **S** in each of the states $\phi_1, \phi_2, \ldots$ pass through unhindered, and which reflects unchanged the system in each of the states $\psi_1, \psi_2, \ldots$. Systems which are in other states may, on the other hand, be changed by collision with the wall.

The system $\phi_1, \phi_2, \ldots, \psi_1, \psi_2, \ldots$ can be assumed to be complete, since otherwise it could be made so by additional $\chi_1, \chi_2, \ldots$ which one could then add to the $\phi_1, \phi_2, \ldots$. We now choose an operator R with a pure discrete spectrum, and only simple eigenvalues $\lambda_1, \lambda_2, \ldots, \mu_1, \mu_2, \ldots$ whose eigenfunctions are $\phi_1, \phi_2, \ldots, \psi_1, \psi_2, \ldots$ respectively. In fact, let the $\lambda_n < 0$ and the $\mu_n > 0$. Let the quantity $\mathfrak{R}$ belong to R. We construct many windows in the wall, each of which is defined as follows: each "molecule" $\mathbf{K}_1, \ldots, \mathbf{K}_N$ of our gas (we are again considering U-gases at the temperature $\mathbf{T} > 0$) is detained there, opened, the quantity $\mathfrak{R}$ measured on the system $\mathbf{S}_1$ or $\mathbf{S}_2$ or ... $\mathbf{S}_N$ contained in it. Then the box is closed again, and according to whether the measured value of $\mathfrak{R}$ is < 0 or > 0, the box, together with its contents, penetrates the window or is reflected, with unchanged momentum. That this contrivance satisfies the desired end is clear -- it remains only to discuss what changes remain in it after such collisions, and how closely it is related to the so-called "Maxwell's demon" of thermodynamics.[194]

In the first place, it must be said that since the measurement (under certain circumstances) changes the

---

[194]Cf. the reference in Note 185. The reader will find a detailed discussion of the difficulties connected with the concept of "Maxwell's demon" in L. Szilard, Z. Physik, 53 (1929).

state of **S**, and perhaps its energy expectation value also, this difference in the mechanical energy must be added or absorbed by the measurement action, in the sense of the first law of thermodynamics (for example, by installing a spring which can be extended or compressed, or something similar). Since it is a case of a purely automatically functioning measuring mechanism, and since only mechanical (not heat!) energies are transformed, certainly no entropy changes occur, and at present, only this is of importance to us. (If **S** is in one of the states $\phi_1, \phi_2, \ldots, \psi_1, \psi_2, \ldots$, then the $\mathfrak{R}$ measurement does not, in general, change **S**, and no compensating changes remain in the measuring apparatus.)

The second point is more doubtful. Our arrangement is rather similar to "Maxwell's demon," i.e., to a semi-permeable wall which transmits molecules coming from the right and reflects those coming from the left. If we insert such a wall in the midst of a container filled with a gas, then all the gas is soon on the left hand side -- i.e., the volume is halved without entropy consumption. This means an uncompensated entropy increase of the gas, and therefore, by the second law of thermodynamics, such a wall cannot exist. Nevertheless, our semi-permeable wall is essentially different from this thermodynamically unacceptable one; because reference is made with it only to the internal properties of the "molecules" $\mathbf{K}_1, \ldots, \mathbf{K}_N$ (i.e., the state of $\mathbf{S}_1$ or ... or $\mathbf{S}_N$ enclosed therein), and not to the exterior (i.e., whether it comes from the right or left, or something similar). This, however, is the decisive circumstance. A thorough going analysis of this question is made possible by the researches of L. Szilard, which clarified the nature of the semi-permeable wall, "Maxwell's demon," and the general role of the "intervention of an intelligent being in thermodynamical systems." We cannot go any further into these things here, especially since the reader can find a treatment of this in the references to Note 194.

In particular, the above treatment shows that two states ϕ, ψ of the system **S** can be certainly divided by a semi-permeable wall if they are orthogonal. We now want to prove the converse: if ϕ, ψ are not orthogonal, then the assumption of such a semi-permeable wall contradicts the second law of thermodynamics. That is, the necessary and sufficient condition for the separability by semi-permeable walls is $(\phi, \psi) = 0$, and not, as in classical theory, $\phi \neq \psi$ (we write ϕ, ψ instead of the I, II used above). This clarifies an old paradox of the classical form of thermodynamics, namely, the uncomfortable discontinuity in the operations with semi-permeable walls: states whose differences are arbitrarily small are always 100% separable, the absolutely equal states are in general not separable! We now have a continuous transition: It will be seen that 100% separability exists only for $(\phi, \psi) = 0$ and for increasing (ϕ, ψ) it becomes steadily worse. Finally, at maximum (ϕ, ψ), i.e., $|(\phi, \psi)| = 1$ (here $||\phi|| = ||\psi|| = 1$, and therefore it follows from $|(\phi, \psi)| = 1$ that $\phi = c\psi$, c constant, $|c| = 1$), the states ϕ, ψ are identical, and the separation is completely impossible.

In order to carry out these considerations, we must anticipate the end result of this section, the value of the entropy of the U-ensemble. Naturally we shall not use this result in its derivation.

Let us then assume that there is a semi-permeable wall separating ϕ and ψ. We shall then prove $(\phi, \psi) = 0$. We consider a $\frac{1}{2}(P_{[\phi]} + P_{[\psi]})$ gas (i.e., of $N/2$ systems in the state ϕ and $N/2$ systems in the state ψ, the trace of this operator is 1), and choose $\mathcal{V}$ (i.e., $\bar{\mathbf{K}}$), and **T** so that the gas is ideal. Let $\bar{\mathbf{K}}$ have the longitudinal cross section shown in Fig. 3: 1 2 3 4 1. We insert a semi-permeable wall at one end aa, and then move it halfway, up to the center bb. The temperature of the gas is kept fixed by contact with a large heat reservoir **W** of temperature **T** at the other end 2 3.

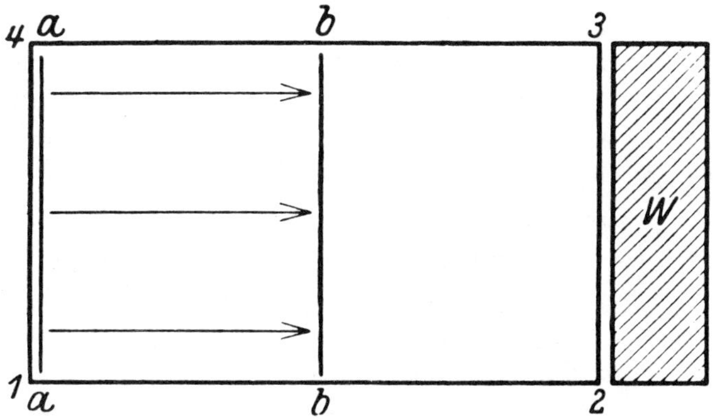

In this process, nothing happens to the ϕ molecules, but the ψ molecules are pushed into the right half of $\overline{\mathbf{K}}$ (between bb and 2 3). That is, the $\frac{1}{2}(P_{[\phi]} + P_{[\psi]})$ gas is a 1:1 mixture of a $P_{[\phi]}$ gas and a $P_{[\psi]}$ gas. Nothing happens to the former, but the latter is isothermally compressed to one half its original volume. From the equation of state of the ideal gas, it follows that in this process the mechanical work $\frac{N}{2} \kappa \mathbf{T} \ln 2$ is performed ($N/2$ is the number of the molecules of the $P_{[\psi]}$ gas, κ is Boltzmann's constant),[195] and since the energy of the gas is not changed (because of the isothermy),[196] this quantity of energy is taken over by the heat reservoir $\mathbf{W}$. The entropy change of the reservoir is then

---

[195]If an ideal gas consists of M molecules, then its pressure is $p = \frac{M\kappa\mathbf{T}}{\mathcal{V}}$. In the compression from the volume $\mathcal{V}_1$ to the volume $\mathcal{V}_2$ therefore, the mechanical work

$$\int_{\mathcal{V}_1}^{\mathcal{V}_2} p\, d\mathcal{V} = M\kappa\mathbf{T}\int_{\mathcal{V}_1}^{\mathcal{V}_2} \frac{d\mathcal{V}}{\mathcal{V}} = M\kappa\mathbf{T} \ln \frac{\mathcal{V}_2}{\mathcal{V}_1}$$

is done. In our case, $M = N/2$, $\mathcal{V}_1 = \mathcal{V}/2$, $\mathcal{V}_2 = \mathcal{V}$.

[196]The energy of an ideal gas, as is well known, depends only on its temperature.

$\mathbf{Q}/\mathbf{T} = N\kappa \cdot \frac{1}{2} \ln 2$ (see Note 186).

After this process, the half of the original gas is present to the left of bb , i.e., $N/4$ molecules. To the right of bb on the other hand, there is the half of the original $P_{[\phi]}$ gas, i.e., $N/4$ molecules, and the entire $P_{[\psi]}$ gas, i.e., $N/2$ molecules -- therefore a total of $3N/4$ molecules of a $\frac{1}{3} P_{[\phi]} + \frac{2}{3} P_{[\psi]}$ gas. We compress or expand these gases to the volumes $\mathcal{V}/4$ and $3\mathcal{V}/4$ respectively, and mechanical work is again taken from or given to the heat reservoir $\mathbf{W}$: this amounts to $\frac{N}{4} \kappa \mathbf{T} \ln 2$ and $\frac{3N}{4} \kappa \mathbf{T} \ln \frac{3}{2}$ respectively (see Note 195), and the entropy increase of the reservoir is then $N\kappa \cdot \frac{1}{4} \ln 2$ and $- N\kappa \cdot \frac{3}{4} \ln \frac{3}{2}$ respectively. Altogether:

$$N\kappa \cdot \left(\frac{1}{2} \ln 2 + \frac{1}{4} \ln 2 - \frac{3}{4} \ln \frac{3}{2}\right) = N\kappa \cdot \frac{3}{4} \ln \frac{4}{3} .$$

Finally, we have a $P_{[\phi]}$ and a $P_{[\psi]}$ gas of $N/4$ and $3N/4$ molecules respectively, with the respective volumes $\mathcal{V}/4$ and $3\mathcal{V}/4$. Originally there was a $\frac{1}{2} P_{[\phi]} + \frac{1}{2} P_{[\psi]}$ gas of N molecules in the volume $\mathcal{V}$ i.e., if we will, two $\frac{1}{2} P_{[\phi]} + \frac{1}{2} P_{[\psi]}$ gases with $N/4$ and $3N/4$ molecules respectively, in the volumes $\mathcal{V}$ and $3\mathcal{V}/4$ respectively. The change effected by the entire process is then this: $N/4$ molecules in volume $\mathcal{V}/4$ changed from a $\frac{1}{2} P_{[\phi]} + \frac{1}{2} P_{[\psi]}$ gas into a $P_{[\phi]}$ gas, $3N/4$ molecules in the volume $3\mathcal{V}/4$ changed from a $\frac{1}{2} P_{[\phi]} + \frac{1}{2} P_{[\psi]}$ gas into a $\frac{1}{3} P_{[\phi]} + \frac{2}{3} P_{[\psi]}$ gas, and the entropy of $\mathbf{W}$ increased by $N\kappa \cdot \frac{3}{4} \ln \frac{4}{3}$. Since the process was reversible, the entire entropy increase must be zero, i.e., the two gas-entropy changes must entirely compensate the change of entropy of $\mathbf{W}$. We must therefore find the entropy changes of the gases.

As we shall see, a U-gas of N molecules has the entropy $- M\kappa \cdot \mathrm{Tr}\,(U \ln U)$ if that of the $P_{[\chi]}$ -gas of equal volume and temperature is taken as zero (see above). If therefore U has a pure discrete spectrum with the eigenvalues $w_1, w_2, \cdots$, then this is

$$- M\kappa \cdot \sum_{n=1}^{\infty} w_n \ln w_n$$

(therefore $x \ln x$ is to be set equal to 0 for $x = 0$) . As may easily be calculated, $P_{[\phi]}$, $\frac{1}{2} P_{[\phi]} + \frac{1}{2} P_{[\psi]}$, $\frac{1}{3} P_{[\phi]} + \frac{2}{3} P_{[\psi]}$ have the respective eigenvalues $1, 0$ and $\frac{1 + \alpha}{2}, \frac{1 - \alpha}{2}, 0$ and

$$\frac{3 + \sqrt{1 + 8\alpha^2}}{6}, \frac{3 - \sqrt{1 - 8\alpha^2}}{6}, 0$$

($\alpha = |(\phi, \psi)|$, therefore $\geq 0, \leq 1$) , in which the multiplicity of the zero is always infinite, but in which the others are simple.[197] Therefore the entropy of the

---

[197] We determine the eigenvalues of $aP_{[\phi]} + bP_{[\psi]}$. The requirement is

$$(aP_{[\phi]} + bP_{[\psi]})f = \lambda f \ .$$

Since the left side is a linear combination of the ϕ, ψ , the right side is also, therefore also f , is too if $\lambda \neq 0$. $\lambda = 0$ is certainly an infinitely multiple eigenvalue, since each f orthogonal to ϕ, ψ belongs to it. It therefore suffices to consider $\lambda \neq 0$ and $f = x\phi + y\psi$ (let ϕ, ψ be linearly independent, otherwise, $\phi = c\psi$, $|c| = 1$, and the two states are identical).

The above equation then becomes

$$a(x + y(\psi, \phi))\cdot\phi + b(x(\phi, \psi) + y)\cdot\psi = \lambda x\cdot\phi + \lambda y\cdot\psi ,$$

i.e.,

$$a\cdot x + a\overline{(\phi, \psi)}\cdot y = \lambda\cdot x, \quad b(\phi, \psi)\cdot x + b\cdot y = \lambda\cdot y \ .$$

gas has increased by

$$-\frac{N}{4}\kappa\cdot 0-\frac{3N}{4}\kappa\cdot\left(\frac{3+\sqrt{1+8\alpha^2}}{2}\ln\frac{3+\sqrt{1+8\alpha^2}}{6}+\frac{3-\sqrt{1+8\alpha^2}}{2}\ln\frac{3-\sqrt{1+8\alpha^2}}{6}\right)$$

$$+N\kappa\cdot\left(\frac{1+\alpha}{2}\ln\frac{1+\alpha}{2}+\frac{1-\alpha}{2}\ln\frac{1-\alpha}{2}\right).$$

This should equal 0 when the entropy increase $N\kappa\cdot\frac{3}{4}\ln\frac{4}{3}$ of **W** is added to it. If we divide by $N\kappa/4$ then we have

$$-\frac{3+\sqrt{1+8\alpha^2}}{2}\ln\frac{3+\sqrt{1+8\alpha^2}}{6}-\frac{3-\sqrt{1+8\alpha^2}}{2}\ln\frac{3-\sqrt{1+8\alpha^2}}{6}$$

$$+2(1+\alpha)\ln\frac{1+\alpha}{2}+2(1-\alpha)\ln\frac{1-\alpha}{2}+3\ln\frac{4}{3}=0$$

Also $0\le\alpha\le 1$.

Now it can easily be seen that the left side increases monotonically as α varies from 0 to 1,[198]

The determinant of these equations must vanish:

$$\begin{vmatrix} a-\lambda, & a\overline{(\phi,\psi)} \\ b(\phi,\psi), & b-\lambda \end{vmatrix}=0,\quad (a-\lambda)(b-\lambda)-ab|(\phi,\psi)|^2=0,$$

$$\lambda^2-(a+b)\lambda+ab(1-\alpha^2)=0,$$

$$\lambda=\frac{a+b\pm\sqrt{(a+b)^2-4ab(1-\alpha^2)}}{2}=\frac{a+b\pm\sqrt{(a-b)^2+4\alpha^2ab}}{2}.$$

If we put $a=1, b=0$ or $a=1/2, b=1/2$ or $a=1/3, b=2/3$ respectively, then the formulas of the text are obtained.

[198]Since $(x\ln x)'=\ln x+1$, therefore

and in fact from 0 to $3 \ln \frac{4}{3}$; therefore α must be zero (for $\alpha \neq 0$ the inverse process to that described

---

$$\left(\frac{1+y}{2} \ln \frac{1+y}{2} + \frac{1-y}{2} \ln \frac{1-y}{2}\right)' = \frac{1}{2}\left(\ln \frac{1+y}{2} + 1\right) - \frac{1}{2}\left(\ln \frac{1-y}{2} + 1\right)$$

$$= \frac{1}{2} \ln \frac{1+y}{1-y}$$

and the derivative of our expression is

$$- 3\cdot\frac{1}{2} \ln \frac{3 + \sqrt{1+8\alpha^2}}{3 - \sqrt{1+8\alpha^2}} \cdot \frac{1}{3} \frac{8\alpha}{\sqrt{1+8\alpha^2}} + 4\cdot\frac{1}{2} \ln \frac{1+\alpha}{1-\alpha}$$

$$= 2\left(\ln \frac{1+\alpha}{1-\alpha} - \frac{2\alpha}{\sqrt{1+8\alpha^2}} \ln \frac{3 + \sqrt{1+8\alpha^2}}{3 - \sqrt{1+8\alpha^2}}\right) .$$

That this is > 0 means that

$$\ln \frac{1+\alpha}{1-\alpha} > \frac{2\alpha}{\sqrt{1+8\alpha^2}} \ln \frac{3 + \sqrt{1+8\alpha^2}}{3 - \sqrt{1+8\alpha^2}} ,$$

i.e.,

$$\frac{1}{2\alpha} \ln \frac{1+\alpha}{1-\alpha} > \frac{2}{3} \cdot \frac{1}{2\beta} \ln \frac{1+\beta}{1-\beta} , \quad \beta = \frac{\sqrt{1+8\alpha^2}}{3} .$$

We shall prove this with $8/9$ in place of $2/3$. Since $1 - \beta^2 = \frac{8}{9}(1 - \alpha^2)$ and $\alpha < \beta$ (which follows from the former, since $\alpha < 1$) , this means that

$$\frac{1-\alpha^2}{2\alpha} \ln \frac{1+\alpha}{1-\alpha} > \frac{1-\beta^2}{2\beta} \ln \frac{1+\beta}{1-\beta}$$

and this is proved if $\frac{1-x^2}{2x} \ln \frac{1+x}{1-x}$ is shown to be monotonically decreasing in $0 < x < 1$. This last property, however, follows, for example, from the power series expansion:

would be entropy decreasing, contrary to the second law). Therefore $(\phi, \psi) = 0$ has been proved.——

After these preparations, we can go on to determine the entropy of a U-gas of N molecules in the volume $\mathcal{V}$ and at temperature **T** -- i.e., more precisely, its entropy excess with respect to a $P_{[\phi]}$ gas under the same conditions. By our earlier remarks and in the sense of the normalization given above, this is the entropy of a U-ensemble of N individual systems. Let $\mathrm{Tr}\, U = 1$, as was done above.

The U, as we know, has a pure discrete spectrum $w_1, w_2, \ldots$ with $w_1 \geq 0, w_2 \geq 0, \ldots, w_1 + w_2 + \ldots = 1$. Let the corresponding eigenfunctions be $\phi_1, \phi_2, \ldots$. Then

$$U = \sum_{n=1}^{\infty} w_n P_{[\phi_n]}$$

(cf. IV.3.). Consequently, our U-gas is composed of a mixture of $P_{[\phi_1]}, P_{[\phi_2]}, \ldots$ gases of $w_1 N, w_2 N, \ldots$ molecules respectively, all in the volume $\mathcal{V}$. Let **T**, $\mathcal{V}$ again be such that all these gases are ideal, and let $\overline{\mathbf{K}}$ be of rectangular cross section. Now we will apply the following reversible interventions in order to separate the $\phi_1, \phi_2, \ldots$ molecules from each other (cf. Fig. 4.). We add an equally large rectangular box $\overline{\mathbf{K}}'$ (1 2 5 6 1) on to $\overline{\mathbf{K}}$ (2 3 4 5 2), and replace the common wall 2 5 by two walls lying next to each other. Let the one (2 5) be fixed and semi-permeable -- transparent for ϕ_1, but opaque for $\phi_2, \phi_3, \ldots$; let the other wall (bb) be movable, but an ordinary, absolutely impenetrable wall. In addition, we insert another semi-permeable wall at dd,

---

$$\frac{1-x^2}{2x} \ln \frac{1+x}{1-x} = (1 - x^2)(1 + \frac{x^2}{3} + \frac{x^4}{5} + \ldots)$$

$$= 1 - (1 - \tfrac{1}{3})x^2 - (\tfrac{1}{3} - \tfrac{1}{5})x^4 - \ldots .$$

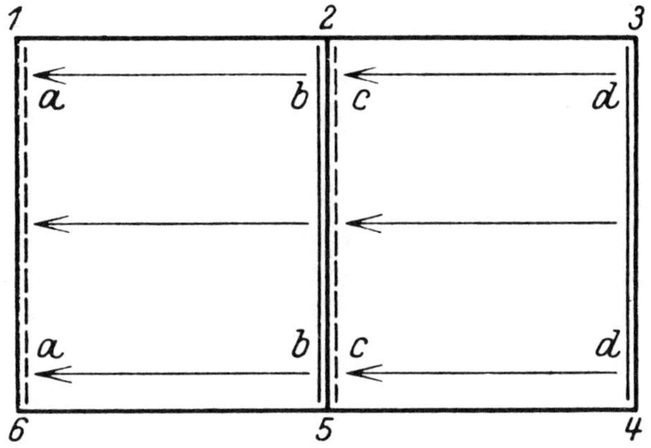

close to 3 4 , which is transparent for $\phi_2, \phi_3, \ldots$ and opaque for ϕ_1 . We then push bb and dd , the distance between them being kept constant, to aa and cc respectively (i.e., close to 1 6 and 2 5 respectively). By this means, the $\phi_2, \phi_3, \ldots$ are not affected, but the ϕ_1 are forced to remain between the moving walls bb, dd . Since the distance between these walls is a constant, no work is done (against the gas pressure), and no heat development takes place. Finally, we replace the walls 2 5, cc by a rigid, absolutely impenetrable wall 2 5 , and remove aa -- in this way the boxes $\overline{\mathbf{K}}$, $\overline{\mathbf{K}}'$ are restored. There is, however, this change. All ϕ_1 molecules are in $\overline{\mathbf{K}}'$, i.e., we have transferred all these from $\overline{\mathbf{K}}$ into the same sized box $\overline{\mathbf{K}}'$, reversibly and without any work being done, without any evolution of heat or tenperature change.[199]

Similarly, we "tap off" the $\phi_2, \phi_3, \ldots$ molecules into the equal boxes $\overline{\mathbf{K}}''$, $\overline{\mathbf{K}}'''$, ... , and have finally, $P_{[\phi_1]}, P_{[\phi_2]}, \ldots$ gases, consisting of $w_1N, w_2N, \ldots$ mole-

[199]Cf., for example, the reference in Note 184 for this artifice which is characteristic of the phenomenological thermodynamical method.

cules, respectively, each in the volume $\mathcal{V}$. We now compress these isothermally to the volumes $w_1\mathcal{V}, w_2\mathcal{V}, \ldots$ respectively. We must therefore add the quantities of heat $w_1 N\kappa T \ln w_1, w_2 N\kappa T \ln w_2, \ldots$, respectively, as compensation, from a large heat reservoir (of temperature T, so that the process may be reversible; the quantities of heat are all less than zero), since the amounts of work done in compressing the individual gases are the negatives of these values (cf. Note 191). Therefore, the entropy increase for this process amounts to

$$\sum_{n=1}^{\infty} w_n N\kappa \cdot \ln w_n \ .$$

Finally, we transform the $P_{[\phi_1]}, P_{[\phi_2]}, \cdots$ gases all into a $P_{[\phi]}$ gas (reversibly, cf. above, ϕ an arbitrarily chosen state). We have then only $P_{[\phi]}$ gases of $w_1 N, w_2 N, \ldots$ molecules respectively, in the volumes $w_1\mathcal{V}, w_2\mathcal{V}, \ldots$. Since all of these are identical and of equal density $(N/\mathcal{V})$, we can mix them, and this is also reversible. We then obtain a $P_{[\phi]}$ gas of N molecules in the volume $\mathcal{V}$ (since

$$\sum_{n=1}^{\infty} w_n = 1 \ ) \ .$$

Consequently, we have carried out the desired reversible process. The entropy has increased by

$$N\kappa \sum_{n=1}^{\infty} w_n \ln w_n$$

and since it is zero in the final state, it was

$$- N\kappa \sum_{n=1}^{\infty} w_n \ln w_n$$

in the initial state.

Since U has the eigenfunctions $\phi_1, \phi_2, \dots$ with the eigenvalues $w_1, w_2, \dots$, $U \ln U$ has the same eigenfunctions, but the eigenvalues $w_1 \ln w_1, w_2 \ln w_2, \dots$. Consequently,

$$\mathrm{Tr}\,(U \ln U) = \sum_{n=1}^{\infty} w_n \ln w_n \ .$$

It may be observed that $w_n \geq 0, \leq 1$, therefore $w_n \ln w_n \leq 0$, and in fact equals zero only for $w_n = 0, 1$. Note that for $w_n = 0$, $w_n \ln w_n$ is to be taken equal to zero -- this follows from the circumstance that in our above considerations, the vanishing w_n are not considered at all. The same conclusion may also be obtained from continuity considerations.

We have then determined the entropy of a U-ensemble, consisting of N individual systems, to be $-N\kappa\,\mathrm{Tr}\,(U \ln U)$. The previous discussion on $w_n \ln w_n$ shows that it is always ≥ 0, and in order that it be 0, all w_n must be zero or 1. Since $\mathrm{Tr}\, U = 1$, exactly one $w_n = 1$, while the others $= 0$, therefore $U = P_{[\phi]}$. That is, the states have an entropy $= 0$, and the other mixtures have entropies > 0.

3. REVERSIBILITY AND EQUILIBRIUM PROBLEMS

We can now prove the irreversibility of the measurement process as asserted in V.1. For example, if U is a state, $U = P_{[\phi]}$, then in the measurement of a quantity $\mathfrak{R}$ whose operator R has the eigenfunctions $\phi_1, \phi_2, \dots$, it goes over into the ensemble

$$U' = \sum_{n=1}^{\infty} (P_{[\phi]}\phi_n, \phi_n) \cdot P_{[\phi_n]} = \sum_{n=1}^{\infty} |(\phi, \phi_n)|^2 P_{[\phi_n]}$$

and if U' is not a state, then an entropy increase has

occurred (the entropy of U was 0, that of U' is > 0), so that the process is irreversible. If U', too, is to be a state, it must be a $P_{[\phi_n]}$, and since the ϕ_n are its eigenfunctions, this means that all $|(\phi, \phi_n)|^2 = 0$ except one (that one $= 1$) i.e., ϕ is orthogonal to all ϕ_n, $n \neq \bar{n}$ -- but then $\phi = c\phi_{\bar{n}}$, where $|c| = 1$, and therefore $P_{[\phi]} = P_{[\phi_{\bar{n}}]}$, $U = U'$. Therefore, each measurement on a state is irreversible, unless the eigenvalue of the measured quantity (i.e., this quantity in the given state) has a sharp value, in which case the measurement does not change the state at all. As we see, the non-causal behavior is thus unambiguously related to a certain concomitant thermodynamical phenomena.

We shall now discuss in complete generality when the process 1.,

$$U \longrightarrow U' = \sum_{n=1}^{\infty} (U\phi_n, \phi_n) \cdot P_{[\phi_n]}$$

increases the entropy.

U has the entropy $-N\kappa \operatorname{Tr}(U \ln U)$. If $w_1, w_2, \ldots$ are its eigenvalues and $\psi_1, \psi_2, \ldots$ its eigenfunctions then this is equal to

$$-N\kappa \sum_{n=1}^{\infty} w_n \ln w_n = -N\kappa \sum_{n=1}^{\infty} (U\psi_n, \psi_n) \ln (U\psi_n, \psi_n) .$$

U' has the eigenvalues $(U\phi_1, \phi_1), (U\phi_2, \phi_2), \ldots,$ and therefore its entropy is

$$-N\kappa \sum_{n=1}^{\infty} (U\phi_n, \phi_n) \ln (U\phi_n, \phi_n) .$$

Consequently the entropy of U is $\geq$ that of U' depending on whether

$$* \quad \sum_{n=1}^{\infty} (U\psi_n, \psi_n) \ln (U\psi_n, \psi_n) \lesseqgtr \sum_{n=1}^{\infty} (U\phi_n, \phi_n) \ln (U\phi_n, \phi_n) .$$

We next show that in *, $\geq$ holds in any case, i.e., that the process $U \rightarrow U'$ is not entropy-diminishing -- this is indeed clear thermodynamically, but it is of importance for our subsequent purposes to have a purely mathematical proof of this fact. We proceed in such a way that U, and with it $\psi_1, \psi_2, \ldots$, are fixed, while the $\phi_1, \phi_2, \ldots$ run through all complete orthonormal sets.

Next, for reasons of continuity, we may limit ourselves to such sets $\phi_1, \phi_2, \ldots$ in which only a finite number of ϕ_n are different from the corresponding ψ_n. Then, for example, let $\phi_n = \psi_n$ for $n > M$. Then the ϕ_n, $n \leq M$ are linear combinations of the ψ_n, $n \leq M$, and conversely -- therefore,

$$\phi_m = \sum_{n=1}^{M} x_{mn}\psi_n \qquad (m = 1, \ldots, M),$$

and the M dimensional matrix $\{x_{mn}\}$ is obviously unitary. We obtain $(U\psi_m, \psi_m) = w_m$ and, as can easily be calculated,

$$(U\phi_m, \phi_m) = \sum_{n=1}^{N} w_n |x_{mn}|^2 \quad (m = 1, \ldots, M)$$

so that

$$\sum_{m=1}^{M} w_m \ln w_m \geq \sum_{m=1}^{M} \left(\sum_{n=1}^{M} w_n |x_{mn}|^2 \right) \ln \left(\sum_{n=1}^{M} w_n |x_{mn}|^2 \right)$$

is to be proved. Since the right side is a continuous function of the M^2 bounded variables x_{mn}, it has a maximum, and it also assumes its maximum value ($\{x_{mn}\}$ unitary); since the left side is its value for

$$x_{mn} \begin{cases} = 1 & \text{for } m = n \\ = 0 & \text{for } m \neq n \end{cases}$$

we must show: the maximum just mentioned occurs at this x_{mn}-complex.

Therefore, let x^0_{mn} $(m, n = 1, \ldots, M)$ be a set of values for which the maximum occurs. If we multiply the matrix $\{x^0_{mn}\}$ by the unitary matrix

$$\left\{\begin{array}{cccc:c} \alpha, & \beta, & 0, & & 0 \\ -\bar{\beta}, & \bar{\alpha}, & 0, & & 0 \\ 0, & 0, & 1, & & 0 \\ \cdots & \cdots & \cdots & & \cdots \\ 0 & 0 & 0 & & 1 \end{array}\right\}, \quad |\alpha|^2 + |\beta|^2 = 1,$$

then we obtain a unitary matrix $\{x'_{mn}\}$, and therefore an acceptable x_{mn}-complex. Now, let $\alpha = \sqrt{1 - \epsilon^2}$, $\beta = \theta\epsilon$ (ϵ real, $|\theta| = 1$). ϵ will be small, and in the following we shall carry in our calculations the $1, \epsilon, \epsilon^2$ terms only, and neglect the $\epsilon^3, \epsilon^4, \ldots$ terms. Then $\alpha \approx 1 - \frac{1}{2}\epsilon^2$, and in the new matrix $\{x'_{mn}\}$,

$$x'_{1n} \approx (1 - \tfrac{1}{2}\epsilon^2)x^0_{1n} + \theta\epsilon x^0_{2n},$$

$$x'_{2n} \approx -\bar{\theta}\epsilon x^0_{1n} + (1 - \tfrac{1}{2}\epsilon^2)x^0_{2n},$$

$$x'_{mn} = x^0_{mn} \quad (m \geq 3),$$

therefore

$$\sum_{n=1}^{M} w_n |x'_{1n}|^2 \approx \sum_{n=1}^{M} w_n |x^0_{1n}|^2 + \sum_{n=1}^{M} 2w_n \,\Re(\bar{\theta} x^0_{1n} \bar{x}^0_{2n})\cdot\epsilon + \sum_{n=1}^{M} w_n(-|x^0_{1n}|^2 + |x^0_{2n}|^2)\cdot\epsilon^2,$$

$$\sum_{n=1}^{M} w_n|x'_{2n}|^2 \approx \sum_{n=1}^{M} w_n|x^o_{2n}|^2 - \sum_{n=1}^{M} 2w_n \,\Re(\bar{\theta}x^o_{1n}\bar{x}^o_{2n})\cdot\epsilon - \sum_{n=1}^{M} w_n(-|x^o_{1n}|^2 + |x^o_{2n}|^2)\cdot\epsilon^2 ,$$

$$\sum_{n=1}^{M} w_n|x'_{mn}|^2 = \sum_{n=1}^{M} w_n|x^o_{mn}|^2 \qquad (m \geq 3) .$$

If we substitute these expressions in $f(x) = x \ln x$, in which

$$f'(x) = \ln x + 1 , \quad f''(x) = \frac{1}{x}$$

and add the resulting expressions together, then

$$\sum_{m=1}^{M}\left(\sum_{n=1}^{M} w_n|x'_{mn}|^2\right)\ln\left(\sum_{n=1}^{M} w_n|x'_{mn}|^2\right)$$

$$\approx \sum_{m=1}^{M}\left(\sum_{n=1}^{M} w_n|x^o_{mn}|^2\right)\ln\left(\sum_{n=1}^{M} w_n|x^o_{mn}|^2\right)$$

$$+\left(\ln\left(\sum_{n=1}^{M} w_n|x^o_{1n}|^2\right) - \ln\left(\sum_{n=1}^{M} w_n|x^o_{2n}|^2\right)\right)\cdot\sum_{n=1}^{M} 2w_n\,\Re(\bar{\theta}x^o_{1n}\bar{x}^o_{2n})\cdot\epsilon$$

$$+\left[-\left(\ln\left(\sum_{n=1}^{M} w_n|x^o_{1n}|^2\right) - \ln\left(\sum_{n=1}^{M} w_n|x^o_{2n}|^2\right)\right)\left(\left(\sum_{n=1}^{M} w_n|x^o_{1n}|^2\right) - \left(\sum_{n=1}^{M} w_n|x^o_{2n}|^2\right)\right)\right.$$

$$\left.+\frac{1}{2}\left(\frac{1}{\sum_{n=1}^{M} w_n|x^o_{1n}|^2} + \frac{1}{\sum_{n=1}^{M} w_n|x^o_{2n}|^2}\right)\left(\sum_{n=1}^{M} 2w_n\Re(\bar{\theta}x^o_{1n}\bar{x}^o_{2n})\right)^2\right]\cdot\epsilon^2 .$$

In order that the first term on the right be the maximum value, the ϵ coefficient must be $= 0$, and the ϵ^2 coefficient ≤ 0. The former has two factors,

$$\ln\left(\sum_{n=1}^{M} w_n |x_{1n}^{\circ}|^2\right) - \ln\left(\sum_{n=1}^{M} w_n |x_{2n}^{\circ}|^2\right)$$

and

$$\sum_{n=1}^{M} 2w_n \,\mathfrak{R}\,(\overline{\theta}x_{1n}^{\circ}\overline{x}_{2n}^{\circ}) \ .$$

If the first is zero, then the first term in the ϵ^2 coefficient $= 0$ (this is always ≤ 0), so that the second term, which is clearly ≥ 0 always, must vanish in order that the entire coefficient be ≤ 0. This means that

$$\sum_{n=1}^{M} 2w_n \,\mathfrak{R}\,(\overline{\theta}x_{1n}^{\circ}\overline{x}_{2n}^{\circ}) = 0 \ .$$

Therefore, the second factor of the ϵ coefficient is $= 0$ in any case, which can also be written

$$2\,\mathfrak{R}\left(\overline{\theta}\sum_{n=1}^{M} w_n x_{1n}^{\circ}\overline{x}_{2n}^{\circ}\right) \ .$$

Since this goes over into the absolute value of the

$$\sum_{n=1}^{M}$$

for appropriate θ, this must disappear:

$$\sum_{n=1}^{M} w_n x_{1n}^{\circ}\overline{x}_{2n}^{\circ} = 0 \ .$$

Since we can replace 1, 2 by any two different

$k, j = 1, \ldots, M$, we have

$$\sum_{n=1}^{M} w_n x^o_{kn} \overline{x}^o_{jn} \quad \text{for} \quad k \neq j \quad .$$

That is, the unitary coordinate transformation with the matrix $\{x^o_{mn}\}$ brings the diagonal matrix with the elements $w_1, \ldots, w_n$ again into diagonal form. Since the diagonal elements are the multipliers (or eigenvalues) of the matrix, they are not changed by the coordinate transformation, and are at most permuted. Before the transformation they were the w_m $(m = 1, \ldots, M)$, afterwards, they are the

$$\sum_{n=1}^{M} w_n |x^o_{mn}|^2$$

$(m = 1, \ldots, N)$. The sums

$$\sum_{n=1}^{M} w_n \ln w_n \, , \quad \sum_{m=1}^{M} \left(\sum_{n=1}^{M} w_n |x^o_{mn}|^2 \right) \ln \left(\sum_{n=1}^{M} w_n |x^o_{mn}|^2 \right)$$

then have the same values. Hence there is at any rate a maximum at

$$x_{mn} \begin{cases} = 1 & \text{for} \quad m = n \\ = 0 & \text{for} \quad m \neq n \end{cases}$$

too, as was asserted.

Let us determine when the equality holds in * . If it does hold, then

$$\sum_{n=1}^{\infty} (U\chi_n, \chi_n) \ln (U\chi_n, \chi_n)$$

takes on its maximum value not only for $\chi_n = \psi_n$ $(n = 1, 2, \ldots)$ (these are the eigenfunctions of U , cf.

above), but also for $x_n = \phi_n$ $(n = 1,2,\dots)$ $(x_1, x_2, \dots$ running through all complete orthonormal sets). This holds in particular if only the first M among the ϕ_n are transformed (i.e., $x_n = \phi_n$ for $n > M$) and hence, of course, transformed unitarily among each other. Let $\mu_{mn} = (U\phi_m, \phi_n)$ $(m, n = 1,\dots,M)$, let $v_1,\dots,v_N$ be the eigenvalues of the finite (and at the same time Hermitian and definite) matrix $\{\mu_{mn}\}$, and $\{\alpha_{mn}\}$ $(m, n = 1,\dots,M)$ the matrix that transforms $\{\mu_{mn}\}$ to the diagonal form. This transforms the $\phi_1,\dots,\phi_M$ into $\omega_1,\dots,\omega_M$,

$$\phi_m = \sum_{n=1}^{M} \alpha_{mn}\omega_n$$

$(m = 1,\dots,M)$, and then

$$U\omega_n = v_n\omega_n , \text{ therefore } (U\omega_m, \omega_n) = \left\{ \begin{array}{ll} v_n , & \text{for } m = n \\ 0 , & \text{for } m \neq n \end{array} \right| .$$

For

$$\xi_m = \sum_{n=1}^{M} x_{mn}\omega_n$$

$(m = 1,\dots,M$, let $\{x_{mn}\}$ also be unitary),

$$(U\xi_k, \xi_j) = \sum_{n=1}^{M} v_n x_{kn}\bar{x}_{jn} .$$

Because of the assumption on the $\phi_1,\dots,\phi_M$,

$$\sum_{n=1}^{M} \left(\sum_{n=1}^{M} v_n|x_{mn}|^2 \right) \ln \left(\sum_{n=1}^{M} v_n|x_{mn}|^2 \right)$$

takes on its maximum for $x_{mn} = \alpha_{mn}$. According to our previous proof, it follows from this that

$$\sum_{n=1}^{M} v_n \alpha_{kn} \overline{\alpha}_{jn} = 0$$

for $k \neq j$, i.e., $(U\phi_k, \phi_j) = 0$ for $k \neq j$, $k, j = 1, \ldots, M$.

This must hold for all M, therefore $U\phi_k$ is orthogonal to all ϕ_j, $k \neq j$ -- hence it is equal to $w'_k\phi_k$ (w'_k a constant). Consequently, the $\phi_1, \phi_2, \ldots$ are the eigenfunctions of U. The corresponding eigenvalues are $w'_1, w'_2, \ldots$ (and therefore a permutation of the $w_1, w_2, \ldots$). But under these circumstances,

$$U' = \sum_{n=1}^{\infty} (U\phi_n, \phi_n) \cdot P_{[\phi_n]} = \sum_{n=1}^{\infty} w'_n \cdot P_{[\phi_n]} = U .$$

We have therefore found.

The process **1.**,

$$U \longrightarrow U' = \sum_{n=1}^{\infty} (U\phi_n, \phi_n) \cdot P_{[\phi_n]}$$

($\phi_1, \phi_2, \ldots$ are the eigenfunctions of the operator R of the measured quantity $\mathfrak{R}$), never diminishes the entropy. It actually increases it, unless all $\phi_1, \phi_2, \ldots$ are eigenfunctions of U, in which case $U = U'$.

In the case mentioned moreover, U commutes with R, and this is actually characteristic for it (because it is equivalent to the existence of the common eigenfunctions $\phi_1, \phi_2, \ldots$, cf. II.10.).

Hence the process **1.** is irreversible in all cases in which it effects a change at all.

The reversibility question should now be treated for the processes **1.**, **2.**, independently of phenomenological thermodynamics, as was announced as the second point of the program in V.2.. The mathematical method with which this can be accomplished we already know: if the second law of thermodynamics holds, the entropy must be equal to

$-N\kappa\,\mathrm{Tr}\,(U \ln U)$, and this may not decrease in any process **1.**, **2.** We must then treat $-N\kappa\,\mathrm{Tr}\,(U \ln U)$ merely as a calculated quantity, independently of its meaning as entropy, and find out what it does in **1.**, **2.**[200]

In **2.**, we obtain

$$U_t = e^{-\frac{2\pi i}{h}tH}\,U e^{\frac{2\pi i}{h}tH}$$

from U, i.e., if we designate the unitary operator

$$e^{-\frac{2\pi i}{h}tH}$$

by A, $U \to U_t = AUA^{-1}$. Since $f \to Af$, because of the unitary nature of A, is an isomorphic mapping of Hilbert space on itself, which transforms each operator P into APA^{-1}, therefore always $F(APA^{-1}) = AF(P)A^{-1}$. Consequently $U_t \ln U_t = A \cdot U \ln U \cdot A^{-1}$. Hence $\mathrm{Tr}\,(U_t \ln U_t) = \mathrm{Tr}\,(U \ln U)$, i.e., our quantity $-N\kappa\,\mathrm{Tr}\,(U \ln U)$ is constant in **2.** We have already ascertained what happens in **1.**, and in fact, without reference to the second law of thermodynamics. If U changes (i.e., $U \neq U'$), then $-N\kappa\,\mathrm{Tr}\,(U \ln U)$ increases, while for unchanged U (i.e., $U = U'$; or $\psi_1, \psi_2, \ldots$ eigenfunctions of U; or U, R commutative), it naturally remains unchanged. In an intervention composed of several process **1.** and **2.** (in arbitrary number and order) $-N\kappa\,\mathrm{Tr}\,(U \ln U)$ remains unchanged if each process **1.** is ineffective (i.e., causes no change), but in all other cases it increases.

Therefore, if only interventions **1.**, **2.** are taken into consideration, then each process **1.**, which effects a change at all, is irreversible.

It is worth noting, there are also other, simpler

---

[200] Naturally, we could neglect the factor $N\kappa$ and consider $-\mathrm{Tr}\,(U \ln U)$. Or, preserving the proportionality with the number of elements N, $-N\,\mathrm{Tr}\,(U \ln U)$.

expressions than $-\mathrm{Tr}\,(U \ln U)$ which do not decrease in **1.**, and are constant in **2.**: for example, the largest eigenvalue of U. Indeed: For **2.**, it is invariant, as are all eigenvalues of U -- while in **1.**, the eigenvalues $w_1, w_2, \ldots$ of U go over into the eigenvalues of U':

$$\sum_{n=1}^{\infty} w_n |x_{1n}|^2, \sum_{n=1}^{\infty} w_n |x_{2n}|^2, \ldots$$

(cf. the earlier considerations of this section), and since, by the unitary nature of the matrix $\{x_{mn}\}$,

$$\sum_{n=1}^{\infty} |x_{1n}|^2 = 1, \sum_{n=1}^{\infty} |x_{2n}|^2 = 1, \ldots$$

all these numbers are $\leq$ than the largest w_n. (A maximum w_n exists, since all $w_n \geq 0$, and since

$$\sum_{n=1}^{\infty} w_n = 1$$

$w_n \longrightarrow 0$.) Now since it is possible so to change U that

$$-\mathrm{Tr}\,(U \ln U) = -\sum_{n=1}^{\infty} w_n \ln w_n$$

remains invariant, but that the largest w_n decreases, we see that these are changes which are possible according to phenomenological thermodynamics -- therefore they are actually possible of execution with our gas processes -- but which can never be brought about by successive applications of **1.**, **2.** alone. This proves that our introduction of gas processes was indeed necessary.

Instead of $-\mathrm{Tr}\,(U \ln U)$ we can also consider $\mathrm{Tr}\,(F(U))$ for appropriate functions $F(x)$. That this increases in **1.** for $U \neq U'$ (for $U = U'$, as well as in **2.**, it is of course invariant), can also be proved, as was done for $F(x) = -x \ln x$, if the special properties of this function, which we used above are also present in $F(x)$. These are: $F''(x) < 0$, and the monotonic decrease of $F'(x)$; but the latter follows from the former. Therefore, for our non-thermodynamical irreversibility considerations, we can use each $\mathrm{Tr}\, F(U)$, if $F(x)$ is a function that is convex from above, i.e., if $F''(x) < 0$ (in $0 \leq x \leq 1$ since all eigenvalues of U lie in that interval).

Finally, it should be shown that the mixing of two ensembles U, V (say in the ratio $\alpha:\beta$; $\alpha > 0$, $\beta > 0$, $\alpha + \beta = 1$) is also not entropy-diminishing, i.e.,

$$- \mathrm{Tr}\,((\alpha U + \beta V) \ln (\alpha U + BV))$$

$$\geq - \alpha\, \mathrm{Tr}\,(U \ln U) - \beta\, \mathrm{Tr}\,(V \ln V)$$

This also holds for each convex $F(x)$ in place of $- x \ln x$. The proof is left to the reader. ——

We shall now investigate the stationary equilibrium superposition, i.e., the mixture of maximum entropy, when the energy is given. The latter is, of course, to be understood to mean that the expectation value of the energy is prescribed -- only this interpretation is admissible, in view of the method indicated in Note 184 for the thermodynamical investigation of statistical ensembles. Consequently, only such mixtures will be allowed, for the U of which $\mathrm{Tr}\, U = 1$, $\mathrm{Tr}\,(UH) = \mathbf{E}$, where H is the energy operator and $\mathbf{E}$ the prescribed energy expectation value. Under these auxiliary conditions, $- N\kappa\, \mathrm{Tr}\,(U \ln U)$ is to be made a maximum. We also make the simplifying assumption that H has a pure discrete spectrum; say the eigenvalues $W_1, W_2, \ldots$ and the eigenfunctions $\phi_1, \phi_2, \ldots$ (there may also be multiple values among these).

Let $\mathfrak{R}$ be a quantity whose operator R has the eigenfunctions (of H) $\phi_1, \phi_2, \ldots$, but only distinct eigenvalues. The measurement of $\mathfrak{R}$ transforms U, by **2.**, into

$$U' = \sum_{n=1}^{\infty} (U\phi_n, \phi_n) P_{[\phi_n]} \quad ,$$

and therefore $-N\kappa\,\mathrm{Tr}\,(U \ln U)$ increases, unless $U = U'$. Also, $\mathrm{Tr}\,(U)$, $\mathrm{Tr}\,(U\,\mathrm{H})$ do not change -- the latter because the ϕ_n are eigenfunctions of H, and therefore $(\mathrm{H}\phi_m, \phi_n)$ vanishes for $m \neq n$:

$$\mathrm{Tr}\,(U'\mathrm{H}) = \sum_{n=1}^{\infty} (U\phi_n, \phi_n)\,\mathrm{Tr}\,(P_{[\phi_n]}\mathrm{H})$$

$$= \sum_{n=1}^{\infty} (U\phi_n, \phi_n)(\mathrm{H}\phi_n, \phi_n)$$

$$= \sum_{m,n=1}^{\infty} (U\phi_m, \phi_n)(\mathrm{H}\phi_n, \phi_m) = \mathrm{Tr}\,(U\,\mathrm{H}) \ .$$

This must also be true because of the commutativity of R, H (i.e., simultaneous measurability of $\mathfrak{R}$ and energy). Consequently, the desired maximum is the same if we limit ourselves to the U', i.e., to statistical operators with eigenfunctions $\phi_1, \phi_2, \cdots$, and, furthermore it is assumed only among these.

Therefore

$$U = \sum_{n=1}^{\infty} w_n P_{[\phi_n]} \quad ,$$

and since U, $U\,\mathrm{H}$, $U \ln U$ all have the eigenfunctions ϕ_n, but the respective eigenvalues w_n, $W_n w_n$, $w_n \ln w_n$, it suffices to make

$$-N\kappa \sum_{n=1}^{\infty} w_n \ln w_n$$

a maximum, with the auxiliary conditions

$$\sum_{n=1}^{\infty} w_n = 1\ , \quad \sum_{n=1}^{\infty} W_n w_n = \mathbf{E}\ .$$

But this is exactly the same problem as that which is obtained for the corresponding equilibrium problem of the ordinary gas theory,[201] and is solved in the same way. According to the well-known rules of extremum calculation, for the set of maximizing $w_1, w_2, \ldots$

$$\frac{\partial}{\partial w_n}\left(\sum_{m=1}^{\infty} w_m \ln w_m\right) + \alpha \frac{\partial}{\partial w_n}\left(\sum_{m=1}^{\infty} w_m\right) + \beta \frac{\partial}{\partial w_m}\left(\sum_{m=1}^{\infty} W_m w_m\right) = 0$$

must hold, in which α, β are suitable constants, and $n = 1, 2, \ldots$, that is,

$$(\ln w_n + 1) + \alpha + \beta W_n = 0\ , \quad w_n = e^{-1-\alpha-\beta W_n} = a e^{-\beta W_n}$$

where the constant $a = e^{-1-\alpha}$ is introduced in place of α . From

$$\sum_{n=1}^{\infty} w_n = 1$$

it follows that

$$a = \frac{1}{\sum_{n=1}^{\infty} e^{-\beta W_n}}\ ,$$

and therefore

---

[201] Cf., for example, Planck, Theorie der Wärmestrahlung, Leipzig, 1913.

$$w_n = \frac{e^{-\beta W_n}}{\sum_{m=1}^{\infty} e^{-\beta W_m}},$$

and because of $\sum_{n=1}^{\infty} W_n w_n = \mathbf{E}$

$$\frac{\sum_{n=1}^{\infty} W_n e^{-\beta W_n}}{\sum_{n=1}^{\infty} e^{-\beta W_n}} = \mathbf{E}$$

which determines β. If, as is customary, we introduce the "partition function,"

$$z(\beta) = \sum_{n=1}^{\infty} e^{-\beta W_n} = \mathrm{Tr}\,(e^{-\beta H})$$

(cf. Notes 183, 184 for this and the following), then

$$z'(\beta) = -\sum_{n=1}^{\infty} W_n e^{-\beta W_n} = -\mathrm{Tr}\,(H e^{-\beta H})$$

and therefore the condition for β is

$$-\frac{z'(\beta)}{z(\beta)} = \mathbf{E}.$$

(We are making the assumption here that

$$\sum_{n=1}^{\infty} e^{-\beta W_n} \quad \text{and} \quad \sum_{n=1}^{\infty} W_n e^{-\beta W_n}$$

converge for all $\beta > 0$, i.e., that $W_n \longrightarrow \infty$ for $n \longrightarrow \infty$, and in fact, with sufficient rapidity. For example, $W_n/\ln n \longrightarrow \infty$ suffices.) We then obtain the following expression for U itself:

$$U = \sum_{n=1}^{\infty} a e^{-\beta W_n} P_{[\phi_n]} = a e^{-\beta H} = \frac{e^{-\beta H}}{\mathrm{Tr}\,(e^{-\beta H})} = \frac{e^{-\beta H}}{z(\beta)} \, .$$

The properties of the equilibrium ensemble U (which is determined by the enumeration of the values of **E** or of β, and which therefore depends on a parameter, as it must) can now be determined with the method customary in gas theory.

The entropy of our ensemble is

$$\mathbf{S} = -N\kappa\,\mathrm{Tr}\,(U \ln U) = -N\kappa\,\mathrm{Tr}\left(\frac{e^{-\beta H}}{z(\beta)} \ln \frac{e^{-\beta H}}{z(\beta)}\right)$$

$$= -\frac{N\kappa}{z(\beta)}\,\mathrm{Tr}\,(e^{-\beta H}(-\beta H - \ln z(\beta)))$$

$$= \frac{\beta N\kappa}{z(\beta)}\,\mathrm{Tr}\,(He^{-\beta H}) + \frac{\ln z(\beta) N\kappa}{z(\beta)}\,\mathrm{Tr}\,(e^{-\beta H})$$

$$= N\kappa\left[-\frac{\beta z'(\beta)}{z(\beta)} + \ln z(\beta)\right] ,$$

and the total energy

$$N\mathbf{E} = -N\frac{z'(\beta)}{z(\beta)}$$

(this, and not **E** itself, is to be considered in conjunction with **S**). Thus U, **S**, N**E** are expressed by β. Instead of inverting the last relationship, i.e., expressing β by **E**, it is more practical to determine the temperature **T** of the equilibrium mixture, and to reduce everything to this. This is done as follows: Our equilibrium mixture is brought into contact with a heat reservoir of temperature $\mathbf{T}'$, and the energy $Nd\mathbf{E}$ is transferred to it from that reservoir. The two laws of thermodynamics imply, then, that the total energy must remain unchanged, and that the entropy must not decrease. Consequently, the heat reservoir loses the energy $Nd\mathbf{E}$, and therefore its entropy increase is $-Nd\mathbf{E}/\mathbf{T}'$, and we must now have

$$d\,\mathbf{S} - \frac{Nd\mathbf{E}}{\mathbf{T}'} = \left(\frac{d\mathbf{S}}{Nd\mathbf{E}} - \frac{1}{\mathbf{T}'}\right)Nd\mathbf{E} \geq 0 \quad .$$

On the other hand, $Nd\mathbf{E} \gtreqless 0$ must hold according to whether $\mathbf{T}' \gtreqless \mathbf{T}$, because the colder body absorbs energy from the warmer -- consequently $\mathbf{T}' \gtreqless \mathbf{T}$ implies $\frac{d\mathbf{S}}{Nd\mathbf{E}} - \frac{1}{\mathbf{T}'} \gtrless 0$ i.e.,

$$\mathbf{T}' \gtrless \frac{Nd\mathbf{E}}{d\mathbf{S}} = \frac{N\frac{d\mathbf{E}}{d\beta}}{\frac{d\mathbf{S}}{d\beta}} \quad .$$

Hence

$$\mathbf{T} = \frac{N\frac{d\mathbf{E}}{d\beta}}{\frac{d\mathbf{S}}{d\beta}} = -\frac{1}{\kappa}\,\frac{\left(\frac{z'(\beta)}{z(\beta)}\right)'}{\left(\ln z(\beta) - \beta\,\frac{z'(\beta)}{z(\beta)}\right)'} = -\frac{1}{\kappa}\,\frac{\left(\frac{z'(\beta)}{z(\beta)}\right)'}{-\beta\left(\frac{z'(\beta)}{z(\beta)}\right)'} = \frac{1}{\kappa\beta} \,,$$

i.e.,

$$\beta = \frac{1}{\kappa\mathbf{T}} \quad .$$

Therefore U, $\mathbf{S}$, $N\mathbf{E}$ are now all expressed as functions of the temperature.

The analogy of the expressions obtained above for the entropy, equilibrium ensemble etc., with the corresponding results of the classical thermodynamical theory is striking. First, the entropy $-N\kappa\,\mathrm{Tr}\,(U \ln U)$.

$$U = \sum_{n=1}^{\infty} w_n P_{[\phi_n]}$$

is a mixture of the ensembles $P_{[\phi_1]}, P_{[\phi_2]}, \dots$ with the relative weights $w_1, w_2, \dots$, i.e., Nw_1 ϕ_1-systems, Nw_2 ϕ_2-systems, ... The Boltzmann entropy of this ensemble is obtained with the aid of the "thermodynamical probability" $N!/(Nw_1)!(Nw_2)!\dots$. It is its κ fold logarithm.[201] Since N is large, we may approximate the factorial by the Stirling formula, $x! \approx \sqrt{2\pi x}\, e^{-x} x^x$ and then $\kappa \ln \frac{N!}{(Nw_1)!(Nw_2)!\dots}$ becomes essentially

$$- N\kappa \sum_{n=1}^{\infty} w_n \ln w_n$$

-- and this is exactly $- N\kappa \operatorname{Tr}(U \ln U)$.

Furthermore, if we had the equilibrium ensemble

$$U = e^{-\frac{H}{\kappa T}}$$

(we neglect the normalization factor $\frac{1}{Z(\beta)}$), this is equal to

$$\sum_{n=1}^{\infty} e^{-\frac{W_n}{\kappa T}} P_{[\phi_n]} ,$$

therefore a mixture of the states $P_{[\phi_1]}, P_{[\phi_2]}, \dots$, i.e., of the stationary states with the energies $W_1, W_2, \dots$, and with the respective (relative) weights

$$e^{-\frac{W_1}{\kappa T}}, e^{-\frac{W_2}{\kappa T}}, \dots .$$

If an energy value is multiple, say $W_{n_1} = \dots = W_{n_\nu} = W$, then $P_{[\phi_{n_1}]} + \dots + P_{[\phi_{n_\nu}]}$ appears in the equilibrium ensemble with the weight

$$e^{-\frac{W}{\kappa T}} ,$$

i.e., the correctly normalized mixture

$$\frac{1}{\nu}\left(P_{[\phi_{n_1}]} + \cdots + P_{[\phi_{n_\nu}]}\right)$$

(cf. the beginning of IV.3.) appears with the weight

$$\nu e^{-\frac{W}{\kappa T}} .$$

But the classical "canonical" ensemble is defined in exactly the same way (aside from the appearance of the specifically quantum mechanical form

$$\frac{1}{\nu}\left(P_{[\phi_{n_1}]} + \cdots + P_{[\phi_{n_\nu}]}\right)):$$

this is known as Boltzmann's Theorem.[201]

For $T \longrightarrow 0$, the weights

$$e^{-\frac{W_n}{\kappa T}}$$

approach 1, therefore our U tends to

$$\sum_{n=1}^{\infty} P_{[\phi_n]} = 1 .$$

Consequently, $U \approx 1$ is the absolute equilibrium state, if no energy limitations apply -- a result that we had already obtained in IV.3. We see that the "a priori equal probability of the quantum orbits" (i.e., of the simple, non-degenerate ones -- in general the multiplicity of the eigenvalues is the "a priori" weight, cf. discussion above) follows automatically from this theory.

It remains to ascertain how much can be said non-thermodynamically about the equilibrium ensemble U of given energy -- i.e., only from the fact that U is stationary (does not change in the course of time, process **2.**), and that it remains unchanged in all measurements which do not affect the energy (i.e., in measurements of quantities that are measurable simultaneously with the energy, process **1.** with commutative R, H, i.e., $\phi_1, \phi_2, \ldots$ eigenfunctions of H).

Because of the differential equation $\frac{\partial}{\partial t} U = \frac{2\pi i}{h}(U H - H U)$ the former means only that H U commute. The latter means that if $\phi_1, \phi_2, \ldots$ are usable, as a complete eigenfunction set of H, then $U = U'$, i.e., $\phi_1, \phi_2, \ldots$ are also eigenfunctions of U. Let the corresponding H-eigenvalues be $W_1, W_2, \ldots$, those of U, $w_1, w_2, \ldots$. If $W_j = W_k$, then we can replace ϕ_j, ϕ_k by

$$\frac{\phi_j + \phi_k}{\sqrt{2}}, \quad \frac{\phi_j - \phi_k}{\sqrt{2}}$$

for H, and therefore these are also eigenfunctions of U, from which it follows that $w_j = w_k$. Therefore, a function $F(x)$ with $F(W_n) = w_n$ $(n = 1, 2, \ldots)$ can be constructed, and $F(H) = U$. It is clear that this is sufficient, and also that it implies the commutativity of H and U.

Hence there results $U = F(H)$, but a determination of $F(x)$ (it is, as we know $F(x) = \frac{1}{Z(\beta)} e^{-\beta x}$, $\beta = \frac{1}{\kappa' T}$) is not accomplished. From $\mathrm{Tr}\, U = 1$, $\mathrm{Tr}\,(U H) = \mathbf{E}$, it follows that

$$\sum_{n=1}^{\infty} F(W_n) = 1, \quad \sum_{n=1}^{\infty} W_n F(W_n) = \mathbf{E}$$

but with this, all that this method can furnish us is exhausted.

4. THE MACROSCOPIC MEASUREMENT

Although our entropy expression, as we saw, is completely analogous to the classical entropy, it is still surprising that it is invariant in the normal evolution in time of the system (process **2.**), and only increases with measurements (process **1.**) -- in the classical theory (where the measurements in general played no role) it increased as a rule even with the ordinary mechanical evolution in time of the system. It is therefore necessary to clear up this apparently paradoxical situation.

The normal classical thermodynamical consideration runs as follows: One could take a container of volume $\mathcal{V}$, in which M molecules of a gas (for simplicity, an ideal gas) of temperature $\mathbf{T}$ are present in the right half (volume $\mathcal{V}/2$, separated by a partition from the other half). If we were to expand this gas isothermally and reversibly to the volume by driving back the partition with the gas pressure, utilizing the mechanical work that this performs, and by keeping the gas temperature constant by means of a large heat reservoir of temperature $\mathbf{T}$), then the entropy outside (in the reservoir) would decrease by $M\kappa \ln 2$ (cf. Note 195), and therefore the gas entropy could increase by the same amount. On the other hand, if we simply remove the partition, the gas diffuses into the free left half, the volume increases to $\mathcal{V}$ -- i.e., the entropy increases by $M\kappa \ln 2$ without the corresponding compensation taking place. The process is consequently irreversible, for the entropy has increased in the course of the simple mechanical evolution in time of the system (namely, in diffusion). Why does our theory give nothing similar?

This situation is best clarified if we set $M = 1$. Thermodynamics is still valid for such a one-molecule gas, and it is true that its entropy increases by $\kappa \ln 2$ if its volume is doubled. Nevertheless, this difference is $\kappa \ln 2$ actually only so long as one knows no more about the molecule than that it is found in the volume $\mathcal{V}/2$ or $\mathcal{V}$, respectively. For example, if the molecule is in the volume $\mathcal{V}$, but it is known whether it is in the right side or left side of the middle of the container, then it suffices to insert a partition in the middle and allow this to be pushed (isothermally and reversibly) by the molecule to the left or right end of the container. In this case, the mechanical work $\kappa T \ln 2$ is performed, i.e., this energy is taken from the heat reservoir. Consequently, at the end of the process, the molecule is again in the volume $\mathcal{V}$, but we no longer know whether it is on the left or right of the middle. Hence there is a compensating entropy decrease of $\kappa \ln 2$ (in the reservoir). That is, we have exchanged our knowledge for the entropy decrease $\kappa \ln 2$.[202] Or, the entropy is the same in the volume $\mathcal{V}$ as in the volume $\mathcal{V}/2$, provided that we know in the first mentioned case, in which half of the container the molecule is to be found. Therefore, if we knew all the properties of the molecule before diffusion (position and momentum), we could calculate for each moment after the diffusion whether it is on the right or left side, i.e., the entropy has not decreased. If, however, the only information at our disposal was the macroscopic one that the volume was initially $\mathcal{V}/2$, then the entropy does increase upon diffusion.

For a classical observer, who knows all coordinates and momenta, the entropy is therefore constant, and is in fact 0, since the Boltzmann "thermodynamical probability" is 1 (cf. the reference in Note 201); just

---

[202] L. Szilard has (see reference in Note 194) shown that one cannot get this "knowledge" without a compensating entropy increase $\kappa \ln 2$. In general, $\kappa \ln 2$ is the "thermodynamic value" of the knowledge, which consists of an alternative of two cases. All attempts to carry out the process described above without the knowledge of the half of the container in which the molecule is located, can be proved to be invalid, although they may occasionally lead to very complicated automatic mechanisms.

as in our theory for states, $U = P_{[\phi]}$, since these again correspond to the highest possible state of knowledge of the observer relative to the system.

The time variations of the entropy are then based on the fact that the observer does not know everything, that he cannot find out (measure) everything which is measurable in principle. His senses allow him to perceive only the so-called macroscopic quantities. But this clarification of the apparent contradiction mentioned at the outset imposes on us the obligation of investigating the precise analog of the classical macroscopic entropy for the quantum mechanical ensemble, i.e., the entropy as seen by an observer who cannot measure all quantities, but only a few special quantities, namely, the macroscopic ones, and even these, under certain circumstances, with only limited accuracy.

In III.3., we learned that all measurements with limited accuracy can be replaced by absolutely accurate measurements of other quantities which are functions of these, and which have discrete spectra. If now $\mathfrak{R}$ is such a quantity, and R is its operator, if $\lambda^{(1)}, \lambda^{(2)}, \dots$ are the distinct eigenvalues, then the measurement of $\mathfrak{R}$ is equivalent to the answering of the following questions: "Is $\mathfrak{R} = \lambda^{(1)}$?" "Is $\mathfrak{R} = \lambda^{(2)}$?", ... In fact, we can also say directly: Assume that $\mathfrak{S}$, with the operator S , is to be measured with limited accuracy -- say one wishes to determine within which interval $c_{n-1} < \lambda \leq c_n$ $(\dots c_{-2} < c_{-1} < c_0 < c_1 < c_2 < \dots)$ it lies. This is then a case of answering all these questions "Does $\mathfrak{S}$ lie in $c_{n-1} < \lambda \leq c_n$?", $n = 0, \pm 1, \pm 2, \dots$.

Such questions now correspond, by III.5., to projections E whose quantities $\mathfrak{E}$ (which have only the two values $0, 1$) are actually to be measured. In our examples, the $\mathfrak{E}$ are the functions $F_n(\mathfrak{R})$, $n = 1, 2, \dots$, in which

$$F_n(\lambda) \begin{cases} = 1 , & \text{for } \lambda = \lambda^{(n)} \\ = 0 , & \text{otherwise} \end{cases}$$

or the functions $G_n(\mathfrak{S})$, $n = 0, \pm 1, \pm 2, \ldots$, in which

$$G_n(\lambda) \begin{cases} = 1, & \text{for } c_{n-1} < \lambda \leq c_n \\ = 0, & \text{otherwise} \end{cases}$$

-- and the corresponding E are the $F_n(R)$ and $G_n(S)$ respectively. Therefore, instead of giving the macroscopically measurable quantities $\mathfrak{S}$ (together with the (macroscopic) measurement precision obtainable, we may equivalently give the questions $\mathfrak{E}$ which are answered by macroscopic measurements, or their projections E (cf. III.5.). This can be viewed as the characterization of a macroscopic observer. The specification of his E. (Thus, classically, one might characterize him by stating that he can measure the temperature and the pressure in each cm^3 of the gas volume [perhaps with certain limitations of precision], but nothing else).[203]

Now it is a fundamental fact with macroscopic measurements that everything which is measurable at all, is also simultaneously measurable, i.e., that all questions which can be answered separately can also be answered simultaneously, i.e., that all the E commute. The reason that the non-simultaneous measurability of quantum mechanical quantities has made such a paradoxical impression is just that this concept is so alien to the macroscopic method of observation. Because of the fundamental importance of this point, it is best to discuss it somewhat more in detail.

Let us consider the method by which two non-simultaneously measurable quantities [e.g., the coordinate q and the momentum p (cf. III.4.)] can be measured simultaneously with limited precision. Let the mean errors

[203] This characterization of the macroscopic observer is due to E. Wigner.

be ϵ, η respectively (according to the uncertainty principle, $\epsilon\eta \sim h$). The discussion in III.4. showed that with such precision requirements simultaneous measurement is indeed possible: the q (position) measurement is performed with light wave lengths which are not too short, the p (momentum) measurement is performed with light wave trains which are not too long. If everything is properly arranged, then the actual measurements consist in detecting two light quanta in some way, e.g., by photographing: one (in the q measurement) is the light quantum scattered by the Compton effect, the other (in the p-measurement by means of the Doppler effect) is reflected, changed in frequency and then, in the determination of this frequency, is deflected by an optical device (prism, diffraction grating). At the end of the experiment therefore, there are two light quanta or two photographic plates, and from the directions of the light quanta, or the blackened places on the plates, we must calculate q and p. But we must emphasize here that nothing prevents us from determining (with arbitrary precision) the two directions mentioned, or the blackened places, because these are obviously simultaneously measurable quantities (they are momenta or coordinates of two different objects). However, excessive precision at this point is not of much help for the measurement of q and p. As was shown in III.4., the connection of these quantities with q and p is such that the uncertainties ϵ, η remain for q and p (even if the above quantities are measured with greater precision), and the apparatus cannot be arranged so that $\epsilon\eta \ll h$.

Therefore, if we introduce the two directions mentioned, or the blackened places themselves as physical quantities (with operators Q', P'), then we see that Q', P' are commutative, but the operators Q, P belonging to q, p can be expressed by means of them with no higher precision than ϵ, η respectively. Let the quantities belonging to Q', P' be q', p'. The interpretation that the actually macroscopically measurable quantities are not the q, p themselves but the q', p' is a very plausible one (indeed the q', p' are in fact measured), and it is in accord with our postulate of the simultaneous measurability of all macroscopic quantities.

It is reasonable to attribute to this result a general significance, and to view it as disclosing a characteristic of the macroscopic method of observation. According to this, the macroscopic procedure consists of the replacing of all possible operators $A,B,C,\ldots$, which as a rule do not commute with each other, by other operators $A',B',C',\ldots$ (of which these are functions to within a certain approximation) which do commute with each other. Since we can just as well denote these functions of $A',B',C',\ldots$ themselves by $A',B',C',\ldots$, we may also say this: $A',B',C',\ldots$ are approximations of the $A,B,C,\ldots$, but commute exactly with one another. If the respective numbers $\epsilon_A, \epsilon_B, \epsilon_C, \ldots$ give a measure for the magnitudes of the operators $A' - A$, $B' - B$, $C' - C, \ldots$, then we see that $\epsilon_A \epsilon_B$ will be of the order of magnitude of $AB - BA$ (that is, $\neq 0$, generally), etc. -- this gives the limit of the approximations which can be achieved. It is, of course, advisable, in enumerating the $A,B,C,\ldots$ to restrict oneself to those operators whose physical quantities are inaccessible to macroscopic observation, at least within a reasonable approximation.

These wholly qualitative developments remain an empty program so long as we cannot show that they require only things which are mathematically practicable. Therefore, for the characteristic case Q, P, we shall discuss further the question of the existence of the above Q', P' on a mathematical basis. For this purpose, let ϵ, η be two positive numbers with $\epsilon\eta = \frac{h}{4\pi}$. We seek two commuting Q', P' such that $Q' - Q$, $P' - P$ (in a sense still to be defined more precisely), have the orders of magnitude ϵ, η respectively.

We do this with quantities q', p' which are measurable with perfect precision, i.e., Q', P' have pure discrete spectra; since they commute, there is a complete orthonormal set consisting of the eigenfunctions common to both, $\phi_1, \phi_2, \ldots$ (cf. II.10.). Let the corresponding eigenvalues of Q', P' be $a_1, a_2, \ldots$ and $b_1, b_2, \ldots$ respectively. Then

$$Q' = \sum_{n=1}^{\infty} a_n P_{[\phi_n]} , \quad P' = \sum_{n=1}^{\infty} b_n P_{[\phi_n]} .$$

Arrange their measurement in such a manner, that it creates one of the states $\phi_1, \phi_2, \ldots$ -- measure a quantity $\mathfrak{R}$ whose operator R has the eigenfunctions $\phi_1, \phi_2, \ldots$ and distinct eigenvalues $c_1, c_2, \ldots$, and then Q', P' are functions of R. That this measurement implies a measurement of Q and P in approximate fashion is clearly implied by this: In the state ϕ_n the values of Q, P are expressed approximately by the respective values of Q', P', i.e., a_n, b_n. That is, their dispersions about these values are small. These dispersions are the expectation values of the quantities $(q - a_n)^2$, $(p - b_n)^2$, i.e.,

$$((Q - a_n 1)^2 \phi_n, \phi_n) = \|(Q - a_n 1)\phi_n\|^2 = \|Q\phi_n - a_n \phi_n\|^2 ,$$

$$((P - b_n 1)^2 \phi_n, \phi_n) = \|(P - b_n 1)\phi_n\|^2 = \|P\phi_n - b_n \phi_n\|^2 .$$

They are the measures for the squares of the differences of Q' and Q, P' and P respectively, i.e., they must be approximately ϵ^2 and η^2 respectively. We therefore require

$$\|Q\phi_n - a_n \phi_n\| \lesssim \epsilon , \quad \|P\phi_n - b_n \phi_n\| \lesssim \eta .$$

Instead of speaking of Q', P', it is then more appropriate only to seek a complete orthonormal set $\phi_1, \phi_2, \ldots$

for which, for suitable choice of $a_1, a_2, \dots$ and $b_1, b_2, \dots$, the above estimates hold.

Individual ϕ (with $\|\phi\| = 1$), for which (for suitable a, b)

$$\|Q\phi - a\phi\| = \epsilon, \quad \|P\phi - b\phi\| = \eta$$

are known from III.4.:

$$\phi_{\rho,\sigma,\gamma} = \phi_{\rho,\sigma,\gamma}(q) = \left(\frac{2\gamma}{h}\right)^{\frac{1}{4}} e^{-\frac{\pi\gamma}{h}(q-\sigma)^2 + \frac{2\pi\rho}{h} iq} .$$

Hence, because of $\epsilon\eta = \frac{h}{4\pi}$ we have again

$$\epsilon = \sqrt{\frac{h\gamma}{4\pi}}, \quad \eta = \sqrt{\frac{h}{4\pi\gamma}}$$

(i.e., $\gamma = \epsilon/\eta$), and we choose $a = \sigma$, $b = \rho$. We now must construct a complete orthonormal set with the help of these $\phi_{\rho,\sigma,\gamma}$. Since σ is the Q- and ρ the P-expectation value, it is plausible that ρ, σ should each run through a set of numbers independently of each other, and in fact, in such a way that the ρ-set has approximately the density ϵ and the σ-set approximately the density η. It proves practical to choose the units

$$2\sqrt{\pi}\cdot\epsilon = \sqrt{h\gamma} \quad \text{and} \quad 2\sqrt{\pi}\cdot\eta = \sqrt{\frac{h}{\gamma}},$$

i.e.,

$$\rho = \sqrt{h\gamma}\,\mu, \quad \sigma = \sqrt{\frac{h}{\gamma}}\,\nu$$

$(\mu, \nu = 0, \pm 1, \pm 2, \dots)$. The

$$\psi_{\mu,\nu} = \phi_{\sqrt{h\gamma}\,\mu,\ \sqrt{\frac{h}{\gamma}}\,\nu,\ \gamma}$$

$(\mu, \nu = 0, \pm 1, \pm 2, \dots)$ ought then to correspond to the

ϕ_n $(n = 1,2,\ldots)$. It is obviously irrelevant that we have two indices μ, ν in place of the one n.

However, these $\psi_{\mu,\nu}$ are not yet orthogonal. (They are normalized, however, and they satisfy

$$||Q\psi_{\mu,\nu} - \sqrt{h\gamma}\ \mu\psi_{\mu,\nu}|| = \epsilon,\ ||P\psi_{\mu,\nu} - \sqrt{\frac{h}{\gamma}}\ \nu\psi_{\mu,\nu}|| = \eta\,.)$$

If we now orthogonalize them by the E. Schmidt process (in order, cf. II.2., proof of THEOREM 8.), then we can prove the completeness of the resulting normalized orthogonal set $\psi'_{\mu,\nu}$ without any particular difficulties, and can also establish the estimates

$$||Q\psi'_{\mu,\nu} - \sqrt{h\gamma}\ \mu\psi'_{\mu,\nu}|| \leq C\epsilon,\ ||P\psi'_{\mu,\nu} - \sqrt{\frac{h}{\gamma}}\ \nu\psi'_{\mu,\nu}|| \leq C\eta$$

with certain fixed C. A value $C \sim 60$ has been obtained in this way, and it could probably be reduced. The proof of this fact leads to rather tedious calculations, which require no new concepts, and we shall omit them. The factors $C \sim 60$ are not important, since $\epsilon\eta = h/4\pi$ measured in macroscopic (CGS) units is exceedingly small (c. 10^{-28}).

Summing up, we can then say that it is justified to assume the commutativity of all macroscopic operators, and in particular the commutativity of the macroscopic projections E introduced above.

The E correspond to all macroscopically answerable questions $\mathfrak{E}$, i.e., to all discriminations of alternatives in the system investigated, that can be carried out macroscopically. They are all commutative. We can conclude from II.5., that $1 - E$ belongs to them along with E, and that $EF,\ E + F - EF,\ E - EF$ belong along with E, F. It is reasonable to assume that there are only a finite number of them: $E_1, \ldots, E_n$. We introduce the notation $E^{(+)} = E,\ E^{(-)} = 1 - E$ and consider all 2^n products $E_1^{(s_1)} \cdots E_n^{(s_n)}$ $(s_1, \ldots, s_n = \pm)$. Any two different ones among these have the product zero: For if $E_1^{(s_1)} \ldots E_n^{(s_n)}$ and $E_1^{(t_1)} \ldots E_n^{(t_n)}$ are two such, and $s_\nu \neq t_\nu$, then there appear in their product the factors $E_\nu^{(s_\nu)}$, $E_\nu^{(t_\nu)}$ i.e., $E_\nu^{(+)} = E$ and $E_\nu^{(-)} = 1 - E$, whose product is zero. Each E_ν is the sum of several such products: Indeed,

$$E_\nu = \sum_{s_1, \ldots, s_{\nu-1}, s_{\nu+1}, \ldots, s_n = \pm} E_1^{(s_1)} \ldots E_{\nu-1}^{(s_{\nu-1})} . E_\nu^{(+)} E_{\nu+1}^{(s_{\nu+1})} \ldots E_n^{(s_n)} .$$

Among these products consider the ones which are different from zero. Call them $E_1', \ldots, E_m'$. (Evidently $m \leq 2^n$, but actually even $m \leq n - 1$, since these must occur among the $E_1, \ldots, E_n$ and be $\neq 0$). Now clearly: $E_\mu' \neq 0$; $E_\mu' E_\nu' = 0$ for $\mu \neq \nu$; each E_μ is the sum of several E_ν': (From the latter it also follows that $n = 2^m$.) It should be noted that $E_\mu + E_\nu = E_\rho'$ can never occur, unless $E_\mu = 0$, $E_\nu = E_\rho'$ or $E_\mu = E_\rho'$, $E_\nu = 0$. Otherwise, E_μ, E_ν would be sums of several E_π', and therefore E_ρ' the sum of ≥ 2 terms E_π' (possibly with repetitions). By II.4., THEOREMS 15, 16., these would all differ from one another, since their number is ≥ 2 and all are $\neq 0$, they also differ from E_ρ' -- therefore their product with E_ρ' would be zero. Hence the product of their sum with E_ρ' would also be zero, but this contradicts the assertion that the sum is $= E_\rho'$.

The properties $\mathfrak{E}_1', \ldots, \mathfrak{E}_m'$ corresponding to the $E_1', \ldots, E_m'$ are then macroscopic properties of the following type: None is absurd. Every two are mutually exclusive. Each macroscopic property obtains by disjunction of several of them. None of them can be resolved by disjunction into two sharper macroscopic properties. $\mathfrak{E}_1', \ldots, \mathfrak{E}_m'$ therefore represent the furthest that we can go in macroscopic discrimination, for they are macroscopically indecomposable.

In the following, we shall not require that their number be finite, but only that there exist macroscopically indecomposable properties $\mathfrak{E}_1', \mathfrak{E}_2', \ldots$. Let their projections be $E_1', E_2', \ldots$, all again different from zero, mutually orthogonal, and each macroscopic E the sum of several of them.

Therefore 1 is also a sum of several of them. If an E_ν' did not occur in this sum, it would be orthogonal to each term and hence to the sum, that is to 1 : $E_\nu' = E_\nu' \cdot 1 = 0$, which is impossible. Therefore $E_1' + E_2' + \ldots = 1$. We drop the prime notation: $\mathfrak{E}_1, \mathfrak{E}_2, \ldots$ and $E_1, E_2, \ldots$. The closed linear manifolds belonging to these will be called $\mathfrak{M}_1, \mathfrak{M}_2, \ldots$, and their dimension numbers $s_1, s_2, \ldots$.

If all the $s_n = 1$, i.e., all $\mathfrak{M}_n$ one dimensional, then $\mathfrak{M}_n = [\phi_n]$, $E_n = P_{[\phi_n]}$ and because $E_1 + E_2 + \ldots = 1$, the $\phi_1, \phi_2, \ldots$ would form a complete orthonormal set. This would mean that macroscopic measurements would themselves make a complete determination of the state of the observed system possible. Since this is ordinarily not the case, we have in general $s_n > 1$, and in fact, $s_n \gg 1$.

In addition, it should be observed that the E_n , which are the elementary building blocks of the macroscopic description of the world, correspond in a certain sense to the ordinary cell division of phase space in the classical theory. We have already seen that they can reproduce the behavior of non-commutative operators in an approximate fashion, in particular, that of Q, P , which are so important for phase space.

Now, what entropy does the mixture U have for a macroscopic observer whose indecomposable projections are $E_1, E_2, \ldots$? Or, more precisely, how much entropy can such an observer maximally obtain by transforming U into V -- i.e., what entropy decrease (under suitable conditions, naturally this decrease may be $\gtrless 0$) can he produce, under the most favorable circumstances, in external objects as compensation for the transition $U \longrightarrow V$?

First, it must be emphasized that he cannot distinguish between each two ensembles U, U', if both give the same expectation value to E_n for each $n = 1, 2, \ldots$, that is, if $\mathrm{Tr}\,(UE_n) = \mathrm{Tr}\,(U'E_n)$ $(n = 1, 2, \ldots)$. After some time, of course, the discrimination may become possible, since U, U' change according to **2.**, and

$$\mathrm{Tr}\,(AUA^{-1}E_n) = \mathrm{Tr}\,(AU'A^{-1}E_n), \quad A = e^{-\frac{2\pi i}{h} tH}$$

must no longer hold.[204] But we considered only measurements which are carried out immediately. Under the above conditions we may therefore regard U, U' as indistinguishable. Furthermore, the observer can also use only such semi-permeable walls which transmit the ϕ of some E_n and reflect the remainder unchanged. This possibility suffices, as can be seen without difficulty. By means of the method of V.2., to transform a

$$U = \sum_{n=1}^{\infty} x_n E_n$$

---

[204] If E_n commutes with H, and therefore with A, the equality still holds because

$$\mathrm{Tr}\,(A.UA^{-1}E_n) = \mathrm{Tr}\,(UA^{-1}E_n.A) = \mathrm{Tr}\,(UA^{-1}AE_n) = \mathrm{Tr}\,(UE_n)\,.$$

But all E_n, i.e., all macroscopically observable quantities, are in no way all commutative with H. Indeed, many such quantities, for example, the center of gravity of a gas in diffusion, change appreciably with t, i.e., $\mathrm{Tr}\,(UE_n)$ is not constant. Since all macroscopic quantities do commute, H is never a macroscopic quantity, i.e., the into a

$$V' = \sum_{n=1}^{\infty} y_n E_n$$

reversibly, so that the entropy difference is still $\kappa \operatorname{Tr} (U' \ln U') - \kappa \operatorname{Tr} (V' \ln V')$, i.e., the entropy of U' equals $- \kappa \operatorname{Tr} (U' \ln U')$. To be sure, in order that such U' with $\operatorname{Tr} U' = 1$ exist in general, the $\operatorname{Tr} E_n$, i.e., the numbers s_n, must be finite. We therefore assume that all s_n are finite. U' has the s_1-fold eigenvalue x_1, the s_2-fold eigenvalue $x_2, \ldots$. Therefore $- U' \ln U'$ has the s_1-fold eigenvalue $- x_1 \ln x_1$, the s_2-fold eigenvalue $- x_2 \ln x_2, \ldots$. Consequently $\operatorname{Tr} U' = 1$ implies

$$\sum_{n=1}^{\infty} s_n x_n = 1$$

and the entropy is equal to

$$- \kappa \sum_{n=1}^{\infty} s_n x_n \ln x_n .$$

Because of

$$U'E_m = \sum_{n=1}^{\infty} x_n E_n E_m = x_m E_m, \quad \operatorname{Tr} (U'E_m) = x_m \operatorname{Tr} E_m = s_m x_m ,$$

$x_m = \dfrac{\operatorname{Tr}(U'E_m)}{s_m}$, therefore the entropy is equal to

$$- \kappa \sum_{n=1}^{\infty} \operatorname{Tr} (U'E_n) \ln \frac{\operatorname{Tr} (U'E_n)}{s_n} .$$

For arbitrary U $(\operatorname{Tr} U = 1)$, the entropy must

---

energy is not measured macroscopically with complete precision. This is plausible without additional comment.

also be equal to

$$- \kappa \sum_{n=1}^{\infty} \mathrm{Tr}\ (UE_n) \ln \frac{\mathrm{Tr}\ (UE_n)}{s_n}$$

because, if we set

$$x_n = \frac{\mathrm{Tr}\ (UE_n)}{s_n}, \quad U' = \sum_{n=1}^{\infty} x_n E_n$$

then $\mathrm{Tr}\ (UE_n) = \mathrm{Tr}\ (U'E_n)$, and since U, U' are indistinguishable, they have the same entropy.

We must also mention the fact that this entropy always exceeds the customary entropy:

$$- \kappa \sum_{n=1}^{\infty} \mathrm{Tr}\ (UE_n) \ln \frac{\mathrm{Tr}\ (UE_n)}{s_n} \geq - \kappa\ \mathrm{Tr}\ (U \ln U)$$

and that the equality holds only for

$$U = \sum_{n=1}^{\infty} x_n E_n \ .$$

By the results of V.3., this is certainly the case if

$$U' = \sum_{n=1}^{\infty} \frac{\mathrm{Tr}\ (UE_n)}{s_n} E_n$$

can be obtained from U by several (not necessarily macroscopic) applications of the process 1. -- because on the left we have $- \kappa\ \mathrm{Tr}\ (U' \ln U')$, and

$$U = \sum_{n=1}^{\infty} x_n E_n$$

means the same as $U = U'$. We take an orthonormal set $\phi_1^{(n)}, \ldots, \phi_{s_n}^{(n)}$ which spans the closed linear manifold $\mathfrak{M}_n$ belonging to E_n. Because of

$$\sum_{n=1}^{\infty} E_n = 1 \quad ,$$

all $\phi_\nu^{(n)}$ $(n = 1,2,\ldots;\ \nu = 1,\ldots,s_n)$ form a complete orthonormal set. Let R be an operator belonging to these eigenfunctions (with only distinct eigenvalues) and $\mathfrak{R}$ its physical quantity. In the measurement of $\mathfrak{R}$, we get from U (by **1.**)

$$U'' = \sum_{n=1}^{\infty} \sum_{\nu=1}^{s_n} (U\phi_\nu^{(n)}, \phi_\nu^{(n)}).P_{[\phi_\nu^{(n)}]} \quad .$$

Then, if we set

$$\psi_\mu^{(n)} = \frac{1}{\sqrt{s_n}} \sum_{\nu=1}^{s_n} e^{\frac{2\pi i}{s_n}\mu\nu} \phi_\nu^{(n)} \qquad (\mu = 1,\ldots,s_n) ,$$

the $\psi_1^{(n)},\ldots,\psi_{s_n}^{(n)}$ form an orthonormal set which spans the same closed linear manifold as the $\phi_1^{(n)},\ldots,\phi_{s_n}^{(n)}$: $\mathfrak{M}_n$. Therefore the $\psi_\nu^{(n)}$ $(n = 1,2,\ldots;\ \nu = 1,2,\ldots,s_n)$ also form a complete orthonormal set, and we form an operator S with these eigenfunctions, and the corresponding physical quantity $\mathfrak{S}$. We must note the validity of the following formulas:

$$\left(P_{[\phi_\nu^{(n)}]}\psi_\mu^{(m)}, \psi_\mu^{(m)}\right) \begin{cases} = 0 & \text{for } m \neq n \\ = \dfrac{1}{s_n} & \text{for } m = n \end{cases} ,$$

$$\sum_{\nu=1}^{s_n} P_{[\phi_\nu^{(n)}]} = \sum_{\nu=1}^{s_n} P_{[\psi_\nu^{(n)}]} = E_n \quad .$$

In the measurement of $\mathfrak{S}$, therefore, U'' becomes (by **1.**)

$$\sum_{m=1}^{\infty} \sum_{\mu=1}^{s_m} (U''\psi_\mu^{(m)}, \psi_\mu^{(m)}) P_{[\psi_\mu^{(m)}]}$$

$$= \sum_{m=1}^{\infty} \sum_{\mu=1}^{s_m} \left[\sum_{n=1}^{\infty} \sum_{\nu=1}^{s_n} (U\phi_\nu^{(n)}, \phi_\nu^{(n)})(P_{[\phi_\nu^{(n)}]} \psi_\mu^{(m)}, \psi_\mu^{(m)}) \right] P_{[\psi_\mu^{(m)}]}$$

$$= \sum_{m=1}^{\infty} \sum_{\mu=1}^{s_m} \left[\sum_{\nu=1}^{s_m} \frac{(U\phi_\nu^{(m)}, \phi_\nu^{(m)})}{s_m} \right] P_{[\psi_\mu^{(m)}]} = \sum_{m=1}^{\infty} \sum_{\nu=1}^{s_m} \frac{\mathrm{Tr}\,(UE_m)}{s_m} P_{[\psi_\mu^{(m)}]}$$

$$= \sum_{m=1}^{\infty} \frac{\mathrm{Tr}\,(UE_m)}{s_m} E_m = U' \quad .$$

Consequently, two processes **1.** suffice to transform U into U' -- and this is all we needed for the proof.

This entropy for states $(U = P_{[\phi]}, \mathrm{Tr}\,(UE_m) = (E_n\phi, \phi) = ||E_n\phi||^2)$,

$$- \kappa \sum_{n=1}^{\infty} ||E_n\phi||^2 \ln \frac{||E_n\phi||^2}{s_n}$$

is no longer subject to the inconveniences of the "macroscopic" entropy: In general, it is not constant in time (i.e., in process **2.**), and not $= 0$ for all states $U = P_{[\phi]}$. In fact: that the $\mathrm{Tr}\,(UE_n)$, from which our entropy is formed, are not time constant in general, was discussed in Note 204. It is easy to determine when the state $U = P_{[\phi]}$ has the entropy 0: Since

$$\frac{||E_n\phi||^2}{s_n} \geq 0, \leq 1$$

all summands

$$||E_n\phi||^2 \ln \frac{||E_n\phi||^2}{s_n}$$

in the entropy expression are ≤ 0. All these must therefore be $= 0$. That is,

$$\frac{||E_n\phi||^2}{s_n} = 0,\ 1 \ .$$

The former means that $E_n\phi = 0$, the latter that $||E_n\phi|| = \sqrt{s_n}$, but since

$$||E_n\phi|| \leq 1,\ s_n \geq 1$$

this implies $s_n = 1$, $||E_n\phi|| = ||\phi||$; i.e., $E_n\phi = \phi$; or: $s_n = 1$, ϕ in $\mathfrak{M}_n$. The latter can certainly not hold for two different n, but also, it cannot hold at all because then $E_n\phi = 0$ would always be true, and therefore $\phi = 0$ since

$$\sum_{n=1}^{\infty} E_n = 1 \ .$$

Hence, for exactly one n, ϕ is in $\mathfrak{M}_n$, and then $s_n = 1$. Since we determined that in general all $s_n \gg 1$, this is impossible. That is, our entropy is always > 0.

Since the macroscopic entropy is time variable, the next question to be answered is this: does it behave like the entropy of the phenomenological thermodynamics in the real world, i.e., does it increase predominantly? This question is answered affirmatively in classical mechanical theory by the so-called Boltzmann H-theorem. In that, however, certain statistical assumptions, the so-called "disorder assumptions" must be made.[205] In quantum

---

[205]For the classical H-theorem, see Boltzmann, Vorlesungen über Gastheorie, Leipzig, 1896, as well as the extremely instructive discussion by P. and T. Ehrenfest in the article cited in Note 185. The "disorder assumptions" which can take the place (in quantum mechanics) of those of Boltzmann

mechanics, it was possible for the author to prove the corresponding theorem without such assumptions.[206] Since the detailed discussion of this subject, as well as of the ergodic theorem closely connected with it (cf. the reference in Note 206, where this theorem is also proved) would go beyond the scope of this volume, we cannot report on these investigations. The reader who is interested in this problem can refer to the treatments in the references.

---

have been formulated by W. Pauli (Sommerfeld-Festschrift, 1928), and the H-theorem is proved there with their help. More recently, the author also succeeded in proving the classical-mechanical ergodic theorem, cf. Proc. Nat. Ac., Jan. and March, 1932, as well as the improved treatment of G. D. Birkhoff, Proc. Nat. Ac., Dec. 1931 and March, 1932.

[206] Z. Physik, 57 (1929).

THE MEASURING PROCESS

1. FORMULATION OF THE PROBLEM

In the discussions so far, we have treated the relation of quantum mechanics to the various causal and statistical methods of describing nature. In the course of this we found a peculiar dual nature of the quantum mechanical procedure which could not be satisfactorily explained. Namely, we found that on the one hand, a state ϕ is transformed into the state ϕ' under the action of an energy operator H in the time interval $0 \leq \tau \leq t$:

$$\frac{\partial}{\partial t} \phi_\tau = -\frac{2\pi i}{h} H \phi_\tau \qquad (0 \leq \tau \leq t),$$

so if we write $\phi_0 = \phi$, $\phi_t = \phi'$, then

$$\phi' = e^{-\frac{2\pi i}{h} tH} \phi$$

which is purely causal. A mixture U is correspondingly transformed into

$$U' = e^{-\frac{2\pi i}{h} tH} U e^{\frac{2\pi i}{h} tH} .$$

Therefore, as a consequence of the causal change of ϕ

into ϕ', the states $U = P_{[\phi]}$ go over into the states $U' = P_{[\phi']}$ (process **2.** in V.1.). On the other hand, the state ϕ -- which may measure a quantity with a pure discrete spectrum, distinct eigenvalues and eigenfunctions $\phi_1, \phi_2, \ldots$ -- undergoes in a measurement a non-causal change in which each of the states $\phi_1, \phi_2, \ldots$ can result, and in fact does result with the respective probabilities $|(\phi, \phi_1)|^2, |(\phi, \phi_2)|^2, \ldots$. That is, the mixture

$$U' = \sum_{n=1}^{\infty} |(\phi, \phi_n)|^2 P_{[\phi_n]}$$

obtains. More generally, the mixture U goes over into

$$U' = \sum_{n=1}^{\infty} (U\phi_n, \phi_n) P_{[\phi_n]}$$

(process **1.** in V.1.). Since the states go over into mixtures, the process is not causal.

The difference between these two processes $U \longrightarrow U'$ is a very fundamental one: aside from the different behaviors in regard to the principle of causality, they are also different in that the former is (thermodynamically) reversible, while the latter is not (cf. V.3.).

Let us now compare these circumstances with those which actually exist in nature or in its observation. First, it is inherently entirely correct that the measurement or the related process of the subjective perception is a new entity relative to the physical environment and is not reducible to the latter. Indeed, subjective perception leads us into the intellectual inner life of the individual, which is extra-observational by its very nature (since it must be taken for granted by any conceivable observation or experiment). (Cf. the discussion above.) Nevertheless, it is a fundamental requirement of the scientific viewpoint -- the so-called principle of the psycho-physical parallelism -- that it must be possible so to describe the extra-physical process of the subjective perception as if it were in reality in the physical world -- i.e., to assign to its parts equivalent physical processes in the objective environment, in ordinary space. (Of course, in this correlating procedure there arises the frequent necessity of localizing some of these processes at points which lie within the portion of space occupied by our own bodies. But this does not alter the fact of their belonging to the "world about us," the objective environment referred to above.) In a simple example, these concepts might be applied about as follows: We wish to measure a temperature. If we want, we can pursue this process numerically until we have the temperature of the environment of the mercury container of the thermometer, and then say: this temperature is measured by the thermometer. But we can carry the calculation further, and from the properties of the mercury, which can be explained in kinetic and molecular terms, we can calculate its heating, expansion, and the resultant length of the mercury column, and then say: this length is seen by the observer. Going still further, and taking the light source into consideration, we could find out the reflection of the light quanta on the opaque mercury column, and the path of the remaining light quanta into the eye of the observer, their refraction in the eye lens, and the formation of an image on the retina, and then we would say: this image is registered by the retina of the observer. And were our physiological knowledge more precise than it is today, we could go still further, tracing the chemical reactions which produce the impression of this image on the retina, in the optic nerve tract and in the brain, and then in the end say: these chemical changes of his brain cells are perceived by the observer. But in any case, no matter how far we calculate -- to the mercury vessel, to the scale of the thermometer, to the retina, or into the brain, at some time we must say: and this is perceived by the observer. That is, we must always divide the world into two parts, the one being the observed system, the other the observer. In the former, we can follow up all physical processes (in principle at least) arbitrarily precisely. In the latter, this is meaningless. The boundary between the two is arbitrary to a very large extent. In particular we saw in the four different possibilities in the example above, that the observer in this sense needs not to become identified with the body of the actual observer: In one instance in the above example, we included even the thermometer in it, while in another instance, even the eyes and optic nerve tract were not included. That this boundary can be pushed arbitrarily deeply into the interior of the body of the actual observer is the content of the principle of the psycho-physical parallelism -- but this does not change the fact that in each method of description the boundary must be put somewhere, if the method is not to proceed vacuously, i.e., if a comparison with experiment is to be possible. Indeed experience only makes statements of this type: an observer has made a certain (subjective) observation; and never any like this: a physical quantity has a certain value.

Now quantum mechanics describes the events which occur in the observed portions of the world, so long as they do not interact with the observing portion, with the aid of the process **2.** (V.1.), but as soon as such an interaction occurs, i.e., a measurement, it requires the application of process **1**. The dual form is therefore justified.[207] However, the danger lies in the fact that

---

[207]N. Bohr, Naturwiss. 17 (1929), was the first to point out that the dual description which is necessitated by the formalism of the quantum mechanical description of nature is fully justified by the physical nature of things that it may be connected with the principle of the psycho-physical parallelism.

the principle of the psycho-physical parallelism is violated, so long as it is not shown that the boundary between the observed system and the observer can be displaced arbitrarily in the sense given above.

In order to discuss this, let us divide the world into three parts: I, II, III. Let I be the system actually observed, II the measuring instrument, and III the actual observer.[208] It is to be shown that the boundary can just as well be drawn between I and II + III as between I + II and III. (In our example above, in the comparison of the first and second cases, I was the system to be observed, II the thermometer, and III the light plus the observer; in the comparison of the second and third cases, I was the system to be observed plus the thermometer, II the light plus the eye of the observer, III the observer, from the retina on; in the comparison of the third and fourth cases, I was everything up to the retina of the observer, II his retina, nerve tracts and brain, III his abstract "ego.") That is, in one case **2.** is to be applied to I, and **1.** to the interaction between I and II + III; and in the other case, **2.** to I + II, and **1.** to the interaction between I + II and III. (In each case, III itself remains outside of the calculation.) The proof of this assertion, that both procedures give the same results regarding I (this and only this belongs to the observed part of the world in both cases), is then our problem.

But in order to be able to accomplish this successfully, we must first investigate more closely the process of forming the union of two physical systems (which leads from I and II to I + II).

---

[208] The discussion which is carried out in the following, as well as that in VI.3., contains essential elements which the author owes to conversations with L. Szilard. Cf. also the similar considerations of Heisenberg, in the reference cited in Note 181.

2. COMPOSITE SYSTEMS

As was stated at the end of the preceding section, we consider two physical systems I, II (which do not necessarily have the meaning of the I, II above), and their combination I + II . In the classical mechanical method of description, I would have k degrees of freedom, and therefore the coordinates $q_1, \ldots, q_k$, in place of which we shall use the one symbol q ; correspondingly, let II have l degrees of freedom, and the coordinates $r_1, \ldots, r_l$ which shall be denoted by r . Therefore, I + II has k + l degrees of freedom and the coordinates $q_1, \ldots, q_k, r_1, \ldots, r_l$, or, more briefly, q, r . In quantum mechanics then, the wave functions of I have the form $\phi(q)$, those of II the form $\xi(r)$ and those of I + II the form $\Phi(q, r)$. In the corresponding Hilbert spaces $\mathfrak{R}^{I}, \mathfrak{R}^{II}, \mathfrak{R}^{I+II}$, the inner product is defined by $\int \phi(q)\overline{\psi(q)}\, dq$, $\int \xi(r)\overline{\eta(r)}\, dr$ and $\iint \Phi(q, r)\overline{\Psi(q, r)}\, dq\, dr$ respectively. The physical quantities of I, II, I + II are correspondingly the (hypermaximal) Hermitian operators A, A, and $\mathcal{A}$ in $\mathfrak{R}^{I}, \mathfrak{R}^{II}$, and $\mathfrak{R}^{I+II}$ respectively.

Each physical quantity in I is naturally also one in I + II , and in fact its $\mathcal{A}$ is to be obtained from its A in this way: to obtain $\mathcal{A}\,\Phi(q, r)$ consider r as a constant and apply A to the q function $\Phi(q, r)$.[209] This rule of transformation is correct in any case for the coordinate and momentum operators $Q_1, \ldots, Q_k$ and $P_1, \ldots, P_k$, i.e.,

$$q_1, \ldots, q_k, \frac{h}{2\pi i}\frac{\partial}{\partial q_1}, \ldots, \frac{h}{2\pi i}\frac{\partial}{\partial q_k}$$

(cf. I.2.), and it conforms with the principles **I.**, **II.** in

---

[209] It can easily be shown that if A is Hermitian or hypermaximal, $\mathcal{A}$ is also.

IV.2.[210] We therefore postulate them generally. (This is the customary procedure in quantum mechanics.)

In the same way, each physical quantity in II is also one in I + II , and its A gives its $\mathcal{A}$ by the same rule: $\mathcal{A}\,\Phi(q, r)$ equals $A\Phi(q, r)$ if in the latter expression, q is taken as constant, and $\Phi(q, r)$ is considered as a function of r .

If $\phi_m(q)$ $(m = 1,2,\ldots)$ is a complete orthonormal set in $\mathfrak{R}^{I}$ and $\xi_n(r)$ $(n = 1,2,\ldots)$ one in $\mathfrak{R}^{II}$, then $\Phi_{m|n}(q, r) = \phi_m(q)\xi_n(r)$ $(m, n = 1,2,\ldots)$ is clearly one in $\mathfrak{R}^{I+II}$. The operators A, A, $\mathcal{A}$ can therefore be represented by matrices $\{a_{m|m'}\}$, $\{a_{n|n'}\}$, and $\{\alpha_{mn|m'n'}\}$ respectively $(m, n', n, n' = 1,2,\ldots)$.[211] We shall make frequent use of this. The matrix representation means that

$$\mathsf{A}\,\phi_m(q) = \sum_{m'=1}^{\infty} a_{m|m'}\phi_{m'}(q),\quad A\xi_n(r) = \sum_{n'=1}^{\infty} a_{n|n'}\xi_{n'}(r)$$

and

$$\mathcal{A}\,\Phi_{mn}(q, r) = \sum_{m',n'=1}^{\infty} \alpha_{mn|m'n'}\Phi_{m'n'}(q, r)\ ,$$

i.e.,

---

[210] For I. this is clear, and for II. also, so long as only polynomials are concerned. For general functions, it can be inferred from the fact that the correspondence of a resolution of the identity and a Hermitian operator is not disturbed in our transition $A \longrightarrow \mathcal{A}$.

[211] Because of the large number and variety of indices, we use this method of denoting the matrices, which differs somewhat from the notation used thus far.

$$\mathcal{A}\, \phi_m(q)\xi_n(r) = \sum_{m'n'=1}^{\infty} \alpha_{mn|m'n'}\, \phi_{m'}(q)\xi_{n'}(r) .$$

In particular the correspondence $A \longrightarrow \mathcal{A}$ means that

$$\mathcal{A}\phi_m(q)\xi_n(r) = (A\, \phi_m(q))\xi_n(r) = \sum_{m'=1}^{\infty} a_{m|m'}\phi_{m'}(q)\xi_n(r) ,$$

i.e.,

$$\alpha_{mn|m'n'} = a_{m|m'}\delta_{n|n'} \left(\delta_{n|n'} \begin{cases} = 1 , \text{ for } n = n' \\ = 0 , \text{ for } n \neq n' \end{cases} \right) .$$

In an analogous fashion, the correspondence $A \longrightarrow \mathcal{A}$ implies that $\alpha_{mn|m'n'} = a_{n|n'}\delta_{m|m'}$.

A statistical ensemble in I + II is characterized by its statistical operator U or by its matrix $\{\upsilon_{mn|m'n'}\}$. This also determines the statistical properties of all quantities in I + II , and therefore the properties of the quantities in I also. Consequently there also corresponds to it a statistical ensemble in I alone. In fact, an observer who could perceive only I , and not II , would view the ensemble of systems I + II as one such of systems I . What is now the statistical operator $\mathcal{U}$ or its matrix $\{u_{m|m'}\}$, which belongs to this I ensemble? We determine it as follows: The I quantity with the matrix $\{a_{m|m'}\}$ has the matrix $\{a_{m|m'}\delta_{n|n'}\}$ as an I + II quantity, and therefore, by reason of a calculation in I , it has the expectation value

$$\sum_{m,m'=1}^{\infty} u_{m|m'}a_{m'|m} ,$$

while the calculation in I + II gives

$$\sum_{m,n,m',n'=1}^{\infty} \upsilon_{mn|m'n'} a_{m'|m} \delta_{n'|n} = \sum_{m,m',n=1}^{\infty} \upsilon_{mn|m'n'} a_{m'|m}$$

$$= \sum_{m,m'=1}^{\infty} \left(\sum_{n=1}^{\infty} \upsilon_{mn|m'n} \right) a_{m'm} \; .$$

In order that both expressions be equal, we must have

$$u_{m|m'} = \sum_{n=1}^{\infty} \upsilon_{mn|m'n} \; .$$

In the same way, our I + II ensemble, if only II is considered and I is ignored, determines a II ensemble, with a statistical operator U and matrix $\{\mu_{n|n'}\}$. By analogy, we obtain

$$u_{n|n'} = \sum_{m=1}^{\infty} \upsilon_{mn|mn'} \; .$$

We have thus established the rules of correspondence for the statistical operators of I, II, I + II , i.e., U , U, *U*. They proved to be essentially different from those which control the correspondence between the operators A , A, *A* of physical quantities.

It should be mentioned that our U , U, *U* correspondence depends only apparently on the choice of the complete orthonormal sets $\phi_m(q)$ and $\xi_n(q)$. Indeed it was derived from an invariant condition (which is satisfied by this arrangement alone): Namely, from the requirement of agreement between the expectation values of A and of *A* , or of those of A and of *A* .

U expresses the statistics in I + II , U and U those statistics restricted to I or II respectively. There now arises the question: do U , U determine *U* uniquely or not? In general one will expect a negative answer because all "probability dependencies" which may exist between the two systems disappear as the information is reduced to the sole knowledge of U and U, i.e., of the separated systems I and II. But if one knows the state of I precisely, as also that of II, "probability questions" do not arise, and then I + II, too, is precisely known. An exact mathematical discussion is, however, preferable to these qualitative considerations, and we shall proceed to this.

The problem is, then: For two given definite matrices $\{u_{m|m'}\}$ and $\{\mu_{n|n'}\}$, find a third definite matrix $\{\upsilon_{mn|m'n'}\}$, such that

$$\sum_{n=1}^{\infty} \upsilon_{mn|m'n} = u_{m|m'}, \quad \sum_{m=1}^{\infty} \upsilon_{mn|mn'} = u_{n|n'} .$$

(From

$$\sum_{m=1}^{\infty} u_{m|m} = 1, \quad \sum_{n=1}^{\infty} u_{n|n} = 1,$$

it then follows directly that

$$\sum_{m,n=1}^{\infty} \upsilon_{mn|mn} = 1,$$

i.e., the correct normalization is obtained.) This problem is always solvable, for example, $\upsilon_{mn|m'n'} = u_{m|m'}u_{m|n'}$ is always a solution (it can easily be seen that this matrix is definite), but the question arises as to whether this is the only solution.

We shall show that this is the case if and only if at least one of the two matrices $\{u_{m|m'}\}$, $\{u_{n|n'}\}$ is a state. First we prove the necessity of this condition, i.e., the existence of several solutions if both matrices correspond to mixtures. In such a case (cf. IV.2.)

$$u_{m|m'} = \alpha v_{m|m'} + \beta w_{m|m'} , \; u_{n|n'} = \gamma v_{n|n'} + \delta w_{n|n'} \; .$$

($v_{m|m'}$, $w_{m|m'}$ definite and $v_{n|n'}$, $w_{n|n'}$ also, differing by more than a constant factor,

$$\sum_{m=1}^{\infty} v_{m|m} = \sum_{m=1}^{\infty} w_{m|m} = \sum_{n=1}^{\infty} v_{n|n} = \sum_{n=1}^{\infty} w_{n|n} = 1$$

$\alpha, \beta, \gamma, \delta > 0, \; \alpha + \beta = 1, \; \gamma + \delta = 1$).

We easily verify that each

$$\upsilon_{mn|m'n} = \pi v_{m|m'} v_{n|n'} + \rho w_{m|m'} v_{n|n'} + \sigma v_{m|m'} w_{n|n'} + \tau w_{m|m'} w_{n|n'}$$

with

$$\pi + \sigma = \alpha, \; \rho + \tau = \beta, \; \pi + \rho = \gamma, \; \sigma + \tau = \delta \; ,$$

$$\pi, \rho, \sigma, \tau > 0 \; ,$$

is a solution. Then π, ρ, σ, τ can be chosen in an infinite number of ways: Because of $\alpha + \beta = \gamma + \delta$ only three of the four equations are independent; therefore, $\rho = \gamma - \pi$, $\sigma = \alpha - \pi$, $\tau = (\delta - \alpha) + \pi$, and in order that all be > 0 , we must require $\alpha - \delta = \gamma - \beta < \pi < \alpha, \gamma$, which is the case for infinitely many π . Now different π, ρ, σ, τ lead to different $\upsilon_{mn|m'n'}$, because the $v_{m|m'} \cdot v_{n|n'}, \ldots, w_{m|m'} \cdot w_{n|n'}$ are linearly independent, since the $v_{m|m'} \cdot w_{m|m'}$ are such, as well as the $v_{n|n'} \cdot w_{n|n'}$.

Next we prove the sufficiency, and here we may assume that $u_{m|m'}$ corresponds to a state (the other case is disposed of in the same way). Then $U = P_{[\phi]}$ and since the complete orthonormal set $\phi_1, \phi_2, \ldots$ was arbitrary, we can assume $\phi_1 = \phi$. $U = P_{[\phi_1]}$ has the matrix

$$u_{m|m'} \begin{cases} = 1 \; , \text{ for } m = m' = 1 \\ = 0 \; , \text{ otherwise} \end{cases} \Bigg| \; .$$

Therefore

$$\sum_{n=1}^{\infty} \upsilon_{mn|m'n} \begin{cases} = 1, & \text{for } m = m' = 1 \\ = 0, & \text{otherwise} \end{cases} .$$

In particular, for $m \neq 1$,

$$\sum_{n=1}^{\infty} \upsilon_{mn|mn} = 0 ,$$

but since all $\upsilon_{mn|mn} \geq 0$ because of the definiteness of $\upsilon_{mn|m'n'}$ $[\upsilon_{mn|mn} = (U \Phi_{mn}, \Phi_{mn})]$, therefore in this case $\upsilon_{mn|mn} = 0$. That is, $(U \Phi_{mn}, \Phi_{mn}) = 0$, and hence, because of the definiteness of U, $(U \Phi_{mn}, \Phi_{m'n'})$ also $= 0$ (cf. II.5., THEOREM 19.), where m', n' are arbitrary. That is, it follows from $m \neq 1$ that $\upsilon_{mn|m'n'} = 0$, and because of the Hermitian nature, this also follows from $m' \neq 1$. For $m = m' = 1$ however, this gives

$$\upsilon_{1n|1n'} = \sum_{m=1}^{\infty} \upsilon_{mn|mn'} = u_{n|n'} .$$

Consequently, as was asserted, the solution $\upsilon_{mn|m'n}$ is determined uniquely.

We can thus summarize our result as follows: A statistical ensemble in I + II with the operator $U = \{\upsilon_{mn|m'n'}\}$ is determined uniquely by the statistical ensembles determined by it in I and II individually, with the respective operators $U = \{u_{m|m'}\}$ and $U = \{u_{n|n'}\}$, if and only if the following two conditions are satisfied:

1. $\upsilon_{mn|m'n'} = v_{m|m'} v_{n|n'}$. (From

$$\mathrm{Tr}\, U = \sum_{m,n=1}^{\infty} \upsilon_{mn|mn} = \sum_{m=1}^{\infty} v_{m|m} \sum_{n=1}^{\infty} v_{n|n} = 1 ,$$

it follows that, by multiplication of $v_{m|m'}$ and $v_{n|n'}$ with two reciprocal constant factors, we can obtain

$$\sum_{m=1}^{\infty} v_{m|m} = 1 , \sum_{n=1}^{\infty} v_{n|n} = 1$$

But then we see that $u_{m|m'} = v_{m|m'}$, $u_{n|n'} = v_{n|n'}$.)

2. Either $v_{m|m'} = \bar{x}_m x_{m'}$ or $v_{n|n'} = \bar{x}_n x_n$. (Indeed $U = P_{[\phi]}$ means that

$$\phi = \sum_{m=1}^{\infty} y_m \phi_m ,$$

and therefore $u_{m|m'} = \bar{y}_m y_{m'}$ and correspondingly for $v_{m|m'}$; by analogy the same is true with $U = P_{[\xi]}$.)

We shall call U and U the projections of U in I and II respectively.[212]

We now apply ourselves to the states of I + II , $U = P_{[\Phi]}$. The corresponding wave functions $\Phi(q, r)$ can be expanded according to the complete orthonormal set $\phi_{mn}(q, r) = \phi_m(q)\xi_n(r)$:

$$\Phi(q, r) = \sum_{m,n=1}^{\infty} f_{mn}\phi_m(q)\xi_n(r) .$$

We can therefore replace them by the coefficients f_{mn} $(m, n = 1,2,\ldots)$ which are subject only to the condition that

$$\sum_{m,n=1}^{\infty} |f_{mn}|^2 = ||\Phi||^2$$

be finite.

---

[212]The projections of a state of I + II are in general mixtures in I or II ; cf. above. This circumstance was discovered by Landau, Z. Physik 45 (1927).

We can define two operators F, F^* by

$$(F.)\qquad \begin{aligned} F\phi(q) &= \int \overline{\Phi(q,\ r)}\phi(q)dq \\ F^*\xi(r) &= \int \Phi(q,\ r)\xi(r)dr \ . \end{aligned}$$

These are linear, but have the peculiarity of being defined in $\mathfrak{R}^{I}$ and $\mathfrak{R}^{II}$ respectively, and of taking on values from $\mathfrak{R}^{II}$ and $\mathfrak{R}^{I}$ respectively. Their relation is that of adjoints, since obviously $(F\phi,\ \xi) = (\phi,\ F^*\xi)$ (the inner product on the left is to be formed in $\mathfrak{R}^{II}$ and that on the right is to be formed in $\mathfrak{R}^{I}$). Since the difference of $\mathfrak{R}^{I}$ and $\mathfrak{R}^{II}$ is mathematically unimportant, we can apply the results of II.11: then, since we are dealing with integral operators, $\Sigma(F)$ and $\Sigma(F^*)$ are equal to

$$\iint |\Phi(q,\ r)|^2 dqdr = ||\Phi||^2 = 1 \quad (||\Phi|| \ \text{ in } \mathfrak{R}^{I+II}!) \ ,$$

and are therefore finite. Consequently F, F^* are continuous, in fact are completely continuous operators, and F^*F as well as FF^* are definite operators, $\mathrm{Tr}\ (F^*F) = \Sigma(F) = 1$, $\mathrm{Tr}\ (FF^*) = \Sigma(F^*) = 1$.

If we again consider the difference between $\mathfrak{R}^{I}$ and $\mathfrak{R}^{II}$ then we see that F^*F is defined and assumes values in $\mathfrak{R}^{I}$, and FF^* similarly in $\mathfrak{R}^{II}$.

Since $F\phi_m(q)$ comes out equal to

$$\sum_{n=1}^{\infty} \overline{f}_{mn}\xi_n(r) \ ,$$

F has the matrix $\{\overline{f}_{mn}\}$ [by use of the complete orthonormal sets $\phi_m(q)$ and $\overline{\xi}_n(r)$ respectively -- note that the latter is a complete orthonormal set along with $\xi_n(r)$], likewise F^* has the matrix $\{f_{mn}\}$ (with the same complete orthonormal systems). Therefore F^*F, FF^*

have the matrices

$$\left\{ \sum_{n=1}^{\infty} \overline{f}_{mn} f_{m'n} \right\}$$

(using the complete orthonormal set $\phi_m(q)$ in $\Re^{I}$) and

$$\left\{ \sum_{n=1}^{\infty} \overline{f}_{mn} f_{mn'} \right\}$$

(using the complete orthonormal set $\overline{\xi_n(r)}$ in $\Re^{II}$).

On the other hand, $U = P_{[\Phi]}$ has the matrix $\{\overline{f}_{mn} f_{m'n'}\}$ (using the complete orthonormal set $\Phi_{mn}(q, r) = \phi_m(q)\xi_n(r)$ in $\Re^{I+II}$), so that its projections in I and II , U and U have the matrices

$$\left\{ \sum_{n=1}^{\infty} \overline{f}_{mn} f_{m'n} \right\}$$

and

$$\left\{ \sum_{m=1}^{\infty} \overline{f}_{mn} f_{mn'} \right\}$$

respectively (with the complete orthonormal sets given above).[213] Consequently

$$(\mathbf{U}.) \qquad \mathsf{U} = F^*F, \quad U = FF^* .$$

Note that the definitions (**F.**) and the equations (**U.**) make no use of the ϕ_m, ξ_n -- hence they are valid independently of these.

The operators U , U are completely continuous, and by II.11. and IV.3., they can be written in the form

---

[213]The mathematical discussion is based on a paper by E. Schmidt, Math. Ann. 83 (1907).

$$U = \sum_{k=1}^{\infty} w_k' P_{[\psi_k]}, \quad U = \sum_{k=1}^{\infty} w_k'' P_{[\eta_k]} ,$$

in which the ψ_k form a complete orthonormal set in $\mathfrak{R}^{I}$, the η_k one in $\mathfrak{R}^{II}$ and all w_k', $w_k'' \geq 0$. We now neglect the terms in each of the two formulas with $w_k' = 0$ or $w_k'' = 0$ respectively, and number the remaining terms with $k = 1, 2, \ldots$. Then the ψ_k and η_k again form orthonormal, but not necessarily complete sets; the sums

$$\sum_{k=1}^{M'}, \quad \sum_{k=1}^{M''}$$

appear in place of the two

$$\sum_{k=1}^{\infty}$$

where M', M'' can be equal to ∞ or finite. Also, all w_k', w_k'' are now > 0.

Let us now consider a ψ_k. $U \psi_k = w_k' \psi_k$ and therefore $F^* F \psi_k = w_k' \psi_k$, $F F^* F \psi_k = w_k' F \psi_k$, $U F \psi_k = w_k' F \psi_k$. Furthermore

$$(F\psi_k, F\psi_l) = (F^* F\psi_k, \psi_l) = (U \psi_k, \psi_l)$$

$$= w_k'(\psi_k, \psi_l) \begin{cases} = w_k' , & \text{for } k = l \\ = 0 , & \text{for } k \neq l \end{cases} ,$$

therefore, in particular, $||F\psi_k||^2 = w'_k$. The $\frac{1}{\sqrt{w_k'}} F\psi_k$ then form an orthonormal set in R^{II} and they are eigenfunctions of U, with the same eigenvalues as the ψ_k for U (i.e., w_k'). That is, each eigenvalue of U is also one of U with at least the same multiplicity. Interchanging U, U shows that they have the same eigenvalues with the same multiplicities. The w'_k and w''_k therefore coincide except for their order. Hence $M' = M'' = M$, and by re-enumeration of the w''_k we can obtain $w'_k = w''_k = w_k$. And if this occurs, then we can clearly choose

$$\eta_k = \frac{1}{\sqrt{w_k}} F\psi_k$$

in general. Then

$$\frac{1}{\sqrt{w_k}} F^* \eta_k = \frac{1}{w_k} F^* F\psi_k = \frac{1}{w_k} \mathsf{U}\psi_k = \psi_k .$$

Therefore

$$(\mathbf{V}.) \qquad \eta_k = \frac{1}{\sqrt{w_k}} F\psi_k, \quad \psi_k = \frac{1}{\sqrt{w_k}} F^* \eta_k .$$ [212]

Let us now extend the orthonormal set $\psi_1, \psi_2, \ldots$ to a complete $\psi_1, \psi_2, \ldots, \psi'_1, \psi'_2, \ldots$ and likewise $\eta_1, \eta_2, \ldots$ to $\eta_1, \eta_2, \ldots, \eta'_1, \eta'_2, \ldots$ (each of the two sets $\psi'_1, \psi'_2, \ldots$ and $\eta'_1, \eta'_2, \ldots$ can be empty, finite or infinite, and in addition each set independently of the other set). We have observed before, that (**F.**), (**U.**) make no reference to the ϕ_m, ξ_n. We may therefore use (**V.**), as well as the above construction, and let them determine the choice of the complete orthonormal sets $\phi_1, \phi_2, \ldots$ and $\xi_1, \xi_2, \ldots$. Specifically we let these coincide with the $\psi_1, \psi_2, \ldots, \psi'_1, \psi'_2, \ldots$ and $\overline{\eta}_1, \overline{\eta}_2, \ldots, \overline{\eta}'_1, \overline{\eta}'_2, \ldots$ respectively. Now let ψ_k correspond to ϕ_{μ_k}, η_k to ξ_{ν_k} $(k = 1, \ldots, M)$ ($\mu_1, \mu_2, \ldots$ different from one another, $\nu_1, \nu_2, \ldots$ likewise). Then

$$F\phi_{\mu_k} = \sqrt{w_k}\, \xi_{\nu_k} ,$$

$$F\phi_m = 0 \quad \text{for} \quad m \neq \mu_1, \mu_2, \ldots .$$

Therefore

$$f_{mn} \begin{cases} = \sqrt{w_k}\ , \text{ for } m = \mu_k,\ n = \nu_k,\ k = 1,2,\ldots \\ \\ = 0\ , \text{ otherwise} \end{cases} ,$$

or equivalently

$$\Phi(q,\ r) = \sum_{k=1}^{M} \sqrt{w_k}\ \phi_{\mu_k}(q)\xi_{\nu_k}(r)\ .$$

By suitable choice of the complete orthonormal sets $\phi_m(q)$ and $\xi_n(r)$ we have thus established that each column of the matrix $\{f_{mn}\}$ contains at most one element $\neq 0$ (that this is real and > 0, namely $\sqrt{w_k}$, is unimportant for what follows). What is the physical meaning of this mathematical statement?

Let A be an operator with the eigenfunctions $\phi_1, \phi_2, \ldots$ and with only distinct eigenvalues, say $a_1, a_2, \ldots$; likewise B with $\xi_1, \xi_2, \ldots$ and $b_1, b_2, \ldots$. A corresponds to a physical quantity in I , B to one in II . They are therefore simultaneously measurable. It is easily seen that the statement "A has the value a_m and B has the value b_n" determines the state $\Phi_{mn}(q,\ r) = \phi_m(q)\xi_n(r)$, and that this has the probability $(P_{[\Phi_{mn}]}\Phi,\ \Phi) = |(\Phi,\ \Phi_{mn})|^2 = |f_{mn}|^2$ in the state $\Phi(q,\ r)$. Consequently, our statement means that A, B are simultaneously measurable, and that if one of them was measured in Φ , then the value of the other is determined by it uniquely. (An a_m with all $f_{mn} = 0$ cannot result, because its total probability

$$\sum_{n=1}^{\infty} |f_{mn}|^2$$

cannot be 0 , if a_m is ever observed -- therefore for

exactly one n, $f_{mn} \neq 0$; likewise for b_n .) That is, there are several possible A values in the state Φ (namely, those a_m for which

$$\sum_{n=1}^{\infty} |f_{mn}|^2 > 0 ,$$

i.e., for which there exists an n with $f_{mn} \neq 0$ -- usually all a_m are such), and an equal number of possible B values (those b_n for which

$$\sum_{n=1}^{\infty} |f_{mn}|^2 > 0 ,$$

i.e., for which there exists an m with $f_{mn} \neq 0$), but Φ establishes a one-to-one correspondence between the possible A values and the possible B values.

If we call the possible m values $\mu_1, \mu_2, \ldots$ and the corresponding possible n values $\nu_1, \nu_2, \ldots$, then

$$f_{mn} \begin{cases} = c_k \neq 0 , & \text{for } m = \mu_k,\ n = \nu_k,\ k = 1,2,\ldots \\ = 0 , & \text{otherwise} \end{cases} ,$$

therefore (M finite or ∞)

$$\Phi(q, r) = \sum_{k=1}^{M} c_k \phi_{\mu_k}(q) \xi_{\nu_k}(r) ,$$

hence

$$u_{mm'} = \sum_{n=1}^{\infty} \bar{f}_{mn} f_{m'n} \begin{cases} = |c_k|^2 , & \text{for } m = m' = \mu_k,\ k = 1,2,\ldots \\ = 0 , & \text{otherwise} \end{cases} ,$$

$$u_{nn'} = \sum_{m=1}^{\infty} \bar{f}_{mn} f_{mn'} \begin{cases} = |c_k|^2, & \text{for } n = n' = \nu_k,\ k = 1, 2, \ldots \\ = 0, & \text{otherwise} \end{cases}$$

and therefore

$$\mathsf{U} = \sum_{k=1}^{M} |c_k|^2 P_{[\phi_{\mu_k}]}, \quad U = \sum_{k=1}^{M} |c_k|^2 P_{[\xi_{\nu_k}]} .$$

Hence, when Φ is projected in I or II , it in general becomes a mixture, while it is a state in I + II only. Indeed, it involves certain information regarding I + II which cannot be made use of in I alone or in II alone, namely the one-to-one correspondence of the A and B values with each other.

For each Φ we can therefore so choose A, B , i.e., the ϕ_m and the ξ_n , that our condition is satisfied; for arbitrary A, B , it may of course be violated. Each state Φ then establishes a particular relation between I and II , while the related quantities A, B depend on Φ . How far Φ determines them, i.e., the ϕ_m and the ξ_n , is not difficult to answer. If all $|c_k|$ are different and $\neq 0$, then U, U (which are determined by Φ) determine the respective ϕ_m, ξ_n uniquely (cf. IV.3.). The general discussion is left to the reader.

Finally, let us mention the fact that for $M \neq 1$ neither U nor U is a state (because all $|c_k|^2 > 0$). For $M = 1$ they both are: $\mathsf{U} = P_{[\phi_{\mu_1}]}$, $U = P_{[\xi_{\nu_1}]}$. Then $\Phi(q, r) = c_1 \phi_{\mu_1}(q) \xi_{\nu_1}(r)$. We can absorb c_1 in $\phi_{\mu_1}(q)$. Therefore U, U are states if and only if $\Phi(q, r)$ has the form $\phi(q)\xi(r)$, and in that case they are equal to $P_{[\phi]}$ and $P_{[\xi]}$ respectively.

On the basis of the above results, we note: If

I is in the state $\phi(q)$ and II in the state $\xi(r)$, then I + II is in the state $\Phi(q, r) = \phi(q)\xi(r)$. If on the other hand I + II is in a state $\Phi(q, r)$ which is not a product $\phi(q)\xi(r)$, then I and II are mixtures and not states, but Φ establishes a one-to-one correspondence between the possible values of certain quantities in I and in II.

3. DISCUSSION OF THE MEASURING PROCESS

Before we complete the discussion of the measuring process in the sense of the ideas developed in VI.1. (with the aid of the formal tools developed in VI.2.), we shall make use of the results of VI.2. to exclude a possible explanation often proposed for the statistical character of the process **1.** (V.1.). This rests on the following idea: Let I be the observed system, II the observer. If I is in a state $U = P_{[\phi]}$ before the measurement, while II on the other hand is in a mixture

$$U = \sum_{n=1}^{\infty} w_n P_{[\xi_n]} ,$$

then I + II is a uniquely determined mixture $\mathcal{U}$, and in fact, as we can easily calculate from VI.2.,

$$\mathcal{U} = \sum_{n=1}^{\infty} w_n P_{[\Phi_n]}, \quad \Phi_n(q, r) = \phi(q)\xi_n(r)$$

If now a measurement of a quantity A takes place in I, then this is to be regarded as an interaction of I and II. This is a process **2.** (V.1.), with an energy operator $\mathcal{H}$. If it has the time duration t, then we obtain

$$\mathcal{U}' = e^{-\frac{2\pi i}{h} t\mathcal{H}} \mathcal{U} e^{\frac{2\pi i}{h} t\mathcal{H}}$$

from U , and in fact,

$$U' = \sum_{n=1}^{\infty} w_n P\left[e^{-\frac{2\pi i}{h} tH} \phi_n \right]$$

If now each

$$e^{-\frac{2\pi i}{h} tH} \Phi_n(q, r)$$

were of the form $\psi_n(q)\eta_n(r)$, where the ψ_n are the eigenfunctions of A , and the η_n any fixed complete orthonormal set, then this intervention would have the character of a measurement. For it transforms each state ϕ of I into a mixture of the eigenfunctions ψ_n of A . The statistical character therefore arises in this way: Before the measurement I was in a (unique) state, but II was a mixture -- and the mixture character of II has, in the course of the interaction, associated itself with I + II , and in particular, it has made a mixture of the projection in I . That is, the result of the measurement is indeterminate, because the state of the observer before the measurement is not known exactly. It is conceivable that such a mechanism might function, because the state of information of the observer regarding his own state could have absolute limitations, by the laws of nature. These limitations would be expressed in the values of the w_n , which are characteristic of the observer alone (and therefore independent of ϕ) .

At this point, the attempted explanation breaks down. For quantum mechanics requires that $w_n = (P_{\psi_n} \phi, \phi) = |(\phi, \psi_n)|^2$, i.e., w_n dependent on ϕ !
There might exist another decomposition

$$U' = \sum_{n=1}^{\infty} w'_n P_{[\Phi'_n]} ,$$

(the $\phi'_n(q, r) = \psi_n(q)\eta_n(r)$ are orthonormal) but this is of no use either; because the w'_n are (except for order) determined uniquely by U' (IV.3.), and are therefore equal to the w_n.[214]

Therefore, the non-causal nature of the process **1.** is not produced by any incomplete knowledge of the state of the observer, and we shall therefore assume in all that follows that this state is completely known.

Let us now apply ourselves again to the problem formulated at the end of VI.1. I, II, III shall have the meanings given there, and, for the quantum mechanical investigation of I, II, we shall use the notation of VI.2., while III remains outside of the calculations (cf. the discussion of this in VI.1.). Let A be the quantity (in I) actually to be measured, $\phi_1(q), \phi_2(q), \ldots$ its eigenfunctions. Let I be in the state $\phi(q)$.

If I is the observed system, II + III the observer, then we must apply the process **1.**, and we find that the measurement transforms I from the state ϕ into one of the states ϕ_n $(n = 1,2,\ldots)$, the probabilities for which are respectively $|(\phi, \phi_n)|^2$ $(n = 1,2,\ldots)$. Now, what is the method of description if I + II is the observed system, and only III the observer?

In this case we must say that II is a measuring instrument which shows on a scale the value of A (in I): the position of the pointer on this scale is a physical quantity B (in II) which is actually observed by III (if II is already within the body of the observer, we have the corresponding physiological concepts in place of the scale and pointer, e.g., retina and image on the retina, etc.) Let A have the values $a_1, a_2, \ldots$, B the values $b_1, b_2, \ldots$, and let the numbering be such that a_n is associated with b_n.

---

[214] This approach is capable of still more variants, which must be rejected for similar reasons.

Initially, I is in the (unknown) state $\phi(q)$ and II in the (known) state $\xi(r)$, therefore I + II is in the state $\Phi(q, r) = \phi(q)\xi(r)$. The measurement (so far as it is performed by II on I) is, as in the earlier example, carried out by an energy operator H (in I + II) in the time t: This is the process **2.**, which transforms the Φ into

$$\Phi' = e^{-\frac{2\pi i}{h} tH} \Phi .$$

Viewed by the observer III, one has a measurement only if the following is the case: If III were to measure (by process **1.**) the simultaneously measurable quantities A, B (in I or II respectively, or both in I + II), then the pair of values a_n, b_n would have the probability 0 for $m \neq n$, and the probability w_n for $m = n$. That is, it suffices "to look at" II, and A is measured in I. Quantum mechanics then requires in addition $w_n = |(\phi, \phi_n)|^2$.

If this is established, then the measuring process so far as it occurs in II, is "explained" theoretically, i.e., the division of I | II + III discussed in VI.1. is shifted to I + II | III.

The mathematical problem is then the following. A complete orthonormal set $\phi_1, \phi_2, \ldots$ is given in I. Such a set $\xi_1, \xi_2, \ldots$ in $\mathfrak{R}^{II}$ as well as a state ξ in R^I, also an (energy) operator H in $\mathfrak{R}^{I+II}$, and a t, are to be found so that the following holds. If ϕ is an arbitrary state in R^I and

$$\Phi(q, r) = \phi(q)\xi(r), \quad \Phi'(q, r) = e^{-\frac{2\pi i}{h} tH} \Phi(q, r) ,$$

then $\Phi'(q, r)$ must have the form

$$\sum_{n=1}^{\infty} c_n \phi_n(q) \xi_n(r)$$

(the c_n are naturally dependent on ϕ). Therefore $|c_n|^2 = |(\phi, \phi_n)|^2$. (That the latter is equivalent to the physical requirement formulated above was discussed in VI.2.)

In the following we shall use a fixed set $\xi_1, \xi_2, \ldots$ and a fixed ξ along with the fixed $\phi_1, \phi_2, \ldots$, and shall investigate the unitary operator

$$\Delta = e^{-\frac{2\pi i}{h} tH}$$

instead of H .

The mathematical problem leads us back to the problem solved in VI.2.: there the quantity corresponding to our present Φ was given, and we showed the existence of c_n, ϕ_n, ξ_n . Now ϕ_n, ξ_n are fixed and Φ, c_n are given dependent on ϕ , and it remains so to determine a fixed Δ that for $\Phi' = \Delta\Phi$ these c_n, ϕ_n, ξ_n result.

We shall show that such a determination of Δ is indeed possible. In this case only the principle is of importance to us, i.e., the existence of any such Δ. The further question, whether the

$$\Delta = e^{-\frac{2\pi i}{h} tH}$$

corresponding to simple and plausible measuring arrangements also have this property, shall not concern us. Indeed, we saw that our requirements coincide with a plausible intuitive criterion of the measurement character in an intervention. Furthermore the arrangements in question are to possess the characteristics of the measurement. Hence quantum mechanics, as applied to observation would be in blatant contradiction with experience, if these Δ did not satisfy the requirements in question (at least approximately).[215] Therefore, in the following, only an abstract

[215]The corresponding calculation for the case of the posi-

Δ which satisfies our conditions exactly, shall be given.

Therefore, let the ϕ_m $(m = 0, \pm 1, \pm 2, \ldots)$ and the ξ_n $(n = 0, \pm 1, \pm 2, \ldots)$ respectively be two given complete orthonormal sets in $\mathfrak{R}^{I}$ and $\mathfrak{R}^{II}$ respectively. (We do not let m, n run over $1, 2, \ldots$, but over $0, \pm 1, \pm 2, \ldots$. This is purely for technical convenience, and is in principle equivalent to the former). Let the state ξ be, for simplicity, ξ_0. We define the operator Δ by

$$\Delta \sum_{m,n=-\infty}^{\infty} x_{mn}\phi_m(q)\xi_n(r) = \sum_{m,n=-\infty}^{\infty} x_{mn}\phi_m(q)\xi_{m+n}(r) ,$$

since the $\phi_m(q)\xi_n(r)$ as well as the $\phi_m(q)\xi_{m+n}(r)$ form a complete orthonormal set in $\mathfrak{R}^{I+II}$, this Δ is unitary. Now

$$\phi(q) = \sum_{m=-\infty}^{\infty} (\phi, \phi_m)\cdot\phi_m(q), \quad \xi(r) = \xi_0(r) ,$$

therefore

$$\Phi(q, r) = \phi(q)\xi(r) = \sum_{m=-\infty}^{\infty} (\phi, \phi_m)\cdot\phi_m(q)\xi_0(r) ,$$

$$\Phi'(q, r) = \Delta\Phi(q, r) = \sum_{m=-\infty}^{\infty} (\phi, \phi_m)\cdot\phi_m(q)\xi_m(r) .$$

Hence our purpose is accomplished. We have in addition $c_n = (\phi, \phi_n)$.

A better overall view of the mechanism of this process can be obtained if we exemplify it by concrete Schrödinger wave functions, and give H in place of Δ.

The observed object, as well as the observer

---

tion measurement discussed in III.4. is contained in a paper by Weizsäcker, Z. Physik 70 (1931).

(i.e., I and II respectively) may be characterized by a single variable q and r respectively, running continuously from $-\infty$ to $+\infty$. That is, let both be thought of as points which can move along a line. Their wave functions then have always the form $\psi(q)$ and $\eta(r)$ respectively. We assume that their masses m_1 and m_2 are so large that the kinetic energy portion of the energy operator (i.e., $\frac{1}{2m_1}(\frac{h}{2\pi i}\frac{\partial}{\partial q})^2 + \frac{1}{2m_2}(\frac{h}{2\pi i}\frac{\partial}{\partial r})^2$) can be neglected. Then there remains of H only the interaction energy part which is decisive for the measurement. For this we choose the particular form $\frac{h}{2\pi i} q \frac{\partial}{\partial r}$.

The Schrödinger time dependent differential equation then is (for the I + II wave functions $\psi_t = \psi_t(q, r)$):

$$\frac{h}{2\pi i}\frac{\partial}{\partial t}\psi_t(q, r) = -\frac{h}{2\pi i} q \frac{\partial}{\partial r}\psi_t(q, r) ,$$

$$(\frac{\partial}{\partial t} + q\frac{\partial}{\partial r})\psi_t(q, r) = 0 ,$$

i.e.,

$$\psi_t(q, r) = f(q, r - tq) .$$

If, for $t = 0$, $\psi_0(q, r) = \Phi(q, r)$, then we have $f(q, r) = \Phi(q, r)$, and therefore

$$\psi_t(q, r) = \Phi(q, r - tq)$$

In particular, if the initial states of I, II are represented by $\phi(q)$ and $\xi(r)$ respectively, then, in the sense of our calculation scheme (if the time t appearing therein is chosen to be 1)

$$\Phi(q, r) = \phi(q)\xi(r) ,$$

$$\Phi'(q, r) = \psi_1(q, r) = \phi(q)\xi(r - q) .$$

We now wish to show that this can be used by II for a position measurement of I, i.e., that the coordinates are tied to each other. (Since q, r have continuous spectra, they are therefore measurable with only arbitrary precision, but not with absolute precision. Hence this can be accomplished only approximately.)

For this purpose, we wish to assume that $\xi(r)$ is different from 0 only in a very small interval $-\epsilon < r < \epsilon$ (i.e., the coordinate r of the observer before the measurement is very accurately known), in addition ξ should of course be normalized:

$$||\xi|| = 1 \text{ , i.e., } \int |\xi(r)|^2 dr = 1 \text{ .}$$

The probability therefore that q lies in the interval $q_0 - \delta < q < q_0 + \delta$, and r in the interval $r_0 - \delta' < r < r_0 + \delta'$ is

$$\int_{q_0-\delta}^{q_0+\delta} \int_{r_0-\delta}^{r_0+\delta} |\Phi'(q, r)|^2 dqdr = \int_{q_0-\delta}^{q_0+\delta} \int_{r_0-\delta}^{r_0+\delta} |\phi(q)|^2 |\xi(r - q)|^2 dqdr \text{ .}$$

If q_0, r_0 are to differ by more than $\delta + \delta' + \epsilon$, then this is 0, i.e., q, r are so very closely tied to each other that the difference can never be greater than $\delta + \delta' + \epsilon$. And for $r_0 = q_0$ this is, equal to

$$\int_{q_0-\delta}^{q_0+\delta} |\phi(q)|^2 dq \text{ ,}$$

if we choose $\delta' \geq \delta + \epsilon$, because of the assumptions on ξ. But since we can choose $\delta, \delta', \epsilon$ arbitrarily small (they must be different from zero, however), this means that q, r are tied to each other with arbitrary closeness, and the probability density has the value furnished by quantum mechanics, $|\phi(q)|^2$.

That is, the relations of the measurement, as we had discussed them in IV.1., and in this section, are realized.

The discussion of more complicated examples, say of an analog to our four-term example of IV.1., or the control determination of the validity of a measurement which II carried out on I , effected by a second observer III , can also be carried out in this fashion. It is left to the reader.

V.3 THE ERGODIC BEHAVIOUR OF QUANTUM MANY-BODY SYSTEMS

LÉON VAN HOVE

Synopsis

By a pertubation technique adapted to the actual properties of gases and solids (and possibly also of liquids) we have established in previous papers that under suitable conditions a quantum many-body system approaches statistical equilibrium as far as those physical quantities are concerned which are diagonal in the unperturbed representation. This result is now extended to non-diagonal quantities of a type broad enough to include all observables of actual interest. A general discussion of the resulting ergodic theorem is given, and its implications for classical statistics are briefly analyzed. The paper ends with a discussion of a recent article by Ingraham on the application of our methods to the case of a very small perturbation. The main arguments of Ingraham are shown to be in error, and the inconsistencies he derives from them are thereby disproved.

1. *Introduction.* The approach of a quantum many-body system to statistical equilibrium has been studied in two previous papers [1]), to be referred to hereafter as S and S', on the basis of a separation of the hamiltonian H' into a main term H describing non-interacting plane wave excitations (like phonons or Bloch electrons in solids, free particles in gases) and a perturbation λV representing their mutual interaction. Using as basic representation the eigenstates $|\alpha\rangle$ of H, each of which describes a set of plane wave excitations, we founded our treatment on the recognition that the matrix elements to be calculated according to perturbation theory exhibit remarkable diagonal singularities, *i.e.* singularities of the form $\delta(\alpha - \alpha')$. A systematic analysis of these singularities made it possible to study the time evolution of certain physical quantities A under the assumption of incoherent phases for the amplitudes $c(\alpha)$ of the initial state φ_0 of the system *)

$$\varphi_0 = \int |\alpha\rangle d\alpha c(\alpha). \tag{1.1}$$

Under proper conditions it could be established that in the course of time the expectation value of A, which is (we put $\hbar = 1$)

$$\langle A \rangle_t = \langle \varphi_0 | U_{-t} A U_t | \varphi_0 \rangle, \quad U_t = \exp[-i(H + \lambda V)t], \tag{1.2}$$

*) A detailed definition of our notation is found in S'.

Originally published in *Physica, 25*, 268-76 (1959).

tends to the equilibrium value $\langle A\rangle_{eq}$ calculated on the basis of microcanonical ensemble theory. This was established for operators A diagonal in the $|\alpha\rangle$-representation, in the case of very small perturbation in S and for finite λV, *i.e.* to general order in the dimensionless parameter λ, in S'.

Our present aim is to extend the general order treatment of S' to a wide class of non-diagonal operators B **). This class is composed of the non-diagonal operators B given by a convergent series, each term of which is a product of creation and destruction operators for individual plane wave excitations. We assume the number of creation and destruction operators in each term of the series to be finite and independent of the large number N of particles in the system. In contrast to the diagonal operators A, the class of operators B just defined is broad enough to contain all quantities of practical interest. Since we will again be able to establish that the expectation value

$$\langle B\rangle_t = \langle \varphi_0|U_{-t}BU_t|\varphi_0\rangle \tag{1.3}$$

tends to the microcanonical equilibrium value $\langle B\rangle_{eq}$ we will have established, for all practical purposes, the ergodic behaviour of our system, using only the properties of the hamiltonian postulated in S' (sections 2 and 7) and the incoherent phase assumption for the initial state (1.1).

The rather broad scope of this result makes it desirable to discuss its significance, also in relation to classical statistics. Whereas the derivation of our main result

$$\langle B\rangle_t \to \langle B\rangle_{eq} \text{ for } t \to \pm\infty \tag{1.4}$$

is presented in the next section, section 3 will be devoted to this general discussion. Section 2 will make free use of the formal technique developed and applied in S'. Although the author is fully aware of the complication of this technique, he believes that the scope of the results may justify at least partially the involved nature of the mathematical methods. Section 3 gives a nontechnical discussion which can be largely followed without knowledge of the detailed formalism of section 2 and S'. Section 4 presents a refutation of a critical discussion of S recently published by Ingraham.

2. *Approach to equilibrium for non-diagonal operators.* For any operator B of the type described above and for any set of diagonal operators $A_1, \ldots, A_n$ the matrix element

$$\langle\alpha|VA_1VA_2V\ldots A_\nu BA_{\nu+1}V\ldots A_nV|\alpha'\rangle \tag{2.1}$$

can have a $\delta(\alpha-\alpha')$-singularity for the same reasons and with the same properties as was the case for the matrix elements $\langle\alpha|VA_1V\ldots A_nV|\alpha'\rangle$

**) As in S and S' the adjectives diagonal and non-diagonal will always refer to the $|\alpha\rangle$-representation.

considered in S' (see S', section 2). The discussion of this singularity requires that (2.1) be written out as a sum over intermediate states α_j. In addition to having one V replaced by B this sum may differ from the sum (S'.2.6) *) by the fact that additional intermediate states may have to be introduced in between various factors of a single term of B. This would for example be the case if $B = V_1V_2$, V_1 and V_2 being individual terms in the expansion of V in products of creation and destruction operators. In this particular case one would obviously insert an additional intermediate state between V_1 and V_2. All considerations presented in S' (section 2) can be repeated for this slightly more general case, however, and we can define in addition the concept of *B-irreducible diagonal part*. It is the diagonal part, *i.e.* the $\delta(\alpha - \alpha')$-singular part, of (2.1) which is obtained when in the calculation of (2.1) one leaves out the diagonal part of each subproduct of the following type

$$\left.\begin{array}{l} VA_jV\ldots A_kV,\ (1 \leq j \leq k < \nu \text{ or } \nu+1 < j \leq k \leq n), \\ VA_jV\ldots A_\nu BA_{\nu+1}V\ldots A_kV,\ (1 \leq j \leq \nu < k \leq n,\ k-j<n-1, \\ VA_jV\ldots A_\nu B,\ (1 \leq j \leq \nu),\quad BA_{\nu+1}V\ldots A_kV,\ (\nu+1 \leq k \leq n). \end{array}\right\} \tag{2.2}$$

We denote by $\{VA_1V\ldots A_\nu BA_{\nu+1}V\ldots A_nV\}_{Bd}$ the diagonal operator defined by the B-irreducible diagonal part of (2.1). This concept may differ from the irreducible diagonal part defined in S' because of the possible occurrence of additional intermediate states inside B (these states are allowed to become equal to each other, to other intermediate states, to α or to α', and the corresponding subproducts, in contrast to the subproducts (2.2), may therefore contribute their diagonal part). The new concept can be extended in an obvious way to products of the type

$$BA_1V\ldots A_nV,\quad VA_1\ldots VA_nB. \tag{2.3}$$

Its importance will appear presently.

In analogy to our analysis in S', which made essential use of the quantity $X_{ll'}(\alpha\alpha')$ defined by (S'.4.14), we must consider the quantity $Y_{ll'}(\alpha)$ defined by

$$\{R_lBR_{l'}\}_d|\alpha\rangle = |\alpha\rangle Y_{ll'}(\alpha) \tag{2.4}$$

where R_l is the resolvent (S'.4.1). Reduction of diagonal parts, using (S'.4.4) and our definition of B-irreducible diagonal part, readily gives

$$\{R_lBR_{l'}\}_d = \{R_lB_{ll'}R_{l'}\}_d \tag{2.5}$$

with

$$B_{ll'} = \{(1 - \lambda VD_l + \lambda^2VD_lVD_l\ldots)B(1 - \lambda D_{l'}V + \lambda^2 D_{l'}VD_{l'}V\ldots)\}_{Bd} \tag{2.6}$$

Hence, using (S'.4.14),

$$Y_{ll'}(\alpha) = \int B_{ll'}(\alpha')d\alpha' X_{ll'}(\alpha'\alpha) \tag{2.7}$$

where $B_{ll'}(\alpha')$ is the eigenvalue of the diagonal operator $B_{ll'}$ for the state α'.

*) This symbol refers to Eq. (2.6) of S'

In analogy to (S'.4.13) we have

$$U_{-t}BU_t = -(2\pi)^{-2}\int_\gamma dl \int_\gamma dl' \exp[i(l-l')t]R_l B R_{l'} \tag{2.8}$$

where the contour γ in the complex plane encircles an interval of the real axis sufficiently large to include the whole energy spectrum of the system and is described counterclockwise. Calculating the expectation value of (2.8) for the initial state φ_0, and using the incoherent phase assumption for the amplitudes $c(\alpha)$ of φ_0 one obtains

$$\langle B\rangle_t = -(2\pi)^{-2}\int_\gamma dl \int_\gamma dl' \exp[i(l-l')t] Y_{ll'}(\alpha) d\alpha |c(\alpha)|^2. \tag{2.9}$$

The long time limit of this quantity is found by repeating for $Y_{ll'}$ the discussion carried our for $X_{ll'}$ in section 6 of S'. No new difficulty occurs in this discussion because the quantity $B_{ll'}(\alpha')$ by virtue of its definition remains finite and has only finite discontinuities for l and l' crossing the real axis. According to (2.7) the pseudopoles of $Y_{E+l,E-l}$ therefore coincide with those of $X_{E+l,E-l}$. In analogy with (S'.6.18) one finds for the limit of (2.9) as $t \to \pm\infty$

$$\langle B\rangle_{\pm\infty} = \pi^{-1}\int_{-\infty}^{\infty} dE \int \lim_{0<\eta\to 0}\{\eta Y_{E\mp i\eta, E\pm i\eta}(\alpha)\} d\alpha |c(\alpha)|^2. \tag{2.10}$$

We go back to (2.7) and remember that $B_{ll'}$ remains finite when l and l' approach the real axis. This gives

$$\lim \eta\, Y_{E\mp i\eta, E\pm i\eta}(\alpha) = \int B_{E\mp i0, E\pm i0}(\alpha') d\alpha' \lim \eta X_{E\mp i\eta, E\pm i\eta}(\alpha'\alpha). \tag{2.11}$$

The limit on the righthand side has been calculated in section 7 of S', with the result (see (S'.7.9) and (S'.7.17) and the third equation thereafter)

$$\pi^{-1}\lim \eta X_{E\pm i\eta, E\mp i\eta}(\alpha'\alpha) = \Delta_E(\alpha')\Delta_E(\alpha)[\int \Delta_E(\alpha'')d\alpha'']^{-1}. \tag{2.12}$$

We obtain by substitution

$$\langle B\rangle_{\pm\infty} = \int_{-\infty}^{\infty} \langle B\rangle_E p_E dE. \tag{2.13}$$

$p_E dE$, defined by (S'.7.22), is the probability that the total energy $H+\lambda V$ be included between E and $E+dE$ for the system in its initial state φ_0, under the incoherent phase assumption. The quantity $\langle B\rangle_E$ is defined by

$$\langle B\rangle_E = [\int \Delta_E(\alpha'')d\alpha'']^{-1}\int B_{E\mp i0, E\pm i0}(\alpha')d\alpha'\Delta_E(\alpha'). \tag{2.14}$$

As will now be shown, it is equal to the microcanonical average of B on the energy shell $H+\lambda V = E$ and its value is independent of the double sign appearing in the definition.

The microcanonical average of B is

$$Sp(BQ_E)/Sp(Q_E), \tag{2.15}$$

Q_E being the projection operator on the energy shell $H+\lambda V = E$. One has

$$Q_E = \delta(H+\lambda V - E) = (2\pi i)^{-1}\lim_{0<\eta\to 0}(R_{E+i\eta} - R_{E-i\eta}).$$

But the resolvent verifies

$$R_{E+i\eta} - R_{E-i\eta} = 2i\eta R_{E+i\eta} R_{E-i\eta},$$

a relation which enables us to write

$$Sp(BQ_E) = \pi^{-1} \lim \eta Sp(R_{E\mp i\eta} B R_{E\pm i\eta}) = \pi^{-1} \int d\alpha \lim \eta Y_{E\mp i\eta, E\pm i\eta}(\alpha).$$

Here we have calculated the trace in the $|\alpha\rangle$-representation. The limit under the integral sign has been found in (2.11) and (2.12). It gives

$$Sp(BQ_E) = \int B_{E\mp i0, E\pm i0}(\alpha') d\alpha' \Delta_E(\alpha'). \tag{2.16}$$

If one now remembers that $Sp(Q_E)$ is $\int \Delta_E(\alpha'')d\alpha''$ as remarked in S', one reaches the announced identity of (2.14) and (2.15) for either value of the double sign appearing in the former expression. Returning to (2.13) we see that the long time limit of $\langle B\rangle_t$ agrees with the equilibrium value

$$\langle B\rangle_{eq} = \int \langle B\rangle_E p_E dE \tag{2.17}$$

of the quantity B as deduced from microcanonical ensemble theory. The ergodic behaviour of our system is thereby established for all observable quantities B of the type described in section 1 and for initial states with incoherent phases. It may be noted that the main difference between the derivation just given and the treatment of diagonal quantities A in S and S' lies in the replacement of A by the diagonal operator $B_{ll'}$.

3. *Discussion.* We have been able to establish that under specified conditions an isolated many-body system approaches microcanonical equilibrium. This has been achieved in the quantum description and we have made essential use of very special properties of the system in this description. In the first place, our analysis is entirely based on the existence of a special orthonormal set of states α, composed of plane wave excitations. In this special representation the total hamiltonian is assumed to split into a diagonal part H and an off-diagonal part λV, and matrix elements of the form (2.1) are supposed to exhibit diagonal singularities with very definite properties. Secondly, we establish the approach to microcanonical equilibrium in a slightly unusual way. What we do is to study physical quantities represented by operators O with definite properties in the α-representation, and establish that their expectation value $\langle O\rangle_t$ tends in the course of time toward the equilibrium average value $\langle O\rangle_{eq}$ calculated from microcanonical theory. Finally our analysis follows the time evolution of the system for $t \to \pm\infty$ beginning with an initial state $\varphi_0 = \int |\alpha\rangle d\alpha c(\alpha)$. We show that $\langle O\rangle_t$ splits into two terms, one depending on the $|c(\alpha)|^2$ only, and the second depending on the relative phases of $c(\alpha)$ for different α's. We simply leave out the phase-dependent term on the ground that it will vanish for all times of practical interest if the initial amplitudes $c(\alpha)$ have incoherent phases.

Our result $\langle O \rangle_t \to \langle O \rangle_{eq}$ is established for the phase-independent part of $\langle O \rangle_t$. This result, being truly non-trivial and involving quite subtle properties of the diagonal singularities mentioned above, seems to us in itself to give additional support to the soundness of the incoherent phase assumption.

Obviously the systems we study are quite special, we use a special representation in expressing their properties and we discuss the approach to equilibrium for special operators O. All these special features, however, are realized in the most common physical examples of ergodic systems, solids, gases and liquids (although for the latter the large size of λV may make our treatment more doubtful). We feel therefore confident that despite its apparent lack of generality our method is quite well adapted to the difficult problem of establishing ergodicity for realistic systems. In fact it is only by complete exploitation of the special conditions assumed that we have been able to study our system in so much greater detail than is the case in the conventional investigations of the ergodic problem *) and to actually carry out a true proof of ergodic behaviour.

Although quantum-mechanical in nature, our analysis applies also to the physical situations one usually calls classical, i.e. when the values of all observable quantities of interest depend on Planck's constant only through negligibly small corrections. For such situations it is important to translate our results into the language of classical theory. The main point in this translation is that the state of the system at any time, although a single quantum state $\varphi_t = U_t \varphi_0$, must be described classically by an ensemble of points in $6N$-dimensional phase space (N being the number of particles). This cannot be otherwise, because the quantum-mechanical wave function corresponding to a single point in classical phase space is a very special type of wave packet and cannot possess the incoherent phases we require φ_t to have for $t = 0$ in our analysis **). This correspondence between single quantum states and classical ensembles leads to certain consequences which we now want to describe.

According to our theory, for sufficiently large times t the expectation value $\langle O \rangle_t$ of a physical quantity O becomes equal to the equilibrium value $\langle O \rangle_{eq}$. This holds equally well for the quantity O^2,

$$\langle O^2 \rangle_t = \langle O^2 \rangle_{eq} \text{ for large } t.$$

In general, of course, $\langle O^2 \rangle_{eq}$ and $\langle O \rangle_{eq}^2$ are different, so that for large t the quantities $\langle O^2 \rangle_t$ and $\langle O \rangle_t^2$ will be both constant in time and have different values, a fact which makes it quite clear that the classical analogue to the

*) The unsatisfactory nature of Von Neumann's approach [2]) to the quantum ergodic problem has now been clearly revealed [3]). It stems from the fact that Von Neumann's form of ergodic property actually imposes no restriction at all on the dynamical system.

**) Starting from a different standpoint [4]) Van Kampen has been led some time ago to take the same view concerning the relation between classical and quantum statistics [5]).

single quantum state φ_t of our system has to be an ensemble. If now measurements of O are made on one and the same system at various times, the measured values q_t will show a time dependence even for large t and, in classical situations (as defined above), the time averages of q_t and $(q_t)^2$ must be equal to $\langle O \rangle_t$ and $\langle O^2 \rangle_t$ respectively, because the measuring process then cannot affect the observable properties of the system.

In our theory, a measurement of O at time t must be described in the conventional quantum-mechanical fashion as giving rise to a reduction of the state vector φ_t to another vector φ_t'. This interpretation holds always, even in classical situations. For the latter case we can translate it by saying that the measurement gives rise to a reduction of the classical ensembe associated with φ_t to the smaller ensemble associated with φ_t'. Although there is nothing wrong with this description, it differs in a non-trivial way from the picture conventionally adopted for a classical many-body system, in which at every instant t the system is regarded as being in one single point of $6N$-dimensional phase space. If our theory is valid, *a gas, liquid or solid in thermal equilibrium can never be said to be in one single point of its classical phase space, even at high temperatures where all measurable quantum effects are numerically negligible. It is in a single quantum state, the classical analogue of which is an ensemble.*

It should be stressed that this unorthodox view cannot lead to any observable discrepancy with the conventional picture of a classical system. Let us verify for example in our quantum description that for a classical situation (as defined above) a measurement carried out at time t does not affect the result of observations at later times $t' = t + \tau$. Let the quantity O be measured at time t, and the quantity O' at time t'. Assuming for simplicity O to have discrete eigenvalue O_n, all we have to establish is the identity

$$\langle \varphi_t | U_{-\tau} O' U_\tau O | \varphi_t \rangle = \sum_n O_n \langle \varphi_t | P_n U_{-\tau} O' U_\tau P_n | \varphi_t \rangle \tag{3.1}$$

where the P_n's are the projection operators verifying $O = \sum_n O_n P_n$. The righthand side of (3.1) includes the reduction of the state vector due to the first measurement, the lefthand side neglects it. Since $P_n^2 = P_n$ the difference between the two is

$$\sum_n O_n \langle \varphi_t | [P_n, U_{-\tau} O' U_\tau] P_n | \varphi_t \rangle .$$

The commutator being proportional to $\hbar$, this difference is indeed negligible in a classical situation.

4. *Refutation of Ingraham's criticism.* In a recent paper [6]) Ingraham has presented a severely critical discussion of our derivation in S of the master equation describing the approach to equilibrium in the case of a very small perturbation (limiting case $\lambda \to 0$, $t \to \infty$, $\lambda^2 t$ finite). In fact, if

this discussion were correct, it would imply complete invalidity of the contents of S, S' and the present paper. Ingraham's analysis, however, is based on a few patently wrong assertions and arguments. If these assertions and arguments are replaced by their corrected versions, the whole criticism of Ingraham becomes groundless and his considerations reduce to those of S and lead to the same conclusions. We would like to devote the last section of the present paper to a refutation of Ingraham's criticism, to which end it will be sufficient to indicate which basic arguments of this author are incorrect. We can concentrate on section 3 of Ingraham's paper, which contains the actual discussion of our work.

The first argument of Ingraham is developed on pp. 107 to 111 of his paper. It tends to show that our method of calculation violates the unitarity of the operator of motion

$$U(t) = \exp[-i(H + \lambda V)t].$$

The essential step is that unitarity of $U(t)$ would imply In (3.20) *), which itself entails In (3.22), i.e. the vanishing of all transition rates. The central error is In (3.20). In deriving this equation, Ingraham tacitly assumes that for $\lambda \to 0$, $t \to \infty$, $\lambda^2 t$ finite, the limit of a product of operators (in the case at hand $U_2{}^* U_2$) is equal to the product of the limits. This assumption is incorrect. Explicit calculation of the limit of the product gives an additional term in the righthand side of In (3.20), removing the inconsistency. Incidentally, while "deriving" In (3.20), Ingraham states that our use of a well known asymptotic formula, In (3.16), would result from a choice. This is patently wrong, because In (3.16) is a *mathematical identity* for $f(\epsilon)$ continuously differentiable and vanishing at least as fast as $|\epsilon|^{\nu}$, $\nu < 0$ for $\epsilon \to \infty$. No other formula would be correct, and no choice can therefore be made.

The second criticism of Ingraham is that ambiguities exist in our calculations, because in products containing 3 or more factors V several prescriptions could be followed in replacing sub-products VAV by their diagonal parts: various choices could be made for these subproducts leading to different results for the total expression (see pp. 112 and 113 of Ingraham's paper). Ingraham has not understood that all choices *have* to be made, and that in the limit $\lambda \to 0$, $t \to \infty$, $\lambda^2 t$ finite, the total expression is the sum of the contributions of all possible choices. Among the latter only a few give non-vanishing contributions in the limiting case considered; all these have been calculated in S, ensuring the correctness of our results in the weak coupling limit.

Misled by the errors just mentioned, Ingraham concludes to the invalidity of the fundamental property on which our work is based, to know the occurrence of diagonal singularities, In (3.1) in his paper. His misunderstanding of this property becomes quite clear when he states it to be

*) By this abbreviation we mean Eq. (3.20) of Ingraham's paper.

equivalent to the property In (3.30), where no summation is carried out on the intermediate state q_1. The summation is absolutely essential for the occurrence of the diagonal singularity. Consequently the operator A in In (3.1) must have eigenvalues $A(E''\alpha'') = A''$ varying smoothly with the parameters E'', α'', and In (3.30) is incorrect. For the same reason, the master equation established in S concerns the *probability density* to find our large system in unperturbed states, i.e. the probability averaged over a very large number of neighboring states. This is a coarse-grained probability distribution. Therefore Ingraham's statement that we "want to obtain irreversible effects while retaining the maximum information permitted by statistical mechanics" (p. 101), according to his introduction the very motivation of his critical study, is incorrect. Concerning the alternative diagonal property In (3.33) proposed by Ingraham, it will be sufficient to say that it does not hold for actual systems, as is easily verified on the example of the electron-phonon system described in the appendix of S'.

Having thus concluded to the invalidity of Ingraham's criticism, we may finally remark that this author seems to be quite confused concerning the order in which the various limiting processes $N \to \infty$, $\lambda \to 0$, $t \to \infty$ must be considered for obtaining the region of validity of a master or transport equation. The number of particles N must go to infinity first, λ and t being finite. When this limiting process, which is necessary e.g. to eliminate surface and shape-dependent effects, is completed, one can consider the limiting case $\lambda \to 0$, $t \to \infty$, $\lambda^2 t$ finite. One then obtains the master equation as derived in S. One can treat also, however, the case of finite λ and t, as was done in S' and the present paper.

The author gratefully acknowledges a very valuable exchange of letters with Prof. M. Fierz on the problems discussed in this paper, especially in section 3. He is indebted to Dr. T. O. Woodruff for a critical reading of the manuscript, and to Prof. W. Opechowski for drawing his attention to recent papers on the quantum ergodic problem.

Received 7-1-59

REFERENCES

1) Van Hove, L., Physica **21** (1955) 517 and **23** (1957) 441.
2) Von Neumann, J., Z. Phys. **57** (1929) 30; Pauli, W. and Fierz, M., Z. Phys. **106** (1937) 572.
3) Fierz, M., Helv. Phys. Acta, **28** (1955) 705; Farquhar, I. E. and Landsberg, P. T., Proc. roy. Soc. **A239** (1957) 134; Bocchieri, P. and Loinger, A., Phys Rev. **111** (1958) 668.
4) Van Kampen, N. G., Physica **20** (1954) 603.
5) Van Kampen, N. G., Proc. Int. Symposium on Transport Processes in Statistical Mechanics (Brussels, Aug. 1956), Interscience, New York and London (1958) p. 239.
6) Ingraham, R., Nuovo Cimento **9** (1958) 99.

V.4 QUANTUM THEORY OF MEASUREMENT AND ERGODICITY CONDITIONS

ADRIANA DANERI, A. LOINGER, AND G. M. PROSPERI

Abstract: Some criticisms to von Neumann's treatment of the measuring process are recalled and discussed.

The problem is reconsidered and reinvestigated, in the spirit of the "philosophy" of Jordan and Ludwig, on the basis of an ergodic theorem recently given.

The measuring apparatus is schematized as a macroscopic system which possesses, besides the energy, at least another macroscopic constant of the motion. The value of this constant characterizes an invariant manifold ("channel"). In each manifold certain ergodicity conditions hold and there exists an equilibrium macro-state towards which the system evolves spontaneously. The apparatus is assumed to be initially in the equilibrium state belonging to a given channel and the interaction with the observed system determines a transition of the apparatus towards a state belonging to another channel, which depends on the initial state of the observed system. Then the apparatus evolves towards a new equilibrium state.

The ergodicity conditions employed are sufficiently realistic, since it has been proved that they are in particular satisfied by that class of Hamiltonians for which Van Hove succeeded in deriving a master equation.

1. Introduction

As is well known, the measuring process plays a central role in quantum mechanics. Its formal mathematical theory was developed by von Neumann [1]) in 1932. In recent years von Neumann's point of view and results have been criticized under different aspects by many authors (Jordan [2]), Bohm [3]), Wigner, Araki and Yanase [4]), Ludwig [5]), Feyerabend [6]), H. S. Green [7]), Durand [8]), Wakita [9])). Various attempts have been made (see refs. [4, 5, 7—10])) to obtain a more satisfactory solution of the problem than that of von Neumann. In our opinion, the treatment given by Ludwig [5, 10]) is the most complete and satisfactory from a physical point of view. However, its mathematical elaboration is still in a preliminary stage.

Originally published in *Nuclear Physics, 33*, 297-319 (1962).

In the present article, following essentially Jordan and Ludwig's "philosophy" †, we give a treatment of the measuring process based on a recent formulation of the ergodic problem [11, 12]). In the course of our exposition we shall also re-examine some points of Ludwig's work, which, in our opinion, deserve a more detailed analysis.

An idea of the point of view adopted can be given in the following terms.

Performing an observation on a microscopic system amounts to putting the system itself in interaction with another large system, which has the property of undergoing, as a result of the interaction, macroscopic modifications which are dependent on the state of the micro-system.

If we assume that an objective meaning can be attributed to the macro-states of a large system, the aim of quantum mechanics is then essentially to make predictions on the trace that our microscopic body will leave in the macroscopic world, when the trace left at a certain time is known. In this line of thought, the measuring process has to be considered — rather than as an "untrennbare Kette zwischen Objekt und Subjekt" (Weizsäcker [13])) — as an "untrennbare Kette zwischen Mikro- und Makrokosmos" (Ludwig).

In order that an objective meaning may be attributed to the macro-states of the large bodies, it is of course necessary that — by virtue of the laws of quantum mechanics and of the structure of the macroscopic bodies — states incompatible with the macroscopic observations be actually impossible. In our opinion, this impossibility should be a consequence of the fact that the result of the interaction of two macro-systems must always be macroscopically determined.

After these premisses it is clear that, in order to build up a satisfactory quantum theory of measurement, it is necessary to have a theory, at least schematic, of the large bodies, i.e. a theory which gives the connection between the macro-properties of these bodies and their microscopic structure described by quantum mechanics.

We notice that in the measuring process the energy of the apparatus is enormously larger than the energy of the micro-object, which should suffer, during the measuring process, the least possible perturbation. Therefore the apparatus must be a macro-system in a thermodynamically metastable state, such that a very small perturbation makes it evolve towards a thermo-dynamically stable state, dependent on the state of the micro-object.

It is therefore clear that the measuring process is closely connected with the problem of the evolution of a large body towards its state of thermodynamic equilibrium and, consequently, with the ergodic problem.

† As will appear presently, this "philosophy" and, consequently, any technical development founded on it are in harmony with the ideas of Bohr. More exactly, as Prof. Rosenfeld has kindly pointed out to us, the situation is the following: "Bohr's epistemological analysis, which is complete in itself, disposes of von Neumann's infinite regress" (cf. sects. 3 and 4) "by appealing to the *logical* necessity of making a sharp distinction between object and measuring instrument. No formal treatment can replace Bohr's argument, for the simple reason that the physical meaning of any mathematical elaboration can only be expressed by means of the classical concepts to which Bohr appeals directly, and which are (in the last resort) not formalizable, but immediately given (as part of common experience)." The significance of our treatment (and of the treatment of Ludwig) is rather to establish rigorously the logical consistency of the way in which the mathematical formalism is connected with he unanalyzed basic concepts.

2. Macro-States of the Measuring Apparatus

It follows from the preceding considerations that a measuring apparatus must be a macroscopic system possessing many states of macroscopic equilibrium compatible with a given value of the macroscopic energy. A system of this kind possesses, besides the energy, other macroscopic constants of the motion. We characterize schematically the macroscopic state of this system in a way analogous to that followed in ref. [12]).

We introduce a suitable basis $\{\Omega_{ak\nu i}\}$ of vectors and assume that the manifolds $\mathbf{C}_{ak\nu}$ spanned by the $\Omega_{ak\nu i}$ corresponding to given sets of values for a, k, ν represent the macro-states of the system. The indices a and k denote, respectively, the value of the macro-energy and the set of values of the other macroscopic constants of the motion which are relative to the macroscopic state considered. These quantities do not change during the free evolution of the system. The index ν denotes the value of the macro-variables which do change during the time evolution. We denote by $\mathbf{C}_{ak}$ the manifold ("channel") spanned by the vectors $\Omega_{ak\nu i}$ corresponding to given values of a and k. We assume that within the manifold $\mathbf{C}_{ak}$ ergodicity conditions are verified, analogous to those formulated for the whole energy shell in refs. [12]); we assume further that for given a and k there exists a manifold $\mathbf{C}_{ake_k}$ having a number of dimensions much larger than that of the other manifolds $\mathbf{C}_{ak\nu}$ $(\nu \neq e_k)$. If the system is initially in a state belonging to the channel $\mathbf{C}_{ak}$, it then evolves spontaneously towards the state $\mathbf{C}_{ake_k}$. The manifolds $\mathbf{C}_{ake_k}$ represent the various states of macroscopic equilibrium.

Let the measuring apparatus be initially in the equilibrium state $\mathbf{C}_{a0e_0}$ of a given channel $\mathbf{C}_{a0}$ and suppose that the interaction of the object system with the apparatus induces a transition of the state of the latter from the channel $\mathbf{C}_{a0}$ to another channel $\mathbf{C}_{ak}$ of the same energy shell, depending upon the particular state of the object system. At the end of the measuring process the apparatus is in a final equilibrium macro-state $\mathbf{C}_{ake_k}$.

3. Basic Assumptions of Quantum Theory

We recall very briefly the assumptions on which quantum theory is founded.

(i) The physical state of a system is represented by a vector (properly by a ray) of a Hilbert space.

(ii) The time evolution of a state is governed, in the Schrödinger picture, by the Schrödinger equation

$$\psi_t = T(t)\psi_0, \qquad i\hbar\frac{\mathrm{d}T}{\mathrm{d}t} = HT, \qquad T(0) = I. \tag{3.1}$$

(iii) An observable A is represented by a Hermitian hypermaximal operator α. The eigenvalues a_r of α give the possible † values of A. An eigenvector $\phi_r^{(s)}$ corres-

† We suppose here that the spectrum of α is discrete. We recall that it is physically impossible to measure with absolute accuracy an observable C having a continuous spectrum. However, it is possible to measure such an observable with an arbitrarily good accuracy. It is in fact sufficient to choose a step function $F(x)$ — which approximates the function x with the wanted accuracy — and measure $F(C)$, which is an observable with a discrete spectrum. Notice that the introduction of $F(C)$ is not a mathematical trick but is essential for reasons of principle.

ponding to the eigenvalue a_r represents a state in which A has the value a_r. If the system is in the state ψ, the probability p_r of finding the value a_r for A, when a measurement is performed, is given by

$$p_r = (\psi, P_{\mathbf{V}_r}\psi) = \sum_s |(\phi_r^{(s)}, \psi)|^2, \tag{3.2}$$

where $P_{\mathbf{V}_r}$ is the projection operator on the eigenmanifold $\mathbf{V}_r$ corresponding to a_r, and the sum $\sum_s$ is taken over a complete orthonormal set $\{\phi_r^{(s)}\}$ $(s = 1, 2, \ldots)$ of $\mathbf{V}_r$.

(iv) Let us suppose we have performed a measurement of a quantity A and that this measurement is a maximal one. If the result is a_r, we postulate that the state of the system immediately after the observation is described by the (unique) eigenvector ϕ_r, corresponding to the eigenvalue a_r. If the measurement of A is not a maximal one — i.e. if the spectrum of α is degenerate — it is commonly assumed (cf. e.g. ref. [14])) that the state after the measurement is

$$N^{-1} \sum_s (\phi_r^{(s)}, \psi)\phi_r^{(s)}, \tag{3.3}$$

where $N^2 \equiv \sum_s |(\phi_r^{(s)}, \psi)|^2$ is a normalization factor. This assumption, however, as it could be derived from what follows, implies a hypothesis on the measuring apparatus which cannot always be considered to be verified. As is well known, the preceding assumptions may be formulated also in the language of the statistical (density) operators. In this way, we have in particular the advantage of obtaining formulae which are valid not only for the pure cases, but also for the mixtures. The statistical operator W of a system changes in time according to the equation (Schrödinger picture)

$$W(t) = T(t)W(0)T^*(t). \tag{3.4}$$

The probability p_r of finding the result a_r for the observable A is

$$p_r = \mathrm{Tr}(P_{\mathbf{V}_r}W) \tag{3.5}$$

and the statistical operator of the system for which the eigenvalue a_r has been found, is

$$\frac{P_{\mathbf{V}_r}WP_{\mathbf{V}_r}}{\mathrm{Tr}(P_{\mathbf{V}_r}W)}. \tag{3.6}$$

4. Consequences of the Preceding Assumptions

Let A and B be two observables, whose operators are respectively α and β. Let us suppose that the spectra of α and β are discrete and non-degenerate and that

$$\alpha\phi_r = a_r\phi_r; \qquad \beta\chi_s = b_s\chi_s,$$

where ϕ_r and χ_s are normalized eigenvectors. If the (normalized) state vector at $t = 0$ is

$$\psi_0 = \sum_r c_r\phi_r, \tag{4.1}$$

the probability of finding the value a_r of A, when a measurement of A is performed at $t = 0$, is, by virtue of postulate (iii),

$$p_{[A=a_r]}(0) = |c_r|^2.$$

If the value a_r has been found, the probability of finding the value b_s of B at a later time t, is

$$p_{[B=b_s]}(t) = \left|\left(\chi_s, \exp\left[-\frac{i}{\hbar}Ht\right]\phi_r\right)\right|^2.$$

If we perform the measurement of A at $t = 0$ without "looking at" the result, the probability of finding at the time t the value b_s for B is evidently given by

$$\sum_r \left|\left(\chi_s, \exp\left[-\frac{i}{\hbar}Ht\right]\phi_r\right)\right|^2 |c_r|^2. \tag{4.2}$$

If no measurement is performed at $t = 0$, the above probability is

$$\begin{aligned}\left|\sum_r c_r \left(\chi_s, \exp\left[-\frac{i}{\hbar}Ht\right]\phi_r\right)\right|^2 &= \sum_r |c_r|^2 \left|\left(\chi_s, \exp\left[-\frac{i}{\hbar}Ht\right]\phi_r\right)\right|^2 \\ &+ \sum_{r\neq r'} c_r^* c_{r'} \left(\chi_s, \exp\left[-\frac{i}{\hbar}Ht\right]\phi_r\right)^* \left(\chi_s, \exp\left[-\frac{i}{\hbar}Ht\right]\phi_{r'}\right).\end{aligned} \tag{4.3}$$

The second term of the right-hand side of (4.3) is the so-called *interference term*; owing to its existence, we are not allowed to think that, when the system is in the state (4.1), the observable A has a value which is well determined even if it cannot be theoretically calculated (this assumption would furnish indeed an expression analogous to (4.2)). We may therefore assert that A has a given value only when it has been actually measured.

Considering that even the measuring apparatus is a physical system describable with the quantum laws, a first consistency problem of the theory consists in trying to prove that the apparatus acts on the object in such a way that the results of a later measurement are given by formula (4.2) rather than by formula (4.3).

5. Von Neumann's Theory of the Measuring Process

Let us denote by I and II the system we are considering and the apparatus for the measurement of A, respectively. Let the Hamiltonian of the total system I + II be

$$H = H_{\mathrm{I}} + H_{\mathrm{II}} + H_{\mathrm{int}}. \tag{5.1}$$

In order that II can actually act as an apparatus for the measurement of A, it is necessary that H_{int} be such that (a) I and II are coupled only during a very small time interval τ; (b) if I is in an eigenstate ϕ_r of α, at the end of the interaction a change depending on ϕ_r has taken place in the apparatus, while the state ϕ_r of I is practically not changed. More precisely, if II is initially in a state Φ_0 belonging to a manifold $\mathbf{V}_0$, we assume that H_{int} is such that

$$\exp\left[-\frac{i}{\hbar}H\tau\right]\phi_r\Phi_0 \approx \exp\left[-\frac{i}{\hbar}H_{\mathrm{int}}\tau\right]\phi_r\Phi_0 = \phi_r\Phi_r, \tag{5.2}$$

where Φ_r belongs to a manifold $\mathbf{V}_r$ and is dependent on the particular Φ_0 of $\mathbf{V}_0$. $\mathbf{V}_0, \mathbf{V}_1, \mathbf{V}_2, \ldots$ are supposed mutually orthogonal in order that the results of the measurement be distinct †.

The measurement of the quantity A of I is thus reduced to the measurement of a quantity $\mathscr{A}$ of II, which has as eigenmanifolds the orthogonal manifolds $\mathbf{V}_0, \mathbf{V}_1, \mathbf{V}_2, \ldots$. If I is at the time $t = 0$ in the state (4.1), one obtains by virtue of the linearity of the time evolution operator

$$\exp\left(-\frac{i}{\hbar} H_{\mathrm{int}}\tau\right) \psi_0 \Phi_0 = \exp\left(-\frac{i}{\hbar} H_{\mathrm{int}}\tau\right) \sum_r c_r \phi_r \Phi_0 = \sum_r c_r \phi_r \Phi_r . \tag{5.3}$$

The probability of finding II in a state belonging to the manifold $\mathbf{V}_r$ is then equal to $|c_r|^2$, i.e. coincides with the probability given by the fundamental assumption (iii), when the measuring apparatus is considered as external to the studied system.

From eq. (5.3) it follows that the probability of finding at a time $t > \tau$ the system I + II in a state $\chi_s X_h$ (where the X_h's are eigenstates of some observable of II) is

$$\begin{aligned}
&\left|\left(\chi_s X_h, \exp\left[-\frac{i}{\hbar}(H_{\mathrm{I}}+H_{\mathrm{II}})t\right] \sum_r c_r \phi_r \Phi_r\right)\right|^2 \\
&= \sum_r |c_r|^2 \left|\left(\chi_s, \exp\left[-\frac{i}{\hbar} H_{\mathrm{I}} t\right] \phi_r\right)\right|^2 \left|\left(X_h, \exp\left[-\frac{i}{\hbar} H_{\mathrm{II}} t\right] \Phi_r\right)\right|^2 \\
&+ \sum_{r \neq r'} c_r^* c_{r'} \left(\chi_s, \exp\left[-\frac{i}{\hbar} H_{\mathrm{I}} t\right] \phi_r\right)^* \left(\chi_s, \exp\left[-\frac{i}{\hbar} H_{\mathrm{I}} t\right] \phi_{r'}\right) \\
&\cdot \left(X_h, \exp\left[-\frac{i}{\hbar} H_{\mathrm{II}} t\right] \Phi_r\right)^* \left(X_h, \exp\left[-\frac{i}{\hbar} H_{\mathrm{II}} t\right] \Phi_{r'}\right) .
\end{aligned} \tag{5.4}$$

Let us now suppose that the second measurement is performed on the system I only. The operator β, considered as an operator in the Hilbert space of I + II, is degenerate and the eigenvectors belonging to an eigenvalue b_s are of the kind $\chi_s X_h$. Owing to postulate (iii) we must sum over the index h both sides of (5.4); since we have obviously

$$\sum_h \left(X_h, \exp\left[-\frac{i}{\hbar} H_{\mathrm{II}} t\right] \Phi_r\right)^* \left(X_h, \exp\left[-\frac{i}{\hbar} H_{\mathrm{II}} t\right] \Phi_{r'}\right) = (\Phi_r, \Phi_{r'}) = \delta_{rr'},$$

† Note that from a purely mathematical point of view, it is always possible to find a unitary operator $T(\tau)$ satisfying the relation

$$T(\tau)\phi_r \Phi_0 = \phi_r \Phi_r .$$

In order that T be unitary it is in fact necessary (and sufficient) that it transforms the vectors forming with the $\phi_r \Phi_0$ a complete orthonormal basis of the Hilbert space of I + II, into vectors orthogonal to each other and to the vectors of the manifold spanned by the $\phi_r \Phi_r$. Now, this is possible in an infinite number of ways. From a physical point of view this result is not significant since the problem consists in ascertaining whether there actually exists in nature a measuring apparatus such that the Hamiltonian H_{int} satisfies the above relation.

the probability of finding for B the value b_s, when no other measurement is performed on II, is

$$\sum_r |c_r|^2 \left| \left(\chi_s, \exp\left[-\frac{i}{\hbar} H_1 t \right] \phi_r \right) \right|^2 . \tag{5.5}$$

This relation coincides with (4.2); the interaction with the measuring apparatus explains the replacement of (4.3) by (4.2).

The interference terms, which have been deleted from (4.3), reappear obviously in (5.4). In order to remove them, it would actually be necessary, according to von Neumann, to perform on the system II a measurement with another apparatus III. The system I + II would then be, with respect to III, in the same situation in which I was with respect to II. In particular, we have consistency between the results given by the theory for the two cases in which (a) I is considered as object and II + III as apparatus, (b) I + II is considered as object and III as apparatus. According to von Neumann, this is sufficient to prove the adequacy of the theory because in von Neumann's opinion any statement on the system II makes sense only when it corresponds to an actual observation made by means of III, any statement concerning III makes sense only when a further observation is performed on it by means of another system IV, and so on. In this chain of instruments we must include, according to von Neumann, besides the apparatus *stricto sensu*, also the sense organs with which the observer observes the apparatus, the nervous system sending to the brain the data of the sense organs, and so on. From the preceding considerations it follows that the point at which one makes the separation between observed system and observer (Heisenberg's "Schnitt") does not matter; in other words, we can put the "boundary" between system and apparatus or between apparatus and sense organs or between sense organs and nervous system, and so on. This is the so-called principle of the psycho-physical parallelism. In this line of thought the transition from one state to another, and consequently the passage from (4.3) to (4.2) would be determined, in the last analysis, by the "abstract ego" becoming conscious of the result.

6. Criticism of von Neumann's Point of View

6.1. SUBJECTIVISM

It is clear that von Neumann's theory is founded on a radically subjectivistic (solipsistic) philosophy. From an objectivistic point of view, the difference between (4.3) and (5.3) ought to be explained by taking explicitly into account the fact that the aim of quantum mechanics is to describe the behaviour of the micro-objects only through their relations with the macroscopic world. But it is then necessary, for obvious consistency reasons, to construct a quantum theory of the large bodies, which explains the objective nature of their macroscopic properties and the changes induced by the micro-objects on the macro-properties of the measuring apparatus.

6.2. INFINITE REGRESS

It is commonly assumed that to any observable there corresponds a given Hermitian operator. However, a real physical meaning can only be given to this mathematical symbol when the procedure by which the quantity studied is measured, is defined at least conceptually. According to von Neumann's theory of measurement, the meaning of a quantity of a given system I is given by the modifications produced by I on some quantity of the system II. The meaning of the latter quantity is given by the modifications determined by II on some quantity of a system III, and so on. Evidently, if von Neumann's chain is never truncated, the physical meaning of an abstract observable of the system I is referred to that of some observable of the system II, the meaning of this observable of II is referred to the meaning of some observable of III and so on indefinitely, the meaning in question being never attained. If, in line with the considerations of subsect. 6.1, we could give an objective character to the macroscopic states of a large system and consider its macro-observables as the quantities to which a meaning is given from the very beginning, a meaning could be given to any other abstract observable by defining the procedure with which it can be measured. This procedure is uniquely defined once the macro-states of a large system have been defined and the nature of the forces between the systems is known.

6.3. WIGNER'S OBJECTION

An objection of a more particular nature to von Neumann's theory was presented by Wigner [4]). He proved that, if the state of the system object + apparatus is described at the end of the measurement by eq. (5.3), results which are in contradiction with conservation principles can be obtained. A way out, however, was found by Wigner himself, who demonstrated that if one takes properly into account the macroscopic nature of the measuring apparatus the difficulty can be overcome.

Wigner considered the following example. Let I be a system describable only with spin variables; let σ_x be the observable to be measured. We shall denote by $\phi_{\sigma'_x}$ and $\phi_{\sigma''_x}$ the eigenstates of σ_x corresponding to the eigenvalues $+1$ and -1 respectively. If we put ψ_0 equal to $\phi_{\sigma'_x}$ or $\phi_{\sigma''_x}$ we get from eq. (5.2)

$$\begin{aligned} \exp\left(-\frac{i}{\hbar}H\tau\right)\phi_{\sigma'_x}\Phi_0 &= \phi_{\sigma'_x}\Phi', \\ \exp\left(-\frac{i}{\hbar}H\tau\right)\phi_{\sigma''_x}\Phi_0 &= \phi_{\sigma''_x}\Phi''. \end{aligned} \tag{6.1}$$

Let us now consider a conservative observable of the total system I + II, not compatible with σ_x: for instance the component M_z of the angular momentum of I + II. We have

$$M_z = L_z + \sigma_z ,$$

where L_z acts in the Hilbert space of II. It is well known that a representation can be chosen in such a way that the eigenstates of σ_x be written as linear combinations of the eigenstates of σ_z, in the following way:

$$\phi_{\sigma'_x} = \frac{1}{\sqrt{2}}(\phi_{\sigma'_z} + \phi_{\sigma''_z}), \qquad \phi_{\sigma''_x} = \frac{1}{\sqrt{2}}(\phi_{\sigma'_z} - \phi_{\sigma''_z}).$$

where $\phi_{\sigma'_z}$ and $\phi_{\sigma''_z}$ are normalized vectors. Adding and subtracting the equations obtained by substituting the above expressions into eqs. (6.1), we get

$$\begin{aligned} 2\exp\left(-\frac{i}{\hbar}H_{\text{int}}\tau\right)\phi_{\sigma'_z}\Phi_0 &= \phi_{\sigma'_z}(\Phi'+\Phi'')+\phi_{\sigma''_z}(\Phi'-\Phi''), \\ 2\exp\left(-\frac{i}{\hbar}H_{\text{int}}\tau\right)\phi_{\sigma''_z}\Phi_0 &= \phi_{\sigma'_z}(\Phi'-\Phi'')+\phi_{\sigma''_z}(\Phi'+\Phi''). \end{aligned} \tag{6.2}$$

We now observe that the expectation values of M_z corresponding to the states described by the right-hand sides of eqs. (6.2) are the same, while the expectation values corresponding to $\phi_{\sigma'_z}\Phi_0$ and to $\phi_{\sigma''_z}\Phi_0$ differ by $2\hbar$. Therefore in this case eq. (5.2) is in contradiction with the conservation theorem of angular momentum. Wigner however showed that the difficulty can be overcome if eqs. (6.1) are replaced by the following:

$$\begin{aligned} \exp\left(-\frac{i}{\hbar}H\tau\right)\phi_{\sigma'_x}\Phi_0 &= \phi_{\sigma'_x}\Phi'+\phi_{\sigma''_x}\Theta, \\ \exp\left(-\frac{i}{\hbar}H\tau\right)\phi_{\sigma''_x}\Phi_0 &= \phi_{\sigma''_x}\Phi''-\phi_{\sigma'_x}\Theta, \end{aligned} \tag{6.3}$$

where $(\Phi', \Theta) = 0 = (\Phi'', \Theta)$, and $\|\Theta\|$ can be supposed to be very small if the states Φ_0, Φ', Φ'' are superpositions of many states corresponding to different eigenvalues of L_z. This fact can be considered as actually occurring since Φ_0, Φ', Φ'' correspond to macroscopic states of the apparatus. We have a similar situation for any observable of the system I not compatible with a conservative observable of the total system I + II (Araki and Yanase [4])).

7. Hints for an Objectivistic Interpretation of the Measuring Process

As emphasized in the preceding section, a satisfactory theory of the measuring process must start from a characterization of the macroscopic properties of a large body. Such properties must have an *objective* character; it is therefore necessary that no interference terms appear in the relations concerning them. Such terms must be absent essentially by virtue of the complexity of the considered system.

In particular, it must be allowed to truncate von Neumann's chain immediately after the first macroscopic system $S^{(i)}$. The interference terms must disappear owing to the nature of the macroscopic observations and to the properties of the Hamiltonian of the system $S^{(1)}+S^{(2)}+\ldots+S^{(i)}$. A first hint for a way out may be obtained from the following rough considerations.

Let us suppose for the sake of simplicity, as it is almost always the case in practice, that the system II is already macroscopic and consider the manifolds $\mathbf{V}_0, \mathbf{V}_1, \ldots \mathbf{V}_r, \ldots$ of II as corresponding to the macroscopic properties. Observe now that in order that the measurement may be performed it is necessary that the changes undergone by the system II owing to its coupling with I persist for a sufficiently long time;

we shall suppose therefore that the above manifolds are invariant with respect to the operator $\exp[-(i/\hbar)H_{\text{II}}t]$, i.e. that for any t

$$\exp\left(-\frac{i}{\hbar}H_{\text{II}}t\right)\Phi_r \in \mathbf{V}_r,$$

and that transitions from one manifold to another occur only by virtue of the interaction with I. Any macroscopic observation on the system II consists then in determining the manifold $\mathbf{V}_r$ to which the state vector belongs. Any observation compatible with the macroscopic observation determines a set of manifolds which may be spanned by vectors X_h ($h = 1, 2, \ldots$) every one of which belongs to a single manifold $\mathbf{V}_r$.

In (5.4) only one of the expressions $(X_h, \exp[-(i/\hbar)H_{\text{II}}t]\Phi_r)$ is different from zero, say that containing Φ_n; eq. (5.4) therefore becomes

$$|c_n|^2 \left|\left(\chi_s, \exp\left[-\frac{i}{\hbar}H_{\text{I}}t\right]\phi_n\right)\right|^2 \left|\left(X_h, \exp\left[-\frac{i}{\hbar}H_{\text{II}}t\right]\Phi_n\right)\right|^2. \tag{7.1}$$

Eq. (7.1) is a relation of the kind (4.2) and is in accordance with the objective character attributed to the macroscopic properties of the system II.

If the considered observation on II is not only compatible with a macroscopic observation, but is itself a macroscopic observation, we must sum in (7.1) over all the states X_h which belong to $\mathbf{V}_n$. We obtain in this way

$$|c_n|^2 \left|\left(\chi_s, \exp\left[-\frac{i}{\hbar}H_{\text{I}}t\right]\phi_n\right)\right|^2; \tag{7.2}$$

this expression is obviously in agreement with (4.2).

The present schematization of the macro-observations and of the measuring apparatus is too rough and will be re-examined in the following; the essential elements to the solution of the measurement problem are, however, already contained in it. The success of this viewpoint rests of course on the fact that the observations which can be *actually* performed on the system II must be compatible with the macroscopic ones. In other words, *it would be impossible to construct an instrument capable of measuring a quantity such that its eigenstates X_h have projections different from zero on distinct manifolds $\mathbf{V}_r$.*

From a mathematical point of view it is of course always possible to find a unitary operator $\exp[-(i/\hbar)H_{\text{int}}\tau]$ corresponding to such a measurement performed on II. From the point of view of physics, however, H_{int} is not arbitrary, but depends upon the nature of the systems; therefore it is likely that the above possibility does not exist. The plausibility of the assumption will be discussed in the following (sect. 5).

8. Analysis of the Physical Characteristics of Some Typical Measuring Apparatus

Let us first suppose that we want to measure the energy of a charged particle with a proportional counter.

The counter is assumed to be initially charged to a given potential; we may assume that there is thermal equilibrium within the gas of the counter, electrostatic and thermal equilibrium within the electrodes. When a charged particle goes through the counter, a practically determined number of ions is produced, which is a function of the particle velocity and consequently, for a given mass, of its energy.

Under the influence of the electric field the electrons migrate. Since the wire is at a positive potential, the electrons go towards it. The gas pressure and the potential are fixed in such a way that in a very small region around the central wire, where the field is stronger, the energy gained by the electron in covering a distance equal to the mean free path is larger than the ionization energy of the gas molecules. In this region secondary ions are created at a rate such that the ratio of their number to the number of initially produced ions is practically constant. When the electrons reach the wire, the electronic instruments connected to the counter record a sudden reduction of the potential, proportional to the number of ions originally created by our particle and therefore a function of its energy. In order to avoid unessential complications, we omit considering the electronic instruments and suppose the counter to be connected to the electrodes of a capacitor. (This is certainly correct for the time intervals of the order of the discharge time.) Then, if there was initially a given voltage between the electrodes of the condenser, and consequently between the electrodes of the counter, after the discharge this voltage decreases and the decrease is a function of the energy of the particle that crossed the counter.

We may distinguish the following phases:

(a) The apparatus is initially in a macroscopic state of metastable equilibrium in which it remains practically for an indefinite time, if no external perturbation intervenes. (Strictly speaking, it is of course possible that some ions are produced by thermal fluctuations inside the gas and it is consequently possible that the apparatus discharges spontaneously; however the interval of time needed for a spontaneous discharge is much larger than that during which a discharge induced by the passage of a particle may occur.)

(b) The particle goes through the counter; if its energy is practically determined, the number of ions generated is practically determined. The passage of the particle removes the system from the state of metastable equilibrium and brings it into a state which can already be considered as *macroscopic* (and characterized by the number of the generated ions and by the voltage between the electrodes); this state is no longer a state of equilibrium and depends essentially on the particle energy.

(c) The apparatus goes spontaneously towards a new equilibrium situation (no ions present, a smaller voltage between the electrodes), which depends on the state corresponding to phase (b) (on the number of the generated ions) and therefore on the energy of the particle.

One gets easily convinced that all the processes that occur in the ordinary measuring apparatus are reducible to the above scheme. Examples: the formation of a trace in a Wilson chamber, in a bubble chamber, in a photographic emulsion.

In order to clarify the question further, we analyze another measuring apparatus. Consider first a Geiger counter. As is well known, it differs from the proportional counter because the potential and the gas pressure are fixed in such a way that the ion multiplication spreads along all the wire; consequently the production even of a small number of ions induces a discharge which is self-sustaining if the voltage does not decrease.

In the apparatus actually employed in practice there is a resistor in series with the counter, or the discharge is interrupted at a given time by some suitable trick. The pulse-height given by the Geiger counter is independent of the number of ions primarily created and therefore of the energy of the incident particle. In our simplified apparatus the passage of a particle through the counter induces a discharge, at the end of which the voltage falls practically to zero. The counter is a "yes or no" apparatus: the particle has passed through, the particle has not passed through.

Let us suppose that a beam of practically monokinetic particles impinges on some target T and let us study the angular distribution of the scattered particles by means of a set of counters. If the particle is scattered within the angle (θ_k', θ_k'') under which the k-th counter is viewed from T (i.e. if the particle is represented by a wave-packet with propagation vectors within this angle), it produces a given number of ions inside the k-th counter; the k-th counter then discharges while the remaining counters stay unaltered. Here the initial metastable state is the state in which all the counters are charged, the intermediate state (depending on the direction in which the particle is scattered) is the state in which ions are created inside the k-th counter, the final state (also depending on the direction of scattering) is a state in which the k-th counter has been triggered while the other counters undergo no change.

9. Outline of a Theory of the Macroscopic Bodies

It is now evident that if we want to build up a theory of the measuring apparatus, we must first of all construct a theory of the macroscopic bodies and give a solution to the problem of the time evolution of a macroscopic body towards its thermodynamic equilibrium state, i.e. give a solution to the ergodic problem.

To this end, we shall recall briefly the characterization of the macroscopic states of large bodies given by statistical mechanics.

Obviously the macroscopic observations performed on a large body cannot define completely its quantal state. In order to characterize formally the macro-observations, one decomposes suitably the state space of the body into a given number of orthogonal manifolds having a very large number of dimensions. The various possible macro-states correspond to these manifolds.

A first interesting quantity is the energy; it is measured macroscopically with a certain inaccuracy. Let us subdivide the set of the possible energy values into intervals $(E_a, E_{a+1}=E_a+\Delta E)$ and denote by $\mathbf{C}_a$ the manifold of the eigenvectors corresponding to the a-th energy interval. Since a macroscopic body is a system spatially bounded, its energy spectrum is a purely discrete one. Let us indicate by S_a the number of dimensions of the manifold $\mathbf{C}_a$. Let us suppose that there exists another macroscopic constant of the motion J, or, more generally, a set of such constants. (Consider, e.g., a system composed of two non-interacting parts, each of which possesses its own energy. In this case the constants of the motion are the total energy and the energy of one of the subsystems.) The spectrum of J can also be decomposed into intervals (J_k, J_{k+1}) and consequently each manifold $\mathbf{C}_a$ can be decomposed into a set of orthogonal manifolds

$\mathbf{C}_{ak}$, which are spanned by the eigenvectors of J belonging to $\mathbf{C}_a$ and to the eigenvalues of (J_k, J_{k+1}). We call $\mathbf{C}_{ak}$ the k-th channel of $\mathbf{C}_a$ and denote by S_{ak} its number of dimensions.

In order to define completely the macroscopic state of the body it is necessary to take account also of other quantities. We represent mathematically this situation by decomposing the manifolds $\mathbf{C}_{ak}$ into submanifolds $\mathbf{C}_{ak\nu}$ $(\nu = 1, 2, \ldots, N_{ak})$. Let $\{\Omega_{ak\nu i}\}$ $(i = 1, 2, \ldots, s_{ak\nu})$ be a basis adapted to the cells $\mathbf{C}_{ak\nu}$.

We omit a detailed discussion of the characterization of the cells $\mathbf{C}_{ak\nu}$, since it would require more precise assumptions on the nature of the system considered. By way of example, we recall here how the manifolds $\mathbf{C}_{ak\nu}$ are defined for an ideal gas or solid.

In a gas one may take as basis $\{\Omega_{ak\nu i}\}$ the set of eigenstates of the unperturbed energy, i.e. the states composed of direct products, suitably symmetrized, of the single particle states. The energy shell $\mathbf{C}_a$ is the manifold spanned by the set of the unperturbed eigenstates corresponding to a given interval of unperturbed energy; in fact, if this interval is sufficiently large, the above manifold coincides practically with the manifold spanned by the corresponding perturbed energy states. The cells $\mathbf{C}_{ak\nu}$ are defined in the following way. One subdivides the single particle energy spectrum into intervals $(\varepsilon_r, \varepsilon_r + \Delta E/n)$ where n is the number of molecules contained in the gas. The number of molecules whose energy lies in the r-th interval, when the system is in a given unperturbed eigenstate, is then denoted by n_r. The energy shell $\mathbf{C}_a$ is the manifold spanned by the set of vectors corresponding to given values of the numbers $n_1, n_2, \ldots$.

In the case of a non-conducting solid, we may take a basis formed by the eigenstates of the phonon unperturbed energy and repeat for the phonon gas the above treatment, taking into account that now the total number of phonons is also allowed to change and must be considered as a dynamical variable.

In the case of a conducting solid, one must characterize analogously, besides the state of the phonon gas, the state of the conducting electron gas.

This characterization of the macro-state of a gas or of a solid suffices to account for several macroscopic properties of the system. For a somewhat more complete characterization, it is necessary to know also the spatial distributions of the particles. To this end it would be sufficient to repeat the preceding arguments taking in lieu of the above single particle states, which correspond to a complete indeterminacy of the particle positions, a set of states corresponding to given position and momentum indeterminacies †.

Going back to the general case, let us refer to a fixed energy shell, dropping accordingly the index a.

We suppose that the Hamiltonian H_{II} of the macro-system and the basis $\{\Omega_{k\nu i}\}$ are such that the following ergodicity conditions are satisfied [12]:

$$\begin{aligned}
&\mathbf{M}\left\{\sum_{i=1}^{s_{k\nu}}\left|\left(\Omega_{k\nu i}, \exp\left[-\frac{i}{\hbar}H_{\text{II}}t\right]\Omega_{k\mu j}\right)\right|^2\right\} = \sum_\rho\sum_i |(\Omega_{k\nu i}, P_\rho \Omega_{k\mu j})|^2 \approx \frac{s_{k\nu}}{S_k}, \\
&\left|\mathbf{M}\left\{\sum_{i=1}^{s_{k\nu}}\left(\Omega_{k\nu i}, \exp\left[-\frac{i}{\hbar}H_{\text{II}}t\right]\Omega_{k\mu j}\right)^*\left(\Omega_{k\nu i}, \exp\left[-\frac{i}{\hbar}H_{\text{II}}t\right]\Omega_{k\mu' j'}\right)\right\}\right| \\
&\qquad = \left|\sum_\rho\sum_i (\Omega_{k\nu i}, P_\rho\Omega_{k\mu j})^*(\Omega_{k\nu i}, P_\rho \Omega_{k\mu' j'})\right| \ll \frac{s_{k\nu}}{S_k} \quad (\mu \neq \mu'),
\end{aligned} \tag{9.1}$$

† Cf. von Neumann [1]), p. 404 and ff.

where $\sum_\rho E_\rho P_\rho$ is the spectral representation of H_{II} and $\mathbf{M}$ denotes the operation of time averaging

$$\lim_{T\to\infty} \frac{1}{T}\int_0^T \mathrm{d}t \ldots .$$

Let us suppose that the initial state Ψ_0 is a superposition of states belonging to different channels $\mathbf{C}_k$:

$$\Psi_0 = \sum_{k,\mu,i} (\Omega_{k\mu i}, \Psi_0)\Omega_{k\mu i}. \tag{9.2}$$

If $u_{k\nu}(t)$ denotes the probability of finding the system in the macrostate $\mathbf{C}_{k\nu}$ at the time t:

$$u_{k\nu}(t) = \sum_{i=1}^{s_{k\nu}} \left|\left(\Omega_{k\nu i}, \exp\left[-\frac{i}{\hbar}H_{\text{II}}t\right]\Psi_0\right)\right|^2, \tag{9.3}$$

we have, by virtue of eqs. (9.1),

$$\mathbf{M}u_{k\nu}(t) \approx p_k \frac{s_{k\nu}}{S_k}, \tag{9.4}$$

where

$$p_k = \sum_{\mu,i} |(\Omega_{k\mu i}, \Psi_0)|^2. \tag{9.5}$$

In particular, if Ψ_0 is assumed to belong to a given cell $\mathbf{C}_{k\mu}$, i.e. if the system is initially assumed to be in a given macro-state, we have

$$\mathbf{M}u_{k\nu}(t) \approx \frac{s_{k\nu}}{S_k}. \tag{9.4'}$$

By virtue of this result, we are allowed to consider $s_{k\nu}/S_k$ as the probability (in the sense of statistical mechanics) of finding the system, at any time, in the macro-state $\mathbf{C}_{k\nu}$.

Let us suppose that, as is the case in practice, there exists in every channel $\mathbf{C}_k$ a cell $\mathbf{C}_{ke_k}$ with a number of dimensions enormously larger than that of the other cells, in such a way that

$$1 - \frac{s_{ke_k}}{S_k} \ll 1, \quad \frac{s_{k\nu}}{S_k} \ll 1 \qquad (\nu \neq e_k). \tag{9.6}$$

(For boson or fermion systems for which the interaction energy between the particles is negligible, as in the examples investigated above, the cells relative to Bose-Einstein or Fermi-Dirac distributions possess this property.)

From (9.4′) it follows, by virtue of the inequalities (9.6), that

$$\begin{aligned}
&\mathbf{M}\left\{\left[\sum_{i=1}^{s_{k\nu}} \left|\left(\Omega_{k\nu i}, \exp\left[-\frac{i}{\hbar}H_{\text{II}}t\right]\Psi_0\right)\right|^2\right]^2\right\} \\
&\quad\leqslant \mathbf{M}\left\{\sum_{i=1}^{s_{k\nu}} \left|\left(\Omega_{k\nu i}, \exp\left[-\frac{i}{\hbar}H_{\text{II}}t\right]\Psi_0\right)\right|^2\right\} \approx \frac{s_{k\nu}}{S_k} \ll 1, \quad \text{for} \quad \nu \neq e_k, \\
&\mathbf{M}\left\{\left[\sum_{i=1}^{s_{ke_k}} \left|\left(\Omega_{ke_k i}, \exp\left[-\frac{i}{\hbar}H_{\text{II}}t\right]\Psi_0\right)\right|^2 - 1\right]^2\right\} \\
&\quad\leqslant \mathbf{M}\left\{1 - \sum_{i=1}^{s_{ke_k}} \left|\left(\Omega_{ke_k i}, \exp\left[-\frac{i}{\hbar}H_{\text{II}}t\right]\Psi_0\right)\right|^2\right\} \approx 1 - \frac{s_{ke_k}}{S_k} \ll 1.
\end{aligned} \tag{9.7}$$

Relations (9.7) entail that, for every initial state $\Psi_0 \in \mathbf{C}_{k\mu}$, the probability $u_{k\nu}(t)$ is, for the overwhelming majority of the time, very small or very near to 1 (for $\nu \neq e_k$ and for $\nu = e_k$, respectively). Consequently, whatever the value of $u_{k\nu}(t_0)$, we may think that, after a time t sufficiently long, $u_{k\nu}(t)$ has assumed the values 0 or 1, i.e.

$$u_{k\nu}(t) \approx \delta_{\nu e_k}, \tag{9.8}$$

and the system is practically in the macro-state $\mathbf{C}_{ke_k}$. The cell $\mathbf{C}_{ke_k}$ represents precisely the state of macroscopic equilibrium relative to the channel $\mathbf{C}_k$.

If Ψ_0 is given by (9.2) we obtain

$$u_{k\nu}(t) \approx p_k \delta_{\nu e_k}. \tag{9.9}$$

In the statistical operator language, after a sufficiently long time the system is macroscopically characterized by the following respective operators:

$$\frac{1}{s_{ke_k}} P_{\mathbf{C}_{ke_k}}, \tag{9.8'}$$

$$\sum_k p_k \frac{1}{s_{ke_k}} P_{\mathbf{C}_{ke_k}}. \tag{9.9'}$$

10. Schematization of the Measuring Apparatus and Mathematical Theory of the Measuring Process

We want now to consider a schematization of the measuring apparatus more accurate than that given in sect. 7.

We consider the apparatus as an isolated macro-system (the energy exchanged with the micro-object during the measurement is certainly negligible compared with the internal energy), having an energy macroscopically determined, i.e. lying in a given interval $(E, E+\Delta E)$. All the considerations of the preceding section are assumed to be applicable to the apparatus; in particular, contrary to the assumptions of sect. 7, we do not suppose here that the channels $\mathbf{C}_k$ correspond to macro-observations, but that each $\mathbf{C}_k$ may be further decomposed into mutually orthogonal submanifolds $\mathbf{C}_{k\nu}$, everyone of which corresponds to a macro-observation.

In the previously considered example of the proportional counter, the macroscopic state of the system is essentially characterized by the average potential of the counter, by the number of ions, by the spatial and energetic distributions of the neutral molecules and ions, by the charge distribution over the electrodes.

The channels are characterized by the values of the constant of the motion

$$U' = U - \frac{e\lambda}{C} N,$$

where U is the voltage between the electrodes, N is the number of ions, e, λ and C are the elementary charge, the ion multiplication coefficient and the electrostatic capacity of the global system counter + condenser respectively. If U_k is the value of U' characteristic of the k-th channel $\mathbf{C}_k$, the equilibrium manifold $\mathbf{C}_{ke_k}$ corresponds to a situation in which there is thermal equilibrium within the gas and within the electrodes, there are no ions within the gas and the voltage is U_k. The non-equilibrium manifolds $\mathbf{C}_{k\nu}$ correspond to situations in which the voltage is still U_k and no ion is present within the gas, but the gas or the electrodes are not in thermal equilibrium, or to situations in which there is a fixed number N of electrons and the voltage has the value $U_k+(e\lambda/C)N$.

In the case of the set of Geiger counters measuring the angular distribution of the scattered particles studied in sect. 8, the channel $\mathbf{C}_0$ corresponds to a situation in which the counters are all charged and no ion is present and the state $\mathbf{C}_{0e_0}$ corresponds to a situation in which gas and electrodes are in thermal equilibrium. The other equilibrium states of interest $\mathbf{C}_{ke_k}$ correspond to situations in which the k-th counter has discharged and there is thermal equilibrium, and the other states of the channel $\mathbf{C}_k$ correspond to situations in which there is no thermal equilibrium or, even if the k-th counter is still charged, there are ions within it.

The macro-system is initially assumed to be in a state Φ_0 belonging to the manifold $\mathbf{C}_{0e_0}$; if the particle is in the state ϕ_k, the interaction brings the apparatus, as supposed in sect. 7, into a state $\Phi_k \in \mathbf{C}_k$; since the system is assumed to be ergodic inside $\mathbf{C}_k$, it then goes spontaneously into $\mathbf{C}_{ke_k}$ (cf. sect. 9). The latter phase is a process of approach towards an equilibrium situation in the thermodynamical sense.

From the considerations developed in sects. 7 and 9, we obtain immediately the following result:

Initial state of the system I + II:

$$\sum_r c_r \phi_r \Phi_0 . \tag{10.1}$$

State after the end of the interaction:

$$\exp\left[-\frac{i}{\hbar} H_{\text{int}}\tau\right] \sum_r c_r \phi_r \Phi_0 = \sum_r c_r \phi_r \Phi_r$$

$$(\Phi_0 \in \mathbf{C}_0;\ \Phi_r \in \mathbf{C}_r). \tag{10.2}$$

Probability of finding the system I *in the state* χ_s *and the system* II *in the macrostate* $\mathbf{C}_{k\nu}$ *at the time* t:

$$\sum_{i=1}^{s_{k\nu}} \left|\left(\chi_s \Omega_{k\nu i}, \exp\left[-\frac{i}{\hbar}(H_{\text{I}}+H_{\text{II}})t\right] \sum_r c_r \phi_r \Phi_r\right)\right|^2$$

$$= |c_k|^2 \left|\left(\chi_s, \exp\left[-\frac{i}{\hbar} H_{\text{I}} t\right] \phi_k\right)\right|^2 \sum_{i=1}^{s_{k\nu}} \left|\left(\Omega_{k\nu i}, \exp\left[-\frac{i}{\hbar} H_{\text{II}} t\right] \Phi_k\right)\right|^2$$

$$\to |c_k|^2 \left|\left(\chi_s, \exp\left[-\frac{i}{\hbar} H_{\text{I}} t\right] \phi_k\right)\right|^2 \delta_{\nu e_k} . \tag{10.3}$$

Eq. (10.3) is of the type desired, the interference terms which appeared in (5.4) are not present in it; their elimination has been performed without the intervention of a system III observing II. In the statistical operator language, eqs. (10.1), (10.2) and (10.3) may be rewritten as follows:

Statistical operator before the measurement:

$$W_0 = P^{\mathrm{I}}_{[\Sigma_r c_r \phi_r]} P^{\mathrm{II}}_{[\Phi_0]} . \tag{10.1'}$$

Statistical operator immediately after the end of the interaction:

$$W' = P_{[\Sigma_r c_r \phi_r \Phi_r]} . \tag{10.2'}$$

The time-evolved of this operator after a large time t can be identified, as far as the macroscopic quantities of II *are concerned, with*

$$\tilde{W}_t = \sum_r |c_r|^2 P^{\mathrm{I}}_{\{\exp[-(i/\hbar)H_{\mathrm{I}}t]\phi_r\}} \frac{1}{s_{re_r}} P^{\mathrm{II}}_{\mathbf{C}_{re_r}} . \tag{10.3'}$$

11. Limitations on the Kind of Quantities Actually Observable; Measurements on Macroscopic Bodies

It follows from the preceding considerations that a given Hermitian operator represents an observable only if it is actually possible to construct an experimental device according to the above scheme. We shall now discuss the various limitations imposed on such a possibility by the structure of the laws of motion. A first restriction is given by the so-called superselection rules [15]). There exist some universal constants of the motion as the electric charge, the baryon number, etc. They determine a decomposition of the Hilbert space of the total system object + apparatus into superselection spaces, which are invariant with respect to the equations of motion. It is never possible therefore to satisfy conditions (5.3) for a Hermitian operator whose eigenstates are superpositions of states corresponding to different values of the above universal constants. Even for the Hermitian operators for which the superselection spaces are invariant, there exist further restrictions. Of this kind are for instance those pointed out by Wigner [4]), which, however, can be overcome, as has been emphasized, by taking into account the fact that the apparatus is a macroscopic system.

Other restrictions are supplied by the following considerations.

If τ is the coupling time between system and apparatus and δt_{I} is the time during which the system I remains in a given eigenstate of α, in order that the quantity A be actually observable, we must obviously have, according to the analysis of the preceding section,

$$\delta t_{\mathrm{I}} \gg \tau . \tag{11.1}$$

If we denote respectively by $\delta\varepsilon_{\mathrm{I}}$ the energy indeterminacy of the system I when it is in an eigenstate of α, and by $\delta\varepsilon$ the energy indeterminacy of I + II corresponding to an unperturbed energy eigenstate, we have

$$\delta t_{\mathrm{I}} \approx \hbar/\delta\varepsilon_{\mathrm{I}} \tag{11.2}$$

and

$$\tau \approx \hbar/\delta\varepsilon . \tag{11.3}$$

From (11.1) it follows that

$$\delta\varepsilon_{\mathrm{I}} \ll \delta\varepsilon \tag{11.4}$$

This relation implies a strong restriction on the nature of the quantities actually observable. In general only quantities quasi-diagonal in the energy representation are observable. An exception is given by the quantities having a very strong degeneracy, as the macroscopic ones.

The indeterminacy $\delta\varepsilon_{\mathrm{I}}$ of the macro-observables is of the order of the width $\Delta\varepsilon_{\mathrm{I}}$ of the energy shell; however, the relaxation time $\tilde{t}_{\mathrm{I}}$ relative to these observables is certainly very large. This is essentially due to the fact that the occupation probability of a macro-state can be written as a sum of a large number of terms, among which some cancellations occur. Consequently the frequency width in the Fourier series expansion of the above probability turns out to be, as a matter of fact, much smaller than $\Delta\varepsilon_{\mathrm{I}}/\hbar$.

As has been previously emphasized, an objective character can be given to the macroscopic quantities only if observations incompatible with these cannot actually be performed, i.e. if quantities whose eigenstates are superpositions of vectors belonging to different $\mathbf{C}_{ak\nu}$ are not observable. (Otherwise, it would not be possible to eliminate the interference terms among the macro-states.)

The above considerations circumscribe the class of the actually observable quantities of a large body, but are not sufficient to exclude the possibility of observations incompatible with macro-observations.

To this end, considerations of a different nature are needed. We notice that in this case the systems I and II of the preceding sections are both macroscopic systems, for every one of which certain ergodicity conditions are satisfied, and it is plausible that similar conditions are satisfied also for the whole system I + II. (Indeed the ergodicity conditions must be a consequence of the very large number of elementary constituents — atoms or molecules — of the systems I and II and of the nature of the forces between these constituents.) If the system I + II satisfies the above ergodicity relations, its final situation is macroscopically represented by a statistical operator (10.3′) which is a mixture of the equilibrium states corresponding to the single channels, with weights equal to the initial occupation weights of these channels. Therefore, a statistical operator at the time $t = 0$ for the system I, which corresponds to a pure state, described by a superposition of vectors belonging to different $\mathbf{C}_{ak\nu}$, is equivalent, so far as the macroscopic observations on II are concerned, to a statistical operator which is a mixture of the above macroscopic states. Consequently it is not possible to put any correspondence between final macrostates of II and initial eigenstates of I corresponding to an observable which is not a macroscopic quantity.

It follows from the preceding considerations that there is a fundamental difference between the characteristics of an apparatus for observations on a microscopic system and those of an apparatus for observations on a macroscopic one.

In the first case, the measuring apparatus has a number of degrees of freedom very large with respect to that of the object system; the fact that the apparatus does not perturb appreciably the system (at least insofar as the quantity considered is concerned) is due to the nature of the interaction system-apparatus and to its short duration. An apparatus for observations on a large body must be, on the contrary, a system small with respect to the body, since in this case the nature of the interaction is very special and the coupling times are long. Moreover it is clear that if one wishes to perform measurements of quantities which change during the time evolution of the system, it is necessary that the relaxation time of the apparatus be small with respect to that of the system.

By way of example, we consider the case of temperature measurements inside a body. The apparatus is given by a set of thermometers distributed inside the body, the thermal capacities of which are very small with respect to that of the body. In order that these thermometers should make it possible to study how the temperature distribution changes with time, one must assume that each thermometer goes towards the equilibrium, together with the region around it, much more rapidly than the temperature changes with time, i.e. in a time small with respect to the time needed in order that a sensible displacement of heat inside the body takes place.

One could think that the following objection can be raised against our point of view. Consider an hyperapparatus composed of as many measuring apparatus as there are particles constituting a given macroscopic body and suppose we perform on these particles a very accurate measurement, for instance of the energy. We should have in this case an observation much more accurate than a macroscopic one, an observation which could even be incompatible with the macro-observations.

Let us observe, however, that in this way the body would be dissolved into its constituents; the ergodicity conditions, by virtue of which we have been able to exhibit the impossibility of non-macroscopic measurements on a large body are, on the contrary, just a consequence of the interactions which exist between the particles of the macroscopic body and which have the effect of lumping the particles into a single whole.

The meaning of our considerations is precisely the following: when the particles of a system do interact so strongly that a body is formed, it is no longer possible to think of observing them independently.

12. Theory of the Measuring Process under "Weak" Ergodicity Conditions

In order to reach the result (9.9) we have used rather restrictive ergodicity conditions. In this section we want to show how, using some plausible assumptions, the same result can be obtained starting from much weaker ergodicity conditions.

We assumed that the initial configuration of the measuring apparatus corresponds to a state belonging to the manifold $\mathbf{C}_{0e_0}$, and that after the end of the interaction the apparatus is in a state belonging to a channel determined by the particular eigenstate of the observable of the micro-object which has been measured. The particular state in this channel depends on the state of the manifold $\mathbf{C}_{0e_0}$ in which the apparatus was before the measurement. Obviously not all the states of the final channel $\mathbf{C}_k$ can actually be reached by the macro-system since the number of dimensions of the channel is different from (in general larger than) the number of dimensions of the manifold $\mathbf{C}_{0e_0}$. We may even add that not all the macro-states of the system can actually be reached. For instance, in a proportional or Geiger counter the positions of the ions which are created when the particle goes through the apparatus are correlated. If we know the positions of some ions the remaining ones must be more or less aligned with these.

We shall then suppose that the state-vector of the apparatus can actually go only into a given manifold **C**, union of a given class of manifolds $\mathbf{C}_{k\nu}$, such that its number of dimensions is equal to the number of dimensions s_{0e_0} of $\mathbf{C}_{0e_0}$. The above class of manifolds corresponds to the macrostates in which the apparatus can actually be found after the end of the interaction. If Ω_{0e_0i} is the initial state of the apparatus, and if the system is in the state ϕ_k, we assume that, owing to the interaction, the apparatus goes into the state

$$U^{(k)}\Omega_{0e_0i} = \sum_\nu{}' \sum_{j=1}^{s_{k\nu}} U^{(k)}_{\nu j,\, e_0 i}\Omega_{k\nu j}, \tag{12.1}$$

$$\sum_\nu{}' s_{k\nu} = s_{0e_0}, \tag{12.2}$$

where $\sum'_\nu$ denotes the sum taken over the above class of cells. The square matrix $\{U^{(k)}_{\nu j,\, e_0 i}\}$ is unitary, i.e. satisfies the conditions

$$\begin{aligned}\sum_\nu{}' \sum_{j=1}^{s_{k\nu}} U^{(k)*}_{\nu j,\, e_0 i'} U^{(k)}_{\nu j,\, e_0 i} &= \delta_{ii'},\\ \sum_{i=1}^{s_{0e_0}} U^{(k)*}_{\nu j,\, e_0 i} U^{(k)}_{\nu' j',\, e_0 i} &= \delta_{\nu\nu'}\delta_{jj'}.\end{aligned} \tag{12.3}$$

If, more generally, the initial state of the apparatus is given by

$$\Phi_0 = \sum_{i=1}^{s_{0e_0}} a_i \Omega_{0e_0 i}, \tag{12.4}$$

owing to the interaction with the system the apparatus will go into the state

$$\Phi_k = U^{(k)}\Phi_0 = \sum_{i=1}^{s_{0e_0}} \sum_\nu{}' \sum_{j=1}^{s_{k\nu}} a_i U^{(k)}_{\nu j,\, e_0 i}\Omega_{k\nu j}. \tag{12.1'}$$

Thus we obtain

$$\begin{aligned}u_{k\nu}(t) &= \sum_{i=1}^{s_{k\nu}} \left| \left(\Omega_{k\nu i}, \exp\left[-\frac{i}{\hbar}H_{\mathrm{II}}t\right]\Phi_k\right)\right|^2\\ &= \sum_{i=1}^{s_{k\nu}} \left| \sum_\mu{}' \sum_{j=1}^{s_{k\mu}} \left(\Omega_{k\nu i}, \exp\left[-\frac{i}{\hbar}H_{\mathrm{II}}t\right]\Omega_{k\mu j}\right) \sum_{l=1}^{s_{0e_0}} a_l U^{(k)}_{\mu j,\, e_0 l}\right|^2\\ &= \sum_{i=1}^{s_{k\nu}} \sum_\mu{}' \sum_{j=1}^{s_{k\mu}} \sum_{\mu'}{}' \sum_{j'=1}^{s_{k\mu'}} \left(\Omega_{k\nu i}, \exp\left[-\frac{i}{\hbar}H_{\mathrm{II}}t\right]\Omega_{k\mu j}\right)^*\\ &\quad\cdot \left(\Omega_{k\nu i}, \exp\left[-\frac{i}{\hbar}H_{\mathrm{II}}t\right]\Omega_{k\mu' j'}\right) \sum_{ll'} U^{(k)*}_{\mu j,\, e_0 l} U^{(k)}_{\mu' j',\, e_0 l'} a_l^* a_{l'}.\end{aligned} \tag{12.5}$$

If we put

$$\begin{aligned}L^{(k\nu)}_{\mu j,\, \mu' j'} &= \frac{S_k}{s_{k\nu}}\left\{\mathbf{M}\sum_{i=1}^{s_{k\nu}}\left[\left(\Omega_{k\nu i}, \exp\left[-\frac{i}{\hbar}H_{\mathrm{II}}t\right]\Omega_{k\mu j}\right)^*\right.\right.\\ &\quad\left.\left.\cdot\left(\Omega_{k\nu i}, \exp\left[-\frac{i}{\hbar}H_{\mathrm{II}}t\right]\Omega_{k\mu' j'}\right)\right] - \frac{s_{k\nu}}{S_k}\delta_{\mu\mu'}\delta_{jj'}\right\}\\ &= \frac{S_k}{s_{k\nu}}\left\{\sum_{i=1}^{s_{k\nu}}\sum_\rho (\Omega_{k\nu i}, P_\rho\Omega_{k\mu j})^*(\Omega_{k\nu i}, P_\rho\Omega_{k\mu' j'}) - \frac{s_{k\nu}}{S_k}\delta_{\mu\mu'}\delta_{jj'}\right\},\end{aligned} \tag{12.6}$$

we have

$$\frac{(\mathbf{M}u_{k\nu}(t)-s_{k\nu}/S_k)^2}{s_{k\nu}^2/S_k^2} = \sum_{\mu\mu'\lambda\lambda'}' \sum_{j=1}^{s_{k\mu}} \sum_{j'=1}^{s_{k\mu'}} \sum_{l=1}^{s_{k\lambda}} \sum_{l'=1}^{s_{k\lambda'}} L^{(k\nu)*}_{\mu j,\,\mu' j'} L^{(k\nu)}_{\lambda l,\,\lambda' l'}$$

$$\cdot \sum_{uu'vv'} U^{(k)}_{\mu j,\,e_0 u} U^{(k)*}_{\mu' j',\,e_0 u'} U^{(k)*}_{\lambda l,\,e_0 v} U^{(k)}_{\lambda' l',\,e_0 v'} a_u a^*_{u'} a^*_v a_{v'}\,. \tag{12.7}$$

We denote now by $\mathfrak{B}_0$ an averaging operation over the initial states Φ_0 performed with the following criterion. We attribute to any set of vectors Φ_0 a weight proportional to the area of the surface determined by them over the complex s_{0e_0}-dimensional spherical surface of equation

$$\sum_{i=1}^{s_{0e_0}} |a_i|^2 = 1. \tag{12.8}$$

One then obtains (cf. Prosperi and Scotti [12]))

$$\mathfrak{B}_0(a_u a^*_{u'} a^*_v a_{v'}) = \frac{\delta_{uu'}\delta_{vv'} + \delta_{uv}\delta_{u'v'}}{s_{0e_0}^2}\,. \tag{12.9}$$

By virtue of eqs. (12.3) it follows that

$$\mathfrak{B}_0\Big(\sum_{uu'vv'} U^{(k)}_{\mu j,\,e_0 u} U^{(k)*}_{\mu' j',\,e_0 u'} U^{(k)*}_{\lambda l,\,e_0 v} U^{(k)}_{\lambda' l',\,e_0 v'} a_u a^*_{u'} a^*_v a_{v'} \Big)$$

$$= \frac{1}{s_{0e_0}^2} \Big(\sum_u U^{(k)}_{\mu j,\,e_0 u} U^{(k)*}_{\mu' j',\,e_0 u} \cdot \sum_v U^{(k)*}_{\lambda l,\,e_0 v} U^{(k)}_{\lambda' l',\,e_0 v}$$

$$+ \sum_u U^{(k)}_{\mu j,\,e_0 u} U^{(k)*}_{\lambda l,\,e_0 u} \cdot \sum_{u'} U^{(k)*}_{\mu' j',\,e_0 u'} U^{(k)}_{\lambda' l',\,e_0 u'} \Big)$$

$$= \frac{1}{s_{0e_0}^2} (\delta_{\mu\mu'}\delta_{jj'}\delta_{\lambda\lambda'}\delta_{ll'} + \delta_{\mu\lambda}\delta_{jl}\delta_{\mu'\lambda'}\delta_{j'l'}), \tag{12.10}$$

from which

$$\sum_{k\nu} \frac{\mathfrak{B}_0(\mathbf{M}u_{k\nu}(t)-s_{k\nu}/S_k)^2}{s_{k\nu}^2/S_k^2} = \frac{1}{s_{0e_0}^2} \sum_{k\nu}' \Big(\sum_{\mu\lambda}' \sum_j \sum_l L^{(k\nu)}_{\mu j,\,\mu j} L^{(k\nu)}_{\lambda l,\,\lambda l}$$

$$+ \sum_{\mu\mu'}' \sum_{jj'} L^{(k\nu)*}_{\mu j,\,\mu' j'} L^{(k\nu)}_{\mu j,\,\mu' j'} \Big) = \sum_{k\nu} \Big[\Big(\frac{1}{s_{0e_0}} \sum_\mu' \sum_j L^{(k\nu)}_{\mu j,\,\mu j} \Big)^2$$

$$+ \frac{1}{s_{0e_0}^2} \sum_{\mu\mu'}' \sum_{jj'} |L^{(k\nu)}_{\mu j,\,\mu' j'}|^2 \Big]. \tag{12.11}$$

We now assume the following ergodicity conditions:

$$\frac{1}{s_{k\mu}} \Big| \sum_{j=1}^{s_{k\mu}} L^{(k\nu)}_{\mu j,\,\mu j} \Big| \ll \frac{1}{\sqrt{N}}\,,$$

$$\frac{1}{s_{k\mu}} \sum_{j=1}^{s_{k\mu}} \frac{1}{s_{k\mu'}} \sum_{j'=1}^{s_{k\mu'}} |L^{(k\nu)}_{\mu j,\,\mu' j'}|^2 \ll \frac{1}{N} \tag{12.12}$$

(where $N = \sum_k N_k$), which are much less strong than conditions (9.1) (cf. Prosperi and Scotti [12])). From these relations and from (12.2) it follows that

$$\left|\frac{1}{s_{0e_0}}\sum_{\mu}{}'\sum_{j=1}^{s_{k\mu}} L_{\mu j,\mu j}^{(k\nu)}\right| \leqslant \frac{1}{s_{0e_0}}\sum_{\mu}{}'\left|\sum_{j=1}^{s_{k\mu}} L_{\mu j,\mu j}^{(k\nu)}\right|$$

$$= \frac{1}{s_{0e_0}}\sum_{\mu}{}' s_{k\mu}\frac{1}{s_{k\mu}}\left|\sum_{j=1}^{s_{k\mu}} L_{\mu j,\mu j}^{(k\nu)}\right| \ll \frac{1}{\sqrt{N}}\frac{1}{s_{0e_0}}\sum_{\mu}{}' s_{k\mu} = \frac{1}{\sqrt{N}},$$

$$\frac{1}{s_{0e_0}^2}\sum_{\mu\mu'}{}'\sum_{j=1}^{s_{k\mu}}\sum_{j'=1}^{s_{k\mu'}}\left|L_{\mu j,\mu' j'}^{(k\nu)}\right|^2$$

$$= \frac{1}{s_{0e_0}^2}\sum_{\mu\mu'}{}' s_{k\mu}s_{k\mu'}\frac{1}{s_{k\mu}}\sum_j\frac{1}{s_{k\mu'}}\sum_{j'}\left|L_{\mu j,\mu' j'}^{(k\nu)}\right|^2$$

$$\ll \frac{1}{N}\frac{1}{s_{0e_0}^2}\left(\sum_{\mu}{}' s_{k\mu}\right)^2 = \frac{1}{N}, \tag{12.13}$$

and in conclusion

$$\sum_{k\nu}\frac{\mathfrak{B}_0(\mathbf{M}u_{k\nu}(t)-s_{k\nu}/S_k)^2}{s_{k\nu}^2/S_k^2} \ll \sum_{k\nu}\frac{2}{N}=2. \tag{12.14}$$

Owing to the replacement of the ergodicity conditions (9.1) by the conditions (12.12), we are no longer allowed to conclude that relations (9.4′) and (9.8) are verified for *every* initial micro-state Φ_0 of the apparatus. However, as follows from relation (12.14), the ergodicity conditions (12.12) allow us to assert that the set of the initial exceptional states for which (9.4′) and (9.8) are not verified has a very small measure. But this is the only result needed in order to be able to draw the conclusion of sect. 10.

We finally remark that in the case in which the energy is the only macroscopic constant of the motion (and hence the channels coincide with the energy shells), the following result can be proved (cf. Prosperi [16])): Ergodicity conditions of the kind (12.12) are satisfied, under suitable restrictive conditions, by the Hamiltonians for which Van Hove [17]) has derived a master equation. These Hamiltonians already represent some physically significant examples and it is very likely that most of the macro-systems belong to this class. Consequently *relations* (12.12) *are much more representative of the physical nature of the systems studied than would appear at first sight.*

We cordially thank our friends P. Bocchieri, A. Scotti and G. Stabilini for helpful discussions.

References

1) J. von Neumann, Mathematical foundations of quantum mechanics (Princeton University Press, Princeton, 1955) chapts. 5 and 6
2) P. Jordan, Phil. Sci. **16** (1949) 269
3) D. Bohm, Quantum theory (Prentice-Hall, New York, 1952), part VI
4) E. P. Wigner, Z. Phys. **133** (1952) 101;
H. Araki and M. M. Yanase, Phys. Rev. **120** (1960) 622
5) G. Ludwig, Z. Phys. **135** (1953) 483; Die Grundlagen der Quantenmechanik (Springer-Verlag, Berlin, 1954) chapt. 5; Festschrift z. 200. Jahrfeier d. Columbia Univ. (Colloquium-Verlag, Berlin) p. 262; Phys. Bl. **11** (1955) 489

6) P. K. Feyerabend, Z. Phys. **148** (1957) 551; Proc. Nineth Symp. Colston Research Soc. (Butterworths Publishers, London, 1957) p. 121
7) H. S. Green, Nuovo Cim. **9** (1958) 880
8) L. Durand III, unpublished preprint (1958)
9) H. Wakita, Prog. Theor. Phys. **23** (1960) 32
10) G. Ludwig, Z. Naturfors. **12** (1957) 662; Z. Phys. **150** (1958) 346; **152** (1958) 98;
J. Schroeter, Z. Naturfors. **14** (1959) 750;
G. Ludwig in Reports of the Varenna Summer School 1960 (to be published);
N. G. van Kampen, Physica **20** (1954) 603; Fortschr. d. Phys. **4** (1956) 405;
H. Kuemmel, Nuovo Cim. **1** (1955) 1057
11) P. Bocchieri and A. Loinger, Phys. Rev. **114** (1959) 948;
G. M. Prosperi and A. Scotti, Nuovo Cim. **13** (1959) 100; **17** (1960) 267
12) G. M. Prosperi and A. Scotti, J. Math. Phys. **1** (1960) 218;
A. Loinger in Reports of the Varenna Summer School 1960 (to be published)
13) C. F. von Weizsäcker, Z. Phys. **118** (1941) 480
14) G. Lueders, Ann. Phys. **8** (1951) 322
15) G. C. Wick, A. S. Wightman and E. P. Wigner, Phys. Rev. **88** (1952) 101
16) G. M. Prosperi, J. of Math. Phys., in course of publication
17) L. Van Hove, Physica **21** (1955) 517; **23** (1957) 441; **25** (1959) 268

V.5 TIME SYMMETRY IN THE QUANTUM PROCESS OF MEASUREMENT*

YAKIR AHARONOV, PETER G. BERGMANN, AND JOEL L. LEBOWITZ

We examine the assertion that the "reduction of the wave packet," implicit in the quantum theory of measurement introduces into the foundations of quantum physics a time-asymmetric element, which in turn leads to irreversibility. We argue that this time asymmetry is actually related to the manner in which statistical ensembles are constructed. If we construct an ensemble time symmetrically by using both initial and final states of the system to delimit the sample, then the resulting probability distribution turns out to be time symmetric as well. The conventional expressions for prediction as well as those for "retrodiction" may be recovered from the time-symmetric expressions formally by separating the final (or the initial) selection procedure from the measurements under consideration by sequences of "coherence destroying" manipulations. We can proceed from this situation, which resembles prediction, to true prediction (which does not involve any postselection) by adding to the time-symmetric theory a postulate which asserts that ensembles with unambiguous probability distributions may be constructed on the basis of preselection only. If, as we believe, the validity of this postulate and the falsity of its time reverse result from the macroscopic irreversibility of our universe as a whole, then the basic laws of quantum physics, including those referring to measurements, are as completely time symmetric as the laws of classical physics. As a by-product of our analysis, we also find that during the time interval between two noncommuting observations, we may assign to a system the quantum state corresponding to the observation that follows with as much justification as we assign, ordinarily, the state corresponding to the preceding measurement.

I. INTRODUCTION

ONE of the perennially challenging problems of theoretical physics is that of the "arrow of time." Everyday experience teaches us that the future is qualitatively different from the past, that our practical powers of prediction differ vastly from those of memory, and that complex physical systems tend to develop in the course of time in patterns distinct from those of their antecedents. On the other hand, all the "microscopic" laws of physics ever seriously propounded and widely accepted are entirely symmetric with respect to the direction of time; they are form-invariant with respect to time reversal.[1,2]

The *de facto* absence of time symmetry in nature enters the formal statement of the laws of nature principally in two areas. One of these is thermodynamics, particularly the second law of thermodynamics; the latter proclaims that the entropy of a thermally isolated system can only increase toward the future. The other area is that of cosmogony; our universe is expanding toward the future. Gold[1] has suggested that these two asymmetric phenomena may well be causally related to each other. A third time-asymmetric effect, the preponderance of outgoing radiation in nature over incoming radiation, may be considered to be a special aspect of the second law.

In quantum theory the dynamical laws of motion, either the Schrödinger or the Heisenberg equations, are time symmetric as are their classical counterparts, Hamilton's equations of motion. It has been suggested, though, that asymmetry in the direction of time, and even thermodynamic irreversibility, enters into quantum theory through the theory of measurement.[3,4] Any measurement performed on a quantum system changes its state discontinuously and in a manner not to be described by the Schrödinger or Heisenberg equations of the isolated system. The performance of a measurement leads to the "reduction of the wave packet." That is to say, if the result of the measurement is known, then the quantum state of the system preceding the measurement has been replaced by the eigenvector of the observable that belongs to the eigenvalue recorded. If the outcome of the measurement is not known, the original state vector must now be replaced by a density matrix diagonal with respect to the eigenvectors of the observables measured, each diagonal element equaling the absolute square of the corresponding component of the original state vector. This density matrix is inequivalent to the original state vector in that all phase relations between the components have been destroyed by the act of measurement, though their norms survive in the density matrix.

Quite aside from entropy considerations, the conventional quantum theory of measurements is concerned exclusively with the prediction of probabilities of specific outcomes of future measurements on the basis of the results of earlier observations. Indeed the reduction of the wave packet has as its operational

* This research was supported by the U. S. Air Force Office of Scientific Research, Aerospace Research Laboratories, the National Science Foundation, and the National Aeronautics and Space Administration.

[1] T. Gold, in Onzième Conseil de l'Institut International de Physique Solvay, *La Structure et l'Evolution de l'Universe* (Edition Stoops, Brussels, 1958), p. 81; Am. J. Phys. **30**, 403 (1962); Proceedings of the Conference on the Arrow of Time, Cornell University, 1963 (to be published).

[2] See, however, O. Penrose and I. C. Percival, Proc. Phys. Soc. (London) **79**, 605 (1962).

[3] J. von Neumann, *Mathematical Foundations of Quantum Mechanics*, transl. by R. T. Beyer (Princeton University Press, Princeton, 1955).

[4] D. Bohm, *Quantum Theory* (Prentice-Hall Inc., Englewood Cliffs, New Jersey, 1951), cf. in particular, p. 608.

Originally published in *Physical Review, 134B*, 1410-16 (1964).

contents nothing but this probabilistic connection between successive observations.[5]

In this paper we propose to examine the nature of the time symmetry in the quantum theory of measurement. Rather than delve into the measurement process itself, which involves a specialized interaction between the atomic system and a macroscopic device,[3–5] we shall simply accept the standard expressions for probabilities of values furnished by the conventional theory. Whereas the conventional theory deals with ensembles of quantum systems that have been "preselected" on the basis of some initial observation, we shall deduce from it probability expressions that refer to ensembles that have been selected from combinations of data favoring neither past nor future. A theory that concerns itself exclusively with such symmetrically selected ensembles (the "time-symmetric theory") will contain only time-symmetric expressions for the probabilities of observations. Logically this time-symmetric theory is contained in the conventional theory but lacks one of the latter's postulates. It will be developed in Sec. II.

In Sec. III we shall consider the case that prior to the final selection some observations are performed that completely destroy coherence of any state previously existing; we shall find that any earlier observations obey probability laws that formally resemble the conventional prediction formula. Likewise, if the initial selection ("preselection") is followed by coherence destroying measurements to be succeeded in turn by some other observations, then these latter observations obey the precise time-reflected expression of the conventional prediction formula. This reflected relationship might be called a "retrodiction" formula. Finally, in Sec. IV we shall return to the true prediction and "retrodiction" situations, i.e., to the consideration of ensembles that have been either strictly preselected or postselected. By adding to the time-symmetric theory one postulate that appears to portray accurately the conditions of our universe (and whose time-reflected proposition does not hold), we are able to recover the conventional asymmetric theory. We present an argument that this asymmetry represents the intrusion of the irreversibility of macroscopic processes into the microscopic domain, so that the totality of the basic (microscopic) laws of nature emerges completely time symmetric.

II. SEQUENCES OF OBSERVATIONS

We shall begin by considering systems which are subjected to sequences of measurements, each of which is individually "complete"; that is to say, that each observation determines a quantum state of the system. We make the conventional assumption about the selection of ensembles of such systems (and of their histories), which is to the effect that initially all systems of the ensemble have yielded a specified nondegenerate eigenvalue of an observable J; no other conditions are imposed. Under these circumstances the conventional quantum theory of measurements states that, given two successive measurements, the probability of a particular outcome of the later observation depends on the outcome of the earlier observation by being the absolute square of the scalar product of the two state-vectors belonging to the two respective eigenvalues. We shall denote the observables to be measured by symbols $A_1, A_2, \cdots, A_k, \cdots$, all of whose eigenvalues are nondegenerate; let the eigenvalues of A_k be denoted by d_k. Only when necessary will distinct eigenvalues of A_k be denoted by Greek superscripts $d_k^{(\alpha)}, d_k^{(\beta)}, \cdots$. For the sake of simplicity we shall work in a Heisenberg representation and assume further that all the A_k are constants of the motion, not necessarily explicitly time-independent. At any rate, between measurements both the quantum states of our systems and the matrix elements of our observables will be constant. If the observables A_k are to be measured in any particular sequence, which, in general, will not correspond to the order of the subscripts $\cdots, k, \cdots$, we shall indicate the sequence of measurements by Latin superscripts, thus: A_k^m.

Suppose now that we perform a sequence of observations, $A_m^{-M}, \cdots, A_i^{j-1}$, yielding the measurements $d_m, \cdots, d_i$; then the probability that the next measurement A_k^j will yield the eigenvalue d_k is

$$p(d_k/d_m,\cdots,d_i)=|\langle d_i|d_k\rangle|^2=\mathrm{Tr}(D_iD_k), \quad (2.1)$$

where the symbol D_k denotes the idempotent operator

$$D_k\equiv|d_k\rangle\langle d_k|, \quad (2.2)$$

etc. If the measurement A_k^j is to be followed by $A_l^{j+1}, A_n^{j+2}, \cdots, A_r^{N-1}, A_s^N$, the probability that the respective outcomes will be $d_k, d_l, \cdots, d_r, d_s$ is

$$p(d_k,\cdots,d_r,d_s/d_m,\cdots,d_i) =\mathrm{Tr}(D_sD_r\cdots D_kD_iD_k\cdots D_r). \quad (2.3)$$

Equations (2.1) and (2.3) hold irrespective of the outcome of the measurements $A_m^{-M}, \cdots, A_h^{j-2}$, and irrespective of the outcome of the members of the ensemble subsequent to the performance of the specified observation(s). These expressions summarize the quantitative content of the conventional theory of measurement in quantum physics.

In passing let us briefly comment on the need in quantum theory for constructing ensembles with well-defined probability characteristics. If, in classical mechanics, we had to deal with a system possessing a phase space with a finite volume Ω, then we could define an *a priori* probability density on that phase space that would be invariant with respect to canonical transformations: the constant probability density Ω^{-1}. One could then modify this density in conformity with any restrictions imposed on the physical system, so as to obtain contingency probabilities by purely deductive methods. In other words, in a finite phase space one might construct statistical mechanics employing a

[5] E. P. Wigner, Am. J. Phys. 31, 6 (1963).

standard ensemble as the point of departure. Because in every realistic physical system the phase space has an infinite volume, a transformation-invariant standard probability density does not exist, and one is led into constructing or conjecturing probability distributions to fit various conditions imposed on the ensemble.

The situation in quantum theory is analogous. If Hilbert space were finite-dimensional, then there would be one density matrix distinguished as being representation-invariant, the normalized multiple of the unit matrix, from which all other density matrices could be derived in response to various contingencies. But again, for all realistic physical systems Hilbert space is infinite-dimensional; hence, there is no "standard ensemble" existing *a priori* and independently of any information about our physical system. Thus, formally, we are forced to construct ensembles of systems having certain restrictive properties. Whether particular classes of restrictions lead to ensembles with unambiguous probability characteristics cannot be decided affirmatively by formal analysis alone, though internal inconsistencies might rule out some conjectures. It is clear that the assumptions underlying the conventional theory of quantum measurements are logically admissible.

Next we shall consider a sequence of measurements $J, A_k^{-M}, \cdots, A_j^{-1}, A_i^0, A_l^1, A_m^2, \cdots, A_n^N, F$, in that order. J and F are to be nondegenerate observables like the others, and their eigenvalues are denoted respectively by a and b. We shall now consider an ensemble of systems whose initial and final states are fixed to correspond to the particular eigenvalues a and b, respectively; we ask for the probability that the outcome of the intervening measurements are $d_j, \cdots, d_n$, respectively. This probability, on the strength of Eq. (2.3), is found to be

$$p(d_j,\cdots,d_n/a,b)=\frac{p(d_j,\cdots,d_n,b/a)}{p(b/a)}=\frac{1}{H(a,b)}\operatorname{Tr}(AD_j\cdots D_nBD_n\cdots D_j), \tag{2.4}$$

where

$$H(a,b)=\sum_{j'}\cdots\sum_{n'}\operatorname{Tr}(AD_{j'}\cdots D_{n'}BD_{n'}\cdots D_{j'}),\quad A=|a\rangle\langle a|,\quad B=|b\rangle\langle b|. \tag{2.5}$$

This expression is manifestly time symmetric. If we change the sequence of measurements to $F, A_n^{-N}, \cdots, A_k^M, J$, Eqs. (2.4), (2.5) remain unchanged. In the exceptional case $H(a,b)=0$ the probability $p(d_j\cdots d_n/a,b)$ is not defined.

The probabilities (2.4), (2.5) refer to a sample that has been selected on the basis of required outcomes of specified initial and final observations. This procedure may appear artificial compared to the usual prescription: "Prepare a system so that the value of J (at the beginning) be a." But from a formal point of view we may legitimately specify any selection that could be performed with physical equipment, however complex.

As a matter of fact, in experimental physics selections are frequently based on combinations of initial and final characteristics. Consider a beam of particles that enters a cloud chamber or similar device controlled by a master pulse. For the device to select an event as belonging to a sample to be evaluated statistically, the particle must enter the chamber and, prior to the onset of any manipulation by magnetic fields, etc., satisfy certain requirements. But in order to be counted the particle must also activate the circuits of counters placed below the chamber; thus, we make the selection on the basis of both the initial and the final state. In some experiments even intermediate specifications may be imposed in addition to initial and final conditions. Thus, our formal treatment of initial and final states on an equivalent footing is not inconsistent with experimental procedures used in some investigations.

Equations (2.4), (2.5) may be thought of as providing the foundation for a time-symmetric theory of measurement. If we assumed the existence of ensembles with well-defined probabilities only if selected on the basis of both initial and final states, we should have a logically closed theory, though one that would never permit extrapolations to time intervals lying outside the interstice between initial and final determination. Given ensembles of any kind with well-defined probability dispersions, we can always form subensembles obeying additional restrictions and hence the time-symmetric ensembles can be obtained from those of the conventional theory by means of a deductive process. The reverse does not hold, i.e., we cannot infer the characteristics of broadly defined ensembles from those of more narrowly defined ensembles.

On the basis of Eqs. (2.4), (2.5) we may calculate probabilities involving only some of the measurements between J and F, or we may calculate contingent probabilities referring to partial samples in which the outcomes of some of these measurements are fixed. In particular we can calculate the contingent probability of the outcome d_l^1, given the outcome d_i^0. To obtain this probability we must, of course, sum over all the possible outcomes of the measurements preceding A_i^0 and over all the possible outcomes following A_l^1, keeping, as before, the outcomes of J and F fixed. The result is

$$p(d_l/d_i;a,b)=\frac{\sum_{m'}\cdots\sum_{n'}\operatorname{Tr}(D_iD_lD_{m'}\cdots D_{n'}BD_{n'}\cdots D_{m'}D_l)}{\sum_{l'}\sum_{m'}\cdots\sum_{n'}\operatorname{Tr}(D_iD_{l'}D_{m'}\cdots D_{n'}BD_{n'}\cdots D_{m'}D_{l'})}=|\langle d_i|d_l\rangle|^2\frac{\sum_{m'}\cdots\sum_{n'}\operatorname{Tr}(D_lD_{m'}\cdots D_{n'}BD_{n'}\cdots D_{m'})}{\sum_{l'}\cdots\sum_{n'}\operatorname{Tr}(D_iD_{l'}D_{m'}\cdots B\cdots D_{m'}D_{l'})}. \tag{2.6}$$

As expected, the history preceding the measurement A_i^0 drops out of our expression, but the coefficient of the squared matrix element of the conventional prediction (2.1) is sensitive to d_i and to d_l as well as to the subsequent history. In other words, postselection will affect the transition probability from d_i to d_l. That this is unavoidable can be understood easily by the consideration of the extreme case in which all observables $A_m, \cdots, A_n, F$ commute with each other as well as with A_l. Depending on the selection of the eigenvalue b, the transition probability in that case will be either 0 or 1.

It is obvious that the time-reflected relationship to (2.6) also holds. That is to say, if we calculate the contingent probability of d_i knowing the outcome d_l of the observation immediately following, we shall obtain an expression that is independent of the whole history subsequent to the measurement A_l^1, but which will depend on the initial selection a as well as observations scheduled prior to A_i^0.

Let us now consider incomplete measurements. The result (2.4), (2.5) can be generalized immediately if we drop the requirement that each intermediate measurement be complete. According to von Neumann,[3] an incomplete observation projects the initial state not on a particular direction but on a particular (multidimensional) linear subspace of the Hilbert space, and may be represented by an idempotent operator D_k. The form of Eqs. (2.4), (2.5) will remain unchanged under this reinterpretation of the symbols D_k. It should be noted, however, that Eq. (2.6) holds only if A_i is nondegenerate.

The replacement of the initial and final states by mixtures is a bit more involved. If we form an ensemble in which histories beginning with state $|a\rangle$ and ending with state $|b\rangle$ form a fraction c_{ab} of the whole

$$c_{ab}\geq 0, \quad \sum_{a'}\sum_{b'} c_{a'b'}=1, \tag{2.7}$$

then the probability $p(d_j,\cdots,d_n/\{c\})$ will be

$$p(d_j,\cdots,d_n/\{c\})=\sum_{a'}\sum_{b'} c_{a'b'}p(d_j,\cdots,d_n/a',b'). \tag{2.8}$$

There exists no simple expression that would depend on the initial and final density matrices. The probabilities (2.8) depend on the fractions of systems within the ensemble passing from specified initial to specified final states, not merely on the initial distribution $\sum_{b'} c_{ab'}$ and the final distribution $\sum_{a'} c_{a'b}$.

III. ASYMPTOTIC PROCEDURES

Whereas we have been able to obtain time-symmetric ensembles from those depending only on initial selection, the reverse procedure is impossible without an additional postulate; that is to say, given a theory of ensembles based on time-symmetric double-selection procedures, we cannot obtain probabilities for ensembles in which the selection is based only on initial (or only on final) observations by deduction alone. In this sense, the time-symmetric theory of Sec. II is more restricted than the conventional theory of measurements.

There is, however, a way to blunt the effects of either pre- or postselection. The method to be described in this section rests on the fact that in quantum theory the type of interference that we call an observation destroys the "coherence" of the state of a system, producing a new situation that is connected with the original situation only by stochastic laws. This stochastic connection, or the lack of a tighter relationship, may be expressed either in terms of the state vector, or its replacement by a density matrix, or purely in terms of probabilistic assertions. Whatever the mode of description, it is possible to sever different portions of the history of a system from each other by the interposition of certain types of measurements. By preceding the final selection in the time-symmetric theory by such "coherence destroying" manipulations, we may formally recover the prediction formula (2.1); by scheduling such procedures following the initial selection of a time-symmetric ensemble, we may obtain the time-reverse of Eq. (2.1), a "retrodiction" formula.

These possibilities are of considerable interest because they present us with a relatively large class of possible procedures all of which lead, asymptotically in most cases, to substantially similar results. Though the interpolation of coherence destroying manipulations, say before the act of final selection within the framework of the time-symmetric theory, does not relieve us of the logical necessity of performing the act of final selection, the particular choice of observable and of its numerical value used for that final selection has no effect on the probabilities of events preceding the coherence destroying acts.

We shall first indicate particular sets of measurements which destroy coherence more or less completely. Such sets of two consecutive measurements may be constructed in closed form if the Hilbert space of a system is finite dimensional, e.g., if the particles in a monochromatic and well-collimated beam can differ only in their states of polarization. Consider, in this case, two observables A_i and A_j whose eigenvectors are related to each other by a unitary matrix U and whose matrix elements all have the same absolute square $1/n$, n being the number of dimensions of the Hilbert space. One possible unitary matrix with this property is, for instance, the following:

$$U_{kl}=[1/(n)^{1/2}]e^{i\theta_{kl}}, \quad \theta_{kl}=(2\pi/n)kl. \tag{3.1}$$

Let us denote the idempotent operators to be constructed from the respective eigenvectors of the two observables by D_1 and D_2, respectively, each of these symbols representing n such different operators. Then the following expression constructed with any density matrix M whatsoever is always a multiple of the unit matrix I:

$$\sum_{d_1'}\sum_{d_2'} D_1'D_2'MD_2'D_1'=(1/n)I. \quad (3.2$$

As for the infinite-dimensional case, the situation is insofar more involved as there exists no density matrix which is precisely a multiple of the unit matrix. We shall assume that the Hilbert space admits a complete set of commuting operators, each having a continuous range of eigenvalues from $-\infty$ to ∞. We shall call these operators x_s and construct by the usual methods a set of operators p_s which satisfy standard canonical commutation relations with each other and with the x_s. In a somewhat symbolic sense the unitary operators leading from the improper joint eigenfunctions of the x_s to the improper joint eigenfunctions of the p_s, i.e., the Fourier integral operators, possess matrix elements all of the same magnitude as in the previous case. In view of the fact that idempotent operators of the type $x(x_0)$, etc., are not really defined, we introduce idempotent operators $X(x,\Delta)$, defined as integral operators whose kernel equals 1 if $x\epsilon\Delta$, and vanishes otherwise. We cover the space of numerical values of the x_s with a denumerable set of domains Δ without overlap. Similarly, we introduce idempotent operators $P(p,E)$, where the domains E cover the momentum space without overlap. The expression constructed in complete analogy to (3.1) will then not equal a multiple of the unit matrix because of the coarseness of the cell structures established in x space and in p space. However, we may establish a sensible limit if we improve the fineness of both cell structures and if we multiply the expression (3.1) on the left by a factor corresponding to the effective n, eventually becoming infinite, so that the right side can actually tend to the identity operator I (whose trace diverges).

We now return to the expression (2.4) and substitute for a certain number of factors centered on F a multiple of I, both in the numerator and the denominator. The constant of proportionality used is immaterial, as it drops out in any case, and we might use I directly. We then see, almost by inspection, that (2.4) reduces to (2.3), the pure prediction formula, and, likewise, that (2.6) reduces to (2.1). We conclude, then, that because of the asymptotic properties of expressions of type (3.2) the prediction formulas may be recovered from the time-symmetric formulas.

We may derive the corresponding "retrodiction" expression by time reversing the procedure that we have just presented. If we follow the initial selection of an ensemble in the time-symmetric theory by a set of coherence-destroying measurements, then the outcome of subsequent observations is related to the final selection as follows:

$$p(d_k,d_l,\cdots,d_s/a,b)=\mathrm{Tr}(D_kD_l\cdots D_sBD_s\cdots D_l). \quad (3.3)$$

If, in particular, we are concerned with the one observation preceding the final selection, then the probability of the outcome d is

$$p(d/b)=|\langle b|d\rangle|^2. \quad (3.4)$$

The coherence destroying properties of the procedure summarized in Eqs. (3.1), (3.2), and of the corresponding asymptotic procedure outlined for the infinite-dimensional Hilbert space may be demonstrated by straightforward computation. It would be of considerable interest if there were a broad range of procedures having the same effect. Generally, sequences of measurements will destroy coherence to a greater or lesser extent provided that they involve all directions of Hilbert space in noncommuting measurements. There are, of course, degrees of noncommutativity: The noncommutavity may involve varying numbers of directions in Hilbert space, and the eigendirections of consecutive operators may differ from each other by various angles. Formally, the extent to which coherence is destroyed by a given sequence may be evaluated in terms of the degree to which matrices of the general form (3.2) approximate a multiple of the unit matrix. That there is some approach to the unit matrix in a sequence of noncommuting measurements is assured by the results to be found in von Neumann.[3] If $D^{(\alpha)}$ is a set of idempotent operators belonging to the same measurement and with properties

$$D^{(\alpha)}D^{(\beta)}=\delta^{\alpha\beta}D^{(\alpha)}, \quad \textstyle\sum_\alpha D^{(\alpha)}=I, \quad (3.5)$$

and if M is an arbitrary density matrix, then

$$M'=\textstyle\sum_\alpha D^{(\alpha)}MD^{(\alpha)} \quad (3.6)$$

is also a density matrix and approximates a multiple of the unit matrix I at least as well as M in the following respects: (a) If we define the entropy of M as usual by the expression

$$S=-k\,\mathrm{Tr}(M\ln M), \quad (3.7)$$

then

$$S'\geqslant S. \quad (3.8)$$

The equality holds only if the idempotent operators commute with M. (b) The range of eigenvalues of M' is not greater than the range of eigenvalues of M; that is to say, the upper limit of its eigenvalues is not larger and the lower limit not smaller. Both entropy and range of eigenvalue spectrum are yardsticks for the approach to λI.

Thus, it appears that we can destroy coherence more or less completely by a wide variety of sequences of measurements and thereby obtain the asymptotic prediction and retrodiction situations within the framework of the time-symmetric theory of measurements.

The existence of the retrodiction formula (3.3), (3.4) suggests that the customary assignment of a state vector to a system on the basis of the most recent preceding observation may be somewhat arbitrary. This assignment is based on the intuitive notion that the measurement is the "cause" and the quantum state the "effect," and that cause must precede effect in time. Also, perhaps, there is the notion that the quantum state of a system embodies the maximum of information available to us about the system at any time; ordinarily, we can know the outcome of all observations in the past but not of those yet in the future.

But, as we have seen, under suitable circumstances the usual prediction formula (2.1) may be replaced by the retrodiction formula (3.4), which bases a probabilistic statement about the outcome of one measurement on the outcome of the measurement next following in time. If the measurement of A (whose eigenvalues are being denoted by d) is preceded by coherence destroying operations as we have assumed in deriving Eqs. (3.3) and (3.4), then we know essentially nothing about the outcome of observations preceding A; that is to say, all possible outcomes of such preceding observations are approximately equally likely. Hence, our probabilistic statement about the outcome of the measurement of A is based primarily on the event immediately following, and the information on which our statement is based ought to be incorporated in an appropriate assignment of quantum state. Thus, we are led into assigning the state $|b\rangle$ to the period of time *preceding* the observation of F yielding the eigenvalue b.

From a purely operational point of view, one might eschew the assignment of quantum states to physical systems altogether and instead rely entirely on probabilistic statements referring to carefully defined ensembles. However, as long as one does assign quantum states to physical systems, it appears defensible to do so either in reliance on the (complete) observation immediately preceding (as is customary) or on the one next following, depending on circumstances. This ambiguity indicates that the quantum state of a system, though undoubtedly containing some elements of "reality" independent of any observer, also has subjective aspects.

We shall conclude this section by pointing out that, in general, the dispersion of probabilities of the outcome of one particular observation will be minimized (i.e., the "negative entropy" associated with this dispersion will be maximized) if we use all information about the system's past and future. This statement is a direct consequence of the properties of the entropy function to be found, e.g., in Khinchin.[6] Hence, if both the initial and final state of a system are known, use of the prediction formulas (2.1) or (2.3) instead of (2.4) will lead to a loss in precision of the probabilistic statement concerning the intermediate observation.

IV. DIRECT PREDICTION

By now we have established that the conventional prediction formulas can be recovered from the time-symmetric expressions (2.4) by means of a model that consists of shielding events close at hand from the terminal selection on which (2.4) is based by the interposition of a series of "coherence destroying" experiments. Each measurement constitutes an interference with the physical system which destroys in a limited and mathematically well-described manner its dynamic behavior as an isolated system.

Normally, the prediction formula (2.1) and its corollary (2.3) are not conceived of as depending on carefully managed follow-up maneuvers, but are assumed to be independent of the subsequent history of the system. That this prediction theory is indeed logically independent of the time-symmetric formula (2.4) may be deduced immediately from the circumstance that in our universe the prediction formula is considered to be universally valid, whereas the time-reflected formula, the retrodiction formula, is not.

Consider an ensemble of similar physical systems of arbitrary provenance and select a sample on the strength of a single complete measurement. The conventional theory of measurement then furnishes us, with respect to this selected subensemble, with relative frequencies of outcomes of a subsequent measurement or of a subsequent series of measurements, regardless of the events that may have preceded the initial selection procedure, as well as of those events that follow on the heels of the specified series of measurements, as long as no further selection is involved. The reverse theory would have to concern itself with the probability of the outcome of certain measurements on an ensemble of similar physical systems, the ensemble to be determined solely on the basis of a pure-state selection immediately following the specified series of measurements; the expression for the probabilities should contain no reference to any events following the terminal selection, nor to the manner in which physically similar systems were collected prior to the onset of the series of measurements. Clearly, in our universe no such "retrodiction theory" would be valid: Suppose we constructed a monochromatic and well-collimated beam of particles possessing nonzero spin, performed some observation referring to the spin distribution of the beam, and then followed up with a Stern-Gerlach experiment singling out those particles in the beam being in a very definite spin state. Suppose we ask for the percentage of particles, from among those passing the postselection test, that had specified outcomes in the antecedent experiment (which should refer to an observable not possessing the specified final state as an eigenstate); surely these probabilities would not be independent of the state of polarization of our beam prior to the performance of the first experiment.

We conclude, therefore, that in order to recover the conventional prediction statement from the time-symmetric formulas of Sec. II, we must adopt a postulate that is logically independent of the time-symmetric theory; the postulate that in our universe ensembles chosen on the basis of an initial complete measurement alone possess unambiguous and reproducible probability characteristics. Once we adopt this

[6] A. I. Khinchin, *Mathematical Foundations of Information Theory*, transl. by R. A. Silverman and M. D. Friedman (Dover Publications, Inc., New York, 1957).

postulate the conventional prediction formulas (2.1), (2.3) follow from the time-symmetric formula (2.4) and from the considerations of Sec. III. We found in that section that there are "coherence destroying" procedures that make the "prediction" expressions (2.6) independent of the particular postselection we choose to perform. But if there are methods by which we can make our probabilities independent of the manner of postselection and if, by our new postulate, there exist unambiguous probabilities even in the absence of any postselection, then these two sets of probabilities should be equal.

Logically, it is conceivable that the time reverse of our new postulate should also hold; this would mean that postselection alone results in an ensemble with well-defined and reproducible probability characteristics. Actually we know that in our universe this proposition is untrue. We are thus confronted with an indubitable asymmetry in time direction. It remains to discuss whether this asymmetry is a property of microphysics proper or whether it represents the intrusion of the macroscopic universe on the microscopic scene. Granting that this question does not lend itself to straightforward logical analysis, it appears to us that the construction of ensembles in the real physical universe is a macroscopic operation and that it depends on the realities of the universe as a whole. Let us return once more to our beam of particles endowed with spin.

If we attempt to analyze the different manner in which past and future histories affect its present characteristics, we find that no matter how we gather our beam, its constituent particles have come from one or several "sources" (e.g., a laboratory device, a distant galaxy, etc.), which determine its properties; there simply is no way of avoiding preselection completely. On the other hand, beams are not collimated toward a "sink," unless we arrange it so in our laboratory. This asymmetry is directly associated with the fact that the origins of all kinds of radiations in the universe are spatially and temporally concentrated, and their destinations are not. The nature of ensembles or beams actually occurring in nature is, in fact, macroscopic, not microscopic; it is determined by the same cause as all macroscopic irreversibility, conceivably by the expansion of the universe.[1]

As for the microscopically determined aspects of quantum measurements, we believe that they can be fairly summarized by the statement that in time-symmetrically constructed ensembles the laws of probability are also time symmetric; further, that to the extent that retrodiction situations may be said to exist, they obey the same laws as the corresponding prediction situations.

ACKNOWLEDGMENTS

The authors wish to acknowledge frequent valuable discussions with O. Penrose and with A. Komar. One of us (P. G. B.) was helped considerably in clarifying his own thinking by the privilege of attending a conference on the arrow of time organized by T. Gold and H. Bondi, which took place in May 1963 at Cornell University. Our postulate as to the sufficiency of preselection in Sec. IV was stimulated by the paper by Penrose and Percival.[2] Their time-asymmetric proposal is, however, concerned with ensembles of classical systems, and not directly related to their construction on the basis of observations. Finally, we wish to acknowledge with thanks receipt of a preprint by W. J. Cocke entitled "Statistical Time-Inversion Invariance." Cocke's paper is, in contrast to ours, primarily concerned with histories of macroscopic systems, which lend themselves to a description in terms of Markov chains. H. Margenau has pointed out that the conventional interpretation of quantum measurements which is associated with the name of von Neumann[3] is currently no longer acceptable to a fair number of physicists.[7] We wish to express our appreciation for his critical reading of our work.

---

[7] H. Margenau, Phil. Sci. **30**, 1 (1963); Ann. Phys. (N. Y.) **23** 469 (1963). Further references are to be found in these two papers

V.6 LYAPOUNOV VARIABLE: ENTROPY AND MEASUREMENT IN QUANTUM MECHANICS

BAIDYANATH MISRA, ILYA PRIGOGINE, AND MAURICE COURBAGE

ABSTRACT We discuss the question of the dynamical meaning of the second law of thermodynamics in the framework of quantum mechanics. Previous discussion of the problem in the framework of classical dynamics has shown that the second law can be given a dynamical meaning in terms of the existence of so-called Lyapounov variables—i.e., dynamical variables varying monotonically in time without becoming contradictory. It has been found that such variables can exist in an extended framework of classical dynamics, provided that the dynamical motion is suitably unstable. In this paper we begin to extend these results to quantum mechanics. It is found that no dynamical variable with the characteristic properties of nonequilibrium entropy can be defined in the *standard* formulation of quantum mechanics. However, if the Hamiltonian has certain well-defined spectral properties, such variables can be defined but only as a *nonfactorizable* superoperator. Necessary nonfactorizability of such entropy operators M has the consequence that they cannot preserve the class of pure states. Physically, this means that the distinguishability between pure states and corresponding mixtures must be lost in the case of a quantal system for which the algebra of observables can be extended to include a new dynamical variable representing nonequilibrium entropy. We discuss how this result leads to a solution of the quantum measurement problem. It is also found that the question of existence of entropy of superoperators M is closely linked to the problem of defining an operator of time in quantum mechanics.

1. Introduction

No other question in theoretical physics seems to have caused as much controversial discussions over as long a period of time as the question of the dynamical meaning of irreversibility expressed in the second law of thermodynamics. With the advent of quantum mechanics and the discovery of the apparently irreversible exponential decay of unstable particles, this question has gained added theoretical importance. Irreversibility is now an essential feature of gross macroscopic phenomena such as the familiar transport processes and it also seems to be intrinsic in such basic processes as the "wave packet reduction" caused by measurement and the decay of unstable particles.

The difficulties encountered by the traditional approach to the problem of the dynamical meaning of the second law are well known (1). In this paper we shall discuss this question in the framework of quantum dynamics from the alternative viewpoint that has emerged from our previous work (1–4). This discussion will lead to the conclusion that the second law can be interpreted as a dynamical principle in an extended framework of quantum dynamics without involving contradictions and that thus interpreted it implies the loss of distinguishability between pure states and mixtures for systems to which the second law applies.

As is well known, a fundamental distinction is made in quantum mechanics between *pure states* (usually represented by unit vectors of a Hilbert space) and the *mixtures* represented by the so-called density operators. The pure states occupy a privileged position in the theory: the quantum superposition principle holds between the pure states; dynamical evolution, as described by the Schrödinger equation, transforms pure states into pure states and the observables of the theory correspond to self-adjoint operators that again map pure states into pure states. The basic laws of quantum mechanics can thus be formulated without ever invoking the notion of mixtures and their representation by density operators. The use of this notion is generally believed to reflect incompleteness of knowledge about the system and it is considered to be only a matter of practical convenience or approximation.

The fundamental distinction between the pure states and mixtures and the privileged position of the pure states are, however, not maintained in measurement processes. As von Neumann's by now "classical" analysis has shown, we have, in addition to the deterministic and reversible evolution of the pure states into pure states described by the Schrödinger equations, the peculiar evolution, called "the reduction of the wave packet," which occurs during measurement processes. This latter evolution is irreversible and typically transforms pure states into mixtures.

Obviously, one can not accept such a dualism of state-evolution as final, and various authors have attempted to overcome it. (We do not intend to survey these attempts here; an excellent account of the subject can be found in ref. 5.) Let us only emphasize here that it is the presumed distinguishability between the pure state (which the Schrödinger equation would predict for the object + apparatus system) and the mixture that arises from the wave packet reduction that is at the root of the conceptual problems posed by quantum theory of measurement. This dualism of state-evolution could be avoided if one could formulate a physical principle that implies the loss of distinguishability between the pure states and the corresponding mixtures in the case of sufficiently complex systems capable of serving as measuring and recording apparatus (6). The main finding of our paper is that the second law of thermodynamics, when suitably interpreted as a dynamical principle, is just the physical principle that leads to the desired loss of distinguishability between the pure states and mixtures.

Before we discuss further our conclusions, it will be useful to consider briefly the problem of irreversibility in classical dynamics. Obviously, the simplest dynamical interpretation of the second law would be to require the existence of a dynamical variable with the characteristic properties of entropy—in particular, the property of monotonic variation with time. However, Poincaré (7) has pointed out that such a dy-

The publication costs of this article were defrayed in part by page charge payment. This article must therefore be hereby marked "*advertisement*" in accordance with 18 U. S. C. §1734 solely to indicate this fact.

† Also at the University of Texas at Austin, Center for Statistical Mechanics and Thermodynamics, Austin, TX 78712.

‡ Postal address: campus plaine ULB, Boulevard du Triomphe, 1050 Bruxelles, Belgique.

Originally published in *Proceedings of the National Academy of Sciences of the United States of America, 76*, 4768-72 (1979).

namical variable can not exist within the class of standard dynamical variables of classical mechanics (i.e., the class of functions on phase space). This seemed to exclude the possibility of interpreting the second law as a dynamical principle.

Our recent work (2–4), however, shows that this difficulty can be bypassed in an extended framework of dynamics. Briefly, our approach links the existence of a dynamical variable having the properties of nonequilibrium entropy to the presence of instabilities of dynamical motion. For systems presenting sufficiently strong instabilities of motion the notion of phase space trajectories loses operational meaning, and dynamics has to be formulated in terms of the motion of distribution functions on phase space. The dynamical variable can correspond now to more general operators (acting on the distribution functions) than just the multiplication operators by phase functions. In this extended framework the second law may be interpreted as requiring the existence of operators M (called Lyapounov variables) such that

$$\langle \rho_t, M\rho_t \rangle \qquad [1.1]$$

varies monotonously with time t as the distribution functions ρ_t evolve according to the Liouville's equation:

$$i\frac{\partial \rho_t}{\partial t} = L\rho_t. \qquad [1.2]$$

Naturally, not all dynamical systems allow the existence of Lyapounov variables M. As expected, it turns out that a suitable degree of dynamical instability (for instance, "mixing instability") is necessary for the existence of M (3). More importantly, it is found that there is a well-defined class of dynamical systems (so-called K-flows) for which the Lyapounov variables exist. In this case one can define a (self-adjoint) operator T representing "internal time" of the system which is "canonically conjugate" to the generator L of dynamical evolution:

$$i[L, T] = I \qquad [1.3]$$

and Lyapounov variables M then can be obtained as monotone positive operator functions of T. The Lyapounov variables M may, under suitable conditions, lead further to (nonunitary) *invertible* similarity transformations Λ defined as

$$\Lambda = M^{1/2} \qquad [1.4]$$

which convert the original deterministic evolution described by the Liouville equation into the stochastic evolution of a *Markov process* satisfying an H theorem of the Boltzmann type. We have exhibited examples of this possibility (4).

One thus begins to see how the second law can be formulated as a dynamical principle in terms of the existence of Lyapounov variables and the important implications of the existence of such variables. In particular, one begins to see the close links among intrinsic irreversibility (expressed in terms of the existence of M), inherent randomness (expressed in terms of the existence of Λ which converts the dynamical evolution to the evolution of a stochastic process), and the instabilities of dynamical motion. In this paper we make a start to extend these results to quantum mechanics.

We show in *Section* 2 that, under fairly general conditions, no entropy operator or Lyapounov variable can be defined in the standard framework of quantum mechanics. This is a consequence of the fact that in the standard formulation the generator of the time evolution group is also the operator representing the energy observable and hence is required to be bounded from below. This result is the quantum analogue of Poincaré's conclusion that entropy cannot be defined as a phase function in classical mechanics.

The limitation of *Section 2* does not apply if the quantum mechanical Liouville equation is taken as the basic equation of motion because the Liouvillian operator need not be semibounded. It is found (*Section* 3) that, if the Liouville operator L (or equivalently the Hamiltonian) has certain well-defined spectral properties, then Lyapounov variables exist as "super operators" (i.e., as operators acting on the space of density operators). They cannot exist, however, even as superoperator for all systems. In particular, they do not exist for quantum systems of finite number of particles enclosed in a finite volume. In this respect the situation in quantum mechanics differs fundamentally from that in classical mechanics. Lyapounov variables exist, for instance, for the classical system of a finite number of hard spheres enclosed in a finite box because this system is known to satisfy the K-flow condition.

As a side remark let us add that this disparity in behavior (with regard to irreversibility) of the classical and quantum systems argues against the plausibility of "hidden variable theories" of quantum mechanics. This point is further developed in a forthcoming publication (8).

The essential point about the entropy superoperators M is that they are necessarily "*nonfactorizable*" and they *cannot* preserve the class of pure states. These results are proved in *Section 4*. Physically this means that, for systems whose algebra of observables can be extended to include a new dynamical variable representing nonequilibrium entropy, the distinguishability between pure states and mixtures must be lost. This opens the way to reconciling the two types of evolution distinguished by von Neumann in his analysis of measurement processes. The measurement process involves always a dynamical system (namely, the measuring + recording apparatus) for which the requirement of the second law—in the existence of an entropy superoperator—is satisfied. But this implies that the distinction between pure states and mixture is lost in measurement processes.

Let us mention that in previous work (2, 9) it has been emphasized that kinetic theory leading to description of irreversible phenomena can only be developed on the level of mixture and it involves nonfactorizable superoperators. We begin to see now the deeper justification for this.

2. Incompatibility between the existence of entropy operator and the semiboundness of generator of time-evolution

Let us start by showing that in order to define an observable with the characteristic property of nonequilibrium entropy it is necessary to go beyond the standard formulation of quantum mechanics in which the Hamiltonian operator H is the generator of dynamical evolution. Suppose that there exists an operator M in the Hilbert space of pure states that represents nonequilibrium entropy. The requirement that M increases monotonically with time then translates into the condition:

$$i[H, M] = D \geq 0. \qquad [2.1]$$

The operator D can now be interpreted as the entropy production operator, and it seems natural to suppose that the measurements of M and D are mutually compatible. This implies

$$[M, D] = 0. \qquad [2.2]$$

The basic reason why conditions **2.1** and **2.2** cannot be satisfied by an operator M (except in the trivial case $D = 0$) is that the Hamiltonian operator H plays a dual role in quantum mechanics: it is the generator of the time-evolution group and it also represents the energy of the system. Hence, it must be bounded from below: $H \geq 0$.

Before proceeding to show the incompatibility of conditions **2.1** and **2.2** (with $D \neq 0$) with the semiboundedness of the Hamiltonian, let us remark that the operators involved in conditions **2.1** and **2.2** are in general unbounded. To obtain reasonably meaningful results it is thus necessary to make the formal commutation conditions **2.1** and **2.2** more precise by making appropriate assumptions about the domains of definition of the involved operators. We need not discuss this question in detail here. Let us only mention that the argument given below is rigorously valid if (*i*) M is bounded and self-adjoint, (*ii*) $M\mathcal{D}_H \subseteq \mathcal{D}_H$, where $\mathcal{D}_H$ denotes the domain of H, and (*iii*) condition **2.2** is interpreted to mean that the operator DM is an extension of MD: $MD \subset DM$. Naturally, these conditions are not necessary but only sufficient to establish the incompatibility of conditions **2.1** and **2.2** with the semiboundedness of H.

For instance, this incompatibility follows also even for unbounded M provided that suitable assumptions are made on the domain of M. But we do not discuss such generalizations here because they do not bring any essentially new point.

Returning to the proof of incompatibility, let us consider the identity

$$\frac{d}{dt}\langle e^{-iMt}\psi, He^{-iMt}\psi\rangle = -i\langle e^{-iMt}\psi, [H, M]e^{-iMt}\psi\rangle \tag{2.3}$$

which can be easily verified to hold for all ψ in the domain $\mathcal{D}_H$ of H. To be more explicit, using the previously stated conditions on the domain of H and condition **2.2**, one can verify the following: (*i*) for all $\psi \in \mathcal{D}_H$, $e^{-iMt}\psi$ also belongs to $\mathcal{D}_H$; (ii) for $\psi \in \mathcal{D}_H$, $\langle e^{-iMt}\psi, He^{-iMt}\psi\rangle$ is differentiable with respect to T with its derivative given by identity **2.3**.

Because the bounded self-adjoint operator M commutes with the self-adjoint operator D, by assumption it follows that $e^{iMt}De^{-iMt} = D$. Thus, identity **2.3** reduces to the equality:

$$\frac{d}{dt}\langle e^{-iMt}\psi, He^{-iMt}\psi\rangle = -\langle\psi, D\psi\rangle. \tag{2.4}$$

Integration of both sides of equality **2.4** from 0 to $t \geq 0$ yields

$$\langle e^{-iMt}\psi, He^{-iMt}\psi\rangle - \langle\psi, H\psi\rangle = -t\langle\psi, D\psi\rangle$$

or,

$$\langle\psi, H\psi\rangle = t\langle\psi, D\psi\rangle + \langle e^{-iMt}\psi, He^{-iMt}\psi\rangle$$
$$\text{for } \psi \in \mathcal{D}_H.$$

Because $H \geq 0$, we have for all $\psi \in \mathcal{D}_H$ and all $t \geq 0$,

$$\langle\psi, H\psi\rangle \geq t\langle\psi, D\psi\rangle \tag{2.5}$$

or $1/t\langle\psi, H\psi\rangle \geq \langle\psi, D\psi\rangle$ for all $\psi \in \mathcal{D}_H$ and all $t \geq 0$. Letting $t \to \infty$ we then obtain $\langle\psi, D\psi\rangle \leq 0$ for a dense set of vectors; or $D \leq 0$.

At this point we may use the assumption that $D \geq 0$ and reach the desired conclusion $D = 0$. However, it is important to note that this conclusion does not depend on the positivity of D but follows only from condition **2.2** and the semiboundedness of H. To see this, we now integrate equality **2.4** from $-t$ to 0 and obtain, as before, the inequality:

$$\langle\psi, H\psi\rangle \geq -t\langle\psi, D\psi\rangle \tag{2.6}$$

for all $\psi \in \mathcal{D}_H$ and all $t \geq 0$.

The inequality **2.6** together with the previously reached conclusion that $D \leq 0$ now proves that $D \equiv 0$. This shows that, in order to be able to define a new dynamical variable M representing nonequilibrium entropy, it is necessary to go beyond the standard formulation of quantum mechanics. Although the property of monotonic increase of entropy (i.e., the condition $D \geq 0$) does not play any role in reaching this conclusion, we shall see that it plays an essential role in establishing the important conclusion that the existence of an entropy "superoperator" implies the loss of distinction between pure and mixed states.

From a purely mathematical point of view, the result just obtained generalizes a result of Putnam (10). In fact, Putnam has shown that condition **2.2** together with the assumption that H is bounded implies that $D = 0$. Here we have generalized this result to semibounded H. Moreover, this generalization is "maximal" in the sense that the theorem fails to hold if the condition of semiboundedness of H is dropped. As discussed below, this fact allows the possibility of constructing Lyapounov variables or entropy operators in the Liouvillian formulation of dynamics.

There is an interesting connection between the nonexistence of entropy operator M in $\mathcal{H}$ and the impossibility [noted by Pauli (11)] of defining an operator of time in the usual formulation of quantum mechanics. Such an operator of time would be, by definition, a self-adjoint operator T which is canonically conjugated to the generator H of the time-evolution group; or

$$[H, T] = iI \tag{2.7}$$

However, as shown in ref. 3, if such a T exists then one can obtain an entropy operator M satisfying conditions **2.1** and **2.2** by simply taking M to be a monotonic operator function of T: $M = f(T)$. One thus obtains yet another proof of Pauli's remark that there can be no operator T in the standard formulation of quantum mechanics. Let us also mention that more recent attempts to define a more general concept of operator T that need not be self-adjoint or satisfy condition **2.7** still conflict with the fact that the generator of motion is the Hamiltonian operator and hence is bounded from below (cf. ref. 12).

The impossibility of defining the entropy operator M, the nonexistence of time operator in the standard formulation of quantum mechanics, and the problem of interpreting and justifying the time–energy uncertainty relationship are thus all linked. Their common origin is the fact that in the usual formulation of quantum mechanics the generator H of the time-translation group is identical with the energy operator of the system. To be able to define the entropy operator M it is thus necessary to overcome this degeneracy. The simplest way of achieving this is to go to the so-called Liouvillian formulation of (quantum) dynamics. The basic object in this formulation is the group describing the time evolution of the density operators. Under dynamical evolution, an initial density operator ρ is transformed in time t to the density operator $e^{-iHt}\rho e^{iHt}$. Therefore the generator of the time-translation group is now the Liouvillian operator L defined by the equation:

$$L\rho = [H, \rho]. \tag{2.8}$$

In the following section we investigate the existence and properties of entropy operator M in the Liouvillian formulation of quantum dynamics.

3. The entropy superoperator

The important advantage gained in going to the Liouvillian formulation of quantum dynamics is that the generator L of the time-translation group is no longer physically required to be bounded from below. In fact if the spectrum of H extends from 0 to $+\infty$ the spectrum of L is the entire real line. The possibility of defining M as a superoperator (i.e., an operator acting on the space spanned by density operators) satisfying the relationships

$$i[L, M] = D \geq 0 \tag{3.1}$$

and

$$[M, D] = 0 \tag{3.2}$$

is thus not excluded by the argument given in the preceding section.

Naturally, one does not expect M to exist, even as a superoperator, for all dynamical systems. Let us recall that in classical mechanics the existence of M is always associated with certain strong instability properties of dynamical motion, which is related to certain well-defined spectral properties of the Liouvillian.

One expects similar conclusions in quantum mechanics. Let us first note that an entropy superoperator (or, equivalently, a Lyapounov variable) cannot exist in either of the following cases: (*i*) the Hamiltonian H has purely discrete spectrum; and (*ii*) H has bounded spectrum.

The first of these statements follows from the fact that, if H has purely discrete spectrum, then so does L (considered as a self-adjoint operator acting on the Hilbert space of the Hilbert–Schmidt operators. On the other hand, it has been shown that, if a pair of self-adjoint operators M and L satisfy relationship **3.1**, then the eigenvectors of L belong to the null space D (3). Thus, if L has a *purely* discrete spectrum its eigenvectors will span the entire space and D will vanish identically. Similarly, the second statement follows from the fact that, if H is bounded, then so is L. The argument of the preceding sections [or the earlier result of Putnam (10)] now shows that, for bounded L, condition **3.2** can not be satisfied unless $D = 0$.

The physical meaning of the statements just proved is that an entropy superoperator can not exist for quantum systems consisting of finite numbers of particles enclosed in a finite volume (because the Hamiltonian has a discrete spectrum in this case) or in theories relying on finite cutoff energy. In quantum mechanics, it is only in the limit of infinite systems that irreversibility expressed in the second law of thermodynamics may become manifest. This is in marked contrast to the situation in classical mechanics. As shown (3), entropy operators can be constructed in classical mechanics for a class of systems which includes, for instance, the system of a finite number of hard spheres enclosed within a finite box. This shows that conceptually the problem of irreversibility in quantum mechanics is more involved than in classical mechanics.

A sufficient condition for the existence of M as a superoperator is that the Hamiltonian H has an absolutely continuous spectrum extending from 0 to $+\infty$. To see this, it may be observed that this condition on H implies that the corresponding Liouvillian has an absolutely continuous spectrum of uniform (actually infinite) multiplicity extending over the entire real line (M. Courbage and B. Misra, unpublished). Under this condition there always exists a self-adjoint operator T which is "canonically conjugate" to L. In fact, in the spectral representation (or direct integral representation) that "diagonalizes" L to make it correspond to the multiplication operator by the real variable λ this operator T can be taken to be the differentiation operator $id/d\lambda$. As mentioned before, monotonically increasing operator functions of T then satisfy defining conditions **3.1** and **3.2** for M.

A consequence of this remark is that the system of a single free particle in infinite volume admits an entropy superoperator, because the Hamiltonian $H = P2/2m$ obviously has an absolutely continuous spectrum extending over the entire interval $[0, +\infty]$. But one does not expect any physical irreversibility for this system. The spurious "irreversibility" associated with the evolution of the free particle in infinite volume is just a reflection of the fact that initially localized wave packets spread out and get rarified with the particle, eventually "escaping to infinity."

Naturally, one would like to distinguish such spurious cases from more physical irreversibility by requiring M to satisfy supplementary physically motivated conditions. We shall not discuss this question in detail here. However, let us mention that we have studied (4) the requirement that the square root $M^{1/2}$ of the Lyapounov variables of classical systems, as well as the semigroup $\Lambda e^{-iLt}\Lambda^{-1}$, $(t \geq 0)$, preserve the positivity of distribution functions on phase space. The physical meaning of this is as follows. Dynamical evolutions admitting the existence of such Lyapounov variables are intrinsically random in the sense that a change of representation effected by this (invertible) transformation Λ converts the original dynamical evolution into the evolution of the stochastic Markov process. In quantum mechanics, one may formulate a corresponding requirement on the entropy operator M or, more accurately, on the Lyapounov variable $M^{-1} = M'$. In other words, one would require of the Lyapounov variable M' that its square root Λ' as well as the semigroup $\Lambda' e^{-itL}\Lambda'^{-1}$ (for $t \geq 0$) map density operators (positive operator with trace 1) to density operators. The physical meaning of the existence of such a Lyapounov variable is the same as in classical mechanics: it means that the quantum Liouvillian evolution in question is equivalent, through an invertible similarity transformation, to the evolution of a quantum stochastic Markov process.

A detailed discussion of how the measurement process can be adequately described in terms of the quantum stochastic processes obtained, through similarity transformation, from the Liouvillian evolution of the object + apparatus system will be presented elsewhere. Here we shall only show that the supposed distinguishability between the pure states and mixtures and the privileged position of the pure states in the usual formulation of quantum mechanics—the two facts that are at the root of the conceptual difficulties posed by the phenomena of wave packet reduction—must necessarily be given up in the case of physical systems, for which entropy superoperators exist. To this end, let us show that the entropy superoperators are necessarily nonfactorizable.

4. Nonfactorizability of entropy superoperators

Nonfactorizability of an entropy superoperator means, by definition, that $M\rho$ cannot be written in the form

$$M\rho = A_1 \rho A_2 \qquad [\mathbf{4.1}]$$

where A_1 and A_2 are two ordinary Hilbert space operators acting on the same Hilbert space on which ρ acts. We shall show that if a positive and hermiticity-preserving superoperator M satisfies conditions **3.1** and **3.2** and is factorizable (i.e., satisfies Eq. **4.1**), then $D = 0$. Let us recall that the positivity- and hermiticity-preserving conditions on superoperator M mean, respectively, that

$$Tr(\rho^* M\rho) \geq 0 \qquad [\mathbf{4.2}]$$

and

$$\rho = \rho^* \text{ implies } M\rho = (M\rho)^*.$$

These are natural requirements to impose on the entropy superoperator M. They lead to the simplification that, if a factorizable superoperator M satisfies them, then $M\rho$ can be written in the form

$$M\rho = A\rho A \qquad [\mathbf{4.3}]$$

where A is a positive operator. It will thus suffice to show that conditions **3.1**, **3.2**, and **4.3** together imply that $D = 0$.

Now it follows from the positivity of D (condition **3.1**) and the assumed form **4.3** for M that

$$i[H, A] \equiv B \geq 0. \qquad [\mathbf{4.4}]$$

To see this, let us compute the action of D on density operators of the form

$$\rho_\phi = |\phi\rangle\langle\phi|$$

with ϕ in the domain $\mathcal{D}_H$ of H. Using definition **2.8** of L and form **4.3** for M, one easily verifies that

$$D\rho_\phi = i[LM - ML]|\phi\rangle\langle\phi| = |B\phi\rangle\langle A\phi| + |A\phi\rangle\langle B\phi| \quad [\mathbf{4.5}]$$

Positivity of D then implies that

$$Tr(\rho_\phi D\rho_\phi) = g\langle\phi, B\phi\rangle\langle\phi, A\phi\rangle \geq 0.$$

Because $A \geq 0$ it follows that $B \geq 0$.

In the same way, computing $[D, M]|\phi\rangle\langle\psi|$ and setting it equal to 0 (condition **3.2**), one finds that

$$|\Big[[H, A], A\Big]\phi\rangle\langle A^2\psi| = |A^2\phi\rangle\langle\Big[[H, A], A\Big]\psi| \quad [\mathbf{4.6}]$$

for all ϕ, ψ in the dense set $\mathcal{D}_H$. But this means

$$\Big[[H, A], A\Big]\phi = CA^2\phi$$

with

$$C = \langle\Big[[H, A], A\Big]\psi, \psi\rangle / \langle A^2\psi, \psi\rangle,$$

a real number independent of ϕ. Thus

$$\Big[[\mathrm{H}, \mathrm{A}], \mathrm{A}\Big] = CA^2$$

with C a real number; or

$$i[B, A] = -CA^2. \quad [\mathbf{4.7}]$$

Since $B \geq 0$ (relationship **3.4**) and $-CA^2$ obviously commutes with A, the argument of *Section 2* applies with B taking the role of H, A that of M, and $-CA^2$ that of D. Thus, we conclude that $-CA^2 = 0$. Now, there are two cases: (*i*) if $C \neq 0$, we conclude that $A^2 = 0$ and hence $A = 0$ with the result that M given by form **4.3** and, *a fortiori*, D vanish; (*ii*) if $C = 0$, then $B = i[H, A]$ and A commutes and again the argument of *Section 2* applies, leading to the conclusion that $B = 0$. This then implies (relationship **4.5**) that $D = 0$. Let us remark that, unlike the argument of *Section 2*, the proof of the nonfactorizability of the entropy operator makes essential use of the positivity, or rather definiteness in sign, of entropy production operator D.

The nonfactorizability of M has the important consequence that M can not preserve the purity of states. In fact, if M were to map pure states $|\phi\rangle\langle\phi|$ into operators of the same form it would follow that M would also map positive operator to positive operators. Such a purity- and positivity-preserving map M is, however, one of the following three forms (13): (*i*) M is factorizable; (*ii*) $M\rho = A'\rho^*A'^*$ with A' an antilinear operation of $\mathcal{H}$; (*iii*) $M\rho = Tr(\rho B).\ |\psi\rangle\langle\psi|$ when B is a fixed positive operator and ψ is a fixed vector. But the entropy superoperator cannot be any of these forms. Case *i* is already ruled out. To rule out case *ii*, let us consider the operator M^2 which, as it can be verified easily, is factorizable. On the other hand, as the entropy superoperator M satisfies conditions **3.1** and **3.2**, M^2 satisfies $i[L, M^2] \equiv D' \equiv 2DM = 2M^{1/2}DM^{1/2} \geq 0$ and $[M^2, D'] = 0$. Thus, M^2 is also an entropy superoperator and hence it can not be factorizable. Finally, this entropy superoperator can not be of the form *iii* because in this case M, being an operator of rank 1, will have a purely discrete spectrum. This, however, contradicts the easily verified fact that entropy superoperators can not have a purely discrete spectrum.

5. Concluding remarks

The preceding considerations lead up to the following conclusions. For an infinite quantal system, there exists the possibility of enlarging the algebra of observables to include an operator M representing nonequilibrium entropy. The operator M can be defined, however, only as a nonfactorizable superoperator. The inclusion of a (necessarily nonfactorizable) entropy operator among the observables thus entails that the pure states lose their privileged position in theory and that the pure and mixed states be treated on equal basis. Physically, this means that, for systems having entropy as an observable, the distinction between the pure and mixed states must cease to be operationally meaningful and there would be limitations on the possibility of realizing coherent superposition of quantum states.

Evidently this conclusion, which is reached here as a logical consequence of our theory of entropy operator, should be further elucidated by an analysis of the physical reason for the loss of the distinction between pure and mixed states.

The corresponding situation for classical systems has been discussed elsewhere (1, 3, 4). In those papers, the existence of an entropy superoperator implies that the phase space trajectories cease to be physically observable. The physical reason for this is linked to the instability of dynamical motion, which is a necessary condition for the existence of an entropy superoperator or a Lyapounov variable.

One would expect a suitable quantum analogue of the instability mechanism to operate for quantal systems (admitting the existence of Lyapounov variables) which is responsible for the loss of physical distinction between pure states and mixtures.

A detailed formulation of the instability mechanism in quantum mechanics is work for the future. It remains, nevertheless, a remarkable fact that, when interpreted as a dynamical principle in terms of the existence of Lyapounov variables or entropy operator M, the second law of thermodynamics implies the loss of distinction between pure states and mixtures. This provides the way to a resolution of the duality of state evolution considered in the "orthodox" theory of measurement by von Neumann.

Finally, the approach to the measurement problem which emerges from this paper is in accord with the often repeated observation by Bohr that any correct account of a measurement process must take into consideration two important facts: irreversibility of measurement processes and the classical nature of measuring apparatus. Our approach links these two features. We show that irreversibility (expressed by the existence of entropy superoperator M for the measuring apparatus) implies the classical nature of apparatus in that the distinction between pure and mixed states is lost.

1. Prigogine, I. (1979) *Order and Fluctuations in Equilibrium and Nonequilibrium Statistical Mechanics*, 17th International Solvay Conference in Physics, Brussels, 20–24 November 1978, in press.
2. Prigogine, I., George, C., Henin, F. & Rosenfeld, L. (1973) *Chem. Scr.* **4,** 5–32.
3. Misra, B. (1978) *Proc. Natl. Acad. Sci. USA* **75,** 1627–1631.
4. Misra, B., Prigogine, I. & Courbage, M. (1979) *Proc. Natl. Acad. Sci. USA* **76,** 3607–3611.
5. d'Espagnat, B. (1976) *Conceptual Foundations of Quantum Mechanics* (Benjamin, Menlo Park, CA), 2nd Ed.
6. Wigner, E. P. (1963) *Am. J. Phys.* **31,** 6–15.
7. Poincaré, H. (1889) *C. R. Hebd. Seances Acad. Sci.* **108,** 550–553.
8. George, C. & Prigogine, I. (1979) *Physica,* in press.
9. George, C., Prigogine, I. & Rosenfeld L. (1972) *K. Dan. Vidensk. Selsk. Mat-fys. Meddel* **38,** 1–44.
10. Putnam, C. R. (1967) *Communication Properties of Hilbert Space Operators* (Springer, New York), Theorem 1.6.3., p. 12.
11. Pauli, W. (1958) in *Encyclopedia of Physics*, ed. Flügge, S. (Springer, Berlin), Vol. 5.
12. Misra, B. & Sudarshan, E. C. G. (1977) *J. Math. Phys.* **18,** 756–763 (see especially p. 762).
13. Davies, E. B. (1976) *Quantum Theory of Open Systems* (Academic, New York), Theorem 3.1, p. 21.

V.7 CAN WE UNDO QUANTUM MEASUREMENTS?

ASHER PERES

The Schrödinger equation cannot convert a pure state into a mixture (just as Newton's equations cannot display irreversibility). However, to observe phase relationships between macroscopically distinguishable states, one has to measure very peculiar operators. An example, constructed explicitly, shows that the *classical* analog of such an operator cannot be measured, because to do so would violate classical irreversibility. This result justifies von Neumann's measurement theory, without any hypothesis on the role of the observer.

The measurement process in quantum physics was analyzed long ago by von Neumann[1] who showed that it could formally be described as the transformation of a pure state $\Psi = \sum c_n \phi_n$ into a mixture $\rho = \sum |c_n|^2 P_n$. Here, the ϕ_n are eigenstates of the dynamical variable being measured, and the P_n are the corresponding projection operators.

This irreversible transformation, commonly called the "collapse of the wave packet," cannot follow from the Schrödinger equation, since the latter generates a unitary mapping of the Hilbert space of states. In fact, the coupling of the eigenstates of the measured system to those of the measuring apparatus is a perfectly reversible process[2-4] as long as we are willing to measure correlations between the two. For these reasons, von Neumann's theory has been considered unsatisfactory, or at least incomplete.

There have been several attempts[5-7] to prove von Neumann's conjecture by supplementing quantum theory with superselection rules forbidding the measurement of operators of a certain type (those which connect macroscopically different states of the apparatus). The purpose of this paper is to show that systems with many degrees of freedom are indeed subject to such superselection rules. A general proof of this assertion would be very difficult, but the following model is typical enough to convey belief in the result.

Consider a macroscopic apparatus designed to measure the z component of the spin of an electron. This apparatus has a pointer (center-of-mass coordinate q, conjugate momentum p) initially localized around $q=0$. The pointer is to move through a macroscopic distance L to the right or the left depending on whether $s_z = \frac{1}{2}$ or $-\frac{1}{2}$. This can be achieved by the coupling $H = 2V(t) s_z p$, where $V(t)$ is a large velocity, so large indeed that we can neglect all the other terms in the Hamiltonian during the brief duration of the coupling.[8]

Before the measurement, the state of the electron is $\binom{\alpha}{\beta}$ and that of the apparatus is $\Psi(q, q_2, q_3, \ldots, q_N)$ where $q_2, q_3, \ldots, q_N$ are the other, "irrelevant," degrees of freedom. Naturally, N is a very large number, say 10^{23}. (It would be more realistic to assume a density matrix instead of the pure state ψ, but this refinement is not needed at the present stage.)

After completion of the coupling, the combined state is

$$\begin{pmatrix}\alpha\\0\end{pmatrix} e^{-iLp}\psi + \begin{pmatrix}0\\\beta\end{pmatrix} e^{iLp}\psi\,, \tag{1}$$

where $L = \int V\,dt$. Since ψ is peaked around $q=0$, $e^{\pm iLp}\psi$ is peaked around $q = \mp L$. Thus, the sign of q is correlated to that of s_z and

$$\langle s_z \rangle = \langle \tfrac{1}{2}\,\mathrm{sign}(q)\rangle = \tfrac{1}{2}(|\alpha|^2 - |\beta|^2)\,. \tag{2}$$

We have performed what von Neumann calls a measurement. (As we shall soon see, a better word would be "premeasurement.")

The question is whether this process is reversible and, in particular, whether the relative phase α/β is still observable. At the present stage, it is, as can be seen by measuring the expectation values of the operators,

$$A_1 = s_x \cos 2Lp + s_y \sin 2Lp \tag{3a}$$

and

$$A_2 = s_x \sin 2Lp - s_y \cos 2Lp\,. \tag{3b}$$

[To measure A_1 and A_2 we divide the electrons in two identical but disjoint ensembles. After each electron passage through the apparatus, we first measure p (modulo π/L) then the component of $\vec{s}$ in the direction of $\tan^{-1}(2Lp)$ or $\cot^{-1}(-2Lp)$. Note that the eigenvalues of A_1 and A_2 are $\pm\frac{1}{2}$.]

These operators can conveniently be combined as

$$A = A_1 + iA_2 = \begin{pmatrix}0&0\\1&0\end{pmatrix} e^{2iLp}\,. \tag{4}$$

A straightforward calculation yields $\langle A\rangle = \alpha\beta^*$

Originally published in *Physical Review, D22*, 879-83 (1980).

which, together with Eq. (2), gives α and β separately, up to a common phase.

However, if we wait some time, the state (1) will evolve into

$$e^{-itH}\left[\begin{pmatrix}\alpha\\0\end{pmatrix}e^{-iLp}\psi+\begin{pmatrix}0\\\beta\end{pmatrix}e^{iLp}\psi\right], \qquad (5)$$

where H is the Hamiltonian of the electron and apparatus. Assuming for simplicity that the two spin states have the same energy, we obtain

$$\langle A\rangle=\alpha\beta^*\int(e^{-itH}e^{iLp}\psi)^*e^{2iLp}e^{-itH}e^{-iLp}\psi d^Nq. \qquad (6)$$

But $e^{iLp}e^{-itH}e^{-iLp}$ is simply $e^{-itH\,(q+L)}$ (i.e., H with the q coordinate shifted by L) and (6) can be written as

$$\langle A\rangle=\alpha\beta^*\langle e^{itH(q-L)}e^{-itH\,(q+L)}\rangle. \qquad (7)$$

The coefficient of $\alpha\beta^*$ would still be one if H did not depend on q, but there is no reason to expect that. As the pointer moves with respect to some fixed scale on the apparatus, its energy may vary somewhat from place to place and the coefficient of $\alpha\beta^*$ may be less than one in absolute value. For small t we get

$$|\langle e^{itH(q-L)}e^{-itH(q+L)}\rangle|^2=1-t^2\delta H^2+\cdots, \qquad (8a)$$

where

$$\delta H^2=\langle[H(q-L)-H(q+L)-\langle H(q-L)-H(q+L)\rangle]^2\rangle. \qquad (8b)$$

Moreover, if the other degrees of freedom of the apparatus are in a mixed state, this coefficient will quickly fall to zero,[9] because of the randomness of the phases. The time needed to erase $\langle A\rangle$ is of the order of $1/\delta H$. It is therefore inversely proportional to the strength of the coupling of the macroscopic degree of freedom q, used for the measurement, with the *other* degrees of freedom of the apparatus. In the present model, this time could be of the order of the size of the pointer divided by the speed of sound (a few microseconds).

This neat distinction between the reversible premeasurement—Eq. (1)—and the ensuing irreversible process is admittedly unrealistic in most instances. In practice, a macroscopic apparatus has almost always an amplification mechanism based on a metastable initial state[10] and irreversibility appears at the very outset of the process. However, the amplification requirement is not essential and it obscures the true nature of the irreversibility of quantum measurements, which is explained below. (The reversal of an idealized premeasurement is illustrated in Fig. 1.)

The above discussion of Eq. (7)—or some similar argument[10,11]—is usually considered as a proof that the relative phase of the two branches of Eq. (5) is "lost" after some finite time. However, such arguments are not convincing, because Eq. (5) represents a pure state (what else could it be?) and this can be shown by measuring the expectation value of *another* operator, namely,

$$A'=e^{-itH}A\,e^{itH}. \qquad (9)$$

Indeed, we trivially have $\langle A'\rangle=\alpha\beta^*$, since the $e^{\pm itH}$ factors in A' cancel those of the wave functions.

However, the operator A' has very peculiar properties. (It is not of course the Heisenberg picture of A, the latter being $e^{itH}Ae^{-itH}$. In fact, we are always working in the Schrödinger picture.) This A' operator is *explicitly time dependent and is also a constant of the motion.*

To verify that it is a constant of the motion, it is enough to observe that its matrix elements between any two Schrödinger states are constant, or simply to go to the Heisenberg picture, where A'_H looks like A_S, without any time dependence.

Now, these explicitly time-dependent constants of the motion are very familiar in classical mechanics. For example, for a free particle, $q-tp/m$ is such a constant. Its meaning simply is the initial value of q. For an harmonic oscillator, such a constant would be $\tan^{-1}(m\omega q/p)-\omega t$, the meaning of which is the initial value of the phase. In general, for a system with N degrees of freedom, there are $2N$ constants of the motion, a few of which may be explicitly time independent (the total energy, momentum, etc.), but almost all

FIG. 1. Idealized premeasurement, using the recoil of a rigid double mirror. If a particle is reflected from the first mirror, a correlation is established between the momentum of the particle and that of the instrument (this is the premeasurement). That correlation is then reversibly destroyed when the particle is reflected from the second mirror. (Note that if we wish to complete the measurement and to observe the recoil of the double mirror *between* the two reflections, the latter must be prepared with $\Delta p\ll h/\lambda$. If this device is part of a double-slit experiment, it allows to determine through which slit the particle passed only at the expense of destroying the interference pattern, because $\Delta q\gg\lambda$. But if we forego observing the recoil, the interference pattern is restored because the *same* Δq is added and subtracted at both reflections.)

of which include the time explicitly. Their physical meaning is to give the $2N$ initial positions and momenta as explicit functions of the positions and momenta at some future time t. The structure of these constants of the motion is of course hopelessly complicated for large N and finite t. It leaves us no choice but to replace Newtonian mechanics by statistical mechanics. It is our inability to make use of these constants of the motion which is the cause of irreversibility.

In the present case, we must measure, instead of A given by Eq. (4),

$$A' = \begin{pmatrix} 0 & 0 \\ 1 & 0 \end{pmatrix} e^{2iLp'},$$

where $p' = e^{-itH} p e^{itH}$ is the value of p immediately after the premeasurement, expressed as a function of p and $2N-1$ other variables at a later time t. In classical physics, we would say that this is so complicated that only a "Maxwell demon" can measure all these variables and then compute p' (assuming H is known). In quantum physics, the task is even more difficult because the $2N$ variables do not commute. Therefore, the Maxwell demon must contrive a *single* measurement[12] for p', which is an incredibly complicated function of $2N$ noncommuting variables and of t.

In other words, we see that not every self-adjoint operator corresponds to an observable, simply because not every classical dynamical variable is observable. It is the inobservability of these operators which makes pure states appear as mixtures and causes the irreversibility of quantum measurements.

In conclusion, let us summarize the assumptions used in the derivation of this result. First, we note that the macroscopic degree of freedom used for the measurement—here, the center of mass of the pointer—is not completely isolated from the other degrees of freedom of the apparatus.[13] (We could, of course, have treated these other degrees of freedom as an external reservoir, but then our result would have been trivial. It is essential that our system be a *closed* one.)

The second assumption is the impossibility of measuring the classical analog of the operator p'. (There is also a tacit assumption that if a classical measurement is impossible, the same is true for the corresponding quantum measurement.) Here, it may be objected that as long as the number of degrees of freedom is finite, it is not impossible to measure p', only very difficult. In principle, a measurement of p' should always be possible at the cost of a great increase of entropy of the rest of the world.[14] From this point of view, as long as we are able to pay the price,[15] we definitely *can* undo quantum measurements,[4] except in the unattainable mathematical limit of an infinite apparatus.[16] However, if we admit that a finite system may appear irreversible (if the time needed for a Poincaré recurrence is longer than the Universe lifetime), the present paper shows how the irreversibility of quantum measurements is rooted in the familiar classical irreversibility.

I am very grateful to J. A. Wheeler for the warm hospitality of the University of Texas and to E. C. G. Sudarshan for many stimulating discussions. The final version of this paper has benefited from comments by J. S. Bell (CERN) and A. Ron (Technion). This work was supported in part by the Center for Theoretical Physics, The University of Texas at Austin, Austin, Texas 78712, and also by NSF Grant No. PHY78-26592.

APPENDIX: A MORE REALISTIC MODEL

The simple model discussed above involves an explicitly time-dependent coupling $V(t)$, supposedly switched on and off by an external agent. This may give the impression that we are dealing with an open (i.e., incompletely described) system, for which the transformation of a pure state into a mixture would be trivial.

In a real-life Stern-Gerlach experiment, this time dependence is of course due to the translational degree of freedom of the electron, which was arbitrarily ignored in our model. A more realistic description of what happens follows.

We write the complete Hamiltonian as

$$H = H_a + H_e + 2Vs_z p u(x_2 - x) u(x - x_1), \tag{A1}$$

where H_a refers to the apparatus, H_e to the free electron (mass m, momentum k, position x), V is a coupling *constant*, u is the unit step function, and x_1 and x_2 are the entrance and exit points of the electron as it passes through the apparatus. The pointer is assumed massive enough so that its velocity p/M is negligible when the electron is outside the apparatus. When it is inside, the pointer velocity is $\pm V$.

The measurement process can be described as a scattering of the electron and the apparatus. Before the "collision," the electron has momentum k. When it reaches the apparatus, it meets an energy barrier of height $\pm Vp$ and thickness $x_2 - x_1$. Inside the barrier, its momentum is $k' = (k^2 \pm 2mVp)^{1/2} \simeq k \pm mVp/k$, where we have assumed that $k^2 \gg 2mVp$, so that most electrons are transmitted (a reflected electron would mean an unsuccessful experiment). The outgoing electron still has momentum k, but has been subject to a phase shift $(k'-k)(x_2 - x_1) = \pm \tau V p$ where $\tau = m(x_2 - x_1)/k$ is the classical time of passage through the apparatus. We now identify $L = \tau V$ and the final state of the combined system is

$$e^{ikx}\left[\begin{pmatrix}\alpha\\0\end{pmatrix}e^{-iLp}\psi+\begin{pmatrix}0\\\beta\end{pmatrix}e^{iLp}\psi\right], \tag{A2}$$

whence the discussion proceeds as before.

However, several remarks are in order. First, we have treated p as a constant during the collision, i.e., we assumed that $\dot{p}=\partial H_a/\partial q=0$. This is of course incompatible with a nontrivial δH [see Eq. (8b)]. However, we can make the change in p arbitrarily small by increasing V and k (keeping their ratio constant, so that L remains unchanged). This does not affect $\dot{p}$, but makes τ arbitrarily small. The condition is easily seen to be $\tau \ll 1/\delta H$, i.e., the premeasurement must be very brief, compared to the time required to make the measurement irreversible.

To avoid a possible misunderstanding, it must be emphasized that δH is *much smaller* than the energy uncertainty $\Delta H=(\langle H^2\rangle-\langle H\rangle^2)^{1/2}$. Indeed there must be *many* different energy eigenstates involved to make the measurement possible.[17,18] In particular, the incoming electron must have $\Delta H \gg \delta H$ because the two branches of the outgoing electron will not interfere if $\delta H > \Delta H$ (that is, we would need an operator much more complicated than A to display their interference[19]).

The above remark is closely related to overall energy conservation. We have assumed hitherto that the outgoing electron had the same energy as the incoming one. This cannot, of course, be rigorously exact if $H(q-L)\neq H(q+L)$. A more correct treatment follows.

First, assume that initially the apparatus is in an eigenstate of energy E_0 and that the electron too is monochromatic with energy $K=k^2/2m$. Then obviously there is no irreversibility since the operator e^{-itH} in Eq. (A2) becomes a phase factor $\exp[-it(E_0+K)]$ and $\langle A\rangle$ is constant. The energy picked up or released by the electron exactly compensates the energy difference in the final state of the apparatus. It is thus important to understand why $|\langle A\rangle|$ may decrease if we have a superposition (or mixture), rather than an energy eigenstate.

Even if $|E_0\rangle$ is an eigenstate of H_a, the states $e^{\pm iLp}|E_0\rangle$ usually are not. We can write

$$e^{\pm iLp}|E_0\rangle=\int c_\pm(E_0,E,K)\,|E\rangle dE\,,$$

where the coefficients $c_\pm$ depend also on K, because $L=m|x_2-x_1|/k$. By virtue of energy conservation, the scattering process can therefore be written as

$$|E_0,K\rangle\to\int c_\pm(E_0,E,K)\,|E,K+E_0-E\rangle dE\,,$$

where the $\pm$ subscripts refer to $s_z=\pm\frac{1}{2}$.

Now let the electron be initially in a superposition $\int g(K)|K\rangle dK$ (the apparatus may still be initially in an energy eigenstate[10]). The outgoing states become

$$\int g(K)c_\pm(E_0,E,K)\,|E,K+E_0-E\rangle dE\,dK\,.$$

In order to compute $\langle A\rangle$, we first note that

$$\langle E',K'+E_0-E'|e^{2iLp}|E,K+E_0-E\rangle$$
$$=\langle E'|e^{2iLp}|E\rangle\delta(K'-E'-K+E)\,,$$

so that

$$\langle A\rangle=\alpha\beta^*\int g(K)g^*(K')c_+(E_0,E,K)c_-^*(E_0,E',K')$$
$$\times\langle E'|e^{2iLp}|E\rangle\delta(K'-E'-K+E)dE\,dE'dK\,dK'\,.$$

We now make two essential physical assumptions. One is that $\Delta K \ll K$ (otherwise, L is ill-defined) so that in $c_\pm$ we can replace K by its average value K_0. The second one is that $\delta H\ll\Delta K$, i.e., $c_\pm(E_0,E,K_0)$ is very small unless $|E-E_0|\ll\Delta K$ (as explained above, $|K_{\text{out}}-K_{\text{in}}|$ must be much smaller than ΔK to allow the two "branches" of the electron to interfere). We can therefore replace $g^*(K+E'-E)$ by $g^*(K)$. Integration over K and K' gives

$$\langle A\rangle=\alpha\beta^*\int c_+(E_0,E,K_0)c_-^*(E_0,E',K_0)$$
$$\times\langle E'|e^{2iLp}|E\rangle dE\,dE'\,.$$

We see that the electron energy no longer appears in the formula (except as an average). The result looks *as if* the final state of the apparatus were $\int c_\pm(E_0,E,K_0)|E\rangle dE$. Therefore, the energy shift of the electron cannot prevent $\langle A\rangle$ from having a nontrivial time dependence, due to the factor $e^{-i(E-E')t}$ in $\langle E'|e^{2iLp}|E\rangle$.

---

*On sabbatical leave from Technion-Israel Institute of Technology, Haifa.

[1]J. von Neumann, *Mathematical Foundations of Quantum Mechanics* (Princeton University Press, Princeton, 1955).

[2]P. A. Moldauer, Phys. Rev. D 5, 1078 (1972).

[3]A. Peres, Am. J. Phys. 42, 886 (1974).

[4]J. S. Bell, Helv. Phys. Acta 48, 93 (1975).

[5]H. Wakita, Prog. Theor. Phys. 23, 32 (1960).

[6]A. Komar, Phys. Rev. 133, B542 (1964).

[7]E. C. G. Sudarshan, Pramana 6, 117 (1976); T. N. Sherry and E. C. G. Sudarshan, Phys. Rev. D 18, 4580 (1978); 20, 857 (1979).

[8]A more realistic model with no explicitly time-depen-

dent coupling is discussed in the Appendix.

[9]It is not necessary that $\langle A\rangle$ be "rigorously" zero because it is an *average* value, not an eigenvalue. The eigenvalues are $\pm\frac{1}{2}$ and the probable error on $\langle A\rangle$ after n measurements is $\sim 1/\sqrt{n}$. Now n cannot exceed e^{2000}, say, because of cosmological limitations. Therefore, after 1000 lifetimes $\langle A\rangle$ is indistinguishable from zero *as a matter of principle*, not only as an approximation.

[10]A. Daneri, A. Loinger, and G. M. Prosperi, Nucl. Phys. 33, 297 (1962); Nuovo Cimento 44B, 119 (1966). These authors assume that the initial state of the apparatus is metastable. This is often true in practice but, as shown in the present paper and in particular in the Appendix, this assumption is not necessary.

[11]K. Hepp, Helv. Phys. Acta 45, 237 (1972).

[12]For example, it is not the same thing to measure $s_x + s_y$ (eigenvalues $\pm 1/\sqrt{2}$) or to measure s_x and s_y separately and sum the results.

[13]A macroscopic superconducting current is *not* acceptable as a measuring instrument, unless it is coupled to some monitoring device with numerous degrees of freedom.

[14]Here again, we may encounter cosmological limitations [A. Peres and N. Rosen, Phys. Rev. 135, B1486 (1964)]. If we assume that the Universe is finite, there must be an upper limit to the amount of entropy which may be generated in it. It is then plausible that if N is large, irreversibility sets in after a finite time t. However, I do not wish to enter deeper into this subject: The purpose of this paper was not to explain classical irreversibility, but only tc clarify its relationship to the quantum measurement problem.

[15]This is the only place where the human observer has any role. He decides (perhaps subjectively) which experiments are feasible. For a fuller discussion, see A. Peres, Found. Phys. (to be published).

[16]B. Whitten-Wolfe and G. G. Emch, Helv. Phys. Acta 49, 45 (1976).

[17]E. P. Wigner, Z. Phys. 133, 101 (1952).

[18]H. Araki and M. M. Yanase, Phys. Rev. 120, 622 (1960); M. M. Yanase, *ibid.* 123, 666 (1961).

[19]If the apparatus or part of it is in thermal equilibrium, we have $\Delta H \simeq kT$ and in most realistic situations $\tau\Delta H \gg 1$. This does not hamper the measurement as long as $\tau\delta H \ll 1$.

VI
Accuracy of Measurements: Quantum Limitations

VI.1 THE UNMEASURABILITY OF THE SPIN OF A FREE ELECTRON

COMMENTARY OF CASIMIR (1967)

Bohr gave much thought to the spin of the electron and to Dirac's theory. I never felt quite at ease about his argument that the spin cannot be observed by classical means although he always succeeded in showing the fallacies in any proposed experimental setup.

COMMENTARY OF ROSENFELD (1971B)

Bohr as a lecturer is a different matter. It is much glossed, but very little written about. Perhaps the only one who has put his view of it in print so far is Larmor; in a speech (later published)* at the Maxwell celebrations in Cambridge in 1931, he commented upon Maxwell's reputation of being 'a poor lecturer' and roundly added: "So perhaps with our friend Bohr: he might want to instruct us about the correlations of too many things at once . . . ." I was sitting near Bohr when the speech was delivered; as this judgment was expressed, Bohr whispered to me: "Imagine, he thinks I am a poor lecturer!" Bohr's lectures, composed with tremendous labour, were indeed masterpieces of allusive evocation of a subtle dialectic; the trouble was that the audience was usually unprepared to catch subtle allusions to conceptions and arguments which were anyhow unfamiliar and hard to grasp.

* *James Clerk Maxwell: A Commemoration Volume 1831–1931*, p. 78, Cambridge University Press, Cambridge (1931).

I am not sure whether Bohr's introductory talk at the conference was really worse than the average; perhaps he had not prepared it so thoroughly, since the idea was to have quite informal discussions: no programme had been set up in advance—Bohr took in turn each of the participants aside and asked him what topic he wished to bring up. At any rate, here is the impression this talk has left in my memory, as I described it (with some hindsight) in 1945: "He had begun with a few general considerations calculated, no doubt, to convey to the audience that peculiar sensation of having the ground suddenly removed from under their feet, which is so effective in promoting receptiveness for complementary thinking. This preliminary result being readily achieved, he had eagerly hastened to his main subject and stunned us all (except Pauli) with the non-observability of the electron spin. I spent the afternoon with Heitler pondering on the scanty fragments of the hidden wisdom which we had been able to jot down in our notebooks."

It was comforting to hear from Klein, when I told him some time ago of our failure to understand what Bohr meant by the impossibility of measuring the spin of the electron, that he had had the the same difficulty when Bohr first discussed the matter with him in the autumn of 1928. Guided by the general correspondence idea, Bohr argued that such a purely quantal concept as the electron spin, vanishing from the theory in the classical limit, could not possibly be brought in direct relation with classical quantities like angular momentum or magnetic moment. It was not immediately clear to Klein, however, how this correspondence argument could be reconciled with the Stern-Gerlach effect, which clearly exhibited a contribution to the magnetic moment of an atom from an electron bound in a ^2S state; but what Bohr demonstrated was precisely that with a free electron a Stern-Gerlach experiment could not succeed, because the effect of the Lorentz force would inevitably blur any Stern-Gerlach pattern. This is the point he ineffectually tried to make in his talk. Fortunately, Mott, during his stay at the Institute, had been engaged in the problem of electron polarization, and in the paper* in which he brilliantly showed how this property could in principle be ascertained by a double scattering experiment, he gave a very clear account of the whole situation. He finished writing this paper shortly after the conference (it was sent off by Bohr on the 25th of April) and we were thus soon able to appreciate at leisure the full force of Bohr's famous argument.

* N. F. Mott, *Proc. Roy. Soc. A124*, 440 (1929).

VI.1 MAGNETIC MOMENT OF THE ELECTRON

NEVILL F. MOTT AND HARRIE S.W. MASSEY

2. Magnetic moment of the electron

We have discussed so far only the magnetic moment of the atom. We shall not review here the evidence, derived from the anomalous Zeeman effect, from the gyromagnetic effect, etc., that the *electron* has a fourth degree of freedom, a magnetic moment $-\epsilon\hbar/2mc$, and a mechanical moment $\frac{1}{2}\hbar$. We shall content ourselves with remarking that according to the Schrödinger theory the ground state of the hydrogen atom is not degenerate, and therefore, in order to account for the splitting in a magnetic field revealed by the Stern–Gerlach experiment, it is necessary to assume that the electron has a fourth degree of freedom.

The first evidence that electrons have a magnetic moment was derived from their behaviour when bound in stationary states in atoms. For the study of collision problems it is necessary to inquire what meaning can be attached to the magnetic moment of a free electron. In the first place, just as in the case of the atom, it is impossible to determine the moment by means of a magnetometer experiment. This can be shown by the following argument, due to Bohr.† Let us suppose that the position of the electron is known with an accuracy Δr and that we wish to determine the magnetic moment at a point distant r from it. It will not be possible to deduce from our measurement anything about the magnetic moment of the electron unless

$$\Delta r \ll r. \tag{5}$$

The field H that we wish to observe will be of order of magnitude

$$H \sim M/r^3.$$

If, however, the electron is in motion with velocity v, there will be a magnetic field due to its motion, of amount $\epsilon v/cr^2$; since we do not know v exactly we cannot allow for this field exactly. From our measurements, therefore, of the magnetic field, it will not be possible to find out anything about the magnetic moment of the electron, unless

$$M/r^3 \gg \epsilon\Delta v/cr^2,$$

where Δv is the uncertainty in our knowledge of v. Since by the

† Cf. Mott, *Proc. Roy. Soc.* A, **124** (1929), 440.

Originally published as section 2, chapter IX of *The Theory of Atomic Collisions* by N. F. Mott and H.S.W. Massey, pp. 214-19 of the third edition, Clarendon Press, Oxford (1965).

uncertainty principle $\Delta r \Delta v > h/m$, this leads to

$$\Delta r \gg r,$$

which contradicts the inequality (5). We conclude therefore that it is not possible to measure the magnetic moment of an electron in this manner.

We shall now show that it is impossible, by means of a Stern–Gerlach experiment, to determine the magnetic moment of a free electron, or to prepare a beam of electrons with the magnetic moments all pointing in the same direction. The argument is also due to Bohr.

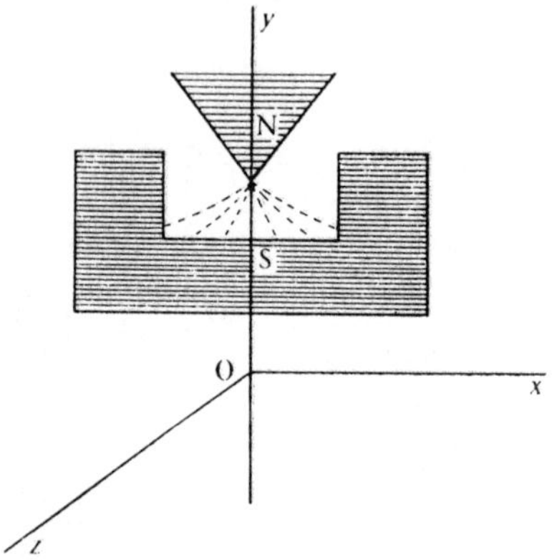

FIG. 35

In Fig. 35 a beam of electrons is supposed to travel parallel to the z-axis (i.e. perpendicular to the plane of the paper). The pole pieces of the magnet are shown, as are also the lines of force. The purpose of the experiment is to observe a splitting in the y-direction. The force on an electron tending to split the beam will be

$$\pm M \frac{\partial H_y}{\partial y}. \tag{6}$$

Now all electrons will experience a force due to their motion through the field. Those moving in the plane Oyz will experience a force in the direction Ox. This force is perpendicular to the direction of the splitting, and its only effect will be to displace the beams to the right or to the left. However, electrons which do not move in the plane Oyz will experience a force in the direction Oy, because the lines of force in an inhomogeneous magnetic field cannot be straight, and there must be a component H_x of H along Ox. This force will have magnitude

$$\epsilon v H_x / c. \tag{7}$$

We can compare (7) with the force (6) tending to produce the splitting. H_x at a point distant Δx from the plane Oyz will be equal to $(\partial H_x/\partial x)\Delta x$, and since div H vanishes, this is equal to $-(\partial H_y/\partial y)\Delta x$. The quantities (6) and (7) therefore stand in the ratio

$$\frac{\epsilon h}{4\pi mc} \frac{\partial H_y}{\partial y} : \frac{\epsilon v}{c} \frac{\partial H_y}{\partial y} \Delta x.$$

Dividing through by common factors this becomes

$$1:4\pi\Delta x/\lambda, \tag{7.1}$$

where λ is the wavelength h/mv of the waves that represent the electrons. Suppose now that $\pm\Delta x$ is the distance from the plane Oyz of the two extremities of the beam. Since Δx must be greater than λ, it is clear that the two extremities of the beam will be deflected in opposite directions through angles greater than the angle of splitting, which we hope to observe.

To see now that it is impossible to observe any splitting, let us consider the trace that the beam would make on a photographic plate. Suppose that it were possible to use finer beams than is allowed by the uncertainty principle, so that the thickness Δy of the beam in the y-direction would be infinitely small. Before passing through the magnetic field, the cross-section of the beam would be as in Fig. 36 (*a*). Afterwards, it would be as in Fig. 36 (*b*), which shows the trace produced on a photographic plate. The tilting of the traces is produced by the Lorentz forces discussed above. If ABC, $A'B'$ are two lines parallel to Oy and distant λ apart, then by (7.1) we see that the tilting is so great that $AB > BC$. If $A\beta\gamma$ is drawn perpendicular to the traces, it follows that $A\beta > \beta\gamma$. But $A\beta < \lambda$, and hence $\beta\gamma$, the distance between the traces, is less than λ. Thus the maximum separation that can be produced is λ. But actually we cannot obtain a trace of breadth comparable with λ. Therefore it is impossible to observe any splitting.

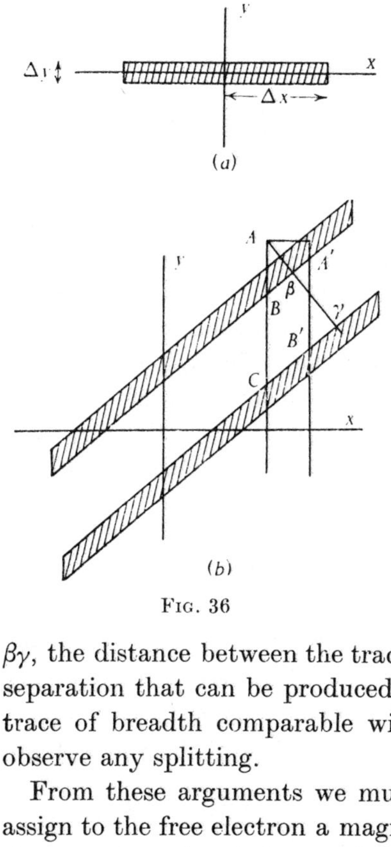

FIG. 36

From these arguments we must conclude that it is meaningless to assign to the free electron a magnetic moment. It is a property of the electron that when it is bound in an S state in an atom, the atom has a magnetic moment. When we consider the relativistic treatment of the electron due to Dirac, we shall see that this magnetic moment is not in general equal to $-\epsilon\hbar/2mc$, unless the velocities of the electron within the atom are small compared with that of light (§ 3.3). A single electron bound in its lowest state in the field of a nucleus of charge $Z\epsilon$ gives, according to Dirac's theory, a magnetic moment†

$$-\tfrac{1}{3}[1+2\sqrt{(1-\gamma^2)}]\epsilon\hbar/2mc \quad (\gamma = Z\epsilon^2/\hbar c). \tag{8}$$

The statement that a free electron has four degrees of freedom is on a different footing, for it is hardly conceivable that an electron in an atom should have four degrees of freedom, and a free electron three. It is interesting to inquire, therefore, whether there is any conceivable experiment by which this fourth degree of freedom can be detected. We wish to know whether it is possible to prepare a beam of electrons that is in some sense 'polarized', and whether it would be possible to detect this polarization.

Although it is now well known that polarization in this sense can be detected (see §§ 4.5, 4.6) it is instructive to consider the following experiment.‡ A beam of atoms is prepared, by means of a Stern–Gerlach experiment, with their axes all pointing in the same direction, say along the z-axis. Electrons are ejected from these atoms by illuminating them with ultra-violet light. The beam of electrons obtained may be said to be polarized, for the following reasons: Assuming for the moment that the electron behaves like a small magnet, let us ask whether forces sufficient to eject the electron would be sufficient to alter appreciably the direction of the magnetic moment. The following purely classical considerations of the order of magnitude of the forces involved show that they are not, and so we may consider that the magnetic moments in the beam of ejected electrons all point in the same direction.§

If an electric field of intensity E acts on an electron for time t, the kinetic energy acquired is $\frac{1}{2}(E\epsilon)^2t^2/m$. The energy that must be given to an electron to remove it from an atom is of order of magnitude $m\epsilon^4/h^2$. Thus to remove an electron from an atom the product of E and t must be of order of magnitude $Et \sim \epsilon m/h$. The average velocity of an electron in an atom is ϵ^2/h. The average couple acting on the electron magnet, due to its motion through the electric field E, will be of order

$$E\frac{\epsilon h}{mc}\frac{\epsilon^2}{h}\frac{1}{c},$$

† This formula is due to Breit, *Nature*, **122** (1928), 649. Cf. § 3.3 of this chapter.

‡ This method of preparing a polarized beam of electrons was first suggested by Fues and Hellmann, *Phys. Zeits.* **31** (1930), 465.

§ There is, of course, a small probability that the direction of the spin-axis is reversed, and the following discussions show this to be of order of magnitude $(1/137)^2$. There is no known method by means of which a completely polarized beam can be produced.

which is equal to $E\epsilon^3/mc^2$. To change the orientation of the electron by an angle comparable with π, this couple must produce a change of angular momentum comparable with h. The time T necessary for this to occur is given by

$$T\frac{E\epsilon^3}{mc^2} \sim h,$$

which gives

$$ET \sim hmc^2/\epsilon^3.$$

We deduce that

$$\frac{ET}{Et} \sim \left(\frac{hc}{\epsilon^2}\right)^2.$$

Thus

$$T \gg t.$$

It would, however, be meaningless to speak of a polarized beam, unless the fact that the beam is polarized could be detected in some way. This could be done if the beam were passed through a gas of ionized atoms, so that some of the electrons were captured. If the neutral atoms formed were shown by means of a Stern–Gerlach experiment to be polarized, then we should have a method of detecting the polarization. The argument used above about the order of magnitude of the forces involved indicates that this should be the case; a proper proof can, however, be given on the basis of Dirac's theory of the electron.

An account of more practical methods of preparing and of detecting a polarized beam is given in §§ 4.2 and 4.6.

We see, then, that the spin of a free electron may be described by the same wave function $\chi_l(s)$ that was used before to describe the magnetic moment of an atom. The function

$$|\chi_l(s)|^2 \quad (s = \pm 1)$$

gives the probability that, if the electron is prepared with its magnetic moment in the direction **l**, then, if the electron be captured by an atom, and if that atom be passed into an inhomogeneous magnetic field, the energy of that atom will be either $\pm MH$. It is necessary to give to the square of the amplitude of the wave function this rather complicated interpretation, because it is not possible to measure the energy of an electron in a magnetic field, unless the electron is captured in an atom. It is further to be noted that, by the statement that an electron is prepared with its magnetic moment in a given direction, it is meant that the electron has been knocked off an atom that has been so prepared.

As in the case of the bound electron, an electron is completely described by a wave function

$$\psi(\mathbf{r}, s).$$

If the forces acting on the electron are so small that the direction of the spin remains constant throughout the experiment considered, then as before this function may be split up into the product

$$\psi(\mathbf{r})\chi(s),$$

where $\psi(\mathbf{r})$ is a solution of Schrödinger's equation. The form of $\psi(\mathbf{r}, s)$ when this is not the case can be found from Dirac's theory.

VI.2 MEASUREMENT OF QUANTUM MECHANICAL OPERATORS

HUZIHIRO ARAKI AND MICHAEL YANASE

The limitation on the measurement of an operator imposed by the presence of a conservation law is studied. It is shown that an operator which does not commute with a conserved (additive) quantity cannot be measured exactly (in the sense of von Neumann). It is also shown for a simple case that an approximate measurement of such an operator is possible to any desired accuracy.

1. INTRODUCTION

IT was pointed out by Wigner[1] that the presence of a conservation law puts a limitation on the measurement of an operator which does not commute with the conserved quantity. The limitation is such that the measurement of such an operator is only approximately possible. An approximate measurement can be done by a measuring apparatus which is large enough in the sense that the apparatus should be a superposition of sufficiently many states with different quantum numbers of the conserved quantity. He has proved these statements for a simple case where the x component of the spin of a spin one-half particle is measured, the z component of the angular momentum being the conserved quantity. The aim of this paper is to present a proof of the above statement for the general case.

In Sec. 2, we will prove that an exact measurement of an operator M which does not commute with a conserved operator L_1 is impossible. In Sec. 3, we will prove that an *approximate* measurement of the operator M is possible if L_1 has discrete eigenvalues and is bounded in the Hilbert space of the measured object.

2. IMPOSSIBILITY OF AN EXACT MEASUREMENT OF AN OPERATOR WHICH DOES NOT COMMUTE WITH A CONSERVED QUANTITY

Suppose we measure a self-adjoint operator M for a system represented by a Hilbert space $\mathfrak{H}_1$. Assume that M has discrete eigenvalues μ and corresponding eigenvectors $\phi_{\mu\rho}$ which are orthonormal and complete in $\mathfrak{H}_1$,

$$M\phi_{\mu\rho}=\mu\phi_{\mu\rho}, \tag{2.1}$$

$$(\phi_{\mu\rho},\phi_{\mu'\rho'})=\delta_{\mu\mu'}\delta_{\rho\rho'}. \tag{2.2}$$

The measuring apparatus is represented by a Hilbert space $\mathfrak{H}_2$. Then a state of the combined system of the measured object and the measuring apparatus is represented by a unit vector in $\mathfrak{H}_1\otimes\mathfrak{H}_2$.

According to von Neumann,[2] the measurement of the operator M in a state ϕ is accomplished by choosing an apparatus in a state ξ (fixed normalized state independent of ϕ) in $\mathfrak{H}_2$ such that the combined system, if it is in the state $\phi_{\mu\rho}\otimes\xi$ before the measurement, goes over after a finite time t into

$$U(t)(\phi_{\mu\rho}\otimes\xi)=\sum_{\rho'}\phi_{\mu\rho'}\otimes X_{\mu\rho\rho'}, \tag{2.3}$$

where $U(t)$ is a unitary operator describing the time-development of the combined system. In order to be able to distinguish the different measured values of the operator M in terms of states of measuring apparatus after the measurement, we require

$$(X_{\mu\rho\rho''},X_{\mu'\rho'\rho'''})=0, \quad \text{if} \quad \mu\neq\mu'. \tag{2.4}$$

We note that, because we are not measuring the degeneracy parameter ρ, we have to allow the possibility that the measuring object remains in any linear combination of $\phi_{\mu\rho'}$, with fixed μ but with arbitrary ρ.[3]

We now assume the existence of a universal conservation law for a self-adjoint operator L which is additive in the sense that

$$L=L_1\otimes 1+1\otimes L_2, \tag{2.5}$$

where L_1 and L_2 are self-adjoint operators in $\mathfrak{H}_1$ and $\mathfrak{H}_2$, respectively. Actually this additivity will be used only before and after the measurement, when the two systems are separated. By universal, we mean that, whatever measuring apparatus we take, $U(t)$ commutes with L,

$$[U(t),L]=0. \tag{2.6}$$

Our claim is that (2.3) is impossible unless

$$[L_1,M]=0. \tag{2.7}$$

For the proof, we first note that, because of the unitarity of $U(t)$ and the conservation law (2.6), we

* Present address: Department of Nuclear Engineering, Kyoto University, Kyoto, Japan.

† On leave of absence from Sophia University, Tokyo, Japan.

[1] E. P. Wigner, Z. Physik **131**, 101 (1952).

[2] J. von Neumann, *Mathematische Grundlagen der Quantenmechanik* (Verlag Julius Springer, Berlin, 1932; English ed.: Princeton University Press, Princeton, 1955).

[3] For any state of the measured object, we can write

$$U(t)(\phi\otimes\xi)=\Sigma_{\mu\rho}\,\phi_{\mu\rho}\otimes X_{\mu\rho}(\phi),$$

where $X_{\mu\rho}(\phi)$ depends on ϕ. The Eqs. (2.3) and (2.4) give the most general form of the above equation satisfying (1) the distinguishability of the measured result,

$$(X_{\mu\rho}(\phi),X_{\mu'\rho'}(\phi))=0, \quad \text{if} \quad \mu\neq\mu',$$

and (2) the requirement that the probability of an eigenvalue μ in the state ϕ, as measured by the state of the measuring apparatus after the measurement, should give the conventionally postulated value,

$$\Sigma_\rho\|X_{\mu\rho}(\phi)\|^2=\Sigma_\rho|(\phi_{\mu\rho},\phi)|^2.$$

Originally published in *Physical Review, 120*, 622-26 (1960).

have

$$(\phi_{\mu'\rho'}\otimes\xi, L(\phi_{\mu\rho}\otimes\xi))$$
$$=(U(t)(\phi_{\mu'\rho'}\otimes\xi), U(t)L(\phi_{\mu\rho}\otimes\xi))$$
$$=(U(t)(\phi_{\mu'\rho'}\otimes\xi), LU(t)(\phi_{\mu\rho}\otimes\xi))$$
$$=(\sum_{\rho'''}\phi_{\mu'\rho'''}\otimes X_{\mu'\rho'\rho'''}, L\sum_{\rho''}\phi_{\mu\rho''}\otimes X_{\mu\rho\rho''}). \quad (2.8a)$$

Hence, as the necessary condition for the conservation law for the operator L, we can write

$$(\phi_{\mu'\rho'}\otimes\xi, L(\phi_{\mu\rho}\otimes\xi))$$
$$=(\sum_{\rho'''}\phi_{\mu'\rho'''}\otimes X_{\mu'\rho'\rho'''}, L\sum_{\rho''}\phi_{\mu\rho''}\otimes X_{\mu\rho\rho''}). \quad (2.8b)$$

Using the additivity of L, (2.5), we obtain

$$(\phi_{\mu'\rho'}\otimes\xi, L(\phi_{\mu\rho}\otimes\xi))$$
$$=(\phi_{\mu'\rho'},L_1\phi_{\mu\rho})(\xi,\xi)+(\phi_{\mu'\rho'},\phi_{\mu\rho})(\xi,L_2\xi)$$
$$=\sum_{\rho''\rho'''}[(\phi_{\mu'\rho'''},L_1\phi_{\mu\rho''})(X_{\mu'\rho'\rho'''},X_{\mu\rho\rho''})$$
$$+(\phi_{\mu'\rho'''},\phi_{\mu\rho''})(X_{\mu'\rho'\rho'''},L_2X_{\mu\rho\rho''})]. \quad (2.8c)$$

Because of the orthogonalities, (2.2) and (2.4), we finally obtain

$$(\phi_{\mu'\rho'},L_1\phi_{\mu\rho})=0, \quad \text{if} \quad \mu\neq\mu', \quad (2.9a)$$

or

$$(\phi_{\mu'\rho'},L_1\phi_{\mu\rho})=\delta_{\mu\mu'}(\phi_{\mu'\rho'},L_1\phi_{\mu\rho}). \quad (2.9b)$$

We are now ready to prove that L_1 commutes with M. For this purpose we decompose M into projection operators

$$M=\sum_\mu \mu P_\mu; \quad P_\mu\phi_{\mu'\rho'}=\delta_{\mu\mu'}\phi_{\mu'\rho'}. \quad (2.10)$$

To prove the commutativity of L_1 and M, (2.7), it is sufficient to prove the commutativity of L_1 and P_μ,

$$P_\mu L_1-L_1P_\mu=0. \quad (2.11)$$

From the self-adjoint nature of P_μ, (2.9b) and (2.10), we see that

$$(\phi_{\mu'\rho'},P_\mu L_1\phi_{\mu''\rho''})=\delta_{\mu\mu'}\delta_{\mu'\mu''}(\phi_{\mu'\rho'},L_1\phi_{\mu''\rho''}),$$
$$(\phi_{\mu'\rho'},L_1P_\mu\phi_{\mu''\rho''})=\delta_{\mu\mu''}\delta_{\mu'\mu''}(\phi_{\mu'\rho'},L_1\phi_{\mu''\rho''}),$$

which manifestly demonstrate (2.11). Thus we have succeeded in proving that (2.3)–(2.6) imply (2.7).[4]

Although we have assumed in the above proof that M has a discrete spectrum, the conclusion holds for any self-adjoint operator M. Namely, suppose

$$M=\int \mu dP(\mu)$$

is a spectral decomposition of M. If M can be measured exactly, $P(\mu)$ for each μ can obviously be measured exactly. Since the projection operator $P(\mu)$ has a discrete eigenvalue 1 or 0, the above proof tells us that $P(\mu)$ for each μ commutes with L_1, which in turn implies (2.7).

3. POSSIBILITY OF AN APPROXIMATE MEASUREMENT

In this section we will discuss the problem of whether the operator M, which does not commute with the conserved operator L_1 of the preceding section, can be measured approximately. We will prove that this is possible if L has a discrete spectrum and L_1 has only a finite number of eigenvalues.

We may assume that the eigenvalues of L_1 are[5] $0, \pm1, \pm2, \cdots \pm l$. We decompose L's into projection operators

$$L=\sum_\lambda \lambda P(\lambda), \quad (3.1a)$$
$$L_i=\sum_\lambda \lambda P_i(\lambda), \quad i=1, 2. \quad (3.1b)$$

The additivity, (2.5), implies

$$P(\lambda)=\sum_{|\lambda'|\leq l} P_1(\lambda')P_2(\lambda-\lambda'). \quad (3.2)$$

As a first step of our proof, we state the following Lemma which will be proved at the end of this section.

Lemma. Given two sets of vectors $\Psi_\alpha{}^i$ and $\Psi_\alpha{}^f$ in a Hilbert space $\mathfrak{H}=\mathfrak{H}_1\otimes\mathfrak{H}_2$ satisfying

$$(\Psi_\alpha{}^i,P(\lambda)\Psi_\beta{}^i)=(\Psi_\alpha{}^f,P(\lambda)\Psi_\beta{}^f), \quad (3.3)$$

for every λ, then there exists a Hilbert space $\mathfrak{H}_2'$ con-

---

[4] If L_2 is unbounded, the above proof does not exclude the possibility that one can measure M, even if M does not commute with L_1, by a measuring apparatus (ξ or $X_{\mu\rho\rho'}$) in a state which is outside the domain of L, because (2.8) would then be meaningless.

However, even if L_2 is unbounded, as long as L_1 is bounded, we can modify the above argument in the following way. We introduce unitary operators

$$V(S)=\exp(iLS); \quad V_j(S)=\exp(iL_jS), \quad j=1, 2. \quad (i)$$

Because of the additivity, (2.5),

$$V(S)=V_1(S)\otimes V_2(S). \quad (ii)$$

Then, by the conservation law (2.6), we have

$$(\phi_{\mu'\rho'}\otimes\xi, V(S)(\phi_{\mu\rho}\otimes\xi))$$
$$=(\sum_{\rho'''}\phi_{\mu'\rho'''}\otimes X_{\mu'\rho'\rho'''}, V(S)(\sum_{\rho''}\phi_{\mu\rho''}\otimes X_{\mu\rho\rho''})). \quad (iii)$$

By the orthogonality, (2.2),

$$(\phi_{\mu'\rho'}\otimes\xi, V_2(S)(\phi_{\mu\rho}\otimes\xi))$$
$$=(\sum_{\rho'''}\phi_{\mu'\rho'''}\otimes X_{\mu'\rho'\rho'''}, V_2(S)(\sum_{\rho''}\phi_{\mu\rho''}\otimes X_{\mu\rho\rho''})) \quad (iv)$$
$$=0 \quad (\text{for } \mu\neq\mu')$$

Combining the two equations above and using the additivity, (ii), we obtain for $\mu\neq\mu'$,

$$(\phi_{\mu'\rho'}\otimes\xi, F(S)(\phi_{\mu\rho}\otimes\xi))$$
$$=(\sum_{\rho'''}\phi_{\mu'\rho'''}\otimes X_{\mu'\rho'\rho'''}, F(S)(\sum_{\rho''}\phi_{\mu\rho''}\otimes X_{\mu\rho\rho''})), \quad (v)$$

where

$$F(S)=(1/iS)[V_1(S)-1]\otimes V_2(S). \quad (vi)$$

Since $F(S)\rightarrow L_1$, as $S\rightarrow 0$, we obtain

$$(\phi_{\mu'\rho'}\otimes\xi, L_1(\phi_{\mu\rho}\otimes\xi))$$
$$=(\sum_{\rho'''}\phi_{\mu'\rho'''}\otimes X_{\mu'\rho'\rho'''}, L_1(\sum_{\rho''}\phi_{\mu\rho''}\otimes X_{\mu\rho\rho''})). \quad (vii)$$

Because of the orthogonality, (2.4), we finally obtain (2.9a) from which we conclude (2.11) as before.

[5] The proof holds without this specification but notations become complicated, especially in dividing various regions of values of λ.

taining $\mathfrak{H}_2$ and a unitary operator U on $\mathfrak{H}'=\mathfrak{H}_1\otimes\mathfrak{H}_2'$ such that (1) a self-adjoint operator L_2' (representing the conserved quantity in $\mathfrak{H}_2'$) is defined on $\mathfrak{H}_2'$ coinciding with L_2 on $\mathfrak{H}_2$, (2) U commutes with the conserved quantity L' on $\mathfrak{H}'$, $L'=L_1\otimes 1+1\otimes L_2'$, and

$$\text{(3)}\quad \Psi_\alpha{}^f=U\Psi_\alpha{}^i. \tag{3.4}$$

If the set of the indices α is finite, $\mathfrak{H}_2'$ can be taken to be $\mathfrak{H}_2$.

This Lemma is used in the following way. We will construct states ξ, $X_{\mu\rho}$, ψ, and $\eta_{\mu\rho}$ satisfying

$$(X_{\mu\rho},X_{\mu'\rho'})=0,\quad \text{if}\quad \mu\neq\mu', \tag{3.5}$$

$$(X_{\mu\rho},\eta_{\mu'\rho'})=0,\quad \text{for any } \mu,\ \rho,\ \mu',\ \rho', \tag{3.6}$$

$$\|\psi\otimes\eta_{\mu\rho}\|^2<\epsilon,$$

$$(\eta_{\mu\rho},\eta_{\mu'\rho'})=0,\quad \text{for } (\mu,\rho)\neq(\mu',\rho') \tag{3.7}$$

in such a way that the two sets of vectors

$$\Psi_\alpha{}^i=\phi_{\mu\rho}\otimes\xi,\quad \Psi_\alpha{}^f=\phi_{\mu\rho}\otimes X_{\mu\rho}+\psi\otimes\eta_{\mu\rho},\quad \alpha=(\mu,\rho), \tag{3.8}$$

fulfill (3.3). If we succeed in constructing such states, then by the Lemma, there exists a unitary operator U in $\mathfrak{H}'$ which conserves L' and for which (3.4) holds. This implies, for any normalized state ϕ in $\mathfrak{H}_1$,

$$U(\phi\otimes\xi)=\sum_{\mu\rho}(\phi_{\mu\rho},\phi)(\phi_{\mu\rho}\otimes X_{\mu\rho})+\eta(\phi), \tag{3.9}$$

$$\eta(\phi)=\sum_{\mu\rho}(\phi_{\mu\rho},\phi)(\psi\otimes\eta_{\mu\rho}), \tag{3.10}$$

and, due to (3.7),

$$\|\eta(\phi)\|^2<\epsilon. \tag{3.11}$$

Thus if we choose the setup of a measurement in such a way that the Hilbert space of the measuring instrument is $\mathfrak{H}_2'$, the initial state of the instrument is ξ, and the time development of the combined system of the measured object and the measuring apparatus in a certain time interval t is described by $U(t)=U$, then we can measure the operator M in terms of the states $X_{\mu\rho}$ of the measuring apparatus after the measurement within the inaccuracy representing by $\eta(\phi)$. This inaccuracy can be made as small as one desires by making ϵ small enough. Because we are only concerned with the effect of the conservation law of L, we have assumed in the above argument that, if U is a unitary operator commuting with the conserved quantity, then there always exists an experimental setup whose time development in a certain time period is described by U. There may be many other conditions on U in addition to that it commutes with L. Hence, our argument does not assure that a system exists whose Hamiltonian leads to U.

We now give an explicit construction of states ξ, $X_{\mu\rho}$, ψ and $\eta_{\mu\rho}$. For this purpose we denote by $\mathfrak{H}_{2\lambda}$ the subspace of $\mathfrak{H}_2$ which is spanned by eigenvectors of L_2 with an eigenvalue λ. ψ is taken to be a normalized eigenstate of L_1 with the eigenvalue 0,

$$L_1\psi=0,\quad (\psi,\psi)=1. \tag{3.12}$$

ξ, $X_{\mu\rho}$, and $\eta_{\mu\rho}$ are given by

$$\xi=\sum\nolimits_\lambda \xi_\lambda,\quad X_{\mu\rho}=\sum\nolimits_\lambda X_{\mu\rho\lambda},\quad \eta_{\mu\rho}=\sum\nolimits_\lambda \eta_{\mu\rho\lambda}, \tag{3.13}$$

where ξ_λ, $X_{\mu\rho\lambda}$, and $\eta_{\mu\rho\lambda}$ are vectors in $\mathfrak{H}_{2\lambda}$ to be specified below.

ξ_λ is any state in $\mathfrak{H}_{2\lambda}$ with the norm given by

$$(\xi_\lambda,\xi_\lambda)=0,\qquad \text{for}\quad |\lambda|>N, \tag{3.14a}$$

$$=(2N+1)^{-1},\quad \text{for}\quad |\lambda|\leq N. \tag{3.14b}$$

N is any integer satisfying

$$N>2l/\epsilon-\tfrac{1}{2}. \tag{3.15}$$

The $X_{\mu\rho\lambda}$ are any states in $\mathfrak{H}_{2\lambda}$, orthogonal to each other and with the norm given by

$$(X_{\mu\rho\lambda},X_{\mu'\rho'\lambda})=0,\qquad \text{for}\quad |\lambda|>N-2l, \tag{3.16a}$$

$$=(2N+1)^{-1}\delta_{\mu\mu'}\delta_{\rho\rho'},\quad \text{for}\quad |\lambda|\leq N-2l. \tag{3.16b}$$

The orthogonal complement of the set $\{X_{\mu\rho\lambda}|\mu,\rho$ varying$\}$ in $\mathfrak{H}_{2\lambda}$ will be denoted by $\mathfrak{H}_{2\lambda}{}^\eta$.

$\eta_{\mu\rho\lambda}$ are taken from $\mathfrak{H}_{2\lambda}{}^\eta$ and defined in the following way

(I) For $|\lambda|>N+l$ or $|\lambda|\leq N-3l$,

$$\eta_{\mu\rho\lambda}=0. \tag{3.17a}$$

(II) For $N+l\geq|\lambda|>N-l$,[6]

$$(\eta_{\mu\rho\lambda},\eta_{\mu'\rho'\lambda})=(2N+1)^{-1}\sum_{\substack{|\lambda'|\leq l\\|\lambda-\lambda'|\leq N}}(\phi_{\mu\rho},P_1(\lambda')\phi_{\mu'\rho'}). \tag{3.17b}$$

(III) For $N-l\geq|\lambda|>N-3l$, $\eta_{\mu\rho\lambda}$ are any states in $\mathfrak{H}_{2\lambda}{}^\eta$ orthogonal to each other and with the norm given by

$$(\eta_{\mu\rho\lambda},\eta_{\mu'\rho'\lambda})=(2N+1)^{-1}(\phi_{\mu\rho},Q_\lambda\phi_{\mu\rho})\delta_{\mu\mu'}\delta_{\rho\rho'}, \tag{3.17c}$$

where Q_λ is a projection operator given by

$$Q_\lambda=\sum_{\substack{|\lambda'|\leq l\\|\lambda-\lambda'|>N-2l}}P_1(\lambda'). \tag{3.18}$$

Note that $(\phi_{\mu\rho},Q_\lambda\phi_{\mu\rho})$ is non-negative (between 0 and 1).

We now show that ξ, $X_{\mu\rho}$, ψ, and $\eta_{\mu\rho}$ thus constructed have the desired properties. ξ is normalized due to (3.14). (3.5) and (3.6) are trivially satisfied by our

[6] This means that $\eta_{\mu\rho\lambda}$ is defined by

$$\eta_{\mu\rho\lambda}=(2N+1)^{-\frac{1}{2}}\sum_{\substack{|\lambda'|\leq l\\|\lambda-\lambda'|\leq N}}P_1{}^{(\lambda)}(\lambda')\phi_{\mu\rho}{}^{(\lambda)}$$

where we have made an isometric linear mapping of $\mathfrak{H}_1$ into $\mathfrak{H}_{2\lambda}{}^\eta$ and $\phi_{\mu\rho}$ and $P_1(\lambda')$ thus mapped are called $\phi_{\mu\rho}{}^{(\lambda)}$ and $P_1{}^{(\lambda)}(\lambda')$.

choice. To prove (3.3), we rewrite (3.3) using (3.2):

$$\sum_{|\lambda'|\le l} (\phi_{\mu\rho},P_1(\lambda')\phi_{\mu'\rho'})(\xi, P_2(\lambda-\lambda')\xi)$$

$$= \sum_{|\lambda'|\le l} (\phi_{\mu\rho},P_1(\lambda')\phi_{\mu'\rho'})(X_{\mu\rho}, P_2(\lambda-\lambda')X_{\mu'\rho'}) + (\eta_{\mu\rho},P_2(\lambda)\eta_{\mu'\rho'}), \quad (3.19)$$

where we have also used (3.12). By (3.13), this is equivalent to

$$\sum_{|\lambda'|\le l} (\phi_{\mu\rho},P_1(\lambda')\phi_{\mu'\rho'}) \times[\|\xi_{\lambda-\lambda'}\|^2-(X_{\mu\rho,\lambda-\lambda'},X_{\mu'\rho',\lambda-\lambda'})] = (\eta_{\mu\rho\lambda},\eta_{\mu'\rho'\lambda}). \quad (3.20)$$

We divide the range of λ into 4 parts and prove (3.20) separately for λ in each of these 4 regions.

(I) If $|\lambda|>N+l$, then $|\lambda-\lambda'|>N$ and (3.20) is trivially satisfied because all terms vanish.

(II) If $N+l\ge|\lambda|>N-l$, then $|\lambda-\lambda'|>N-2l$, and hence the term containing X still vanishes. Due to (3.14b), the left-hand side of (3.20) becomes

$$\sum_{|\lambda'|\le l} (\phi_{\mu\rho},P_1(\lambda')\phi_{\mu'\rho'})\|\xi_{\lambda-\lambda'}\|^2 = (2N+1)^{-1} \sum_{\substack{|\lambda'|\le l\\ |\lambda-\lambda'|\le N}} (\phi_{\mu\rho},P_1(\lambda')\phi_{\mu'\rho'}),$$

which is equal to the right-hand side of (3.20) calculated by (3.17b).

(III) If $N-l\ge|\lambda|>N-3l$, then $|\lambda-\lambda'|\le N$ and hence $\|\xi_{\lambda-\lambda'}\|^2$ is always $(2N+1)^{-1}$. By the orthogonality, (2.2), the definition (3.16b) and the equation

$$\sum_{|\lambda'|\le l} P_1(\lambda)=1, \quad (3.21)$$

the left-hand side of (3.20) becomes

$$\sum_{|\lambda'|\le l} (\phi_{\mu\rho},P_1(\lambda')\phi_{\mu'\rho'})[(2N+1)^{-1}-(X_{\mu\rho,\lambda-\lambda'},X_{\mu'\rho',\lambda-\lambda'})]$$

$$=(2N+1)^{-1}\delta_{\mu\mu'}\delta_{\rho\rho'} \sum_{|\lambda'|\le l} (\phi_{\mu\rho},P_1(\lambda')\phi_{\mu\rho}) \times[1-(2N+1)(X_{\mu\rho,\lambda-\lambda'},X_{\mu'\rho',\lambda-\lambda'})]. \quad (3.22)$$

Because of (3.16b), the inside of the square bracket of (3.22) vanishes for $|\lambda-\lambda'|\le N-2l$ and is unity for $|\lambda-\lambda'|>N-2l$. Thus, due to (3.17c) and (3.18), (3.22) is equal to the right-hand side of (3.20).

(IV) If $N-3l\ge|\lambda|$, then $|\lambda-\lambda'|\le N-2l$ and the left-hand side of (3.20) becomes

$$\sum_{|\lambda'|\le l} (\phi_{\mu\rho},P_1(\lambda)\phi_{\mu'\rho'})(1-\delta_{\mu\mu'}\delta_{\rho\rho'}).$$

Because of (3.21) and the orthogonality, (2.2), this expression vanishes and hence is equal to the right-hand side of (3.20) which also vanishes due to (3.17a). This completes the proof of (3.4).

Finally, we will prove (3.7). Since ψ is normalized, $\|\psi\otimes\eta_{\mu\rho}\|$ is equal to $\|\eta_{\mu\rho}\|$. By (3.13), we have

$$(\eta_{\mu\rho},\eta_{\mu'\rho'})=\sum_\lambda (\eta_{\mu\rho\lambda},\eta_{\mu'\rho'\lambda}).$$

By (3.20), we get

$$\sum_\lambda (\eta_{\mu\rho\lambda},\eta_{\mu\rho'\lambda'})$$

$$= \sum_{\substack{\lambda\lambda'\\|\lambda'|\le l}} (\phi_{\mu\rho},P_1(\lambda')\phi_{\mu'\rho'})[\|\xi_{\lambda-\lambda'}\|^2-\|X_{\mu\rho,\lambda-\lambda'}\|^2\delta_{\mu\mu'}\delta_{\rho\rho'}]$$

$$= \sum_{\substack{\lambda\lambda'\\|\lambda'|\le l}} (\phi_{\mu\rho},P_1(\lambda')\phi_{\mu'\rho'})[\|\xi_{\lambda}\|^2-\|X_{\mu\rho,\lambda}\|^2\delta_{\mu\mu'}\delta_{\rho\rho'}].$$

By (3.21) and (2.2),

$$\sum_{|\lambda'|\le l} (\phi_{\mu\rho},P_1(\lambda')\phi_{\mu'\rho'})=\delta_{\mu\mu'\rho\rho'}.$$

By (3.14) and (3.16)

$$\sum_\lambda [\|\xi_\lambda\|^2-\|X_{\mu\rho\lambda}\|^2]=4l(2N+1)^{-1}.$$

Combining these, and using (3.15), we obtain

$$\|\psi\otimes\eta_{\lambda\rho}\|^2=\frac{4l}{2N+1}<\epsilon.$$

$$(\eta_{\mu\rho},\eta_{\mu'\rho'})=0, \quad \text{for } (\mu,\rho)\neq(\mu',\rho')$$

In the above construction, $\mathfrak{H}_{2\lambda}$ for $|\lambda|\le N-3l$ should have at least the dimension of $\mathfrak{H}_1$. We need higher dimension for $\mathfrak{H}_{2\lambda}$ with $N-3l<|\lambda|\le N$.[7]

Finally we will give a proof of our Lemma. For this purpose, we denote the subspace of $\mathfrak{H}$ spanned by eigenvectors of L with eigenvalue λ by $\mathfrak{H}_\lambda$, the subspace spanned by $P(\lambda)\Psi_\alpha{}^i$ with varying α by $\mathfrak{H}_\lambda{}^i$, the subspace spanned by $P(\lambda)\Psi_\alpha{}^f$ with varying α by $\mathfrak{H}_\lambda{}^f$, the orthogonal complement of $\mathfrak{H}_\lambda{}^i$ in $\mathfrak{H}_\lambda$ by $\mathfrak{H}_\lambda{}^i{}_\perp$, and the orthogonal complement of $\mathfrak{H}_\lambda{}^f$ in $\mathfrak{H}_\lambda$ by $\mathfrak{H}_\lambda{}^f{}_\perp$. Obviously

$$\mathfrak{H}=\oplus_\lambda(\mathfrak{H}_\lambda{}^i\oplus\mathfrak{H}_\lambda{}^i{}_\perp)=\oplus_\lambda(\mathfrak{H}_\lambda{}^f\oplus\mathfrak{H}_\lambda{}^f{}_\perp). \quad (3.23)$$

We will first show that

$$U_\lambda(\sum_\alpha C_\alpha P(\lambda)\Psi_\alpha{}^i)=\sum_\alpha C_\alpha P(\lambda)\Psi_\alpha{}^f, \quad (3.24)$$

defines a unitary mapping U_λ of $\mathfrak{H}_\lambda{}^i$ onto $\mathfrak{H}_\lambda{}^f$, where $\{C_\alpha\}$ is a set of arbitrary complex numbers. To see this, we note that, due to (3.3), $\sum_\alpha C_\alpha P(\lambda)\Psi_\alpha{}^i$ and

[7] In the above construction, the measuring apparatus is a superposition of eigenstates of L_2 with different eigenvalues λ varying over the range of the order $1/\epsilon$. However, if one counts the number of equations to be satisfied, one finds a possibility of constructing a similar measuring apparatus which is a superposition of eigenstates of L_2 with eigenvalues, near a certain large value of the order $1/\epsilon$, but varying only over the range of the order of the dimension of $\mathfrak{H}_1$, provided that the latter is finite. Here we will not pursue the problem of such minimization, but we will only note that, if we do minimize the number of eigenvalues of L_2 to be used in the measuring apparatus, then $X_{\mu\rho}$ will be nearly strictly determined and if that is the case, there is a fair chance that $X_{\mu\rho}$ cannot be made macroscopically distinguishable any better than $\phi_{\mu\rho}$.

$\sum_\alpha C_\alpha P(\lambda)\Psi_\alpha^f$ converge, diverge, or vanish simultaneously. Hence, U_λ is a one-to-one mapping of $\mathfrak{H}_\lambda^i$ onto $\mathfrak{H}_\lambda^f$. Since this mapping is linear and, due to (3.3), isometric, U_λ is a unitary mapping of $\mathfrak{H}_\lambda^i$ onto $\mathfrak{H}_\lambda^f$ as was to be proved. This also proves that the dimensions of $\mathfrak{H}_\lambda^i$ and $\mathfrak{H}_\lambda^f$ are the same.

If this dimension is finite, the dimensions of $\mathfrak{H}_\lambda{}^i{}_\perp$ and $\mathfrak{H}_\lambda{}^f{}_\perp$ are the same. Then there always exists a unitary mapping $U_{\lambda\perp}$ of $\mathfrak{H}_\lambda{}^i{}_\perp$ onto $\mathfrak{H}_\lambda{}^f{}_\perp$. Now we define an operator U in $\mathfrak{H}$.

$$U=\oplus_\lambda(U_\lambda\oplus U_{\lambda\perp}). \qquad (3.25)$$

Because of the unitarity of U_λ and $U_{\lambda\perp}$ and the decomposition, (3.23), U is obviously unitary. For any $\Psi\epsilon\mathfrak{H}$,

$$U\Psi=\sum_\lambda (U_\lambda\Psi_\lambda{}^i+U_{\lambda\perp}\Psi_\lambda{}^i{}_\perp), \qquad (3.26)$$

where

$$\Psi=\sum_\lambda(\Psi_\lambda{}^i+\Psi_\lambda{}^i{}_\perp),\ \Psi_\lambda{}^i\epsilon\mathfrak{H}_\lambda{}^i,\ \Psi_\lambda{}^i{}_\perp\epsilon\mathfrak{H}_\lambda{}^i{}_\perp, \qquad (3.27)$$

is a unique decomposition of Ψ according to the first equation of (3.23). Since the subspace $\mathfrak{H}_\lambda$ of $\mathfrak{H}$ spanned by eigenvectors of $L=L_1\otimes 1+1\otimes L_2$ with the eigenvalue λ is mapped onto itself by U, U commutes with L. This completes the proof for the case where the dimension of $\mathfrak{H}_\lambda^i$ and $\mathfrak{H}_\lambda^f$ is finite.

If this dimension is infinite, then the dimensions of $\mathfrak{H}_\lambda{}^i{}_\perp$ and $\mathfrak{H}_\lambda{}^f{}_\perp$ can be different. In such a case we introduce a new Hilbert space $\mathfrak{H}_2{}^r$ (on which the conserved quantity $L_2{}^r$ is defined) in such a way that the dimension of $\mathfrak{H}_\lambda{}^r$ is at least the number of indices α where $\mathfrak{H}_\lambda{}^r$ is the subspace of $\mathfrak{H}^r\equiv\mathfrak{H}_1\otimes\mathfrak{H}_2{}^r$ spanned by the eigenstates of $L^r=L_1\otimes 1+1\otimes L_2{}^r$ with eigenvalues λ. Then since the dimension of $\mathfrak{H}_\lambda^i$ and $\mathfrak{H}_\lambda^f$ does not exceed the cardinal number of the set of the indices α, $\mathfrak{H}_\lambda{}^{ir}\equiv\mathfrak{H}_\lambda{}^i{}_\perp\oplus\mathfrak{H}_\lambda{}^r$ and $\mathfrak{H}_\lambda{}^{fr}\equiv\mathfrak{H}_\lambda{}^f{}_\perp\oplus\mathfrak{H}_\lambda{}^r$ have the same dimension. Hence, there always exists a unitary mapping $U_{\lambda\perp}$ of $\mathfrak{H}_\lambda{}^{ir}$ onto $\mathfrak{H}_\lambda{}^{fr}$.

We are now in the position to construct the Hilbert space $\mathfrak{H}_2{}'$ and the unitary operator U for this case. $\mathfrak{H}_2{}'$ is taken to be $\mathfrak{H}_2\oplus\mathfrak{H}_2{}^r$. $L_2{}'$ on $\mathfrak{H}_2{}'$ is taken to be $L_2\oplus L_2{}^r$. $\mathfrak{H}'$ can be decomposed as

$$\mathfrak{H}'=\oplus_\lambda(\mathfrak{H}_\lambda{}^i\oplus\mathfrak{H}_\lambda{}^{ir})=\oplus_\lambda(\mathfrak{H}_\lambda{}^f\oplus\mathfrak{H}_\lambda{}^{fr}). \qquad (3.28)$$

U is defined as unitary mapping

$$U=\oplus_\lambda(U_\lambda\oplus U_{\lambda\perp}). \qquad (3.29)$$

Instead of (3.26), (3.27), we have, for any $\Psi\epsilon\mathfrak{H}'$

$$U\Psi=\sum_\lambda (U_\lambda\Psi_\lambda{}^i+U_{\lambda\perp}\Psi_\lambda{}^{ir}), \qquad (3.30)$$

$$\Psi=\sum_\lambda (\Psi_\lambda{}^i+\Psi_\lambda{}^{ir}),\ \Psi_\lambda{}^i\epsilon\mathfrak{H}_\lambda{}^i,\ \Psi_\lambda{}^{ir}\epsilon\mathfrak{H}_\lambda{}^{ir}. \qquad (3.31)$$

Then by the same argument as in the previous case, we can show the unitarity of U, and commutativity with L', where L' is defined as $L'\equiv L_1\otimes 1+1\otimes L_2{}'$.

We note that in our application of the Lemma, the number of the indices α is the same as the dimension of $\mathfrak{H}_1$.

ACKNOWLEDGMENT

The authors are very much indebted to Professor E. P. Wigner for many helpful comments.

One of the authors (M. M. Y.) wishes to express his sincere gratitude to the Physics Department of Princeton University for its hospitality.

VI.3 OPTIMAL MEASURING APPARATUS

MICHAEL YANASE

An upper limit for the accuracy of the measurement of a simple quantity which does not commute with a conserved quantity is obtained in terms of the "size" of the apparatus. The "size" of the apparatus is defined as the mean square value $\hbar^2 M^2$ of the conserved quantity for the apparatus which is, in the example chosen, the z component of the angular momentum. The measured quantity is the projection of a spin in a perpendicular direction. It is found that the probability of an unsuccessful measurement is at least $1/8M^2$.

1. INTRODUCTION

IT was shown recently that a quantum mechanical operator which does not commute with the operator of a conserved quantity can be measured only approximately. There is a finite probability that the measurement is unsuccessful, but this probability can be very small if the measuring apparatus contains a large amount of the conserved quantity.[1] It was shown, in particular, that if the product of the probability of an unsuccessful measurement and of the maximum value of the conserved quantity which is present in the measuring apparatus exceeds a certain value, no contradiction with the conservation law occurs. The objective of the present article is to find the "best" measuring apparatus for a given "size." The conserved quantity will not have an upper limit in the initial state of this; rather, we specify the mean-square value of the conserved quantity and ask for the minimum probability for an unsuccessful measurement, consistent with the prescribed mean square of the conserved quantity and, of course, the validity of the conservation law.

We require that the operator of the conserved quantity for the apparatus commute with the operator, by which the final state of the apparatus is measured. This condition is necessary because otherwise—as a consequence of the result of our previous work—we cannot ascertain the result of the measurement exactly.

The condition to be obtained will be only a necessary one. In other words, for the given mean square of the conserved quantity, the probability of a malfunctioning of the apparatus cannot be smaller than the value to be calculated. Whether an apparatus with the specified properties is actually possible will not be decided. All that can be claimed is that the existence of such an apparatus is not in conflict with the conservation law considered.

The quantity to be measured and the conservation law to be considered will be the same as in the first publication on this subject: The quantity is the component of the spin of a particle in a given direction; the conserved quantity, the angular momentum about a direction perpendicular to the aforementioned direction. It will be shown in Sec. 2 that the minimum probability for the malfunctioning of the apparatus is inversely proportional to the mean square of the conserved quantity, and the proportionality constant will be determined. Section 3 will contain a discussion of the results.

2. MINIMIZATION OF THE MALFUNCTIONING PROBABILITY

We measure the x component of the spin of a particle with spin $\frac{1}{2}$; the z component of the spin of this particle is the additive conserved quantity. To make the compu-

* On leave of absence from Sophia University, Tokyo, Japan.

[1] E. P. Wigner, Z. Physik 133, 101 (1952); H. Araki and M. M. Yanase, Phys. Rev. 120, 622 (1960).

Originally published in *Physical Review, 123*, 666-68 (1961).

tation simpler, we choose the eigenvalues of the z component as 0 and 1 and denote the corresponding eigenstates ψ_0 and ψ_1, respectively. Then eigenstates of the x component are $2^{-\frac{1}{2}}(\psi_0+\psi_1)$ and $2^{-\frac{1}{2}}(\psi_0-\psi_1)$. The state of the measuring apparatus before the measurement will be denoted by ξ. The measurement will result in a unitary transformation U in the Hilbert space of the combined system of the object and the measuring apparatus:

$$U[2^{-\frac{1}{2}}(\psi_0+\psi_1)\xi]=2^{-\frac{1}{2}}(\psi_0+\psi_1)X+2^{-\frac{1}{2}}(\psi_0-\psi_1)\eta, \quad (2.1)$$

$$U[2^{-\frac{1}{2}}(\psi_0-\psi_1)\xi]=2^{-\frac{1}{2}}(\psi_0-\psi_1)X'+2^{-\frac{1}{2}}(\psi_0+\psi_1)\eta', \quad (2.2)$$

where X, X', η, η' are the states of the measuring apparatus after the measurement.

From (2.1) and (2.2) we have

$$U(\psi_0\xi)=\psi_0 u+\psi_1 v, \quad (2.3)$$

$$U(\psi_1\xi)=\psi_0 w+\psi_1 z, \quad (2.4)$$

where

$$\begin{aligned} u&=\tfrac{1}{2}(X+X'+\eta+\eta'),\\ v&=\tfrac{1}{2}(X-X'-\eta+\eta'),\\ w&=\tfrac{1}{2}(X-X'+\eta-\eta'),\\ z&=\tfrac{1}{2}(X+X'-\eta-\eta'). \end{aligned} \quad (2.4a)$$

It follows from the conservation law for the z component of the angular momentum that

$$U(\psi_0\xi_\mu)=\psi_0 u_\mu+\psi_1 v_{\mu-1}, \quad (2.5)$$

$$U(\psi_1\xi_{\mu-1})=\psi_0 w_\mu+\psi_1 z_{\mu-1}, \quad (2.6)$$

where we decomposed ξ, u, v, w, z, into $\xi=\sum \xi_\mu$, $u=\sum u_\mu$, $v=\sum v_\mu$, $w=\sum w_\mu$, $z=\sum z_\mu$, the index μ being the eigenvalues of the z component of the angular momentum of the apparatus, and the ξ_μ, u_μ, etc., corresponding eigenstates. We normalize

$$(\xi,\xi)=1. \quad (2.7)$$

After the measurement, the apparatus is separated from the object, and subject to the second measurement to distinguish the states corresponding to the states of the measured object. For this second measurement we require the following conditions. First, the two states X and X' should be the two orthogonal eigenstates of the operator N of the second measurement, i.e.,

$$(X,X')=0. \quad (2.8)$$

Secondly, N should commute with the operator L_z of the conserved quantity for the apparatus, otherwise we cannot measure N exactly,[1] i.e.,

$$[N,L_z]=0. \quad (2.9)$$

This leads to the relation

$$(X_\mu,X_\mu')=0, \quad (2.10)$$

the X_μ being eigenfunctions of both N and L_z. Because of (2.4a) this can be written also as

$$(u_\mu+z_\mu,\ u_\mu+z_\mu)=(v_\mu+w_\mu,\ v_\mu+w_\mu). \quad (2.10a)$$

In an ideal measurement, both η and η' would be zero. It is not possible to accomplish this, but we shall try to find the minimum ϵ of the probability that the measurement was unsuccessful, i.e., the minimum of

$$\begin{aligned} \epsilon&=(\eta,\eta)+(\eta',\eta')\\ &=\tfrac{1}{2}[(u-z)^2+(v-w)^2]\\ &=\tfrac{1}{2}[(\textstyle\sum u_\mu-\sum z_\mu)^2+(\sum v_\mu-\sum w_\mu)^2], \end{aligned} \quad (2.11)$$

consistent with a definite "size" of the measuring apparatus, to be defined below.

The conservation law, together with (2.5), (2.6), (2.7), and the unitary nature of U gives

$$\begin{aligned} (\xi_\mu,\xi_\mu)&=(u_\mu,u_\mu)+(v_{\mu-1},v_{\mu-1})\\ &=(w_{\mu+1},w_{\mu+1})+(z_\mu,z_\mu); \end{aligned} \quad (2.12)$$

$$\begin{aligned} 1&=(u,u)+(v,v)=\textstyle\sum (u_\mu,u_\mu)+\sum (v_\mu,v_\mu)\\ &=(w,w)+(z,z)=\textstyle\sum (w_\mu,w_\mu)+\sum (z_\mu,z_\mu); \end{aligned} \quad (2.13)$$

and

$$(u_\mu,w_\mu)+(v_{\mu-1},z_{\mu-1})=0. \quad (2.14)$$

We define the "size" of the measuring apparatus as the mean square of the additive conserved quantity which we denote by M^2

$$(\xi,L_z^2\xi)=\sum \mu^2\xi_\mu^2=M^2. \quad (2.15)$$

The problem is, therefore, to obtain the smallest value of ϵ consistent with Eqs. (2.12) to (2.15), the ξ_μ, u_μ, v_μ, w_μ, z_μ, being otherwise arbitrary eigenstates of L_z to the eigenvalue μ. The ξ_μ can be eliminated from (2.13) by means of (2.10) and one obtains

$$\sum \mu^2(u_\mu^2+v_{\mu-1}^2)=\sum \mu^2(w_{\mu-1}^2+z_\mu^2)=M^2. \quad (2.16)$$

We note that (2.14) can be always satisfied because no equation depends on the angles between the u_μ and w_μ, or the v_μ and z_μ, except (2.14). It is easy to see that, in order to obtain the smallest possible ϵ at given $\|u_\mu\|$, $\|v_\mu\|$, $\|w_\mu\|$, $\|z_\mu\|$, the Hilbert vectors u_μ and z_μ must be parallel for all μ and the same applies for the vectors v_μ and w_μ.

To solve the preceding equations, we note that, because of (2.10a) and (2.12),

$$u_\mu+z_\mu=v_\mu+w_\mu, \quad (2.17)$$

$$u_\mu-z_\mu=(w_{\mu+1}^2-v_{\mu-1}^2)/(v_\mu+w_\mu), \quad (2.18)$$

where the u_μ, z_μ are now the *lengths* of the corresponding vectors (which are parallel), and the same applies to v_μ and z_μ.

We introduce new variables s_μ, t_μ such that

$$w_\mu+v_\mu=s_\mu, \quad (2.19)$$

$$w_\mu-v_\mu=t_\mu. \quad (2.20)$$

If ϵ is small, M is large and the s_μ, t_μ can be considered to be continuous functions of μ. It can be verified from the solutions to be obtained that

$$s_\mu = O(M^{-\frac{1}{2}}), \quad \partial s_\mu/\partial\mu = \dot{s}_\mu = O(M^{-\frac{3}{2}}), \quad \text{etc.}; \tag{2.21}$$

$$t_\mu = O(M^{-\frac{3}{2}}), \quad \partial t_\mu/\partial\mu = \dot{t}_\mu = O(M^{-\frac{5}{2}}), \quad \text{etc.} \tag{2.22}$$

Using these estimates, we obtain from (2.18), (2.19), (2.20)

$$u_\mu - z_\mu = \dot{s}_\mu + t_\mu + O(M^{-7/2}). \tag{2.23}$$

whereas (2.11) becomes

$$\begin{aligned}\epsilon &= \tfrac{1}{2}\{\sum [\dot{s}_\mu + t_\mu + O(M^{-7/2})]^2 + \sum t_\mu^2\} \\ &= \tfrac{1}{2}\sum [\dot{s}_\mu^2 + 2t_\mu^2 + 2\dot{s}_\mu t_\mu + O(M^{-5})].\end{aligned} \tag{2.24}$$

The s and t are still subject to the conditions (2.13) and (2.15) which now read

$$\tfrac{1}{2}\sum [s_\mu^2 + O(M^{-3})] = 1, \tag{2.25}$$

$$\tfrac{1}{2}\sum \mu^2[s_\mu^2 + O(M^{-3})] = M^2. \tag{2.26}$$

Since the t do not occur in these equations, the derivative of (2.24) with respect to t_μ will be zero at the minimum for ϵ:

$$2t_\mu = -\dot{s}_\mu + O(M^{-7/2}). \tag{2.27}$$

Hence ϵ becomes

$$\epsilon = \sum [\tfrac{1}{4}\dot{s}_\mu^2 + O(M^{-5})]. \tag{2.28}$$

Using $-\lambda$ and $2m^2$ as the Lagrange multipliers for (2.25) and (2.26), we obtain the Euler equation for s_μ

$$-\ddot{s}_\mu + (4m^2\mu^2 - 2\lambda)s_\mu = 0, \tag{2.29}$$

where terms of lower order of magnitude have been omitted. The solution of (2.29) is

$$s_\mu = c\exp(-m\mu^2); \quad c^2 = 2(m/\pi)^{\frac{1}{2}}. \tag{2.30}$$

This already satisfies the normalization condition (2.25). The condition (2.26) gives

$$M^2 = 1/4m, \tag{2.31}$$

and this gives, finally, for the probability (2.28) of an unsuccessful measurement

$$\epsilon = 1/8M^2, \tag{2.32}$$

for a given M. Tracing through the omitted low-order terms, one finds that (2.32) is accurate up to terms of the order $1/M^3$.

3. CONCLUSIONS AND DISCUSSIONS

(1) The relation (2.32) tells us that the probability of an unsuccessful measurement is $1/8M^2$, if one uses the "best possible" apparatus with given M^2. Therefore, neglecting terms of the order $1/M^3$, the inequality

$$\epsilon > 1/8M^2 \tag{3.1}$$

holds. In other words, with given M^2, we cannot make the probability of unsuccessful measurement smaller than $\hbar^2/8M^2$ (in dimensionless units). This limitation of the measurement is not the consequence of an uncertainty relation for the simultaneous measurement of *two* noncommuting operators, but the consequence of the existence of the additive conserved quantity whose operator does not commute with the *single* operator to be measured.

(2) We required the condition $(X_\mu, X_\mu') = 0$, (2.10), so that the conservation law does not interfere with the possibility of distinguishing X and X'. However, if we loosen this condition and require only $(X, X') = 0$, we obtain the same value for ϵ except for terms of order $1/M^3$ or lower. However, in this case the second measurement (which distinguishes X and X') can be carried out only approximately, and the total probability for an unsuccessful measurement becomes larger.

(3) The theory of measurement, as described by von Neumann,[2] contains no limitation for the size of the measuring apparatus. As we have seen, there always are such limitations unless the operator to be measured commutes with all the operators of additive conserved quantities.[3] Recently the macroscopic character of the measuring apparatus has been studied in detail by several authors,[4] but no quantitative conditions have been established before.

ACKNOWLEDGMENTS

The author expresses his sincere gratitude to Professor E. P. Wigner for his encouragement and help, without which this work would not have been completed. He expresses his gratitude also to the Physics Department of Princeton University for its kind and warm hospitality. He also benefited from the helpful comments of Dr. H. Araki.

---

[2] J. von Neumann, *Mathematical Foundation of Quantum Mechanics* (Princeton University Press, Princeton, New Jersey, 1955).

[3] See reference 1. We will discuss this problem, "exactly measurable operators," in another article.

[4] G. Ludwig, *Die Grundlagen der Quantenmechanik* (Springer-Verlag, Berlin, 1954), Chap. V; H. S. Green, Nuovo cimento **9**, 880 (1958); P. K. Feyerabend, *Observation and Interpretation*, edited by S. Koerner (Butterworth's Scientific Publications, Ltd., London, 1957), p. 121; H. Wakita, Progr. Theoret. Phys. (Kyoto) **23**, 32 (1960); H. Margenau (to be published); S. Watanabe, Revs. Modern Phys. **27**, 179 (1955).

VI.4 TIME IN THE QUANTUM THEORY AND THE UNCERTAINTY RELATION FOR TIME AND ENERGY

YAKIR AHARONOV AND DAVID BOHM

Because time does not appear in Schrödinger's equation as an operator but only as a parameter, the time-energy uncertainty relation must be formulated in a special way. This problem has in fact been studied by many authors and we give a summary of their treatments. We then criticize the main conclusion of these treatments; viz., that in a measurement of energy carried out in a time interval, Δt, there must be a minimum uncertainty in the transfer of energy to the observed system, given by $\Delta(E'-E)\geqslant h/\Delta t$. We show that this conclusion is erroneous in two respects. First, it is not consistent with the general principles of the quantum theory, which require that all uncertainty relations be expressible in terms of the mathematical formalism, i.e., by means of operators, wave functions, etc. Secondly, the examples of measurement processes that were used to derive the above uncertainty relation are not general enough. We then develop a systematic presentation of our own point of view, with regard to the role of time in the quantum theory, and give a concrete example of a measurement process not satisfying the above uncertainty relation.

1. HISTORICAL SUMMARY OF THE STATE OF THE PROBLEM OF TIME MEASUREMENT IN THE QUANTUM THEORY

As is well known, the uncertainty relations in quantum mechanics can be regarded in two closely related ways. First of all, they are a direct mathematical consequence of the replacement of classical numbers by operators, and of adding the basic principle that the statistical distributions of the corresponding observables can be obtained by means of the usual formulas from the wave function and its probability interpretation.[1] Secondly, however, it can be shown by analyses such as that of the Heisenberg microscope experiment that they are also limitations on the possible accuracy of measurements.[2]

These considerations apply to observables such as x, p, and H. With regard to the measurement of time, however, a further problem appears, because time enters into Schrödinger's equation, not as an operator (i.e., and "observable") but rather, as a parameter, which is a "c" number that has a well defined value. Nevertheless, the uncertainty principle, $\Delta E\,\Delta t\geqslant h$, is generally accepted as valid, even though it is not deduced directly from commutation relations in the way described above.

The justification of the time-energy uncertainty relationship has been attempted in several ways. (We shall restrict ourselves here entirely to a discussion of the nonrelativistic case, since the theory of relativity has no essential relationship to the measurement problems that we are going to treat in this paper.)

First, one can begin with the wave function

$$\psi(x,t)=\sum_E C_E\psi_E(\mathbf{x})e^{-iEt/\hbar}, \tag{1}$$

where $\psi_E(x)$ is the eigenfunction of the Hamiltonian H of the system belonging to the eigenvalue E, and C_E is an arbitrary coefficient. If we consider a wave packet of width ΔE in energy space (i.e., ΔE is the range in which $|C_E|$ is appreciable), it immediately follows from the properties of Fourier analysis that $(\Delta E)\tau\geqslant h$, where τ is the time during which the wave packet does not change significantly (τ may be regarded as the mean life of the state in question[3]).

The above is a discussion in terms of the Schrödinger representation. Mandelstamm and Tamm[4] have formulated what is, in essence, the same point of view but it is expressed in the Heisenberg representation. They consider a dynamical variable A, which is a function of the time (e.g., the location of the needle on a clock dial or the position of a free particle in motion) and which can therefore be used to indicate time. If ΔA is the uncertainty in A, then the uncertainty in time is

$$\Delta t=\Delta A/|\langle\dot{A}\rangle_{\text{av}}|,$$

provided that $\dot{A}$ does not change significantly during the time, Δt, and that $\Delta\dot{A}/|\langle\dot{A}\rangle_{\text{av}}|$ is negligible. From the relation

$$\Delta A\;\Delta H\geqslant|\langle(A,H)\rangle_{\text{av}}|=\hbar|\langle\dot{A}\rangle_{\text{av}}|,$$

we obtain

$$\Delta A\;\Delta H/\langle\dot{A}\rangle_{\text{av}}=\Delta t\;\Delta H\geqslant h. \tag{2}$$

Since H represents the Hamiltonian of the isolated system, ΔH is also equal to ΔE, the uncertainty in energy of that system.

It should be noted that the method proposed by Mandelstamm and Tamm can actually lead to a determination of time, only when the system is not in a stationary state; i.e., only when the wave function takes the form of a packet, consisting of a linear superposition of stationary states. In other words, the ΔH appearing in equation (2) is determined by the range of energies

* Now at Brandeis University, Waltham, Massachusetts.

[1] The uncertainty relations are obtained in this way using Schwarz's inequality with the expressions for

$$\langle(A-\bar{A})^2\rangle_{\text{av}}\langle(B-\bar{B})^2\rangle_{\text{av}}=(\Delta A)^2(\Delta B)^2.$$

See, for example, D. Bohm, *Quantum Theory* (Prentice-Hall, Inc., Englewood Cliffs, New Jersey, 1951), pp. 205–206.

[2] W. Heisenberg, *The Physical Principles of the Quantum Theory* (Dover Publications, New York, 1930), Chap. 2.

[3] V. Fock and N. Krylov, J. Phys. (U.S.S.R.) **11**, 112 (1947), present a more detailed account of the lifetime of a state.

[4] L. Mandelstamm and I. Tamm, J. Phys. (U.S.S.R.) **9**, 249 (1945).

Originally published in *Physical Review, 122*, 1649-58 (1961).

in the wave packet. In this way, the relation of the Mandelstamm and Tamm treatment to the Schrödinger representation is made clear.

The above is then a discussion of the relation $\Delta E\,\Delta t \gtrsim h$ insofar as this has been obtained from the mathematical formalism of the quantum theory (i.e., the wave function, operators, and probability interpretation). Naturally, as is necessary in the case of observables such as x and p, this uncertainty relation must also be analyzed in terms of the interaction of the measuring apparatus with the observed system. Landau and Peierls,[5,6] for example, do this by considering a special example, in which the momentum of a free particle is measured by means of a collision with a heavy test particle (also free). To simplify the problem they consider a case in which the measuring particle is a perfectly reflecting mirror, and discuss only the movement in one dimension (perpendicular to the mirror). They then apply the laws of conservation of energy and momentum, which are

$$p'+P'-(p+P)=0, \tag{3a}$$

$$E'+\epsilon'-(E+\epsilon)=0, \tag{3b}$$

where lower case letters refer to the observed particle, capitals to the test particle, unprimed quantities to values before collision, and primed quantities to values after collision. Because $E=P^2/2M$, $\epsilon=p^2/2m$, one can solve for the momentum of the observed particle before and after collision, in terms of the corresponding momenta of the test particle.

In order to define the time of measurement, Landau and Peierls[5] (and also Landau and Lifshitz)[6] consider the case of a time-dependent interaction between the particle and the mirror, which lasts for some known period of time, Δt. This period Δt, which is the uncertainty in the time of measurement, then implies (e.g., according to perturbation theory) an uncertainty in the energy of the combined system consisting of observed particle and mirror, of magnitude $h/\Delta t$, resulting from the time-dependent interaction. Instead of Eq. (3b) for the exact conservation of energy, we must therefore write

$$|\epsilon'+E'-(\epsilon+E)| \gtrsim h/\Delta t. \tag{3c}$$

Evidently the momentum of the test particle before and after collision can be measured with arbitrary accuracy, so that $\Delta P=\Delta P'=0$. As a result, we obtain from Eq. (3a), $\Delta p=\Delta p'$; and from (3c), we have

$$\Delta(\epsilon'-\epsilon) \gtrsim h/\Delta t. \tag{4}$$

Since $\epsilon=p^2/2m$, we can also write the above result as

$$(v'-v)\Delta p \gtrsim h/\Delta t. \tag{5}$$

(Note that although Δp itself may be very small, there is still a minimum uncertainty in energy transfer, because the change of velocity will then become very large, if Δt is finite.)

Landau and Peierls therefore conclude that there is an uncertainty relation between the energy transferred to the system and the time at which the energy is measured. This means that the energy of the observed system cannot be measured in a short time, without changing it in an unpredictable and uncontrollable way. In other words, *energy measurements carried out in short periods of time are not reproducible.*

Fock and Krylov[3] criticize the derivation of the above results, but come to essentially the same conclusion. In effect, they do not accept the definition of the time of measurement by means of a time-dependent potential of interaction between the two particles. This is because in a real collision, there is no such time-dependent potential. Rather, the time of collision is determined by the movement of the particles themselves, in such a way that one of them serves as a clock. Let us suppose that it is the test particle which fulfills this function. This particle defines the time, t, as that at which it passes a definite point, X, by the equation, $t=X/V$. The time, as defined in this way, has an uncertainty $\Delta t=\Delta X/V$ (provided, as will actually be the case in our example, that $\Delta V/V \ll 1$).

In order to define the time of collision, we must have some information about the initial location of the observed particle, as well as that of the test particle. For simplicity, let us suppose that the initial velocity of the test particle is so much higher than that of the observed particle that the latter can be regarded as essentially at rest until the collision. The mean initial position x of the observed particle will be taken to be at the origin, while the uncertainty in this position is represented by Δx. Evidently we must choose $\Delta x \lessgtr \Delta X$, if the location of the test particle is to serve as a definition of the time of collision. Therefore

$$\Delta t=\Delta X/V \gtrsim \hbar/(\Delta P)V=\hbar/\Delta E.$$

(This is just the well-known uncertainty relation between the energy of the test particle and the time that is defined by the movement of its coordinate.)

Fock and Krylov then point out that in this case, the laws of conservation of energy and momentum are both satisfied *exactly*, so that Eqs. (3a) and (3b) can be used directly, while the approximate form (3c) for the conservation of energy is not to be applied here (since perturbation theory no longer has any relevance to the problem.) From (3a), we obtain

$$\Delta(p-p')=\Delta(P-P').$$

If V is chosen large enough, we can for a given $\Delta t \gtrsim \hbar/V\Delta P$ make ΔP and $\Delta P'$ arbitrarily small, and, as a result, we can likewise make $\Delta(p-p')$ as small as

[5] L. Landau and R. Peierls, Z. Physik **69**, 56 (1931).

[6] See also L. Landau and E. Lifschitz, *Quantum Mechanics* (Pergamon Press, New York, 1958), pp. 150–153.

we please. From Eq. (3b), it then follows that

$$\Delta(\epsilon'-\epsilon)=\frac{(p'-p)}{m}\frac{\Delta(p'+p)}{2}=(v'-v)\Delta p=\Delta(\epsilon'-\epsilon)\geqslant\frac{\hbar}{\Delta t},$$

where we have used the result that $\Delta(p'-p)$ is negligible. The above is exactly the same Eq. (5) as that obtained by Landau and Peierls, but the uncertainty in energy transfer to the observed system is now deduced on the basis of the fact that the (time-dependent) position of one of the particles is used to define the time of collision.

Fock and Krylov then go on to criticize the approach of Mandelstamm and Tamm, suggesting that it is incomplete. They assert that by means of the wave function and the operators of the observed system, one can discuss only the statistical features of any measurement. In order to discuss an *individual* measurement process, they refer to what they call "Bohr's uncertainty relation," $\Delta(\epsilon'-\epsilon)\Delta t\geqslant h$, where ϵ and ϵ' are the actual values of the energy of an individual observed system before and after measurement.

To clarify this distinction between the statistical uncertainty relations discussed by Mandelstamm and Tamm, and the Bohr relation, they point out that, for example, in observation of a state with lifetime τ (as described by its wave function), one can make measurements in times much shorter than τ. Therefore, it is necessary to distinguish between the time intervals defined by the wave function of the observed system, and the time interval representing the actual duration of an individual measurement. The time interval defined by the wave function has in measurements generally only a *statistical* significance.

Even if one treated the measurement process by means of a many-body Schrödinger equation, including the apparatus coordinates, the same distinction would arise. For it would the necessary to observe the combined system by means of additional apparatus; and here too, there will be a "Bohr uncertainty principle" for the individual observation and a statistical uncertainty principle following from the wave function, which applies to an ensemble of cases. To treat the apparatus by quantum theory is, in effect, to push back the well known "cut" between classical and quantum sides another stage. While it is always permissible and sometimes convenient to do this, it cannot change the content of the theory.

Let us now sum up the problem. Mandelstamm and Tamm propose a mathematical operator uncertainty relation between energy and time, as determined by the wave function. Fock and Krylov regard such a treatment as incomplete, because it applies only statistically to a large number of measurements, and because within it one cannot even consider the question of the interval of time needed to carry out an individual measurement. To complete the treatment, they call attention to the "Bohr uncertainty relation" (discussed also by Landau and Peierls,[5] as well as by Landau and Lifshitz[6]) which applies to individual measurements, and which refers to the relation between the error in the measurement of energy and the duration of the measurement process. While criticizing some of the methods of Landau and Peierls, they agree with the essential conclusion that energy cannot be measured in arbitrarily short periods of time, without introducing uncertainties, according to the relation $\Delta(\epsilon'-\epsilon)\geqslant h/\Delta t$.

2. CRITICISM OF COMMONLY ACCEPTED INTERPRETATIONS OF THE TIME-ENERGY UNCERTAINTY RELATION

The main conclusion of Landau and Peierls,[5] Landau and Lifschitz,[6] and Fock and Krylov[3] (given in Sec. 1), viz., that energy cannot be measured in a short time without introducing an uncertainty in its value, represents a very widely accepted interpretation of the time-energy uncertainty relation. This conclusion is, as we shall show, erroneous, the error being based in part on an inadequate formulation of Bohr's point of view concerning measurement, and in part on the use of an illustrative example of a measurement process, that was not sufficiently typical of the general case of such a measurement. (In fact, in Sec. 4 we shall give a counter-example, in which the energy of a particle is measured to arbitrary accuracy in as short a time as we please).

With regard to Bohr's point of view concerning measurements, it is important to stress here his continual insistence that the minimum ambiguities in the results of any individual measurement process (in the sense of what Fock and Krylov called the "Bohr uncertainty relations") are always exactly the same as the minimum ambiguities in the possibility of definition in the mathematical theory of the observables that are being measured.[7] (These latter ambiguities are, of course, the "statistical" uncertainty relations referred to by Fock and Krylov.)

The ambiguities in the results of individual measurements are regarded as originating in the indivisible quantum connections of the object under investigation to the apparatus (and indeed to the whole universe), which give rise to a minimum ambiguity in the degree to which well-defined classical properties (e.g., position and momentum) can be assigned to the object as a result of any such measurement.[8] In addition, however, the result of each measurement defines a "quantum state" of the observed system, specified by its wave function. This wave function is, of course, not a representation of an individual system, but it implies, in general, only a set of statistical predictions concerning the results of possible measurements. Nevertheless, if these predictions were such that the minimum ambiguity in the definition of the results of an individual measurement were less than the minimum statistical

[7] N. Bohr, Die Naturwissenschaften, 251 (1928).
[8] W. Heisenberg, *Physics and Philosophy* (Allen and Unwin, London, 1959), Chap. 3.

fluctuation implied by the mathematical theory, then there would be a contradiction. Vice versa, if they were such that the minimum ambiguity in the result of an individual measurement were greater than the minimum statistical fluctuation as described above, then this would lead to an arbitrary restriction, not related in any general way to the mathematical formalism, a restriction that evidently has no place in a coherent over-all framework of theory. Moreover, one could, in general, expect that with sufficient effort, it would be possible to find an example of an individual measurement process with the same minimum ambiguity as that implied by the formalism; and if such a process is found, then the supposedly greater minimum ambiguity in the results of an individual measurement will be contradicted. For these reasons, it is necessary to consider the statistical and individual uncertainty relations as two equally essential sides of what is basically the same limitation on the precise definability and measurability of the state of any system. In other words, as Bohr[9] has stressed, *there can be no limitation on individual measurements that cannot also be obtained from the mathematical formalism and the statistical interpretation.*

There is no question that all the above considerations apply for common examples of the uncertainty principle (e.g., x and p). However, as we pointed out in Sec. 1, time enters into Schrödinger's equation only as a parameter, so that there is no straightforward way to apply these ideas to the time-energy uncertainty relation. Of course, we can, with Mandelstamm and Tamm, obtain an uncertainty relation between the lifetime of a state of the observed system and its energy. But let us recall here that (as pointed out by Fock and Krylov), the operators of the observed system have no connection whatsoever with the duration of the measurement process (which is evidently determined, in general, by the apparatus). Keeping this fact in mind, let us now raise the question of whether there can be a genuine uncertainty relation between the energy transferred to the observed system and the time at which the measurement took place (as has been suggested by Landau and Peierls, Fock and Krylov, and other authors).

In accordance with Bohr's point of view on the subject, as we have described it above, we are led to point out that one cannot safely regard any given uncertainty relations as representing a real limitation on the accuracy of all possible measurements of the quantities under discussion unless the relationship has been shown to follow from the mathematical formalism. On the other hand, all of the authors referred to above seem to be satisfied to establish the time-energy uncertainty relations as applying to individual measurements by what Fock and Krylov called "illustrative examples." Such a point of view would imply, of course, that the uncertainty relations applying to individual measurements could in principle, have a basis that is independent of the statistical relations obtainable from the mathematical formalism. As we have already pointed out, however, such a procedure is arbitrary and therefore subject to the continual danger of being contradicted by the development of new examples of measurement processes, which reduce the ambiguity down to the minimum allowed by the formalism. (For, as is quite evident, there is no way to be sure that conclusions obtained from an illustrative example have universal validity).

It follows from the above discussion that to complete the treatment of the time-energy uncertainty relation, it is necessary to develop a method of showing how the time of measurement and the energy transferred in this measurement are to be expressed in terms of suitable operators. The method that we shall use in this paper starts from our discussion of the example first treated by Landau and Peierls, and then by Fock and Krylov; viz., the one in which the energy of a free particle is measured by collision with another particle. As we saw in Sec. 1, in such an interaction, the time of collision is determined physically by the state of some system which serves as a clock. In the example of Fock and Krylov, it is determined by the position of the test particle (which was taken to be free). Now, for any system, one can define a Hermitian operator representing such a time. In the case of a free particle, this is

$$t_c=\frac{y}{v_y}=\tfrac{1}{2}M\left(y\frac{1}{p_y}+\frac{1}{p_y}y\right), \tag{7}$$

where y and p_y are respectively position and momentum of the particle in question.[10]

The commutation relations between the above operator and the Hamiltonian, H_c, of the "clock" in question are (as can easily be verified for the case of a free particle, for which $H_c=p_y^2/2M$),

$$[H_c,t_c]=i\hbar. \tag{8}$$

This procedure is evidently very similar to that of Mandelstamm and Tamm. However, they discussed only the operators of the observed system, and obtained an uncertainty relation (2) between the energy of this system and the "inner" time as defined by dynamical variables in this system (e.g., the lifetime of a state). On the other hand, we are applying the relations (8) to the energy, E_c, of the "clock" in the apparatus, and to the time, t_c, of measurement as determined by this clock.

Since the time of measurement can be represented by an operator, t_c, belonging only to the observing apparatus, it follows that this *time must commute with every operator of the observed system and, in particular, with its Hamiltonian.* There is therefore no reason inherent in the principles of the quantum theory why the energy of a system cannot be measured in as short a time as we

[9] N. Bohr (private communication).

[10] There is a singularity for $p_y=0$, but it is easily shown that this will be unimportant if p_y is large enough, as will be the case in our applications.

please. (Recall, however, that in accordance with the treatment of Mandelstamm and Tamm, any such a measurement of the energy of the observed system to an accuracy ΔE must leave the "inner" time undefined to $\Delta t \geqslant \hbar/\Delta E$).

In view of the above discussion, it is evident that the usual treatment of the energy-time uncertainty relation (e.g., as discussed by Fock and Krylov) must be in some way erroneous. Since the particular illustrative example chosen by all the authors cited here (which is, in fact, the one usually given) has in fact been treated correctly, it follows that the mistake must be that this example is not sufficiently typical of the general case. And, indeed, as we shall see in Sec. 4, one can suggest more general methods of measurement of energy, which do not lead to the above limitation. In this way, we confirm our conclusion, based on general considerations regarding the principles of the quantum theory, that it is always possible to obtain true limitations on the measurability of any observable from the mathematical formalism, and that any other limitations that are added to these are arbitrary restrictions, which can eventually be contradicted, if further examples of measurement processes are sought.[11]

Finally, it is instructive to point out that problems similar to those connected with the time-energy uncertainty relations arise in the more familiar example of the position momentum relationship, $\Delta p_x \, \Delta x \geqslant \hbar$. To bring out the analogy, we can ask ourselves whether the momentum of a particle can be measured to arbitrary accuracy by means of an apparatus, which is localized in space. (Here, p_x takes the place of E, while the region of space in which the apparatus is located takes the place of the duration, Δt, of the measurement.) At first sight, it may seem that if the apparatus is localized in a very small region of size ΔX, the momentum of a particle cannot be measured to an accuracy greater than $\Delta p_x = h/\Delta X$. This, however, is not the case, because, what is defined here is the coordinate, X, of the apparatus, and not that of the measured particle, x. Since p_x commutes with X, there is no inherent limitation on how accurately it can be measured, if X is defined.

To illustrate such a possibility, consider the measurement of the momentum of a photon, by measuring its energy and using the relation $p = E/c$. (This is analogous to the measurement of energy of a particle by measuring its momentum and using the relation $E = p^2/2M$.) We can do this by means of an atom which is very highly localized, provided that this atom has a sharp level, excited above the ground state by the amount $E = pc$. If the photon has the appropriate energy, it will be absorbed and eventually reemitted (being delayed and perhaps scattered). It is observable whether this happens or not. If it does happen, then this provides a measurement of the energy, and through this, of the momentum. It is evident that the uncertainty in this momentum has no essential relation to the size of the atom, but only to the lifetime of the excited state. The momentum has therefore been measured by an apparatus, which is as localized in space as we please. (Of course, the position of the photon after the measurement is over is indeterminate, just as happens with "inner" time variables in the analogous case of time measurement.)

[11] This conclusion, the validity of which is fairly evident, will be obtained again in Sec. 3 from a more detailed discussion of the mathematical formalism.

3. TREATMENT OF TIME OF MEASUREMENT IN TERMS OF THE MATHEMATICAL FORMALISM OF THE QUANTUM THEORY

We saw in Sec. 1 that (as pointed out by Fock and Krylov[3]) there is a need to make a careful distinction between the time at which a measurement takes place and the time as defined by the wave function and operators of the observed system (e.g., the lifetime of an excited state). In Sec. 2, we showed how such a distinction can be represented within the mathematical formalism of the quantum theory by considering as operators certain variables that have hitherto usually been associated with the observing apparatus; viz., those variables determining the time at which interaction between the apparatus and the observed system takes place. This implies, of course, that the wave function must now be extended, so as to depend on these latter variables, It is equivalent to placing the "cut" between observing apparatus and observed system at a different point.

It is well known that while there is a certain kind of arbitrariness in the location of this cut, it is not completely arbitrary. For example, in the treatment of the energy levels of a hydrogen atom, one can, in a certain approximation, regard the nucleus as a classical particle in a well-defined position. If, however, the treatment aims at being accurate enough to take the reduced mass into account, both electron and nucleus must be treated quantum-mechanically, and the cut is introduced instead between the atom as a whole and its environment. The place of the cut therefore depends, in general, on how accurate a treatment is required for the problem under discussion. Of course, it follows that once a given place of the cut is justified, then it can always be moved further toward the classical side without changing the results significantly.

If we are interested only in discussing what Fock and Krylov called the "Bohr uncertainty relation" (the one referring to an *individual* measurement of energy and time), then we are justified in placing the physical variables that determine the time of the measurement on the classical side of the cut. For, as is quite evident, in this aspect of the uncertainty relations, these variables are by definition regarded as classical, in the sense that their uncertainty represents only an inherent ambiguity in the possibility of defining the state of an *individual* system. We have seen in Sec. 2, however, that according to the Bohr point of view, every uncertainty relation that appears in this way must also be able to appear as a statistical fluctuation in a corresponding operator, which must, of course, be calculated from an appropriate wave function. To discuss this side of the uncertainty relations, it is clear that we must change the position of the cut, so that the corresponding variables now fall on the quantum-mechanical side.

In the subsequent discussion of how the uncertainty relations appear in the mathematical formalism, we shall begin with the case in which the time determining variables are placed on the classical side of the cut. In this case, the time variables can be reflected in the Schrödinger equation only in the time parameter t which can, of course, have an arbitrarily well-defined value. This time parameter is related to measurement in several ways.

First of all such a relation comes about in the preparation of a system in a definite quantum state, and in observations carried out later on that system. Consider, for example, a quantum state prepared at a time determined by means of a shutter (which we are, of course, now regarding as being on the classical side of the cut). There must be some relationship between the time, t_s, at which the shutter functions and the time parameter, t appearing the Schrödinger's equation. Indeed, if the Hamiltonian of the observed system does not depend on time, then it is easily seen that the wave function takes the form

$$\psi=\psi(\mathbf{x},\, t-t_s), \tag{9}$$

where ψ is a solution of Schrödinger's equation for the system in question. The form of ψ is determined by choosing that solution which at $t=t_s$ becomes equal to the function, $\psi_0(\mathbf{x})$, representing the quantum state in which the "preparing" measurement leaves the system. Then, when an observation is made, the time t_m, of the measurement is likewise determined by suitable variables on the classical side of the cut. The probability of any given result is, of course, computed in the well-known way from the wave function, $\psi=\psi(\mathbf{x},\, t_m-t_s)$.

It is clear that as far as this one-body treatment is concerned, there is certainly nothing in the formalism which would prevent the system under discussion from being either prepared or observed in a state of definite energy, when t_m and t_s are as well defined as we please. Thus, if the system is in a state of definite energy, E (so that the uncertainty, ΔE, in its energy is zero, while the lifetime $\tau\gg\hbar/\Delta E$ of the state is infinite), its wave function, $\psi=\psi_E(\mathbf{x})e^{-iEt/\hbar}$ [where $\psi_E(\mathbf{x})$ is the eigenfunction of the Hamiltonian operator belonging to the eigenvalue, E] is evidently able to represent such a state, no matter what value is given to t. A wave function of this kind is therefore evidently compatible with the statement that at some time, $t=t_s$, the system was prepared in the eigenstate of the energy represented by $\psi_E(\mathbf{x})$. Thus, *in the one-body treatment alone*, no reason for an uncertainty relation between the energy of the system and the time of measurement can be found. And this is indeed basically the reason why Fock and Krylov were led to postulate such an uncertainty relation independently, and to try to justify this relationship by means of illustrative examples of measurement processes (see Sec. 1).

Let us now go on to consider the time-energy uncertainty relation from the other aspect, in which the variables determining the time of measurement are placed on the quantum-mechanical side of the cut. In this case, we must introduce these variables into the wave function, so that we are in this way led to a many-body Schrödinger equation. Let us recall, however, that the "cut" has not been abolished, but merely pushed back another stage. Thus, as was pointed out in the discussion of the treatment of Fock and Krylov given in Sec. 1, there is implied an additional observing apparatus on the classical side of the cut, with the aid of which the many-body system under discussion can be observed. The probabilities for the results of such observations are determined by the wave functions, which take the form

$$\Psi=\Psi(\mathbf{x},y_i,t), \tag{10}$$

where y_i, represents the apparatus variables on the quantum-mechanical side of the cut (which include those that describe the time of measurement). The time parameter t here plays a role similar to that which it had in the one-body problem; viz., through it the time frame on the large-scale classical side of the cut is brought into relationship with the quantum-mechanical formalism by means of suitable observations.

We shall now consider as an example of the approach described above, the measurement of energy and time by means of a collision of two particles, as treated in Sec. 1. The initial wave function of the combined system is, for this case, a product of two packet functions, one representing the test particle coming in with a very high velocity, V, and the other representing the observed particle, essentially at rest (with a velocity that is negligible in comparison to V) and with its center at the origin. After the collision, it is well known[12,13] that the wave function becomes a sum of products of such packets, correlated in such a way that an observation of the properties of the test particle can yield information about the particle under discussion.

As far as this particular example is concerned, it will not be relevant here to go into a more detailed discussion of the problem of solving Schrödinger's equation. All that is important here is that as we saw in Sec. 2, the time of collision is given essentially by the operator,

$$t_c=\tfrac{1}{2}M\left(y\frac{1}{p_y}+\frac{1}{p_y}y\right),$$

(where y and p_y refer to the position and momentum of

[12] See reference 1, Chap. 22.

[13] J. von Neumann, *Mathematical Foundations of Quantum Mechanics* (Princeton University Press, Princeton, New Jersey, 1955), Chap. 6.

the test particle, respectively), so that the operator, t_c, commutes with the Hamiltonian, H_s, of the observed system. As a result, there is, as we have already stated in Sec. 2, no uncertainty relation between the time of measurement and the energy of the observed system. But in the treatment that we are now using, we have obtained this result by allowing the variables determining the time of measurement to fall on the quantum-mechanical side of the cut. (As was to be expected, of course, the actual physical consequences of the theory do not depend on which side of the cut these variables are placed.)

In the example given above, the same apparatus was used to determine both the time of measurement and the momentum of the particle. However, it is possible, and in fact frequently advantageous, to consider a more general situation, in which the time determining variables are separated from those which are used to measure other quantities (such as momentum). This, in fact, would be the correct way of completing the treatment of the example given by Landau and Peierls[5,6] (see Sec. 1), in which the time of measurement was determined by an interaction between the test particle and the observed particle which was assumed to last for some interval Δt.

If there is a time-dependent interaction between apparatus and observed system which lasts for an interval Δt, then the Schrödinger equation will have to have a corresponding potential, which represents this interaction. The form of this potential will depend on where we place the cut.

If the apparatus determining the time of interaction is taken to be on the classical side, then the potential will be a certain well defined function of time, which is nonzero only in the specified interval of length Δt. We may write this potential as

$$V = V(\mathbf{x},y,t), \tag{11}$$

where $\mathbf{x}$ represents the coordinate of the observed particle, and y that of the test particle. This is indeed the usual way by which measurements are represented in the mathematical formalism.[12,13]

In the next section we shall apply this method in order to treat a specific example, in which it will be shown in detail that the energy of a system can be measured in an arbitrarily short time.

If, on the other hand, the variables determining the time of interaction are placed on the quantum mechanical side of the cut, then we cannot regard the potential as a well-defined function of time. Instead, we must write

$$V = V(\mathbf{x},y,z), \tag{12}$$

where z is the variable that determines the time of interaction.

If the particles determining the time of interaction are heavy enough, then they will move in an essentially classical way, very nearly following a definite orbit, $Z=Z(t)$. To the extent that this happens, we obtain, as a good approximation,

$$V(\mathbf{x},y,z) = V(\mathbf{x},y,z(t)). \tag{13}$$

To treat this problem mathematically, we begin with the Schrödinger equation for the whole system.

$$i\hbar\frac{\partial\Psi}{\partial t}(\mathbf{x},y,z,t) = [H_p+H_y+H_z+V(\mathbf{x},y,z)]\Psi(\mathbf{x},y,z,t), \tag{14}$$

where H_p represents the Hamiltonian of the observed particle, H_y that of the apparatus, H_z that of the time-determining variable, z. We must then show the equivalence of this treatment with that obtained by placing the time-determining variable on the classical side of the cut. To do this, it is sufficient to demonstrate that in a suitable approximation, equation (14) leads to the time-dependent Schrödinger equation for the $\mathbf{x}$ and y variables alone, viz.,

$$i\hbar\frac{\partial\psi}{\partial t}(\mathbf{x},y,t) = [H_p+H_y+V(\mathbf{x},y,z(t))]\psi(\mathbf{x},y,t). \tag{15}$$

We shall simplify the problem[14] by letting the time determining variable be represented by a heavy free particle mass M, for which we have $H_z=p_z^2/2M$. We suppose that the initial state of the time-determining variable can be represented by a wave packet narrow enough in z space, so that $\Delta t = \Delta z/|\dot{z}|$ can be made as small as is necessary. This packet is

$$\Phi_0(z,t) = \sum_{p_z} C_{p_z} \exp\left\{\frac{i}{\hbar}\left[zp_z - \frac{p_z^2}{2M}t\right]\right\}. \tag{16}$$

Because M is very large, the wave packet will spread very slowly, and to a good approximation, we shall have

$$\Phi_0(z,t) = \Phi(z-v_zt)\exp\left\{\frac{i}{\hbar}\left[\bar{p}_z z - \frac{(\bar{p}_z)^2}{2M}t\right]\right\}, \tag{17}$$

where $v_z=\bar{p}_z/M$ is the mean velocity, and $\Phi(z-v_zt)$ is just a form factor for the wave packet which is, in general, a fairly regular function which varies slowly in comparison with the wavelength, $\lambda = h/\bar{p}_z$.

If the interaction, $V(\mathbf{x},y,z)$ is neglected, a solution for the whole problem will be

$$\Psi(\mathbf{x},y,z,t) = \Phi_0(z,t)\psi_0(\mathbf{x},y,t), \tag{18}$$

where $\psi_0(\mathbf{x},y,t)$ is a solution of the equation

$$i\hbar\partial\psi_0(\mathbf{x},y,t)/\partial t = (H_p+H_y)\psi_0(\mathbf{x},y,t). \tag{19}$$

When this interaction is taken into account, the solution will, in general, take the form

$$\Psi(\mathbf{x},y,z,t) = \sum\nolimits_n \Phi_n(z,t)\psi_n(\mathbf{x},y,t)C_n,$$

[14] Our procedure is along lines similar to those developed by H. L. Armstrong, Am. J. Phys. 22, 195 (1957).

where the sum is taken over the respective eigenfunctions, $\Phi_n(z,t)$ and $\psi_n(\mathbf{x},y,t)$ of H_z and (H_p+H_y). If such a sum is necessary, there will be correlations between the time-determining variable and the other variables, with the result that there will be no valid approximation in which an equation such as (15) involving only $\mathbf{x}$ and y can be separated out. However, if the mass, M, of the time determining particle is great enough, so that the potential $V(\mathbf{x},y,z)$ does not vary significantly in a wavelength, $\lambda=h/p_z$, then, as is well known, the adiabatic approximation will apply. In this case, one can obtain a simple solution, consisting of a single product, even when interaction is taken into account. To show in more detail how this comes about, we first write the solution in the form

$$\Psi=\Phi_0(z,t)\psi(\mathbf{x},y,z,t).$$

When this function is substituted into Schrödinger's equation (15), the result is

$$i\hbar\frac{\partial\psi}{\partial t}=\left[H_p+H_y+V(\mathbf{x},y,z)-\frac{\hbar^2}{M}\frac{\partial\ln\Phi_0}{\partial z}\frac{\partial}{\partial z}-\frac{\hbar^2}{2M}\frac{\partial^2}{\partial z^2}\right]\psi. \quad (20)$$

If M is large, and if the potential does not vary too rapidly as a function of z, the last term on the right-hand side of (20) can be neglected.[15] Moreover,

$$\frac{\partial\ln\Phi_0}{\partial z}=\frac{i}{\hbar}\left[\bar{p}_z+\hbar\frac{\partial\ln\Phi(z-v_zt)}{\partial z}\right].$$

Because $\Phi(z-v_zt)$ does not vary significantly in a wavelength, this term too can be neglected in the above equation, and we obtain

$$i\hbar\frac{\partial\psi(\mathbf{x},y,z,t)}{\partial t}=\left[H_p+H_y+V(\mathbf{x},y,z)-i\hbar v_z\frac{\partial}{\partial z}\right]\psi(\mathbf{x},y,z,t).$$

We then make the substitution, $z-v_zt=u$, and

$$\psi'(\mathbf{x},y,u,t)=\psi(\mathbf{x},y,z,t)=\psi(\mathbf{x},y,u+v_zt,t).$$

With the relation

$$\partial\psi'/\partial t=(\phi\psi/\partial t)+(v_z\partial\psi/\partial z),$$

we have

$$i\hbar\partial\psi'/\partial t=[H_p+H_y+V(\mathbf{x},y,u+v_zt)]\psi'(\mathbf{x},y,u,t). \quad (21)$$

Note that this equation does not contain derivatives of u, so that u can be given a definite value in it.

The complete wave function is, of course, obtained by multiplying $\psi'(\mathbf{x},y,u,t)$ by $\Phi_0(z-v_zt)=\Phi_0(u)$. Now, this was assumed to be a narrow packet centering at $u=0$, such that the spread of u can be neglected. As a result, we can write $u=0$ in the above equation. The result is

$$i\hbar\frac{\partial\psi'(\mathbf{x},y,0,t)}{\partial t}=[H_p+H_y+V(\mathbf{x},y,v_zt)]\psi'(\mathbf{x},y,0,t). \quad (22)$$

In this way, we have obtained the Schrödinger equation for x, y, with the appropriate time-dependent potential $V(\mathbf{x},y,v_zt)$, the relationship between the time parameter t and the time determining variable z being, in this case, $t=z/v_z$. We have therefore completed the demonstration of the equivalence of the two treatments, in which the time-determining variables are placed on different sides of the cut.

4. EXAMPLE OF A REPRODUCIBLE MEASUREMENT OF ENERGY IN A WELL-DEFINED TIME

We saw in Secs. 2 and 3 that there is no reason inherent in the principles of the quantum theory why a reproducible and exact measurement of energy cannot be made in an arbitrarily short period of time. Since Landau and Peierls,[5] Fock and Krylov,[6] and many others have considered examples leading to a contrary conclusion, it is necessary to complete the discussion by giving a specific example of a method of measuring energy precisely in as short a time as we please. This we shall do in the present section. Following the development of our example, it will become clear in what way the previous treatments of this problem were inadequate.

As a preliminary step, we discuss the treatment of the measurement of energy by means of the Schrödinger equation for the apparatus and the observed system together. The Hamiltonian of the combined system is

$$H=H_0(p_x,x)+H_0'(p_y,y)+H_I(p_x,x;p_y,y,t), \quad (23)$$

where $H_0(p_x,x)$ is the Hamiltonian of the observed system, $H_0'(p_y,y)$, that of the apparatus, and $H_I(p_x,x;p_y,y,t)$ is the interaction, which is zero except during a certain interval of time between t_0 and $t_0+\Delta t$. (Here we are adopting the point of view described in Sec. 3, in which we regard the time determining variables as being on the classical side of the cut, so that they do not appear explicitly in Schrödinger's equation.) It will be adequate for our purposes to assume that both the observed system and the apparatus are free particles, with respective Hamiltonians

$$H_0=p_x^2/2M;\quad H_0'=p_y^2/2M. \quad (24)$$

To simplify the problem, we consider the ideal case of a measurement of p_x which does not change p_x. This will happen if H_I is not a function of x. (The satisfaction of this condition will evidently guarantee that reproducible measurements of p_x, and therefore $H_0=p_x^2/2M$ will be possible). The Hamiltonian of the whole system will then be taken as

$$H=(p_x^2/2M)+(p_y^2/2M)+yp_xg(t), \quad (25)$$

[15] If $V(z,y,z)$ varies too rapidly, then $(\hbar^2/2M)\partial^2\psi/\partial z^2$ will not be negligible, even when M is large.

where $g(t)$ is everywhere zero, except between t_0 and $t_0+\Delta t$, where it is constant. (The interaction Hamiltonian is similar to a vector potential in its effects).

With the Hamiltonian (25), p_x is, of course, a constant of the motion. The equations of motion for the remaining variables are then

$$\dot{x}=(p_x/m)+yg(t), \quad \dot{p}=-p_x g(t), \quad \dot{y}=p_y/M. \tag{26}$$

On solving for p_y, we obtain (using p_x=constant)

$$p_y-p_y^0=-p_x g(t)\Delta t. \tag{27}$$

This equation implies a correlation between $p_y-p_y^0$ and p_x, such that if $p_y-p_y^0$ is observed, we can calculate p_x.

It is important also to consider the behavior of $\dot{x}$. Although p_x is constant, $\dot{x}$ shifts suddenly at $t=t_0$ from p_x/m to $p_x/m+g(t)y$, and remains at this value until $t=t_i$, after which it returns to its initial value. (In a similar way, the velocity and momentum differ in the case of a vector potential.)

The above behavior of the velocity is, as a simple calculation shows, just what is needed to produce the uncertainty in position, which is required by the improved definition of the momentum resulting from the measurement.

It is easily seen that if $g(t)$ is large enough, the measurement described above can be carried out in as short a time as we please. In order that a given accuracy, Δp_x, be possible, the change of deflection of the apparatus, $\Delta(p_y-p_y^0)$ due to the shift Δp_x, must be greater than the uncertainty, $\Delta(p_y^0)$ in the initial state of the apparatus. This means that we must have

$$\Delta p_x g(t)\Delta t \cong \Delta p_y^0,$$

and if $g(t)$ is large enough, both Δt and Δp_x can be made arbitrarily small for a given Δp_y^0.

This hypothetical example confirms our conclusion once again that accurate energy measurements can be reproduced in an arbitrarily short time. We shall now show how to carry out such a measurement by means of a concrete experiment. To guide us in the choice of this experiment, we note that the essential feature of the interaction described in Eq. (25) is that it implies a force that is independent of the x coordinate of the particle, and which alters the velocity suddenly at $t=t_0$ to bring it back to its original value at $t=t_1$.[16] This force is therefore equivalent in its effect to a pair of equal and opposite pulses in a uniform electric field, the first at t_0 and the second at t_1. In order to approximate such pulses, we shall consider two condensers, the fields of which cross the observed particle at the times t_0 and t_1. The condensers are assumed to have a length, l, in the Y direction which is much greater than their thickness, d, in the X direction. Therefore, they will produce a uniform electric field in the X direction, except for edge effects which can be neglected when $l \gg d$. Each condenser will go by the particle at a velocity, V_y in the Y direction, which is assumed to be so great that the electric field acts for a very short time, l/V_y, with the result that the field approximates the one cited in our mathematical example, where the period of action was infinitesimal. If the two condensers follow each other, one at $t=t_0$, the other at $t=t_1$, then we shall approach the case treated in Eq. (25).

As in the case of the collision treated in Secs. 1 and 3, the time of measurement is defined as the time at which the condenser passes the observed particle. (This means that we are now shifting to the point of view in which the time-determining variables are on the quantum-mechanical side of the cut, but as we saw in Sec. 3, both both points of view are equivalent and can be used interchangeably). As in the case of collision, the uncertainty Δt in this time will be given by $\Delta y/V_y$, provided that the observed particle is initially localized in the Y direction, with a velocity v_y much smaller than V_y. This will imply an uncertainty in the energy of the condenser, $\Delta E_c=V_y\Delta P_y \gtrsim \hbar/\Delta t$.

In the interaction between particle and condenser, the transfer of X component of the momentum (neglecting edge effects) is

$$\Delta p_x=p_x^0=F\tau=e\mathcal{E}\tau=e\mathcal{E}l/V_y, \tag{28}$$

where τ is the time taken by the condenser to pass the particle. (Note that τ and Δt are different quantities). This transfer is independent of initial conditions, and is calculable to arbitrary accuracy. (Since V_y is as large as you please, the Y component of the velocity of the particle can be neglected in the above calculation.) The above transfer of momentum implies a transfer of energy to the particle,

$$\epsilon-\epsilon_0=\frac{(p_x^0+\Delta p_x)^2}{2m}-\frac{p_x^2}{2m}=\frac{p_x^0}{m}\Delta p_x+\frac{(\Delta p_x)^2}{2m}. \tag{29}$$

By conservation, this transfer must be equal to the energy loss E_0-E_1 of the condenser. Since the initial momentum P_x of the condenser in the X direction is zero, and since the mass M of the condenser is large, the term $(\Delta p_x)^2/2M$ which represents the energy of the condenser due to momentum transfer in the X direction, will be negligible. The energy change must therefore be the result of alteration in the Y component of the condenser momentum, so that it will be equal to $E_0-E_1\cong V_y\Delta P_y$.

Equation (29) can now be used to permit p_x^0 to be measured if $E_0-E=\epsilon-\epsilon_0$ is known. For since $\epsilon-\epsilon_0$ depends on p_x^0 and since Δp_x can be obtained from Eq. (28), p_x^0 can be calculated in terms of E_0-E_1.

There is, however, a limitation on how accurately E_0-E_1 can be measured because we require that the

[16] In the hypothetical example of Eq. (25), this force resulted from the time-dependent interaction, which was equivalent to a corresponding vector potential, which would produce a field, $\mathcal{E}=-(1/c)\partial\mathbf{A}/\partial t$, that is nonzero only when $\mathbf{A}(t)$ changes; i.e., at the beginning and the end of the interval.

Y coordinate of the condenser shall serve as a clock to an accuracy, δt, with the result that the uncertainty relation, $\delta E \gtrsim \hbar/\delta t$ will hold for the energy of the condenser. By conservation, the same uncertainty relation must hold for the energy $\epsilon - \epsilon_0$ transferred to the particle.

By evaluating $\delta(\epsilon - \epsilon_0)$ from Eq. (29), we obtain

$$\delta(\epsilon_0 - \epsilon)\delta t = \delta p_x^0 \frac{\Delta p_x}{m} \delta t = \delta p_x^0 \Delta v_x \delta t = \delta p_x'(v_x' - v_x)\delta t \gtrsim h.$$

This is exactly the same relation as was obtained in the collision example given in Sec. 1. In other words, the measurement on the first condenser alone, must satisfy the condition that if it is carried out in a time defined as δt, there will be an uncertain energy transfer, $\delta E \gtrsim h/\delta t$. It is at this point, however, that the second condenser plays an essential role. For immediately after the interaction with the first condenser is over, it will bring about a transfer of X component of the momentum, which is equal and opposite to that transferred to the first condenser. As a result, the velocity of the observed particle will return to its initial value, just as happened in the mathematical example discussed in the beginning of this section. Thus, the momentum and the energy have been measured without their being changed. There is, therefore, no limitation on the accuracy with which the energy of the particle can be measured, regardless of the value of δt, which can be made as small as we please by making V_y very large.

A similar two-stage interaction can be carried out in the collision example described in Sec. 1. To do this, we recall that the uncertainty in energy transfer, $\delta(\epsilon' - \epsilon) = |v - v'|\delta p$ is large because $|v - v'|$ is large. Nevertheless $v - v'$ can be determined with arbitrary accuracy from the results of the measurement. After this is done, one can then send in a second test particle, with initial momentum calculated to be such as to change v' back to v. After the two collisions, there will be, as in the case of the condensers, no uncertainty in energy of the observed particle. In the collision experiment, the change of velocity depends on the value of the momentum of the observed particle, so that the initial conditions in the second collision must be arranged, in accordance with this value, which is learned from the first collision. On the other hand, in the condenser experiment, $v_x' - v_x$ is independent of initial conditions so that the second condenser can be prearranged to cancel out this shift of velocity without any information from the results of the first interaction.

At first sight, one might raise the question as to whether our conclusions could be invalidated by effects of radiation, or by currents which might be induced in the condenser. Since we are discussing only the problem of nonrelativistic quantum mechanics, we can assume that the velocity of light is infinite. In this case, radiation and relativistic effects, in general, can be made negligible, no matter how sudden the shift of potential is. As for currents induced in the condenser, these can be avoided by charging an insulator instead of a metal plate. The field will still be uniform, but the charges will not be mobile, so that no currents will be induced in the condenser.

The error in the treatments of Landau and Peierls, Fock, and Krylov, and others, as discussed in Sec. 1, is now evident. For in all of these treatments, the example used was that of a single collision of a pair of particles. For this case, our own treatment also gave the result that energy transfer in a short time must be uncertain. But as shown in our general canonical treatment of the problem [see Eqs. (25)–(28)], it is clear that this is not the correct way to measure momentum and energy without changing them. To accomplish this purpose we need an interaction of the kind described in the above equations, which changes the velocity only while interaction is taking place, but which brings it back to the initial value after interaction is over. And, as we have seen, it is possible to realize such a measurement in a concrete example.

5. SUMMARY AND CONCLUSIONS

There has been an erroneous interpretation of uncertainty relations of energy and time. It is commonly realized, of course, that the "inner" times of the observed system (defined as, for example, by Mandelstamm and Tamm[4]) do obey an uncertainty relation $\Delta E\, \Delta t \gtrsim h$, where ΔE is the uncertainty of the energy of the system, and Δt is, in effect, a lifetime of states in that system. It goes without saying that whenever the energy of any system is measured, these "inner" times must become uncertain in accordance with the above relation, and that this uncertainty will follow in any treatment of the measurement process. In addition, however, there has been a widespread impression that there is a further uncertainty relation between the *duration* of measurement and the *energy transfer* to the observed system. Since this cannot be deduced directly from the operators of the observed system and their statistical fluctuation, it was regarded as an additional principle that had to be postulated independently and justified by suitable illustrative examples. As was shown by us, however, this procedure is not consistent with the general principles of the quantum theory, and its justification was based on examples that are not general enough.

Our conclusion is then that there are no limitations on measurability which are not obtainable from the mathematical formalism by considering the appropriate operators and their statistical fluctuation; and as a special case we see that energy can be measured reproducibly in an arbitrarily short time.

ACKNOWLEDGMENTS

We are grateful to Professor M. H. L. Pryce and to Dr. G. Carmi for helpful discussions, and one of us (Y. Aharonov) wishes to acknowledge aid from a grant provided by the D.S.I.R. at the University of Bristol, while this work was being done.

VI.5 THE FUNDAMENTAL NOISE LIMIT OF LINEAR AMPLIFIERS

H. Heffner

Summary—If the uncertainty principle of quantum mechanics is applied to the process of signal measurement, two theorems relating to amplifier noise performance can be deduced. First, it can be shown that it is impossible to construct a linear noiseless amplifier. Second, if the amplifier is characterized as having additive white Gaussian noise, it can be shown that the minimum possible noise temperature of any linear amplifier is

$$T_n = \left[\ln \frac{2 - 1/G}{1 - 1/G}\right]^{-1} \frac{h\nu}{k}.$$

In the limit of high gain G this expression reduces to that previously derived for the ideal maser and parametric amplifier. It is shown that the minimum noise amplifier does not degrade the signal but rather allows the use of an inaccurate detector to make measurements on an incoming signal to the greatest accuracy consistent with the uncertainty principle.

Introduction

SINCE THE advent of the maser, there have been a number of treatments of the noise figure or noise temperature of this and other potentially low noise devices such as the parametric amplifier.[1–4] Most of these have treated each specific device as a quantum system and have determined a limiting noise temperature arising because of amplified spontaneous emission. Although the details of the calculations differ, investigations of the minimum noise temperature due to this effect yield values of the order of $h\nu/k$ for both the maser and the parametric amplifier.

The maser and the parametric amplifiers are phase preserving amplifiers, or, in the terminology employed here, linear amplifiers. They have been characterized

* Received January 8, 1962; revised manuscript received, April 30, 1962.

† Stanford Electronics Laboratories, Stanford University, Calif.

[1] K. Shimoda, H. Takehashi, and C. H. Townes, "Fluctuations in amplification of quanta with application to maser amplifiers," *J. Phys. Soc. Japan*, vol. 12, pp. 686–700; June, 1957.

[2] J. Weber, "Maser noise considerations," *Phys. Rev.*, vol. 108, pp. 537–541; November, 1957.

[3] M. W. Muller, "Noise in a molecular amplifier," *Phys. Rev.*, vol. 106, pp. 8–12; April, 1957.

[4] W. H. Louisell, A. Yariv, and A. E. Siegman, "Quantum fluctuations and noise in parametric processes," *Phys. Rev.*, vol. 124, pp. 1646–1654; December, 1961.

Originally published in *Proceedings of the IRE, 50*, 1604-08 (1962).

broadly as voltage amplifiers. There is another type of amplifier which does not preserve the signal phase which can be classed as a quantum counter. Weber[2,5] has proposed two forms of such amplifiers and has pointed out that they have no spontaneous emission noise. Thus these phase-insensitive counters have a zero-limiting noise temperature. This result opens the question of whether there are possible forms of linear amplifiers—linear in the sense of phase preserving—which also have a limiting noise temperature approaching zero.

In order to settle this question one needs to look for a general physical principle which will apply to all amplifiers regardless of the details of the specific amplification process. An appropriate one is the uncertainty principle of quantum mechanics which forms one of the basic postulates of quantum theory. Although it has resulted in important limitations on the accuracy of measurements possible in atomic systems, its corresponding limitations on the accuracy of signal measurements have until recently gone unnoticed. At the 1959 Quantum Electronics Conference, Serber and Townes[6] investigated the role of the uncertainty principle in maser noise and Friedburg[7] considered its implication on the noise figure of a general amplifier. Each of these papers is open to some criticism. First, the uncertainty principle as generally interpreted is a statement about the results obtained in a physical measurement. As such it can be applied to a signal detector but not to an amplifier. An amplifier is not a measuring instrument which produces a set of data. Amplification is rather a process, a transformation of the signal. The uncertainty principle can be applied to the measurement of the results of that processing but not directly to the processing itself. A second objection to both papers concerns their lack of rigor in problems of statistical averaging, particularly in regard to the phase of a signal. Finally, in the case of Friedburg, the wrong constant was used in the statement

[5] J. Weber, "Masers," *Rev. Mod. Phys.*, vol. 31, pp. 681–710; July, 1959. See also N. Bloembergen, "Solid-state infrared quantum counter," *Phys. Rev. Lett.*, vol. 2, pp. 84–85; February, 1959.

[6] R. Serber and C. H. Townes, "Amplification and Complementarity," in "Quantum Electronics," C. H. Townes, Ed.; Columbia University Press, New York, N. Y., pp. 233–255; 1960.

[7] H. Friedburg, "General amplifier noise limit," in "Quantum Electronics," C. H. Townes, Ed.; Columbia University Press, New York, N. Y., pp. 228–232; 1960.

of the uncertainty principle. As implied before, Friedburg's conclusions strictly speaking apply only to a detector, not to an amplifier, and as such do not indicate what class of amplifier falls under the uncertainty limitation, nor how this limitation is affected by the gain.

This purpose of this paper is to develop in a simple but rigorous fashion the limitations on the noise performance of linear amplifiers which are implied by the uncertainty principle. First it will be shown that there is no such thing as a noiseless linear amplifier. Next, by characterizing the amplifier as adding white noise, the minimum possible noise temperature is derived. The resulting expression when the gain is large is exactly that derived for the limiting noise performance of the maser and the parametric amplifier.

The Uncertainty Principle

The uncertainty principle, first formulated by Heisenberg in 1927, claims that complete accuracy is impossible to obtain in the simultaneous measurements of certain physical quantities. In its most familiar form, it asserts the fundamental inaccuracy which must result in the simultaneous measurement of a particle's momentum p and position x. If we define the uncertainty in measurement to be the rms deviation from the mean in the distribution obtained from an ensemble of measurements, then the uncertainty in the measurement of momentum Δp and the uncertainty in the measurement of position Δx are related by

$$\Delta p \Delta x \geq h/4\pi. \qquad (1)$$

This relation can be interpreted in the following way. The process of measuring cannot be divorced from the physical process being measured. Not only does the act of observing affect the system being observed, but it does this in a way which cannot be precisely predicted. It is this quality of unpredictability which formed the new content of the uncertainty principle.

In its most general form, the uncertainty principle applies to measurements of any two canonically conjugate quantities[8] such as, for example, the energy of

[8] In mathematical form, the uncertainty principle states that if

a system and the precise time at which the system possesses this energy,

$$\Delta E \Delta t \geq h/4\pi. \tag{2}$$

Still another form of the principle applies to the measurement of the number of quanta in an oscillation and its phase

$$\Delta n \Delta \phi \geq \tfrac{1}{2}. \tag{3}$$

This latter statement of the principle may be made plausible by substituting the relations $E = nh\nu$ and $\phi = 2\pi\nu t$ into the preceding equation. We shall use this last formulation of the uncertainty principle to derive two basic theorems on amplifier noise performance.

The Unavoidable Noisiness of Linear Amplifiers

The first conclusion which emerges from the uncertainty principle is: *It is impossible to construct a noiseless linear amplifier*. We can prove this statement by postulating the existence of a noiseless linear amplifier and then showing that it violates the uncertainty principle.

Suppose we have a perfect linear amplifier by which we mean the following. If during any given interval we measure the number of photons n_2 produced at the output, we find that it is related to the number of input photons n_1 by a constant G, the gain of the amplifier.

$$n_2 = Gn_1. \tag{4}$$

Secondly, if we measure the phase ϕ_2 of the output, we find that it is equivalent to the input phase ϕ_1 with perhaps the inclusion of an additive phase shift θ.[9]

$$\phi_2 = \phi_1 + \theta. \tag{5}$$

---

the operators A and B which represent physical observables a and b satisfy the commutation relation

$$AB - BA = iC,$$

then the uncertainties in the measurement of a and b satisfy the relation

$$\Delta a \Delta b = \tfrac{1}{2}\langle C \rangle,$$

where the term uncertainty stands for the root mean square deviation from expectation value, *e.g.*,

$$\Delta a = [\langle A^2 \rangle - \langle A \rangle^2]^{1/2}.$$

[9] These are operationally valid constructs since we can prepare two signal sources, one of which has an accurately known photon output and the other of which has an accurately known phase. If

Such an amplifier is linear in the sense that the phase is preserved and the output quanta are linearly related to the input quanta. It is perfect in that no noise is added. Note that frequency converters which derive their gain solely by the frequency conversion factor do not fall under this definition of a linear *amplifier* in that the photon gain G is unity.

Let us now attach to the perfect linear amplifier an ideal detector, ideal in the sense that it is capable of detecting the number n_2 of output photons and the output phase ϕ_2 within an uncertainty,

$$\Delta n_2 \Delta \phi_2 = \tfrac{1}{2}, \tag{6}$$

the minimum value allowed by the uncertainty principle. (See Fig. 1.) Thus we imagine that we make a measurement of the output photons and phase which together with the uncertainties introduced by the detector we write symbolically as $(n_2 \pm \Delta n_2)$ and $(\phi_2 \pm \Delta \phi_2)$.

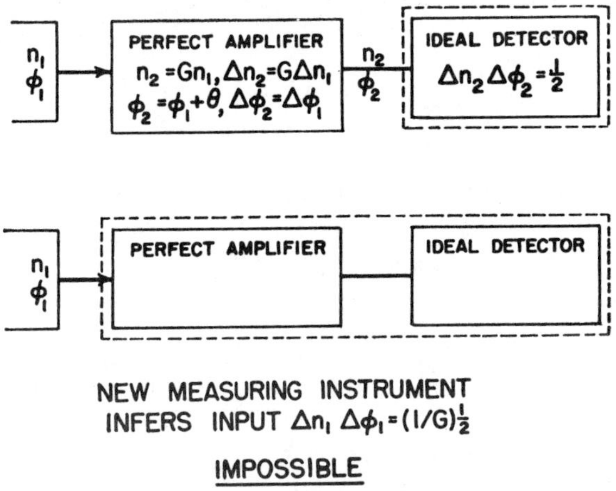

Fig. 1—Thought experiment to show the nonexistence of a perfect amplifier.

We now change our point of view and look upon the combination of perfect amplifier and ideal detector as a

---

each of these is applied in succession to the amplifier input and if in succession two detectors are applied to the amplifier output, one of which measures number of photons without giving phase information and another which measures phase without giving photon number information, we can determine the constants of the amplifier to arbitrary precision. Such a determination can, in fact, be made with a single signal source and a single detector if a sufficiently large signal level is used since the accuracy in the determination of phase and the relative accuracy in the simultaneous determination of photon number are each proportional to $1/n^{1/2}$.

single measuring instrument. The measurement of $(n_2 \pm \Delta n_2)$ and $(\phi_2 \pm \Delta\phi_2)$ when referred to the input of the amplifier implies an input number of photons $(n_1 \pm \Delta n_1) = 1/G(n_2 \pm \Delta n_2)$ and an input phase $(\phi_1 \pm \Delta\phi_1) = (\phi_2 \pm \Delta\phi_2) - \theta$. Thus the uncertainty in the measurement of input photons and phase is such that

$$\Delta n_1 \Delta\phi_1 = \frac{1}{G} \cdot \frac{1}{2}. \tag{7}$$

This conclusion is clearly impossible since it violates the uncertainty principle. Therefore our postulated perfect linear amplifier cannot exist. It must add some uncertainty, that is, noise.

Minimum Noise Temperature of a Linear Amplifier

We can pursue this argument even further to prove the following result. *The minimum noise temperature of a linear amplifier characterized by additive white Gaussian noise is*

$$T_n = \left[\ln \frac{2 - 1/G}{1 - 1/G}\right]^{-1} \frac{h\nu}{k}, \tag{8}$$

and the minimum mean square phase fluctuation is

$$\Delta\phi_a{}^2 = \frac{(G - 1)h\nu B}{2P}. \tag{9}$$

Here h is Planck's constant, ν is the frequency, k is Boltzmann's constant, B is the bandwidth and P is the signal power.

The proof of this assertion employs the same conceptual scheme used previously, a linear amplifier (no longer considered noiseless) followed by an ideal detector. Again a measurement is made by the detector of the output phase ϕ_2 and photon number n_2. This measurement will include as before the same uncertainties introduced by the detection process but will also include the uncertainties which we have seen must be added by the amplifier. Let us label the uncertainties introduced by the detector Δn_d and $\Delta\phi_d$ and those added by the amplifier Δn_a and $\Delta\phi_a$. If we assume the processes which give rise to these two sets of uncertainties are uncorrelated, then the total uncertainties as actually measured, Δn_2 and $\Delta\phi_2$, can be obtained from

$$\Delta n_2^2 = \Delta n_a^2 + \Delta n_d^2$$

$$\Delta\phi_2^2 = \Delta\phi_a^2 + \Delta\phi_d^2. \tag{10}$$

These equations merely assert that the variances of two uncorrelated random processes add.

Let us again shift our point of view and look upon the combination of amplifier and detector as a single measuring instrument. The measured uncertainties Δn_2, $\Delta\phi_2$ now imply an uncertainty in the measurement of the input phase and photon number given by

$$\Delta n_1^2 = \frac{1}{G^2}(\Delta n_a^2 + \Delta n_d^2)$$

$$\Delta\phi_1^2 = \Delta\phi_a^2 + \Delta\phi_d^2. \tag{11}$$

Our proof proceeds by first demanding that the inferred input uncertainties be the least allowed by the uncertainty principle, *i.e.*, $\Delta n_1 \Delta\phi_1 = \frac{1}{2}$. This condition insures that the amplifier uncertainty is the smallest possible. We then characterize this amplifier uncertainty by white Gaussian noise and determine the noise temperature corresponding to this minimum uncertainty.

First, however, we must make sure that the detector is matched to the amplifier, for although the product of the uncertainties $\Delta n_d \Delta\phi_d$ is set, their ratio is not. We can assure the best detection performance by minimizing the product $\Delta n_1 \Delta\phi_1$ given by (11) with respect to the ratio $(\Delta n_d/\Delta\phi_d)$ while still demanding that

$$\Delta n_d \Delta\phi_d = \tfrac{1}{2}. \tag{12}$$

This process results in the relation

$$\frac{\Delta n_d}{\Delta\phi_d} = \frac{\Delta n_a}{\Delta\phi_a} \tag{13}$$

which simply states that the minimum over-all uncertainty comes about when the detector measures number and phase with the same relative uncertainties as those introduced by the amplifier.

We must make sure that the detector is matched to the amplifier in another sense. Let us assume that the amplifier has a bandwidth B. This characteristic implies that the detector should sample the output at intervals of $\tau = \frac{1}{2}B$. If the interval is made longer, the full information transmission capabilities of the amplifier are not being used, while if it is shorter, not only are some of the data redundant, but also, because of the fluctuations in the output, the uncertainties are greater than necessary. Thus the matching of the detector to the amplifier implies two things, first that the ratios of the uncertainties are made equal and second that the time interval over which the number of output photons and the phase are detected is one half of the reciprocal bandwidth.

Let us now multiply (11) together, introduce the conditions of (12) and (13), and demand that the uncertainty in the measurements referred to the input be the minimum allowed by the uncertainty principle, that is

$$\Delta n_1 \Delta\phi_1 = \tfrac{1}{2}. \tag{14}$$

The resulting equation is

$$\frac{G^2 - 1}{4} = \Delta n_a{}^2 \Delta\phi_a{}^2 + \Delta n_a \Delta\phi_a. \tag{15}$$

If we put this equation in the form

$$\left(\frac{\Delta\phi_a}{\Delta n_a}\right)^2 \Delta n_a{}^4 + \left(\frac{\Delta\phi_a}{\Delta n_a}\right) \Delta n_a{}^2 - \left(\frac{G^2 - 1}{4}\right) = 0 \tag{16}$$

we can solve for $\Delta n_a{}^2$ in terms of the ratio $(\Delta n_a/\Delta\phi_a)$ to give

$$\Delta n_a{}^2 = \frac{G - 1}{2}\left(\frac{\Delta n_a}{\Delta\phi_a}\right). \tag{17}$$

Little more can be done unless we specify the nature of the amplifier uncertainty. Let us characterize the uncertainty as additive white noise. The statistical properties of a signal contaminated by white Gaussian noise have been extensively studied.[10] One result is that the probability density function for the output phase approaches a Gaussian distribution for large SNR's and has a variance given by

$$\Delta\phi_a{}^2 = \frac{\Delta P}{2P}. \tag{18}$$

Here ΔP is the noise power and P is the signal power. For large SNR's the variance in the power distribution is given by

[10] S. O. Rice, "Mathematical analysis of random noise," *Bell Sys. Tech. J.*, vol. 23, pp. 282–332, January, 1944; vol. 24, pp. 46–156, January, 1945; "Statistical properties of a sine-wave plus random noise," *Bell Sys. Tech. J.*, vol. 27, pp. 109–157, January, 1948.

$$\Delta P^2 = 2P\Delta P \tag{19}$$

so that (18) becomes

$$\Delta\phi_a{}^2 = \frac{\Delta P^2}{4P^2}\cdot \tag{20}$$

Since the integration time of the amplifier is $\tau = \frac{1}{2}B$, (20) can be put in terms of the variance of the number of photons $\Delta n_a{}^2$ detected during this interval

$$\frac{\Delta n_a}{\Delta\phi_a} = \frac{2P\tau}{h\nu}\cdot \tag{21}$$

This result inserted in (17) together with the relation

$$\Delta n_a{}^2 = \frac{\Delta P^2\tau^2}{(h\nu)^2} = \frac{2P\Delta P\tau^2}{(h\nu)^2} \tag{22}$$

gives for the minimum noise power introduced in the output of the amplifier

$$\Delta P = (G - 1)h\nu B. \tag{23}$$

The effective noise temperature T_n is obtained by dividing by the amplifier gain to refer the noise power to the input and then determining what temperature is required for a black body to radiate the same power. That is, we must find the value T_n for which

$$\frac{h\nu B}{e^{h\nu/kT_n} - 1} = \left(1 - \frac{1}{G}\right) h\nu B. \tag{24}$$

The result is

$$T_n = \left[\ln\frac{2 - 1/G}{1 - 1/G}\right]^{-1} \frac{h\nu}{k}\cdot \tag{8}$$

According to (19) and (23), the minimum mean square phase fluctuations are

$$\Delta\phi_a{}^2 = \frac{(G - 1)h\nu B}{2P}\cdot \tag{9}$$

These last two equations give relations for the best possible noise performance of any linear amplifier whose uncertainty is characterized by white Gaussian noise. It is interesting to note that in the limit of high gain, the minimum noise temperature becomes

$$T_n = \frac{1}{\ln 2} \frac{h\nu}{k}, \tag{25}$$

which is precisely the value obtained for the minimum noise temperature of the maser and the parametric amplifier derived from detailed consideration of the amplification mechanisms in the two cases. Thus we can say that both the maser and the parametric amplifier represent ideal amplifiers in so far as their ultimate noise performance is concerned.

We should also remark in passing, (8) and (9) indicate that the parametric up-converter which has power gain by virtue of the change in frequency but has unity photon gain ($G=1$), possesses a limiting noise temperature and phase uncertainty of zero. This result is in agreement with the detailed calculations of Louisell, Yariv and Siegman.[4] This sort of amplifier which does not multiply the *number* of photons, however, does not improve the capability of detecting a signal. The accuracy of the detection process at the output of the up-converter is no better than if it were performed at the input. The change in photon frequency and hence energy is immaterial since the limiting detector uncertainty is dependent solely on the number of photons arriving, not upon their energy.

The Implications of the Minimum Noise Limit

At conventional communications frequencies, the minimum noise temperature given by (8) is entirely negligible. At optical frequencies though, it can amount to several tens of thousands of degrees. Such a number is misleading, however, for it implies that the insertion of even the best amplifier has seriously degraded our ability to detect a signal. We can show that such is not the case by recasting the results of the previous section in slightly different form. Eqs. (22) and (23) can be combined to give the uncertainty produced by the amplifier in the number of photons as

$$\Delta n_a^2 = (G - 1)n_a, \tag{26}$$

and (9) can be rewritten in the form

$$\Delta\phi_a^2 = \frac{(G - 1)}{4n_a} \tag{27}$$

to give the uncertainty in phase produced by the amplifier. If we refer these quantities to the input we have

$$\Delta n_{ai}^2 = \frac{1}{G^2} \Delta n_a^2 = (1 - 1/G)n_1 \tag{28}$$

and

$$\Delta\phi_a^2 = \frac{(1 - 1/G)}{4n_1} \cdot \tag{29}$$

From (26) and (27) we see that if the amplifier gain is high, the output uncertainties introduced by the amplifier are considerably larger than those of even a poor matched detector. In this case, the total uncertainty Δn_1 in the inferred measurement of the input photon number and $\Delta\phi_1$ in the inferred measurement of input phase are closely given by (28) and (29), which relate to the amplifier alone. In the limit of high gain, the product of these uncertainties is

$$\Delta n_1 \Delta\phi_1 \cong \Delta n_{ai} \Delta\phi_a \cong \tfrac{1}{2}. \tag{30}$$

This is, of course, the minimum value allowed by the uncertainty principle. Thus the minimum noise amplifier allows us to use a poor detector, one which introduces uncertainties considerably larger than the minimum necessary, and still measure an incoming signal with an accuracy approaching the best allowed by the uncertainty principle. There still remains a question of what limitation is put on the rate of information transmission by this maximum allowable accuracy of detection. The answer to this question, however, must await the development of a quantum theory of communication.

VI.6 QUANTUM NOISE IN LINEAR AMPLIFIERS

H. A. HAUS AND JAMES A. MULLEN

The classical definition of noise figure, based on signal-to-noise ratio, is adapted to the case when quantum noise is predominant. The noise figure is normalized to "uncertainty noise." General quantum mechanical equations for linear amplifiers are set up using the condition of linearity and the requirement that the commutator brackets of the pertinent operators are conserved in the amplification. These equations include as special cases the maser, the parametric amplifier, and the parametric up-converter. Using these equations the noise figure of a general amplifier is derived. The minimum value of this noise figure is equal to 2. The significance of the result with regard to a simultaneous phase and amplitude measurement is explored.

INTRODUCTION

THE availability of coherent signals at optical frequencies has stimulated research in their use for communication purposes. Ways of processing optical frequencies are considered that are similar to those of the low end of the coherent frequency spectrum. With the use of classical communication techniques, classical performance criteria will be applied. One purpose of this paper is to extend classical noise performance criteria to linear quantum amplifiers in which *the predominant noise is quantum mechanical* in nature. These criteria will be applied to a wide class of linear quantum mechanical amplifiers.

The purpose of a sensitive linear amplifier is to increase the power, or photon flux, of an incoming signal with as small a noise contamination as possible so that the signal may be conveniently detected at high power levels. The incoming signal, if used for communication purposes, carries amplitude modulation, phase modulation, frequency modulation, or some other type of modulation. Here we shall discuss noise problems mainly in the context of amplifiers processing sinusoidal carriers with narrow band amplitude and/or phase modulation. In this connection it must be noted that the presence of a modulation of bandwidth B calls for a minimum rate of detection. The received signal must be detected within a succession of observation times each of duration τ, where $\tau=\frac{1}{2}B$, in order to utilize the information contained in the modulation.

Noise in masers, including sponteneous emission noise, has been analyzed in many papers including the classical papers by Shimoda, Takahasi, and Townes,[1] and Serber and Townes.[2] A quantum mechanical treatment of the parametric amplifier and up-converter has been presented in a paper by Louisell *et al.*[3] We shall develop a unified set of equations for all "linear" amplifiers, special cases of which are the maser, the parametric amplifier, and the parametric up-converter. On the basis of these equations and the criteria of noise performance, it will be possible to present a proof on the limiting noise performance achievable by any one of these amplifiers used singly or in combination with other linear amplifiers. The connection of the fundamental noise of these amplifiers with the uncertainty principle will be studied.

I. NOISE FIGURE

In the noise theory of classical amplifiers (i.e., amplifiers operating with a very large number of quanta) the deterioration of the signal-to-noise ratio as the signal passes the amplifier is used as a measure of amplifier noise performance. The signal-to-noise ratio (SNR) is defined in the classical limit as the ratio of signal power to noise power. Mathematically one may describe a phase and amplitude modulated signal in the presence of noise by

$$A(t)=A_0(t)\cos[\omega_0 t+\phi_0(t)]+\delta A_p\cos[\omega_0 t+\phi_0(t)] +\delta A_q\sin[\omega_0 t+\phi_0(t)]. \quad (1.1)$$

The first term in this equation represents the signal in the absence of noise. The remaining two terms are the inphase and quadrature perturbations of the amplitude due to the noise. These are slowly varying with time if the noise is narrowband. We envisage an ensemble of identical signal waveforms with accompanying noise. The signal part may be extracted from the waveform by taking an ensemble average indicated by the brackets $\langle\ \rangle$

$$\langle A(t)\rangle=A_0(t)\cos[\omega_0 t+\phi_0(t)]. \quad (1.2)$$

[1] K. Shimoda, H. Takahasi, and C. H. Townes, J. Phys. Soc. Japan **12**, 686 (1957).

[2] R. Serber and C. H. Townes, in *Quantum Electronics*, edited by C. H. Townes (Columbia University Press, New York, 1960), pp. 233–255.

[3] W. H. Louisell, A. Yariv, and A. E. Siegman, Phys. Rev. **124**, 1646 (1961).

Originally published in *Physical Review, 128*, 2407-13 (1962).

The noise part is extracted as follows:

$$\langle A(t)^2\rangle - \langle A(t)\rangle^2$$
$$= \langle \delta A_p{}^2\rangle \cos^2[\omega_0 t + \phi_0(t)] + \langle \delta A_q{}^2\rangle \sin^2[\omega_0 t + \phi_0(t)] + \langle \delta A_p \delta A_q\rangle \sin 2[\omega_0 t + \phi_0(t)]. \quad (1.3)$$

Additive stationary noise is characterized by

$$\langle \delta A_p{}^2\rangle = \langle \delta A_q{}^2\rangle \quad \text{and} \quad \langle \delta A_p \delta A_q\rangle = 0. \quad (1.4)$$

The noise as defined here measures the mean square deviation from the signal of the measured ensemble of waveforms all containing the same signal. If the noise is stationary, this mean square deviation becomes

$$\langle A(t)^2\rangle - \langle A(t)\rangle^2 = \langle \delta A_p{}^2\rangle. \quad (1.5)$$

A signal-to-noise ratio may be defined, based on a time average, over an observation time T, long compared to the inverse bandwidth of signal and noise. The signal power is

$$S = \frac{1}{T}\int_0^T A_0{}^2(t)\cos^2(\omega_0 t + \phi_0)dt = \tfrac{1}{2}[A_0{}^2(t)]_{\text{av}}, \quad (1.6)$$

where the square bracket indicates a time average.

The noise is defined correspondingly as

$$N = \frac{1}{T}\int_0^T \{\langle A^2(t)\rangle - \langle A(t)\rangle^2\}dt$$
$$= \tfrac{1}{2}\{\langle \delta A_p{}^2\rangle + \langle \delta A_q{}^2\rangle\}_{\text{av}} = \langle \delta A_p{}^2\rangle. \quad (1.7)$$

The last equality holds for stationary noise. The noise figure of an amplifier is defined as the signal-to-noise ratio at the input of the amplifier divided by the signal-to-noise ratio at the output[4]

$$F = (S_i/N_i)/(S_0/N_0). \quad (1.8)$$

In defining S_i/N_i one envisages measurements of the signal and noise at the amplifier input and output by a noise-free measurement apparatus. The noise figure is usually defined by choosing a standard input noise corresponding to thermal noise of the input source of 290°K. If the amplifier is linear,

$$S_0 = GS_i, \quad (1.9)$$

and the amplifier noise is additive,

$$N_0 = GN_i + N_a, \quad (1.10)$$

then, one has for the "excess noise figure," $F-1$,

$$F - 1 = N_a/GN_i. \quad (1.11)$$

The excess noise figure measures the increase of the mean square deviation of the normalized signal as caused by the amplifier noise.

When adapting the noise figure of linear amplifiers for the quantum case, one faces two problems. First of all, one must establish that linear quantum mechanical amplifiers, like linear classical amplifiers, introduce additive noise. Secondly, in the limit when quantum effects are of importance, physical measurements, in general, introduce uncertainties, i.e., noise, and it is not clear that the concept of a noise-free measurement as envisaged in the classical noise figure definition can be applied. Thus, for example, simultaneous measurements of amplitude and phase of an electric field cannot be carried out with arbitrarily high precision but must obey the principle of complementarity. It is clear, therefore, that special measures have to be taken if the uncertainty introduced by the measurement is to be negligible compared to the noise in the system.

In quantum theory, a physical quantity is described by an operator. The expectation value of the operator represents the result of a set of measurements on an ensemble of identically prepared systems. In the evaluation of quantum noise we shall consider field amplitude measurements on an ensemble of amplifiers all of which are fed by a transmitter signal that is nonthermally attenuated (such as the attenuation of a radiation field by the inverse square law). At the transmitter the signal has a classical power level and, thus, can be accurately phase and amplitude controlled. If a measurement of the electric field $E(t)$ at the receiver inputs is performed at the time t, the signal is defined by $\langle E(t)\rangle$, and the noise by[5] $\langle E(t)^2\rangle - \langle E(t)\rangle^2$. By performing many such measurements at random time instants delayed with respect to each other by times long compared to the inverse bandwidth of the receivers, one finds the signal power

$$S = [\langle E(t)\rangle^2]_{\text{av}}, \quad (1.12)$$

and the noise power

$$N = [\langle E(t)^2\rangle - \langle E(t)\rangle^2]_{\text{av}}. \quad (1.13)$$

Such measurements do not mutually intefere, and, therefore, are not accompanied by a fundamental uncertainty. These measurements are analogous to the "noise-free" measurements implied in the classical signal-to-noise ratio definition. The signal-to-noise ratio is thus

$$S/N = [\langle E(t)\rangle^2]_{\text{av}}/[\langle E(t)^2\rangle - \langle E(t)\rangle^2]_{\text{av}}. \quad (1.14)$$

With the aid of definition (1.14), Eq. (1.8) may be used as the definition for the quantum noise figure. In the case of linear quantum amplifiers, as discussed in this paper, Eqs. (1.9), (1.10), and (1.11) are also made valid. It is not expedient to normalize the quantum noise figure to thermal noise. It is more appropriate to normalize it to the minimum noise within the observation time τ of $h\nu/2$ that is associated with attenuated signals that before attenuation were classically phase and amplitude controlled (see Appendix). Then, if we express the amplifier noise power multiplied by the observation time τ in terms of a photon number n_a at

[4] H. T. Friis, Proc. Inst. Radio Engrs. 33, 458 (1945).

[5] R. Senitzky, Phys. Rev. 111, 3 (1958).

the amplifier output,

$$N_a = h\nu n_a, \tag{1.15}$$

one has

$$F-1=2n_a/G. \tag{1.16}$$

Since the basic noise energy to which we compare the amplifier noise corresponds to half a photon within the observation time, it is natural to measure the output noise content, referred to the input by division through G, in terms of the energy of half a photon *at the output frequency*. In this sense, Eq. (1.16) can be applied to the general case for different input and output frequencies if G is interpreted as the *photon number gain*.

II. THE EQUATIONS OF LINEAR AMPLIFIERS

Consider a system of weakly interacting particles that, in the absence of radiation, has N available energy levels. If the particles obey Bose-Einstein statistics, the entry of a particle into a particular level (i) and its exit from this level may be described by creation and annihilation operators, $b_i^\dagger$ and b_i that obey commutator relations.

$$[b_i, b_j^\dagger] = \delta_{ij}, \tag{2.1}$$

$$[b_i, b_j] = 0. \tag{2.2}$$

The operation of b_i on a wave function of the system with n_i particles in level i produces a wave function with $n_i - 1$ particles in that level, the operation of $b_i^\dagger$ produces one with $n_i + 1$ particles in the level i. The number of particles in level i is given by

$$n_i = b_i^\dagger b_i. \tag{2.3}$$

If the particles obey Fermi-Dirac statistics, a Bose-Einstein description is possible,[2] provided that the number of available states in a particular level is much greater than the number of particles occupying it. If the states within a particular energy level are all equivalent for the physical processes envisaged, wave functions may be constructed that correspond to a discrete number of particles in a particular energy level with no distinction made between different states within that level.[6] Creation operators $b_i^\dagger$ and annihilation operators b_i that increase or decrease the number of particles in a particular energy level upon operation on a wave function can be shown to obey the commutation relations (2.1) and (2.2), provided the number of particles in the energy level is much less than the number of states of the level.[7] Linear quantum amplifiers must of necessity operate by means of transitions from and to weakly occupied levels since nonlinear effects occur as soon as the deviation from Bose-Einstein behavior of these levels becomes appreciable.

[6] For example, if one is interested only in the interaction with an electromagnetic field of the spin of weakly interacting particles, all states of the particles that have the same spin are equivalent.

[7] H. A. Haus, Internal Memorandum, Massachusetts Institute of Technology Research Laboratory of Electronics, 1962 (unpublished).

The electromagnetic field is represented by the creation and annihilation operators $a_i^\dagger$ and a_i for photons in the modes i, j, etc., of the electromagnetic system. These obey the usual commutation relations

$$a_i a_j^\dagger - a_j^\dagger a_i = \delta_{ij}, \tag{2.4}$$

$$a_i a_j - a_j a_i = 0. \tag{2.5}$$

When the fields are made to interact with matter, coupling results among the various operators of the system. In the interaction representation, differential equations in time are obtained relating the time rate of change of every operator to all the other operators of energy levels and electromagnetic modes partaking in the interaction. These equations are, in general, nonlinear. However, under proper "biasing" or "pumping" conditions linear interactions may result among some of the energy-level and photon operators of the system. Thus, consider a system like the three-level maser in which a pumping excitation establishes a steady state in the occupation of two energy levels. If a small signal is applied with a frequency corresponding to the energy difference between one of the pumped levels and an intermediate level, the small perturbations produced in the steady state of the strongly excited levels may be disregarded, and linear equations are obtained for the operators of the weakly excited level and the photons of the small applied signal. In more complicated devices, more level and photon operators may be particpating in the interaction initiated by the applied small signal. In any case, if the device is to act as a linear amplifier for some photons it is a necessary requirement that the equations of motion permit a linearization of the equations for the signal photon operators and the operators directly involved in the small signal interaction. If the amplifier is to be time independent, a further requirement is that the coefficients of the differential equations be time independent. These two requirements strongly restrict the form of the equations. Integration of the differential equations in time leads to linear relations among the operators at an initial time t_0, and final time t_1.

Taking as an example a system within which a photon operator interacts with an energy level operator, one has two possible cases:

$$\begin{aligned} a(t_1) &= M_{aa}a(t_0) + M_{ab}b^\dagger(t_0), \\ b(t_1) &= M_{ba}a(t_0) + M_{bb}b^\dagger(t_0), \end{aligned} \tag{2.6}$$

or

$$\begin{aligned} a(t_1) &= M_{aa}a(t_0) + M_{ab}b(t_0), \\ b(t_1) &= M_{ba}a(t_0) + M_{bb}b(t_0). \end{aligned} \tag{2.7}$$

The first system of equations is of the type obtained by Serber and Townes[2] for the ideal maser. In this system, the creation operator $b^\dagger$ of the energy level couples to the annihilation operator a of the photons. In the second system the annihilation operators of the energy level b couples to a. We shall later see the significance of this latter type of coupling. An analysis of the parametric amplifier leads to equations of the form (2.6) where $b^\dagger$ is replaced by a creation operator representing the photons at the idler frequency.[3]

The general form of the linear equations is immediately apparent for a linear system of arbitrary degree of complexity. If one comprises all operators representing "output" quantities at $t=t_1$ in a column matrix $\mathbf{v}$, those representing "input" quantities at $t=t_0$ in a column matrix $\mathbf{u}$, one may write the general linear relations in matrix form as

$$\mathbf{v}=\mathbf{M}\mathbf{u}. \tag{2.8}$$

In the example of Eq. (2.6)

$$\mathbf{u}=\begin{bmatrix} a(t_0) \\ b^\dagger(t_0) \end{bmatrix}, \quad \text{and} \quad \mathbf{v}=\begin{bmatrix} a(t_1) \\ b^\dagger(t_1) \end{bmatrix}.$$

The requirements of linearity and time independence have led to an equation of the form of (2.8). There is another important condition that has to be met by Eq. (2.8) which restricts the matrix $\mathbf{M}$: The commutator brackets of the operators must be conserved. Indeed, considering a system with photons and Bose-Einstein particles, the commutator brackets are invariants of the complete (nonlinear) equations of motion. The linearized equations that describe the small signal interactions, must conserve commutator brackets insofar as they are good approximations to the behavior of the same set of operators in the complete equations. Consider next a system with photons and Fermi-Dirac particles. One notes that the complete equations of motion preserve the commutator brackets of the photon operators and anticommutator brackets of the particle operators. As mentioned before, one may construct effectively Bose-Einstein operators from the input Fermi-Dirac operators. Corresponding operators may be constructed for the output. These will also be effectively Bose-Einstein provided the number of particles in the level of interest has remained small compared to the number of states in the level (as needs be if the amplification is to be linear). Therefore, the commutator brackets of the constructed operators are also conserved in the amplification. Let us study the matrix $\mathbf{C}$ consisting of the commutator brackets of the operators contained in the input matrix $\mathbf{u}$. It is easily seen that this matrix is constructed by

$$\mathbf{C}=\mathbf{u}\mathbf{u}^\dagger-(\mathbf{u}_t{}^\dagger\mathbf{u}_t)_t. \tag{2.9}$$

Here the subscripts t indicate the transpose. The operators obey initially the commutation relations (2.1), (2.2), (2.4), and (2.5). Photon operators and level operators commute. The commutator relations can be conveniently summarized by defining a matrix $\mathbf{P}$ of the same order as $\mathbf{u}$ by

$$\mathbf{P}=\text{diag}(\pm1,\pm1,\pm1,\cdots\pm1), \tag{2.10}$$

where for the ith diagonal element, the plus sign is chosen when an annihilation operator appears in the ith row of the matrix $\mathbf{u}$, and the minus sign if a creation operator appears. Using this definition for $\mathbf{P}$, one has from Eq. (2.9)

$$\mathbf{C}=\mathbf{u}\mathbf{u}^\dagger-(\mathbf{u}_t{}^\dagger\mathbf{u}_t)_t=\mathbf{P} \tag{2.11}$$

at $t=t_0$. Because

$$\mathbf{C}(t_1)=\mathbf{C}(t_0), \tag{2.12}$$

one has from Eqs. (2.8) and (2.12)

$$\begin{aligned}\mathbf{v}\mathbf{v}^\dagger-(\mathbf{v}_t{}^\dagger\mathbf{v}_t)_t=\mathbf{P}&=\mathbf{M}\mathbf{u}\mathbf{u}^\dagger\mathbf{M}^\dagger-[(\mathbf{u}^\dagger\mathbf{M}^\dagger)_t(\mathbf{M}\mathbf{u})_t]_t\\&=\mathbf{M}(\mathbf{u}\mathbf{u}^\dagger-\mathbf{u}_t{}^\dagger\mathbf{u}_t)\mathbf{M}^\dagger=\mathbf{M}\mathbf{P}\mathbf{M}^\dagger.\end{aligned}$$

It follows that

$$\mathbf{M}\mathbf{P}\mathbf{M}^\dagger=\mathbf{P}. \tag{2.13}$$

One may easily check that in the special cases of an ideal maser and parametric amplifier mentioned previously, the expressions as derived in references 2 and 3 obey the relation (2.13). Here this relation has been derived in general as a consequence of the requirement of conservation of commutator brackets. It is this relation that imposes a fundamental limit on the noise performance of all linear amplifiers. It is of interest to show that in an amplifier with photon gain greater than unity the signal creation (annihilation) operator must couple to at least one annihilation (creation) operator of a molecular state or of a photon of frequency different from the signal frequency. Suppose the output photons pertain to the operator a_j, the input photons to a_k. Let these, in turn, correspond to v_m or $v_m{}^\dagger$ and u_n or $u_n{}^\dagger$. If the signal level is sufficiently large, one may disregard the uncertainty contained in the commutator relations, i.e., one may set

$$\langle v_m{}^\dagger v_m\rangle=n_m\cong\langle v_m v_m{}^\dagger\rangle, \tag{2.14}$$

where n_m is the number of output photons,

$$\langle u_n{}^\dagger u_n\rangle=n_n\cong\langle u_n u_n{}^\dagger\rangle, \tag{2.15}$$

with n_n, the number of input photons. All the other input quantities are assumed unexcited

$$\langle u_i{}^\dagger u_i\rangle=0,\quad \langle u_i u_i{}^\dagger\rangle=1,\quad i\neq n. \tag{2.16}$$

We thus have from Eqs. (2.8), (2.14)–(2.16),

$$n_m\cong|M_{mn}|^2 n_n. \tag{2.17}$$

$|M_{mn}|^2$ is the photon gain of the amplifier. But, from Eq. (2.13)

$$\textstyle\sum_i P_{ii}|M_{mi}|^2=P_{mm}. \tag{2.18}$$

We recall that $P_{ii}=\pm1$, depending upon whether u_i stands for an annihilation operator or creation operator. Thus, this equation shows that $|M_{mn}|^2$ is necessarily less than unity, unless at least one of the P_{ii}'s is of opposite sign to P_{mm}. A simple situation exists when u_m interacts with only one other operator u_l,

$$|M_{mm}|^2+(P_{ll}/P_{mm})|M_{ml}|^2=1. \tag{2.19}$$

The ideal maser, Eq. (2.6), uses the same input and output frequency, the photons of which couple to a single level operator $b^\dagger$ $(=u_l)$. The gain $|M_{mm}|^2$ is greater than unity, because P_{ll}/P_{mm} is equal to -1.

The parametric amplifier using the same input and output frequency couples the signal photon operator $a_1=u_m$ to the idler photon operator $a_2{}^\dagger=u_l$. Again $P_{ll}/P_{mm}=-1$ and the gain $|M_{mm}|^2>1$.

The ideal parametric up-converter couples the input signal frequency operator $a_1=u_m$ to the output frequency operator $a_2=u_l$. $P_{ll}/P_{mm}=1$ and the gain $|M_{ml}|^2\leqslant 1$.

A system within which a photon operator $a=u_m$ couples to a level operator $b=u_l$ exhibits gain less than, or at most equal to, unity since $P_{ll}/P_{mm}=1$. [Compare Eq. (2.7).]

An interesting special case is the "lossless scatterer," with the P_{ii}'s all of the same sign. The coupling between a transmitting antenna and a receiver may be represented in this way. Here all u's represent annihilation operators a_i pertaining to the same frequency, but different spatial modes. The relation (2.13) assures conservation of the total photon number, but only a fraction $|M_{jk}|^2$ of the input (transmitter) photons represented by, say $\langle u_k{}^\dagger u_k\rangle$ reach the receiver as output photons $\langle v_j{}^\dagger v_j\rangle$.

III. AMPLIFIER NOISE

It should be recalled that the u's contain the a's and b's, or their Hermitian conjugates. In the study of signal and noise we are interested in averages of expectation values of electric field components, whose operators are given by

$$E_j(t)=k_j i(a_j{}^\dagger e^{i\omega_j t}-a_j e^{-i\omega_j t}), \tag{3.1}$$

where k_j is some real constant dependent upon the geometry of the system.

Let the photon operator a_j, or $a_j{}^\dagger$, be represented by u_j, so that

$$\langle E_j(t)\rangle=\pm k_j i\langle u_j{}^\dagger e^{\pm i\omega_j t}-u_j e^{\mp i\omega_j t}\rangle$$
$$=\pm 2k_j|\langle u_j\rangle|\cos(\omega_j t+\Phi), \tag{3.2}$$

and

$$[\langle E_j(t)\rangle^2]_{\text{av}}=2k_j{}^2[|\langle u_j\rangle|^2]_{\text{av}}=2k_j{}^2[|\langle u_j{}^\dagger\rangle|^2]_{\text{av}}, \tag{3.3}$$

because the average indicated by the square bracket is equivalent to a time average. Similarly,

$$[\langle E_j(t)^2\rangle]_{\text{av}}=k_j{}^2[\langle u_j{}^\dagger u_j+u_j u_j{}^\dagger\rangle]_{\text{av}}. \tag{3.4}$$

Since the final expressions are symmetric in u and $u^\dagger$ and do not depend on time, both cases $u_j=a_j$ and $u_j=a_j{}^\dagger$ lead to the same final expressions for $[\langle E(t)\rangle^2]_{\text{av}}$ and $[\langle E(t)^2\rangle]_{\text{av}}$. Thus one may use the u's directly to obtain the required averages.

Suppose the matrix $\mathbf{M}$ of Eq. (2.8) represents an amplifier with its input described by the operator u_n and its output by v_m. Then the output signal-to-noise ratio is given by [compare Eqs. (1.14) and (3.2)–(3.4)]

$$\frac{S_0}{N_0}=\frac{[2|\langle v_m\rangle|^2]_{\text{av}}}{[\langle v_m{}^\dagger v_m\rangle+\langle v_m v_m{}^\dagger\rangle-2|\langle v_m\rangle|^2]_{\text{av}}}. \tag{3.5}$$

Now, from Eq. (2.8)

$$v_m=\sum_i M_{mi}u_i. \tag{3.6}$$

Because all inputs but the signal input m are fed by incoherent noise, we have for i, $j\neq m$, n

$$\langle u_i\rangle=\langle u_i{}^\dagger\rangle=0, \tag{3.7}$$

$$\begin{aligned}\langle u_i{}^\dagger u_j\rangle&=0,\ \text{if } u_i \text{ stands for an annihilation operator,}\\ &=\delta_{ij},\ \text{if } u_i \text{ stands for a creation operator,}\end{aligned} \tag{3.8}$$

$$\begin{aligned}\langle u_i u_j{}^\dagger\rangle&=0,\ \text{if } u_i \text{ stands for a creation operator,}\\ &=\delta_{ij},\ \text{if } u_i \text{ stands for an annihilation operator.}\end{aligned}$$

Thus

$$\langle v_m\rangle=M_{mn}\langle u_n\rangle, \tag{3.9}$$

and

$$\langle v_m{}^\dagger v_m\rangle+\langle v_m v_m{}^\dagger\rangle-2|\langle v_m\rangle|^2$$
$$=\sum_{i\neq n}|M_{mi}|^2+|M_{mn}|^2\{\langle u_n{}^\dagger u_n+u_n u_n{}^\dagger\rangle-2|\langle u_n\rangle|^2\}. \tag{3.10}$$

But the input noise is taken as the noise accompanying a signal after a large radiative attenuation. This noise expressed in terms of a photon number is equal to the uncertainty noise of $\frac{1}{2}$ photon as shown in the Appendix. The input signal-to-noise ratio to which the noise figure will be normalized has the same form as Eq. (3.5) except that u_n replaces v_m and is equal to

$$S_i/N_i=2[n_n]_{\text{av}}, \tag{3.11}$$

where $[n_n]_{\text{av}}$ is the average number of input photons. Introducing Eqs. (3.9)–(3.11) into Eq. (3.5) and using the noise figure definition (1.8), we have

$$F=(\sum_i|M_{mi}|^2)/|M_{mn}|^2. \tag{3.12}$$

The input signal has dropped out as is characteristic of a linear amplifier. The noise figure F of (3.12) can be made unity. It is found, however, that the gain $|M_{mn}|^2$ of the "amplifier" is then also necessarily equal to unity. If one wants gain, F must be optimized for fixed gain. Another way of accomplishing this is to use the definition of "noise measure"[8] M,

$$M=(F-1)/(1-1/G). \tag{3.13}$$

Here G is the photon number gain, in the present case $|M_{mn}|^2$. The noise measure M has a nontrivial minimum as we now proceed to show. We obtain

$$M=(\sum_{i\neq n}|M_{mi}|^2)/(|M_{mn}|^2-1). \tag{3.14}$$

[8] H. A. Haus and R. B. Adler, *Circuit Theory of Linear Noisy Networks* (Technology Press, Cambridge, Massachusetts, 1959).

According to Eq. (2.13)

$$\sum_i P_{ii}|M_{mi}|^2 = P_{mm}.$$

Thus

$$|M_{mn}|^2 - 1 = -\sum_{i\neq n}(P_{ii}/P_{nn})|M_{mi}|^2 + (P_{mm}/P_{nn}) - 1. \quad (3.15)$$

Since we are, in general, interested in photon gain, $|M_{mn}|^2 > 1$, it follows that the right-hand side of (3.15) must be positive. In such a case, since $P_{ii}/P_{nn} = \pm 1$, it follows immediately that

$$\sum_{i\neq n}|M_{mi}|^2 \geq -\sum_{i\neq n}(P_{ii}/P_{nn})|M_{mi}|^2 + (P_{mm}/P_{nn}) - 1. \quad (3.16)$$

Thus, from (3.14)–(3.16) we find

$$M > 1. \quad (3.17)$$

At high gain $M = F - 1$. It follows that in that case

$$F \geq 2. \quad (3.18)$$

The quantum noise introduced in an amplifier leads to a noise-to-signal ratio at its output double of that at its input, if the input signal-to-noise ratio is assumed to be that of uncertainty noise. Considering specific devices, we note that the ideal maser characterized by Eq. (2.6) has a noise measure of unity, because the equality sign in Eq. (3.18) applies, as one can easily see. The same is true for the parametric amplifier. In general, one can see that coupling of the signal photon annihilation operator to other annihilation operators tends to increase the noise figure by virtue of the fact that negative terms appear in the sum on the right-hand side of Eq. (3.16).

We have introduced the concept of noise measure because it possessed a nontrivial minimum, whereas the noise figure could be made equal to unity at a complete sacrifice of gain. There are other advantages in the use of "noise measure" that have been successfully employed in the analysis of noise in classical linear amplifiers.[8] Using the concept of noise measure one may state in simple terms the limit on the optimum noise performance of a passive interconnection of linear two-port amplifiers (amplifiers with one input terminal pair and one output terminal pair). In the classical circuit-theoretical application for which this theorem was originally derived it states[8] that any passive interconnection of two-port amplifiers resulting in an overall two-port amplifier (amplifier with one input terminal pair and one output terminal pair) leads to an optimum (minimum) noise measure that cannot be lower than that of the best amplifier, namely, the amplifier with the lowest value of optimum noise measure. This proof can now be easily extended to the case of an interconnection of linear quantum amplifiers. In the quantum case, two-port amplifiers are those that are described by one input signal photon operator and one output signal photon operator as discussed in this section. A passive interconnection is one with a net positive (or zero) internal loss of photons. Systems with linear frequency transformations are not ruled out. Thus the proof in the present section, extended along the lines of reference 8, in effect shows that the optimum noise measure of a linear quantum mechanical system with one input and one output cannot be better than unity ($F = 2$ at high gain).

IV. SIMULTANEOUS PHASE AND AMPLITUDE MEASUREMENTS

The technical purpose of amplifiers is to raise signal levels so that signal processing may be effected at conveniently high power levels. In the process, amplifiers must introduce as little noise as possible. Since we have found that all linear quantum amplifiers introduce noise, one may ask the question whether, in those cases in which ultimate sensitivity is desired, one should not dispense with linear amplification. We shall discuss this question as one of principle, although it is clear that technical requirements may call for linear amplification for reasons other than those considered here.

We have defined the quantum noise figure on the basis of a noise-free measurement. In this context it is immediately apparent that the use of an amplifier is not desired in those cases in which a "noise-free" measurement can be administered to the incoming signal, namely, an instantaneous amplitude measurement alone.

If one envisages simultaneous amplitude and phase measurement the situation is not that simple and deserves further study.[2,9] Suppose we intend to measure with an ideal measuring apparatus the phase ϕ_0 and amplitude A_0 of an incoming wave. The measurement introduces an rms uncertainty in phase $\Delta\phi$ and in photon number Δn such that at best

$$\Delta n \Delta\phi = \tfrac{1}{2}. \quad (4.1)$$

But if A_0^2 is measured in units of power, Δn is related to the inphase uncertainty of amplitude ΔA_p by (cf. Eq. (1.1)]

$$\tau A_0(\langle\delta A_p^2\rangle)^{1/2} = \tau A_0 \Delta A_p = h\nu\Delta n, \quad (4.2)$$

where ΔA_p is defined as $(\langle\delta A_p^2\rangle)^{1/2}$. Further

$$\Delta\phi = \Delta A_q / A_0. \quad (4.3)$$

We thus have

$$\tau \Delta A_p \Delta A_q = \tfrac{1}{2} h\nu \quad (4.4)$$

for an ideal measurement apparatus. If the measurement apparatus is preceded by an ideal linear amplifier of high gain then the measurement does not have to introduce an uncertainty beyond that introduced by the amplifier noise. The minimum amplifier noise at high gain referred to the input is equal to that caused by quantum attenuation and, when measured in terms of energy, is according to Eq. (3.18) equal to half a

[9] H. Heffner, Proc. Inst. Radio Engrs. 50, 1604 (1962).

photon within an observation time. Further, the noise is stationary. It follows from (1.4)

$$\langle \delta A_p^2 \rangle = \langle \delta A_q^2 \rangle. \tag{4.5}$$

Using Eq. (1.7),

$$\tau \langle \delta A_p^2 \rangle = \tau (\Delta A_p)^2 = \tfrac{1}{2} h\nu. \tag{4.6}$$

Using (4.5) and (4.6) we find that the uncertainty introduced into the final measurement by the amplifier noise is just equal to that introduced by an ideal detector not preceded by an amplifier, Eq. (4.4). It is characteristic of linear amplifiers that the final measurement results in equal inphase and quadrature uncertainties whereas a measurement performed without the use of a preamplifier may choose the relative magnitudes of each, subject only to Eq. (4.1).

ACKNOWLEDGMENTS

The authors gratefully acknowledge helpful discussions with Professor F. M. H. Villars of MIT and Dr. F. Horrigan, presently of Saclay.

APPENDIX

Radiative Attenuation Noise

Using Eq. (2.8) to represent radiative attenuation, we interpret u_l as the annihilation operator of the input photons at the transmitter, v_k as the annihilation operator of the output photons at the receiver. All the other operators in **u** represent spatial modes not used for the transmission process and are unexcited, $\langle u_i \rangle = 0$, $\langle u_i^\dagger u_i \rangle = 0$, $\langle u_i u_i^\dagger \rangle = 1$ for $i \neq l$. We have for the signal-to noise ratio at the receiver, as in Eq. (3.5),

$$\left(\frac{S}{N}\right)_{\text{rec}} = \frac{2[|\langle v_k \rangle|^2]_{\text{av}}}{[\langle v_k^\dagger v_k + v_k v_k^\dagger \rangle - 2|\langle v_k \rangle|^2]_{\text{av}}}. \tag{A1}$$

Introducing Eq. (2.8), where according to Eq. (2.13), with $\mathbf{P} = \mathbf{1}$,

$$\mathbf{MM}^+ = \mathbf{1} \tag{A2}$$

one has

$$\left(\frac{S}{N}\right)_{\text{rec}} = \frac{2|M_{kl}|^2[|\langle u_l \rangle|^2]_{\text{av}}}{(|M_{kl}|^2[u_l^\dagger u_l + u_l u_l^\dagger - 2|\langle u_l \rangle|^2]_{\text{av}} + \sum_{i \neq k} |M_{ki}|^2)}. \tag{A3}$$

The first term in the denominator represents the attenuated noise of the transmitted signal. The additive noise introduced in the transmission process is represented by $\sum_{i \neq k} |M_{kl}|^2$ and is stationary. If the attenuation is very large, $|M_{kl}|^2 \ll 1$, and if the noise accompanying the signal at the input is not inordinately larger than that imposed by the uncertainty relation, then this term is negligible compared to the second one. Then the noise at the receiver is entirely determined by the zero-point fluctuations of the modes other than the one used for transmission regardless of the input noise. Further, using Eq. (A2)

$$\sum_{i \neq l} |M_{ki}|^2 \cong \sum |M_{ki}|^2 = 1. \tag{A4}$$

Finally,

$$|M_{kl}|^2[|\langle u_l \rangle|^2]_{\text{av}} \cong [n]_{\text{av}},$$

the average number of received photons as long as this number is much larger than unity. We find

$$(S/N)_{\text{rec}} \cong 2[n]_{\text{av}}.$$

VI.7 OPTICAL CHANNELS: PRACTICAL LIMITS WITH PHOTON COUNTING

JOHN R. PIERCE

Abstract—**In optical communication, ideal amplification of the received signal leads to a limiting signaling rate of 1 nat per photon. This is much inferior to the optimum limit of kT joules/nat, which we can theoretically approach by counting photons. Practically, the rates we can attain by photon counting will be limited by how elaborate codes we can instrument rather than by thermal photons.**

Consider a free-space path such as we might have between space vehicles. For a wavelength λ, a distance L and transmitting antennas of effective areas A_T and A_R the ratio of received power P_R to transmitted power P_T is[1]

$$P_R/P_T = A_T A_R/\lambda^2 L^2. \tag{1}$$

This suggests the use of a short wavelength.

Going to optical wavelengths requires very smooth antenna surfaces and very precise pointing. Further, at optical frequencies we encounter quantum effects. Here we disregard antenna and pointing problems and consider how quantum effects will limit a communication system.

In receiving a signal mixed with Johnson noise, we have the option of amplifying the signal with an ideal amplifier of power gain G and bandwidth B. The Gaussian noise power P_n in the output of the amplifier will be[2,3,6]

$$P_n = \frac{hf}{e^{hf/hT} - 1} B + (G - 1)hfB. \tag{2}$$

When $hf \ll kT$, the noise power density at the input of the amplifier is nearly kT. According to Shannon[4,5] when noise of this power density is added to a signal, the limiting information rate R in nats per joule of transmitted power, which is attained as B approaches infinity, is

$$R = 1/kT \text{ nats/joule}. \tag{3}$$

Paper approved by the Editor for Communication Theory of the IEEE Communications Society for publication without oral presentation. Manuscript received April 7, 1978; revised June 14, 1978.

The author is with the Department of Electrical Engineering, California Institute of Technology, Pasadena, CA 91125.

Originally published in *IEEE Transactions on Communications, COM-26*, 1819-21 (1978).

When $hf \gg kT$ the second term on the right of (2) dominates. We see that this second term is *not* amplified noise because hfB is multiplied by $(G - 1)$, not G. However, when the gain G is very large, the second term is very nearly equal to a Gaussian noise density hf multiplied by the gain G. Thus, following Shannon the limiting information rate in nats per joule of transmitted power will be

$$R = 1/hf \text{ nats/joule.} \tag{4}$$

The energy per photon is hf. Thus, the limiting rate given by (4) is 1 nat per photon. This limit holds only if we amplify

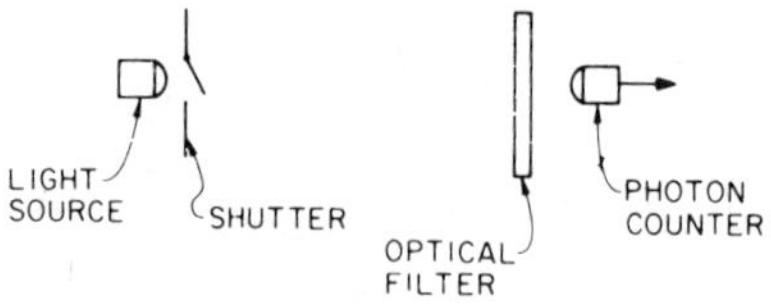

Figure 1. A Photon Channel. A light source transmits photons during intervals when the shutter is open. Received signal and thermal photons pass through an optical filter intended to eliminate out-of-band thermal photons. Each photon that passes the optical filter is counted by a photon counter. In a real system, the light source and shutter could be replaced by a pulsed laser.

the received signal with an ideal amplifier. It *does not* hold for simple photon counting or for various other forms of transmission and reception.[6-9]

Indeed, if no noise is added to the signal, the number of nats we can transmit per photon is unbounded.[6-9] We can see this by a very simple argument.

Consider the signaling system shown in Figure 1. The transmitter consists of a light source and a shutter.* The receiver consists of an optical filter and a photon counter, which we will assume to emit a pulse when one or more photons strike it. The purpose of the optical filter is to exclude thermal photons. Initially, we will assume that there are no thermal photons and we will dispense with the optical filter.

In signaling, we assume a code word whose length is N time intervals, each of duration t. Each of these code words has a "pulse" in one time interval only, so that the form of signaling is quantized pulse position modulation.

In signaling, we open the shutter during only one time interval out of the N in the code word. If we impose no bandwidth limitation in the transmission path, so that classically the path has a constant loss and delay, we can never receive a

* In a "practical" system we could send a pulse by pulsing a semiconductor laser rather than by opening a shutter.

photon during any time interval except the one corresponding to the time interval during which the shutter was open**.

Let us assume that the average number of photons received in the time interval corresponding to the opening of the shutter is M. If M is fairly large, we will almost always receive at least one photon, and thus we will almost always receive a proper code word. The information per code word will then be

$\ln N$ nats.

The average number of photons per code word is

M photons.

Hence, the number of nats per photon is

$$(\ln N)/M \text{ nats/photon.} \qquad (5)$$

By making N large, we can make the information rate in nats/photon as large as we wish. But growth is only as the logarithm of N, and is impractical to make $\ln N$ very large.

What about the effects of thermal photons in limiting the rate of signaling? Oddly enough, the theoretical limit for very small signal strength and unlimited bandwidth is the classical limit given by (3), that is, kT joules/nat.[6,7] I have included in Appendix A a simple demonstration that for a coherent signal source, kT joules/nat can be approached when $hf \gg kT$.

Let us now ask, how many nats per photon do we need in order to attain the limiting kT joules/nat?

A photon has an energy hf. If we require kT joules/nat, the number Q of nats per photon must be

$$Q = hf/kT = 4.80 \times 10^{-11}(f/T) \text{ nats/photon.} \qquad (6)$$

How big will Q be for optical frequencies and low temperatures? Let us assume a wavelength of 5,000 Å, corresponding to a frequency of 6×10^{14} Hz, and a temperature of 6K. For these values,

$Q = 4{,}800$ nats/photon.

** One can quibble here or later that P_R/P_T in (1) changes with frequency or that propagation near the source may be slightly dispersive, or that the actual energy of the emitted photon cannot be known. But, for a signaling rate far less than the frequency of the light used, transmission is practically the same for all frequency components of the signal, and the fractional range in photon energies is extremely small.

If we tried to attain this by means of quantized pulse position modulation, length N of our code words would have to be at least

$$N = e^{4,800}.$$

This, of course, is ridiculous.

The practical conclusion is that in optical signaling at low temperatures we encounter insuperable problems of encoding long before we approach the theoretical limit of kT joules/nat.

This seems plausible from another point of view. At what rate do we receive thermal photons? P_n of (2) gives the thermal photon power in one mode of propagation if we set $G = 1$. The rate p at which we receive thermal photons in a coherent (one-mode) system is thus

$$p = (e^{hf/kT} - 1)^{-1} B \text{ photons/second.} \tag{7}$$

Let us set $B = 1$, which allows time intervals t of around 1 s. For $B = 1, f = 6 \times 10^{14}$ and $T = 6$,

$$p = e^{-4,800}.$$

This is almost no noise photons per time interval.

With codes of any reasonable length and elaborateness, we will fall far short of kT joules/nat, so far short that we can afford to ignore thermal photons. For optical frequencies and low temperatures, the rate at which we can signal, measured in nats per photon, will be limited by our ability to implement codes, not by thermal photons.

APPENDIX

Here we will consider a single-mode photon communication system in which there is a noise source of photons with an average of n_0 photons per time interval, and a signal source of photons with an average of n photons per time interval when the transmitter is *on* (when the shutter is open).

According to Shannon,[4] the rate of transmission R measured in nats per symbol (that is, per time interval) is

$$R = H(Y) - H(Y \mid X). \tag{A-1}$$

Here X represents shutter position (0 = closed; 1 = open) and Y represents photons received (0 = none; 1 = 1 or more). R is a function of n, n_0 and the probability α that the shutter will be open.

In order to compute $H(Y)$ we need to know the probabilities $p(0)$ (that will receive no photons in an interval) and $p(1)$ (that will receive some photons in an interval), assuming that we have no knowledge of X. We see that

$$p(0) = \alpha p(0 \mid 1) + (1 - \alpha)p(0 \mid 0) \tag{A-2}$$

$$p(1) = 1 - p(0). \tag{A-3}$$

It turns out that the number of nats per photon is greatest when the number n of signal photons and the number n_0 of noise photons are much smaller than unity. When this is so, we can use the following approximate expressions for the case in which the shutter is closed, so that there are thermal photons only, and no signal photons

$$\begin{aligned} p(0 \mid 0) &= 1 - n_0 \\ p(1 \mid 0) &= n_0. \end{aligned} \tag{A-4}$$

We have no expression for the case in which the shutter is open and we receive *both* coherent signal photons and noise photons. However, if we assume that we will receive more photons than the number of signal photons alone, we will *underestimate* R by disregarding the effect of noise photons when the shutter is open and taking

$$\begin{aligned} p(0 \mid 1) &= 1 - n \\ p(1 \mid 1) &= n. \end{aligned} \tag{A-5}$$

Making these assumptions, and assuming that $n_0 \ll n$ and $\alpha \ll 1$, we find

$$\frac{R}{\alpha n} = -\ln n_0. \tag{A-6}$$

This is the number of nats per photon transmitted. Equation (A-6) leads to the classical limit of kT joules/nat when the frequency is high enough so that quantum effects are very strong.

From (2) we see that for a single-mode transmission system, when $hf \gg kT$ the number n_0 of thermal photons in a time t can be taken as

$$n_0 = Bt\ e^{-hf/kT}. \tag{A-7}$$

From (A-6) and (A-7)

$$\frac{R}{\alpha n} = \frac{hf}{kT} - \ln Bt. \tag{A-8}$$

Because we have assumed that $hf \gg kT$, the first term on the right is much larger than unity. What about ln Bt?

B is really a sort of mean bandwidth of the optical filter in Fig. 1. This must be made wide enough so as not to lose many signal photons when the shutter is open, but narrow enough not to let in too many noise photons. This means that Bt should be around unity, but probably somewhat larger.

Thus, when quantum effects are most pronounced, we can disregard the second term in (A-9). Then

$$\frac{R}{\alpha n} = \frac{hf}{kT} \text{ nats/photon.} \qquad \text{(A-9)}$$

This corresponds to

$$\frac{1}{kT} \text{ nats/joule.} \qquad \text{(A-10)}$$

Gordon[6] makes a somewhat similar calculation of channel capacity but does not carry the argument to (A-10) above. Equation (A-10) appears to be implicit in Helstrom *et al*[7] but is not stated explicitly.

ACKNOWLEDGMENT

The author wishes to thank J. P. Gordon, Carl W. Helstrom and Jon Mathews for helpful correspondence and comments.

REFERENCES

1. Schelkunoff, S. A. and Friis, H. T., *Antenna Theory and Practice,* Wiley and Sons, New York, 1952, p. 43.
2. Heffner, H., The Fundamental Noise Limit for Linear Amplifiers, *Proc. IRE,* 50, 1604-1608, 1962.
3. Gordon, J. P., Louisell, W. H., and Walker, L. R., Quantum Fluctuations and Noise in Parametric Processes, II, *Phys. Rev.,* 129, January, 1963, pp. 481-485.
4. Shannon, C. E. and Weaver, W., *The Mathematical Theory of Communication,* The University of Illinois Press, 1959.
5. Pierce, J. R., *Symbols, Signals and Noise,* Harper and Row, 1961, Chapter 10.
6. Gordon, J. P., Quantum Effects in Communication Systems, *Proc. Inst. Radio Eng.,* 50, 1898-1908, September, 1962.
7. Helstrom, C. W., Liu, J. W. S., and Gordon, P., Quantum Mechanical Communication Theory, *Proc. IEEE,* 58, October, 1970, pp. 1578-1598.
8. Helstrom, C. W., Capacity of the Pure-State Quantum Channel, *Proc. IEEE,* 62, January, 1974, pp. 140-141.
9. Yuen, Horace P., Kennedy, Robert S. and Lax, Melvin, Optimum Testing of Multiple Hypotheses in Quantum Detection Theory, *IEEE Trans. Information Theory,* IT-21, March, 1975, pp. 125-134.

.VI.8 QUANTUM NONDEMOLITION MEASUREMENTS

VLADIMIR B. BRAGINSKY, YURI I. VORONTSOV, AND KIP S. THORNE

Scientists have understood since the 1920's that the physical laws which govern atoms, molecules, and elementary particles are very different from the laws of everyday experience. The special laws of the atomic and molecular "microworld" are called quantum mechanics; those of everyday experience are classical mechanics. The laws of quantum mechanics were forced on physicists and chemists in the 1920's as the only possible way to understand the spectral properties of the light emitted by atoms and molecules.

Quantum mechanics tells us that, whenever a person measures some property of an electron (or of any other object in the microworld), his measurement inevitably will disturb the electron in a somewhat unpredictable way. The more accurate the measurement, the bigger and more unpredictable the disturbance. The disturbance is not due to the person's incompetence; rather, it is an intrinsic and inevitable feature of the laws of quantum mechanics.

Summary. Some future gravitational-wave antennas will be cylinders of mass ~100 kilograms, whose end-to-end vibrations must be measured so accurately (10^{-19} centimeter) that they behave quantum mechanically. Moreover, the vibration amplitude must be measured over and over again without perturbing it (quantum nondemolition measurement). This contrasts with quantum chemistry, quantum optics, or atomic, nuclear, and elementary particle physics, where one usually makes measurements on an ensemble of identical objects and does not care whether any single object is perturbed or destroyed by the measurement. This article describes the new electronic techniques required for quantum nondemolition measurements and the theory underlying them. Quantum nondemolition measurements may find application elsewhere in science and technology.

As an example, if a person measures the position of an electron in space with complete accuracy, his measurement inevitably will kick the electron with a totally unpredictable force. A second measurement of the electron's position, immediately after the first one, will give the same position as the first one did; but a measurement of the electron's momentum will give a completely unexpected result.

If, nevertheless, the momentum is measured very carefully and some definite result is obtained, that momentum measurement inevitably will disturb the electron's position by an unpredictable amount. If a second momentum measurement is made, the result is known in advance: it will be the same as just obtained. But if the next measurement is of

Vladimir B. Braginsky is professor of physics and Yuri I. Vorontsov is associate professor of physics at the Physics Faculty, Moscow University, Moscow, U.S.S.R. 117234. Kip S. Thorne is professor of theoretical physics in the W. K. Kellogg Radiation Laboratory, California Institute of Technology, Pasadena 91125.

Originally published in *Science, 209*, 547-57 (1980).

position, nobody can know the result in advance.

It matters not at all how the person makes his measurements—with the best technology of the 1920's, or the best of the 1980's, or the best of the 23rd century: an accurate position measurement must completely disturb the momentum; an accurate momentum measurement must completely disturb the position.

As bizarre as this situation may seem, it is even more bizarre when studied in greater depth—as was done theoretically by Niels Bohr, Werner Heisenberg, John von Neumann, Wolfgang Pauli, and others in the 1920's and 1930's. [See the textbook of Bohm (*1*) for details; see (*2*) for a detailed illustrative example.] It turns out that the unpredictable disturbance is a direct result of the extraction of information about the particle's position or momentum. It matters not how the information is extracted, nor where it is stored—in a person's brain, on magnetic tape, or in some minute change of the state of some other particle. So long as the information exists somewhere in the universe outside the original particle (more precisely, "outside the particle's wave function"), future measurements of the particle will reveal that the disturbance has occurred. The only way to undo the disturbance is to "run the measuring apparatus perfectly backward" and thereby reinsert all the information back into the particle. Only if no trace of the information remains anywhere, not even in the experimenter's brain, can the particle return to its original undisturbed state.

The quantum theory of measurement (*1*), which tells us these things and more, is very widely but not universally accepted by physicists. Einstein never fully accepted it (*3*); Lamb, a Nobel Prize winner for his experimental work in quantum physics, does not fully accept it (*4*). The authors of this article do accept it, and will presume it to be correct throughout this article.

The quantum theory of measurement tells us that, if a measurement is somewhat imprecise, then the magnitude of its disturbance is somewhat but not entirely predictable. For example, a very careful measurement of the east-west position of an electron, with an imprecision Δx, can be guaranteed to disturb its east-west momentum by not much more than $\Delta p = \hbar/(2\Delta x)$, where $\hbar$ ($\equiv 1.054 \times 10^{-27}$ cm g cm/sec) is Planck's constant. However, no matter how careful the measurement may be, the momentum uncertainty afterward will be at least $\hbar/(2\Delta x)$. Similarly, a momentum measurement of precision Δp will leave the position uncertain by at least $\Delta x = \hbar/(2\Delta p)$—but if the measurement is very careful, the position disturbance need not be much larger than this. The limit $\Delta x \Delta p \geq \hbar/2$, which holds for either type of measurement, is called the Heisenberg uncertainty principle.

The ultimate limits imposed by the uncertainty principle have been explored in great detail during the past decade by C. W. Helstrom, R. L. Stratanovich, J. P. Gordon, and others. They have developed a beautiful, mathematical theory of optimum quantum mechanical measurements (quantum detection and estimation theory) (*5*). Unfortunately, this theory assumes one can make a precise measurement of one observable or another, or of some combination of observables; but it does not spell out how such precise measurements can be realized technically—even in principle.

This gap in the theory is being confronted today in the effort to detect cosmic gravitational waves (*6*). Gravity-wave detectors consist of aluminum (or sapphire or silicon or niobium) bars, weighing between 10 kilograms and 10 tons, which are driven into motion by passing waves of gravity. The motions are very tiny: for the gravity waves that theorists predict are bathing the earth, a displacement $\delta x \cong 10^{-19}$ centimeter might be typical (*6*). And this displacement may oscillate, due to oscillations of the gravity wave, with a period $P \sim 10^{-3}$ second. To see the details of the gravity wave, one must thus make repeated measurements of the bar's position with precision $\Delta x \lesssim 10^{-19}$ cm, and with time intervals between measurements of $\bar{\tau} \lesssim 10^{-3}$ second.

For all measurements ever made in the past on a heavy bar, the effects of quantum mechanics were totally negligible; the classical mechanics of everyday experience gave a perfectly adequate description of the bar's behavior. But one never before tried to make measurements of such enormous precision as 10^{-19} cm. If the bar is suspended freely like a pendulum, as it is in some detectors (*6*), then over time intervals $\bar{\tau} \sim 10^{-3}$ second it will behave as though it were not suspended at all. It will be as free to move horizontally as the electron described above—and like the electron it will be subject to the laws of quantum mechanics: an "initial" measurement of the bar's east-west position with precision $\Delta x_i \cong 10^{-19}$ cm will inevitably disturb the bar's east-west momentum by $\Delta p \geq (\hbar/2\Delta x_i)$, and correspondingly will disturb its velocity by $\Delta v = \Delta p/m \geq (\hbar/2m\Delta x_i)$, where m is the bar's mass. During the time interval $\bar{\tau} \sim 10^{-3}$ second between measurements, the mass will move away from its initial position by an amount, $\Delta x_m = \Delta v\ \bar{\tau} \geq (\hbar\bar{\tau}/2m\Delta x_i)$, which is unpredictable because Δv is unpredictable. Putting in numbers ($\bar{\tau} = 10^{-3}$ second, $m = 10$ tons, $\Delta x_i = 10^{-19}$ cm), we find $\Delta x_m \geq 5 \times 10^{-19}$ cm—which is somewhat larger than the desired precision of our sequence of measurements. If the next measurement reveals a position changed by as much as 5×10^{-19} cm, we have no way of knowing whether the change was due to a passing gravity wave or to the unpredictable, quantum mechanical disturbance made by our first measurement. In effect, our first measurement plus subsequent free motion of the bar has "demolished" all possibility of making a second measurement of the same precision, $\Delta x \sim 10^{-19}$ cm, as the first, and of thereby monitoring the bar and detecting the expected gravity waves.

In principle one can circumvent this problem by making the bar much heavier than 10 tons (recall that Δx_m is inversely proportional to the mass). However, this is impractical. In principle another solution is to shorten the time between measurements (recall that Δx_m is directly proportional to $\bar{\tau}$). However, this will weaken the gravitational-wave signal ($\delta x_{GW} \propto \bar{\tau}^2$ for $\bar{\tau} \lesssim 10^{-3}$ second) even more than it reduces the unpredictable movement of the bar ($\Delta x_m \propto \bar{\tau}$).

The best solution is cleverness: find some way to make the gravity-wave effect stronger; this is being done in laser-interferometer gravity-wave detectors (*6*), but only at the price of having to make 10^{-16} cm measurements of the relative displacement of two bars as far apart as several kilometers. Alternatively, find some way to circumvent the effects of the Heisenberg uncertainty principle—that is, some way to prevent the inevitable disturbance due to the first measurement, plus subsequent free motion, from demolishing the possibility of a second accurate measurement: a quantum nondemolition (QND) method.

One QND method which could work in principle is this: instead of measuring the position of the 10-ton bar, measure its momentum with a small enough initial error, $\Delta p_i \sim 10^{-9}$ g cm/sec, to detect the expected gravity waves. Thereby inevitably disturb the bar's position by an unknown amount $\Delta x \gtrsim \hbar/2\Delta p_i \sim 5 \times 10^{-19}$ cm. Wait a time $\bar{\tau} \sim 10^{-3}$ second and then make another momentum measurement. As the bar moves freely between the measurements, its momentum remains fixed. The uncertainty Δx in the bar's position does not by free evolution produce a new uncertainty Δp_m in the momentum. Consequently the second measurement can have as good accuracy, 10^{-9} g cm/sec, as the initial measurement; and a momentum change of (a few) $\times 10^{-9}$ g cm/sec due to a passing gravity wave can be seen.

Momentum measurements can be quantum nondemolition, but position measurements cannot be, for this simple reason: in its free motion between measurements the bar keeps its momentum constant, but it changes its position by an amount $\delta x = (p/m)\hat{\tau}$ that depends on the momentum, and that therefore is uncertain because of measurement-induced uncertainties of the momentum. We say that momentum is a QND variable, but position is not.

Unfortunately, however, it is far easier to measure position than momentum. Nobody has yet invented a technically realizable way of making momentum measurements with the required precision.

The problem of inventing a technically realizable QND measurement scheme was first posed in 1974 (*8*). This reference also formalized the concept of QND measurements. Subsequent developments in the theory of QND are due largely to Unruh (*9, 10*); Braginsky, Vorontsov, and Khalili (*11, 12*); Caves, Thorne, Drever, Zimmermann, and Sandberg (*13-15*); and Hollenhorst (*16*). All of this work has been theoretical: it has shown that in principle QND measurement schemes can completely circumvent the disturbing back-action effects of one's measurements, and it has led to several tentative designs for practical QND measurements in gravity-wave detectors—designs which do not involve measuring momentum.

Actual laboratory work on QND measurement schemes is only now beginning to get under way, and the levels of sensitivity required are so great that we cannot hope for any laboratory results until several years from now. Nevertheless, it is reasonable to expect QND measurements to be a routine part of gravity-wave technology by the late 1980's.

The purpose of this article is to make as wide an audience as possible aware of these developments, so that people can begin to ask whether the QND idea might be useful in other areas of science and technology. To achieve this purpose effectively, we feel it necessary to write the rest of this article at a somewhat technical level. We hope thereby to convey to physicists, engineers, chemists, and others familiar with elementary quantum mechanics and elementary electronics, a deep enough understanding of the QND idea that they can begin to think creatively about it themselves.

Resonant-Bar Gravitational-Wave Antennas

Although the QND idea is explained most easily, as we have done, in terms of bars which move freely (free masses), QND measurements are most needed for a different type of gravity-wave antenna: one made of a bar which oscillates mechanically in its fundamental mode (bar mass, $m \simeq 10$ to 10,000 kg; oscillation frequency, $\omega/2\pi \simeq 500$ to 10,000 hertz) (*6*). The expected gravity waves should produce changes of oscillation amplitude $\delta x \simeq 10^{-18}$ to 10^{-19} cm, which are less than or of order the width of the quantum mechanical wave packet of the oscillator $\delta x_{QM} = (\hbar/2m\omega)^{1/2}$, if the oscillator is in its ground state or in a coherent (minimal-wave-packet) state. Here, by contrast with nuclear, atomic, and elementary-particle physics, there is only one quantum mechanical system being measured (the oscillator), rather than an ensemble of systems; and we must make a continuous sequence of measurements on this one system.

Such an oscillator will actually behave quantum mechanically even in the presence of thermal Brownian motion and at bar temperatures $kT >> \hbar\omega$, so long as its quality factor Q is sufficiently high—that is, so long as the fundamental mode of the bar is coupled sufficiently weakly to the other, thermalized modes (*7*). In particular, when one is making energy measurements which put the oscillator in an energy eigenstate, Brownian motion during one cycle will change the number of quanta n in the oscillator by less than unity if (*7*)

$$(n + \tfrac{1}{2})kT/Q < \hbar\omega/4\pi \qquad (1a)$$

and if one is making amplitude measurements, Brownian motion during the measurement time $\bar{\tau}$ will change the amplitude by less than the coherent-state wave-packet width $(\hbar/2m\omega)^{1/2}$ if (*7*)

$$2kT\bar{\tau}/Q < \hbar \qquad (1b)$$

(see Eqs. 40 and 41 below).

In order to monitor the effects of a weak gravity-wave force on such an oscillator, one must use a measurement technique whose back action on the oscillator, together with subsequent free evolution, does not substantially disturb the probability distributions of the observables being measured—a QND technique. In the following sections we shall describe the theory of such QND measurement techniques as applied to oscillators, to free masses, and more generally to any quantum mechanical system. Throughout our description we shall try to give short, elementary proofs of most of the results quoted. More elegant and rigorous proofs will be found in the primary literature. To understand our discussion, the reader must be familiar with elementary quantum mechanics and elementary electronic circuit theory, but little other specialized knowledge should be needed.

General Theory of Quantum Nondemolition Measurements

Consider a system, such as an oscillator, that has some observable $\hat{A}$ which an experimenter wishes to monitor. For the moment, assume that the system's only coupling to the external world is through the experimenter's measuring apparatus. We define a QND measurement of $\hat{A}$ as a sequence of precise measurements of $\hat{A}$ such that the result of each measurement is completely predictable from the result of the first measurement—plus, perhaps, other information about the initial state of the system. This definition, and the ramifications which follow, are a refinement by Caves [in (*15*)] of Braginsky and Vorontsov's (*8*) original concept of quantum nondemolition. A similar refinement has been developed independently by Unruh (*10*).

Quantum nondemolition measurements are ideal tools for use in the detection of weak external forces (such as gravity waves) that act on the system. One need only perform a QND monitoring of the evolution of $\hat{A}$ and watch for deviations from the predicted evolution.

Most observables cannot, even in principle, be monitored in a QND way. In any precise measurement of an observable $\hat{A}$, the back action of the measuring apparatus uncontrollably and unpredictably kicks all observables $\hat{C}$ which fail to commute with $\hat{A}$; and then, in the subsequent free evolution of the system, the contamination in $\hat{C}$ may be fed into $\hat{A}$, making the results of future measurements of $\hat{A}$ unpredictable. Only very special observables can be immune to such feedback contamination; they are called QND observables [or sometimes generalized QND observables (*15*)]. Mathematically, $\hat{A}$ is a QND observable if and only if, when the system is evolving freely in the Heisenberg picture, $\hat{A}$ commutes with itself at the different moments of time t_j, t_k when one makes one's measurements (*10, 15*)

$$[\hat{A}(t_j), \hat{A}(t_k)] = 0 \qquad (2)$$

If this condition is satisfied at all times t_j and t_k, then $\hat{A}$ is called a continuous QND observable; if it is satisfied only at special times, then $\hat{A}$ is a stroboscopic

QND observable. If $\hat{A}$ is conserved during free evolution ($d\hat{A}/dt = 0$), then it is guaranteed to satisfy Eq. 2 for all t_j, t_k and therefore to be a continuous QND observable.

In the case of a free particle, the energy and momentum are conserved and are continuous QND observables, but the position is not: $\hat{x}(t + \tau) = \hat{x}(t) + \hat{p}\tau/m$, so

$$[\hat{x}(t), \hat{x}(t + \tau)] = i\hbar\tau/m \tag{3}$$

Precise measurements of $\hat{x}$ perturb $\hat{p}$ uncontrollably, and the contamination in $\hat{p}$ subsequently feeds back into $\hat{x}$ as the particle moves freely.

For a harmonic oscillator the position and momentum satisfy the commutation relations

$$[\hat{x}(t), \hat{x}(t + \tau)] = (i\hbar/m\omega) \sin \omega\tau \tag{4a}$$

$$[\hat{p}(t), \hat{p}(t + \tau)] = i\hbar m\omega \sin \omega\tau \tag{4b}$$

These imply that $\hat{x}$ and $\hat{p}$ are not continuous QND observables. However, if one makes one's measurements stroboscopically at times separated by an integral number of half-periods ($\tau = k\pi/\omega$; $\sin \omega\tau = 0$), then the commutators in Eqs. 4a and 4b vanish. This means that $\hat{x}$ and $\hat{p}$ are stroboscopic QND observables (*12, 13*). Stroboscopic QND measurements (*12, 13*) of $\hat{x}$ or $\hat{p}$ drive the oscillator into a state where $\hat{x}$ is known precisely—for example, at moments $t = k\pi/\omega$ and $\hat{p}$ is known precisely at $t = (k + \frac{1}{2})\pi/\omega$; but at other times $\hat{x}$ and $\hat{p}$ are highly uncertain. For an oscillator the conserved quantities, which are guaranteed to be QND observables at any and all times, include the energy (*8*) and the real and imaginary parts of the complex amplitude (*13*)

$$\hat{X}_1 = \hat{x}(t) \cos \omega t - [\hat{p}(t)/m\omega] \sin \omega t \tag{5a}$$

$$\hat{X}_2 = \hat{x}(t) \sin \omega t + [\hat{p}(t)/m\omega] \cos \omega t \tag{5b}$$

High-precision measurements of $\hat{X}_1$ or $\hat{X}_2$ (whether fully QND or not) are called back-action-evading measurements (*14, 15*) because they enable the measured component of the amplitude (for example, $\hat{X}_1$) to avoid back-action contamination by the measuring device, at the price of strongly contaminating the other component ($\hat{X}_2$). (The uncertainty relation

$$\Delta X_1 \Delta X_2 \geq \hbar/2m\omega \tag{5c}$$

is enforced by the commutation relations $[\hat{X}_1, \hat{X}_2] = i\hbar/m\omega$, which follow from $[\hat{x}, \hat{p}] = i\hbar$.)

Let $\hat{A}$ be a QND observable which is to be monitored by a sequence of perfect QND measurements at times t_0, t_1, t_2, ... Since $\hat{A}(t_0)$ and $\hat{A}(t_j)$ commute (QND assumption; Heisenberg picture), one can perform a perfect "state-preparation measurement" at time t_0, which puts the system into a simultaneous eigenstate $|\psi_0\rangle$ of the observables $\hat{A}(t_0)$, $\hat{A}(t_1)$, $\hat{A}(t_2)$, . . . with some (not previously predictable) eigenvalues $A(t_0)$, $A(t_1)$, $A(t_2)$, ... From the results of this first measurement one can compute the eigenvalues $A(t_0)$, $A(t_1)$, $A(t_2)$, ... Later, as the system evolves freely, its state $|\psi_0\rangle$ remains fixed in time, while its observable $\hat{A}$ evolves through the values $\hat{A}(t_1)$, $\hat{A}(t_2)$, ... Subsequent perfect measurements of $\hat{A}$ at times t_1, t_2, . . . must give the known eigenvalues $A(t_1)$, $A(t_2)$ and must leave the state of the system $|\psi_0\rangle$ unchanged. If $\hat{A}$ is a continuous QND observable, then the QND measurements can be made continuously, and each measurement can last as long or as short a time as one wishes. If $\hat{A}$ is a stroboscopic QND observable, then each measurement must be made very quickly (stroboscopically) to avoid contamination. Examples will be given below, and further detail will be found in section IV of (*15*).

The apparatus used in any measurement consists of a sequence of stages, through which information flows toward the experimenter's eyes and brain. Measurement theory asserts that, although the early stages of the apparatus may behave quantum mechanically, the late stages must be classical. There is no universally accepted definition of classical. We shall regard a stage as classical if the quantum mechanical uncertainties of it and of subsequent stages have no significant influence on the overall accuracy of the measurement. If the system being studied interacts directly with a classical stage, the measurement is called direct. When, between the system and the first classical stage, there is a quantum stage (quantum mechanical readout system, QRS), the measurement is called indirect (*17*). For example, the measurement of the position of a particle by its blackening of a photographic plate is direct. The measurement of position by the particle's scattering of light or of an electron is indirect. The vast majority of measurements are indirect. In electronic measuring systems the first classical stage is often the first amplifier.

In deducing quantum limitations on the sensitivity of a specific measuring scheme, one must analyze quantum mechanically everything that precedes the first classical stage. The overall accuracy of the measurement is governed not only by quantum fluctuations in the quantum stages, but also by the details of the couplings between those stages and between them and the measured system. These all influence the signal which enters the first classical stage, and that signal ultimately determines the quantum errors of measurement.

In practice, if not in principle, the reduction of the wave function occurs when the signal enters the first classical stage. If that signal carries information not only about the observable $\hat{A}$ which interests us, but also about observables $\hat{C}$ that fail to commute with $\hat{A}$, an exact measurement is impossible. This is because any flow of information about $\hat{C}$ into the first classical stage will, according to the uncertainty principle, be accompanied by unpredictable back-action forces into the quantum stages—back-action forces which must ultimately contaminate all observables that fail to commute with $\hat{C}$, including $\hat{A}$.

Because of this back action, the measurement error must always exceed an ultimate quantum limit. We shall derive that limit under the special assumption that in the Heisenberg picture $\hat{A}$ and $\hat{C}$ are time-independent—either because they are constants of the motion such as $\hat{X}_1$ and $\hat{X}_2$, or because they are time-evolving observables [such as $\hat{x}(t)$ and $\hat{p}(t)$] evaluated at some fixed moment of time [such as $\hat{A} = \hat{x}(0)$, $\hat{C} = \hat{p}(0)$]. We assume that the "readout observable" of the last quantum stage, $\hat{Q}_R$, which couples into the first classical stage, is expressible as

$$\hat{Q}_R = f(\alpha\hat{A} + \beta\hat{C}) \tag{6a}$$

where

$$[\hat{A},\hat{C}] = 2i\gamma\hbar \quad \text{so} \quad \Delta A \Delta C \geq \gamma\hbar \tag{6b}$$

with γ a real number. The time evolution of the readout observable $\hat{Q}_R$ is embodied in the function f and/or in the real parameters α and β. Typically, α and β will be sinusoidal functions of time, which are used to code and separate the $\hat{A}$ and $\hat{C}$ signals. We assume that the first classical stage (usually an amplifier) is equally sensitive to signals at the $\hat{A}$ and $\hat{C}$ frequencies. Then no matter how accurately the first classical stage monitors $\hat{Q}_R$, it must give errors in $\hat{A}$ and $\hat{C}$ related by $\Delta A = (\bar{\beta}/\bar{\alpha})\Delta C$, where $\bar{\alpha}$ and $\bar{\beta}$ are the root-mean-square (r.m.s.) values of α and β. These relative errors, combined with the uncertainty relation (Eq. 6b), imply the ultimate quantum limit

$$\Delta A \geq [(\bar{\beta}/\bar{\alpha})\gamma\hbar]^{1/2} \tag{6c}$$

Return now to the general situation where $\hat{A}$ and $\hat{C}$ might be time-dependent. In order that the instantaneous signal at time t not contain any contaminant information about observables $\hat{C}(t)$ which fail to commute with $\hat{A}(t)$, it is necessary and sufficient that $\hat{A}(t)$ commute with that part of the Hamiltonian $\hat{H}_I(t)$ which describes the interaction of the system with the measuring apparatus (*1*)

$$[\hat{A}(t),\hat{H}_I(t)] = 0 \tag{7}$$

In order that information about $\hat{A}$ enter the measuring apparatus, $\hat{H}_I$ must depend upon $\hat{A}$. Usually one achieves these conditions by direct coupling of $\hat{A}$ to some observable $\hat{M}$ of the measuring apparatus

$$\hat{H}_I = K\hat{A}\hat{M} \tag{8}$$

In summary, the condition in Eq. 7 guarantees no direct, instantaneous back action of the measuring apparatus on the quantity $\hat{A}$ being measured; and the condition in Eq. 2 guarantees that variables $\hat{C}$ which have been contaminated by back action will not subsequently, by free evolution (with $\hat{H}_I$ turned off), feed their contamination into $\hat{A}$. Often, however, $\hat{H}_I$ is turned on for a long time—even for all time. Then there is danger that $\hat{H}_I$ may catalyze an evolutionary feeding of $\hat{C}$ into $\hat{A}$. One can be sure this does not happen if an analysis of the system plus measuring apparatus, including all couplings, reveals that $[\hat{A}(t_1), \hat{A}(t_2)] = 0$ for all times t_1 and t_2 at which signals enter the classical stage. However, such a full analysis may be prohibitively difficult.

Fortunately, in one common situation a full analysis is not necessary: Caves [in (*15*)] has shown that, if $\hat{A}$ is a continuous QND observable and $\hat{H}_I$ contains no system observables except $\hat{A}$, then the Heisenberg picture evolution of $\hat{A}$ with couplings turned on is identical to its free evolution, and consequently $\hat{A}$ is fully isolated from back action—both direct and indirect. Caves (*15*) has also proved "full isolation of $\hat{A}$ with $\hat{H}_I$ turned on" under more general circumstances.

Just as $\hat{H}_I$ might catalyze an indirect feeding of contaminated variables $\hat{C}$ into $\hat{A}$, so also such feeding might be catalyzed by the coupling of the system to a classical external force $F(t)$ (for example, to a gravitational wave). This coupling is embodied in a piece of the Hamiltonian

$$\hat{H}_F = \mu F\hat{x} \tag{9}$$

where μ is a coupling constant and $\hat{x}$ is a dynamical variable of the system ($\hat{x}$ is position in the case of a gravitational-wave antenna). If $\hat{A}$ satisfies the self-commutation condition (Eq. 2) even in the presence of $\hat{H}_F$, then $\hat{A}$ can remain free from contamination. If, in addition, free evolution with $\hat{H}_F$ turned on causes an eigenvalue $A(t)$ of $\hat{A}(t)$ to evolve in such a way that, from a precise knowledge of $A(t)$, one can deduce $F(t)$, then $\hat{A}$ is called a QNDF observable [(*15*); see also Unruh (*10*), where this is denoted QNDD]. QNDF observables are ideal tools for monitoring weak, classical forces.

If, on the other hand, the term $\hat{H}_F$ in the Hamiltonian catalyzes an evolutionary feeding of contaminated observables $\hat{C}$ into $\hat{A}$ (that is, if $[\hat{A}(t_j), \hat{A}(t_k)] \neq 0$ in the presence of $\hat{H}_F$), then although $\hat{A}$ may be highly sensitive to the presence of an external force, one cannot hope to monitor the details of the force by measurements of $\hat{A}$.

In the case of an oscillator with position $\hat{x}$ coupled to the force (for example, a gravitational-wave detector), $\hat{X}_1$ and $\hat{X}_2$ (Eqs. 5) are QNDF observables and thus can be used for perfect monitoring of the forces (*13, 15*). By contrast, the oscillator's energy, although a QND observable, is not QNDF. As a result, precise measurements of the energy can reveal the presence of an arbitrarily weak force; but they cannot determine the strength of the force with a precision better than a factor of 3 (*13, 10, 15*)—unless the force is so strong that it increases the energy by an amount large compared to the initial energy. Examples and proofs will be sketched later.

When one is using a quantum system to monitor a classical force F, one can increase one's sensitivity by increasing the response of the measured quantity $\hat{A}$ to F. In (*8*) and (*18*) it is shown that, if $\hat{A}$ is the energy of an oscillator, F produces a change of $\hat{A}$ which is larger, the larger the oscillator's initial energy. Formally, such a measuring scheme satisfies Unruh's (*10*) general condition for the dependence of $\hat{A}$'s response on the initial state of the detector (detector-dependent response, or DDR)

$$[\hat{A}, \hat{x}] \neq \text{a } C\text{-number} \qquad (10)$$

In the case of (*8*) and (*18*) $\hat{A}$ is the detector's (oscillator's) Hamiltonian, $\hat{x}$ is its position, and $[\hat{A},\hat{x}] = -(i\hbar/m)\hat{p}$. The larger the initial energy of oscillation, the larger will be $\langle \hat{p}^2 \rangle$, and the larger will be $d\hat{A}/dt$. Further details will be given later.

This completes our sketch of the general theory of QND measurements. This theory will now be applied to various types of measurements of harmonic oscillators, with emphasis on issues relevant to gravitational-wave detection.

Position Measurements

A resonant-bar gravity-wave antenna is an oscillator with mass m, frequency ω, position $\hat{x}$, and momentum $\hat{p}$, which couples to a gravitational wave (classical external force F) with a coupling energy $\hat{H}_F = -\hat{x}F(t)$. In most experiments the antenna's position $\hat{x}$ is coupled by a transducer ($\hat{H}_I = K\hat{x}\hat{q}$; $K \equiv$ coupling constant) to an electromagnetic circuit (quantum readout system), which we shall describe as an oscillator with capacitance C, inductance L, generalized coordinate (equal to charge on the capacitor) $\hat{q}$, and generalized momentum (equal to flux in the inductor) $\hat{\pi}$. More complicated QRS's can be used; but this is the typical case. The voltage on the capacitor, which is proportional to $\hat{q}$, is monitored by an amplifier—the first classical stage of the measuring system. Thus $\hat{q}$ is the readout observable $\hat{Q}_R$ (see Eqs. 6).

The coupled antenna, force, and QRS are governed by the Hamiltonian

$$\hat{H} = \frac{\hat{p}^2}{2m} + \frac{1}{2} m\omega^2\hat{x}^2 + \frac{\hat{\pi}^2}{2L} + \frac{\hat{q}^2}{2C} + \hat{H}_F + \hat{H}_I$$

$$\hat{H}_F = -\hat{x}F(t) \qquad \hat{H}_I = K\hat{x}\hat{q} \qquad (11)$$

for which the Heisenberg evolution equations are

$$d\hat{x}/dt = \hat{p}/m$$
$$d\hat{p}/dt = -m\omega^2\hat{x} + F(t) - K\hat{q}$$
$$d\hat{q}/dt = \hat{\pi}/L$$
$$d\hat{\pi}/dt = -\hat{q}/C - K\hat{x} \qquad (12)$$

Because these equations ignore the first classical stage (amplifier) and its detailed back action on the QRS, they cannot tell us the actual sensitivity of the measuring system. On the other hand, they can tell us the ultimate quantum mechanical limit on the sensitivity.

Suppose, as a first case, that the signal $\hat{Q}_R = \hat{q}$ is fed continuously into the amplifier for a time much longer than a quarter-cycle of the antenna, and that one's goal is to measure $\hat{x}_0$, the initial value of the oscillator's position. During the measurement $\hat{x}(t)$, which feeds $\hat{\pi}$ and thence $\hat{Q}_R \equiv \hat{q}$, oscillates between $\hat{x}_0$ and $\hat{p}_0$. [$\hat{x}(t) = \hat{x}_0 \cos \omega t + (\hat{p}_0/m\omega) \sin \omega t$, aside from minor modifications due to the couplings. Note that $\hat{x}_0 \equiv \hat{X}_1$, $\hat{p}_0/m\omega \equiv \hat{X}_2$; Eqs. 5.] Consequently, the signal $\hat{Q}_R$ entering the amplifier contains not only $\hat{x}_0$ but also, unavoidably, $\hat{p}_0$. Since their relative strengths in the signal are $p_0/x_0 = m\omega$, the measurement determines them with relative precisions $\Delta p_0 = m\omega\Delta x_0$. Taking account of the uncertainty relation $\Delta x_0 \Delta p_0 \geq \hbar/2$, we find (*19, 18, 12, 15*)

$$\Delta x_0 = \Delta p_0/m\omega \geq (\hbar/2m\omega)^{1/2} \quad (13)$$

(This is a specific example of the general quantum limit of Eqs. 6.) Such a measurement is called an amplitude-and-phase measurement because it gives information about both the amplitude $[x_0^2 + (p_0/m\omega)^2]^{1/2}$ and the phase $\psi_0 = \tan^{-1}(p_0/m\omega x_0)$ of the antenna's motions. An ideal amplitude-and-phase measurement with the limiting sensitivity in Eq. 13 drives the antenna into a coherent (minimal-wave-packet) state. If such a measurement (state preparation) has put the antenna into a coherent state with $\langle \hat{x}(t) \rangle = x_0 \cos \omega t + (p_0/m\omega) \sin \omega t$, then a classical force $F = F_0 \cos(\omega t + \varphi)$ acting for a time $\bar{\tau} >> 2\pi/\omega$ will leave the state coherent but change its amplitudes by $\delta x_0 = (F_0\bar{\tau}/2m\omega) \sin \varphi$, $\delta p_0/m\omega = (F_0\bar{\tau}/2m\omega) \cos \varphi$ (*15*). A subsequent ideal amplitude-and-phase measurement can reveal this change if the force F_0 exceeds the quantum limit (*18*)

$$F_0 \simeq (2/\bar{\tau})(m\omega\hbar)^{1/2} \quad (14)$$

No amplitude-and-phase measurement can do better than this.

An alternative derivation of the quantum limits in Eqs. 13 and 14, due to Giffard (*19*), takes detailed account of quantum fluctuations in the amplifier and their back action on the QRS.

The quantum limits in Eqs. 13 and 14 are traceable to the fact that $\hat{x}$ is not a continuous QND observable; a continuous measurement of $\hat{x}$ produces direct back action on $\hat{p}$, which then contaminates $\hat{x}$ through free evolution. On the other hand, $\hat{x}$ is a stroboscopic QND observable (see Eq. 4a). Consequently, by stroboscopic measurements (*12, 13*) at times $t = 0, \pi/\omega, 2\pi/\omega, \ldots$ one can monitor $\hat{x}$ with perfect precision, in principle (except for the ridiculous limit from relativistic quantum theory, $\Delta x \geq \hbar/mc \simeq 10^{-41}$ cm for $m \simeq 10$ kg). Stroboscopic measurements can be achieved with the system of Eq. 11 by pulsing on and off the transducer's coupling constant K. By a sequence of perfect stroboscopic measurements one can monitor an arbitrarily weak force F_0.

Perfect stroboscopic measurements require that $\hat{x}$ be coupled to the QRS for arbitrarily short time intervals τ at $t = 0, \pi/\omega, \ldots$ (and also that the QRS transfer its information to the first classical stage in a time less than π/ω). If τ is finite then the momentum spread $\Delta p \geq \hbar/2\Delta x$, produced by a measurement of precision Δx, causes a mean position spread $(\Delta x)_s \simeq (\Delta p/m)\tau \geq \hbar\tau/2m\Delta x$ during the next measurement. The resulting r.m.s. error is (*18, 12, 13*)

$$\Delta x \gtrsim (\hbar\tau/m)^{1/2} \quad (15)$$

The shorter the measurement time τ, the more accurate the measurement can be.

Unfortunately, short measurements

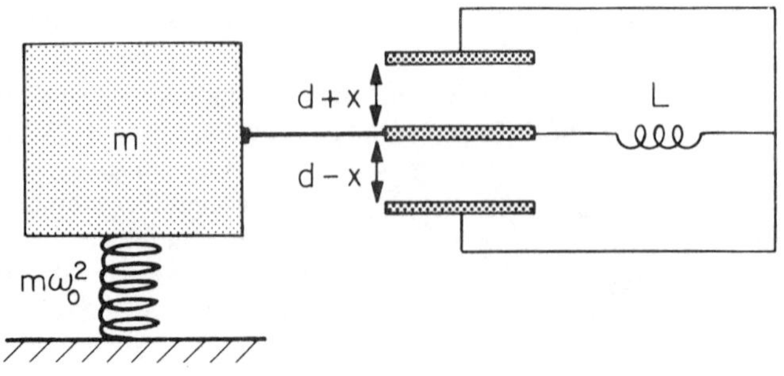

Fig. 1. Scheme for coupling a mechanical oscillator's (position)$^2 \equiv x^2$ to an electromagnetic QRS.

require very strong coupling of the antenna to the QRS in order to surmount the quantum mechanical zero-point energy that accompanies the signal through the QRS and into the amplifier. This is quantified in (*20, 15, 14*) for the case of a mechanical oscillator with transducer and QRS that feed the amplifier a sinusoidal voltage signal

$$V_s = \sqrt{2}\, Kx|g_{21}|\cos \Omega t \qquad (16a)$$

Here Ω (assumed $>> 1/\tau$) is the signal frequency, K is the transducer's coupling constant, and g_{21} is the transfer function of the QRS. This signal carries an r.m.s. power $P_s = (Kx|g_{21}|)^2/4\mathrm{Re}(g_{22})$, where $\mathrm{Re}(g_{22})$ is the real part of the output impedance of the QRS. Accompanying this signal in the experimental bandwidth $1/2\tau$ is a fluctuating quantum mechanical zero-point power $P_{fl} = \frac{1}{2}\hbar\Omega/2\tau$—half of it at the known phase of the signal, the other half at the other phase. This zero-point noise leads to $\Delta x \gtrsim (\hbar/2m\omega)^{1/2}\,(1/\beta\omega\tau)^{1/2}$, where

$$\beta \equiv \frac{(K|g_{21}|)^2}{m\omega^2 \mathrm{Re}(g_{22}\Omega)} \qquad (16b)$$

is a dimensionless coupling constant (*21, 20*). If one averages over N successive stroboscopic measurements (total bandwidth $1/2N\tau$), then the accuracy improves as $N^{-1/2}$

$$\Delta x \gtrsim (\hbar/2m\omega)^{1/2}\,(1/\beta N\omega\tau)^{1/2} \qquad (17)$$

Optimization of the measurement time τ in Eqs. 15 and 17 leads to an ultimate quantum limit for stroboscopic measurements with finite coupling (*14, 15, 20*):

$$\tau_{\text{optimal}} = (\beta N\omega^2)^{-1/2} \qquad (18a)$$

$$\Delta x \gtrsim (\hbar/m\omega)^{1/2}(\beta N)^{-1/4} \qquad (18b)$$

A coupling as large as $\beta = 1$ is difficult to achieve. Therefore, to beat the amplitude-and-phase quantum limit $(\hbar/2m\omega)^{1/2}$, one will probably have to average over a large number N of measurements.

By a sequence of stroboscopic measurements at the quantum limit of Eq. 18b, one can monitor a classical force $F = F_0 \cos(\omega t + \varphi)$. If the phase φ is near $\pi/2$ or $3\pi/2$, then the optimal times for the stroboscopic measurements are $t = 0,\ \pi/\omega,\ 2\pi/\omega, \ldots$; the force produces during N half-cycles $\delta x = (\pi/2)(NF_0/m\omega^2)$, and the force is measurable if

$$F_0 \gtrsim (2/\pi)(\hbar m\omega^3)^{1/2}\beta^{-1/4}N^{-5/4} \qquad (19)$$

If the phase φ is near 0 or π, then the precision of Eq. 19 is achieved by stroboscopic measurements at $t = \pi/2\omega$, $3\pi/2\omega, \ldots$. Since the phase of a gravitational wave is not predictable in advance, two antennas are needed; one to be monitored at $t = 0,\ \pi/\omega, \ldots$, the other at $t = \pi/2\omega,\ 3\pi/2\omega, \ldots$ (*12, 13*).

The stroboscopic limits of Eqs. 15, 17, 18, and 19 strictly speaking refer to a harmonic oscillator with only one degree of freedom. Unfortunately, a resonant-bar gravitational-wave antenna has many normal modes which can all be simultaneously perturbed by the back action of each measurement. However, if the strongly perturbed modes have commensurate eigenfrequencies, then stroboscopic QND measurements on the fundamental mode are also QND for the other modes (*12*), and results near the limits of Eqs. 15 to 18 may be achievable.

Stroboscopic measurements can be carried out on electromagnetic oscillators [such as inductance-capacitance (*LC*) circuits] as well as on mechanical oscillators. For example, one could send a collimated pulse of electrons through the capacitor so quickly that it spends a time $\tau << 2\pi/\omega$ between the capacitor plates. The electrons will be deflected by the electric field of the capacitor, which is proportional to the oscillator's generalized coordinate q (≡ charge on plates); and by measuring the deflection one can infer q. A stroboscopic sequence of such measurements can reveal q, in principle, to an accuracy $(\hbar\tau/L)^{1/2} = (\hbar C\omega^2\tau)^{1/2}$ (see Eq. 15) and can reveal the corresponding voltage in the capacitor to $\Delta V = (\hbar\omega/C)^{1/2}(\omega\tau)^{1/2}$—which is a factor $(\omega\tau)^{1/2}$ better than the standard amplitude-and-phase quantum limit.

Energy Measurements

Suppose that one has developed a method for making accurate measurements of a harmonic oscillator (antenna), and that an initial "state-preparation" measurement has put the oscillator into an eigenstate with energy E_0. A force $F_0 \cos(\omega t + \varphi)$ acting for time $\bar{\tau}$ will change the oscillator's state. Because the phase of the initial state is completely indeterminate, no interference terms show up in the new state's energy expectation value (*22*)

$$\langle E\rangle = E_0 + W \qquad W \equiv F_0^2\bar{\tau}^2/8m \quad (20a)$$

However, interference is a dominant effect in the variance of the new state's energy (*22*)

$$\sigma(E) = (2E_0 W)^{1/2} \quad (20b)$$

The next measurement is likely to reveal a changed energy, and thereby tell us that a force has acted, if $\sigma(E) \gtrsim \hbar\omega$. (Here we assume the force to be weak, $W < E_0$.) Rewritten in terms of F_0, this detection criterion is

$$F_0 \gtrsim \frac{2}{\bar{\tau}}\left(\frac{m\omega\hbar}{n_0 + \frac{1}{2}}\right)^{1/2} \quad (21)$$

where $n_0 = E_0/\hbar\omega - 1/2$ is the number of quanta in the initial state. This force-detection method can be arbitrarily sensitive if n_0 is made arbitrarily large (*8, 18*). However, because there is no unique relationship between the measured energy and F_0 $[\sigma(E) >> \langle E\rangle - E_0]$, this method cannot tell us the precise magnitude of F_0. In other words, the energy is not a QNDF observable (*13, 10, 15*); see the discussion following Eq. 9.

A perfect energy measurement (perfect up to one quantum) is possible only if (i) the interaction Hamiltonian $\hat{H}_I$ for the oscillator and QRS involves the oscillator energy $\hat{H}_0$, and (ii) $\hat{H}_I$ commutes with $\hat{H}_0$; see Eq. 7 and associated discussion.

If instead $\hat{H}_I = K\hat{x}\hat{q}$, as occurs in most measuring systems, then the directly measured quantity is $\hat{x}(t)$—or the amplitudes $\hat{x}_0$ and $\hat{p}_0$—and the measurement is of the amplitude-and-phase type with quantum limits $\Delta x_0 = \Delta p_0/m\omega \geq (\hbar/2m\omega)^{1/2}$ (Eq. 13). From the measured x_0 and p_0 one can compute the oscillator's energy

$$E_0 = p_0^2/2m + m\omega^2 x_0^2/2 = (n_0 + 1/2)\hbar\omega$$

up to an accuracy, for $n_0 >> 1$

$$\Delta E = p_0\Delta p_0/m + m\omega^2 x_0\Delta x_0 \gtrsim n_0^{1/2}\hbar\omega \quad (22)$$

The effect of the force will be discernible if this error is less than $\sigma(E)$ (Eqs. 20), which implies the same force-detection criterion (Eq. 14) as we derived from our original amplitude-and-phase discussion.

One way to achieve an $\hat{H}_I$ which involves $\hat{H}_0$ rather than $\hat{x}$—and to thereby beat the amplitude-and-phase limit (Eq. 14)—is to make the oscillator's mass m and spring constant $m\omega^2$ depend weakly on a variable $\hat{q}$ of the QRS: $m = m_0/(1 + K\hat{q})$, $m\omega^2 = m_0\omega_0^2(1 + K\hat{q})$ (*10*). Then the total Hamiltonian becomes

$$\hat{H} = \hat{H}_0 + \hat{H}_I + \hat{H}_{QRS}$$

$$\hat{H}_0 = \hat{p}^2/2m_0 + \frac{1}{2}m\omega_0^2\hat{x}^2 \quad (23)$$

$$\hat{H}_I = K\hat{q}\hat{H}_0$$

where $\hat{H}_{QRS}$ is the Hamiltonian of the free quantum readout system. Unruh (*10*) has given a pedagogical example of this for an electromagnetic oscillator: the "mass" m is an inductance; the "spring constant" $m\omega^2$ is 1/(a capacitance); and the inductor coil and capacitor plates are attached to a mechanicai pivot with angular position $\hat{q}$, which varies the inductance and capacitance in the required manner. One can also vary the inductance and capacitance by letting the mechanical QRS move appropriate materials in and out of the inductor and capacitor (*11, 13*).

For a mechanical oscillator with electromagnetic QRS there are also several ways to make the mass and spring constant depend on a QRS variable. In Fig. 1 the oscillator's mass is attached to a movable capacitor plate of the QRS; the energy in the capacitor's electric field is

$$E_c = [1 - (\hat{x}/d)^2]\hat{q}^2/2C \qquad (24)$$

where C is the capacitance when $x = 0$; and consequently the charge $\hat{q}$ on the central plate of the QRS renormalizes the spring constant to $m\omega^2 = m_0\omega_0^2 - \hat{q}^2/2Cd^2$. The spring constant can also be renormalized by the QRS "momentum" $\hat{\pi}$ ($\equiv$ flux in inductor) by attaching the oscillator's mass to a current-carrying coil that resides between two oppositely wound coils of the QRS inductor. To renormalize the oscillator's mass m one might attach to it a conducting plate that resides in the inductor's magnetic field. The velocity of the plate through the magnetic field would induce an electric dipole moment on the plate, which in turn would couple by its velocity to the magnetic field, giving an interaction energy proportional to $\hat{p}^2\hat{\pi}^2$ and thence a mass renormalization.

Unfortunately, these various ideas have not yet produced a viable design for clean coupling of a mechanical oscillator's energy $\hat{H}_0$ to a QRS. On the other hand, designs without clean coupling can still yield measurements of $\hat{H}_0$ more accurate than the amplitude-and-phase quantum limit $(n_0 + 1/2)^{1/2}\hbar\omega$. An example is a QRS that couples only to $\hat{x}^2$, but that averages $\hat{x}^2$ over a number of cycles before sending it into the first classical stage (amplifier) (*9, 11*). The measurement scheme of Fig. 1 will do this if the period of the circuit's (QRS) oscillations is much longer than the period of the mechanical oscillations. Then the circuit's capacitance (Eq. 24) and resonant frequency will be sensitive to the time average of $\hat{x}^2$ and thence to $\hat{H}_0$, with only small admixtures of sensitivity to the time-varying part of $\hat{x}^2$ and thence to the oscillator's phase $\hat{\psi}$. This is equivalent to the statement in Eq. 6 that $\hat{Q}_R = f(\hat{H}_0 + \alpha\langle\hat{H}_0\rangle\hat{\psi})$ with $\bar{\alpha} << 1$, which in turn permits accuracies much better than $\Delta E = (n_0 + 1/2)^{1/2}\hbar\omega$. A detailed analysis of this type of scheme is given in (*11*), but for an electromagnetic oscillator with a mechanical QRS and with $\hat{H}_I = K\hat{x}^2\hat{q}$ rather than $K\hat{x}^2\hat{q}^2$ as in Fig. 1 and Eq. 24. That analysis reveals a limiting sensitivity

$$\Delta E \gtrsim \left(n_0 + \frac{1}{2}\right)^{1/2} (\Omega/\omega)^{1/2}\hbar\omega \qquad (25a)$$

where $E = (n_0 + 1/2)\hbar\omega$ is the oscillator's energy, ω is its frequency, Ω is the frequency of the QRS, and $\Omega << \omega$. The corresponding limit on the detection of a classical force $F_0 \cos(\omega t + \varphi)$, which drives changes in the oscillator's energy, is

$$F_0 \gtrsim \frac{2}{\hat{\tau}}\left(\frac{m\omega\hbar}{\omega/\Omega}\right)^{1/2} \qquad (25b)$$

if $\omega/\Omega < n_0 + 1/2$. If $\omega/\Omega > n_0 + 1/2$, then the limit on ΔE in Eq. 25a is replaced by $\hbar\omega$, the ultimate precision with which one can ever measure energy changes; correspondingly, the force limit in Eq. 25b is replaced by Eq. 21.

In measurements of the time average of $\hat{x}^2$ and thence $\hat{H}_0$, it is not essential that the interaction Hamiltonian $\hat{H}_I$ involve $\hat{x}^2$. Instead $\hat{H}_I$ can be proportional to $\hat{x}$, and then the internal workings of the QRS can produce the average of $\hat{x}^2$ at the entrance to the first classical stage.

Back-Action-Evading Measurements of $\hat{X}_1$

The QNDF observable $\hat{X}_1 = \hat{x}\cos\omega t - (\hat{p}/m\omega)\sin\omega t$ (real part of complex amplitude; Eq. 5), like the position $\hat{x}$, has a continuous spectrum of eigenvalues; and in principle it can be measured arbitrarily quickly and accurately (13, 15). Suppose that an initial state-preparation measurement at $t = 0$ has put the oscillator into an eigenstate $|\xi_0\rangle$ of $\hat{X}_1(0)$ with eigenvalue ξ_0. A classical force $F(t)$ [total Hamiltonian $\hat{H} = \hat{H}_0 - \hat{x}F(t)$] will change $\hat{X}_1$ as seen in the Heisenberg picture

$$\hat{X}_1(t) = \hat{X}_1(0) - \int_0^t [F(t')/m\omega]\sin\omega t'dt' \tag{26}$$

In the Heisenberg picture the oscillator's state remains fixed in time at $|\xi_0\rangle$, but this is an eigenstate of $\hat{X}_1(t)$ with eigenvalue

$$\xi(t) = \xi_0 - \int_0^t [F(t')/m\omega]\sin\omega t'dt' \tag{27a}$$

Subsequent perfect measurements of $\hat{X}_1$ must yield this eigenvalue and will reveal the full details of its evolution. It evolves in exactly the same manner as X_1 would evolve for a classical oscillator (13, 15).

One pays the price, in these measurements, of not knowing anything about the imaginary part of the complex amplitude $\hat{X}_2$ (Eq. 5c). However, if one has a second oscillator coupled to the same force $F(t)$, one can measure the imaginary part $\hat{Y}_2$ of its complex amplitude, giving up all information about the real part $\hat{Y}_1$. One's measurements must give the eigenvalue

$$\eta(t) = \eta_0 + \int_0^t [F(t')/m\omega]\cos\omega t'dt' \tag{27b}$$

which evolves in exactly the same manner as the X_2 or Y_2 of a classical oscillator. From the output of either oscillator, or better from the two outputs, one can deduce all details of the evolution of $F(t)$, no matter how weak $F(t)$ may be (13, 15). Thus $\hat{X}_1$ and $\hat{Y}_2$ are QNDF observables.

A perfect measurement of $\hat{X}_1$ (or $\hat{Y}_2$) requires (i) that the interaction Hamiltonian $\hat{H}_I$ depend on $\hat{X}_1$ and (ii) that $\hat{H}_I$ commute with $\hat{X}_1$ (Eq. 7 and associated discussion). The simplest example is

$$\hat{H}_I = K\hat{X}_1\hat{q} = K\hat{x}\hat{q}\cos\omega t - (K/m\omega)\hat{p}\,\hat{q}\sin\omega t \tag{28}$$

A coupling of this type can be achieved, for a mechanical oscillator, by using a capacitive position transducer with sinusoidally modulated coupling constant ($\hat{H}_I = K\hat{x}\hat{q}\cos\omega t$), followed by an inductive momentum transducer with modulated coupling constant [$H_I = -(K/m\omega)\hat{p}\hat{q}\sin\omega t$]. The two transducers together produce a voltage output

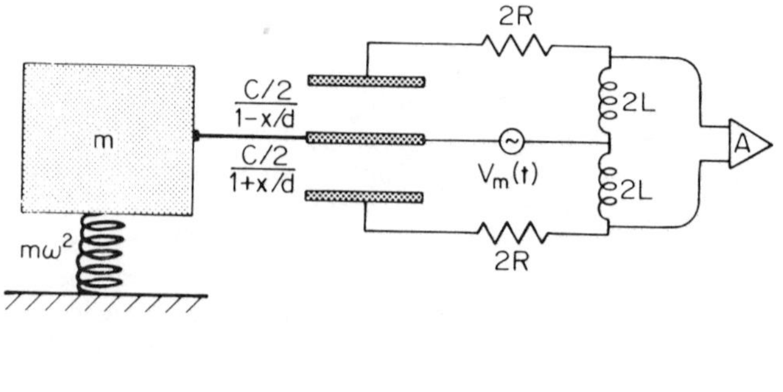

Fig. 2. Scheme for stroboscopic or continuous back-action-evading measurements of a mechanical oscillator. This scheme was devised independently in 1978 by V. B. Braginsky and by R. W. P. Drever, but has not previously been published.

$$\begin{aligned}\hat{V} &= \partial\hat{H}_1/\partial\hat{q} \\ &= K\hat{x}\cos\omega t - (K/m\omega)\hat{p}\sin\omega t \\ &= K\hat{X}_1 \end{aligned} \tag{29}$$

which drives an electromagnetic circuit, the QRS, in which the charge $\hat{q}$ flows (*15*). While capacitive position transducers and inductive velocity transducers are easy to construct, inductive momentum transducers are not. The momentum and velocity of the oscillator are related by

$$\begin{aligned}\dot{x} &= \partial(H_0 + H_1)/\partial p \\ &= p/m - (K/m\omega)q\sin\omega t \end{aligned} \tag{30a}$$

which means that the classical Lagrangian $L = p\dot{x} - (H_0 + H_1)$ for oscillator plus transducers is

$$L = \frac{1}{2}m\dot{x}^2 - \frac{1}{2}m\omega^2x^2 - Kxq\cos\omega t + (K/\omega)\dot{x}q\sin\omega t + \frac{1}{2}m(K\sin\omega t/m\omega)^2q^2 \tag{30b}$$

The first two terms represent the oscillator, the third is the capacitive position transducer, the fourth is an inductive velocity transducer (wire, physically attached to oscillator, moves through external magnetic field), and the last is a negative capacitor in the QRS circuit. Thus, an inductive momentum transducer is equivalent to an inductive velocity transducer (easy to construct) plus a negative capacitor (hard) (*15*). Although negative capacitors are not standard electronic components, they can be constructed in principle, and in principle they can be noise-free (*15*).

For an electromagnetic oscillator with mechanical QRS, one can achieve the desired $\hat{H}_1 = K\hat{X}_1\hat{q}$ using a capacitive transducer for the oscillator's position $\hat{x}$ ($\equiv$ charge in oscillator's capacitor) and an inductive transducer for its momentum $\hat{p}$ ($\equiv$ flux in oscillator's inductor). The momentum transducer turns out to involve a standard mechanical current transducer (current $\equiv \dot{x}$) plus a negative spring in the QRS (*15*). In principle negative springs can be noise-free (*15*).

The sinusoidal modulations required in the transducers must be regulated by an external, classical clock, which has the same frequency ω as one's oscillator. One cannot use the oscillator itself as the clock because in extracting the required oscillatory information from the oscillator one will produce an unacceptably large back action on $\hat{X}_1$. However, before the experiment begins one can check the frequency of the clock against that of the oscillator. In principle they can be made to agree perfectly, and in principle the clock can be made fully classical so its outputs are real numbers, $\cos\omega t$ and $\sin\omega t$, rather than operators (*10, 15*). In practice, frequency drifts and quantum features of the clock need not cause serious experimental problems (*15, 20*).

A perfect measurement of $\hat{X}_1$, which lasts a finite time $\bar{\tau}$, requires infinitely strong coupling in the transducers ($K \to \infty$) in order to give a signal that overwhelms zero-point noise in the QRS. If one has only finite coupling, then the zero-point noise accompanying the signal gives rise to a limit (Eqs. 16 and 17 with $x \to X_1$ and $N\tau \to \bar{\tau}$) (*13–15*)

$$\Delta X_1 \gtrsim (\hbar/2m\omega)^{1/2}(\beta\omega\bar{\tau})^{-1/2} \tag{31}$$

Here β is the dimensionless coupling constant (Eqs. 16). Thus, whereas stroboscopic measurements with limited coupling can beat the amplitude-and-phase limit by a factor of only $(\beta\omega\bar{\tau})^{-1/4}$ (Eq. 18b with $N = \omega\bar{\tau}/\pi$), continuous back-action-evading measurements of X_1 can beat it by $(\beta\omega\bar{\tau})^{-1/2}$. Stroboscopic measurements are worse because of their smaller duty cycle.

In the realistic case of weak coupling, $\beta < 1$, one must average over many cycles ($\omega\bar{\tau} >> 1/\beta$) in order to substantially beat the amplitude-and-phase limit. In this case one can make use of a trick analogous to measuring the energy by coupling to $\hat{x}^2$ and averaging: one can perform a "single-transducer, back-action-evading measurement" (*13–15*) by coupling to

$$\hat{x} \cos \omega t \cos \Omega t = \frac{1}{2} (\hat{X}_1 + \hat{X}_1 \cos 2\omega t + \hat{X}_2 \sin 2\omega t) \cos \Omega t \qquad (32)$$

(that is, $\hat{H}_1 = 2K\hat{x}\hat{q} \cos \omega t \cos \Omega t$) and then sending the signal through a filter (the QRS) with band pass at frequency $\Omega >> \omega$ and bandwidth $\Delta f = 1/2\tau_*$ $<< \omega/2\pi$. The filter will "average the X_2 signal away" until its amplitude has fallen by $1/2\omega\tau_*$ relative to that of the X_1 signal. Since the initial r.m.s. X_2 signal strength is $1/\sqrt{2}$ that of the X_1, this corresponds to $\hat{Q}_{\rm R} = f(\hat{X}_1 + \hat{X}_2/2\sqrt{2}\omega\tau_*)$ in Eq. 6a, which together with the uncertainty relation in Eq. 5c and the argument of Eqs. 6 tells us that (*20*)

$$\Delta X_1 \gtrsim (\hbar/2m\omega)^{1/2} (2\sqrt{2}\omega\tau_*)^{-1/2} \qquad (33)$$

This is the error in X_1 due to back action from measurement of X_2. The additional error due to zero-point noise accompanying the X_1 signal into the amplifier is (Eqs. 16 and 17 with $x \rightarrow X_1$ and $N\tau \rightarrow \bar{\tau}$) (*14, 15, 20*)

$$\Delta X_1 \gtrsim (\hbar/2m\omega)^{1/2} (\beta\omega\bar{\tau})^{-1/2} \qquad (34)$$

Here $1/2\bar{\tau}$ is the bandwidth of the experiment ($\bar{\tau}$ is the larger of the QRS averaging time τ_*, and the averaging time in subsequent electronics). The ultimate quantum limit on the sensitivity is Eq. 33 if $\beta > 2\sqrt{2}\tau_*/\bar{\tau}$, and Eq. 34 if $\beta < 2\sqrt{2}\tau_*/\bar{\tau}$. Note that Eq. 34 is the same limit (to within factors of order unity) as in the case of exact coupling to $\hat{X}_1 = \hat{x} \cos \omega t - (\hat{p}/m\omega) \sin \omega t$. Thus, when $\beta < 2\sqrt{2}\tau_*/\bar{\tau}$ and $\omega\bar{\tau} >> 1$, one can abandon the momentum transducer without any serious loss of accuracy.

This type of single-transducer, time-averaged, back-action-evading measurement of X_1 appears today to be the most viable technique for beating the amplitude-and-phase limit (Eq. 14) in gravitational-wave detection. In place of Eq. 14 one will face the limiting measurable force

$$F_0 \gtrsim (2/\bar{\tau})(m\omega\hbar)^{1/2} \times \text{Max}\,[(\beta\omega\bar{\tau})^{-1/2}, (2\sqrt{2}\omega\tau_*)^{-1/2}] \qquad (35)$$

Thermal Noise in the Oscillator and Amplifier

The quantum limits derived above are not achievable in the laboratory today because thermal noise exceeds quantum mechanical noise.

Ignore for the moment thermal (Nyquist) noise in the oscillator. Then if the resistors in the QRS are cooled sufficiently, the dominant nonquantum noise will be that in the amplifier (first classical stage). The amplifier, which we assume to be linear, can be characterized by its power gain G and its noise temperature $T_{\rm n}$. The QRS feeds the amplifier a signal at frequency $f = \Omega/2\pi$, to which the amplifier adds a noise power per unit bandwidth

$$\frac{dP_{\rm n}}{df} = \frac{\hbar\Omega}{\exp(\hbar\Omega/kT_{\rm n}) - 1} \qquad (36)$$

Here k is Boltzmann's constant. If the incoming signal has power $P_{\rm s}$, then the amplified signal and noise have power (*23*)

$$GP_{\rm s} + \left(G \frac{\hbar\Omega}{\exp(\hbar\Omega/kT_{\rm n}) - 1} + \frac{\hbar\Omega}{2}\right) \Delta f \qquad (37)$$

Here Δf is the bandwidth, and the $\hbar\Omega/2$ is a zero-point energy that accompanies the signal throughout its trek through the amplifier and other electronics, but does not get amplified (*23*). The quantum limits of previous sections of this article are attributable to this zero-point energy. In the presence of a real linear amplifier, with nonnegligible noise temperature $T_n >> \hbar\Omega/k$ and large gain $G >> 1$, the signal power P_s must fight not $(\hbar\Omega/2)\Delta f$, but rather

$$\left(\frac{\hbar\Omega}{\exp(\hbar\Omega/kT_n) - 1} + \frac{1}{G}\frac{\hbar\Omega}{2}\right)\Delta f \simeq kT_n\,\Delta f \quad (38)$$

(see Eq. 37). Consequently, it is reasonable to expect that the amplifier noise will modify our quantum limits (Eqs. 6c, 13-15, 17-19, 21, 22, 25, 31, and 33-35) by replacing $\hbar$ with (*12*)

$$\hbar \rightarrow \frac{2kT_n}{\Omega} \quad (39)$$

These modified quantum limits are sometimes called amplifier limits. It will never be possible, even in principle, to reduce these amplifier limits below the corresponding quantum limits (*19, 24, 25*).

The best linear amplifiers that have been built are parametric amplifiers and maser amplifiers, which operate at microwave frequencies and have (kT_n/Ω) as small as $\sim 10\hbar$. With such amplifiers one can only hope to get within a factor 20 or $\sqrt{20}$ of our quantum limits ($\hbar \rightarrow 20\hbar$). And even to achieve this one must design a QRS which upconverts the oscillator's signal frequency (kilohertz in the gravitational-wave case) to the microwave (gigahertz) region.

Any physical oscillator (such as the fundamental mode of a gravitational-wave bar antenna) is weakly coupled to a thermal bath of dynamical systems (sound waves in the bar). This coupling produces a frictional damping of large-amplitude motions, and it also produces a thermal-buffeting random walk of the oscillator's amplitude (Nyquist noise). The r.m.s. random-walk change of the oscillator's amplitude during time $\bar{\tau}$ is

$$(\Delta x_0)_{\mathrm{Nyq}} = (\Delta p_0/m\omega)_{\mathrm{Nyq}} = (\Delta X_1)_{\mathrm{Nyq}} = (\Delta X_2)_{\mathrm{Nyq}} \simeq (kT/m\omega^2)^{1/2}(\omega\bar{\tau}/Q)^{1/2} \quad (40)$$

Here T is the temperature of the thermal bath (the bar's temperature), and Q is the oscillator's quality factor (number of radians of oscillation required for frictional damping of large-amplitude oscillations by a factor e in energy). The corresponding r.m.s. energy change is

$$(\Delta E)_{\mathrm{Nyq}} \simeq (2E_0kT)^{1/2}(\omega\bar{\tau}/Q)^{1/2} \quad (41)$$

These Nyquist noises must not exceed the amplifier limits (quantum limits with $\hbar \rightarrow 2kT_n/\Omega$) if one is to achieve the amplifier limits in real experiments. Some numbers will be given below.

Prospects for Stroboscopic Measurements

One possible scheme for stroboscopic measurements of a mechanical oscillator (gravitational-wave antenna with mass $m \simeq 10$ kg and frequency $\omega \simeq 3 \times 10^4$ sec^{-1}) is shown schematically in Fig. 2. The mass of the oscillator is physically attached to the central, movable plate of a capacitor (capacitance between outer plates = C), which plays the role of transducer. The capacitor resides in the QRS—a high-frequency LC circuit [frequency $\Omega = (LC)^{-1/2} \simeq 10^{10}$ sec^{-1}], which has small losses [amplitude damping time $\tau = 2(RC\Omega^2)^{-1} << 0.1/\omega$] and which is driven at its resonant frequency Ω by an external generator. In practice this circuit would be a microwave cavity (*26*). At the measurement times $\omega t = 0, \pi, 2\pi, \ldots$ the generator is turned on for a time $\tau/2$ and then turned off, and in an additional time $\tau/2$ the excitations in the circuit die out. During the brief on-time τ, the amplifier sees a voltage signal $V_s = (V_0/d)\,\Omega\tau x \cos \Omega t$, where V_0/d is the amplitude of the oscillating electric field between the capacitor plates. The experimenter averages the amplitude of this signal (with alternating sign) over N measurements to determine the position x of the oscillator.

It is straightforward to analyze the noise performance of this system by standard circuit theory. Alternatively, one can invoke the general formulas of Eqs. 15 to 19 for stroboscopic measurement schemes. Assuming that the resistor's physical temperature is less than the amplifier's noise temperature $T_n \simeq 10$ K, the amplifier noise dominates and in Eq. 18b we must replace $\hbar \to 2kT_n/\Omega$. Assuming that the amplifier is properly impedance-matched to the circuit, the measurement will achieve the limiting precision (Eq. 18b)

$$\Delta x \simeq \left(\frac{2kT_n/\Omega}{m\omega} \right)^{1/2} (\beta N)^{-1/4} \qquad (42)$$

Comparison of the voltage signal with Eq. 16a reveals that $K|g_{21}| = (V_0/\sqrt{2}d)(\Omega\tau)$; scrutiny of Fig. 2 reveals that the QRS output impedance, as seen by the amplifier, is $g_{22} = 2\tau/C$; consequently the dimensionless coupling constant of Eq. 16b is $\beta = (V_0/d)^2 C\Omega\tau/(4m\omega^2)$. Combining this with the required pulse time $\tau = (\beta N\omega^2)^{-1/2}$ (Eq. 18a), we find

$$\beta N = \left[\frac{(V_0/d)^2 CN\Omega}{4m\omega^3} \right]^{2/3} \qquad (43)$$

To avoid voltage breakdown in the capacitor, its electric field amplitude should not exceed $(V_0/d) \simeq 10^6$ volt/cm. Assuming other reasonable parameters $C \simeq 1$ pf, $\Omega \simeq 10^{10}$ sec^{-1}, $N \simeq 1000$, $T_n \simeq 10$ K, $m \simeq 10$ kg, and $\omega \simeq 3 \times 10^4$ sec^{-1}, we find

$$\beta N \simeq 20 \qquad \Delta x \simeq 1 \times 10^{-17} \text{ cm} \qquad (44)$$

Thus this system can achieve a sensitivity that is a factor $(20)^{1/4} \simeq 2.1$ below the amplitude-and-phase amplifier limit; but this is still an order of magnitude worse than the amplitude-and-phase quantum limit $(\hbar/2m\omega)^{1/2} \simeq 1 \times 10^{-18}$ cm.

Nyquist noise in the antenna (Eq. 40 with $\omega\hat{\tau} = \pi N$) will be less than the measurement precision $\Delta x \simeq 1 \times 10^{-17}$ cm if the antenna is cooled to 4 K and has a quality factor $Q \simeq 4 \times 10^9$. This is comparable to the best mechanical Q that has been achieved (*27*) for a sapphire crystal at 4 K.

Prospects for Single-Transducer Back-Action-Evading Measurements

The configuration of Fig. 2 can also be used in a single-transducer, back-action-evading measurement of $\hat{X}_1$. In this case the circuit's amplitude damping time $2(RC\Omega^2)^{-1}$ becomes the averaging time τ_* of the QRS filter (previously it was the stroboscopic pulse length), and we require $\tau_* >> 1/\omega$ (previously it was $<< 1/\omega$). Instead of being pulsed, the generator's modulating voltage has the steady-state form $V_m = U_0 \sin \Omega t \sin \omega t$, which produces an electric field (V_0/d) $\cos \Omega \cos \omega t$ in the capacitors ($V_0 = U_0\Omega/2\omega$). That electric field, interacting with the motions $x = X_1 \cos \omega t + X_2 \sin \omega t$ of the mechanical oscillator, produces a signal voltage

$$\begin{aligned} V_s = (V_0/d)(\Omega\tau_*/2)\, [X_1 \sin \Omega t \\ + (2\omega\tau_*)^{-1} X_1 \sin \Omega t \sin 2\omega t \\ - (2\omega\tau_*)^{-1} X_2 \sin \Omega t \cos 2\omega t] \end{aligned} \qquad (45)$$

at the output of the QRS. Amplification of this signal produces information about X_1 and X_2 with relative accuracies $\Delta X_1 = (2\sqrt{2}\omega\tau_*)^{-1}\, \Delta X_2$.

Assuming that the resistor noise is negligible compared to amplifier noise (which it will be if the physical temperature of the resistors is somewhat less than the noise temperature $T_n = 10\ K$ of the amplifier), we can compute the noise performance of this system from Eqs.

16, 33, and 34 with $\hbar \rightarrow 2kT_n/\Omega$. The best quality factor that has been achieved (*26*) for a superconducting microwave resonator (our QRS circuit) with a narrow capacitive gap is $Q_e = \Omega\tau_*/2 \simeq 10^7$, corresponding to $\tau_* \simeq 10^{-3}$ second. Consequently back-action forces (Eq. 33) limit the sensitivity to

$$\Delta X_1 \simeq \left(\frac{kT_n/\Omega}{m\omega}\right)^{1/2} \frac{1}{(2\sqrt{2}\omega\tau_*)^{1/2}}$$
$$\simeq 2 \times 10^{-18} \text{ cm} \qquad (46)$$

a factor 9 below the amplitude-and-phase amplifier limit and approximately twice the amplitude-and-phase quantum limit. (Here we use $T_n = 10$ K, $\Omega = 10^{10}$ sec^{-1}, $m = 10$ kg, and $\omega = 3 \times 10^4$ sec^{-1} as before.) In order that Nyquist noise in the mechanical oscillator (Eq. 40) not exceed this sensitivity, the averaging time must not exceed $\bar{\tau} \simeq 0.01$ sec. (Here we use the same oscillator temperature and Q as before, $T = 4$ K and $Q = 4 \times 10^9$.) To achieve the limit in Eq. 46 we also require a coupling constant $\beta \simeq 2\sqrt{2}\tau_*/\bar{\tau} \simeq 0.3$ (Eq. 34). To compute β, first derive $K|g_{21}| = (V_0/d)(\Omega\tau_*/2\sqrt{2})$ from Eqs. 16a and 45 with $x \rightarrow X_1$; then evaluate the impedance seen by the amplifier in Fig. 2 at the X_1 signal frequency $\Omega = (LC)^{-1/2}$, $g_{22} = 2\tau_*/C$; then evaluate Eq. 16b

$$\beta = (V_0/d)^2 C\Omega\tau_*/(16m\omega^2) \qquad (47)$$

The required β of 0.3 can be achieved with the same electric field in the capacitive gap as we used before: $V_0/d = 1 \times 10^6$ V/cm.

This example and that of the last section confirm that it is easier to achieve a given level of sensitivity by continuous, single-sensor, back-action evasion than by stroboscopic techniques. However, along the route toward realization of such experiments there remain a series of difficult experimental problems—not least of which is the frequency stability of the clock that regulates the voltage generator.

On the Limiting Frequency Stability of a Generator

Although current technology can achieve the frequency stability required by the above examples, it is of longer term interest to know ultimate quantum mechanical limits on the stabilities of clocks.

At present the world's most stable clocks are the superconducting cavity stabilized oscillator (SCSO) (*28*) and the hydrogen maser (*29*). Both involve self-excited electromagnetic oscillations inside a cavity. In the SCSO the clock frequency Ω is regulated by the cavity's normal mode, and a change Δl of a typical dimension l of the cavity will produce a frequency change

$$\Delta\Omega/\Omega \simeq \Delta l/l \qquad (48)$$

In the maser, if the electromagnetic quality factory Q_e of the cavity (Teflon bubble) exceeds $\frac{1}{2}\Omega \times$ (mean time hydrogen atoms spend in cavity) $\equiv \Omega_a$, then Eq. 48 will be true. Otherwise, $\Delta\Omega/\Omega \simeq (\Delta l/l)(Q_e/Q_a)$, and the limit derived below is correspondingly modified.

A quantum limit on the frequency stability of any electromagnetic oscillator satisfying Eq. 48 is derived in (*30*). The source of the limit is quantum fluctuations in the deformation of the cavity walls by electromagnetic stresses. Since the stresses in the electromagnetic field are equal to its energy density $\hat{H}_e/l^3$ (with $\hat{H}_e$ the Hamiltonian of the electromagnetic oscillator), the force on the walls is $\hat{H}_e/l$, and this deforms the walls by $\delta l = \hat{H}_e/lk$, where k is the mechanical spring constant of the walls. The electromagnetic field is in a thermalized coherent state with n_0 quanta, which possesses quantum fluctuations $\Delta H_e \geq n_0^{1/2}\hbar\Omega$;

consequently, $\Delta l \geq n_0^{1/2}\hbar\Omega/kl$, which leads to frequency fluctuations (Eq. 48)

$$\Delta\Omega/\Omega \gtrsim n_0^{1/2}\hbar\Omega/kl^2 \qquad (49)$$

This electromagnetic back-action limit must be contrasted with the limiting precision for measurements of Ω during an averaging time $\bar{\tau}$: $\Delta\Omega \gtrsim \Delta\psi/\bar{\tau}$, where $\Delta\psi \gtrsim n_0^{-1/2}$ is the quantum uncertainty in the phase of the oscillator's coherent state

$$\Delta\Omega/\Omega \gtrsim n_0^{-1/2}(\Omega\bar{\tau})^{-1} \qquad (50)$$

(Townes-Schawlow limit). These two limits lead to an optimal number of quanta n_0 and an ultimate quantum limit

$$\frac{\Delta\Omega}{\Omega} \gtrsim \left(\frac{\hbar}{kl^2\bar{\tau}}\right)^{1/2} \qquad (51)$$

For a cavity with wall thickness comparable to cavity dimensions l, or for a "cavity" made by coating the outside of a dielectric crystal with superconducting material (*31*), the spring constant k is related to the Young's modulus E_M of the cavity walls by $k \simeq E_M V/l^2$, where V is the cavity volume. Then

$$\Delta\Omega/\Omega \gtrsim (\hbar/E_M V\bar{\tau})^{1/2} \qquad (51')$$

In practice $E_M \lesssim 10^{13}$ dyne/cm^2, $V \simeq 1$ cm^3, so $\Delta\Omega/\Omega \gtrsim 10^{-20}$ $(\bar{\tau}/1 \text{ sec})^{-1/2}$. This limit is achievable in principle, but current technology is far from it.

References and Notes

1. D. Bohm, *Quantum Theory* (Prentice-Hall, Englewood Cliffs, N.J., i951), especially chapters 6 and 22.
2. M. O. Scully, R. Shea, J. D. McCullen, *Phys. Rep.* **43**, 486 (1978).
3. A. Einstein, "Autobiographical notes" and "Reply to criticisms," in *Albert Einstein: Philosopher-Scientist*, P. A. Shilpp, Ed. (Banta, Menasha, Wis., 1949).
4. W. Lamb, *Phys. Today* **22**, 23 (1969).
5. C. W. Helstrom, *Quantum Detection and Estimation Theory* (Academic Press, New York, 1976).
6. V. B. Braginsky and V. N. Rudenko, *Phys. Rep.* **46**, 165 (1978); J. A. Tyson and R. P. Giffard, *Annu. Rev. Astron. Astrophys.* **16**, 521 (1978); R. Weiss, in *Sources of Gravitational Radiation*, L. Smarr, Ed. (Cambridge Univ. Press, Cambridge, 1979); D. H. Douglass and V. B. Braginsky, in *General Relativity: An Einstein Centenary Survey*, S. W. Hawking and W. Israel, Eds. (Cambridge Univ. Press, Cambridge, 1979); K. S. Thorne, *Rev. Mod. Phys.* **52**, 285 (1980).
7. V. B. Braginsky and V. S. Nazarenko, *Zh. Eksp. Teor. Fiz.* **57**, 1421 (1969); English translation in *Sov. Phys. JETP* **30**, 770 (1970).
8. V. B. Braginsky and Yu. I. Vorontsov, *Usp. Fiz. Nauk.* **114**, 41 (1974); English translation in *Sov. Phys. Usp.* **17**, 644 (1975).
9. W. G. Unruh, *Phys. Rev. D* **18**, 1764 (1978).
10. ______, *ibid.* **19**, 2888 (1979).
11. V. B. Braginsky, Yu. I. Vorontsov, F. Ya. Khalili, *Zh. Eksp. Teor. Fiz.* **73**, 1340 (1977); English translation in *Sov. Phys. JETP* **46**, 705 (1977).
12. ______, *Pis'ma Zh. Eksp. Teor. Fiz.* **27**, 296 (1978); English translation in *Sov. Phys. JETP Lett.* **27**, 276 (1978).
13. K. S. Thorne, R. W. P. Drever, C. M. Caves, M. Zimmermann, V. D. Sandberg, *Phys. Rev. Lett.* **40**, 667 (1978).
14. K. S. Thorne, C. M. Caves, V. D. Sandberg, M. Zimmermann, R. W. P. Drever, in *Sources of Gravitational Radiation*, L. Smarr, Ed. (Cambridge Univ. Press, Cambridge, 1979).
15. C. M. Caves, K. S. Thorne, R. W. P. Drever, V. D. Sandberg, M. Zimmermann, *Rev. Mod. Phys.* **52**, 341 (1980).
16. J. N. Hollenhorst, *Phys. Rev. D* **19**, 1669 (1979).
17. L. I. Mandelstam, *Complete Collected Works* (1950), vol. 5.
18. V. B. Braginsky, *Physical Experiments with Test Bodies* (Nauka, Moscow, 1970); English translation published as NASA-TT F762, National Technical Information Service, Springfield, Va.
19. R. P. Giffard, *Phys. Rev. D* **14**, 2478 (1976).
20. R. W. P. Drever, K. S. Thorne, C. M. Caves, M. Zimmermann, V. D. Sandberg, in preparation.
21. G. W. Gibbons and S. W. Hawking, *Phys. Rev. D* **4**, 2191 (1971).
22. I. I. Gol'dman and V. D. Krivchenkov, *Problems in Quantum Mechanics* (Pergamon, London, 1961).
23. J. P. Gordon, in *Quantum Electronics and Coherent Light*, P. A. Miles, Ed. (Academic Press, New York, 1964), pp. 156–169.
24. J. Weber, *Rev. Mod. Phys.* **31**, 681 (1959).
25. H. Heffner, *Proc. IRE* **50**, 1604 (1962).
26. V. B. Braginsky, V. I. Panov, V. G. Petnikov, V. D. Popel'nyuk, *Prib. Tekh. Eksp.* **20**, 234 (1977); English translation in *Instrum. Exp. Tech. (USSR)* **20**, 269 (1977).
27. Kh. S. Bagdasarov, V. B. Braginsky, V. P. Mitrofonov, V. S. Shiyan, *Vestn. Mosk. Univ. Fiz. Astronomiya* **18**, 98 (1977).
28. S. R. Stein and J. P. Turneaure, *Proc. IEEE* **63**, 1249 (1975).
29. R. F. C. Vessot, in *Experimental Gravitation*, B. Bertotti, Ed. (Academic Press, New York, 1974), p. 111.
30. V. B. Braginsky and S. P. Vyatchanin, *Zh. Eksp. Teor. Fiz.* **74**, 828 (1978); English translation in *Sov. Phys. JETP* **47**, 433 (1978).
31. Kh. S. Bagdasarov, V. B. Braginsky, P. I. Zubietov, *Pis'ma Zh. Tekh. Fiz.* **3**, 991 (1977); English translation in *Sov. Tech. Phys. Lett.* **3**, 406 (1977).

32. For valuable discussions we thank C. M. Caves, R. W. P. Drever, and F. Ya. Khalili. For financial support we thank: (i) at Moscow University, the Ministry of Higher Education of the U.S.S.R.; (ii) at Caltech, the U.S. National Aeronautics and Space Administration (NGR 05-002-256 and a grant from PACE) and the U.S. National Science Foundation (AST76-80801 A02); (iii) for collaborative aspects of this work, the Cooperative Program in Physics between the U.S. National Academy of Sciences and the U.S.S.R. Academy of Sciences under the auspices of the U.S.-U.S.S.R. Joint Commission on Scientific and Technological Cooperation (contract NSF-C310, Task Order 379).

GUIDE TO SOME FURTHER LITERATURE

The following papers are cited, not with any thought of providing a comprehensive bibliography, but rather with the aim of providing a few points of entry to an enormous literature. The traditional detective novel is loaded with clues, most of them irrelevant, distractive or downright deceptive. The few truly decisive items have no star to distinguish them. That is how the author conceals his plot. If we have distinguished no items with a star in our bibliography, it is because we do not *know* the plot!

I. Questions of Principle

Consciousness: Wigner reasons that an observation is only then an observation when it becomes part of "the consciousness of the observer" (Wigner, 1969) and points to "the impressions which the observer receives as the basic entities between which quantum mechanics postulates correlations" (Wigner, 1971). For Bohr, the central point is not "consciousness" (Kalckar, 1967), not even an "observer," but an experimental device—grain of silver bromide, Geiger counter—capable of an "irreversible act of amplification" (Bohr, 1958, p. 88) that brings the measuring process to a "close" (Bohr, 1958, p. 73). Only then, he emphasized, is one person able to describe the result of the measurement to the other "in plain language" (Bohr, 1963, p. 3). For more on consciousness, see Fuller and Putnam (1966), Popper and Eccles (1977), Pugh (1976), and Wheeler (1981a,c) as a few items in an enormous literature. Moreover, "consciousness" may not be the unique faculty that thinkers of the past assumed, feature as it is of the brain. Thus, the eye, the "window of the mind" has evolved independently in at least forty different places and times, according to Salvini-Plawen and Mayr (1978). In addition, the very mechanism of that evolution, beautiful as it is, would appear (see, for example, Cooper, 1973, Nass and Cooper, 1975, and Cooper, 1975) a natural spelling out of the principles of computers and automata sketched out in a preliminary way by von Neumann (1958).

Is *meaning*—a term appropriate for a community—more relevant than the individual consciousness for understanding where the "individual quantum phenomenon" links up with human perception? Is this why it is so important that the result of a quantum phenomenon, being "closed by irreversible amplification" (Bohr, 1958, p. 73), is "unambiguously communicable" from one to another in "plain language"? (Bohr, 1963, pp. 5, 6, and 3). If so, a direct link would seem to be established with a central theme of American and British philosophical investigations in recent decades, work most briefly summarized in the phrase of Føllesdal (1975), "Meaning is the joint product of all the evidence that is available to those who communicate." It reflects the spread of views on a topic so elusive as this to recall the words of David

Hume (1779): "What peculiar privilege has this little agitation of the brain that we call *thought* that we must make it a model for the entire universe?"

Bohr's *phenomenon* and philosophers' *phenomenalism* are to be distinguished. Phenomenon is an *individual quantum process* brought to a close by an irreversible act of amplification. Phenomenalism is "the view that the existence of a physical object is dependent upon its being perceived by some percipient, that a physical object is nothing but a construct made up of the percepts that are the immediate objects of perception. It follows directly from this that the physical world is completely dependent on some perceiver(s) for its existence and that if all perceivers were annihilated, the physical world would accordingly cease to exist" (Halverson, 1967). This view of physics has ancient antecedents. Parmenides of Elea argued that "what is . . . is identical with the thought that recognizes it" (Parmenides, ~550 B.C.). Bishop George Berkeley advocated the position "*Esse est percipere aut percipi*," "to be is to perceive or to be perceived." (Berkeley in Calkins, ed., 1929).

E. E. Thomas (1921), analyzing R. H. Lotze's *philosophical system of teleological idealism* (1874–1879), distinguishes it from Berkeley's and similar ideas with these words: "Many indeed have maintained that we create the world of being—that out of our minds, or out of some universal mind in whom we exist and have our being, we introduce into sense the individualising element and so exercise a creative activity in relation to the things which we perceive around us. Lotze tells us, however, that we do no such thing, but that the sense qualities themselves are the creators of their own individuality; that indeed sense qualities are the constitutive elements in a continuously creative activity, and that the individual whole which lives in and through such activity is what we understand by a real thing . . . that a 'real thing is nothing but the realised individual law of its procedure.'" Husserl (1913) provides a still more widely read and widely influential account of phenomenology.

Ernst Mach, patron saint of *positivism*, and inspiration in Einstein's early thinking about relativity, special and general, opposes any "dualism of feeling and motion" as arising "from an improper formulation of the questions involved." He goes on, "I maintain that every physical concept is nothing but a certain definite connexion of the sensory *elements* which I denote by *ABC* . . . , and that every physical fact rests therefore on such a connexion. These *elements*—elements in the sense that no further resolution has as yet been made of them—are the simplest building stones of the physical world that we have yet been able to reach" (Mach, 1897).

For an up-to-date account of the many steps by which the eye processes entering photons into visual perception, reference may be made to Hubel and Wiesel (1977) and Hubel (1978).

How concepts are formed from sensory experience has been elucidated in part by the famous and extensive studies of learning in children by Jean Piaget and his school, summarized by Richmond (1971). How decisive the difference is between Berkeley's "*tree that falls in the forest*"—a many quantum process—and the individual quantum phenomenon is stressed in Wheeler (1981a,c). He notes that in the transition from individual quantum effects to classical concepts, "Many a chance 'yes, no' gives rise to the substantiality 'how much.'"

Popper (1982a,b; 1983) argues for an almost *classical interpretation of Heisenberg's indeterminacy* relations; he hopes for a "way in which the interference of the subject with the object may be visualized as *something of no fundamental importance*—a *consequence* of the prevailing laws of interaction, and therefore in no way capable of explaining these laws, or their indeterminist character'' (1982b, p. 191).

The vastness of the literature on "reality" is very inadequately indicated by the foregoing sampling; and as to the great *scope* of that literature, it may be enough to give one tiny taste by citing the article of Heath (1972) on "nothing."

From these links between quantum theory and epistemology, we turn in the following five sections of this bibliography to issues that lie more clearly in the traditional purview of physics. For surveys of quantum theory additional to those cited in the Preface, the following, among others, may be mentioned: d'Abro (1939, 1951), DeWitt and Graham (1971), Dirac (1971), Eisenbud (1971), Feyerabend (1962), Flato (1976), Gamow (1966), Heisenberg (1930), Hermann (1979), Hoffmann (1959), Hund (1974), Jammer (1966), Jauch (1973), Klein (1959, 1970), Kramers (1956), Landé (1955, 1960, 1965), Mehra (1974), Pais (1979), Piron (1976), Price and Chissick (1976). Puligandla (1966), Ruark and Urey (1930), Whittaker (1953), Yourgrau and van der Merwe (1971), and Bohr's *Collected Works* (Rosenfeld and Nielsen, 1972-1976); Nielsen, 1977).

II. Interpretations of the Act of Measurement

As early as 1909, G. I. Taylor showed that one obtains *interference fringes even with feeble light*; and other experiments with similar findings have been reported by Dempster and Batho (1927), Janossy and Naray (1957), Frisch (1965), Donstov and Baz' (1967), and Reynolds, Spartalian, and Scarl (1969).

An irreversible act of amplification is an essential feature of the elementary quantum phenomenon as conceived by Bohr; but not one mention of that term occurs in the formulation of the act of measurement as presented by von Neumann (1932), London and Bauer (1939; printed here in translation as II.1), and Wigner (1976). This difference in outlook is far from having been resolved today, central though it is. Many issues have been and continue to be explored in the attempt to clarify the situation. One of them is *irreversibility*, on which we cite, for example, Davydov (1947), the discussions by Frisch, von Weizäcker, Bohm, Aharonov and Petersen, Prosperi, and others in Bastin (1971) and Belinfante (1975).

Consider *amplification* as we know it from Bohr's discussion, as an "irreversible process" that brings the elementary quantum phenomenon to a close, and *redundancy* as that term is understood in information theory

(Pierce, 1961): What do they have in common? Can we construct a macroscopic redundant record of a certain observable by making sufficiently many microscopic copies of the same observable? Several papers pursue this line of investigation (Green, 1958; Bell, 1975; Machida and Namiki, 1980a,b; Araki, 1980; Zurek, 1982a,b). In particular, Whitten-Wolf and Emch (1976), updating Hepp (1972) and Emch (1972), propose a model of an amplification device (two infinite non-interacting chains of $\frac{1}{2}$-spin), which, though rather idealized and unrealistic, nevertheless provides, they propose, an exactly solvable model in which one can see both the "reduction of the wavepacket" and the transfer of information from a microscopic to a macroscopic level.

Relativity of quantum observables is ignored in the usual treatments of the process of measurement. There observables are assumed to be absolute. A preferred direction—z-axis—in the measurement of the spin is assumed to be "God-given." Zurek (1981, 1982a,b) points out that in fact physical observables are always defined in a relative manner, with respect to other physical systems. This observation alone may be significant for reconciling distinct and definite outcomes of measurements as perceived by the individual observer with the indefinite superpositions of outcomes which follow from Schrödinger evolution of the combined apparatus-system wave function. In particular he shows that only the "open" apparatus interacting with the environment will "know" what observable of the measured system it is supposed to record. The very interaction which will define the preferred "pointer observable" of the apparatus will also remove spurious correlation elements from the apparatus-system density matrix, thus accomplishing von Neumann's "second stage of the measurement," reduction of the wavepacket. In the absence of the environment, on Zurek's view, the observable measured by the quantum apparatus will be uncertain, but the outcome of the measurement may appear definite (Zurek, 1982a,b). This at first sight preposterous statement may be illustrated by an EPR-correlated pair of spin-$\frac{1}{2}$ particles, with their wave function given by

$$|\sigma\rangle = \{|\uparrow\rangle_1 \otimes |\downarrow\rangle_2 - |\downarrow\rangle_1 \otimes |\uparrow\rangle_2\}/2^{1/2}.$$

Zurek notes that "Nothing can keep one from thinking of the first spin as of the measured system and of the second spin as of a quantum apparatus. After the first stage of measurement, which has resulted in the wave function $|\sigma\rangle$, the state of the spin-system with respect to the spin-apparatus is quite definite: The (no. 1) spin-system always points in the direction which is opposite to the orientation of the (no. 2) spin-apparatus. This is a definite, 'coordinate-independent' statement. Thus, from the point of view of the spin-apparatus, the measurement has already yielded a definite result. However, from the point of view of an external observer the measurement has not been completed: the state of the spin remains indefinite with respect to the state of the coordinate system in which this observer describes physics."

The *non-linear Schrödinger equation* foreshadowed in Wigner's lecture notes (1976, printed here as section II.2) appears in a concrete form in Wigner (in Meystre and Scully, 1983). This is different from the non-linear wave equation proposed by Bialynicki-Birula and Mycielski (1976), and discussed by Shimony (1979), an upper limit to the critical parameter in which has been set by the observations of Shull, Atwood, Arthur, and Horne (1980).

Irreversible increase of entropy and the problem of measurement in quantum physics are intimately connected, as pointed out by von Neumann (1929, 1932): Irreversible evolution could get rid of spurious correlations represented by off-diagonal elements in the apparatus-system density matrix. Assuming that irreversible evolutions are possible, one can accomplish reduction of the wavepacket. This conclusion stimulated many to seek the resolution of the measurement problem by invoking irreversibility and the second law of thermodynamics, among them Green (1958); Daneri, Loinger, and Prosperi (1962); Prigogine, George, Henin, and Rosenfeld (1973); and Penrose (1981). The unitary and reversible evolution given and demanded by quantum theory cannot, however, dissipate any information. This circumstance leads to the fundamental difficulty discussed by Zeh (1971) among others. However, it may be possible to transfer the information rather than dissipate it (Davies, 1974, p. 168; Zeh, 1971; Zurek, 1982a). Then the forgetting of the off-diagonal correlation terms would not be due to the irreversible decay of information, but rather due to its transfer. This scenario ties in with the process of amplification, where the information about the chosen observable of the measured system is recorded in many separate copies at the expense of the information about complementary observables.

The relative state or "many-worlds" approach to measurement described by Everett (1957; reprinted here as II.3) and independently put forward by Cooper and Van Vechten (1969) and other early papers on this line of thought are brought together in the book of Dewitt and Graham (1973); and Cooper (1976) has written a further paper. This outlook, originally defended by Wheeler (1957) has subsequently been judged wanting by him (1977), Bell (1971b), Zeh (1975), and Wigner (1976, printed here as II.2).

Interpretation basis is the term used by Deutsch (1981) to describe a preferred basis system in which the splitting of the universes in Everett's many-world interpretation takes place. His work aims to solve the problem of the non-uniqueness of the relative states as defined by Everett and exemplified by the many options open (Zurek, 1981) for the polarizations of the two photons produced in the annihilation of positronium. On the view of Deutsch, "At the instant of completion of a measurement, the interpretation basis is determined by the requirement that a measurement has indeed taken place"; and he provides a system of equations designed actually to calculate this interpretation basis.

Does *Gödel's undecidability have any connection with quantum indeterminism*? A few items of literature bearing on this issue will be found in Section V, below.

In connection with the theme of

"*law without law*" (Wheeler, 1979–1981, reprinted here as I.13), special reference should be made to the philosophical arguments already put forward for indeterminacy before the advent of quantum theory by Fritz Exner, described in recent papers of Forman (1971) and Hanle (1979). No guide would seem more valuable in the further exploration of the thesis of "law without law" than the well-known motto, "More is different" (P. W. Anderson, 1972). Charles S. Peirce (1897) asks, "May these forces of nature not be somehow amenable to reason? May they not have naturally grown up?" In contrast, James Clerk Maxwell (in Campbell and Garnett, 1969) argues that "No theory of evolution can be formed to account for this similarity of molecules, for evolution necessarily implies continual change, and the molecule is incapable of growth, or decay, or generation, or destruction."

"*Quantum logic*" is a term occasionally used for one of the several mathematically equivalent ways available to state the content of quantum theory. It differs as much from wave mechanics in the mathematical machinery that it uses for this purpose, as wave mechanics differs from matrix mechanics or both from Dirac's abstract operator algebra. Its focus of attention is propositions, such as the proposition that the spin of this silver atom points in the $+z$ direction. The totality of all conceivable propositions forms a "lattice." There are subsets of this lattice which are "Boolean"; and if classical theory is applied, the lattice as a whole would be Boolean. In quantum theory one asks questions such as (1) Given a certain proposition, tell what Boolean sub-lattices of the total lattice are compatible with that proposition; (2) Derive the existence of the Hilbert space of quantum theory from the structure of the lattice of propositions; and conversely, derive the structure of the lattice of propositions from quantum mechanics in its Hilbert space formulation. There was a time when a few thought that "quantum logic" meant a new kind of logic. That mistaken illusion being by now demolished, one translates the phrase "quantum logic" today as the logic of the relations between each proposition and all the others. Thus, if the proposition is true that the spin of the silver atom is in the $+z$ direction, then the proposition is false that the spin of the silver atom is in the $-z$ direction; and the proposition that the spin of the silver atom is in the $+x$ direction is neither true nor false—from which one goes on to discuss probabilities. Jammer (1974) gives, in chapter VIII of his book, an extensive account of the history and literature of the subject, from the path-breaking paper of Birkhoff and von Neumann (1936) to the extended development in such books as Mackey (1963, pp. 61–85), Jauch (1968, chapters 5 and 8), Ludwig (1970), Hooker (1975), Mittelstaedt (1978, 1979), and Suppes (1976). It is a fortunate circumstance for the history and further development of this subject that the hitherto unpublished lecture notes of von Neumann have recently been edited and published in book form (von Neumann, 1981). Related to this outlook is a theorem of Gleason (1957) which simplified the axiomatics of quantum mechanics. In a separable real or complex

Hilbert space H of more than two dimensions, every measure $\mu(A)$ on the subspaces of H has the form $\mu(A) = \text{trace}\ (TP_A)$, where P_A denotes the orthogonal projection on the subspace A and T is a fixed operator of trace class.

When is a "measurement of nothing" a measurement? This question, considered by Epstein (1945) and Renninger (1960), has been illuminated by a recent paper of Dicke (1982).

It may give a final impression of the scope of the topic of "Interpretations of the Act of Measurement" to list a sampling of other items: a book of Audi (1973) on the interpretation of quantum mechanics, a review by Ballentine (1970) on the statistical interpretation of quantum mechanics, a book of Whyte (1974) on the "universe of experience," a paper of DeWitt (1970) on "quantum mechanics and reality," a discussion by Aharonov and Vardi (1980) on the meaning of an individual Feynman history, and papers of Aharonov and Albert (1980, 1981) on measurement in relativistic quantum mechanics.

Other general references include Bohm (1953), Ludwig (1953), Körner and Pryce (1957), d'Espagnat (1971a,b), Belinfante (1978), de Broglie (1941, 1953, 1962), Feynman (1951), Furry (1936), Moyal (1949), and Wigner (1961, 1963).

III. "Hidden Variables" versus "Phenomenon" and Complementarity

Bell's inequality (1966; III.4 in this book) was extended by Clauser, Horne, Shimony, and Holt (1969) to cover actual systems (see also Fehrs and Shimony, 1974, and Shimony, 1974), providing an experimental test for all "local hidden-variable theories"; it has been shown also to apply to "objective local theories" (Clauser and Horne, 1974) and to "realistic local theories" (Clauser and Shimony, 1978). The latter paper contains an extensive review of the subject of Bell's inequality and hidden variables.

The *polarizations of the two annihilation photons* given out when positronium disappears have a correlation in angle that can be measured as suggested by Wheeler (1946; III.1 here) and that has been calculated by Pryce and Ward (1947) and by Snyder, Pasternack, and Hornbostel (1948).

Maximum and minimum rates of coincidence between counters detecting annihilation quanta *were measured* for perpendicular and parallel orientations of the two polarizations by Bleuler and Bradt (1948), Hanna (1948), Vlasov and Dzeljepov (1949), Wu and Shakhov (1950), Hereford (1951), and by Bertolino, Beltoni, and Lazarini (1955). All of these investigators found qualitative agreement with the predictions of quantum theory, though less than perfect angular resolution made quantitative comparisons difficult.

The *correlation in polarization of annihilation photons* was measured anew by Langhoff (1960) with improved angular resolution, using both ^{22}Na and ^{64}Cu as radioactive sources of the positrons to be slowed down and annihilated in the sample. The dependence of coincidence rate upon angle between the two polarizations agreed well with quantum theory.

Still more precise measurements of coincidence rate in its dependence on angle between the two polarizations

(Kasday, Ullman, and Wu, 1970; Kasday, 1971; Kasday, Ullman, and Wu, 1975; Bruno, d'Agostino, and Maconi, 1977; Mesenheimer, 1979), supplemented by reasonable additional symmetry assumptions, appear to indicate disagreement with Bell's inequality, agreement with quantum theory; they also to rule out the *Bohm-Aharonov theory* (1957, 1960) in which the state vector is assumed to change at random into one of the members of an ensemble of product states.

Agreement with Bell's inequality and therefore disagreement with quantum theory was reported in measurements of the correlation of the polarizations of annihilation quanta by Faraci, Gutkowski, Nottarigo, and Penisi (1974). Moreover, their measurements made on annihilation photons, some of which had a coherence length of 7 cm, others 47 cm, suggest that there is a decrease of correlation with increase in the difference between the flight paths of the two photons.

Disagreement with the results of Faraci et al., agreement with the predictions of quantum theory, was found by Wilson, Lowe, and Butt (1976). They used a ^{64}Cu source and were able to vary the separation between photon source and polarizers by as much as 2.5 m, and the difference in separation by as much as 1 m, as compared to a 12 cm coherence length for the annihilation photons. They found no dependence of polarization correlation upon either separation.

Correlation of spins in low-energy (p, p) scattering has been observed by Lamehi-Rachti and Mittig (1976). Departures from Rutherford scattering arise at low energies almost exclusively from the interaction of protons in a state of zero orbital angular momentum. That state is symmetric in proton coordinates. In order to satisfy the Pauli exclusion principle, it must be antisymmetric in proton spins; that is, it must possess total spin zero. Such a state gives rise to a maximal anticorrelation of the spin directions of the two particles. The measurements, as discussed by the observers and by Pipkin (1978), indicate that the hypothesis of agreement with quantum theory, disagreement with the Bell inequality, has only 7 chances in 10^4 to be wrong. The experiment was not unambiguous because the efficiency of the analyzers of polarization was low.

Correlation of the polarization of the photons emitted in a 2-photon *atomic cascade* was investigated—to discriminate between quantum theory and Bell's inequality—by Kocher and Commins, 1967; Freedman and Clauser, 1972; Holt, 1973 (still unpublished; the only such experiment to disagree with quantum theory); Freedman and Holt, 1975; Clauser, 1976; Fry and Thompson, 1976 (III.8 in this book); and Aspect, Grangier, and Roger (1981). An extensive review has been given by Pipkin (1978). In the experiment of Aspect, Grangier, and Roger, the relevant quantity δ,

$$\delta = \left| \frac{\begin{pmatrix} \textit{coincidence rate for} \\ \textit{polarizers separated by } 22.5^\circ \end{pmatrix} - \begin{pmatrix} \textit{same, for} \\ 67.5^\circ \end{pmatrix}}{\begin{pmatrix} \textit{coincidence rate with} \\ \textit{both polarizers removed} \end{pmatrix}} \right| - \tfrac{1}{4},$$

is found to be $(5.72 \pm 0.43) \times 10^{-2}$, violating by more than 13 standard deviations the condition $\delta \leq 0$ required by Freedman's transcription of the Bell inequality, but agreeing with the prediction of quantum theory, $\delta = (5.8 \pm 0.2) \times 10^{-2}$. Moreover, these investigators report, "moving each polarizer up to 6.5 m from the source, that is, to four coherence lengths of the wave packet associated with the lifetime of the intermediate state of the cascade (5 ns), we observed no change in the results." These experiments provide the clearest evidence to date for that violation of the Bell inequality that is predicted by quantum theory.

A "pilot-wave theory" or "theory of the double solution," proposed by de Broglie (1928), can be regarded as a first attempt to cast quantum theory in a deterministic mold, and thus achieve the goal of the modern-day "hidden variable" theories. According to de Broglie, the usual wave function ψ is to play a twofold role: apart from being a "probability wave" it is also to represent and determine the trajectory of the particle as a singular solution of the wave equation. Pauli (1928a) objected to de Broglie's ideas. He showed in a specific example that the motion of the singularity-particle will take place in a many-dimensional, and hence fictitious, configuration space.

Exploration of any hidden-variable approach to quantum theory was abandoned for many years after von Neumann (1932) published his much-discussed proof, later reformulated by Jauch and Piron (1963), that hidden variables of a rather restricted class directly contradict the experimentally confirmed predictions of quantum theory.

Theories of hidden variables were resuscitated by Bohm (1952a,b), first to point out that von Neumann's theorem did not have the generality commonly attributed to it. Bell (1966) identified the specific assumption of von Neumann that was not satisfied by so-called "realistic hidden variable theories."

The history and status of theories of hidden variables are reviewed by Belinfante (1973), Jammer (1974), and Pipkin (1978). Apart from the papers of Bohm and Bell included in the present collection there is a wide-ranging literature, not all of it free from speculation, on hidden variables. A few points of entry are provided by the papers of Bohm and Aharonov, 1957, 1960; Bub, 1969; Bohm and Bub, 1966a,b; Bohm and Vigier, 1954; Vigier, 1951, 1956; and Rietdijk, 1971.

An inequality, an upper bound on the strength of correlations of polarization which can be allowed by a deterministic and local theory of hidden variables, was derived by Bell (1964) in the course of reexamining the arguments of Einstein, Podolsky, and Rosen (1935). Extensions of Bell's inequality have been given by Bell, 1971a; Clauser and Horne, 1974; McGuire and Fry, 1973; and Garuci and Selleri, 1976.

The so-called "non-locality" of quantum theory as it shows in the Einstein-Podolsky-Rosen and other experiments is discussed in some detail by d'Espagnat, 1976, part 3, and 1979; Bohm and Hiley, 1975, 1976; and Hiley, 1977; and in other papers too numerous to cite. Several interesting papers of John S. Bell consider the bearing of so-called "non-locality" on what shall be understood by the term "reality." Bell (1976) is "particularly concerned with *local* "*beables*," those which (unlike, for example, the total energy) can be assigned to some bounded space-time region." He formulates one concept of "local causality," and shows that "ordinary relativistic quantum field theory is not locally causal in this sense." Bell (1980a), on "Bertlmann's socks and the nature of reality," presents the EPR experiment in a particularly vivid form, and outlines four alternative ways to come to terms with its implications for "reality." In particular he discusses the idea that the settings of the polarization analyzers (in Bohm's version of the EPR experiment) might not be independent in the sense assumed by apparently free-willed experimental physicists, and adds, "But this way of arranging quantum mechanical correlations would be even more mind-boggling than one in which causal chains go faster than light. Apparently separate parts of the world would be deeply and conspiratorially entangled, and our apparent free will would be entangled with them." Bell (1980b) defends the de Broglie-Bohm version of quantum theory "as sharp where the usual one is fuzzy, and general where the usual one is special" and illustrates this view by analyzing the delayed-choice double-slit experiment. Henry P. Stapp (1972, 1981) advocates a "psychophysical theory" in which "the physical world described by the laws of physics is a structure of tendencies in the world of mind."

"*Delayed choice experiments*" were prefigured by Einstein, Tolman, and Podolsky (1931, reprinted here as 1.7), as well as by Bohr's solitary and pregnant sentence (1949, p. 230), ". . . it . . . can make no difference, as regards observable effects obtainable by a definite experimental arrangement, whether our plans for constructing or handling the instruments are fixed beforehand or whether we prefer to postpone the completion of our planning until a later moment when the particle is already on its way from one instrument to another." They were foreshadowed in an analysis of the uncertainty principle, as it shows up in the operation of an electron microscope, done by eighteen-year-old Karl Friedrich von Weizsäcker (1931) at the instance of Heisenberg. On page 128 of this paper appears this sentence (in our translation), "Physically one would reason in this case [atom not in the image plane but at the focal distance] that a photon absorbed at the focal distance must, already before it went through the lens, have become equivalent to a plane wave whose direction of propagation can be calculated from the position of the absorbing atom." The general concept of delayed-choice experiments was formulated, and seven types of such experiments were spelled out, in Wheeler (1978). Scully and Drühl (1982) "propose and analyze an experiment designed to probe the extent to which information accessible to an observer—and the 'eraser' of this information—affect measured results." They point out that their new experiment, too, lets itself be operated in the delayed-choice mode.

Would a world be conceivable in

which there is *no quantum indeterminism*? Peres and Zurek (1982) investigate conceivable alternatives and conclude that "Theories where the observed world is deterministic but the observer is not (whatever the reason for that) lead to Bell's nonseparability theorem. If, on the other hand, the observer too is deterministic, the theory is not verifiable. It follows that quantum theory must be the logically preferred option. Its inability to completely describe the measurement process is not a flaw of the theory but a *logical necessity* and is analogous to Gödel's undecidability theorem."

"*Stochastic quantum mechanics*" is a name often given to the view that (1) the Schrödinger equation is essentially a diffusion equation, though an equation for the diffusion of probability amplitude rather than probability density itself; and (2) that this diffusion is driven by a force arising from other dynamical entities, that is, from a special class of hidden variables: random impacts of particles postulated *ad hoc*, fluctuating fields, electromagnetic or otherwise, etc.

This outlook was first introduced by Fürth (1933) and is expounded at length in Fényes (1952). In their work quantum mechanics is described as a "time-symmetric stochastic process," with little said about the origin of the random external force. In contrast, Kalitsin (1953), inspired by Einstein and Hopf (1910) and by Einstein and Stern (1913), likewise takes the mechanics of the electron to be classical at bottom, but views the fluctuating force on it that makes it behave "quantum mechanically" as originating from the fluctuating electromagnetic field of the vacuum. Well though this treatment reproduces the ground state of the harmonic oscillator, it fails for other systems. (1) It predicts that the electron of the hydrogen atom, originally in its ground state, will gain energy from the vacuum fluctuations. (2) When a system breaks into two parts, as in the Einstein-Podolsky-Rosen experiment, the fluctuations at the location of the one system and the other would have to have a quite artificial correlation to reproduce the well-tested predictions of standard quantum mechanics. Boyer (1980b) surveys "stochastic electrodynamics."

A bibliography of over 200 *papers on "stochastic quantum mechanics"* has been prepared by Robert Hobart of Middle Georgia College, Cochran, Georgia. We are most indebted to him for making it available to us. Space limitations have made it impossible for us to include more than an arbitrary but representative sampling of the literature: Bergia, Lugli, and Zamboni, 1980; Boyer, 1968, 1974, 1975, 1980a,b; Claverie and Diner, 1977; Comisar, 1965; Davidson, 1979; De La Peña, 1969, 1977; Einstein and Hopf, 1910; Einstein and Stern, 1913; Fényes, 1952; Fürth, 1933; Kalitsin, 1953; Kracklauer, 1974; Marshall, 1965; Nelson, 1966; Santos, 1979; Vigier, 1979. The subject of each paper is sufficiently indicated in most cases by its title as listed in the bibliography at the end of this book.

IV. Field Measurements

The paper of Landau and Peierls (1931, reprinted here as IV.1) stimulated Bohr and Rosenfeld (1933, reprinted here as IV.2; and 1950, reprinted here as IV.3) to do a thoroughgoing analysis of the *measurability of electromagnetic field quantities*. Stages in the evolution of Bohr's thinking on this topic are described by Rosenfeld (1955, 1967).

Corinaldesi (1951), pursuing a line of development opened up by Heisenberg twenty years earlier, analyzes *charge fluctuations in quantum electrodynamics*.

Bryce DeWitt (1962, 1965) extends to the *dynamics of geometry* considerations on measurability analogous to those worked out by Bohr and Rosenfeld for the electromagnetic field. Salecker and Wigner (1958) show that there is one way of measuring geometry by use of light rays and test particles which falls far short of the accuracy expected from the considerations of DeWitt. Bohr and Rosenfeld had initially encountered a similar difficulty when they were trying to find an apparatus that would measure with the accuracy allowed by theory. In the end, however, they found that it was more reasonable to improve the apparatus than to question the theory.

V. Irreversibility and Quantum Theory

Maxwell's demon "was conceived in discussions among Maxwell, Tate, and Thomson; he was first described in a letter from Maxwell to Tate in 1867, and publicly introduced in Maxwell's *Theory of Heat* (1871)," according to Brush (1966), who also reprints papers of Thomson (1874) and Poincaré (1893) analyzing implications of the Maxwell demon. Szilard's pathbreaking paper (1929, here V.1) has led to an extensive literature, one of the more significant treatises in which is Brillouin (1962).

An attempt to define *thermodynamic entropy in terms of information content* has been made by Jaynes (1957a,b, 1979). Whether his work can serve as a foundation for statistical mechanics remains a subject of controversy (Friedman and Shimony, 1971; Dias and Shimony, 1981). Szilard's paper has been criticized by Jauch and Baron (1972) for suggesting that the connection between entropy and information is straightforward. They have also criticized Szilard's discussion of the "one-particle-gas" model on physical grounds.

John von Neumann's 1932 formulation of the quantum *theory of measurement* has been critically discussed by Feyerabend (1957), Margenau (1963), Sneed (1966), and Furry (1966) as both too idealized and too restrictive.

Ergodicity was a central topic of classical mechanics from early days. It was the subject of a paper by twenty-two-year-old Enrico Fermi (1901–1954; Fermi, 1923). It is the focus of numerous treatises: Hopf (1937), Khinchin (1957), Jacobs (1962–1963), Billingsley (1965), Rohlin (1967), Arnold and Avez (1968), Moser (1973), Ornstein (1974), Sinai (1976), and Cornfeld, Fomin, and Sinai (1981). It took time to see how to extend these considerations to the new framework of quantum theory. Einstein (1917b) discussed the rule of Bohr and Sommerfeld for quantization in the case when the Hamilton-Jacobi equation is not separable, and showed that if tori exist in phase space, it is possible to define action and angle variables. Percival (1973) reviews the literature of classical mechanics which shows that "bound systems which have an analytic Hamiltonian, and which are sufficiently close to being separable, have a significant fraction of their phase space full of trajectories which behave very much like those of separable systems." Associated with such "regular" regions of phase space, Percival shows by semiclassical methods, is a quantum energy spectrum with "regular" properties; but he shows that there is another part of the spectrum, an "irregular part," with strongly contrasting properties. McDonald and Kaufmann (1979) and Casati, Valz-Gris, and Guarnieri (1980) present numerical work on irregular spectra. Korsch and Berry (1981) use the Wigner phase-space density function to analyze what happens in the course of time to a distribution initially confined to a well-circumscribed region of phase space. In the beginning it spreads out much as expected from classical mechanics. Classically, however, the distribution, for the case of a non-integrable Hamiltonian, develops an intricate structure of whorls and tendrils. Other workers had already pointed out that these tendrils can no longer evolve according to classical expectations when their dimensions grow so small as not even to include a single quantum cell in phase space. Korsch and Berry illustrate this result by numerical computations.

General ergodic properties of quantum systems, discussed by van Hove (1959, reprinted here as V.3) have also been investigated by, among others, von Neumann (1929), Pauli and Fierz (1937), Krylov (1979), and van Hove (1955, 1957). For a penetrating discussion of these problems and many additional references see Davies (1976).

The relationship between "*quantum ergodicity*" and so-called "*collapse of the wavepacket*" proposed by Daneri, Loinger, and Prosperi (1962, reprinted here as V.4; also 1966 and Prosperi, 1967) has been endorsed by Rosenfeld (1965, 1968b), criticized by Jauch, Wigner, and Yanase (1967) and Bub (1968), and defended by Loinger (1968). Additional considerations on *ergodicity fluctuations and entropy* as they relate to quantum theory will be found in Pauli (1928b); Kemble (1939a,b); Vonsovsky (1946); Stueckelberg (1952); Groenewold (1952); Klein (1952, 1956); Van Kampen (1954, 1958); Fierz (1955); Jancel (1955); Gamba (1955); Farinelli and Gamba (1956); Farquhar and Landsberg (1957); Ludwig (1958a,b); Prosperi and Scotti (1959, 1960); Wannier (1965).

Further discussion of *time reversal* (Aharonov, Bergmann, and Lebowitz, 1964, reprinted here as V.5) and its relevance to the quantum theory of measurement can be found in Belinfante (1975). See also Rankin (1965).

The *Lyapounov superoperator formalism* mentioned by Misra, Prigogine, and Courbage (1979a, reprinted here as V.6) has been described in much greater detail by Prigogine, George, Henin, and Rosenfeld (1973), and by Misra, Prigogine, and Courbage (1979b) and applied to problems of the quantum theory of measurement by George, Prigogine, and Rosenfeld (1972).

The relation between *irreversibility*,

amplification and "reduction of the wavepacket" has been a subject of many papers. For example, Green (1958) has stressed metastability of the initial state of the measuring apparatus. In his model the detector consists of two sets of oscillators kept at different temperatures, coupled by a quantum system, the wave function of which is to be "collapsed." A critical assessment of Green's paper has been given by Furry (1966). Hepp (1972) has argued that the generation of probabilities from probability amplitudes can take place if the apparatus is an infinite system. Though Hepp thus concludes that in a certain mathematical limit "rigorous reduction of the wavepacket" can be obtained, Bell (1975) has shown that at least in one of the models investigated by Hepp, such reduction does not take place. The non-reduction of the wavepacket becomes particularly apparent in the Heisenberg picture. Haake and Weidlich (1968) have argued that the reduction of the wavepacket can be achieved if the apparatus possesses "channel structure." They have investigated in detail a particular model, one where the "apparatus" is a field mode. More on this topic will be found in Weidlich (1967) and Weidlich and Haake (1969). Machida and Namiki (1980a,b) stress that any reasonable model of a macroscopic apparatus must subsume a large number of distinct Hilbert spaces. Their reasoning has been given a detailed mathematical development by Araki (1980) in terms of continuous superselection rules. Zurek (1981, 1982b) has argued that it is the coupling between the Hilbert space of the apparatus pointer and the rest of the apparatus which is responsible for the effective superselection rules. According to this reasoning, the pointer of the apparatus cannot be regarded as an isolated system.

The *thermodynamics of black holes* (Bekenstein, 1973, 1974, 1975; Hawking, 1974, 1975) has suggested to some that irreversibility and noncausality may be inherent in the laws of nature (Hawking, 1976). Penrose (1973) proposes that such a finding might allow one to understand evolution from pure states to mixtures and, consequently, the process of measurement itself, in a completely new way.

For a long time it was thought that one could make a simple-minded *application of the considerations of Szilard in the theory of computers*, and that each stage in the computation would require the dissipation of an energy of the order of magnitude of kT. By 1961, however, this doctrine was no longer beyond question (Landauer, 1961); and twelve years later, Bennett (1973) in an epoch-making paper, proved that a computation can, in principle, be operated as close to reversibility as desired. In the words of Landauer, the process is not "lossless, but only as reversible as automobiles or locomotives which can back up along their path. We also assume that the frictional forces are proportional to velocity and thus, the accompanying energy losses, per logic step, can be made as small as desired, by sufficiently slow computation." Landauer (1981a) comments on the implications of Szilard's work: "It taught us that measurement requires energy expenditure. But it does not seem to be adequately recognized that coupling a meter to an object to be measured, and letting the object thus influence the meter, does not require dissipation. The dissipation arises from the need to reset the meter after subsequent decoupling, in order to prepare the meter for further use. Again it is the *destruction* of information which is associated with the real need for energy dissipation. This relationship to information destruction has been emphasized only in our own field of computation, and not equally in the related analyses of the measurement process and of channel capacity requirements."

Distinguishability is the first requirement for measurement. Measurements in general, and measurements at the quantum level in particular, deal with information, often information of the "yes, no" type rather than the "how much" type. How much information of the "yes, no" type does one need to distinguish two nearby values of the angle of polarization of a beam of photons? And how should the probability of a yes or no count depend upon the angle of polarization to give—if nature were in the mood to give—the maximum number of distinguishable directions of polarization? Just the way nature says it does, according to contributions of Wootters (1980, 1981), who also goes on to show that "distance in Hilbert space measures distinguishability." This result raises the still unanswered question whether all of quantum theory can be derived out of appropriate information-theoretic considerations.

That *probability amplitudes are complex numbers*, not reals or quaternions, has been connected by Stueckelberg (1960) with the principle of complementarity or indeterminism; and some of the extensive subsequent discussion of these issues is summarized in Jammer (1974, pp. 358–359).

This question about the ties between quantum theory and information theory leads to interest in two other intellectual partners of information theory, memory theory and undecidability theory. Out of the vast literature on *memory theory*, it may be appropriate to cite as representative contributions Ovenden and Roy (1973) and Pearl and Crolotte (1980). As regards *undecidability*, Chaitin (1975, 1981) notes that "Gödel's theorem (1931; see also Nagel and Newman, 1958; and Hofstadter, 1979) may be demonstrated using arguments having an information-theoretic flavor. In such an approach it is possible to argue that if a theorem contains more information than a given set of axioms, then it is impossible for the theorem to be derived from the axioms. In contrast with the traditional proof based on the paradox of the liar, this new viewpoint suggests that the incompleteness phenomenon discovered by Gödel is natural and widespread rather than pathological and unusual."

Even further from any immediately obvious tie to quantum theory, but also an incentive to think about the deeper meaning of quantum theory, is the *Austrian theory of capital* (von Hayek, 1948, 1975; von Mises, 1966) as discussed in this connection by Tipler (1981). Tipler considers Friedrich von Hayek's definition of capital (an extension of that originally given by Eugen von Böhm-Bawerk) as analogous to the Feynman sum over histories, but more complex in depending in general on all the other histories.

VI. Accuracy of Measurements: Quantum Limitations

To measure the direction of the spin of a free electron is impossible, as Niels Bohr explained, Mott and Massey (1965, VI.1) demonstrate, and Pauli (1932) illustrates by numerous examples; but for a bound electron the direction of the spin can be measured. Experiments of the Stern-Gerlach type measure, for example, the direction of the spin of the valence electron of a neutral silver atom. More recently, the direction of the spin has been measured for an otherwise free electron trapped in a manmade mixture of electric and magnetic fields (Van Dyck, Ekstrom, and Dehmelt, 1976; Dehmelt, 1981; Van Dyck, Schwinberg, and Dehmelt, 1981).

Uncertainties of quantum origin arise in the typical Stern-Gerlach experiment when the measuring apparatus has a finite size, as elucidated in an extensive literature summarized by Wigner (1976, II.2), Araki and Yanase (1960, VI.2), and Yanase (1961, VI.3).

The principle of *indeterminism as applied to time and energy*, analyzed by Aharonov and Bohm (1961, VI.4), has been discussed in additional contexts by Allcock (1969a,b,c), who also gives extensive citations of the literature (1969a). Wigner (1972) works out the wave function that minimizes the product of the uncertainties in time and in energy.

All kinds of observing devices and amplifiers are affected by *Brownian motion*, as spelled out in a charming review article by Barnes and Silverman (1934). See Wax, ed. (1954) for a collection of papers on the theory of Brownian motion. A brief and elementary review of the subject of *amplifiers* will be found in Halkias and Alley (1977). A more advanced discussion can be found in Caves (1982b).

More on the *uncertainty principle as it affects the operation of the ideal amplifier*, the subject of two contributions reprinted here (Heffner, 1962, VI.5, and Haus and Mullen, 1962, VI.6), will be found in the book of Pierce and Posner (1980, pp. 285–292), the book edited by Townes (1960), and in articles of Helstrom (1976) and Serber and Townes (1960).

The capacity of a communication channel (Shannon and Weaver, 1949; Pierce, 1981a) is analyzed as it depends on quantum limitations in the paper of Pierce (1978, VI.7) reprinted here.

A *computer* has some similarity in principle to an *information channel*. The process of computation, far from requiring a minimum energy of the order of kT per logic step, is in principle a reversible process, a fascinating conclusion proved by Bennett (1973, 1982).

On the basis of this result, Pierce (1981b), in an unpublished letter, draws up the following assessment:

(1) "Need it take energy to compute? Not in the following sense: A reversible finite state machine can act as a computer. In principle it can go from the initial condition to the final condition and back again without any net loss of energy.

(2) "What about (1) and information theory? As (1) is stated, nothing. If we know the data, program and machine, in principle we know the 'answer'. There is no uncertainty and hence there is no information involved in the operation of the reversible computer.

(3) "What about getting output from the computer of (1)? If we don't know the state of the computer output device, which is a part of the reversible finite-state computer, we can find the state of this output device only through an expenditure of kT joules per nat.

(4) "What about inputting to the reversible computer? If we know the state, this costs us nothing. See (5) below.

(5) "What about 'writing on a tape'? Ideally, if we know the initial state of the tape, this costs us nothing. Think of the tape as a lot of frictionlessly hinged bars that can lie left or right on a flat surface. If we know that a bar lies left, in principle we can raise it to vertical and lower it to right without loss of energy.*

* "But we might need a different special-purpose machine for each initial and/or output state."

(6) "If we don't know the state, what will writing cost? If we know nothing at all about the state it will cost kT joules per nat to find the state and nothing to write thereafter.*

(7) "What about 'duplicating a tape'? It costs nothing if we know the states of both tapes. If we don't know anything about the state of one tape it costs kT joules per nat to find it.*

(8) "What about writing into our idealized computer? Ideally, it costs nothing if we know the state. If we don't know the state we can learn the state before we can write in for nothing at a cost of kT joules per nat. How much is this in total? That depends both on the complexity of the computer and on how little or how much we do know about its state."

Landauer (1981a,b) notes that "*the energies* required by channel capacity considerations represent energy *dissipation* only if the message is destroyed at the receiving end and a computer need not do that. Consider, for example, a high density *reel* of storage tape sent physically through space, at high velocity . . . [constituting] transmission of information [with a] required dissipation [of] at most a few kT." Further discussion will be found in Louisell (1964).

How much mass-*energy* does it take *to dispose of* a bit of *information*? How much is the minimum mass increment experienced by a black hole in the process? This question is considered in the framework of quantum theory by Bekenstein (1981) and—with nonidentical conclusions—by Unruh and Wald (1982) and by Deutsch (1982).

One *acquires information* with high effectiveness, without violating the uncertainty principle that connects frequency and time, *by* use of "*chirp radar*," a scheme of great consequence for bats and men (Griffin, 1950, 1958; Klauder, Price, Darlington, and Albersheim, 1960; Farnett et al., 1970; Casasent, 1978; Leith, 1978). See also Cutrona (1970) and Jensen et al. (1977) for radar in a broader context.

To *count individual atoms* is to approach one quantum limit to the measuring process. This feat has been accomplished by Hurst, Payne, Kramer, and Young (1979) and their Oak Ridge colleagues, surely with great consequences for the future of detecting and measuring devices of many kinds.

The idea of *quantum nondemolition measurements*, also pregnant with incalculable consequences for the future of measuring devices, reviewed here in Braginsky, Vorontsov, and Thorne (1980, VI.8), is also the subject of a rapidly increasing and instructive literature (Braginsky and Manukin, 1977; Caves, Thorne, Drever, Sandberg, and Zimmermann, 1980; Braginsky and Viatchanin, 1981; and Caves, 1982).

A superfluid helium interferometer is proposed by Chiao (1982) as a detector of the Lense-Thirring effect and of gravitational radiation.

BIBLIOGRAPHY

This bibliography does not attempt to include references cited in papers photographically reproduced in the present collection. It does contain all references cited in papers freshly set in type, and also all references cited in the Preface and other introductory material, and in the section just prior to this one, the *Guide to Some Further Literature*. In the square bracket following each reference appears the place where it is cited: [III.2] means "cited in paper 2 of Section III"; [III] means "cited in Section III of *Guide to Some Further Literature*"; and [Refs.] typically means some edited collection or other book, one or more chapters of which are more specifically listed by author elsewhere in the bibliography.

AAAS, 1961, 1970, 1980, *Guide to Scientific Instruments*, special numbers of *Science*. [Preface]

Ahmad, S.M.W. and Wigner, E. P., 1975, Invariant theoretic derivation of the connection between momentum and velocity, *Nuovo Cimento 28A*, 1–11. [II.2]

Aharonov, Y. and Albert, D. Z., 1980, States and observables in relativistic quantum field theories, *Phys. Rev. D21*, 3316–3324. [II]

Aharonov, Y. and Albert, D. Z., 1981, Can we make sense out of the measurement process in relativistic quantum mechanics? *Phys. Rev. D24*, 359–370. [II]

Aharonov, Y. and Bohm, D., 1961, Time in the quantum theory and the uncertainty relation for time and energy, *Phys. Rev. 122*, 1649–1658; reprinted here, VI.4. [VI, VI.4]

Aharonov, Y. and Vardi, M., 1980, Meaning of an individual "Feynman Path," *Phys. Rev. D21*, 2235–2240. [II]

Aharonov, Y., Bergmann, P. G., and Lebowitz, J. L., 1964, Time symmetry in the quantum process of measurement, *Phys. Rev. 134B*, 1410–1416; reprinted here, V.5. [V, V.5]

Allcock, G. R., 1969a, The time of arrival in quantum mechanics, 1. Formal considerations, *Ann. Phys. 53*, 253–285. [VI]

Allcock, G. R., 1969b, The time of arrival in quantum mechanics, 2. The individual measurement, *Ann. Phys. 53*, 286–310. [VI]

Allcock, G. R., 1969c, The time of arrival in quantum mechanics, 3. The measurement ensemble, *Ann. Phys. 53*, 311–348. [VI]

Alley, C. O., 1972, Apollo 11 laser ranging retro-reflector (LR^3) experiment, in Maglic, ed., 1972, pp. 127–156. [III.13]

Anderson, P. W., 1972, More is different: broken symmetry and the nature of the hierarchical structure of science, *Science 177*, 393. [II]

Araki, H., 1980, On a characterization of the state space of quantum mechanics, *Commun. Math. Phys.* (Germany), *75*, No. 1, 1–24. [II, V]

Araki, H. and Yanase, M. M., 1960, Measurement of quantum mechanical operators, *Phys. Rev. 120*, 622–

626; reprinted here, VI.2. [II.2, VI, VI.2]

Arnold, V. I. and Avez, A., 1968, *Ergodic Problems of Classical Mechanics*, Benjamin, New York. [V]

Aspect, A., 1975, Proposed experiment to test separable hidden-variable theories, *Phys. Lett. 54A*, 117–118. [III.13]

Aspect, A., 1976, Proposed experiment to test the non-separability of quantum mechanics, *Phys. Rev. D14*, 1944–1951; reprinted here, III.10. [III.10, III.13]

Aspect, A., Grangier, P., and Roger, G., 1981, Experimental tests of realistic local theories via Bell's theorem, *Phys. Rev. Lett. 47*, 460–463. [III]

Audi, M., 1973, *The Interpretation of Quantum Mechanics*, University of Chicago Press. [II]

Ballentine, E., 1970, The statistical interpretation of quantum mechanics, *Rev. Mod. Phys. 42*, 358–381. [II]

Barnes, R. B. and Silverman, S., 1934, Brownian motion as a limit to measuring processes, *Rev. Mod. Phys. 6*, 162–192. [VI]

Bartell, L. S., 1980, Complementarity in the double-slit experiment: On simple realizable systems for observing intermediate particle-wave behavior, *Phys. Rev. D21*, 1698–1699; reprinted here, III.12. [III.12]

Barut, A. O., ed., 1980, *Foundations of Radiation Theory and Quantum Electrodynamics*, Plenum, New York. [Refs.]

Bastin, T., ed., 1971, *Quantum Theory and Beyond*, Cambridge University Press. [II]

Bauer, E., *see* London, F. W.

Bekenstein, J., 1973, Black holes and entropy, *Phys. Rev. D7*, 2333–2346. [V]

Bekenstein, J., 1974, Generalized second law of thermodynamics in black-hole physics, *Phys. Rev. D19*, 3292–3300. [V]

Bekenstein, J., 1975, The quantum mass spectrum of the Kerr black hole, *Lett. Nuovo Cimento 11*, 467–470. [V]

Bekenstein, J., 1981, Energy cost of information transfer, *Phys. Rev. Lett. 46*, 623–625. [VI]

Belinfante, F. J., 1973, *A Survey of Hidden-Variable Theories*, Pergamon, Oxford. [II.2, III]

Belinfante, F. J., 1975, *Measurements and Time Reversal in Objective Quantum Theory*, Pergamon, Oxford; see page 39 for "indelible." [II, V]

Belinfante, F. J., 1978, Can individual elementary particles have individual properties?, *Am. J. Phys. 46*, 329–336. [II]

Bell, J. S., 1964, On the Einstein Podolsky Rosen paradox, *Physics 1*, 195–200; reprinted here, III.5 [II.2, III]

Bell, J. S., 1966, On the problem of hidden variables in quantum mechanics, *Rev. Mod. Phys. 38*, 447–452; reprinted here, III.4. [III]

Bell, J. S., 1971a, Introduction to the hidden variable question, in d'Espagnat, B., *Foundations of Quantum Mechanics*, Academic Press, New York and London. [III]

Bell, J. S., 1971b, On the hypothesis that the Schroedinger equation is exact. Contribution to the international colloquium on issues in contemporary physics and philosophy of science, and their relevance for our society, Pennsylvania State University, September 1971. [II]

Bell, J. S., 1975, On wave packet reduction in the Coleman-Hepp model, *Helv. Phys. Acta 48*, 93–98. [II, V]

Bell, J. S., 1976, The theory of local beables, *Epistemological Letters 9*,

11; originally appeared as CERN preprint TH-2053, 1975. [III]

Bell, J. S., 1980a, Bertlmann's socks and the nature of reality, *J. de Physique, Colloque C2, Supplement*, tome *42*, C2 41–62. [III]

Bell, J. S., 1980b, De Broglie-Bohm, delayed-choice, double-slit experiment, and density matrix, *Int. J. Quantum Chem.: A Symposium*, *14*, 155–159. [III]

Beltrametti, E. G. and Cassinelli, G., 1981, *The Logic of Quantum Mechanics*, Addison-Wesley, Reading, Mass. [Preface]

Bender, P. L., Currie, D. G., Dicke, R. H., Eckhardt, D. H., Faller, J. E., Kaula, W. M., Mulholland, J. D., Plotkin, H. H., Poultney, S. K., Silverberg, E. C., Wilkinson, D. T., Williams, J. G., and Alley, C. O., 1973, The lunar laser ranging experiment, *Science 182*, 229–238. [III.13]

Bennett, C. H., 1973, Logical reversibility of computation, *IBM J. Res. Dev. 17*, 525. [V, VI]

Bennett, C. H., 1982, The thermodynamics of computation—a review, presented at MIT Conference on Physics of Computation, *Int. J. Theoret. Phys. 21*, 905–940. [VI]

Bergia, S., Lugli, P., and Zamboni, N., 1980, Zero-point energy, Planck's Law, and the pre-history of stochastic electrodynamics, part II: Einstein and Stern's paper of 1913, *Ann. Fond. L. de Broglie 5*, 39–62. [III]

Berkeley, G., 1710, *Treatise Concerning the Principles of Understanding* (2nd ed., 1734) Dublin. [I]

Bertolini, G., Bettoni, M., Lazzarini, E., 1955, Angular correlation of scattered annihilation radiation, *Nuovo Cimento 2*, 661–662. [III]

Bialynicki-Birula, I., and Mycielski, J., 1976, Nonlinear wave mechanics, *Ann. of Phys.* (N.Y.) *100*, 62–93. [II]

Billingsley, P., 1965, *Ergodic Theory and Information*, Wiley, New York. [V]

Birkoff, G. and von Neumann, J., 1936, The logic of quantum mechanics, *Ann. Math. 37*, reprinted in J. von Neumann, *Collected Works*, Vol. 4, 105–125. [II]

Bleuler, E. and Bradt, H. L., 1948, Correlation between the states of polarization of the two quanta of annihilation radiation. *Phys. Rev. 73*, 1398. [III]

Bloch, E., 1930, *L'ancienne et Nouvelle Théorie des Quanta*, Hermann, Paris. [II.1]

Bohm, D., 1951, The paradox of Einstein, Rosen and Podolsky, originally published as sections 15–19, Chapter 22 of Bohm, D., 1951, *Quantum Theory*, Prentice-Hall, Englewood Cliffs, N.J.; reprinted here, III.2.

Bohm, D., 1952a, A suggested interpretation of the quantum theory in terms of "hidden variables": Part I, *Phys. Rev. 85*, 166–179, reprinted here, III.3. [II.2., III]

Bohm, D., 1952b, A suggested interpretation of the quantum theory in terms of "hidden variables": Part II, *Phys. Rev. 85*, 180–193, reprinted here, III.3. [II.2, III, III.3]

Bohm, D., 1953, Proof that probability density approaches $|\psi|^2$ in causal interpretation of quantum theory, *Phys. Rev. 89*, 458–466. [II]

Bohm, D. and Aharanov, Y., 1957, Discussion of experimental proof for the paradox of Einstein, Rosen and Podolsky, *Phys. Rev. 108*, 1070–1076. [II.2, III]

Bohm, D., and Aharonov, Y., 1960, Further discussion of possible experimental tests for the paradox of

Einstein, Podolsky and Rosen, *Nuovo Cimento 17*, 964–976. [III]

Bohm, D. and Bub, J., 1966a, A proposed solution of the measurement problem in quantum mechanics by a hidden variable theory, *Rev. Mod. Phys. 38*, 453–469. [II.2, III]

Bohm, D. and Bub, J., 1966b, A refutation of the proof by Jauch and Piron that hidden variables can be excluded in quantum mechanics, *Rev. Mod. Phys. 38*, 470–475. [II.2, III]

Bohm, D., and Hiley, B. J., 1975, On the intuitive understanding of nonlocality as implied by quantum theory, *Found. Phys. 5*, 93–109. [III]

Bohm, D. and Hiley, B. J., 1976, Nonlocality and polarization correlations of annihilation quanta, *Nuovo Cimento 35B*, 137–144. [III]

Bohm, D. and Vigier, J.-P., 1954, Model of the causal interpretation of quantum theory in terms of a fluid with irregular fluctuations, *Phys. Rev. 96*, 208–216. [II.2, III]

Bohr, N., 1913a, On the constitution of atoms and molecules: Introduction and Part I—binding of electrons by positive nuclei, *Phil. Mag. 26*, 1–25. [II.2]

Bohr, N., 1913b, On the constitution of atoms and molecules: Part II—systems containing only a single nucleus, *Phil. Mag. 26*, 476–508. [II.2]

Bohr, N., 1913c, On the constitution of atoms and molecules: Part III—systems containing several nuclei, *Phil. Mag. 26*, 857–871. [II.2]

Bohr, N., 1914, On the effect of electric and magnetic fields on spectral lines, *Phil. Mag. 27*, 506–524. [II.2]

Bohr, N., 1915a, On the series spectrum of hydrogen and the structure of the atom, *Phil. Mag. 29*, 332–335. [II.2]

Bohr, N. 1915b, On the quantum theory of radiation and the structure of the atom, *Phil. Mag. 30*, 394–415. [II.2]

Bohr, N., 1928, The quantum postulate and the recent development of atomic theory, *Nature, 121*, 580–590; reprinted here, I.4. [I.4]

Bohr, N., 1934, *Atomic Theory and the Description of Nature*, Cambridge University Press. [I.1]

Bohr, N., 1935a, Quantum mechanics and physical reality, *Nature 121*, 65; reprinted here, I.9. [I.9]

Bohr, N., 1935b, Can quantum-mechanical description of physical reality be considered complete? *Phys. Rev. 48*, 696–702, reprinted here, I.10. [I.10]

Bohr, N., 1939, The causality problem in atomic physics, in *New Theories in Physics*, International Institute of Intellectual Co-operation, Paris. [I.1]

Bohr, N., 1958, *Atomic Physics and Human Knowledge*, Wiley, New York. [I, I.1]

Bohr, N., 1949, Discussion with Einstein on epistemological problems in atomic physics, in Schilpp, P. A., ed., 1949, 200–241; reprinted here. I.1. [III, III.13]

Bohr, N., 1963, *Essays 1958-1962 on Atomic Physics and Human Knowledge*, Wiley, New York. [I]

Bohr, N., Kramers, H. A., and Slater, J. C., 1924, The quantum theory of radiation, *Phil. Mag. 47*, 785–802. [II.2]

Bohr, N. and Rosenfeld, L., 1933, On the question of the measurability of the electromagnetic field quantities, original in German, *Mat.-Fys. Medd. Dan. Vidensk. Selsk. 12*, no. 8; English translation by A. Petersen, originally published in Cohen and Stachel, eds., 1979, 357–412; reprinted here, IV.2. [IV, II.2]

Bohr, N. and Rosenfeld, L., 1950, Field and charge measurements in quantum electrodynamics, *Phys. Rev. 78*, 794–798; reprinted here, IV.3. [II.2, IV, IV.3]

Boorse, H. A. and Motz, L., eds., 1966, *The World of the Atom*, Basic Books. New York. [Refs.]

Born, M. and Jordan, P., 1925, Zur Quantenmechanik, *Z. Phys. 34*, 858–888; English translation, On quantum mechanics, in van der Waerden, 1967. [II.2]

Born, M., 1926, Zur Quantenmechanik der Stossvorgänge, *Z. Phys. 37*, 863–867; reprinted in *Dokumente der Naturwissenschaft, 1*, 48–52, 1962; reprinted here, I.2. English translation by J. A. Wheeler and W. H. Zurek, 1981. [II.1]

Born, M., Heisenberg, W., and Jordan, P., 1926, Zur Quantenmechanik II, *Z. Phys. 35*, 557–615; English translation, On quantum mechanics II, in van der Waerden, 1967. [II.2]

Bothe, W. and Geiger, H., 1924, Ein Weg zur experimentellen Nachprüfung der Theorie von Bohr, Kramers und Slater, *Z, Phys. 26*, 44. [II.2]

Boyer, T. H., 1968, Quantum electromagnetic zero-point energy of a conducting spherical shell and the Casimir model for a charged particle, *Phys. Rev. 174*, 1764–1776. [III]

Boyer, T. H., 1974, Van der Waals forces and zero-point energy for dielectric and permeable materials, *Phys. Rev. A9*, 2078–2084. [III]

Boyer, T. H., 1975, Random electrodynamics: The theory of classical electrodynamics with classical electromagnetic zero-point radiation, *Phys. Rev. D11*, 790–808. [III]

Boyer, T. H., 1980a, Thermal effects of acceleration through random classical radiation, *Phys. Rev. D21*, 2137–2148. [III]

Boyer, T. H., 1980b, A brief survey of stochastic electrodynamics, in Barut, 1980, pp. 49–63. [III]

Braginsky, V. B. and Viatchanin, S. P., 1981, On the quantum-non-demolition measurement of the energy of optical quanta; preprint, Physical Dept., Moscow State University. [VI]

Braginsky, V. B. and Manukin, A. B., 1977, *Measurement of Weak Forces in Physics Experiments*, ed. D. H. Douglass, University of Chicago Press. [VI]

Braginsky, V. B., Vorontsov, Y. I., and Thorne, K. S., 1980, Quantum nondemolition measurements, *Science 209*, 547–557; reprinted here, VI.8. [VI, VI.8]

Brillouin, L., 1962, *Science and Information Theory*, 2nd ed., Academic Press, New York. [V]

Brittin, W. E., ed., 1966, *Lectures in Theoretical Physics*, vol. VIII A, *Statistical Physics and Solid State Physics*, University of Colorado Press, Boulder. [Refs.]

Bruno, M., d'Agostino, M., and Maroni, C., 1977, Measurement of linear polarization of positron annihilation photons, *Nuovo Cimento 40B*, 142–152. [III]

Brush, S. G., 1966, *Kinetic Theory: Vol. 2. Irreversible Processes*, Pergamon, Oxford. [V]

Bub, J., 1968, The Daneri-Loinger-Prosperi quantum theory of measurement, *Nuovo Cimento 57B*, 503–520. [V]

Bub, J., 1969, What is a hidden variable theory of quantum phenomena?, *Int. J. Theoret. Phys. 2*, 101–123. [II.2, III]

Calkins, M. W., ed., 1929, *Berkeley: Essays, Principles, Dialogs, with Selections from Other Writings*, Scribner, New York; reprinted in 1957, pp. 125–126. [I]

Campbell, L. and Garnett, W., 1969, *The Life of James Clerk Maxwell*, Johnson Reprint Corporation, N. Y., p. 359. [II]

Casasent, D., ed., 1978, *Optical Data Processing*, in Vol. 23 of *Topics in Applied Physics*, Springer, New York. [VI]

Casati, G., Valz-Gris, F., and Guarnieri, I., 1980, On the connection between quantization of non-integrable systems and statistical theory of spectra, *Lett. Nuovo Cimento, 28*, ser. 2, no. 8, 279–282. [V]

Cassirer, E., 1910, *Substanzbegriff und Funktionsbegriff*, Berlin; English translation, 1923, *Substance and Function and Einstein's Theory of Relativity*, Open Court, Chicago. [II.1]

Cassirer, E., 1936, *Determinismus und Indeterminismus in der modernen Physik*, Göteborg; English translation, 1956, *Determinism and Indeterminism in Modern Physics*, Yale University Press. [II.1]

Caves, C. M., 1982a, Quantum nondemolition measurements, in Meystre and Scully, eds., 1982. [VI]

Caves, C. M., 1982b, Quantum limits on noise in linear amplifiers, *Phys. Rev. D.*, to appear. [VI]

Caves, C. M., Thorne, K. S., Drever, R.W.P., Sandberg, V. D., and Zimmermann, M., 1980, On the measurement of a weak classical force coupled to a quantum-mechanical oscillator., *Rev. Mod. Phys. 52*, 341–392. [VI]

Chaitin, G. J., 1975, Randomness and mathematical proof, *Sci. Am. 232*, No. 5, 47–52. [V]

Chaitin, G. J., 1981, Algorithmic information theory, in *Encyclopedia of Statistical Sciences*, S. Kotz and N. L. Johnson, eds., Wiley, New York. [V]

Chiao, R. Y., 1982, Interference and inertia: A superfluid helium interferometer using an internally porous powder and its inertial interactions, *Phys. Rev. B25*, 1655–1662. [VI]

Clauser, J. F., 1976, Experimental investigation of a polarization correlation anomaly, *Phys. Rev. Lett. 36*, 1223–1226. [Refs., II.2, III]

Clauser, J. F. and Horne, M. A., 1974, Experimental consequences of objective local theories, *Phys. Rev. 10*, 526–535. [III]

Clauser, J. F., Horne, M. A., Shimony, A., and Holt R. A., 1969, Proposed experiment to test local hidden-variable theories, *Phys. Rev. Lett. 23*, 880–884; reprinted here, III.6. [III, III.6, III.13]

Clauser, J. F. and Shimony, A., 1978, Bell's theorem. Experimental tests and implications *Rep. Prog. Phys. 41*, 1881–1927. [III]

Claverie, P. and Diner, S., 1977, Stochastic electrodynamics and quantum theory, *Int. J. Quantum Chem. 12, Sup. 1*, 41–82. [III]

Cohen, R. S. and Stachel, J. J., eds., 1979, *Selected Papers of Léon Rosenfeld*, Reidel, Dordrecht. [Refs.]

Colodny, R. G., ed., 1962, *Frontiers of Science and Philosophy*, University of Pittsburgh Press. [Refs.]

Comisar, G. G., 1965, Brownian-motion model of non-relativistic quantum mechanics, *Phys. Rev. B138*, 1332–1337. [III]

Compton, A. H. and Simon, A. W., 1925a, Measurements of the beta rays excited by hard x-rays, *Phys. Rev. 25*, 107. [II.2]

Compton, A. H. and Simon, A. W., 1925b, Measurement of beta rays associated with scattered x-rays, *Phys. Rev. 25*, 306–313. [II.2]

Compton, A. H. and Simon, A. W., 1925c, Directed quanta of scattered x-rays, *Phys. Rev. 26*, 289–299. [II.2]

Cooper, L. N., 1973, A possible organization of animal memory and learning, in Lundqvist and Lundqvist, eds., 1973, pp. 252–264. [I]

Cooper, L. N., 1975, A theory for the acquisition of animal memory, in Zichichi, ed., 1975, pp. 808–839. [I]

Cooper, L. N., 1976, How possible becomes actual in the quantum theory, *Proc. Am. Phil. Soc. 120*, 37–45. [II]

Cooper, L. N., and Van Vechten, D., 1969, On the interpretation of measurement within the quantum theory, *Am. J. Phys. 37*, No. 12, 1212–1220. [II]

Corinaldesi, E., 1951, Charge fluctuations in quantum electrodynamics, *Nuovo Cimento 8*, 494–497. [IV]

Cornfeld, I. P., Fomin, S. V., and Sinai, Ya. G., 1981, *Ergodic Theory*, Springer, Berlin and New York. [V]

Currie, D. G., Steggerda, C. A., Rayner, J., and Buennagel, A., 1972, Second generation timing system for Apollo lunar laser ranging experiment, in *Proceedings of the Fourth NASA and Department of Defense Precise Time and Time Interval Meeting*, U.S. Government Printing Office, Washington, D.C., 41–45. [III.13]

Cutrona, L. J., 1970, Synthetic aperture radar, Chapter 23 in Skolnik, ed., 1970. [VI]

d'Abro, A., 1939, *The Decline of Mechanism*, Van Nostrand, New York. [I]

d'Abro, A., 1951, *The Rise of the New Physics; Its Mathematical and Physical Theories*, Dover, New York. [I]

Daneri, A., Loinger, A., and Prosperi, G. M., 1962, Quantum theory of measurement and ergodicity conditions, *Nucl. Phys. 33*, 297–319; reprinted here, V.4. [II, V, V.4]

Daneri, A., Loinger, A. and Prosperi, G. M., 1966, Further remarks on the relations between statistical mechanics and quantum theory of measurement, *Nuovo Cimento 44B*, 119–128. [V]

Davidson, M., 1979, A model for the stochastic origins of Schrödinger's equation, *J. Math. Phys. 20*, 1865–1869. [III]

Davies, E. B., 1976, *Quantum Theory of Open Systems*, Academic Press, London. [V]

Davies, P.C.W., 1974, *The Physics of Time Asymmetry*, University of California Press. [II]

Davis, J., and Tango, W. J., eds., *High Angular Resolution Stellar Interferometry*, IAU *Colloquium No. 50*, Chatterton Astronomy Department, School of Physics, University of Sydney, N.S.W. 2006, Australia, pp. 8–14. [Refs., III.13]

Davisson, C. and Germer, L., 1927, The scattering of electrons by a single crystal of nickel, *Nature 119*, 558–560. [II.2]

Davydov, B., 1947, Quantum mechanics and thermodynamic irreversibility, *J. Phys. Moscow 11*, 33. [II]

de Broglie, L., 1923a, Ondes et quanta, *C. R. Acad. Sci. Paris 177*, 517. [II.2]

de Broglie, L., 1923b, Quanta de lumière, diffraction et interférences, *C. R. Acad. Sci. Paris 177*, 548. [II.2]

de Broglie, L., 1923c, Les quanta, la théorie cinétique des gaz et le principe de Fermat, *C. R. Acad. Sci. Paris 177*, 630. [II.2]

de Broglie, L., 1924a, *Thèse*, Masson et Cie., Paris. [II.2]

de Broglie, L., 1924b, A tentative theory of light quanta, *Phil. Mag. 47*, 446–458. [II.2]

de Broglie, L., 1924c, Sur la définition générale de la correspondance entre onde et mouvement, *C. R. Acad. Sci. Paris 179*, 39. [II.2]

de Broglie, L., 1924d, Sur un théorème de Bohr, *C. R. Acad. Sci. Paris 179*, 676. [II.2]

de Broglie, L., 1924e, Sur la dynamique du quantum de lumière et les interférence, *C. R. Acad. Sci. Paris 179*, 1029. [II.2]

de Broglie, L., 1925, Recherches sur la théorie des quanta, *Ann. de Phys. 3*, 22–128. [II.2]

de Broglie, L., 1926, *Ondes et mouvements*, Gauthier-Villars, Paris. [II.2]

de Broglie, L., 1928, La nouvelle dynamique des quanta, in *Electrons et Photons* -Rapports et Discussions du Cinquième Conseil de Physique tenu à Bruxelles du 24 au 29 Octobre 1927 sous les Auspices de l'Institut International de Physique Solvay, Gauthier-Villars, Paris. [III]

de Broglie, L., 1930, *Introduction à l'Étude de la Mécanique Ondulatoire*, Hermann, Paris. [II.1]

de Broglie, L., 1941, *Continu et discontinu en physique moderne*, Michel, Paris. [II]

de Broglie, L., 1953, *La physique quantique restera-t-elle indéterministe?*, Gauthier-Villars, Paris. [II]

de Broglie, L., 1962, *New Perspectives in Physics*, Basic Books, New York. [II]

Dehmelt, H., 1981, Invariant frequency ratios in electron and positron geonium spectra yield refined data on electron structure, pp. 1–3 in D. Kleppner and F. Pipkin, eds., *Atomic Physics 7*, Plenum, New York. [VI]

De La Peña-Auerbach, L., 1969, New formulation of stochastic theory and quantum mechanics, *J. Math. Phys. 10*, 1620–1630. [III]

De La Peña, L., and Cetto, A. M., 1977, Derivation of quantum mechanics from stochastic electrodynamics, *J. Math. Phys. 18*, 1612–1622. [III]

d'Espagnat, B., ed., 1971a, *Foundations of Quantum Mechanics*, Proceedings of the International School of Physics "Enrico Fermi," Course 49, Academic Press, New York and London. [II, Refs.]

d'Espagnat, B., 1971b, *Conceptual Foundations of Quantum Mechanics*, W. A. Benjamin, Menlo Park, Calif. (now Addison-Wesley, Reading, Mass.). [II]

d'Espagnat, B., 1979, The quantum theory and reality, *Sci. Amer. 241* (Nov.), 158–181. [III, III.13]

Deutsch, D., 1981, Quantum theory as a universal physical theory; working paper, Center for Theoretical Physics, University of Texas at Austin and Department of Astrophysics, Oxford University. [II]

Deutsch, D., 1982, Is there a fundamental bound on the rate at which information can be processed? *Phys. Rev. Lett. 48*, 286–288. [VI]

DeWitt, B. S., 1962, The quantization of geometry, in Witten, 1962, pp. 266–381. [IV]

DeWitt, B. S., 1965, *Dynamical Theory of Groups and Fields*, Gordon and Breach, New York. [IV]

DeWitt, B. S., 1970, Quantum mechanics and reality, *Physics Today 23*, Sept., 30–35. [II, II.2, IV]

DeWitt, B. S. and Graham, R. N., 1971, Resource letter IQM-1 on the interpretation of quantum mechanics, *Am. J. Phys. 39*, 724–738. [Preface]

DeWitt, B. S., and Graham, N., eds., 1973, *The Many-Worlds Interpretation of Quantum Mechanics*, Princeton University Press. [II, II.2]

Dias, T. and Shimony, A., 1981, A critique of Jaynes' maximum entropy principle, *Adv. Appl. Math. 2*, 172–211. [V]

Dicke, R. H., 1981, Interaction-free quantum measurements, a paradox?, *Am. J. Phys. 49*, 925–929. [II]

Dirac, P.A.M., 1927, The physical interpretation of quantum dynamics, *Proc. Roy. Soc. A113*, 621–641. [II.1]

Dirac, P.A.M., 1971, *The Development of Quantum Theory*, Gordon and Breach, New York. [I]

Dontsov, Y. P. and Baz', A. I., 1967, Interference experiments with statistically independent photons, *Sov. Phys.-JETP 25*, 1–5. [II]

Dushman, S., 1938, *The Elements of Quantum Mechanics*, Wiley, New York, and Chapman and Hall, London. [II.1]

Edwards, P., ed., 1972, *The Encyclopedia of Philosophy*, Macmillan Publishing Co. and The Free Press, New York. [Refs.]

Einstein, A., 1905, Über einen die Enzeugung und Verwandlung des Lichtes betreffenden heuristischen Gesichtspunkt, *Ann. Physik 17*, 132–148; English translation, On a heuristic point of view about the creation and conversion of light, in Boorse and Motz, 1966, and in ter Haar, 1967. [II.2]

Einstein, A., 1916a, Quantentheorie des Strahlung, *Physik. Gesell. Zurich Mitt. 16*, 47–62; reprinted in *Phys. Z. 19*, 121–128, 1917; English translation, Quantum theory of radiation, in van der Waerden, 1967. [II.2]

Einstein, A., 1916b, Strahlungsemission und Absorption nach der Quantentheorie, *Verh. D. Phys. Ges. 18*, 318–323. [II.2]

Einstein, A., 1917a, Zur Quantentheorie der Strahlung, *Phys. Z. 18*, 121; English translation, On the quantum theory of radiation, in van der Waerden, 1967. [II.2]

Einstein, A., 1917b, Zum Quantensatz von Sommerfeld und Epstein, *Verh. D. Phys. Ges. 19*, 82–92. [V]

Einstein, A., 1920, Letter to Niels Bohr, May 2, 1920.

Einstein, A., 1936, Physics and reality, *J. Franklin Inst. 221*, 349–382. [I.1]

Einstein, A., 1949, Autobiographical notes, in Schilpp, 1949, pp. 2–95.

Einstein, A., before 1953, as quoted in Rosenfeld, 1953; reprinted in Cohen and Stachel, eds., 1979, p. 469. [I.1]

Einstein, A. and Ehrenfest, P., 1922, Quantentheoretische Bemerkungen zum Experiment von Stern und Gerlach, *Z. Phys. 11*, 31–34. [II.2]

Einstein, A. and Hopf., L., 1910, 2. Über einen Satz der Wahrscheinlichkeitsrechnung und seine Anwendung in der Strahlungstheorie, *Ann. de Phys. 33*, 1096–1105. [III]

Einstein, A. and Infeld, L., 1939, *The Evolution of Physics: The Growth of Ideas from Early Concepts to Relativity and Quanta*, Simon and Schuster, New York, [I]

Einstein, A., Podolsky, B., and Rosen, W., 1935, Can quantum mechanical description of physical reality be considered complete?, *Phys. Rev. 47*, 777–780; reprinted here, I.8. [III, I.8]

Einstein, A. and Stern, O., 1913, Einige Argumente für die Annahme einer molekularen Agitation beim absoluten Nullpunkt, *Ann. Physik 40*, 551–560; English translation in Bergia, Lugli, and Zamboni, 1980. [III]

Einstein, A., Tolman, R. C., and Podolsky, B., 1931, Knowledge of past and future in quantum mechanics, *Phys. Rev. 37*, 780–781; reprinted here, I.7 [III]

Eisenbud, L., 1971, *The Conceptual Foundations of Quantum Mechanics*, Van Nostrand-Reinhold, New York. [I]

Emch, G. G., 1972, On quantum measuring processes, *Helv. Phys. Acta, 45*, 1049–1056. [II]

Epstein, P. S., 1945, The reality problem in quantum mechanics, *Am. J. Phys. 13*, 127–136. [II]

Everett, H., III, 1957, Relative state formulation of quantum mechanics, *Rev. Mod. Phys. 29*, 454–462; reprinted here, II.3. [II, II.2, II.3]

Faraci, G., Gutkowski, D., Notarrigo, S., and Penisi, A. R., 1974, An experimental test of the EPR paradox, *Lett. Nuovo Cimento 9*, 607–611. [II.2, III, III.13]

Farinelli, U. and Gamba, A., 1956, Entropy in quantum mechanics, *Nuovo Cimento 3*, 1033. [V]

Farnett, E. C., et. al., 1970, Pulse compression radar, Chapter 20 in Skolnik, ed., 1970. [VI]

Farquhar, I. E. and Landsberg, P. T., 1957, On the quantum statistical ergodic and H-theorems, *Proc. Roy. Soc. A239*, 134. [V]

Fényes, I., 1952, Eine wahrscheinlichkeits-theoretische Begründung und Interpretation der Quantenmechanik, *Z. Phys. 132*, 81–106. [III]

Fehrs, M. H. and Shimony, A., 1974, Approximate measurement in quantum mechanics, I, *Phys. Rev. D9*, 2317–2320. [III]

Fermi, E., 1923, Beweis, dass ein mechanisches Normalsystem im allgemeinen quasi-ergodisch ist, *Phys. Z. 24*, 261–264. [V]

Feyerabend, P. K., 1957, On the quantum theory of measurement, in Körner and Pryce, 1957, pp. 121–130. [V]

Feyerabend, P. K., 1962, Problems of microphysics, in Colodny, ed., 1962. [I, II]

Feynman, R. P., 1951, The concept of probability in quantum mechanics, *Proceedings of the Second Berkeley Symposium on Mathematical Statistics and Probability*, University of California Press, pp. 533–541. [II]

Fierz, M., 1955, Der ergodensatz in der Quantenmechanik, *Helv. Phys. Acta 28*, 705–715. [V]

Flato, M., ed., 1976, *Quantum Mechanics, Determinism, Causality and Particles*, Reidel, Dordrecht. [I]

Fleming, G. N., 1965, Nonlocal propperties of stable particles, *Phys. Rev. 139B*, 963–968. [II.2]

Føllesdal, D., 1975, Meaning and experience, in Guttenplan, ed., 1975, pp. 25–44. [I]

Forman, P., 1971, Weimar culture, causality and quantum theory 1918–1927, *Historical Studies in the Physical Sciences 3*, 1–115. [II]

Forsee, A., 1963, *Albert Einstein, Theoretical Physicist*, Macmillan, New York.

Franck, J. and Hertz, G., 1914, Über die Erregung der Quecksilberresonanzlinie 253.6 $\mu\mu$ durch Elektronenstösse, *Verh. D. Phys. Ges. 16*, 512–517. [II.2]

Freedman, S. J. and Clauser, J. F., 1972, Experimental test of local hidden variable theories, *Phys. Rev. Lett. 28*, 938–941; reprinted here, III.7. [II.2, III, III.7, III.13]

Freedman, S. J. and Holt, R. A., 1975, Tests of local hidden-variable theories in atomic physics, *Comments Atom. Molec. Phys. 5*, 55–62. [III]

Friedman, K. and Shimony, A., 1971,

Jaynes' maximum entropy prescription and probability theory, *J. Stat. Phys.*, *3*, 381–384. [V]

Frisch, O., 1965, Take a photon . . . , *Contemp. Phys.* *7*, 45–53. [II]

Fry, E. S. and Thompson, R. C., 1976, Experimental test of local hidden-variable theories, *Phys. Rev. Lett.* *37*, 465–468; reprinted here, III.8. [III, III.8]

Fuller, R. W. and Putnam, P., 1966, On the origin of order in behavior, *Gen. Sys.* *11*, 99–112. [I]

Furry, W. H., 1936, Note on the quantum-mechanical theory of measurement, *Phys. Rev.* *49*, 393–399. [I]

Furry, W. H., 1966, Some aspects of the quantum theory of measurement, in Brittin, ed., 1966, pp. 1–64. [V, V.2]

Fürth, R., 1933, Über einige Beziehungen zwichen klassischer Statistik und Quantenmechanik, *Z. Phys.* *81*, 143–162. [III]

Gamba, A., 1955, Thermodynamics and quantum mechanics, *Nuovo Cimento* *1*, 358–360. [V]

Gamow, G., 1966, *Thirty Years That Shook Physics*, Anchor Books, New York. [I]

Garuccio, A. and Selleri, F., 1976, Nonlocal interactions and Bell's inequality, *Nuovo Cimento* *B36*, 176–185. [III]

George, C., Prigogine, I., and Rosenfeld, L., 1972, The macroscopic level of quantum mechanics, *Mat. -Fys. Medd. Dan. Vidensk. Selsk.* *38*, 1–44. [V]

Gerlach, W. and Stern, O., 1921, Der experimentelle Nachweis des magnetischen Moments des Silberatoms, *Z. Phys.* *8*, 110. [II.2]

Gerlach, W. and Stern, O., 1922, Der experimentelle Nachweis der Richtungsquantelung im Magnetfeld, *Z. Phys.* *9*, 349–352. [II.2]

Gleason, A. M., 1957, Measures on the closed subspaces of a Hilbert space, *J. Math. and Mech.* *6*, 885–893. [II, II.2]

Gödel, K., 1931, Über formal unentscheidbare Sätze der Principia Mathematica und verwandter Systeme I, *Monatshefte für Mathematik und Physik*, *38*, 173–198; English translation by Meltzer, B., 1962, *On Formally Undecidable Propositions of Principia Mathematica and Related Systems*, with an introduction by R. B. Braithwaite, Basic Books, New York. [V]

Good, I. J., ed., 1961, *The Scientist Speculates*, Heinemann, London, and Basic Books, New York. [Refs.]

Green, H. S., 1958, Observation in quantum mechanics, *Nuovo Cimento* *9*, 880–889. [II, V]

Griffin, D. R., 1950, The navigation of bats, *Sci, Amer.* *183* (Aug.), 52–55. [VI]

Griffin, D. R., 1958, More about bat radar, *Sci. Amer.* *199* (July), 40–44. [VI]

Groenewold, H. J., 1952, Information in quantum measurements, *Proc. Kon. Ned. Akad. v. Wet.* *B55*, 219–227.[V]

Guttenplan, S., ed., 1975, *Mind and Language*, Clarendon Press, Oxford. [Refs.]

Haake, F. and Weidlich, W., 1968, A model for the measuring process in quantum theory, *Z. Phys.* *213*, 451–465. [V]

Halkias, C. C. and Alley, C. L., 1977, Amplifier, in *McGraw-Hill Encyclopedia of Science and Technology*, McGraw-Hill, New York. [VI]

Halverson, W. H., 1967, Phenomenalism, Chapter 22 in *A Concise Introduction to Philosophy*, Random House, New York, 200–208. [I]

Hanle, P. A., 1979, Indeterminism before Heisenberg: the case of Franz Exner and Erwin Schrödinger, *Historical Studies in the Physical Sciences 10*, 225–269. [II]

Hanna, R. C., 1948, Polarization of annihilation radiation, *Nature 102*, 332. [III]

Hartshorne, C., and Weiss, P., 1931–1935, *Collected Papers of Charles Sanders Peirce*, Vol. 1, paragraphs 170, 172–175, Harvard University. [Refs.]

Haus, H. A. and Mullen, J. A., 1962, Quantum noise in linear amplifiers, *Phys. Rev. 128*, 2407–2413; reprinted here, VI.6. [VI]

Hawking, S. W., 1974, A variational principle for black holes, *Commun. Math. Phys.* (Germany) *33*, No. 4, 323–334. [V]

Hawking, S. W., 1975, Particle creation by black holes, *Commun. Math. Phys.* (Germany) *43*, 199–220. [V]

Hawking, S. W., 1976, Black holes and thermodynamics, *Phys. Rev. D13*, 191–197. [V]

Hawking, S. W. and Israel, W., 1979, *General Relativity—An Einstein Centenary Survey*, Cambridge University Press. [Refs.]

Heath, P. L., 1972, Nothing, in Edwards, ed., 1972, Vol. 5, pp. 524–525. [I]

Heffner, H., 1962, The fundamental noise limit of linear amplifiers, *Proc. IRE 50*, 1604–1608. Reprinted here, VI.5. [VI, VI.5]

Hegerfeldt, G. C., 1974, Remark on causality and particle localization, *Phys. Rev. D10*, 3320–3321. [II.2]

Heisenberg, W., 1925, Über quantentheoretische Umdeutung kinematischer und mechanischer Beziehungen, *Z. Phys. 33*, 879–893; English translation, Quantumtheoretical reinterpretation of kinematic and mechanical relations, in van der Waerden, 1967, or, The interpretation of kinematic and mechanical relationships according to quantum theory, in Ludwig, 1968. [II.2].

Heisenberg, W., 1927, Über den anschaulichen Inhalt der quantentheoretischen Kinematik und Mechanik, *Z. Phys. 43*, 172–198; reprinted in *Dokumente der Naturwissenschaft 4*, 9–35 (1935); English translation by J. A. Wheeler and W. H. Zurek printed here, I.3. [I.3, II.1, II.2]

Heisenberg, W., 1930, *The Physical Principles of the Quantum Theory*, University of Chicago Press; reprinted by Dover, New York (1949, 1967). [I]

Heisenberg, W., 1967, Quantum theory and its interpretation, in Rozental, 1967, pp. 94–108. [I.1, I.3, I.4]

Helstrom, C. W., 1976, *Quantum Detection and Estimation*, Academic Press, New York. [VI]

Hepp, K., 1972, Quantum theory of measurement and macroscopic observables, *Helv. Phys. Acta 45*, 237–248. [II, V]

Hereford, F., 1951, The angular correlation of photoelectrons ejected by annihilation quanta, *Phys. Rev. 81*, 482; see also 627–628. [III]

Hermann, A., 1979, *The New Physics*, Heinz Moos Verlag, Munich. [I]

Hiley, B. J., 1977, A review of Bernard d'Espagnat's "Conceptual Foundations of Quantum Mechanics," *Contemp. Phys. 18*, 411–414. [III]

Hoffmann, B., 1959, *The Strange Story of the Quantum*, Dover, New York. [I]

Hofstadter, D. R., 1979, *Gödel, Escher, Bach: An Eternal Golden Braid*, Random House, New York. [V]

Holt, R. A., 1973, Atomic cascade experiments, doctoral dissertation, Harvard University, Cambridge, Mass.; available from University Microfilms, Inc., Ann Arbor, Michigan 48106. [III]

Holt, R. A. and Pipkin, F. M., 1974, Quantum mechanics vs. hidden variables: polarization correlation measurement on an atomic mercury cascade, Harvard University preprint, unpublished. [II.2]

Hooker, C. A., ed., 1975, *The Logico-Algebraic Approach to Quantum Mechanics*, Reidel, Dordrecht. [II]

Hopf, E., 1937, *Ergodentheorie*, Springer, Berlin. [V]

Houtappel, R.M.F., Van Dam, H., and Wigner, E. P., 1965, The conceptual basis and use of the geometric invariance principles, *Rev. Mod. Phys. 37*, 595–632. [II.2]

Hubel, 1978, Effects of deprivation on the visual cortex of cat and monkey, Harvey Lectures, series 72, Academic Press, New York. [I]

Hubel, D. and Wiesel, T., 1977, Functional architecture of Macaque monkey visual cortex, Ferrier Lectures, *Proc. Roy. Soc. London B198*, 1–59. [I]

Hume, D., 1779, *Dialogues Concerning Natural Religion*, Robinson, London; 1947 reprint by Thomas Nelson, London; U.S. edition thereof, Bobbs-Merrill, Indianapolis and New York; p. 148 in latter edition [I].

Hund, F., 1974, *The History of Quantum Theory*, Harrap, London. [I]

Hurst, G. S., Payne, M. G., Kramer, S. D., and Young, J. P., 1979, Resonance ionization spectroscopy and one-atom detection, *Rev. Mod. Phys. 51*, 767–819. [VI]

Husserl, E., 1900–1901, *Logische Untersuchungen*, 2 vols., N. Niemeyer, Halle; English translation of 2nd edition, *Logical Investigations*, 1970, trans. J. N. Findlay, Humanities Press, New York and Routledge and Kegan Paul, London. [II.1]

Husserl, E., 1913, *Ideen zu einer reinen Phänomenologie*, Halle; English translation, *Ideas: General Introduction to Pure Phenomenology*, 1931, trans. W. R. Boyce, Gibson, London. [I, II.1]

Isham, C. J., Penrose, R., and Sciama, D. W., eds., 1981, *Quantum Gravity 2*, Oxford University Press, Oxford. [Refs.]

Jacobs, K., 1962–1963, *Lecture Notes on Ergodic Theory*, Vols. 1 and 2, University of Aarhus, Denmark. [V]

Jahn, R. G., ed., 1981, *The Role of Consciousness in the Physical World*, Westview, Boulder, Colorado. [Refs.]

Jammer, M., 1966, *The Conceptual Development of Quantum Mechanics*, McGraw-Hill, New York. [I]

Jammer, M., 1974, *The Philosophy of Quantum Mechanics*, Wiley, New York. [Preface, I, II, II.2, III, V]

Jancel, R., 1955, Sur la théorie ergodique en mécanique quantique, *C. R. Acad. Sci. Paris*, 1693–1695. [V]

Janossy, L. and Naray, Z., 1957, The interference phenomena of light at very low intensities, *Acta Phys. Acad. Sci. Hung. 7*, 403–424. [II]

Jauch, J. M., 1968, *Foundations of Quantum Mechanics*, Addison-Wesley, Reading, Mass. [Preface, I, II]

Jauch, J. M., 1973, *Are Quanta Real? A Galilean Dialogue*, Indiana University Press. [I]

Jauch, J. M. and Baron, J. G., 1972, Entropy, information and Szilard's paradox, *Helv. Phys. Acta 45*, 220–232. [V]

Jauch, J. M. and Piron, C., 1963, Can

hidden variables be excluded in quantum mechanics? *Helv. Phys. Acta 36*, 827–837. [III]

Jauch, J. M., Wigner, E. P., and Yanase, M. M., 1967, Some comments concerning measurements in quantum mechanics, *Nuovo Cimento 48B*, 144–151. [V]

Jaynes, E. T., 1957a, b, Information theory and statistical mechanics, Part I—*Phys. Rev. 106*, 620–630; Part II—*Phys. Rev. 108*, 171–190. [V]

Jaynes, E. T., 1979, Where do we stand on maximum entropy? in Levine and Tribus, eds., 1979, pp. 15–118. [V]

Jensen, H., Graham, L. C., Porcello, L. J., and Leith, E. N., 1977, Side looking radar, *Sci. Amer. 237* (Oct.), 84–94. [VI]

Kalckar, J., 1967, Niels Bohr and his youngest disciples, in Rozental, 1967, pp. 227–239. [I, I.1]

Kalitsin, N. S., 1953, On the interaction of an electron with the fluctuating electromagnetic field of the vacuum, in Russian, *Sov. Phys. -JETP 25*, 407–409. [III]

Kangro, H., ed., 1972, *Original Papers in Quantum Physics*, trans. D. ter Haar and S. G. Brush, Taylor and Francis, London. [II.2]

Kasday, L., Experimental test of quantum predictions for widely separated photons, in d'Espagnat, ed., 1971. [III.13]

Kasday, L. R., Ullman, J., and Wu, C. S., 1970, The Einstein-Podolsky-Rosen argument: positron annihilation experiment, *Bull. Am. Phys. Soc. 15*, 586. [II.2, III]

Kasday, L. R., Ullman, J. D., and Wu, C. S., 1975, Angular correlation of Compton-scattered annihilation photons of hidden variables, *Nuovo Cimento B25*, 633–661. [III]

Kemble, E. C., 1937, *The Fundamental Principles of Quantum Mechanics*, McGraw Hill, New York, London. [II.1]

Kemble, E. C., 1939a, Fluctuations, thermodynamic equilibrium and entropy, *Phys. Rev. 56*, 1013–1023. [V]

Kemble, E. C., 1939b, The quantum mechanical basis of statistical mechanics, *Phys. Rev. 56*, 1146–1164. [V]

Khinchin, A. I., 1957, *Mathematical Foundations of Information Theory*, Dover, New York. [V]

Klauder, J. R., Price, A. C., Darlington, S., and Albersheim, W. J., 1960, The theory and design of chirp radar, *Bell Sys. Tech. J. 39*, 809. [VI]

Klein, M. J., 1952, The ergodic theorem in quantum statistical mechanics, *Phys. Rev. 87*, 111–115. [V]

Klein, M. J., 1956, Entropy and the Ehrenfest urn model, *Physica 22*, 569–575. [V]

Klein, M. J., ed., 1959, *Paul Ehrenfest. Collected Scientific Papers*, North Holland, Amsterdam. [I]

Klein, M. J., 1970, *Paul Ehrenfest. Volume 1: The Making of a Theoretical Physicist*, North Holland, Amsterdam. [I]

Kochen, S. and Specker, E. P., 1967, The problem of hidden variables in quantum mechanics, *J. Math. and Mech. 17*, 59–87. [II.2]

Kocher, C. A. and Commins, E. D., 1967, Polarization correlation of photons emitted in an atomic cascade, *Phys. Rev. Lett. 18*, 575–577. [III]

Körner, S. and Pryce, M.H.L., eds., 1957, *Observation and Interpretation in the Philosophy of Physics—With Special Reference to Quantum Mechanics*, Constable and Co., London. [II, V]

Korsch, H. J. and Berry, M. V., 1981, Evolution of Wigner's phase-space density under a nonintegrable quantum map, *Physica 3D*, 627. [V]

Kracklauer, A. F., 1974, Comment on the derivation of the Schrödinger equation from Newtonian mechanics, *Phys. Rev. 10*, 1358–1360. [III]

Kramers, H. A., 1956, *Collected Scientific Papers*, North Holland, Amsterdam. [I]

Krylov, N. S., 1979, *Works on the Foundations of Statistical Physics*, Princeton University Press. [V]

Kuhn, T. S., 1978, *Black-Body Theory and the Quantum Discontinuity, 1894–1912*, Clarendon Press, Oxford. [II.2]

Kuhn, T. S., Heilbron, J. L., Forman, P. L., and Allen, L., 1967, *Sources for History of Quantum Physics*, American Philosophical Society, Philadelphia, Pennsylvania. [Preface]

Lamb, W., 1969, An operational interpretation of nonrelativistic quantum mechanics, *Physics Today 22* (April), 23–28. [Preface]

Lamehi-Rachti, M. and Mittig, W., 1976, Quantum mechanics and hidden variables: A test of Bell's inequality by the measurement of the spin correlation in low-energy proton-proton scattering, *Phys. Rev. D14*, 2543–2555; reprinted here, III.9. [II.2, III, III.9]

Landau, L. and Peierls, R., 1931, Erweiterung des Unbestimmtheitsprinzips für die relativistiche Quantentheorie, *Z. Phys.*, *69*, 56; English translation by D. ter Haar, ed., 1965, Extension of the uncertainty principle to relativistic quantum theory, *Collected Papers of L. D. Landau*, Gordon and Breach, New York, 40–51. Reprinted here, IV.1. [IV, IV.1]

Landauer, R., 1961, Irreversibility and heat generation in the computing process, *IBM J. Res. Dev. 5*, 183–191. [V]

Landauer, R., 1981a, Fundamental physical limitations of the computational process, in Meijer, Mountain, and Soulen, 1981. [V, VI]

Landauer, R., 1981b, Uncertainty principle and minimal energy dissipation in computers, presented at the Massachusetts Institute of Technology Conference on Physics of Computation, May 6–8, 1981, and published in *Int. J. Theoret. Phys. 21*, 283–297. [VI]

Landé, A., 1955, *Foundations of Quantum Theory: A Study in Continuity and Symmetry*, Yale University Press. [I]

Landé, A., 1960, *From Dualism to Unity in Quantum Physics*, Cambridge University Press. [I]

Landé, A., 1965, *New Foundations of Quantum Mechanics*, Cambridge University Press. [I]

Langhoff, H., 1960, Die Linearpolarisation der Vernichtungsstrahlung von Positronen, *Z. Phys. 160*, 186–193. [III]

Leite Lopes, J. and Paty, M., eds., 1977, *Quantum Mechanics, a Half Century Later*, Reidel, Dordrecht, Holland. [Refs.]

Leith, E. N., 1978, Synthetic aperture radar, in Casasent, 1978, pp. 89–117. [VI]

Levine, R. D., and Tribus, M., eds., 1979, *The Maximum Entropy Formalism*, A Conference held at the Massachusetts Institute of Technology on May 2–4, 1978, MIT Press, Cambridge, Mass. [Refs.]

Liewer, K. M., 1979, The prototype very long base line amplitude interferometer, in Davis and Tango, eds., 1979. [III.13]

Loinger, A., 1968, Comments on a recent paper concerning the quantum theory of measurement, *Nucl. Phys. A108*, 245–249. [V]

London, F. W. and Bauer, E., 1939, *La théorie de l'observation en mécanique quantique*. (No. 775 of Actualités scientifiques et industrielles: Exposés de physique générale, publiés sous la direction de Paul Langevin; Hermann, Paris.) English translations done independently by Abner Shimony, by J. A. Wheeler and W. H. Zurek, and by J. McGrath and S. McLean McGrath; reconciled, published for the first time in the present collection, II.1. [II, II.1, II.2]

Lotze, R. H., 1874–1879, Systeme der Philosophie, 2 vols. Hirzel, Leipzig. [I]

Louisell, W. H., 1964, *Radiation and Noise in Quantum Electronics*, McGraw-Hill, New York. [VI]

Lüders, G., 1951, Über die Zustandsänderung durch den Messprozess, *Ann. Physik 8*, 322–328. [V.2]

Ludwig, G., 1953, Der messprozess, *Z. Phys. 135*, 483–511. [II]

Ludwig, G., 1958a,b, Zum ergodensatz und zum begriff der makroskopischen Observablen, I and II, *Z. Phys. 150*, 346–374 and *152*, 98–115. [V]

Ludwig, G., ed., 1968, *Wave Mechanics*, Pergamon, Oxford. [Preface, Refs.]

Ludwig, G., 1970, *Deutung des Begriffs "physikalische Theorie" und axiomatische Grundlegung der Hilbertraumstruktur der Quantenmechanik durch Hauptsätze des Messens*, Springer, Berlin. [II]

Mach, E., 1897, *Contributions to the Analysis of the Sensations*, pp. 191–192, trans. C. M. Williams, Open Court, Chicago; a second German edition appeared in 1900. [I]

Machida, S. and Namiki, M. 1980a, Theory of measurement in quantum mechanics: Mechanism of reduction of wave packet I, *Prog. Theor. Phys. 63*, No. 5, 1457–1473. [II,V]

Machida, S. and Namiki, M., 1980b, Theory of measurement in quantum mechanics: Mechanism of reduction of wave packet II, *Prog. Theor. Phys. 63*, No. 6, 1833–1847. [II,V]

Mackey, G. W., 1963, *Mathematical Foundations of Quantum Mechanics*, Benjamin, New York. [Preface, II, II.2]

Maglic, B., ed., 1972, *Adventures in Experimental Physics*, Alpha 1972, World Science Publications, Princeton, N.J. [III.13 references]

Margenau, H., 1963, Measurements in quantum mechanics, *Ann. Phys. 23*, 469–485. [V]

Marshall, T. W., 1965, Statistical electrodynamics, *Proc. Camb. Phil. Soc. 61*, 537–546. [III]

Maxwell, J. C., 1871, *Theory of Heat*, Longmans, London, Chapter 22. [V]

McDonald, S. W. and Kaufman, A. N., 1979, Spectrum and eigenfunctions for a Hamiltonian with stochastic trajectories, *Phys. Rev. Lett. 42*, 1189–1191. [V]

McGuire, J. H. and Fry, E. S., 1973, Restrictions on nonlocal hidden-variable theory. *Phys. Rev. D7*, 555–557. [III]

Mehra, J., 1974, *The Quantum Principle: Its Interpretation and Epistemology*, Reidel, Dordrecht. [I]

Meijer, P. H. E., Mountain, R. D., and Soulen, R. J., Jr., eds., 1981, *Sixth International Conference on Noise in Physical Systems*, NBS Spec. Publ. 614, Washington D. C. [Refs.]

Mesenheimer, K., 1979, Thesis, University of Freiburg, unpublished. [III]

Meystre, P. and Scully, M. O., 1983, *Quantum Optics, Experimental Gravitation, and Measurement Theory*, Plenum, New York. [II, Refs.]

Misra, B., Prigogine, I., and Courbage, M., 1979a, Lyapounov variable: entropy and measurement in quantum mechanics, *Proc. Natl. Acad. Sci. U.S.A. 76*, 4768–4772; reprinted here, V.6. [V, V.6]

Misra, B., Prigogine, I., Courbage, M., 1979b, From deterministic dynamics to probabilistic descriptions, *Physica A98*, 1–26. [V]

Mittelstaedt, P., 1978, *Quantum Logic.* Reidel, Dordrecht [Preface, II]

Mittelstaedt, P., 1979, Quantum logic, in Toraldo di Francia, ed., 1979, pp. 264–299. [II]

Møller, C., 1972, *The Theory of Relativity*, 2nd ed., Clarendon, Oxford, pp. 176–180. [II.2]

Moser, J., 1973, *Stable and Random Motions in Dynamical Systems with Special Emphasis on Celestial Dynamics*, Princeton University Press and University of Tokyo Press. [V]

Mott, N. F., 1929, The wave-mechanics of α-ray tracks, *Proc. Roy. Soc. London A126*, 79–84; reprinted here, I.6. [I.6]

Mott, N. F. and Massey, H.S.W., 1965, Magnetic moment of the electron, section 2, chapter IX of *The Theory of Atomic Collisions*, Clarendon, Oxford, pp. 214–219. Reprinted here, VI.1. [VI, VI.1]

Moyal, J. E., 1949, Quantum mechanics as a statistical theory, *Proc. Camb. Phil. Soc. 45*, 99. [II]

Nagel, E. and Newman, J. R., 1958, *Gödel's Proof*, New York University Press. [V]

Nass, M. M and Cooper, L. N., 1975, A theory for the development of feature-detecting cells in visual cortex, *Bio. Cyb. 19*, 1–18. [I]

Nelson, E., 1966, Derivation of Schrödinger equation from Newtonian mechanics, *Phys. Rev. 150*, 1079–1085. [III]

Nielsen, J. R., ed., 1977, *Niels Bohr Collected Works 4: The Periodic System (1920–1923)*, North Holland, Amsterdam. [I]

O'Connell, R. F. and Wigner, E. P., 1977, On the relation between momentum and velocity in elementary systems, *Phys. Lett. 61A*, 353–354. [II.2]

O'Connell, R. F. and Wigner, E. P., 1978, Position operators for systems exhibiting the special relativistic relation between momentum and velocity, *Phys. Lett. 67A*, 319–321. [II.2]

Ornstein, D., 1974, *Ergodic Theory, Randomness and Dynamical Systems*, Yale University Press. [V]

Ovenden, M. W. and Roy, A. E., 1973, Disseminated storage of information with sequential coding, *Bull. Math. Biol. 35*, 663–688. [V]

Pais, A., 1979, Einstein and the Quantum Theory, *Rev. Mod. Phys. 51*, 863–914. [I]

Parmenides of Elea, ~550 B.C., quoted from his poem, *Nature*, section on Truth, as summarized in article "Parmenides" in *Encyclopedia Britannica*, Vol. 17, Chicago, 1973, p. 394 [I].

Pauli, W., 1928a, discussion in *Electrons et Photons—Rapports et Discussions du Cinquième Conseil de Physique tenu à Bruxelles du 24 au 29 Octobre 1927 sous les Auspices de l'Institut International de Physique*

Solvay, Gauthier-Villars, Paris, pp. 280–282. [III]

Pauli, W., 1928b, Über das H-theorem vom Anwachsen der Entropie vom standpunkt der neuen Quantenmechanik, in *Probleme der modernen Physik, Arnold Sommerfeld zum 60. Geburtstage, gewidmet von seinen Schulern*, Hirzel, Leipzig, pp. 30–45. [V]

Pauli, W., ed., 1955, *Niels Bohr and the Development of Physics*, Pergamon, New York. [Refs.]

Pauli, W., 1932, Les théories quantiques du magnétisme, l'electron magnétique, in *Le Magnétisme*, Rapports et Discussions du Sixième Conseil de Physique Solvay, Bruxelles, 1930, pp. 175–238; discussion, pp. 239–280, Paris. [VI]

Pauli, W., ed., 1955, *Niels Bohr and the Development of Physics*, McGraw-Hill, New York. [Refs.]

Pauli, W. and Fierz, M., 1937, Über das H-Theorem in der Quantenmechanik, *Z. Phys. 106*, 572–587. [V]

Pearl, J. and A. Crolotte, 1980, Storage space versus validity of answers in probabilistic question-answering systems, *IEEE Trans. Inf. The. IT-26*, 633–640. [V]

Peirce, C. S., 1897, in Hartshorne and Weiss, pp. 1931–1935. [II]

Penrose, R., 1979, Singularities and time asymmetry, in Hawking and Israel, 1979, pp. 581–638. [V]

Penrose, R., 1981, Time asymmetry and quantum gravity, in Isham, Penrose, and Sciama, 1981. [II]

Percival, I. C., 1973, Regular and irregular spectra, *J. Phys. B6*, L229. [V]

Peres, A., 1980, Can we undo quantum measurements?, *Phys. Rev. D22*, 879–883; reprinted here, V.7. [V.7]

Peres, A. and Zurek, W. H., 1982, Is quantum theory universally valid? *Am. J. Phys. 50*, 807–809. [Refs.]

Peterson, A., 1968, *Quantum Physics and the Philosophical Tradition*, M.I.T. Press, Cambridge, Mass. [I.1]

Pierce, J. R., 1961, *Symbols, Signals and Noise: The Nature and Process of Communication*, Harper, New York. [II]

Pierce, J. R., 1978, Optical channels: practical limits with photon counting, *IEEE Trans. Com. COM-26*, 1819–1821; reprinted here, VI.7. [VI.7]

Pierce, J. R., 1981a, *An Introduction to Information Theory: Symbols, Signals and Noise*, 2nd rev. ed., Peter Smith, Magnolia, Mass. [VI]

Pierce, J. R., 1981b, unpublished letter of June 17 to J. A. Wheeler (except for its opening and closing sentences). [VI]

Pierce, J. R. and Posner, E. C., 1980, *Introduction to Communication Science and Systems*, Plenum Press, New York. [VI]

Pipkin, F. M., 1978, Atomic physics tests of the basic concepts in quantum mechanics, *Advances in Atomic and Molecular Physics 14*, 281–340. [II.2, III]

Piron, C., 1976, *Foundations of Quantum Physics*, Benjamin, Reading, Mass. [I]

Planck, M., 1900a, Über eine Verbesserung der Wien'schen Spektralgleichung, *Verh. D. Phys. Ges. 2*, 202–204; English translation, On an improvement of Wien's equation for the spectrum, in ter Haar, 1967, and Kangro, 1972. [II.2]

Planck, M., 1900b, Zur Theorie des Gesetzes der Energieverteilung im Normalspectrum, *Verh. D. Phys. Ges. 2*, 237–245; English translation, On the theory of the energy distribution law of the normal spectrum, in ter Haar, 1967, and Kangro, 1972. [II.2]

Poincaré, H., 1893, Le mécanisme et l'expérience, *Revue de Metaphysique et de Morale, 1*, 534–537; English translation in Brush, 1966, pp. 203–207. [V]

Polanyi, M. and Wigner, E., 1925, Bildung und Zerfall von Molekülen, *Z. Phys. 38*, 429–434. [II.2]

Popper, K. R., 1982a, *The Open Universe: An Argument for Indeterminism*, ed. W. W. Bartley, III, Rowan and Littlefield, Totowa, New Jersey (uncorrected advance page proof received January 1982). [I]

Popper, K. R., 1982b, *Quantum Theory and the Schism in Physics*, ed. W. W. Bartley, III, Rowan and Littlefield, Totowa, New Jersey (uncorrected advance page proof received January 1982). [I].

Popper, K. R., 1983, *Realism and the Aim of Science*, ed. W. W. Bartley, III, Rowan and Littlefield, Totowa, New Jersey (uncorrected advance page proof received January 1982). [I].

Popper, K. R., and Eccles, J. C., 1977, *The Self and Its Brain*, Springer, Berlin. [I]

Price, W. C. and Chissick, S. S., 1976, *The Uncertainty Principle and Foundations of Quantum Mechanics*, Wiley, New York. [I]

Prigogine, I., George C., Henin, F., and Rosenfeld, L., 1973, Unified approach to dynamics and thermodynamics, *Chemica Scripta 4*, 5–32. [II, V]

Prosperi, G. M., 1967, Quantum theory of measurement, in *Encyclopedic Dictionary of Physics*, supplementary volume 2, pp. 275–280, Pergamon, Oxford. [V]

Prosperi, G. M. and Scotti, A., 1959, Ergodicity theorem in quantum mechanics; evaluation of the probability of an exceptional initial condition, *Nuovo Cimento 13*, 1007–1012. [V]

Prosperi, G. M. and Scotti, A., 1960, Ergodicity conditions in quantum mechanics, *J. Math. Phys. 1*, 218–221. [V]

Pryce, M.H.L., 1948, The mass-centre in the restricted theory of relativity and its connexion with the quantum theory of elementary particles, *Proc. Roy. Soc. London A*195, 62–81. [II.2]

Pryce, M.H.L., and Ward, J. C., 1947, Angular correlation effects with annihilation radiation, *Nature 160*, 435. [III]

Pugh, G. E., 1976, *On the Origin of Human Values* (chapter, Human values, free will, and the conscious mind), Basic Books, New York; preprinted in *Zygon*, 11, 2–24, 1976. [I]

Puligandla, R., 1966, *Quantum Theory: An Examination of the Copenhagen Interpretation*, Sterling Publishers, New Delhi. [I]

Rankin, B., 1965, Quantum mechanical time, *J. Math. Phys. 6*, 1057–1071. [V]

Renninger, M., 1960, Messungen ohne Störung des "Messobjekts," *Z. Phys. 158*, 417–421. [II]

Reynolds, G. T., Spartalian, K., and Scarl, D. B., 1969, Interference effects produced by single photons, *Nuovo Cimento 10: 61B*, 355–364. [II]

Richmond, P. G., 1971, *An Introduction to Piaget*, Basic Books, New York. [I]

Rietdijk, C. W., 1971, *On Waves, Particles and Hidden Variables: A New Approach*, Van Gorcum, Assen. [III]

Robertson, H. P., 1929, The uncertainty principle, *Phys. Rev. 34*, 163–164; reprinted here, I.5. [II.2]

Rohlin, A., 1967, Lectures on the entropy theory of measure-preserving transformations (Russian), *Uspekhi Mat. Nauk 22*, No. 5, 3–56; for an early survey, by the author, of part of the material covered, translated into English, see *Russian Maths. Surveys 15*, No. 4, 1–22, 1960. [V]

Rosenfeld, L., 1953, Strife about complementarity, *Science Progress 163*, 393–410; reprinted in Cohen and Stachel, 1979. [Refs.]

Rosenfeld, L., 1955, On quantum electrodynamics, in Pauli, ed., 1955, pp. 114–136. [IV.2, IV]

Rosenfeld, L., 1963, Niels Bohr's contribution to epistemology, *Physics Today 16*, 47–54; reprinted in Cohen and Stachel, 1979, [I.4]

Rosenfeld, L., 1965, The measuring process in quantum mechanics, *Supp. Prog. Theo. Phys.*, Commemoration Issue for the 30th Anniversary of the Meson Theory by Dr. H. Yukawa, pp. 222–231. [V]

Rosenfeld, L., 1967, Niels Bohr in the thirties: consolidation and extension of the conception of complementarity, in Rozental, 1967, pp. 114–136. [I.8, I.9, IV]

Rosenfeld, L., 1968a, in *Fundamental Problems in Elementary Particle Physics*, Proceedings of the Fourteeth Solvay Conference, Interscience, New York.

Rosenfeld, L., 1968b, Questions of method in the consistency problem of quantum mechanics, *Nucl. Phys. A108*, 241–244. [V]

Rosenfeld, L., 1971a, Men and ideas in the history of atomic theory, *Arch. Hist. Exact Sci. 7*, 69–90; reprinted in Cohen and Stachel, 1979. [I.2, I.3, I.4]

Rosenfeld, L., 1971b, Quantum theory in 1929: recollections from the first Copenhagen conference, Rhodos, Copenhagen; reprinted in Cohen and Stachel, 1979. [VI.1]

Rosenfeld, L. and Nielsen, J. R., eds., 1972, *Niels Bohr. Collected Works 1: Early Work (1905–1911)*, North Holland, Amsterdam. [I]

Rosenfeld, L. and Nielsen, J. R., eds., 1976, *Niels Bohr. Collected Works 3: The Correspondence Principle (1918–1923)*, North Holland, Amsterdam. [I]

Rozental, S., ed., 1967, *Niels Bohr: His life and work as seen by his friends and colleagues*, North-Holland, Amsterdam. [Refs.]

Ruark, A. E. and Urey, H. C., 1930, *Atoms, Molecules and Quanta*, McGraw-Hill, New York. [I]

Salecker, H. and Wigner, E. P., 1958, Quantum limitations of the measurement of spacetime distances, *Phys. Rev. 109*, 571–577. [IV]

Salvini-Plawen, V. and Mayr, E., 1978, On the evolution of photoreceptors and eyes, *Evol. Biol. 10*, 207–263. [I]

Santos, E., 1979, Coupled harmonic oscillators in stochastic electrodynamics, *Lett. Nuovo Cimento, 25*, 360–364. [III]

Schilpp, P. A., 1949, *Albert Einstein: Philosopher-Scientist*, Library of Living Philosophers, Evanston, Illinois.

Schrödinger, E., 1926, Quantisierung als Eigenwert problem, *Ann. Physik 79*, 361–376; English translation,

Quantization as a problem of proper values, in Schrödinger, 1928, or Quantization as an eigenvalue problem, in Ludwig, 1968. [II.2]

Schrödinger, E., 1928, *Collected Papers on Wave Mechanics*, Blackie and Son, London. [II.2, Refs.]

Schrödinger, E., 1935, Die gegenwärtige Situation in der Quantenmechanik, *Naturwissenschaften 23*, 807–812, 823–828, and 844–849; English translation by J. D. Trimmer, 1980, The present situation in quantum mechanics: a translation of Schrödinger's "cat paradox" paper, *Proc. Am. Phil. Soc. 124*, 323–338; reprinted here. I.11.

Scully, M. O. and Drühl, K., 1982, The quantum eraser: a proposed photon correlation experiment concerning observation and "delayed choice" in quantum mechanics, preprint to appear in Meystre, P. and Scully, M. O., 1983, *Quantum Optics*, Plenum, New York. [III]

Serber, R. and Townes, C. H., 1960, Limits on electromagnetic amplification due to complementarity, in Townes, ed., 1960. [VI]

Shannon, C. E. and Weaver, W., 1949, *The Mathematical Theory of Communication*, University of Illinois Press, Urbana. [VI]

Shimony, A., 1974, Approximate measurement in quantum mechanics, II, *Phys. Rev. D9*, 2321–2323. [III]

Shimony, A., 1979, Proposed neutron interferometer test of some nonlinear variants of wave mechanics, *Phys. Rev. A20*, 394–396. [II]

Shull, C. G., Atwood, D. K., Arthur, J., and Horne, M. A., 1980, Search for a nonlinear variant of the Schrödinger equation by neutron interferometry, *Phys. Rev. Lett. 44*, 765–768. [II]

Silverberg, E. C., 1974, Operation and performance of a lunar ranging station, *Applied Optics 13*, 565–574. [III.13]

Sinai, Ya. G., 1976, *Introduction to Ergodic Theory*, Princeton University Press, trans. V. Scheffer from the 1973 book published by Erevan State University. [V]

Skolnik, M. I., ed., 1970, *Radar Handbook*, McGraw Hill, New York. [Refs.]

Sneed, J. S., 1966, Von Neumann's argument for the projection postulate, *Phil. Sci. 33*, 22–39. [V]

Snyder, H. S., Pasternack, S., and Hornbostel, J., 1948, Angular correlation of scattered annihilation radiation, *Phys. Rev. 73*, 440–448. [III]

Stapp, H. P., The Copenhagen interpretation, 1972, *Am. J. Phys. 40*, 1098–1116. [III]

Stapp, H. P., 1977, Are superluminal connections necessary?, *Nuovo Cimento 40B*, 191–204. [III.13]

Stapp, H. P., 1981, Mind, matter and quantum mechanics, preprint LBL-12631, Lawrence Berkeley Laboratory, Berkeley, California. [III]

Steggerda, C. A., 1973, A Precision Event Timer for Lunar Ranging, *Technical Report 74–038*, University of Maryland Department of Physics and Astronomy, College Park, Maryland. [III.13]

Stueckelberg, E.C.G., 1952, Théorème H et unitarité de S, *Helv. Phys. Acta 25*, 577–580. [V]

Stueckelberg, E.C.G., 1960, Quantum theory in real Hilbert space, *Helv. Phys. Acta 33*, 727–752. [V]

Suppes, P., ed., 1976, *Logic and Probability in Quantum Mechanics*, Reidel, Dordrecht. [II]

Szilard, L., 1929, Über die Entropieverminderung in einen thermodynamischen System bei Eingriffen intelligenter Wesen, *Z. Phys. 53*, 840–856; English translation, 1964, On the decrease of entropy in a thermodynamic system by the intervention of intelligent beings, *Behav. Sci. 9*, 301–310; reprinted here, V.1. [V, V.1]

Taylor, G. I., 1909, Interference fringes with feeble light, *Proc. Camb. Phil. Soc. 15*, 114–115. [II]

ter Haar, D., ed., 1967, *The Old Quantum Theory*, Pergamon, Oxford. [II.2, Refs.]

Thomas, E. E., 1921, *Lotze's Theory of Reality*, Longmans Green, London. [I]

Thomson, W., 1874, The kinetic theory of the dissipation of energy, *Proc. Roy. Soc. Edinburgh*, *8*, 325–334; reprinted in Brush, ed., 1966, pp. 176–187. [V]

Tipler, F. J., 1981, Some thoughts on the analogy between quantum mechanics and the Austrian theory of capital; draft preprint, Center for Theoretical Physics, University of Texas, Austin and Depts. of Mathematics and Physics, Tulane University, New Orleans. [V]

Toraldo di Francia, N., ed., 1979, *Problems in the Foundations of Physics*, Proceedings of the International School of Physics "Enrico Fermi," Course 72, North-Holland, Amsterdam. [Refs.]

Townes, C. H., 1960, *Quantum Electronics, A Symposium*, Columbia University Press, New York. [VI]

Unruh, W. G. and Wald, R. M., 1982, Acceleration radiation and the generalized second law of thermodynamics; scheduled to appear in *Phys. Rev. D25*, 942–958. [VI]

Van der Waerden, B. L., ed., 1967, *Sources of Quantum Mechanics*, North-Holland, Amsterdam; paperback reprint, Dover, New York, 1968. [Preface, II.2]

Van Dyck, Jr., R. S., Ekstrom, P., and Dehmelt, H. G., 1976, Axial, magnetron and spin-cyclotron-beat frequencies measured on single electron almost at rest in free space (geonium), *Nature 262*, 776–777. [VI]

Van Dyck, Jr., R. S., Schwinberg, P. B., and Dehmelt, H. G., 1981, Electron magnetic moment from geonium spectra, pp. 4–36 in D. Kleppner and F. Pipkin, eds., *Atomic Physics 7*, Plenum, New York. [VI]

van Hove, L., 1955, Quantum-mechanical perturbations giving rise to a statistical transport equation, *Physica 21*, 517–540. [V]

van Hove, L., 1957, The approach to equilibrium in quantum statistics, *Physica 23*, 441–480. [V]

van Hove, L., 1959, The ergodic behaviour of quantum many-body systems, *Physica 25*, 268–276; reprinted here, V.3. [V, V.3]

Van Kampen, N. G., 1954, Quantum statistics of irreversible processes, *Physica 20*, 603–622. [V]

Van Kampen, N. G., 1958, Correspondence principle in quantum statistics, *Proc. Int. Symposium on Transport Processes in Statistical Mechanics* (Brussels, 1956), Interscience, New York. [V]

Vigier, J. P., 1951, Introduction géometrique de l'onde pilote en théorie unitaire affine, *C. R. Acad. Sci. Paris 233*, 1010–1012. [II.2, III]

Vigier, J. P., 1954, Structure des micro-objets dans l'interprétation causale de la théorie des quanta, Thése, Paris. [II.2]

Vigier, J. P., 1956, *Stucture des micro-objets dans l'interprétation causale de la théorie des quanta*, Gauthier-Villars, Paris. [II.2, III]

Vigier, J. P., 1979, Model of quantum statistics in terms of a fluid with irregular stochastic fluctuations propagating at the velocity of light: A derivation of Nelson's equations, *Lett. Nuovo Cimento 24*, 265–272. [III]

Vlasov, N. A. and Dzeljepov, B. S., 1949, Poljarizatzija annigiljatzionnikh gammakvantov, *Dokl. Akad. Nauk U.S.S.R. 69*, 777–779. [III]

von Hayek, F. A., 1948, *Individualism and Economic Order*, University of Chicago Press, especially Essay 4, The use of knowledge in society. [V]

von Hayek, F. A., 1975, *Pure Theory of Capital*, University of Chicago Press, paperback reprint. [V]

von Mises, L., 1966, *Human Action*, 3rd revised ed., Contemporary Books, Chicago. [V]

von Neumann, J., 1927, Wahrscheinlichkeits-theoretischer Aufbau der Quantenmechanik, *Göttinger Nachrichten I*, No. 10, 245–272. [II.1]

von Neumann, J., 1929, Beweis des Ergodensatzes und des H-Theorems in der neuen Mechanik, *Z. Phys. 57*, 30–70. [II, V]

von Neumann, J., 1932, "Measurement and reversibility" and "The measuring process," chapters V and VI of *Mathematische Grundlagen der Quantenmechanik*, Springer, Berlin; translation into English by R. T. Beyer, 1955, *Mathematical Foundations of Quantum Mechanics*, Princeton University Press, pp. 347–445; reprinted here, V.2. [II, II.1, II.2, III, V, V.2]

von Neumann, J., 1958, *The Computer and the Brain*, Yale University Press, pp. 60–61. [I]

von Neumann, J., 1981, Continuous geometries with a transition probability, ed. Israel Halpern, *Memoirs of the American Mathematical Society 34*, No. 262, November 1981. [II]

Vonsovsky, S., 1946, Derivation of fundamental kinetic equations in quantum mechanics, *J. Phys. Moscow 10*, 367–376. [V]

Wannier, G. H., 1965, Quantum-mechanical proof of the second law, *Am. J. Phys. 33*, 222–225. [V]

Wax, N., ed., 1954, *Selected Papers on Noise and Stochastic Processes*, Dover, New York. [VI]

Weidlich, W., 1967, Problems of the quantum theory of measurement, *Z. Phys. 205*, No. 3, 199–220. [V]

Weidlich, W. and Haake, F., 1969, On the quantumstatistical theory of the measuring process, *J. Phys. Soc. Japan 26*, Supplement, 231–232. [V]

Weizsäcker, K. F. v., 1931, Ortbestimmung eines Elektrons durch ein Mikroskop, *Z. Phys. 70*, 114–130. [III]

Weyl, H., 1927, Quantenmechanik und Gruppentheorie, *Z. Phys. 46*, 1–46. [II.2]

Wheeler, J. A., 1946, Polyelectrons, *Ann. N.Y. Acad. Sci. 48*, 219–238; excerpt reprinted here, III.1. [II.2, III, III.1]

Wheeler, J. A., 1957, Assessment of Everett's "relative state" formulation of quantum theory, *Rev. Mod. Phys. 29*, No. 3, 463–465. [II, II.2]

Wheeler, J. A., 1977, Include the observer in the wave function? in Leite Lopes and Paty, eds., 1977, pp. 1–18. [II, II.2, III.13]

Wheeler, J. A., 1978, The "past" and the "delayed-choice" double-slit experiment, in A. R. Marlow, ed., 1978, *Mathematical Foundations of Quantum Theory*, Academic Press, New York. [III, III.13]

Wheeler, J. A. 1979, Frontiers of time, in Toraldo di Francia, ed., 1979, pp. 395–497. [I.13, II, III.13]

Wheeler, J. A., 1980, Beyond the black hole, in Woolf, ed., 1980, pp. 341–375. [I.13, II]

Wheeler, J. A., 1981a, Delayed-choice experiments and the Bohr-Einstein dialog, in *The American Philosophical Society and The Royal Society: Papers read at a meeting, June 5, 1980*, American Philosophical Society, Philadelphia. [I, I.13, II]

Wheeler, J. A., 1981b, This participatory universe, in Overbye, D., 1981, Messenger at the gates of time, *Science '81, 2*, no. 5, 60–67; last paragraph reprinted here, I.13.

Wheeler, J. A., 1981c, Not consciousness, but the distinction between the probe and the probed, as central to the elemental quantum act of observation, in Jahn, ed., 1981, pp. 87–111. [I, II]

Wheeler, J. A., 1979–1981, Law without law, excerpts reprinted here as section I.13 appeared in Wheeler, 1981a, 1979, 1980, 1981b. [I.13, II]

Whittaker, E. T., 1953, *History of the Theories of Aether and Electricity*, 1900–1926, Nelson, London. [I]

Whitten-Wolfe, B. and Emch, G. G., 1976, A mechanical quantum measuring process, *Helv. Phys. Acta*, *49*, 45–55. [II]

Whyte, L. L., 1974, *The Universe of Experience*, Harper and Row, New York. [II]

Wickes, W. C., Alley, C. O., and Jakubowicz, O., 1981, A delayed-choice quantum mechanics experiment, University of Maryland working paper; reprinted here, III.13. [III.13]

Wightman, A. S., 1962, On the localizability of quantum mechanical systems, *Rev. Mod. Phys. 34*, 845–872. [II.2]

Wigner, E. P., 1952, Die Messung quantenmechanischer Operatoren, *Z. Phys. 133*, 101–108. [II.2]

Wigner, E. P., 1959, *Group Theory and Its Applications to the Quantum Mechanics of Atomic Spectra*, Academic Press, New York. [II.2]

Wigner, E. P., 1961, Remarks on the mind-body question, in Good, 1961, pp. 284–302; reprinted in Wigner, 1967, pp. 171–184, from which this paper is reprinted here, I.12. [I.12, II]

Wigner, E. P., 1963, The problem of measurement, *Am. J. Phys. 31*, 6-15; reprinted here II.4. [II, II.2, II.4]

Wigner, E. P., 1967, *Symmetries and Reflections*, Indiana University Press. [I.12, Refs.]

Wigner, E. P., 1969, Are we machines?, *Proc. Amer. Phil. Soc. 113*, 95–101. [I]

Wigner, E. P., 1971, The philosophical problem, in d'Espagnat, ed., 1971, pp. 1–3. [I]

Wigner, E. P., 1972, On the time-energy uncertainty relation, pp. 237–247 in A. Salam and E. P. Wigner, eds., *Aspects of Quantum Theory*, Cambridge University Press, 1972. [VI]

Wigner, E. P., 1976, Interpretation of quantum mechanics, mimeographed notes of E. P. Wigner, revised and printed here for the first time, II.2. [II, II.2, VI]

Williams, J. G., Dicke, R. H., Bender, P. L., Alley, C. O., Carter, W. E., Currie, D. G., Eckhardt, D. H., Faller, J. E., Kaula, W. M., Mulholland, J. D., Plotkin, H. H., Poultney, S. K., Shelas, P. J., Silverberg, E. C., Sinclair, W. S., Slade, M. A., and Wilkinson, D. T., 1976, New test of the equivalence principle from lunar laser ranging, *Phys. Rev. Lett. 36*, 551–554. [III.13]

Wilson, A. R., Lowe, J., and Butt, D. K., 1976, Measurement of the relative planes of polarization of annihilation quanta as a function of separation distance, *J. Phys. G2*, 613–624. [III]

Witten, L., 1962, *Gravitation*, Wiley, New York. [Refs.]

Woolf, H., ed., 1980, *Some Strangeness in the Proportion: A Centennial Symposium to Celebrate the Achievements of Albert Einstein*, Addison-Wesley, Reading, Mass. [Refs.]

Wootters, W. K., 1980, The acquisition of information from quantum measurements, Ph.D. dissertation, University of Texas at Austin. [V]

Wootters, W. K., 1981, Statistical distribution and Hilbert space, *Phys. Rev. 23*, 357–362. [V]

Wootters, W. K. and Zurek, W. H., 1979, Complementarity in the double-slit experiment: quantum non-separability and a quantitative statement of Bohr's principle, *Phys. Rev. D19*, 473–484; reprinted here, III.11. [III.13]

Wu, C. S. and Shaknov, I., 1950, The angular correlation of scattered annihilation radiation, *Phys. Rev. 77*, 136. [II.2, III]

Yanase, M. M., 1961, Optimal measuring apparatus, *Phys. Rev. 123*, 666–668; reprinted here, VI.3. [II.2, VI, VI.3]

Yourgrau, W. and van der Merve, A., eds., 1971, *Perspectives in Quantum Theory, Essays in Honor of Alfred Landé*, M.I.T. Press, Cambridge, Mass. [I]

Zeh, H. D., 1970, On the interpretation of measurement in quantum theory, *Found. Phys. 1*, 69–76; reprinted here, II.5. [II.5]

Zeh, H. D., 1971, On the irreversibility of time and observation in quantum theory, in d'Espagnat, ed., 1971, pp. 263–273. [II]

Zeh, H. D., 1975, Connections between quantum theory, thermodynamics, cosmology and subjective perception, paper presented at Fourth International Conference on the Unity of Sciences, New York. [II]

Zichichi, A., ed., 1975, *Lepton and Hadron Structure*, 1974 International School of Subnuclear Physics, Erice, Trapani, Sicily, 14–21 July. [Refs.]

Zurek, W. H., 1981, Pointer basis of quantum apparatus: Into what mixture does the wave packet collapse? *Phys. Rev. D24*, 1516–1525. [II, V]

Zurek, W. H., 1982a, Information transfer in quantum measurements: Irreversibility and amplification, California Institute of Technology preprint; to appear in Meystre and Scully, 1983. [II]

Zurek, W. H., 1982b, Environment-induced superselection rules, California Institute of Technology preprint, to appear in *Phys. Rev. D*. [II, V]

Library of Congress Cataloging in Publication Data
Main entry under title:

Quantum theory and measurement.

(Princeton series in physics)
Bibliography: pp. 787-811
1. Quantum theory—Addresses, essays, lectures.
2. Physical measurements—Addresses, essays, lectures.
I. Wheeler, John Archibald, 1911- . II. Zurek,
Wojciech Hubert, 1951- . III. Series.
QC174.125.Q38 1983 530.1′2 82-47620
ISBN 0-691-08315-0
ISBN 0-691-08316-9 (pbk.)